FIFTH EDITION *Ebbing*

GENERAL CHEMISTRY

SOLUTIONS MANUAL

DARRELL D. EBBING
WAYNE STATE UNIVERSITY

GEORGE H. SCHENK
WAYNE STATE UNIVERSITY

DAVID BOOKIN
MT. SAN JACINTO COLLEGE

HOUGHTON MIFFLIN COMPANY ■ BOSTON ■ TORONTO
■ GENEVA, ILLINOIS ■ PALO ALTO ■ PRINCETON, NEW JERSEY

Sponsoring Editor: Kathi Prancan
Project Editor: Peggy Flanagan
Senior Production/Design Coordinator: Jill Haber
Senior Manufacturing Coordinator: Priscilla Bailey
Marketing Manager: Clint Crockett

The cover photograph, taken by Brian Coleman, depicts a piece of twisted blown crystal glass pumped with argon, xenon, and krypton. The gases are pumped into the glass at 20 millimeters of pressure. The red rod is a tube of soda lime glass filled with argon, and the blue rod is a second tube filled with argon and xenon.

Printed in the U. S. A.

ISBN: 0-395-75932-3

123456789-CS-00 99 98 97 96

CONTENTS

PREFACE

This complete Solutions Manual provides worked-out answers to all of the problems that appear in *General Chemistry* by Darrell D. Ebbing. This includes detailed, step-by-step solutions for all in-chapter Exercises as well as for the Practice Problems, Unclassified Problems, Cumulative-Skills Problems, and Conceptual Problems that appear at the end of the chapters. Also provided are answers to all the Review Questions.

Please note the following:

Significant figures: The answer is first shown with 1 to 2 nonsignificant figures and no units, and the least significant digit is underlined. The answer is then rounded off to the correct number of significant figures with units.

Equilibrium calculations: The answer is first shown at the beginning of each problem. This will help students gauge whether they are on track through the complicated working-out of the solution. The answer is then restated at the end of the problem.

Great effort and care have gone into the preparation of this manual. The solutions have been checked and rechecked for accuracy and completeness several times. I would like to thank John Goodinow who helped check the solutions for accuracy.

D. B.

FIFTH EDITION *Ebbing*

GENERAL CHEMISTRY

SOLUTIONS MANUAL

1. CHEMISTRY AND MEASUREMENT

■ Solutions to Exercises

Note on significant figures: The final answer to each mathematical solution is given first with one nonsignificant figure (the significant figure is underlined) and then is rounded to the correct number of figures. Intermediate answers usually also have at least one nonsignificant figure.

1.1 In words: Mass gases from wood = mass wood - mass ash. We calculate:

Mass gases from wood = 1.85 g - 0.28 g = 1.57 g

Assuming that all oxygen in the air is converted to gas in the vessel, we can write that the mass of gases in the vessel = mass gases from wood + mass air:

Mass of gases in vessel = 1.57 g + 9.45 g = 11.02 g

Thus the mass of gases in the vessel after burning is 11.02 g.

1.2 Physical properties: soft; silver color; melting point of 64°C.
Chemical properties: metal; reacts with water, oxygen, and chlorine.

1.3 a. $\frac{5.61 \times 7.891}{9.1} = 4.\underline{8}6 = 4.9$

(Two significant figures because 9.1 has only two.)

b. $8.91 - 6.435 = 2.4\underline{7}5 = 2.48$

(Two decimal places because 8.91 has only two decimal places.

c. $6.81 - 6.730 = 0.0\underline{8}0 = 0.08$

(Two decimal places because 6.81 has only two decimal places.)

d. $38.91 \times (6.81 - 6.730) = 38.91 \times 0.0\underline{8}0 = \underline{3}.113 = 3$

(The calculated difference of $0.0\underline{8}0$ has two significant figures because 6.81 has two decimal places. When 0.080 is used in multiplying, it has only one significant figure, so the product of 0.080 and 38.91 can have only one significant figure.)

1.4 a. 1.84×10^{-9} m = 1.84 nm
b. 5.67×10^{-12} s = 5.67 ps
c. 7.85×10^{-3} g = 7.85 mg
d. 9.7×10^{3} m = 9.7 km
e. 0.000732 s = 732 μs
f. 0.000000000154 m = 154 pm

1.5 a. $°C = \frac{°F - 32}{1.8} = \frac{[102.5 - 32]}{1.8} = 39.\underline{1}6 = 39.2\ °C$

(Three significant figures from the numerator of [70.5]. (Note that 32 and 1.8 are both exact numbers and do not affect significant figures.)

b. $K = °C + 273.15 = -78 + 273.15 = 19\underline{5}.15 = 195$

(78 has no decimal places.)

1.6 $d = \frac{m}{V} = \frac{159\ g}{20.2\ cm^3} = 7.8\underline{7}1 = 7.87\ g/cm^3$

(The metal is iron.)

1.7 $V = \frac{m}{d} = 30.3\ g \times \frac{1\ cm^3}{0.789\ g} = 38.\underline{4}03 = 38.4\ cm^3$

1.8 $121\ pm \times \frac{10^{-12}\ m}{1\ pm} \times \frac{10^{3}\ mm}{1\ m} = 1.21 \times 10^{-7}\ mm$

1.9 $67.6\ Å^3 \times \frac{10^{-30}\ m^3}{1\ Å^3} \times \frac{10^{3}\ dm}{1\ m^3} = 6.76 \times 10^{-26}\ dm^3$

1.10 $\frac{36 \text{ in}}{1 \text{ yd}} \times \frac{2.54 \text{ cm}}{1 \text{ in}} \times \frac{1 \text{ m}}{10^2 \text{ cm}} = \frac{9.144 \times 10^{-1} \text{ m}}{1 \text{ yd}}$

$$3.54 \text{ yd} \times \frac{9.144 \times 10^{-1}\text{m}}{\text{yd}} = 3.2\underline{3}69 = 3.24 \text{ m}$$

■ Answers to Review Questions

1.1 Chemistry has changed technology in the areas of communications and computers. In communications, optical fibers of extremely pure glass have replaced telephone wires. Computers have been made possible by the invention of computer chips made of certain pure chemicals. Chemistry has changed biology in the areas of DNA research and in the area of therapy for cystic fibrosis.

1.2 An experiment is an observation of natural phenomena carried out in a controlled manner so that the results can be duplicated and rational conclusions obtained. A theory is a tested explanation of natural phenomena. Experiments are performed and some regularity is observed; this leads to more experiments and a theory. A hypothesis is a tentative explanation of some regularity.

1.3 Rosenberg conducted controlled experiments, noting a basic relationship that could be stated as a scientific law. He formed a hypothesis that platinum compounds are responsible for preventing cell division. After more testing, the hypothesis became a theory involving cisplatin.

1.4 Matter is any material thing that occupies space. Mass is the quantity of matter in a material. The difference between mass and weight is that mass remains the same wherever it is measured, but weight varies with the distance from the center of the earth.

1.5 This law states that mass remains constant during a chemical change (reaction). Weighing a flash bulb before and after flashing illustrates this law. The bulb is found to have the same mass after the reaction that occurs during flashing, showing that mass is conserved during chemical changes.

1.6 Liquid mercury reacts with oxygen gas to form solid red mercury(II) oxide.

1.7 Gases are easily compressible and fluid as compared to solids, which are rigid. A gas fits into a container of any size and shape. Liquids are relatively incompressible fluids. A liquid has a fixed volume but no fixed shape. Solids are relatively incompressible and rigid. A solid has fixed shape and volume.

1.8 An example of a substance is the element sodium. Among its physical properties: It is a solid, it has a white color, and it melts at 98°C. Among its chemical properties: It reacts vigorously with water, and it burns in chlorine gas to form sodium chloride.

1.9 An example of an element: sodium; of a compound: sodium chloride, or table salt; of a heterogeneous mixture: salt and sugar; of a homogeneous mixture: sodium chloride dissolved in water to form a solution.

1.10 a. Two phases: liquid and solid.
b. Phases: liquid water, solid quartz, and solid seashells.

1.11 If the material is a pure compound, all samples should have the same melting point, the same color, and the same elemental composition. If it is a mixture, there should be a varying difference in these properties among the various samples.

1.12 The precision of a set of measurements refers to how close the values are to each other. The precision of one measurement is indicated by using significant figures.

1.13 Mul./div. rule calculation: 3.00 x 2.0 = 6.0 (2.0 has only two significant figures).
Add/sub rule calculation: 3.000 + 7.00 = 10.00 (7.00 has two decimal places).

1.14 An exact number arises when we count a small number of items, or when we define a number rather than measure it. Examples: *exact:* 0 K (absolute zero); *measured:* 11 K.

1.15 The SI system starts with seven base units and uses prefixes to obtain units of different sizes. Units for other quantities are obtained by deriving them from any of the seven base units.

1.16 SI length unit = meter (39.37 in). SI mass unit = kilogram (2.2 U.S. pounds).

1.17 An absolute temperature scale is a scale in which the lowest temperature that can be attained theoretically is zero degrees. Degrees Celsius are related to kelvins by $^{o}C = (K - 273.15)$.

1.18 Density is the mass of an object per unit volume. It can be used to check purity and to help identify substances.

1.19 Units should be carried along in a calculation to obtain units for the answer and to avoid errors.

1.20 The conversion factor for converting liters to U.S. gallons is obtained by multiplying the conversion factor for liters to quarts by that for converting quarts to U.S. gallons:

$$\frac{1\ \text{qt}}{0.9464\ \text{L}} \times \frac{1\ \text{U.S. gal}}{4\ \text{qt}} = 0.264\underline{1}58 = \frac{0.2642\ \text{U.S. gal}}{\text{L}}$$

■ Solutions to Practice Problems

Note on significant figures: The final answer is given first with one nonsignificant figure (the rightmost significant figure is underlined) to each problem and then is rounded to the correct number of significant figures. Intermediate answers usually also have at least one nonsignificant figure. Atomic weights are rounded to two decimal places, except for that of hydrogen.

1.21 By law of conservation of mass:

Mass(sodium carbonate) + mass(acetic acid solution)

= mass(contents of reaction vessel) + mass(carbon dioxide)

15.9 g + 20.0 g = 29.3 g + mass(carbon dioxide)

Mass of carbon dioxide = 20.0 g + 15.9 g - 29.3 g = $6.\underline{6}0$ = 6.6 g

1.22 By law of conservation of mass:

Mass(iron) + mass(acid) = mass(beaker contents) + mass(hydrogen)

5.6 g + 15.0 = 20.4 g + mass(hydrogen)

Mass(hydrogen) = 5.6 g + 15.0 g - 20.4 g = $0.\underline{2}0$ = 0.2 g

1.23 By law of conservation of mass:

Mass(zinc) + mass(sulfur) = mass(zinc sulfide)

65.4 g Zn + 32.1 g S must = (65.4 + 32.1) g ZnS

20.0 g Zn + unknown mass of S = x g ZnS

Use this to write a proportion from the masses of zinc and zinc sulfide in the last two equations:

$$\frac{x}{20.0 \text{ g Zn}} = \frac{(65.4 + 32.1) \text{ g ZnS}}{65.4 \text{ g Zn}}$$

Solving gives x = $29.\underline{8}1$ = 29.8 g

1.24 Using the law of conservation of mass:

Mass(bromine) = mass(aluminum bromide) - mass(aluminum)

Mass(bromine) = (266.7 g aluminum bromide - 27.0 g Al) g Br

15.0 g Al + x g Br = unknown mass of aluminum bromide

Now write a proportion from the masses of Al and Br in the last two equations:

$$\frac{x}{15.0 \text{ g Al}} = \frac{(266.7 - 27.0) \text{ g Br}}{27.0 \text{ g Al}}$$

Solving gives x = $13\underline{3}.1$ = 133 g Br

1.25 a. Solid b. Liquid c. Gas d. Solid

1.26 a. Solid b. Gas c. Solid d. Liquid

1.27 a. Physical change b. Physical change c. Chemical change d. Physical change

1.28 a. Physical change b. Chemical change c. Chemical change d. Physical change

1.29 Physical change: Liquid mercury is cooled to solid mercury.
Chemical changes: (1) Solid mercury oxide forms liquid mercury metal and gaseous oxygen; (2) glowing wood and oxygen forms burning wood (forms ash and gaseous products).

1.30 Physical changes: (1) Solid iodine is heated to gaseous iodine; (2) gaseous iodine is cooled to form solid iodine.
Chemical change: Solid iodine and zinc metal are ignited to form a white powder.

1.31 a. Physical property
b. Chemical property
c. Physical property
d. Physical property
e. Chemical property

1.32 a. Physical property
b. Physical property
c. Chemical property
d. Chemical property
e. Physical property

1.33 Physical properties: (1) Iodine is solid at room temperature; (2) the solid has a lustrous blue-black color; (3) the solid vaporizes to a gas (without forming a liquid); (4) the gas is violet colored.
Chemical properties: (1) Iodine combines chemically with many metals; (2) iodine combines chemically with aluminum to form the compound aluminum iodide.

1.34 Physical properties: (1) is solid at room temperature; (2) has an orange-red color; (3) has a density of 11.1 g/cm^3; (4) is insoluble in water.
Chemical property: Mercury(II) oxide decomposes when heated to form mercury and oxygen.

1.35 a. Physical process
b. Chemical reaction
c. Physical process
d. Chemical reaction
e. Physical process

1.36 a. Chemical reaction
b. Physical process
c. Physical process
d. Physical process
e. Chemical reaction

1.37 a. Solution b. Substance c. Substance d. Heterogeneous mixture

1.38 a. Heterogeneous mixture b. Substance c. Solution d. Substance

1.39 a. Liquid and gaseous forms of a pure substance
b. Liquid and solid forms of a mixture
c. Liquid and solid forms of a pure substance
d. Solid form of a mixture

1.40 a. Liquid and solid forms of a mixture
b. Liquid and solid forms of a mixture
c. Solid form of a mixture
d. Liquid and gaseous forms of a pure substance

1.41 All significant figures (s.f.) are underlined:

a. $\underline{73.0000}$ (6 s.f.)
b. $0.0\underline{503}$ (3 s.f.)
c. $\underline{6.300}$ (4 s.f.)
d. $0.\underline{80090}$ (5 s.f.)
e. $\underline{5.10} \times 10^{-7}$ (3 s.f.)
f. $\underline{2.001}$ (4 s.f.)

1.42 All significant figures (s.f.) are underlined:

a. $\underline{130.0}$ (4 s.f.)
b. $0.0\underline{738}$ (3 s.f.)
c. $0.\underline{224800}$ (6 s.f.)
d. $\underline{1008}$ (4 s.f.)
e. $\underline{4.380} \times 10^{-8}$ (4 s.f.)
f. $\underline{9.100} \times 10^{4}$ (4 s.f.)

1.43 $4\underline{0}{,}000 \text{ km} = 4.\underline{0} \times 10^4 \text{ km}$

1.44 $15\underline{0}{,}000{,}000 \text{ km} = 1.5\underline{0} \times 10^8 \text{ km}$

1.45 a. $(8.71 \times 0.0301) \div 0.056 = 4.\underline{6}8 = 4.7$

b. $0.71 + 81.8 = 82.\underline{5}1 = 82.5$

c. $(934 \times 0.00435) + 107 = 11\underline{1}.06 = 111$

d. $(847.89 - 847.73) \times 14673 = 2.\underline{3}4 \times 10^3 = 2.3 \times 10^3$

1.46 a. $(0.871 \times 0.23) \div 5.871 = 0.03\underline{4}1 = 0.034$

b. $8.937 - 8.930 = 0.00\underline{7}0 = 0.007$

c. $8.937 + 8.930 = 17.86\underline{7}0 = 17.867$

d. $(0.00015 \times 54.6) + 1.002 = 1.01\underline{0}1 = 1.010$

1.47 Volume of 5.10 cm $= (4/3) \times 3.1416 \times (5.10 \text{ cm})^3 = 5.5\underline{5}64 \times 10^2 \text{ cm}^3$

Volume of 5.00 cm $= (4/3) \times 3.1416 \times (5.00 \text{ cm})^3 = 5.2\underline{3}60 \times 10^2 \text{ cm}^3$

Volume difference $= (5.5\underline{5}64 \times 10^2 - 5.2\underline{3}60 \times 10^2) = 0.3\underline{2}04 \times 10^2 = 32 \text{ cm}^3$

1.48 Volume of 3.50 cm $= 3.1416 \times (1.500 \text{ cm})^2 \times 3.50 \text{ cm} = 24.\underline{7}40 \text{ cm}^3$

Volume of 3.10 cm $= 3.1416 \times (1.500 \text{ cm})^2 \times 3.10 \text{ cm} = 21.\underline{9}12 \text{ cm}^3$

Volume difference $= (24.\underline{7}40 - 21.\underline{9}12) = 2.\underline{8}28 = 2.8 \text{ cm}^3$

1.49 a. 5.89×10^{-12} s = 5.89 ps

b. 2.130×10^{-9} m = 2.130 nm

c. 0.00721 g = 7.21 mg

d. 6.05×10^3 m = 6.05 km

1.50 a. 4.851×10^{-6} g = 4.851 µg

b. 3.16×10^{-2} m = 3.16 cm

c. 2.591×10^{-9} s = 2.591 ns

d. 8.93×10^{-12} g = 8.93 pg

1.51 a. 6.15 ps = 6.15×10^{-12} s
b. 3.781 µm = 3.781×10^{-6} m
c. 1.546 Å = 1.546×10^{-10} m
d. 9.7 mg = 9.7×10^{-3} g

1.52 a. 8.55 km = 8.55×10^{3} m
b. 1.98 ns = 1.98×10^{-9} s
c. 2.54 cm = 2.54×10^{-2} m
d. 6.923 µg = 6.923×10^{-6} g

1.53 a. $\frac{(68^\circ F - 32)}{1.8} = 2\underline{0}.0 = 20.^\circ C$

b. $\frac{(-11^\circ F - 32)}{1.8} = -2\underline{3}.8 = -24^\circ C$

c. $(37^\circ C \times 1.8) + 32 = 9\underline{8}.6 = 99^\circ F$

d. $(-70^\circ C \times 1.8) + 32 = -9\underline{4}.0 = -94^\circ F$

1.54 a. $\frac{(121^\circ F - 32)}{1.8} = 4\underline{9}.4 = 49^\circ C$

b. $\frac{(-15^\circ F - 32)}{1.8} = -2\underline{6}.1 = -26^\circ C$

c. $(-45^\circ C \times 1.8) + 32 = -4\underline{9}.0 = -49^\circ F$

d. $(65^\circ C \times 1.8) + 32 = 149^\circ F = 149^\circ F$

1.55 $(-21.1^\circ C \times 1.8) + 32 = -5.\underline{9}8 = -6.0^\circ F$

1.56 $(-196^\circ C \times 1.8) + 32 = -32\underline{0}.8 = -321^\circ F$

1.57 Density = mass ÷ volume = 12.4 g ÷ 1.64 cm^3 = 7.5$\underline{6}$09 = 7.56 g/cm^3

1.58 Density = mass ÷ volume = 17.84 g ÷ 25.0 mL = 0.71$\underline{3}$6 = 0.714 g/mL

1.59 Calculate density: Density = 9.42 g ÷ 10.7 mL = 0.88$\underline{0}$3 = 0.880 g/mL
The density is closest to that of benzene (0.879 g/cm^3), so the liquid is benzene.

1.60 Calculate density: Density = 16.3 g ÷ 2.3 cm^3 = 7.$\underline{0}$8 = 7.1 g/cm^3
The density is closest to that of cassiterite (6.99 g/cm^3), so the mineral is cassiterite.

1.61 $$\text{Mass(Pt)} = d \times V = \frac{21.4\text{ g}}{\text{cm}^3} \times 5.9\text{ cm}^3 = 1.\underline{2}6 \times 10^2 = 1.3 \times 10^2\text{ g}$$

1.62 $$\text{Mass(gas)} = d \times V = \frac{0.70\text{ g}}{\text{cm}^3} \times \frac{1\text{ cm}^3}{1\text{ mL}} \times 51.6\text{ mL} = 3\underline{6}.1 = 36\text{ g}$$

1.63 $$\text{Volume(alc)} = 19.8\text{ g alc} \times 1\text{ cm}^3\text{ alc}/0.789\text{ g alc} = 25.\underline{0}9 = 25.1\text{ cm}^3\text{ alc}$$

1.64 $$\text{Volume(Br)} = 88.5\text{ g Br} \times 1\text{ mL Br}/3.10\text{ g Br} \times 1\text{ cm}^3/\text{mL} = 28.\underline{5}4 = 28.5\text{ cm}^3$$

1.65 $$0.348\text{ kg} \times (10^3\text{ g}/1\text{ kg}) \times (10^3\text{ mg}/1\text{ g}) = 3.48 \times 10^5\text{ mg}$$

1.66 $$483\text{ mg} \times (1\text{ g}/10^3\text{ mg}) \times (10^6\ \mu\text{g}/1\text{ g}) = 4.83 \times 10^5\ \mu\text{g}$$

1.67 $$555\text{ nm} \times \frac{1\text{ m}}{10^9\text{ nm}} \times \frac{10^3\text{ mm}}{1\text{ m}} = 5.55 \times 10^{-4}\text{ mm}$$

1.68 $$0.96\text{Å} \times \frac{1\text{ m}}{10^{10}\text{ Å}} \times \frac{10^3\text{ mm}}{1\text{ m}} = 9.6 \times 10^{-8}\text{ mm}$$

1.69 $$3.73 \times 10^8\text{ km}^3 \times \frac{10^9\text{ m}^3}{1\text{ km}^3} \times \frac{10^3\text{ dm}^3}{1\text{ m}^3} = 3.73 \times 10^{20}\text{ dm}^3\text{(L), or } 3.73 \times 10^{17}\text{ m}^3$$

1.70 $$1.4\ \mu\text{m}^3 \times \frac{(1\text{ m})^3}{(10^6\ \mu\text{m})^3} \times \frac{(10\text{ dm})^3}{(1\text{ m})^3} = 1.4 \times 10^{-15}\text{ dm}^3\text{(L)}$$

1.71 $$3.58\text{ ton} \times \frac{2000\text{ lb}}{1\text{ ton}} \times \frac{16\text{ oz}}{1\text{ lb}} \times \frac{1\text{ g}}{0.03527\text{ oz}} = 3.2\underline{4}8 \times 10^6 = 3.25 \times 10^6\text{ g}$$

1.72 $$2.45\text{ Btu} \times \frac{252.0\text{ cal}}{1\text{ Btu}} \times \frac{4.184\text{ J}}{1\text{ cal}} = 2.5\underline{8}3 \times 10^3 = 2.58 \times 10^3\text{ J}$$

1.73 $2425 \text{ fath} \times \frac{6 \text{ ft}}{1 \text{ fath}} \times \frac{12 \text{ in}}{1 \text{ ft}} \times \frac{2.54 \times 10^{-2} \text{ m}}{1 \text{ in}} = 4.43\underline{4}8 \times 10^3 = 4.435 \times 10^3 \text{ m}$

1.74 $9.6 \times 10^9 \text{ bbl} \times \frac{42 \text{ gal}}{1 \text{ bbl}} \times \frac{4 \text{ qt}}{1 \text{ gal}} \times \frac{9.46 \times 10^{-4} \text{ m}^3}{1 \text{ qt}} = 1.\underline{5}2 \times 10^9 = 1.5 \times 10^9 \text{ m}^3$

■ Solutions to Unclassified Problems

1.75 9.85 g Na + 63.11 g H_2O - 72.53 g soln = 0.43 g hydrogen

1.76 0.853 g tablet + 56.519 g acid - 57.152 soln = 0.220 g carbon dioxide

1.77 From the law of conservation of mass:

5.40 g Al + 18.50 g Fe_2O_3 = 11.17 g Fe + 10.20 g Al_2O_3 + g Fe_2O_3 unreacted

g Fe_2O_3 unreacted = 5.40 g + 18.50 g - (11.17 g + 10.20 g) = 2.53 g

1.78 From the law of conservation of mass:

20.6 g NaBr + mass Cl_2 reacted = 16.0 g Br_2 + 11.7 g NaCl

Mass Cl_2 reacted = 16.0 g + 11.7 g - 20.6 g = 7.1 g

1.79 53.10 g + 5.348 g + 56.1 g = $114.\underline{5}4$ = 114.5 g total mass

1.80 68.1 g + 48.7 g + 5.318 g = $122.\underline{1}18$ = 122.1 g total mass

1.81 a. Bromine b. Phosphorus c. Gold d. Carbon (as graphite)

1.82 a. Copper b. Sulfur c. Chlorine d. Mercury

1.83 The fraction of iron (Fe) in each sample is calculated as follows:

A. $\dfrac{1.094\text{ g Fe}}{1.518\text{ g}} = 0.720\underline{6}8$; B. $\dfrac{1.449\text{ g Fe}}{2.056\text{ g}} = 0.704\underline{8}$

C. $\dfrac{1.335\text{ g Fe}}{1.873\text{ g}} = 0.712\underline{8}$

Each sample has a slightly different composition, indicating a mixture, not a compound.

1.84 The fraction of mercury (Hg) in each sample is calculated as follows:

A. $\dfrac{0.9641\text{ g Hg}}{1.0410\text{ g}} = 0.926\underline{1}3$; B. $\dfrac{1.4293\text{ g Hg}}{1.5434\text{ g}} = 0.9260\underline{7}$

C. $\dfrac{1.1283\text{ g Hg}}{1.2183\text{ g}} = 0.9261\underline{3}$

Each compound has the same fraction of mercury to four significant figures (and should for O), so the data are consistent with the hypothesis that the material is a compound.

1.85 $(60.8\text{ cm})^3 = 2.2\underline{4}7 \times 10^5 = 2.25 \times 10^5\text{ cm}^3$

1.86 $(3.1416) \times (2.13\text{ cm})^2 \times (56.32\text{ cm}) = 8.0\underline{2}7 \times 10^2 = 8.03 \times 10^2\text{ cm}^3$

1.87 $47.8\text{ in} \times 12.5\text{ in} \times 19.5\text{ in} \times \dfrac{1\text{ gal}}{231\text{ in}^3} = 50.\underline{4}3 = 50.4\text{ gal}$

1.88 $\dfrac{4}{3} \times 3.1416 \times (150.0\text{ in})^3 \times \dfrac{1\text{ gal}}{277.4\text{ in}^3} = 5.09\underline{6}3 \times 10^4$

$= 5.096 \times 10^4\text{ gal};\ 1.41\underline{4} \times 10^7\text{ in}^3$

1.89 5.85 cm sphere volume $= (4/3)(3.1416)(5.85\text{ cm})^3 = 8.3\underline{8}60 \times 10^2\text{ cm}^3$

5.61 cm sphere volume $= (4/3)(3.1416)(5.61\text{ cm})^3 = 7.3\underline{9}56 \times 10^2\text{ cm}^3$

Difference $= 0.9\underline{9}04 \times 10^2 = 9.9 \times 10^1\text{ cm}^3$

1.90 8.50 cm circle area = $(3.1416)(8.50\text{ cm})^2 = 2.2\underline{6}98 \times 10^2\text{ cm}^2$

7.98 cm circle area = $(3.1416)(7.98\text{ cm})^2 = 2.0\underline{0}06 \times 10^2\text{ cm}^2$

Difference = $0.2\underline{6}92 \times 10^2 = 2.7 \times 10^1\text{ cm}^2$

1.91 a. $0.7\underline{5}98 = 0.76$ b. $16.\underline{2}9 = 16.3$ c. $47\underline{6}.3 = 476$ d. $0.11\underline{1}9 = 0.112$

1.92 a. $0.034\underline{6}55 = 0.0347$ b. $1.87\underline{0}3 = 1.870$ c. $58.\underline{0}6 = 58.1$ d. $17.6\underline{1}56 = 17.62$

1.93 a. 9.12 cg/mL b. 66 pm c. 7.1 μm d. 56 nm

1.94 a. 1.86 cg/mL b. 77 pm c. 6.5 nm d. 0.85 μm

1.95 a. 1.07×10^{-12} s b. 5.8×10^{-6} m c. 3.19×10^{-7} m d. 1.53×10^{-2} s

1.96 a. 7.3×10^{-3} K b. 2.75×10^{-10} m c. 1.96×10^{-2} s d. 4.5×10^{-5} m

1.97 $(3410^oC \times 1.8) + 32 = 6170^oF$

1.98 $(1677^oC \times 1.8) + 32 = 305\underline{0}.6 = 3051^oF$

1.99 $(825^oC \times 1.8) + 32 = 15\underline{1}7 = 1.52 \times 10^3\ ^oF$

1.100 $(50^oC \times 1.8) + 32 = 12\underline{2}.0 = 122^oF$

1.101 $K = 273.15 + 29.8 = 302.\underline{9}5 = 303.0$; $(29.8^oC \times 1.8) + 32 = 85.\underline{6}4 = 85.6^oF$

1.102 $K = 273.15 + (-38.9) = 234.\underline{2}5 = 234.3$; $(-38.9^oC \times 1.8) + 32 = -38.\underline{0}2 = -38.0^oF$

1.103 $(1666^oF - 32) \div 1.8 = 907.\underline{7}7 = 907.8^oC$; $273.15 + 907.77 = 1180.\underline{9}2 = 1180.9$ K

1.104 $(236^oF - 32) \div 1.8 = 11\underline{3}.33 = 113^oC$; $273.15 + 113.33 = 38\underline{6}.48 = 386$ K

1.105 Density = $(22.5\ g/cm^3)(1\ kg/10^3\ g)(10^2\ cm/1\ m)^3 = 2.25 \times 10^4\ kg/m^3$

1.106 Density = $(5.96\ g/cm^3)(1\ kg/10^3\ g)(10^2\ cm/1\ m)^3 = 5.96 \times 10^3\ kg/m^3$

1.107 Density = $38.4\ g \div (65.7\ mL - 51.2\ mL) = 2.6\underline{4}8 = 2.65\ g/mL$

1.108 Volume(water) = $(109.3 - 70.7)\ g \times (1\ cm^3/0.997\ g) = 38.\underline{7}1\ cm^3$, or $38.\underline{7}1$ mL

Density(ore) = $70.7\ g \div (53.2 - 38.71)\ mL = 4.8\underline{7}9 = 4.88\ g/mL$

1.109 Density(unknown) = $22.3\ g/15.0\ mL = 1.4\underline{8}6\ g/mL = 1.49\ g/cm^3$

This density is closest to that of chloroform ($1.489\ g/cm^3$), so the unknown is chloroform.

1.110 Density(unknown) = $35.6\ g/12.9\ cm^3 = 2.7\underline{5}9 = 2.76\ g/cm^3$

The calcite will float on the liquids with greater densities: tetrabromoethane ($2.96\ g/cm^3$) and methylene iodide ($3.33\ g/cm^3$).

1.111 Mass = $21.4\ g/cm^3 \times (2.20\ cm)^3 = 2.2\underline{7}8 \times 10^2 = 2.28 \times 10^2\ g = 0.228\ kg$

1.112 Mass = $2.33\ g/cm^3 \times (3.50\ cm)^2 \times 3.1416 \times 12.40\ cm = 1.1\underline{1}1 \times 10^3\ g = 1.11\ kg$

1.113 Volume = $(1\ mL/1.053\ g) \times 35.00\ g = 33.2\underline{3}8 = 33.24\ mL\ (33.24\ cm^3)$

1.114 Volume = $(1\ mL/0.902\ g) \times 0.985\ kg \times (10^3\ g/1\ kg) = 10\underline{9}2\ mL$

$1092\ mL \times (1\ L/10^3\ mL) = 1.0\underline{9}2 = 1.09\ L$

1.115 a. $8.45\ kg \times (10^3\ g/1\ kg) \times (10^3 mg/1\ g) = 8.45 \times 10^6\ mg$

b. $318\ \mu s \times (1\ s/10^6\ \mu s) \times (10^3\ ms/1\ s) = 0.318\ ms$

c. $93\ km \times (10^3 m/1\ km) \times (10^9\ nm/1\ m) = 9.3 \times 10^{13}\ nm$

d. $37.1\ mm \times (1\ m/10^3\ mm) \times (10^2\ cm/1\ mm) = 3.71\ cm$

1.116 a. 239 Å x (1 m/10^{10} Å) x (10^6µm/1 m) = 2.39 x 10^{-2} µm

b. 19.6 kg x (10^3 g/1 kg) x (10^3mg/1 g) = 1.96 x 10^7 mg

c. 24.8 cm x (1 m/10^2 cm) x (10^3 mm/1 m) = 248 mm

d. 4.3 ns x (1 s/10^9 ns) x (10^6 µs/1 s) = 4.3 x 10^{-3} µs

1.117 a. 5.91 kg x (10^3 g/1 kg) x (10^3 mg/1 g) = 5.91 x 10^6 mg

b. 753 mg x (1 g/10^3 mg) x (10^6µg/1 g) = 7.53 x 10^5 µg

c. 90.1 MHz x (1 x 10^6 Hz/1 MHz) x (1 kHz/1 x 10^3 Hz) = 9.01 x 10^4 kHz

d. 498 mJ x (1 J/10^3 mJ) x (1 kJ/10^3 J) = 4.98 x 10^{-4} kJ

1.118 a. 7.19 µg x (1 g/10^6 µg) x (10^3 mg/1 g) = 7.19 x 10^{-3} mg

b. 104 pm x (1 m/10^{12} pm) x (10^{10} Å /1 m) = 1.04 Å

c. 0.010 mm x (1 m/10^3 mm) x (10^2 mm/1 m) = 1.0 x 10^{-3} cm

d. 0.0605 kPa x (10^3 Pa/1 kPa) x (10^2 cPa/1 Pa) = 6.05 x 10^3 cPa

1.119 Volume = 12,230 km^3 x [(10^4 dm)/(1 km)]^3 x (1 L/1 dm^3) = 1.2230 x 10^{16} L

(Assuming the final zero in 12,230 km^3 is significant.)

1.120 Volume = 0.477 km^3 x [(10^4 dm)/(1 km)]^3 x (1 L/1 dm^3) = 4.77 x 10^{11} L

1.121 Volume = 10.0 ft x 12.0 ft x 9.0 ft x (12 in/1 ft)^3 x (2.54 cm/1 in)^3 x (1 dm/10 cm)^3

x (1 L/1 dm^3) = 3.$\underline{0}$58 x 10^4 = 3.1 x 10^4 L

1.122 Volume = 3.1416 x 6.0 ft x (15.0 ft)^2 x $\frac{(12\text{ in})^3}{(1\text{ ft})^3}$ x $\frac{(2.54\text{ cm})^3}{(1\text{ in})^3}$ x $\frac{(1\text{ dm})^3}{(10\text{ cm})^3}$

x $\frac{1\text{ L}}{1\text{ dm}^3}$ = 1.$\underline{2}$009 x 10^5 = 1.2 x 10^5 L

1.123 Mass (wt) = 563 carats x (200 mg/1 carat) x (1 g/10^3 mg) = $11\underline{2}.6$ = 113 g

1.124 Mass (ton) = 49.6 x 10^6 tr oz x (31.10 g/1 tr oz) x (1 ton/10^6 g) = $1.5\underline{4}2 \times 10^3$

= 1. 54 x 10^3 ton

■ Cumulative Skills Problems

(require two or more operational skills learned in this chapter)

1.125 From the law of conservation of mass:

10.0 g marble + (50.0 mL x 1.096 g/mL) HCl = 60.4 g soln + mass CO_2

Mass CO_2 = 10.0 g + 54.8 g - 60.4 g = 4.4 g; Vol CO_2 = 4.4 g/(1.798 g/L) = 2.4 L

1.126 From the law of conservation of mass:

10.8 g zinc ore + (50.0 mL x 1.153 g/mL) H_2SO_4 = 65.1 g soln + mass H_2S

Mass H_2S = 10.8 g + 57.65 - 65.1 g = $3.\underline{3}5$ g

Liter H_2S = 3.35 g ÷ (1.393 g/L) = $2.\underline{4}04$ = 2.4 L

1.127 Volume of sphere = $(4/3)(3.1416)(1.58\text{ in})^3(2.54\text{ cm/in})^3$ = $27\underline{0}.746\text{ cm}^3$

Mass of sphere = 270.746 cm^3 x 7.88 g/cm^3 = $2.1\underline{3}3 \times 10^3$ = 2.13 x 10^3 g

1.128 Assume the balloon is a perfect sphere and calculate the volume as done in the previous problem, using radius = 3.00 ÷ 2 = 1.50 ft.

Volume = $(4/3)(3.1416)[1.50\text{ ft} \times (12\text{ in/1 ft}) \times (2.54\text{ cm/in})]^3$

Volume = $4.0\underline{0}32 \times 10^5\text{ cm}^3$ = $4.0\underline{0}32 \times 10^2$ L of helium

Mass = 4.0032 x 10^2 L x 0.166 g/L = $66.\underline{4}54$ = 66.5 g

1.129 Mass $= (840{,}000\ mi^2 - 132{,}000\ mi^2) \times (5280\ ft/1\ mi)^2 \times 5000\ ft \times (12\ in/1\ ft)^3$

$\times (2.54\ cm/1\ in)^3 \times (0.917\ g/cm^3)$

$= 2.\underline{5}6 \times 10^{21} = 2.6 \times 10^{21}\ g$

1.130 Mass $= 5.5 \times 10^6\ mi^2 \times (5280\ ft/1\ mi)^2 \times (7500 - 1500)\ ft \times (12\ in/1\ ft)^3$

$\times (2.54\ cm/1\ in)^3 \times (0.917\ g/1\ cm^3)$

$= 2.\underline{3}8 \times 10^{22} = 2.4 \times 10^{22}\ g$

1.131 Let x = mass of ethanol and y = mass of water. From the law of conservation of mass:

Total mass = x + y = 49.6 g

So y = 49.6 g - x, and

Total volume = volume of ethanol + volume of water

Express the volumes of ethanol and water using both densities (given) and masses x and y:

$54.2\ cm^3 = (x/0.789\ g/cm^3) + [(49.6\ g - x)/(0.998\ g/cm^3)]$

Solve the above equation for x:

$54.2\ cm^3 = (x/0.789\ g/cm^3) + [49.6\ g/(0.998\ g/cm^3)] - [x/(0.998\ g/cm^3)]$

$(x/0.789) - (x/0.998) = [54.2 - (49.6/0.998)]\ g$

Multiply both sides by (0.789)(0.998):

$0.998x - 0.789x = (0.789)(0.998) \times [54.2 - (49.6/0.998)]\ g$

$0.2090x = 3.\underline{5}438\ g$

$x = 1\underline{6}.95\ g$

Answer 1: Mass % ethanol = (16.95 g/49.6 g soln) x 100% = $3\underline{4}.18$ = 34%

To calculate the proof, first find the volume of ethanol from its mass of 16.95 g:

Volume $= 16.95\ g/(0.789\ g/cm^3) = 21.\underline{4}8\ cm^3$

Volume % $= (21.48\ cm^3/54.2\ cm^3) \times 100\% = 39.\underline{6}3\%$

Answer 2: Proof = 2 x volume % = 2 x 39.63 = $79.\underline{2}7$ = 79.3 proof

1.132 Let x = mass of silver and y = mass of gold. From the law of conservation of mass:

Total mass = 9.35 g = x + y

So x = 9.35 g - y, and

Total volume = volume of silver + volume of gold

Express the volumes of silver and gold using both the densities (given) and masses x and y:

$$0.654\ cm^3 = [x/10.5\ g/cm^3)] + [y/(19.3\ g/cm^3)]$$

Substituting (9.35g - y) for x gives:

$$0.654\ cm^3 = [(9.35\ g - y)/(10.5\ g/cm^3)] + [y/(19.3\ g/cm^3)]$$

Solving the above equation for y:

$$0.654\ cm^3 = (9.35/10.5)\ cm^3 - [y/(10.5\ g/cm^3)] + [y/(19.3\ g/cm^3)]$$

Isolate the y terms on the left side:

$$(y/10.5) - (y/19.3) = [(9.35/10.5) - 0.654]\ g$$

Multiply both sides by (10.5)(19.3):

$$19.3y - 10.5y = (10.5)(19.3) \times [(9.35/10.5) - 0.654]\ g$$

$$8.80\ y = 47.92\ g$$

$$y = 5.446\ g$$

Answer 1: % gold = (5.446 g/9.35 g) x 100% = 58.$\underline{2}$4 = 58.2%

To calculate the proportion (karats), multiply the fraction of gold by 24 karats/fraction of 1:

Answer 2: Karats = 0.5824 x (24 karats/fraction of 1.00) = 13.$\underline{9}$7 = 14.0 karats

1.133 Mass water displaced = 18.49 - 16.21 = 2.28 g

Volume mineral = 2.28 g x (1 cm^3/0.9982 g) = 2.2$\underline{8}$4 cm^3

Mass air displaced = 2.284 cm^3 x 1.205 x 10^{-3} g/cm^3 = 2.7$\underline{5}$2 x 10^{-3} g

$$\text{Density} = \frac{18.49 + 2.752 \times 10^{-3}\ g}{2.284\ cm^3} = 8.0\underline{9}6 = 8.10\ g/cm^3$$

1.134 Mass water displaced = 7.35 - 5.40 = 1.95 g

Volume mineral = 1.95 g x (1 cm^3/0.9982 g) = 1.9$\underline{5}$3 cm^3

Mass air displaced = 1.953 cm^3 x 1.205 x 10^{-3} g/cm^3 = 2.3$\underline{5}$3 x 10^{-3} g

$$\text{Density} = \frac{7.35 + 2.353 \times 10^{-3}\ \text{g}}{1.953\ \text{cm}^3} = 3.7\underline{6}4 = 3.76\ \text{g/cm}^3$$

1.135 Mass of ethanol displaced = 15.8 - 10.5 = 5.3 g

Volume object = 5.3 g x (1 cm^3/0.789 g) = 6.$\underline{7}$17 cm^3

$$\text{Density} = \frac{15.8\ \text{g}}{6.717\ \text{cm}^3} = 2.\underline{3}52 = 2.4\ \text{g/cm}^3$$

1.136 Mass of mercury displaced = 255 - 101 = 154 g

Volume metal = 154 g x (1 cm^3/13.6 g) = 11.$\underline{3}$2 cm^3

$$\text{Density} = \frac{255\ \text{g}}{11.32\ \text{cm}^3} = 22.\underline{5}2 = 22.5\ \text{g/cm}^3$$

2. ATOMS, MOLECULES, AND IONS

Solutions to Exercises

Note on significant figures: The final answer to each mathematical solution is given first with one nonsignificant figure (the rightmost significant figure is underlined) and then is rounded to the correct number of figures. Intermediate answers usually also have at least one nonsignificant figure.

2.1 The element whose nucleus has 17 protons has an atomic number of 17, and is therefore chlorine (whose symbol is Cl). The mass number is 17 + 18 = 35. The symbol is: $^{35}_{17}\text{Cl}$.

2.2 Multiply each isotopic mass by its fractional abundance, and then add both abundances:

$$34.96885 \text{ amu} \times 0.75771 = 26.49\underline{6}247$$
$$36.96590 \text{ amu} \times 0.24229 = \underline{\ \ 8.956467}$$
$$\text{Atomic weight of chlorine} = 35.45\underline{2}714 = 35.453 \text{ amu}$$

2.3 a. Se: Group VIA, Period 4; nonmetal
b. Cs: Group IA, Period 6; metal
c. Fe: Group VIIIB, Period 4; metal
d. Cu: Group IB, Period 4; metal
e. Br: Group VIIA, Period 4; nonmetal

2.4 The formula is K_2CrO_4.

2.5 a. CaO: calcium oxide (Group IIA forms only 2+ cations.)
b. $PbCrO_4$: lead(II) chromate (CrO_4^{2-} is the chromate anion [Table 2.5], so this is Pb^{2+}.)

2.6 The formula is $Tl(NO_3)_2$.

2.7 a. Dichlorine hex(a)oxide b. Phosphorus trichloride c. Phosphorus pentachloride

2.8 a. CS_2 b. SO_3

2.9 Perbromate ion: BrO_4^-

2.10 Sodium carbonate decahydrate

2.11 $Na_2S_2O_3{\bullet}5H_2O$

2.12 Balance O first in parts (a) and (b) because it occurs in only one product. Balance S first in part (c) because it appears in only one product. Balance H first in part (d) because it appears in just one reactant, as well as in the product.

a. Write a 2 in front of $POCl_3$ for O; this requires a 2 in front of PCl_3 for final balance:

$$2PCl_3 + O_2 \rightarrow 2POCl_3$$

b. Write a 6 in front of N_2O to balance O; this requires a 6 in front of N_2 for final balance:

$$6N_2O + P_4 \rightarrow 6N_2 + P_4O_6$$

c. Write $2As_2S_3$ and $6SO_2$ to achieve ultimately an even number of oxygens on the right to balance what will always be an even number of oxygens on the left. The $2As_2S_3$ then requires $2As_2O_3$. Finally, to balance (6 + 12) O's on the right, write $9O_2$.

$$2As_2S_3 + 9O_2 \rightarrow 2As_2O_3 + 6SO_2$$

d. Write a 4 in front of H_3PO_4; this requires a 3 in front of $Ca(H_2PO_4)_2$ for 12 H's.

$$4H_3PO_4 + Ca_3(PO_4)_2 \rightarrow 3Ca(H_2PO_4)_2$$

■ Answers to Review Questions

2.1 Atomic theory states that matter is composed of small particles called atoms. There are more than 100 different kinds of atoms, from which an infinite number of different kinds of compounds can be made. A chemical reaction consists of rearranging the combinations of atoms present in certain kinds of matter to form new kinds of matter.

2.2 Apply the law of multiple proportions: The masses of chlorine for a fixed mass of iron are in the ratio of small whole numbers. Divide each amount of chlorine (1.270 g and 1.904 g) by 1.270 (the lower amount); this gives 1.000 g and 1.499 g, respectively. Convert these to whole numbers by multiplying by the appropriate number (2), giving 2.000 g and 2.998 g. The ratio of these amounts of chlorine is essentially 2:3.

2.3 A cathode-ray tube consists of a negative electrode, or cathode, and a positive electrode, or anode, in an evacuated tube. Cathode rays travel from the cathode to the anode when a high voltage is connected to the cathode and anode. Some of the rays pass through the hole in the anode to form a beam, which is then bent toward positively charged electric plates in the tube. This implies that the cathode rays are negatively charged.

2.4 Thomson concluded that a cathode ray consists of a beam of electrons, which are constituents of all matter. Thus the evidence that cathode rays are a part of all matter is based on Thomson's experiments. He found that cathode rays have the same characteristics no matter which gas occupies the tube and no matter what the electrodes are made of. This indicates that the particles making up the cathode rays (electrons) are found in all matter.

2.5 Millikan sprayed oil into a chamber and watched the motion of a single droplet of oil. Often the droplet would have an electric charge, which it picked up from the friction of forming the oil mist. After measuring the mass and the mass-to-charge ratio of a droplet, Millikan could calculate the charge on the droplet. After many experiments, Millikan selected the smallest charge as the charge of one electron.

2.6 The nuclear model of the atom is based on experiments of Geiger, Marsden, and Rutherford. Rutherford stated that most of the mass of an atom is concentrated in a positively charged center called the nucleus, around which negatively charged electrons move. The nucleus, although containing most of the mass, occupies only a very small portion of the space of the atom. Alpha particles mostly pass through this space of an atom, being undeflected by the very lightweight electrons. If an alpha particle happens to strike a nucleus, it is scattered through a wide angle because it is repelled by the positively charged nucleus.

2.7 The plum-pudding model predicts that alpha particles will be deflected through small angles. What is actually observed, however, is that occasionally alpha particles are deflected through a wide angle, perhaps even backward. These results are correctly explained by the nuclear model (see the previous question).

2.8 The atomic nucleus consists of two kinds of particles, protons and neutrons. The mass of each is about the same, the order of 1.67×10^{-27} kg. The neutron is electrically neutral but the proton is positively charged. An electron has a much smaller mass, on the order of 9×10^{-31} kg, and it is negatively charged.

2.9 Protons (hydrogen nuclei) were discovered as products of experiments involving the collision of alpha particles (helium nuclei) with nitrogen atoms. Experiments with other nuclei also produced protons. Neutrons were discovered as the radiation product of collisions of alpha particles with beryllium atoms. The resulting radiation was discovered to consist of particles having a mass approximately equal to that of a proton and having no charge (neutral). Experiments on other elements also showed they contained neutrons.

2.10 Oxygen consists of three different *isotopes*.

2.11 Dalton obtained the relative atomic masses in a compound by measuring the relative masses of elements required to form the compound. From this he deduced the relative *atomic* masses. To do this, he had to know the formula of the compound.

2.12 A mass spectrometer works by bombarding atoms or molecules with a beam of electrons (cathode rays). In a mass spectrometer, gaseous atoms or molecules pass through an inlet tube into the ionization chamber. Inside this chamber, electrons (cathode rays) traveling from a heated filament toward a positively charged plate strike the atoms or molecules. This displaces electrons from the atoms or molecules, producing positively charged ions. The positively charged ions are pulled toward the negative grid of the spectrometer, where some of them pass through slits to form a positive beam of electricity. This beam then travels through the magnetic field in the spectrometer, wherein the positively charged ions are deflected according to their mass-to-charge ratios. From measurements of these beams, one can obtain the masses of the ions and their relative numbers, or abundances.

2.13 The atomic weight of an element is the *average* atomic mass for the naturally occurring element. The atomic weight would be different elsewhere in the universe if the percentages of isotopes in the element were different from those on earth.

2.14 The element is tin.

2.15 Characteristic metal properties: lustrous appearance; good conductor of heat and electricity.

2.16 The formula is C_2H_6.

2.17 A molecular formula indicates how many of certain kinds of atoms are in the molecule. A structural formula also shows how the atoms are bonded together.

2.18 The mixture and water molecules have the same number of atoms, but the atoms are combined in different ways.

2.19 An ionic binary compound: NaCl; a molecular binary compound: H_2O.

2.20 The two hydrogen atoms and one oxygen atom in water form bonds with a definite spatial arrangement. The calcium and chloride ions in calcium chloride form ions and exist in a crystal with a regular repeating arrangement of many ions, not just three ions.

2.21 CuCl: copper(I) chloride; $CuCl_2$: copper(II) chloride.

Advantages of the Stock system: More than two different ions of the same metal can be named with the Stock system. In the former (older) system, a new suffix other than *-ic* and *-ous* must be established and/or memorized.

2.22 In words, the equation means that one N_2 molecule, or 2 N atoms, reacts with three H_2 molecules, or 6 H atoms, to form two NH_3 molecules (a total of 2 N and 6 H atoms). There are the same number of N atoms (2) and the same number of H atoms (6) on both sides.

■ Solutions to Practice Problems

Note on significant figures: The final answer to each problem is given first with one nonsignificant figure (the rightmost significant figure is underlined), and then is rounded to the correct number of significant figures. Intermediate answers usually also have at least one nonsignificant figure. Atomic weights are rounded to two decimal places, except for that of hydrogen.

2.23 a. Neon b. Zinc c. Silver d. Magnesium

2.24 a. Calcium b. Lead c. Mercury d. Tin

2.25 a. K b. S c. Fe d. Mn

2.26 a. Se b. P c. Na d. Au

2.27 The mass of the electron is found by multiplying the two values:

$$1.605 \times 10^{-19}\,\text{C} \times \frac{5.64 \times 10^{-12}\,\text{kg}}{\text{C}} = 9.0\underline{5}2 \times 10^{-31} = 9.05 \times 10^{-31}\,\text{kg}$$

2.28 The mass of the fluorine atom is found by multiplying the two values:

$$1.602 \times 10^{-19}\ \text{C} \times \frac{1.97 \times 10^{-7}\ \text{kg}}{\text{C}} = 3.1\underline{5}59 \times 10^{-26} = 3.16 \times 10^{-26}\ \text{kg}$$

2.29 Isotope of A: atom C; atom with same mass number: atom B.

2.30 Isotope of A: atom D; atom with same mass number: atom B.

2.31 The number of neutrons = mass number - atomic number, so

Cl-35: No. of neutrons = 35 - 17 = 18; No. protons = no. electrons = 17

Cl-37: No. of neutrons = 37 - 17 = 20; No. protons = no. electrons = 17

2.32 The number of neutrons = mass number - atomic number, so

Li-6: No. of neutrons = 6 - 3 = 3; No. protons = no. electrons = 3

Li-7: No. of neutrons = 7 - 3 = 4; No. protons = no. electrons = 3

2.33 The element selenium has a nucleus with 34 protons. The mass number = 34 + 36 = 70. The notation for the nucleus is ${}^{70}_{34}\text{Se}$.

2.34 The element boron has a nucleus with 5 protons. The mass number = 5 + 6 = 11. The notation for the nucleus is ${}^{11}_{5}\text{B}$.

2.35 Divide the mass of N by one-third of the mass of hydrogen to find the relative mass of N:

$$\frac{\text{Atomic mass of N}}{\text{Atomic mass of H}} = \frac{7.933\ \text{g N}}{1/3 \times 1.712\ \text{g H}} = \frac{13.9\underline{0}1\ \text{g N}}{1\ \text{g H}} = \frac{13.90}{1}$$

2.36 Divide the mass of S by one-half of the mass of hydrogen to find the relative mass of S:

$$\frac{\text{Atomic mass of S}}{\text{Atomic mass of H}} = \frac{9.330\ \text{g S}}{1/2 \times 0.587\ \text{g H}} = \frac{31.\underline{7}8\ \text{g S}}{1\ \text{g H}} = \frac{31.8}{1}$$

2.37 B-10: 10.013 x 0.1978 = 1.98$\underline{0}$5
B-11: 11.009 x 0.8022 = $\underline{8.8314}$
Average = 10.81$\underline{1}$9 = 10.812 amu (= atomic weight).

2.38 Element-84: $84.9118 \times 0.7215 = 61.2\underline{6}3$
Element-86: $86.9092 \times 0.2785 = \underline{24.204}$
Average $= 85.4\underline{6}7 = 85.47$ amu (= atomic weight).
Identity: rubidium (Rb).

2.39 Mg-24: $23.985 \times 0.7870 = 18.8\underline{7}6$
Mg-25: $24.986 \times 0.1013 = 2.53\underline{1}1$
Mg-26: $25.983 \times 0.1117 = \underline{2.9023}$
Average $= 24.3\underline{0}9 = 24.31$ amu (= atomic weight).

2.40 Element-27: $27.977 \times 0.9221 = 25.7\underline{9}8$
Element-28: $28.976 \times 0.0470 = 1.3\underline{6}2$
Element-29: $29.974 \times 0.0309 = \underline{0.9262}$
Average $= 28.0\underline{8}6 = 28.09$ amu (= atomic weight).
Identity: silicon (Si).

2.41 a. C: IVA, Period 2; nonmetal
b. Po: VIA, Period 6; metal
c. Cr: VIB, Period 4; metal
d. Mg: IIA, Period 3; metal
e. B: IIIA, Period 2; metalloid

2.42 a. Si: IVA, Period 3; metalloid
b. F: VIIA, Period 2; nonmetal
c. Ca: IIA, Period 4; metal
d. Co: VIIIB, Period 4; metal
e. Xe: VIIIA, Period 5; nonmetal

2.43 a. Tellurium b. Aluminum

2.44 a. Bismuth b. Calcium

2.45 Examples are: a. O (oxygen) b. Na (sodium) c. Fe (iron) d. Ce (cerium)

2.46 Examples are: a. Ag (silver) b. Cl (chlorine) c. Al (aluminum) d. U (uranium)

2.47 The solid sulfur consists of all S_8 molecules, which are four times as heavy as S_2 molecules. Hot sulfur is a mixture of S_2 and S_8 just above the boiling point, but at high temperatures only S_2 molecules are formed. Both hot sulfur and solid sulfur consist of molecules *all* with S-S bonds.

2.48 The solid phosphorus consists of all P_4 molecules, which are twice as heavy as P_2 molecules. Hot phosphorus is a mixture of P_4 and P_2 molecules above the boiling point, but at high temperatures, only P_2 molecules are formed. Both solid phosphorus and phosphorus vapor consist of molecules *all* with P-P bonds.

2.49 $2.05 \times 10^{22}\ N_2O\text{ molec.} \times \dfrac{2\text{ N atoms}}{1\ N_2O\text{ molec.}} = 4.10 \times 10^{22}\text{ N atoms}$

$$1.00\text{ g} \times \frac{2.05 \times 10^{22}\ N_2O\text{ molec.}}{1.50\text{ g}} \times \frac{2\text{ N atoms}}{1\ N_2O\text{ molec.}} = 2.7\underline{3}3 \times 10^{22}$$

$$= 2.73 \times 10^{22}\text{ N atoms}$$

2.50 In 3.34×10^{22} HNO_3 molecules, there are 3.34×10^{22} N atoms.

$$1.50\text{ g} \times \frac{3.34 \times 10^{22}\ HNO_3\text{ molec.}}{3.50\text{ g}} \times \frac{3\text{ O atoms}}{1\ HNO_3\text{ molec.}} = 4.2\underline{9}4 \times 10^{22}$$

$$= 4.29 \times 10^{22}\text{ O atoms}$$

2.51 $3.3 \times 10^{21}\text{ H atoms} \times \dfrac{1\ NH_3\text{ molec}}{3\text{ H atoms}} = 1.1 \times 10^{21}\ NH_3\text{ molec}$

2.52 $4.2 \times 10^{23}\text{ H atoms} \times \dfrac{1\ C_2H_5OH\text{ molec.}}{6\text{ H atoms}} = 7.0 \times 10^{22}\ C_2H_5OH\text{ molec.}$

2.53 a. N_2H_4 b. H_2O_2 c. C_3H_8O d. PCl_3

2.54 a. $C_3H_8O_3$ b. Si_2H_6 c. NH_3O d. SF_4

2.55 $\dfrac{2Fe}{1Fe_2(SO_4)_3} \times \dfrac{1Fe_2(SO_4)_3}{3SO_4^{\ 2-}} \times \dfrac{1SO_4^{\ 2-}}{4\text{ O}} = \dfrac{2Fe}{12\text{ O}} = \dfrac{1Fe}{6\text{ O}}$

The ratio is 2 Fe to 12 O, as written; this simplifies to 1 Fe to 6 O.

2.56 $\dfrac{4H}{1NH_4} \times \dfrac{3NH_4}{1(NH_4)_3PO_4} \times \dfrac{1(NH_4)_3PO_4}{4O} = \dfrac{12H}{4\text{ O}} = \dfrac{3H}{1\text{ O}}$

The ratio is 12 H to 4 O, as written; this simplifies to 3 H to 1 O.

2.57 a. $Fe(CN)_3$ b. K_2SO_4 c. Li_3N d. $SrCl_2$

2.58 a. $CoBr_2$ b. $(NH_4)_2SO_4$ c. Na_3PO_4 d. $Fe(NO_3)_3$

2.59 a. Na_2SO_4: sodium sulfate (Group IA forms only 1+ cations.)
b. CaO: calcium oxide (Group IIA forms only 2+ cations.)
c. CuCl: copper(I) chloride (Group IB forms 1+ and 2+ cations.)
d. Cr_2O_3: chromium(III) oxide (Group VIB forms numerous oxidation states.)

2.60 a. Na_2O: sodium oxide (Group IA forms only 1+ cations.)
b. Mn_2O_3: manganese(III) oxide (Group VIIB forms numerous oxidation states.)
c. NH_4HCO_3: ammonium bicarbonate or ammonium hydrogen carbonate
d. $Cu(NO_3)_2$: copper(II) nitrate (Group IB forms 1+ and 2+ cations.)

2.61 a. Lead(II) dichromate: $PbCr_2O_7$ (Dichromate is in Table 2.5.)
b. Barium hydrogen carbonate: $Ba(HCO_3)_2$ (The HCO_3^- ion is in Table 2.5.)
c. Cesium oxide: Cs_2O (Group 1A ions form 1+ cations.)
d. Iron(II) acetate: $Fe(C_2H_3O_2)_2$ (The acetate ion = 1- [Table 2.5]; for the sum of charges to be 0, two must be used.)

2.62 a. Sodium thiosulfate: $Na_2S_2O_3$ (The $S_2O_3^{2-}$ is in Table 2.5.)
b. Copper(I) oxide: Cu_2O (Each Cu must be 1+ to balance O^{2-}.)
c. Calcium hydrogen carbonate: $Ca(HCO_3)_2$ (The HCO_3^- ion is in Table 2.5.)
d. Tin(II) fluoride: SnF_2 (Two F^- must be used to balance Sn^{2+}.)

2.63 a. Molecular b. Ionic c. Molecular d. Ionic

2.64 a. Ionic b. Molecular c. Molecular d. Ionic

2.65 a. Dinitrogen monoxide
b. Tetraphosphorus dec(a)oxide
c. Arsenic trichloride
d. Dichlorine hept(a)oxide

2.66 a. Dinitrogen difluoride
b. Nitrogen trichloride
c. Dinitrogen pent(a)oxide
d. Tetr(a)arsenic hex(a)oxide

2.67 a. NBr_3 b. XeO_4 c. OF_2 d. Cl_2O_5

2.68 a. ClF_3 b. Cl_2O c. N_2F_4 d. PF_5

2.69 a. Bromic acid: $HBrO_3$ c. Disulfurous acid: $H_2S_2O_5$
b. Hyponitrous acid: $H_2N_2O_2$ d. Arsenic acid: H_3AsO_4

2.70 a. Selenous acid: H_2SeO_3 c. Hypoiodous acid: HIO
b. Bromous acid: $HBrO_2$ d. Diphosphoric acid: $H_4P_2O_7$

2.71 $Na_2SO_4{\bullet}10H_2O$ is sodium sulfate decahydrate.

2.72 $NiSO_4{\bullet}6H_2O$ is nickel(II) sulfate hexahydrate.

2.73 Iron(II) sulfate heptahydrate is $FeSO_4{\bullet}7H_2O$.

2.74 Cobalt(II) chloride hexahydrate is $CoCl_2{\bullet}6H_2O$.

2.75 $$1\ As_4O_6 \times \frac{6\ \text{O atoms}}{1\ As_4O_6} + 6\ H_2O \times \frac{1\ \text{O atom}}{1\ H_2O} = 12\ \text{O atoms}$$

2.76 $$2\ PbO \times \frac{1\ \text{O atom}}{1\ PbO} + 2\ SO_2 \times \frac{2\ \text{O atoms}}{1\ SO_2} = 6\ \text{O atoms}$$

The equation is not balanced

2.77 a. Balance: $Sn + NaOH \rightarrow Na_2SnO_2 + H_2$

If Na is balanced first by writing a 2 in front of NaOH, the entire equation is balanced.

$$Sn + 2NaOH \rightarrow Na_2SnO_2 + H_2$$

b. Balance: $Al + Fe_3O_4 \rightarrow Al_2O_3 + Fe$

First balance O (it appears once on each side) by writing a 3 in front of Fe_3O_4 and a 4 in front of Al_2O_3:

$$Al + 3Fe_3O_4 \rightarrow 4Al_2O_3 + Fe$$

Now balance Al against the 8 Al's on the right and Fe against the 9 Fe's on the left:

$$8Al + 3Fe_3O_4 \rightarrow 4Al_2O_3 + 9Fe$$

c. Balance: $CH_3OH + O_2 \rightarrow CO_2 + H_2O$

First balance H (it appears once on each side) by writing a 2 in front of H_2O:

$$CH_3OH + O_2 \rightarrow CO_2 + 2H_2O$$

To avoid fractional coefficients for O, multiply the equation by 2:

$$2CH_3OH + 2O_2 \rightarrow 2CO_2 + 4H_2O$$

Finally, balance O by changing $2O_2$ to "$3O_2$"; this balances the entire equation:

$$2CH_3OH + 3O_2 \rightarrow 2CO_2 + 4H_2O$$

d. Balance: $P_4O_{10} + H_2O \rightarrow H_3PO_4$

First balance P (it appears once on each side) by writing a 4 in front of H_3PO_4:

$$P_4O_{10} + H_2O \rightarrow 4H_3PO_4$$

Finally, balance H by writing a 6 in front of H_2O; this balances the entire equation:

$$P_4O_{10} + 6H_2O \rightarrow 4H_3PO_4$$

e. Balance: $PCl_5 + H_2O \rightarrow H_3PO_4 + HCl$

First balance Cl (it appears once on each side) by writing a 5 in front of HCl:

$$PCl_5 + H_2O \rightarrow H_3PO_4 + 5HCl$$

Finally, balance H by writing a 4 in front of H_2O; this balances the entire equation:

$$PCl_5 + 4H_2O \rightarrow H_3PO_4 + 5HCl$$

2.78 a. Balance: $Cl_2O_7 + H_2O \rightarrow HClO_4$

Balance Cl (appears only once on each side) by writing a 2 in front of $HClO_4$; this balances the entire equation:

$$Cl_2O_7 + H_2O \rightarrow 2HClO_4$$

b. Balance: $MnO_2 + HCl \rightarrow MnCl_2 + Cl_2 + H_2O$

First balance O (appears only once on each side) by writing a 2 in front of H_2O:

$$MnO_2 + HCl \rightarrow MnCl_2 + Cl_2 + 2H_2O$$

Finally, balance H and Cl by writing a 4 in front of HCl to balance the entire equation:

$$MnO_2 + 4HCl \rightarrow MnCl_2 + Cl_2 + 2H_2O$$

c. Balance: $Na_2S_2O_3 + I_2 \rightarrow NaI + Na_2S_4O_6$

First balance S by writing a 2 in front of $Na_2S_2O_3$:

$$2Na_2S_2O_3 + I_2 \rightarrow NaI + Na_2S_4O_6$$

Finally, balance Na by writing a 2 in front of NaI; this balances the entire equation:

$$2Na_2S_2O_3 + I_2 \rightarrow 2NaI + Na_2S_4O_6$$

d. Balance: $Al_4C_3 + H_2O \rightarrow Al(OH)_3 + CH_4$

First balance Al with a 4 in front of $Al(OH)_3$ and balance C with a 3 in front of CH_4:

$$Al_4C_3 + H_2O \rightarrow 4Al(OH)_3 + 3CH_4$$

Finally, balance H and O with a 12 in front of H_2O; this balances the entire equation:

$$Al_4C_3 + 12H_2O \rightarrow 4Al(OH)_3 + 3CH_4$$

e. Balance: $NO_2 + H_2O \rightarrow HNO_3 + NO$

First balance H with a 2 in front of HNO_3:

$$NO_2 + H_2O \rightarrow 2HNO_3 + NO$$

Finally, balance N with a 3 in front of NO_2; this balances the entire equation:

$$3NO_2 + H_2O \rightarrow 2HNO_3 + NO$$

2.79 a. Balance Cl with a 2 in front of HCl; this balances the entire equation:

$$SbCl_5 + H_2O \rightarrow SbOCl_3 + 2HCl$$

b. Balance O with a 2 in front of MgO:

$$Mg + SiO_2 \rightarrow 2MgO + Si$$

Finally, balance Mg with a 2 in front of Mg; this balances the entire equation:

$$2Mg + SiO_2 \rightarrow 2MgO + Si$$

c. Balance Cl with a 2 in front of NaCl; this balances the entire equation:

$$CaCl_2 + Na_2CO_3 \rightarrow CaCO_3 + 2NaCl$$

d. Tentatively balance C with a 6 in front of CO_2:

$$C_6H_6 + O_2 \rightarrow 6CO_2 + H_2O$$

Next balance H; use 6 (not 3) H_2O's and double the C_6H_6 and CO_2 coefficients to avoid an odd number of O's:

$$2C_6H_6 + O_2 \rightarrow 12CO_2 + 6H_2O$$

Finally, balance O with a 15 in front of O_2 to balance the entire equation:

$$2C_6H_6 + 15O_2 \rightarrow 12CO_2 + 6H_2O$$

e. Balance Al by writing a 2 in front of $Al(OH)_3$:

$$Al_2S_3 + H_2O \rightarrow 2Al(OH)_3 + H_2S$$

Next balance S by writing a 3 in front of H_2S:

$$Al_2S_3 + H_2O \rightarrow 2Al(OH)_3 + 3H_2S$$

Finally, balance O and H by writing a 6 in front of H_2O to balance the entire equation:

$$Al_2S_3 + 6H_2O \rightarrow 2Al(OH)_3 + 3H_2S$$

2.80 a. Balance Cl with a 4 in front of HCl:

$$TiCl_4 + H_2O \rightarrow TiO_2 + 4HCl$$

Finally, balance H and O with a 2 in front of H_2O:

$$TiCl_4 + 2H_2O \rightarrow TiO_2 + 4HCl$$

b. Balance O with a 4 in front of H_2O:

$$Fe_3O_4 + H_2 \rightarrow Fe + 4H_2O$$

Balance Fe with a 3 in front of Fe:

$$Fe_3O_4 + H_2 \rightarrow 3Fe + 4H_2O$$

Finally, balance H with a 4 in front of H_2:

$$Fe_3O_4 + 4H_2 \rightarrow 3Fe + 4H_2O$$

c. Balance O by writing a 2 in front of H_2O:

$$V_2O_5 + H_2 \rightarrow V_2O_3 + 2H_2O$$

Finally, balance H with a 2 in front of H_2:

$$V_2O_5 + 2H_2 \rightarrow V_2O_3 + 2H_2O$$

d. Balance H with a 4 in front of H_2O; this also balances O and the equation:

$$(NH_4)_2Cr_2O_7 \rightarrow Cr_2O_3 + 4H_2O + N_2$$

e. Tentatively balance the C with a 4 in front of CO_2:

$$C_4H_{10} + O_2 \rightarrow 4CO_2 + H_2O$$

Next balance H; use 10 (not 5) H_2O's and double the C_4H_{10} and CO_2 coefficients to avoid an odd number of O's:

$$2C_4H_{10} + O_2 \rightarrow 8CO_2 + 10H_2O$$

Finally, balance O with a 13; this also balances the equation:

$$2C_4H_{10} + 13O_2 \rightarrow 8CO_2 + 10H_2O$$

2.81 Balance: $Ca_3(PO_4)_2(s) + H_2SO_4(aq) \rightarrow CaSO_4(s) + H_3PO_4(aq)$

Balance Ca first with a 3 in front of $CaSO_4$:

$$Ca_3(PO_4)_2 + H_2SO_4 \rightarrow 3CaSO_4 + H_3PO_4$$

Next balance the P with a 2 in front of H_3PO_4:

$$Ca_3(PO_4)_2 + H_2SO_4 \rightarrow 3CaSO_4 + 2H_3PO_4$$

Finally, balance the S with a 3 in front of H_2SO_4; this balances the equation:

$$Ca_3(PO_4)_2(s) + 3H_2SO_4(aq) \rightarrow 3CaSO_4(s) + 2H_3PO_4(aq)$$

2.82 Balance: $K(s) + H_2O(l) \rightarrow KOH(aq) + H_2(g)$

Balance H first with a 2 in front of H_2O and KOH:

$$K + 2H_2O \rightarrow 2KOH + H_2$$

Finally, balance K with a 2 in front of K; this balances the equation:

$$2K(s) + 2H_2O(l) \rightarrow 2KOH(aq) + H_2(g)$$

2.83 Balance: $NH_4Cl(aq) + Ba(OH)_2(aq) \rightarrow NH_3(g) + BaCl_2(aq) + H_2O(l)$

Balance O first with a 2 in front of H_2O:

$$NH_4Cl + Ba(OH)_2 \rightarrow NH_3 + BaCl_2 + 2H_2O$$

Balance H with a 2 in front of NH_4Cl and a 2 in front of NH_3; this balances the equation:

$$2NH_4Cl(aq) + Ba(OH)_2(aq) \xrightarrow{\Delta} 2NH_3(g) + BaCl_2(aq) + 2H_2O(l)$$

2.84 Balance: $PbS(s) + PbSO_4(s) \rightarrow Pb(l) + SO_2(g)$

Balance S first with a 2 in front of SO_2:

$$PbS + PbSO_4 \rightarrow Pb + 2SO_2$$

Balance Pb with a 2 in front of Pb; this balances the equation:

$$PbS(s) + PbSO_4(s) \xrightarrow{\Delta} 2P(l) + 2SO_2(g)$$

■ Solutions to Unclassified Problems

2.85 Calculate the ratio of oxygen for 1 g (fixed amount) of nitrogen in both compounds:

$$\text{A:}\ \frac{2.755\ \text{g O}}{1.206\ \text{g N}} = \frac{2.28\underline{4}4\ \text{g O}}{1\ \text{g N}} \qquad \text{B:}\ \frac{4.714\ \text{g O}}{1.651\ \text{g N}} = \frac{2.85\underline{5}2\ \text{g O}}{1\ \text{g N}}$$

$$\frac{\text{g O in B/1 g N}}{\text{g O in A/1 g N}} = \frac{2.85\underline{5}2\ \text{g O}}{2.28\underline{4}4\ \text{g O}} = \frac{1.24\underline{9}8\ \text{g O}}{1\ \text{g O}}$$

B contains 1.25 times as many O atoms as A does (there are 5 O's in B for every 4 O's in A).

2.86 Calculate the ratio of oxygen for 1 g (fixed amount) of sulfur in both compounds:

$$\text{A:}\ \frac{1.811\ \text{g O}}{1.210\ \text{g S}} = \frac{1.49\underline{6}6\ \text{g O}}{1\ \text{g S}} \qquad \text{B:}\ \frac{1.779\ \text{g O}}{1.783\ \text{g S}} = \frac{0.997\underline{7}5\ \text{g O}}{1\ \text{g S}}$$

$$\frac{\text{g O in A/1 g S}}{\text{g O in B/1 g S}} = \frac{1.49\underline{6}6\ \text{g O}}{0.997\underline{7}5\ \text{g O}} = \frac{1.49\underline{9}9\ \text{g O}}{1\ \text{g O}}$$

A contains 1.50 times as many O atoms as B does (there are 3 O's in A for every 2 O's in B).

2.87 $$\text{A: Fraction P} = \frac{1.156\ \text{g}}{(1.156 + 3.971)\ \text{g}} = 0.225\underline{4}7$$

$$\text{B: Fraction P} = \frac{1.542\ \text{g}}{(1.542 + 5.297)\ \text{g}} = 0.225\underline{4}7$$

A and B are the same compound because the fractions of P (and Cl) are the same.

2.88 $$\text{A: Fraction P} = \frac{2.581\ \text{g}}{(2.581 + 3.332)\ \text{g}} = 0.436\underline{4}9$$

$$\text{B: Fraction P} = \frac{3.718\ \text{g}}{(3.718 + 2.881)\ \text{g}} = 0.563\underline{4}1$$

A and B are not the same compound because the fractions of P (and O) aren't the same. The ratios of O atoms in the compounds for a fixed amount of P are calculated as follows:

(continued)

$$A: \text{ g O per 1 g P} = \frac{3.332 \text{ g O}}{2.581 \text{ g P}} = \frac{1.2909 \text{ g O}}{1 \text{ g P}}$$

$$B: \text{ g O per 1 g P} = \frac{2.881 \text{ g O}}{3.718 \text{ g P}} = \frac{0.77487 \text{ g O}}{1 \text{ g P}}$$

$$\frac{\text{g O in A per 1 g P}}{\text{g O in B per 1 g P}} = \frac{1.2909 \text{ g O}}{0.77487 \text{ g O}} = \frac{1.6659 \text{ g or atoms O in A}}{1 \text{ g or atoms O in B}}$$

A contains 1.66 or 1 2/3 as many O's as B (there are 5 O's in A for every 3 O's in B).

2.89 The difference between -1.12×10^{-18} C and 9.60×10^{-19} C is -1.6×10^{-19} C. If this charge is equivalent to 1 electron, the number of excess electrons on a drop may be found by dividing the negative charge by the charge of 1 electron.

$$\text{Drop 1: } \frac{-3.20 \times 10^{-19} \text{ C}}{-1.6 \times 10^{-19} \text{ C}} = 2.0 \cong 2 \text{ electrons}$$

$$\text{Drop 2: } \frac{-6.40 \times 10^{-19} \text{ C}}{-1.6 \times 10^{-19} \text{ C}} = 4.0 \cong 4 \text{ electrons}$$

$$\text{Drop 3: } \frac{-9.60 \times 10^{-19} \text{ C}}{-1.6 \times 10^{-19} \text{ C}} = 6.0 \cong 6 \text{ electrons}$$

$$\text{Drop 4: } \frac{-1.12 \times 10^{-18} \text{ C}}{-1.6 \times 10^{-19} \text{ C}} = 7.0 \cong 7 \text{ electrons}$$

2.90 The smallest difference in charge for the oil drop is -1.85×10^{-19}; assume this is the fundamental unit of negative charge. Use this to divide into each drop's charge:

$$\text{Drop 1: } \frac{-5.55 \times 10^{-19} \text{ C}}{-1.85 \times 10^{-19} \text{ C}} = 3.0 \cong 3 \text{ charge units}$$

$$\text{Drop 2: } \frac{-9.25 \times 10^{-19} \text{ C}}{-1.85 \times 10^{-19} \text{ C}} = 5.0 \cong 5 \text{ charge units}$$

$$\text{Drop 3: } \frac{-1.11 \times 10^{-18} \text{ C}}{-1.85 \times 10^{-19} \text{ C}} = 6.0 \cong 6 \text{ charge units}$$

$$\text{Drop 4: } \frac{-1.48 \times 10^{-18} \text{ C}}{-1.85 \times 10^{-19} \text{ C}} = 8.0 \cong 8 \text{ charge units}$$

2.91 For the Eu atom to be neutral, the number of electrons must equal the number of protons, so a neutral europium atom has 63 electrons. The +3 charge on the Eu^{3+} indicates there are 3 more protons than electrons, so the number of electrons = 63 - 3 = 60.

2.92 For the Cs atom to be neutral, the number of electrons must equal the number of protons, so a neutral cesium atom has 55 electrons. The +1 charge on the Cs^{+} indicates there is 1 more proton than electrons, so the number of electrons = 55 - 1 = 54.

2.93 The number of protons = mass number - number of neutrons = 69 - 38 = 31. The element with Z = 31 is gallium (Ga).
The ionic charge = number of protons - number of electrons = 31 - 28 = +3.
Symbol: ${}^{69}_{31}Ga^{3+}$.

2.94 The number of protons = mass number - number of neutrons = 119 - 69 = 50. The element with Z = 50 is tin (Sn).
The ionic charge = number of protons - number of electrons = 50 - 48 = +2.
Symbol: ${}^{119}_{50}Sn^{2+}$.

2.95 The sum of the fractional abundances must equal 1. Thus, the abundance of one isotope can be expressed in terms of the other. Let y = the fractional abundance of Ag-107. Then the fractional abundance of Ag-109 = (1 - y). We can write one equation in one unknown:

$$\text{Atomic weight} = 107.87 = 106.91y + 108.90(1 - y)$$

$$107.87 = 108.90 - 1.99y$$

$$y = \frac{108.90 - 107.87}{1.99} = 0.51\underline{7}58 = 0.518$$

The fractional abundance of Ag-107 = 0.518.
The fractional abundance of Ag-109 = 1 - 0.51758 = 0.48$\underline{2}$41 = 0.482.

2.96 As in the previous problem, the sum of the fractional abundances must equal 1. Let y = the fractional abundance of ^{63}Cu. Then the fractional abundance of ^{65}Cu = (1 - y). We write one equation in one unknown:

$$\text{Atomic weight} = 63.546 = 62.9298y + 64.9278(1 - y)$$

$$63.546 = 64.9278 - 1.9980y$$

$$y = \frac{64.9278 - 63.546}{1.9980} = 0.691\underline{5}9$$

The fractional abundance of ^{63}Cu = 0.691$\underline{5}$9 = 0.6916.
The fractional abundance of ^{65}Cu = 1 - 0.69159 = 0.308$\underline{4}$1 = 0.3084.

2.97 a. Bromine, Br b. Oxygen, O c. Niobium, Nb d. Fluorine, F

2.98 a. Oxygen, O b. Mercury, Hg c. Boron, B d. Potassium, K

2.99 a. Chromium(III) ion b. Chromium(II) ion c. Copper(I) ion d. Copper(II) ion

2.100 a. Manganese(II) ion b. Nickel(II) ion c. Cobalt(II) ion d. Cobalt(III) ion

2.101 All possible ionic compounds: Na_2SO_4, NaCl, $NiSO_4$, and $NiCl_2$.

2.102 All possible ionic compounds: CaO, $Ca(NO_3)_2$, Cr_2O_3, and $Cr(NO_3)_3$.

2.103 a. Tin(II) phosphate
b. Ammonium nitrite
c. Magnesium hydroxide
d. Chromium(II) sulfate

2.104 a. Ammonium carbonate
b. Nickel(II) nitrate
c. Barium sulfate
d. Mercury(II) chloride

2.105 a. Hg_2Cl_2 [Mercury(I) exists as the polyatomic Hg_2^{2+} ion (Table 2.5).]
b. CuO
c. $(NH_4)_2Cr_2O_7$
d. ZnS

2.106 a. H_2O_2 b. $Fe(NO_3)_3$ c. $Ni_3(PO_4)_2$ d. SF_4

2.107 a. Arsenic trichloride
b. Selenium dioxide
c. Dinitrogen trioxide
d. Silicon tetrafluoride

2.108 a. Chlorine tetrafluoride
b. Carbon disulfide
c. Nitrogen trifluoride
d. Sulfur hexafluoride

2.109 a. Balance the C and H first:

$$C_2H_6 + O_2 \rightarrow 2CO_2 + 3H_2O$$

Avoid a fractional coefficient for O on the left by doubling all coefficients except O_2's, and then balance the O's:

$$2C_2H_6 + 7O_2 \rightarrow 4CO_2 + 6H_2O$$

b. Balance the P first:

$$P_4O_6 + H_2O \rightarrow 4H_3PO_3$$

Then balance the O (or H), which also gives the H (or O) balance:

$$P_4O_6 + 6H_2O \rightarrow 4H_3PO_3$$

c. Balancing the O first is the simplest approach. (Starting with K and Cl and then O will cause the initial coefficient for $KClO_3$ to be changed in balancing O last.)

$$4KClO_3 \rightarrow KCl + 3KClO_4$$

d. Balance the N first:

$$(NH_4)_2SO_4 + NaOH \rightarrow 2NH_3 + H_2O + Na_2SO_4$$

Then balance the Na, followed by O; this also balances the H:

$$(NH_4)_2SO_4 + 2NaOH \rightarrow 2NH_3 + 2H_2O + Na_2SO_4$$

e. Balance the N first:

$$2NBr_3 + NaOH \rightarrow N_2 + NaBr + HOBr$$

Note that NaOH and HOBr each have 1 O, and that NaOH and NaBr each have 1 Na; thus the coefficients of all three must be equal; from $2NBr_3$ this coefficient must = 6Br/2 = 3:

$$2NBr_3 + 3NaOH \rightarrow N_2 + 3NaBr + 3HOBr$$

2.110 a. Balance the Na first:

$$3NaOH + H_3PO_4 \rightarrow Na_3PO_4 + H_2O$$

Then balance the O; this also balances the H:

$$3NaOH + H_3PO_4 \rightarrow Na_3PO_4 + 3H_2O$$

b. Balance the Cl with a 4 in front of the HCl; then balance the O's with a 2 in front of H_2O:

$$SiCl_4 + 2H_2O \rightarrow SiO_2 + 4HCl$$

c. Balance the O first with an 8 in front of CO; then balance the C with an 8 in front of C:

$$Ca_3(PO_4)_2 + 8C \rightarrow Ca_3P_2 + 8CO$$

d. Balance the O by multiplying O_2 by 3 and doubling both products to give a total of 6 O's on both sides of the equation:

$$H_2S + 3O_2 \rightarrow 2SO_2 + 2H_2O$$

Then balance H and S with a 2 in front of H_2S:

$$2H_2S + 3O_2 \rightarrow 2SO_2 + 2H_2O$$

e. Since $N_2O_5 \rightarrow NO_2 + O_2$ has 2 N's on the left and 1 N on the right, try a tentative N-balancing by writing a 2 in front of NO_2:

$$N_2O_5 \rightarrow 2NO_2 + O_2$$

Now there are 5 O's on the left and 6 O's on the right. Begin to balance the O's with a 6 in front of N_2O_5; this gives

$$6N_2O_5 \rightarrow 2NO_2 + O_2$$

Because changing NO_2 will change the oxygen balance, first balance the N's using $12NO_2$:

$$6N_2O_5 \rightarrow 12NO_2 + O_2$$

Now there are only (24 + 2) O's on the right, so balance the O's by writing $3O_2$:

$$6N_2O_5 \rightarrow 12NO_2 + 3O_2$$

This can be reduced to

$$2N_2O_5 \rightarrow 4NO_2 + O_2$$

2.111 Let x = number of protons. Since the mass number is 62, you get

$$62 = x + 1.21\,x = 2.21\,x$$

Thus, x = $2\underline{8}.054$, or 28. The element is nickel (Ni). Since the ion has a +2 charge, there are 26 electrons.

2.112 Let x = number of protons. Since the mass number is 45, you get

$$45 = x + 1.14\,x = 2.14\,x$$

Thus, x = $2\underline{1}.03$, or 21. The element is scandium (Sc). Since the ion has a +3 charge, there are 18 electrons.

2.113 The average atomic mass would be

Natural carbon:	12.011 x 1/2 =	$6.00\underline{5}500$
Carbon-13:	13.00335 x 1/2 =	6.501675
	Average =	$12.50\underline{7}175$ = 12.507 amu (= atomic mass)

2.114 The average atomic mass would be

Natural chlorine:	35.4527 x 1/2 =	$17.726\underline{3}500$
Chlorine-35:	34.96885 x 1/2 =	17.4844250
	Average =	$35.210\underline{7}750$ = 35.2108 amu (= atomic mass)

■ Cumulative-Skills Problems

(require skills from previous chapters)

2.115 The spheres occupy a diameter of 2 x 1.86 Å = 3.72 Å.

$$\text{The line of Na atoms} = \frac{3.72\text{ Å}}{1\text{ Na atom}} \times 2.619 \times 10^{22}\text{ Na atoms}$$

$$= 9.7\underline{4}27 \times 10^{22}\text{ Å}$$

$$9.7\underline{4}27 \times 10^{22}\text{ Å} \times \frac{1 \times 10^{-10}\text{ m}}{1\text{ Å}} \times \frac{1\text{ mile}}{1.609 \times 10^{3}\text{ m}} = 6.0\underline{5}5 \times 10^{9}$$

$$= 6.06 \times 10^{9}\text{ miles}$$

2.116 The spheres occupy a diameter of 2 x 0.99 Å = 1.98 Å.

$$\text{The line of Cl atoms} = \frac{1.98\ \text{Å}}{1\ \text{Cl atom}} \times 1.699 \times 10^{22}\ \text{Cl atoms}$$

$$= 3.3\underline{6}4 \times 10^{22}\ \text{Å}$$

$$3.3\underline{6}4 \times 10^{22}\ \text{Å} \times \frac{1 \times 10^{-10}\ \text{m}}{1\ \text{Å}} \times \frac{1\ \text{mile}}{1.609 \times 10^{3}\ \text{m}} = 2.0\underline{9}07 \times 10^{9}$$

$$= 2.09 \times 10^{9}\ \text{miles}$$

2.117 $NiSO_4 \bullet 7H_2O(s) \rightarrow NiSO_4 \bullet 6H_2O(s) + H_2O(g)$

[8.753] = [8.192 g + (8.753 - 8.192 = 0.561 g)]

The 8.192 g of $NiSO_4 \bullet 6H_2O$ must contain 6 x 0.561 = 3.366 g H_2O.

Mass of anhydrous $NiSO_4$ = 8.192 g $NiSO_4 \bullet 6H_2O$ - 3.366 g $6H_2O$ = 4.826 g

2.118 $CoSO_4 \bullet 7H_2O(s) \rightarrow CoSO_4 \bullet H_2O(s) + 6H_2O(g)$

[3.548 g] = [2.184 g + (3.548 - 2.184 = 1.364 g)]

Mass of one H_2O per 3.548 g of $CoSO_4 \bullet 7H_2O$ = 1.364 g ÷ 6 = 0.22733 g

Mass of anhydrous $CoSO_4$ = 2.184 g $CoSO_4 \bullet H_2O$ - 0.22733 g H_2O

$$= 1.95\underline{6}6 = 1.957\ \text{g}$$

2.119 $$\text{Mass of O} = 0.6015\ \text{L} \times \frac{1.330\ \text{g O}}{1\ \text{L}} = 0.799\underline{9}95\ \text{g O}$$

$$15.9994\ \text{amu O} \times \frac{3.177\ \text{g X}}{0.799995\ \text{g O}} = 63.5\underline{3}8\ \text{amu X} = 63.54\ \text{amu}$$

$$= \text{atomic wt of X}$$

X is copper, which has an atomic number of 29.

2.120 $$\text{Mass of Cl} = 0.4810\ \text{L} \times \frac{2.948\ \text{g Cl}}{1\ \text{L}} = 1.41\underline{7}99\ \text{g Cl}$$

$$35.453\ \text{amuCl} \times \frac{4.315\ \text{g X}}{1.41799\ \text{g Cl}} = 107.\underline{8}8\ \text{amu X} = \text{atomic wt of X}$$

X is silver, which has an atomic number of 47.

3. CHEMICAL REACTIONS: AN INTRODUCTION

■ Solutions to Exercises

Note on significant figures: The final answer to each mathematical solution is given first with one nonsignificant figure (the rightmost significant figure is underlined) and then is rounded to the correct number of figures. Intermediate answers usually also have at least one nonsignificant figure.

3.1 a. Ionic equation:

$$2H^+(aq) + 2NO_3^-(aq) + Mg(OH)_2(s) \rightarrow 2H_2O(l) + Mg^{2+}(aq) + 2NO_3^-(aq)$$

Net ionic equation:

$$2H^+(aq) + Mg(OH)_2(s) \rightarrow 2H_2O(l) + Mg^{2+}(aq)$$

b. Ionic equation:

$$Pb^{2+}(aq) + 2NO_3^-(aq) + 2Na^+(aq) + SO_4^{2-}(aq) \rightarrow$$
$$PbSO_4(s) + 2Na^+(aq) + 2NO_3^-(aq)$$

Net ionic equation:

$$Pb^{2+}(aq) + SO_4^{2-}(aq) \rightarrow PbSO_4(s)$$

3.2 Molecular equation showing the formation of insoluble PbI_2:

$$Pb(C_2H_3O_2)_2(aq) + 2NaI(aq) \rightarrow PbI_2(s) + 2NaC_2H_3O_2(aq)$$

Ionic equation:

$$Pb^{2+}(aq) + 2I^-(aq) \rightarrow PbI_2(s)$$

3.3 a. Weak acid b. Weak acid c. Strong acid d. Strong base

3.4 Molecular equation showing the formation of the water molecule:

$$HCN(aq) + LiOH(aq) \rightarrow H_2O(l) + LiCN(aq)$$

Net ionic equation (HCN is a weak acid, but Li^+ is omitted because LiOH is a strong base):

$$HCN(aq) + OH^-(aq) \rightarrow H_2O(l) + CN^-(aq)$$

3.5 Molecular equations:

$$H_2SO_4(aq) + KOH(aq) \rightarrow KHSO_4(aq) + H_2O(l)$$

$$KHSO_4(aq) + KOH(aq) \rightarrow K_2SO_4(aq) + H_2O(l)$$

Net ionic equations:

$$H^+(aq) + OH^-(aq) \rightarrow H_2O(l)$$

$$HSO_4^-(aq) + OH^-(aq) \rightarrow H_2O(l) + SO_4^{2-}(aq)$$

3.6 Molecular equation showing insoluble $CaCO_3$ and formation of CO_2 gas and molecular water:

$$CaCO_3(s) + 2HNO_3(aq) \rightarrow CO_2(g) + H_2O(l) + Ca(NO_3)_2(aq)$$

Net ionic equation:

$$CaCO_3(s) + 2H^+(aq) \rightarrow CO_2(g) + H_2O(l) + Ca^{2+}(aq)$$

3.7 a. The oxidation number of K, x_K, = +1 and that of O, x_O, = -2. Letting x_{Cr} equal the oxidation number of the chromium atom, write an equation and solve for x_{Cr}:

$$2x_K + x_{Cr} + 7x_O = 0$$

$$+2 + 2x_{Cr} - 14 = 0$$

$$x_{Cr} = +6$$

b. The oxidation number of O, x_O, = -2. Letting x_{Mn} equal the oxidation number of the manganese atom, write an equation and solve for x_{Mn}:

$$x_{Mn} + 4x_O = -1$$

$$x_{Mn} - 8 = -1$$

$$x_{Mn} = +7$$

3.8 Assign oxidation numbers:

$$\overset{0}{I_2} + \overset{+5}{NO_3^-} \rightarrow \overset{+5}{IO_3^-} + \overset{+4}{NO_2}$$

Separate into half-reactions:

$$I_2 \rightarrow IO_3^- \;;\; NO_3^- \rightarrow NO_2$$

Balance each half-reaction by adding electrons:

$$6\,H_2O + I_2 \rightarrow 2\,IO_3^- + 12\,H^+ + 10\,e^-$$

$$e^- + 2\,H^+ + NO_3^- \rightarrow NO_2 + H_2O$$

Multiply by factors so that when added, the electrons cancel.

$$6\,H_2O + I_2 \rightarrow 2\,IO_3^- + 12\,H^+ + 10\,e^-$$

$$10\,e^- + 20\,H^+ + 10\,NO_3^- \rightarrow 10\,NO_2 + 10\,H_2O$$

Add and simplify to get

$$I_2\,(s) + 10\,NO_3^-\,(aq) + 8\,H^+\,(aq) \rightarrow 2\,IO_3^-\,(aq) + 10\,NO_2\,(g) + 4\,H_2O\,(l)$$

3.9 The skeleton equation is $H_2O_2 + ClO_2 \rightarrow ClO_2^- + O_2$. Splitting into half-reactions gives

$$ClO_2 \rightarrow ClO_2^- \;; \qquad H_2O_2 \rightarrow O_2.$$

The first half-reaction is balanced by adding $1e^-$ to the left to balance the charges.

$$ClO_2 + e^- \rightarrow ClO_2^-$$

The second half-reaction is balanced with H atoms.

$$H_2O_2 \rightarrow O_2 + 2\,H^+$$

Finally, because the right side has a net charge of +2 and the left side a charge of 0, the half-reaction is balanced by adding $2e^-$ to the right to balance the charges.

$$H_2O_2 \rightarrow O_2 + 2\,H^+ + 2\,e^-$$

To combine, multiply the first half-reaction by 2 so that the e's cancel on adding the half-reactions.

$$2\,ClO_2 + 2\,e^- \rightarrow 2\,ClO_2^-$$

$$H_2O_2 \rightarrow O_2 + 2\,H^+ + 2\,e^-$$

$$H_2O_2 + 2\,ClO_2 \rightarrow 2\,ClO_2^- + O_2 + 2\,H^+$$

Add 2 OH^- to both sides of the equation, and replace the 2 H^+ and 2 OH^- on the right side with 2 H_2O. The balanced equation in basic solution is

$$H_2O_2 + 2\,ClO_2 + 2\,OH^- \rightarrow 2\,ClO_2^- + O_2 + 2\,H_2O$$

■ Answers to Review Questions

3.1 An electrolyte solution exhibits electrical conductivity because of its ionization to positively and negatively charged ions. Such a solution conducts electricity by means of the movement of the positive and negative ions formed by ionization. In a sodium chloride solution, for example, the negative chloride ions carry a negative charge to the positive electrode and the positive sodium ions carry a positive charge to the negative electrode. This movement of ions, or electric charge, is responsible for the electric current that flows in the solution.

3.2 Some electrolyte solutions are strongly conducting and others are weakly conducting because the former are almost completely ionized whereas the others are weakly ionized. Thus, the former solutions will have many more ions to conduct electricity than will the latter solutions, if both are present at the same concentrations.

3.3 A strong electrolyte is an electrolyte that exists in solution almost entirely as ions. An example is HCl. When HCl dissolves in water, it reacts almost completely to give H_3O^+ (or H^+) and Cl^- ions; almost no unreacted HCl molecules remain. A weak electrolyte is an electrolyte that dissolves in water to give an equilibrium between a relatively low level of ions and a relatively high level of a molecular substance. An example is NH_3. When NH_3 dissolves in water, it reacts very little with the water, so the level of NH_3 is relatively high and the level of the NH_4^+ and OH^- ions is relatively low.

3.4 The advantage of using a molecular equation to represent an ionic equation is that it states explicitly what chemical species have been added and what chemical species are obtained as products. It also makes stoichiometric calculations easy to perform. The disadvantages are (1) the molecular equation does not represent the fact that the reaction actually involves ions, and (2) the molecular equation does not indicate which species exist as ions and which exist as molecular solids or molecular gases.

3.5 A spectator ion is an ion that does not take part in the reaction; it remains unchanged after the reaction is over. In the following ionic reaction, the Na^+ and Cl^- are spectator ions:

$$Na^+(aq) + OH^-(aq) + H^+(aq) + Cl^-(aq) \rightarrow Na^+(aq) + Cl^-(aq) + H_2O(l)$$

3.6 A net ionic equation is an ionic equation from which spectator ions have been canceled. The value of such an equation is that it shows the essential reaction that occurs. An example is the ionic equation representing the reaction of $CaCl_2(aq) + K_2CO_3(aq) \rightarrow CaCO_3(s) + 2\,KCl(aq)$:

$$Ca^{2+}(aq) + CO_3^{2-}(aq) \rightarrow CaCO_3(s)$$

3.7 The major types of chemical reactions are combination reactions, decomposition reactions, displacement (single replacement) reactions, precipitation reactions, and combustion reactions.

A combination reaction is a reaction in which two substances combine to form a third substance.

An example is: $2\,Na(s) + Cl_2(g) \rightarrow 2\,NaCl(s)$

A decomposition reaction is a reaction in which a single compound reacts to give two or more substances.

An example is: $2\,HgO(s) \rightarrow 2\,Hg(l) + O_2(g)$

(continued)

A displacement reaction is a reaction in which an element reacts with a compound displacing an element from it.

An example is: $Cu (s) + 2 AgNO_3 (aq) \rightarrow 2 Ag (s) + Cu(NO_3)_2 (aq)$

A precipitation reaction is a reaction that appears to involve the exchange of parts of the reaction.

An example is: $2 KCl (aq) + Pb(NO_3)_2 (aq) \rightarrow 2 KNO_3 (aq) + PbI_2 (s)$

A combustion reaction is a reaction of a substance with oxygen, usually with rapid release of heat to produce a flame.

An example is: $CH_4 (g) + 2 O_2 (g) \rightarrow CO_2 (g) + 2 H_2O (l)$

3.8 To prepare AgCl and $NaNO_3$, first make solutions of $AgNO_3$ and NaCl by weighing equivalent amounts of both solid compounds. Then mix the two solutions together, forming a precipitate of silver chloride and a solution of soluble sodium nitrate. Filter off the silver chloride and wash it with water to remove the sodium nitrate solution. Then allow it to dry to obtain pure crystalline silver chloride. Finally, take the filtrate containing the sodium nitrate and evaporate it, leaving pure crystalline sodium nitrate.

3.9 An example of a neutralization reaction is

$$HBr(acid) + KOH(base) \rightarrow KBr(salt) + H_2O(l)$$

3.10 An example of polyprotic acid is carbonic acid, H_2CO_3. The successive neutralization is given by the following molecular equations:

$$H_2CO_3(aq) + NaOH(aq) \rightarrow NaHCO_3(aq) + H_2O(l)$$

$$NaHCO_3(aq) + NaOH(aq) \rightarrow Na_2CO_3(aq) + H_2O(l)$$

3.11 Magnesite must evolve carbon dioxide, the only odorless gas in Table 3.4. Thus, magnesite is probably a metal (M) carbonate salt. It is possible only to write a net ionic equation from the observations, using $MCO_3(s)$ to represent the mineral (which may be insoluble in water):

$$MCO_3(s) + 2H^+(aq) \rightarrow CO_2(g) + H_2O(l) + M^{2+}(aq)$$

(In fact, magnesite is $MgCO_3$, so it would be insoluble in water.)

3.12 In oxidation, at least one atom in a substance loses electrons, so that its oxidation number increases. In reduction, at least one atom in a substance gains electrons, so that its oxidation number decreases.

3.13 Oxidation and reduction must occur together because the electrons lost from oxidation must be absorbed by the substance reduced.

3.14 A displacement reaction is an oxidation-reduction reaction in which a free element appears to displace another element in a compound.

$Fe(s)$ +	$2AgNO_3(aq) \rightarrow 2Ag(s) + Fe(NO_3)_2(aq)$	
reducing agent	oxidizing agent	

■ Solutions to Practice Problems

Note on significant figures: The final answer to each problem is given first with one nonsignificant figure (the rightmost significant figure is underlined) and then is rounded to the correct number of significant figures. Intermediate answers usually also have at least one nonsignificant figure. Atomic weights are rounded to two decimal places, except for that of hydrogen.

3.15 a. $H^+(aq) + OH^-(aq) \rightarrow H_2O(l)$

b. $Ag^+(aq) + Br^-(aq) \rightarrow AgBr(s)$

c. $S^{2-}(aq) + 2H^+(aq) \rightarrow H_2S(g)$

d. $OH^-(aq) + NH_4^+(aq) \rightarrow NH_3(g) + H_2O(l)$

3.16 a. $H^+(aq) + NH_3(aq) \rightarrow NH_4^+(aq)$

b. $H^+(aq) + OH^-(aq) \rightarrow H_2O(l)$

c. $Pb^{2+}(aq) + 2Br^-(aq) \rightarrow PbBr_2(s)$

d. $MgCO_3(s) + 2H^+(aq) \rightarrow Mg^{2+}(aq) + H_2O(l) + CO_2(g)$

3.17 Molecular equation: $Pb(NO_3)_2(aq) + Na_2SO_4(aq) \rightarrow PbSO_4(s) + 2NaNO_3(aq)$

Net ionic equation: $Pb^{2+}(aq) + SO_4^{2-}(aq) \rightarrow PbSO_4(s)$

3.18 Molecular equation: $K_2CO_3(aq) + 2HBr(aq) \rightarrow CO_2(g) + H_2O(l) + 2KBr(aq)$

Ionic equation: $CO_3^{2-}(aq) + 2H^+(aq) \rightarrow CO_2(g) + H_2O(l)$

3.19 a. AgBr is insoluble (silver halides are insoluble).

b. $Pb(NO_3)_2$ is soluble (all nitrates are soluble).

c. $SrSO_4$ is insoluble (in Group IIA, the sulfates of Ba, Sr, and Ca are insoluble).

d. Na_2CO_3 is soluble (all sodium salts are soluble).

3.20 a. $(NH_4)_2SO_4$ is soluble (all ammonium salts are soluble).

b. $Ca(NO_3)_2$ is soluble (all nitrates are soluble).

c. $BaCO_3$ is insoluble (carbonates are insoluble except for Group IA).

d. $PbSO_4$ is insoluble (lead sulfate is insoluble).

3.21 a. $FeSO_4(aq) + NaCl(aq) \rightarrow NR$

b. $Na_2CO_3(aq) + MgBr_2(aq) \rightarrow MgCO_3(s) + 2NaBr(aq)$

$CO_3^{2-}(aq) + Mg^{2+}(aq) \rightarrow MgCO_3(s)$

c. $MgSO_4(aq) + 2NaOH(aq) \rightarrow Mg(OH)_2(s) + Na_2SO_4(aq)$

$Mg^{2+}(aq) + 2OH^-(aq) \rightarrow Mg(OH)_2(s)$

d. $NiCl_2(aq) + NaBr(aq) \rightarrow NR$

3.22 a. $AgNO_3(aq) + NaI(aq) \rightarrow AgI(s) + NaNO_3(aq)$

$Ag^+(aq) + I^-(aq) \rightarrow AgI(s)$

b. $Ba(NO_3)_2(aq) + K_2SO_4(aq) \rightarrow BaSO_4(s) + 2KNO_3(aq)$

$Ba^{2+}(aq) + SO_4^{2-}(aq) \rightarrow BaSO_4(s)$

c. $Mg(NO_3)_2(aq) + K_2SO_4(aq) \rightarrow NR$

d. $CaCl_2(aq) + Al(NO_3)_3(aq) \rightarrow NR$

3.23 a. $Ba(NO_3)_2(aq) + LiSO_4(aq) \rightarrow BaSO_4(s) + 2LiNO_3(aq)$

$Ba^{2+}(aq) + SO_4^{2-}(aq) \rightarrow BaSO_4(s)$

b. $Ca(NO_3)_2(aq) + NaBr(aq) \rightarrow NR$

c. $Al_2(SO_4)_3(aq) + 6NaOH(aq) \rightarrow 2Al(OH)_3(s) + 3Na_2SO_4(aq)$

$Al^{3+}(aq) + 3OH^-(aq) \rightarrow Al(OH)_3(s)$

d. $3CaBr_2(aq) + 2Na_3PO_4(aq) \rightarrow Ca_3(PO_4)_2(s) + 6NaBr(aq)$

$3Ca^{2+}(aq) + 2PO_4^{3-}(aq) \rightarrow Ca_3(PO_4)_2(s)$

3.24 a. $ZnCl_2(aq) + Na_2S(aq) \rightarrow ZnS(s) + 2NaCl(aq)$

$Zn^{2+}(aq) + S^{2-}(aq) \rightarrow ZnS(s)$

b. $CaCl_2(aq) + Na_2S(aq) \rightarrow NR$

c. $MgSO_4(aq) + KI(aq) \rightarrow NR$

d. $MgSO_4(aq) + K_2CO_3(aq) \rightarrow MgCO_3(s) + K_2SO_4(aq)$

$Mg^{2+}(aq) + CO_3^{2-}(aq) \rightarrow MgCO_3(s)$

3.25 a. Weak acid b. Strong base c. Strong acid d. Weak acid

3.26 a. Weak acid b. Weak acid c. Strong base d. Strong acid

3.27 a. $NaOH(aq) + HNO_3(aq) \rightarrow H_2O(l) + NaNO_3(aq)$

$H^+(aq) + OH^-(aq) \rightarrow H_2O(l)$

b. $2HCl(aq) + Ba(OH)_2(aq) \rightarrow 2H_2O(l) + BaCl_2(aq)$

$H^+(aq) + OH^-(aq) \rightarrow H_2O(l)$

c. $2HC_2H_3O_2(aq) + Ca(OH)_2(aq) \rightarrow 2H_2O(l) + Ca(C_2H_3O_2)_2(aq)$

$HC_2H_3O_2(aq) + OH^-(aq) \rightarrow H_2O(l) + C_2H_3O_2^-(aq)$

d. $NH_3(aq) + HNO_3(aq) \rightarrow NH_4NO_3(aq)$

$NH_3(aq) + H^+(aq) \rightarrow NH_4^+(aq)$

3.28 a. $Al(OH)_3(s) + 3HCl(aq) \rightarrow AlCl_3(aq) + 3H_2O(l)$

$Al(OH)_3(s) + 3H^+(aq) \rightarrow Al^{3+}(aq) + 3H_2O(l)$

b. $2HBr(aq) + Sr(OH)_2(s) \rightarrow 2H_2O(l) + SrBr_2(aq)$

$2H^+(aq) + Sr(OH)_2(s) \rightarrow 2H_2O(l) + Sr^{2+}(aq)$

c. $Ba(OH)_2(aq) + 2HC_2H_3O_2(aq) \rightarrow Ba(C_2H_3O_2)_2(aq) + 2H_2O(l)$

$OH^-(aq) + HC_2H_3O_2(aq) \rightarrow C_2H_3O_2^-(aq) + H_2O(l)$

d. $HNO_3(aq) + KOH(aq) \rightarrow H_2O(l) + KNO_3(aq)$

$H^+(aq) + OH^-(aq) \rightarrow H_2O(l)$

3.29 a. $2HBr(aq) + Ca(OH)_2(aq) \rightarrow 2H_2O(l) + CaBr_2(aq)$

$H^+(aq) + OH^-(aq) \rightarrow H_2O(l)$

b. $3HNO_3(aq) + Al(OH)_3(s) \rightarrow 3H_2O(l) + Al(NO_3)_3(aq)$

$3H^+(aq) + Al(OH)_3(s) \rightarrow 3H_2O(l) + Al^{3+}(aq)$

c. $2HCN(aq) + Ca(OH)_2(aq) \rightarrow 2H_2O(l) + Ca(CN)_2(aq)$

$HCN(aq) + OH^-(aq) \rightarrow H_2O(l) + CN^-(aq)$

d. $HCN(aq) + LiOH(aq) \rightarrow H_2O(l) + LiCN(aq)$

$HCN(aq) + OH^-(aq) \rightarrow H_2O(l) + CN^-(aq)$

3.30 a. $HClO_4(aq) + LiOH(aq) \rightarrow H_2O(l) + LiClO_4(aq)$

$H^+(aq) + OH^-(aq) \rightarrow H_2O(l)$

b. $2HNO_2(aq) + Ba(OH)_2(aq) \rightarrow 2H_2O(l) + Ba(NO_2)_2(aq)$

$HNO_2(aq) + OH^-(aq) \rightarrow H_2O(l) + NO_2^-(aq)$

c. $HNO_2(aq) + NaOH(aq) \rightarrow H_2O(l) + NaNO_2(aq)$

$HNO_2(aq) + OH^-(aq) \rightarrow H_2O(l) + NO_2^-(aq)$

d. $2HCN(aq) + Sr(OH)_2(aq) \rightarrow 2H_2O(l) + Sr(CN)_2(aq)$

$HCN(aq) + OH^-(aq) \rightarrow H_2O(l) + CN^-(aq)$

3.31 a. $2KOH(aq) + H_3PO_4(aq) \rightarrow K_2HPO_4(aq) + 2H_2O(l)$

$2OH^-(aq) + H_3PO_4(aq) \rightarrow HPO_4^{2-}(aq) + 2H_2O(l)$

b. $3H_2SO_4(aq) + 2Al(OH)_3(s) \rightarrow 6H_2O(l) + Al_2(SO_4)_3(aq)$

$3H^+(aq) + Al(OH)_3(s) \rightarrow 3H_2O(l) + Al^{3+}(aq)$

c. $2HC_2H_3O_2(aq) + Ca(OH)_2(aq) \rightarrow 2H_2O(l) + Ca(C_2H_3O_2)_2(aq)$

$HC_2H_3O_2(aq) + OH^-(aq) \rightarrow H_2O(l) + C_2H_3O_2^-(aq)$

d. $H_2SO_3(aq) + NaOH(aq) \rightarrow H_2O(l) + NaHSO_3(aq)$

$H_2SO_3(aq) + OH^-(aq) \rightarrow HSO_3^-(aq) + H_2O(l)$

3.32 a. $Ca(OH)_2(aq) + 2H_2SO_4(aq) \rightarrow 2H_2O(l) + Ca(HSO_4)_2(aq)$

$OH^-(aq) + 2H^+ + SO_4^-(aq) \rightarrow H_2O(l) + HSO_4^-(aq)$ (ionizes to $H^+ + SO_4^{2-}$)

b. $2H_3PO_4(aq) + Ca(OH)_2(aq) \rightarrow H_2O(l) + Ca(H_2PO_4)_2(aq)$

$H_3PO_4(aq) + OH^-(aq) \rightarrow H_2O(l) + H_2PO_4^-(aq)$

c. $NaOH(aq) + H_2SO_4(aq) \rightarrow NaHSO_4(aq) + H_2O(l)$

$OH^-(aq) + H^+(aq) \rightarrow H_2O(l)$

d. $Sr(OH)_2(aq) + 2H_2CO_3(aq) \rightarrow Sr(HCO_3)_2(aq) + 2H_2O(l)$

$OH^-(aq) + H_2CO_3(aq) \rightarrow HCO_3^-(aq) + H_2O(l)$

3.33 Molecular equations: $2H_2SO_3(aq) + Ca(OH)_2(aq) \rightarrow 2H_2O(l) + Ca(HSO_3)_2(aq)$

$Ca(HSO_3)_2(aq) + Ca(OH)_2(aq) \rightarrow 2H_2O(l) + 2CaSO_3(s)$

Ionic equations: $H_2SO_3(aq) + OH^-(aq) \rightarrow H_2O(l) + HSO_3^-(aq)$

$Ca^{2+}(aq) + HSO_3^-(aq) + OH^-(aq) \rightarrow CaSO_3(s) + H_2O(l)$

3.34 Molecular equations: $2H_3PO_4(aq) + Ca(OH)_2(aq) \rightarrow 2H_2O(l) + Ca(H_2PO_4)_2(aq)$

$Ca(H_2PO_4)_2(aq) + Ca(OH)_2(aq) \rightarrow 2H_2O(l) + 2CaHPO_4(aq)$

$2CaHPO_4(aq) + Ca(OH)_2(aq) \rightarrow 2H_2O(l) + Ca_3(PO_4)_2(s)$

Ionic equations: $H_3PO_4(aq) + OH^-(aq) \rightarrow H_2O(l) + H_2PO_4^-(aq)$

$H_2PO_4^-(aq) + OH^-(aq) \rightarrow H_2O(l) + HPO_4^{2-}(aq)$

$2HPO_4^{2-}(aq) + 2OH^-(aq) + 3Ca^{2+}(aq) \rightarrow 2H_2O(l) + Ca_3(PO_4)_2(s)$

3.35 a. Molecular equation: $CaS(aq) + 2HBr(aq) \rightarrow CaBr_2(aq) + H_2S(g)$

Ionic equation: $S^{2-}(aq) + 2H^+(aq) \rightarrow H_2S(g)$

b. Molecular equation: $MgCO_3(s) + 2HNO_3(aq) \rightarrow$

$Mg(NO_3)_2(aq) + CO_2(g) + H_2O(l)$

Ionic equation: $MgCO_3(s) + 2H^+(aq) \rightarrow Mg^{2+}(aq) + CO_2(g) + H_2O(l)$

c. Molecular equation: $K_2SO_3(aq) + H_2SO_4(aq) \rightarrow K_2SO_4(aq) + SO_2(g) + H_2O(l)$

Ionic equation: $SO_3^{2-}(aq) + 2H^+(aq) \rightarrow SO_2(g) + H_2O(l)$

3.36 a. Molecular equation: $BaCO_3(s) + 2HNO_3(aq) \rightarrow CO_2(g) + H_2O(l) + Ba(NO_3)_2(aq)$

Ionic equation: $BaCO_3(s) + 2H^+(aq) \rightarrow CO_2(g) + H_2O(l) + Ba^{2+}(aq)$

b. Molecular equation: $K_2S(aq) + 2HCl(aq) \rightarrow H_2S(g) + 2KCl(aq)$

Ionic equation: $S^{2-}(aq) + 2H^+(aq) \rightarrow H_2S(g)$

c. Molecular equation: $CaSO_3(s) + 2HI(aq) \rightarrow SO_2(g) + H_2O(l) + CaI_2(aq)$

Ionic equation: $CaSO_3(s) + 2H^+(aq) \rightarrow SO_2(g) + H_2O(l) + Ca^{2+}(aq)$

3.37 Molecular equation: $FeS(s) + 2HCl(aq) \rightarrow H_2S(g) + FeCl_2(aq)$

Ionic equation: $FeS(s) + 2H^+(aq) \rightarrow H_2S(g) + Fe^{2+}(aq)$

3.38 Molecular equation: $BaCO_3(s) + 2HBr(aq) \rightarrow BaBr_2(aq) + CO_2(g) + H_2O(l)$

Ionic equation: $BaCO_3(s) + 2H^+(aq) \rightarrow Ba^{2+}(aq) + CO_2(g) + H_2O(l)$

3.39 a. Because all 3 O's = a total of -6, both Ga's = +6; thus, the oxidation number of Ga = +3.

b. Because both O's = a total of -4, the oxidation number of Pb = +4.

c. Because the 4 O's = a total of -8 and K = +1, the oxidation number of Br = +7.

d. Because the 4 O's = a total of -8 and the 2 K's = +2, the oxidation number of Mn = +6.

3.40 a. Because the 3 O's = a total of -6, the oxidation number of Cr = +6.

b. Because the 2 Cl's = a total of -2, both Hg's = +2; thus, the oxidation number of Hg = +1.

c. Because the 3 O's = a total of -6 and the 3 H's = a total of +3, the oxidation number of Ga = +3.

d. Because the 4 O's = a total of -8 and the 3 Na's = a total of +3, the oxidation number of P = +5.

3.41 a. Because the charge of -1 = [x_N + 2 (from 2 H's)], x_N must equal -3.

b. Because the charge of -1 = [x_I - 6 (from 3 O's)], x_I must equal +5.

c. Because the charge of -1 = [x_{Al} - 8 (4 O's) + 4 (4 H's)], x_{Al} must equal +3.

d. Because the charge of -1 = [x_P - 8 (4 O's) + 2 (2 H's)], x_P must equal +5.

3.42 a. Because the charge of -1 = [x_N - 6 (from 3 O's)], x_N must equal +5.

b. Because the charge of -2 = [x_{Cr} - 8 (from 4 O's)], x_{Cr} must equal +6.

c. Because the charge of -2 = [x_{Zn} - 8 (4 O's) + 4 (4 H's)], x_{Zn} must equal +2.

d. Because the charge of -1 = [x_{As} - 6 (3 O's) + 2 (2 H's)], x_{As} must equal +3.

3.43 a. From the list of common polyatomic anions in Table 2.5, the formula of the ClO_3 anion must be ClO_3^-. Thus, the oxidation state of Mn is Mn^{2+} (see also Table 3.5). For the oxidation state of Cl in ClO_3^-, the charge of -1 must equal x_{Cl} - 6, so x_{Cl} must equal +5.

b. From the list of common polyatomic anions in Table 2.5, the formula of the CrO_4 anion must be CrO_4^{2-}. Thus, the oxidation state of Fe is Fe^{3+}. For the oxidation state of Cr in CrO_4^{2-}, the charge of -2 must equal x_{Cr} - 8, so x_{Cr} must equal +6.

c. From the list of common polyatomic anions in Table 2.5, the formula of the Cr_2O_7 anion must be $Cr_2O_7^{2-}$. Thus, the oxidation state of Hg is Hg^{2+}. For the oxidation state of Cr in $Cr_2O_7^{2-}$, the charge of -2 must equal $2x_{Cr}$ - 14, so x_{Cr} must equal +6.

d. From the list of common polyatomic anions in Table 2.5, the formula of the PO_4 anion must be PO_4^{3-}. Thus, the oxidation state of Co is Co^{2+}. For the oxidation state of P in PO_4^{3-}, the charge of -3 must equal x_P - 8, so x_P must equal +5.

3.44 a. From the formula of ClO_3^- in the list of common polyatomic anions in Table 2.5, the formula of the BrO_3 anion must be BrO_3^-. Thus, the oxidation state of Hg is +1, and its formula must be Hg_2^{2+} (from Table 3.5). For the oxidation state of Br in BrO_3^-, the charge of -1 must equal x_{Br} - 6, so x_{Br} must equal +5.

b. From the list of common polyatomic anions in Table 2.5, the formula of the SO_4 anion must be SO_4^{2-}. Thus, the oxidation state of Cr is +3. For the oxidation state of S in SO_4^{2-}, the charge of -2 must equal x_S - 8, so x_S must equal +6.

c. From the formula of SO_4^{2-} in the list of common polyatomic anions in Table 2.5, the formula of the SeO_4 anion must be SO_4^{2-}. Thus, the oxidation state of Co is +2. For the oxidation state of Se in SeO_4^{2-}, the charge of -2 must equal x_{Se} - 8, so x_{Se} must be +6.

d. From the list of common polyatomic anions in Table 2.5, the formula of the SO_3 anion must be SO_3^{2-}. Thus, the oxidation state of Cu is +1. For the oxidation state of S in SO_3^{2-}, the charge of -2 must equal x_S - 6, so x_S must equal +4.

3.45 a. Phosphorus changes from an oxidation number of 0 in P_4 to +5 in P_4O_{10}, losing electrons and acting as a reducing agent. Oxygen changes from an oxidation number of 0 in O_2 to -2 in P_4O_{10}, gaining electrons and acting as an oxidizing agent.

b. Cobalt changes from an oxidation number of 0 in Co(s) to +2 in $CoCl_2$, losing electrons and acting as a reducing agent. Chlorine changes from an oxidation number of 0 in Cl_2 to -1 in $CoCl_2$, gaining electrons and acting as an oxidizing agent.

3.46 a. Carbon changes from an oxidation number of 0 in C to +2 in CO, losing electrons and acting as a reducing agent. Zinc changes from an oxidation number of +2 in ZnO to 0 in Zn, gaining electrons and acting as an oxidizing agent.

b. Iron changes from an oxidation number of 0 in Fe(s) to +2 in FeS, losing electrons and acting as a reducing agent. Sulfur changes from an oxidation number of 0 in S_8 to -2 in FeS, gaining electrons and acting as an oxidizing agent.

3.47 a. Al changes from oxidation number 0 to +3; Al is the reducing agent.
F changes from oxidation number 0 to -1; F_2 is the oxidizing agent.

b. Hg changes from oxidation state +2 to 0; Hg^{2+} is the oxidizing agent.
N changes from oxidation state +3 to +5; NO_2^- is the reducing agent.

3.48 a. C changes from oxidation number +2 to +4; the CO is the reducing agent.
Fe changes from oxidation number +3 to 0; the Fe_2O_3 is the oxidizing agent.

b. S changes from oxidation number -2 to +6; the PbS is the reducing agent.
O changes from oxidation number -1 to -2; the H_2O_2 is the oxidizing agent.

3.49 The half-reaction method will be used. The balanced half-reactions of the oxidizing agent and reducing agent will be given as obtained by this method. The next equation will be the sum of the half-reactions after each has been adjusted to have the same number of electrons (the electrons will not appear). If this is not the *ultimate (final) equation,* it will be labeled the *penultimate equation,* and it will then be simplified to the ultimate (final) equation.

a. $Cr_2O_7^{2-} + 14H^+ + 6e^- \rightarrow 2Cr^{3+} + 7H_2O$

$C_2O_4^{2-} \rightarrow 2CO_2 + 2e^-$

Ultimate (final) equation—no simplification needed:

$Cr_2O_7^{2-} + 3C_2O_4^{2-} + 14H^+ \rightarrow 2Cr^{3+} + 6CO_2 + 7H_2O$

b. $NO_3^- + 4H^+ + 3e^- \rightarrow NO + 2H_2O$

$Cu \rightarrow Cu^{2+} + 2e^-$

Ultimate (final) equation—no simplification needed:

$2NO_3^- + 8H^+ + 3Cu \rightarrow 2NO + 3Cu^{2+} + 4H_2O$

c. $MnO_2 + 4H^+ + 2e^- \rightarrow Mn^{2+} + 2H_2O$

$HNO_2 + H_2O \rightarrow NO_3^- + 3H^+ + 2e^-$

Penultimate equation:

$MnO_2 + HNO_2 + H_2O + 4H^+ \rightarrow Mn^{2+} + NO_3^- + 2H_2O + 3H^+$

Ultimate (final) equation:

$MnO_2 + HNO_2 + H^+ \rightarrow Mn^{2+} + NO_3^- + H_2O$

d. $PbO_2 + SO_4^{2-} + 4H^+ + 2e^- \rightarrow PbSO_4 + 2H_2O$

$Mn^{2+} + 4H_2O \rightarrow MnO_4^- + 8H^+ + 5e^-$

Penultimate equation:

$5PbO_2 + 2Mn^{2+} + 5SO_4^{2-} + 20H^+ + 8H_2O \rightarrow$

$5PbSO_4 + 2MnO_4^- + 16H^+ + 10H_2O$

Ultimate (final) equation:

$5PbO_2 + 2Mn^{2+} + 5SO_4^{2-} + 4H^+ \rightarrow 5PbSO_4 + 2MnO_4^- + 2H_2O$

e. $Cr_2O_7^{2-} + 14H^+ + 6e^- \rightarrow 2Cr^{3+} + 7H_2O$

$HNO_2 + H_2O \rightarrow NO_3^- + 3H^+ + 2e^-$

Penultimate equation:

$Cr_2O_7^{2-} + 3HNO_2 + 14H^+ + 3H_2O \rightarrow 2Cr^{3+} + 3NO_3^- + 7H_2O + 9H^+$

Ultimate (final) equation:

$Cr_2O_7^{2-} + 3HNO_2 + 5H^+ \rightarrow 2Cr^{3+} + 3NO_3^- + 4H_2O$

3.50 The half-reaction method will be used. The balanced half-reactions of the oxidizing agent and reducing agent will be given as obtained by this method. The next equation will be the sum of the half-reactions after each has been adjusted to have the same number of electrons (the electrons will not appear). If this is not the *ultimate (final) equation,* it will be labeled the *penultimate equation,* and it will then be simplified to the ultimate (final) equation.

a. $BiO_3^- + 6H^+ + 2e^- \rightarrow Bi^{3+} + 3H_2O$

$Mn^{2+} + 4H_2O \rightarrow MnO_4^- + 8H^+ + 5e^-$

Penultimate equation:

$5BiO_3^- + 2Mn^{2+} + 30H^+ + 8H_2O \rightarrow 5Bi^{3+} + 2MnO_4^- + 15H_2O + 16H^+$

Ultimate (final) equation:

$5BiO_3^- + 2Mn^{2+} + 14H^+ \rightarrow 5Bi^{3+} + 2MnO_4^- + 7H_2O$

b. $Cr_2O_7^{2-} + 14H^+ + 6e^- \rightarrow 2Cr^{3+} + 7H_2O$

$I^- + 3H_2O \rightarrow IO_3^- + 6H^+ + 6e^-$

Penultimate equation:

$Cr_2O_7^{2-} + I^- + 14H^+ + 3H_2O \rightarrow 2Cr^{3+} + IO_3^- + 7H_2O + 6H^+$

Ultimate (final) equation:

$Cr_2O_7^{2-} + I^- + 8H^+ \rightarrow 2Cr^{3+} + IO_3^- + 4H_2O$

c. $MnO_4^- + 8H^+ + 5e^- \rightarrow Mn^{2+} + 4H_2O$

$H_2SO_3 + H_2O \rightarrow SO_4^{2-} + 4H^+ + 2e^-$

Penultimate equation:

$2MnO_4^- + 5H_2SO_3 + 16H^+ + 5H_2O \rightarrow 2Mn^{2+} + 5SO_4^{2-} + 8H_2O + 20H^+$

Ultimate (final) equation:

$2MnO_4^- + 5H_2SO_3 \rightarrow 2Mn^{2+} + 5SO_4^{2-} + 3H_2O + 4H^+$

d. $Cr_2O_7^{2-} + 14H^+ + 6e^- \rightarrow 2Cr^{3+} + 7H_2O$

$Fe^{2+} \rightarrow Fe^{3+} + e^-$

Ultimate (final) equation:

$Cr_2O_7^{2-} + 6Fe^{2+} + 14H^+ \rightarrow 2Cr^{3+} + 6Fe^{3+} + 7H_2O$

e. $ClO_3^- + 5H^+ + 4e^- \rightarrow HClO + 2H_2O$

$As + 3H_2O \rightarrow H_3AsO_3 + 3H^+ + 3e^-$

Penultimate equation:

$3ClO_3^- + 4As + 15H^+ + 12H_2O \rightarrow 3HClO + 4H_3AsO_3 + 6H_2O + 12H^+$

Ultimate (final) equation:

$3ClO_3^- + 4As + 3H^+ + 6H_2O \rightarrow 3HClO + 4H_3AsO_3$

3.51 The balanced half-reactions of the oxidizing agent and reducing agent will be given as obtained by the half-reaction method, *as if the reaction occurred in acid solution.* The next equation will be the sum of the half-reactions after adjusting the number of electrons to be equal (the electrons will not appear). If this is not the *ultimate (final) equation,* it will be labeled the *penultimate equation,* which will be simplified to the ultimate (final) equation. Then OH^- ions will be added to both sides of the equation so that OH^- will react with the H^+ to give H_2O.

a. $H_2O_2 + 2H^+ + 2e^- \rightarrow 2H_2O$

$Mn^{2+} + 2H_2O \rightarrow MnO_2 + 4H^+ + 2e^-$

Penultimate equation:

$H_2O_2 + Mn^{2+} + 2H^+ + 2H_2O \rightarrow MnO_2 + 2H_2O + 4H^+$

(continued)

Ultimate (final) equation (in acid):

$H_2O_2 + Mn^{2+} \rightarrow MnO_2 + 2H^+$

Ultimate (final) equation (after adding $2OH^-$ to both sides):

$H_2O_2 + Mn^{2+} + 2OH^- \rightarrow MnO_2 + 2H_2O$

b. $MnO_4^- + 4H^+ + 3e^- \rightarrow MnO_2 + 2H_2O$

$NO_2^- + H_2O \rightarrow NO_3^- + 2H^+ + 2e^-$

Penultimate equation:

$2MnO_4^- + 3NO_2^- + 8H^+ + 3H_2O \rightarrow 2MnO_2 + 3NO_3^- + 4H_2O + 6H^+$

Ultimate (final) equation (in acid):

$2MnO_4^- + 3NO_2^- + 2H^+ \rightarrow 2MnO_2 + 3NO_3^- + H_2O$

Ultimate (final) equation (after adding $2OH^-$ to both sides and canceling H_2O):

$2MnO_4^- + 3NO_2^- + H_2O \rightarrow 2MnO_2 + 3NO_3^- + 2OH^-$

c. $ClO_3^- + 2H^+ + e^- \rightarrow ClO_2 + H_2O$

$Mn^{2+} + 2H_2O \rightarrow MnO_2 + 4H^+ + 2e^-$

Penultimate equation:

$2ClO_3^- + Mn^{2+} + 4H^+ + 2H_2O \rightarrow 2ClO_2 + MnO_2 + 2H_2O + 4H^+$

Ultimate (final) equation (same in acid or in base):

$2ClO_3^- + Mn^{2+} \rightarrow 2ClO_2 + MnO_2$

d. $MnO_4^- + 4H^+ + 3e^- \rightarrow MnO_2 + 2H_2O$

$NO_2 + H_2O \rightarrow NO_3^- + 2H^+ + e^-$

Penultimate equation:

$MnO_4^- + 3NO_2 + 4H^+ + 3H_2O \rightarrow MnO_2 + 3NO_3^- + 2H_2O + 6H^+$

Ultimate (final) equation (in acid):

$MnO_4^- + 3NO_2 + H_2O \rightarrow MnO_2 + 3NO_3^- + 2H^+$

Ultimate (final) equation (after adding $2OH^-$ to both sides and canceling H_2O):

$MnO_4^- + 3NO_2 + 2OH^- \rightarrow MnO_2 + 3NO_3^- + H_2O$

e. $Cl_2 + 2e^- \rightarrow 2Cl^-$

$Cl_2 + 6H_2O \rightarrow 2ClO_3^- + 12H^+ + 10e^-$

Ultimate (final) equation (in acid after dividing coefficients by 2):

$3Cl_2 + 3H_2O \rightarrow 5Cl^- + ClO_3^- + 6H^+$

Ultimate (final) equation (after adding $6OH^-$ to both sides and canceling $3H_2O$):

$3Cl_2 + 6OH^- \rightarrow 5Cl^- + ClO_3^- + 3H_2O$

3.52 The balanced half-reactions of the oxidizing agent and reducing agent will be given as obtained by the half-reaction method, *as if the reaction occurred in acid solution.* The next equation will be the sum of the half-reactions after adjusting the number of electrons to be equal (the electrons will not appear). If this is not the *ultimate (final) equation,* it will be labeled the *penultimate equation,* which will be simplified to the ultimate (final) equation. Then OH^- ions will be added to both sides of the equation so that OH^- will react with the H^+ to give H_2O.

a. $H_2O_2 + 2H^+ + 2e^- \rightarrow 2H_2O$

$Cr(OH)_4^- \rightarrow CrO_4^{2-} + 4H^+ + 3e^-$

Penultimate equation:

$3H_2O_2 + 2Cr(OH)_4^- + 6H^+ \rightarrow 6H_2O + 2CrO_4^{2-} + 8H^+$

Ultimate (final) equation (in acid):

$3H_2O_2 + 2Cr(OH)_4^- \rightarrow 6H_2O + 2CrO_4^{2-} + 2H^+$

Ultimate (final) equation (after adding $2OH^-$ to both sides and combining H_2O's):

$3H_2O_2 + 2Cr(OH)_4^- + 2OH^- \rightarrow 8H_2O + 2CrO_4^{2-}$

b. $MnO_4^- + 4H^+ + 3e^- \rightarrow MnO_2 + 2H_2O$

$Br^- + 3H_2O \rightarrow BrO_3^- + 6H^+ + 6e^-$

Ultimate (final) equation (in acid):

$2MnO_4^- + Br^- + 2H^+ \rightarrow 2MnO_2 + BrO_3^- + H_2O$

Ultimate (final) equation (after adding $2OH^-$ and canceling H_2O):

$2MnO_4^- + Br^- + H_2O \rightarrow 2MnO_2 + BrO_3^- + 2OH^-$

c. $H_2O_2 + 2H^+ + 2e^- \rightarrow 2H_2O$

$Co^{2+} + 3OH^- \rightarrow Co(OH)_3 + e^-$

Penultimate equation (after replacing $2H^+ + 2OH^-$ with $2H_2O$):

$H_2O_2 + 2Co^{2+} + 2H_2O + 4OH^- \rightarrow 2H_2O + 2Co(OH)_3$

Ultimate (final) equation (turns out to be basic):

$H_2O_2 + 2Co^{2+} + 4OH^- \rightarrow 2Co(OH)_3$

d. $ClO^- + 2H^+ + 2e^- \rightarrow Cl^- + H_2O$

$Pb(OH)_4^{2-} \rightarrow PbO_2 + 2H_2O + 2e^-$

Ultimate (final) equation (in acid):

$ClO^- + Pb(OH)_4^{2-} + 2H^+ \rightarrow Cl^- + PbO_2 + 3H_2O$

Ultimate (final) equation (after adding $2OH^-$ to both sides and canceling $2H_2O$):

$ClO^- + Pb(OH)_4^{2-} \rightarrow Cl^- + PbO_2 + H_2O + 2OH^-$

e. $NO_3^- + 9H^+ + 8e^- \rightarrow NH_3 + 3H_2O$

$Zn + 4OH^- \rightarrow Zn(OH)_4^{2-} + 2e^-$

Penultimate equation:

$NO_3^- + 4Zn + 16OH^- + 9H^+ \rightarrow NH_3 + 4Zn(OH)_4^{2-} + 3H_2O$

Ultimate (final) equation (after combining $9OH^- + 9H^+$ as 9 H_2O and canceling $3H_2O$):

$NO_3^- + 4Zn + 6H_2O + 7OH^- \rightarrow NH_3 + 4Zn(OH)_4^{2-}$

3.53 The half-reaction method will be used. The balanced half-reactions of the oxidizing agent and reducing agent will be given as obtained by this method. The next equation will be the sum of the half-reactions after each has been adjusted to have the same number of electrons (the electrons will not appear). If this is not the *ultimate (final) equation,* it will be labeled the *penultimate equation,* and it will then be simplified to the ultimate (final) equation.

a. $NO_3^- + 2H^+ + e^- \rightarrow NO_2 + H_2O$

$8H_2S \rightarrow S_8 + 16H^+ + 16e^-$

Penultimate equation:

$16NO_3^- + 8H_2S + 32H^+ \rightarrow 16NO_2 + S_8 + 16H_2O + 16H^+$

Ultimate (final) equation:

$16NO_3^- + 8H_2S + 16H^+ \rightarrow 16NO_2 + S_8 + 16H_2O$

b. $NO_3^- + 4H^+ + 3e^- \rightarrow NO + 2H_2O$

$Cu \rightarrow Cu^{2+} + 2e^-$

Ultimate (final) equation—no simplification needed:

$2NO_3^- + 8H^+ + 3Cu \rightarrow 2NO + 3Cu^{2+} + 4H_2O$

c. $MnO_4^- + 8H^+ + 5e^- \rightarrow Mn^{2+} + 4H_2O$

$SO_2 + 2H_2O \rightarrow SO_4^{2-} + 4H^+ + 2e^-$

Penultimate equation:

$2MnO_4^- + 5SO_2 + 16H^+ + 10H_2O \rightarrow 2Mn^{2+} + 5SO_4^{2-} + 8H_2O + 20H^+$

Ultimate (final) equation:

$2MnO_4^- + 5SO_2 + 2H_2O \rightarrow 2Mn^{2+} + 5SO_4^{2-} + 4H^+$

d. $Bi(OH)_3 + 3e^- \rightarrow Bi + 3OH^-$

$Sn(OH)_3^- + 3OH^- \rightarrow Sn(OH)_6^{2-} + 2e^-$

Penultimate equation:

$2Bi(OH)_3 + 3Sn(OH)_3^- + 9OH^- \rightarrow 2Bi + 3Sn(OH)_6^{2-} + 6OH^-$

Ultimate (final) equation:

$2Bi(OH)_3 + 3Sn(OH)_3^- + 3OH^- \rightarrow 2Bi + 3Sn(OH)_6^{2-}$

3.54 The balanced half-reactions of the oxidizing agent and reducing agent will be given as obtained by the half-reaction method, *as if the reaction occurred in acid solution.* The next equation will be the sum of the half-reactions after adjusting the number of electrons to be equal (the electrons will not appear). If this is not the *ultimate (final) equation,* it will be labeled the *penultimate equation,* which will be simplified to the ultimate (final) equation. For basic solution, OH^- will be added to both sides of the equation to react with the H^+ to give H_2O.

a. $Hg_2^{2+} + 2e^- \rightarrow 2Hg$

$8H_2S \rightarrow S_8 + 16H^+ + 16e^-$

Ultimate (final) equation:

$8Hg_2^{2+} + 8H_2S \rightarrow 16Hg + S_8 + 16H^+$

b. $I_2 + 2e^- \rightarrow 2I^-$

$S^{2-} + 4H_2O \rightarrow SO_4^{2-} + 8H^+ + 8e^-$

Ultimate (final) equation (in acid solution):

$4I_2 + S^{2-} + 4H_2O \rightarrow 8I^- + SO_4^{2-} + 8H^+$

Ultimate (final) equation (after adding $8OH^-$ to both sides and canceling $4H_2O$):

$4I_2 + S^{2-} + 8OH^- \rightarrow 8I^- + SO_4^{2-} + 4H_2O$

c. $NO_3^- + 9H^+ + 8e^- \rightarrow NH_3 + 3H_2O$

$Al + 4OH^- \rightarrow Al(OH)_4^- + 3e^-$

Penultimate equation:

$3NO_3^- + 8Al + 27H^+ + 32OH^- \rightarrow 3NH_3 + 8Al(OH)_4^- + 9H_2O$

Ultimate (final) equation (after replacing $27H^+ + 27OH^-$ with $27H_2O$ and canceling $9H_2O$):

$3NO_3^- + 8Al + 18H_2O + 5OH^- \rightarrow 3NH_3 + 8Al(OH)_4^-$

d. $MnO_4^- + 4H^+ + 3e^- \rightarrow MnO_2 + 2H_2O$

$C_2O_4^{2-} \rightarrow 2CO_2 + 2e^-$

Ultimate (final) equation (in acid solution):

$2MnO_4^- + 3C_2O_4^{2-} + 8H^+ \rightarrow 2MnO_2 + 6CO_2 + 4H_2O$

Ultimate (final) equation (after adding $2OH^-$ to both sides and canceling H_2O):

$2MnO_4^- + 3C_2O_4^{2-} + 4H_2O \rightarrow 2MnO_2 + 6CO_2 + 8OH^-$

3.55 The balanced half-reactions of the oxidizing agent and reducing agent will be given as obtained by the half-reaction method, *as if the reaction occurred in acid solution.* The next equation will be the sum of the half-reactions after adjusting the number of electrons to be equal (the electrons will not appear). If this is not the *ultimate (final) equation,* it will be labeled the *penultimate equation,* which will be simplified to the ultimate (final) equation. For basic solution, OH^- will be added to both sides of the equation to react with the H^+ to give H_2O.

a. $MnO_4^- + 4H^+ + 3e^- \rightarrow MnO_2 + 2H_2O$

$I^- + 3H_2O \rightarrow IO_3^- + 6H^+ + 6e^-$

Ultimate (final) equation (in acid):

$2MnO_4^- + I^- + 2H^+ \rightarrow 2MnO_2 + IO_3^- + H_2O$

Ultimate (final) equation (after adding $2OH^-$ and canceling H_2O):

$2MnO_4^- + I^- + H_2O \rightarrow 2MnO_2 + IO_3^- + 2OH^-$

b. $Cr_2O_7^{2-} + 14H^+ + 6e^- \rightarrow 2Cr^{3+} + 7H_2O$

$2Cl^- \rightarrow Cl_2 + 2e^-$

Ultimate (final) equation:

$Cr_2O_7^{2-} + 6Cl^- + 14H^+ \rightarrow 2Cr^{3+} + 3Cl_2 + 7H_2O$

c. $NO_3^- + 4H^+ + 3e^- \rightarrow NO + 2H_2O$

$S_8 + 16H_2O \rightarrow 8SO_2 + 32H^+ + 32e^-$

Penultimate equation:

$32NO_3^- + 3S_8 + 128H^+ + 48H_2O \rightarrow 32NO + 24SO_2 + 64H_2O + 96H^+$

Ultimate (final) equation:

$32NO_3^- + 3S_8 + 32H^+ \rightarrow 32NO + 24SO_2 + 16H_2O$

d. $MnO_4^- + 4H^+ + 3e^- \rightarrow MnO_2 + 2H_2O$

$H_2O_2 \rightarrow O_2 + 2H^+ + 2e^-$

Penultimate equation:

$2MnO_4^- + 3H_2O_2 + 2H^+ \rightarrow 2MnO_2 + 3O_2 + 4H_2O$

Ultimate (final) equation:

$2MnO_4^- + 3H_2O_2 \rightarrow 2MnO_2 + 3O_2 + 2H_2O + 2OH^-$

e. $2NO_3^- + 12H^+ + 10e^- \rightarrow N_2 + 6H_2O$

$Zn \rightarrow Zn^{2+} + 2e^-$

Ultimate (final) equation:

$2NO_3^- + 5Zn + 12H^+ \rightarrow N_2 + 5Zn^{2+} + 6H_2O$

3.56 The balanced half-reactions of the oxidizing agent and reducing agent will be given as obtained by the half-reaction method, *as if the reaction occurred in acid solution.* The next equation will be the sum of the half-reactions after adjusting the number of electrons to be equal (the electrons will not appear). If this is not the *ultimate (final) equation,* it will be labeled the *penultimate equation,* which will be simplified to the ultimate (final) equation. For basic solution, OH^- will be added to both sides of the equation to react with the H^+ to give H_2O.

a. $Cr_2O_7^{2-} + 14H^+ + 6e^- \rightarrow 2Cr^{3+} + 7H_2O$

$H_2O_2 \rightarrow O_2 + 2H^+ + 2e^-$

Penultimate equation:

$Cr_2O_7^{2-} + 3H_2O_2 + 14H^+ \rightarrow 2Cr^{3+} + 3O_2 + 7H_2O + 6H^+$

(continued)

Ultimate (final) equation:

$Cr_2O_7^{2-} + 3H_2O_2 + 8H^+ \rightarrow 2Cr^{3+} + 3O_2 + 7H_2O$

b. $MnO_4^- + 4H^+ + 3e^- \rightarrow MnO_2 + 2H_2O$

$CN^- + H_2O \rightarrow CNO^- + 2H^+ + 2e^-$

Penultimate equation:

$2MnO_4^- + 3CN^- + 8H^+ + 3H_2O \rightarrow 2MnO_2 + 3CNO^- + 4H_2O + 6H^+$

Ultimate (final) equation (in acid):

$2MnO_4^- + 3CN^- + 2H^+ \rightarrow 2MnO_2 + 3CNO^- + H_2O$

Ultimate (final) equation (after adding $2OH^-$ to both sides and canceling H_2O):

$2MnO_4^- + 3CN^- + H_2O \rightarrow 2MnO_2 + 3CNO^- + 2OH^-$

c. $OCl^- + 2H^+ + 2e^- \rightarrow Cl^- + H_2O$

$Cr(OH)_4^- \rightarrow CrO_4^{2-} + 4H^+ + 3e^-$

Penultimate equation:

$3OCl^- + 2Cr(OH)_4^- + 6H^+ \rightarrow 3Cl^- + 2CrO_4^{2-} + 3H_2O + 8H^+$

Ultimate (final) equation (in acid):

$3OCl^- + 2Cr(OH)_4^- \rightarrow 3Cl^- + 2CrO_4^{2-} + 3H_2O + 2H^+$

Ultimate (final) equation (after adding $2OH^-$ to both sides and combining H_2O's):

$3OCl^- + 2Cr(OH)_4^- + 2OH^- \rightarrow 3Cl^- + 2CrO_4^{2-} + 5H_2O$

d. $Br_2 + 2e^- \rightarrow 2Br^-$

$SO_2 + 2H_2O \rightarrow SO_4^{2-} + 4H^+ + 2e^-$

Ultimate (final) equation:

$Br_2 + SO_2 + 2H_2O \rightarrow 2Br^- + SO_4^{2-} + 4H^+$

e. $NO_3^- + 4H^+ + 3e^- \rightarrow NO + 2H_2O$

$8CuS \rightarrow S_8 + 8Cu^{2+} + 16e^-$

Ultimate (final) equation:

$16NO_3^- + 24CuS + 64H^+ \rightarrow 16NO + 3S_8 + 24Cu^{2+} + 32H_2O$

■ Solutions to Unclassified Problems

3.57 For the reaction of magnesium metal and hydrobromic acid, the equations are as follows.

Molecular equation: $Mg(s) + 2HBr(aq) \rightarrow H_2(g) + MgBr_2(aq)$

Ionic equation: $Mg(s) + 2H^+(aq) \rightarrow H_2(g) + Mg^{2+}(aq)$

3.58 For the reaction of aluminum metal and perchloric acid, the equations are as follows.

Molecular equation: $2Al(s) + 6HClO_4(aq) \rightarrow 3H_2(g) + 2Al(ClO_4)_3(aq)$

Ionic equation: $2Al(s) + 6H^+(aq) \rightarrow 3H_2(g) + 2Al^{3+}(aq)$

3.59 For the reaction of nickel(II) sulfate and lithium hydroxide, the equations are as follows.

Molecular equation: $NiSO_4(aq) + 2LiOH(aq) \rightarrow Ni(OH)_2(s) + Li_2SO_4(aq)$

Ionic equation: $Ni^{2+}(aq) + 2OH^-(aq) \rightarrow Ni(OH)_2(s)$

3.60 For the reaction of potassium sulfate and barium bromide, the equations are as follows.

Molecular equation: $BaBr_2(aq) + K_2SO_4(aq) \rightarrow BaSO_4(s) + 2KBr(aq)$

Ionic equation: $Ba^{2+}(aq) + SO_4^{2-}(aq) \rightarrow BaSO_4(s)$

3.61 a. Molecular equation: $LiOH(aq) + HCN(aq) \rightarrow LiCN(aq) + H_2O(l)$

Ionic equation: $OH^-(aq) + HCN(aq) \rightarrow CN^-(aq) + H_2O(l)$

b. Molecular equation: $Li_2CO_3(aq) + 2HNO_3(aq) \rightarrow 2LiNO_3(aq) + CO_2(g) + H_2O(l)$

Ionic equation: $CO_3^{2-}(aq) + 2H^+(aq) \rightarrow CO_2(g) + H_2O(l)$

c. Molecular equation: $LiCl(aq) + AgNO_3(aq) \rightarrow LiNO_3(aq) + AgCl(s)$

Ionic equation: $Cl^-(aq) + Ag^+(aq) \rightarrow AgCl(s)$

d. Molecular equation: $MgSO_4(aq) + LiCl(aq) \rightarrow$ NR (Li_2SO_4 and $MgCl_2$ are soluble.)

3.62 a. Molecular equation: $Al(OH)_3(s) + 3HNO_3(aq) \rightarrow Al(NO_3)_3(aq) + 3H_2O(l)$

Ionic equation: $Al(OH)_3(s) + 3H^+(aq) \rightarrow Al^{3+}(aq) + 3H_2O(l)$

b. Molecular equation: $FeS(s) + 2HClO_4(aq) \rightarrow Fe(ClO_4)_2(aq) + H_2S(g)$

Ionic equation: $FeS(s) + 2H^+(aq) \rightarrow Fe^{2+}(aq) + H_2S(g)$

c. Molecular equation: $CaCl_2(aq) + NaNO_3(aq) \rightarrow$ NR [$Ca(NO_3)_2$ and NaCl are soluble.]

d. Molecular equation: $MgSO_4(aq) + Ba(NO_3)_2(aq) \rightarrow Mg(NO_3)_2(aq) + BaSO_4(s)$

Ionic equation: $SO_4^{2-}(aq) + Ba^{2+}(aq) \rightarrow BaSO_4(s)$

3.63 a. Molecular equation: $Sr(OH)_2(aq) + 2HC_2H_3O_2(aq) \rightarrow$
$Sr(C_2H_3O_2)_2(aq) + 2\,H_2O(l)$

Ionic equation: $HC_2H_3O_2(aq) + OH^-(aq) \rightarrow C_2H_3O_2^-(aq) + H_2O(l)$

b. Molecular equation: $NH_4I(aq) + CsCl(aq) \rightarrow$ NR (NH_4Cl and CsI are soluble.)

c. Molecular equation: $NaNO_3(aq) + CsCl(aq) \rightarrow$ NR (NaCl and $CsNO_3$ are soluble.)

d. Molecular equation: $NH_4I(aq) + AgNO_3(aq) \rightarrow NH_4NO_3(aq) + AgI(s)$

Ionic equation: $I^-(aq) + Ag^+(aq) \rightarrow AgI(s)$

3.64 a. Molecular equation: $2HClO_4(aq) + BaCO_3(s) \rightarrow$

$$Ba(ClO_4)_2(aq) + H_2CO_3(aq)\ [CO_2(g) + H_2O(l)]$$

Ionic equation: $2H^+(aq) + BaCO_3(s) \rightarrow Ba^{2+}(aq) + CO_2(g) + H_2O(l)$

b. Molecular equation: $H_2CO_3(aq) + Sr(OH)_2(aq) \rightarrow 2H_2O(l) + SrCO_3(s)$

Ionic equation: $H_2CO_3(aq) + Sr^{2+}(aq) + 2\ OH^-(aq) \rightarrow 2\ H_2O(l) + SrCO_3(s)$

c. Molecular equation: $2K_3PO_4(aq) + 3MgCl_2(aq) \rightarrow 6KCl(aq) + Mg_3(PO_4)_2(s)$

Ionic equation: $2PO_4^{3-}(aq) + 3Mg^{2+}(aq) \rightarrow Mg_3(PO_4)_2(s)$

d. Molecular equation: $FeSO_4(aq) + MgCl_2(aq) \rightarrow$ NR
($FeCl_2$ and $MgSO_4$ are soluble.)

3.65 For each preparation, the compound to be prepared is given first, followed by the compound from which it is to be prepared. Then the method of preparation is given, followed by the molecular equation for the preparation reaction. Steps such as evaporation, etc., are not given in the molecular equation.

a. To prepare $CuCl_2$ from $CuSO_4$, add a solution of $BaCl_2$ to a solution of the $CuSO_4$, precipitating $BaSO_4$. The $BaSO_4$ can be filtered off, leaving aqueous $CuCl_2$, which can be obtained in solid form by evaporation of the solution. Molecular equation:

$$CuSO_4(aq) + BaCl_2(aq) \rightarrow BaSO_4(s) + CuCl_2(aq)$$

b. To prepare $Ca(C_2H_3O_2)_2$ from $CaCO_3$, add a solution of acetic acid, $HC_2H_3O_2$, to the solid $CaCO_3$, forming CO_2, H_2O, and aqueous $Ca(C_2H_3O_2)_2$. The aqueous $Ca(C_2H_3O_2)_2$ can be converted to the solid form by evaporation of the solution, which also removes the CO_2 and H_2O products. Molecular equation:

$$CaCO_3(s) + 2HC_2H_3O_2(aq) \rightarrow Ca(C_2H_3O_2)_2(aq) + CO_2(g) + H_2O(l)$$

c. To prepare $NaNO_3$ from Na_2SO_3, add a solution of nitric acid, HNO_3, to the solid Na_2SO_3, forming SO_2, H_2O, and aqueous $NaNO_3$. The aqueous $NaNO_3$ can be converted to the solid by evaporation of the solution, which also removes the SO_2 and H_2O products. Molecular equation:

$$Na_2SO_3(s) + 2HNO_3(aq) \rightarrow 2NaNO_3(aq) + SO_2(g) + H_2O(l)$$

d. To prepare $MgCl_2$ from $Mg(OH)_2$, add a solution of hydrochloric acid (HCl) to the solid $Mg(OH)_2$, forming H_2O and aqueous $MgCl_2$. The aqueous $MgCl_2$ can be converted to the solid form by evaporation of the solution. Molecular equation:

$$Mg(OH)_2(s) + 2HCl(aq) \rightarrow MgCl_2(aq) + 2H_2O(l)$$

3.66 For each preparation, the compound to be prepared is given first, followed by the compound from which it is to be prepared. Then the method of preparation is given, followed by the molecular equation for the preparation reaction. Steps such as evaporation, etc., are not given in the molecular equation.

a. To prepare $MgCl_2$ from $MgCO_3$, add a solution of hydrochloric acid (HCl) to the solid $MgCO_3$, forming CO_2, H_2O, and aqueous $MgCl_2$. The aqueous $MgCl_2$ can be converted to the solid form by evaporation of the solution. Molecular equation:

$$MgCO_3(s) + 2HCl(aq) \rightarrow MgCl_2(aq) + CO_2(g) + H_2O(l)$$

b. To prepare $NaNO_3$ from NaCl, add a solution of $AgNO_3$ to a solution of the NaCl, precipitating AgCl. The AgCl can be filtered off, leaving aqueous $NaNO_3$, which can be obtained in solid form by evaporation of the solution. Molecular equation:

$$NaCl(aq) + AgNO_3(aq) \rightarrow AgCl(s) + NaNO_3(aq)$$

c. To prepare $Al(OH)_3$ from $Al(NO_3)_3$, add a solution of NaOH to a solution of $Al(NO_3)_3$, precipitating $Al(OH)_3$. The $Al(OH)_3$ can be filtered off and dried to remove any water adhering to it. Molecular equation:

$$Al(NO_3)_3(aq) + 3NaOH(aq) \rightarrow Al(OH)_3(s) + 3NaNO_3(aq)$$

d. To prepare HCl from H_2SO_4, add a solution of $BaCl_2$ to the solution of H_2SO_4, precipitating $BaSO_4$. The $BaSO_4$ can be filtered off, leaving the desired solution of aqueous HCl. Molecular equation:

$$H_2SO_4(aq) + BaCl_2(aq) \rightarrow BaSO_4(s) + 2HCl(aq)$$

3.67 a. Decomposition b. Decomposition c. Combination d. Displacement

3.68 a. Combination b. Displacement c. Decomposition d. Combination

3.69 a. $Pb(NO_3)_2 + H_2SO_4$ [$\rightarrow PbSO_4(s) + HNO_3(aq)$];
$Pb(NO_3)_2 + MgSO_4$ [$\rightarrow PbSO_4(s) + Mg(NO_3)_2(aq)$]
$Pb(NO_3)_2 + Ba(OH)_2$ [$\rightarrow Pb(OH)_2(s) + Ba(NO_3)_2(aq)$
b. $Ba(OH))_2 + MgSO_4$ [$\rightarrow BaSO_4(s) + Mg(OH)_2(s)$]
c. $Ba(OH))_2 + H_2SO_4$ [$\rightarrow BaSO_4(s) + H_2O(l)$]

3.70 a. $AgNO_3 + H_3PO_4$ [$\rightarrow Ag_3PO_4(s) + HNO_3(aq)$]
$AgNO_3 + Sr(OH)_2$ [$\rightarrow AgOH(s) + Sr(NO_3)_2(aq)$]
b. $Sr(OH)_2 + CuSO_4$ [$\rightarrow SrSO_4(s) + Cu(OH)_2(s)$]
c. $Sr(OH)_2 + H_3PO_4$ [$\rightarrow Sr_3(PO_4)_2(s) + H_2O(l)$]

3.71 a. $ZnS + 2HCl$ [$\rightarrow H_2S(g) + ZnCl_2(aq)$]
$2HCl + Na_2SO_3$ [$\rightarrow 2NaCl(aq) + H_2O(l) + SO_2(g)$]
$2HCl + CaCO_3(s)$ [$\rightarrow 2NaCl(aq) + H_2O(l) + CO_2(g)$]
b. $CaCO_3(s) + 2HCl$ [$\rightarrow CO_2(g) + CaCl_2(aq)$ (+ H_2O)]
$CaCO_3(s) + H_2SO_4$ [$\rightarrow CO_2(g) + CaSO_4(s)$ (+ H_2O)]
$ZnS + H_2SO_4$ [$\rightarrow H_2S(g) + ZnSO_4(aq)$]
$ZnS + 2HCl$ [$\rightarrow H_2S(g) + ZnCl_2(aq)$]
c. $CaCO_3(s) + H_2SO_4$ [$\rightarrow CaSO_4(s) + H_2O(l)$ (+ CO_2)]

3.72 a. $ZnS + 2HBr$ [$\rightarrow H_2S(g) + ZnBr_2(aq)$]
b. $SrCO_3(s) + 2HBr$ [$\rightarrow CO_2(g) + SrBr_2(aq)$ (+ H_2O)]
$SrCO_3(s) + H_2SO_4$ [$\rightarrow CO_2(g) + SrSO_4(s)$ (+ H_2O)]
$ZnS(s) + H_2SO_4$ [$\rightarrow H_2S(g) + ZnSO_4(aq)$]
c. $SrCO_3(s) + H_2SO_4$ [$\rightarrow SrSO_4(s) + H_2O(l)$ (+ CO_2)]

3.73 a. $Fe(s) \rightarrow Fe^{2+} + 2e^-$

$NO_3^- + 4H^+ + 3e^- \rightarrow NO + 2H_2O$

$3 \times (Fe(s) \rightarrow Fe^{2+} + 2e^-)$

$2 \times (NO_3^- + 4H^+ + 3e^- \rightarrow NO + 2H_2O)$

$3Fe(s) + 2NO_3^- + 8H^+ + \cancel{6e^-} \rightarrow 3Fe^{2+} + 2NO + 4H_2O + \cancel{6e^-}$

b. $Fe^{2+} \rightarrow Fe^{3+} + 1\,e^-$

$NO_3^- + 4\,H^+ + 3\,e^- \rightarrow NO + 2\,H_2O$

$3 \times (Fe^{2+} \rightarrow Fe^{3+} + 1\,e^-)$

$1 \times (NO_3^- + 4\,H^+ + 3\,e^- \rightarrow NO + 2\,H_2O)$

$3\,Fe^{2+} + NO_3^- + 4H^+ + \cancel{3e^-} \rightarrow 3Fe^{3+} + NO + 2\,H_2O + \cancel{3e^-}$

c. $3\,Fe(s) + 2\,NO_3^- + 8\,H^+ + \cancel{6e^-} \rightarrow 3Fe^{2+} + 2\,NO + 4\,H_2O + \cancel{6e^-}$

$3\,Fe^{2+} + NO_3^- + 4H^+ + \cancel{3e^-} \rightarrow 3Fe^{3+} + NO + 2\,H_2O + \cancel{3e^-}$

$Fe(s) + NO_3^- + 4\,H^+ + \rightarrow Fe^{3+} + NO + 2\,H_2O$

3.74 a. $Tl(s) \rightarrow Tl^+ + 1\,e^-$

$NO_3^- + 4\,H^+ + 3\,e^- \rightarrow NO + 2\,H_2O$

$3 \times (Tl(s) \rightarrow Tl^+ + 1\,e^-)$

$1 \times (NO_3^- + 4\,H^+ + 3\,e^- \rightarrow NO + 2\,H_2O)$

$3\,Tl(s) + NO_3^- + 4H^+ + \cancel{3e^-} \rightarrow 3Tl^+ + NO + 2\,H_2O + \cancel{3e^-}$

b. $Tl^+ \rightarrow Tl^{3+} + 2\,e^-$

$NO_3^- + 4\,H^+ + 3\,e^- \rightarrow NO + 2\,H_2O$

$3 \times (Tl^+ \rightarrow Tl^{3+} + 2\,e^-)$

$2 \times (NO_3^- + 4\,H^+ + 3\,e^- \rightarrow NO + 2\,H_2O)$

$3\,Tl^+ + 2\,NO_3^- + 8\,H^+ + \cancel{6e^-} \rightarrow 3Tl^{3+} + 2\,NO + 4\,H_2O + \cancel{6e^-}$

c. $3\,Tl(s) + NO_3^- + 4H^+ + \cancel{3e^-} \rightarrow \cancel{3Tl^+} + NO + 2\,H_2O + \cancel{3e^-}$

$\cancel{3Tl^+} + 2\,NO_3^- + 8\,H^+ + \cancel{6e^-} \rightarrow 3Tl^{3+} + 2\,NO + 4\,H_2O + \cancel{6e^-}$

$Tl(s) + NO_3^- + 4H^+ \rightarrow Tl^{3+} + NO + 2\,H_2O$

3.75 The balanced half-reactions of the oxidizing agent and reducing agent will be given as obtained by the half-reaction method. The next equation will be the sum of the half-reactions after adjusting the number of electrons to be equal. If this is not the *ultimate (final) equation,* it will be labeled the *penultimate equation,* which will be simplified to the ultimate equation. For basic solution, OH^- will be added to both sides of the equation to react with the H^+ to give H_2O.

a. $MnO_4^- + 4H^+ + 3e^- \rightarrow MnO_2 + 2H_2O$

$8S^{2-} \rightarrow S_8 + 16e^-$

Ultimate (final) equation (in acid):

$16MnO_4^- + 24S^{2-} + 64H^+ \rightarrow 16MnO_2 + 3S_8 + 32H_2O$

Ultimate (final) equation (after adding $64OH^-$ to both sides and canceling $32H_2O$):

$16MnO_4^- + 24S^{2-} + 32H_2O \rightarrow 16MnO_2 + 3S_8 + 64OH^-$

b. $IO_3^- + 6H^+ + 6e^- \rightarrow I^- + 3H_2O$

$HSO_3^- + H_2O \rightarrow SO_4^{2-} + 3H^+ + 2e^-$

Penultimate equation:

$IO_3^- + 3HSO_3^- + 6H^+ + 3H_2O \rightarrow I^- + 3SO_4^{2-} + 3H_2O + 9H^+$

Ultimate (final) equation:

$IO_3^- + 3HSO_3^- \rightarrow I^- + 3SO_4^{2-} + 3H^+$

c. $CrO_4^{2-} + 4H^+ + 3e^- \rightarrow Cr(OH)_4^-$

$Fe(OH)_2 + OH^- \rightarrow Fe(OH)_3 + e^-$

Penultimate equation:

$CrO_4^{2-} + 3Fe(OH)_2 + 4H^+ + 3OH^- \rightarrow Cr(OH)_4^- + 3Fe(OH)_3$

Ultimate (final) equation (in acid):

$CrO_4^{2-} + 3Fe(OH)_2 + H^+ + 3H_2O \rightarrow Cr(OH)_4^- + 3Fe(OH)_3$

Ultimate (final) equation (after adding OH^- to both sides and combining H_2O's):

$CrO_4^{2-} + 3Fe(OH)_2 + 4H_2O \rightarrow Cr(OH)_4^- + 3Fe(OH)_3 + OH^-$

d. $Cl_2 + 2e^- \rightarrow 2Cl^-$

$Cl_2 + 2H_2O \rightarrow 2ClO^- + 4H^+ + 2e^-$

Ultimate (final) equation (in acid after dividing coefficients by 2):

$Cl_2 + H_2O \rightarrow Cl^- + ClO^- + 2H^+$

Ultimate (final) equation (after adding $2OH^-$ to both sides and canceling H_2O):

$Cl_2 + 2OH^- \rightarrow Cl^- + ClO^- + H_2O$

3.76 The balanced half-reactions of the oxidizing agent and reducing agent will be given as obtained by the half-reaction method. The next equation will be the sum of the half-reactions after adjusting the number of electrons to be equal. If this is not the *ultimate (final) equation,* it will be labeled the *penultimate equation,* which will be simplified to the ultimate equation. For basic solution, OH^- will be added to both sides of the equation to react with the H^+ to give H_2O.

a. $MnO_4^- + 8H^+ + 5e^- \rightarrow Mn^{2+} + 4H_2O$

$8H_2S \rightarrow S_8 + 16H^+ + 16e^-$

Penultimate equation:

$16MnO_4^- + 40H_2S + 128H^+ \rightarrow 16Mn^{2+} + 5S_8 + 64H_2O + 80H^+$

Ultimate (final) equation:

$16MnO_4^- + 40H_2S + 48H^+ \rightarrow 16Mn^{2+} + 5S_8 + 64H_2O$

b. $2NO_3^- + 10H^+ + 8e^- \rightarrow N_2O + 5H_2O$

$Zn \rightarrow Zn^{2+} + 2e^-$

Ultimate (final) equation:

$2NO_3^- + 4Zn + 10H^+ \rightarrow N_2O + 4Zn^{2+} + 5H_2O$

c. $MnO_4^{2-} + 4H^+ + 2e^- \rightarrow MnO_2 + 2H_2O$

$MnO_4^{2-} \rightarrow MnO_4^- + e^-$

Ultimate (final) equation (in acid solution):

$3MnO_4^{2-} + 4H^+ \rightarrow MnO_2 + 2MnO_4^- + 2H_2O$

(continued)

Ultimate (final) equation (after adding $4OH^-$ to both sides and canceling $2H_2O$):

$$3MnO_4^{2-} + 2H_2O \rightarrow MnO_2 + 2MnO_4^- + 4OH^-$$

d. $Br_2 + 2e^- \rightarrow 2Br^-$

$$Br_2 + 6H_2O \rightarrow 2BrO_3^- + 12H^+ + 10e^-$$

Ultimate (final) equation (in acid after dividing coefficients by 2):

$$3Br_2 + 3H_2O \rightarrow 5Br^- + BrO_3^- + 6H^+$$

Ultimate (final) equation (after adding $6OH^-$ to both sides and canceling $3H_2O$):

$$3Br_2 + 6OH^- \rightarrow 5Br^- + BrO_3^- + 3H_2O$$

3.77 The half-reaction method can be used to find the electron change for each half-reaction in acid solution:

$$O_2 + 4H^+ + 4e^- \rightarrow 2H_2O$$

$$Fe(OH)_2 + H_2O \rightarrow Fe(OH)_3 + H^+ + e^-$$

Penultimate equation:

$$O_2 + 4Fe(OH)_2 + 4H^+ + 4H_2O \rightarrow 2H_2O + 4Fe(OH)_3 + 4H^+$$

Ultimate (final) equation (for either acid or base):

$$O_2 + 4Fe(OH)_2 + 2H_2O \rightarrow 4Fe(OH)_3$$

3.78 The half-reaction method can be used to find the electron change for each half-reaction in acid solution; OH^- can then be added to the ultimate equation for basic solution.

$$Bi^{3+} + 3e^- \rightarrow Bi$$

$$SnO_2^{2-} + H_2O \rightarrow SnO_3^{2-} + 2H^+ + 2e^-$$

Ultimate (final) equation (in acid solution):

$$2Bi^{3+} + 3SnO_2^{2-} + 3H_2O \rightarrow 2Bi + 3SnO_3^{2-} + 6H^+$$

Ultimate (final) equation (after adding $6OH^-$ to each side and canceling $3H_2O$):

$$2Bi^{3+} + 3SnO_2^{2-} + 6OH^- \rightarrow 2Bi + 3SnO_3^{2-} + 3H_2O$$

■ Cumulative-Skills Problems
(requires skills from previous chapters)

3.79 For this reaction, the formulas are listed first, followed by the molecular and net ionic equations, the names of the products, and the molecular equation for another reaction giving the same precipitate.

Lead(II) nitrate is $Pb(NO_3)_2$, and cesium sulfate is Cs_2SO_4.

Molecular equation: $Pb(NO_3)_2(aq) + Cs_2SO_4(aq) \rightarrow PbSO_4(s) + 2CsNO_3(aq)$

Net ionic equation: $Pb^{2+}(aq) + SO_4^{2-}(aq) \rightarrow PbSO_4(s)$

$PbSO_4$ is lead(II) sulfate, and $CsNO_3$ is cesium nitrate.

Molecular equation: $Pb(NO_3)_2(aq) + Na_2SO_4(aq) \rightarrow PbSO_4(s) + 2NaNO_3\,(aq)$

3.80 For this reaction, the formulas are listed first, followed by the molecular and net ionic equations, the names of the products, and the molecular equation for another reaction giving the same precipitate.

Silver nitrate is $AgNO_3$, and strontium chloride is $SrCl_2$.

Molecular equation: $2AgNO_3(aq) + SrCl_2(aq) \rightarrow 2AgCl(s) + Sr(NO_3)_2(aq)$

Net ionic equation: $Ag^+(aq) + Cl^-(aq) \rightarrow AgCl(s)$

AgCl is silver(I) chloride, and $Sr(NO_3)_2$ is strontium nitrate.

Molecular equation: $AgClO_4(aq) + NaCl(aq) \rightarrow AgCl(s) + NaClO_4(aq)$

3.81 Net ionic equation: $2Br^-(aq) + Cl_2(g) \rightarrow 2Cl^-(aq) + Br_2(l)$

Molecular equation:	$CaBr_2(aq)$ +	$Cl_2(g)$	$\rightarrow CaCl_2(aq)$ +	$Br_2(l)$
Masses:	40.0 g	14.2 g	22.2 g	(40.0 + 14.2 - 22.2) g

Combining the three known masses gives the unknown mass of Br_2. Now use a ratio of the known masses of $CaBr_2$ to Br_2 to convert pounds of Br_2 to grams of $CaBr_2$:

$$10.0\text{ lb } Br_2 \times \frac{40.0\text{ g } CaBr_2}{(40.0 + 14.2 - 22.2)\text{ g } Br_2} \times \frac{453.6\text{ g}}{1\text{ lb}} = 56\underline{7}0$$

$$= 5.67 \times 10^3\text{ g } CaBr_2$$

3.82 Net ionic equation: $Ba^{2+}(aq) + CO_3^{2-}(aq) \rightarrow BaCO_3(s)$

Molecular equation: $BaS(aq) + Na_2CO_3(aq) \rightarrow Na_2S(aq) + BaCO_3(s)$

Masses: 33.9 g 21.2 g 15.6 g (33.9 + 21.2 - 15.6) g

Combining the three known masses gives the unknown mass of $BaCO_3$. Now use a ratio of the known masses of BaS to $BaCO_3$ to ultimately obtain grams of BaS:

$$10.0 \text{ ton } BaCO_3 \times \frac{33.9 \text{ g BaS}}{(33.9 + 21.2 - 15.6) \text{ g } BaCO_3} = 8.5\underline{8}2 \text{ ton BaS}$$

$$8.5\underline{8}2 \text{ ton BaS} \times \frac{2000 \text{ lb}}{1 \text{ ton}} \times \frac{453.6 \text{ g}}{1 \text{ lb}} = 7.7\underline{8}5 \times 10^6 = 7.79 \times 10^6 \text{ g BaS}$$

3.83 $Hg(NO_3)_2 + H_2S \rightarrow HgS\ (s) + 2\ HNO_3\ (aq)$

$Hg^{2+} + H_2S \rightarrow HgS\ (s) + 2\ H^+\ (aq)$

The acid formed is nitric acid, a strong acid. The other product is mercury(II) sulfide.

Mass HNO_3 = (81.15 g + 8.52 g) - 58.16 g = $31.5\underline{1}$ g

Mass of solution = $550.\underline{0}$ g H_2O + $31.5\underline{1}$ g HNO_3 = $581.\underline{5}1$ = 581.5 g

3.84 $Hg(NO_3)_2 + H_2S \rightarrow HgS\ (s) + 2\ HNO_3\ (aq)$

$Hg^{2+} + H_2S \rightarrow HgS\ (s) + 2\ H^+\ (aq)$

The acid formed is nitric acid, a strong acid. The other product is mercury(II) sulfide.

Mass HNO_3 = (65.65 g + 4.26 g) - 54.16 g = $15.7\underline{5}$ g

Mass of solution = $490.\underline{0}$ g H_2O + $15.7\underline{5}$ g HNO_3 = $505.\underline{7}5$ = 505.8 g

3.85 Let the number of Fe^{3+} ions = y; then the number of Fe^{2+} ions = (7 - y).

Since Fe_7S_8 is neutral, the number of positive charges must equal the number of negative charges. If the signs are omitted, then:

Total charge on both Fe^{2+} and Fe^{3+} = total charge on all 8 sulfide ions

$3y + 2(7 - y) = 8 \times 2$

$y + 14 = 16$

$y = 2$

So the ratio of Fe^{2+} to Fe^{3+} = 5/2

3.86 To define the problem in terms of percentages, use 100 X_2O_3 oxides. Then the number of X^{3+} ions = 100, and the sum of the X^{2+} and X^{5+} ions also = 100. Let the number of X^{2+} ions = y; then the number of X^{5+} ions = (100 - y). Since X_2O_3 is neutral, the number of positive charges must be equal the number of negative charges. If the signs are omitted, then;

Total charge on X^{2+}, X^{3+}, and X^{5+} = total charge of all 300 oxide ions

$2y + 100 \times 3 + 5(100 - y) = 300 \times 2$

$-3y + 800 = 600$

$y = (200 \div 3) = 66.67$

The percentage of X^{2+} in the 100 X_2O_3 's = $(66.67 \div 200) \times 100 = 33.3\%$

4. CALCULATIONS WITH CHEMICAL FORMULAS AND EQUATIONS

Solutions to Exercises

Note on significant figures: The final answer to each mathematical solution is given first with one nonsignificant figure (the rightmost significant figure is underlined) and then is rounded to the correct number of figures. Intermediate answers usually also have at least one nonsignificant figure.

4.1 a. NO_2

1 x AW of N		=	14.0067 amu
2 x AW of O	= 2 x 15.9994 amu	=	31.9988 amu
	MW of NO_2	=	46.0055 = 46.0 amu (3 s.f.)

b. $C_6H_{12}O_6$

6 x AW of C	= 6 x 12.011	=	72.066 amu
12 x AW of H	= 12 x 1.0079	=	12.0948 amu
6 x AW of O	= 6 x 15.9994	=	95.9964 amu
	MW of $C_6H_{12}O_6$	=	180.1572 amu = 180 amu (3 s.f.)

c. NaOH

1 x AW of Na	=	22.98977 amu
1 x AW of O	=	15.9994 amu
1 x AW of H	=	1.0079 amu
MW of NaOH	=	39.9971 amu = 40.0 amu (3 s.f.)

d. $Mg(OH)_2$

1 x AW of Mg		=	24.305 amu
2 x AW of O	= 2 x 15.9994	=	31.9988 amu
2 x AW of H	= 2 x 1.0079	=	2.0158 amu
MW of $Mg(OH)_2$		=	58.3196 amu = 58.3 amu (3 s.f.)

4.2 a. The atomic weight of Ca = 40.08 amu; thus, the molar mass = 40.08 g/mol, and 1 mol of Ca = 6.022×10^{23} Ca atoms.

$$\text{Mass of 1 Ca} = \frac{40.08 \text{ g}}{1 \text{ mol Ca}} = \frac{1 \text{ mol}}{6.022 \times 10^{23} \text{ atoms}} = 6.65\underline{5}6 \times 10^{-23}$$

$$= 6.656 \times 10^{-23} \text{ g/atom}$$

b. The molecular weight of C_2H_5OH, or C_2H_6O, = (2 x 12.01) + (6 x 1.008) + 16.00 = 46.0$\underline{6}$8. Its molar mass = 46.07 g/mol, and 1 mol = 6.022×10^{23} molecules of C_2H_6O.

$$\text{Mass of 1 } C_2H_6O = \frac{46.07 \text{ g}}{1 \text{ mol } C_2H_6O} = \frac{1 \text{ mol}}{6.022 \times 10^{23} \text{ atoms}}$$

$$= 7.65\underline{0}3 \times 10^{-23} = 7.650 \times 10^{-23} \text{ g/molecule}$$

4.3 The molar mass of H_2O_2 is 34.02 g/mol. Therefore,

$$0.909 \text{ mol } H_2O_2 \times \frac{34.02 \text{ g } H_2O_2}{1 \text{ mol } H_2O_2} = 30.\underline{9}2 = 30.9 \text{ g } H_2O_2$$

4.4 The molar mass of HNO_3 is 63.01 g/mol. Therefore,

$$28.5 \text{ g } HNO_3 \times \frac{1 \text{ mol } HNO_3}{63.01 \text{ g } HNO_3} = 0.45\underline{2}3 = 0.452 \text{ mol } HNO_3$$

4.5 Convert the mass of HCN from milligrams to grams. Then convert grams of HCN to moles of HCN. Finally, convert moles of HCN to the number of HCN molecules.

$$56 \text{ mg HCN} \times \frac{1 \text{ g}}{1000 \text{ mg}} \times \frac{1 \text{ mol HCN}}{27.02 \text{ g HCN}} \times \frac{6.022 \times 10^{23} \text{ HCN molec}}{1 \text{ mol HCN}}$$

$$= 1.\underline{2}48 \times 10^{21} = 1.2 \times 10^{21} \text{ HCN molec.}$$

4.6 The molecular weight of NH_4NO_3 = 80.05; thus, its molar mass = 80.05 g/mol. Hence

$$\%N = \frac{28.02\text{ g}}{80.05\text{ g}} \times 100\% = 35.\underline{0}0 = 35.0\%$$

$$\%H = \frac{4.032\text{ g}}{80.05\text{ g}} \times 100\% = 5.0\underline{3}6 = 5.04\%$$

$$\%N = \frac{48.00\text{ g}}{80.05\text{ g}} \times 100\% = 59.\underline{9}6 = 60.0\%$$

4.7 From the previous exercise, NH_4NO_3 is 35.0% N (fraction N = 0.350), so the mass of N in 48.5 g of NH_4NO_3 is

$$48.5\text{ g } NH_4NO_3 \times (0.350\text{ g N/1 g } NH_4NO_3) = 16.\underline{9}75 = 17.0\text{ g N}$$

4.8 First convert the mass of CO_2 to moles of CO_2. Next convert this to moles of C (1 mol of CO_2 is equivalent to 1 mol C). Finally, convert to mass of carbon, changing mg to g first:

$$5.80 \times 10^{-3}\text{g } CO_2 \times \frac{1\text{ mol } CO_2}{44.01\text{ g } CO_2} \times \frac{1\text{ mol C}}{1\text{ mol } CO_2} \times \frac{12.01\text{ g C}}{1\text{ mol C}}$$

$$= 1\ 5\underline{8}3 \times 10^{-3}\text{ g C}$$

Do the same series of calculations for water, noting that 1 mol H_2O contains 2 mol H.

$$1.58 \times 10^{-3}\text{g } H_2O \times \frac{1\text{ mol } H_2O}{18.02\text{ g } H_2O} \times \frac{2\text{ mol H}}{1\text{ mol } H_2O} \times \frac{1.008\text{ g H}}{1\text{ mol H}}$$

$$= 1.7\underline{6}7 \times 10^{-4}\text{ g H}$$

The mass percentages of C and H can be calculated using the masses from the previous calculations:

$$\%\ C = \frac{1.583\text{ mg}}{3.87\text{ mg}} \times 100\% = 40.\underline{9}0 = 40.9\%\text{ C}$$

$$\%\ H = \frac{0.1767\text{ mg}}{3.87\text{ mg}} \times 100\% = 4.5\underline{6}58 = 4.57\%\text{ H}$$

The mass percentage of O can be determined by subtracting the sum of the above percentages from 100%:

$$\%\ O = 100.000\% - (40.90 + 4.5658) = 54.\underline{5}342 = 54.5\%\text{ O}$$

4.9 Convert the masses to moles, which are proportional to subscripts in the empirical formula:

$$33.4 \text{ g S} \times \frac{1 \text{ mol S}}{32.07 \text{ g S}} = 1.0\underline{4}14 \text{ mol S}$$

$$(83.5 - 33.4) \text{ g O} \times \frac{1 \text{ mol O}}{16.00 \text{ g O}} = 3.1\underline{3}12 \text{ mol O}$$

Next obtain the smallest integers from the moles by dividing each by the smallest number of moles:

$$\text{For O: } \frac{3.1312}{1.0414} = 3.01 \qquad \text{For S: } \frac{1.0414}{1.0414} = 1.00$$

The empirical formula is SO_3.

4.10 For a 100.0-g sample of benzoic acid, 68.8 g are C, 5.0 g are H, and 26.2 g are O. Using the molar masses, convert these masses to moles:

$$68.8 \text{ g C} = 68.8 \text{ g C} \times \frac{1 \text{ mol C}}{12.01 \text{ g C}} = 5.7\underline{2}9 \text{ mol C}$$

$$5.0 \text{ g H} = 5.0 \text{ g H} \times \frac{1 \text{ mol H}}{1.008 \text{ g H}} = 4.\underline{9}6 \text{ mol H}$$

$$26.2 \text{ g O} = 26.2 \text{ g O} \times \frac{1 \text{ mol O}}{16.00 \text{ g O}} = 1.6\underline{3}8 \text{ mol O}$$

These numbers are in the same ratio as the subscripts in the empirical formula. They must be changed to integers. First divide each one by the smallest mol number:

$$\text{For C: } \frac{5.729}{1.638} = 3.497 \qquad \text{For H: } \frac{4.96}{1.638} = 3.03$$

$$\text{For O: } \frac{1.638}{1.638} = 1.000$$

Rounding off, we obtain $C_{3.5}H_{3.0}O_{1.0}$. Multiplying the numbers by 2 gives whole numbers, for an empirical formula of $C_7H_6O_2$.

4.11 For a 100.0-g sample of acetaldehyde, 54.5 g are C, 9.2 g are H, and 36.3 g are O. Using the molar masses, convert these masses to moles:

$$54.5 \text{ g C} = 54.5 \text{ g C} \times \frac{1 \text{ mol C}}{12.01 \text{ g C}} = 4.5\underline{3}7 \text{ mol C}$$

$$9.2 \text{ g H} = 9.2 \text{ g H} \times \frac{1 \text{ mol H}}{1.008 \text{ g H}} = 9.\underline{1}2 \text{ mol H}$$

$$36.3 \text{ g O} = 36.3 \text{ g O} \times \frac{1 \text{ mol O}}{16.00 \text{ g O}} = 2.2\underline{6}8 \text{ mol O}$$

These numbers are in the same ratio as the subscripts in the empirical formula. They must be changed to integers. First divide each one by the smallest mol number:

$$\text{For C: } \frac{4.537}{2.268} = 2.000 \qquad \text{For H: } \frac{9.12}{2.268} = 4.02$$

$$\text{For O: } \frac{2.268}{2.268} = 1.000$$

Rounding off, we obtain C_2H_4O, the empirical formula, which is also the molecular formula.

4.12 H_2 + Cl_2 → 2HCl
1 molec. (mol) H_2 + 1 molec. (mol) Cl_2 → 2 molec. (mol) HCl (molec., mole interp.)
2.016 g H_2 + 70.9 g Cl_2 → 2 x 36.5 g HCl (mass interp.)

4.13 Equation: Na + H_2O → $1/2H_2$ + NaOH, or 2Na + $2H_2O$ → H_2 + 2NaOH

From the above equation, 1 mol of Na corresponds to 1/2 mol of H_2, or 2 mol of Na corresponds to 1 mol of H_2. Therefore,

$$7.81 \text{ g } H_2 \times \frac{1 \text{ mol } H_2}{2.016 \text{ g } H_2} \times \frac{2 \text{ mol Na}}{1 \text{ mol } H_2} \times \frac{22.99 \text{ g Na}}{1 \text{ mol Na}} = 17\underline{8}.1 = 178 \text{ g Na}$$

4.14 Balanced equation: 2ZnS + $3O_2$ → 2ZnO + $2SO_2$

Convert grams of ZnS to moles of ZnS. Then determine the relationship between ZnS and O_2 (2ZnS is equivalent to $3O_2$). Finally, convert to mass O_2.

$$5.00 \times 10^3 \text{ g ZnS} \times \frac{1 \text{ mol ZnS}}{97.46 \text{ g ZnS}} \times \frac{3 \text{ mol } O_2}{2 \text{ mol ZnS}} \times \frac{32.00 \text{ g } O_2}{1 \text{ mol } O_2} \times \frac{1 \text{ kg}}{1000 \text{ g}}$$

$$= 2.4\underline{6}3 = 2.46 \text{ kg } O_2$$

4.15 Balanced equation: $2HgO \rightarrow 2Hg + O_2$

Convert the mass of O_2 to mol of O_2. Using the fact that 1 mol of O_2 is equivalent to 2 mol of Hg, determine the number of mol of Hg, and convert to mass of Hg.

$$6.47 \text{ g } O_2 \times \frac{1 \text{ mol } O_2}{32.00 \text{ g } O_2} \times \frac{2 \text{ mol Hg}}{1 \text{ mol } O_2} \times \frac{200.59 \text{ g Hg}}{1 \text{ mol Hg}} = 81.\underline{1}1$$

$$= 81.1 \text{ g Hg}$$

4.16 First determine the limiting reactant by calculating the moles of $AlCl_3$ that would be obtained if Al and HCl were totally consumed:

$$0.15 \text{ mol Al} \times \frac{2 \text{ mol } AlCl_3}{2 \text{ mol Al}} = 0.1\underline{5}0 \text{ mol } AlCl_3$$

$$0.35 \text{ mol HCl} \times \frac{2 \text{ mol } AlCl_3}{6 \text{ mol HCl}} = 0.1\underline{1}66 \text{ mol } AlCl_3$$

Because the HCl produces the smaller amount of $AlCl_3$, the reaction will stop when HCl is totally consumed but before the Al is consumed. The limiting reactant is therefore HCl. The amount of $AlCl_3$ produced must be 0.1$\underline{1}$66, or 0.12 mol.

4.17 First determine the limiting reactant by calculating the moles of ZnS produced by totally consuming Zn and S_8:

$$7.36 \text{ g Zn} \times \frac{1 \text{ mol Zn}}{65.39 \text{ g Zn}} \times \frac{8 \text{ mol ZnS}}{8 \text{ mol Zn}} = 0.11\underline{2}56 \text{ mol ZnS}$$

$$6.45 \text{ s } S_8 \times \frac{1 \text{ mol } S_8}{256.56 \text{ g } S_8} \times \frac{8 \text{ mol ZnS}}{1 \text{ mol } S_8} \; 0.20\underline{1}1 \text{ mol ZnS}$$

The reaction will stop when Zn is totally consumed; S_8 is in excess and not all of it is converted to ZnS. The limiting reactant is therefore Zn. Now convert the moles of ZnS obtained from the Zn to grams of ZnS:

$$0.11256 \text{ mol Zn} \times \frac{97.46 \text{ g ZnS}}{1 \text{ mol ZnS}} = 10.\underline{9}7 = 11.0 \text{ g ZnS}$$

4.18 First write the balanced equation:

$$CH_3OH + CO \rightarrow HC_2H_3O_2$$

Convert grams of each reactant to moles of acetic acid:

$$15.0 \text{ g } CH_3OH \times \frac{1 \text{ mol } CH_3OH}{32.04 \text{ g } CH_3OH} \times \frac{1 \text{ mol } HC_2H_3O_2}{1 \text{ mol } CH_3OH} = 0.46\underline{8}1 \text{ mol } HC_2H_3O_2$$

$$10.0 \text{ g CO} \times \frac{1 \text{ mol CO}}{28.01 \text{ g CO}} \times \frac{1 \text{ mol } HC_2H_3O_2}{1 \text{ mol } CH_3OH} = 0.35\underline{7}0 \text{ mol } HC_2H_3O_2$$

Thus, CO is the limiting reactant, and 0.03570 mol $HC_2H_3O_2$ is obtained. The mass of product is

$$0.3570 \text{ mol } HC_2H_3O_2 \times \frac{60.05 \text{ g } HC_2H_3O_2}{1 \text{ mol } HC_2H_3O_2} = 21.\underline{4}4 \text{ g } HC_2H_3O_2$$

The percentage yield is:

$$\frac{19.1 \text{ g actual yield}}{21.44 \text{ g theo. yield}} \times 100\% = 89.\underline{0}8 = 89.1\%$$

4.19 Convert mass of NaCl to moles of NaCl. Then divide moles of solute by liters of solution.

$$0.0678 \text{ g NaCl} \times \frac{1 \text{ mol NaCl}}{58.44 \text{ g NaCl}} = 1.1\underline{6}0 \times 10^{-3} \text{ mol NaCl}$$

$$25.0 \text{ mL} \times \frac{1 \text{ L}}{1000 \text{ mL}} = 0.0250 \text{ L}$$

$$\text{Molarity} = \frac{1.160 \times 10^{-3} \text{ mol NaCl}}{0.0250 \text{ L soln}} = 0.046\underline{4}1 = 0.0464M$$

4.20 Convert grams of NaCl to moles NaCl to volume of NaCl solution.

$$0.0958 \text{ g NaCl} \times \frac{1 \text{ mol NaCl}}{58.44 \text{ g NaCl}} \times \frac{1 \text{ L soln}}{0.163 \text{ mol NaCl}} \times \frac{1000 \text{ mL}}{1 \text{ L}}$$

$$= 10.\underline{0}6 = 10.1 \text{ mL NaCl}$$

4.21 One (1) liter of solution is equivalent to 0.15 mol NaCl. The amount of NaCl in 50.0 mL of solution is

$$50.0 \text{ mL} \times \frac{1 \text{ L}}{1000 \text{ mL}} \times \frac{0.15 \text{ mol NaCl}}{1 \text{ L soln}} = 0.007\underline{5}0 \text{ mol NaCl}$$

$$0.00750 \text{ mol NaCl} \times \frac{58.4 \text{ g NaCl}}{1 \text{ mol NaCl}} = 0.4\underline{3}8 = 0.44 \text{ g NaCl}$$

4.22 Use the rearranged version of the dilution formula from the text to calculate the initial volume of 1.5 M sulfuric acid required:

$$V_i = \frac{M_f V_f}{M_i} = \frac{0.18 \text{ M} \times 100.0 \text{ mL}}{1.5 \text{ M}} = 1\underline{2}.0 = 12 \text{ mL}$$

4.23 Follow the method of Example 4.12 to convert the mass of CaC_2O_4 to mass of Ca^{2+}:

$$0.1402 \text{ g } CaC_2O_4 \times \frac{1 \text{ mol } CaC_2O_4}{128.10 \text{ g } CaC_2O_4} \times \frac{1 \text{ mol Ca}}{1 \text{ mol } CaC_2O_4} \times \frac{40.08 \text{ g Ca}}{1 \text{ mol Ca}}$$

$$= 0.04386\underline{6}6 \text{ g Ca}$$

Now calculate the percentage of calcium in the 128.3 mg (0.1283 g) limestone:

$$\frac{0.04386\underline{6}6 \text{ g Ca}}{0.1283 \text{ g limestone}} \times 100\% = 34.1\underline{9}0 = 34.19\%$$

4.24 Convert the volume of Na_3PO_4 to moles using the molarity of Na_3PO_4.

$$45.7 \times 10^{-3} \text{ L } Na_3PO_4 \times \frac{0.265 \text{ mol } Na_3PO_4}{1 \text{ L}} = 0.012\underline{1}1 \text{ mol } Na_3PO_4$$

Finally, calculate the amount of $NiSO_4$ required to react with this amount of Na_3PO_4:

$$0.1211 \text{ mol } Na_3PO_4 \times \frac{3 \text{ mol } NiSO_4}{2 \text{ mol } Na_3PO_4} \times \frac{1 \text{ L } NiSO_4}{0.375 \text{ M } NiSO_4}$$

$$= 0.048\underline{4}4 \text{ L } (48.4 \text{ mL } NiSO_4)$$

4.25 Determine the moles of NaOH needed for the titration.

$$0.0391 \text{ L NaOH} \times \frac{0.108 \text{ mol NaOH}}{1 \text{ L}} = 0.0042\underline{2}3 \text{ mol NaOH}$$

$$= 0.0042\underline{2}3 \text{ mol } HC_2H_3O_2$$

Because the moles of NaOH are equal to the moles of $HC_2H_3O_2$, the latter can be converted to grams of $HC_2H_3O_2$ and then to the mass percentage.

$$0.004223 \text{ mol } HC_2H_3O_2 \times \frac{60.05 \text{ g } HC_2H_3O_2}{1 \text{ mol } HC_2H_3O_2} = 0.25\underline{3}59 \text{ g}$$

$$\text{Mass percentage} = \frac{0.25359 \text{ g } HC_2H_3O_2}{5.00 \text{ g vinegar}} \times 100\% = 5.0\underline{7}1 = 5.07\%$$

■ Answers to Review Questions

4.1 The molecular weight is the sum of the atomic weights of all the atoms in a *molecule,* whereas the formula weight is the sum of the atomic weights of all the atoms in one *formula unit* of the compound, whether the compound is molecular or not. A given substance could have both a molecular weight and a formula weight if it existed as discrete molecules.

4.2 A formula weight equals the sum of the atomic weights of all atoms in a formula.

4.3 A mole of N_2 contains Avogadro's number of N_2 molecules and 2 x Avogadro's number of N atoms. One mole of $Fe_2(SO_4)_3$ contains 3 moles of SO_4^{2-} ions; it contains 12 moles of O.

4.4 Because the ratio of the masses of N and O in N_2O_4 is the same as that in NO_2, the N_2O_4 will have the same percentage composition as NO_2: 30.5% N and 69.5% O.

4.5 A sample of the compound of known mass is burned. The percentage composition of the compound is determined from the masses of CO_2 and H_2O obtained.

4.6 The empirical formula is obtained from the percentage composition by considering a sample of 100 g of the pure substance. Then multiply the 100 g by the fraction of each element to find the masses of each element in the 100 g. Convert the masses of the elements to moles of the elements using the atomic mass of each element. Divide the moles of each by the smallest number to obtain the smallest ratio of each atom. If necessary, multiply by a whole number to obtain integers to obtain the subscripts in the empirical formula.

4.7 The number, *n*, of empirical formula units in the molecule is unknown. On inspecting the molecular formula of $C_6H_{12}O_2$, we see that there are two O's in the formula and that all subscripts are divisible by 2. Hence, the empirical formula is C_3H_6O (and $n = 2$).

4.8 Because *n*, the number of empirical formula units, equals the molecular weight divided by the empirical formula weight, the molecular formula equals (34.0 ÷ 17.0) x HO = H_2O_2.

4.9 Using one molecule, one mole, and the molar mass of 16.0 g for CH_4, we can summarize the three interpretations in three lines below the balanced equation:

CH_4	+	$2O_2$	→	CO_2	+	$2H_2O$
1 molecule	+	2 molecules	→	1 molecule	+	2 molecules
1 mole	+	2 moles	→	1 mole	+	2 moles
16.0 g	+	2 x 32.0 g	→	44.0 g	+	2 x 18.0 g

4.10 A chemical equation yields the mole ratio of a reactant to a second reactant or product. Once the mass of a reactant is converted to moles, this can be multiplied by the appropriate mole ratio to give the moles of a second reactant or product. Multiplying this number of moles by the appropriate molar mass gives mass. Thus, the masses of two different substances are related by a chemical equation.

4.11 The limiting reactant is the reactant that is entirely consumed when the reaction is complete. Because the reaction stops when the limiting reactant is used up, the moles of product are always determined by the starting number of moles of the limiting reactant.

4.12 By definition,

$$\text{Molar concentration} = \frac{\text{moles of solute}}{\text{liters of solution}}$$

Therefore,

$$\text{Moles of solute} = \text{molar concentration} \times \text{liters of solution}$$

4.13 In dilution, only additional water, not the solute, is added to the container. Thus, the moles of solute does not change during dilution. In mathematical terms, the product of molar concentration and liters of solution remains constant.

4.14 The number of moles of hydrochloric acid reacted equals the product of molar concentration and volume of acid used in the titration. From the chemical equation, we see that the number of moles of hydrochloric acid equals the number of moles of sodium hydroxide.

■ Solutions to Practice Problems

Note on significant figures: The final answer to each problem is given first with one nonsignificant figure (the rightmost significant figure is underlined), and then is rounded to the correct number of significant figures. Intermediate answers usually also have at least one nonsignificant figure. Atomic weights are rounded to two decimal places, except for that of hydrogen.

4.15 a. Formula weight of CH_3OH = AW of C + 4(AW of H) + AW of O. Using the values of atomic weights in the periodic table (inside front cover) rounded to four significant figures, we have

$$FW = 12.01 \text{ amu} + (4 \times 1.008 \text{ amu}) + 16.00 \text{ amu} = 32.0\underline{4}2 = 32.0 \text{ amu (3 s.f.)}$$

b. FW of PCl_3 = AW of P + 3(AW of Cl)

$$= 30.97 \text{ amu} + (3 \times 35.45 \text{ amu}) = 137.3\underline{2}0 = 137 \text{ amu (3 s.f.)}$$

c. FW of K_2CO_3 = 2(AW of K) + AW of C + 3(AW of O)

$$= (2 \times 39.10 \text{ amu}) + 12.01 \text{ amu} + (3 \times 16.00 \text{ amu})$$

$$= 138.2\underline{1}0 = 138 \text{ amu (3 s.f.)}$$

d. FW of $Ni_3(PO_4)_2$ = 3(AW of Ni) + 2(AW of P) + 8(AW of O)

$$= (3 \times 58.70 \text{ amu}) + (2 \times 30.97 \text{ amu}) + (8 \times 16.00 \text{ amu})$$

$$= 366.0\underline{4}0 = 366 \text{ amu (3 s.f.)}$$

4.16 a. FW of $HC_2H_3O_2$ = 2(AW of C) + 4(AW of H) + 2(AW of O)

$$= (2 \times 12.01) + (4 \times 1.008) + (2 \times 16.00)$$

$$= 60.0\underline{5}2 = 60.1 \text{ amu (3 s.f.)}$$

b. FW of PCl_5 = AW of P + 5(AW of Cl)

$$= 30.97 \text{ amu} + (5 \times 35.45 \text{ amu})$$

$$= 208.2\underline{2}0 = 208 \text{ amu (3 s.f.)}$$

c. FW of K_2SO_4 = 2(AW of K) + AW of S + 4(AW of O)

$$= (2 \times 39.10 \text{ amu}) + 32.07 \text{ amu} + (4 \times 16.00 \text{ amu})$$

$$= 174.2\underline{7}0 = 174 \text{ amu (3 s.f.)}$$

d. FW of $Ca(OH)_2$ = AW of Ca + 2(AW of H) + 2(AW of O)

= 40.08 amu + (2 x 1.008 amu) + (2 x 16.00 amu)

= $74.0\underline{9}6$ = 74.1 amu (3 s.f.)

4.17 First find the formula weight of NH_4NO_3 by adding the respective atomic weights. Then convert it to the molar mass:

FW of NH_4NO_3 = 2(AW of N) + 4(AW of H) + 3(AW of O)

= (2 x 14.01 amu) + (4 x 1.008 amu) + (3 x 16.00 amu)

= $80.0\underline{5}2$ amu

The molar mass of NH_4NO_3 = 80.05 g/mol.

4.18 First find the formula weight of H_3PO_4 by adding the respective atomic weights. Then convert it to the molar mass:

FW of H_3PO_4 = 3(AW of H) + AW of P + 4(AW of O)

= (3 x 1.008) + 30.97 + (4 x 16.00)

= $97.9\underline{9}4$ amu

The molar mass of H_3PO_4 = 97.99 g/mol.

4.19 a. The atomic weight of Na = 22.99 amu; thus, the molar mass = 22.99 g/mol. Because 1 mol of Na atoms = 6.022×10^{23} Na atoms, we calculate

$$\text{Mass of one Na atom} = \frac{22.99 \text{ g/mol}}{6.022 \times 10^{23} \text{ atom/mol}} = 3.81\underline{7}7 \times 10^{-23} \text{ g/atom}$$

b. The atomic weight of S = 32.07 amu; thus, the molar mass = 32.07 g/mol. Because 1 mol of S atoms = 6.022×10^{23} S atoms, we calculate

$$\text{Mass of one S atom} = \frac{32.07 \text{ g/mol}}{6.022 \times 10^{23} \text{ atom/mol}} = 5.32\underline{5}47 \times 10^{-23} \text{ g/atom}$$

c. The formula weight of CH_3Cl = [12.01 + (3 x 1.008) + 35.45] = 50.48 amu; thus, the molar mass = 50.48 g/mol. Because 1 mol of CH_3Cl molecules = 6.022×10^{23} CH_3Cl molecules, we calculate

$$\text{Mass of one } CH_3Cl \text{ molec} = \frac{50.48 \text{ g/mol}}{6.022 \times 10^{23} \text{ molec./mol}}$$

$$= 8.38\underline{2}6 \times 10^{-23} \text{ g/molec}$$

d. The formula weight of Na_2SO_3 = [(2 x 22.99) + 32.07 + (3 x 16.00)] = 126.05 amu; thus, the molar mass = 126.05 g/mol. Because 1 formula weight of Na_2SO_3 = 6.022×10^{23} Na_2SO_3 formula units, we calculate

$$\text{Mass of one } Na_2SO_3 = \frac{126.05 \text{ g/mol}}{6.022 \times 10^{23} \text{ units/mol}} = 2.09\underline{3}2 \times 10^{-22} \text{ g/unit}$$

4.20 As in the previous problem, the atomic weights are used with units of g/mol. In addition, the formula weights have been found by addition of the atomic weights and expressed in g/mol.

a. $$\text{Mass of one Fe atom} = \frac{55.85 \text{ g/mol}}{6.022 \times 10^{23} \text{ atom/mol}} = 9.27\underline{4}3 \times 10^{-23} \text{ g/atom}$$

b. $$\text{Mass of one F atom} = \frac{19.00 \text{ g/mol}}{6.022 \times 10^{23} \text{ atom/mol}} = 3.15\underline{5}1 \times 10^{-23} \text{ g/atom}$$

c. $$\text{Mass of one } N_2O \text{ molec.} = \frac{44.02 \text{ g/mol}}{6.022 \times 10^{23} \text{ atom/mol}} = 7.30\underline{9}9 \times 10^{-23} \text{ g/molec.}$$

d. $$\text{Mass of one } K_2CrO_4 = \frac{194.20 \text{ g/mol}}{6.022 \times 10^{23} \text{ atom/mol}} = 3.22\underline{4}8 \times 10^{-22} \text{ g/unit}$$

4.21 First find the formula weight (in amu) using the periodic table (inside front cover):

$$\text{FW of } (CH_3CH_2)_2O = (4 \times 12.01 \text{ amu}) + (10 \times 1.008 \text{ amu}) + 16.00 \text{ amu}$$

$$= 74.12 \text{ amu}$$

$$\text{Mass of } (CH_3CH_2)_2O \text{ molec.} = \frac{74.12 \text{ g/mol}}{6.022 \times 10^{23} \text{ atom/mol}}$$

$$= 1.23\underline{0}8 \times 10^{-22} \text{ g/molec.}$$

4.22 First find the formula weight (in amu) using the periodic table (inside front cover):

$$\text{FW of glycerol} = (3 \times 12.01\text{ amu}) + (8 \times 1.008\text{ amu}) + (3 \times 16.00\text{ amu})$$

$$= 92.09\text{ amu}$$

$$\text{Mass of glycerol molec} = \frac{92.09\text{ g/mol}}{6.022 \times 10^{23}\text{ atom/mol}} = 1.5\underline{2}9 \times 10^{-22}\text{ g/molec.}$$

4.23 From the table of atomic weights, we obtain the following molar masses for parts a through d: Na = 22.99 g/mol; S = 32.07 g/mol; C = 12.01 g/mol; H = 1.008 g/mol; Cl = 35.45 g/mol; and O = 16.00 g/mol.

a. $0.15\text{ mol Na} \times \frac{22.99\text{ g Na}}{1\text{ mol Na}} = 3.\underline{4}48 = 3.4\text{ g Na}$

b. $0.594\text{ mol S} \times \frac{32.07\text{ g S}}{1\text{ mol S}} = 19.\underline{0}4 = 19.0\text{ g S}$

c. Using molar mass = 50.48 g/mol for CH_3Cl, we obtain

$$2.78\text{ mol }CH_3Cl \times \frac{50.48\text{ g }CH_3Cl}{1\text{ mol }CH_3Cl} = 14\underline{0}.3 = 140.\text{ g }CH_3Cl$$

d. Using molar mass = 126.05 g/mol for Na_2SO_3, we obtain

$$38\text{ mol }Na_2SO_3 \times \frac{126.05\text{ g }Na_2SO_3}{1\text{ mol }Na_2SO_3} = 4.\underline{7}90 \times 10^3$$

$$= 4.8 \times 10^3\text{ g }Na_2SO_3$$

4.24 From the table of atomic weights, we obtain the following molar masses for parts a through d: N = 14.01 g/mol; O = 16.00 g/mol; K = 39.10 g/mol; and Cr = 52.00 g/mol.

a. $0.205\text{ mol Fe} \times \frac{55.85\text{ g Fe}}{1\text{ mol Fe}} = 11.\underline{4}4 = 11.4\text{ g Fe}$

b. $0.83\text{ mol F} \times \frac{19.00\text{ g F}}{1\text{ mol F}} = 1\underline{5}.77 = 16\text{ g F}$

c. Using molar mass = 44.02 g/mol for N_2O, we obtain

$$5.8\text{ mol }N_2O \times \frac{44.02\text{ g }N_2O}{1\text{ mol }N_2O} = 2\underline{5}5.3 = 2.6 \times 10^2\text{ g }N_2O$$

d. Using molar mass 194.20 g/mol for K_2CrO_4, we obtain

$$48.1 \text{ mol } K_2CrO_4 \times \frac{194.20 \text{ g } K_2CrO_4}{1 \text{ mol } K_2CrO_4} = 93\underline{4}1.02$$

$$= 9.34 \times 10^3 \text{ g } K_2CrO_4$$

4.25 First find the molar mass of H_3BO_3: (3 x 1.008 amu) + 10.81 amu + (3 x 16.00 amu) = 61.83. Therefore, the molar mass of H_3BO_3 = 61.83 g/mol. The mass of H_3BO_3 is calculated as follows:

$$0.543 \text{ mol } H_3BO_3 \times \frac{61.83 \text{ g } H_3BO_3}{1 \text{ mol } H_3BO_3} = 33.\underline{5}7 = 33.6 \text{ g } H_3BO_3$$

4.26 First find the molar mass of CS_2: 12.01 amu + (2 x 32.07 amu) = 76.15 amu. Therefore, the molar mass of CS_2 = 76.15 g/mol. The mass of CS_2 is calculated as follows:

$$0.0205 \text{ mol } CS_2 \times \frac{76.15 \text{ g } CS_2}{1 \text{ mol } CS_2} = 1.5\underline{6}10 = 1.56 \text{ g } CS_2$$

4.27 From the table of atomic weights, we obtain the following rounded molar masses for parts a through d: C = 12.01 g/mol; Br = 79.90 g/mol; H = 1.008 g/mol; Li = 6.94 g/mol; and O = 16.00 g/mol.

a. $$3.43 \text{ g C} \times \frac{1 \text{ mol C}}{12.01 \text{ g C}} = 0.28\underline{5}59 = 0.286 \text{ mol C}$$

b. $$7.05 \text{ g } Br_2 \times \frac{1 \text{ mol } Br_2}{159.80 \text{ g } Br_2} = 0.044\underline{1}1 = 0.0441 \text{ mol } Br_2$$

c. The molar mass of C_4H_{10} = (4 x 12.01) + (10 x 1.008) = 58.12 g C_4H_{10}/mol C_4H_{10}. The mass of C_4H_{10} is calculated as follows:

$$76 \text{ g } C_4H_{10} \times \frac{1 \text{ mol } C_4H_{10}}{58.12 \text{ g } C_4H_{10}} = 1.\underline{3}07 = 1.3 \text{ mol } C_4H_{10}$$

d. The molar mass of Li_2CO_3 = (2 x 6.94) + 12.01 + (3 x 16.00 g) = 73.89 g Li_2CO_3/mol Li_2CO_3. The mass of Li_2CO_3 is calculated as follows:

$$35.4 \text{ g } Li_2CO_3 \times \frac{1 \text{ mol } Li_2CO_3}{73.89 \text{ g } Li_2CO_3} = 0.47\underline{9}09 = 0.479 \text{ mol } Li_2CO_3$$

4.28 From the table of atomic weights, we obtain the following rounded molar masses for parts a through d: As = 74.92 g/mol; P = 30.97 g/mol; N = 14.01 g/mol; H = 1.008 g/mol; Al = 26.98 g/mol; and O = 16.00 g/mol.

a. $$2.57 \text{ g As} \times \frac{1 \text{ mol As}}{74.92 \text{ g As}} = 0.034\underline{3}03 = 0.0343 \text{ mol As}$$

b. $$7.83 \text{ g } P_4 \times \frac{1 \text{ mol } P_4}{123.88 \text{ g } P_4} = 0.063\underline{2}06 = 0.0632 \text{ mol } P_4$$

c. The molar mass of N_2H_4 = (2 x 14.01) + (4 x 1.008) = 32.052 g N_2H_4/mol N_2H_4. The mass of N_2H_4 is calculated as follows:

$$41.4 \text{ g } N_2H_4 \times \frac{1 \text{ mol } N_2H_4}{32.052 \text{ g } N_2H_4} = 1.29\underline{1}6 = 1.29 \text{ mol } N_2H_4$$

d. The molar mass of $Al_2(SO_4)_3$ = (2 x 26.98) + (3 x 32.07) + (12 x 16.00) = 342.17 g $Al_2(SO_4)_3$/mol $Al_2(SO_4)_3$. The mass of $Al_2(SO_4)_3$ is calculated as follows:

$$153 \text{ g } Al_2(SO_4)_3 \times \frac{1 \text{ mol } Al_2(SO_4)_3}{342.17 \text{ g } Al_2(SO_4)_3} = 0.44\underline{7}15$$

$$= 0.447 \text{ mol } Al_2(SO_4)_3$$

4.29 Calculate the formula weight of calcium sulfate: 40.08 amu + 32.07 amu + (4 x 16.00 amu) = 136.15 amu. Therefore, the molar mass of $CaSO_4$ is 136.15 g/mol. Use this to convert the mass of $CaSO_4$ to moles:

$$0.791 \text{ g } CaSO_4 \times \frac{1 \text{ mol } CaSO_4}{136.15 \text{ g } CaSO_4} = 0.0058\underline{1}1 \text{ mol } CaSO_4$$

Calculate the molecular weight of water: (2 x 1.008 amu) + 16.00 amu = 18.02 amu. Therefore, the molar mass of H_2O = 18.02 g/mol. Use this to convert the rest of the sample to moles of water:

$$0.209 \text{ g } H_2O \times \frac{1 \text{ mol } H_2O}{18.02 \text{ g } H_2O} = 0.011\underline{5}9 \text{ mol } H_2O$$

Because 0.01159 mol is about twice 0.005811 mol, both numbers of moles are consistent with the formula $CaSO_4 \cdot 2H_2O$.

4.30 Calculate the formula weight of copper(II) sulfate: 63.55 amu + 32.07 amu + (4 x 16.00 amu) = 159.62 amu. Thus, the molar mass of $CuSO_4$ is 159.62 g/mol. From the previous problem, the molar mass of H_2O is 18.02 g/mol. Use this to convert the rest of the sample to moles of water:

$$0.558 \text{ g } H_2O \times \frac{1 \text{ mol } H_2O}{18.02 \text{ g } H_2O} = 0.030\underline{9}6 \text{ mol } H_2O$$

Calculate the moles of $CuSO_4$ to be able to compare relative molar amounts of $CuSO_4$ and H_2O. Then divide the moles of H_2O by the moles of $CuSO_4$:

$$0.989 \text{ g } CuSO_4 \times \frac{1 \text{ mol } CuSO_4}{159.62 \text{ g } CuSO_4} = 0.0061\underline{9}6 \text{ mol } CuSO_4$$

$$0.03096 \text{ mol } H_2O \div 0.006196 \text{ mol } CuSO_4 = 4.99/1, \text{ or about } 5{:}1$$

(consistent with $CuSO_4{\bullet}5H_2O$)

4.31 The following rounded atomic weights are used: Li = 6.94 g/mol; Br = 79.90 g/mol; N = 14.01 g/mol; H = 1.008 g/mol; Pb = 207.2 g/mol; Cr = 52.00 g/N_A; O = 16.00 g/mol; and S = 32.07 g/mol. Also, Avogadro's number is 6.022×10^{23} atoms, so

a. $$\text{No. Li atoms} = 7.46 \text{ g Li} \times \frac{6.022 \times 10^{23} \text{ atoms}}{6.941 \text{ g Li}} = 6.4\underline{7}3 \times 10^{23} \text{ atoms}$$

b. $$\text{No. Br atoms} = 32.0 \text{ g } Br_2 \times \frac{2 \times 6.022 \times 10^{23} \text{ atoms}}{(2 \times 79.90) \text{ g } Br_2}$$

$$= 2.4\underline{1}2 \times 10^{23} \text{ atoms}$$

c. $$\text{No. } NH_3 \text{ molec.} = 43 \text{ g } NH_3 \times \frac{6.022 \times 10^{23} \text{ molec.}}{17.03 \text{ g } NH_3}$$

$$= 1.\underline{5}2 \times 10^{24} \text{ molec.}$$

d. $$\text{No. } PbCrO_4 \text{ units} = 159 \text{ g } PbCrO_4 \times \frac{6.022 \times 10^{23} \text{ units}}{323.2 \text{ g } PbCrO_4}$$

$$= 2.9\underline{6} \times 10^{23} \text{ units}$$

e. $$\text{No. } SO_4^{2-} \text{ ions} = 14.3 \text{ g } Cr_2(SO_4)_3 \times \frac{3 \times 6.022 \times 10^{23} \text{ ions}}{392.21 \text{ g } Cr_2(SO_4)_3}$$

$$= 6.5\underline{8}7 \times 10^{22} \text{ ions}$$

4.32 These rounded atomic weights are used: Al = 26.98 g/mol; I = 126.90 g/mol; N = 14.01 g/mol; O = 16.00 g/mol; Na = 22.99 g/mol; Cl = 35.45 g/mol; Ca = 40.08 g/mol; and P = 30.97 g/mol. Also, Avogadro's number is 6.022×10^{23} atoms, so

a. $$\text{No. Al atoms} = 25.7 \text{ g Al} \times \frac{6.022 \times 10^{23} \text{ atoms}}{26.98 \text{ g Al}} = 5.7\underline{3}6 \times 10^{23} \text{ atoms}$$

b. $$\text{No. I atoms} = 7.92 \text{ g } I_2 \times \frac{2 \times 6.022 \times 10^{23} \text{ atoms}}{2 \times 126.90 \text{ g } I_2} = 3.7\underline{5}8 \times 10^{22} \text{ atoms}$$

c. $$\text{No. } N_2O_5 \text{ molec.} = 38.1 \text{ g } N_2O_5 \times \frac{6.022 \times 10^{23} \text{ molec.}}{108.02 \text{ g } N_2O_5}$$

$$= 2.1\underline{2}4 \times 10^{23} \text{ molec.}$$

d. $$\text{No. } NaClO_4 \text{ units} = 3.31 \text{ g } NaClO_4 \times \frac{6.022 \times 10^{23} \text{ units}}{122.44 \text{ g } NaClO_4}$$

$$= 1.6\underline{2}8 \times 10^{22} \text{ units}$$

e. $$\text{No. } Ca^{2+} \text{ ions} = 6.54 \text{ g } Ca_3(PO_4)_2 \times \frac{3 \times 6.022 \times 10^{23} \text{ ions}}{310.18 \text{ g } Ca_3(PO_4)_2}$$

$$= 3.8\underline{0}9 \times 10^{22} \text{ ions}$$

4.33 Calculate the molecular weight of CCl_4: 12.01 amu + (4 x 35.45 amu) = 153.81 amu. Use this and Avogadro's number to express it as 153.81 g/N_A to calculate the number of molecules:

$$7.58 \text{ mg } CCl_4 \times \frac{1 \text{ g}}{1000 \text{ mg}} \times \frac{6.022 \times 10^{23} \text{ molec.}}{153.81 \text{ g } CCl_4} = 2.9\underline{6}8 \times 10^{19}$$

$$= 2.97 \times 10^{19} \text{ molec.}$$

4.34 Calculate the molecular weight of ClF_3: 35.45 amu + (3 x 19.00 amu) = 92.45 amu. Use this and Avogadro's number to express it as 92.45 g/N_A to calculate the number of molecules:

$$5.88 \text{ mg } ClF_3 \times \frac{1 \text{ g}}{1000 \text{ mg}} \times \frac{6.022 \times 10^{23} \text{ molec.}}{92.45 \text{ g } ClF_3} = 3.8\underline{3}0 \times 10^{19}$$

$$= 3.83 \times 10^{19} \text{ molec.}$$

4.35 $\text{Mass \% carbon} = \frac{\text{mass of C in sample}}{\text{mass of sample}} \times 100\%$

$$\text{\% carbon} = \frac{1.584 \text{ g}}{1.836 \text{ g}} \times 100\% = 86.2\underline{7}4 = 86.27\%$$

4.36 $\text{Mass \% alcohol} = \frac{\text{mass of alcohol in soln}}{\text{mass of soln}} \times 100\%$

$$\text{\% alcohol} = \frac{4.87 \text{ g}}{5.69 \text{ g}} \times 100\% = 85.\underline{5}8 = 85.6\%$$

4.37 $\text{Mass \% phosphorus oxychloride} = \frac{\text{mass of } POCl_3 \text{ in sample}}{\text{mass of sample}} \times 100\%$

$$\text{\% } POCl_3 = \frac{1.72 \text{ mg}}{8.53 \text{ mg}} = 100\% = 20.\underline{1}6 = 20.2\%$$

4.38 $\text{Mass \% sulfur} = \frac{\text{mass of S in sample}}{\text{mass of sample}} \times 100\%$

$$\text{\% S} = \frac{1.64 \text{ mg}}{3.17 \text{ mg}} \times 100\% = 51.\underline{7}3 = 51.7\%$$

4.39 Start with the definition for percentage nitrogen and rearrange this equation to find the mass of N in the fertilizer.

$$\text{Mass \% N} = \frac{\text{mass of N in fert.}}{\text{mass of fert.}} \times 100\%$$

$$\text{Mass N} = \frac{\text{mass \% N}}{100\%} \times \text{mass of fert.} = \frac{15.8\%}{100\%} \times 4.15 \text{ kg} = 0.65\underline{5}7$$

$$= 0.656 \text{ kg N}$$

4.40 Start by finding the mass of 1.000 L of seawater, using the density of 1.025 g/cm^3.

$$\text{Mass seawater} = 1.00 \text{ L} \times \frac{10^3 \text{ cm}^3}{1 \text{ L}} \times \frac{1.025 \text{ g}}{1 \text{ cm}^3} = 1.025 \times 10^3 \text{ g}$$

Continue with the definition for percentage of bromine in the seawater and rearrange this equation to find the mass of Br in the seawater.

$$\text{Mass \% Br} = \frac{\text{mass of Br in seaw.}}{\text{mass of seaw.}} \times 100\%$$

$$\text{Mass Br} = \frac{\text{mass \% Br}}{100\%} \times \text{mass seaw.} = \frac{0.0065\%}{100\%} \times 1025 \text{ g} = 0.06\underline{6}6$$

$$= 0.067 \text{ g Br}$$

4.41 Convert moles to mass, using the molar masses from the respective atomic weights. Then calculate the mass percentages from the respective masses.

$$0.0972 \text{ mol Al} \times \frac{26.98 \text{ g Al}}{1 \text{ mol Al}} = 2.6\underline{2}2 \text{ g Al}$$

$$0.0381 \text{ mol Mg} \times \frac{24.31 \text{ g Mg}}{1 \text{ mol Mg}} = 0.92\underline{6}2 \text{ g Mg}$$

$$\text{\% Al} = \frac{\text{mass of Al}}{\text{mass of alloy}} = \frac{2.622 \text{ g Al}}{3.548 \text{ g alloy}} \times 100\% = 73.\underline{9}007 = 73.9\% \text{ Al}$$

$$\text{\% Mg} = \frac{\text{mass of Mg}}{\text{mass of alloy}} = \frac{0.9262 \text{ g Mg}}{3.548 \text{ g alloy}} \times 100\% = 26.\underline{1}05 = 26.1\% \text{ Mg}$$

4.42 Convert moles to mass, using the molar masses from the respective atomic weights of 20.18 g/mol for Ne and 83.80 g/mol for Kr. Then calculate the mass percentages from the respective masses.

$$0.0856 \text{ mol Ne} \times \frac{20.18 \text{ g Ne}}{1 \text{ mol Ne}} = 1.7\underline{2}7 \text{ g Ne}$$

$$0.0254 \text{ mol Kr} \times \frac{83.80 \text{ g Kr}}{1 \text{ mol Kr}} = 2.1\underline{2}9 \text{ g Kr}$$

$$\% \text{ Ne} = \frac{\text{mass of Ne}}{\text{mass of mix}} = \frac{1.727 \text{ g Ne}}{3.856 \text{ g mix}} \times 100\% = 44.\underline{7}9 = 44.8\% \text{ Ne}$$

$$\% \text{ Kr} = \frac{\text{mass of Kr}}{\text{mass of mix}} = \frac{2.129 \text{ g Ne}}{3.856 \text{ g mix}} \times 100\% = 55.\underline{2}1 = 55.2\% \text{ Kr}$$

4.43 In each part, the numerator consists of the mass of the element in one mole of the compound; the denominator is the mass of one mole of the compound. Use the atomic weights of C = 12.01 g/mol; O = 16.00 g/mol; K = 39.10 g/mol; Mn = 54.94 g/mol; Co = 58.93 g/mol; and N = 14.01 g/mol. (100% can be written as 100% or 100.000% because it is an exact number.)

a. $$\% \text{ C} = \frac{\text{mass of C}}{\text{mass of CO}} = \frac{12.01 \text{ g C}}{28.01 \text{ g CO}} \times 100\% = 42.878 = 42.9\% \text{ (3 s.f.)}$$

$$\% \text{ O} = 100.000\% - 42.8\underline{7}8\%\text{C} = 57.122 = 57.1\%$$

b. $$\% \text{ C} = \frac{\text{mass of C}}{\text{mass of } CO_2} = \frac{12.01 \text{ g C}}{44.01 \text{ g } CO_2} \times 100\% = 27.289 = 27.3\%$$

$$\% \text{ O} = 100.000\% - 27.2\underline{8}9\% \text{ C} = 72.7\underline{1}1 = 72.7\%$$

c. $$\% \text{ K} = \frac{\text{mass of K}}{\text{mass of } KMnO_4} = \frac{39.10 \text{ g K}}{158.04 \text{ g } KMnO_4} \times 100\% = 24.741$$

$$= 24.7\%$$

$$\% \text{ Mn} = \frac{\text{mass of Mn}}{\text{mass of } KMnO_4} = \frac{54.94 \text{ g Mn}}{158.04 \text{ g } KMnO_4} \times 100\% = 34.763$$

$$= 34.8\%$$

$$\% \text{ O} = 100.000\% - (24.7\underline{4}1 + 34.7\underline{6}3) = 40.496 = 40.5\%$$

d. $\% \text{Co} = \dfrac{\text{mass of Co}}{\text{mass of Co(NO}_3)_2} = \dfrac{58.93 \text{ g Mn}}{182.95 \text{ g Co(NO}_3)_2} \times 100\% = 32.211$

$= 32.2\%$

$\% \text{N} = \dfrac{\text{mass of N}}{\text{mass of Co(NO}_3)_2} = \dfrac{2 \times 14.01 \text{ g N}}{182.95 \text{ g Co(NO}_3)_2} \times 100\% = 15.316$

$= 15.3\%$

$\% \text{O} = 100.000\% - (32.2\underline{1}1 + 15.3\underline{1}6) = 52.473 = 52.5\%$

4.44 In each part, the numerator consists of the mass of the element in one mole of the compound; the denominator is the mass of one mole of the compound. For atomic weights, use Cl = 35.45 g/mol; the other atomic weights are given in the previous problem.

a. $\% \text{N} = \dfrac{\text{mass of N}}{\text{mass of NO}} = \dfrac{14.01 \text{ g N}}{30.01 \text{ g NO}} \times 100\% = 46.684 = 46.7\% \text{ (3 s.f.)}$

$\% \text{O} = 100.000\% - 46.684\% \text{ N} = 53.316 = 53.3\%$

b. $\% \text{N} = \dfrac{\text{mass of N}}{\text{mass of N}_2\text{O}} = \dfrac{2 \times 14.01 \text{ g N}}{44.02 \text{ g N}_2\text{O}} \times 100\% = 63.653 = 63.7\%$

$\% \text{O} = 100.000\% - 63.653 = 36.347 = 36.3\%$

c. $\% \text{K} = \dfrac{\text{mass of K}}{\text{mass of KClO}_4} = \dfrac{39.10 \text{ g K}}{138.55 \text{ g KClO}_4} \times 100\% = 28.221 = 28.2\%$

$\% \text{Cl} = \dfrac{\text{mass of Cl}}{\text{mass of KClO}_4} = \dfrac{35.45 \text{ g Cl}}{138.55 \text{ g KClO}_4} \times 100\% = 25.586 = 25.6\%$

$\% \text{O} = 100.000\% - (28.221 + 25.586) = 46.193 = 46.2\%$

d. $\% \text{Mn} = \dfrac{\text{mass of Mn}}{\text{mass of Mn(NO}_3)_2} = \dfrac{54.94 \text{ g Mn}}{178.95 \text{ g Mn(NO}_3)_2} \times 100\% = 30.701$

$= 30.7\%$

$\% \text{N} = \dfrac{\text{mass of N}}{\text{mass of Mn(NO}_3)_2} = \dfrac{2 \times 14.01 \text{ g N}}{182.95 \text{ g Mn(NO}_3)_2} \times 100\% = 15.652$

$= 15.7\%$

$\% \text{O} = 100.000\% - (30.701 + 15.652) = 53.646 = 53.6\%$

4.45 One mol of $CF_3CHBrCl$ contains 2 mol of C (AW = 12.01), 3 mol of F (AW = 19.00), one mol of H (AW = 1.008), one mol of Br (AW = 79.90), and one mol of Cl (AW = 35.45).

$$\% C = \frac{\text{mass of C}}{\text{mass of } C_2F_3HBrCl} = \frac{2 \times 12.01 \text{ g}}{197.38 \text{ g}} \times 100\% = 12.169 = 12.2\%$$

$$\% F = \frac{\text{mass of F}}{\text{mass of } C_2F_3HBrCl} = \frac{3 \times 19.0 \text{ g}}{197.38 \text{ g}} \times 100\% = 28.878 = 28.9\%$$

$$\% Br = \frac{\text{mass of Br}}{\text{mass of } C_2F_3HBrCl} = \frac{79.90 \text{ g}}{197.38 \text{ g}} \times 100\% = 40.4802 = 40.5\%$$

$$\% Cl = \frac{\text{mass of Cl}}{\text{mass of } C_2F_3HBrCl} = \frac{35.45 \text{ g}}{197.38 \text{ g}} \times 100\% = 17.9602 = 18.0\%$$

$$\% H = \frac{\text{mass of H}}{\text{mass of } C_2F_3HBrCl} = \frac{1.008 \text{ g}}{197.38 \text{ g}} \times 100\% = 0.51069 = 0.511\%$$

4.46 One mol of $C_{13}H_{21}ClN_2O_2$ contains 13 mol of C (AW = 12.01), 21 mol of H (AW = 1.008), one mol of Cl (AW = 35.45), two mol of N (AW = 14.01), and 2 mol of O (AW = 16.00).

$$\% C = \frac{\text{mass of C}}{\text{mass of } C_{13}H_{21}ClN_2O_2} = \frac{13 \times 12.01 \text{ g}}{272.77 \text{ g}} \times 100\% = 57.238$$

$$= 57.2\%$$

$$\% Cl = \frac{\text{mass of Cl}}{\text{mass of } C_{13}H_{21}ClN_2O_2} = \frac{35.45 \text{ g}}{272.77 \text{ g}} \times 100\% = 12.996 = 13.0\%$$

$$\% N = \frac{\text{mass of N}}{\text{mass of } C_{13}H_{21}ClN_2O_2} = \frac{2 \times 14.01 \text{ g}}{272.77 \text{ g}} \times 100\% = 10.272$$

$$= 10.3\%$$

$$\% O = \frac{\text{mass of O}}{\text{mass of } C_{13}H_{21}ClN_2O_2} = \frac{2 \times 16.00 \text{ g}}{272.77 \text{ g}} \times 100\% = 11.731$$

$$= 11.7\%$$

$$\% H = 100.000\% - (57.238 + 12.996 + 10.272 + 11.731) = 7.763 = 7.76\%$$

4.47 Find the moles of C in each amount in one-step operations: Calculate the moles of each compound using the molar mass; then multiply by the number of moles of C per mole of compound:

$$\text{Mol C (glucose)} = 4.71\text{ g} \times \frac{1\text{ mol}}{180.2\text{ g}} \times \frac{6\text{ mol C}}{1\text{ mol glucose}} = 0.15\underline{7}\text{ mol}$$

$$\text{Mol C (ethanol)} = 5.85\text{ g} \times \frac{1\text{ mol}}{46.07\text{ g}} \times \frac{2\text{ mol C}}{1\text{ mol ethanol}} = 0.25\underline{4}\text{ mol (more C)}$$

4.48 Find the moles of S in each amount in one-step operations: Calculate the moles of each compound using the molar mass; then multiply by the number of moles of S per mole of compound.

$$\text{Mol S (CaSO}_4\text{)} = 40.8\text{ g} \times \frac{1\text{ mol CaSO}_4}{136.15\text{ g}} \times \frac{1\text{ mol S}}{1\text{ mol CaSO}_4}$$

$$= 0.29\underline{9}7\text{ mol (more S)}$$

$$\text{Mol S (Na}_2\text{SO}_3\text{)} = 35.2\text{ g} \times \frac{1\text{ mol Na}_2\text{SO}_3}{126.05\text{ g}} \times \frac{1\text{ mol S}}{1\text{ mol Na}_2\text{SO}_3}$$

$$= 0.27\underline{9}3\text{ mol}$$

4.49 First calculate the mass of C in the glycol by multiplying the mass of CO_2 by the molar mass of C and the reciprocal of the molar mass of CO_2. Then calculate the mass of H in the glycol by multiplying the mass of H_2O by the molar mass of 2H and the reciprocal of the molar mass of H_2O. Then use the masses to calculate the mass percentages. Calculate O by difference.

$$9.06\text{ mg CO}_2 \times \frac{1\text{ mol CO}_2}{44.01\text{ g CO}_2} \times \frac{12.01\text{ g C}}{1\text{ mol C}} = 2.4\underline{7}2\text{ mg C}$$

$$5.58\text{ mg H}_2\text{O} \times \frac{1\text{ mol H}_2\text{O}}{18.02\text{ g H}_2\text{O}} \times \frac{2\text{ H}}{1\text{ H}_2\text{O}} \times \frac{1.008\text{ g H}}{1\text{ mol H}} = 0.62\underline{4}3\text{ mg H}$$

$$\text{Mass O} = 6.38\text{ mg} - (2.472 + 0.6243) = 3.2\underline{8}4\text{ mg O}$$

$$\%\text{ C} = (2.472\text{ mg C}/6.38\text{ mg glycol}) \times 100\% = 38.\underline{7}4 = 38.7\%$$

$$\%\text{ H} = (0.6243\text{ mg H}/6.38\text{ mg glycol}) \times 100\% = 9.7\underline{8}5 = 9.79\%$$

$$\%\text{ O} = (3.284\text{ mg O}/6.38\text{ mg glycol}) \times 100\% = 51.\underline{4}7 = 51.5\%$$

4.50 First calculate the mass of C in the phenol by multiplying the mass of CO_2 by the molar mass of C and the reciprocal of the molar mass of CO_2. Then calculate the mass of H in the phenol by multiplying the mass of H_2O by the molar mass of 2H and the reciprocal of the molar mass of H_2O. Then use the masses to calculate the mass percentages. Calculate O by difference.

$$14.67\text{ mg } CO_2 \times \frac{1\text{ mol } CO_2}{44.01\text{ g } CO_2} \times \frac{12.01\text{ g C}}{1\text{ mol C}} = 4.00\underline{3}3\text{ mg C}$$

$$3.01\text{ mg } H_2O \times \frac{1\text{ mol } H_2O}{18.02\text{ g } CO_2} \times \frac{2\text{ H}}{1\text{ } H_2O} \times \frac{1.008\text{ g H}}{1\text{ mol H}} = 0.33\underline{6}8\text{ mg H}$$

$$\text{Mass O} = 5.23\text{ mg} - (4.0033 + 0.3368) = 0.8\underline{8}99\text{ mg O}$$

$$\%\text{ C} = (4.0033\text{ mg}/5.23\text{ mg}) \times 100\% = 76.\underline{5}4 = 76.5\%$$

$$\%\text{ H} = (0.3368\text{ mg}/5.23\text{ mg}) \times 100\% = 6.4\underline{3}9 = 6.44\%$$

$$\%\text{ O} = (0.8899\text{ mg}/5.23\text{ mg}) \times 100\% = 1\underline{7}.0 = 17\%$$

4.51 Start by calculating the moles of Os and O; then divide each by the smaller number of moles to obtain integers for the empirical formula.

$$\text{Mol Os} = 2.16\text{ g Os} \times \frac{1\text{ mol}}{190.2\text{ g Os}} = 0.011\underline{3}6\text{ mol (smaller no.)}$$

$$\text{Mol O} = (2.89 - 2.16)\text{ g O} \times \frac{1\text{ mol}}{16.00\text{ g Os}} = 0.04\underline{5}6\text{ mol}$$

$$\text{Integer for Os} = 0.01136 \div 0.01136 = 1.0\underline{0}0$$

$$\text{Integer for O} = 0.0456 \div 0.01136 = 4.0\underline{1}$$

Because 4.01 = 4.0 within experimental error, the empirical formula is OsO_4.

4.52 Start by calculating the moles of W and O; then divide each by the smaller number of moles to obtain integers for the empirical formula.

$$\text{Mol W} = 4.23\text{ g W} \times \frac{1\text{ mol}}{183.85\text{ g W}} = 0.023\underline{0}1\text{ mol (smaller no.)}$$

$$\text{Mol O} = (5.34 - 4.23)\text{ g O} \times \frac{1\text{ mol}}{16.00\text{ g O}} = 0.069\underline{3}8\text{ mol}$$

(continued)

Integer for W = 0.02301 ÷ 0.02301 = 1.0$\underline{0}$0

Integer for O = 0.06938 ÷ 0.02301 = 3.0$\underline{1}$5

Because 3.015 = 3.0 within experimental error, the empirical formula is WO_3.

4.53 Assume a sample of 100.0 g of potassium manganate. By multiplying this by the percentage composition, we obtain 39.6 g of K, 27.9 g of Mn, and 32.5 g of O. Convert each of these masses to moles by dividing by molar mass.

$$\text{Mol K} = 39.6 \text{ g K} \times \frac{1 \text{ mol}}{39.10 \text{ g K}} = 1.0\underline{1}3 \text{ mol}$$

$$\text{Mol Mn} = 27.9 \text{ g Mn} \times \frac{1 \text{ mol}}{54.94 \text{ g Mn}} = 0.50\underline{7}8 \text{ mol (smallest no.)}$$

$$\text{Mol O} = 32.5 \text{ g O} \times \frac{1 \text{ mol}}{16.00 \text{ g Mn}} = 2.0\underline{3}1 \text{ mol}$$

Now divide each number of moles by the smallest number to obtain the smallest set of integers for the empirical formula.

Integer for K = 1.013 ÷ 0.5078 = 1.998, or 2

Integer for Mn = 0.5078 ÷ 0.5078 = 1.000, or 1

Integer for O = 2.031 ÷ 0.5078 = 3.999, or 4

The empirical formula is thus K_2MnO_4.

4.54 Assume a sample of 100.0 g of hydroquinone. By multiplying this by the percentage composition, we obtain 65.4 g of C, 5.5 g of H, and 29.1 g of O. Convert each of these masses to moles by dividing by the molar mass.

$$\text{Mol C} = 65.4 \text{ g C} \times \frac{1 \text{ mol}}{12.01 \text{ g C}} = 5.4\underline{4}5 \text{ mol}$$

$$\text{Mol H} = 5.5 \text{ g H} \times \frac{1 \text{ mol}}{1.008 \text{ g H}} = 5.\underline{4}6 \text{ mol}$$

$$\text{Mol O} = 29.1 \text{ g O} \times \frac{1 \text{ mol}}{16.00 \text{ g O}} = 1.8\underline{1}9 \text{ mol (smallest no.)}$$

Now divide each number of moles by the smallest number to obtain the smallest set of integers for the empirical formula.

(continued)

Integer for C = 5.445 ÷ 1.819 = 2.99, or 3

Integer for H = 5.46 ÷ 1.819 = 3.0, or 3

Integer for O = 1.819 ÷ 1.819 = 1.00, or 1

The empirical formula is thus C_3H_3O.

4.55 Assume a sample of 100.0 g of acrylic acid. By multiplying this by the percentage composition, we obtain 50.0 g C, 5.6 g H, and 44.4 g O. Convert each of these masses to moles by dividing by the molar mass.

$$\text{Mol C} = 50.0 \text{ g C} \times \frac{1 \text{ mol}}{12.01 \text{ g C}} = 4.1\underline{6}3 \text{ mol}$$

$$\text{Mol H} = 5.6 \text{ g H} \times \frac{1 \text{ mol}}{1.008 \text{ g H}} = 5.\underline{5}6 \text{ mol}$$

$$\text{Mol O} = 44.4 \text{ g O} \times \frac{1 \text{ mol}}{16.00 \text{ g O}} = 2.7\underline{7}5 \text{ mol (smallest no.)}$$

Now divide each number of moles by the smallest number to obtain the smallest number of moles and the tentative integers for the empirical formula.

Tentative integer for C = 4.163 ÷ 2.775 = 1.50, or 1.5

Tentative integer for H = 5.56 ÷ 2.775 = 2.00, or 2

Tentative integer for O = 2.775 ÷ 2.775 = 1.00, or 1

Because 1.5 is not a whole number, multiply each tentative integer by 2 to obtain the final integer for the empirical formula:

C: 2 x 1.5 = 3

H: 2 x 2 = 4

O: 2 x 1 = 2

The empirical formula is thus $C_3H_4O_2$.

4.56 Assume a sample of 100.0 g of malonic acid. By multiplying this by the percentage composition, we obtain 34.6 g C, 3.9 g H, and 61.5 g O. Convert each of these masses to moles by dividing by the molar mass.

$$\text{Mol C} = 34.6 \text{ g C} \times \frac{1 \text{ mol}}{12.01 \text{ g C}} = 2.8\underline{8}1 \text{ mol (smallest no.)}$$

$$\text{Mol H} = 3.9 \text{ g H} \times \frac{1 \text{ mol}}{1.008 \text{ g H}} = 3.\underline{8}7 \text{ mol}$$

$$\text{Mol O} = 61.5 \text{ g O} \times \frac{1 \text{ mol}}{16.00 \text{ g O}} = 3.8\underline{4}4 \text{ mol}$$

Now divide each number of moles by the smallest number to obtain the smallest number of moles and the tentative integers for the empirical formula.

Tentative integer for C = 2.881 ÷ 2.881 = 1.00, or 1

Tentative integer for H = 3.87 ÷ 2.881 = 1.34, or 1-1/3

Tentative integer for O = 3.884 ÷ 2.881 = 1.334, or 1-1/3

Because 1-1/3 is not a whole number, multiply each tentative integer by 3 to give the final integer for the empirical formula:

C: 3 x 1 = 3

H: 3 x 1-1/3 = 4

O: 3 x 1-1/3 = 4

The empirical formula is thus $C_3H_4O_4$.

4.57 a. Assume for the calculation that you have 100.0 g; of this quantity, 92.25 g is C and 7.75 g is H. Now, convert these masses to moles:

$$92.25 \text{ g C} \times \frac{1 \text{ mol C}}{12.01 \text{ g C}} = 7.68\underline{1}09 \text{ mol C}$$

$$7.75 \text{ g H} \times \frac{1 \text{ mol H}}{1.008 \text{ g H}} = 7.6\underline{8}8 \text{ mol H}$$

Usually you divide all the mole numbers by the smaller one, but in this case both are equal so the ratio of number of C atoms to the number of H atoms is 1:1. Thus the empirical formula for both compounds is CH.

b. Obtain n, the number of empirical formula units in the molecule, by dividing the molecular weight of 52.03 amu and 78.05 amu, by the empirical formula weight of 13.018 amu:

$$\text{For 52.03:}\quad n \;=\; \frac{52.03\text{ amu}}{13.018\text{ amu}} \;=\; 3.99\underline{6}8\text{, or 4}$$

$$\text{For 78.05:}\quad n \;=\; \frac{78.05\text{ amu}}{13.018\text{ amu}} \;=\; 5.99\underline{5}5\text{, or 6}$$

The molecular formulas are: for 52.03, $(CH)_4$ or C_4H_4; and for 78.05, $(CH)_6$ or C_6H_6.

4.58 a. Assume for the calculation that you have 100.0 g; of this quantity, 85.62 g is C and 14.38 g is H. Now convert these masses to moles:

$$85.62\text{ g C} \;=\; \times\; \frac{1\text{ mol C}}{12.01\text{ g C}} \;=\; 7.12\underline{9}05\text{ mol C}$$

$$14.38\text{ g H} \times \frac{1\text{ mol H}}{1.008\text{ g H}} \;=\; 14.\underline{2}6\text{ mol H}$$

Divide both mole numbers by the smaller one:

$$\text{For C:}\quad \frac{7.129\text{ mol}}{7.129\text{ mol}} \;=\; 1.0\underline{0}$$

$$\text{For H:}\quad \frac{14.26\text{ mol}}{7.129\text{ mol}} \;=\; 2.0\underline{0}02$$

The empirical formula is obviously CH_2.

b. Obtain n, the number of empirical formula units in the molecule, by dividing the molecular weights of 28.03 amu and 56.06 amu by the empirical formula weight of 14.026 amu:

$$\text{For 28.03:}\quad n \;=\; \frac{28.03\text{ amu}}{14.026\text{ amu}} \;=\; 1.99\underline{8}4\text{, or 2}$$

$$\text{For 56.06:}\quad n \;=\; \frac{56.06\text{ amu}}{14.026\text{ amu}} \;=\; 3.99\underline{6}8\text{, or 4}$$

The molecular formulas are: for 28.03, $(CH_2)_2$ or C_2H_4; and for 56.06, $(CH_2)_4$ or C_4H_8.

4.59 The formula weight corresponding to the empirical formula C_2H_6N may be found by adding the respective atomic weights.

$$\text{Formula weight} = (2 \times 12.01 \text{ amu}) + (6 \times 1.008 \text{ amu}) + 14.01 \text{ amu} = 44.08 \text{ amu}$$

Dividing the molecular weight by the formula weight gives the number of times the C_2H_6N unit occurs in the molecule. Because the molecular weight is an average of 88.5 ([90 + 87] ÷ 2), this quotient is

$$88.5 \text{ amu} \div 44.1 \text{ amu} = 2.\underline{0}06, \text{ or } 2$$

Therefore, the molecular formula is $(C_2H_6N)_2$, or $C_4H_{12}N_2$.

4.60 The formula weight corresponding to the empirical formula BH_3 may be found by adding the respective atomic weights.

$$\text{Formula weight} = 10.81 \text{ amu} + (3 \times 1.008 \text{ amu}) = 13.83 \text{ amu}$$

Dividing the molecular weight by the formula weight gives the number of times the BH_3 unit occurs in the molecule. Because the molecular weight is 28 amu, this quotient is

$$28 \text{ amu} \div 13.83 \text{ amu} = 2.\underline{0}2$$

Therefore, the molecular formula is $(BH_3)_2$, or B_2H_6.

4.61 Assume a sample of 100.0 g of oxalic acid. By multiplying this by the percentage composition, we obtain 26.7 g C, 2.2 g H, and 71.1 g O. Convert each of these masses to moles by dividing by the molar mass.

$$\text{Mol C} = 26.7 \text{ g C} \times \frac{1 \text{ mol}}{12.01 \text{ g C}} = 2.2\underline{2}3 \text{ mol}$$

$$\text{Mol H} = 2.2 \text{ g H} \times \frac{1 \text{ mol}}{1.008 \text{ g H}} = 2.\underline{1}8 \text{ mol (smallest no.)}$$

$$\text{Mol O} = 71.1 \text{ g O} \times \frac{1 \text{ mol}}{16.00 \text{ g O}} = 4.4\underline{4}3 \text{ mol}$$

Now divide each number of moles by the smallest number to obtain the smallest set of integers for the empirical formula.

$$\text{Integer for C} = 2.223 \div 2.18 = 1.02, \text{ or } 1$$

$$\text{Integer for H} = 2.18 \div 2.18 = 1.00, \text{ or } 1$$

$$\text{Integer for O} = 4.443 \div 2.18 = 2.0\underline{3}8, \text{ or } 2$$

(continued)

The empirical formula is thus CHO_2. The formula weight corresponding to this formula may be found by adding the respective atomic weights:

$$\text{Formula weight} = 12.01 \text{ amu} + 1.008 \text{ amu} + (2 \times 16.00 \text{ amu}) = 45.02 \text{ amu}$$

Dividing the molecular weight by the formula weight gives the number of times the CHO_2 unit occurs in the molecule. Because the molecular weight is 90 amu, this quotient is

$$90 \text{ amu} \div 45.02 \text{ amu} = 2.\underline{0}0, \text{ or } 2$$

The molecular formula is thus $(CHO_2)_2$, or $C_2H_2O_4$.

4.62 Assume a sample of 100.0 g of adipic acid. By multiplying this by the percentage composition, we obtain 49.3 g C, 6.9 g H, and 43.8 g O. Convert each of these masses to moles by dividing by the molar mass.

$$\text{Mol C} = 49.3 \text{ g C} \times \frac{1 \text{ mol}}{12.01 \text{ g C}} = 4.1\underline{0}5 \text{ mol}$$

$$\text{Mol H} = 6.9 \text{ g H} \times \frac{1 \text{ mol}}{1.008 \text{ g H}} = 6.\underline{8}5 \text{ mol}$$

$$\text{Mol O} = 43.8 \text{ g O} \times \frac{1 \text{ mol}}{16.00 \text{ g O}} = 2.7\underline{3}8 \text{ mol (smallest no.)}$$

Now divide each number of moles by the smallest number to obtain the smallest set of integers for the empirical formula.

$$\text{Tentative integer for C} = 4.105 \div 2.738 = 1.4\underline{9}9, \text{ or } 1.5$$

$$\text{Tentative integer for H} = 6.85 \div 2.738 = 2.\underline{5}0, \text{ or } 2.5$$

$$\text{Tentative integer for O} = 2.738 \div 2.738 = 1.0\underline{0}0, \text{ or } 1$$

Because 1.5 and 2.5 are not whole numbers, multiply each tentative integer by 2 to give the final integers for the empirical formula:

C: $2 \times 1.5 = 3$

H: $2 \times 2.5 = 5$

O: $2 \times 1 = 2$

The empirical formula is thus $C_3H_5O_2$. The formula weight corresponding to this formula may be found by adding the respective atomic weights:

(continued)

Formula weight = (3 x 12.01 amu) + (5 x 1.008 amu) + (2 x 16.00 amu) = 73.1 amu

Dividing the molecular weight by the formula weight gives the number of times the CHO_2 unit occurs in the molecule. Because the molecular weight is 146 amu, this quotient is

146 amu ÷ 73.1 amu = 2.$\underline{0}$0, or 2

The molecular formula is thus $(C_3H_5O_2)_2$, or $C_6H_{10}O_4$.

4.63

C_2H_4	+	$3O_2$	→	$2CO_2$	+	$2H_2O$
1 molecule C_2H_4	+	3 molecules O_2	→	2 molecules CO_2	+	2 molecules H_2O
1 mole C_2H_4	+	3 moles O_2	→	2 moles CO_2	+	2 moles H_2O
28.05 g C_2H_4	+	3 x 32.00 g O_2	→	2 x 44.01 g CO_2	+	2 x 18.016 g H_2O

4.64

$2H_2S$	+	$3O_2$	→	$2SO_2$	+	$2H_2O$
2 molecules H_2S	+	3 molecules O_2	→	2 molecules SO_2	+	2 molecules H_2O
2 moles H_2S	+	3 moles O_2	→	2 moles SO_2	+	2 moles H_2O
2 x 34.06 g H_2S	+	3 x 32.00 g O_2	→	2 x 64.06 g CO_2	+	2 x 18.016 g H_2O

4.65 By inspecting the balanced equation, obtain a conversion factor of 8 mol CO_2 to 2 mol C_4H_{10}. Multiply the given amount of 0.15 moles of C_4H_{10} by the conversion factor to obtain the moles of H_2O.

$$0.15 \text{ mol } C_4H_{10} \times \frac{8 \text{ mol } CO_2}{2 \text{ mol } C_4H_{10}} = 0.6\underline{0}0 = 0.60 \text{ mol } CO_2$$

4.66 By inspecting the balanced equation, obtain a conversion factor of 3 mol H_2O to 1 mol C_2H_5OH. Multiply the given amount of 0.25 mol of C_2H_5OH by the conversion factor to obtain the moles of H_2O.

$$0.25 \text{ mol } C_2H_5OH \times \frac{3 \text{ mol } H_2O}{1 \text{ mol } C_2H_5OH} = 0.7\underline{5}0 = 0.75 \text{ mol } H_2O$$

4.67 By inspecting the balanced equation, obtain a conversion factor of 3 mol O_2 to 2 mol Fe_2O_3. Multiply the given amount of 5.21 mol Fe_2O_3 by the conversion factor to obtain moles of O_2.

$$5.21 \text{ mol } Fe_2O_3 \times \frac{3 \text{ mol } O_2}{2 \text{ mol } Fe_2O_3} = 7.8\underline{1}50 = 7.82 \text{ mol } O_2$$

4.68 By inspecting the balanced equation, obtain a conversion factor of 3 mol $NiCl_2$ to 1 mol $Ni_3(PO_4)_2$. Multiply the given amount of 0.479 mol $Ni_3(PO_4)_2$ by the conversion factor to obtain moles of $NiCl_2$.

$$0.479 \text{ mol } Ni_3(PO_4)_2 \times \frac{3 \text{ mol } NiCl_2}{1 \text{ mol } Ni_3(PO_4)_2} = 1.4\underline{3}7 = 1.44 \text{ mol } NiCl_2$$

4.69 $3NO_2 + H_2O \rightarrow 2HNO_3 + NO$

3 mol of NO_2 is equivalent to 2 mol of HNO_3 (from equation).

1 mol of NO_2 is equivalent to 46.01 g NO_2 (from molecular weight of NO_2).

1 mol of HNO_3 is equivalent to 63.02 g HNO_3 (from molecular weight of HNO_3).

$$5.00 \text{ g } HNO_3 \times \frac{1 \text{ mol } HNO_3}{63.02 \text{ g } HNO_3} \times \frac{3 \text{ mol } NO_2}{2 \text{ mol } HNO_3} \times \frac{46.01 \text{ g } NO_2}{1 \text{ mol } NO_2} = 5.4\underline{7}6$$

$$= 5.48 \text{ g } NO_2$$

4.70 $2Ca_3(PO_4)_2 + 6SiO_2 + 10C \rightarrow P_4 + 6CaSiO_3 + 10CO$

2 mol of $Ca_3(PO_4)_2$ is equivalent to 1 mol of P_4 (from equation).

1 mol of P_4 is equivalent to 123.9 g P_4 (from molecular weight of P_4).

1 mol of $Ca_3(PO_4)_2$ is equivalent to 310.2 g $Ca_3(PO_4)_2$ [from molecular weight of $Ca_3(PO_4)_2$].

$$5.00 \text{ g } P_4 \times \frac{1 \text{ mol } P_4}{123.88 \text{ g } P_4} \times \frac{2 \text{ mol } Ca_3(PO_4)_2}{1 \text{ mol } P_4} \times \frac{310.18 \text{ g } Ca_3(PO_4)_2}{1 \text{ mol } Ca_3(PO_4)_2}$$

$$= 25.\underline{0}4 = 25.0 \text{ g } Ca_3(PO_4)_2$$

4.71 $WO_3 + 3H_2 \rightarrow W + 3H_2O$

1 mol of W is equivalent to 3 moles of H_2 (from equation).

1 mol of H_2 is equivalent to 2.016 g H_2 (from molecular weight of H_2).

1 mol of W is equivalent to 183.8 g W (from atomic weight of W).

4.81 kg of H_2 is equivalent to 4.81×10^3 g of H_2.

$$4.81 \times 10^3 \text{ g } H_2 \times \frac{1 \text{ mol } H_2}{2.016 \text{ g } H_2} \times \frac{1 \text{ mol W}}{3 \text{ mol } H_2} \times \frac{183.85 \text{ g W}}{1 \text{ mol W}} = 1.4\underline{6}2 \times 10^5$$

$$= 1.46 \times 10^5 \text{ g W}$$

4.72 $4C_3H_6 + 6NO \rightarrow 4C_3H_3N + 6H_2O + N_2$

4 mol of C_3H_6 is equivalent to 4 mol of C_3H_3N (from equation).

1 mol of C_3H_6 is equivalent to 42.08 g C_3H_6 (from molecular weight of C_3H_6).

1 mol of C_3H_3N is equivalent to 53.06 g C_3H_3N (from molecular weight of C_3H_3N).

651 kg of C_3H_6 is equivalent to 6.51×10^5 g C_3H_6.

$$6.51 \times 10^5 \text{ g } C_3H_6 \times \frac{1 \text{ mol } C_3H_6}{42.08 \text{ g } C_3H_6} \times \frac{4 \text{ mol } C_3H_3N}{4 \text{ mol } C_3H_6} \times \frac{53.06 \text{ g } C_3H_3N}{1 \text{ mol } C_3H_3N}$$

$$= 8.2\underline{0}8 \times 10^5 = 8.21 \times 10^5 \text{ g } C_3H_3N$$

4.73 Using the approach of the previous problem, we write the equation and set up the calculation below the equation (after calculating the two molecular weights):

$CS_2 + 3Cl_2 \rightarrow CCl_4 + S_2Cl_2$

$$62.7 \text{ g } Cl_2 \times \frac{1 \text{ mol } Cl_2}{70.90 \text{ g } Cl_2} \times \frac{1 \text{ mol } CS_2}{3 \text{ mol } Cl_2} \times \frac{76.15 \text{ g } CS_2}{1 \text{ mol } CS_2} = 22.\underline{4}48$$

$$= 22.4 \text{ g } CS_2$$

4.74 Using the approach of, say, problem 4.66, we write the equation and set up the calculation below the equation (after calculating the two formula weights):

$$2NaOH + Cl_2 \rightarrow NaCl + NaClO + H_2O$$

$$54.2 \text{ g NaOH} \times \frac{1 \text{ mol NaOH}}{40.00 \text{ g NaOH}} \times \frac{1 \text{ mol } Cl_2}{2 \text{ mol NaOH}} \times \frac{70.90 \text{ g } Cl_2}{1 \text{ mol } Cl_2} = 48.\underline{0}4$$

$$= 48.0 \text{ g } Cl_2$$

4.75 Using the approach of, say, problem 4.66, we write the equation and set up the calculation below the equation (after calculating the two molecular weights):

$$2N_2O_5 \rightarrow 4NO_2 + O_2$$

$$1.618 \text{ g } O_2 \times \frac{1 \text{ mol } O_2}{32.00 \text{ g } O_2} \times \frac{4 \text{ mol } NO_2}{1 \text{ mol } O_2} \times \frac{46.01 \text{ g } NO_2}{1 \text{ mol } NO_2} = 9.30\underline{5}52$$

$$= 9.306 \text{ g } NO_2$$

4.76 Using the approach of, say, problem 4.66, we write the equation and set up the calculation below the equation (after calculating the two formula weights):

$$3Cu + 8HNO_3 \rightarrow 3Cu(NO_3)_2 + 2NO + 4H_2O$$

$$5.92 \text{ g } Cu(NO_3)_2 \times \frac{1 \text{ mol } Cu(NO_3)_2}{187.56 \text{ g } Cu(NO_3)_2} \times \frac{2 \text{ mol NO}}{3 \text{ mol } Cu(NO_3)_2} \times \frac{30.01 \text{ g NO}}{1 \text{ mol NO}}$$

$$= 0.63\underline{1}4 = 0.631 \text{ g NO}$$

4.77 First determine whether KO_2 or H_2O is the limiting reagent by calculating the moles of O_2 that each would form if each were the limiting reagent. Identify the limiting reactant by the smaller number of moles of O_2 formed.

$$0.10 \text{ mol } H_2O \times \frac{3 \text{ mol } O_2}{2 \text{ mol } H_2O} = 0.1\underline{5}0 \text{ mol } O_2$$

$$0.15 \text{ mol } KO_2 \times \frac{3 \text{ mol } O_2}{4 \text{ mol } KO_2} = 0.1\underline{1}25 \text{ mol } O_2 \quad (KO_2 \text{ is the limiting reactant.})$$

The moles of O_2 produced = 0.11 mol.

4.78 First determine whether NH_3 or O_2 is the limiting reagent by calculating the moles of NO that each would form if each were the limiting reagent. Identify the limiting reactant by the smaller number of moles of NO formed.

$$0.120 \text{ mol } NH_3 \times \frac{4 \text{ mol NO}}{4 \text{ mol } NH_3} = 0.1\underline{2}00 \text{ mol NO}$$

$$0.140 \text{ mol } NH_3 \times \frac{4 \text{ mol NO}}{5 \text{ mol } O_2} = 0.11\underline{2}0 \text{ mol NO} \quad (O_2 \text{ is the limiting reactant.})$$

The moles of NO produced = 0.112 mol.

4.79 First determine whether CO or H_2 is the limiting reagent by calculating the moles of CH_3OH that each would form if each were the limiting reagent. Identify the limiting reactant by the smaller number of moles of CH_3OH formed. Use the molar mass of CH_3OH to calculate the mass of CH_3OH formed. Then calculate the mass of the unconsumed reactant.

$$CO + 2H_2 \rightarrow CH_3OH$$

$$10.2 \text{ g } H_2 \times \frac{1 \text{ mol } H_2}{2.016 \text{ g } H_2} \times \frac{1 \text{ mol } CH_3OH}{2 \text{ mol } H_2} = 2.5\underline{2}9 \text{ mol } CH_3OH$$

$$35.4 \text{ g CO} \times \frac{1 \text{ mol CO}}{28.01 \text{ g CO}} \times \frac{1 \text{ mol } CH_3OH}{1 \text{ mol CO}} = 1.2\underline{6}3 \text{ mol } CH_3OH$$

(CO is the limiting reagent.)

$$\text{Mass } CH_3OH \text{ formed} = 1.2\underline{6}3 \text{ mol } CH_3OH \times \frac{32.042 \text{ g } CH_3OH}{1 \text{ mol } CH_3OH}$$

$$= 40.\underline{4}7 = 40.5\text{g } CH_3OH$$

Hydrogen is left unconsumed at the end of the reaction. The mass of H_2 that reacts can be calculated from the moles of product obtained:

$$1.2\underline{6}3 \text{ mol } CH_3OH \times \frac{2 \text{ mol } H_2}{1 \text{ mol } CH_3OH} \times \frac{2.016 \text{ g } H_2}{1 \text{ mol } H_2} = 5.0\underline{9}2 \text{ g } H_2$$

The unreacted H_2 = 10.2 g total H_2 - 5.092 g reacted H_2 = 5.$\underline{1}$08 = 5.1 g H_2.

4.80 First determine whether CS_2 or O_2 is the limiting reagent by calculating the moles of SO_2 that each would form if each were the limiting reagent. Identify the limiting reagent by the smaller number of moles of SO_2 formed. Use the molar mass of SO_2 to calculate the mass of SO_2 formed. Then calculate the mass of the unconsumed reactant.

$$CS_2 + 3O_2 \rightarrow CO_2 + 2SO_2$$

$$35.0 \text{ g } O_2 \times \frac{1 \text{ mol } O_2}{32.00 \text{ g } O_2} \times \frac{2 \text{ mol } SO_2}{3 \text{ mol } O_2} = 0.72\underline{9}1 \text{ mol } SO_2$$

$$15.0 \text{ g } CS_2 \times \frac{1 \text{ mol } CS_2}{76.15 \text{ g } CS_2} \times \frac{2 \text{ mol } SO_2}{1 \text{ mol } CS_2} = 0.39\underline{4}0 \text{ mol } SO_2$$

(CS_2 is the limiting reagent.)

$$\text{Mass } SO_2 \text{ formed} = 0.39\underline{4}0 \text{ mol } SO_2 \times \frac{64.07 \text{ g } SO_2}{1 \text{ mol } SO_2} = 25.\underline{2}4 \text{ g } SO_2$$

Oxygen is left unconsumed at the end of the reaction. The mass of O_2 that reacts can be calculated from the moles of product obtained:

$$0.39\underline{4}0 \text{ mol } SO_2 \times \frac{3 \text{ mol } O_2}{2 \text{ mol } SO_2} \times \frac{32.00 \text{ g } O_2}{1 \text{ mol } O_2} = 18.\underline{9}1 \text{ g } O_2$$

The unreacted O_2 = 35.0 g total O_2 - 18.91 g reacted O_2 = 16.$\underline{0}$9 = 16.1 g O_2.

4.81 First determine which of the three reactants is the limiting reagent by calculating the moles of $TiCl_4$ that each would form if each were the limiting reagent. Identify the limiting reagent by the smallest number of moles of $TiCl_4$ formed. Use the molar mass of $TiCl_4$ to calculate the mass of $TiCl_4$ formed.

$$3TiO_2 + 4C + 6Cl_2 \rightarrow 3TiCl_4 + 2CO_2 + 2CO$$

$$4.15 \text{ g } TiO_2 \times \frac{1 \text{ mol } TiO_2}{79.88 \text{ g } TiO_2} \times \frac{3 \text{ mol } TiCl_4}{3 \text{ mol } TiO_2} = 0.051\underline{9}5 \text{ mol } TiCl_4$$

$$5.67 \text{ g C} \times \frac{1 \text{ mol C}}{12.01 \text{ g C}} \times \frac{3 \text{ mol } TiCl_4}{4 \text{ mol C}} = 0.35\underline{4}07 \text{ mol } TiCl_4$$

$$6.78 \text{ g } Cl_2 \times \frac{1 \text{ mol } Cl_2}{70.90 \text{ g } Cl_2} \times \frac{3 \text{ mol } TiCl_4}{6 \text{ mol } Cl_2} = 0.047\underline{8}1 \text{ mol } TiCl_4$$

(Cl_2 is the limiting reagent.)

(continued)

$$\text{Mass TiCl}_4\text{ formed} = 0.47\underline{8}1\text{ mol TiCl}_4 \times \frac{189.68\text{ g TiCl}_4}{1\text{ mol TiCl}_4} = 9.0\underline{6}8$$

$$= 90.7\text{ g TiCl}_4$$

4.82 First determine which of the three reactants is the limiting reagent by calculating the moles of HCN that each would form if each were the limiting reagent. Identify the limiting reagent by the smallest number of moles of HCN formed. Use the molar mass of HCN to calculate the mass of HCN formed.

$$2NH_3 + 3O_2 + 2CH_4 \rightarrow 2HCN + 6H_2O$$

$$11.5\text{ g NH}_3 \times \frac{1\text{ mol NH}_3}{17.03\text{ g NH}_3} \times \frac{2\text{ mol HCN}}{2\text{ mol NH}_3} = 0.67\underline{5}\text{ mol HCN}$$

$$10.5\text{ g CH}_4 \times \frac{1\text{ mol CH}_4}{16.04\text{ g CH}_4} \times \frac{2\text{ mol HCN}}{2\text{ mol CH}_4} = 0.65\underline{4}\text{ mol HCN}$$

$$10.0\text{ g O}_2 \times \frac{1\text{ mol O}_2}{32.00\text{ g O}_2} \times \frac{2\text{ mol HCN}}{3\text{ mol O}_2} = 0.20\underline{8}3\text{ mol HCN}$$

(O_2 is the limiting reagent.)

$$\text{Mass HCN formed} = 0.20\underline{8}3\text{ mol HCN} \times \frac{27.03\text{ g HCN}}{1\text{ mol HCN}} = 5.6\underline{3}03$$

$$= 5.63\text{ g HCN}$$

4.83 First determine which of the two reactants is the limiting reagent by calculating the moles of aspirin that each would form if each were the limiting reagent. Identify the limiting reagent by the smallest number of moles of aspirin formed. Use the molar mass of aspirin to calculate the theoretical yield in grams of aspirin. Then calculate the percentage yield.

$$C_7H_6O_3 + C_4H_6O_3 \rightarrow C_9H_8O_4 + C_2H_4O_2$$

$$4.00\text{ g C}_4\text{H}_6\text{O}_3 \times \frac{1\text{ mol C}_4\text{H}_6\text{O}_3}{102.09\text{ g C}_4\text{H}_6\text{O}_3} \times \frac{1\text{ mol C}_9\text{H}_8\text{O}_4}{1\text{ mol C}_4\text{H}_6\text{O}_3}$$

$$= 0.039\underline{1}8\text{ mol C}_9\text{H}_8\text{O}_4$$

(continued)

$$2.00 \text{ g } C_7H_6O_3 \times \frac{1 \text{ mol } C_7H_6O_3}{138.12 \text{ g } C_7H_6O_3} \times \frac{1 \text{ mol } C_9H_8O_4}{1 \text{ mol } C_7H_6O_3}$$

$$= 0.014\underline{4}8 \text{ mol } C_9H_8O_4$$

Thus, $C_7H_6O_3$ is the limiting reagent. The theoretical yield of $C_9H_8O_4$ is

$$0.014\underline{4}8 \text{ mol } C_9H_8O_4 \times \frac{180.15 \text{ g } C_9H_8O_4}{1 \text{ mol } C_9H_8O_4} = 2.6\underline{0}9 \text{ g } C_9H_8O_4$$

The percentage yield is

$$\% \text{ yield} = \frac{\text{actual yield}}{\text{theoretical yield}} \times 100\% = \frac{2.10 \text{ g}}{2.609 \text{ g}} \times 100\% = 80.\underline{4}9 = 80.5\%$$

4.84 First determine which of the two reactants is the limiting reagent by calculating the moles of methyl salicylate that each would form if each were the limiting reagent. Identify the limiting reagent by the smallest number of moles of methyl salicylate formed. Use the molar mass of methyl salicylate to calculate the theoretical yield in grams of methyl salicylate. Then calculate the percentage yield.

$$C_7H_6O_3 + CH_3OH \rightarrow C_8H_8O_3 + H_2O$$

$$11.20 \text{ g } CH_3OH \times \frac{1 \text{ mol } CH_3OH}{32.04 \text{ g } CH_3OH} \times \frac{1 \text{ mol } C_8H_8O_3}{1 \text{ mol } CH_3OH} = 0.34\underline{9}6 \text{ mol } C_8H_8O_3$$

$$1.50 \text{ g } C_7H_6O_3 \times \frac{1 \text{ mol } C_7H_6O_3}{138.12 \text{ g } C_7H_6O_3} \times \frac{1 \text{ mol } C_8H_8O_3}{1 \text{ mol } C_7H_6O_3}$$

$$= 0.010\underline{8}6 \text{ mol } C_8H_8O_3$$

Thus, $C_7H_6O_3$ is the limiting reagent. The theoretical yield of $C_8H_8O_3$ is

$$0.010\underline{8}6 \text{ mol } C_8H_8O_3 \times \frac{152.14 \text{ g } C_8H_8O_3}{1 \text{ mol } C_8H_8O_3} = 1.6\underline{5}2 \text{ g } C_8H_8O_3$$

The percentage yield is

$$\% \text{ yield} = \frac{\text{actual yield}}{\text{theoretical yield}} \times 100\% = \frac{1.31 \text{ g}}{1.651 \text{ g}} \times 100\% = 79.\underline{3}4 = 79.3\%$$

4.85 $$\text{Molarity} = \frac{\text{moles solute}}{\text{liters of solution}} = \frac{0.0341 \text{ mol}}{0.0250 \text{ L}} = 1.3\underline{6}4 = 1.36 \text{ M}$$

4.86 $$\text{Molarity} = \frac{\text{moles solute}}{\text{liters of solution}} = \frac{0.0285 \text{ mol}}{0.0500 \text{ L}} = 0.57\underline{0}0 = 0.570 \text{ M}$$

4.87 Find the number of moles of solute ($KMnO_4$), using the molar mass of 158.03 g $KMnO_4$ per 1 mol $KMnO_4$:

$$0.798 \text{ g } KMnO_4 \times \frac{1 \text{ mol } KMnO_4}{158.03 \text{ g } KMnO_4} = 5.0\underline{4}97 \times 10^{-3} \text{ mol } KMnO_4$$

$$\text{Molarity} = \frac{\text{moles solute}}{\text{liters of solution}} = \frac{5.0497 \times 10^{-3} \text{ mol}}{0.0500 \text{ L}} = 0.10\underline{0}99 = 0.101 \text{ M}$$

4.88 Find the number of moles of solute ($H_2C_2O_4$), using the molar mass of 90.04 g $H_2C_2O_4$ per 1 mol $H_2C_2O_4$:

$$1.274 \text{ g } H_2C_2O_4 \times \frac{1 \text{ mol } H_2C_2O_4}{90.04 \text{ g } H_2C_2O_4} = 1.41\underline{4}9 \times 10^{-2} \text{ mol } H_2C_2O_4$$

$$\text{Molarity} = \frac{\text{moles solute}}{\text{liters of solution}} = \frac{1.4149 \times 10^{-2} \text{ mol}}{0.1000 \text{ L}} = 0.141\underline{4}9 = 0.1415 \text{ M}$$

4.89 $$0.150 \text{ mol } CuSO_4 \times \frac{1 \text{ L solution}}{0.120 \text{ mol } CuSO_4} = 1.2\underline{5}0 = 1.25 \text{ L soln}$$

4.90 $$0.00752 \text{ mol } HClO_4 \times \frac{1 \text{ L solution}}{0.126 \text{ mol } HClO_4} = 0.059\underline{6}8 \text{ L (59.7 mL soln)}$$

4.91 $$0.0353 \text{ g KOH} \times \frac{1 \text{ mol KOH}}{56.10 \text{ g KOH}} \times \frac{1 \text{ L soln}}{0.0176 \text{ mol KOH}} = 0.035\underline{7}51 \text{ L (35.8 mL soln)}$$

4.92 $$0.949 \text{ g } H_2SO_4 \times \frac{1 \text{ mol } H_2SO_4}{98.09 \text{ g } H_2SO_4} \times \frac{1 \text{ L soln}}{0.215 \text{ mol } H_2SO_4}$$

$$= 0.044\underline{9}99 \text{ L (45.0 mL soln)}$$

4.93 From the molarity, 1 L of heme solution is equivalent to 0.0019 mol of heme solute. Before starting the calculation, note that 25 mL of solution is equivalent to 25 x 10^{-3} L of solution:

$$25 \times 10^{-3}\text{ L soln} \times \frac{0.0019\text{ mol heme}}{1\text{ L soln}} = 4.\underline{7}50 \times 10^{-5} = 4.8 \times 10^{-5}\text{ mol heme}$$

4.94 From the molarity, 1L of insulin solution is equivalent to 0.0048 mol of insulin solute. Before starting the calculation, note that 28 mL of solution is equivalent to 28 x 10^{-3} L of solution:

$$28 \times 10^{-3}\text{ L soln} \times \frac{0.0048\text{ mol insulin}}{1\text{ L soln}} = 1.\underline{3}44 \times 10^{-4}$$

$$= 1.3 \times 10^{-4}\text{ mol insulin}$$

4.95 Multiply the volume of solution by molarity to convert it to moles; then convert to mass of solute by multiplying by the molar mass:

$$50 \times 10^{-3}\text{ L soln} \times \frac{0.025\text{ mol }Na_2Cr_2O_7}{1\text{ L soln}} \times \frac{262.0\text{ g }Na_2Cr_2O_7}{1\text{ mol }Na_2Cr_2O_7}$$

$$= 0.3\underline{2}75 = 0.33\text{ g }Na_2Cr_2O_7$$

4.96 Multiply the desired volume of solution by the molarity to convert it to moles; then convert to mass of solute by multiplying by the molar mass:

$$0.250\text{ L soln} \times \frac{0.10\text{ mol }Na_2SO_4}{1\text{ L soln}} \times \frac{142.05\text{ g }Na_2SO_4}{1\text{ mol }Na_2SO_4} = 3.\underline{5}51$$

$$= 3.6\text{ g }Na_2SO_4$$

Weigh out 3.6 g of pure Na_2SO_4 and place it in a 250-mL volumetric flask. Add enough water to fill the flask to the mark on the neck.

4.97 Following Example 4.18, use the rearranged version of the dilution formula to calculate the initial volume of 15.8 M HNO_3 required:

$$V_i = \frac{M_f V_f}{M_i} = \frac{0.12\text{ M} \times 1000\text{ mL}}{15.8\text{ M}} = 7.\underline{5}9 = 7.6\text{ mL}$$

4.98 Following Example 4.18, use the rearranged version of the dilution formula to calculate the initial volume of 12.4 M HCl required:

$$V_i = \frac{M_f\ V_f}{M_i} = \frac{0.25\ M\ \times\ 1500\ mL}{12.4\ M} = 3\underline{0}.2 = 30.\ mL$$

4.99 Following Example 4.12, use the appropriate conversion factors to convert the mass of $BaSO_4$ to the mass of Ba^{2+} ions:

$$0.513\ g\ BaSO_4 \times \frac{1\ mol\ BaSO_4}{233.40\ g\ BaSO_4} \times \frac{1\ mol\ Ba^{2+}}{1\ mol\ BaSO_4} \times \frac{137.33\ g\ Ba^{2+}}{1\ mol\ Ba^{2+}}$$

$$= 0.30\underline{1}84\ g\ Ba^{2+}$$

Then calculate the percentage of barium in the 458-mg (0.458-g) compound:

$$\frac{0.30184\ g\ Ba^{2+}}{0.458\ g} \times 100\% = 65.\underline{9}039 = 65.9\%\ Ba^{2+}$$

4.100 Following Example 4.12, use the appropriate conversion factors to convert the mass of AgI to the mass of the I^- ion:

$$2.185\ g\ AgI \times \frac{1\ mol\ AgI}{234.77\ g\ AgI} \times \frac{1\ mol\ I^-}{1\ mol\ AgI} \times \frac{126.90\ g\ I^-}{1\ mol\ I^-} = 1.18\underline{1}1\ g\ I^-$$

Then calculate the percentage of iodine in the 1.545-g compound:

$$\frac{1.1811\ g\ I^-}{1.545\ g} \times 100\% = 76.4\underline{4}7 = 76.45\%\ I^-$$

4.101 a. The mass of chloride ion in the AgCl from the copper chloride compound is:

$$86.00\ mg\ AgCl \times \frac{35.45\ mg\ Cl^-}{143.32\ mg\ AgCl} = 21.2\underline{7}1\ mg\ Cl^-$$

The percentage of chlorine in the 59.40 mg sample is:

$$\frac{21.2\underline{7}1\ mg\ Cl^-}{59.40\ mg\ sample} \times 100\% = 35.8\underline{0}9 = 35.81\%\ Cl^-$$

b. Of the various approaches, it is as easy to calculate the theoretical %Cl^- in both CuCl and $CuCl_2$ as it is to use another approach:

$$\text{CuCl:}\quad \frac{35.45\text{ mg Cl}^-}{99.00\text{ mg CuCl}} \times 100\% = 35.8\underline{0}8\%$$

$$\text{CuCl}_2\text{:}\quad \frac{70.90\text{ mg Cl}^-}{134.45\text{ mg CuCl}_2} \times 100\% = 52.7\underline{3}3\%$$

The compound is obviously CuCl.

4.102 a. The mass of chloride ion in the AgCl from the gold chloride compound is:

$$100.3\text{ mg AgCl} \times \frac{35.45\text{ mg Cl}^-}{143.32\text{ mg AgCl}} = 24.8\underline{0}9\text{ mg Cl}^-$$

The percentage of chloride in the 162.7 mg sample is:

$$\frac{24.8\underline{0}9\text{ mg Cl}^-}{162.7\text{ mg sample}} \times 100\% = 15.2\underline{4}8 = 15.25\%\text{ Cl}^-$$

b. Of the various approaches, it is as easy to calculate the theoretical %Cl^- in both AuCl and $AuCl_3$ as it is to use another approach:

$$\text{AuCl:}\quad \frac{35.45\text{ mg Cl}^-}{232.4\text{ mg AuCl}} \times 100\% = 15.2\underline{5}3\%$$

$$\text{AuCl}_3\text{:}\quad \frac{106.35\text{ mg Cl}^-}{303.32\text{ mg AuCl}_3} \times 100\% = 35.06\underline{2}0\%$$

The compound is obviously AuCl.

4.103 Following Example 4.19, first calculate the moles of chlorine in the compound:

$$0.3048\text{ g AgCl} \times \frac{1\text{ mol AgCl}}{143.32\text{ g AgCl}} \times \frac{1\text{ mol Cl}^-}{1\text{ mol AgCl}} = 0.00212\underline{7}\text{ mol Cl}^-$$

(continued)

Then, following Example 4.10, calculate the g Fe^{X+} from the g Cl^-:

$$g\ Fe^{X+} = 0.1348\ g\ comp - (0.002127\ mol\ Cl^- \times \frac{35.45\ g\ Cl^-}{1\ mol\ Cl^-})$$

$$= 0.0593\underline{9}7\ g\ Fe^{X+}$$

Now calculate the moles of Fe^{X+} using the molar mass:

$$0.059397\ g\ Fe^{X+} \times \frac{1\ mol\ Fe^{X+}}{55.85\ g\ Fe^{X+}} = 0.00106\underline{3}5\ mol\ Fe^{X+}$$

Finally, divide the mole numbers by the smallest mole number:

$$\text{For Cl: } \frac{0.002127\ mol\ Cl^-}{0.0010635\ mol} = 2.00;\ \text{for } Fe^{X+}: \frac{0.0010635\ mol\ Fe^{X+}}{0.0010635\ mol} = 1.00$$

Thus, the formula is $FeCl_2$.

4.104 Following Example 4.19, first calculate the moles of Ba^{2+} in the compound:

$$2.012\ g\ BaCrO_4 \times \frac{1\ mol\ BaCrO_4}{253.33\ g\ BaCrO_4} \times \frac{1\ mol\ Ba^{2+}}{1\ mol\ BaCrO_4} = 0.00794\underline{2}2\ mol\ Ba^{2+}$$

Then, following Example 4.10, calculate the g O from the g Ba^{2+}:

$$g\ O = 1.345\ g\ comp - (0.0079422\ mol\ Ba^{2+} \times \frac{137.33\ g\ Ba^{2+}}{1\ mol\ Ba^{2+}})$$

$$= 0.254\underline{3}0\ mol\ O$$

Now calculate the moles of O using the molar mass:

$$0.25430\ g\ O \times \frac{1\ mol\ O}{16.00\ g\ O} = 0.0158\underline{9}4\ mol\ O$$

Finally, divide the mole numbers by the smallest mole number:

$$\text{For O: } \frac{0.015894\ mol\ O}{0.0079422\ mol} = 2.00;\ \text{for } Ba^{2+}: \frac{0.0079422\ mol\ Ba^{2+}}{0.0079422\ mol} = 1.00$$

Thus, the formula is BaO_2 (barium peroxide).

4.105 Using molarity, convert the volume of Na_2CO_3 to moles of Na_2CO_3; then use the equation to convert to moles of HNO_3, and finally to volume:

$$2HNO_3 + Na_2CO_3 \rightarrow 2NaNO_3 + H_2O + CO_2$$

$$42.4 \times 10^{-3}\ \text{L}\ Na_2CO_3 \times \frac{0.150\ \text{mol}\ Na_2CO_3}{1\ \text{L soln}} \times \frac{2\ \text{mol}\ HNO_3}{1\ \text{mol}\ Na_2CO_3}$$

$$\times \frac{1\ \text{L}\ HNO_3}{0.250\ \text{mol}\ HNO_3} = 0.050\underline{8}8\ \text{L}\ (50.9\ \text{mL})\ HNO_3$$

4.106 As in the previous problem, use molarity to convert to volume; then use the equation to convert to moles of the other reactant, and finally to volume:

$$Na_2CO_3 + Ca(OH)_2 \rightarrow CaCO_3 + 2NaOH$$

$$53.1 \times 10^{-3}\ \text{L}\ Ca(OH)_2 \times \frac{0.150\ \text{mol}\ Ca(OH)_2}{1\ \text{L soln}} \times \frac{2\ \text{mol}\ Na_2CO_3}{1\ \text{mol}\ Ca(OH)_2}$$

$$\times \frac{1\ \text{L}\ Na_2CO_3}{0.350\ \text{mol}\ Na_2CO_3} = 0.022\underline{7}57\ \text{L}\ (22.8\ \text{mL})\ Na_2CO_3$$

4.107 The reaction is $H_2SO_4 + 2NaHCO_3 \rightarrow Na_2SO_4 + 2H_2O + CO_2$.

$$2.05\ \text{g}\ NaHCO_3 \times \frac{1\ \text{mol}\ NaHCO_3}{84.01\ \text{g}\ NaHCO_3} \times \frac{1\ \text{mol}\ H_2SO_4}{2\ \text{mol}\ NaHCO_3} \times \frac{1\ \text{L soln}}{0.150\ \text{mol}\ H_2SO_4}$$

$$= 0.081\underline{3}40\ \text{L}\ (81.3\ \text{mL})\ \text{soln}$$

4.108 The reaction is:

$$10FeSO_4 + 2KMnO_4 + 8H_2SO_4 \rightarrow 5Fe_2(SO_4)_3 + 2MnSO_4 + K_2SO_4 + 4H_2O$$

$$3.36\ \text{g}\ FeSO_4 \times \frac{1\ \text{mol}\ FeSO_4}{151.92\ \text{g}\ FeSO_4} \times \frac{2\ \text{mol}\ KMnO_4}{10\ \text{mol}\ FeSO_4} \times \frac{1\ \text{L soln}}{0.238\ \text{mol}\ KMnO_4}$$

$$= 0.018\underline{5}9\ \text{L}\ (18.6\ \text{mL})\ \text{soln}$$

4.109 First find the mass of H_2O_2 required to react with $KMnO_4$.

$$5H_2O_2 + 2KMnO_4 + 3H_2SO_4 \rightarrow 5O_2 + 2MnSO_4 + K_2SO_4 + 8H_2O$$

$$46.9 \times 10^{-3}\ \text{L}\ KMnO_4 \times \frac{0.145\ \text{mol}\ KMnO_4}{1\ \text{L soln}} \times \frac{5\ \text{mol}\ H_2O_2}{2\ \text{mol}\ KMnO_4} \times \frac{34.02\ \text{g}\ H_2O_2}{1\ \text{mol}\ H_2O_2}$$

$$= 0.57\underline{8}4\ \text{g}\ H_2O_2$$

$$\%\ H_2O_2 = (\text{mass}\ H_2O_2 \div \text{mass sample}) \times 100$$

$$= (0.5784\ \text{g} \div 20.0\ \text{g}) \times 100 = 2.8\underline{9}2 = 2.89\%$$

■ Solutions to Unclassified Problems

4.110 First find the mass of Fe^{2+} required to react with the $K_2Cr_2O_7$.

$$6FeSO_4 + K_2Cr_2O_7 + 7H_2SO_4 \rightarrow 3Fe_2(SO_4)_3 + Cr_2(SO_4)_3 + 7H_2O + K_2SO_4$$

$$41.4 \times 10^{-3}\ \text{L}\ KMnO_4 \times \frac{0.150\ \text{mol}\ K_2Cr_2O_7}{1\ \text{L soln}} \times \frac{6\ \text{mol}\ FeSO_4}{1\ \text{mol}\ K_2Cr_2O_7} \times \frac{1\ \text{mol}\ Fe^{2+}}{1\ \text{mol}\ FeSO_4}$$

$$\times \frac{55.85\ \text{g}\ Fe^{2+}}{1\ \text{mol}\ Fe^{2+}} = 2.0\underline{8}1\ \text{g}\ Fe^{2+}$$

2.081 g Fe^{2+} in reaction = 2.081 g Fe^{2+} in ore, so

$$\%\ Fe = (\text{mass}\ Fe^{2+} \div \text{mass ore}) \times 100 = (2.081\ \text{g} \div 3.33\ \text{g}) \times 100\%$$

$$= 62.\underline{4}9 = 62.5\%$$

4.111 For 1 mol of caffeine, there are 8 mol of C, 10 mol of H, 4 mol of N, and 2 mol of O. Convert these amounts to masses by multiplying by the respective molar masses:

8 mol C x 12.01 g C/1 mol C =	96.08 g C
10 mol H x 1.008 g H/1 mol H =	10.08 g H
4 mol N x 14.01 g N/1 mol N =	56.04 g N
2 mol O x 16.00 g O/1 mol O =	32.00 g O
1 mol of caffeine (total) =	194.20 g (molar mass)

(continued)

Each mass % is calculated by dividing the mass of the element by the molar mass of caffeine and multiplying by 100%: Mass % = (mass element ÷ mass caffeine) x 100%.

Mass % C = (96.08 g ÷ 194.20 g) x 100% = 49.5% (3 s.f.)

Mass % H = (10.08 g ÷ 194.20 g) x 100% = 5.19% (3 s.f.)

Mass % N = (56.04 g ÷ 194.20 g) x 100% = 28.9% (3 s.f.)

Mass % O = (32.00 g ÷ 194.20 g) x 100% = 16.5% (3 s.f.)

4.112 For 1 mol of morphine, there are 17 mol of C, 19 mol of H, 1 mol of N, and 3 mol of O. Convert these amounts to masses by multiplying by the respective molar masses:

17 mol C x 12.01 g C/1 mol C =	204.17 g C
19 mol H x 1.008 g H/1 mol H =	19.15 g H
1 mol N x 14.01 g N/1 mol N =	14.01 g N
3 mol O x 16.00 g O/1 mol O =	<u>48.00 g O</u>
1 mol of morphine (total) =	285.33 g (molar mass)

Each mass % is calculated by dividing the mass of the element by the molar mass of morphine and multiplying by 100%: Mass % = (mass element ÷ mass morphine) x 100%.

Mass % C = (204.17 g ÷ 285.33 g) x 100% = 71.6% (3 s.f.)

Mass % H = (19.15 g ÷ 285.33 g) x 100% = 6.71% (3 s.f.)

Mass % N = (14.01 g ÷ 285.33 g) x 100% = 4.91% (3 s.f.)

Mass % O = (48.00 g ÷ 285.33 g) x 100% = 16.8% (3 s.f.)

4.113 Assume a sample of 100.0 g of dichlorobenzene. By multiplying this by the percentage composition, we obtain 49.0 g C, 2.7 g of H, and 48.2 g of Cl. Convert each mass to moles by dividing by the molar mass:

$$49.0\ \text{g C} \times \frac{1\ \text{mol C}}{12.01\ \text{g C}} = 4.0\underline{8}0\ \text{mol C}$$

$$2.7\ \text{g H} \times \frac{1\ \text{mol H}}{1.008\ \text{g H}} = 2.\underline{6}8\ \text{mol H}$$

$$48.2\ \text{g Cl} \times \frac{1\ \text{mol Cl}}{35.45\ \text{g Cl}} = 1.3\underline{6}0\ \text{mol Cl}$$

(continued)

Divide each number of moles by the smallest number to obtain the smallest set of integers for the empirical formula.

Integer for C = 4.080 mol ÷ 1.360 mol = 3.00, or 3

Integer for H = 2.68 mol ÷ 1.360 mol = 1.97, or 2

Integer for Cl = 1.360 mol ÷ 1.360 mol = 1.00, or 1

The empirical formula is thus C_3H_2Cl. Find the formula weight by adding the atomic weights:

Formula weight = (3 x 12.01 amu) + (2 x 1.008 amu) + 35.45 amu

= 73.4$\underline{9}$6 = 73.50 amu

Divide the molecular weight by the formula weight to find the number of times the C_3H_2Cl unit occurs in the molecule. Because the molecular weight is 147 amu, this quotient is

147 amu ÷ 73.50 amu = 2.00, or 2

The molecular formula is $(C_3H_2Cl)_2$, or $C_6H_4Cl_2$.

4.114 Assume a sample of 100.0 g of sorbic acid. By multiplying this by the percentage composition, we obtain 64.3 g C, 7.2 g H, and 28.5 g O. Convert each mass to moles by dividing by the molar mass.

$$64.3 \text{ g C} \times \frac{1 \text{ mol C}}{12.01 \text{ g C}} = 5.3\underline{5}3 \text{ mol C}$$

$$7.2 \text{ g H} \times \frac{1 \text{ mol H}}{1.008 \text{ g H}} = 7.\underline{1}4 \text{ mol H}$$

$$28.5 \text{ g O} \times \frac{1 \text{ mol O}}{16.00 \text{ g O}} = 1.7\underline{8}1 \text{ mol O}$$

Divide each number of moles by the smallest number to obtain the smallest set of integers for the empirical formula.

Integer for C = 5.353 mol ÷ 1.781 mol = 3.01, or 3

Integer for H = 7.14 mol ÷ 1.781 mol = 4.0, or 4

Integer for O = 1.781 mol ÷ 1.781 mol = 1.00, or 1

(continued)

The empirical formula is thus C_3H_4O. Find the formula weight by adding the atomic weights:

$$\text{Formula weight} = (3 \times 12.01 \text{ amu}) + (4 \times 1.008 \text{ amu}) + 16.00 \text{ amu}$$

$$= 56.0\underline{6}2 = 56.06 \text{ amu}$$

Divide the molecular weight by the formula weight to find the number of times the C_3H_4O unit occurs in the molecule. Because the molecular weight was given as 112 amu, this quotient is

$$112 \text{ amu} \div 56.06 \text{ amu} = 2.00, \text{ or } 2$$

The molecular formula is $(C_3H_4O)_2$, or $C_6H_8O_2$.

4.115 For these calculations, the relative numbers of moles of gold and chlorine must be determined. These can be found from the masses of the two elements in the sample:

$$\text{Total mass} = \text{mass of Au} + \text{mass of Cl} = 328 \text{ mg}$$

The mass of chlorine in the precipitated AgCl = the mass of chlorine in the compound of gold and chlorine. The mass of Cl in the 0.464 g of AgCl is

$$0.464 \text{ g AgCl} \times \frac{1 \text{ mol AgCl}}{143.32 \text{ g AgCl}} \times \frac{1 \text{ mol Cl}}{1 \text{ mol AgCl}} \times \frac{35.45 \text{ g Cl}}{1 \text{ mol Cl}}$$

$$= 0.11\underline{4}8 \text{ g Cl (114.8 mg Cl)}$$

$$\text{Mass \% Cl} = \frac{\text{mass Cl}}{\text{mass comp.}} \times 100\% = \frac{114.8 \text{ mg}}{328 \text{ mg}} \times 100\% = 35.0\% \text{ Cl}$$

To find the empirical formula, convert each mass to moles:

$$\text{Mass Au} = 328 \text{ mg} - 114.8 \text{ mg Cl} = 213.2 \text{ mg Au}$$

$$0.1148 \text{ g Cl} \times \frac{1 \text{ mol Cl}}{35.45 \text{ g Cl}} = 0.0032\underline{3}8 \text{ mol Cl}$$

$$0.2132 \text{ g Au} \times \frac{1 \text{ mol Au}}{196.97 \text{ g Au}} = 0.001082 \text{ mol}$$

Divide both numbers of moles by the smaller number (.001082) to find the integers:

Integer for Cl: $0.003238 \text{ mol} \div 0.001082 \text{ mol} = 2.99$, or 3

Integer for Au: $0.001082 \text{ mol} \div 0.001082 \text{ mol} = 1.00$, or 1

The empirical formula is $AuCl_3$.

4.116 To find the mass percentages of Cl and Sc, find the mass of the chlorine in the precipitated AgCl, which is equal to the mass of the chlorine in the compound of scandium and chlorine. The mass of Cl in the 167.4 mg (0.1674 g) of AgCl is

$$0.1674 \text{ g AgCl} \times \frac{1 \text{ mol AgCl}}{143.32 \text{ g AgCl}} \times \frac{1 \text{ mol Cl}}{1 \text{ mol AgCl}} \times \frac{35.45 \text{ g Cl}}{1 \text{ mol Cl}} = 0.041\underline{4}1 \text{ g Cl}$$

The mass of Sc equals the difference between the sample mass and the mass of Cl:

$$\text{Mass Sc} = 0.0589 \text{ g} - 0.04141 \text{ g} = 0.017\underline{4}9 \text{ g Sc}$$

The mass percentage of each element is found by dividing the mass of the element by the mass of the compound and multiplying by 100%: Mass % = (mass elem. ÷ mass comp.) x 100%.

$$\text{Mass \% Cl} = (0.04141 \text{ g} \div 0.0589 \text{ g}) \times 100\% = 70.\underline{3}05 = 70.3\%$$

$$\text{Mass \% Sc} = (0.01749 \text{ g} \div 0.0589 \text{ g}) \times 100\% = 29.\underline{6}9 = 29.7\%$$

To find the empirical formula, convert each mass to moles:

$$0.04141 \text{ g Cl} = \frac{1 \text{ mol Cl}}{35.45 \text{ g Cl}} = 0.00116\underline{8}0 \text{ mol Cl}$$

$$0.01749 \text{ g Sc} = \frac{1 \text{ mol Sc}}{44.96 \text{ g Sc}} = 0.00038\underline{9}0 \text{ mol Sc}$$

Divide both numbers of moles by the smaller number (0.0003890) to find the integers:

Integer for Cl = 0.0011680 mol ÷ 0.0003890 mol = 3.00, or 3

Integer for Sc = 0.0003890 mol ÷ 0.0003890 mol = 1.00, or 1

The empirical formula is $ScCl_3$.

4.117 Find the % composition of C and S from the analysis:

$$0.01665 \text{ g } CO_2 \times \frac{1 \text{ mol } CO_2}{44.01 \text{ g } CO_2} \times \frac{1 \text{ mol C}}{1 \text{ mol } CO_2} \times \frac{12.01 \text{ g C}}{1 \text{ mol C}}$$

$$= 0.00454\underline{4} \text{ g C}$$

$$\% \text{ C} = (0.004544 \text{ g C} \div 0.00796 \text{ g comp.}) \times 100\% = 57.\underline{0}9\%$$

(continued)

$$0.01196 \text{ g BaSO}_4 \times \frac{1 \text{ mol BaSO}_4}{233.39 \text{ g BaSO}_4} \times \frac{1 \text{ mol S}}{1 \text{ mol BaSO}_4} \times \frac{32.07 \text{ g S}}{1 \text{ mol S}}$$

$$= 0.0016\underline{4}3 \text{ g S}$$

$$\% \text{ S} = (0.001643 \text{ g S} \div 0.00431 \text{ g comp.}) \times 100\% = 38.\underline{1}2\%$$

$$\% \text{ H} = 100.00\% - (57.09 + 38.12)\% = 4.\underline{7}9\%$$

We now obtain the empirical formula by calculating moles from the grams corresponding to each mass percentage of element:

$$57.09 \text{ g C} \times \frac{1 \text{ mol C}}{12.01 \text{ g C}} = 4.75\underline{4} \text{ mol C}$$

$$38.12 \text{ g S} \times \frac{1 \text{ mol S}}{32.07 \text{ g C}} = 1.18\underline{9} \text{ mol S}$$

$$4.79 \text{ g H} \times \frac{1 \text{ mol H}}{1.008 \text{ g H}} = 4.7\underline{5}2 \text{ mol H}$$

Dividing the moles of the elements by the smallest number (1.189), we obtain for C: 3.997, or 4; for S: 1.000, or 1; and for H: 3.996, or 4. Thus, the empirical formula is C_4H_4S (formula weight = 84). Because the formula weight was given as 84 amu, the molecular formula is also C_4H_4S.

4.118 Find the % composition of H and N from the analysis:

$$0.00663 \text{ g H}_2\text{O} \times \frac{1 \text{ mol H}_2\text{O}}{18.02 \text{ g H}_2\text{O}} \times \frac{2 \text{ mol H}}{1 \text{ mol H}_2\text{O}} \times \frac{1.008 \text{ g H}}{1 \text{ mol H}} = 0.00074\underline{1}7 \text{ g H}$$

1.46 mg N_2 from the analysis is equivalent to 0.00146 g N.

$$\% \text{ H} = (0.0007417 \text{ g H} \div 0.00971 \text{ g comp.}) \times 100\% = 7.6\underline{4}\%$$

$$\% \text{ N} = (0.00146 \text{ g N} \div 0.00971 \text{ g comp.}) \times 100\% = 15.\underline{0}\%$$

$$\% \text{ C} = 100.00\% - (7.64 + 15.0)\% = 77.\underline{4}\%$$

Calculate the moles from grams to obtain the empirical formula:

$$77.4 \text{ g C} \times \frac{1 \text{ mol C}}{12.01 \text{ g C}} = 6.4\underline{4} \text{ mol C}$$

(continued)

$$7.64 \text{ g H} \times \frac{1 \text{ mol H}}{1.008 \text{ g H}} = 7.5\underline{8} \text{ mol H}$$

$$15.0 \text{ g N} \times \frac{1 \text{ mol N}}{14.01 \text{ g N}} = 1.0\underline{7} \text{ mol N}$$

Dividing the moles of elements by the smallest number (1.07) gives for C: 6.02, or 6; for H: 7.08, or 7; and for N: 1.00, or 1. The empirical formula is thus C_6H_7N (formula weight = 93 amu). Because the molecular weight was given as 93 amu, the molecular formula is also C_6H_7N.

4.119 If one heme molecule contains one iron atom, then the number of moles of heme in 35.2 mg heme must be the same as the number of moles of iron in 3.19 mg of iron. Start by calculating the moles of Fe (= moles heme):

$$3.19 \times 10^{-3} \text{ g Fe} \times \frac{1 \text{ mol Fe}}{55.85 \text{ g Fe}} = 5.7\underline{1}2 \times 10^{-5} \text{ mol Fe or heme}$$

$$\text{Molar mass of heme} = \frac{35.2 \times 10^{-3} \text{ g}}{5.712 \times 10^{-5} \text{ mol}} = 61\underline{6}.2 = 616 \text{ g/mol}$$

The molecular weight of heme is 616 amu.

4.120 Convert the mass of $BaSO_4$ to mass of S to find the percentage of sulfur:

$$0.00546 \text{ g BaSO}_4 \times \frac{1 \text{ mol BaSO}_4}{233.40 \text{ g BaSO}_4} \times \frac{1 \text{ mol S}}{1 \text{ mol BaSO}_4} \times \frac{32.07 \text{ g S}}{1 \text{ mol S}}$$

$$= 0.7502 \times 10^{-3} \text{ g S}$$

$$\text{Mass \% S} = \frac{0.7502 \times 10^{-3} \text{ g S}}{8.19 \times 10^{-3} \text{ g pen. V}} \times 100\% = 9.1\underline{6}0 = 9.16\%$$

Convert the mass of S to moles; then recognizing that moles of S = moles of pen. V, use that number of moles to calculate the molar mass:

$$0.7502 \times 10^{-3} \text{ g S} \times \frac{1 \text{ mol S}}{32.07 \text{ g S}} = 2.3\underline{3}9 \times 10^{-5} \text{ mol S}$$

$$\text{Molar mass pen. V} = \frac{8.19 \times 10^{-3} \text{ g}}{2.339 \times 10^{-5} \text{ mol}} = 35\underline{0}.1 = 350 \text{ g/mol}$$

The molecular weight of penicillin V is 350 amu.

4.121 For g $CaCO_3$, use this equation: $CaCO_3 + H_2C_2O_4 \rightarrow CaC_2O_4 + H_2O + CO_2$.

$$0.472 \text{ g } CaC_2O_4 \times \frac{1 \text{ mol } CaC_2O_4}{128.10 \text{ g } CaC_2O_4} \times \frac{1 \text{ mol } CaCO_3}{1 \text{ mol } CaC_2O_4} \times \frac{100.09 \text{ g } CaCO_3}{1 \text{ mol } CaCO_3}$$

$$= 0.36\underline{8}8 \text{ g } CaCO_3$$

$$\text{Mass \% } CaCO_3 = \frac{\text{mass } CaCO_3}{\text{mass limestone}} \times 100\% = \frac{0.36\underline{8}8 \text{ g}}{0.413 \text{ g}} \times 100\%$$

$$= 89.\underline{2}9 = 89.3\%$$

4.122 For the mass of TiO_2, use this equation: $TiO_2 + C + 2Cl_2 \rightarrow TiCl_4 + CO_2$.

$$35.4 \text{ g } TiCl_4 \times \frac{1 \text{ mol } TiCl_4}{189.68 \text{ g } TiCl_4} \times \frac{1 \text{ mol } TiO_2}{1 \text{ mol } TiCl_4} \times \frac{79.88 \text{ g } TiO_2}{1 \text{ mol } TiO_2} = 14.\underline{9}1 \text{ g } TiO_2$$

$$\text{Mass \% } TiO_2 = \frac{\text{mass } TiO_2}{\text{mass rutile}} \times 100\% = \frac{14.\underline{9}1 \text{ g}}{17.4 \text{ g}} \times 100\% = 85.\underline{6}8 = 85.7\%$$

4.123 Calculate the theoretical yield using this equation: $2C_2H_4 + O_2 \rightarrow 2C_2H_4O$.

$$10.6 \text{ g } C_2H_4 \times \frac{1 \text{ mol } C_2H_4}{28.05 \text{ g } C_2H_4} \times \frac{2 \text{ mol } C_2H_4O}{2 \text{ mol } C_2H_4} \times \frac{44.05 \text{ g } C_2H_4O}{1 \text{ mol } C_2H_4O}$$

$$= 16.\underline{6}5 \text{ g } C_2H_4O$$

$$\text{\% yield} = \frac{\text{actual yield}}{\text{theo. yield}} \times 100\% = \frac{9.69 \text{ g}}{16.65 \text{ g}} = 58.\underline{1}9 \text{ g} = 58.2\%$$

4.124 Calculate the theoretical yield using this equation: $C_6H_6 + HNO_3 \rightarrow C_6H_5NO_2 + H_2O$.

$$20.3 \text{ g } C_6H_6 \times \frac{1 \text{ mol } C_6H_6}{78.11 \text{ g } C_6H_6} \times \frac{1 \text{ mol } C_6H_5NO_2}{1 \text{ mol } C_6H_6} \times \frac{123.11 \text{ g } C_6H_5NO_2}{1 \text{ mol } C_6H_5NO_2}$$

$$= 32.\underline{0}0 \text{ g } C_6H_5NO_2$$

$$\text{\% yield} = \frac{\text{actual yield}}{\text{theo. yield}} \times 100\% = \frac{28.7 \text{ g}}{32.00 \text{ g}} \times 100\% = 89.\underline{6}9 = 89.7\%$$

4.125 To find Zn, use these equations:

$$2C + O_2 \rightarrow 2CO \text{ and } ZnO + CO \rightarrow Zn + CO_2$$

Two mol C produces 2 mol CO; because 1 mol ZnO reacts with 1 mol CO, 2 mol ZnO will react with 2 mol CO. Thus, 2 mol C is equivalent to 2 mol ZnO, or 1 mol C is equivalent to 1 mol ZnO. Using this to calculate mass of C from mass of ZnO, we have

$$75.0 \text{ g ZnO} \times \frac{1 \text{ mol ZnO}}{81.39 \text{ g ZnO}} \times \frac{1 \text{ mol C}}{1 \text{ mol ZnO}} \times \frac{12.01 \text{ g C}}{1 \text{ mol C}} = 11.\underline{0}7 \text{ g C}$$

Thus, all of the ZnO is used up in reacting with just 11.07 g of C, making ZnO the limiting reagent. Use the mass of ZnO to calculate the mass of Zn formed:

$$75.0 \text{ g ZnO} \times \frac{1 \text{ mol ZnO}}{81.39 \text{ g ZnO}} \times \frac{1 \text{ mol Zn}}{1 \text{ mol ZnO}} \times \frac{65.39 \text{ g Zn}}{1 \text{ mol Zn}} = 60.\underline{2}56 = 60.3 \text{ g Zn}$$

4.126 To find CH_4, use these equations:

$$4NH_3 + 5O_2 \rightarrow 4NO + 6H_2O \text{ and } 2NO + 2CH_4 \rightarrow 2HCN + 2H_2O + H_2$$

Four mol NH_3 produces 4 mol NO; because 2 mol CH_4 reacts with 2 mol NO, 4 mol CH_4 will react with 4 mol NO. Thus, 4 mol NH_3 is equivalent to 4 mol CH_4. Using this to calculate the mass of CH_4 from the mass of NH_3, we have

$$24.2 \text{ g } NH_3 \times \frac{1 \text{ mol } NH_3}{17.03 \text{ g } NH_3} \times \frac{4 \text{ mol } CH_4}{4 \text{ mol } NH_3} \times \frac{16.04 \text{ g } CH_4}{1 \text{ mol } CH_4} = 22.\underline{8} \text{ g } CH_4$$

Thus, all of the NH_3 is used up in reacting with just 22.8 g of CH_4, making NH_3 the limiting reagent. Use the mass of NH_3 to calculate the mass of HCN formed:

$$24.2 \text{ g } NH_3 \times \frac{1 \text{ mol } NH_3}{17.03 \text{ g } NH_3} \times \frac{2 \text{ mol } CH_4}{2 \text{ mol } NH_3} \times \frac{27.03 \text{ g HCN}}{1 \text{ mol HCN}}$$

$$= 38.\underline{4}1 = 38.4 \text{ g HCN}$$

4.127 Divide the mass of $CaCl_2$ by its molar mass and volume to find molarity:

$$2.25 \text{ g } CaCl_2 \times \frac{1 \text{ mol } CaCl_2}{110.98 \text{ g } CaCl_2} \times \frac{1}{1.000 \text{ L soln}} = 0.020\underline{2}7 = 0.0203 \text{ M } CaCl_2$$

The $CaCl_2$ dissolves to form Ca^{2+} and $2Cl^-$ ions. Therefore, the molarities of the ions are 0.0203 M Ca^{2+} and 2 x 0.0203, or 0.0406, M Cl^- ions.

4.128 Divide the mass of $Fe_2(SO_4)_3$ by its molar mass and volume to find molarity:

$$3.45\ g\ Fe_2(SO_4)_3 \times \frac{1\ mol\ Fe_2(SO_4)_3}{399.91\ g\ Fe_2(SO_4)_3} \times \frac{1}{1.000\ L\ soln} = 0.0086\underline{2}7$$

$$= 0.00863\ M\ Fe_2(SO_4)_3$$

The $Fe_2(SO_4)_3$ dissolves to form $2Fe^{3+}$ and $3SO_4^{2-}$ ions. Therefore, the molarities of the ions are 2 x 0.00863, or 0.0173, M Fe^{3+} and 3 x 0.00863, or 0.0259, M SO_4^{2-}.

4.129 Divide the mass of $K_2Cr_2O_7$ by its molar mass and volume to find molarity. Then calculate the volume needed to prepare 1.00L of a 0.100 M solution.

$$89.3\ g\ K_2Cr_2O_7 \times \frac{1\ mol\ K_2Cr_2O_7}{294.20\ g\ K_2Cr_2O_7} = 0.30\underline{3}5\ mol\ K_2Cr_2O_7$$

$$\text{Molarity} = \frac{0.30\underline{3}5\ mol\ K_2Cr_2O_7}{1.00\ L} = 0.30\underline{3}5\ M$$

$$V_i = \frac{V_f \times M_f}{M_i} = \frac{1.00\ L \times 0.100\ M}{0.3035\ M} = 0.32\underline{9}4\ L\ (329\ mL)$$

4.130 Divide the mass by the molar mass and then by volume to find molarity. Then calculate the volume needed to prepare 1.00 L of 0.1150 M.

$$69.3\ g\ H_2C_2O_4 \times \frac{1\ mol\ H_2C_2O_4}{90.04\ g\ H_2C_2O_4} = 0.76\underline{9}7\ mol\ H_2C_2O_4$$

$$\text{Molarity} = \frac{0.7697\ mol\ H_2C_2O_4}{1.00\ L} = 0.76\underline{9}7\ M$$

$$V_i = \frac{V_f \times M_f}{M_i} = \frac{1.00\ L \times 0.150\ M}{0.7697\ M} = 0.1\underline{948}\ L\ (195\ mL)$$

Place 195 mL of the 0.7697-M solution in a 1-L volumetric flask and dilute to 1.00 L.

4.131 Assume a volume of 1.000 L (1000 cm^3) for the 6.00% NaBr solution, and convert to moles and then to molarity.

$$1000\ cm^3 \times \frac{1.046\ g\ soln}{1\ cm^3} \times \frac{6.00\ g\ NaBr}{100\ g\ soln} \times \frac{1\ mol\ NaBr}{102.89\ g\ NaBr} = 0.60\underline{9}97\ mol$$

$$\text{Molarity NaBr} = \frac{0.60997\ mol}{1.000\ L} = 0.60\underline{9}97 = 0.610\ M$$

4.132 Assume a volume of 1.000 L (1000 mL) for the 4.50% NH_3 solution, and convert to moles and then to molarity.

$$1000 \text{ mL} \times \frac{0.979 \text{ g soln}}{1 \text{ mL}} \times \frac{4.50 \text{ g } NH_3}{100 \text{ g soln}} \times \frac{1 \text{ mol } NH_3}{17.03 \text{ g } NH_3} = 2.5\underline{8}7 \text{ mol } NH_3$$

$$\text{Molarity } NH_3 = \frac{2.5\underline{8}7 \text{ mol}}{1.000 \text{ L}} = 2.5\underline{8}7 = 2.59 \text{ M } NH_3$$

4.133 Following example 4.20, first calculate the moles of $BaCl_2$:

$$1.028 \text{ g } BaSO_4 \times \frac{1 \text{ mol } BaSO_4}{233.40 \text{ g } BaSO_4} \times \frac{1 \text{ mol } BaCl_2}{1 \text{ mol } BaSO_4} = 0.00440\underline{4}4 \text{ mol } BaCl_2$$

Then calculate the molarity from the moles and volume (0.0500 L):

$$\text{Molarity} = \frac{0.00440\underline{4}4 \text{ mol } BaCl_2}{0.0500 \text{ L}} = 0.088\underline{0}89 = 0.0881 \text{ M}$$

4.134 Following Example 4.20, first calculate the moles of $CaCl_2$:

$$1.437 \text{ g } CaC_2O_4 \times \frac{1 \text{ mol } CaC_2O_4}{128.10 \text{ g } CaC_2O_4} \times \frac{1 \text{ mol } CaCl_2}{1 \text{ mol } CaC_2O_4}$$

$$= 0.0112\underline{1}77 \text{ mol } CaCl_2$$

Then calculate the molarity from the moles and volume (0.0500 L):

$$\text{Molarity} = \frac{0.0112\underline{1}77 \text{ mol } CaCl_2}{0.0500 \text{ L}} = 0.22\underline{4}35 = 0.224 \text{ M}$$

4.135 First calculate the g of thallium(I) sulfate:

$$0.2122 \text{ g TlI} \times \frac{1 \text{ mol TlI}}{331.28 \text{ g TlI}} \times \frac{1 \text{ mol } Tl_2SO_4}{2 \text{ mol TlI}} \times \frac{504.83 \text{ g } Tl_2SO_4}{1 \text{ mol } Tl_2SO_4}$$

$$= 0.161\underline{6}8 \text{ g } Tl_2SO_4$$

Then calculate the % Tl_2SO_4 in the rat poison:

$$\% \, Tl_2SO_4 = \frac{0.16168 \text{ g}}{0.7590 \text{ g}} \times 100\% = 21.3\underline{0}1 = 21.30\%$$

4.136 First calculate the g of $CaCO_3$:

$$0.6332 \text{ g } CaC_2O_4 \times \frac{1 \text{ mol } CaC_2O_4}{128.10 \text{ g } CaC_2O_4} \times \frac{1 \text{ mol } CaCO_3}{1 \text{ mol } CaC_2O_4} \times \frac{100.1 \text{ g } CaCO_3}{1 \text{ mol } CaCO_3}$$

$$= 0.494\underline{7}9 \text{ g } CaCO_3$$

Then calculate the % $CaCO_3$ in the antacid:

$$\% \ CaCO_3 = \frac{0.49479 \text{ g}}{0.680 \text{ g}} \times 100\% = 72.\underline{7}64 = 72.8\%$$

4.137 The mass of copper(II) ion and mass of sulfate ion in the 98.77 mg sample is:

$$0.09877 \text{ g} \times 0.3250 = 0.0321\underline{0}0 \text{ g } Cu^{2+} \text{ ion}$$

$$0.11666 \text{ g } BaSO_4 \times \frac{96.07 \text{ g } SO_4^{2-}}{233.40 \text{ g } BaSO_4} = 0.0480\underline{1}9 = 0.04802 \text{ g } SO_4^{2-}$$

The mass of water left = 0.09877 g - (0.03210 g Cu^{2+} + 0.04802 g SO_4^{2-})

= 0.01865 g H_2O

The moles of water left = 0.01865 g ÷ 18.02 g/mol = 1.035×10^{-3} mol

The moles of Cu^{2+} = 0.03210 g ÷ 63.55 g/mol = $5.05\underline{1}1 \times 10^{-4}$ mol

The ratio of water to Cu^{2+}, or $CuSO_4$ = $1.035 \times 10^{-3} \div 5.0511 \times 10^{-4}$ = 2.05 or 2

The formula is thus $CuSO_4 \bullet 2\ H_2O$

4.138 The mass of copper(II) ion and mass of sulfate ion in the 85.42 mg sample is:

$$0.08542 \text{ g} \times 0.2976 = 0.025\underline{4}21 \text{ g } Cu^{2+} \text{ ion}$$

$$0.09333 \text{ g } BaSO_4 \times \frac{96.07 \text{ g } SO_4^{2-}}{233.40 \text{ g } BaSO_4} = 0.0384\underline{1}6 = 0.03842 \text{ g } SO_4^{2-}$$

The mass of water left = 0.08542 g - (0.025421 g Cu^{2+} + 0.038416 g SO_4^{2-})

= 0.021583 g H_2O

(continued)

The moles of water left = 0.021583 ÷ 18.02 g/mol = 1.1977×10^{-3} mol

The moles of Cu^{2+} = 0.025421 g ÷ 63.55 g/mol = 4.0002×10^{-4} mol

The ratio of water to Cu^{2+}, or $CuSO_4$ = $1.1977 \times 10^{-3} \div 4.000 \times 10^{-4}$ = 2.99 or 3

The formula is $CuSO_4{\bullet}3\ H_2O$

4.139 From the equations $NH_3 + HCl \rightarrow NH_4Cl$ and $NaOH + HCl \rightarrow NaCl + H_2O$, we write

Mol NH_3 = mol $HCl(NH_3)$

Mol NaOH = mol HCl(NaOH)

We can calculate the mol NaOH and the sum [mol $HCl(NH_3)$ + mol HCl(NaOH)] from the titration data. Because the sum = mol NH_3 + mol NaOH, we can calculate the unknown mol of NH_3 from the difference: Mol NH_3 = sum - mol NaOH.

$$\text{Mol HCl (NaOH)} + \text{mol HCl (NH}_3) = 0.0463\ \text{L} \times \frac{0.0213\ \text{mol HCl}}{1.000\ \text{L}}$$

$$= 0.0098\underline{6}2\ \text{mol HCl}$$

$$\text{Mol NaOH} = 0.0443\ \text{L} \times \frac{0.128\ \text{mol NaOH}}{1.000\ \text{L}} = 0.0056\underline{7}0\ \text{mol NaOH}$$

$$\text{Mol HCl(NH}_3) = 0.009862\ \text{mol} - 0.005670\ \text{mol} = 0.004192\ \text{mol}$$

$$\text{Mol NH}_3 = \text{mol HCl(NH}_3) = 0.004192\ \text{mol NH}_3$$

Because all of the N in the $(NH_4)_2SO_4$ was liberated as, and titrated as, NH_3, the amount of N in the fertilizer is equal to the amount of N in the NH_3. Thus, the moles of NH_3 can be used to calculate the mass percentage of N in the fertilizer:

$$0.004192\ \text{mol NH}_3 \times \frac{1\ \text{mol N}}{1\ \text{mol NH}_3} \times \frac{14.01\ \text{g N}}{1\ \text{mol N}} = 0.058\underline{7}3\ \text{g N}$$

$$\%\ \text{N} = \frac{\text{mass N}}{\text{mass fert.}} \times 100\% = \frac{0.058\underline{7}3\ \text{g N}}{0.608\ \text{g}} \times 100\% = 9.6\underline{5}9 = 9.66\%$$

4.140 From $NaHCO_3$ + HCl → NaCl + H_2O + CO_2 and HCl + NaOH → NaCl + H_2O, we write

$$\text{Mol NaHCO}_3 = \text{mol HCl(NaHCO}_3)$$

$$\text{Mol NaOH} = \text{mol HCl(NaOH)}$$

We can calculate the mol NaOH and the sum [mol HCl($NaHCO_3$) + mol HCl(NaOH)] from the titration data. Because the sum = mol $NaHCO_3$ + mol NaOH, we can calculate the unknown mol of $NaHCO_3$ from the difference: Mol $NaHCO_3$ = sum - mol NaOH.

$$\text{Mol HCl (NaOH)} + \text{mol HCl (NaHCO}_3) = 0.0500\text{ L} \times \frac{0.190\text{ mol HCl}}{1.000\text{ L}}$$

$$= 0.0095\underline{0}0\text{ mol HCl}$$

$$\text{Mol NaOH} = 0.0471\text{ L} \times \frac{0.128\text{ mol NaOH}}{1.000\text{ L}} = 0.0060\underline{2}9\text{ mol NaOH}$$

$$\text{Mol HCl(NaHCO}_3) = 0.009500\text{ mol} - 0.006029\text{ mol} = 0.003471\text{ mol}$$

$$\text{Mol NaHCO}_3 = \text{mol HCl(NaHCO}_3) = 0.003471\text{ mol}$$

Because all of the $NaHCO_3$ in the tablet was titrated as $NaHCO_3$, the moles of $NaHCO_3$ can be used to calculate the mass percentage of $NaHCO_3$ in the tablet:

$$0.003471\text{ mol NaHCO}_3 \times \frac{84.01\text{ g NaHCO}_3}{1\text{ mol NaHCO}_3} = 0.29\underline{1}6\text{ g NaHCO}_3$$

$$\%\text{ NaHCO}_3 = \frac{\text{mass NaHCO}_3}{\text{mass tab.}} \times 1\ 100\% = \frac{0.29\underline{1}6\text{ g N}}{0.500\text{ g}} \times 100\%$$

$$= 58.\underline{3}2 = 58.3\%$$

4.141 For CaO + 3C → CaC_2 + CO, find the limiting reactant in terms of moles of CaC_2 obtainable:

$$\text{Mol CaC}_2 = 1.15 \times 10^3\text{ g C} \times \frac{1\text{ mol C}}{12.01\text{ g C}} \times \frac{1\text{ mol CaC}_2}{3\text{ mol C}} = 31.\underline{9}1\text{ mol}$$

$$\text{Mol CaC}_2 = 1.15 \times 10^3\text{ g CaO} \times \frac{1\text{ mol CaO}}{56.08\text{ g CaO}} \times \frac{1\text{ mol CaC}_2}{1\text{ mol CaO}} = 20.\underline{5}06\text{ mol}$$

(continued)

Because CaO is the limiting reactant, calculate the mass of CaC_2 from it:

$$\text{Mass } CaC_2 = 20.506 \text{ mol } CaC_2 \times \frac{64.10 \text{ g } CaC_2}{1 \text{ mol } CaC_2} = 1.3\underline{1}4 \times 10^3$$

$$= 1.31 \times 10^3 \text{ g } CaC_2$$

4.142 For $CaF_2 + H_2SO_4 \rightarrow 2HF + CaSO_4$, find the limiting reactant in terms of moles of HF obtainable:

$$\text{Mol HF} = 12.8 \text{ g } CaF_2 \times \frac{1 \text{ mol } CaF_2}{78.08 \text{ g } CaF_2} \times \frac{2 \text{ mol HF}}{1 \text{ mol } CaF_2} = 0.32\underline{7}9 \text{ mol}$$

$$\text{Mol HF} = 13.2 \text{ g } H_2SO_4 \times \frac{1 \text{ mol } H_2SO_4}{98.09 \text{ g } H_2SO_4} \times \frac{2 \text{ mol HF}}{1 \text{ mol } H_2SO_4} = 0.26\underline{9}1 \text{ mol}$$

Because H_2SO_4 is the limiting reagent, calculate the mass of HF from it:

$$\text{Mass HF} = 0.2691 \text{ mol HF} \times \frac{20.01 \text{ g HF}}{1 \text{ mol HF}} = 5.3\underline{8}4 = 5.38 \text{ g HF}$$

4.143 From the equation $2Na + H_2O \rightarrow 2NaOH + H_2$, convert the mass of H_2 to mass of Na, and then use the mass to calculate the percentage:

$$0.108 \text{ g } H_2 \times \frac{1 \text{ mol } H_2}{2.016 \text{ g } H_2} \times \frac{2 \text{ mol Na}}{1 \text{ mol } H_2} \times \frac{22.99 \text{ g Na}}{1 \text{ mol Na}} = 2.4\underline{6}3 \text{ g Na}$$

$$\% \text{ Na} = \frac{\text{mass Na}}{\text{mass amalgam}} \times 100\% = \frac{2.463 \text{ g}}{15.23 \text{ g}} \times 100\% = 16.\underline{1}7 = 16.2\%$$

4.144 From the equation $CaCO_3 \rightarrow CaO + CO_2$, convert the mass of CO_2 to mass of $CaCO_3$, and then use the mass to calculate the percentage:

$$0.00401 \text{ g } CO_2 \times \frac{1 \text{ mol } CO_2}{44.01 \text{ g } CO_2} \times \frac{2 \text{ mol } CaCO_3}{1 \text{ mol } CO_2} \times \frac{100.09 \text{ g } CaCO_3}{1 \text{ mol } CaCO_3}$$

$$= 0.0091\underline{2}0 \text{ g } CaCO_3$$

$$\% \text{ } CaCO_3 = \frac{\text{mass } CaCO_3}{\text{mass sandst.}} \times 100\% = \frac{0.0091\underline{2}0 \text{ g}}{0.0218 \text{ g}} \times 100\% = 41.\underline{8}3\%$$

(continued)

Because the sandstone contains only SiO_2 and $CaCO_3$, the difference between 100% and the percentage of $CaCO_3$ is the percentage of SiO_2:

$$\% \ SiO_2\text{(silica)} = 100.00\% - 41.83\% = 58.\underline{1}7 = 58.2\%$$

■ Cumulative-Skills Problems

(require skills from previous chapters)

4.145 Let y = the mass of CuO in the mixture. Then 0.500 g - y = the mass of Cu_2O in the mixture. Multiplying the appropriate conversion factors for Cu times the mass of each oxide will give one equation in one unknown for the mass of 0.425 g Cu:

$$0.425 = y\left[\frac{63.55 \text{ g Cu}}{79.55 \text{ g CuO}}\right] + (0.500 - y)\left[\frac{127.10 \text{ g Cu}}{143.10 \text{ g } Cu_2O}\right]$$

Simplifying the equation by dividing the conversion factors and combining terms gives:

$$0.425 = 0.798\underline{8}7\ y + 0.8881\underline{9}0\ (0.500 - y)$$

$$0.08932\ y = 0.01\underline{9}095$$

$$y = 0.2\underline{1}39 = 0.21 \text{ g} = \text{mass of CuO}$$

4.146 Let y = the mass of Fe_2O_3 in the mixture. Then 0.500 g - y = the mass of FeO in the mixture. The mass of Fe in the mixture is 0.720 x 0.500 g = 0.360 g. Multiplying the appropriate conversion factors for Fe times the mass of each oxide will give one equation in one unknown for the mass of 0.360 g Fe in the mixture:

$$0.360 = y\left[\frac{111.70 \text{ g Fe}}{159.70 \text{ g } Fe_2O_3}\right] + (0.500 - y)\left[\frac{55.85 \text{ g Fe}}{71.85 \text{ g FeO}}\right]$$

Simplifying the equation by dividing the conversion factors and combining terms gives:

$$0.360 = 0.6994\ y + 0.7773\ (0.500 - y)$$

$$0.07790\ y = 0.02865$$

$$y = 0.36\underline{7}7 = 0.368 \text{ g} = \text{mass of } Fe_2O_3$$

4.147 Use the data to find the molar mass of the metal and anion. Start with X_2.

Mass X_2 in MX = 4.52 g - 3.41 g = 1.11 g

Molar mass X_2 = 1.11 g ÷ 0.0158 mol = 70.$\underline{2}$5 g/mol

Molar mass X = 70.25 ÷ 2 = 35.$\underline{1}$4 = 35.1 g/mol

Thus X is Cl, chlorine.

Moles of M in 4.52 g MX = 0.0158 x 2 = 0.0316 mol

Molar mass of M = 3.41 g ÷ 0.0316 mol = 10$\underline{7}$.9 = 108 g/mol

Thus M is Ag, silver.

4.148 Use the data to find the molar mass of the metal and the anion. Start with M^{2+}.

Mol M^{2+} in MX_2 = 0.158 mol ÷ 2 = 0.0790 mol

Molar mass of M^{2+} = 1.92 g ÷ 0.0790 mol = 24.$\underline{3}$0 = 24.3 g/mol

Thus M^{2+} is Mg^{2+}.

Now find the mass of MX_2 and then find the molar mass of X:

(1 - 0.868) mass MX_2 = 1.92 g, or mass MX_2 = 14.$\underline{5}$45 g

Mass of X in MX_2 = 14.$\underline{5}$45 g - 1.92 g = 12.$\underline{6}$25 g

The moles of X in MX_2 = 0.158 mol

Molar mass of X^- = 12.$\underline{6}$25 g ÷ 0.158 mol = 79.$\underline{9}$05 = 79.9 g/mol

Thus X^- is Br^- (molar mass 79.90).

4.149 After finding the volume of the alloy, convert it to mass Fe using density and % Fe. Then use Avogadro's number and the atomic weight for the number of atoms.

$$\text{Vol.} = 10.0 \text{ cm} \times 20.0 \text{ cm} \times 15.0 \text{ cm} = 3.00 \times 10^3 \text{ cm}^3$$

$$\text{Mass Fe} = 3.00 \times 10^3 \text{ cm}^3 \times \frac{8.17 \text{ g alloy}}{1 \text{ cm}^3} \times \frac{54.7 \text{ g Fe}}{100.0 \text{ g alloy}} = 1.3\underline{4}07 \times 10^4 \text{ g}$$

$$\text{No. of Fe atoms} = 1.3\underline{4}07 \times 10^4 \text{ g Fe} \times \frac{1 \text{ mol Fe}}{55.85 \text{ g Fe}} \times \frac{6.022 \times 10^{23} \text{ Fe atoms}}{1 \text{ mol Fe}}$$

$$= 1.4\underline{4}56 \times 10^{26} = 1.45 \times 10^{26} \text{ Fe atoms}$$

4.150 After finding the volume of the cylinder, convert it to mass Co using density and % Co. Then use Avogadro's number and the atomic weight for the number of atoms.

$$\text{Vol.} = 3.1416 \times (2.50 \text{ cm})^2 \times 10 \text{ cm} = 196.35 \text{ cm}^3$$

$$\text{Mass Co} = 1.9635 \text{ cm}^3 \times \frac{8.20 \text{ g alloy}}{1 \text{ cm}^3} \times \frac{12.0 \text{ g Co}}{100.0 \text{ g alloy}} = 1.9\underline{3}2 \times 10^2 \text{ g}$$

$$\text{No. of Co atoms} = 1.9\underline{3}2 \times 10^2 \text{ g Co} \times \frac{1 \text{ mol Co}}{58.93 \text{ g Co}} \times \frac{6.022 \times 10^{23} \text{ Co atoms}}{1 \text{ mol Co}}$$

$$\text{No. of Co atoms} = 1.9\underline{7}4 \times 10^{24} = 1.97 \times 10^{24} \text{ Co atoms}$$

4.151 Use the density, formula weight, and percentage to convert to molarity. Then combine the 0.200 mol with mol/L to obtain the volume in liters.

$$\frac{0.807 \text{ g soln}}{1 \text{ mL}} \times \frac{0.940 \text{ g ethanol}}{1.00 \text{ g soln}} \times \frac{1 \text{ mol ethanol}}{46.07 \text{ g ethanol}} \times \frac{1000 \text{ mL}}{1 \text{ L}}$$

$$= \frac{16.46 \text{ mol ethanol}}{1 \text{ L ethanol}}$$

$$\text{L ethanol} = 0.200 \text{ mol ethanol} \times \frac{\text{L ethanol}}{16.46 \text{ mol ethanol}} = 0.012\underline{1}506 = 0.0122 \text{ L}$$

4.152 Use the density, formula weight, and percentage to convert to molarity. Then combine the 0.350 mol with mol/L to obtain the volume in liters.

$$\frac{1.072 \text{ g soln}}{1 \text{ mL}} \times \frac{0.560 \text{ g glycol}}{1.00 \text{ g soln}} \times \frac{1 \text{ mol glycol}}{62.07 \text{ g ethanol}} \times \frac{1000 \text{ mL}}{1 \text{ L}}$$

$$= = \frac{9.6\underline{7}2 \text{ mol glycol}}{1 \text{ L glycol}}$$

$$\text{L glycol} = 0.350 \text{ mol glycol} \times \frac{\text{L glycol}}{9.6\underline{7}2 \text{ mol glycol}} = 0.036\underline{1}8 = 0.0362 \text{ L}$$

4.153 Convert the 2.290 g of Ag to mol AgI, which is chemically equivalent to moles of KI. Use that to calculate the molarity of the KI.

$$2.290 \text{ g AgI} \times \frac{1 \text{ mol AgI}}{234.77 \text{ g AgI}} = 9.75\underline{4}2 \times 10^{-3} \text{ mol AgI (eq. to mol KI)}$$

$$\text{Molarity} = \frac{9.75\underline{4}2 \times 10^{-3} \text{ mol KI}}{0.0100 \text{ L}} = 0.975\underline{4}2 = 0.975 \text{ M}$$

4.154 Convert the 5.483 g of $BaSO_4$ to mol $BaSO_4$, which is chemically equivalent to moles of Na_2SO_4. Use that to calculate the molarity of the Na_2SO_4.

$$5.483 \text{ g } BaSO_4 \times \frac{1 \text{ mol } BaSO_4}{233.40 \text{ g } BaSO_4} = 0.0234\underline{9}2 \text{ mol } BaSO_4 \text{ (eq. to mol } Na_2SO_4)$$

$$\text{Molarity} = \frac{0.0234\underline{9}2 \text{ mol } Na_2SO_4}{0.0250 \text{ L}} = 0.939\underline{7} = 0.940 \text{ M}$$

4.155 Convert the 6.026 g of $BaSO_4$ to mol $BaSO_4$; then from the equation deduce that 3 mol $BaSO_4$ is equivalent to 1 mol $M_2(SO_4)_3$ and is equivalent to 2 mol of M. Use that with 1.200 g of the metal M to calculate the atomic weight of M.

$$6.026 \text{ g } BaSO_4 \times \frac{1 \text{ mol } BaSO_4}{233.40 \text{ g } BaSO_4} \times \frac{2 \text{ mol M}}{3 \text{ mol } BaSO_4} = 0.0172\underline{1}2 \text{ mol M}$$

$$\text{Atomic wt of M in g/mol} = \frac{1.200 \text{ g M}}{0.017212 \text{ mol M}} = 69.7\underline{1}9 \text{ g/mol (= gallium)}$$

4.156 Convert the 7.964 g of AgCl to mol AgCl; then, from the equation, deduce that 2 mol AgCl is equivalent to 1 mol MCl_2 and is equivalent to 1 mol M. Use that with 2.434 g of the metal M to calculate the atomic weight of M.

$$7.964 \text{ g AgCl} \times \frac{1 \text{ mol AgCl}}{143.32 \text{ g AgCl}} \times \frac{1 \text{ mol M}}{2 \text{ mol AgCl}} = 0.0277\underline{8}4 \text{ mol M}$$

$$\text{Atomic wt of M in g/mol} = \frac{2.434 \text{ g M}}{0.027784 \text{ mol M}} = 87.6\underline{0}4 \text{ g/mol (= strontium)}$$

4.157 Use the density, formula weight, percentage, and volume to convert to mol H_3PO_4. Then, from the equation $P_4O_{10} + 6H_2O \rightarrow 4H_3PO_4$, deduce that 4 mol H_3PO_4 is equivalent to 1 mol of P_4O_{10}, and use that to convert to mol P_4O_{10}.

$$15\underline{0}0 \text{ mL} \times \frac{1.025 \text{ g soln}}{1 \text{ mL}} \times \frac{0.0500 \text{ g } H_3PO_4}{1 \text{ g soln}} \times \frac{1 \text{ mol } H_3PO_4}{98.00 \text{ g } H_3PO_4}$$

$$= 0.78\underline{4}4 \text{ mol } H_3PO_4$$

$$0.7844 \text{ mol } H_3PO_4 \times \frac{1 \text{ mol } P_4O_{10}}{4 \text{ mol } H_3PO_4} = 0.19\underline{6}1 \text{ mol } P_4O_{10}$$

$$\text{Mass } P_4O_{10} = 0.1961 \text{ mol } P_4O_{10} \times \frac{283.92 \text{ g } P_4O_{10}}{1 \text{ mol } P_4O_{10}}$$

$$= 55.\underline{6}77 = 55.7 \text{ g } P_4O_{10}$$

4.158 Use the density, formula weight, percentage, and volume to convert to mol $FeCl_3$, which is chemically equivalent to mol Fe from the equation $2Fe + 3Cl_2(g) \rightarrow 2FeCl_3$.

$$\frac{1.067 \text{ g soln}}{1 \text{ mL}} \times \frac{0.0800 \text{ g } FeCl_3}{1 \text{ g soln}} \times \frac{1 \text{ mol } H_3PO_4}{162.20 \text{ g } FeCl_3} \times 2500 \text{ mL}$$

$$1.3\underline{1}57 \text{ mol } FeCl_3$$

$$1.3157 \text{ mol } FeCl_3 = 1.3157 \text{ mol Fe}$$

$$\text{Mass Fe} = 1.3157 \text{ mol Fe} \times \frac{55.85 \text{ g Fe}}{\text{mol Fe}} = 73.\underline{4}8 = 73.5 \text{ g Fe}$$

4.159 Convert the 0.1068 g of hydrogen to mol H_2; then deduce from the equation that 3 mol H_2 is equivalent to 2 mol Al. Use the moles of Al to calculate mass Al and the percentage Al.

$$0.1068\text{ g }H_2 \times \frac{1\text{ mol }H_2}{2.016\text{ g }H_2} \times \frac{2\text{ mol Al}}{3\text{ mol }H_2} = 0.0353\underline{1}74\text{ mol Al}$$

$$\%\text{ Al} = \frac{0.0353\underline{1}74\text{ mol Al} \times \frac{26.98\text{ g Al}}{\text{mol Al}}}{1.118\text{ g alloy}} \times 100\% = 85.2\underline{2}9 = 85.23\%$$

4.160 Convert the 0.1228 g of hydrogen to mol H_2; then deduce from the equation that 3 mol H_2 is equivalent to 2 mol Fe. Use the moles of Fe to calculate mass Fe and the percentage Fe.

$$0.1228\text{ g }H_2 \times \frac{1\text{ mol }H_2}{2.016\text{ g }H_2} \times \frac{2\text{ mol Fe}}{3\text{ mol }H_2} = 0.0406\underline{0}85\text{ mol Fe}$$

$$\%\text{ Fe} = \frac{0.0406\underline{0}85\text{ mol Fe} \times \frac{55.85\text{ g Fe}}{\text{mol Fe}}}{2.358\text{ g alloy}} \times 100\% = 96.1\underline{8}2 = 96.18\%$$

4.161 Use the formula weight of $Al_2(SO_4)_3$ to convert to mol $Al_2(SO_4)_3$. Then deduce from the equation that 1 mol $Al_2(SO_4)_3$ is equivalent to 3 mol H_2SO_4, and calculate the moles of H_2SO_4 needed. Combine density, percentage, and formula weight to obtain molarity of H_2SO_4. Then combine molarity and moles to obtain volume.

$$18.7\text{ g }Al_2(SO_4)_3 \times \frac{1\text{ mol }Al_2(SO_4)_3}{342.17\text{ g }Al_2(SO_4)_3} \times \frac{3\text{ mol }H_2SO_4}{1\text{ mol }Al_2(SO_4)_3}$$

$$= 0.16\underline{3}9\text{ mol }H_2SO_4$$

$$\frac{1.104\text{ g soln}}{1\text{ mL}} \times \frac{0.150\text{ g }H_2SO_4}{1\text{ g soln}} \times \frac{1\text{ mol }H_2SO_4}{98.09\text{ g }H_2SO_4} \times \frac{1000\text{ mL}}{1\text{ L}}$$

$$= 1.6\underline{8}8\text{ mol }H_2SO_4\text{/L}$$

$$0.1639\text{ mol }H_2SO_4 \times \frac{1\text{ L }H_2SO_4}{1.688\text{ mol }H_2SO_4} = 0.097\underline{1}0\text{ L (97.1 mL)}$$

4.162 Use the formula weight of Na_3PO_4 to convert to mol Na_3PO_4. Then deduce from the equation that 1 mol Na_3PO_4 is equivalent to 3 mol NaOH, and calculate the moles of NaOH needed. Combine density, percentage, and formula weight to obtain molarity of NaOH. Then combine molarity and moles to obtain the volume.

$$26.2 \text{ g } Na_3PO_4 \times \frac{1 \text{ mol } Na_3PO_4}{163.94 \text{ g } Na_3PO_4} \times \frac{3 \text{ mol NaOH}}{1 \text{ mol } Na_3PO_4} = 0.47\underline{9}4 \text{ mol NaOH}$$

$$\frac{1.133 \text{ g soln}}{1 \text{ mL}} \times \frac{0.120 \text{ g NaOH}}{1 \text{ g soln}} \times \frac{1 \text{ mol NaOH}}{40.00 \text{ g NaOH}} \times \frac{1000 \text{ mL}}{1 \text{ L}}$$

$$= 3.3\underline{9}9 \text{ mol NaOH}$$

$$0.4794 \text{ mol NaOH} \times \frac{\text{L NaOH}}{3.399 \text{ mol NaOH}} = 0.14\underline{1}04 \text{ L } (141 \text{ mL})$$

4.163 The equations for the neutralization are $2HCl + Mg(OH)_2 \rightarrow MgCl_2 + 2H_2O$ and $3HCl + Al(OH)_3 \rightarrow AlCl_3 + 3H_2O$. Calculate the moles of HCl and write two equations in two unknowns using the relations that mol $Mg(OH)_2$ = mol $MgCl_2$, and mol $Al(OH)_3$ = mol $AlCl_3$

$$0.0485 \text{ L HCl} \times \frac{0.187 \text{ mol HCl}}{\text{L HCl}} = 0.0090\underline{6}95 \text{ mol HCl}$$

Rearrange the equation 0.0090695 mol HCl = 2 [mol $Mg(OH)_2$] + 3 [mol $Al(OH)_3$] to:

(1) $0.0090695 - 2 [\text{mol } Mg(OH)_2] = 3 [\text{mol } Al(OH)_3]$, or

or

(1a) $0.0030231 - 2/3 [\text{mol } Mg(OH)_2] = \text{mol } Al(OH)_3$

Substitute mol $Mg(OH)_2$ for mol $MgCl_2$ (molar mass = 95.21), and mol $Al(OH)_3$ for mol $AlCl_3$ (molar mass = 133.33) in the second equation for the sum of the weights of the two chlorides:

(2) $[95.21 \text{ g/mol} \times \text{mol } Mg(OH)_2] + [133.33 \text{ g/mol} \times \text{mol } Al(OH)_3] = 0.4200 \text{ g}$

Substitute equation 1a into equation 2 for the mol of $Al(OH)_3$:

$$[95.21 \times \text{mol } Mg(OH)_2] + [133.33 \times (0.0030231 - 2/3 \text{ mol } Mg(OH)_2)] = 0.4200$$

$$6.323 \text{ mol } Mg(OH)_2 + 0.4030\underline{7}0 = 0.4200$$

(continued)

Mol $Mg(OH)_2$ = 0.01693 ÷ 6.323 = 0.0026728 mol $Mg(OH)_2$

0.0026728 mol $Mg(OH)_2$ x 58.33 g $Mg(OH)_2$/mol $Mg(OH)_2$ = 0.15590 g $Mg(OH)_2$

Mol $Al(OH)_3$ = 0.0030231 - 2/3(0.0026728 mol Mg(OH)

= 0.0012412 mol $Al(OH)_3$

0.0012412 mol $Al(OH)_3$ x 78.00 g $Al(OH)_3$/mol $Al(OH)_3$ = 0.096681 g $Al(OH)_3$

% $Mg(OH)_2$ = [0.15590 g ÷ (0.15590 + 0.096681) g] x 100%

= 61.72 = 61.7%

4.164 The equations for the neutralization are 2HCl + $MgCO_3$ → $MgCl_2$ + H_2O + CO_2 and 2HCl + $CaCO_3$ → $CaCl_2$ + H_2O + CO_2. Calculate the moles of HCl and write two equations in two unknowns, using the relations that mol $MgCO_3$ = mol $MgCl_2$, and mol $CaCO_3$ = mol $CaCl_2$:

$$0.04133 \text{ L HCl} \times \frac{0.08750 \text{ mol HCl}}{\text{L HCl}} = 0.0036164 \text{ mol HCl}$$

Rearrange the equation 0.0036164 mol HCl = 2 [mol $CaCO_3$] + 2 [mol $MgCO_3$] to

(1) 0.0036164 - 2 [mol $CaCO_3$] = 2 [mol $MgCO_3$],

or

(1a) 0.0018082 - mol $CaCO_3$ = mol $MgCO_3$

Substitute mol $CaCO_3$ for mol $CaCl_2$ (molar mass = 110.98), and mol $MgCO_3$ for mol $MgCl_2$ (molar mass = 95.21) in the second equation for the sum of the weights of the two chlorides:

(2) [110.98 g/mol x mol $CaCO_3$] + [95.21 g/mol x mol $MgCO_3$] = 0.1900 g

Substitute equation 1a into equation 2 for the mol of $MgCO_3$:

[110.98 x mol $CaCO_3$] + [95.21 x (0.0018082 - mol $CaCO_3$)] = 0.1900

15.77 mol $CaCO_3$ + 0.172159 = 0.1900

Mol $CaCO_3$ = 0.017841 ÷ 15.77 = 0.0011313

(continued)

$0.0011313 \text{ mol } CaCO_3 \times 100.09 \text{ g } CaCO_3/\text{mol } CaCO_3 = 0.11\underline{3}234 \text{ g } CaCO_3$

$\text{Mol } MgCO_3 = 0.0018082 \text{ mol} - 0.0011313 \text{ mol } CaCO_3$

$= 0.0006769 \text{ mol } MgCO_3$

$0.0006769 \text{ mol } MgCO_3 \times 84.32 \text{ g } MgCO_3/\text{mol } MgCO_3 = 0.057076 \text{ g } MgCO_3$

$\% \ CaCO_3 = [0.113234 \text{ g} \div (0.113234 + 0.057076) \text{ g}] \times 100\%$

$= 66.\underline{4}87 = 66.5\%$

5. THE GASEOUS STATE

Solutions to Exercises

Note on significant figures: The final answer to each mathematical solution is given first with one nonsignificant figure (the rightmost significant figure is underlined) and then is rounded to the correct number of figures. Intermediate answers usually also have at least one nonsignificant figure.

5.1 Rearrange the equation $gd_{Hg}h_{Hg} = gd_oh_o$ to solve for h_{Hg}:

$$h_{Hg} = \frac{d_oh_o}{d_{Hg}} = \frac{0.775\ g/cm^3 \times 7.68\ cm\ oil}{13.596\ g/cm^3} = 0.43\underline{7}7\ cm\ Hg = 4.38\ mmHg$$

5.2 Rearrange $P_fV_f = P_iV_i$ to solve for V_f (at constant T and n):

$$V_f = V_i \times \frac{P_i}{P_f} = 20.0\ L \times \frac{1.00\ atm}{0.830\ atm} = 24.\underline{0}96 = 24.1\ L$$

5.3 Rearrange $V_f/T_f = V_i/T_i$ to solve for V_f (at constant P and n). Use T_i as the sum of 19 + 273 (= 292 K) and T_f as the sum of 25 + 273 (= 298 K) in the equation.

$$V_f = V_i \times \frac{T_f}{T_i} = 4.38\ dm^3 \times \frac{298\ K}{292\ K} = 4.4\underline{7}0 = 4.47\ dm^3$$

5.4 Use the combined gas law to solve for V_f with the usual form of the equation. Use T_i as the sum of 24 + 273 (= 297) and T_f as the sum of 35 + 273 (= 308 K) in the equation.

$$V_f = V_i \times \frac{P_i}{P_f} \times \frac{T_f}{T_i} = 5.41\ dm^3 \times \frac{101.5\ kPa}{102.8\ kPa} \times \frac{308\ K}{297\ K}$$

$$= 5.5\underline{3}9 = 5.54\ dm^3$$

5.5 Use the ideal gas law, PV = nRT, and solve for n:

$$n = \frac{PV}{RT} = P \times \frac{V}{RT}$$

If V and T are constant, n is proportional to P (R is a constant).

5.6 Convert kg of O_2 to moles O_2 and convert temperature to kelvins. Then use the ideal gas law.

$$3.03\ kg\ O_2 \times \frac{1000\ g}{1\ kg} \times \frac{1\ mol\ O_2}{32.00\ g\ O_2} = 94.\underline{6}88\ mol\ O_2$$

23 + 273 = 296 K

$$P = \frac{nRT}{V} = \frac{(94.688)(0.08206\ L \bullet atm/K \bullet mol)(296\ K)}{50.0\ L} = 46.\underline{0}0 = 46.0\ atm$$

5.7 Because density equals mass per unit volume, calculating the mass of 1 L (an exact number) of helium will yield the density of helium. Tablulate the values of the variables:

Variable	Value
P	752 mmHg x (1 atm/760 mmHg) = 0.98$\underline{9}$47 atm
V	1 L (exact number)
T	21 + 273 = 294 K
n	unknown

Using the ideal gas law, solve for n, and convert moles to grams of helium:

$$n = \frac{PV}{RT} = \frac{(0.98947\ atm)(1\ L)}{(0.08206\ L \bullet atm/K \bullet mol)(294\ K)} = 0.041\underline{0}1\ mol$$

$$0.041\underline{0}1\ mol\ He \times \frac{4.00\ g\ He}{1\ mol\ He} = 0.16\underline{4}04\ g\ He/1\ L\ (= density)$$

Mass air - mass He = 1.188 g - 0.16404 g = 1.02$\underline{3}$96 = 1.024 g difference

5.8 Calculate the moles of the vapor from the ideal gas law and then calculate the molar mass of the vapor. Tabulate the values of the variables:

Variable	Value
P	0.862 atm
V	1 L (exact number)
T	25 + 273 = 298 K
n	unknown

$$n = \frac{PV}{RT} = \frac{(0.862 \text{ atm})(1 \text{ L})}{(0.08206 \text{ L} \bullet \text{atm/K} \bullet \text{mol})(298 \text{ K})} = 0.035\underline{2}5 \text{ mol}$$

$$\text{Molar mass} = \frac{\text{grams vapor}}{\text{moles vapor}} = \frac{2.26 \text{ g}}{0.03525 \text{ mol}} = 64.\underline{1}14 \text{ g/mol}$$

Therefore, the molecular weight is 64.1 amu.

5.9 First determine the number of moles of Cl_2 from the mass of HCl:

$$9.41 \text{ g HCl} \times \frac{1 \text{ mol HCl}}{36.46 \text{ HCl}} \times \frac{5 \text{ mol } Cl_2}{16 \text{ mol HCl}} = 0.080\underline{6}53 \text{ mol } Cl_2$$

Next calculate the moles of the vapor from the moles of Cl_2 and use the ideal gas law to calculate the volume of the vapor. Tabulate the values of the variables:

Variable	Value
P	787 mmHg x (1 atm/760 mm Hg) = 1.0355 atm
T	(40 + 273) K = 313 K
n	$0.080\underline{6}53$ mol
V	unknown

Rearrange the ideal gas law to obtain V:

$$V = \frac{nRT}{P} = \frac{(0.080653 \text{ mol})(0.08206 \text{ L} \bullet \text{atm/K} \bullet \text{mol})(313 \text{ K})}{1.0355 \text{ atm}}$$

$$= 2.0\underline{0}1 = 2.00 \text{ L}$$

5.10 Each gas obeys the ideal gas law. In each case, convert grams to moles and substitute into the ideal gas law to determine the partial pressure of each.

$$1.031 \text{ g } O_2 \times \frac{1 \text{ mol } O_2}{32.00 \text{ g } O_2} = 0.0322\underline{1}88 \text{ mol } O_2$$

$$P = \frac{nRT}{V} = \frac{(0.0322188)(0.08206 \text{ L} \bullet \text{atm/K} \bullet \text{mol})(291 \text{ K})}{10.0 \text{ L}} = 0.076\underline{9}4 \text{ atm}$$

$$0.572 \text{ g } CO_2 \times \frac{1 \text{ mol } CO_2}{44.01 \text{ g } CO_2} = 0.012\underline{9}97 \text{ mol } CO_2$$

$$P = \frac{nRT}{V} = \frac{(0.012997)(0.08206 \text{ L} \bullet \text{atm/K} \bullet \text{mol})(291 \text{ K})}{10.0 \text{ L}} = 0.031\underline{0}4 \text{ atm}$$

The total pressure is equal to the sum of the partial pressures:

$$P = P_{O_2} + P_{CO_2} = 0.076\underline{9}4 + 0.031\underline{0}4 = 0.108\underline{0}2 = 0.1080 \text{ atm}$$

$$\text{Mole frac. } O_2 = 0.0322188 \div (0.0322188 + 0.012997) = 0.71\underline{2}55 = 0.713$$

5.11 Determine the number of moles of O_2 from the mass of $KClO_3$:

$$1.300 \text{ g } KClO_3 \times \frac{1 \text{ mol } KClO_3}{122.5 \text{ g } KClO_3} \times \frac{3 \text{ mol } O_2}{2 \text{ mol } KClO_3} = 0.0159\underline{1}84 \text{ mol } O_2$$

Find the partial pressure of O_2 using Dalton's law:

$$P = P_{O_2} + P_{H_2O}$$

$$P_{O_2} = P - P_{H_2O} = (745 - 21.1) \text{ mmHg} = 72\underline{3}.9 \text{ mmHg}$$

Solve for the volume using the ideal gas law.

Variable	Value
P	723.9 mmHg x 1 atm/760.0 mmHg = 0.95$\underline{2}$5 atm
V	unknown
T	296 K
n	0.0159184

$$V = \frac{nRT}{P} = \frac{(0.0159184 \text{ mol})(0.08206 \text{ L} \bullet \text{atm/K} \bullet \text{mol})(296 \text{ K})}{0.9525 \text{ atm}}$$

$$= 0.40\underline{5}9 = 0.406 \text{ L}$$

5.12 $u = \sqrt{\dfrac{3RT}{M}} = \sqrt{\dfrac{3 \times 8.31 \text{ kg} \bullet \text{m}^2/(\text{s}^2 \bullet \text{K} \bullet \text{mol}) \times 295 \text{ K}}{153.8 \times 10^{-3} \text{ kg/mol}}}$

$= 21\underline{8}.7 = 219 \text{ m/s}$

5.13 Determine the rms molecular speed for N_2 at 455°C (728 K):

$$u = \sqrt{\frac{3RT}{M}} = \sqrt{\frac{3 \times 8.31 \text{ kg} \bullet \text{m}^2/(\text{s}^2 \bullet \text{K} \bullet \text{mol}) \times 728 \text{ K}}{28.02 \times 10^{-3} \text{ kg/mol}}} = 80\underline{4}.81 \text{ m/s}$$

After writing this equation with the same speed for H_2, square both sides and solve for T:

$$u = \frac{3RT}{M_m}$$

$$T = \frac{u^2 M_m}{3R} = \frac{(804.81)^2(2.016 \times 10^{-3} \text{ kg/mol}}{(3)(8.31 \text{ kg} \bullet \text{m}^2/\text{s}^2 \bullet \text{K} \bullet \text{mol})} = 52.\underline{3}79 = 52.4 \text{ K}$$

Because the average kinetic energy of a molecule is proportional to T, the absolute temperature at which an H_2 molecule has the same average kinetic energy as an N_2 molecule at 728 K is 728 K.

5.14 The rate of effusion for He = 10.0 mL/3.52 s; that of O_2, rate(O_2), is unknown.

$$\frac{\text{rate}(O_2)}{\text{rate(He)}} = \sqrt{\frac{M_m(\text{He})}{M_m(O_2)}}$$

$$\frac{\text{rate}(O_2)}{10.0 \text{ mL}/3.52 \text{ s}} = \sqrt{\frac{4.00}{32.00}} = 0.35\underline{3}55$$

$$\text{Rate}(O_2) = 0.35\underline{3}55 \times \frac{10.0 \text{ mL}}{3.52 \text{ s}} = 1.0\underline{0}44 \text{ mL/s}$$

$$\text{Time for } O_2 \text{ to diffuse} = \frac{10.0 \text{ mL}}{1.0044 \text{ mL/s}} = 9.9\underline{5}6 = 9.96 \text{ s}$$

5.15 The rate of effusion is inversely proportional to the molar mass. To solve for the molar mass of the unknown, square both sides of the equation:

$$\frac{\text{rate}(H_2)}{4.67 \times \text{rate}(H_2)} = \sqrt{\frac{2.016\ \text{g/mol}}{M_m(\text{gas})}}$$

$$\frac{1}{(4.67)^2} = \frac{2.016\ \text{g/mol}}{M_m(\text{gas})}$$

$$M_m\ (\text{gas}) = (4.67)^2(2.016\ \text{g/mol}) = 43.\underline{9}6 = 44.0\ \text{g/mol}\ \ (\text{molec. wt} = 44.0\ \text{amu})$$

5.16 From Table 5.7, a = 5.489 L^2•atm/mol^2, and b = 0.06380 L/mol. Into the van der Waals equation, substitute R = 0.08206 L•atm/K•mol, T = 273.2 K, and V = 22.41 L.

$$P = \frac{nRT}{(V - nb)} - \frac{n^2a}{V^2}$$

$$P = \frac{1.000\ \text{mol} \times 0.08206\ \text{L•atm/(K•mol)} \times 273.2\ \text{K}}{22.41\ \text{L} - (1.000\ \text{mol} \times 0.06499\ \text{L/mol})} -$$

$$= \frac{(1.000\ \text{mol})^2 \times 5.570\ \text{L}^2\text{•atm/mol}^2}{(22.41\ \text{L})^2}$$

$$P = 1.00\underline{3}3 - 0.0110\underline{9}1 = 0.99\underline{2}21 = 0.992\ \text{atm}$$

Using the ideal gas law, P = 1.00$\underline{0}$4 atm (larger).

■ Answers to Review Questions

5.1 Pressure is the force exerted per unit area of surface. The SI unit of pressure is obtained as follows:

$$P = \frac{\text{force}}{\text{area}} = \frac{\text{kg•m/s}^2}{\text{m}^2} = \text{kg/(m•s}^2) = \text{pascals}$$

5.2 A manometer measures the pressure of a gas within a vessel. A mercury manometer balances the pressure from a column of mercury against the gas pressure; the mercury column height is proportional to the gas pressure. Liquids other than mercury can be used.

5.3 The height of the liquid in a manometer depends on the density of the liquid; the acceleration of gravity, g; and the pressure of the gas being measured.

5.4 From Boyle's law, PV = constant. Because this is true for conditions P_i and V_i as well as conditions P_f and V_f, we can write

$$P_fV_f = P_iV_i$$

Dividing both sides of this equation by P_f gives

$$V_f = V_i \times (P_i/P_f)$$

5.5 A linear relationship between variables such as x and y is given by the mathematical relation

$$y = a + bx$$

The variable y is directly proportional to x only if a = 0.

5.6 When we convert -273.15°C to degrees Fahrenheit, we obtain -459.67°F. The relationship between volume, V, and temperature, T_F, in °F, is $V = a + bT_F$. At absolute zero, V = 0, so that

$$0 = a + b(-459.67), \text{ or } a = 459.67b$$

Thus, we obtain:

$$V = 459.67b + bT_F, \text{ or } V = b(T_F + 459.67)$$

Thus, an absolute temperature scale with °F degrees uses absolute temperature = T_F + 459.67.

5.7 From Charles's law, V/T = constant. Because this is true for conditions T_i and V_i as well as conditions T_f and V_f, we can write

$$V_f/T_f = V_i/T_i$$

Multiplying both sides of the equation by T_f gives

$$V_f = V_i \times (T_f/T_i)$$

5.8 Avogadro's law states that equal volumes of any two gases at the same temperature and pressure contain the same number of molecules. The law of combining volumes states that the volumes of reactant gases at a given pressure and temperature are in ratios of small whole numbers. The combining-volume law may be explained from Avogadro's law using the reaction $N_2 + 3H_2 \rightarrow 2NH_3$ as follows: In Avogadro's terms, this equation says that Avogadro's number of N_2 molecules reacts with three times Avogadro's number of H_2 molecules to form two times Avogadro's number of NH_3 molecules. From Avogadro's law, it follows that one volume of N_2 reacts with three volumes of H_2 to form two volumes of NH_3. This result is true for all gas reactions.

5.9 The standard conditions are 0^oC and 1 atm pressure.

5.10 The molar gas volume is the volume of one mole of gas at any given temperature and pressure. At standard conditions (STP), the molar gas volume equals 22.4 L.

5.11 Boyle's law states that V is proportional to 1/P; Charles's law states that V is proportional to T. These two laws can be combined into one law:

$$V = \text{constant} \times T/P \quad \text{(for a given amount of gas)}$$

For any amount of gas, the constant will have a specific value we can specify as R. For exactly one mole of gas, the volume can be symbolized as V_m, the molar volume of gas. Thus, the above equation can be written as

$$V_m = R \times T/P$$

Because V_m has the same value for all gases, we can write this equation for n moles of gas if we multiply both sides by n. This yields the equation

$$nV_m = nRT/PV \text{ or } PV = nRT$$

5.12 The variables in the ideal gas law are P, V, and T. The SI units of these variables are pascals (P), cubic meters (V), and kelvins (T).

5.13 Find the value of R in units of (liter•atmospheres) per (kelvin•mole) by multiplying the value of R in (liter•mmHg) per (kelvin•mole) by the conversion factor of 760 mmHg/1 atm. To three significant figures, this is

$$0.0821 \frac{\text{L•atm}}{\text{K•mol}} \times \frac{760 \text{ mmHg}}{1 \text{ atm}} = 62.4 \frac{\text{L•mmHg}}{\text{K•mol}}$$

5.14 Six empirical gas laws can be obtained. They can be stated as follows:

$P \times V$ = constant (T and n constant)
P/T = constant (V and n constant)
P/n = constant (T and V constant)
V/T = constant (P and n constant)
V/n = constant (P and T constant)
$n \times T$ = constant (P and V constant)

5.15 The postulates and supporting evidence are the following: (1) Gases are composed of molecules, whose sizes are negligible compared with the distance between the molecules. The fact that gases are compressible is in agreement with this postulate. (2) Molecules move randomly in linear motion. The random motion of molecules explains Brownian motion. (3) The forces of attraction or repulsion between molecules in a gas are very weak. This explains why a gas fills any container. (4) The collisions of gas molecules are elastic. If this was not true, we might expect the average speed of the molecules to continue to decrease. In that case, we would see a steady drop in pressure. No such thing is observed. (5) The average kinetic energy of a molecule in a gas is proportional to the absolute temperature. This explains why Brownian motion increases with temperature.

5.16 Boyle's law requires that the temperature be constant. Postulate 5 of the kinetic theory holds that the average kinetic energy of a molecule (or many molecules) is proportional to the absolute temperature. At a constant temperature, the average molecular force from all collisions is constant. If we increase the volume of a gas, this decreases the number of molecules per unit volume and so decreases the frequency of collisions per unit wall area.

5.17 Kinetic theory says that gas pressure on a container wall results from the bombardment of the wall by the gas molecules.

5.18 The rms speed of a molecule equals $(3RT/M_m)^{1/2}$, where M_m is the molar mass of the gas. The rms speed does not depend on the molar volume.

5.19 A gas appears to diffuse more slowly because it never travels very far in one direction before it collides with another molecule and moves in another direction. Thus, it must travel a very long, crooked path as the result of collisions.

5.20 Effusion is the passage of a gas through a very small hole in a container. It results from the gas molecules encountering the hole by chance, rather than colliding with the walls of the container. The faster the molecules move, the more likely they are to encounter the hole. Thus, the rate of effusion depends on the average molecular speed, which depends inversely on molecular mass.

5.21 The behavior of a gas begins to deviate significantly from that predicted by the ideal gas law at high pressures and relatively low temperatures.

5.22 The a constant in the van der Waals equation is related to intermolecular forces. The b constant is related to the molecular volume.

■ Solutions to Practice Problems

Note on significant figures: The final answer to each problem is given first with one nonsignificant figure (the rightmost significant figure is underlined) and then is rounded to the correct number of figures. Intermediate answers usually also have at least one nonsignificant figure. Atomic weights are rounded to two decimal places, except for that of hydrogen.

5.23 Let d_{dp} and h_{dp} symbolize the density and height, respectively, of dibutyl phthalate. Then rearrange the equation $gd_{Hg}h_{Hg} = gd_{dp}h_{dp}$, and solve for h_{Hg} (the pressure):

$$h_{Hg} = \frac{d_{db}}{d_{Hg}} \times h_{db} = 64.3 \text{ mm} \times \frac{1.046 \text{ g dp/cm}^3 \text{ dp}}{13.596 \text{ g Hg/cm}^3 \text{ Hg}}$$

$$= 4.9\underline{4}68 = 4.95 \text{ mmHg}$$

5.24 Let d_{oil} and h_{oil} symbolize the density and height, respectively, of mineral oil. Then substitute into the equation from the previous problem.

$$h_{Hg} = \frac{d_{oil}}{d_{Hg}} \times h_{oil} = 79.3 \text{ mm} \times \frac{0.857 \text{ g oil/cm}^3 \text{ oil}}{13.596 \text{ g Hg/cm}^3 \text{ Hg}}$$

$$= 4.9\underline{9}85 = 5.00 \text{ mmHg}$$

5.25 Rearrange the equation $gd_{Hg}h_{Hg} = gd_{liq}h_{liq}$, and solve for d_{liq}.

$$d_{liq} = \frac{h_{Hg}}{h_{liq}} \times d_{Hg} = 13.596 \text{ g Hg/mL Hg} \times \frac{8.56 \text{ mmHg}}{95.6 \text{ mmliq}}$$

$$1.2\underline{1}77 = 1.22 \text{ g liq/ mL liq}$$

5.26 Rearrange the equation $gd_{Hg}h_{Hg} = gd_{liq}h_{liq}$, and solve for d_{liq}.

$$d_{liq} = \frac{h_{Hg}}{h_{liq}} \times d_{Hg} = 13.596 \text{ g Hg/mL Hg} \times \frac{6.85 \text{ mmHg}}{65.7 \text{ mmliq}}$$

$$1.4\underline{1}7 = 1.42 \text{ g liq/ mL liq}$$

5.27 Using Boyle's law, solve for V_f of the neon gas at 1.564 atm pressure.

$$V_f = V_i \times \frac{P_i}{P_f} = 3.15 \text{ L} \times \frac{0.951 \text{ atm}}{1.564 \text{ atm}} = 1.9\underline{1}53 = 1.92 \text{ L}$$

5.28 Using Boyle's law, solve for V_f of the helium gas at 543 mmHg.

$$V_f = V_i \times \frac{P_i}{P_f} = 2.68 \text{ L} \times \frac{789 \text{ mmHg}}{543 \text{ mmHg}} = 3.8\underline{9}4 = 3.89 \text{ L}$$

5.29 Using Boyle's law, let V_f = volume at 0.974 atm (P_f), V_i = 50.0 L, and P_i = 19.8 atm.

$$V_f = V_i \times \frac{P_i}{P_f} = 50.0 \text{ L} \times \frac{19.8 \text{ atm}}{0.974 \text{ atm}} = 10\underline{1}6.4 = 1.02 \times 10^3 \text{ L}$$

5.30 Using Boyle's law, let V_f = volume at 1.584 atm (P_f), $V_i = 8.58\ m^3$, and P_i = 1.020 atm.

$$V_f = V_i \times \frac{P_i}{P_f} = 8.58\ m3 \times \frac{1.020\ atm}{1.584\ atm} = 5.5\underline{2}5 = 5.53\ L$$

5.31 Using Boyle's law, let P_i = pressure of 345 cm^3 of gas, and solve for it:

$$P_i = P_f \times \frac{V_f}{V_i} = 2.51\ kPa \times \frac{0.0457\ cm^3}{345\ cm^3} = 3.3\underline{2}4 \times 10^{-4} = 3.32 \times 10^{-4}\ kPa$$

5.32 Using Boyle's law, let P_f = final pressure of 25.0 dm^3 of gas, and solve for it:

$$P_f = P_i \times \frac{V_i}{V_f} = 101\ kPa \times \frac{456\ dm^3}{25\ dm^3} = 1.8\underline{4}2 \times 10^3 = 1.84 \times 10^3\ kPa$$

5.33 Use Charles's law: $T_i = 18^oC + 273 = 291\ K$, and $T_f = 0^oC + 273 = 273\ K$.

$$V_f = V_i \times \frac{T_f}{T_i} = 2.67\ mL \times \frac{273\ K}{291\ K} = 2.5\underline{0}48 = 2.50\ mL$$

5.34 Use Charles's law: $T_i = 0^oC + 273 = 273\ K$, and $T_f = 20^oC + 273 = 293\ K$.

$$V_f = V_i \times \frac{T_f}{T_i} = 22.41\ L \times \frac{293\ K}{273\ K} = 24.\underline{0}51 = 24.1\ L$$

5.35 Use Charles's law: $T_i = 22^oC + 273 = 295\ K$, and $T_f = -197^oC + 273 = 76\ K$.

$$V_f = V_i \times \frac{T_f}{T_i} = 2.54\ L \times \frac{76\ K}{295\ K} = 0.6\underline{5}4 = 0.65\ L$$

5.36 Use Charles's law: $T_i = 0^oC + 273 = 273\ K$, and $T_f = 25^oC + 273 = 298\ K$.

$$V_f = V_i \times \frac{T_f}{T_i} = 5.83\ L \times \frac{298\ K}{273\ K} = 6.3\underline{6}3 = 6.36\ L$$

5.37 Use Charles's law: $T_i = 25^oC + 273 = 298$ K, and V_f is the difference between the vessel's volume of 39.5 cm^3 and the 18.8 cm^3 of ethanol that is forced into the vessel.

$$T_f \; T_i \times \frac{V_f}{Vi} = 298 \text{ K} \times \frac{(39.5 - 18.8) \text{ cm}^3}{39.5 \text{ cm}^3} = 15\underline{6}.1 \text{ K} (-11\underline{6}.9 \text{ or } -117^oC$$

5.38 Use Charles's law: $T_i = 18^oC + 273 = 291$ K, $V_i = 58.2$ cm^3, and $V_f = 46.7$ cm^3.

$$T_f = T_i \times \frac{V_f}{V_i} = 291 \text{ K} \times \frac{46.7 \text{ cm}^3}{58.2 \text{ cm}^3} = 23\underline{3}.5 \text{ K} (-3\underline{9}.5 \text{ or } -40°C)$$

5.39 Use the combined law: $T_i = 31^oC + 273 = 304$ K, and $T_f = 0^oC + 273 = 273$ K.

$$V_f = V_i \times \frac{P_i}{P_f} \times \frac{T_f}{T_i} = 41.3 \text{ mL} \times \frac{753 \text{ mmHg}}{760 \text{ mmHg}} \times \frac{273 \text{ K}}{304 \text{ K}} = 36.\underline{7}4 = 36.7 \text{ mL}$$

5.40 Use the combined law: $T_i = 23^oC + 273 = 296$ K, and $T_f = 0^oC + 273 = 273$ K.

$$V_f = V_i \times \frac{P_i}{P_f} \times \frac{T_f}{T_i} = 3.84 \text{ mL} \times \frac{785 \text{ mmHg}}{760 \text{ mmHg}} \times \frac{273 \text{ K}}{296 \text{ K}} = 3.6\underline{5}8 = 3.66 \text{ mL}$$

5.41 The balanced equation is

$$4NH_3 + 5O_2 \rightarrow 4NO + 6H_2O$$

The ratio of moles of NH_3 to moles of NO = 4 to 4, or 1 to 1, so 1 volume of NH_3 will produce 1 volume of NO at the same temperature and pressure.

5.42 The balanced equation is

$$CO_2 + 3H_2 \rightarrow CH_3OH + H_2O$$

The ratio of moles of H_2 to moles of CO_2 = 3 to 1, so 3 volumes of H_2 are required to react with 1 volume of CO_2 at the same temperature and pressure.

5.43 Solve the ideal gas law for V:

$$V = \frac{nRT}{P} = nRT\left(\frac{1}{P}\right)$$

If the temperature and number of moles are held constant, then the product nRT is constant, and volume is inversely proportional to pressure:

$$V = \text{constant} \times \frac{1}{P}$$

5.44 Solve the ideal gas law for V:

$$V = \frac{nR}{P} \times (T)$$

If n and P are held constant, the nR/P quotient is a constant. The equation may now be written

$$V = \text{constant} \times T$$

5.45 Calculate the moles of oxygen, and then solve the ideal gas law for P:

$$n = 89.6 \text{ g} \times \frac{1 \text{ mol}}{32.0 \text{ g}} = 2.8\underline{0}0 \text{ mol}$$

$$P = \frac{nRT}{V} = \frac{(2.80 \text{ mol})(0.08206 \text{ L•atm/K•mol})(294 \text{ K})}{8.58 \text{ L}} = 7.8\underline{7}3 = 7.87 \text{ atm}$$

5.46 Calculate the moles of methane, and then solve the ideal gas law for P:

$$n = 7.68 \text{ g} \times \frac{1 \text{ mol}}{16.0 \text{ g}} = 0.48\underline{0}0 \text{ mol}$$

$$P = \frac{nRT}{V} = \frac{(0.480 \text{ mol})(0.08206 \text{ L} \bullet \text{atm/K} \bullet \text{mol})(292 \text{ K})}{5.00 \text{ L}}$$

$$= 2.3\underline{0}03 = 2.30 \text{ atm}$$

5.47 Using the moles of chlorine, solve the ideal gas law for V:

$$V = \frac{nRT}{P} = \frac{(3.50 \text{ mol})(0.08206 \text{ L} \bullet \text{atm/K} \bullet \text{mol})(307 \text{ K})}{2.45 \text{ atm}} = 35.\underline{9}9 = 36.0 \text{ L}$$

5.48 Calculate the moles of oxygen, and then solve the ideal gas law for V:

$$n = 5.67 \text{ g} \times \frac{1 \text{ mol}}{32.0 \text{ g}} = 0.17\underline{7}18 \text{ mol}$$

$$V = \frac{nRT}{P} = \frac{(0.17718 \text{ mol})(0.08206 \text{ L} \bullet \text{atm/K} \bullet \text{mol})(296 \text{ K})}{0.985 \text{ atm}}$$

$$= 4.3\underline{6}9 = 4.37 \text{ L}$$

5.49 Solve the ideal gas law for temperature in K, and convert to ^{o}C:

$$T = \frac{(3.50 \text{ atm})(4.00 \text{ L})}{(0.410 \text{ mol})(0.08206 \text{ L}\bullet\text{atm/K}\bullet\text{mol})} = 41\underline{6}.1 = 416 \text{ K}$$

$$^{o}\text{C} = 416 - 273 = 143^{o}\text{C}$$

5.50 Find the moles of C_3H_8, solve the ideal gas law for temperature in K, and convert to ^{o}C:

$$n = 5.65 \text{ g} \times \frac{1 \text{ mol}}{44.09 \text{ g}} = 0.12\underline{8}1 \text{ mol}$$

$$T = \frac{(956/760 \text{ atm})(2.50 \text{ L})}{(0.1281 \text{ mol})(0.08206 \text{ L}\bullet\text{atm/K}\bullet\text{mol})} = 29\underline{9}.1 \text{ K}$$

$$^{o}\text{C} = 299.1 - 273 = 2\underline{6}.1 = 26^{o}\text{C}$$

5.51 Because density equals mass per unit volume, calculating the mass of 1 L (exact number) of gas will give the density of the gas. Start from the ideal gas law and calculate n; then convert the moles of gas to grams using the molar mass.

$$n = \frac{(751/760 \text{ atm})(1 \text{ L})}{(295 \text{ K})(0.08206 \text{ L}\bullet\text{atm/K}\bullet\text{mol})} = 0.040\underline{8}19 \text{ mol}$$

$$0.040819 \text{ mol} \times \frac{17.03 \text{ g}}{1 \text{ mol}} = 0.69\underline{5}1 \text{ g}$$

Therefore, the density of NH_3 at 22°C is 0.695 g/L.

5.52 As in the previous problem, calculate the mass of 1 L (exact number) of gas to obtain the density of the gas. Start from the ideal gas law and calculate n; then convert to grams.

$$n = \frac{(967/760 \text{ atm})(1 \text{ L})}{(329 \text{ K})(0.08206 \text{ L}\bullet\text{atm/K}\bullet\text{mol})} = 0.047\underline{1}2 \text{ mol}$$

$$0.04712 \text{ mol} \times \frac{34.08 \text{ g}}{1 \text{ mol}} = 1.6\underline{0}58 \text{ g}$$

Therefore, the density of H_2S at 56°C is 1.61 g/L.

5.53 As in Problem 5.51, calculate the mass of 1 L (exact number) using the ideal gas law; convert to mass.

$$n = \frac{(0.897 \text{ atm})(1 \text{ L})}{(296 \text{ K})(0.08206 \text{ L}\bullet\text{atm/K}\bullet\text{mol})} = 0.036\underline{9}29 \text{ mo}$$

$$0.036929 \text{ mol} \times \frac{58.12 \text{ g}}{1 \text{ mol}} = 2.1\underline{4}6 = 2.15 \text{ g}$$

Therefore, the density of C_4H_{10} is 2.15 g/L.

5.54 As in Problem 5.51, calculate the mass of 1 L (exact number) using the ideal gas law; convert to mass.

$$n = \frac{(781/760 \text{ atm})(1 \text{ L})}{(371 \text{ K})(0.08206 \text{ L}\bullet\text{atm/K}\bullet\text{mol})} = 0.033\underline{7}5 \text{ mol}$$

$$0.033\underline{7}5 \text{ mol} \times \frac{119.5 \text{ g}}{1 \text{ mol}} = 4.0\underline{3}3 = 4.03 \text{ g}$$

Therefore, the density of $CHCl_3$ at 99°C is 4.03 g/L.

5.55 The ideal gas law gives n moles, which then is divided into the mass of 1.585 g for molar mass.

$$n = \frac{(753/760 \text{ atm})(1 \text{ L})}{(363 \text{ K})(0.08206 \text{ L}\bullet\text{atm/K}\bullet\text{mol})} = 0.033\underline{2}62 \text{ mol}$$

$$\text{Molar mass} = \frac{1.585 \text{ g}}{0.033262 \text{ mol}} = 47.\underline{6}51 \text{ g/mol}$$

The molecular weight is 47.7 amu.

5.56 The moles in 237 mL (0.237 L) of the compound is obtained from the ideal gas law. Dividing the mass (0.548 g) of the gas by the moles gives the molar mass and molecular weight.

$$n = \frac{(755/760 \text{ atm})(0.237 \text{ L})}{(373 \text{ K})(0.08206 \text{ L•atm/K•mol})} = 0.0076\underline{9}2 \text{ mol}$$

$$\text{Molar mass} = \frac{0.548 \text{ g}}{0.007692 \text{ mol}} = 71.\underline{2}4 \text{ g/mol}$$

The molecular weight is 71.2 amu.

5.57 The moles in 250 mL (0.250 L) of the compound is obtained from the ideal gas law. The 1.28-g mass of the gas then is divided by the moles to obtain the molar mass and molecular weight.

$$n = \frac{(786/760 \text{ atm})(0.250 \text{ L})}{(394 \text{ K})(0.08206 \text{ L•atm/K•mol})} = 0.0079\underline{9}6 \text{ mol}$$

$$\text{Molar mass} = \frac{1.28 \text{ g}}{0.007996 \text{ mol}} = 16\underline{0}.08 \text{ g/mol}$$

The molecular weight is 160 amu.

5.58 The moles in 345 mL (0.345 L) of the compound is obtained from the ideal gas law. The 2.30-g mass of the gas then is divided by the moles to obtain the molar mass and molecular weight.

$$n = \frac{(985/760 \text{ atm})(0.345 \text{ L})}{(421 \text{ K})(0.08206 \text{ L•atm/K•mol})} = 0.012\underline{9}4 \text{ mol}$$

$$\text{Molar mass} = \frac{2.30 \text{ g}}{0.01294 \text{ mol}} = 17\underline{7}.7 \text{ g/mol}$$

The molecular weight is 178 amu.

5.59 For a gas at a given temperature and pressure, the density depends on molecular weight (or for a mixture, the *average* molecular weight). Thus, at the same temperature and pressure, the density of NH_4Cl gas would be greater than that of a mixture of NH_3 and HCl because the average molecular weight of NH_3 and HCl would be lower than that of NH_4Cl.

5.60 For a gas at a given temperature and pressure, the density depends on molecular weight (or for a mixture, the *average* molecular weight). Thus, at the same temperature and pressure, the density of PCl_5 gas would be greater than that of a mixture of PCl_3 and Cl_2 formed by the decomposition of PCl_5 because the average molecular weight of PCl_3 and Cl_2 would be lower than that of PCl_5.

5.61 The 0.025 mol CaC_2 will form 0.025 mol C_2H_2. The volume is found from the ideal gas law:

$$\text{Vol} = \frac{(0.025 \text{ mol})(0.08206 \text{ L} \bullet \text{atm/K} \bullet \text{mol})(299 \text{ K})}{684/760 \text{ atm}} = 0.6\underline{8}15 = 0.68 \text{ L}$$

5.62 The 0.0420 mol Mg will form 0.420 mol of H_2. The volume is found from the ideal gas law:

$$\text{Vol} = \frac{(0.0420 \text{ mol})(0.08206 \text{ L} \bullet \text{atm/K} \bullet \text{mol})(301 \text{ K})}{665/760 \text{ atm}} = 1.1\underline{8}56 = 1.19 \text{ L}$$

5.63 Use the equation to obtain the moles of CO_2 and then the ideal gas law to obtain the volume.

$$2LiOH(s) + CO_2(g) \rightarrow Li_2CO_3(aq) + H_2O(l)$$

$$348 \text{ g LiOH} \times \frac{1 \text{ mol LiOH}}{23.95 \text{ g LiOH}} \times \frac{1 \text{ mol } CO_2}{2 \text{ mol LiOH}} = 7.2\underline{6}51 \text{ mol } CO_2$$

$$V = \frac{(7.2651 \text{ mol})(0.08206 \text{ L} \bullet \text{atm/K} \bullet \text{mol})(294 \text{ K})}{(781/760) \text{ atm}} = 17\underline{0}.6 = 171 \text{ L}$$

5.64 Use the equation to obtain the moles of ammonia and then the ideal gas law to obtain the volume.

$$Mg_3N_2(s) + 6H_2O(l) \rightarrow 3Mg(OH)_2(s) + 2NH_3(g)$$

$$4.56 \text{ g } Mg_3N_2 \times \frac{1 \text{ mol } Mg_3N_2}{100.9 \text{ g } Mg_3N_2} \times \frac{2 \text{ mol } NH_3}{1 \text{ mol } Mg_3N_2} = 0.090\underline{3}86 \text{ mol } NH_3$$

$$V = \frac{(0.090386 \text{ mol})(0.08206 \text{ L} \bullet \text{atm/K} \bullet \text{mol})(297 \text{ K})}{(743/760) \text{ atm}} = 2.2\underline{5}3 = 2.25 \text{ L}$$

5.65 Use the equation to obtain the moles of ammonia and then the ideal gas law to obtain the volume.

$$2NH_3(g) + CO_2(g) \rightarrow NH_2CONH_2(aq) + H_2O(l)$$

$$454 \text{ g urea} \times \frac{1 \text{ mol urea}}{60.06 \text{ g urea}} \times \frac{2 \text{ mol } NH_3}{1 \text{ mol urea}} = 15.\underline{1}2 \text{ mol } NH_3$$

$$V = \frac{(15.12 \text{ mol})(0.08206 \text{ L•atm/K•mol})(298 \text{ K})}{1.50 \text{ atm}} = 24\underline{6}.4 = 246 \text{ L}$$

5.66 Use the equation to obtain the moles of ammonia and then the ideal gas law to obtain the volume.

$$4NH_3(g) + 5O_2(g) \rightarrow 4NO(g) + 6H_2O(g)$$

$$50.0 \text{ g NO} \times \frac{1 \text{ mol NO}}{30.0 \text{ g NO}} \times \frac{5 \text{ mol } O_2}{4 \text{ mol NO}} = 2.0\underline{8}3 \text{ mol } O_2$$

$$V = \frac{(2.083 \text{ mol})(0.08206 \text{ L•atm/K•mol})(308 \text{ K})}{2.15 \text{ atm}} = 24.\underline{4}87 = 24.5 \text{ L}$$

5.67 Use the equation to obtain the moles of ammonia and then the ideal gas law to obtain the volume.

$$2NH_3(g) + H_2SO_4(aq) \rightarrow (NH_4)_2SO_4(aq)$$

$$75.0 \text{ g } (NH_4)_2SO_4 \times \frac{1 \text{ mol } (NH_4)_2SO_4}{132.1 \text{ g } (NH_4)_2SO_4} \times \frac{2 \text{ mol } NH_3}{1 \text{ mol } (NH_4)_2SO_4}$$

$$= 1.1\underline{3}55 \text{ mol } NH_3$$

$$V = \frac{nRT}{P} = \frac{(1.1355 \text{ mol})(0.08206 \text{ L • atm/K • mol})(288 \text{ K})}{1.15 \text{ atm}}$$

$$= 23.\underline{3}3 = 23.3 \text{ L}$$

5.68 Use the equation to obtain the moles of ammonia, and then the ideal gas law to obtain the volume.

$$2NaHCO_3(s) \rightarrow Na_2CO_3(s) + CO_2(g) + H_2O(l)$$

(continued)

$$26.8\text{ g NaHCO}_3 \times \frac{1\text{ mol NaHCO}_3}{84.0\text{ g NaHCO}_3} \times \frac{1\text{ mol CO}_2}{2\text{ mol NaHCO}_3} = 0.15\underline{9}52\text{ mol CO}_2$$

$$V = \frac{nRT}{P} = \frac{(0.15952\text{ mol})(0.08206\text{ L}\bullet\text{atm/K}\bullet\text{mol})(348\text{ K})}{(756/760)\text{ atm}}$$

$$= 4.5\underline{7}9 = 4.58\text{ L}$$

5.69 Calculate the partial pressure of each gas; then add the pressures since the total pressure is equal to the sum of the partial pressures:

$$P(\text{He}) = \frac{(0.0200\text{ mol})(0.08206\text{ L}\bullet\text{atm/K}\bullet\text{mol})(283\text{ K})}{5.00\text{ L}} = 0.092\underline{8}9\text{ atm}$$

$$P(\text{H}_2) = \frac{(0.0100\text{ mol})(0.08206\text{ L}\bullet\text{atm/K}\bullet\text{mol})(283\text{ K})}{5.00\text{ L}} = 0.046\underline{4}4\text{ atm}$$

The total pressure $= 0.092\underline{8}9 + 0.046\underline{4}4 = 0.139\underline{3}3 = 0.1393$ atm

5.70 Calculate the partial pressure of each gas; then add the pressures since the total pressure is equal to the sum of the partial pressures:

$$P(\text{He}) = \frac{(0.0300\text{ mol})(0.08206\text{ L}\bullet\text{atm/K}\bullet\text{mol})(293\text{ K})}{4.00\text{ L}} = 0.18\underline{0}3\text{ atm}$$

$$P(\text{O}_2) = \frac{(0.0200\text{ mol})(0.08206\text{ L}\bullet\text{atm/K}\bullet\text{mol})(293\text{ K})}{4.00\text{ L}} = 0.12\underline{0}2\text{ atm}$$

The total pressure $= 0.18\underline{0}3 + 0.12\underline{0}2 = 0.30\underline{0}5 = 0.301$ atm

5.71 Convert mass of O_2 and mass of He to moles. Use the ideal gas law to calculate the partial pressures, and then add to obtain the total pressures.

$$0.00103\text{ g O}_2 \times \frac{1\text{ mol O}_2}{32.00\text{ g O}_2} = 3.2\underline{1}9 \times 10^{-5}\text{ mol O}_2$$

$$0.00041\text{ g He} \times \frac{1\text{ mol He}}{4.00\text{ g He}} = 1.\underline{0}25 \times 10^{-4}\text{ mol He}$$

(continued)

$$P = \frac{nRT}{V} = \frac{(3.219 \times 10^{-5}\ \text{mol})(0.08206\ \text{L} \bullet \text{atm/K} \bullet \text{mol})(288\ \text{K})}{0.2000\ \text{L}}$$

$$= 0.0038\underline{0}4\ \text{atm}\ O_2$$

$$P = \frac{nRT}{V} = \frac{(1.025 \times 10^{-4}\ \text{mol})(0.08206\ \text{L} \bullet \text{atm/K} \bullet \text{mol})(288\ \text{K})}{0.2000\ \text{L}}$$

$$= 0.01\underline{2}11\ \text{atm He}$$

$$P = P_{O_2} + P_{He} = (0.003804 + 0.01211)\ \text{atm} = 0.01\underline{5}91 = 0.016\ \text{atm}$$

5.72 Obtain the partial pressure of helium using Dalton's law; then use the ideal gas law to obtain the mass of helium.

$$P_{He} = P - P_{O_2} = 3.00\ \text{atm} - 0.200\ \text{atm} = 2.8\underline{0}0\ \text{atm}$$

$$n_{He} = \frac{P_{He}V}{RT} = \frac{(2.80\ \text{atm})(1.00\text{L})}{(0.08206\ \text{L}\bullet\text{atm/K}\bullet\text{mol})(293\ \text{K})} = 0.11\underline{6}46\ \text{mol He}$$

$$\text{Mass}_{He} = 0.11646\ \text{mol He} \times \frac{4.003\ \text{g He}}{1\ \text{mol He}} = 0.46\underline{6}1 = 0.466\ \text{g He}$$

5.73 For each gas, Pgas = P x (mole fraction of gas).

$$P_{H_2} = 760\ \text{mmHg} \times 0.250 = 19\underline{0}.0 = 190\ \text{mmHg}$$

$$P_{CO_2} = 760\ \text{mmHg} \times 0.650 = 49\underline{4}.0 = 494\ \text{mmHg}$$

$$P_{HCl} = 760\ \text{mmHg} \times 0.054 = 4\underline{1}.04 = 41\ \text{mmHg}$$

$$P_{HF} = 760\ \text{mmHg} \times 0.028 = 2\underline{1}.28 = 21\ \text{mmHg}$$

$$P_{SO_2} = 760\ \text{mmHg} \times 0.017 = 1\underline{2}.92 = 13\ \text{mmHg}$$

$$P_{H_2S} = 760\ \text{mmHg} \times 0.001 = 0.\underline{7}6 = 0.8\ \text{mmHg}$$

5.74 For each gas, Pgas = P x (mole fraction of gas).

$$P_{He} = 6.91\ \text{atm} \times 0.790 = 5.4\underline{5}89 = 5.46\ \text{atm}$$

$$P_{N_2} = 6.91\ \text{atm} \times 0.170 = 1.1\underline{7}47 = 1.17\ \text{atm}$$

$$P_{O_2} = 6.91\ \text{atm} \times 0.040 = 0.2\underline{7}64 = 0.28\ \text{atm}$$

5.75 The total pressure is the sum of the partial pressures of CO and H_2O, so

$$P_{CO} = P - P_{water} = 689 \text{ mmHg} - 23.8 \text{ mmHg} = 66\underline{5}.2 \text{ mmHg}$$

$$n_{CO} = \frac{P_{CO}V}{RT} = \frac{(665/760 \text{ atm})(3.85 \text{L})}{(0.08206 \text{ L•atm/K•mol})(298 \text{ K})} = 0.13\underline{7}8 \text{ mol CO}$$

$$0.13\underline{7}8 \text{ g CO} \times \frac{1 \text{ mol HCOOH}}{1 \text{ mol CO}} \times \frac{46.03 \text{ g HCOOH}}{1 \text{ mol HCOOH}}$$

$$= 6.3\underline{4}2 = 6.34 \text{ g HCOOH}$$

5.76 The total pressure is the sum of the partial pressures of N_2 and H_2O, so

$$P_{N_2} = P - P_{H_2O} = 97.8 \text{ kPa}\left[\frac{1 \text{ atm}}{101.3 \text{ kPa}}\right] - 16.5 \text{ mmHg}\left[\frac{1 \text{ atm}}{760 \text{ mmHg}}\right]$$

$$= 0.94\underline{3}7 \text{ atm}$$

$$n_{N_2} = \frac{P_{N_2}V}{RT} = \frac{(0.9437 \text{ atm})(4.16 \text{ L})}{(0.08206 \text{ L•atm/K•mol})(292 \text{ K})} = 0.16\underline{3}8 \text{ mol } N_2$$

$$0.16\underline{3}8 \text{ g } N_2 \times \frac{1 \text{ mol } NH_4NO_2}{1 \text{ mol } N_2} \times \frac{64.04 \text{ g } NH_4NO_2}{1 \text{ mol } NH_4NO_2}$$

$$= 10.\underline{4}9 = 10.5 \text{ g } NH_4NO_2$$

5.77 Substitute 298 K (25°C) and 398 K (125°C) into Maxwell's distribution:

$$u_{25} = \sqrt{\frac{3RT}{M_m}} = \sqrt{\frac{3 \times 8.31 \text{ kg}\bullet\text{m}^2/(\text{s}^2\bullet\text{K}\bullet\text{mol}) \times 298 \text{ K}}{28.02 \times 10^{-3} \text{ kg/mol}}}$$

$$= 51\underline{4}.9 = 515 \text{ m/s}$$

$$u_{125} = \sqrt{\frac{3RT}{M_m}} = \sqrt{\frac{3 \times 8.31 \text{ kg}\bullet\text{m}^2/(\text{s}^2\bullet\text{K}\bullet\text{mol}) \times 398 \text{ K}}{28.02 \times 10^{-3} \text{ kg/mol}}}$$

$$= 59\underline{5}.07 = 595 \text{ m/s}$$

Graph as in Figure 5.19.

5.78 Substitute 296 K (23°C) into Maxwell's distribution:

$$u_{23} = \sqrt{\frac{3RT}{M_m}} = \sqrt{\frac{3 \times 8.31 \text{ kg}\bullet\text{m}^2/(\text{s}^2\bullet\text{K}\bullet\text{mol}) \times 296 \text{ K}}{159.8 \times 10^{-3} \text{ kg/mol}}}$$

$$= 21\underline{4}.8 = 215 \text{ m/s}$$

The rms speed is the same at 1.50 atm as at 1.00 atm because the temperature is the same.

5.79 Substitute 330 K (57°C) into Maxwell's distribution:

$$u_{330\text{ K}} = \sqrt{\frac{3RT}{M_m}} = \sqrt{\frac{3 \times 8.31 \text{ kg}\bullet\text{m}^2/(\text{s}^2\bullet\text{K}\bullet\text{mol}) \times 330 \text{ K}}{352 \times 10^{-3} \text{ kg/mol}}}$$

$$= 15\underline{2}.8 = 153 \text{ m/s}$$

5.80 Substitute 365 K into Maxwell's distribution:

$$u_{365\text{ K}} = \sqrt{\frac{3RT}{M_m}} = \sqrt{\frac{3 \times 8.31 \text{ kg}\bullet\text{m}^2/(\text{s}^2\bullet\text{K}\bullet\text{mol}) \times 365 \text{ K}}{2.016 \times 10^{-3} \text{ kg/mol}}}$$

$$= 2.1\underline{2}4 \times 10^3 = 2.12 \times 10^3 \text{ m/s}$$

(2.12 km/s, which is less than the escape velocity)

5.81 Because $u(CO_2) = u(H_2)$, we can equate the two right-hand sides of the Maxwell distributions:

$$\sqrt{\frac{3RT(CO_2)}{M_m(CO_2)}} = \sqrt{\frac{3RT(H_2)}{M_m(H_2)}}$$

Squaring both sides, rearranging to solve for $T(CO_2)$, and substituting numerical values, we have

$$T(CO_2) = T(H_2) \times \frac{M_m(CO_2)}{M_m(H_2)} = 293 \text{ K} \times \frac{44.01 \text{ g/mol}}{2.016 \text{ g/mol}} = 63\underline{9}6$$

$$= 6.40 \times 10^3 \text{ K}$$

5.82 Substitute 375 m/s into the Maxwell distribution, square both sides, and solve for T:

$$u = 375 \text{ m/s} = \sqrt{\frac{3 \times 8.31 \text{ kg} \bullet \text{m}^2/(\text{s}^2 \bullet \text{K} \bullet \text{mol}) \times T}{32.00 \times 10^{-3} \text{ kg/mol}}}$$

$$140625 = \frac{3 \times 8.31 \times T}{32.00 \times 10^{-3}}; \quad T = 18\underline{0}.505 \text{ K } (-9\underline{2}.49 \text{ or } -92°\text{C})$$

5.83 Because the ratio is the same at any temperature, $T(N_2) = T(O_2)$. Write a ratio of two Maxwell distributions after omitting $T(O_2)$ and $T(H_2)$ in each distribution:

$$\frac{u_{N_2}}{u_{O_2}} = \frac{\sqrt{\frac{3RT(N_2)}{M_m}}}{\sqrt{\frac{3RT(O_2)}{M_m}}} = \frac{\sqrt{\frac{3 \times 8.31 \text{ kg} \bullet \text{m}^2/(\text{s}^2 \bullet \text{K} \bullet \text{mol})}{28.02 \times 10^{-3} \text{ kg/mol}}}}{\sqrt{\frac{3 \times 8.31 \text{ kg} \bullet \text{m}^2/(\text{s}^2 \bullet \text{K} \bullet \text{mol})}{32.00 \times 10^{-3} \text{ kg/mol}}}}$$

$$\frac{u_{N_2}}{u_{O_2}} = \sqrt{\frac{32.00}{28.02}} = \frac{1.06\underline{8}6}{1} = \frac{1.069}{1}$$

5.84 Because the ratio is the same at any temperature, $T(H_2) = T(H_2S)$. Write a ratio of two Maxwell distributions after omitting $T(H_2)$ and $T(H_2S)$ in each distribution:

$$\frac{u_{H_2}}{u_{H_2S}} = \frac{\sqrt{\frac{3RT(H_2)}{M_m}}}{\sqrt{\frac{3RT(H_2S)}{M_m}}} = \frac{\sqrt{\frac{3 \times 8.31 \text{ kg} \bullet \text{m}^2/(\text{s}^2 \bullet \text{K} \bullet \text{mol})}{2.016 \times 10^{-3} \text{ kg/mol}}}}{\sqrt{\frac{3 \times 8.31 \text{ kg} \bullet \text{m}^2/(\text{s}^2 \bullet \text{K} \bullet \text{mol})}{34.08 \times 10^{-3} \text{ kg/mol}}}}$$

$$\frac{u_{H_2}}{u_{H_2S}} = \sqrt{\frac{34.08}{2.016}} = \frac{4.11\underline{1}51}{1} = \frac{4.11}{1}$$

5.85 Because the ratio is the same at any temperature, $T(H_2) = T(I_2)$. A ratio of two Maxwell distributions can be written as in the previous two problems, but this also can be simplified by canceling the 3 x 8.31 terms and rearranging the denominators to give

$$\frac{u_{H_2}}{u_{I_2}} = \frac{\sqrt{M_m(I_2)}}{\sqrt{M_m(H_2)}} = \sqrt{\frac{253.8}{2.016}} = \frac{11.2\underline{2}05}{1} = \frac{11.22}{1}$$

(continued)

Because hydrogen diffuses 11.22 times as fast as iodine, the time it would take would be 1/11.22 of the time required for iodine:

$$t(H_2) = 52 \text{ s} \times (1/11.22) = 4.\underline{6}3 = 4.6 \text{ s}$$

5.86 Because the ratio is the same at any temperature, T(He) = $T(N_2)$. A ratio of two Maxwell distributions can be written as in Problems 5.79 and 5.80, but this also can be simplified by canceling the 3 x 8.31 terms and rearranging the denominators to give

$$\frac{u_{He}}{u_{N_2}} = \frac{\sqrt{M_m(N_2)}}{\sqrt{M_m(He)}} = \sqrt{\frac{28.02}{4.00}} = \frac{2.6\underline{4}7}{1} = \frac{2.65}{1}$$

Because helium diffuses 2.65 times as fast as nitrogen, the time it would take would be 1/2.65 of the time required for nitrogen:

$$t(He) = 10.6 \text{ hr} \times (1/2.65) = 4.00 \text{ hr}$$

5.87 Because the diffusion occurs at the same temperature, T(gas) = T(Ar). A ratio of two Maxwell distributions can be written, but it can be simplified by canceling the 3 x 8.31 terms and rearranging the denominators. To simplify the definition of the rates, we assume the time is 1 second, and define the rate(Ar) as 9.23 mL/1 s and the rate(gas) as 4.83 mL/1 s. Then we write a ratio of two Maxwell distributions:

$$\frac{u_{gas}}{u_{Ar}} = \frac{4.83 \text{ mL}/1 \text{ s}}{9.23 \text{ mL}/1 \text{ s}} = \frac{\sqrt{M_m(Ar)}}{\sqrt{M_m(gas)}} = \frac{\sqrt{39.95}}{\sqrt{M_m(gas)}}$$

$$\sqrt{M_m(gas)} = \frac{9.23 \text{ mL}/1 \text{ s}}{4.83 \text{ ml}/1 \text{ s}} \times \sqrt{39.95} = 12.\underline{0}7$$

$$M_m(gas) = 14\underline{5}.8 \text{ g/mol; molecular weight} = 146 \text{ amu}$$

5.88 As in the previous problem, the temperatures are the same and a ratio of two Maxwell distributions can be written and simplified. In contrast, here we assume the volume is 1 mL, and we define the rate(N_2) as 1 mL/68.3 s and the rate(gas) as 1 mL/85.6 s. Then we write

$$\frac{u_{gas}}{u_{N_2}} = \frac{1 \text{ mL}/85.6 \text{ s}}{1 \text{ mL}/68.3 \text{ s}} = \frac{\sqrt{M_m(N_2)}}{\sqrt{M_m(gas)}} = \frac{\sqrt{28.02}}{\sqrt{M_m(gas)}}$$

We now rearrange and solve for M_m(gas):

$$\sqrt{M_m(gas)} = \frac{1 \text{ mL}/68.3 \text{ s}}{1 \text{ ml}/85.6 \text{ s}} \times \sqrt{28.02} = 6.6\underline{3}4$$

$$M_m(gas) = 44.\underline{0}1 \text{ g/mol; molecular weight} = 44.0 \text{ amu}$$

5.89 Solving the van der Waals equation for n = 1 and T = 355.2 K for P gives

$$P = \frac{RT}{(V - b)} - \frac{a}{V^2} = \frac{(0.08206\ L{\bullet}atm/K{\bullet}mol) \times 355.2\ K}{(35.00\ L - 0.08710\ L)} - \frac{12.56\ L^2{\bullet}atm}{(35.00\ L)^2}$$

$$P = 0.834\underline{8}7 - 0.0102\underline{5}3 = 0.824\underline{6}2 = 0.8246\ atm$$

$$P(\text{ideal gas law}) = 0.832\underline{8}\ atm$$

5.90 Solving the van der Waals equation for n = 1 and T = 393.2 K for P gives

$$P = \frac{RT}{(V - b)} - \frac{a}{V^2} = \frac{(0.08206\ L{\bullet}atm/K{\bullet}mol) \times 393.2\ K}{(32.50\ L - 0.03049\ L)} - \frac{5.537\ L^2{\bullet}atm}{(32.50\ L)^2}$$

$$P = 0.993\underline{7}3 - 0.00524\underline{2}1 = 0.988\underline{4}90 = 0.9885\ atm$$

$$P(\text{ideal gas law}) = 0.992\underline{7}9\ atm$$

5.91 To calculate a/V^2 in the van der Waals equation, we obtain V from the ideal gas law at 1.00 atm:

$$V = \frac{RT}{P} = \frac{(0.08206\ L{\bullet}atm/K{\bullet}mol) \times 273\ K}{1.00\ atm} = 22.\underline{4}0\ L$$

$$\frac{a}{V^2} = \frac{5.570}{(22.4\ L)^2} = 1.1\underline{1}0 \times 10^{-2}$$

At 1.00 atm, V = 22.4 L, and $a/V^2 = 1.110 \times 10^{-2}$. Substituting into the van der Waals equation:

$$V = \frac{RT}{P + \frac{a}{V^2}} + b = \frac{(0.08206\ L \bullet atm/K \bullet mol)(273\ K)}{(1.00 + 1.110 \times 10^{-2})\ atm} + 0.6499$$

$$= 22.\underline{2}2 = 22.2\ L$$

At 10.0 atm, the van der Waals equation gives 2.08 L. The ideal gas law gives 22.4 L for 1.00 atm and 2.24 L for 10.0 atm.

5.92 To substitute for a/V^2 in the van der Waals equation, we obtain V from the ideal gas law at 1.00 atm:

$$V = \frac{RT}{P} = \frac{(0.08206\ L{\bullet}atm/K{\bullet}mol) \times 273\ K}{1.00\ atm} = 22.\underline{4}0\ L$$

$$\frac{a}{V^2} = \frac{1.382}{(22.4\ L)^2} = 2.7\underline{5}4 \times 10^{-3}$$

At 1.00 atm, V = 22.4 L, and $a/V^2 = 2.75 \times 10^{-3}$. Substituting into the van der Waals equation:

$$V = \frac{RT}{P + \frac{a}{V^2}} + b = \frac{(0.08206\ L \bullet atm/K \bullet mol)(273\ K)}{(1.00 + 2.75 \times 10^{-3})\,atm} + 0.03186$$

$$= 22.\underline{3}7 = 22.4\ L$$

At 10.0 atm, the van der Waals equation gives 2.21 L. The ideal gas law gives 22.4 L for 1.00 atm and 2.24 L for 10.0 atm.

■ Solutions to Unclassified Problems

5.93 Calculate the mass of 1 cm^2 of the 20.5 m of water above the air in the glass. The volume is the product of the area of 1 cm^2 and the height of 20.5×10^2 cm (20.5 m) of water. The density of 1.00 g/cm^3 must be used to convert volume to mass:

$$m = d \times V$$

$$m = 1.00\ g/cm^3 \times (1.00\ cm^2 \times 20.5 \times 10^2\ cm) = 2.05 \times 10^3\ g, \text{ or } 2.05\ kg$$

The pressure exerted on an object at the bottom of the column of water is

$$P = \frac{force}{area} = \frac{(m)(g)}{area} = \frac{(2.05\ kg)(9.807\ m/s^2)}{1.00\ cm^2 \times \left[\frac{10^{-2}m}{1\ cm}\right]^2} = 2.01 \times 10^5\ kg/ms^2$$

$$= 2.01 \times 10^5\ Pa$$

The total pressure on the air in the tumbler = the barometric pressure and the water pressure:

$$P = 1.00 \times 10^2\ kPa + 2.01 \times 10^2\ kPa = 3.01 \times 10^2\ kPa$$

(continued)

Multiply the initial volume by a factor accounting for the change in pressure to find V_f:

$$V_f = V_i\left[\frac{P_i}{P_f}\right] = 243\ cm^3\left[\frac{1.00 \times 10^2\ kPa}{3.01 \times 10^2\ kPa}\right] = 80.\underline{7}3 = 80.7\ cm^3$$

5.94 The volume of 1 m^2 of the 30.0 m of water above the air is 30.0 m^3. The mass of water above this area is:

$$m = d \times V = \frac{1.025\ g}{cm^3} \times \left[\frac{1\ cm}{10^{-2}\ m}\right]^3 \times 30.0\ m^3$$

$$= 3.0\underline{7}5 \times 10^7 g\ \ (3.0\underline{7}5 \times 10^4\ kg)$$

The pressure on this area is

$$P = \frac{force}{area} = \frac{(m)(g)}{area} = \frac{(3.075 \times 10^4\ kg)(9.807\ m/s)}{1\ m^2}$$

$$= 3.0\underline{1}56 \times 10^5\ kg/\ ms^2\ \text{ or Pa}$$

$$P = 3.0156 \times 10^5\ Pa \times \frac{1\ atm}{1.013 \times 10^5\ Pa} = 2.9\underline{7}69\ atm$$

The total pressure at 30.0 m is

$$P = P_{air} + P_{water} = 1.00\ atm + 2.9769\ atm = 3.9\underline{7}69\ atm$$

The density of a gas is directly proportional to the pressure, so

$$d_2 = d_1\left[\frac{P_2}{P_1}\right] = (1.205\ g/L)\left[\frac{3.9769\ atm}{1.00\ atm}\right] = 4.7\underline{9}2 = 4.79\ g/L$$

5.95 Use the combined gas law and solve for V_f:

$$V_f = V_i \times \frac{P_i}{P_f} \times \frac{T_f}{T_i} = 183\ mL \times \frac{738\ mmHg}{760\ mmHg} \times \frac{273\ K}{294\ K} = 16\underline{5}.009 = 165\ mL$$

5.96 Use the combined gas law and solve for V_f:

$$V_f = V_i \times \frac{P_i}{P_f} \times \frac{T_f}{T_i} = 12.0\ L \times \frac{10.0\ atm}{1.00\ atm} \times \frac{273\ K}{293\ K} = 11\underline{1}.8 = 112\ L$$

5.97 Use the combined gas law and solve for V_f:

$$V_f = V_i \times \frac{P_i}{P_f} \times \frac{T_f}{T_i} = 5.0\ dm^3 \times \frac{100.0\ kPa}{79.0\ kPa} \times \frac{293\ K}{287\ K} = 6.\underline{4}6 = 6.5\ dm^3$$

5.98 Let V_i = 1 volume (vol) in the combined gas law; solving for V_f will give the relative final volume, or factor for the increase:

$$V_f = V_i \times \frac{P_i}{P_f} \times \frac{T_f}{T_i} = 1\ vol \times \frac{1.00\ atm}{1.00 \times 10^{-3}\ atm} \times \frac{253\ K}{288\ K}$$

$$= 87\underline{8}.4\ vol,\ \text{or } 878\ \text{times}$$

5.99 Use the ideal gas law to calculate the moles of helium and combine this with Avogadro's number to obtain the number of helium atoms:

$$n = \frac{PV}{RT} = \frac{(765/760\ atm)(0.01205\ L)}{(0.08206\ L\bullet atm/K\bullet mol)(296\ K)} = 4.9\underline{9}3 \times 10^{-4}\ mol$$

$$4.993 \times 10^{-4}\ mol\ He \times \frac{6.022 \times 10^{23}\ He^{2+}\ ions}{1\ mol\ He} \times \frac{1\ atom}{1\ He^{2+}\ ion} = 3.0\underline{0}67 \times 10^{20}$$

$$= 3.01 \times 10^{20}\ atoms$$

5.100 Use the ideal gas law to calculate the moles of nitrogen and combine this with the molar mass to obtain the mass of nitrogen. Then calculate the mass percentage:

$$n = \frac{PV}{RT} = \frac{(749/760\ atm)(0.00177\ L)}{(0.08206\ L\bullet atm/K\bullet mol)(298\ K)} = 7.1\underline{3}3 \times 10^{-5}\ mol$$

$$7.1\underline{3}3 \times 10^{-5}\ mol\ N_2 \times \frac{28.02\ g\ N}{1\ mol\ N_2} = 1.998 \times 10^{-3}\ g\ N\ \ (1.9\underline{9}8\ mg\ N)$$

$$\%N = \frac{mass\ N}{mass\ comp.} \times 100\% = \frac{1.9\underline{9}8\ mg\ N}{8.75\ mg} \times 100\% = 22.\underline{8}3 = 22.8\%$$

5.101 Calculate the molar mass, M_m, by dividing the mass of 1 L of air by the moles of the gas from the ideal gas equation:

$$M_m = \frac{mass}{n} = 1.2929\ g\ air \times \frac{(0.082058\ L\bullet atm/K\bullet mol)(273.15\ K)}{(1\ atm)(1\ L)}$$

$$= 28.97\underline{9}2\ g/mol = 28.979\ g/mol\ (amu)$$

5.102 First calculate the molar mass, M_m, by dividing the mass of 1 L of air by the moles of the gas from the ideal gas equation:

$$M_m = \frac{\text{mass}}{n} = 1.22 \text{ g gas} \times \frac{(0.08206 \text{ L•atm/K•mol})(293 \text{ K})}{(1 \text{ atm})(1 \text{ L})} = 29.\underline{3}3 \text{ g/mol}$$

Find the empirical formula from the 80.0% C and 20% O by assuming 1 gram of compound and calculating the moles:

$$0.800 \text{ C} \times \frac{1 \text{ mol C}}{12.01 \text{ g C}} = 0.06661 \text{ mol C};\ 0.200 \text{ g H} \times \frac{1 \text{ mol H}}{1.008 \text{ g H}}$$

$$= 0.1984 \text{ mol H}$$

$$\frac{0.06661 \text{ mol C}}{0.06661} = 1.00, \text{ or } 1 \text{ mol C};\ \frac{0.1984 \text{ mol H}}{0.06661} = 2.98, \text{ or } 3 \text{ mol H}$$

The simplest formula is CH_3, whose empirical formula weight is 15.03 g/mol. The number of CH_3 units contained in the molecular weight of 29.33 is

$$\frac{\text{molecular weight}}{\text{formula weight}} = \frac{29.33}{15.03} = 1.95, \text{ or } 2$$

The molecular formula is $(CH_3)_2$, or C_2H_6.

5.103 Use the ideal gas law to calculate the moles of CO_2. Then convert to mass of LiOH.

$$n = \frac{PV}{RT} = \frac{(1.00 \text{ atm})(5.8 \times 10^2 \text{ L})}{(0.08206 \text{ L•atm/K•mol})(273 \text{ K})} = 2\underline{5}.89 \text{ mol } CO_2$$

$$25.89 \text{ mol } CO_2 \times \frac{2 \text{ mol LiOH}}{1 \text{ mol } CO_2} \times \frac{23.95 \text{ g LiOH}}{1 \text{ mol LiOH}}$$

$$= 1.\underline{2}4 \times 10^3 = 1.2 \times 10^3 \text{ g LiOH}$$

5.104 Use the ideal gas law to calculate moles of CO_2. Then convert to mass of pyruvic acid.

$$n = \frac{PV}{RT} = \frac{(343/760 \text{ atm})(0.0203 \text{ L})}{(0.08206 \text{ L•atm/K•mol})(303 \text{ K})} = 3.6\underline{8}47 \times 10^{-4} \text{ mol } CO_2$$

$$3.6\underline{8}47 \times 10^{-4} \text{ mol } CO_2 \times \frac{1 \text{ mol pyr. a.}}{1 \text{ mol } CO_2} \times \frac{88.06 \text{ g pyr. a.}}{1 \text{ mol pyr. a.}}$$

$$= 0.032\underline{4}4 = 0.0324 \text{ g pyr. a}$$

5.105 Convert mass to moles of $KClO_3$ and then use the equation below to convert to moles of O_2. Use the ideal gas law to convert moles of O_2 to pressure at 21°C (294 K).

$$2KClO_3(s) \rightarrow 2KCl(s) + 3O_2(g)$$

$$85.0 \text{ g } KClO_3 \times \frac{1 \text{ mol } KClO_3}{122.55 \text{ g } KClO_3} = 0.69\underline{3}5 \text{ mol } KClO_3$$

$$0.69\underline{3}5 \text{ mol } KClO_3 \times \frac{3 \text{ mol } O_2}{2 \text{ mol } KClO_3} = 1.0\underline{4}02 \text{ mol } O_2$$

$$P = \frac{nRT}{V} = \frac{(1.0\underline{4}02 \text{ mol } O_2)(0.08206 \text{ L} \bullet \text{atm/K} \bullet \text{mol})(294 \text{ K})}{2.50 \text{ L}}$$

$$= 10.\underline{0}3 = 10.0 \text{ atm } O_2$$

5.106 Convert mass to moles of $KCHO_2$ and then use the equation below to convert to moles of H_2. Use the ideal gas law to convert moles of H_2 to pressure at 21°C (294 K).

$$KCHO_2(s) + KOH(s) \rightarrow K_2CO_3(s) + H_2(g)$$

$$75.0 \text{ g } KCHO_2 \times \frac{1 \text{ mol } KCHO_2}{84.12 \text{ g } KCHO_2} = 0.89\underline{1}58 \text{ mol } KCHO_2$$

From the equation, 0.89158 mol of $KCHO_2$ produces 0.89158 mol of H_2.

$$P = \frac{nRT}{V} = \frac{(0.89\underline{1}58 \text{ mol } O_2)(0.08206 \text{ L} \bullet \text{atm/K} \bullet \text{mol})(294 \text{ K})}{2.50 \text{ L}}$$

$$= 8.6\underline{0}3 = 8.60 \text{ atm } H_2$$

5.107 Find the number of moles of CO_2 first. Then convert this to moles HCl and molarity of HCl.

$$n = \frac{742/760 \text{ atm} \times 0.141 \text{ L}}{0.08206 \text{ L} \bullet \text{atm/K} \bullet \text{mol} \times 300 \text{ K}} = 0.0055\underline{9}1 \text{ mol}$$

$$\text{mol HCl} = 0.005591 \text{ mol } CO_2 \times (2 \text{ mol HCl}/1 \text{ mol } CO_2) = 0.011\underline{1}8 \text{ mol}$$

$$\text{M HCl} = 0.011\underline{1}8 \text{ mol HCl} \div 0.0249 \text{ L HCl} = 0.44\underline{8}9 = 0.449 \text{ mol/L HCl}$$

5.108 Find the number of moles of CO_2 first. Then convert this to moles HCl and molarity of HCl.

$$n = \frac{731/760 \text{ atm} \times 0.154 \text{ L}}{0.08206 \text{ L} \bullet \text{atm/K} \bullet \text{mol} \times 296 \text{ K}} = 0.0060\underline{9}8 \text{ mol}$$

$$\text{mol HCl} = 0.006098 \text{ mol } CO_2 \times (2 \text{ mol HCl}/1 \text{ mol } CO_2) = 0.012\underline{1}9 \text{ mol}$$

$$\text{M HCl} = 0.012\underline{1}9 \text{ mol HCl} \div 0.0235 \text{ L HCl} = 0.51\underline{8}7 = 0.519 \text{ mol/L HCl}$$

5.109 The number of moles of carbon dioxide is:

$$n = \frac{646/760 \text{ atm} \times 0.1500 \text{ L}}{0.08206 \text{ L} \bullet \text{atm/K} \bullet \text{mol} \times 300 \text{ K}} = 0.0051\underline{7}9 \text{ mol}$$

The number of moles of molecular acid used is:

$$0.1250 \text{ mol/L} \times 0.04141 \text{ L} = 0.005176 \text{ mol acid}$$

Thus the acid is H_2SO_4 since one mole of H_2SO_4 reacts to form one mole of CO_2.

5.110 The number of moles of carbon dioxide is:

$$n = \frac{722/760 \text{ atm} \times 0.1250 \text{ L}}{0.08206 \text{ L} \bullet \text{atm/K} \bullet \text{mol} \times 290 \text{ K}} = 0.0049\underline{9}0 \text{ mol}$$

The number of moles of molecular acid used is:

$$0.2040 \text{ mol/L} \times 0.04890 \text{ L} = 0.0099756 \text{ mol acid}$$

Thus the acid is HCl since the ratio of acid to CO_2 = 0.0099756 ÷ 0.004990 = 2.00 to 1.

5.111 Use Maxwell's distribution to calculate the temperature in kelvins; then convert to °C.

$$T = \frac{u^2 M_m}{3R} = \frac{(0.510 \times 10^3 \text{ m/s})^2(17.03 \times 10^{-3} \text{ kg/mol})}{3(8.31 \text{ kg} \bullet \text{m}^2/\text{s}^2 \bullet \text{K} \bullet \text{mol})}$$

$$= 17\underline{7}.6 = 178 \text{ K } (-95\ ^\circ\text{C})$$

5.112 Use Maxwell's distribution to calculate the temperature in kelvins.

$$T = \frac{u^2 M_m}{3R} = \frac{(3.53 \times 10^3 \text{ m/s})^2 (4.003 \times 10^{-3} \text{ kg/mol})}{3(8.31 \text{ kg•m}^2/\text{s}^2\text{•K•mol})}$$

$$= 2.0\underline{0}08 \times 10^3 = 2.00 \times 10^3 \text{ K}$$

5.113 Calculate the ratio of the root-mean-square molecular speeds, which is the same as the ratio of the rates of effusion:

$$\frac{\text{u of U(235)F}_6}{\text{u of U(238)F}_6} = \frac{\sqrt{M_m[\text{U(238)F}_6]}}{\sqrt{M_m[\text{U(235)F}_6]}} = \sqrt{\frac{352.04 \text{ g/ mol}}{349.03 \text{ g/ mol}}}$$

$$= \frac{1.004\underline{3}02}{1} = \frac{1.0043}{1}$$

5.114 Calculate the ratio of the root-mean-square molecular speeds, which is the same as the ratio of the rates of effusion through the barrier:

$$\frac{\text{u of H}_2(1,1)}{\text{u of H}_2(1,2)} = \frac{\sqrt{M_m[\text{H}_2(1,2)]}}{\sqrt{M_m[\text{H}_2(1,1)]}} = \sqrt{\frac{3.0219 \text{ g/ mol}}{2.0156 \text{ g/ mol}}} = \frac{1.224\underline{4}4}{1} = \frac{1.1224}{1}$$

5.115 First calculate the apparent molar masses at each pressure, using the ideal gas law. Only the calculation of the apparent molar mass for 0.2500 atm will be shown; the other values will be summarized in a table.

$$n = \frac{PV}{RT} = \frac{(0.2500 \text{ atm})(3.1908 \text{ L})}{(0.082057 \text{ L•atm/K•mol})(273.15 \text{ K})} = 3.55\underline{8}96 \times 10^{-2} \text{ mol}$$

$$\text{Apparent molar mass} = \frac{1.000 \text{ g}}{3.55896 \times 10^{-2} \text{ mol}} = 28.0\underline{9}8 = 28.10 \text{ g/mol}$$

(continued)

The following table summarizes the apparent molar masses calculated as above for all P's; these data are plotted in the graph to the right of the table.

P (atm)	App. Molar Mass (g/mol)
0.2500	28.10
0.5000	28.14
0.7500	28.19
1.0000	28.26

Extrapolation back to P = 0 gives 28.07 g/mol for the molar mass of the unknown gas (CO).

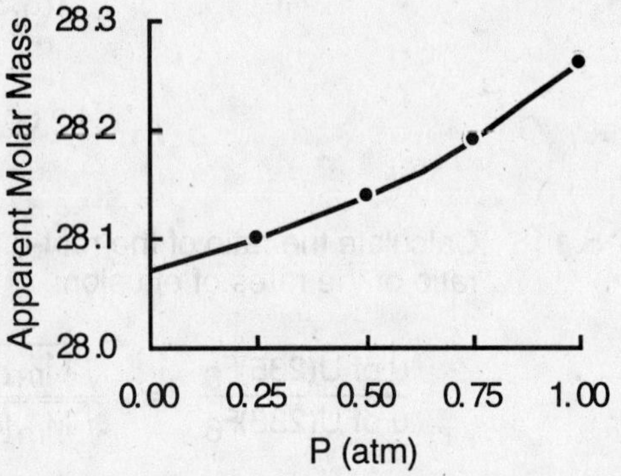

5.116 First calculate the apparent molar mass at each pressure, using the ideal gas law. Only the calculation of the molar mass at 0.2500 atm will be shown; the other values will be summarized in a table.

$$n = \frac{PV}{RT} = \frac{(0.2500\ \text{atm})(2.801\ \text{L})}{(0.082057\ \text{L•atm/K•mol})(273.15\ \text{K})} = 3.12\underline{4}2 \times 10^{-2}\ \text{mol}$$

$$\text{Apparent molar mass} = \frac{1.000\ \text{g}}{3.1242 \times 10^{-2}\ \text{mol}} = 32.0\underline{0}8\ \text{g/mol}$$

The following table summarizes the apparent molar masses calculated as above for all P's; these data are plotted in the graph to the right of the table.

P (atm)	App. Molar Mass (g/mol)
0.2500	32.008
0.5000	32.020
0.7500	32.021
1.0000	32.029

Extrapolation back to P = 0 gives close to 32.00 g/mol for the molar mass of O_2.

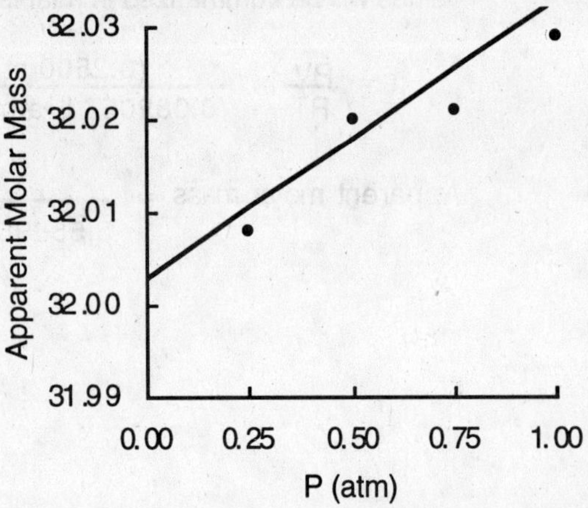

5.117 Use $CO + 1/2O_2 \rightarrow CO_2$, instead of 2CO. First find the moles of CO and O_2 by using the ideal gas law.

$$n_{CO} = \frac{PV}{RT} = \frac{(0.500\ atm)(2.00\ L)}{(0.08206\ L{\bullet}atm/K{\bullet}mol)(300\ K)} = 0.040\underline{6}2\ mol$$

$$n_{O_2} = \frac{PV}{RT} = \frac{(1.00\ atm)(1.00\ L)}{(0.08206\ L{\bullet}atm/K{\bullet}mol)(300\ K)} = 0.040\underline{6}2\ mol$$

There are equal amounts of CO and O_2, but (from the equation) only half as many moles of O_2 as CO are required for the reaction. Therefore, when 0.04062 moles of CO have been consumed, only 0.04062/2 moles of O_2 will have been used up. Then 0.04062/2 mol O_2 will remain and 0.04062 mol CO_2 will have been produced. At the end,

$$n_{CO} = 0\ mol;\ n_{O_2} = 0.0203\ mol;\ and\ n_{CO_2} = 0.04062\ mol$$

However, the total volume with the valve open is 3.00 L, so the partial pressures of O_2 and CO_2 must be calculated from the ideal gas law for each:

$$\frac{nRT}{V} = \frac{(0.0203\ mol\ O_2)(0.08206\ L \bullet atm/K \bullet mol)(300\ K)}{3.00\ L}$$

$$= 0.16\underline{6}58 = 0.167\ atm\ O_2$$

$$\frac{nRT}{V} = \frac{(0.04062\ mol\ CO_2)(0.08206\ L \bullet atm/K \bullet mol)(300\ K)}{3.00\ L}$$

$$= 0.33\underline{3}32 = 0.333\ atm\ CO_2$$

5.118 Use $H_2 + 1/2O_2 \rightarrow H_2O$. First find the moles of H_2 and O_2 by using the ideal gas law.

$$n_{H_2} = \frac{PV}{RT} = \frac{(0.500\ atm)(2.00\ L)}{(0.08206\ L{\bullet}atm/K{\bullet}mol)(533\ K)} = 0.022\underline{8}6\ mol$$

$$n_{O_2} = \frac{PV}{RT} = \frac{(1.00\ atm)(1.00\ L)}{(0.08206\ L{\bullet}atm/K{\bullet}mol)(533\ K)} = 0.022\underline{8}6\ mol$$

When all the H_2 has been consumed, half of the O_2 will remain. At the end, there will be 0.02886 mol of H_2O formed. However, the final volume is 3.00 L, so the partial pressures of H_2O and O_2 must be calculated from the ideal gas law for each:

(continued)

$$\frac{nRT}{V} = \frac{(0.02286 \text{ mol } H_2O)(0.08206 \text{ L} \bullet \text{atm/K} \bullet \text{mol})(533 \text{ K})}{3.00 \text{ L}}$$

$$= 0.33\underline{3}28 = 0.333 \text{ atm } H_2O$$

$$\frac{nRT}{V} = \frac{(0.01143 \text{ mol } O_2)(0.08206 \text{ L} \bullet \text{atm/K} \bullet \text{mol})(533 \text{ K})}{3.00 \text{ L}}$$

$$= 0.16\underline{6}64 = 0.167 \text{ atm } O_2$$

■ Cumulative-Skills Problems

(require skills from previous chapters)

5.119 Assume a 100.0-g sample, giving 85.2 g CH_4 and 14.8 g C_2H_6. Convert each to moles:

$$85.2 \text{ g } CH_4 \times \frac{1 \text{ mol } CH_4}{16.04 \text{ g } CH_4} = 5.3\underline{1}1 \text{ mol } CH_4$$

$$14.8 \text{ g } C_2H_6 \times \frac{1 \text{ mol } C_2H_6}{30.07 \text{ g } C_2H_6} = 0.49\underline{2}2 \text{ mol } C_2H_6$$

$$V_{CH_4} = \frac{(5.311 \text{ mol})(0.08206 \text{ L•atm/K•mol})(291 \text{ K})}{(748/760) \text{ atm}} = 12\underline{8}.9 \text{ L}$$

$$V_{C_2H_6} = \frac{(0.4922 \text{ mol})(0.08206 \text{ L•atm/K•mol})(291 \text{ K})}{(748/760) \text{ atm}} = 11.\underline{9}4 \text{ L}$$

The density is calculated as follows:

$$d = \frac{85.2 \text{ g } CH_4 + 14.8 \text{ g } C_2H_6}{(128.9 + 11.94) \text{ L}} = 0.71\underline{0}02 = 0.710 \text{ g/L}$$

5.120 Assume a 100.0-g sample, giving 34.3 g He, 51.7 g N_2, and 14.0 g O_2. Convert to moles:

$$34.3 \text{ g He} \times \frac{1 \text{ mol He}}{4.00 \text{ g}} = 8.5\underline{7}50 \text{ mol He}$$

$$51.7 \text{ g } N_2 \times \frac{1 \text{ mol } N_2}{28.02 \text{ g}} = 1.8\underline{4}57 \text{ mol } N_2$$

(continued)

$$14.0 \text{ g } O_2 \times \frac{1 \text{ mol } O_2}{32.00 \text{ g } O_2} = 0.43\underline{7}50 \text{ mol } O_2$$

$$V_{He} = \frac{(8.575 \text{ mol})(0.08206 \text{ L•atm/K•mol})(295 \text{ K})}{(755/760) \text{ atm}} = 20\underline{8}.9 \text{ L}$$

$$V_{N_2} = \frac{(1.8457 \text{ mol})(0.08206 \text{ L•atm/K•mol})(295 \text{ K})}{(755/760) \text{ atm}} = 44.\underline{9}7 \text{ L}$$

$$V_{O_2} = \frac{(0.4375 \text{ mol})(0.08206 \text{ L•atm/K•mol})(295 \text{ K})}{(755/760) \text{ atm}} = 10.\underline{6}6 \text{ L}$$

The density is calculated as follows:

$$d = \frac{34.3 \text{ g He} + 51.7 \text{ g } N_2 + 14.0 \text{ g } O_2}{(208.9 + 44.97 + 10.66) \text{ L}} = 0.37\underline{8}02 = 0.378 \text{ g/L}$$

5.121 First subtract the height of mercury equivalent to the 25.00 cm (250 mm) of water inside the tube from 771 mmHg to get P_{gas}. Then subtract the vapor pressure of water, 18.7 mmHg, from P_{gas} to get P_{O_2}.

$$h_{Hg} = \frac{(h_W)(d_W)}{d_{Hg}} = \frac{250 \text{ mm} \times 0.99987 \text{ g/cm}^3}{13.596 \text{ g/cm}^3} = 18.\underline{3}8 \text{ mmHg}$$

$$P_{gas} = P - P_{25 \text{ cm water}} = 771 \text{ mmHg} - 18.38 \text{ mmHg} = 75\underline{2}.62 \text{ mmHg}$$

$$P_{O_2} = 75\underline{2}.62 \text{ mmHg} - 18.7 \text{ mmHg} = 73\underline{3}.92 \text{ mmHg}$$

$$n = \frac{PV}{RT} = \frac{(733.92/760 \text{ atm})(0.0310 \text{ L})}{(0.08206 \text{ L•atm/K•mol})(294 \text{ K})} = 0.0012\underline{4}08 \text{ mol } O_2$$

$$\text{Mass} = (2 \times 0.0012408) \text{ mol } Na_2O_2 \times \frac{77.98 \text{ g } Na_2O_2}{1 \text{ mol } Na_2O_2}$$

$$= 0.19\underline{3}51 = 0.194 \text{ g } Na_2O_2$$

5.122 Proceed as in the previous problem, subtracting the height of water and then vapor pressure to find P_{H2}.

$$h_{Hg} = \frac{(h_W)(d_W)}{d_{Hg}} = \frac{310 \text{ mm} \times 0.99987 \text{ g/cm}^3}{13.596 \text{ g/cm}^3} = 22.\underline{8}0 \text{ mmHg}$$

(continued)

$P_{gas} = P - P_{31\text{ cm water}} = 751\text{ mmHg} - 22.80\text{ mmHg} = 72\underline{8}.2\text{ mmHg}$

$P_{H_2} = (728.2 - 15.5)\text{ mmHg} = 71\underline{2}.7\text{ mmHg}$

$$n = \frac{PV}{RT} = \frac{(712.7/760\text{ atm})(0.0240\text{ L})}{(0.08206\text{ L•atm/K•mol})(291\text{ K})} = 0.00094\underline{2}4\text{ mol }H_2$$

$$\text{Mass} = 0.0009424\text{ mol }H_2 \times \frac{1\text{ mol Zn}}{1\text{ mol }H_2} \times \frac{65.39\text{ g Zn}}{1\text{ mol Zn}}$$

$$= 0.061\underline{6}29 = 0.0616\text{ g Zn}$$

5.123 First find the moles of CO_2:

$$n = \frac{PV}{RT} = \frac{(785/760\text{ atm})(1.94\text{ L})}{(0.08206\text{ L•atm/K•mol})(298\text{ K})} = 0.081\underline{9}4\text{ mol }CO_2$$

Set up one equation in one unknown: x = mol $CaCO_3$; (0.08194 - x) = mol $MgCO_3$.

$7.85\text{ g} = (100.1\text{ g/mol})x + (84.32\text{ g/mol})(0.08194 - x)$

$$x = \frac{(7.85 - 6.9092)}{(100.1 - 84.32)} = 0.05\underline{9}62\text{ mol }CaCO_3$$

$(0.08194 - x) = 0.02\underline{2}32\text{ mol }MgCO_3$

$$\%\ CaCO_3 = \frac{0.05962\text{ mol }CaCO_3 \times 100.1\text{ g/mol}}{7.85\text{ g}} \times 100\% = 7\underline{6}.02 = 76\%$$

$\%\ MgCO_3 = 100.00\% - 76.02\% = 2\underline{3}.98 = 24\%$

5.124 First find the moles of H_2S:

$$n = \frac{PV}{RT} = \frac{(745/760\text{ atm})(1.049\text{ L})}{(0.08206\text{ L•atm/K•mol})(296\text{ K})} = 0.042\underline{3}3\text{ mol }H_2S$$

Set up one equation in one unknown: x = mol ZnS; (0.04233 - x) = mol PbS.

$6.12\text{ g} = (97.46\text{ g/mol})x + (239.25\text{ g/mol})(0.04233 - x)$

$$x = \frac{(6.12 - 10.127)}{(97.46 - 239.25)} = 0.028\underline{2}6\text{ mol ZnS}$$

$(0.04233 - x) = 0.0104\underline{0}70\text{ mol PbS}$

(continued)

$$\% \text{ ZnS} = \frac{0.02826 \text{ mol ZnS} \times 97.46 \text{ g/mol}}{6.12 \text{ g}} \times 100\% = 45.\underline{0}0 = 45.0\%$$

$$\% \text{ PbS} = 55.0\%$$

5.125 Write a mass-balance equation to solve for the moles of each gas, using 28.01 g/mol for the molar mass of N_2 and 20.18 g/mol for the molar mass of Ne. Let y = mol of each gas:

$$28.01 \text{ y} + 20.18 \text{ y} = 10.0 \text{ g}$$

$$\text{y} = (10.0 \div 48.19) = 0.20\underline{7}51 \text{ mol}$$

Total moles (n) = 0.41502 mol (use below)

Use the ideal gas law to calculate the volume, which can then be used to calculate density.

$$\text{Vol} = \frac{(0.41502 \text{ mol})(0.08206 \text{ L} \bullet \text{atm/ K} \bullet \text{mol})(500 \text{ K})}{10.00 \text{ atm}} = 1.7\underline{0}2 \text{ L}$$

$$\text{d} = 10.0 \text{ g} \div 1.702 \text{ L} = 5.8\underline{7}5 = 5.88 \text{ g/ L}$$

5.126 Write a mass-balance equation to solve for moles of each, using 39.95 g/mol for the molar mass of Ar and 20.18 g/mol for the molar mass of Ne. Let y = mol of Ar and 2y = mol of Ne:

$$39.95 \text{ y} + 20.18 \text{ (2y)} = 50.0 \text{ g}$$

$$\text{y} = (50.0 \div 80.31) = 0.62\underline{2}5 \text{ mol}$$

Total moles (n) = mol Ar + mol Ne = 0.6225 + (2)(0.6225) = $1.86\underline{7}7$ mol (use below)

The volume of the mixture = 50.0 g ÷ 4.00 g/L = 12.5 L

Use the ideal gas law to calculate the total pressure

$$\text{P(total)} = \frac{(1.8677 \text{ mol})(0.08206 \text{ L} \bullet \text{atm/ K} \bullet \text{mol})(350 \text{ K})}{12.5 \text{ L}} = 4.2\underline{9}2 \text{ atm}$$

The partial pressure of Ne = 2/3 the total P = 2/3(4.292 atm) = $2.8\underline{6}1$ = 2.86 atm.

5.127 Rearrange the equation PM_m = dRT to find the quantity RT/M_W. .Then plug into the equation for the root-mean-square speed.

$$\frac{RT}{M_W} = \frac{P}{d}$$

$$u = \sqrt{\frac{3RT}{M_W}} = \sqrt{\frac{3P}{d}} = \sqrt{\frac{3\left(\frac{675}{700}\right)\text{ atm x } (8.31\text{ kg}\bullet\text{m}^2/\text{s}^2\bullet\text{K}\bullet\text{mol})}{(3.00 \times 10^{-3}\text{ kg/L})(0.08206\text{ L}\bullet\text{atm/K}\bullet\text{mol})}}$$

$$= 29\underline{9}.90 = 300.\text{ m/s}$$

5.128 Rearrange the root-mean-square speed equation to get $M_W = 3RT/u^2$.

$$M_W = \frac{(3)(8.31\text{ kg}\bullet\text{m}^2/(\text{s}^2\bullet\text{mol}\bullet\text{K})(298.2\text{ K})}{(5.00 \times 10^2\text{ m/s})^2} = 0.029\underline{7}37\text{ kg/mol}$$

$$= 29.\underline{7}4\text{ g/mol}$$

Now, rearrange the equation PM_m = dRT to find the density, d = PM_W / RT.

$$d = \frac{PM_W}{RT} = \frac{(2.50\text{ atm})(29.74\text{ g/mol})}{(0.08206\text{ L}\bullet\text{atm/K}\bullet\text{mol})(298.2\text{ K})} = 3.0\underline{3}8 = 3.04\text{ g/L}$$

■ Solution to Conceptual Problem

5.129 Liquid nitrogen, which boils at -196°C, is often transported in evacuated Dewar flasks similar to thermos bottles. In some cases, this could include transportation between floors using a common passenger elevator. Assuming an elevator with dimensions of 2 meters by 2 meters by 2 meters and a density of liquid nitrogen of 0.808 g/mL, what would you predict would be the outcome of accidentally smashing a 4-L Dewar in the elevator? Is transporting liquid nitrogen in an elevator a safe practice?

Discussion

The boiling point of liquid nitrogen is well below room temperature, so the liquid in the broken flask should quickly vaporize. The elevator car has a volume of 8×10^3 L, whereas the 4 L of liquid nitrogen will occupy 2.6×10^3 L at STP:

$$4 \text{ L } N_2 \text{ (liquid)} \times \frac{1000 \text{ mL } N_2}{1 \text{ L } N_2} \times \frac{0.808 \text{ g } N_2}{1 \text{ mL } N_2} \times \frac{1 \text{ mol } N_2}{28.02 \text{ g } N_2} \times \frac{22.4 \text{ L } N_2}{1 \text{ mol } N_2}$$

$$= 2.6 \times 10^3 \text{ L } N_2 \text{ (gas)}$$

You might expect the nitrogen to partially fill the elevator. Although our atmosphere is mostly nitrogen (78% by volume), we require the oxygen in the air to survive. Even partial replacement of this oxygen by nitrogen could lead to suffocation of the elevator passengers. Thus, transportation of liquid nitrogen in a passenger elevator is not a safe practice.

6. THERMOCHEMISTRY

Solutions to Exercises

Note on significant figures: The final answer to each mathematical solution is given first with one nonsignificant figure (the rightmost significant figure is underlined) and then is rounded to the correct number of figures. Intermediate answers usually also have at least one nonsignificant figure.

6.1 Substitute into the formula $E_k = 1/2\ mv^2$ using SI units:

$$E_k = 1/2 \times 9.11 \times 10^{-31}\ \text{kg} \times (5.0 \times 10^6\ \text{m/s})^2 = 1.\underline{1}3 \times 10^{-17} = 1.1 \times 10^{-17}\ \text{J}$$

$$1.13 \times 10^{-17}\ \text{J} \times \frac{1\ \text{cal}}{4.184\ \text{J}} = 2.\underline{7}2 \times 10^{-18} = 2.7 \times 10^{-18}\ \text{cal}$$

6.2 Heat is evolved; therefore, the reaction is exothermic. The value of q is -1170 kJ.

6.3 The thermochemical equation is

$$2N_2H_4(l) + N_2O_4(l) \rightarrow 3N_2(g) + 4H_2O(g);\ \Delta H = -1.049 \times 10^3\ \text{kJ}$$

6.4 a. $N_2H_4(l) + 1/2N_2O_4(l) \rightarrow 3/2N_2(g) + 2H_2O(g);\ \Delta H = -5.245 \times 10^2\ \text{kJ}$

b. $4H_2O(g) + 3N_2(g) \rightarrow 2N_2H_4(l) + N_2O_4(l);\ \Delta H = 1.049 \times 10^3\ \text{kJ}$

6.5 The reaction is

$$2N_2H_4(l) + N_2O_4(l) \rightarrow 3N_2(g) + 4H_2O(g);\ \Delta H = -1.049 \times 10^3\text{ kJ}$$

$$10.0\text{ g } N_2H_4 \times \frac{1\text{ mol } N_2H_4}{32.02\text{ g}} \times \frac{1\text{ mol } N_2O_4}{2\text{ mol } N_2H_4} \times \frac{1.049 \times 10^3\text{ kJ}}{1\text{ mol } N_2H_4}$$

$$= 16\underline{3}.80\text{ kJ } (1.64 \times 10^2\text{ kJ})$$

6.6 Substitute into the equation q = s x m x Δt to obtain the heat transferred.

$$q = s \times m \times \Delta t = \frac{0.449\text{ J}}{g{\bullet}°C} \times (100.0°C - 20.0°C) \times 5.00\text{ g}$$

$$= 17\underline{9}.6 = 180.\text{ J}$$

6.7 The heat released by the reaction, q_{rxn}, equals the negative value of the heat absorbed by the solution. Divide this by the moles of HCl to find ΔH for the reaction:

$$q = s \times m \times \Delta t = \frac{-4.184\text{ J}}{g{\bullet}°C} \times (31.8°C - 25.0°C) \times (33 + 42)\text{ g}$$

$$= 2\underline{1}33.8\text{ J}$$

$$\text{Mol HCl} = 1.20\text{ mol/L} \times 0.033\text{ L} = 0.03\underline{9}6\text{ mol}$$

$$\Delta H = \frac{-2133.8\text{ J}}{0.0396\text{ mol}} = -5\underline{3}884\text{ J/mol} = -54\text{ kJ/mol}$$

$$HCl(aq) + NaOH(aq) \rightarrow NaCl(aq) + H_2O(l);\ \Delta H = -54\text{ kJ}$$

6.8 Use Hess's law to find ΔH for $4Al(s) + 3MnO_2(s) \rightarrow 2Al_2O_3 + 3Mn(s)$ from the following data for reactions 2 and 3:

$$2Al(s) + 3/2O_2(g) \rightarrow Al_2O_3(s);\quad \Delta H = -1676\text{ kJ}$$

$$Mn(s) + O_2(g) \rightarrow MnO_2(s);\quad \Delta H = -521\text{ kJ}$$

If you take reaction 2 and multiply it by 2 and add the reverse of three times reaction 3, you obtain reaction 1 (you cancel as you would in algebraic equations). If you add the corresponding enthalpy changes, you obtain the enthalpy change of reaction 1. The layout follows:

(continued)

$4Al(s) + 3O_2(g)$	$\rightarrow$	$2Al_2O_3(s)$;	ΔH = (-1676 kJ) x (2)
$3MnO_2(s)$	$\rightarrow$	$3Mn(s) + 3O_2(g)$;	ΔH = (-521 kJ) x (-3)
$4Al(s) + 3MnO_2(s)$	$\rightarrow$	$2Al_2O_3(s) + 3Mn(s)$	ΔH = -1789 kJ

6.9 The reaction is

$H_2O(l)$	$\rightarrow$	$H_2O(g)$
-285.8		-241.8

ΔH°_{vap} = [-241.8 - (-285.8)] kJ = 44.0 kJ

6.10 The reaction is

$3NO_2(g)$	+	$H_2O(l)$	$\rightarrow$	$2HNO_3(aq)$	+	$NO(g)$
3 x 33.2		-285.8		2 x -206.6		90.3

ΔH° = [(2 x -206.6) + 90.3] - [(3 x 33.2) + (-285.8)] = -136.$\underline{7}$0 = -136.7 kJ

6.11 The net chemical equation is

$2NH_4^+(aq)$	+	$2OH^-(aq)$	$\rightarrow$	$2NH_3(g)$	+	$2H_2O(l)$
2(-132.8)		2(-229.9)		2(-45.9)		2(-285.8)

ΔH° = [2(-45.9) + 2(-285.8)] - [2(-132.8) + 2(-229.9)] = 62.$\underline{0}$0 = 62.0 kJ

■ Answers to Review Questions

6.1 *Energy* is the capacity or potential to move matter. *Kinetic energy* is the energy associated with an object by virtue of its motion. *Potential energy* is the energy an object has by virtue of its position in a field of force. *Internal energy* is the sum of the kinetic and potential energies of the particles making up a substance.

6.2 A joule is the SI unit of energy. One joule is the product of a force of one newton (kg•m/s^2) times a distance of 1 m. One joule is also the product of a mass of 1 kg times an acceleration of 1 m per s^2 times a distance of 1 m.

6.3 Originally a calorie was defined as the amount of energy required to raise the temperature of one gram of water by one degree Celsius. At present, the calorie is defined as 4.184 J.

6.4 At either of the two highest points above the earth in a pendulum's cycle, the energy of the pendulum is all potential energy and is equal to the product mgh (m = mass of pendulum, g = constant acceleration of gravity, and h = height of pendulum). As the pendulum moves downward, its potential energy decreases from mgh to near zero, depending on how close it comes to the earth's surface. During the downward motion, its potential energy is converted to kinetic energy. When it reaches the lowest point (middle) of its cycle, the pendulum has its maximum kinetic energy and minimum potential energy. As it rises above the lowest point, its kinetic energy begins to be converted to potential energy. When it reaches the other high point in its cycle, the energy of the pendulum is again all potential energy. At rest, the energy of the pendulum has been transferred to the surroundings.

6.5 As the heat flows into the gas, the gas molecules gain energy and move at a faster average speed. The internal energy of the gas increases.

6.6 An *exothermic* reaction evolves heat. For example, the burning of one mol of methane, $CH_4(g)$, yields carbon dioxide, water, and 890 kJ of heat. An *endothermic* reaction absorbs heat. For example, the reaction of one mol of barium hydroxide with ammonium nitrate absorbs 170.4 kJ of heat in order to form ammonia, water, and barium nitrate.

6.7 Changes in internal energy depend only on the initial and final states of the system, which are determined by variables such as temperature and pressure. Such changes do not depend on how the changes were made.

6.8 The enthalpy change equals the heat of reaction only at constant pressure (changes in temperature do not alter this equality).

6.9 The enthalpy change is positive (the enthalpy increases) for an endothermic reaction at constant pressure.

6.10 It is important to give the states when writing an equation for ΔH because ΔH depends on the states of all reactants and products. If any state changes, ΔH changes.

6.11 The value of ΔH doubles and its sign changes when the equation is doubled and reversed.

6.12 First convert the 10.0-g mass of water to moles of water, using its molar mass. Then, using the equation, multiply the moles of water by the ratio (1 mol CH_4/2 mol H_2O). Finally, multiply the moles of CH_4 by -890.2 kJ/mol CH_4 (as done in Section 6.5).

6.13 The *heat capacity* of a substance is the quantity of heat needed to raise the temperature of the sample of substance one degree Celsius (or one kelvin). The *specific heat* of a substance is the quantity of heat required to raise the temperature of one gram of a substance by one degree Celsius (or one kelvin) at constant pressure.

6.14 A simple calorimeter consists of an insulated container, such as a polystyrene cup, with a thermometer positioned in the liquid held by the container. The heat of the reaction is obtained by conducting the reaction in the calorimeter. The temperature of the mixture is measured before and after the reaction. The heat capacity of the calorimeter and its contents also must be measured.

6.15 Hess's law is based on the fact that if a chemical equation is the sum of multiples of other reactions, the ΔH of this equation equals a similar sum of multiples of the ΔH's of the other reactions. In other words, Hess's law is based on the additivity property of ΔH.

6.16 The heat of sublimation, $\Delta H_{sub} = \Delta H_{fus} + \Delta H_{vap}$.

6.17 The thermodynamic standard state refers to the standard thermodynamic conditions chosen for substances when listing or comparing thermodynamic data: 1 atm pressure and the specified temperature (usually 25°C).

6.18 The reference form of an element is the stablest form (physical state and allotrope) of the element under standard thermodynamic conditions. The standard enthalpy of formation of an element in its reference form is zero.

6.19 The standard enthalpy of formation of a substance is the enthalpy change for the formation of one mole of the substance from its elements at standard pressure (1 atm) and a specified temperature (25°C unless otherwise noted).

6.20 The equation for the formation of $H_2S(g)$ is

$$H_2(g) + 1/8S_8(s) \rightarrow H_2S(g)$$

6.21 The reaction of $C(g) + 4H(g) \rightarrow CH_4(g)$ is not an appropriate equation for calculating the ΔH_f of methane because the most stable form of each element is not used. Both $H_2(g)$ and C(s), as graphite, should be used instead of H(g) and C(g).

6.22 A fuel is any substance that is burned or similarly reacted to provide heat and other forms of energy. The fossil fuels are petroleum, gas, and coal. They were formed long before our present civilization by the decomposition and compression of aquatic plants and animals beneath the sediment of swamps and seas.

6.23 One two-step set of reactions for the formation of methane from coal is

$$C(s) + H_2O(g) \rightarrow CO(g) + H_2(g)$$

$$CO(g) + 3H_2(g) \rightarrow CH_4(g) + H_2O(g)$$

6.24 Some possible rocket fuel/oxidizer combinations are H_2/O_2 and hydrazine/dinitrogen tetroxide. The chemical equations for their reactions are

$$H_2(g) + 1/2O_2(g) \rightarrow H_2O(g); \quad \Delta H^\circ = -242 \text{ kJ}$$

$$2N_2H_4(l) + N_2O_4(l) \rightarrow 3N_2(g) + 4H_2O(g); \quad \Delta H^\circ = -1049 \text{ kJ}$$

■ Solutions to Practice Problems

Note on significant figures: The final answer to each problem is given first with one nonsignificant figure (the rightmost significant figure is underlined) and then is rounded to the correct number of figures. Intermediate answers usually also have at least one nonsignificant figure. Atomic weights are rounded to two decimal places, except for that of hydrogen.

6.25 The SI units of force must be kg•m/s^2 (= newton, N) to be consistent with the joule, the SI unit of energy:

$$\frac{\text{kg}\bullet\text{m}}{\text{s}^2} \times \text{m} = \frac{\text{kg}\bullet\text{m}^2}{\text{s}^2} = \text{joule, J}$$

6.26 Solving E_p = mgh, you obtain g = E_p/mh. Thus, the SI unit of g is (kg•m^2/s^2)/(kg•m) = m/s^2.

6.27 The heat in kcal released is

$$-890.3 \text{ kJ} \times \frac{1000 \text{ J}}{1 \text{ kJ}} \times \frac{1 \text{ cal}}{4.184 \text{ J}} \times \frac{1 \text{ kcal}}{1000 \text{ cal}} = -212.\underline{7}8 = -212.8 \text{ kcal}$$

6.28 The heat in kcal released is

$$-541.4 \text{ kJ} \times \frac{1000 \text{ J}}{1 \text{ kJ}} \times \frac{1 \text{ cal}}{4.184 \text{ J}} \times \frac{1 \text{ kcal}}{1000 \text{ cal}} = -129.\underline{3}9 = -129.4 \text{ kcal}$$

6.29 The kinetic energy in J and in cal is

$$E_k = 1/2 \times 4.85 \times 10^3 \text{ lb} \times \frac{0.4536 \text{ kg}}{1 \text{ lb}} \times \left[\frac{48 \text{ mi}}{1 \text{ h}}\right]^2 \times \left[\frac{1609 \text{ m}}{1 \text{ mi}}\right]^2$$

$$\times \left[\frac{1 \text{ h}}{3600 \text{ s}}\right]^2 = 5.\underline{0}62 \times 10^5 = 5.1 \times 10^5 \text{ J}$$

$$E_k = 5.\underline{0}62 \times 10^5 \times \frac{1 \text{ cal}}{4.184 \text{ J}} = 1.\underline{2}09 \times 10^5 = 1.2 \times 10^5 \text{ cal}$$

6.30 The kinetic energy in J and in cal is

$$E_k = 1/2 \times 228 \text{ gr} \times \frac{0.0000648 \text{ kg}}{1 \text{ gr}} \times \left[\frac{2520 \text{ ft}}{1 \text{ s}}\right]^2 \times \left[\frac{1 \text{ yd}}{3 \text{ ft}}\right]^2$$

$$\times \left[\frac{0.9144 \text{ m}}{1 \text{ yd}}\right]^2 = 4.3\underline{5}8 \times 10^3 = 4.36 \times 10^3 \text{ J}$$

$$E_k = 4.3\underline{5}8 \times 10^3 \times \frac{1 \text{ cal}}{4.184 \text{ J}} = 1.0\underline{4}1 \times 10^3 = 1.04 \times 10^3 \text{ cal}$$

6.31 To insert the mass of 1 molecule of N_2O in the formula, multiply the molar mass by the reciprocal of Avogadro's number. The kinetic energy in J is

$$E_k = 1/2 \times \frac{44.02 \text{ g}}{1 \text{ mol}} \times \frac{1 \text{ mol}}{6.022 \times 10^{23} \text{ molec.}} \times \frac{1 \text{ kg}}{1000 \text{ g}} \times \left[\frac{379 \text{ m}}{1 \text{ s}}\right]^2$$

$$= 5.2\underline{4}9 \times 10^{-21} = 5.25 \times 10^{-21} \text{ J/molec.}$$

6.32 To insert the mass of 1 molecule of ClO_2 in the formula, multiply the molar mass by the reciprocal of Avogadro's number. The kinetic energy in J is

$$E_k = 1/2 \times \frac{67.45 \text{ g}}{1 \text{ mol}} \times \frac{1 \text{ mol}}{6.022 \times 10^{23} \text{ molec.}} \times \frac{1 \text{ kg}}{1000 \text{ g}} \times \left[\frac{306 \text{ m}}{1 \text{ s}}\right]^2$$

$$= 5.2\underline{4}3 \times 10^{-21} = 5.24 \times 10^{-21} \text{ J/molec.}$$

6.33 Exothermic reactions evolve heat, so the sign of q will be negative because energy is lost from the system to the surroundings. The flask will feel hot to the touch.

6.34 Endothermic reactions absorb heat, so the sign of q will be positive because energy must be gained by the system from the surroundings. The flask will feel cold to the touch.

6.35 The release of 939 kJ of heat per 2 mol HCN means that the reaction is exothermic. Because energy is lost from the system to the surroundings, q is negative, and because the reaction involves 2 mol HCN, q for the reaction is -939 kJ.

6.36 The gain of 66.4 kJ of heat per 2 mol NO_2 means that the reaction is endothermic. Because energy is gained by the system from the surroundings, q is positive, and because the reaction involves 2 mol NO_2, q for the reaction is +66.4 kJ.

6.37 The decomposition of 2 mol $KClO_3$ to KCl and O_2 must yield 2 mol KCl to balance the potassium and chlorine. To balance the oxygen, 3 mol O_2 must be written as a product:

$$2KClO_3(s) \rightarrow 2KCl(s) + 3O_2(g)$$

To write a thermochemical equation, the sign of ΔH must be negative because heat is evolved:

$$2KClO_3(s) \rightarrow 2KCl(s) + 3O_2(g); \ \Delta H = -44.7 \text{ kJ}$$

6.38 The reaction of Fe(s) with HCl must yield H_2 and $FeCl_3$. To balance the hydrogen, 2HCl must be written first as a reactant. However, to balance the chlorine, 3 x 2HCl must be written finally as a reactant and $2FeCl_3$ must be written as a product:

$$2Fe(s) + 6HCl(aq) \rightarrow 2FeCl_3(aq) + 3H_2(g)$$

To write a thermochemical equation, the sign of ΔH must be negative because heat is evolved:

$$2Fe(s) + 6HCl(aq) \rightarrow 2FeCl_3(aq) + 3H_2(g); \ \Delta H = 2 \times -87.9 \text{ kJ} = -175.\underline{8} \text{ kJ}$$

6.39 The first equation is

$$P_4(s) + 5O_2(g) \rightarrow P_4O_{10}(s); \ \Delta H = -2940 \text{ kJ}$$

The second equation is

$$P_4O_{10}(s) \rightarrow P_4(s) + 5O_2(g); \ \Delta H = ?$$

The second equation has been obtained by reversing the first equation. Therefore, to obtain ΔH for the second equation, ΔH for the first equation must be reversed in sign:

$$-(-2940) = +2940 \text{ kJ}$$

6.40 The first equation is

$$CS_2(l) + 3O_2(g) \rightarrow CO_2(g) + 2SO_2(g); \ \Delta H = -1{,}075 \text{ kJ}$$

(continued)

The second equation is

$$1/2CS_2(l) + 3/2O_2(g) \rightarrow 1/2CO_2(g) + SO_2(g); \quad \Delta H = ?$$

The second equation has been obtained by dividing each coefficient in the first equation by 2. Therefore, to obtain ΔH for the second equation, ΔH for the first equation must be divided by 2:

$$(-1{,}075) \div 2 = -53\underline{7}.5 = -538 \text{ kJ}$$

6.41 The first equation is

$$1/4P_4O_{10}(s) + 3/2H_2O(l) \rightarrow H_3PO_4(aq); \quad \Delta H = -113.2 \text{ kJ}$$

The second equation is

$$P_4O_{10}(s) + 6H_2O(l) \rightarrow 4H_3PO_4(aq); \quad \Delta H = ?$$

The second equation has been obtained from the first by multiplying each coefficient by 4. Therefore, to obtain the ΔH for the second equation, the ΔH for the first equation must be multiplied by 4: $-113.2 \times 4 = -452.8$ kJ

6.42 The first equation is

$$4NH_3(g) + 5O_2(g) \rightarrow 4NO(g) + 6H_2O(g); \quad \Delta H = -906 \text{ kJ}$$

The second equation is

$$NO(g) + 3/2H_2O(g) \rightarrow NH_3(g) + 5/4O_2(g); \quad \Delta H = ?$$

The second equation has been obtained by reversing the first equation and dividing each coefficient by 4. Therefore, to obtain ΔH for the second equation, the ΔH for the first equation must be reversed in sign and divided by 4: $+906 \div 4 = 22\underline{6}.50 = 227$ kJ.

6.43 Because hydrogen is written as H_2 in the equation, the molar mass of H_2 = 2.016 g per mol H_2. From the equation, 2 mol H_2 evolve 484 kJ heat. Divide this by the 2.016 g/mol H_2 to obtain the amount of heat evolved per gram of hydrogen:

$$\frac{-484 \text{ kJ}}{2 \text{ mol } H_2} \times \frac{1 \text{ mol } H_2}{2.016 \text{ g } H_2} = -12\underline{0}.03 = \frac{-120 \text{ kJ}}{\text{g } H_2}$$

6.44 Because nitric oxide is written as NO in the equation, the molar mass of NO = 30.01 g per mol NO. From the equation, 2 mol NO evolve 114 kJ heat. Divide the 114 kJ by the 30.01 g/mol NO to obtain the amount of heat evolved per gram of NO:

$$\frac{-114 \text{ kJ}}{2 \text{ mol NO}} \times \frac{1 \text{ mol NO}}{30.01 \text{ g NO}} = -1.8\underline{9}9 = \frac{-1.90 \text{ kJ}}{\text{g NO}}$$

6.45 The molar mass of ammonia is 17.03 g/mol. From the equation, 4 moles of NH_3 evolve 1267 kJ of heat. Divide 25.6 g NH_3 by its molar mass and the 4 moles of NH_3 in the equation to obtain the amount of heat evolved:

$$25.6 \text{ g } NH_3 \times \frac{1 \text{ mol } NH_3}{17.03 \text{ g } NH_3} \times \frac{-1267 \text{ kJ}}{1 \text{ mol } NH_3} = -47\underline{6}.1 = -476 \text{ kJ}$$

6.46 The molar mass of H_2S is 34.08 g/mol. From the equation, 2 moles of H_2S evolve 1037 kJ of heat. Divide 36.9 g H_2S by its molar mass and the 2 moles of H_2S in the equation to obtain the amount of heat evolved:

$$36.9 \text{ g } H_2S \times \frac{1 \text{ mol } H_2S}{34.09 \text{ g } H_2S} \times \frac{-1037 \text{ kJ}}{1 \text{ mol } H_2S} = -56\underline{1}.2 = -561 \text{ kJ}$$

6.47 The molar mass of C_3H_8 is 44.06 g/mol. From the equation, 1 mol C_3H_8 evolves 2044 kJ heat. Divide 255 kJ by 2044 kJ and multiply by the molar mass to obtain the mass needed:

$$255 \text{ kJ} \times \frac{1 \text{ mol } C_3H_8}{2044 \text{ kJ}} \times \frac{44.06 \text{ g } C_3H_8}{1 \text{ mol } C_3H_8} = 5.4\underline{9}6 = 5.50 \text{ g } C_3H_8$$

6.48 The molar mass of C_2H_5OH is 46.05 g/mol. From the equation, 1 mol C_2H_5OH evolves 1235 kJ heat. Divide 345 kJ by 1235 kJ and multiply by the molar mass to obtain the mass needed:

$$345 \text{ kJ} \times \frac{1 \text{ mol } C_2H_5OH}{1235 \text{ kJ}} \times \frac{46.05 \text{ g } C_2H_5OH}{1 \text{ mol } C_2H_5OH} = 12.\underline{8}6 = 12.9 \text{ g } C_2H_5OH$$

6.49 Multiply the 180 g (0.180 kg) of water by the specific heat of 4.18 J/(g•°C) and by Δt, to obtain heat in joules:

$$180 \text{ g} \times (96°\text{C} - 15°\text{C}) \times \frac{4.18 \text{ J}}{\text{g}•°\text{C}} = 6.\underline{0}94 \times 10^4 = 6.1 \times 10^4 \text{ J}$$

6.50 Multiply the 1.51×10^3 g (1.51 kg) iron by the specific heat of 0.450 J/(g•°C) and by Δt, to obtain heat in joules:

$$1.51 \times 10^3 \text{g} \times (178^\circ\text{C} - 21^\circ\text{C}) \times \frac{0.449 \text{ J}}{1 \text{ g}\bullet^\circ\text{C}} = 1.0\underline{6}4 \times 10^5 = 1.06 \times 10^5 \text{ J}$$

6.51 Use the 2.26×10^3 J/g (2.26 kJ/g) heat of vaporization to calculate the heat of condensation. Then use it to calculate Δt, the temperature change.

$$\text{Heat of condens.} = \frac{2.26 \times 10^3 \text{ J}}{1 \text{ g}} \times 124 \text{ g} = 2.8\underline{0}24 \times 10^5 \text{ J}$$

$$\text{Temp. change} = \Delta t = \frac{2.8024 \times 10^5 \text{ J}}{6.44 \times 10^4 \text{ g}} \times \frac{\text{g}\bullet^\circ\text{C}}{1.015 \text{ J}} = +4.2\underline{8}7 = +4.29^\circ\text{C}$$

6.52 Use the 334 J/g (0.334 kJ/g) heat of fusion to calculate the heat of melting. Then use the heat of melting to calculate Δt, the temperature change.

$$\text{Heat of melting} = \frac{334 \text{ J}}{1 \text{ g}} = 35.0 \text{ g} = 11,\underline{6}90 \text{ J}$$

Heat melting + heat warming = heat lost by 210 g H_2O

Let T = the final temperature, and substitute into the above equation:

$$11,690 \text{ J} + \left[\frac{4.18 \text{ J}}{1 \text{ g}\bullet^\circ\text{C}} \times 35 \text{ g} \times (\text{T} - 0^\circ\text{C})\right]$$

$$= \left[\frac{4.18 \text{ J}}{1 \text{ g}\bullet^\circ\text{C}} \times 210 \text{ g} \times (21.0^\circ\text{C} - \text{T})\right]$$

$$\frac{18,433.8 \text{ J} - 11,690 \text{ J}}{(146.3 + 877.8) \text{ J/}^\circ\text{C}} = 6.5\underline{8}509 = 6.59^\circ\text{C}$$

6.53 The enthalpy change for the reaction is equal in magnitude and opposite in sign to the heat-energy change occurring from the cooling of the solution and calorimeter.

$$q_{calorimeter} = (1071 \text{ J/}^\circ\text{C})(21.56^\circ\text{C} - 25.00^\circ\text{C}) = -36\underline{8}4.2 \text{ J}$$

Thus, 15.3 g $NaNO_3$ is equivalent to 3684.2 J heat energy. The amount of heat absorbed by 1.000 mol $NaNO_3$ is calculated from +3684.2 J (opposite sign):

(continued)

$$1.000 \text{ mol NaNO}_3 \times \frac{85.00 \text{ g NaNO}_3}{1 \text{ mol NaNO}_3} \times \frac{3684.2 \text{ J}}{15.3 \text{ g NaNO}_3} = 2.0\underline{4}6 \times 10^4 \text{ J}$$

Thus, the enthalpy change, ΔH, for the reaction is 2.05×10^4 J, or 20.5 kJ, per mol $NaNO_3$.

6.54 The enthalpy change for the reaction is equal in magnitude and opposite in sign to the heat energy produced from the warming of the solution and the calorimeter.

$$q_{calorimeter} = (1258 \text{ J/}^\circ\text{C})(38.7^\circ\text{C} - 25.0^\circ\text{C}) = +1.7\underline{2}3 \times 10^4 \text{ J}$$

Thus, 23.6 g $CaCl_2$ is equivalent to 3.141×10^4 J heat energy. The amount of heat released by 1.000 mol $CaCl_2$ is calculated from -3.141×10^4 J (opposite sign):

$$1.000 \text{ mol CaCl}_2 \times \frac{110.98 \text{ g CaCl}_2}{1 \text{ mol CaCl}_2} \times \frac{-1.723 \times 10^4 \text{ J}}{23.6 \text{ g CaCl}_2} = -8.1\underline{0}2 \times 10^4 \text{ J}$$

Thus, the enthalpy change, ΔH, for the reaction is -8.10×10^4 J, or -81.0 kJ, per mol $CaCl_2$.

6.55 The energy change for the reaction is equal in magnitude and opposite in sign to the heat energy produced from the warming of the solution and the calorimeter.

$$q_{calorimeter} = (9.63 \text{ kJ/}^\circ\text{C})(33.73 - 25.00^\circ\text{C}) = +84.\underline{0}6 \text{ kJ}$$

Thus, 2.84 g C_2H_5OH is equivalent to 84.06 kJ heat energy. The amount of heat released by 1.000 mol C_2H_5OH is calculated from -84.06 kJ (opposite sign):

$$1.000 \text{ mol C}_2\text{H}_5\text{OH} \times \frac{46.07 \text{ g C}_2\text{H}_5\text{OH}}{1 \text{ mol C}_2\text{H}_5\text{OH}} \times \frac{-84.06 \text{ kJ}}{2.84 \text{ g C}_2\text{H}_5\text{OH}} = -13\underline{6}3.6 \text{ kJ}$$

Thus, the enthalpy change, ΔH, for the reaction is -1.36×10^3 kJ/mol ethanol.

6.56 The energy change for the reaction is equal in magnitude and opposite in sign to the heat energy produced from the warming of the solution and the calorimeter.

$$q_{calorimeter} = (12.05 \text{ kJ/}^\circ\text{C})(37.18^\circ\text{C} - 25.00^\circ\text{C}) = +146.\underline{7}6 \text{ kJ}$$

Thus, 3.51 g C_6H_6 is equivalent to 146.76 kJ heat energy. The amount of heat released by 1.000 mol C_6H_6 is calculated from -146.76 kJ (opposite sign):

$$1.000 \text{ mol C}_6\text{H}_6 \times \frac{78.11 \text{ g C}_6\text{H}_6}{1 \text{ mol C}_6\text{H}_6} \times \frac{-146.76 \text{ kJ}}{3.51 \text{ g NaNO}_3} = -32\underline{6}5.9 \text{ kJ}$$

Thus, the enthalpy change, ΔH, for the reaction is -3.27×10^3 kJ/mol benzene.

6.57 Using the equations in the data, reverse the direction of the first reaction and reverse the sign of its ΔH. Then divide the second equation by 2, divide its ΔH by 2, and add. Setup:

$$H_2O(l) + 1/2O_2(g) \rightarrow H_2O_2(l); \quad \Delta H = (-98.0\ kJ) \times (-1)$$

$$H_2(g) + 1/2O_2(g) \rightarrow H_2O(l); \quad \Delta H = (-571.6\ kJ) \div (2)$$

$$H_2(g) + O_2(g) \rightarrow H_2O_2(l); \quad \Delta H = -187.8\ kJ$$

6.58 Using the equations in the data, reverse the direction of the first reaction and reverse the sign of its ΔH. Then multiply the second equation by 2, multiply its ΔH by 2, and add. Setup:

$$N_2(g) + 2H_2O(l) \rightarrow N_2H_4(l) + O_2(g); \quad \Delta H = (-622.2\ kJ) \times (-1)$$

$$2H_2(g) + O_2(g) \rightarrow 2H_2O(l); \quad \Delta H = (-285.8\ kJ) \times (2)$$

$$N_2(g) + 2H_2(g) \rightarrow N_2H_4(l); \quad \Delta H = 50.6\ kJ$$

6.59 Using the equations in the data, multiply the second equation by 2 and reverse its direction; do the same to its ΔH. Then multiply the first equation by 2 and its ΔH by 2. Finally, multiply the third equation by 3 and its ΔH by 3. Then add. Setup:

$$4NH_3(g) \rightarrow 2N_2(g) + 6H_2(g); \quad \Delta H = (-91.8\ kJ) \times (-2)$$

$$2N_2(g) + 2O_2(g) \rightarrow 4NO(g); \quad \Delta H = (180.6\ kJ) \times (2)$$

$$6H_2(g) + 3O_2(g) \rightarrow 6H_2O(g); \quad \Delta H = (-483.7\ kJ) \times (3)$$

$$4NH_3(g) + 5O_2(g) \rightarrow 4NO(g) + 6H_2O(g); \quad \Delta H = -906.3\ kJ$$

6.60 Using the equations in the data, reverse the direction of the second equation and reverse the sign of its ΔH. Then reverse the direction of the first equation and multiply by 1/2; do the same to its ΔH. Finally, multiply the third equation and its ΔH by 1/2. Add all three equations. Setup:

$$CH_4(g) \rightarrow C(graph) + 2H_2(g); \quad \Delta H = (-74.9\ kJ) \times (-1)$$

$$NH_3(g) \rightarrow 1/2N_2(g) + 3/2H_2(g); \quad \Delta H = (-91.8\ kJ) \times (-1/2)$$

$$1/2H_2(g) + C(graph) + 1/2N_2(g) \rightarrow HCN(g); \quad \Delta H = (270.3\ kJ) \times (1/2)$$

$$CH_4(g) + NH_3(g) \rightarrow HCN(g) + 3H_2(g); \quad \Delta H = 255.\underline{9}5 = 256.0\ kJ$$

6.61 After reversing the first equation in the data, add all the equations. Setup:

$$C_2H_4(g) + 3O_2(g) \rightarrow 2CO_2(g) + 2H_2O(l); \quad \Delta H = (-1401\ kJ)$$
$$2CO_2(g) + 3H_2O(l) \rightarrow C_2H_6(g) + 7/2O_2(g); \quad \Delta H = (-1550) \times (-1)$$
$$H_2(g) + 1/2O_2(g) \rightarrow H_2O(l); \quad \Delta H = (-286\ kJ)$$

$$C_2H_4(g) + H_2(g) \rightarrow C_2H_6(g); \quad \Delta H = -137\ kJ$$

6.62 After reversing the first equation, double the second and third equations, and then add. Setup:

$$2CO_2(g) + 2H_2O(l) \rightarrow CH_3COOH(l) + 2O_2(g); \quad \Delta H = (-871\ kJ) \times (-1)$$
$$2C(graph) + 2O_2(g) \rightarrow 2CO_2(g); \quad \Delta H = (-394\ kJ) \times (2)$$
$$2H_2(g) + O_2(g) \rightarrow 2H_2O(l); \quad \Delta H = (-286\ kJ) \times (2)$$

$$2C(graph) + 2H_2(g) + O_2(g) \rightarrow CH_3COOH(l); \quad \Delta H = -489\ kJ$$

6.63 Write the ΔH^o values (Table 6.2) underneath each compound in the balanced equation:

$$CCl_4(l) \rightarrow CCl_4(g)$$
$$-139 \qquad -96.0\ (kJ)$$

$$\Delta H^o_{vap} = [\Delta H^o_f(CCl_4)] - [\Delta H^o_f(CCl_4)] = [-96] - [-139]\ kJ = +43\ kJ$$

6.64 Write the ΔH^o values (Appendix C) underneath each compound in the balanced equation:

$$C_2H_5OH(l) \rightarrow C_2H_5OH(g)$$
$$-277.6 \qquad -235.4\ (kJ)$$

$$\Delta H^o_{vap} = [\Delta H^o_f(C_2H_5OH)] - [\Delta H^o_f(C_2H_5OH)] = [-235.4] - [-277.6] = +42.2\ kJ$$

6.65 Write the ΔH^o values (Table 6.2) underneath each compound in the balanced equation:

$$2H_2S(g) + 3O_2(g) \rightarrow 2H_2O(l) + 2SO_2(g)$$
$$2(-20) \qquad 3(0) \qquad 2(-285.8) \qquad 2(-296.8)\ (kJ)$$

$$\Delta H^o = \Sigma n\ \Delta H^o(products) - \Sigma m\ \Delta H^o(reactants)$$
$$= [2(-285.8) + 2(-296.8)] - [2(-20) + 3(0)]\ kJ = -112\underline{5}.2 = -1125\ kJ$$

6.66 Write the ΔH^o values (Table 6.2) underneath each compound in the balanced equation:

$CS_2(l)$	+	$3O_2(g)$	$\rightarrow$	$CO_2(g)$	+	$2SO_2(g)$	
87.9		3(0)		-393.5		2(-296.8)	(kJ)

$$\Delta H^o = \Sigma n\ \Delta H^o(\text{products}) - \Sigma m\ \Delta H^o(\text{reactants})$$
$$= [(-393.5) + 2(-296.8)] - [(87.9) + 3(0)] = -1075.0 \text{ kJ}$$

6.67 Write the ΔH^o values (Appendix C) underneath each compound in the balanced equation:

$2PbS(s)$	+	$3O_2(g)$	$\rightarrow$	$2SO_2(g)$	+	$2PbO(s)$	
2(-98.3)		3(0)		2(-296.8)		2(-219.0)	(kJ)

$$\Delta H^o = \Sigma n\ \Delta H^o(\text{products}) - \Sigma m\ \Delta H^o(\text{reactants})$$
$$= [2(-296.8) + 2(-219)] - [2(-98.3) + 3(0)] = -835.0 \text{ kJ}$$

6.68 Write the ΔH^o values (Appendix C) underneath each compound in the balanced equation:

$Fe_2O_3(s)$	+	$3CO(g)$	$\rightarrow$	$2Fe(s)$	+	$3CO_2(g)$	
-825.5		3(-110.5)		2(0)		3(-393.5)	(kJ)

$$\Delta H^o = \Sigma n\ \Delta H^o(\text{products}) - \Sigma m\ \Delta H^o(\text{reactants})$$
$$= [2(0) + 3(-393.5)] - [(-825.5) + 3(-110.5)] = -23.5 \text{ kJ}$$

6.69 Write the ΔH^o values (Table 6.2) underneath each compound in the balanced equation:

$HCl(g)$	$\rightarrow$	$H^+(aq)$	+	$Cl^-(aq)$	
-92.3		0		-167.5	(kJ)

$$\Delta H^o = \Sigma n\ \Delta H^o(\text{products}) - \Sigma m\ \Delta H^o(\text{reactant})$$
$$= [(0) + (-167.5)] - [-92.3] = -75.2 \text{ kJ}$$

6.70 Write the ΔH^o values (Table 6.2) underneath each compound in the balanced equation:

$CaCO_3(s)$ +	$CO_2(g)$ +	$H_2O(l)$ →		$Ca^{2+}(aq)$ +	$2HCO_3^-(aq)$
-1206.9	-393.5	-285.8		-543.0	2(-691.1) (kJ)

$$\Delta H^o = \Sigma n\ \Delta H^o(\text{products}) - \Sigma m\ \Delta H^o(\text{reactants})$$

$$= [(-543.0) + 2(-691.1)] - [(-1206.9) + (-393.5) + (-285.8)] = -39.0 \text{ kJ}$$

6.71 Calculate the molar heat of formation from the equation with the ΔH_f^o values below each substance; then convert to the heat for 10.0 g of $MgCO_3$ using the molar mass of 84.3.

$MgCO_3(s)$	→	$MgO(s)$ +	$CO_2(g)$
-1112 kJ		-601.2 kJ	-393.5 kJ

$$\Delta H^o = \Sigma n\ \Delta H^o(\text{products}) - \Sigma m\ \Delta H^o(\text{reactants})$$

$$\Delta H^o = -601.2 \text{ kJ} + (-393.5 \text{ kJ}) - (-1112 \text{ kJ}) = 11\underline{7}.3 \text{ kJ}$$

$$\text{Heat} = 10.0 \text{ g} \times \frac{1 \text{ mol}}{84.3 \text{ g}} \times \frac{117.3 \text{ kJ}}{\text{mol}} = 13.\underline{9}1 = 13.9 \text{ kJ}$$

6.72 Calculate the molar heat of formation from the equation with the ΔH_f^o values below each substance; then convert to the heat for 10.0 g of $BaCO_3$ using the molar mass of 197.3.

$BaCO_3(s)$	→	$BaO(s)$ +	$CO_2(g)$
-1219 kJ		-548.1 kJ	-393.5 kJ

$$\Delta H^o = \Sigma n\ \Delta H^o(\text{products}) - \Sigma m\ \Delta H^o(\text{reactants})$$

$$\Delta H^o = -548.1 \text{ kJ} + (-393.5 \text{ kJ}) - (-1219 \text{ kJ}) = 27\underline{7}.4 \text{ kJ}$$

$$\text{Heat} = 10.0 \text{ g} \times \frac{1 \text{ mol}}{197.3 \text{ g}} \times \frac{277.4 \text{ kJ}}{\text{mol}} = 14.\underline{0}59 = 14.1 \text{ kJ}$$

■ Solutions to Unclassified Problems

6.73 Using Table 1.5 and 4.184 J/cal, convert the 686 Btu/lb to J/g:

$$\frac{686\ \text{Btu}}{1\ \text{lb}} \times \frac{252\ \text{cal}}{1\ \text{Btu}} \times \frac{4.184\ \text{J}}{1\ \text{cal}} \times \frac{1\ \text{lb}}{0.4536\ \text{kg}} \times \frac{1\ \text{kg}}{10^3\ \text{g}}$$

$$= 1.5\underline{9}4 \times 10^3 = 1.59 \times 10^3\ \text{J/g}$$

6.74 Using Table 1.5 and 4.184 J/cal, convert the 51600 Btu/lb to J/g:

$$\frac{51600\ \text{Btu}}{1\ \text{lb}} \times \frac{252\ \text{cal}}{1\ \text{Btu}} \times \frac{4.184\ \text{J}}{1\ \text{cal}} \times \frac{1\ \text{lb}}{0.4536\ \text{kg}} \times \frac{1\ \text{kg}}{10^3\ \text{g}}$$

$$= 1.1\underline{9}9 \times 10^5 = 1.20 \times 10^5\ \text{J/g}$$

6.75 Substitute into the equation $E_p = mgh$, and convert to SI units.

$$E_p = 1.00\ \text{lb} \times \frac{9.807\ \text{m}}{\text{s}^2} \times 167\ \text{ft} \times \frac{0.4536\ \text{kg}}{1\ \text{lb}} \times \frac{0.9144\ \text{m}}{3\ \text{ft}}$$

$$= \frac{22\underline{6}.433\ \text{kg} \bullet \text{m}^2}{\text{s}^2} = 226\ \text{J}$$

At the bottom, all the potential energy is converted to kinetic energy, so $E_k = 225.8$ kg•m^2/s^2. Because $E_k = 1/2mv^2$, solve for v, the speed (velocity):

$$\text{Speed} = \sqrt{\frac{E_k}{1/2 \times m}} = \sqrt{\frac{226.433\ \text{kg} \bullet \text{m}^2/\text{s}^2}{1/2 \times 1.00\ \text{lb} \times 0.4536\ \text{kg/lb}}} = 31.\underline{6}0 = 31.6\ \text{m/s}$$

6.76 Substitute into the equation $E_k = 1/2mv^2$, and convert to SI units.

$$E_k = 1/2 \times 2354\ \text{lb} \times \frac{0.4536\ \text{kg}}{1\ \text{lb}} \times \left[\frac{11.2\ \text{km}}{\text{s}} \times \frac{10^3\ \text{m}}{1\ \text{km}}\right]^2$$

$$= 6.6\underline{9}7 \times 10^{10}\ \text{kg} \bullet \text{m}^2/\text{s}^2 = 6.70 \times 10^{10}\ \text{J}$$

6.77 The equation is

$$CaCO_3(s) \rightarrow CaO(s) + CO_2(g); \quad \Delta H = 178.3 \text{ kJ}$$

Use the molar mass of 100.08 g/mol to convert the heat per mol to heat per 12.0 g.

$$12.0 \text{ g } CaCO_3 \times \frac{1 \text{ mol } CaCO_3}{100.08 \text{ g } CaCO_3} \times \frac{178.3 \text{ kJ}}{1 \text{ mol } CaCO_3}$$

$$= 21.\underline{3}8 \text{ kJ (21.4 kJ absorbed by 12.0 g)}$$

6.78 The equation is

$$CaO(s) + H_2O(l) \rightarrow Ca(OH)_2(s); \quad \Delta H = -65.2 \text{ kJ}$$

Use the molar mass in the conversion.

$$28.4 \text{ g CaO} \times \frac{1 \text{ mol CaO}}{56.08 \text{ g CaO}} \times \frac{-65.2 \text{ kJ}}{1 \text{ mol CaO}}$$

$$= -33.\underline{0}1 \text{ kJ (33.0 kJ released per 28.4 g)}$$

6.79 The equation is

$$2HCHO_2(l) + O_2(g) \rightarrow 2CO_2(g) + 2H_2O(l)$$

Use the molar mass of 46.03 g/mol to convert -30.3 kJ/5.48 g to ΔH per mol of acid.

$$\frac{-30.3 \text{ kJ}}{5.48 \text{ g } HCHO_2} \times \frac{46.03 \text{ g } HCHO_2}{1 \text{ mol } HCHO_2} = -25\underline{4}.508 = -255 \text{ kJ/mol}$$

6.80 The equation is

$$HC_2H_3O_2(l) + 2O_2(g) \rightarrow 2CO_2(g) + 2H_2O(l)$$

Use the molar mass of 60.05 g/mol to convert -52.0 kJ/3.58 g to ΔH per mol of acid.

$$\frac{-52.0 \text{ kJ}}{3.58 \text{ g } HC_2H_3O_2} \times \frac{60.05 \text{ g } HC_2H_3O_2}{1 \text{ mol } HC_2H_3O_2} = -87\underline{2}.2 = -872 \text{ kJ/mol}$$

6.81 Divide the 235 J heat by the mass of lead and the Δt to obtain the specific heat.

$$\text{Specific heat} = \frac{235 \text{ J}}{121.6 \text{ g } (35.5°C - 20.4°C)} = 0.12\underline{7}98 = \frac{0.128 \text{ J}}{(g \cdot °C)}$$

6.82 Divide the 47.0 J heat by the mass of copper and the Δt to obtain the specific heat.

$$\text{Specific heat} = \frac{47.0\ \text{J}}{35.4\ \text{g}\ (3.45°\text{C})} = 0.38\underline{4}8 = \frac{0.385\ \text{J}}{(\text{g}\bullet°\text{C})}$$

6.83 The energy used to heat the Zn comes from cooling the water. Calculate q for water:

$$q_{wat} = \text{specific heat x mass x } \Delta t$$

$$q_{wat} = \frac{4.18\ \text{J}}{\text{g}\bullet°\text{C}} \times 50.0\ \text{g} \times (96.68°\text{C} - 100.00°\text{C}) = -69\underline{3}.88\ \text{J}$$

The sign of q for the Zn is the reverse of the sign of q for water because the Zn is absorbing heat:

$$q_{met} = -(q_{wat}) = -(-693.88) = 69\underline{3}.88\ \text{J}$$

$$\text{Sp. heat} = \frac{693.88\ \text{J}}{(25.3\ \text{g})(96.68 - 25.00)\ °\text{C}} = 0.38\underline{2}6 = \frac{0.383\ \text{J}}{(\text{g}\bullet°\text{C})}$$

6.84 The energy given up by the metal is used to heat the water. Calculate q for water:

$$q_{wat} = \text{specific heat x mass x } \Delta t$$

$$q_{wat} = \frac{4.18\ \text{J}}{\text{g}\bullet°\text{C}} \times 26.7\ \text{g} \times (30.00°\text{C} - 25.00°\text{C}) = 55\underline{8}.03\ \text{J}$$

The sign of q for the metal is the reverse of the sign of q for water because the metal is giving up heat:

$$q_{met} = -(q_{wat}) = -55\underline{8}.03\ \text{J}$$

$$\text{Sp. heat} = \frac{-558.03\ \text{J}}{(19.6\ \text{g})(30.00 - 61.67)°\text{C}} = 0.89\underline{8}9 = \frac{0.899\ \text{J}}{(\text{g}\bullet°\text{C})}$$

6.85 Use Δt and the heat capacity of 547 J/°C to calculate q:

$$q = C\,\Delta t = (547\ \text{J/°C})(36.66 - 25.00)°\text{C} = 6.3\underline{7}8 \times 10^3\ \text{J}\ (6.3\underline{7}8\ \text{kJ})$$

Energy is released in the solution process in raising the temperature, so ΔH is negative:

$$\Delta H = \frac{-6.378\ \text{kJ}}{6.48\ \text{g LiOH}} \times \frac{23.95\ \text{g LiOH}}{1\ \text{mol LiOH}} = -23.\underline{5}7 = -23.6\ \text{kJ/mol}$$

6.86 Use Δt and the heat capacity of 682 J/^oC to calculate q:

$$q = C\,\Delta t = (682\ J/^oC)(14.14 - 25.00)^oC = -7.4\underline{0}65 \times 10^3\ J\ (-7.4\underline{0}65\ kJ)$$

Energy is absorbed in the solution process in lowering the temperature, so the sign of ΔH must be reversed, making the heat positive:

$$\Delta H = \frac{7.4065\ kJ}{21.45\ g\ KNO_3} \times \frac{101.1\ g\ KNO_3}{1\ mol\ KNO_3} = 34.\underline{9}1 = 34.9\ kJ/mol$$

6.87 Use Δt and the heat capacity of 13.43 kJ/^oC to calculate q:

$$q = C\,\Delta t = (13.43\ kJ/^oC)(35.81 - 25.00)^oC = 145.\underline{1}7\ kJ$$

As in the previous two problems, the sign of ΔH must be reversed, making the heat negative:

$$\Delta H = \frac{-145.17\ kJ}{10.00\ g\ HC_2H_3O_2} \times \frac{60.05\ g\ HC_2H_3O_2}{1\ mol\ HC_2H_3O_2} = -871.\underline{7}4 = -871.7\ kJ/mol$$

6.88 Use Δt and the heat capacity of 15.8 kJ/^oC to calculate q:

$$q = C\,\Delta t = (15.8\ kJ/^oC)(20.54 - 20.00)^oC = 8.\underline{5}3\ kJ$$

As in the previous three problems, the sign of ΔH must be reversed, making the heat negative:

$$\Delta H = \frac{-8.53\ kJ}{0.548\ g\ sugar} \times \frac{150.1\ g\ sugar}{1\ mol\ sugar} = -2.\underline{3}36 \times 10^3 = -2.3 \times 10^3\ kJ/mol$$

6.89 Using the equations in the data, reverse the direction of the first reaction, and reverse the sign of its ΔH. Then add the second and third equations and their ΔH's.

$H_2O(g)$	$+\ SO_2(g)$	$\rightarrow$	$H_2S(g) + 3/2O_2(g)$;	$\Delta H = (-519\ kJ) \times (-1)$
$H_2(g)$	$+\ 1/2O_2(g)$	$\rightarrow$	$H_2O(g)$;	$\Delta H = (-242\ kJ)$
$1/8S_8(rh.)$	$+\ O_2(g)$	$\rightarrow$	$SO_2(g)$;	$\Delta H = (-297\ kJ)$
$H_2(g)$	$+\ 1/8S_8(rh.)$	$\rightarrow$	$H_2S(g)$;	$\Delta H = -20\ kJ$

6.90 Using the equations in the data, reverse the direction of the reaction involving oxidation of the glycol, and reverse the sign of its ΔH. Then add half of the reaction involving oxidation of C_2H_4O, and half of its ΔH.

$$2CO_2(g) + 3H_2O(l) \rightarrow HOC_2H_4OH(l) + 5/2O_2(g); \quad \Delta H = (-1189.8 \text{ kJ}) \times (-1)$$

$$C_2H_4O(g) + 5/2O_2(g) \rightarrow 2CO_{2(g)} + 2H_2O(l); \quad \Delta H = (-2612.2 \text{ kJ}) \times (1/2)$$

$$C_2H_4O(g) + H_2O(l) \rightarrow HOC_2H_4OH(l); \quad \Delta H = -116.3 \text{ kJ}$$

6.91 Write the ΔH^o values (Table 6.2) underneath each compound in the balanced equation.

$CH_4(g)$ +	$H_2O(g)$ →	$CO(g)$ +	$3H_2(g)$
-74.9	-241.8	-110.5	3(0) (kJ)

$$\Delta H^o = [-110.5 + 3(0)] - [(-74.9) + (-241.8)] = 206.2 \text{ kJ}$$

6.92 Write the ΔH^o values (Table 6.2) underneath each compound in the balanced equation.

$2CH_4(g)$ +	$O_2(g)$ →	$2CO(g)$ +	$4H_2(g)$
2(-74.9)	0	2(-110.5)	4(0) (kJ)

$$\Delta H^o = [2(-110.5) + 4(0)] - [2(-74.9) + 0] = -71.2 \text{ kJ}$$

6.93 Write the ΔH^o values (Table 6.2) underneath each compound in the balanced equation. The ΔH_f^o of -635 kJ/mol CaO is given in the problem.

$CaCO_3(s)$ →	$CaO(s)$ +	$CO_2(g)$
-1206.9	-635	-393.5 (kJ)

$$\Delta H^o = [(-635) + (-393.5)] - [-1206.9] = 17\underline{8}.4 = 178 \text{ kJ}$$

6.94 Write the ΔH^o values (Table 6.2) underneath each compound in the balanced equation.

$2NaHCO_3(s)$ →	$Na_2CO_3(s)$ +	$H_2O(g)$ +	$CO_2(g)$
2(-947.7)	-1130.8	-241.8	-393.5 (kJ)

$$\Delta H^o = [(-1130.8) + (-241.8) + (-393.5)] - [2(-947.7)] = 129.3 \text{ kJ}$$

6.95 Calculate the molar heat of reaction from the equation with the ΔH_f° values below each substance; then convert to the heat for the reaction at 25°C.

$$2\,H_2(g) + O_2(g) \rightarrow 2\,H_2O(l)$$

0.0 kJ 0.0 kJ -285.84 kJ

$$\Delta H^{\circ} = 2 \times (-285.84\text{ kJ}) - 0 - 0 = -571.68\text{ kJ}$$

The moles of oxygen in 2.000 L with a density of 1.11 g/L is:

$$2.000\text{ L } O_2 \times \frac{1.11\text{ g } O_2}{\text{L } O_2} \times \frac{1\text{ mol } O_2}{32.0\text{ g } O_2} = 0.069\underline{3}7\text{ mol } O_2$$

The heat of reaction from 0.06937 mol O_2 is:

$$0.069\underline{3}7\text{ mol } O_2 \times \frac{-571.68\text{ kJ}}{\text{mol } O_2} = -39.\underline{6}57 = -39.7\text{ kJ}$$

6.96 Calculate the molar heat of reaction from the equation with the ΔH_f° values below each substance; then convert to the heat for the reaction at 25°C.

$$Cl_2(g) + 2\,Na(s) \rightarrow 2\,NaCl(s)$$

0.0 kJ 0.0 kJ - 411.1 kJ

$$\Delta H^{\circ} = 2 \times (-411.1\text{ kJ}) - 0 - 0 = -822.2\text{ kJ}$$

The moles of chlorine in 4.000 L with a density of 2.46 g/L is:

$$4.000\text{ L } Cl_2 \times \frac{2.46\text{ g } Cl_2}{\text{L } Cl_2} \times \frac{1\text{ mol } Cl_2}{70.9\text{ g } Cl_2} = 0.13\underline{8}7\text{ mol } Cl_2$$

The heat of reaction from 0.1387 mol Cl_2 is:

$$0.13\underline{8}7\text{ mol } Cl_2 \times \frac{-822.2\text{ kJ}}{\text{mol } Cl_2} = -11\underline{4}.03 = -114\text{ kJ}$$

6.97 Let ΔH_f^o = the unknown standard enthalpy of formation for sucrose. Then write this symbol and the other ΔH_f^o values underneath each compound in the balanced equation, and solve for ΔH_f^o, using the ΔH^o of -5641 kJ for the reaction.

$C_{12}H_{22}O_{11}(s)$ +	$12O_2(g)$ →	$12CO_2(g)$ +	$11H_2O(l)$	
ΔH_f^o	12(0)	12(-393.5)	11(-285.8)	(kJ)

ΔH^o = -5641 = [12(-393.5) + 11(-285.8)] - [ΔH_f^o + 12(0)]

ΔH_f^o = -2225 kJ/mol sucrose

6.98 Let ΔH_f^o = the unknown standard enthalpy of formation for acetone. Then write this symbol and the other ΔH_f^o values underneath each compound in the balanced equation, and solve for ΔH_f^o, using the ΔH^o of -1791 kJ for the reaction.

$CH_3COCH_3(l)$ +	$4O_2(g)$ →	$3CO_2(g)$ +	$3H_2O(l)$	
ΔH_f^o	4(0)	3(-393.5)	3(-285.8)	(kJ)

ΔH^o = -1791 = [3(-393.5) + 3(-285.8)] - [ΔH_f^o + 4(0)]

ΔH_f^o = $-24\underline{6}.9$ = -247 kJ/mol acetone

6.99 First calculate the heat evolved from the molar amounts represented by the balanced equation. From Appendix C, obtain the individual heats of formation, and write those below the reactants and products in the balanced equation. Then multiply the molar heats of formation by the number of moles in the balanced equation, and write those products below the molar heats of formation.

$2Al(s)$ +	$3NH_4NO_3(s)$ →	$3N_2(g)$ +	$6H_2O(g)$ +	$Al_2O_3(s)$	
0.0 kJ	-365.6 kJ	0.0 kJ	-241.826	-1676 kJ	(kJ/mol)
0.0 kJ	-1096.8 kJ	0.0 kJ	-1450.96	-1676 kJ	(kJ/eqn.)

The total heat of reaction of 2 moles of Al and 3 moles of NH_4NO_3 is

ΔH = -1450.96 - 1676 - (-1096.8 kJ) = $-203\underline{0}.16$ kJ

Now 245 kJ represents the following fraction of the total heat of reaction:

$$\frac{245 \text{ kJ}}{2030.16 \text{ kJ}} = 0.12\underline{0}68$$

(continued)

Thus, 245 kJ requires the fraction 0.12068 of the moles of each reactant: 0.12068 x 2, or 0.24136, mol of Al and 0.12068 x 3, or 0.36204, mol of $(NH_4)NO_3$. The mass of each reactant and the mass of the mixture are as follows:

$$0.24136 \text{ mol Al} \times 26.98 \text{ g/mol Al} = 6.5\underline{1}19 \text{ g of Al}$$

$$0.36204 \text{ mol } (NH_4)NO_3 \times 80.04 \text{ g/mol } (NH_4)NO_3 = 28.\underline{9}776 \text{ g } (NH_4)NO_3$$

$$6.5119 \text{ g Al} + 28.9776 \text{ g } (NH_4)NO_3 = 35.\underline{4}8 = 35.5 \text{ g of the mixture}$$

6.100 First calculate the heat evolved from the molar amounts represented by the balanced equation. From Appendix C, obtain the individual heats of formation, and write those below the reactants and products in the balanced equation. Then multiply the molar heats of formation by the number of moles in the balanced equation, and write those products below the molar heats of formation.

$2Al(s)$ +	$Fe_2O_3(s)$ →	$2Fe(l)$ +	$Al_2O_3(s)$	
0.0 kJ	-825.5 kJ	+13.13 kJ	-1676 kJ	(kJ/mol)
0.0 kJ	-825.5 kJ	+26.26 kJ	-1676 kJ	(kJ/eqn.)

The total heat of reaction of 2 moles of Al and 1 mole of Fe_2O_3 is

$$\Delta H = +26.26 - 1676 - (-825.5 \text{ kJ}) = -824.24 \text{ kJ}$$

Now 348 kJ represents the following fraction of the total heat of reaction:

$$\frac{348 \text{ kJ}}{824.24 \text{ kJ}} = 0.42\underline{2}2$$

Thus, 348 kJ requires the fraction 0.4222 of the moles of each reactant: 0.4222 x 2, or 0.8444, mol of Al and 0.4222 x 1, or 0.4222, mol of Fe_2O_3. The mass of each reactant and the mass of the mixture are as follows:

$$0.8444 \text{ mol Al} \times 26.98 \text{ g/mol Al} = 22.\underline{7}82 \text{ g Al}$$

$$0.4222 \text{ mol } Fe_2O_3 \times 159.7 \text{ g/mol } Fe_2O_3 = 67.\underline{4}25 \text{ g } Fe_2O_3$$

$$22.782 \text{ g Al} + 67.425 \text{ g } Fe_2O_3 = 90.\underline{2}07 = 90.2 \text{ g of the mixture}$$

■ Cumulative-Skills Problems
(require skills from previous chapters)

6.101 The heat lost, q, by the water(s) with a temperature higher than the final temperature must be equal to the heat gained, q, by the water(s) with a temperature lower than the final temperature.

q(lost by water at higher temp) = q(gained by water at lower temp)

Σ(s x m x Δt) = Σ(s x m x Δt)

Divide both sides of the equation by the specific heat, s, to eliminate this term, and substitute the other values. Since the mass of the water at 50.0°C is greater than the sum of the masses of the other two waters, assume that the final temperature will be greater than 37°C and greater than 15°C. Use this to set up the three Δt expressions, and write one equation in one unknown, letting t equal the final temperature. Simplify by omitting the "grams" from 45.0 g, 25.0 g, and 15.0 g.

Σ(m x Δt) = Σ(m x Δt)

45.0 x (50.0°C - t) = 25.0 x (t - 15.0 °C) + 15.0 x (t - 37°C)

2250°C - 45.0 t = 25.0 t - 375°C + 15.0 t - 555.0°C

-85.0 t = -3180°C

t = 37.$\underline{4}$1 = 37.4°C (the final temperature)

6.102 The heat lost, q, by the substances(s) with a temperature higher than the final temperature must be equal to the heat gained, q, by the substances(s) with a temperature lower than the final temperature.

q(lost) = q(gained)

Σ(s x m x Δt) = Σ(s x m x Δt)

Since the temperature of the iron, 95.0°C is far larger than the temperatures of the other two substances, assume that the final temperature will be greater than 35.5°C and greater than 25°C. Then write the heat term for the iron on the left and the sum of the heat terms for water and ethanol on the right.

(continued)

Since the masses, m, are equal, divide both sides by m to eliminate this term, and solve one equation in one unknown, letting t equal the final temperature. (To simplify the setup, only the temperature unit is retained.)

$$\Sigma(s \times \Delta t) = \Sigma(s \times \Delta t)$$

$$0.449 \times (95.0^{\circ}C - t) = 2.43 \times (t - 35.5^{\circ}C) + 4.18 \times (t - 25.0^{\circ}C)$$

$$42.655^{\circ}C - 0.449\,t = 2.43\,t - 86.265^{\circ}C + 4.18\,t - 104.5^{\circ}C$$

$$-7.059\,t = -233.42^{\circ}C$$

$$t = 33.\underline{0}6 = 33.1^{\circ}C \quad \text{(the final temperature)}$$

The final temperature is actually less than that of the ethanol, but rewriting the equation to account for this still gives the same final temperature.

6.103 First calculate the mole fraction of each gas in the product, assuming 100 g product:

$$\text{Mol CO} = 33 \text{ g CO} \times 1 \text{ mol CO}/28.01 \text{ g CO} = 1.\underline{1}78 \text{ mol CO}$$

$$\text{Mol } CO_2 = 67 \text{ g } CO_2 \times 1 \text{ mol } CO_2/44.01 \text{ g } CO_2 = 1.\underline{5}22 \text{ mol } CO_2$$

$$\text{Mol frac. CO} = \frac{1.178 \text{ mol CO}}{(1.178 + 1.522) \text{ mol}} = 0.4\underline{3}63$$

$$\text{Mol frac. } CO_2 = \frac{1.522 \text{ mol } CO_2}{(1.178 + 1.522) \text{ mol}} = 0.5\underline{6}37$$

Now calculate the starting moles of C, which equals the total moles of CO and CO_2:

$$\text{Starting mol C(s)} = 1.00 \text{ g C} \times 1 \text{ mol C}/12.01 \text{ g C}$$

$$= 0.083\underline{2}6 \text{ mol C} = \text{mol CO} + CO_2$$

Use the mol fractions to convert mol CO + CO_2 to mol CO and mol CO_2:

$$\text{Mol CO} = \frac{0.4363 \text{ mol CO}}{1 \text{ mol total}} \times 0.08326 \text{ mol total} = 0.03\underline{6}33 \text{ mol CO}$$

$$\text{Mol } CO_2 = \frac{0.5637 \text{ mol } CO_2}{1 \text{ mol total}} \times 0.08326 \text{ mol total} = 0.04\underline{6}93 \text{ mol } CO_2$$

Now use enthalpies of formation (Table 6.2) to calculate the heat of combustion for both:

(continued)

C(s) +	$1/2O_2(g)$	→	CO(g)
0.03633	excess		0.03633 (mol)
0	0		-110.5 (kJ/mol)
0	0		-4.014 (kJ/0.03633 mol)

C(s) +	$O_2(g)$	→	$CO_2(g)$
0.04693	excess		0.04693 (mol)
0	0		-393.5 (kJ/mol)
0	0		-18.47 (kJ/0.04693 mol)

Total $\Delta H = -4.014 + (-18.47) = -2\underline{2}.48 = -22$ kJ

Heat released = 22 kJ

6.104 The balanced equations are

$CH_4(g) + 2O_2(g) \rightarrow CO_2(g) + 2H_2O(l)$

$C_2H_6(g) + 3.5O_2(g) \rightarrow 2CO_2(g) + 3H_2O(l)$

First calculate the moles of CH_4 and C_2H_6 (let 80.0% = 0.800 g, and 20.0% = 0.200 g).

$$\text{Mol } CH_4 = 0.800 \text{ g } CH_4 \times \frac{1 \text{ mol } CH_4}{16.04 \text{ g } CH_4} = 0.049\underline{8}7 \text{ mol } CH_4$$

$$\text{Mol } C_2H_6 = 0.200 \text{ g } C_2H_6 \times \frac{1 \text{ mol } C_2H_6}{30.07 \text{ g } C_2H_6} = 0.0066\underline{5}1 \text{ mol } C_2H_6$$

Now use the enthalpies of formation (Table 6.2) to calculate the total heat:

$CH_4(g)$ +	$2O_2(g)$	→	$CO_2(g)$	+	$2H_2O(l)$
0.04987	excess		0.04987		2(0.04987) (mol)
-74.9	0		-393.5		-285.8 (kJ/mol)
$-3.7\underline{3}5$	0		$-19.\underline{6}2$		$-28.\underline{5}1$ (kJ/eqn.)

$C_2H_6(g)$ +	$3.5O_2$	→	$2CO_2(g)$	+	$3H_2O(l)$
0.006651	excess		2(0.006651)		3(0.006651) (mol)
-84.7	0		-393.5		-285.8 (kJ/mol)
$-0.56\underline{3}3$	0		$-5.2\underline{3}4$		$-5.7\underline{0}3$ (kJ/eqn.)

For the combustion of CH_4:

$\Delta H = [(-28.51) + (-19.62)] - [(-3.735) + 0] = -44.\underline{3}95$ kJ

(continued)

For the combustion of C_2H_6:

$$\Delta H = [(-5.234) + (-5.703)] - [(-0.5633) + 0] = -10.3\underline{7}4 \text{ kJ}$$

For the combustion of the total (1.00 g) of CH_4 and C_2H_6:

$$\Delta H = -44.395 + (-10.374) = -54.\underline{7}69 = -54.8 \text{ kJ}$$

Heat evolved = 54.8 kJ

6.105 The ΔH° for the reactions to produce CO and CO_2 in each case is equal to the ΔH_f° of the respective gases. Thus ΔH° to produce CO is -110.5 kJ/mol and ΔH° to produce CO_2 is -393.5 kJ/mol. Since one mol of graphite is needed to produce one mol of either CO or CO_2, one equation in one unknown can be written for the heat produced, using m for the moles of CO and (2.00 - m) for moles of CO_2.

$$-481 \text{ kJ} = m \times (-110.5 \text{ kJ}) + (2.00 - m) \times (-393.5 \text{ kJ})$$

$$-481 = -110.5\,m - 787 + 393.5\,m$$

$$\text{mol CO} = m = 1.0\underline{8}1; \quad \text{mass CO} = 1.081 \times 28 \text{ g/mol} = 30.\underline{2}6 = 30.3 \text{ g}$$

$$\text{mol } CO_2 = (2.00 - m) = 0.9\underline{1}9; \quad \text{mass } CO_2 = 0.9\underline{1}9 \text{ mol} \times 44.0 \text{ g/mol} = 4\underline{0}.4 = 40. \text{ g}$$

6.106 The ΔH° values for each reaction are as follows:

$$CH_4(g) + 2\,O_2(g) \rightarrow CO_2(g) + 2\,H_2O(l)$$

$$\Delta H^\circ = -393.5 \text{ kJ} + 2(-285.84 \text{ kJ}) - (-74.87 \text{ kJ}) - 0 = -890.\underline{3}1 \text{ kJ}$$

$$C_2H_4(g) + 3\,O_2(g) \rightarrow 2\,CO_2(g) + 2\,H_2O(l)$$

$$\Delta H^\circ = 2(-393.5 \text{ kJ}) + 2(-285.84 \text{ kJ}) - (-52.47 \text{ kJ}) - 0 = -1411.\underline{1}5 \text{ kJ}$$

Since 10.0 g of CH_4 and C_2H_4 produce 520 kJ of heat, one equation in one unknown can be written for the heat produced, using m for the mass of CH_4 and (10.00 - m) for the mass of C_2H_4. Use m /16.0 for the moles of CH_4 and (10.00 - m) /28.0 for the moles of C_2H_4.

(continued)

$-520\ kJ = (m/16.0) \times (-890.31\ kJ) + [(10.00 - m)/28.0] \times (-1411.15\ kJ)$

$-520 = -55.\underline{6}3\ m + (-503.\underline{9}8) + 50.4\underline{0}\ m$

$\text{mass } CH_4 = m = 1\underline{6}.02 \div 5.23 = 3.\underline{0}63 = 3.1\ g$

$\text{mass \% } CH_4 = (3.06\ g \div 10.0\ g) \times 100\% = 3.06 = 3.1\%$

6.107 The equation is

$4NH_3(g) + 5O_2(g) \rightarrow 4NO(g) + 6H_2O(g);\ \Delta H^o = -906\ kJ$

First determine the limiting reactant by calculating the moles of NH_3 and of O_2; then, assuming one of the reactants is totally consumed, calculate the moles of the other reactant needed for the reaction.

$$10.0\ g\ NH_3 \times \frac{1\ mol\ NH_3}{17.03\ g\ NH_3} = 0.58\underline{7}2\ NH_3$$

$$20.0\ g\ O_2 \times \frac{1\ mol\ O_2}{32.00\ g\ O_2} = 0.62\underline{5}0\ mol\ O_2$$

$$0.62\underline{5}0\ mol\ O_2 \times \frac{4 mol\ NH_3}{5\ mol\ O_2} = 0.500\ mol\ NH_3 \text{ needed if } O_2 \text{ is totally consumed}$$

Because NH_3 is present in excess of what is needed, O_2 must be the limiting reactant. Now calculate the heat released on the basis of the complete reaction of 0.622250 mol O_2:

$$\Delta H = \frac{-906\ kJ}{5\ mol\ O_2} \times 0.6250\ mol\ O_2 = -11\underline{3}.25 = -113\ kJ$$

The heat released by the complete reaction of the 20.0 g (0.6250 mol) of $O_2(g)$ is 113 kJ.

6.108 The equation is

$CS_2(g) + 3Cl_2(g) \rightarrow S_2Cl_2(g) + CCl_4(g);\ \Delta H^o = -232\ kJ$

First determine the limiting reactant by calculating the moles of CS_2 and of Cl_2; then, assuming one of the reactants is totally consumed, calculate the moles of the other reactant needed for the reaction.

(continued)

$$10.0 \text{ g } CS_2 \times \frac{1 \text{ mol } CS_2}{76.13 \text{ g } CS_2} = 0.13\underline{1}4\ CS_2$$

$$10.0 \text{ g } Cl_2 \times \frac{1 \text{ mol } Cl_2}{70.91 \text{ g } Cl_2} = 0.14\underline{1}0\ Cl_2$$

$$0.1410 \text{ mol } Cl_2 \times \frac{1 \text{ mol } CS_2}{3 \text{ mol } Cl_2} = 0.047\underline{0}0\ CS_2$$

Because CS_2 is present in excess of what is needed, Cl_2 must be the limiting reactant. Now calculate the heat released on the basis of the complete reaction of 0.1410 mol Cl_2:

$$\Delta H = \frac{-232 \text{ kJ}}{3 \text{ mol } Cl_2} \times 0.1410 \text{ mol } Cl_2 = -10.\underline{9}04 = -10.9 \text{ kJ}$$

The heat released by the complete reaction of 10.0 g of $Cl_2(g)$ is 10.9 kJ.

6.109 The equation is

$$N_2(g) + 3H_2(g) \rightarrow 2NH_3(g);\ \Delta H^o = -91.8 \text{ kJ}$$

a. To find the heat evolved from the production of 1.00 L of NH_3, convert the 1.00 L to mol NH_3, using the density and molar mass (17.03 g/mol). Then convert the moles to heat (ΔH) using ΔH^o:

$$\text{moles } NH_3 = \frac{d \times V}{Mwt} = \frac{0.696 \text{ g/L} \times 1 \text{ L}}{17.03 \text{ g/mol}} = 0.040\underline{8}7 \text{ mol } NH_3$$

$$\Delta H = \frac{-91.8 \text{ kJ}}{2 \text{ mol } NH_3} \times 0.40\underline{8}7 \text{ mol } NH_3 = -1.8\underline{7}6 \text{ (1.88 kJ heat evolved)}$$

b. First find the moles of N_2 using the density and molar mass (28.02 g/mol). Then convert to the heat needed to raise the N_2 from 25°C to 400°C:

$$\text{moles } N_2 = \frac{d \times V}{Mwt} = \frac{1.145 \text{ g/L} \times 0.500 \text{ L}}{28.02 \text{ g/mol}} = 0.020\underline{4}3 \text{ mol } N_2$$

$$0.020\underline{4}3 \text{ mol } N_2 \times \frac{29.12 \text{ J}}{(\text{mol}\bullet°\text{C})} \times (400 - 25)°\text{C} = 22\underline{3}.2 \text{ J } (0.22\underline{3}2 \text{ kJ})$$

$$\% \text{ heat for } N_2 = \frac{0.22\underline{3}2 \text{ kJ}}{1.876 \text{ kJ}} \times 100\% = 11.\underline{8}9 = 11.9\%$$

6.110 The equation is

$$2H_2S(g) + 3O_2(g) \rightarrow 2H_2O(l) + 2SO_2(g); \quad \Delta H^o = -1125 \text{ kJ}$$

a. First calculate the moles of SO_2 using the density and molar mass (64.07 g/mol). Then convert the moles to heat (ΔH) using ΔH^o:

$$\text{moles } SO_2 = \frac{d \times V}{\text{Mwt}} = \frac{2.62 \text{ g/L} \times 1 \text{ L}}{64.07 \text{ g/mol}} = 0.040\underline{8}9 \text{ mol } SO_2$$

$$\Delta H = \frac{-1125 \text{ kJ}}{2 \text{ mol } SO_2} \times 0.40\underline{8}8 \text{ mol } SO_2 = -22.\underline{9}9 \text{ (23.0 kJ heat evolved)}$$

b. Recall that 1.000 L of SO_2 = 0.04088 mol SO_2. Use this with the molar heat capacity of 30.2 J/(mol•°C) to calculate the heat needed to raise the temperature of SO_2 from 25°C to 500°C:

$$0.040\underline{8}9 \text{ mol } SO_2 \times \frac{30.2 \text{ J}}{(\text{mol}\bullet{}^\circ\text{C})} \times (500 - 25)^\circ\text{C} = 58\underline{6}.4 \text{ J } (0.58\underline{6}4 \text{ kJ})$$

$$\%\text{ heat for } SO_2 = \frac{0.5864 \text{ kJ}}{22.99 \text{ kJ}} \times 100\% = 2.5\underline{5}06 = 2.55\%$$

6.111 The glucose equation is

$$C_6H_{12}O_6 + 6O_2 \rightarrow 6CO_2 + 6H_2O; \quad \Delta H^o = -280\underline{2}.8 \text{ kJ}$$

Convert the 2.50×10^3 kcal to mol of glucose, using the ΔH^o of -2802.8 kJ for the reaction and the conversion factor of 4.184 kJ/kcal:

$$2.50 \times 10^3 \text{ kcal} \times \frac{4.184 \text{ kJ}}{1.000 \text{ kcal}} \times \frac{1 \text{ mol glucose}}{2802.8 \text{ kJ}} = 3.7\underline{3}1 \text{ mol glucose}$$

Next convert mol glucose to mol LiOH using the above equation for glucose and the equation for LiOH: $2LiOH(s) + CO_2(g) \rightarrow Li_2CO_3(s) + H_2O(l)$.

$$3.7\underline{3}1 \text{ mol glucose} \times \frac{6 \text{ mol } CO_2}{1 \text{ mol glucose}} \times \frac{2 \text{ mol LiOH}}{1 \text{ mol } CO_2} = 44.\underline{7}7 \text{ mol LiOH}$$

Finally, use the molar mass of LiOH to convert moles to mass:

$$44.77 \text{ mol LiOH} \times \frac{23.95 \text{ g LiOH}}{1 \text{ mol LiOH}} = 1.0\underline{7}2 \times 10^3 \text{ g (1.07 kg) LiOH}$$

6.112 Using ΔH_f^o = -1273 kJ for glucose, the conversion of glucose to heat, and ΔH^o, in the body is

$$C_6H_{12}O_6(s) + 6O_2(g) \rightarrow 6CO_2(g) + 6H_2O(l); \quad \Delta H^o = -280\underline{2}.8 \text{ kJ}$$

Convert the 1.00 x 10^2 kcal to mol of glucose, using the ΔH^o of -2802.8 kJ:

$$100 \text{ kcal} \times \frac{4.184 \text{ kJ}}{1.000 \text{ kcal}} \times \frac{1 \text{ mol glucose}}{2803 \text{ kJ}} = 0.14\underline{9}26 \text{ mol glucose}$$

Next convert mol glucose to mol KO_2 using the above equation for glucose and the equation for KO_2: $4KO_2(s) + 2H_2O(l) \rightarrow 4KOH(s) + 3O_2(g)$.

$$0.14\underline{9}26 \text{ mol glucose} \times \frac{6 \text{ mol } O_2}{1 \text{ mol glucose}} \times \frac{4 \text{ mol } KO_2}{3 \text{ mol } O_2} = 1.1\underline{9}4 \text{ mol } KO_2$$

$$1.1\underline{9}4 \text{ mol } KO_2 \times \frac{71.1 \text{ g } KO_2}{1 \text{ mol } KO_2} = 84.\underline{8}9 = 84.9 \text{ g } KO_2$$

■ Solution to Conceptual Problem

6.113 Hess's law allows the determination of the enthalpy of a reaction by looking at other, related reactions. For example, we can use the enthalpies of formation in Table 6.2 to determine the enthalpy of reaction for the reaction of methane and chlorine (page 235) or to determine the heat of vaporization.

If we are interested in creating an extremely exothermic reaction, we can make a very good estimate to what the best reactants and products would be based on enthalpy. Based on information in Table 6.2 and Appendix C, design the most exothermic reaction. Remember that the reaction must be balanced.

Discussion

Using standard enthalpies of formation, the heat of reaction is obtained by summing enthalpies of formation of products (multiplied by coefficients) minus a similar term for the reactants. Therefore, the student should be looking for large negative values for the products and large positive numbers for the reactants. Many reactions are possible. Based only on Table 6.2, the formation calcite ($CaCO_3(s)$) from the elements C(g), O(g), and Ca(s) gives a heat of reaction of -2171.1 kJ/mol.

Students should note the state of the reactants. In the example given here, C and O are not in their normal states and large quantities of energy would be required to get them to these atomic states. As a follow-up, students could be encouraged to design a highly exothermic reaction with all reactants and products in their normal states.

7. QUANTUM THEORY OF THE ATOM

Solutions to Exercises

Note on significant figures: The final answer to each mathematical solution is given first with one nonsignificant figure (the rightmost significant figure is underlined) and then is rounded to the correct number of figures. Intermediate answers usually also have at least one nonsignificant figure.

7.1 Rearrange the equation $c = \nu\lambda$, which relates wavelength to frequency and the speed of light (3.00×10^8 m/s):

$$\lambda = \frac{c}{\nu} = \frac{3.00 \times 10^8 \text{ m/s}}{3.91 \times 10^{14}/\text{s}} = 7.6\underline{7}2 \times 10^{-7} = 7.67 \times 10^{-7} \text{ m, or } 767 \text{ nm}$$

7.2 Rearrange the equation $c = \nu\lambda$, which relates frequency to wavelength and the speed of light (3.00×10^8 m/s). Recognize that 456 nm = 4.56×10^{-7} m.

$$\nu = \frac{c}{\lambda} = \frac{3.00 \times 10^8 \text{ m/s}}{4.56 \times 10^{-7} \text{ m}} = 6.5\underline{7}8 \times 10^{14} = 6.58 \times 10^{14}/\text{s}$$

7.3 First use the wavelength to calculate the frequency from $c = \nu\lambda$. Then calculate the energy using $E = h\nu$.

$$\nu = \frac{c}{\lambda} = \frac{3.00 \times 10^8 \text{ m/s}}{1.0 \times 10^{-6} \text{ m}} = 3.\underline{0}0 \times 10^{14}/\text{s}$$

$$\nu = \frac{c}{\lambda} = \frac{3.00 \times 10^8 \text{ m/s}}{1.0 \times 10^{-8} \text{ m}} = 3.\underline{0}0 \times 10^{16}/\text{s}$$

(continued)

$$\nu = \frac{c}{\lambda} = \frac{3.00 \times 10^{8} \text{ m/s}}{1.0 \times 10^{-10} \text{ m}} = 3.\underline{0}0 \times 10^{18}/\text{s}$$

$$E = h\nu = 6.63 \times 10^{-34} \text{ J•s} \times 3.00 \times 10^{14}/\text{s} = 1.\underline{9}8 \times 10^{-19} = 2.0 \times 10^{-19} \text{ J (IR)}$$

$$E = h\nu = 6.63 \times 10^{-34} \text{ J•s} \times 3.00 \times 10^{16}/\text{s} = 1.\underline{9}8 \times 10^{-17} = 2.0 \times 10^{-17} \text{ J (UV)}$$

$$E = h\nu = 6.63 \times 10^{-34} \text{ J•s} \times 3.00 \times 10^{18}/\text{s} = 1.\underline{9}8 \times 10^{-15} = 2.0 \times 10^{-15} \text{ J (x ray)}$$

The x-ray photon (shortest wavelength) has the greatest amount of energy; the infrared photon (longest wavelength) has the least amount of energy.

7.4 Solve the equation $E = -R_H/n^2$ for both E_i and E_f, and calculate the energy change for the transition from n = 3 to n = 1. Convert E to wavelength using $E = h\nu$ and $c = \nu\lambda$.

$$\left[\frac{-R_H}{9}\right] - \left[\frac{-R_H}{1}\right] = \frac{8\,R_H}{9} = h\nu\ ?$$

The frequency of the emitted radiation is

$$\nu = \frac{8R_H}{9h} = \frac{8}{9} \times \frac{2.179 \times 10^{-18} \text{ J}}{6.63 \times 10^{-34} \text{ J•s}} = 2.9\underline{2}1 \times 10^{15} = 2.92 \times 10^{15}/\text{s}$$

$$\lambda = \frac{c}{\nu} = \frac{3.00 \times 10^{8} \text{ m/s}}{2.92 \times 10^{15}/\text{s}} = 1.0\underline{2}7 \times 10^{-7} = 1.03 \times 10^{-7} \text{ m, or } 103 \text{ nm}$$

7.5 Calculate the frequency from $c = \nu\lambda$, recognizing that 589 nm is 5.89×10^{-7} m.

$$\nu = \frac{c}{\lambda} = \frac{3.00 \times 10^{8} \text{ m/s}}{5.89 \times 10^{-7} \text{ m}} = 5.0\underline{9}3 \times 10^{14} = 5.09 \times 10^{14}/\text{s}$$

Finally, calculate the energy difference.

$$E = h\nu = 6.63 \times 10^{-34} \text{ J•s} \times 5.093 \times 10^{14}/\text{s} = 3.3\underline{7}6 \times 10^{-19} = 3.38 \times 10^{-19} \text{ J}$$

7.6 To calculate wavelength, use the mass of an electron, m = 9.11 x 10^{-31} kg, and Planck's constant (h = 6.63 x 10^{-34} J•s, or 6.63 x 10^{-34} kg•m^2/s^2).

$$\lambda = \frac{h}{m\nu} = \frac{6.63 \times 10^{-34}\ kg \bullet m^2/s}{9.11 \times 10^{-31}\ kg \times 2.19 \times 10^{6}\ m/s} = 3.32\underline{3} \times 10^{-10}\ m\ (332\ pm)$$

7.7 a. The value of n must be a positive whole number.
b. The values for l can range only from 0 to (n - 1). Here, l has a value greater than n.
c. The values for m_l range from -l to +l. Here, m_l has a value greater than l.
d. The values for m_s are either + 1/2 or -1/2, not 0.

■ Answers to Review Questions

7.1 Light is a form of electromagnetic radiation. In terms of waves, light can be described as a continuously repeating change or oscillation in electric and magnetic fields traveling through space. Two characteristics of light are frequency and wavelength (given in nanometers, nm). The eye can see the light of wavelengths from about 400 nm (violet) to less than 800 nm (red).

7.2 The relationship among the different characteristics of light waves is $\nu = c/\lambda$, where ν is the frequency, c is the speed of light, and λ is the wavelength.

7.3 Starting with the shortest wavelengths, the electromagnetic spectrum consists of gamma rays, x rays, far ultraviolet (UV), near UV, visible light, near infrared (IR), far IR, microwaves, radar, and TV/FM radio waves (longest wavelengths).

7.4 The term *quantized* means that the possible values of the oscillations or vibrations of a solid atom are limited to only certain values. Planck was trying to explain the colors of light emitted by glowing hot solids. The formula he arrived at was $E = nh\nu$, where E is energy, n is the whole (quantum) number, h is Planck's constant, and ν is frequency.

7.5 *Photoelectric effect* is the term applied to the ejection of electrons from the surface of a metal or from other materials when light shines on it. Electrons are only ejected when the frequency (or energy) of light is larger than a certain minimum, or threshold, value that is constant for each metal. If a photon has a frequency equal to or greater than this minimum value, then it will eject one electron from the metal surface.

7.6 The wave-particle picture of light regards the wave and particle depictions of light as complementary views of the same physical entity—hence the phrase "wave-particle duality of light." Neither view alone is a complete description of all the properties of light. The wave picture characterizes light only by wavelength and frequency. The particle picture characterizes light only as having an energy equal to $h\nu$.

7.7 The equation that relates the particle properties of light is $E = h\nu$. The symbol E is energy, h is Planck's constant, and ν is the frequency of the light.

7.8 According to physical theory at Rutherford's time, an electrically charged particle revolving around a center will continuously radiate energy. This implies that the electron in the atom would continuously lose energy, spiraling into smaller and smaller orbits in the process and ultimately spiraling into the nucleus. Thus, the theory apparently could not explain the stability of the electrons around the atom and the atom itself.

7.9 According to Bohr, an electron in an atom can have only specific energy values. An electron in an atom can change energy only by going from one energy level (of allowed energy) to another energy level (of allowed energy). An electron in a higher energy level can go to a lower energy level by emitting a photon of an energy equal to the difference in energy. However, when an electron is in its lowest energy level, no further changes in energy can occur. Thus, the electron does not continuously radiate energy, as thought at Rutherford's time. These features solve the difficulty alluded to in 7.8

7.10 Emission of a photon occurs when an electron in a higher energy level undergoes a transition to a lower energy level. The energy lost is emitted as a photon.

7.11 Absorption of a photon occurs when a photon of a certain required energy is absorbed by a certain electron in an atom. The energy of the photon must be equal to the energy necessary to excite the electron of the atom from a lower energy level, usually the lowest, to a higher energy level. The photon's energy is converted into electronic energy.

7.12 The diffraction of an electron beam is evidence for electron waves. A practical example of diffraction is the operation of the electron microscope.

7.13 The square of a wave function, rather than the value of the function itself, equals the probability for finding an electron at a given point in space. More complex mathematical manipulations of the wave function yield values for other parameters.

7.14 The uncertainty principle says that we can no longer think of the electron as having a precise orbit in an atom similar to the orbit of the planets around the sun. This principle says that it is impossible to know with absolute precision both the position and the speed of a particle such as an electron. Application of Heisenberg's uncertainty principle shows that the uncertainties of position and speed are significant when the principle is applied to electrons.

7.15 Quantum mechanics vastly changes Bohr's original picture of the hydrogen atom in that we can no longer think of the electron as having a precise orbit around the nucleus in this atom. Recall that Bohr's theory depended on the hydrogen electron having specific energy values and thus specific positions and speeds around the nucleus. But quantum mechanics and the uncertainty principle say that it is impossible to know with absolute precision both the speed and the position of an electron. So Bohr's energy levels are only the most probable paths of the electrons.

7.16 a. The principal quantum number can have an integer value between 1 and infinity.
b. The angular momentum quantum number can have any integer value between 0 and (n - 1).
c. The magnetic quantum number can have any integer value between -l and +l.
d. The spin quantum number can be either +1/2 or -1/2.

7.17 The notation is 4f. This subshell contains 7 orbitals.

7.18 An s orbital has a spherical shape. A p orbital has two lobes positioned along a straight line through the nucleus at the center of the line (a dumbbell shape).

■ Solutions to Practice Problems

Note on significant figures: The final answer to each problem is given first with one nonsignificant figure (the rightmost significant figure is underlined), and then is rounded to the correct number of figures. Intermediate answers usually also have at least one nonsignificant figure. Atomic weights are rounded to two decimal places, except for that of hydrogen.

7.19 Using $c = 2.998 \times 10^8$ m/s for four significant figures, solve $c = \lambda\nu$ for λ:

$$\lambda = \frac{c}{\nu} = \frac{2.998 \times 10^8 \text{ m/s}}{1.255 \times 10^6\text{/s}} = 2.38\underline{8}8 \times 10^2 = 238.9 \text{ m}$$

7.20 Using $c = 2.998 \times 10^8$ m/s for four significant figures, solve $c = \lambda\nu$ for λ:

$$\lambda = \frac{c}{\nu} = \frac{2.998 \times 10^8 \text{ m/s}}{1.145 \times 10^{10}\text{/s}} = 2.61\underline{8}3 \times 10^{-2} \text{ m (2.618 cm)}$$

7.21 Solve $c = \lambda\nu$ for ν. Recognize that 465 nm = 465×10^{-9} m, or 4.65×10^{-7} m.

$$\nu = \frac{c}{\lambda} = \frac{3.00 \times 10^8 \text{ m/s}}{4.65 \times 10^{-7} \text{ m}} = 6.4\underline{5}1 \times 10^{14} = 6.45 \times 10^{14}\text{/s}$$

7.22 Solve $c = \lambda\nu$ for ν. Recognize that 656 nm = 656×10^{-9} m = 6.56×10^{-7} m.

$$\nu = \frac{c}{\lambda} = \frac{3.00 \times 10^8 \text{ m/s}}{6.56 \times 10^{-7} \text{ m}} = 4.5\underline{7}3 \times 10^{14} = 4.57 \times 10^{14}\text{/s}$$

7.23 Radio waves travel at the speed of light, so divide the distance by c:

$$56 \times 10^9 \text{ m} \times \frac{1 \text{ s}}{3.00 \times 10^8 \text{ m}} = 1.\underline{8}6 \times 10^2 = 1.9 \times 10^2 \text{ s}$$

7.24 Electromagnetic signals travel at the speed of light, so divide the distance by c:

$$998 \times 10^9 \text{ m} \times \frac{1 \text{ s}}{3.00 \times 10^8 \text{ m}} = 3.3\underline{2}6 \times 10^3 = 3.33 \times 10^3 \text{ s (0.925 hr)}$$

7.25 To do the calculation, divide 1 meter by the number of wavelengths in 1 meter to find the wavelength of this transition. Then, use the speed of light to calculate the frequency:

$$\lambda = \frac{1 \text{ m}}{1{,}650{,}763.73} = 6.05802106 \times 10^{-7} \text{ m}$$

$$\nu = \frac{c}{\lambda} = \frac{2.99792458 \times 10^8 \text{ m/s}}{6.05802106 \times 10^{-7} \text{ m}} = 4.948865162 \times 10^{14} \text{ /s}$$

7.26 To find the wavelength, divide the speed of light by the frequency:

$$\lambda = \frac{c}{\nu} = \frac{2.99792458 \times 10^8 \text{ m/s}}{9{,}192{,}631{,}770 \text{ /s}} = 3.26122557 \times 10^{-2} \text{ m}$$

7.27 Solve for E, using E = hν, and four significant figures for h:

$$E = h\nu = (6.626 \times 10^{-34} \text{ J}\bullet\text{s}) \times (1.255 \times 10^6\text{/s}) = 8.31\underline{5}6 \times 10^{-28} = 8.316 \times 10^{-28} \text{ J}$$

7.28 Solve for E, using E = hν, and four significant figures for h:

$$E = h\nu = (6.626 \times 10^{-34} \text{ J}\bullet\text{s}) \times (1.145 \times 10^{10}\text{/s}) = 7.58\underline{6}7 \times 10^{-24} = 7.587 \times 10^{-24} \text{ J}$$

7.29 Recognize that 535 nm = 535 x 10^{-9} m = 5.35 x 10^{-7} m. Then calculate ν and E.

$$\nu = \frac{c}{\lambda} = \frac{3.00 \times 10^8 \text{ m/s}}{5.35 \times 10^{-7} \text{ m}} = 5.6\underline{0}7 \times 10^{14}\text{/s}$$

$$E = h\nu = (6.63 \times 10^{-34} \text{ J}\bullet\text{s}) \times (5.607 \times 10^{-14}\text{/s}) = 3.7\underline{1}7 \times 10^{-19} = 3.72 \times 10^{-19} \text{ J}$$

7.30 Recognize that 451 nm = 451 x 10^{-9} m = 4.51 x 10^{-7} m. Then calculate ν and E.

$$\nu = \frac{c}{\lambda} = \frac{3.00 \times 10^{8}\ \text{m/s}}{4.51 \times 10^{-7}\ \text{m}} = 6.6\underline{5}1 \times 10^{14}/\text{s}$$

$$E = h\nu = (6.63 \times 10^{-34}\ \text{J•s}) \times (6.651 \times 10^{14}/\text{s}) = 4.4\underline{0}96 \times 10^{-19} = 4.41 \times 10^{-19}\ \text{J}$$

7.31 First calculate the wavelength of this transition from the frequency, using the speed of light:

$$\lambda = \frac{c}{\nu} = \frac{3.00 \times 10^{8}\ \text{m/s}}{3.84 \times 10^{14}/\text{s}} = 7.8\underline{1}2 \times 10^{-7} = 7.81 \times 10^{-7}\ \text{m (781 nm)}$$

Using Figure 7.4, note that 781 nm is just on the edge of the red end of the spectrum and is barely visible to the eye.

7.32 First calculate the wavelength of this transition from the frequency, using the speed of light:

$$\lambda = \frac{c}{\nu} = \frac{3.00 \times 10^{8}\ \text{m/s}}{5.41 \times 10^{14}/\text{s}} = 5.5\underline{4}52 \times 10^{-7} = 5.55 \times 10^{-7}\ \text{m (555 nm)}$$

Using Figure 7.4, note that 555 nm is in the yellow-green region of the spectrum and is visible to the eye.

7.33 Solve the equation $E = -R_H/n^2$ for both E_4 and E_3; equate to hν and solve for ν.

$$E_4 = \frac{-R_H}{4^2} = \frac{-R_H}{16}; \qquad E_3 = \frac{-R_H}{3^2} = \frac{-R_H}{9}$$

$$\left[\frac{-R_H}{16}\right] - \left[\frac{-R_H}{9}\right] = \frac{7\,R_H}{144} = h\nu$$

The frequency of the emitted radiation is:

$$\nu = \frac{7R_H}{144h} = \frac{7}{144} \times \frac{2.179 \times 10^{-18}\ \text{J}}{6.63 \times 10^{-34}\ \text{J•s}} = 1.5\underline{9}8 \times 10^{14} = 1.60 \times 10^{14}/\text{s}$$

7.34 Solve the equation $E = -R_H/n^2$ for both E_5 and E_3; equate to $h\nu$ and solve for ν.

$$E_5 = \frac{-R_H}{5^2} = \frac{-R_H}{25}; \qquad E_3 = \frac{-R_H}{3^2} = \frac{-R_H}{9}$$

$$\left[\frac{-R_H}{25}\right] - \left[\frac{-R_H}{9}\right] = \frac{16\,R_H}{225} = h\nu$$

$$\nu = \frac{16R_H}{225h} = \frac{16}{225} \times \frac{2.179 \times 10^{-18}\text{ J}}{6.63 \times 10^{-34}\text{ J•s}} = 2.3\underline{3}7 \times 10^{14} = 2.34 \times 10^{14}/\text{s}$$

7.35 Solve the equation $E = -R_H/n^2$ for both E_2 and E_1; solve for ν and convert to λ.

$$E_2 = \frac{-R_H}{2^2} = \frac{-R_H}{4}; \qquad E_1 = \frac{-R_H}{1^2} = \frac{-R_H}{1}$$

$$\left[\frac{-R_H}{4}\right] - \left[\frac{-R_H}{1}\right] = \frac{3\,R_H}{4} = h\nu$$

$$\nu = \frac{3R_H}{4h} = \frac{3}{4} \times \frac{2.179 \times 10^{-18}\text{ J}}{6.63 \times 10^{-34}\text{ J•s}} = 2.4\underline{6}5 \times 10^{15}/\text{s}$$

$$\lambda = \frac{c}{\nu} \times \frac{3.00 \times 10^{8}\text{ m/s}}{2.465 \times 10^{15}/\text{s}} = 1.2\underline{1}6 \times 10^{-7} = 1.22 \times 10^{-7}\text{ m (near UV)}$$

7.36 Solve the equation $E = -R_H/n^2$ for both E_5 and E_4; solve for ν and convert to λ.

$$E_2 = \frac{-R_H}{5^2} = \frac{-R_H}{25}; \qquad E_1 = \frac{-R_H}{4^2} = \frac{-R_H}{16}$$

$$\left[\frac{-R_H}{25}\right] - \left[\frac{-R_H}{16}\right] = \frac{9\,R_H}{400} = h\nu$$

$$\nu = \frac{9R_H}{400h} = \frac{9}{400} \times \frac{2.179 \times 10^{-18}\text{ J}}{6.63 \times 10^{-34}\text{ J•s}} = 7.3\underline{9}5 \times 10^{13}/\text{s}$$

$$\lambda = \frac{c}{\nu} \times \frac{3.00 \times 10^{8}\text{ m/s}}{7.395 \times 10^{13}/\text{s}} = 4.0\underline{5}69 \times 10^{-6} = 4.06 \times 10^{-6}\text{ m (near IR)}$$

7.37 This is the lowest energy transition from the n = 6 level, so the electron must undergo a transition to the n = 5 level. Solve the Balmer equation using Bohr's approach:

$$E_i = \frac{-R_H}{6^2} = \frac{-R_H}{36}; \qquad E_f = \frac{-R_H}{5^2} = \frac{-R_H}{25}$$

$$h\nu = \left[\frac{-R_H}{36}\right] - \left[\frac{-R_H}{25}\right] = \frac{11\,R_H}{900}$$

$$\lambda = \frac{11R_H}{900h} = \frac{11 \times (2.179 \times 10^{-18}\ J)}{900 \times (6.63 \times 10^{-34}\ J{\bullet}s)} = 4\,.0\underline{1}7 \times 10^{13}/s$$

$$\lambda = \frac{c}{\nu} = \frac{3.00 \times 10^8\ m/s}{4.017 \times 10^{13}/s} = 7.4\underline{6}6\ \times\ 10^{-6}\ m\ (7470\ nm)$$

7.38 This is the highest energy transition from the n = 6 level, so the electron must undergo a transition to the n = 1 level. Solve the Balmer equation using Bohr's approach:

$$E_i = \frac{-R_H}{6^2} = \frac{-R_H}{36}; \quad E_f = \frac{-R_H}{1^2} = \frac{-R_H}{1}$$

$$h\nu = \left[\frac{-R_H}{36}\right] - \left[\frac{-R_H}{1}\right] = \frac{35\,R_H}{36}$$

$$\nu = \frac{35R_H}{36h} = \frac{35 \times (2.179 \times 10^{-18}\ J)}{36 \times (6.63 \times 10^{-34}\ J{\bullet}s)} = 3.1\underline{9}5 \times 10^{15}/s$$

$$\lambda = \frac{c}{\nu} = \frac{3.00 \times 10^8\ m/s}{3.195 \times 10^{15}/s} = 9.3\underline{8}9 \times 10^{-8}\ m\ (93.9\ nm)$$

7.39 Noting that 285.2 nm = 2.852 x 10^{-7} m, convert the 285.2-nm wavelength to frequency. Then convert the frequency to energy using E = hν. Use these constants to four significant figures: h = 6.626 x 10^{-34} J•s and c = 2.998 x 10^8 m/s.

$$\nu = \frac{c}{\lambda} = \frac{2.998 \times 10^{\,8}\ m/s}{2.852 \times 10^{-7}\ m} = 1.05\underline{1}1 \times 10^{15}/s$$

$$E = h\nu = (6.626 \times 10^{-34}\ J{\bullet}s) \times (1.05\underline{1}1 \times 10^{15}/s) = 6.96\underline{4}58 \times 10^{-19}$$

$$= 6.965 \times 10^{-19}\ J$$

7.40 Noting that 422.7 nm = 4.227×10^{-7} m, convert the 422.7-nm to frequency. Then convert the frequency to energy using E = hν. Use these constants to four significant figures: h = 6.626×10^{-34} J•s and c = 2.998×10^{8} m/s.

$$\nu = \frac{c}{\lambda} = \frac{2.998 \times 10^{8}\ \text{m/s}}{4.227 \times 10^{-7}\ \text{m}} = 7.09\underline{2}5 \times 10^{14}/\text{s}$$

$$E = h\nu = (6.626 \times 10^{-34}\ \text{J}\bullet\text{s}) \times (7.09\underline{2}5 \times 10^{14}/\text{s}) = 4.69\underline{9}4 \times 10^{-19}$$

$$= 4.699 \times 10^{-19}\ \text{J}$$

7.41 The mass of a neutron = 1.67493×10^{-27} kg. Its speed or velocity, v, of 3.90 km/s = 3.90×10^{3} m/s. Substitute these parameters into the de Broglie relation and solve for λ:

$$\lambda = \frac{h}{m\nu} = \frac{6.63 \times 10^{-34}\ \text{kg}\bullet\text{m}^2/\text{s}}{1.67493 \times 10^{-27}\ \text{kg} \times 3.90 \times 10^{3}\ \text{m/s}}$$

$$= 1.0\underline{1}4 \times 10^{-10}\ \text{m, or 101 pm}$$

7.42 The mass of a proton = 1.67262×10^{-27}. Its speed or velocity, v, of 5.80 km/s = 5.80×10^{3} m/s. Substitute these parameters into the de Broglie relation and solve for λ:

$$\lambda = \frac{h}{m\nu} = \frac{6.63 \times 10^{-34}\ \text{kg}\bullet\text{m}^2/\text{s}}{1.67262 \times 10^{-27}\ \text{kg} \times 5.80 \times 10^{3}\ \text{m/s}} = 6.8\underline{3}4 \times 10^{-11}\ \text{m (x rays)}$$

7.43 The mass of an electron = 9.10953×10^{-31} kg. The wavelength, λ, given as 10.0 pm, is equivalent to 1.00×10^{-11} m. Substitute these parameters into the de Broglie relation and solve for the frequency, ν:

$$\nu = \frac{h}{m\lambda} = \frac{6.63 \times 10^{-34}\ \text{kg}\bullet\text{m}^2/\text{s}}{9.10953 \times 10^{-31}\ \text{kg} \times 1.00 \times 10^{-11}\ \text{m}} = 7.2\underline{7}8 \times 10^{7}$$

$$= 7.28 \times 10^{7}\ \text{m/s}$$

7.44 The mass of a neutron = 1.67493×10^{-27} kg. The wavelength, λ, given as 10.0 pm, is equivalent to 1.00×10^{-11} m. Substitute these parameters into the de Broglie relation and solve for the frequency, ν:

$$\nu = \frac{h}{m\lambda} = \frac{6.63 \times 10^{-34}\ \text{kg}\bullet\text{m}^2/\text{s}}{1.67493 \times 10^{-27}\ \text{kg} \times 1.00 \times 10^{-11}\ \text{m}} = 3.9\underline{5}8 \times 10^4$$

$$= 3.96 \times 10^4\ \text{m/s}$$

7.45 Substitute the 1.45×10^{-1} kg mass of the baseball and the 20.0 m/s velocity, v, into the de Broglie relation and solve for wavelength (recall that 1 pm = 10^{-12} m).

$$\lambda = \frac{h}{m\nu} = \frac{6.63 \times 10^{-34}\ \text{kg}\bullet\text{m}^2/\text{s}}{1.45 \times 10^{-1}\ \text{kg} \times 20.0\ \text{m/s}} = 2.2\underline{8}6 \times 10^{-34}\ \text{m}$$

$$= 2.29 \times 10^{-22}\ \text{pm}$$

Because this is much smaller than 100 pm, the wavelength is much smaller than the diameter of one atom.

7.46 The mass of O_2 to three significant figures is 32 ÷ 6.022×10^{23} = 5.313×10^{-23} g = 5.313×10^{-26} kg. Substitute this mass and the 479 m/s velocity, v, into the de Broglie relation and solve for wavelength (recall that 1 pm = 10^{-12} m).

$$\lambda = \frac{h}{m\nu} = \frac{6.63 \times 10^{-34}\ \text{kg}\bullet\text{m}^2/\text{s}}{(5.313 \times 10^{-26}\ \text{kg}) \times (479\ \text{m/s})} = 2.6\underline{0}5 \times 10^{-11}\ \text{m} = 26.1\ \text{pm}$$

Because this is on the order of 100 pm, the wavelength is on the order of an atomic diameter.

7.47 The possible values of l range from 0 to (n - 1), so l may be 0, 1, 2, or 3. The possible values of m_l range from - l to + l, so m_l may be -3, -2, -1, 0, +1, +2, or +3.

7.48 The possible l values are 0, 1, 2, 3, and 4. The possible m_l values are -4, -3, -2, -1, 0, 1, 2, 3, or 4.

7.49 For the M shell, n = 3; there are three subshells in this shell (l = 0, 1, and 2). An f subshell has l = 3; the number of orbitals in this subshell is 2(3) + 1 = 7 (m_l = -3, -2, -1, 0, 1, 2, and 3).

7.50 For the N shell, $n = 4$; there are four subshells in this shell ($l = 0, 1, 2,$ and 3). For a g subshell, the value of $l = 4$; the number of orbitals in this subshell is $2(4) + 1 = 9$ ($m_l = -4, -3, -2, -1, 0, 1, 2, 3,$ and 4).

7.51 a. 3p b. 4d c. 4s d. 5f

7.52 a. 6d b. 5g c. 4f d. 6p

7.53
a. Impossible; n starts at 1, not at zero.
b. Impossible; l may only be as large as (n - 1).
c. Possible.
d. Impossible; m_l may not exceed l in magnitude.
e. Possible.

7.54
a. Not permissible; m_s may be only + 1/2 or -1/2.
b. Not permissible; l can only be as large as (n - 1).
c. Not permissible; m_l may not exceed +2 in magnitude.
d. Not permissible; n may not be zero.
e. Not permissible; m_s may only be + 1/2 or -1/2.

■ Solutions to Unclassified Problems

7.55 Use $c = \nu\lambda$ to calculate frequency; then use $E = h\nu$ to calculate energy.

$$\nu = \frac{c}{\lambda} \times \frac{3.00 \times 10^{8}\ \text{m/s}}{4.61 \times 10^{-7}\ \text{m}} = 6.5\underline{0}7 \times 10^{14} = 6.51 \times 10^{14}/\text{s}$$

$$E = h\nu = (6.63 \times 10^{-34}\ \text{J•s}) \times (6.507 \times 10^{14}/\text{s}) = 4.3\underline{1}4 \times 10^{-19} = 4.31 \times 10^{-19}\ \text{J}$$

7.56 Use $c = \nu\lambda$ to calculate frequency; then use $E = h\nu$ to calculate energy.

$$\nu = \frac{c}{\lambda} \times \frac{3.00 \times 10^{8}\ \text{m/s}}{5.54 \times 10^{-7}\ \text{m}} = 5.4\underline{1}52 \times 10^{14} = 5.42 \times 10^{14}/\text{s}$$

$$E = h\nu = (6.63 \times 10^{-34}\ \text{J•s}) \times (5.4152 \times 10^{14}/\text{s}) = 3.5\underline{9}02 \times 10^{-19} = 3.59 \times 10^{-19}\ \text{J}$$

7.57 Calculate the frequency corresponding to 4.10×10^{-19} J. Then convert that to wavelength.

$$\nu = \frac{E}{h} = \frac{4.10 \times 10^{-19}\ \text{J}}{6.63 \times 10^{-34}\ \text{J}\bullet\text{s}} = 6.1\underline{8}4 \times 10^{14}/\text{s}$$

$$\lambda = \frac{c}{\nu} = \frac{3.00 \times 10^{8}\ \text{m/s}}{6.184 \times 10^{14}/\text{s}} = 4.8\underline{5}1 \times 10^{-7} = 4.85 \times 10^{-7}\text{m (blue)} = 485\ \text{nm}$$

7.58 Calculate the frequency corresponding to 3.34×10^{-19} J. Then convert that to wavelength.

$$\nu = \frac{E}{h} = \frac{3.34 \times 10^{-19}\ \text{J}}{6.63 \times 10^{-34}\ \text{J}\bullet\text{s}} = 5.0\underline{3}7 \times 10^{14}/\text{s}$$

$$\lambda = \frac{c}{\nu} = \frac{3.00 \times 10^{8}\ \text{m/s}}{5.037 \times 10^{14}/\text{s}} = 5.9\underline{5}59 \times 10^{-7}\ \text{m (= 596 nm = yellow)}$$

7.59 Solve for frequency, using E = hν.

$$\nu = \frac{E}{h} = \frac{4.34 \times 10^{-19}\ \text{J}}{6.63 \times 10^{-34}\ \text{J}\bullet\text{s}} = 6.5\underline{4}6 \times 10^{14} = 6.55 \times 10^{14}/\text{s}$$

7.60 Solve for frequency, using E = hν.

$$\nu = \frac{E}{h} = \frac{5.90 \times 10^{-19}\ \text{J}}{6.63 \times 10^{-34}\ \text{J}\bullet\text{s}} = 8.8\underline{9}8 \times 10^{14} = 8.90 \times 10^{14}/\text{s}$$

7.61 First calculate E_p, the energy of the 345-nm photon, noting that this is equivalent to 3.45×10^{-7} m.

$$E_p = \frac{hc}{\lambda} = \frac{(6.63 \times 10^{-34}\ \text{J}\bullet\text{s}) \times (3.00 \times 10^{8}\ \text{m/s})}{(3.45 \times 10^{-7}\ \text{m})} = 5.7\underline{6}52 \times 10^{-19}\ \text{J}$$

Now subtract the work function of Ca = 4.34×10^{-19} J (Problem 7.59) from E_p:

$$5.7652 \times 10^{-19}\ \text{J} - 4.34 \times 10^{-19}\ \text{J} = 1.4\underline{2}52 \times 10^{-19}\ \text{J}$$

(continued)

From Section 6.1 and Example 6.1, we note that for this situation, E = 1/2mv^2. Recall that the mass of the electron is 9.11 x 10^{-31} kg. Now calculate speed, v:

$$v = \sqrt{\frac{2E}{m}} = \sqrt{\frac{2 \times 1.4252 \times 10^{-19}\ J}{9.11 \times 10^{-31}\ kg}} = 5.5\underline{9}36 \times 10^5$$

$$= 5.59 \times 10^5\ m/s$$

7.62 First calculate E_p, the energy of the 285-nm photon, noting that this is equivalent to 2.85 x 10^{-7} m.

$$E_p = \frac{hc}{\lambda} = \frac{(6.63 \times 10^{-34}\ J{\bullet}s) \times (3.00 \times 10^8\ m/s)}{(2.85 \times 10^{-7}\ m)} = 6.9\underline{7}8 \times 10^{-19}\ J$$

Now subtract the work function of Mg = 5.90 x 10^{-19} J (Problem 7.60) from E_p:

$$6.978 \times 10^{-19}\ J - 5.90 \times 10^{-19}\ J = 1.078 \times 10^{-19}\ J$$

From Section 6.1 and Example 6.1, we note that for this situation, E = 1/2mv^2. Recall that the mass of the electron is 9.11 x 10^{-31} kg. Now calculate speed, v:

$$v = \sqrt{\frac{2E}{m}} = \sqrt{\frac{2 \times 1.078 \times 10^{-19}\ J}{9.11 \times 10^{-31}\ kg}} = 4.8\underline{6}4 \times 10^5 = 4.86 \times 10^5\ m/s$$

7.63 This is a transition from the n = 5 level to the n = 2 level. Solve the Balmer equation using Bohr's approach.

$$E_i = \frac{-R_H}{5^2} = \frac{-R_H}{25}; \qquad E_f = \frac{-R_H}{2^2} = \frac{-R_H}{4}$$

$$h\nu = \left[\frac{-R_H}{25}\right] - \left[\frac{-R_H}{4}\right] = \frac{21\ R_H}{100}$$

$$\nu = \frac{21\ R_H}{100h} = \frac{21 \times (2.179 \times 10^{-18}\ J)}{100 \times (6.63 \times 10^{-34}\ J{\bullet}s)} = 6.9\underline{0}18 \times 10^{14}/s$$

$$\lambda = \frac{c}{\nu} = \frac{3.00 \times 10^8\ m/s}{6.9018 \times 10^{14}/s} = 4.3\underline{4}7 \times 10^{-7}\ m\ (435\ nm)$$

7.64 This is a transition from the n = 3 level to the n = 2 level. Solve the Balmer equation using Bohr's approach.

$$E_i = \frac{-R_H}{3^2} = \frac{-R_H}{9}; \qquad E_f = \frac{-R_H}{2^2} = \frac{-R_H}{4}$$

$$h\nu = \left[\frac{-R_H}{9}\right] - \left[\frac{-R_H}{4}\right] = \frac{5\ R_H}{36}$$

$$\nu = \frac{5\ R_H}{36} = \frac{5 \times (2.179 \times 10^{-18}\ J)}{36 \times (6.63 \times 10^{-34}\ J{\bullet}s)} = 4.5\underline{6}46\ \times\ 10^{14}/s$$

$$\lambda = \frac{c}{\nu} = \frac{3.00 \times 10^{8}\ m/s}{4.5657 \times 10^{14}/s} = 6.5\underline{7}2 \times 10^{-7}\ m\ (657\ nm)$$

7.65 Use 397 nm = 3.97×10^{-7} m, and convert to frequency and then to energy.

$$\nu = \frac{c}{\lambda} = \frac{3.00 \times 10^{8}\ m/s}{3.97 \times 10^{-7}\ m} = 7.5\underline{5}6 \times 10^{14}/s$$

$$E = h\nu = (6.63 \times 10^{-34}\ J{\bullet}s) \times (7.556 \times 10^{14}) = 5.0\underline{0}9 \times 10^{-19}\ J$$

Substitute this energy into the Balmer formula, recalling that the Balmer series is an emission spectrum, so ΔE is negative:

$$E_2 = \frac{-R_H}{2^2} = \frac{-R_H}{4}; \qquad E_i = \frac{-R_H}{n_i^2}$$

$$\left[\frac{-R_H}{4}\right] - \left[\frac{-R_H}{n_i^2}\right] = -R_H\left[\frac{1}{4} - \frac{1}{n_i^2}\right] = E\ \text{(of line)}$$

$$\left[\frac{1}{4} - \frac{1}{n_i^2}\right] = \frac{5.009 \times 10^{-19}\ J}{2.179 \times 10^{-18}\ J} = 0.22\underline{9}9$$

$$\frac{1}{n_i^2} = \frac{1}{4} - 0.22\underline{9}9 = 0.02\underline{0}1$$

$$n_i = \left[\frac{1}{0.0201}\right]^{1/2} = 7.\underline{0}53 = 7.1\ (= n)$$

7.66 Convert 9.50×10^{-8} m into frequency and then to energy.

$$\nu = \frac{c}{\lambda} \times \frac{3.00 \times 10^{8}\ \text{m/s}}{9.50 \times 10^{-8}\ \text{m}} = 3.1\underline{5}8 \times 10^{15}/\text{s}$$

$$E = h\nu = (6.63 \times 10^{-34}\ \text{J•s}) \times (3.158 \times 10^{15}) = 2.0\underline{9}4 \times 10^{-18}\ \text{J}$$

Substitute this energy into the Balmer formula, recalling that the Lyman series is an emission spectrum, so ΔE is negative, and $n_f = 1$:

$$E_1 = \frac{-R_H}{1^2} = \frac{-R_H}{1}; \quad E_i = \frac{-R_H}{n_i^2}$$

$$\left[\frac{-R_H}{1}\right] - \left[\frac{-R_H}{n_i^2}\right] = -R_H\left[\frac{1}{1} - \frac{1}{n_i^2}\right] = E \text{ (of line)}$$

$$\left[1 - \frac{1}{n_i^2}\right] = \frac{2.094 \times 10^{-18}\ \text{J}}{2.179 \times 10^{-18}\ \text{J}} = 0.96\underline{0}49$$

$$\frac{1}{n_i^2} = 1 - 0.96\underline{0}99 = 0.03\underline{9}01$$

$$n_i = \left[\frac{1}{0.03901}\right]^{1/2} = 5.\underline{0}6 = 5\ (= n)$$

7.67 Employ the Balmer formula, using $Z = 2$ for the He^+ ion.

$$E_3 = (2)^2 \frac{-R_H}{3^2} = (4) \frac{-R_H}{9}; \quad E_2 = (2)^2 \frac{-R_H}{2^2} = (4) \frac{-R_H}{4}$$

$$4\left[\frac{-R_H}{9}\right] - 4\left[\frac{-R_H}{4}\right] = 4 \times \frac{5R_H}{36} = h\nu$$

The frequency of the radiation is

$$\nu = \frac{4 \times 5R_H}{36h} = \frac{20}{36} \times \frac{2.179 \times 10^{-18}\ \text{J}}{6.63 \times 10^{-34}\ \text{J•s}} = 1.8\underline{2}59 \times 10^{15}/\text{s}$$

$$\lambda = \frac{c}{\nu} = \frac{3.00 \times 10^{8}\ \text{m/s}}{1.8259 \times 10^{15}/\text{s}} = 1.6\underline{4}3 \times 10^{-7}\ \text{m (164 nm = near UV)}$$

7.68 Employ the Balmer formula, using Z = 3 for the Li^{2+} ion.

$$E_4 = (3)^2 \frac{-R_H}{4^2} = (9)\frac{-R_H}{16}; \quad E_3 = (3)^2 \frac{-R_H}{3^2} = (9)\frac{-R_H}{9}$$

$$9\left[\frac{-R_H}{16}\right] - 9\left[\frac{-R_H}{9}\right] = 9 \times \frac{7R_H}{144} = h\nu$$

$$\text{Frequency} = \nu = \frac{9 \times 7R_H}{144h} = \frac{63}{144} \times \frac{2.179 \times 10^{-18}\ J}{6.63 \times 10^{-34}\ J{\bullet}s} = 1.4\underline{3}788 \times 10^{15}/s$$

$$\lambda = \frac{c}{\nu} = \frac{3.00 \times 10^8\ m/s}{1.438 \times 10^{15}/s} = 2.0\underline{8}6 \times 10^{-7}\ m\ (209\ nm = UV)$$

7.69 First use the wavelength of 10.0 pm (1.00×10^{-11} m) and the mass of 9.11×10^{-31} kg to calculate the velocity, v. Then use the kinetic energy equation to calculate kinetic energy from velocity.

$$v = \frac{h}{\lambda m} = \frac{6.626 \times 10^{-34}\ J{\bullet}s}{(1.00 \times 10^{-11}\ m) \times (9.11 \times 10^{-31}\ kg)} = 7.2\underline{7}3 \times 10^7\ m/s$$

$$E = 1/2mv^2 = 1/2 \times (9.11 \times 10^{-31}\ kg) \times (7.273 \times 10^7\ m/s)^2 = 2.4\underline{0}9 \times 10^{-15}\ J$$

$$E_{eV} = 2.409 \times 10^{-15}\ J \times \frac{1\ eV}{1.602 \times 10^{-19}\ J} = 1.5\underline{0}3 \times 10^4$$

$$= 1.50 \times 10^4\ ev$$

7.70 First use the wavelength of 10.0 pm (1.00×10^{-11} m) and the mass of 1.675×10^{-27} kg to calculate the velocity, v. Then use the kinetic energy equation to calculate kinetic energy from velocity.

$$v = \frac{h}{\lambda m} = \frac{6.626 \times 10^{-34}\ J{\bullet}s}{(1.00 \times 10^{-11}\ m) \times (1.675 \times 10^{-27}\ kg)} = 3.9\underline{5}58 \times 10^4\ m/s$$

$$E = 1/2mv^2 = 1/2 \times (1.675 \times 10^{-27}\ kg) \times (3.9558 \times 10^4\ m/s)^2 = 1.3\underline{1}05 \times 10^{-18}\ J$$

$$E_{eV} = 1.3105 \times 10^{-18}\ J \times \frac{1\ eV}{1.602 \times 10^{-19}\ J} = 8.1\underline{8}03 = 8.18\ eV$$

7.71 a. Five b. Seven c. Three d. One

7.72 a. Seven b. Nine c. One d. Three

7.73 The possible subshells for the n = 6 shell are 6s, 6p, 6d, 6f, 6g, and 6h.

7.74 The possible subshells for the n = 7 shell are 7s, 7p, 7d, 7f, 7g, 7h, and 7i.

■ Cumulative-Skills Problems
(require skills from previous chapters)

7.75 First use Avogadro's number to calculate the energy for one Cl_2 molecule.

$$\frac{239 \text{ kJ}}{1 \text{ mol}} \times \frac{1000 \text{ J}}{1 \text{ kJ}} \times \frac{1 \text{ mol}}{6.02 \times 10^{23} \text{ molec.}} = 3.9\underline{7}01 \times 10^{-19} \text{ J/molec.}$$

Then convert energy to frequency and finally to wavelength.

$$\nu = \frac{E}{h} = \frac{3.9701 \times 10^{-19} \text{ J}}{6.63 \times 10^{-34} \text{ J•s}} = 5.9\underline{8}8 \times 10^{14}\text{/s}$$

$$\lambda = \frac{c}{\nu} = \frac{3.00 \times 10^{8} \text{ m/s}}{5.988 \times 10^{14}\text{/s}} = 5.0\underline{1}0 \times 10^{-7} \text{ m}$$

= 501 nm, in the visible spectrum

7.76 First use Avogadro's number to calculate the energy per H_2 molecule.

$$\frac{432 \text{ kJ}}{1 \text{ mol}} \times \frac{1000 \text{ J}}{1 \text{ kJ}} \times \frac{1 \text{ mol}}{6.02 \times 10^{23} \text{ molec.}} = 7.1\underline{7}6 \times 10^{-19} \text{ J/molec.}$$

Then convert energy to frequency and finally to wavelength.

$$\nu = \frac{E}{h} = \frac{7.176 \times 10^{-19} \text{ J}}{6.63 \times 10^{-34} \text{ J•s}} = 1.0\underline{8}2 \times 10^{15}\text{/s}$$

$$\lambda = \frac{c}{\nu} = \frac{3.00 \times 10^{8} \text{ m/s}}{1.082 \times 10^{15}\text{/s}} = 2.7\underline{7}2 \times 10^{-7} \text{ m (= 277 nm, in the UV)}$$

7.77 First calculate the energy needed to heat the 0.250 L of water from 20.0°C to 100.0°C.

$$0.250\ \text{L} \times \frac{1000\ \text{g}}{1\ \text{L}} \times \frac{4.184\ \text{J}}{(\text{g}\bullet°\text{C})} \times (100.0 - 20.0)°\text{C} = 8.3\underline{6}8 \times 10^4\ \text{J}$$

Then calculate the frequency, the energy of one photon, and the number of photons.

$$\nu = \frac{c}{\lambda} = \frac{3.00 \times 10^8\ \text{m/s}}{0.125\ \text{m}} = 2.4\underline{0}0 \times 10^9/\text{s}$$

$$\text{E of 1 photon} = h\nu = (6.63 \times 10^{-34}\ \text{J}\bullet\text{s}) \times (2.40 \times 10^9/\text{s}) = 1.5\underline{9}1 \times 10^{-24}\ \text{J}$$

$$\text{No. photons} = h\nu = 8.368 \times 10^4\ \text{J} \times \frac{1\ \text{photon}}{1.591 \times 10^{-24}\ \text{J}} = 5.2\underline{5}9 \times 10^{28}$$

$$= 5.26 \times 10^{28}\ \text{photons}$$

7.78 First calculate the energy needed to heat the 1.00 L of water from 20.0°C to 30.0°C.

$$1.00\ \text{L} \times \frac{1000\ \text{g}}{1\ \text{L}} \times \frac{4.184\ \text{J}}{(\text{g}\bullet°\text{C})} \times (30.0 - 20.0)°\text{C} = 4.1\underline{8}4 \times 10^4\ \text{J}$$

Then calculate the frequency, the energy of one photon, and the number of photons.

$$\nu = \frac{c}{\lambda} = \frac{3.00 \times 10^8\ \text{m/s}}{2.80 \times 10^{-6}\ \text{m}} = 1.0\underline{7}1 \times 10^{14}/\text{s}$$

$$\text{E of 1 photon} = h\nu = (6.63 \times 10^{-34}\ \text{J}\bullet\text{s}) \times (1.071 \times 10^{14}/\text{s}) = 7.1\underline{0}07 \times 10^{-20}\ \text{J}$$

$$\text{No. photons} = h\nu = 4.184 \times 10^4\ \text{J} \times \frac{1\ \text{photon}}{7.1007 \times 10^{-20}\ \text{J}} = 5.8\underline{9}2 \times 10^{23}$$

$$= 5.89 \times 10^{23}\ \text{photons}$$

7.79 First write the following equality for the energy to remove one electron, $E_{removal}$:

$$E_{removal} = E_{425\ nm} - E_k \text{ of ejected photon}$$

Use $E = h\nu$ to calculate the energy of the photon. Then recall from Chapter 6 that E_k, kinetic energy, = $1/2mv^2$. Use this to calculate E_k.

(continued)

$$E_{405\,nm} = \frac{hc}{\lambda} = \frac{(6.63 \times 10^{-34}\ J\bullet s) \times (3.00 \times 10^{8}\ m/s)}{4.25 \times 10^{-7}\ m} = 4.6\underline{8}0 \times 10^{-19}\ J$$

$$E_k = 1/2mv^2 = \frac{1}{2} \times (9.10953 \times 10^{-31}\ kg) \times (4.88 \times 10^5\ m/s)^2 = 1.0\underline{8}47 \times 10^{-19}\ J$$

Subtract to find $E_{removal}$ and convert it to kJ/mol:

$$E_{removal} = 4.680 \times 10^{-19}\ J - (1.0847 \times 10^{-19}\ J) = 3.5\underline{9}53 \times 10^{-19}$$

$$= 3.60 \times 10^{-19}\ J/electron$$

$$E_{removal} = \frac{3.5953 \times 10^{-19}\ J}{e^-} \times \frac{6.02 \times 10^{23}\ e^-}{1\ mol} \times \frac{1\ kJ}{1000\ J}$$

$$= 21\underline{6}.4 = 216\ kJ/mol$$

7.80 First write the following equality for the energy to remove one electron, $E_{removal}$:

$$E_{removal} = E_{405\ nm} - E_k \text{ of ejected electron}$$

Use E = hν to calculate the energy of the photon. Then recall from Chapter 6 that E_k, kinetic energy, = $1/2mv^2$. Use this to calculate E_k.

$$E_{405\,nm} = \frac{hc}{\lambda} = \frac{(6.63 \times 10^{-34}\ J\bullet s) \times (3.00 \times 10^{8}\ m/s)}{4.05 \times 10^{-7}\ m} = 4.9\underline{1}1 \times 10^{-19}\ J$$

$$E_k = 1/2mv^2 = \frac{1}{2} \times (9.10953 \times 10^{-31}\ kg) \times (3.36 \times 10^5\ m/s)^2 = 5.1\underline{4}2 \times 10^{-20}\ J$$

Subtract to find $E_{removal}$ and convert it to kJ/mol:

$$E_{removal} = 4.911 \times 10^{-19} - 5.142 \times 10^{-20} = 4.3\underline{9}69 \times 10^{-19}$$

$$= 4.40 \times 10^{-19}\ J/electron$$

$$E_{removal} = \frac{4.3969 \times 10^{-19}\ J}{e^-} \times \frac{6.02 \times 10^{23}\ e^-}{1\ mol} \times \frac{1\ kJ}{1000\ J}$$

$$= 26\underline{4}.69 = 265\ kJ/mol$$

7.81 First calculate the energy, E, in joules using the product of voltage and charge:

$$E = (4.00 \times 10^{3}\ V) \times (1.602 \times 10^{-19}\ C) = 6.4\underline{0}8 \times 10^{-16}\ J$$

Now use the kinetic energy equation $E_k = 1/2mv^2$ and solve for velocity:

$$v = \sqrt{\frac{2E_k}{m}} = \sqrt{\frac{2 \times 6.408 \times 10^{-16}\ J}{9.11 \times 10^{-31}\ kg}} = 3.7\underline{5}07 \times 10^{7}\ m/s$$

$$\lambda = \frac{h}{mv} = \frac{6.626 \times 10^{-34}\ J{\bullet}s}{(9.11 \times 10^{-31}\ kg) \times (3.7507 \times 10^{7}\ m/s)}$$

$$= 1.9\underline{3}9 \times 10^{-11}\ m\ (19.4\ pm)$$

7.82 First calculate the energy, E, in joules using the product of voltage and charge:

$$E = (1.00 \times 10^{4}\ V) \times (1.602 \times 10^{-19}\ C) = 1.6\underline{0}2 \times 10^{-15}\ J$$

Now use the kinetic energy equation $E_k = 1/2mv^2$ and solve for velocity:

$$v = \sqrt{\frac{2E_k}{m}} = \sqrt{\frac{2 \times 1.602 \times 10^{-16}\ J}{9.11 \times 10^{-31}\ kg}} = 5.9\underline{3}04 \times 10^{7}\ m/s$$

$$\lambda = \frac{h}{mv} = \frac{6.626 \times 10^{-34}\ J{\bullet}s}{(9.11 \times 10^{-31}\ kg) \times (5.9304 \times 10^{7}\ m/s)}$$

$$= 1.2\underline{2}6 \times 10^{-11}\ m\ (12.3\ pm)$$

■ Solution to Conceptual Problem

7.83 The brightly colored fireworks that are observed during many celebrations are the result of several exothermic reactions. Based on your knowledge of electronic structure, propose how different colors are generated. Can you give an example of a reactant for fireworks that would generate your favorite color.

Discussion

The colors observed during fireworks displays are the result of highly exothermic reactions that occur from the propellant and gun powder mixtures. The exothermicity of these reactions cause the decomposition of the compounds in the mixture to form atoms. The hot mixture causes these atoms to become electronically excited. As the electrons in these atoms undergo transitions back to lower states, and eventually back to their ground states, electromagnetic radiation is emitted, giving the line spectrum of the elements in the fireworks. Some of this radiation is in the visible region of the spectrum.

Selection of an appropriate element to reproduce a particular color requires two steps. Students must first determine the wavelength region corresponding to their favorite color. Then, they must find an element that emits in that wavelength region. The line spectra of several elements are shown in Figure 7.2; others can be found in the CRC Handbook of Chemistry and Physics or Lange's Handbook of Chemistry. The video on "Flame Tests" accompanying the text gives a demonstration of colors observed from some elements. Thus, if a student were to choose yellow as a favorite color, Figure 7.4 shows that the yellow region is from about 570 nm to 610 nm. Sodium emits at 590 nm, so its compounds in fireworks should emit yellow light. You could use flame tests during class discussion to test students' conclusions.

8. ELECTRON CONFIGURATIONS AND PERIODICITY

■ Solutions to Exercises

Note on significant figures: The final answer to each mathematical solution is given first with one nonsignificant figure (the rightmost significant figure is underlined) and then is rounded to the correct number of figures. Intermediate answers usually also have at least one nonsignificant figure.

8.1 a. Possible orbital diagram.
b. Possible orbital diagram.
c. Impossible orbital diagram; there are two electrons in a 2p orbital with the same spin.
d. Possible electron configuration.
e. Impossible electron configuration; only two electrons are allowed in an s subshell.
f. Impossible electron configuration; only six electrons are allowed in a p subshell.

8.2 Look at the periodic table. Start with hydrogen and go through the periods, writing down the subshells being filled, stopping with manganese (Z = 25). You obtain the following order:

Order:	1s	2s2p	3s3p	4s3d
Period:	first	second	third	fourth

Now fill the subshells with electrons, remembering that you have a total of 25 electrons to distribute. You obtain

$1s^2 2s^2 2p^6 3s^2 3p^6 4s^2 3d^5$

or

$1s^2 2s^2 2p^6 3s^2 3p^6 3d^5 4s^2$

8.3 Arsenic is in Group VA and in Period 4, so the five outer electrons should occupy the 4s and 4p subshells; the five valence electrons have the configuration $4s^24p^3$.

8.4 Because the sum of the $6s^2$ and $6p^2$ electrons gives four outer (valence) electrons, lead should be in Group IVA, and it is. Looking at the table, you find lead in Period 6. From its position, it would be classified as a main-group element.

8.5 The electronic configuration of phosphorus is $1s^22s^22p^63s^23p^3$. The orbital diagram is:

8.6 The radius tends to decrease across a row of the periodic table from left to right, and it tends to increase from the top of a column to the bottom. Therefore, in order of increasing radius,

Be < Mg < Na

8.7 It is more likely that 1000 kJ/mol is the ionization energy for iodine because ionization energies tend to decrease with atomic number in a group (I is below Cl in Group VIIA).

8.8 Fluorine should have a more negative electron affinity because (1) carbon has only two electrons in the p subshell, (2) the -1 fluoride ion has a stable noble-gas configuration, and (3) the electron can approach the fluorine nucleus more closely than the carbon nucleus.

8.9 a. The formula is H_2Se. (Group VIA elements form hydrogen compounds of the form H_2X.)
b. The formula is $CaSeO_4$.

■ Answers to Review Questions

8.1 In the original Stern-Gerlach experiment, a beam of silver atoms is directed into the field of a specially designed magnet. (The same can be done for other atoms.) The beam of silver atoms, like the hydrogen atoms described in the first section, is split into two by the magnetic field; half are bent toward one magnetic pole face, and the other half toward the other magnetic pole face. This effect shows that the atoms themselves act as magnets with a positive or a negative component, as indicated by the positive or negative spin quantum numbers.

8.2 In effect, the electron acts as though it were a sphere of spinning charge (Figure 8.3). Like any circulating electric charge, it creates a magnetic field with a spin axis that has more than one possible direction relative to a magnetic field. Electron spin is subject to a quantum restriction to one of two directions, corresponding to the m_s quantum numbers +1/2 and -1/2.

8.3 The Pauli exclusion principle limits the configurations of an atom by excluding configurations in which two or more electrons have the same four quantum numbers. For example, each electron in the same orbital must have different m_s values. This also implies that only two electrons occupy one orbital.

8.4 According to the principles discussed in Section 7.5, the number of orbitals in the g subshell ($l = 4$) is given by $2l + 1$ and is thus equal to 9. Because each orbital can hold a maximum of 2 electrons, the g subshell can hold a maximum of 18 electrons.

8.5 The orbitals in order of increasing energy up to and including the 3p orbitals (but not including the 3d orbitals) are as follows: 1s, 2s, 2p, 3s, and 3p.

8.6 The *noble-gas core* is an inner-shell configuration corresponding to one of the noble gases. The *pseudo-noble-gas-core* is an inner-shell configuration corresponding to one of the noble gases together with $(n-1)d^{10}$ electrons. Like the noble-gas core electrons, the d^{10} electrons are not involved in chemical reactions. The *valence electron* is an electron (of an atom) located outside the noble-gas core or pseudo-noble-gas core. It is an electron primarily involved in chemical reactions.

8.7 The orbital diagram for the $1s^22s^22p^4$ ground state of oxygen is

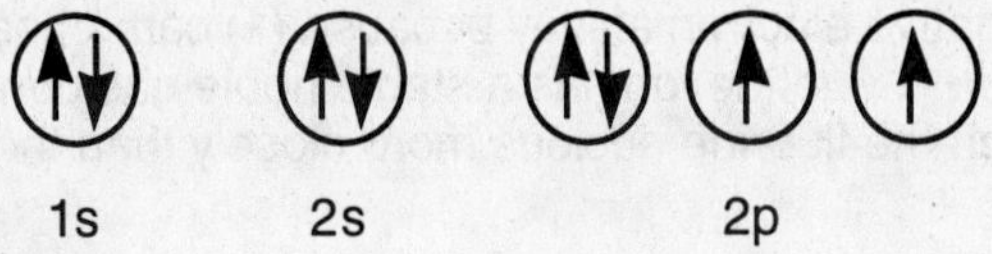

Another possible oxygen orbital diagram, but not a ground state, is

8.8 A *diamagnetic substance* is a substance that is not attracted by a magnetic field or is very slightly repelled by such a field. This property generally indicates that the substance has only paired electrons. A *paramagnetic substance* is a substance that is weakly attracted by a magnetic field. This property generally indicates that the substance has one or more unpaired electrons. *Ground-state oxygen* has two unpaired 2p electrons and is therefore paramagnetic.

8.9 In Groups IA and IIA, the outer s subshell is being filled: s^1 for Group IA and s^2 for Group IIA. In Groups IIIA to VIIIA, the outer p subshell is being filled: p^1 for IIIA, p^2 for IVA, p^3 for VA, p^4 for VIA, p^5 for VIIA, and p^6 for VIIIA. In the transition elements, the (n - 1)d subshell is being filled, from d^1 to d^{10} electrons. In the lanthanides and actinides, the f subshell is being filled, from f^1 to f^{14} electrons.

8.10 Mendeleev arranged the elements in increasing order of atomic weight, an arrangement that was later changed to atomic numbers. His periodic table was divided into rows (periods) and columns (groups). In his first attempt, he left spaces for what he believed to be undiscovered elements. In row 5, under aluminum and above indium in Group III he left a blank space. This Group III element he called eka-aluminum, and he predicted its properties from those of aluminum and indium. Later, the French chemist de Boisbaudran discovered this element and named it gallium.

8.11 In a plot of atomic radii versus atomic number (Figure 8.15), the major trends that emerge are the following: (1) Within each period (horizontal row), the atomic radius tends to decrease with increasing atomic number or nuclear charge. The largest atom in a period is thus the Group 1A atom, and the smallest atom in a period is thus the noble-gas atom. (2) Within each group (vertical column), the atomic radius tends to increase with the period number.

In a plot of ionization energy versus atomic number (Figure 8.17), the major trends are (1) the ionization energy within a period increases with atomic number, and (2) the ionization energy within a group tends to decrease going down the group.

8.12 The alkaline earth element with the smallest radius is beryllium (Be).

8.13 Group VIIA (halogens) is the main group with the most negative electron affinities. Configurations with filled subshells (ground states of the noble-gas elements) would form unstable negative ions when adding one electron per atom.

8.14 The Na^+ and Mg^{2+} ions are stable because they are isoelectronic with the noble gas neon. If Na^{2+} and Mg^{3+} ions were to exist, they would be very unstable because they would not be isoelectronic with any noble-gas structure and because of the energy needed to remove an electron from an inner shell.

8.15 The elements tend to increase in metallic character from right to left in any period. They also tend to increase in metallic character down any column (group) of elements.

8.16 A *basic oxide* is an oxide that reacts with acids and gives basic solutions with water. An example is calcium oxide, CaO. An *acidic oxide* is an oxide that reacts with bases; it gives acidic solutions with water. An example is carbon dioxide, CO_2.

8.17 Rubidium is the alkali metal atom with a $5s^1$ configuration.

8.18 Atomic number = 117 (protons in last known element + those needed to reach Group VIIA).

8.19 The following elements are in Groups IIIA to VIA:

Group IIIA	Group IVA	Group VA	Group VIA
B: metalloid	C: nonmetal	N: nonmetal	O: nonmetal
Al: metal	Si: metalloid	P: nonmetal	S: nonmetal
Ga: metal	Ge: metalloid	As: metalloid	Se: nonmetal
In: metal	Sn: metal	Sb: metalloid	Te: metalloid
Tl: metal	Pb: metal	Bi: metal	Po: metal

Yes, each column displays the expected increasing metallic character.

8.20 The oxides of the following elements are listed as either acidic, basic, amphoteric, or t.g.n.i. (text gives no information):

Group IIIA	Group IVA	Group VA	Group VIA
B: acidic	C: acidic	N: acidic	O: amphoteric (H_2O)
Al: amphoteric	Si: acidic	P: acidic	S: acidic
Ga: amphoteric	Ge: t.g.n.i.	As: t.g.n.i.	Se: acidic
In: basic	Sn: amphoteric	Sb: t.g.n.i.	Te: t.g.n.i.
Tl: basic	Pb: amphoteric	Bi: t.g.n.i.	Po: t.g.n.i.

8.21 $2K(s) + 2H_2O(l) \rightarrow 2KOH(aq) + H_2(g)$

8.22 Barium should be a soft, reactive metal. Barium should form the basic oxide, BaO. Barium metal, for example, would be expected to react with water according to the equation

$$Ba(s) + 2H_2O(l) \rightarrow Ba(OH)_2(aq) + H_2(g)$$

8.23 The two common allotropes of carbon are graphite, a soft black substance, and diamond, a very hard, clear, crystalline substance.

8.24 a. White phosphorus b. Sulfur c. Bromine d. Sodium

■ Solutions to Practice Problems

Note on significant figures: The final answer to each problem is given first with one nonsignificant figure (the rightmost significant figure is underlined) and then is rounded to the correct number of figures. Intermediate answers usually also have at least one nonsignificant figure. Atomic weights are rounded to two decimal places, except for that of hydrogen.

8.25 a. Allowed; $1s^2 2s^1 2p^3$.
b. Not allowed; the two electrons in the first 2p orbital must have opposite spins.

c. Allowed; $1s^22s^22p^4$.
d. Not allowed; the 2s orbital should have just the first two electrons, not the third.

8.26 a. Not allowed; the electrons in the 2s orbital should have opposite spins.
b. Allowed; electron configuration is $1s^22s^22p^3$.
c. Not allowed; the electrons in the second 2p orbital must have opposite spins.
d. Allowed; electron configuration is $1s^22s^22p^4$.

8.27 a. Impossible; the 2s orbital can only hold two electrons.
b. Possible.
c. Impossible; the 2p subshell can only hold six electrons.
d. Possible; however, the 3s and 3p orbitals should be filled before the 3d orbital.

8.28 a. Possible.
b. Possible; however, the 3s and 3p orbitals should be filled before the 3d orbital .
c. Impossible state; the 2p subshell can hold no more than six electrons.
d. Impossible state; the 3d subshell can hold no more than ten electrons.

8.29 The six possible orbital diagrams for $1s^22p^1$ are

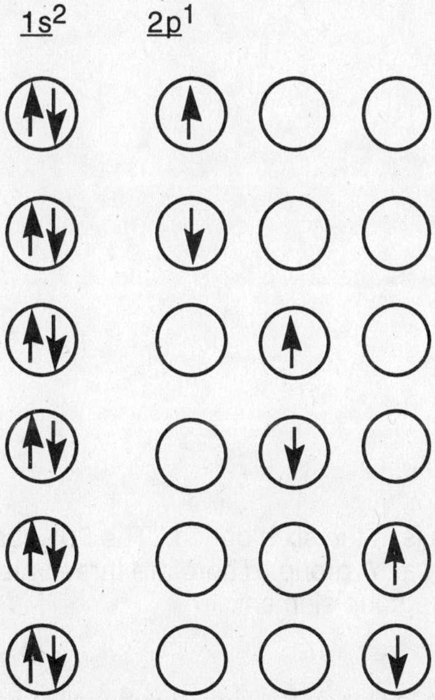

8.30 The twelve possible orbital diagrams for $1s^12p^1$ are

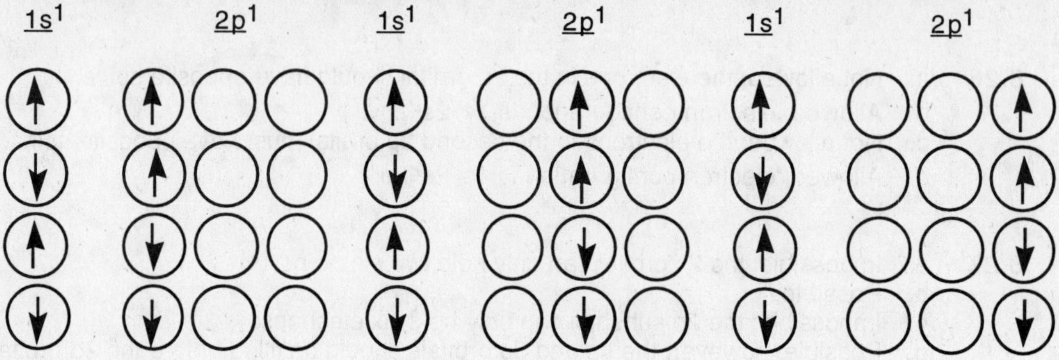

8.31 Phosphorus (Z = 15): $1s^22s^22p^63s^23p^3$

8.32 Chlorine (Z = 17): $1s^22s^22p^63s^23p^5$

8.33 Vanadium (Z = 23): $1s^22s^22p^63s^23p^63d^34s^2$

8.34 Cobalt (Z = 27): $1s^22s^22p^63s^23p^63d^74s^2$

8.35 Bromine (Z = 35): $4s^24p^5$

8.36 Bismuth (Z = 83): $6s^26p^3$

8.37 Titanium (Z = 22): $3d^24s^2$

8.38 Cadmium (Z = 48): $4d^{10}5s^2$

8.39 The highest value of n is 6, so thallium (Tl) is in the sixth period. The 5d subshell is filled and there is a 6p electron, so Tl belongs in an A group. There are three valence electrons, so Tl is in Group IIIA. It is a main-group element.

8.40 The highest value of n is 6, so iridium (Ir) is in Period 6. The 5d subshell is not completely filled, so Ir belongs to a B group. There are nine valence electrons, so Ir is in Group VIIIB. Iridium is a d-transition element.

8.41 Nickel (Z = 28): [Ar]

3d 4s

8.42 Terbium (Z = 65): [Xe]

4f 6s

8.43 Potassium (Z = 19): [Ar]

4s

All the subshells are filled in the argon core; however, the 4s electron is unpaired, causing the ground state of the potassium atom to be a paramagnetic substance.

8.44 Zinc (Z = 30): [Ar]

3d 4s

All of the subshells through the 4s subshell are filled, so the ground state of a zinc atom is a diamagnetic substance.

8.45 Atomic radius increases going down a column (group), from F to Cl, and increases going from right to left in a row, from S to Cl. Thus, the order by increasing atomic radius is F, Cl, S.

8.46 Atomic radius increases going down a column (group), from S to Se, and increases going from right to left in a row, from Se to As. Thus, the order by increasing atomic radius is S, Se, As.

8.47 Ionization energy increases going up a column (group), from Ca to Mg, and increases going left to right in a row, from Mg to S. Thus, the order by increasing ionization energy is Ca, Mg, S.

8.48 Ionization energy increases going left to right in a row. Thus, the order by increasing ionization energy is Na, Al, Cl, Ar.

8.49 a. In general, the electron affinity becomes more negative in going from left to right within a period. Thus, Cl has a more negative electron affinity than S.

b. In general, the electron affinity of a nonmetal is more negative than that of a metal. Thus, Se has a more negative electron affinity than K.

8.50 a. In general, the electron affinity becomes more negative in going from left to right within a period. Thus, Br has a more negative electron affinity than As.

b. In general, a nonmetal has a more negative electron affinity than a metal. Thus, F has a more negative electron affinity than Li.

8.51 The expected positive oxidation states of tellurium are +4 and +6. The corresponding oxides have the simplest formulas of TeO_2 and TeO_3.

8.52 Because chlorine forms the ClO_4^- ion, bromine should form the BrO_4^- ion, and the expected formula of lithium perbromate is $LiBrO_4$.

■ Solutions to Unclassified Problems

8.53 Strontium: $1s^22s^22p^63s^23p^63d^{10}4s^24p^65s^2$

8.54 Tin: $1s^22s^22p^63s^23p^63d^{10}4s^24p^64d^{10}5s^25p^2$

8.55 Polonium: $6s^26p^4$

8.56 Thallium: $6s^26p^1$

8.57 The orbital diagram for arsenic is:

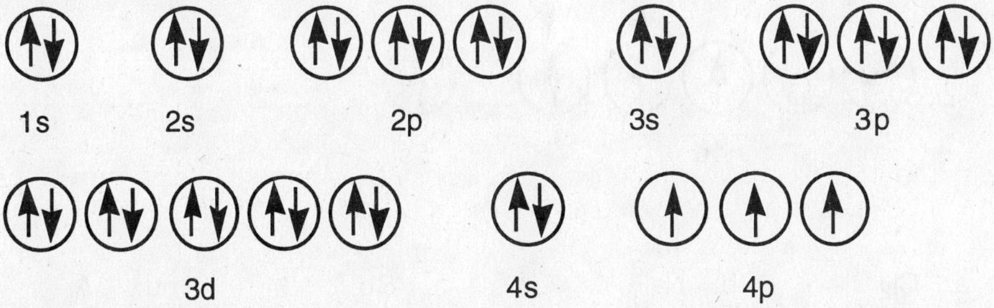

8.58 The orbital diagram for germanium is

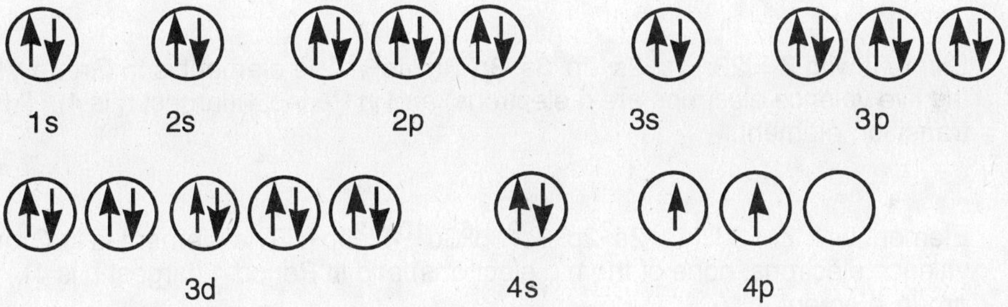

8.59 For eka-lead: [Rn] $5f^{14}6d^{10}7s^27p^2$. It is a metal; the oxide is eka-PbO or eka-PbO_2.

8.60 For eka-bismuth: [Rn] $5f^{14}6d^{10}7s^27p^3$. It is a metal; the oxide is eka-Bi_2O_3 or eka-Bi_2O_5.

8.61 The ionization energy of Fr is ~370 kJ/mol (slightly less than that of Cs).

8.62 The ionization energy of At is <1000 kJ/mol (slightly less than that of I).

8.63 Niobium: [Kr]

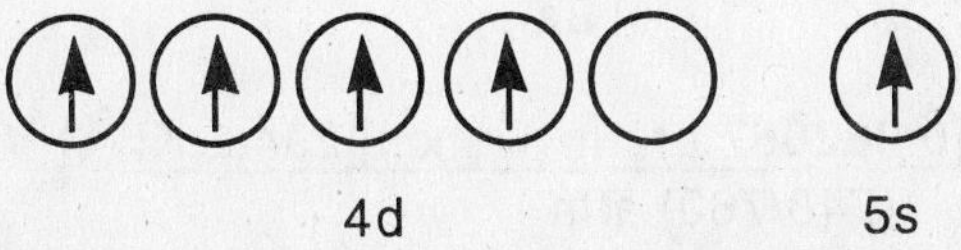

8.64 Ruthenium: [Kr]

8.65 a. Cl_2 b. Na c. Sb d. Ar

8.66 a. F_2 b. Ba c. Ga d. O_2

8.67 Element with Z = 23: $1s^2 2s^2 2p^6 3s^2 3p^6 3d^3 4s^2$. The element is in Group VB (three of the five valence electrons are d electrons) and in Period 4 (largest n is 4). It is a d-block transition element.

8.68 Element with Z = 33: $1s^2 2s^2 2p^6 3s^2 3p^6 3d^{10} 4s^2 4p^3$. The element is in Group VA (five valence electrons, none of them d electrons) and in Period 4 (largest n is 4). It is a main-group element.

■ Cumulative-Skills Problems
(require skills from previous chapters)

8.69 The equation is:

$$Ba(s) + 2H_2O(l) \rightarrow Ba(OH)_2(aq) + H_2(g)$$

Using the equation, calculate the moles of H_2; then use the ideal gas law to convert to volume.

$$\text{mol } H_2 = 2.50 \text{ g Ba} \times \frac{1 \text{ mol Ba}}{137.33} \times \frac{1 \text{ mol } H_2}{1 \text{ mol Ba}} = 0.018\underline{2}04 \text{ mol } H_2$$

$$V = \frac{nRT}{P} = \frac{[0.018204 \text{ mol}][0.082057 \text{ L•atm/(K•mol)}][294.2 \text{ K}]}{(748/760) \text{ atm}}$$

$$= 0.44\underline{6}51 \text{ L } (447 \text{ mL})$$

8.70 The equation is:

$$2Cs(s) + 2H_2O(l) \rightarrow 2CsOH(aq) + H_2(g)$$

Use the ideal gas law to calculate moles of H_2 from 48.1 mL (0.0481 L). Then use the equation to convert moles of H_2 to moles and mass of Cs.

$$\text{mol } H_2 = n = \frac{PV}{RT} = \frac{[(768/760)\text{ atm}][0.0481\text{ L}]}{[0.082057\text{ L•atm/(K•mol)}][292.2\text{ K}]} = 0.0020\underline{2}7\text{ mol } H_2$$

$$\text{mol Cs} = 0.002027\text{ mol } H_2 \times \frac{2\text{ mol Cs}}{1\text{ mol } H_2} = 0.004\underline{0}54\text{ mol Cs}$$

$$\text{Mass Cs} = 0.004054\text{ mol Cs} \times \frac{132.9\text{ g Cs}}{1\text{ mol Cs}} = 0.53\underline{8}7 = 0.539\text{ g Cs}$$

8.71 Radium is in Group IIA; hence, the radium cation is Ra^{2+}, and its oxide is RaO. Use the atomic weights to calculate the percentage of Ra in RaO.

$$\%\text{ Ra} = \frac{226\text{ amu Ra}}{226\text{ amu Ra} + 16.00\text{ amu O}} \times 100\% = 93.\underline{3}8 = 93.4\%\text{ Ra}$$

8.72 Tellurium is in Group VIA; hence, its anion is Te^{2-}, and its hydrogen compound is H_2Te. Use the atomic weights to calculate the percentage of Te in H_2Te.

$$\%\text{ Te} = \frac{127.60\text{ amu Te}}{127.60\text{ amu Te} + (2)(1.00794\text{ amu H})} \times 100\%$$

$$= 98.44\underline{4}7 = 98.445\%\text{ Te}$$

8.73 Convert 5.00 mg (0.00500 g) Na to moles of Na; then convert to energy using the first ionization energy of 496 kJ/mol Na.

$$\text{mol Na} = 0.00500\text{ g Na} \times \frac{1\text{ mol Na}}{22.99\text{ g Na}} = 2.1\underline{7}4 \times 10^{-4}\text{ mol Na}$$

$$2.1\underline{7}4 \times 10^{-4}\text{ mol Na} \times \frac{496\text{ kJ}}{1\text{ mol Na}} = 0.10\underline{7}8 = 0.108\text{ kJ}$$

8.74 Convert 2.65 mg (0.00265 g) Cl atoms to moles of Cl(g); then convert to energy using the electron affinity of 349 kg/mol Cl(g).

$$\text{mol Cl} = 0.00265 \text{ g Cl} \times \frac{1 \text{ mol Cl}}{35.45 \text{ g Cl}} = 7.4\underline{7}53 \times 10^{-5} \text{ mol Cl}$$

$$7.4\underline{7}53 \times 10^{-5} \text{ mol Cl} \times \frac{-349 \text{ kJ}}{1 \text{ mol Cl}} = -2.6\underline{0}8 \times 10^{-2} = -2.61 \times 10^{-2} \text{ kJ}$$

8.75 Use the Bohr formula, where $n_f = \infty$ and $n_i = 1$.

$$\Delta E = -R_H \left[\frac{1}{\infty^2} - \frac{1}{1^2} \right] = -R_H[-1] = R_H = \frac{2.179 \times 10^{-18} \text{ J}}{1 \text{ H atom}}$$

$$\text{I.E.} = \frac{2.179 \times 10^{-18} \text{ J}}{1 \text{ H atom}} \times \frac{6.022 \times 10^{23} \text{ H atoms}}{1 \text{ mol H}} = \frac{1.3\underline{1}29 \times 10^{6} \text{ J}}{1 \text{ mol H}}$$

$$= 1.313 \times 10^{3} \text{ kJ/mol H}$$

8.76 Use the Bohr formula, where $n_f = \infty$, $n_i = 1$, and $Z = 2$.

$$\Delta E = -R_H \left[\frac{2^2}{\infty^2} - \frac{2^2}{1^2} \right] = R_H[-4] = 4R_H = \frac{4 \times 2.179 \times 10^{-18} \text{ J}}{1 \text{ He}^+ \text{ ion}}$$

$$\text{I.E.} = \frac{4 \times 2.179 \times 10^{-18} \text{ J}}{1 \text{ He}^+ \text{ ion}} \times \frac{6.022 \times 10^{23} \text{ He}^+ \text{ ions}}{1 \text{ mol He}^+} = \frac{5.2\underline{4}7 \times 10^{6} \text{ J}}{1 \text{ mol He}^+}$$

$$= 5.25 \times 10^{3} \text{ kJ/mol He}^+$$

8.77 Add the three equations after reversing the equation for the lattice energy and its ΔH:

$Na(g) \rightarrow Na^+(g) + e^-$		$\Delta H = +496$ kJ/mol
$Cl(g) + e^- \rightarrow Cl^-(g)$		$\Delta H = -349$ kJ/mol
$Na^+(g) + Cl^-(g) \rightarrow NaCl(s)$		$-1(\Delta H = 786$ kJ/mol)
$Na(g) + Cl(g) \rightarrow NaCl(s)$		$\Delta H = -639$ kJ/mol

8.78 Add the three equations after reversing the equation for the lattice energy and its ΔH:

$$K(g) \rightarrow K^+(g) + e^- \qquad \Delta H = +419 \text{ kJ/mol}$$

$$Br(g) + e^- \rightarrow Br^-(g) \qquad \Delta H = -325 \text{ kJ/mol}$$

$$K^+(g) + Br^-(g) \rightarrow KBr(s) \qquad -1(\Delta H = 689 \text{ kJ/mol})$$

$$K(g) + Br(g) \rightarrow KBr(s) \qquad \Delta H = -595 \text{ kJ/mol}$$

■ Solution to Conceptual Problem

8.79 The majority of the electrical wiring in homes and businesses is base on copper metal. Copper is not the only element in the periodic table with suitable chemical and physical properties, however. Based on your knowledge of periodic relationships in chemistry, propose a reasonable alternative conductor to be used in power-line and household applications. Support your conclusion with information on key physical and chemical properties for your chosen element.

Discussion

Many elements could meet specific properties that would be desirable for electrical wiring. The periodic table can be useful in locating some elements to look at. Thus, elements with unpaired electrons might be possible candidates. However, other properties need to be considered. Some that the student should consider are electrical conductivity, ductility (how easily is it drawn into wire?), abundance and cost of the element, melting point, and chemical reactivity. Students might use a periodic-table database, such as KC? Discoverer (JCE Software), or one of the chemistry handbooks to locate properties of the elements they could use to argue their case for alternative conductors.

9. IONIC AND COVALENT BONDING

■ Solutions to Exercises

9.1 The Lewis symbol for oxygen is :Ö· and the Lewis symbol for magnesium is ·Mg·. The magnesium atom loses two electrons and the oxygen atom accepts two electrons. You can represent this electron transfer as follows:

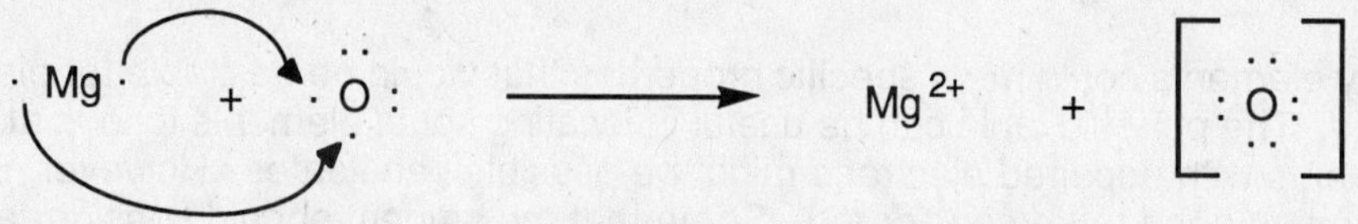

9.2 The electron configuration of the Ca atom is [Ar]$4s^2$. By losing two electrons, the atom assumes a 2+ charge and the argon configuration. The Lewis symbol is Ca^{2+}. The S atom has the configuration [Ne]$3s^2 3p^4$. By gaining two electrons, the atom assumes a 2- charge and the argon configuration. The Lewis symbol is

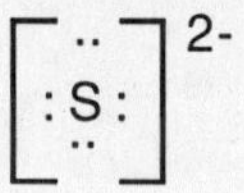

9.3 The electron configuration of lead is [Xe]$4f^{14}5d^{10}6s^2 6p^2$. The electron configuration of Pb^{2+} is [Xe]$4f^{14}5d^{10}6s^2$.

9.4 The electron configuration of manganese is [Ar]$3d^5 4s^2$. To find the ion configuration, first remove the 4s electrons, then the 3d electrons. In this case, only two electrons need to be removed. The electron configuration of Mn^{2+} is [Ar]$3d^5$.

9.5 S^{2-} has a larger radius than S. The anion has more electrons than the atom. The electron-electron repulsion is greater; hence, the valence orbitals expand. The anion radius is larger than the atomic radius.

9.6 The ionic radii increase down any column because of the addition of electron shells. All of these ions are from the IIA family; therefore, $Mg^{2+} < Ca^{2+} < Sr^{2+}$.

9.7 Cl^-, Ca^{2+}, and P^{3-} are isoelectronic. In an isoelectronic sequence, the ionic radius decreases with increasing atomic number. Therefore, in order of increasing ionic radius, we have Ca^{2+}, Cl^-, and P^{3-}.

9.8 The absolute value of the electronegativity differences are C-O, 1.0; C-S, 0.0; and H-Br, 0.7. C-O is the most polar bond.

9.9 First, calculate the total number of valence electrons. C has 4, Cl has 7, and F has 7. The total number is $4 + (2 \times 7) + (2 \times 7) = 32$. From rule 1, the expected skeleton consists of a carbon atom surrounded by Cl and F atoms. Distribute the electron pairs to the surrounding atoms to satisfy the octet rule. All 32 electrons (16 pairs) are accounted for.

$$\begin{array}{c} ..\\ :Cl:\\ ..\ \ ..\ \ ..\\ :F:C:F:\\ ..\ \ ..\ \ ..\\ :Cl:\\ .. \end{array}$$

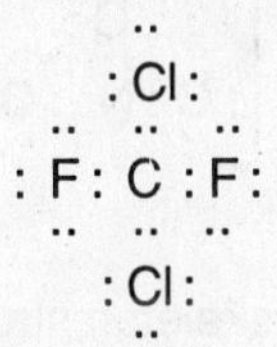

9.10 The total number of electrons in CO_2 is $4 + (2 \times 6) = 16$. Because carbon is more electropositive than oxygen, it is expected to be the central atom. Distribute the electrons to the surrounding atoms to satisfy the octet rule.

$$:\overset{..}{\underset{..}{O}}:C:\overset{..}{\underset{..}{O}}:$$

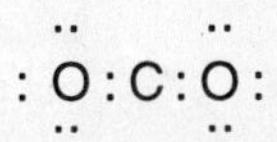

All 16 electrons have been used, but notice there are only 4 electrons on carbon. This is 4 electrons short of a complete octet, suggesting the existence of double bonds. Move a pair of electrons from each oxygen to the carbon-oxygen bonds.

$$\overset{..}{\underset{..}{O}}::C::\overset{..}{\underset{..}{O}}$$

or

$$\overset{..}{\underset{..}{O}}=C=\overset{..}{\underset{..}{O}}$$

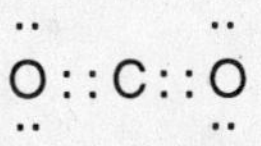

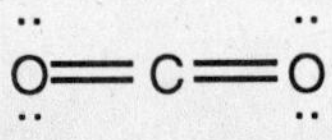

9.11 a. There are (3 x 1) + 6 = 9 valence electrons in H_3O. The H_3O^+ ion has one less electron than is provided by the neutral atoms because the charge on the ion is +1. Hence, there are 8 valence electrons in H_3O^+. The electron-dot formula is

$$\left[\begin{array}{c} H \\ \ddot{} \\ H:O:H \\ \ddot{} \end{array} \right]^+$$

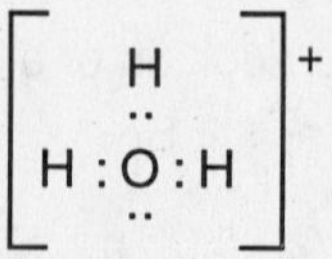

b. Cl has 7 valence electrons and O has 6 valence electrons. The total number of valence electrons from the neutral atoms is 7 + (2 x 6) = 19. The charge on the ClO_2^- is -1, which provides one more electron than the neutral atoms. This makes a total of 20 valence electrons. The electron-dot formula for ClO_2^- is

$$\left[:\ddot{O}:\ddot{Cl}:\ddot{O}: \right]^-$$

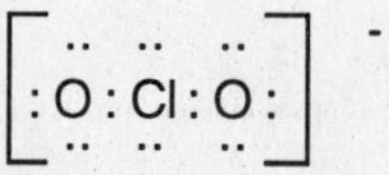

9.12 The resonance formulas for NO_3^- are

$$\left[\begin{array}{c} :\ddot{O}: \\ \| \\ :\ddot{O}:N:\ddot{O}: \end{array} \right]^- \quad \left[\begin{array}{c} :\ddot{O}: \\ \\ :\ddot{O}::N:\ddot{O}: \end{array} \right]^- \quad \left[\begin{array}{c} :\ddot{O}: \\ \\ :\ddot{O}:N::\ddot{O}: \end{array} \right]^-$$

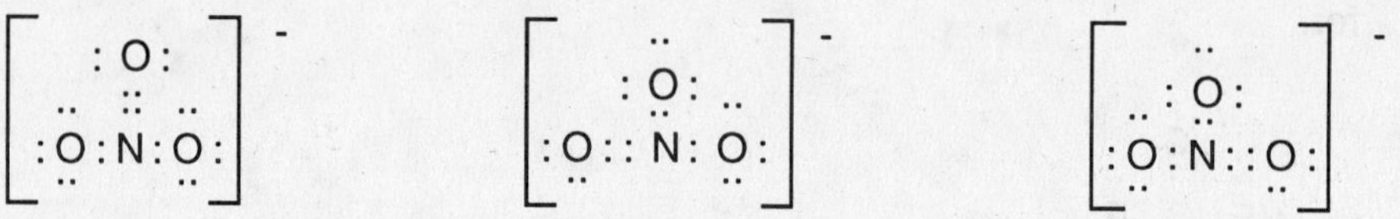

9.13 The number of valence electrons in SF_4 is 6 + (4 x 7) = 34. The skeleton structure is a sulfur atom surrounded by fluorine atoms. After the electron pairs are placed on the F atoms to satisfy the octet rule, two electrons remain.

$$\begin{array}{ccc} :\ddot{F} & & \ddot{F}: \\ & \diagdown \diagup & \\ & S & \\ & \diagup \diagdown & \\ :\ddot{F} & & \ddot{F}: \end{array}$$

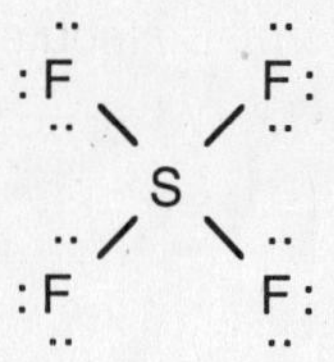

These additional two electrons are put on the sulfur atom because it can expand its octet.

$$\begin{array}{ccc} :\ddot{F} & & \ddot{F}: \\ & \diagdown \diagup & \\ & :S & \\ & \diagup \diagdown & \\ :\ddot{F} & & \ddot{F}: \end{array}$$

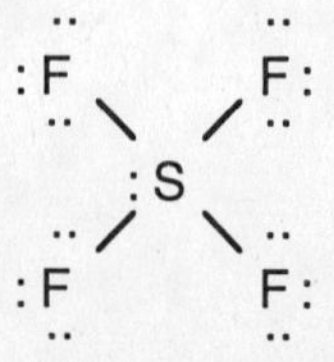

9.14 Be has two valence electrons and Cl has seven valence electrons. The total number of valence electrons is 2 + (2 x 7) = 16 in the BeCl$_2$ molecule. Be, a Group IIA element, can have fewer than eight electrons around it. The electron-dot formula of BeCl$_2$ is

$$:\!\ddot{\underset{..}{Cl}}\!:Be:\!\ddot{\underset{..}{Cl}}\!:$$

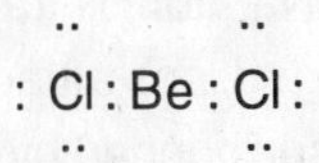

9.15 The total number of electrons in H$_3$PO$_4$ is 3 + 5 + 24 = 32. Assume a skeleton structure in which the phosphorus atom is surrounded by the more electronegative four oxygen atoms. The hydrogen atoms are then attached to the oxygen atoms. Distribute the electron pairs to the surrounding atoms to satisfy the octet rule. If you assume all single bonds (structure on the left), the formal charge on the phosphorus is +1 and the formal charge on the top oxygen is -1. Using the principle of forming a double bond with a pair of electrons on the atom with the negative formal charge, you obtain the structure on the right. The formal charge on all oxygens in this structure is zero; the formal charge on phosphorus is 5 - 5 = 0. This is the better structure.

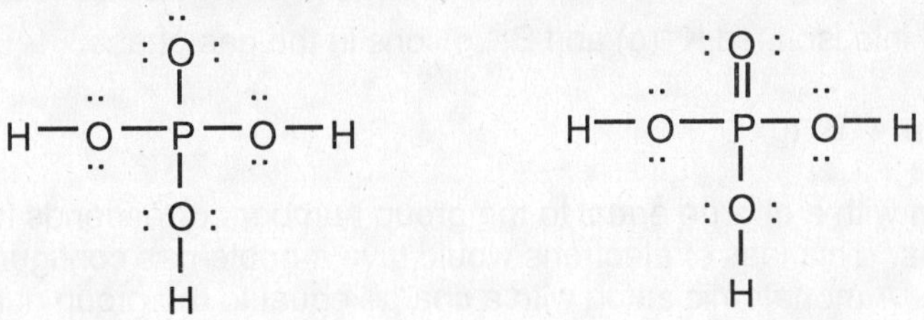

9.16 The bond length can be predicted by adding the covalent radii of the two atoms. For O-H, we have 66 pm + 37 pm = 103 pm.

9.17 As the bond order increases, the bond length decreases. Since the C=O is a double bond, we would expect it to be the shorter one, 123 pm.

9.18

$$H_2C\!=\!CH_2 + 3\,O_2 \longrightarrow 2\,O\!=\!C\!=\!O + 2\,H_2O$$

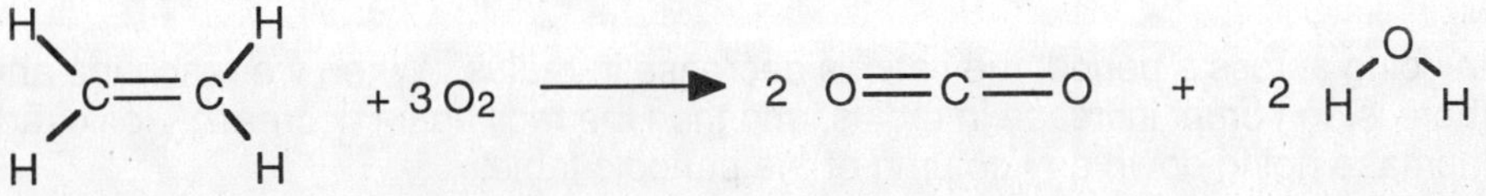

One C=C bond, four C-H bonds, and three O$_2$ bonds are broken. There are four C=O bonds and four O-H bonds formed.

$$\Delta H = \{[602 + (4 \times 411) + (3 \times 494)] - [(4 \times 799) + (4 \times 459)]\}\text{ kJ}$$
$$= -1304 \text{ kJ}$$

■ Answers to Review Questions

9.1 As an Na atom approaches a Cl atom, the outer electron of the Na atom is transferred to the Cl atom. The result is an Na^+ and a Cl^- ion. Positively charged ions attract negatively charged ions, so, finally, the NaCl crystal consists of Na^+ ions surrounded by six Cl^- ions surrounded by six Na^+ ions.

9.2 Ions tend to attract as many ions of opposite charge about them as possible. The result is that ions tend to form crystalline solids rather than molecular substances.

9.3 The energy terms involved in the formation of an ionic solid from atoms are the ionization energy of the metal atom, the electron affinity of the nonmetal atom, and the energy of the attraction of the ions forming the ionic solid. The energy for the solid will be low if the ionization energy of the metal is low, the electron affinity of the nonmetal is high, and the energy of the attraction of the ions is large.

9.4 The lattice energy for potassium bromide is the change in energy that occurs when KBr(s) is separated into isolated K^+(g) and Br^-(g) ions in the gas phase.

$$KB(s) \rightarrow K^+(g) + Br^-(g)$$

9.5 A monatomic cation with a charge equal to the group number corresponds to the loss of all valence electrons. This loss of electrons would give a noble-gas configuration, which is especially stable. A monatomic anion with a charge equal to the group number minus eight would have a noble-gas configuration.

9.6 Most of the transition elements have configurations in which the outer s subshell is doubly occupied. These electrons will be lost first, and we might expect each to be lost with almost equal ease, resulting in 2+ ions.

9.7 If we assume that the ions are spheres that are just touching, the distances between centers of the spheres will be related to the radii of the spheres. For example, in LiI, we assume that the 1- ions are large spheres that are touching. The distance between centers of the I^- ions equals two times the radius of the I^- ion.

9.8 In going across a period, the cations decrease in radius. When we reach the anions, there is an abrupt increase in radius, and then the radii again decrease. Ionic radii increase going down any column of the periodic table.

9.9 As the H atoms approach one another, their 1s orbitals begin to overlap. Each electron can then occupy the space around both atoms; that is, the two electrons are shared by the atoms.

9.10

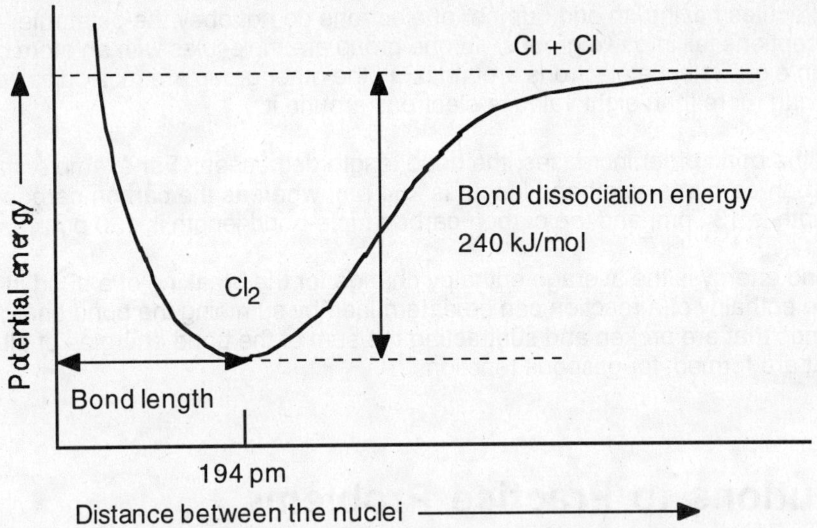

9.11 An example is thionyl chloride, $SOCl_2$:

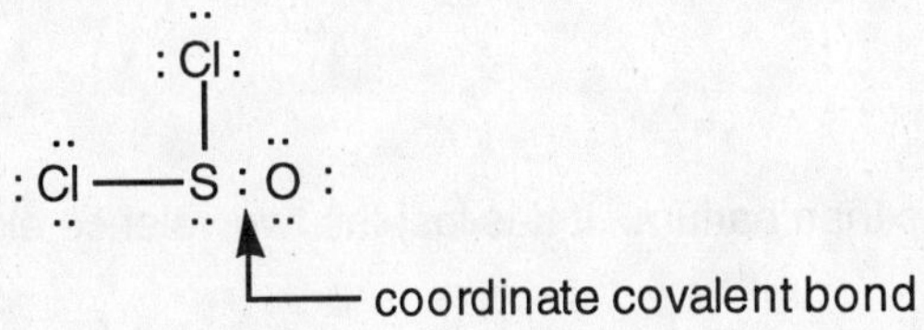

Note that the O atom has eight electrons around it; that is, it has two more electrons than the neutral atom. These two electrons must have come from the S atom. Thus, this bond is a coordinate covalent bond.

9.12 In many atoms of the main-group elements, bonding uses an s orbital and the three p orbitals of the valence shell. These four orbitals are filled with eight electrons, thus accounting for the octet rule.

9.13 Electronegativity increases from left to right (with the exception of the noble gases) and decreases from top to bottom in the periodic table.

9.14 The absolute difference in the electronegativities of the two atoms in a bond gives a rough measure of the polarity of the bond.

9.15 Resonance is used to describe the electron structure of a molecule in which bonding electrons are delocalized. In a resonance description, the molecule is described in terms of two or more Lewis formulas. If we want to retain Lewis formulas, resonance is required because each Lewis formula assumes that a bonding pair of electrons occupies the region between two atoms. We must imagine that the actual electron structure of the molecule is a composite of all resonance formulas.

9.16 Molecules having an odd number of electrons do not obey the octet rule. The other exceptions fall into two groups. In one group are molecules with an atom having fewer than eight valence electrons around it. In the other group are molecules with an atom having more than eight valence electrons around it.

9.17 As the bond order increases, the bond length decreases. For example, the average carbon-carbon single-bond length is 154 pm, whereas the carbon-carbon double-bond length is 134 pm, and the carbon-carbon triple-bond length is 120 pm.

9.18 Bond energy is the average enthalpy change for the breaking of a bond in a molecule. The enthalpy of a reaction can be determined by summing the bond energies of all the bonds that are broken and subtracting the sum of the bond energies of all the bonds that are formed, for gaseous reactions.

■ Solutions to Practice Problems

9.19 a. Ba has the valence configuration [Xe]$6s^2$. It has two valence electrons ($6s^2$), giving the Lewis formula

· Ba ·

b. Ba^{2+} has two fewer electrons than barium. It has lost the two valence electrons. The Lewis formula is

Ba^{2+}

c. Iodine has the electron configuration [Kr]$4d^{10}5s^25p^5$. It has seven valence electrons ($5s^25p^5$), giving the Lewis formula

```
 ..
: I ·
 ..
```

d. I^- has one more electron than iodine. It has gained one electron in its valence shell. The Lewis formula is

```
[  ..  ]-
[ : I : ]
[  ..  ]
```

9.20 a. In has the electron configuration [Kr]$4d^{10}5s^25p^1$. It has three valence electrons, giving the Lewis formula

```
  ·
· In ·
```

b. In^{3+} has three fewer electrons than In. It has lost the three valence electrons. The Lewis formula is

In^{3+}

c. P has the electron configuration $[Ne]3s^2 3p^3$. It has five valence electrons, giving the Lewis formula

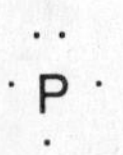

d. P^{3-} has three more valence electrons than P. It has gained three electrons in its valence shell. The Lewis formula is

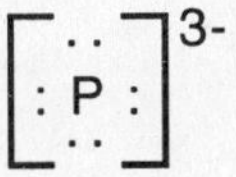

9.21 a. The Lewis symbols for Na and I are Na · and :I· (with dot pairs above and below). If the sodium atom loses one electron and the iodine atom gains one electron, both will assume noble-gas configurations. This can be represented as follows:

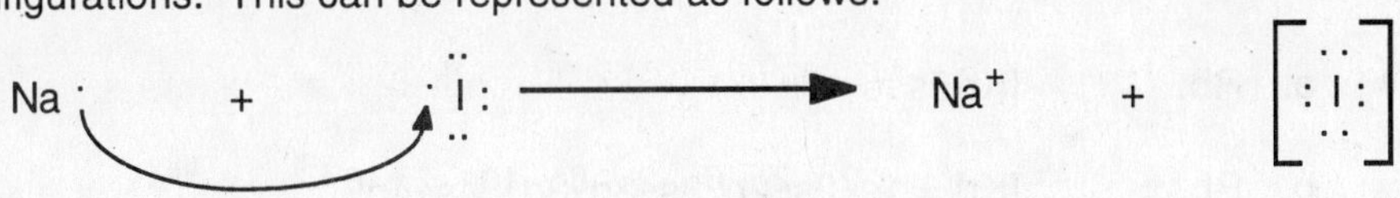

b. The Lewis symbols for Na and S are Na · and :S· (with a dot above and a dot pair below). The S atom gains two electrons to assume a noble-gas configuration. Since the sodium atom has only one valence electron to lose, two sodium atoms must take part in the electron transfer. This can be represented as follows:

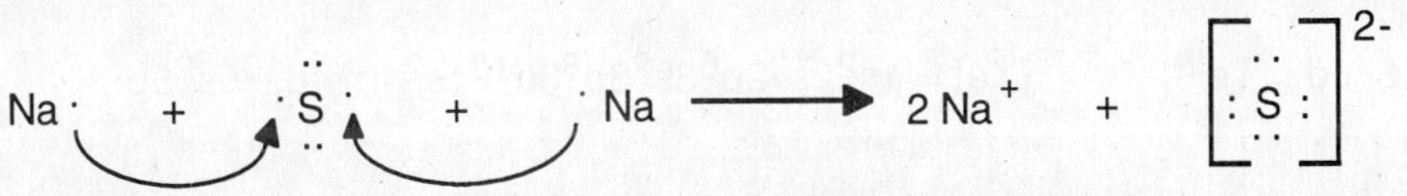

9.22 a. The Lewis symbols for Mg and I are · Mg · and :I· (with dot pairs above and below). The magnesium atom loses two electrons from its valence shell to assume a noble-gas configuration. Since an iodine can accept only one electron into its valence shell, two iodine atoms must participate in the electron transfer. The electron transfer can be represented as follows:

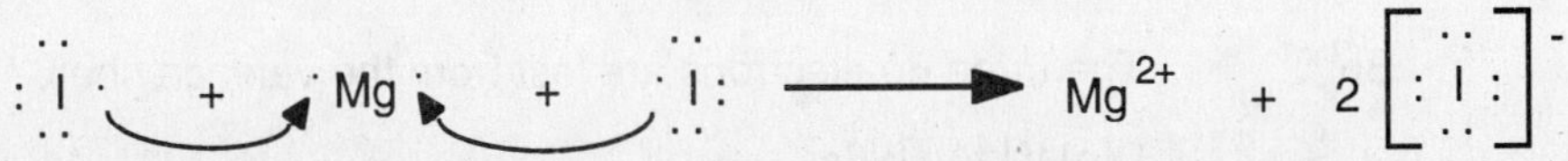

b. The Lewis symbols for Mg and Se are ·Mg· and :S̈ë· . If the magnesium atom loses two electrons and the selenium atom gains two electrons, both will have noble-gas configurations.

$$\cdot Mg\cdot \; + \; \cdot\ddot{S}\ddot{e}: \; \longrightarrow \; Mg^{2+} \; + \; \left[:\ddot{S}\ddot{e}:\right]^{2-}$$

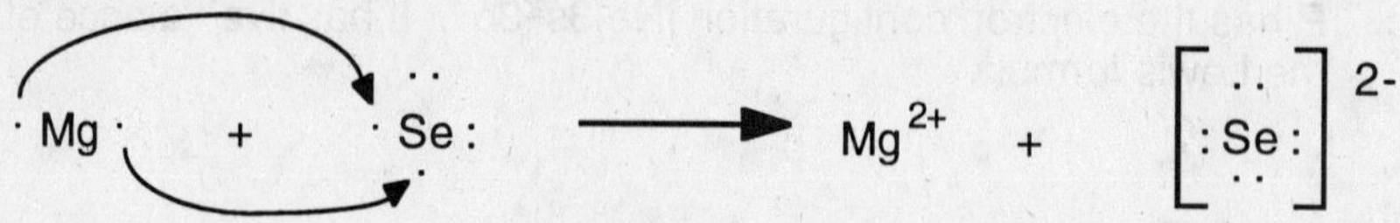

9.23 a. Mg: [Ne]$3s^2$ ·Mg·

b. Mg^{2+}:   [Ne] = $1s^2 2s^2 2p^6$ Mg^{2+}

c. Se^{2-}:   [Kr] = $1s^2 2s^2 2p^6 3s^2 3p^6 3d^{10} 4s^2 4p^6$ $\left[:\ddot{S}\ddot{e}:\right]^{2-}$

d. Br^-:   [Kr] = $1s^2 2s^2 2p^6 3s^2 3p^6 3d^{10} 4s^2 4p^6$ $\left[:\ddot{B}\ddot{r}:\right]^{-}$

9.24 a. Rb: [Kr]$5s^1$ Rb·

b. Rb^+:   [Kr] = $1s^2 2s^2 2p^6 3s^2 3p^6 3d^{10} 4s^2 4p^6$ Rb^+

c. I^-:   [Xe] = $1s^2 2s^2 2p^6 3s^2 3p^6 3d^{10} 4s^2 4p^6 4d^{10} 5s^2 5p^6$ $\left[:\ddot{I}:\right]^{-}$

d. Te^{2-}:   [Xe] = $1s^2 2s^2 2p^6 3s^2 3p^6 3d^{10} 4s^2 4p^6 4d^{10} 5s^2 5p^6$ $\left[:\ddot{T}\ddot{e}:\right]^{2-}$

9.25 a. Sn: [Kr]$4d^{10} 5s^2 5p^2$

b. Sn^{2+}:   The two 5p electrons are lost from the valence shell.
[Kr]$4d^{10} 5s^2$

9.26 a. Bi: [Xe]$4f^{14} 5d^{10} 6s^2 6p^3$

b. Bi^{3+}:   The three 6p electrons are lost from the valence shell.
[Xe]$4f^{14} 5d^{10} 6s^2$

9.27 The 2+ ion is formed by the loss of electrons from the 4s subshell.

Ni^{2+}: [Ar]$3d^8$

The 3+ ion is formed by the loss of electrons from the 4s and 3d subshells.

Ni^{3+}: [Ar]$3d^7$

9.28 The 1+ ion is formed by the loss of the one electron in the 4s subshell.

Cu^+: [Ar]$3d^{10}$

The 2+ ion is formed by the loss of a 4s and a 3d electron.

Cu^{2+}: [Ar]$3d^9$

9.29 a. $Rb^+ < Rb$

The cation is smaller than the neutral atom because it has lost all its valence electrons; hence, it has one less shell of electrons. The electron-electron repulsion is reduced, so the orbitals shrink from the increased attraction of the electrons to the nucleus.

b. $Se < Se^{2-}$

The anion is larger than the neutral atom because it has more electrons. The electron-electron repulsion is greater, so the valence orbitals expand to give a larger radius.

9.30 a. $I < I^-$

The anion is larger than the neutral atom because it has more electrons. The electron-electron repulsion is greater, so the valence orbitals expand to give a larger radius.

b. $Ca^{2+} < Ca$

The cation is smaller than the neutral atom because it has lost all its valence electrons; hence, it has one less shell of electrons. The electron-electron repulsion is reduced, so the orbitals shrink from the increased attraction of the electrons to the nucleus.

9.31 $S^{2-} < Se^{2-} < Te^{2-}$

All have the same number of electrons in the valence shell. The radius increases with the increasing number of filled shells.

9.32 P^{3-} is larger. It has the same number of valence electrons as N^{3-}, but the valence shell has n = 3 for P^{3-} and n = 2 for N^{3-}. The radius increases with the increasing number of filled shells.

9.33 Smallest Na^+ (Z = 11), F^- (Z = 9), N^{3-} (Z = 7) Largest

These ions are isoelectronic. The atomic radius increases with the decreasing nuclear charge (Z).

9.34 Smallest Cs^+ (Z = 55), I^- (Z = 53), Te^{2-} (Z = 52) Largest

These ions are isoelectronic. The atomic radius increases with the decreasing nuclear charge (Z).

9.35

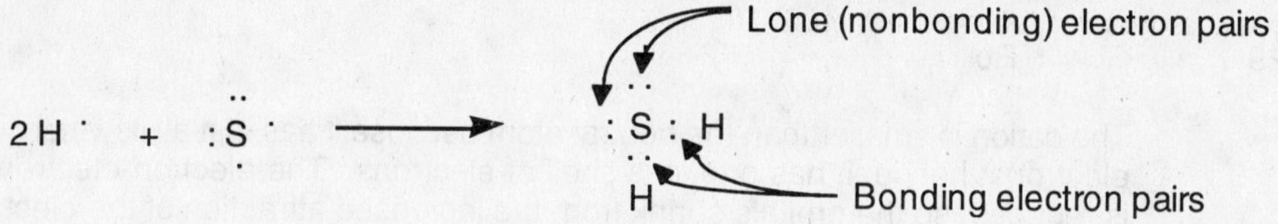

9.36

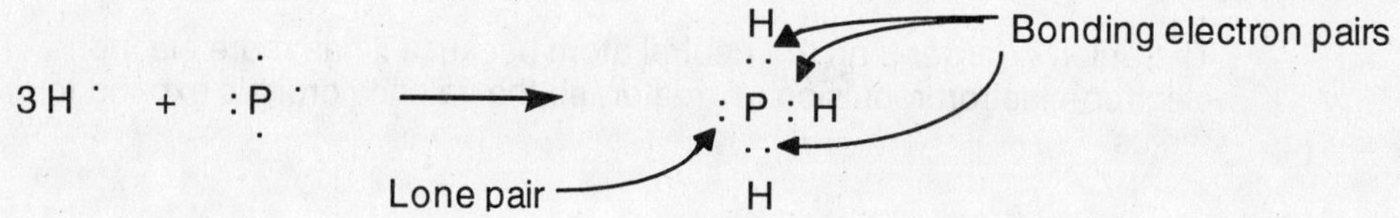

9.37 The Lewis symbols for the atoms are :Cl· and ·Si· . The silicon atom has four valence electrons. Each of these can be paired with the unpaired electron of the chlorine atom. The expected formula is $SiCl_4$.

9.38 The Lewis symbols for the atoms are H· and :As· . There are three unpaired electrons in the valence shell of arsenic. Each can be paired with the electron on the hydrogen atom to form a bond. The expected formula is AsH_3.

9.39 a. Cs, Ba, Sr

Electronegativity increases from left to right and bottom to top in the periodic table.

b. Ca, Ga, Ge

Electronegativity increases from left to right within a period.

c. As, P, S

Electronegativity increases from left to right and bottom to top in the periodic table.

9.40 a. P, N, O

Electronegativity increases from left to right and bottom to top in the periodic table.

b. Na, Mg, Al

Electronegativity increases from left to right within a period.

c. Al, Si, C

Electronegativity increases from left to right and bottom to top in the periodic table.

9.41 $X_{Se} - X_H = 2.4 - 2.1 = 0.3$

$X_{Cl} - X_P = 3.0 - 2.1 = 0.9$

$X_{Cl} - X_N = 3.0 - 3.0 = 0.0$

The difference in electronegativity is smallest for the N-Cl bond; hence, it is the least polar.

9.42 $X_O - X_{Si} = 3.5 - 1.8 = 1.7$

$X_{Br} - X_C = 2.8 - 2.5 = 0.3$

$X_{Br} - X_{As} = 2.8 - 2.0 = 0.8$

The bonds arranged by increasing difference in electronegativity are C-Br, As-Br, Si-O.

9.43 The atom with the greater electronegativity has the partial negative charge.

a. $\overset{\partial+}{\text{H}}\text{—}\overset{\partial-}{\text{Se}}$ b. $\overset{\partial+}{\text{P}}\text{—}\overset{\partial-}{\text{Cl}}$ c. Nonpolar

9.44 a. $\overset{\partial+}{\text{Si}}\text{—}\overset{\partial-}{\text{O}}$ b. $\overset{\partial+}{\text{C}}\text{—}\overset{\partial-}{\text{Br}}$ c. $\overset{\partial+}{\text{As}}\text{—}\overset{\partial-}{\text{Br}}$

9.45 a. Total number of valence electrons = 7 + 7 = 14. Br-Br is the skeleton. Distribute the remaining 12 electrons.

$$:\!\ddot{\underset{..}{Br}}\!-\!\ddot{\underset{..}{Br}}\!:$$

b. Total valence electrons = (2 x 1) + 6 = 8. By rule 1, the skeleton is H-Se-H. Distribute the remaining four electrons.

$$H\!-\!\ddot{\underset{..}{Se}}\!-\!H$$

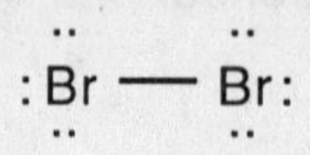

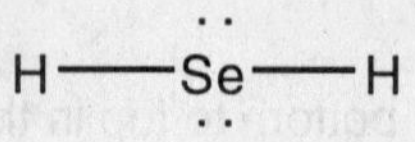

c. Total valence electrons = (2 x 7) + 4 + 6 = 24. By rule 1, the skeleton is

```
        O
        |
   F — C — F
```

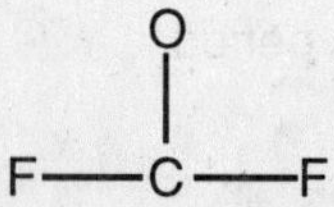

After distributing the remaining electrons, it is necessary to form a C=O bond.

```
         :O:
         ‖
    :F — C — F:
```

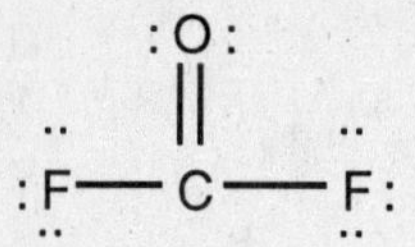

9.46 a. Total valence electrons = 7 + 7 = 14. F-Br is the skeleton. Distribute the remaining 12 electrons.

$$:\!\ddot{\underset{..}{F}}\!-\!\ddot{\underset{..}{Br}}\!:$$

b. Total valence electrons = 5 + (3 x 7) = 26. By rule 1, the skeleton is

```
         Br
         |
   Br — P — Br
```

Distribute the remaining 20 electrons.

```
          :Br:
           |
    :Br — P — Br:
```

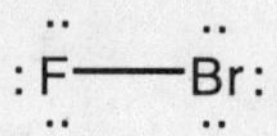

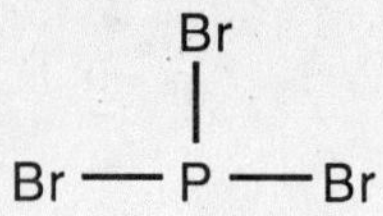

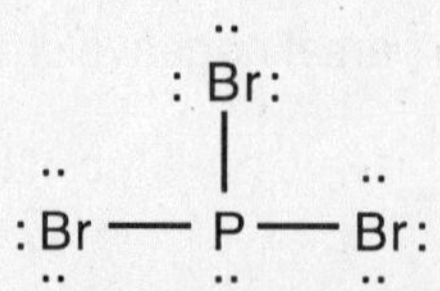

c. Total valence electrons = 7 + 5 + 6 = 18. From rule 1, the skeleton is

O — N — F

Distribute the remaining 14 electrons so that the O and F atoms have an octet.

:Ö — N — F̈:

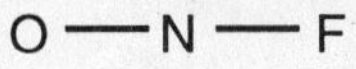

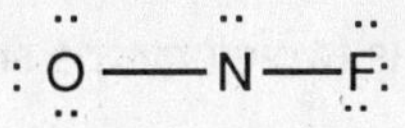

Draw a N-O double bond to achieve an N octet.

Ö = N — F̈:

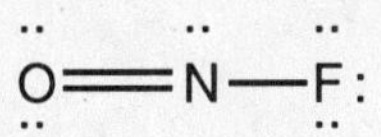

9.47 a. Total valence electrons = 2 x 5 = 10. The skeleton is P-P. Distribute the remaining electrons symmetrically:

:P — P:

Neither P atom has an octet. There are four fewer electrons than needed. This suggests the presence of a triple bond. Make one lone pair from each P a bonding pair.

:P≡P:

b. Total valence electrons = 4 + 6 + (2 x 7) = 24. By rule 1, the skeleton is

```
       Br
       |
   O — C — Br
```

Distribute the remaining 18 electrons.

```
       :Br:
        |
   :Ö — C — Br:
```

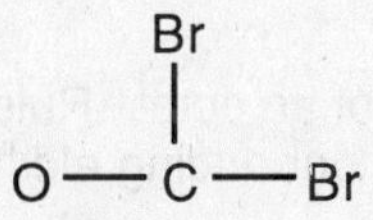

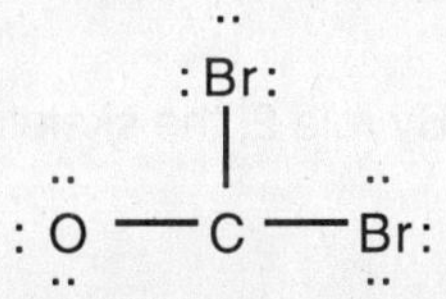

Notice that carbon is two electrons short of an octet. This suggests the presence of a double bond. By rule 4, the most likely double bond is between C and O.

```
       :Br:
        |
   :O = C — Br:
```

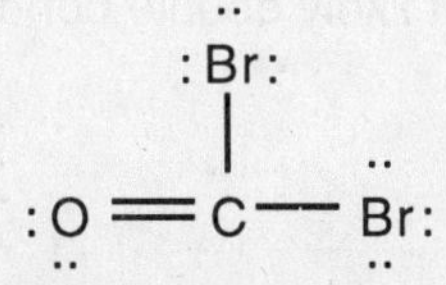

c. Total valence electrons = 1 + 5 + (2 x 6) = 18. The skeleton is most likely H-O-N-O. Distribute the remaining 12 electrons:

$$H - \ddot{\underset{..}{O}} - \ddot{N} - \ddot{\underset{..}{O}}:$$

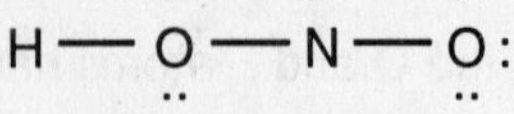

Notice that the N atom does not have an octet. It is two electrons short. By rule 4, the most likely double bond is between N and O.

$$H - \ddot{\underset{..}{O}} - \ddot{N} = \ddot{O}:$$

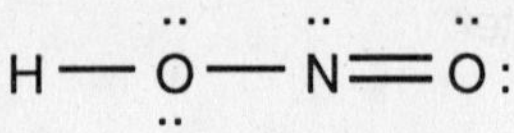

9.48 a. Total valence electrons = 4 + 6 = 10. The skeleton is C-O. Distribute the remaining eight electrons:

$$:\ddot{C} - \ddot{O}:$$

Neither atom has an octet. There are four fewer electrons than needed. This suggests a triple bond.

$$:C \equiv O:$$

b. Total valence electrons = 7 + 4 + 5 = 16. The skeleton is Br-C-N. Distribute the remaining 12 electrons:

$$:\ddot{\underset{..}{Br}} - \ddot{C} - \ddot{N}:$$

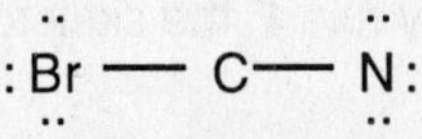

Notice that the carbon atom is four electrons short of an octet. Rule 4 suggests a triple bond between C and N made from the four nonbonding electrons on nitrogen.

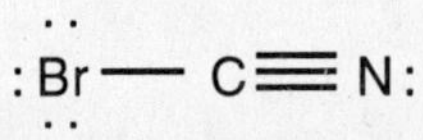

$$:\ddot{\underset{..}{Br}} - C \equiv N:$$

c. Total valence electrons = (2 x 5) + (2 x 7) = 24. By rule 2, the skeleton is F-N-N-F. Distribute the remaining 18 electrons.

$$:\ddot{\underset{..}{F}} - \ddot{N} - \ddot{N} - \ddot{\underset{..}{F}}:$$

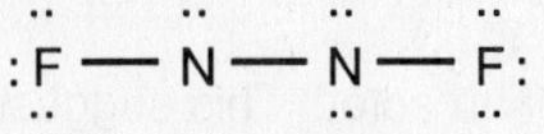

Notice that one of the nitrogens is two electrons short of an octet. This suggests the presence of a double bond. By rule 4, the most likely double bond is between the nitrogens.

$$:\ddot{\underset{..}{F}} - \ddot{N} = \ddot{N} - \ddot{\underset{..}{F}}:$$

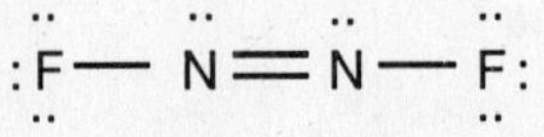

9.49 a. Total valence electrons = 7 + 6 + 1 = 14. The skeleton is Cl-O. Distribute the remaining 12 electrons.

$$\left[:\overset{..}{\underset{..}{Cl}} - \overset{..}{\underset{..}{O}}: \right]^{-}$$

b. Total valence electrons = 4 + (3 x 7) + 1 = 26. By rule 1, the skeleton is

$$\begin{array}{c} Cl \\ | \\ Cl-Sn-Cl \end{array}$$

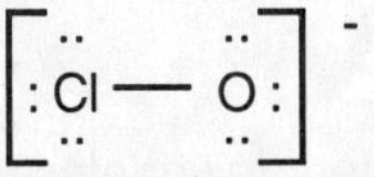

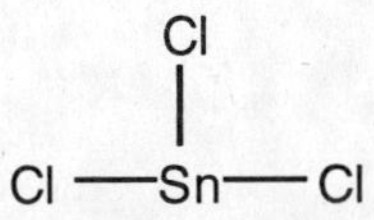

Distribute the remaining 20 electrons so that each atom has an octet.

$$\left[\begin{array}{c} :\overset{..}{Cl}: \\ | \\ :\overset{..}{\underset{..}{Cl}} - \overset{..}{\underset{..}{Sn}} - \overset{..}{\underset{..}{Cl}}: \end{array} \right]^{-}$$

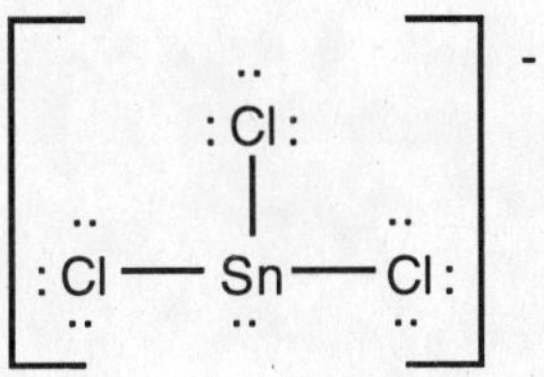

c. Total valence electrons = (2 x 6) + 2 = 14. The skeleton is S-S. Distribute the remaining 12 electrons.

$$\left[:\overset{..}{\underset{..}{S}} - \overset{..}{\underset{..}{S}}: \right]^{2-}$$

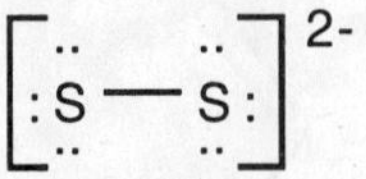

9.50 a. Valence electrons = 7 + (2 x 7) - 1 = 20. The skeleton is Br-I-Br. Distribute the remaining 16 electrons.

$$\left[:\overset{..}{\underset{..}{Br}} - \overset{..}{\underset{..}{I}} - \overset{..}{\underset{..}{Br}}: \right]^{+}$$

b. Total valence electrons = 7 + (2 x 7) - 1 = 20. By rule 2, the skeleton is F-Cl-F. Distribute the remaining 16 electrons.

$$\left[:\overset{..}{\underset{..}{F}} - \overset{..}{\underset{..}{Cl}} - \overset{..}{\underset{..}{F}}: \right]^{+}$$

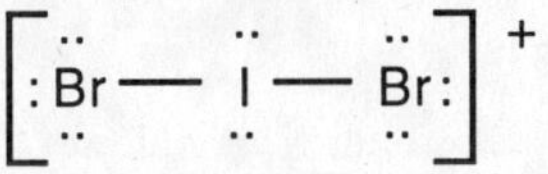

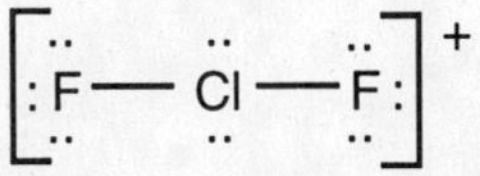

c. Total valence electrons = 4 + 5 + 1 = 10. The skeleton is C-N. Distribute the remaining eight electrons.

Notice that neither atom has an octet. There are four fewer electrons than needed, suggesting a triple bond:

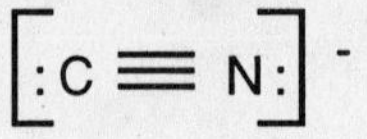

9.51 a. One possible electron-dot formula for FNO_2 is

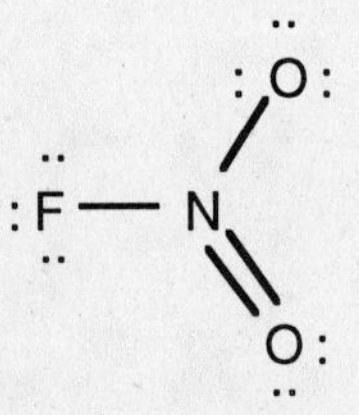

Because the nitrogen-oxygen bonds are expected to be equivalent, the structure must be described in resonance terms.

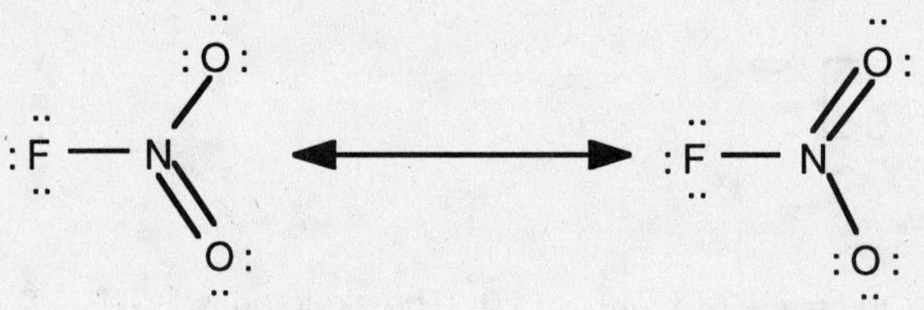

One electron pair is delocalized over the nitrogen atom and the two oxygen atoms.

b. One possible electron-dot formula for SO_3 is

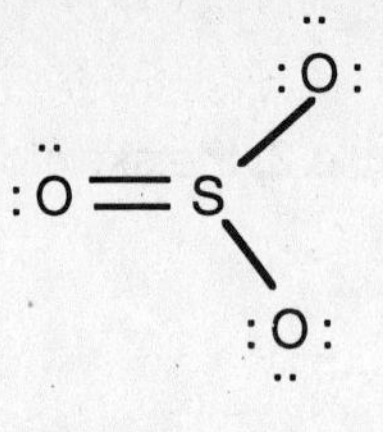

(continued)

Because the sulfur-oxygen bonds are expected to be equivalent, the structure must be described in resonance terms.

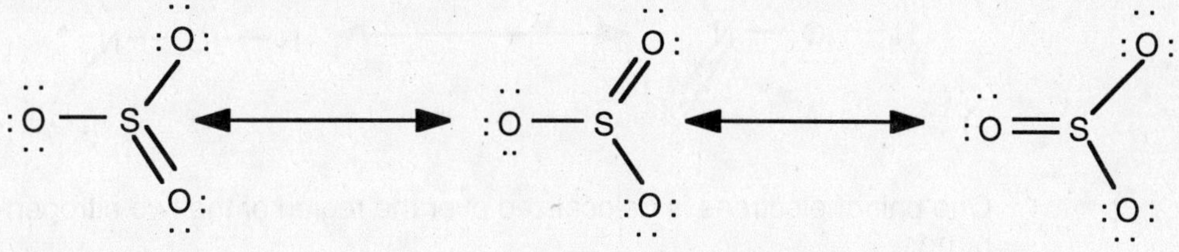

One electron pair is delocalized over the region of all three sulfur-oxygen bonds.

9.52 a. One possible electron-dot formula for NO_2^- is

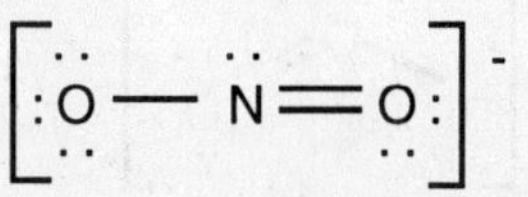

Because the nitrogen-oxygen bonds are expected to be equivalent, the structure must be described in resonance terms. One electron pair is delocalized over the region of both nitrogen-oxygen bonds.

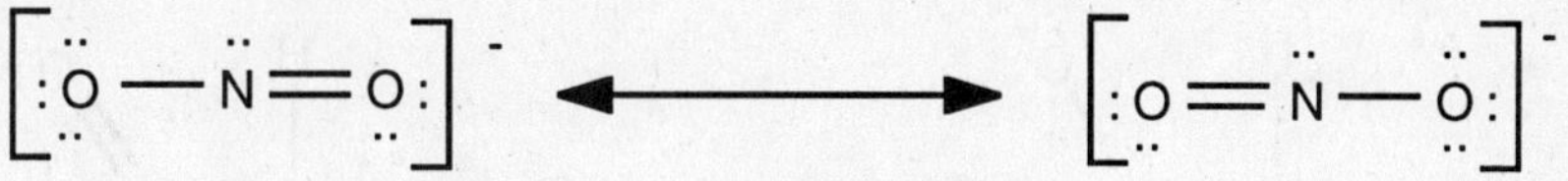

b. One possible electron-dot structure for HNO_3 is

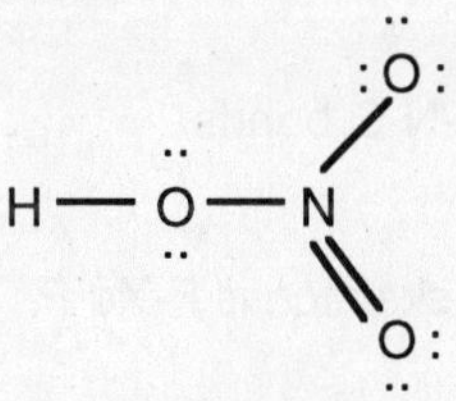

Because two of the nitrogen-oxygen bonds are expected to be equivalent, the structure must be described in terms of resonance. The oxygen bonded to the hydrogen atom is distinguishable from the other two.

(continued)

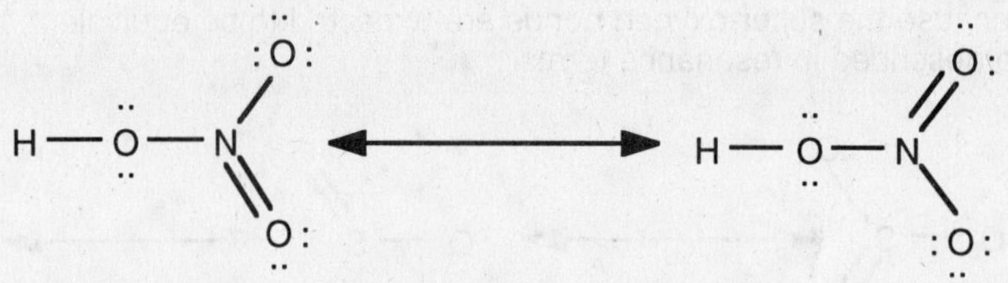

One pair of electrons is delocalized over the region of the two nitrogen-oxygen bonds.

9.53

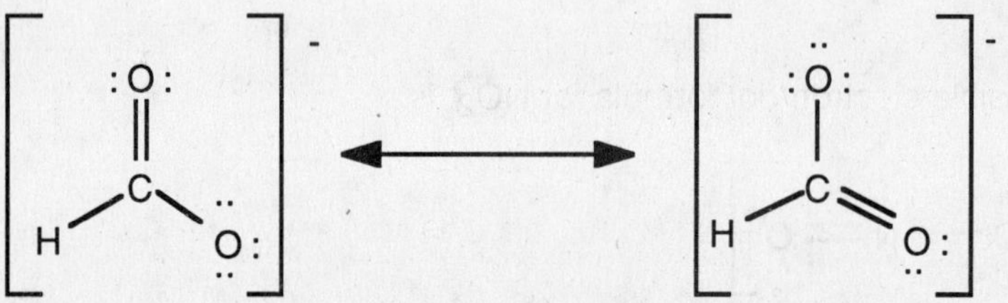

The two carbon-oxygen bonds are expected to be the same. This means that one pair of electrons is delocalized over the region of the O-C-O bonds.

9.54

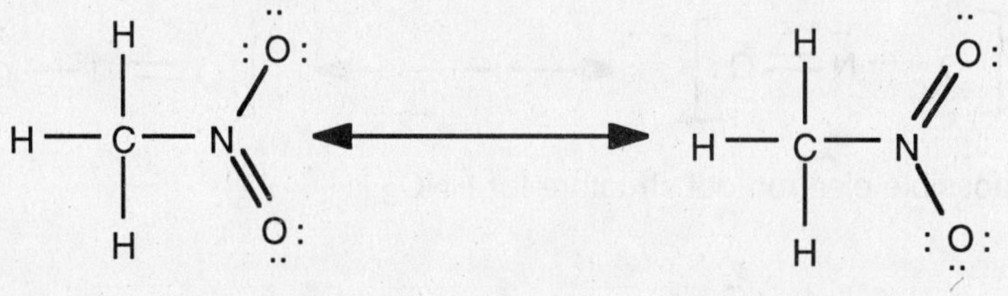

One pair of electrons is delocalized over the O-N-O bonds.

9.55 a. Total valence electrons = 8 + (2 x 7) = 22. The skeleton is F-Xe-F. Place six electrons around each fluorine atom to satisfy its octet.

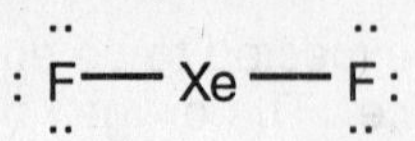

There are three electron pairs remaining. Place them on the xenon atom.

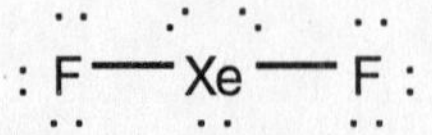

b. Total valence electrons = 6 + (4 x 7) = 34. The skeleton is

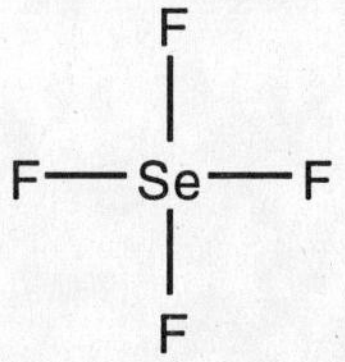

Distribute 24 of the remaining 26 electrons on the fluorine atoms. The remaining pair of electrons is placed on the selenium atom.

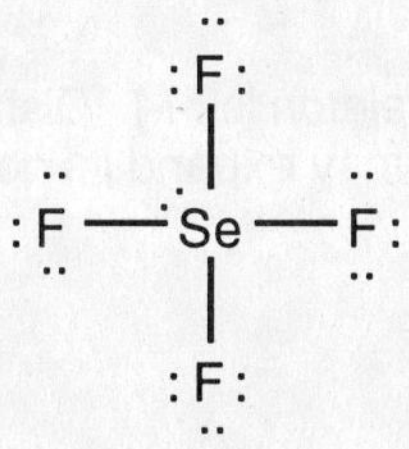

c. Total valence electrons = 6 + (6 x 7) = 48. The skeleton is

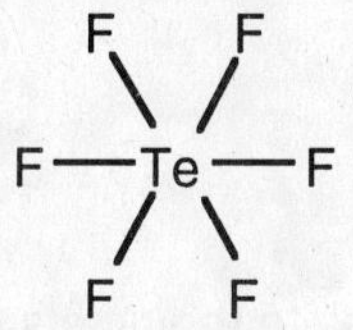

Distribute the remaining 36 electrons on the fluorine atoms.

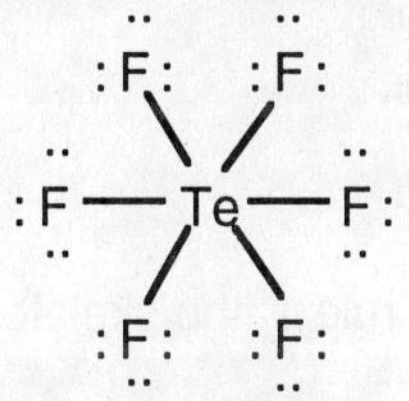

d. Total valence electrons = 8 + (5 x 7) - 1 = 42. The skeleton is

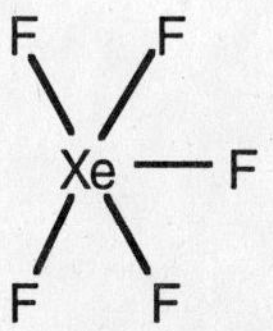

(continued)

Use 30 of the remaining 32 electrons on the fluorine atoms to complete their octets. The remaining two electrons form a lone pair on the xenon atom.

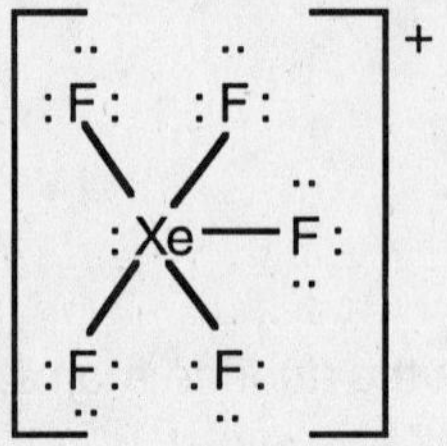

9.56 a. Total valence electrons = (3 x 7) + 1 = 22. The skeleton is I-I-I. Distribute the remaining 18 electrons. The central iodine atom may expand its octet.

$$[\ddot{\underset{\cdot\cdot}{:}\ddot{I}} - \ddot{\underset{\cdot\cdot}{I}} - \ddot{\underset{\cdot\cdot}{I}:}]^-$$

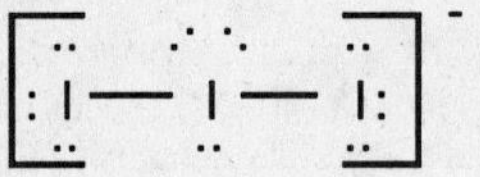

b. Total valence electrons = 7 + (3 x 7) = 28. By rule 1, the skeleton is

```
F — Cl — F
     |
     F
```

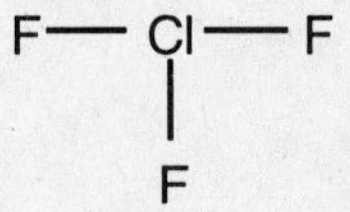

Distribute 18 of the remaining 22 electrons to complete the octets of the fluorine atoms. The four remaining electrons form two sets of lone pairs on the chlorine atom.

```
:F — Cl — F:
     |
    :F:
```

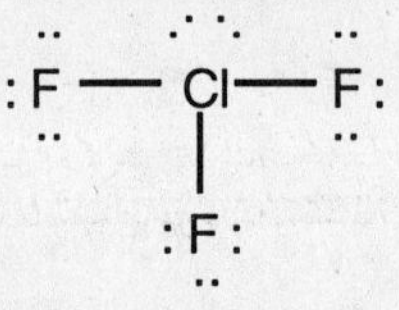

c. Total valence electrons = 7 + (4 x 7) + 1 = 36. By rule 1, the skeleton is

```
      F
      |
F — I — F
      |
      F
```

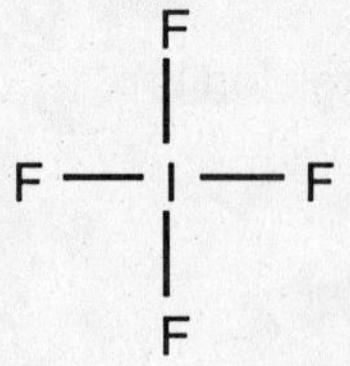

(continued)

Distribute 24 of the 28 remaining electrons to complete the octets of the fluorine atoms. The four electrons remaining form two sets of lone pairs on the iodine atom.

$$\left[\begin{array}{c} :\ddot{F}: \\ | \\ :\ddot{F}-\ddot{I}-\ddot{F}: \\ | \\ :\ddot{F}: \end{array}\right]^{-}$$

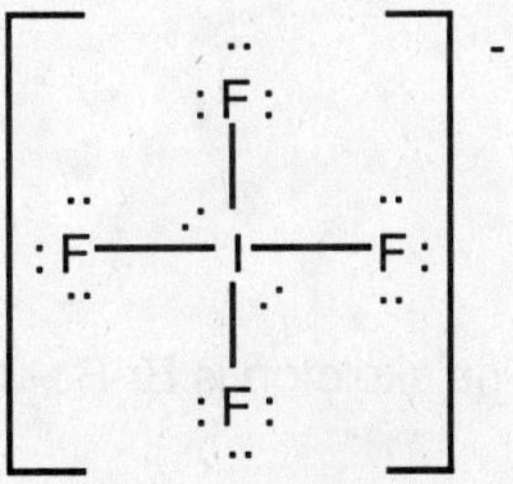

d. Total valence electrons = 7 + (5 x 7) = 42. By rule 1, the skeleton is

$$\begin{array}{c} F \quad F \\ \diagdown \diagup \\ Br - F \\ \diagup \diagdown \\ F \quad F \end{array}$$

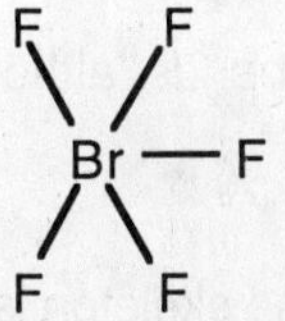

Use 30 of the remaining 32 electrons to complete the octets of the fluorine atoms. The two electrons remaining form a lone pair on the bromine atom.

$$\begin{array}{c} :\ddot{F}: \quad :\ddot{F}: \\ \diagdown \diagup \\ :\ddot{Br}-\ddot{F}: \\ \diagup \diagdown \\ :\ddot{F}: \quad :\ddot{F}: \end{array}$$

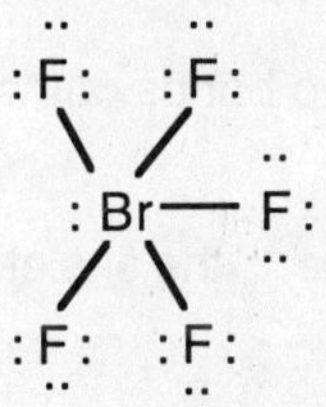

9.57 a. Total valence electrons = 3 + (3 x 7) = 24. By rule 1, the skeleton is

$$\begin{array}{c} Cl \\ | \\ Cl-B-Cl \end{array}$$

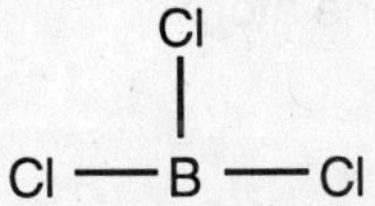

Distribute the remaining 18 electrons.

$$\begin{array}{c} :\ddot{Cl}: \\ | \\ :\ddot{Cl}-B-\ddot{Cl}: \end{array}$$

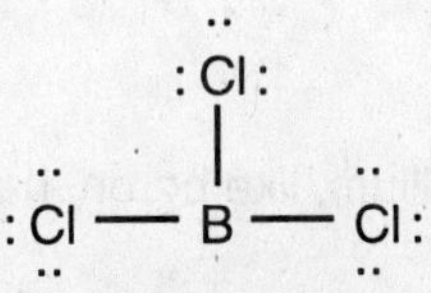

Although boron has only six electrons, it has the normal number of covalent bonds.

b. Total valence electrons = 3 + (2 x 7) - 1 = 16. The skeleton is Cl-Tl-Cl. Distribute the remaining 12 electrons.

$$\left[:\overset{..}{\underset{..}{Cl}}\!-\!Tl\!-\!\overset{..}{\underset{..}{Cl}}: \right]^+$$

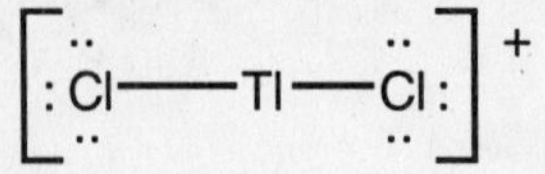

Tl has only four electrons around it.

c. Total valence electrons = 2 + (2 x 7) = 16. The skeleton is Br-Be-Br. Distribute the remaining 12 electrons.

$$:\overset{..}{\underset{..}{Br}}\!-\!Be\!-\!\overset{..}{\underset{..}{Br}}:$$

In covalent compounds, beryllium frequently has two bonds, although it does not have an octet.

9.58 a. Total valence electrons = 2 + (2 x 7) = 16. By rule 1, the skeleton is F-Be-F. Distribute the remaining 12 electrons.

$$:\overset{..}{\underset{..}{F}}\!-\!Be\!-\!\overset{..}{\underset{..}{F}}:$$

In covalent compounds, beryllium commonly has two bonds, although it does not have an octet.

b. Total valence electrons = 2 + (3 x 7) + 1 = 24. By rule 1, the skeleton is

$$\begin{array}{c} F \\ | \\ F\!-\!Be\!-\!F \end{array}$$

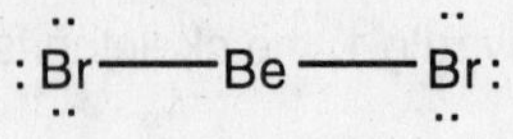

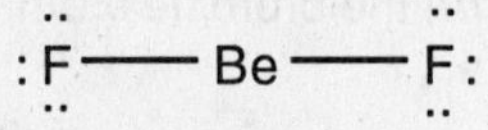

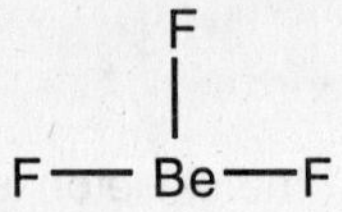

Distribute the remaining 18 electrons on the fluorine atoms.

$$\left[\begin{array}{c} :\overset{..}{F}: \\ | \\ :\overset{..}{\underset{..}{F}}\!-\!Be\!-\!\overset{..}{\underset{..}{F}}: \end{array} \right]^-$$

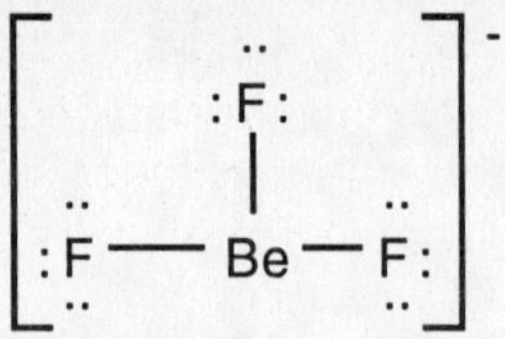

This ion is isoelectronic with BF_3, and the beryllium, like boron, has only six electrons around it.

c. Total valence electrons = 3 + (3 x 7) = 24. By rule 1, the skeleton is

$$\begin{array}{c} Br \\ | \\ Br - Al - Br \end{array}$$

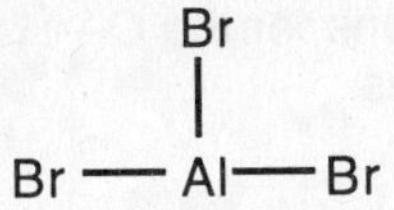

Distribute the remaining 18 electrons to the bromine atoms.

$$\begin{array}{c} : \ddot{Br} : \\ | \\ : \ddot{Br} - Al - \ddot{Br} : \end{array}$$

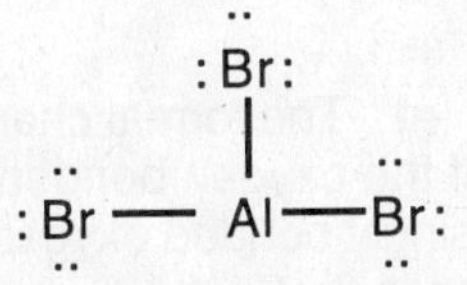

9.59 a. The total number of electrons in O_3 is 3 x 6 = 18. Assume a skeleton structure in which one oxygen atom is singly bonded to the other two oxygen atoms. This requires 6 electrons for the three single bonds, leaving 12 electrons to be used. It is impossible to fill the outer octets of all three oxygens by writing three electron pairs around each, so a double bond must be written between the central oxygen and one of the other oxygens. Then distribute the electron pairs to the oxygen atoms to satisfy the octet rule. As shown below, there are three bonds and six lone pairs of electrons, or 18 electrons, in the structure. Thus, all 18 electrons are accounted for. One of the possible resonance structures is shown below; the other structure would have the double bond written between the left and central oxygens.

$$: \ddot{O} - \ddot{O} = \ddot{O}$$

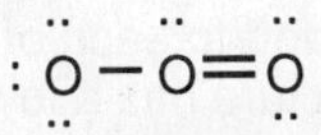

Starting with the left oxygen, the formal charge of this oxygen is 6 - 1 - 6 = -1. The formal charge of just the central oxygen is 6 - 3 - 2 = +1. The formal charge of the right oxygen is 6 - 2 - 4 = 0. The sum of all three = 0.

b. The total number of electrons in CO is 4 + 6 = 10. Assume a skeleton structure in which the oxygen atom is singly bonded to carbon. This requires two electrons for the single bond, leaving eight electrons to be used. It is impossible to fill the outer octets of the carbon and oxygen by writing four electron pairs around each, so a triple bond must be written between the carbon and oxygen. Then distribute the electron pairs to both atoms to satisfy the octet rule. As shown below, there is one triple bond and two lone pairs of electrons, or ten electrons, in the structure. Thus, all ten electrons are accounted for. The structure is

$$: C \equiv O :$$

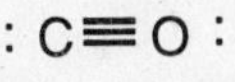

The formal charge of the carbon is 4 - 3 - 2 = -1. The formal charge of the oxygen is 6 - 3 - 2 = +1. The sum of both = 0.

c. The total number of electrons in HNO₃ is 1 + 5 + 18 = 24. Assume a skeleton structure in which the nitrogen atom is singly bonded to two oxygens and doubly bonded to one oxygen. This requires two electrons for the O—H single bond, leaving eight electrons to be used for the N bonds.

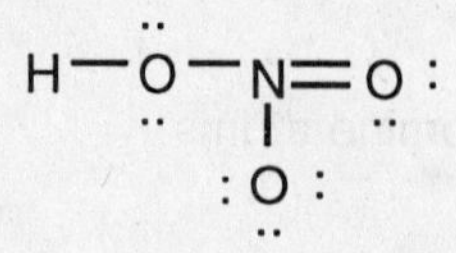

$$H-\ddot{O}-N=\ddot{O}:$$
$$\quad\quad\quad |$$
$$\quad\quad\quad :\ddot{O}:$$

The formal charge of the nitrogen is 5 - 4 - 0 = +1. The formal charge of the hydrogen is 1 - 1 - 0 = 0. The formal charge of the oxygen bonded to the hydrogen is 6 - 2 - 4 = 0. The formal charge of the other singly bonded oxygen is 6 - 1 - 6 = -1. The formal charge of the doubly bonded oxygen is 6 - 2 - 4 = 0.

9.60 a. The total number of electrons in ClNO is 7 + 5 + 6 = 18. Assume a skeleton structure in which one chlorine atom is singly bonded to the nitrogen atom. If a N-O single bond is assumed, 20 electrons are needed to fill the outer octets. Hence, a N=O double bond must be used, the structure being

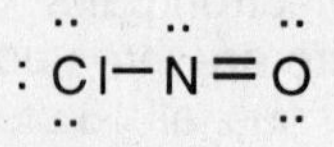

$$:\ddot{C}l-\ddot{N}=\ddot{O}$$

The formal charge of the chlorine is 7 - 1 - 6 = 0. The formal charge of the nitrogen is 5 - 3 - 2 = 0. The formal charge of the oxygen is 6 - 2 - 4 = 0.

b. The total number of electrons in POCl₃ is 5 + 6 + 21 = 32. Assume a skeleton structure in which the phosphorus atom is singly bonded to each of the three chlorines and doubly bonded to the oxygen. This structure has a lower formal charge than one with a single P-O bond.

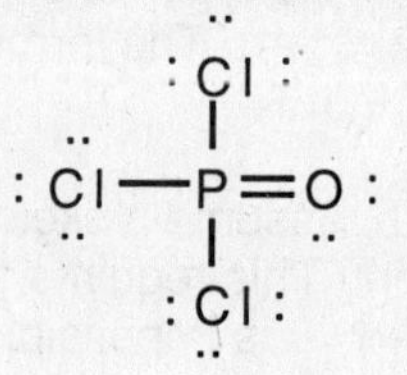

$$\quad\quad :\ddot{C}l:$$
$$\quad\quad\quad |$$
$$:\ddot{C}l-P=\ddot{O}:$$
$$\quad\quad\quad |$$
$$\quad\quad :\ddot{C}l:$$

The formal charge of phosphorus is 5 - 5 - 0 = 0. The formal charge on each chlorine is 7 - 1 - 6 = 0. The formal charge of oxygen is 6 - 2 - 4 = 0.

c. The total number of electrons in N₂O is 2 x 5 + 6 = 16. Assume a skeleton structure in which the two nitrogen atoms are double bonded to each other. The oxygen atom can be singly bonded or doubly bonded to the central nitrogen. Using a N=O double bond gives the lowest formal charge.

$$:\ddot{N}=N=\ddot{O}$$

(continued)

The formal charge of the end nitrogen is 5 - 2 - 4 = -1. The formal charge of the nitrogen between the end nitrogen and oxygen is 5 - 4 = +1. The formal charge of the oxygen is 6 - 2 - 4 = 0.

9.61 a. The total number of electrons in SOF_2 is 6 + 6 + 14 = 26. Assume a skeleton structure in which the sulfur atom is singly bonded to the two fluorine atoms. If a S-O single bond is assumed, 26 electrons are needed. However, there would be formal charges of +1 on the sulfur and -1 on the oxygen. Using a S=O double bond requires only 26 electrons and results in zero formal charge.

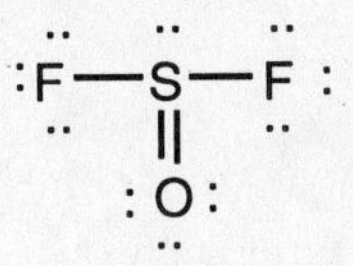

The formal charge on each of the two fluorines is 7 - 1 - 6 = 0. The formal charge on the oxygen is 6 - 2 - 4 = 0. The formal charge on the sulfur is 6 - 4 - 2 = 0.

b. The total number of electrons in H_2SO_3 is 2 + 6 + 18 = 26. Assume a skeleton structure in which the sulfur atom is singly bonded to the three oxygen atoms. Then form single bonds from the two hydrogens to each of two oxygens. If a S-O single bond is assumed, 26 electrons are needed. However, there would be formal charges of +1 on the sulfur and -1 on the oxygen. A S=O double bond also requires 26 electrons but results in zero formal charge on all atoms.

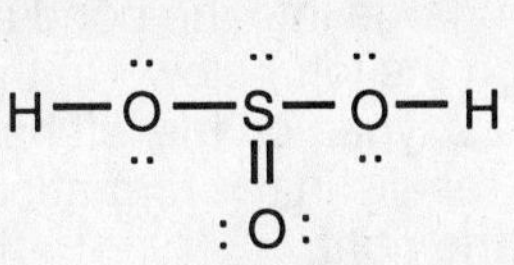

The formal charge of each hydrogen is 1 - 1 = 0. The formal charge of each oxygen bonded to a hydrogen is 6 - 2 - 4 = 0. The formal charge of the oxygen doubled bonded to the sulfur is 6 - 2 - 4 = 0. The formal charge of the sulfur is 6 - 4 - 2 = 0.

c. The total number of electrons in $HClO_2$ is 1 + 7 + 12 = 20. Assume a skeleton structure in which the chlorine atom is singly bonded to the two oxygen atoms and the hydrogen is singly bonded to one of the oxygens. If all Cl-O single bonds are assumed, 20 electrons are needed, but the chlorine exhibits a formal charge of +1 and one oxygen exhibits a formal charge of -1. Hence, a pair of electrons on the oxygen without the hydrogen is used to form a Cl=O double bond.

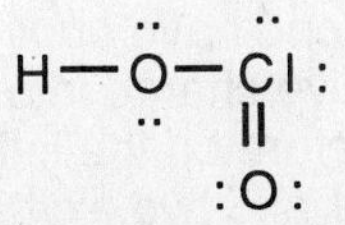

The formal charge of hydrogen is 1 - 1 = 0. The formal charge of the oxygen bonded to hydrogen is 6 - 2 - 4 = 0. The formal charge of the oxygen that is not bonded to the hydrogen is 6 - 2 - 4 = 0. The formal charge of the chlorine is 7 - 3 - 4 = 0.

9.62 a. The total number of electrons in ClO_2F is $7 + 12 + 7 = 26$. Assume a skeleton structure in which the chlorine atom is singly bonded to the two oxygen atoms and the fluorine atom. If all single bonds are assumed, the chlorine has a formal charge of +2 and each oxygen has a formal charge of -1. The general principle to be followed whenever two atoms on a bond have opposite charges is to move an electron pair on the atom with the negative charge into the bond and form a double bond. Doing this once gives the structure on the left below. In this structure, the chlorine now has a formal charge of +1 and the oxygen on the left has a formal charge of -1. Following the general principle again results in the formation of a second double bond, as shown in the structure on the right.

$$:\overset{..}{\underset{..}{F}}: \qquad :\overset{..}{\underset{..}{F}}:$$
$$:\overset{..}{\underset{..}{O}}-\overset{..}{Cl}=\overset{..}{\underset{..}{O}} \qquad \overset{..}{\underset{..}{O}}=\overset{..}{Cl}=\overset{..}{\underset{..}{O}}$$

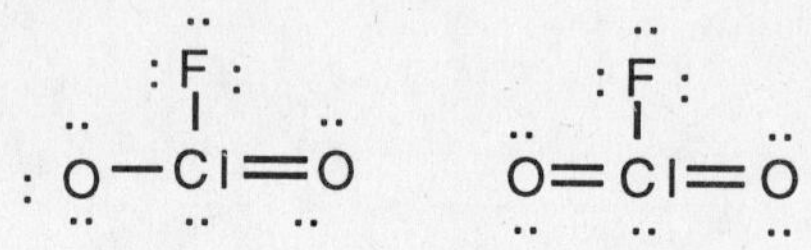

In the structure on the right, the formal charge of each oxygen is $6 - 2 - 4 = 0$. The formal charge of the fluorine is $7 - 1 - 6 = 0$. The formal charge of the chlorine is $7 - 5 - 2 = 0$.

b. The total number of electrons in SO_2 is $6 + 12 = 18$. Assume a skeleton structure in which the sulfur atom is singly bonded to the two oxygen atoms. If all single bonds are assumed, the sulfur atom has a formal charge of +2 and each of the oxygen atoms has a formal charge of -1. As in part a, the general principle to be followed whenever two atoms on a bond have opposite charges is to move an electron pair on the atom with the negative charge into the bond and form a double bond. Doing this once gives the structure on the left below. In this structure, the sulfur now has a formal charge of +1 and the oxygen on the left has a formal charge of -1. Following the general principle again results in the formation of a second double bond, as shown in the structure on the right.

$$:\overset{..}{\underset{..}{O}}-\overset{..}{S}=\overset{..}{\underset{..}{O}} \qquad \overset{..}{\underset{..}{O}}=\overset{..}{S}=\overset{..}{\underset{..}{O}}$$

The formal charge of both oxygens in the structure on the right is $6 - 2 - 4 = 0$. The formal charge of the sulfur is $6 - 4 - 2 = 0$.

c. The total number of electrons in ClO_3^- including the charge is 26. Assume a skeleton structure in which the chlorine atom is singly bonded to the three oxygen atoms. If all single bonds are assumed, the chlorine atom has a formal charge of +2 and each of the oxygen atoms has a formal charge of -1. As in part a, the general principle to be followed whenever two atoms on a bond have opposite charges is to move an electron pair on the atom with the negative charge into the bond and form a double bond. Doing this once gives the structure on the left below. In this structure, the chlorine now has a formal charge of +1 and two oxygens have a formal charge of -1. Following the general principle again results in the formation of a second double bond, as shown in the structure on the right.

(continued)

$$\left[\begin{array}{c}:\ddot{O}:\\|\\:\ddot{O}-Cl=\ddot{O}:\end{array}\right]^- \quad \left[\begin{array}{c}:\ddot{O}:\\|\\\ddot{O}=Cl=\ddot{O}\end{array}\right]^-$$

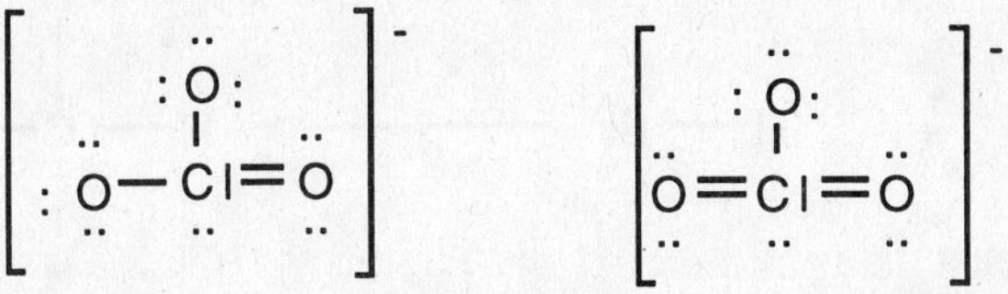

In the right-hand structure, the formal charge of the two doubly bonded oxygens is 6 - 2 - 4 = 0. The formal charge of the singly bonded oxygen is 6 - 1 - 6 = -1 (= negative charge of -1). The formal charge of the chlorine is 7 - 5 - 2 = 0.

9.63 r_F = 64 pm

r_P = 110 pm

$d_{P-F} = r_F + r_P$ = 64 pm + 110 pm = 174 pm

9.64 r_B = 88 pm

r_{Cl} = 99 pm

$d_{B-Cl} = r_B + r_{Cl}$ = 88 pm + 99 pm = 187 pm

9.65 a. $d_{C-H} = r_C + r_H$ = 77 pm + 37 pm = 114 pm
b. $d_{S-Cl} = r_S + r_{Cl}$ = 104 pm + 99 pm = 203 pm
c. $d_{Br-Cl} = r_{Br} + r_{Cl}$ = 114 pm + 99 pm = 213 pm
d. $d_{Si-O} = r_{Si} + r_O$ = 117 pm + 66 pm = 183 pm

9.66 $d_{C-H} = r_C + r_H$ = 77 pm + 37 pm = 114 pm d_{exp} = 107 pm

$d_{C-Cl} = r_C + r_{Cl}$ = 77 pm + 99 pm = 176 pm d_{exp} = 177 pm

The calculated bond distances agree very well with the experimental values.

9.67 Methylamine   147 pm   Single C-N bond is longer.

Acetonitrile   116 pm   Triple C-N bond is shorter.

9.68 Formaldehyde has a shorter carbon-oxygen bond than methanol. In methanol the carbon-oxygen bond is a single bond, and in formaldehyde it is a double bond.

9.69

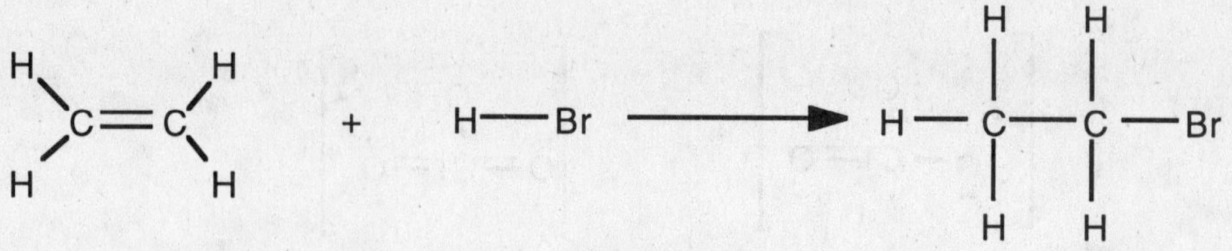

$$\text{H}_2\text{C}=\text{CH}_2 + \text{H}-\text{Br} \longrightarrow \text{H}_3\text{C}-\text{CH}_2-\text{Br}$$

In the reaction, a C=C double bond is converted to a C-C single bond. An H-Br bond is broken, and one C-H bond and one C-Br bond are formed.

$\Delta H \cong$ BE(C=C) + BE(H-Br) - BE(C-C) - BE(C-H) - BE(C-Br)

$= (602 + 362 - 346 - 411 - 285)$ kJ $= -78$ kJ

9.70

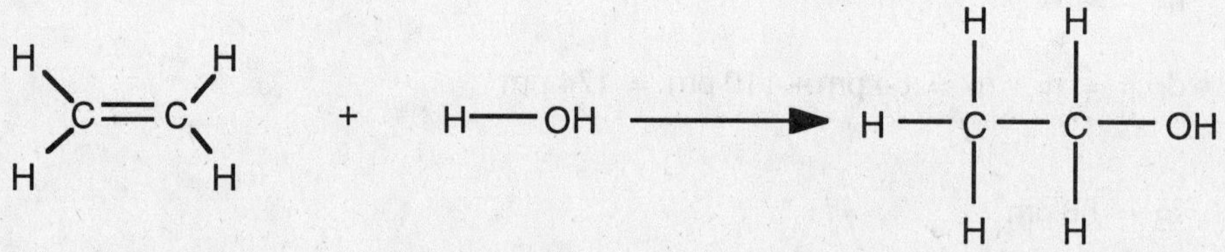

$$\text{H}_2\text{C}=\text{CH}_2 + \text{H}-\text{OH} \longrightarrow \text{H}_3\text{C}-\text{CH}_2-\text{OH}$$

In the reaction, a C=C bond and an O-H bond are converted to a C-C single bond, a C-O bond, and a C-H bond.

$\Delta H \cong [\text{BE(C=C)} + \text{BE(O-H)}] - [\text{BE(C-C)} + \text{BE(C-O)} + \text{BE(C-H)}]$

$= [(602 + 459) - (346 + 358 + 411)]$ kJ $= -54$ kJ

■ Solutions to Unclassified Problems

9.71 a. Strontium is a metal and oxygen is a nonmetal. The binary compound is likely to be ionic. Strontium, in Group IIA, forms Sr^{2+} ions; oxygen, from Group VIA, forms O^{2-} ions. The binary compound has the formula SrO and is named strontium oxide.

b. Carbon and bromine are both nonmetals; hence, the binary compound is likely to be covalent. Carbon usually forms four bonds and bromine usually forms one bond. The formula for the binary compound is CBr_4. It is called carbon tetrabromide.

c. Gallium is a metal and fluorine is a nonmetal. The binary compound is likely to be ionic. Gallium is in Group IIIA and forms Ga^{3+} ions. Fluorine is in Group VIIA and forms F^- ions. The binary compound is GaF_3 and is named gallium(III) fluoride.

d. Nitrogen and bromine are both nonmetals; hence, the binary compound is likely to be covalent. Nitrogen usually forms three bonds and bromine usually forms one bond. The formula for the binary compound is NBr_3. It is called nitrogen tribromide.

9.72 a. Sodium is a metal and sulfur is a nonmetal. The binary compound is likely to be ionic. Sodium, in Group IA, forms Na^+ ions. Sulfur, in Group VIA, forms S^{2-} ions. The binary compound is Na_2S. It is named sodium sulfide.

b. Aluminum is a metal and fluorine is a nonmetal. The binary compound is likely to be ionic. Aluminum, in Group IIIA, forms Al^{3+} ions. Fluorine, in Group VIIA, forms F^- ions. The binary compound is AlF_3. It is named aluminum fluoride.

c. Calcium is a metal and chlorine is a nonmetal. The binary compound is likely to be ionic. Calcium, in Group IIA, forms Ca^{2+} ions. Chlorine, in Group VIIA, forms Cl^- ions. The binary compound is $CaCl_2$. It is named calcium chloride.

d. Silicon and bromine are both nonmetals; hence, the binary compound is likely to be covalent. Silicon usually forms four bonds and bromine usually forms one bond. The formula for the binary compound is $SiBr_4$. It is called silicon tetrabromide.

9.73 Total valence electrons = 5 + (4 x 6) + 3 = 32. By rule 1, the skeleton is

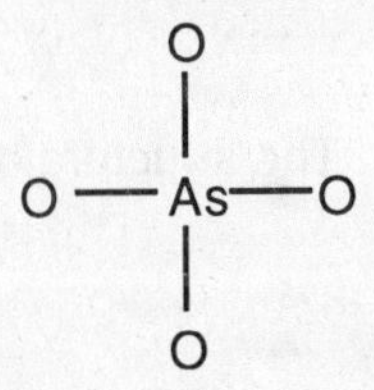

Distribute the remaining 24 electrons to complete the octets around the oxygen atoms.

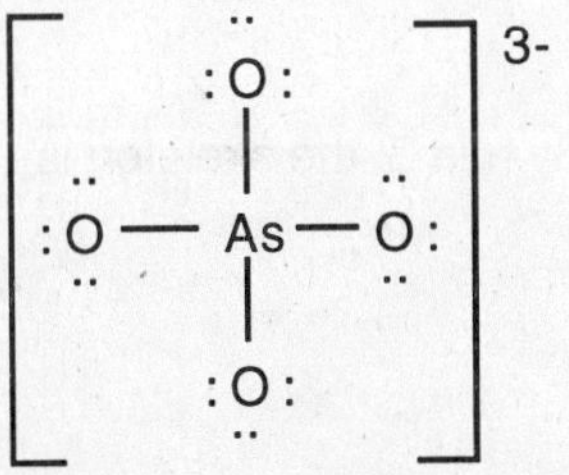

or

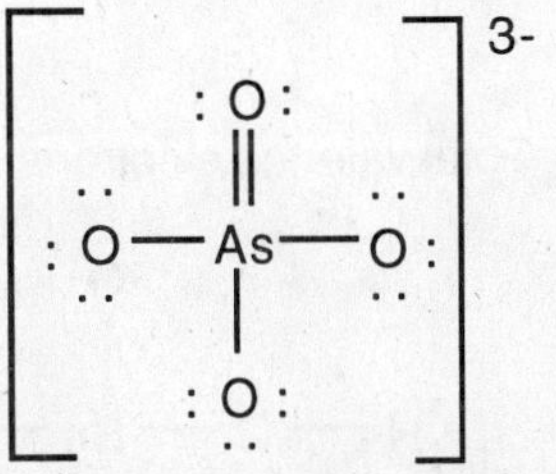

The formula for lead(II) arsenate is $Pb_3(AsO_4)_2$. The structure on the right has no formal charges.

9.74 Total valence electrons = 6 + (3 x 6) + 2 = 26. By rule 1, the skeleton is

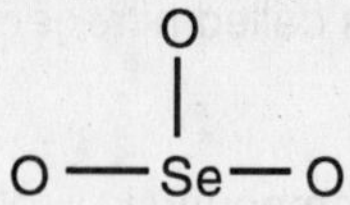

```
      O
      |
  O — Se — O
```

Distribute the remaining 20 electrons to complete the octets of oxygen and selenium atoms.

$$\left[\begin{array}{c} :\ddot{O}: \\ | \\ :\ddot{O} - Se - \ddot{O}: \end{array}\right]^{2-} \quad \text{or} \quad \left[\begin{array}{c} :\ddot{O}: \\ \| \\ :\ddot{O} - Se - \ddot{O}: \end{array}\right]^{2-}$$

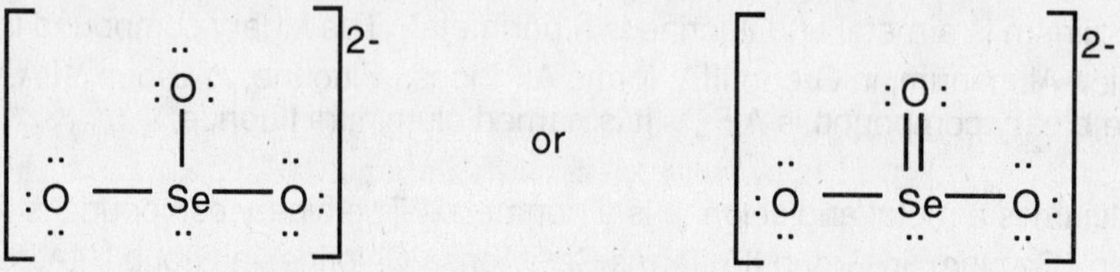

The formula for aluminum selenite is $Al_2(SeO_3)_3$. The structure on the right has no formal charges.

9.75 Total valence electrons = 1 + 7 + (3 x 6) = 26. By rule 3, the skeleton is

```
  H — O — I — O
          |
          O
```

Distribute the remaining 18 electrons to satisfy the octet rule. The structure on the right has no formal charges.

$$H - \ddot{O} - \overset{..}{\underset{|}{I}} - \ddot{O}: \quad \text{or} \quad H - \ddot{O} - \overset{..}{\underset{\|}{I}} = \ddot{O}$$
$$\qquad\qquad :\ddot{O}: \qquad\qquad\qquad :\ddot{O}:$$

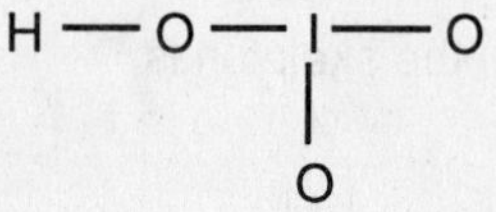

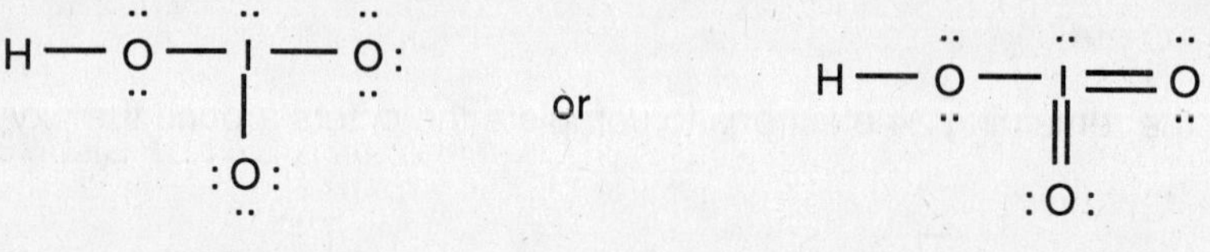

9.76 Total valence electrons = (2 x 1) + 6 + (4 x 6) = 32. By rule 3, the skeleton is

```
          O
          |
  H — O — Se — O — H
          |
          O
```

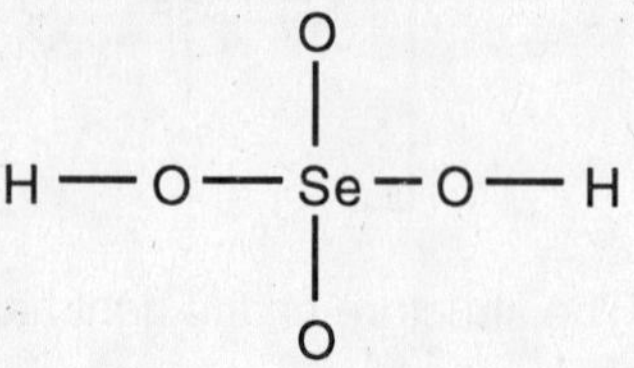

(continued)

Distribute the remaining 20 electrons to satisfy the octet rule. The structure on the right has no formal charges.

$$\begin{array}{c} \ddot{\text{:O:}} \\ | \\ \text{H}-\ddot{\text{O}}-\text{Se}-\ddot{\text{O}}-\text{H} \\ | \\ \ddot{\text{:O:}} \end{array} \quad \text{or} \quad \begin{array}{c} \ddot{\text{:O:}} \\ \| \\ \text{H}-\ddot{\text{O}}-\text{Se}-\ddot{\text{O}}-\text{H} \\ \| \\ \ddot{\text{:O:}} \end{array}$$

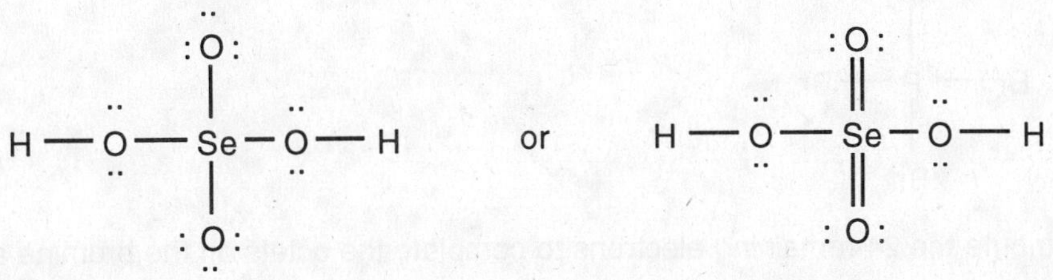

9.77 Total valence electrons = 5 + (2 x 1) + 1 = 8. By rule 1, the skeleton is H-N-H. Distribute the remaining four electrons to complete the octet of the nitrogen atom.

$$\left[\text{H}-\ddot{\underset{\cdot\cdot}{\text{N}}}-\text{H} \right]^-$$

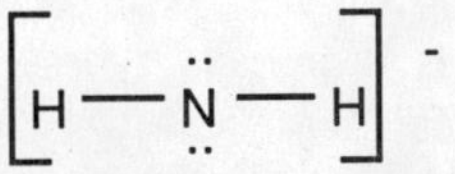

9.78 Total valence electrons = 3 + (4 x 1) + 1 = 8. The skeleton structure will use all the electrons. Because there are no electrons left, the Lewis structure is

$$\left[\begin{array}{c} \text{H} \\ | \\ \text{H}-\text{Al}-\text{H} \\ | \\ \text{H} \end{array} \right]^-$$

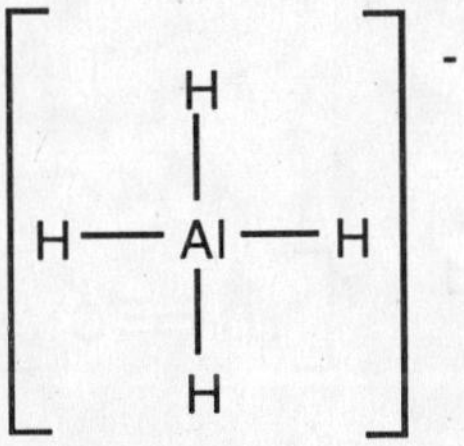

9.79 Total valence electrons = 5 + (2 x 6) - 1 = 16. The skeleton is O-N-O. Distribute the remaining electrons on the oxygen atoms.

$$\ddot{\text{:}}\ddot{\text{O}}-\text{N}-\ddot{\text{O}}\text{:}$$

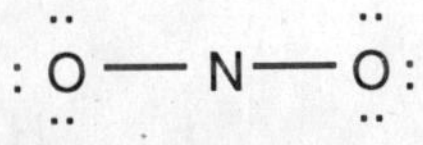

The nitrogen atom is short four electrons. Rule 2 allows the use of two double bonds, one with each oxygen.

$$\left[\ddot{\text{:O}}=\text{N}=\ddot{\text{O:}} \right]^+$$

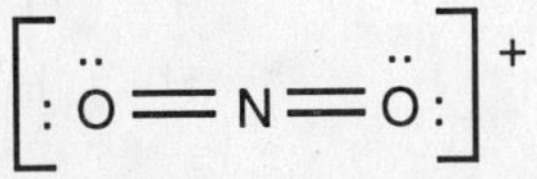

9.80 Total valence electrons = 5 + (4 × 7) − 1 = 32. The skeleton is

```
         Br
         |
   Br — P — Br
         |
         Br
```

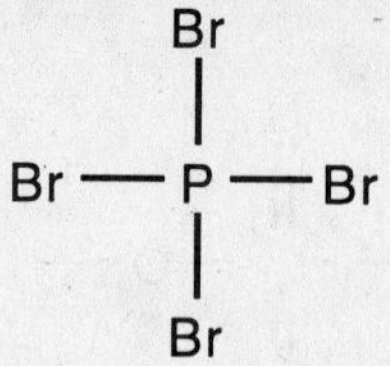

Distribute the 24 remaining electrons to complete the octets on the bromine atoms.

$$\left[\begin{array}{c} :\ddot{B}r: \\ | \\ :\ddot{B}r - P - \ddot{B}r: \\ | \\ :\ddot{B}r: \end{array}\right]^+$$

9.81 a.

```
       :Ö:                        :O:
        |                          ||
 :Cl — Se — Cl:     or     :Cl — Se — Cl:
```

b. :Se̤=C=S̤e:

c.
$$\left[\begin{array}{c} :\ddot{C}l: \\ | \\ :\ddot{C}l - Ga - \ddot{C}l: \\ | \\ :\ddot{C}l: \end{array}\right]^-$$

d. $\left[:C \equiv C: \right]^{2-}$

9.82 a.

```
       :Ö:                        :O:
        |                          ||
 :Br — P — Br:     or      :Br — P — Br:
        |                          |
       :B̈r:                       :B̈r:
```

b.c. $\left[:\ddot{F} - \ddot{I} - \ddot{F}: \right]^+$

d. $\left[:N \equiv O: \right]^+$

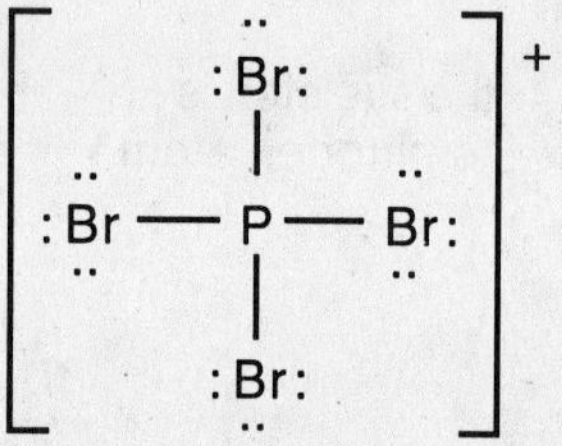

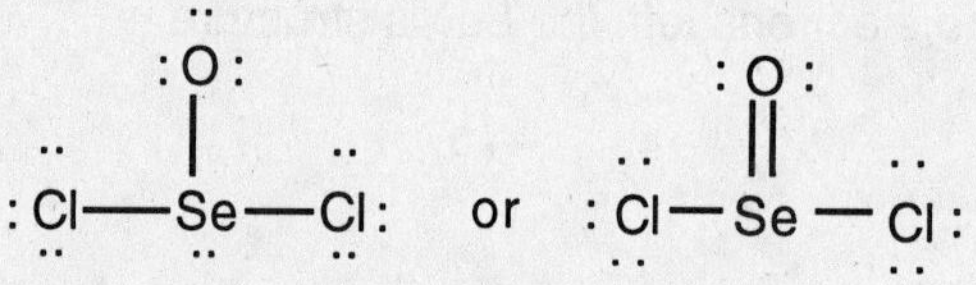

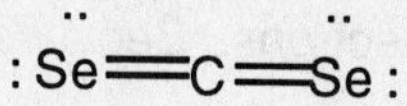

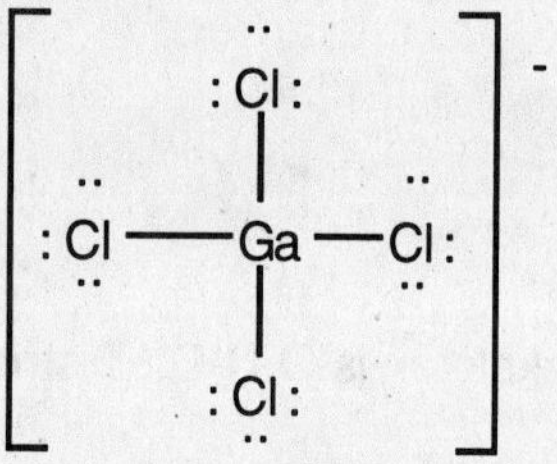

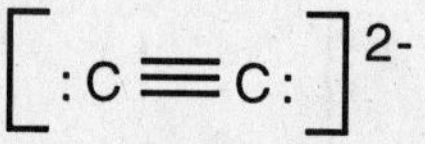

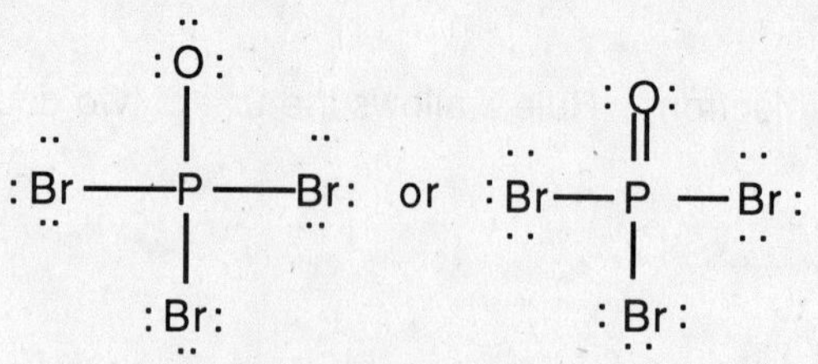

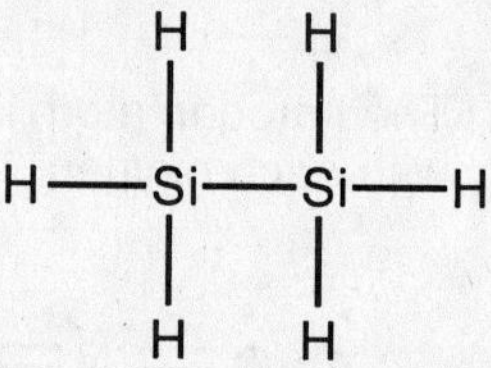

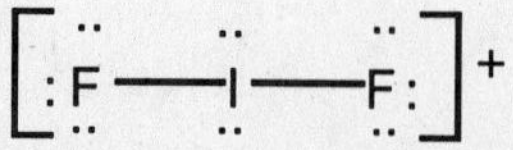

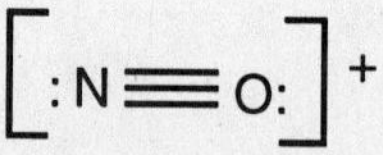

9.83 a. Total valence electrons = 5 + (3 x 7) = 26. By rule 1, the skeleton is

$$\text{Cl}-\text{Sb}-\text{Cl}$$
$$|$$
$$\text{Cl}$$

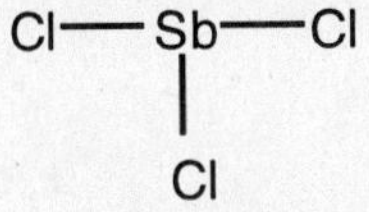

Distribute the remaining 20 electrons to the chlorine atoms and the antimony atom to complete their octets.

$$:\ddot{\text{Cl}}-\ddot{\text{Sb}}-\ddot{\text{Cl}}:$$
$$|$$
$$:\ddot{\text{Cl}}:$$

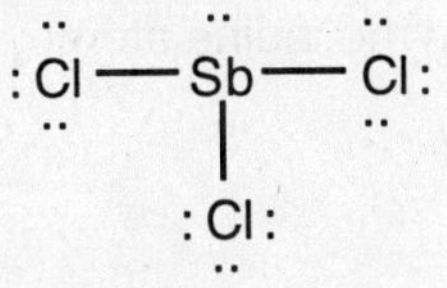

b. Total valence electrons = 7 + 4 + 5 = 16. The skeleton is I-C-N. Distribute the remaining 12 electrons.

$$:\ddot{\text{I}}-\ddot{\text{C}}-\ddot{\text{N}}:$$

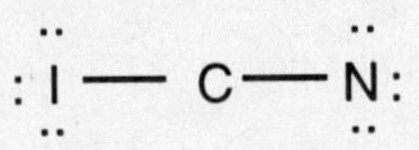

Notice that the carbon atom is four electrons short of an octet. Rule 4 suggests a triple bond between C and N made from the four nonbonding electrons on the nitrogen.

$$:\ddot{\text{I}}-\text{C}\equiv\text{N}:$$

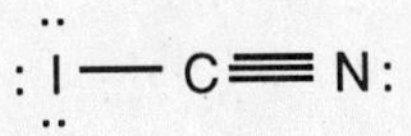

c. Total valence electrons = 7 + (3 x 7) = 28. By rule 1, the skeleton is

$$\text{Cl}-\text{I}-\text{Cl}$$
$$|$$
$$\text{Cl}$$

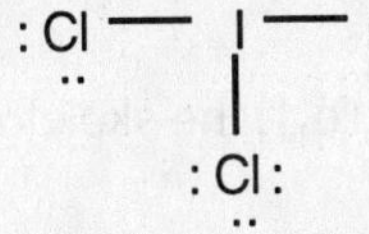

Distribute 18 of the remaining 22 electrons to complete the octets of the chlorine atoms. The four remaining electrons form two sets of lone pairs on the iodine atom.

$$:\ddot{\text{Cl}}-\ddot{\text{I}}-\ddot{\text{Cl}}:$$
$$|$$
$$:\ddot{\text{Cl}}:$$

d. Total valence electrons = 7 + (5 x 7) = 42. By rule 1, the skeleton is

$$\begin{array}{c} F \diagdown F \\ I - F \\ F \diagup F \end{array}$$

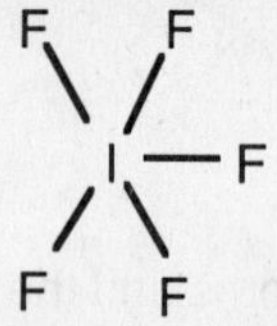

Use 30 of the remaining 32 electrons to complete the octets of the fluorine atoms. The two electrons remaining form a lone pair on the iodine atom:

$$\begin{array}{c} :\!\ddot{F}\!: :\!\ddot{F}\!: \\ :\!\ddot{I} - \ddot{F}\!: \\ :\!\ddot{F}\!: :\!\ddot{F}\!: \end{array}$$

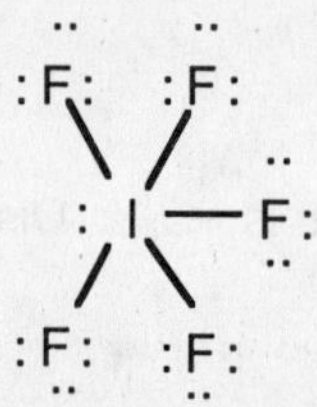

9.84 a. Total valence electrons = 3 + (4 x 7) + 1 = 32. By rule 1, the skeleton is

$$\begin{array}{c} Cl \\ | \\ Cl - Al - Cl \\ | \\ Cl \end{array}$$

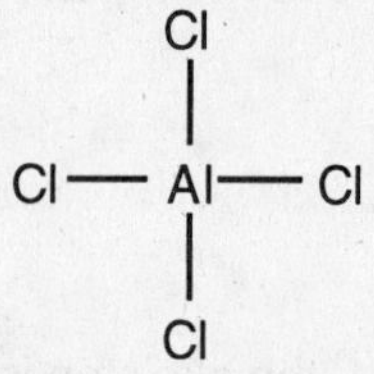

Distribute the remaining 24 electrons to complete the octets of the Cl atoms.

$$\left[\begin{array}{c} :\!\ddot{Cl}\!: \\ | \\ :\!\ddot{Cl} - Al - \ddot{Cl}\!: \\ | \\ :\!\ddot{Cl}\!: \end{array} \right]$$

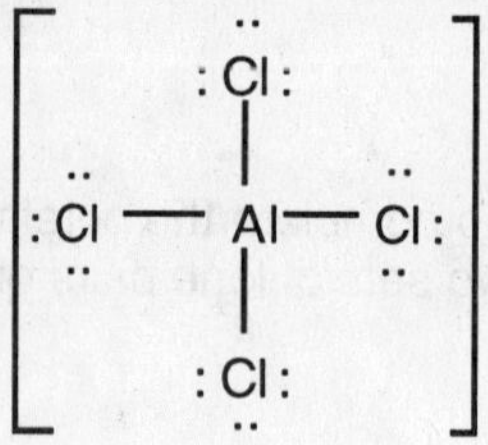

b. Total valence electrons = 3 + (6 x 7) + 3 = 48. By rule 1, the skeleton is

$$\begin{array}{c} F \diagdown F \\ F - Al - F \\ F \diagup F \end{array}$$

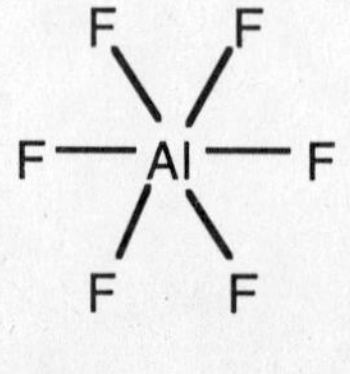

(continued)

Distribute the remaining 36 electrons to complete the octets of the F atoms.

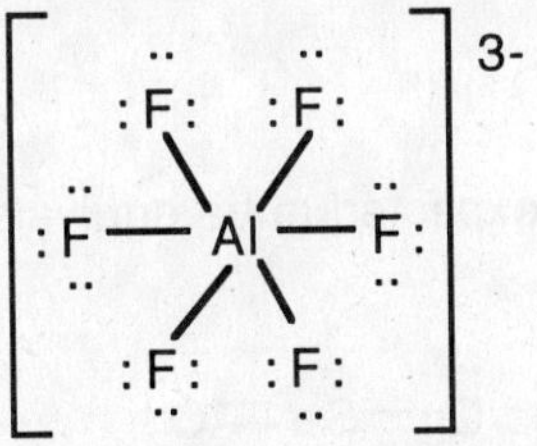

c. Total valence electrons = 7 + (3 x 7) = 28. By rule 1, the skeleton is

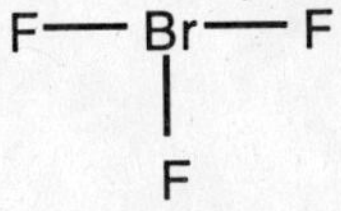

Distribute 18 of the remaining 22 electrons to complete the octets of the fluorine atoms. The four remaining electrons form two sets of lone pairs on the bromine atom.

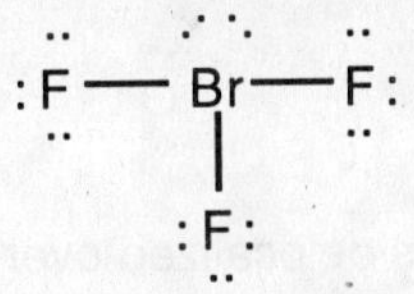

d. Total valence electrons = 7 + (6 x 7) - 1 = 48. The skeleton is

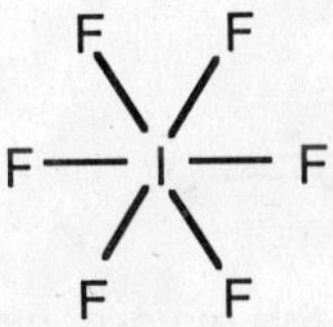

Distribute the remaining 36 electrons to complete the octets of the fluorine atoms.

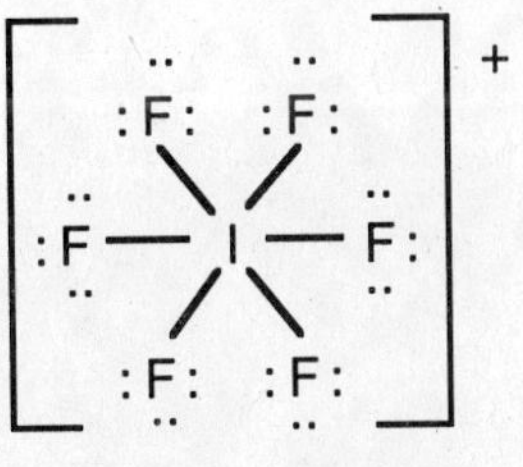

9.85 a. One possible electron-dot structure is

$$:\ddot{O}=\ddot{S}e-\ddot{O}:$$

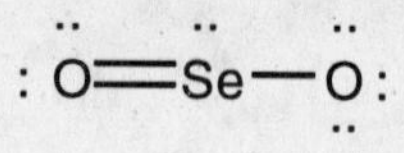

Because the selenium-oxygen bonds are expected to be equivalent, the structure must be described in resonance terms.

$$:\ddot{O}=\ddot{S}e-\ddot{O}: \longleftrightarrow :\ddot{O}-\ddot{S}e=\ddot{O}:$$

One electron pair is delocalized over the selenium atom and the two oxygen atoms.

b. The possible electron-dot structures are

[resonance structures of N₂O₄ showing four resonance forms connected by double-headed arrows]

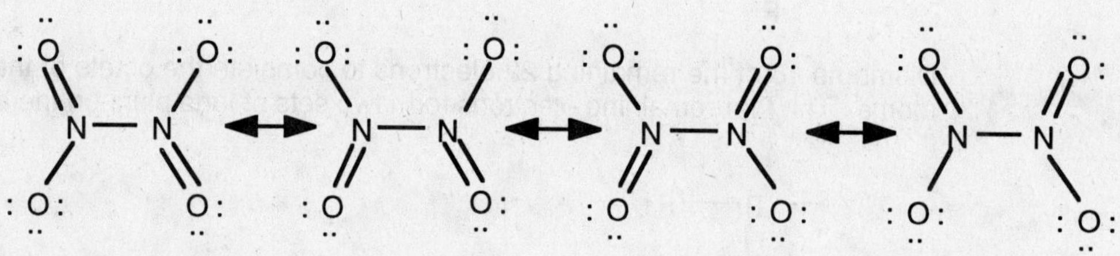

At each end of the molecule, a pair of electrons is delocalized over the region of the nitrogen atom and the two oxygen atoms.

9.86 a. The possible electron-dot structures are

$$:\ddot{O}=\ddot{O}-\ddot{O}: \longleftrightarrow :\ddot{O}-\ddot{O}=\ddot{O}:$$

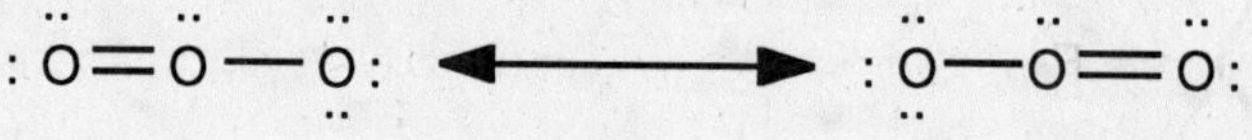

A pair of electrons is delocalized over the region of the oxygen-oxygen bonds.

b. The possible electron-dot structures are

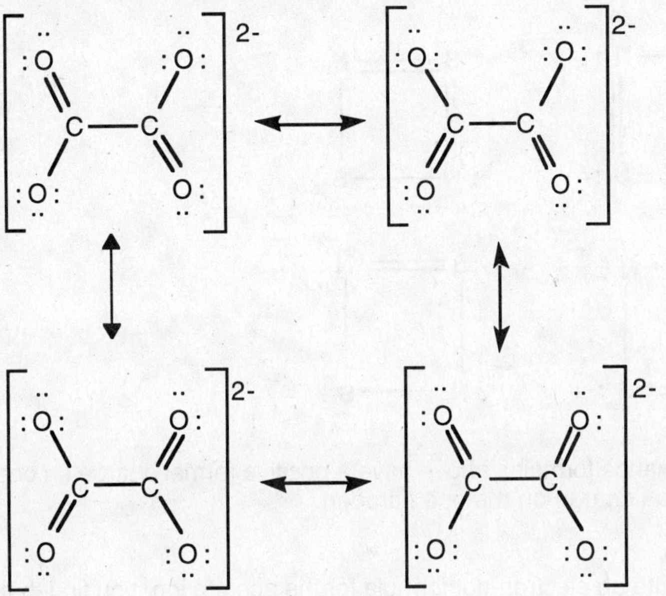

At each end of the molecule, a pair of electrons is delocalized over the region of the two carbon-oxygen bonds.

9.87 The compound S_2N_2 will have a four-membered ring structure. Following the rules in Sec 9.6, we calculate the total number of valence electrons of sulfur and nitrogen, which will be 6 + 6 + 5 + 5 = 22. Writing the skeleton of the four-membered ring with single S-N bonds will use up eight electrons, 14 electrons for electron pairs. After writing an electron pair on each atom, there will be 6 electrons left for 3 electron pairs. No matter where these electrons are written, either a nitrogen or a sulfur will be left with less than eight electrons. This suggests that one or more double bonds are needed. If only one S=N double bond is written, writing the remaining 12 electrons as 6 electron pairs will give the sulfur in the S=N double bond a formal charge of +1 (see second line of formulas below). Writing two S=N double bonds using the same sulfur for both double bonds, will give a formal charge of zero for all four atoms (see first line of formulas).

(continued)

The two resonance formulas below have a zero formal charge on all atoms:

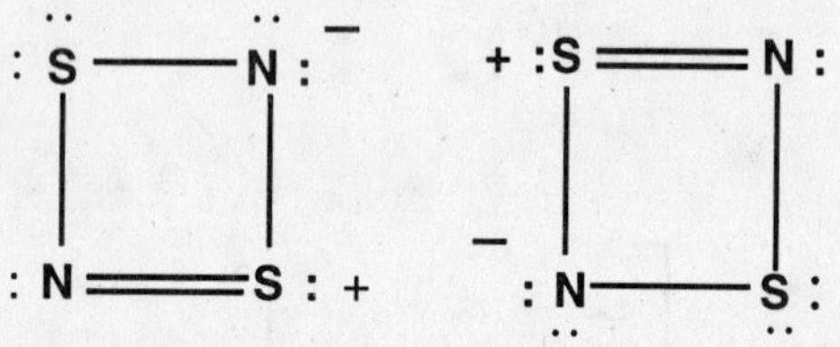

The two resonance formulas above have a positive formal charge on one sulfur and a negative formal charge on the one nitrogen.

9.88 If you try to write an electron-dot formula for the acetate ion, you find that you can write two formulas.

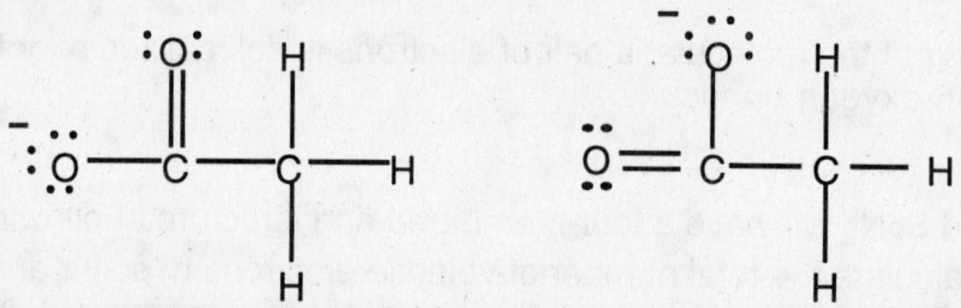

According to theory, the C=O double bond is delocalized; that is, a bonding pair of electrons is spread over the carbon atom and the two oxygen atoms rather than localized between a carbon and one oxygen. This results in both bonds being of identical length. See the discussion of ozone and the discussion of the carbonate ion in Section 9.7.

9.89 The possible electron-dot structures are

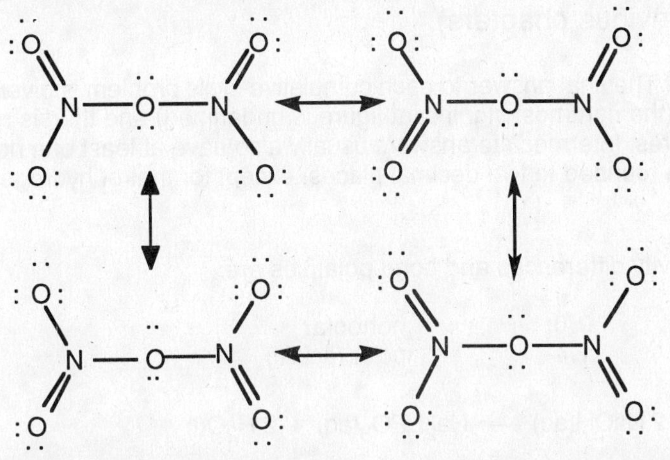

Because double bonds are shorter, the terminal N-O bonds are 118 pm and the central N-O bonds are 136pm.

9.90 The only possible electron-dot structure is

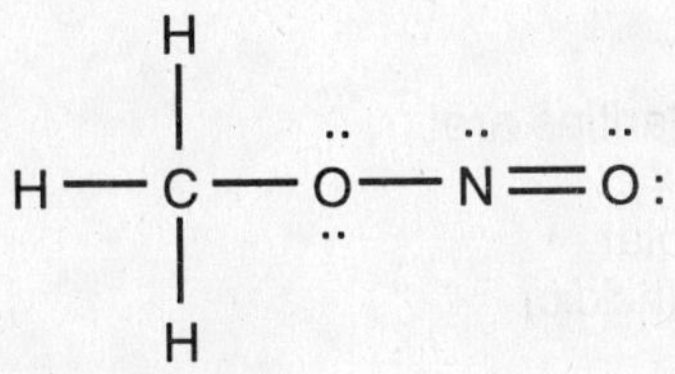

Because double bonds are shorter, the terminal N=O bond is 122pm and the inner N-O bond is 137pm.

9.91 $\Delta H = BE(H\text{-}H) + BE(O{=}O) - 2BE(H\text{-}O) - BE(O\text{-}O)$

$= (432 + 494 - 2 \times 459 - 142)\text{ kJ} = -134\text{ kJ}$

9.92 $\Delta H = 2BE(H\text{-}H) + BE(N{\equiv}N) - BE(N\text{-}N) - 4BE(N\text{-}H)$

$= (2 \times 432 + 942 - 167 - 4 \times 386)\text{ kJ} = +95\text{ kJ}$

9.93 $\Delta H = BE(N{=}N) + BE(F\text{-}F) - BE(N\text{-}N) - 2BE(N\text{-}F)$

$= (418 + 155 - 167 - 2 \times 283)\text{ kJ} = -160\text{ kJ}$

9.94 $\Delta H = BE(C{\equiv}N) + 2BE(H\text{-}H) - BE(C\text{-}N) - 2BE(C\text{-}H) - 2BE(N\text{-}H)$

$= (887 + 2 \times 432 - 305 - 2 \times 411 - 2 \times 386)\text{ kJ} = -148\text{ kJ}$

■ Cumulative-Skills Problems
(require skills from previous chapters)

Note on significant figures: The final answer to each cumulative-skills problem is given first with one nonsignificant figure (the rightmost significant figure is underlined) and then is rounded to the correct number of figures. Intermediate answers usually also have at least one nonsignificant figure. Atomic weights are rounded to two decimal places, except for that of hydrogen.

9.95 The electronegativity differences and bond polarities are:

P- H	0.0	nonpolar
O- H	1.4	polar (acidic)

$$H_3PO_3(aq) + 2\ NaOH(aq) \rightarrow Na_2HPO_3(aq) + 2\ H_2O(l)$$

$$\frac{mol}{L} = \frac{0.1250\ mol\ NaOH}{L\ NaOH} \times 0.02250\ L\ NaOH \times \frac{1}{0.2000\ L\ H_3PO_3}$$

$$\times \frac{1\ mol\ H_3PO_3}{2\ mol\ NaOH} = 0.00703\underline{1}2 = 0.007031\ M\ H_3PO_3.$$

9.96 The electronegativity differences and bond polarities are:

P- H	0.0	nonpolar
O- H	1.4	polar (acidic)

$$H_3PO_2(aq) + NaOH(aq) \rightarrow NaH_2PO_2(aq) + H_2O(l)$$

$$\frac{mol}{L} = \frac{0.1250\ mol\ NaOH}{L\ NaOH} \times 0.02250\ L\ NaOH \times \frac{1}{0.2000\ L\ H_3PO_2}$$

$$\times \frac{1\ mol\ H_3PO_2}{1\ mol\ NaOH} = 0.0140\underline{6}2 = 0.01406\ M\ H_3PO_2.$$

9.97 After assuming a 100.0-g sample, convert to moles:

$$10.9\ g\ Mg \times \frac{1\ mol\ Mg}{24.3\ g\ Mg} = 0.44\underline{8}5\ mol\ Mg$$

$$31.8\ g\ Cl \times \frac{1\ mol\ Cl}{35.453\ g\ Cl} = 0.89\underline{6}96\ mol\ Cl$$

$$57.3\ g\ O \times \frac{1\ mol\ O}{16.00\ g\ O} = 3.5\underline{8}1\ mol\ O$$

(continued)

Divide by 0.4485:

$$\text{Mg: } \frac{0.4485}{0.4485} = 1; \quad \text{Cl: } \frac{0.89696}{0.4485} = 2.00; \quad \text{O: } \frac{3.581}{0.4485} = 7.98$$

The simplest formula is $Mg(ClO_4)_2$, magnesium perchlorate. The Lewis formulas are Mg^{2+} and

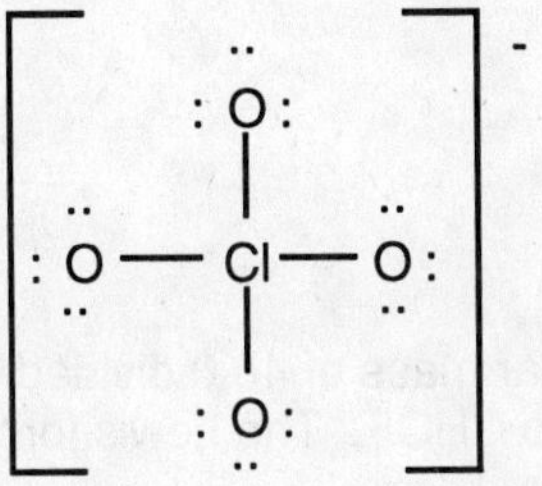

or

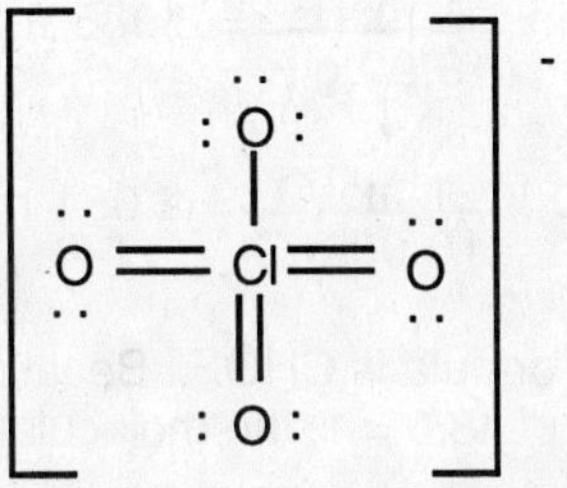

9.98 After assuming a 100.0-g sample, convert to moles:

$$30.3 \text{ g Ca} \times \frac{1 \text{ mol Ca}}{40.08 \text{ g Ca}} = 0.75\underline{5}9 \text{ mol Ca}$$

$$21.2 \text{ g N} \times \frac{1 \text{ mol N}}{14.007 \text{ g N}} = 1.5\underline{1}3 \text{ mol N}$$

$$48.5 \text{ g O} \times \frac{1 \text{ mol O}}{16.00 \text{ g O}} = 3.0\underline{3}1 \text{ mol O}$$

Divide by 0.7559:

$$\text{Ca: } \frac{0.7559}{0.7559} = 1; \quad \text{N: } \frac{1.513}{0.7559} = 2.00; \quad \text{O: } \frac{3.031}{0.7559} = 4.01$$

The simplest formula is $Ca(NO_2)_2$, calcium nitrite. The Lewis formulas are Ca^{2+} and

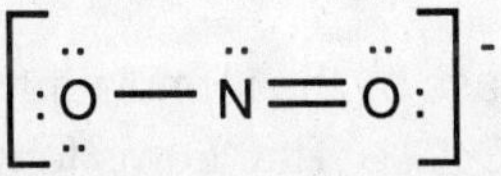

9.99 After assuming a 100.0-g sample, convert to moles:

$$25.0 \text{ g C} \times \frac{1 \text{ mol C}}{12.01 \text{ g C}} = 2.0\underline{8}1 \text{ mol C}$$

$$2.1 \text{ g H} \times \frac{1 \text{ mol H}}{1.008 \text{ g H}} = 2.\underline{0}8 \text{ mol H}$$

$$39.6 \text{ g F} \times \frac{1 \text{ mol F}}{18.99 \text{ g F}} = 2.0\underline{8}5 \text{ mol F}$$

$$33.3 \text{ g O} \times \frac{1 \text{ mol O}}{16.00 \text{ g O}} = 2.0\underline{8}1 \text{ mol O}$$

The simplest formula is CHOF. Because the molecular mass of 48.0 divided by the formula mass of 48.0 = 1, the molecular formula is also CHOF. The Lewis formula is

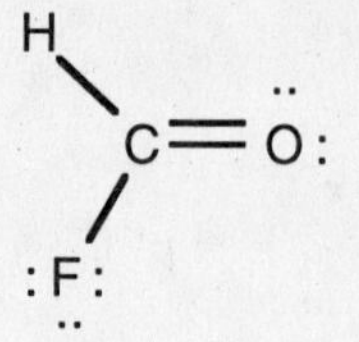

9.100 After assuming a 100.0-g sample, convert to moles:

$$14.5 \text{ g C} \times \frac{1 \text{ mol C}}{12.01 \text{ g C}} = 1.2\underline{0}7 \text{ mol C}$$

$$85.5 \text{ g Cl} \times \frac{1 \text{ mol Cl}}{35.453 \text{ g Cl}} = 2.4\underline{1}16 \text{ mol Cl}$$

Divide by 1.207:

$$\text{C: } \frac{1.207}{1.207} = 1; \quad \text{Cl: } \frac{2.4116}{1.207} = 1.998$$

The simplest formula is CCl_2. Dividing the molecular mass by the formula mass gives 166 amu ÷ 82.92 amu = 2.0. The molecular formula is C_2Cl_4. The Lewis formula is

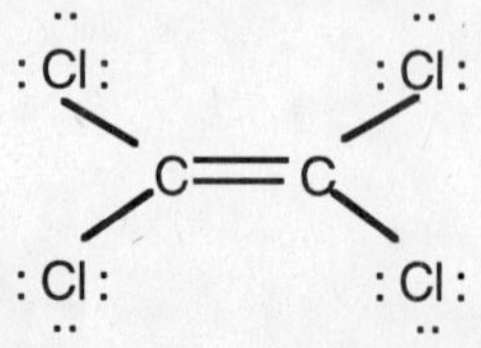

9.101 First calculate the number of moles in one liter:

$$n = \frac{PV}{RT} = \frac{1.00 \text{ atm} \times 1.00 \text{ L}}{0.0821 \text{ L} \bullet \text{atm}/(\text{K} \bullet \text{mol}) \times 424 \text{ K}} = 0.028\underline{7}27 \text{ mol}$$

$$MW = \frac{g}{mol} = \frac{7.49 \text{ g}}{0.028727 \text{ mol}} = 26\underline{0}.7 \text{ g/mol}$$

MW = 260.7 amu = 118.69 amu Sn + n(35.453 amu Cl)

$$n = \frac{260.7 - 118.69}{35.453} = 4.0\underline{0}55$$

The formula is $SnCl_4$, which is molecular because the electronegativity difference = 1.2. The Lewis formula is

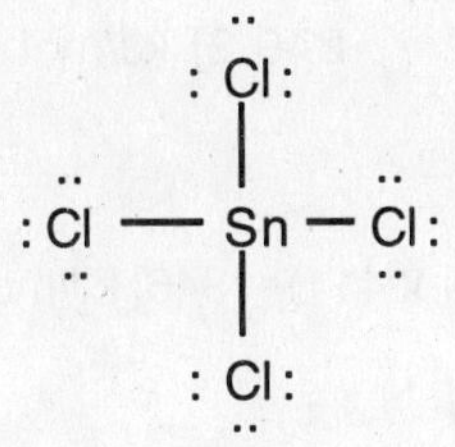

9.102 First calculate the number of moles present:

$$n = \frac{PV}{RT} = \frac{(765/760) \text{ atm} \times 0.0142 \text{ L}}{0.0821 \text{ L} \bullet \text{atm}/(\text{K} \bullet \text{mol}) \times 296 \text{ K}} = 5.8\underline{8}16 \times 10^{-4} \text{ mol}$$

$$MW = \frac{g}{mol} = \frac{0.100 \text{ g}}{5.8816 \times 10^{-4} \text{ mol}} = 17\underline{0}.02 \text{ g/mol}$$

MW = 170.02 amu = 74.92 amu As + n(18.998 amu F)

$$n = \frac{170.02 - 74.92}{18.998} = 5.0\underline{0}57$$

The formula is AsF_5. The Lewis formula is

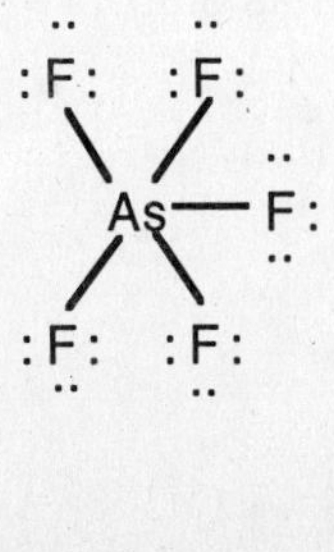

9.103 $HCN(g) \longrightarrow H(g) + C(g) + N(g)$

[135 218.0 715.0 473] kJ/mol

$\Delta H = 1271$ kJ/mol

$BE(C{\equiv}N) = \Delta H - BE(C\text{-}H) = [1271 - 411]$ kJ/mol

$= 860$ kJ/mol (Table 9.5 has 887 kJ/mol.)

9.104 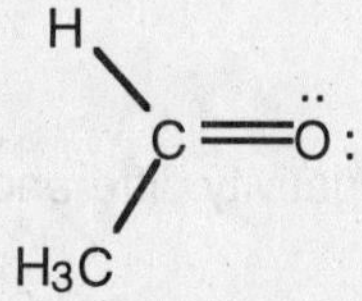 $\longrightarrow 4\,H(g) + 2\,C(g) + O(g)$

[-166 4 x 218 2 x 715 249.2] kJ/mol

$\Delta H = 2717.2$ kJ/mol

$BE(C{=}O) = \Delta H - [4 \times BE(C\text{-}H)] - BE(C\text{-}C) = [2717.2 - (4 \times 411) - 346]$ kJ/mol

$= 72\underline{7}.2$ kJ/mol (Table 9.5 has 727 kJ/mol.)

9.105 Use the O-H bond and its bond energy of 459 kJ/mol to calculate X_O.

$$BE(O\text{-}H) = 1/2\,[BE(H\text{-}H) + BE(O\text{-}O)] + k(X_O - X_H)^2$$

$$459 \text{ kJ/mol} = 1/2(432 \text{ kJ/mol} + 142 \text{ kJ/mol}) + 98.6 \text{ kJ } (X_O - X_H)^2$$

Collecting the terms gives

$$\frac{459 - 287}{98.6} = (X_O - X_H)^2$$

Taking the square root of both sides gives

$$1.3\underline{2}07 = 1.32 = (X_O - X_H)$$

Because $X_H = 2.1$, $X_O = 2.1 + 1.32 = 3.42 = 3.4$. (Figure 9.11 has 3.5.)

9.106 $BE(N\text{-}I) = 1/2[BE(N\text{-}N) + BE(I\text{-}I)] + k(X_N - X_I)^2$

$BE(N\text{-}I) = 1/2(167 + 149) + 98.6(3.0 - 2.5)^2$

$BE(N\text{-}I) = 158 + 24.65 = 18\underline{2}.65 = 183 \text{ kJ/mol}$

9.107 $X = \dfrac{\text{I.E. - E.A.}}{2} = \dfrac{[1250 - (-349)] \text{ kJ/mol}}{2} = 79\underline{9}.5 \text{ kJ/mol}$

$\dfrac{799.5 \text{ kJ/mol}}{230 \text{ kJ/mol}} = 3.47$ (Pauling's X = 3.0.)

9.108 $X = \dfrac{\text{I.E. - E.A.}}{2} = \dfrac{[1314 - (-141)] \text{ kJ/mol}}{2} = 72\underline{7}.5 \text{ kJ/mol}$

$\dfrac{727.5 \text{ kJ/mol}}{230 \text{ kJ/mol}} = 3.16$ (Pauling's X = 3.5.)

■ Solution to Conceptual Problem

9.109 Bond energies are often tabulated to estimate unknown bond energies, other thermodynamic data, and even qualitative identification of compounds by infrared spectroscopy (see Figure 9.21). However, it is not always possible to represent a given bond with a specific energy. Consider the data for the S=O bond in Table 9.5. Why is there a difference between the S=O bond energies in SO_2 vs. SO_3? Give another example of bonds between the same atoms that have different energies in different compounds. Explain.

Discussion

The concept of bond energy is presented in Section 9.11. Predictions of bond energies based on the type of bonding are difficult, so this question does not have a simple answer. Rather, this question should be used to get students to think more broadly about bonding concepts.

We can assume that as long as an A–B bond is similar in different compounds that the A–B bond energy is about the same in these compounds. Conversely, if the A–B bond energy in two compounds is different, we conclude that the bonding situations are different. Thus, we conclude that the bonding situations in SO_2 and SO_3 are substantially different.

(continued)

Several points might be discussed: the relationship between bond energy and bond order, the delocalization (resonance) stabilization of a molecule, and how surrounding atoms affect the electron density of a bond and therefore its energy.

As discussed in Section 9.11, bond order affects bond energy. Thus, the bond energy for C–C is given as 346 kJ/mol, whereas the bond energy for C=C is 602 kJ/mol. Bonding in sulfur compounds is still a matter of some debate. Assuming that the octet rule holds (bonding involving only s and p orbitals), we obtain the following resonance pictures of SO_2 and SO_3.

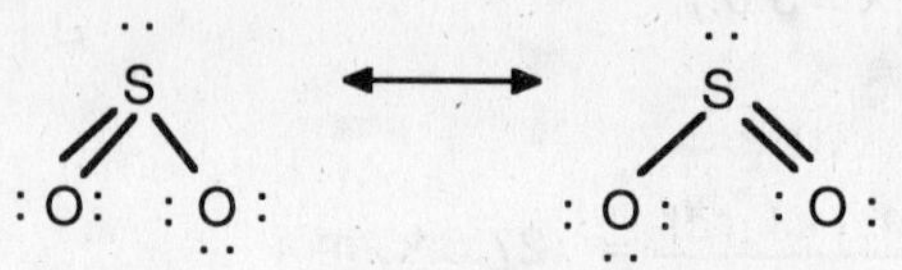

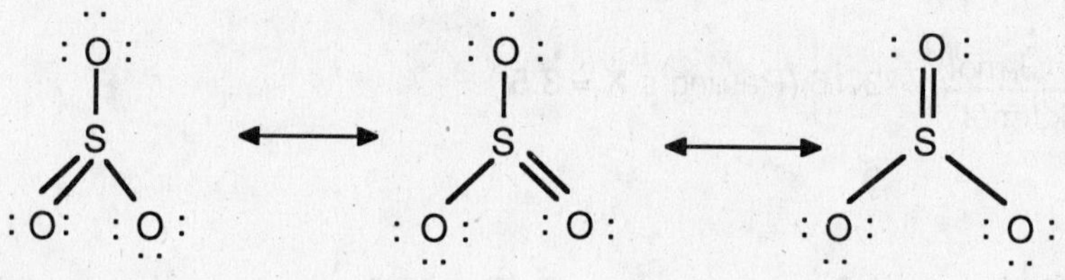

There is some evidence for resonance involving two double bonds, in which d orbitals participate:

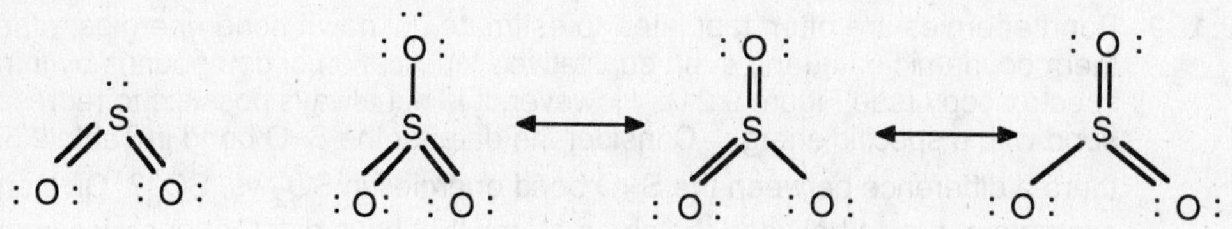

Note that the formula for SO_2 has zero formal charges for the sulfur and oxygen atoms. These bonding pictures would predict that SO_3 has greater single-bond character and therefore the S–O bond in SO_3 would have a lower bond energy.

Table 9.5 also shows that there is a difference in C=O bond energies in compounds such as acetone, CH_3COCH_3, where the bond energy is about 745 kJ/mol, and in CO_2, where the bond energy is 799 kJ/mol. The greater bond energy in CO_2 might be attributed to delocalization energy. That is, in CO_2, the pi electrons are free to move between two C=O bonds, whereas in acetone, there is only one such bond. The classical "textbook" discussion of delocalization compares benzene to cyclohexene. See, for example, John McMurray, Organic Chemistry, Brooks/Cole: Pacific Grove, CA, 1992, p. 532–534.

10. MOLECULAR GEOMETRY AND CHEMICAL BONDING THEORY

■ Solutions to Exercises

10.1 a. A Lewis structure of ClO_3^- is

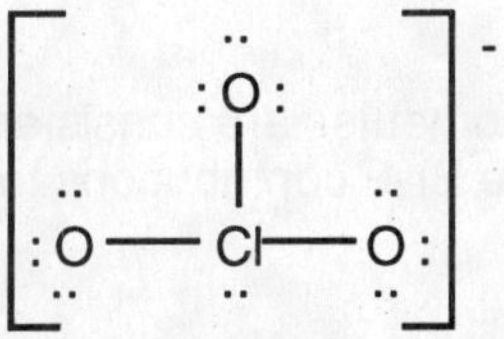

There are four electron pairs arranged tetrahedrally about the central atom. Three pairs are bonding and one pair is nonbonding. The expected geometry is trigonal pyramidal.

b. The Lewis structure of OF_2 is

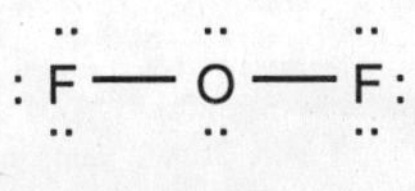

There are four electron pairs arranged tetrahedrally about the central atom. Two pairs are bonding and two pairs are nonbonding. The expected geometry is bent.

c. The Lewis structure of SiF_4 is

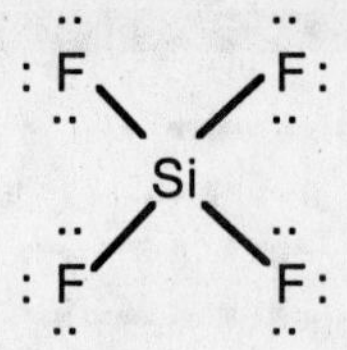

There are four bonding electron pairs arranged tetrahedrally around the central atom. The expected geometry is tetrahedral.

10.2 First, distribute the valence electrons to the bonds and the chlorine atoms. Then distribute the remaining electrons to iodine.

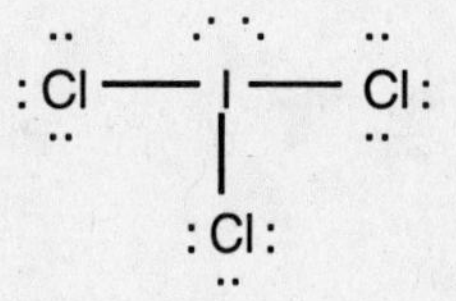

The five electron pairs around iodine should have a trigonal bipyramidal arrangement, with two lone pairs occupying equatorial positions. The molecule is T-shaped.

10.3 Both (b) trigonal pyramidal and (c) T-shaped geometries are consistent with a nonzero dipole moment. In trigonal planar geometry, the Br-F contributions to the dipole moment would cancel.

10.4 On the basis of symmetry, (b) SiF_4 would be expected to have a dipole moment of zero. The bonds are all symmetric about the central atom.

10.5 The Lewis structure for ammonia, NH_3, is

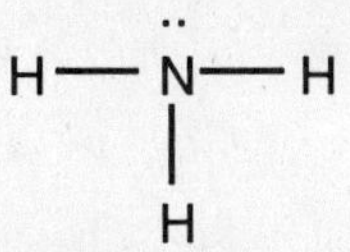

There are four pairs of electrons around the nitrogen atom; According to the VSEPR model, these are arranged tetrahedrally around the nitrogen atom and sp^3 hybridization is expected. This gives the following bonding description:

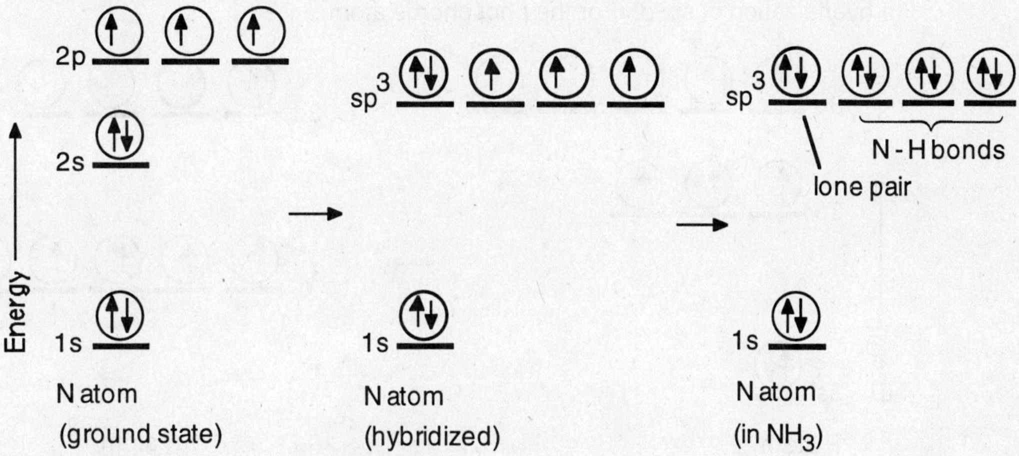

Each N-H bond is formed by the overlap of a 1s orbital of a hydrogen atom with one of the singly occupied sp^3 hybrid orbitals of the nitrogen atom.

10.6 The Lewis structure for PCl_5 is

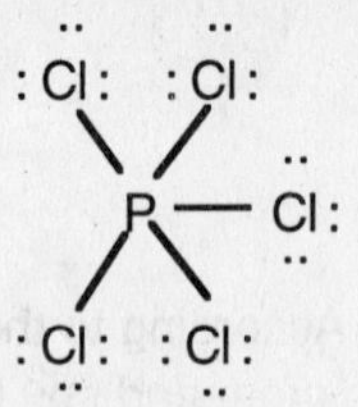

The phosphorus atom has five single bonds and no lone pairs around it. This suggests a hybridization of sp^3d. For the phosphorus atom,

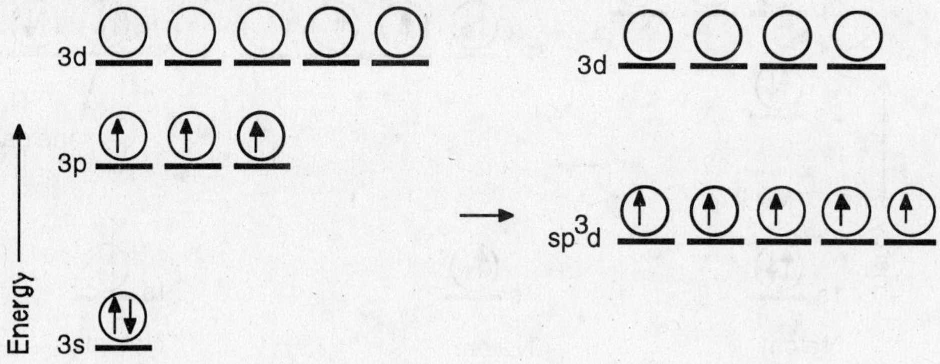

P atom (ground state)

P atom (hybridized)

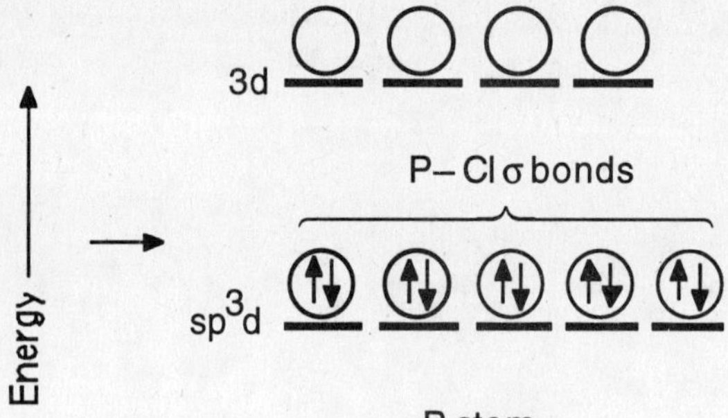

P atom (in PCl_5)

Each Cl atom has one singly occupied 3p orbital (recall the valence-shell configuration is $3s^23p^5$). Each P-Cl σ bond is formed by the overlap of a phosphorus sp^3d hybrid orbital with a singly occupied chlorine 3p atomic orbital.

10.7 The Lewis structure of CO_2 is

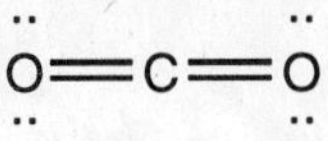

Because there are two σ bonds to the carbon atom, sp hybridization is suggested. The changes on this atom are

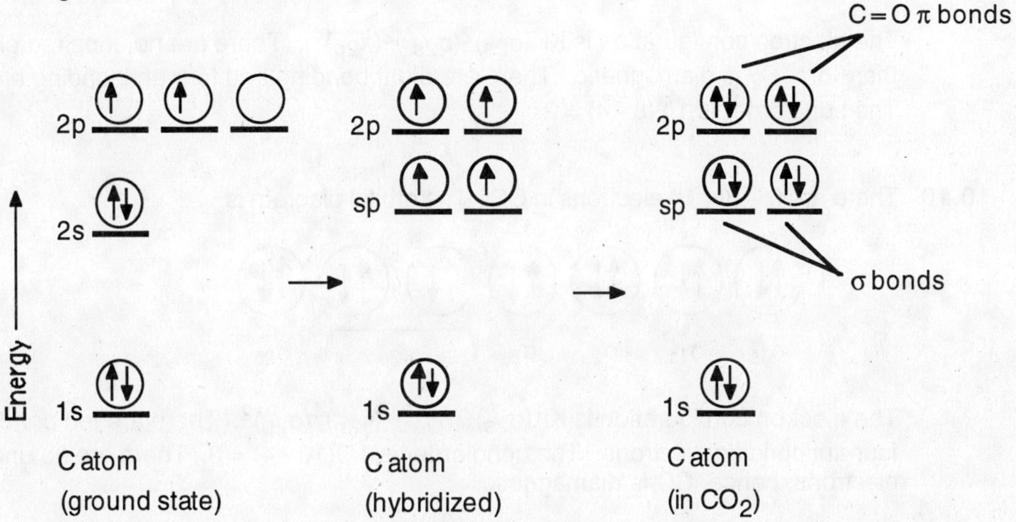

The hybrid orbitals on C are used to describe σ bonds. Each hybrid orbital on the C atom overlaps with a 2p orbital that is pointing along the bond axis on the O atoms to form two σ bonds. Each C atom 2p orbital that is perpendicular to the bond axis overlaps another 2p orbital that is parallel to it on each O atom. The result is a π bond to each O atom. Each C=O consists of one σ bond and one π bond.

10.8 The structural formulas for the isomers are as follows:

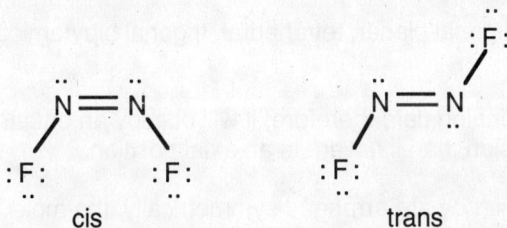

cis trans

These compounds exist as separate isomers. For these to interconvert, one end of the molecule would have to rotate with respect to the other end. This would require breaking the π bond and expending considerable energy.

10.9 There are 2 x 6 = 12 electrons in C_2. They occupy the orbitals as shown below.

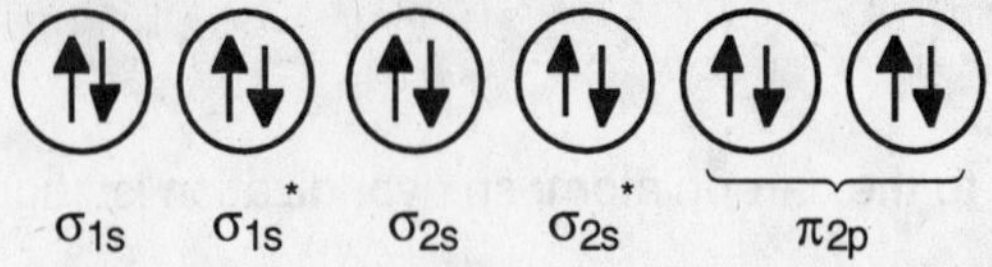

The electron configuration is $KK(\sigma_{2s})^2(\sigma_{2s}^*)^2(\pi_{2p})^4$. There are no unpaired electrons; therefore, C_2 is diamagnetic. There are eight bonding and four antibonding electrons. The bond order is 1/2(8 - 4) = 2.

10.10 There are 6 + 8 = 14 electrons in CO. The orbital diagram is

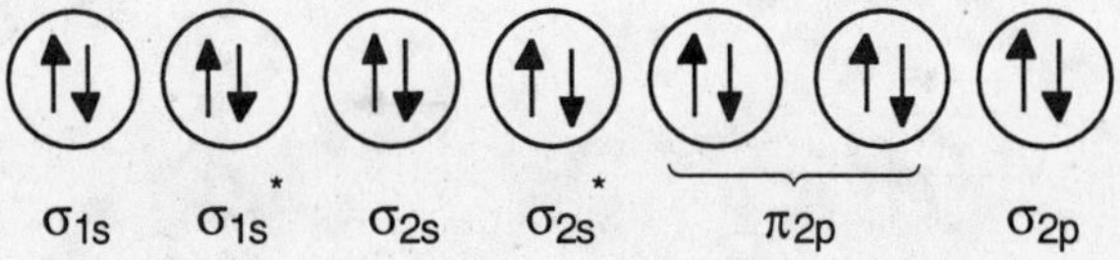

The electron configuration is $KK(\sigma_{2s})^2(\sigma_{2s}^*)^2(\pi_{2p})^4(\sigma_{2p})^2$. There are ten bonding and four antibonding electrons. The bond order is 1/2(10 - 4) = 3. There are no unpaired electrons; hence, CO is diamagnetic.

■ Answers to Review Questions

10.1 The VSEPR model is used to predict the geometry of molecules. The electron pairs around an atom are assumed to arrange themselves to reduce electron repulsion. The molecular geometry is determined by the positions of the bonding electron pairs.

10.2 The arrangements are linear, trigonal planar, tetrahedral, trigonal bipyramidal, and octahedral.

10.3 A lone pair is "larger" than a bonding pair; therefore, it will occupy an equatorial position, where it encounters less repulsion than if it were in an axial position.

10.4 The bonds could be polar, but if they are arranged symmetrically, the molecule will be nonpolar. The bond dipoles will cancel.

10.5 Nitrogen trifluoride has three N-F bonds arranged to form a trigonal pyramid. These bonds are polar and would give a polar molecule with partial negative charges on the fluorine atoms and a partial positive charge on the nitrogen atom. However, there is also a lone pair of electrons on nitrogen that is directed away from the bonds. The result is that the lone pair nearly cancels the polarity of the bonds and gives a molecule with a very small dipole moment.

10.6 Certain orbitals, such as p orbitals and hybrid orbitals, have lobes in given directions. Bonding to these orbitals is directional; that is, the bonding is in preferred directions. This explains why the bonding gives a particular molecular geometry.

10.7 The angle is 109.5°.

10.8 A sigma bond has a cylindrical shape about the bond axis. A pi bond has a distribution of electrons above and below the bond axis.

10.9 In ethylene, C_2H_4, the changes on a given carbon atom may be described as follows:

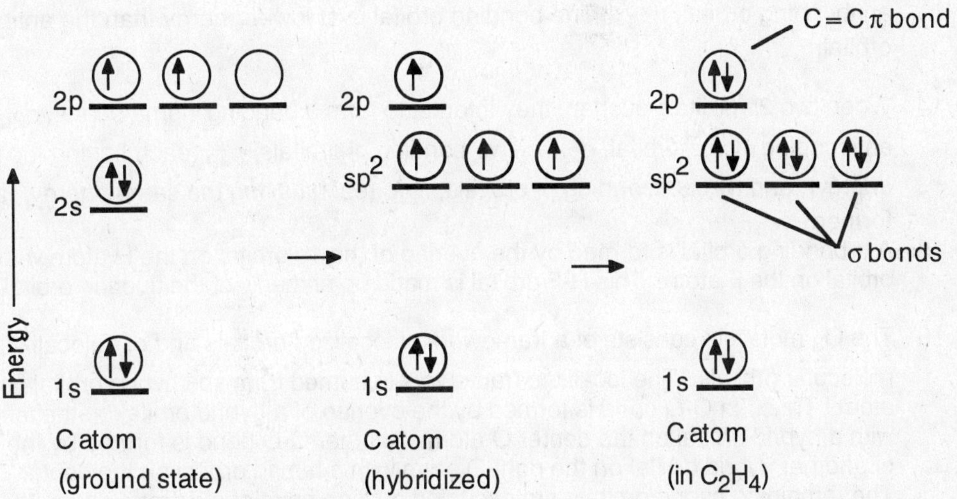

An sp^2 hybrid orbital on one carbon atom overlaps a similar hybrid orbital on the other carbon atom to form a σ bond. The remaining hybrid orbitals on the two carbon atoms overlap 1s orbitals from the hydrogen atoms to form four C-H bonds. The unhybridized 2p orbital on one carbon atom overlaps the unhybridized 2p orbital on the other carbon atom to form a π bond. The σ and π bonds together constitute a double bond.

10.10 Both of the unhybridized 2p orbitals, one from each carbon atom, are perpendicular to their CH_2 planes. When these orbitals overlap each other, they fix both planes to be in the same plane. The two ends of the molecule cannot twist around without breaking the π bond, which requires considerable energy. Therefore, it is possible to have stable molecules with the following structures:

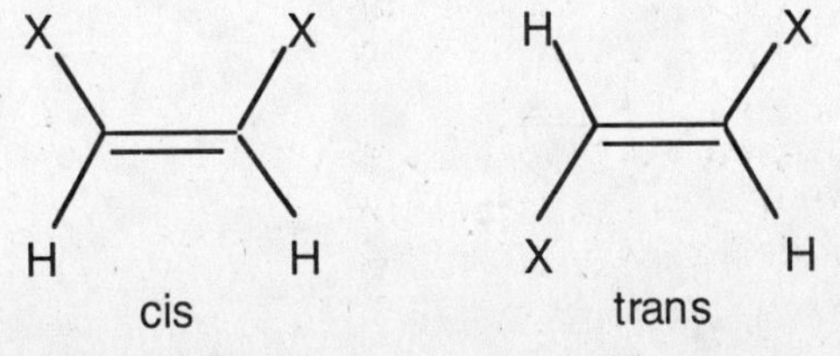

Because these have the same molecular formulas, they are isomers. In this case, they are called cis-trans isomers, or geometrical isomers.

10.11 In a bonding orbital, the probability of finding electrons between the two nuclei is high. For this reason, the energy of the bonding orbital is lower than that of the separate atomic orbitals. In an antibonding orbital, the probability of finding electrons between the two nuclei is low. For this reason, the energy of the antibonding orbital is higher than that of the separate atomic orbitals.

10.12 The factors determining the strength of interaction of two atomic orbitals are (1) the energy difference between the interacting orbitals and (2) the magnitude of their overlap.

10.13 When two 2s orbitals overlap, they interact to form a bonding orbital, σ_{2s}, and an antibonding orbital, σ_{2s}^*. The bonding orbital is at lower energy than the antibonding orbital.

10.14 When two 2p orbitals overlap, they interact to form a bonding sigma orbital, σ_{2p}, and an antibonding sigma orbital, σ_{2p}^*. Two bonding pi orbitals, π_{2p} (each having the same energy), and two antibonding pi orbitals, π_{2p}^* (each having the same energy), are also formed.

10.15 A σ bonding orbital is formed by the overlap of the 1s orbital on the H atom with the 2p orbital on the F atom. This H-F orbital is made up primarily of the fluorine orbital.

10.16 The O_3 molecule consists of a framework of localized orbitals and of delocalized pi molecular orbitals. The localized framework is formed from sp^2 hybrid orbitals on each atom. Thus, an O-O bond is formed by the overlap of a hybrid orbital on the left O atom with a hybrid orbital on the center O atom. Another O-O bond is formed by the overlap of another hybrid orbital on the right O atom with a hybrid orbital on the center O atom. The remaining hybrid orbitals are occupied by lone pairs of electrons. Also, there is one unhybridized p orbital on each of the atoms. These p orbitals are perpendicular to the plane of the molecule and overlap sidewise to give three pi molecular orbitals that are delocalized. The two orbitals of lowest energy are occupied by pairs of electrons.

Solutions to Practice Problems

10.17 The number of electron pairs, lone pairs and geometry are for the central atom.

	Electron-Dot Structure	Number of Electron Pairs	Number of Lone Pairs	Geometry
a.	:Cl—C(—Cl:)(—Cl:)—Cl:	4	0	Tetrahedral
b.	H—Se—H	4	2	Bent (angular)
c.	:F—As(—F:)—F:	4	1	Trigonal pyramidal
d.	:Cl—Al(—Cl:)—Cl:	3	0	Trigonal planar

10.18 The number of electron pairs, lone pairs and geometry are for the central atom.

	Electron-Dot Structure	Number of Electron Pairs	Number of Lone Pairs	Geometry
a.	:S=C=S:	2	0	Linear
b.	H—Si(H)(H)—H	4	0	Tetrahedral

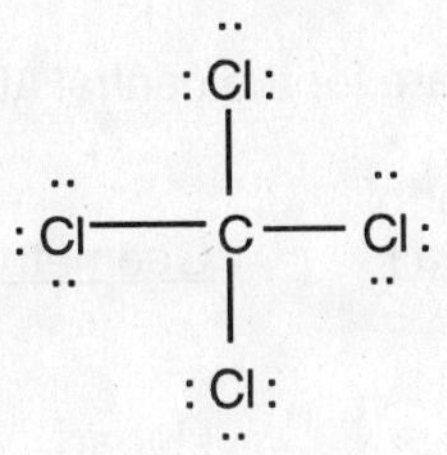

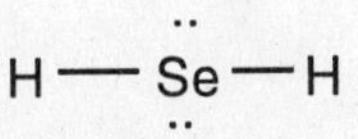

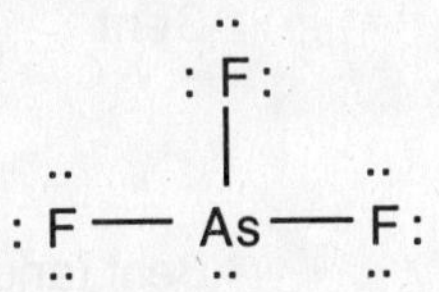

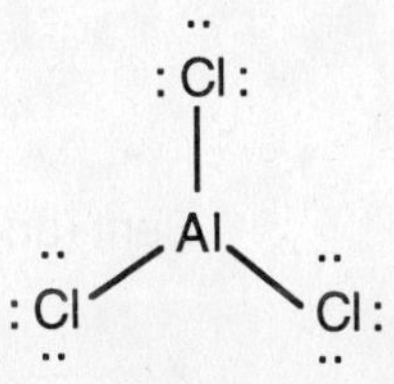

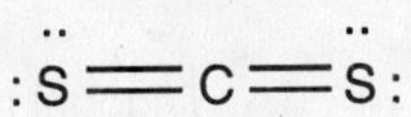

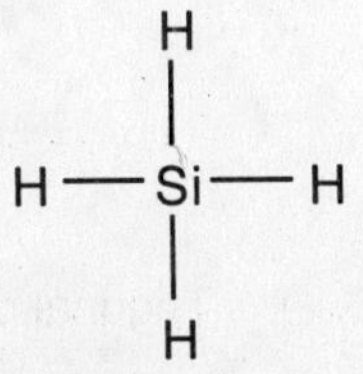

c. $\overset{..}{:}\overset{..}{Cl}:$
    $:\overset{..}{Cl}-\overset{|}{P}-\overset{..}{Cl}:$ 4   1   Trigonal pyramidal

d. $:\overset{..}{Br}-Be-\overset{..}{Br}:$ 2   0   Linear

10.19 The number of electron pairs, lone pairs and geometry are for the central atom.

	Electron-Dot Structure	Number of Electron Pairs	Number of Lone Pairs	Geometry
a.	$[:\overset{..}{Cl}-Tl-\overset{..}{Cl}:]^+$	2	0	Linear
b.	$[:\overset{..}{O}-N=\overset{..}{O}:]^-$	3	1	Bent
c.	$[H-\overset{..}{N}-H]^-$	4	2	Bent (angular)
d.	$[:\overset{..}{O}-\overset{..}{Cl}-\overset{..}{O}:]^-$	4	2	Bent (angular)

10.20 The number of electron pairs, lone pairs and geometry are for the central atom.

	Electron-Dot Structure	Number of Electron Pairs	Number of Lone Pairs	Geometry
a.	$\begin{bmatrix} H \\ H:\overset{..}{O}:H \end{bmatrix}^+$	4	1	Trigonal pyramidal
b.	$\begin{bmatrix} :\overset{..}{F}: \\ :\overset{..}{F}-Be-\overset{..}{F}: \end{bmatrix}^-$	3	0	Trigonal planar

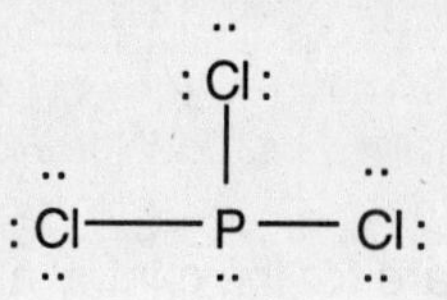

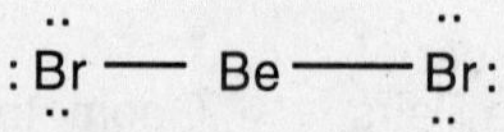

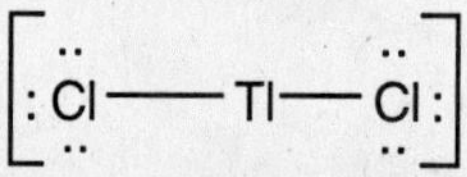

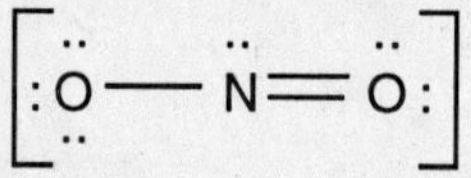

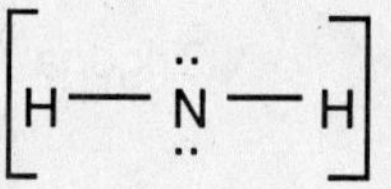

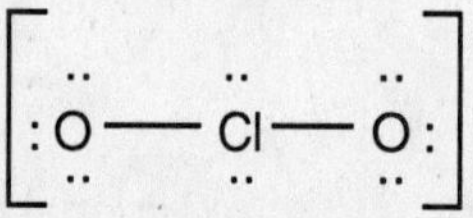

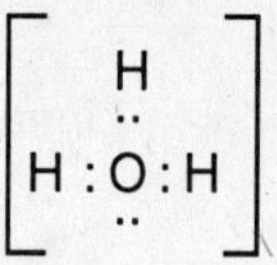

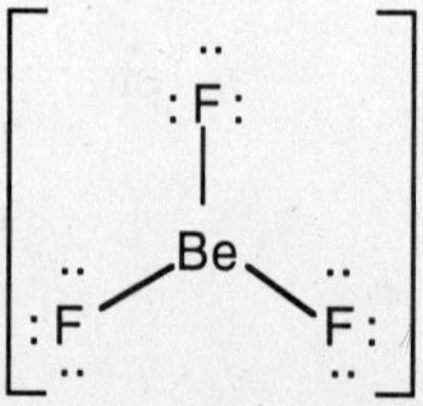

c. 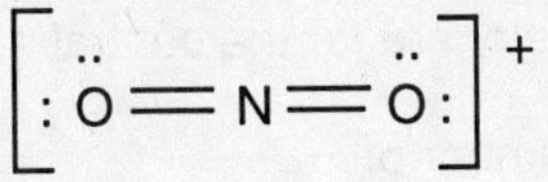	2	0	Linear
d. 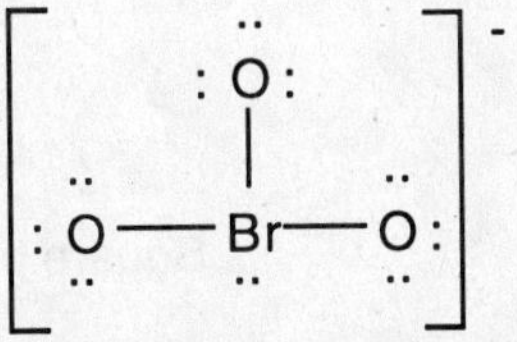	4	1	Trigonal pyramidal

10.21 The number of electron pairs, lone pairs and geometry are for the central atom.

Electron-Dot Structure	Number of Electron Pairs	Number of Lone Pairs	Geometry
a. 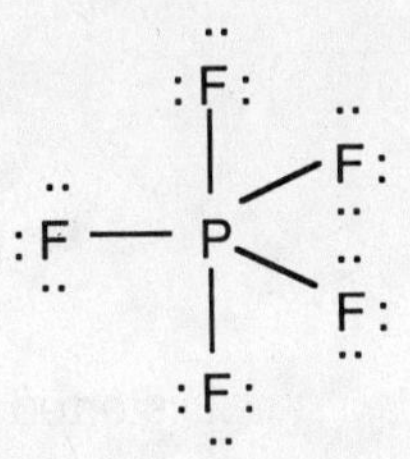	5	0	Trigonal bipyramidal
b. 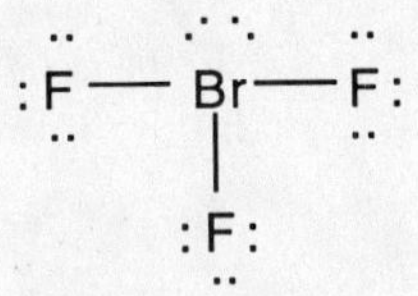	5	2	T-shaped
c. 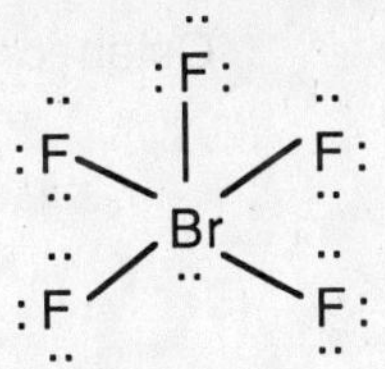	6	1	Square pyramidal
d. 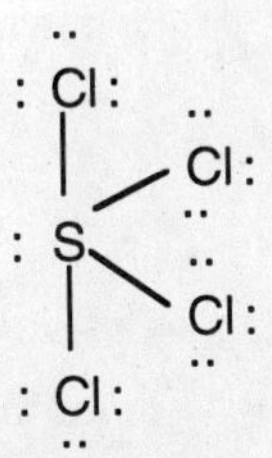	5	1	Distorted tetrahedral

10.22 The number of electron pairs, lone pairs and geometry are for the central atom.

	Electron-Dot Structure	Number of Electron Pairs	Number of Lone Pairs	Geometry
a.	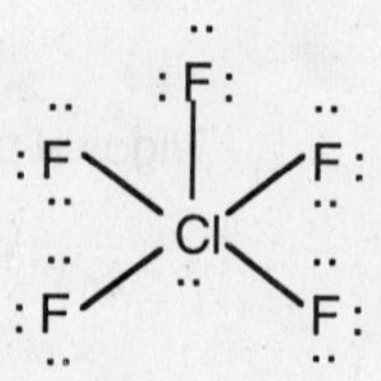	6	1	Square pyramidal
b.	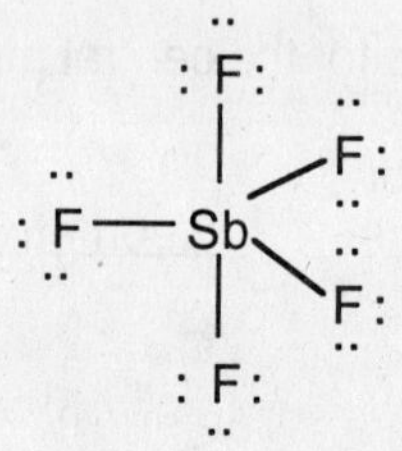	5	0	Trigonal bipyramidal
c.	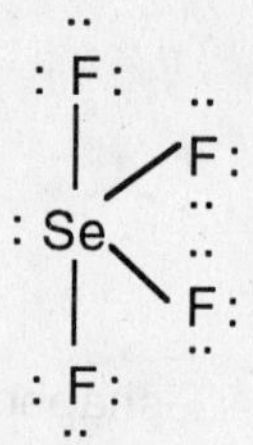	5	1	Distorted tetrahedral
d.	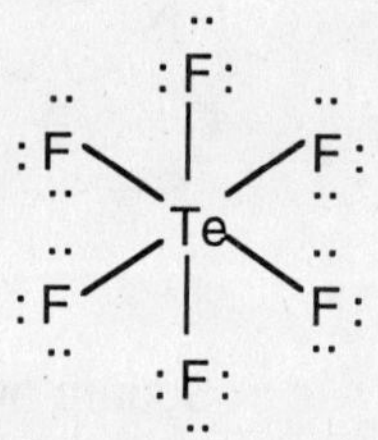	6	0	Octahedral

10.23 The number of electron pairs, lone pairs and geometry are for the central atom.

	Electron Dot Structure	Number of Electron Pairs	Number of Lone Pairs	Geometry
a.	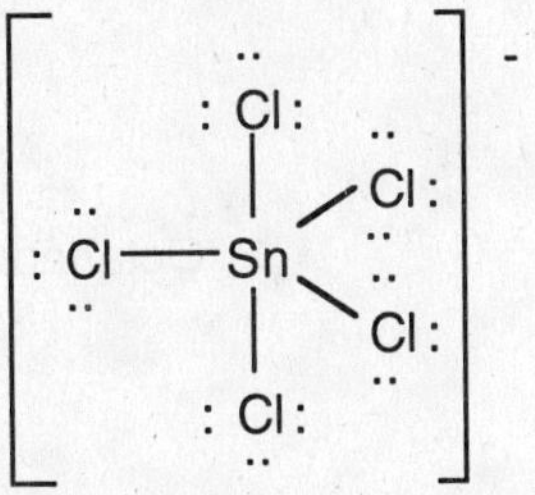	5	0	Trigonal bipyramidal
b.	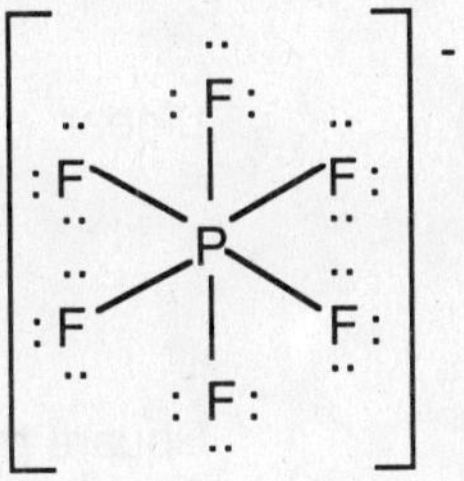	6	0	Octahedral
c.	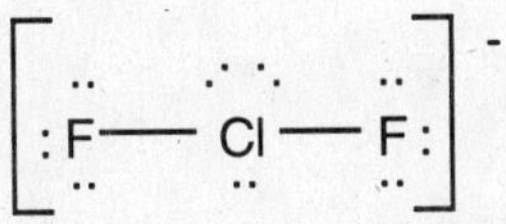	5	3	Linear
d.	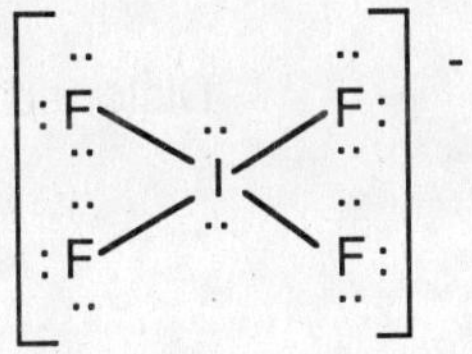	6	2	Square planar

10.24 The number of electron pairs, lone pairs and geometry are for the central atom.

	Electron Dot Structure	Number of Electron Pairs	Number of Lone Pairs	Geometry
a.	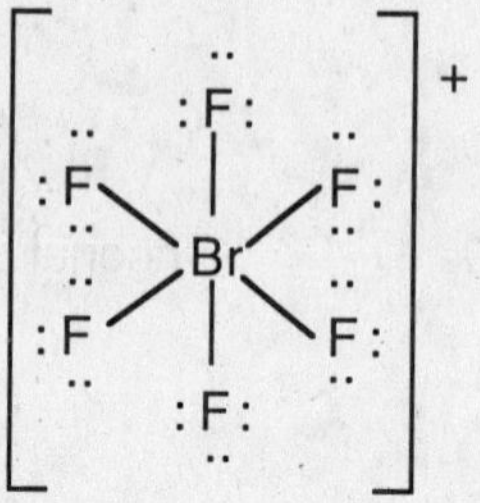	6	0	Octahedral
b.		5	3	Linear
c.	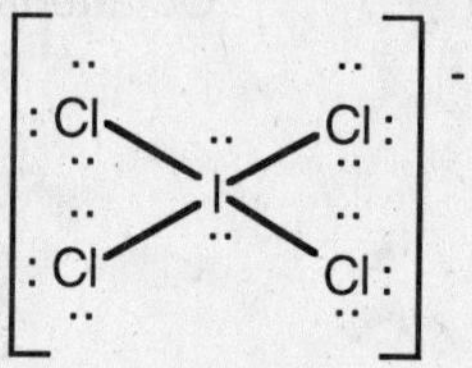	6	2	Square planar
d.	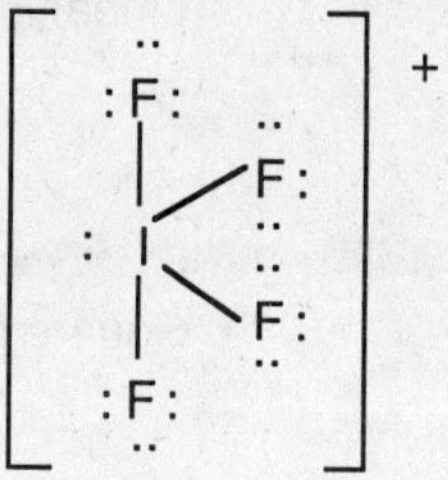	5	1	Distorted tetrahedral

10.25 a. Trigonal pyramidal and T-shaped. Trigonal planar would have a dipole moment of zero.

b. Bent. Linear would have a dipole moment of zero.

10.26 a. Trigonal pyramidal and T-shaped. Trigonal planar would have a dipole moment of zero.

b. Distorted tetrahedral. Others would have a dipole moment of zero.

10.27 TeF_6 (octahedral) and BeF_2 (linear) have zero dipole moments.

10.28 $AlCl_3$ (trigonal planar) and $SiCl_4$ (tetrahedral) have zero dipole moments.

10.29 a. Aluminum has three valence electrons. Each bromine contributes one electron; hence, the Al has a total of six valence electrons, or three electron pairs. The hybridization of Al is sp^2.

b. Beryllium has two valence electrons. The chlorine atoms each donate one electron to their bonds. There are four electrons, or two electron pairs, around Be. The bonding can be described in terms of sp hybridization of the beryllium.

c. Silicon has four valence electrons. Each Cl atom contributes an electron to give a total of eight electrons, or four electron pairs, around the Si atom. The hybridization of Si is sp^3.

d. Beryllium has two valence electrons, to which may be added one electron from each of the three fluorine atoms and one electron to account for the charge on the ion. This gives a total of six electrons, or three electron pairs, around the Be atom. The hybridization of Be is sp^2.

10.30 a. Nitrogen has five valence electrons. Each chlorine atom donates one electron to give a total of eight electrons, or four electron pairs, around the N atom. The hybridization is sp^3.

b. Boron has three valence electrons. Each chlorine atom contributes one electron. The B atom has a total of six electrons, or three electron pairs, around it. The hybridization is sp^2.

c. Beryllium has two valence electrons. The Br atoms each donate one electron to their bonds. There are four electrons, or two electron pairs, around Be. The hybridization is sp.

d. Aluminum has three valence electrons, to which may be added one electron from each of the fluorine atoms and one electron to account for the negative charge on the ion. This gives a total of eight electrons, or four electron pairs, around Al. The hybridization is sp^3.

10.31 a. The Lewis structure is

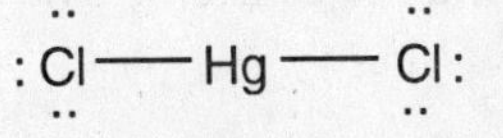

The presence of two single bonds and no lone pairs suggests sp hybridization. Thus, an Hg atom with the configuration $[Xe]4f^{14}5d^{10}6s^2$ is promoted to $[Xe]4f^{14}5d^{10}6s^16p^1$, then hybridized. An Hg-Cl bond is formed by overlapping an Hg hybrid orbital with a 3p orbital of Cl.

b. The Lewis structure is

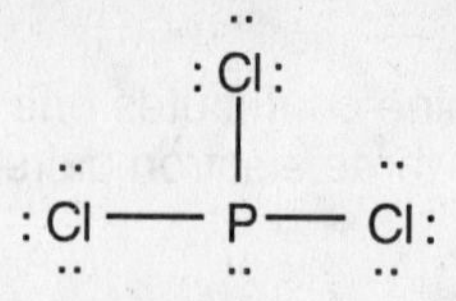

The presence of three single bonds and one lone pair suggests sp^3 hybridization of the P atom. Three hybrid orbitals each overlap a 3p orbital of a Cl atom to form a P-Cl bond. The fourth hybrid orbital contains the lone pair.

10.32 a. The Lewis structure is

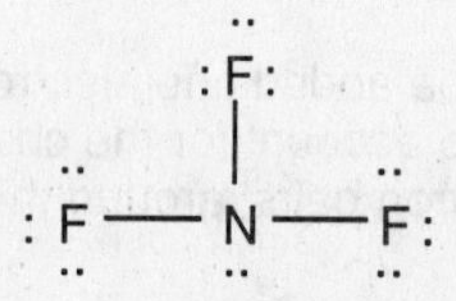

The presence of three single bonds and one lone pair suggests sp^3 hybridization of the N atom. Three hybrid orbitals each overlap a 2p orbital of an F atom to form an N-F bond. The fourth hybrid orbital contains the lone pair.

b. The Lewis structure is

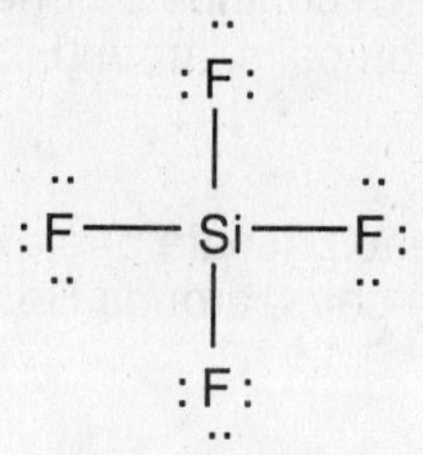

The presence of four single bonds and no lone pairs suggests sp^3 hybridization of the Si atom. Each hybrid orbital overlaps a 2p orbital of an F atom to form an Si-F bond.

10.33 a. Bromine has seven valence electrons. Each F atom donates one electron to give a total of twelve electrons, or six electron pairs, around the Br atom. The hybridization is sp^3d^2.

b. Bromine has seven valence electrons. Each F atom donates one electron to give a total of ten electrons, or five electron pairs, around the Br atom. The hybridization is sp^3d.

c. Arsenic has five valence electrons. The Cl atoms each donate one electron to give a total of ten electrons, or five electron pairs, around the As atom. The hybridization is sp^3d.

d. Chlorine has seven valence electrons, to which may be added one electron from each F atom minus one electron for the charge on the ion. This gives a total of ten electrons, or five electron pairs, around chlorine. The hybridization is sp^3d.

10.34 a. Phosphorus has five valence electrons. The Cl atoms each donate one electron to give a total of ten electrons, or five electron pairs, around the P atom. The hybridization is sp^3d.

b. Selenium has six valence electrons. Each F atom donates one electron to give a total of twelve electrons, or six electron pairs, around the Se atom. The hybridization is sp^3d^2.

c. Tellurium has six valence electrons. The F atoms each donate one electron to give a total of ten electrons, or five electron pairs, around the Te atom. The hybridization is sp^3d.

d. Iodine has seven valence electrons, to which may be added one electron from each of the F atoms and one electron to account for the charge on the ion. This gives a total of twelve electrons, or six electron pairs, around the I atom. The hybridization is sp^3d^2.

10.35 The P atom in PCl_6^- has six single bonds around it and no lone pairs. This suggests sp^3d^2 hybridization. Each bond in this ion is a σ bond formed by overlap of an sp^3d^2 hybrid orbital on P with a 3p orbital on Cl.

10.36 The central I atom in I_3^- has two single bonds and three lone pairs around it. This suggests sp^3d hybridization. Each I-I bond is formed by the overlap of an sp^3d hybrid orbital on the central I with a 5p orbital from a terminal I.

10.37 a. The structural formula of formaldehyde is

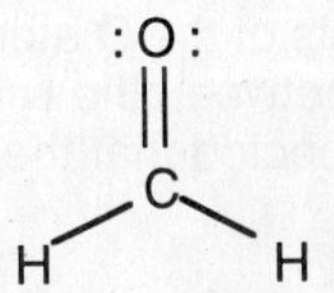

Because the C is bonded to three other atoms, it is assumed to be sp^2 hybridized. One 2p orbital remains unhybridized. The carbon-hydrogen bonds are σ bonds formed by the overlap of an sp^2 hybrid orbital on C with a 1s orbital on H. The remaining sp^2 hybrid orbital on C overlaps with a 2p orbital on O to form a σ bond. The unhybridized 2p orbital on C overlaps with a parallel 2p orbital on O to form a π bond. Together the σ and π bonds constitute a double bond.

b. The nitrogen atoms are sp hybridized. A σ bond is formed by the overlap of an sp hybrid orbital from each N. The remaining sp hybrid orbitals contain lone pairs of electrons. The two unhybridized 2p orbitals on one N overlap with the parallel unhybridized 2p orbitals on the other N to form two π bonds.

10.38 a. Each nitrogen atom is bonded to two other atoms and also has a lone pair of electrons in its valence shell. Therefore, the nitrogen atoms are expected to be sp^2 hybridized. The N-H bonds are σ bonds formed by the overlap of an sp^2 hybrid orbital of an N atom with a 1s orbital on an H atom. Each N also contains a lone pair of electrons in an sp^2 hybrid orbital. The remaining sp^2 hybrid orbitals, one on each N, overlap with each other to form a σ bond. The two 2p orbitals, one on each N, are oriented parallel to each other so that they may overlap to form a π bond.

b. The carbon atom is bonded to two atoms and has no lone pairs; hence, it is expected to be sp hybridized. The C-H bond is a σ bond formed by the overlap of an sp hybrid orbital on carbon with a 1s orbital on hydrogen. The other sp hybrid orbital overlaps with an sp orbital on the nitrogen atom to form a σ bond. The remaining sp hybrid orbital on N contains a lone pair. There remain two unhybridized 2p orbitals on the carbon and two 2p orbitals on the nitrogen atom. One 2p orbital from the C and a parallel 2p orbital on the N overlap to form a π bond. The remaining two 2p orbitals overlap to form a second π bond between the carbon and nitrogen.

10.39

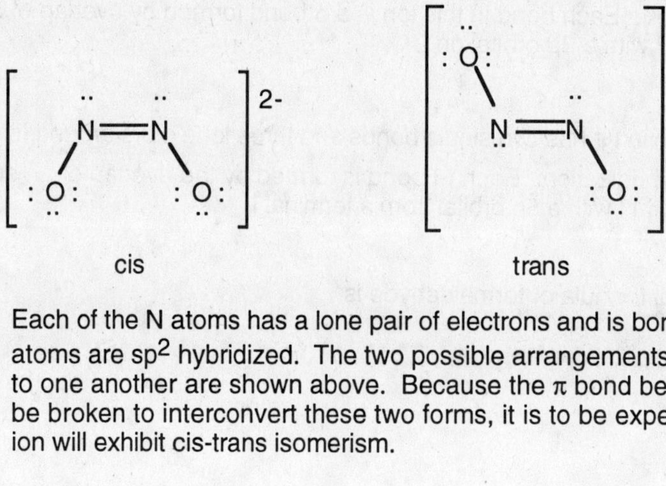

cis trans

Each of the N atoms has a lone pair of electrons and is bonded to two atoms. The N atoms are sp^2 hybridized. The two possible arrangements of the O atoms relative to one another are shown above. Because the π bond between the N atoms must be broken to interconvert these two forms, it is to be expected that the hyponitrite ion will exhibit cis-trans isomerism.

10.40

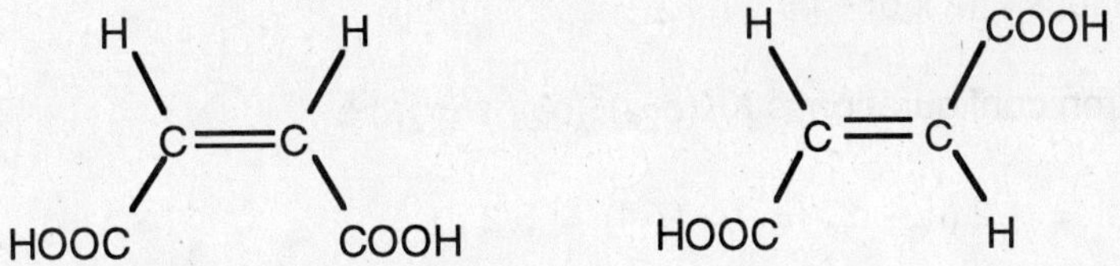

The C=C bond consists of a σ bond and a π bond. The π bond depends on the parallel orientation of the two p orbitals that overlap to form the bond. In order to interconvert the cis and trans forms, one end of the molecule would have to be rotated relative to the other end, thus changing the orientation of the p orbitals and breaking the bond. Because this requires considerable energy, the cis and trans forms are stable relative to interconversion.

10.41 a. Total electrons = 2 x 5 = 10.

The electron configuration is $KK(\sigma_{2s})^2(\sigma_{2s}^*)^2(\pi_{2p})^2$.

Bond order = $1/2\ (n_b - n_a)$ = 1/2 (6-4) = 1

The B_2 molecule is stable. It is paramagnetic because the two electrons in the π_{2p} subshell occupy separate orbitals.

b. Total electrons = (2 x 5) - 1 = 9.

The electron configuration is $KK(\sigma_{2s})^2(\sigma_{2s}^*)^2(\pi_{2p})^1$.

Bond order = $1/2\ (n_b - n_a)$ = 1/2 (5-4) = 1/2

The B_2^+ molecule should be stable and is paramagnetic because there is one unpaired electron in the π_{2p} subshell.

c. Total electrons = (2 x 8) + 1 = 17.

The electron configuration is $KK(\sigma_{2s})^2(\sigma_{2s}^*)^2(\pi_{2p})^4(\sigma_{2p})^2(\pi_{2p}^*)^3$.

Bond order = $1/2\ (n_b - n_a)$ = 1/2 (10-7) = 3/2

The O_2^- molecule should be stable and is paramagnetic because there is one unpaired electron in the π_{2p}^* subshell.

10.42 a. Total electrons = (2 x 6) - 1 = 11.

The electron configuration is $KK(\sigma_{2s})^2(\sigma_{2s}^*)^2(\pi_{2p})^3$.

Bond order = $1/2\ (n_b - n_a)$ = 1/2 (7-4) = 3/2

The C_2^+ molecule is expected to be stable and paramagnetic.

b. Total electrons = 2 x 10 = 20.

The electron configuration is $KK(\sigma_{2s})^2(\sigma_{2s}^*)^2(\pi_{2p})^4(\sigma_{2p})^2(\pi_{2p}^*)^4(\sigma_{2p}^*)^2$.

Bond order = $1/2\ (n_b - n_a)$ = 1/2 (10-10) = 0

The Ne_2 molecule is expected to be unstable and diamagnetic.

c. Total electrons = (2 x 6) + 1 = 13.

The electron configuration is $KK(\sigma_{2s})^2(\sigma_{2s}^*)^2(\pi_{2p})^4(\sigma_{2p})^1$.

Bond order = $1/2\ (n_b - n_a)$ = 1/2 (9-4) = 5/2

The C_2^- molecule is expected to be stable. It is paramagnetic because there is one unpaired electron in the σ_{2p} subshell.

10.43 Total electrons = 6 + 7 + 1 = 14.

The electron configuration is $KK(\sigma_{2s})^2(\sigma_{2s}^*)^2(\pi_{2p})^4(\sigma_{2p})^2$.

Bond order = $1/2\ (n_b - n_a)$ = 1/2 (10-4) = 3

The CN^- ion is diamagnetic.

10.44 Total electrons = 5 + 7 = 12.

The electron configuration is $KK(\sigma_{2s})^2(\sigma_{2s}^*)^2(\pi_{2p})^4$.

Bond order = $1/2\ (n_b - n_a)$ = 1/2 (8-4) = 2

The BN molecule is diamagnetic.

Solutions to Unclassified Problems

10.45 The number of electron pairs, lone pairs and geometry are for the central atom.

	Electron-Dot Structure	Number of Electron Pairs	Number of Lone Pairs	Geometry
a.	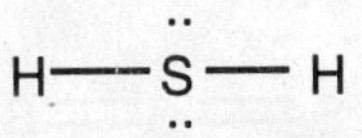	4	2	Bent
b.	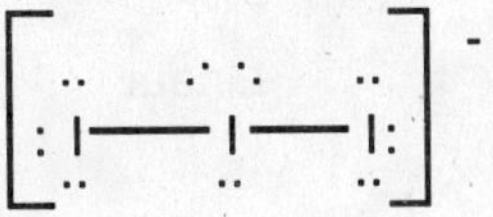	5	3	Linear
c.	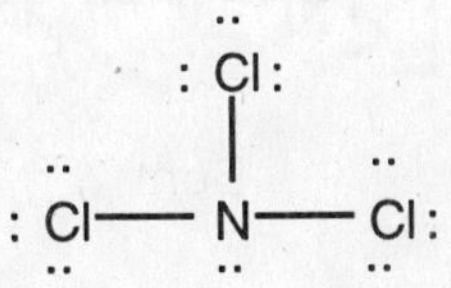	4	1	Trigonal pyramidal
d.	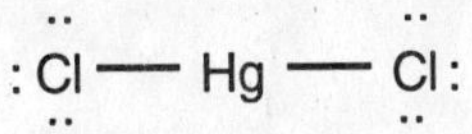	2	0	Linear

10.46

	Electron-Dot Structure	Number of Electron Pairs	Number of Lone Pairs	Geometry
a.	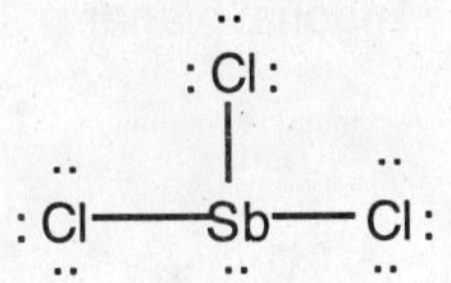	4	1	Trigonal pyramidal
b.	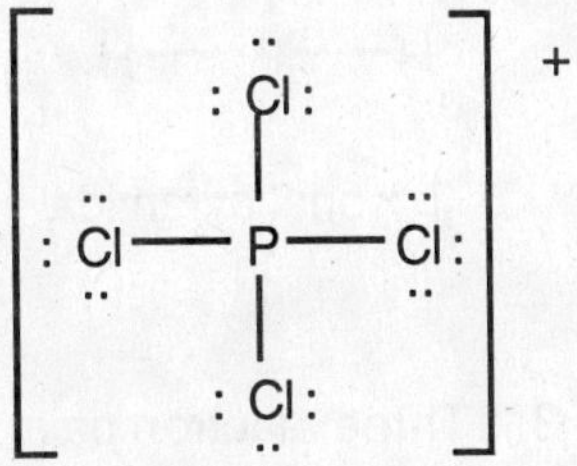	4	0	Tetrahedral
c.	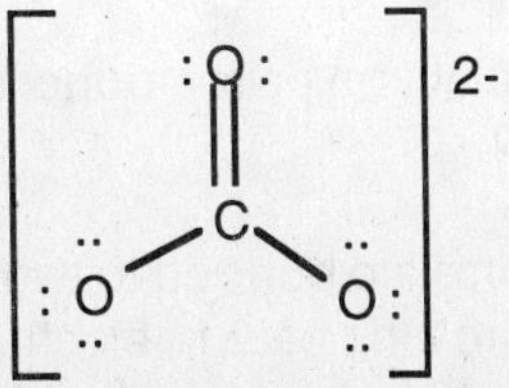	3	0	Trigonal planar

d. 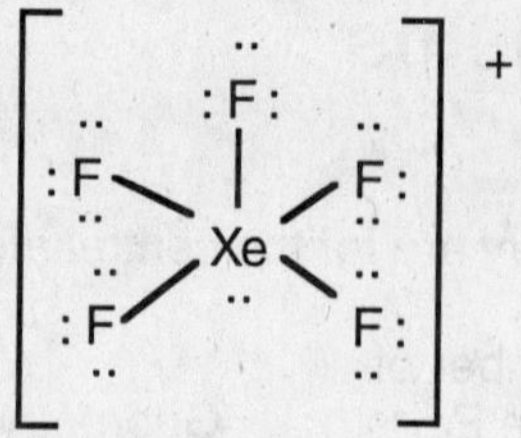 6 1 Square pyramidal

10.47 a. 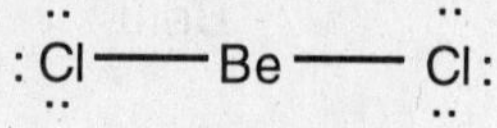

Linear

c. 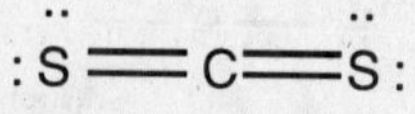

Linear

b. 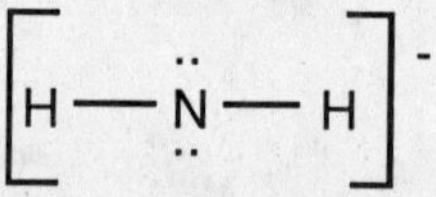

Bent

d. 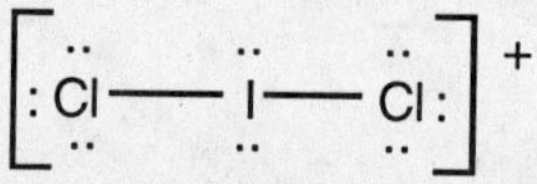

Bent

10.48 a. 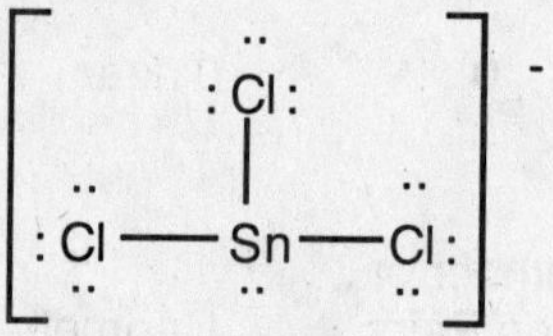

Trigonal pyramidal

c. 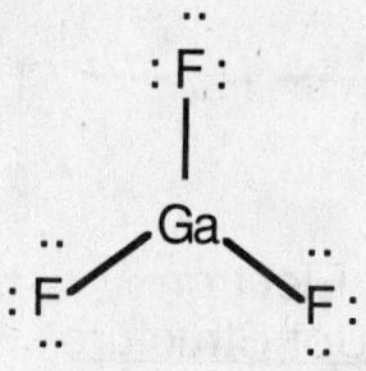

Trigonal planar

b. 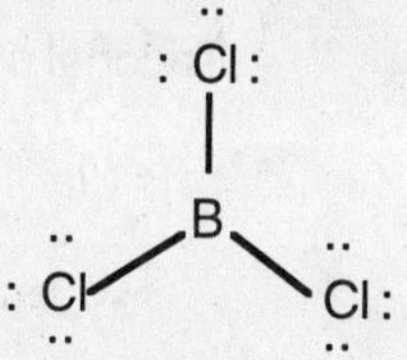

Trigonal planar

d. 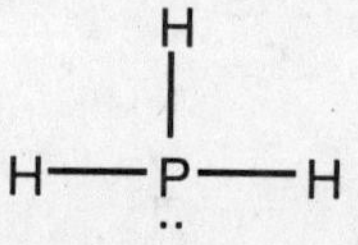

Trigonal pyramidal

10.49 a. 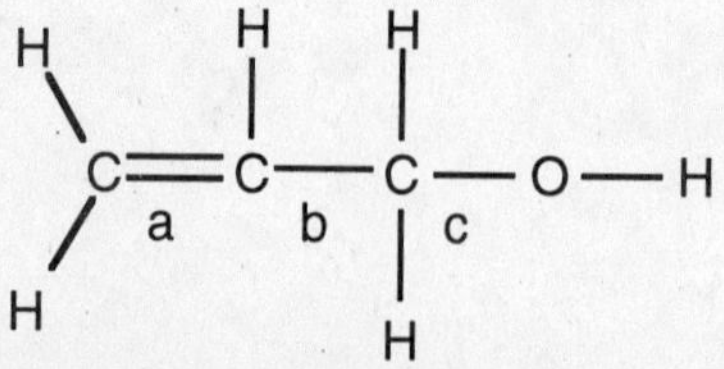

C_a and C_b: Three electron pairs around each. They are sp^2 hybridized.

C_c: Four electron pairs around it. It is sp^3 hybridized

b. 

Both C atoms are bonded to two other atoms and have no lone pairs of electrons. They are sp hybridized

10.50 a. 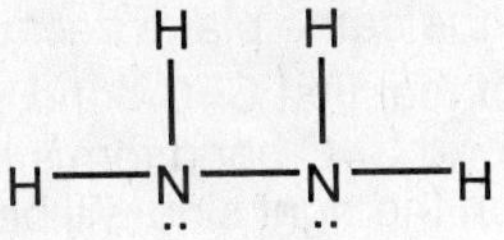

Each N has three atoms and one lone pair of electrons. They are sp^3 hybridized.

b. $:N\equiv C-C\equiv N:$

Each N as a σ bond to one atom and a lone pair of electrons. They are sp hybridized.

c. 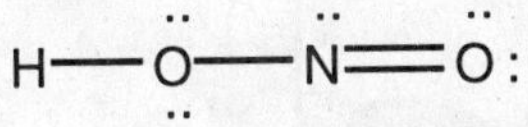

N has σ bonds to two atoms and a lone pair of electrons. It is sp^2 hybridized.

10.51 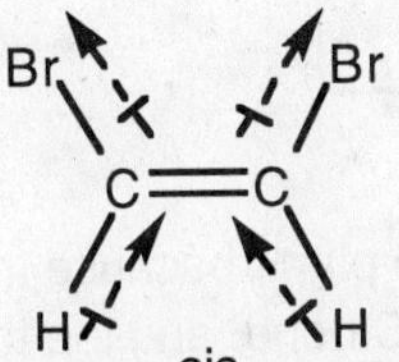

cis

has a net dipole

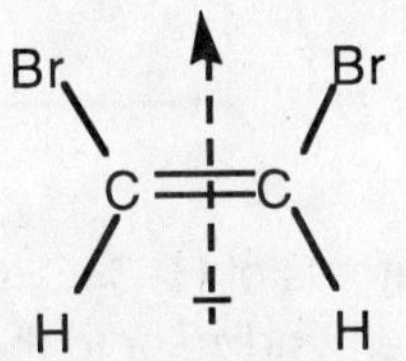

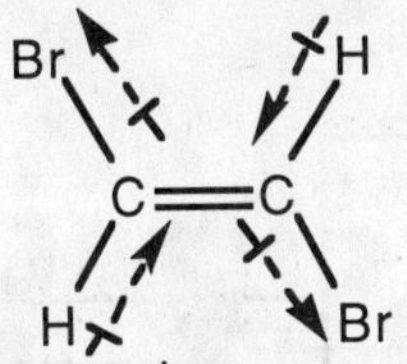

trans

has no net dipole. The two C-Br bond dipoles cancel and the two C-H bond dipoles cancel.

10.52 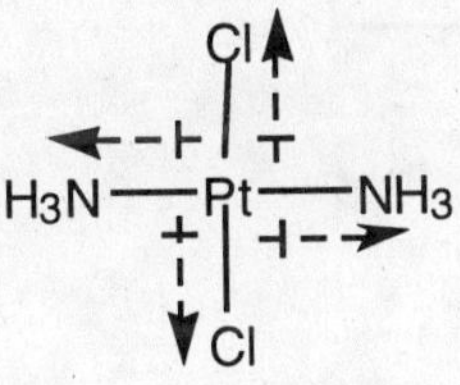

zero

Identical bond dipoles cancel each other.

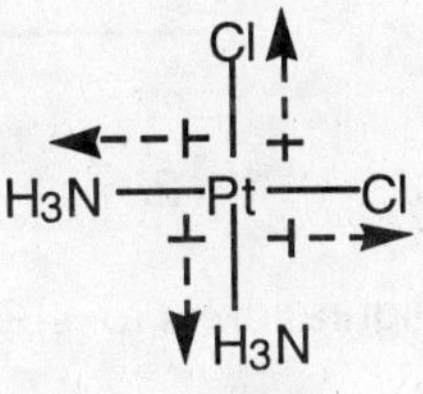

nonzero

Bond dipoles cannot cancel. Dissimilar bonds across from each other.

10.53 All four hydrogen atoms of $H_2C{=}C{=}CH_2$ cannot lie in the same plane because the second C=C bond forms perpendicular to the plane of the first C=C bond. By looking at Figure 10.24, you can see how this is so. The second C=C bond forms in the plane of the C-H bonds. Thus the plane of the C-H bonds on the right side will be perpendicular to the plane of the C-H bonds on the left side. This is shown below using the π orbitals of the two C=C double bonds.

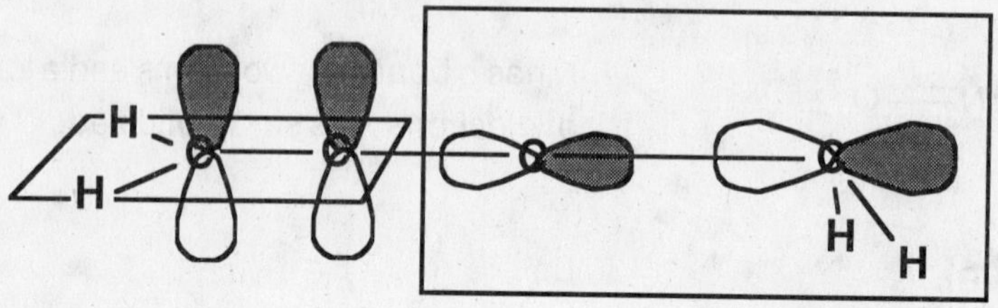

10.54 All four hydrogen atoms of $H_2C{=}C{=}C{=}CH_2$ lie in the same plane because the second C=C bond forms perpendicular to the plane of the first C=C bond, and the third C=C bond forms perpendicular to the plane of the second C=C bond. Thus the right-hand CH_2 is in the same plane as the left-hand H_2C group.

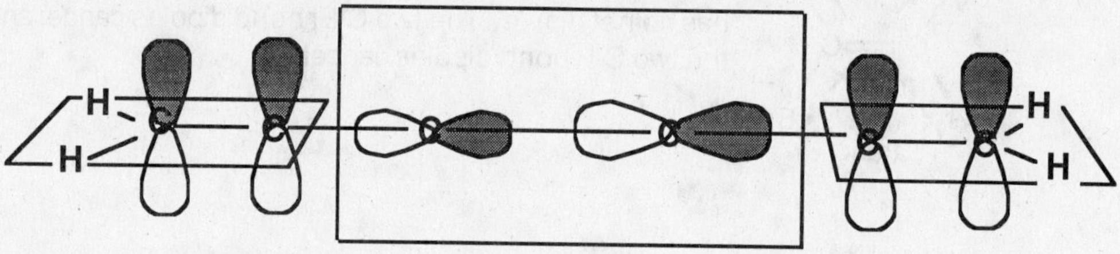

10.55 Total electrons = 2 + 1 - 1 = 2.

The electron configuration is $(\sigma_{1s})^2$.

The two electrons in the HeH^+ ion reside in the σ_{1s} molecular orbital. There are no electrons in any antibonding orbitals. The ion is expected to be stable.

10.56 Total electrons = (2 x 2) - 1 = 3.

The electron configuration is $(\sigma_{1s})^2(\sigma_{1s}^*)^1$.

$$\text{Bond order} = 1/2\,(n_b - n_a) = 1/2\,(2\text{-}1) = 1/2$$

The He_2^+ ion is expected to be stable.

10.57 Total electrons = (2 x 6) + 2 = 14.

The electron configuration is $KK(\sigma_{2s})^2(\sigma_{2s}^*)^2(\pi_{2p})^4(\sigma_{2p})^2$.

Bond order = $1/2\ (n_b - n_a)$ = 1/2 (10-4) = 3

10.58 Total electrons = (2 x 8) + 2 = 18.

The electron configuration is $KK(\sigma_{2s})^2(\sigma_{2s}^*)^2(\pi_{2p})^4(\sigma_{2p})^2(\pi_{2p}^*)^4$.

Bond order = $1/2\ (n_b - n_a)$ = 1/2 (10-8) = 1

10.59 The molecular orbital configuration of O_2 is $KK(\sigma_{2s})^2(\sigma_{2s}^*)^2(\pi_{2p})^4(\sigma_{2p})^2(\pi_{2p}^*)^2$. O_2^+ has one electron less than O_2. The difference is in the number of electrons in the π_{2p}^* antibonding orbital. This means that the bond order is larger for O_2^+ than for O_2.

O_2: Bond order = $1/2\ (n_b - n_a)$ = 1/2 (10-6) = 2

O_2^+: Bond order = $1/2\ (n_b - n_a)$ = 1/2 (10-5) = 5/2

It is expected that the species with the higher bond order, O_2^+, has the shorter bond length. In O_2^-, there is one more electron than in O_2. This additional electron occupies a π_{2p}^* orbital. Increasing the number of electrons in antibonding orbitals decreases the bond order; hence, O_2^- should have a longer bond length than O_2.

10.60 The molecular orbital configuration of N_2 is $KK(\sigma_{2s})^2(\sigma_{2s}^*)^2(\pi_{2p})^4(\sigma_{2p})^2$. N_2^+ has one electron less than N_2. The difference is in the number of electrons in the σ_{2p} bonding orbital. Because N_2^+ has fewer electrons in bonding orbitals than N_2 and the same number in antibonding orbitals, N_2^+ has a lower bond order and a longer bond distance. In N_2^-, there is one more electron than in N_2. This additional electron occupies a π_{2p}^* antibonding orbital. Because N_2^- has more electrons in antibonding orbitals and the same number in bonding orbitals, the bond order is lower; hence, N_2^- has a longer bond length than N_2.

10.61 As shown in Figure 10.32, the occupation of molecular orbitals by the N_2 valence electrons is:

$$KK(\sigma_{2s})^2(\sigma_{2s}^*)^2(\pi_{2p})^4(\sigma_{2p})^2(\pi_{2p}^*)^0$$

To form the first excited state of N_2, a σ_{2p} electron is promoted to the π_{2p}^* orbital, giving:

$$KK(\sigma_{2s})^2(\sigma_{2s}^*)^2(\pi_{2p})^4(\sigma_{2p})^1(\pi_{2p}^*)^1$$

The differences in properties are as follows:

Magnetic character: The ground state is diamagnetic (all electrons paired), and the excited state is paramagnetic.
Bond order: ground state order = 1/2 (8-2) = 3
excited state order = 1/2 (7-3) = 2
Bond-dissociation energy: ground state energy = 942 kJ; excited state energy is less than 942 kJ (excited state does not possess a stable triple bond like the ground state)
Bond length: ground state length = 110 pm; excited state length is more than 110 pm.

10.62 For N_2, refer to Figure 10.32. Note that the energy of the highest occupied orbital of atomic nitrogen is greater than the energy of the highest occupied orbital of N_2. Thus the outer electrons of atomic nitrogen are farther from the nucleus and have a smaller ionization energy than that of N_2.

The occupation of the molecular orbitals of O_2 are almost the same as shown in Figure 10.32; the electronic configuration given in Example 10.7 is below with that of atomic oxygen:

$$O_2\text{: } KK(\sigma_{2s})^2(\sigma_{2s}^*)^2(\pi_{2p})^4(\sigma_{2p})^2(\pi_{2p}^*)^2$$

$$O\text{: } 1s^2 2s^2 2p^4$$

Referring again to Figure 10.32 (after inserting two π_{2p}^* electrons for O_2), we see that the energy of the highest occupied orbital of O_2 is greater than the energy of the highest occupied orbital of atomic oxygen. Thus the outer electrons of O_2 are farther from the nucleus and have a smaller ionization energy than that of atomic oxygen.

■ Cumulative-Skills Problems

(require skills from previous chapters)

Note on significant figures: The final answer to each cumulative-skills problem is given first with one nonsignificant figure (the rightmost significant figure is underlined), and then is rounded to the correct number of figures. Intermediate answers usually have at least one nonsignificant figure. Atomic masses are rounded to two decimal places, except for that of hydrogen.

10.63 After assuming a 100.0-g sample, convert to moles:

$$60.4 \text{ g Xe} \times \frac{1 \text{ mol Xe}}{131.29 \text{ g Xe}} = 0.46\underline{0}05 \text{ mol Xe}$$

$$22.1 \text{ g O} \times \frac{1 \text{ mol O}}{16.00 \text{ g O}} = 1.3\underline{8}1 \text{ mol O}$$

$$17.5 \text{ g F} \times \frac{1 \text{ mol F}}{18.99 \text{ g F}} = 0.92\underline{1}5 \text{ mol F}$$

Divide by 0.460 :

$$\text{Xe: } \frac{0.460}{0.460} = 1; \qquad \text{O: } \frac{1.381}{0.460} = 3.00; \qquad \text{F: } \frac{0.9215}{0.460} = 2.00$$

The simplest formula is XeO_3F_2. This is also the molecular formula. The Lewis formula is

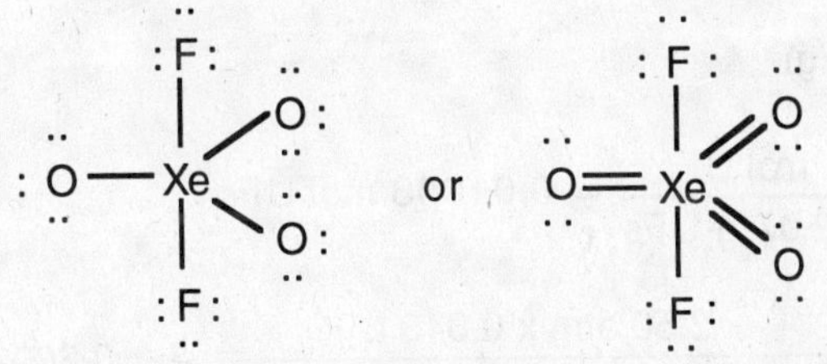

Number of electron pairs = 5, number of lone pairs = 0; hence, the geometry is trigonal bipyramidal. Because xenon has five single bonds, it will require five orbitals to describe the bonding. This suggests sp^3d hybridization.

10.64 After assuming a 100.0-g sample, convert to moles:

$$58.8 \text{ g Xe} \times \frac{1 \text{ mol Xe}}{131.29 \text{ g Xe}} = 0.44\underline{7}8 \text{ mol Xe}$$

$$7.2 \text{ g O} \times \frac{1 \text{ mol O}}{16.00 \text{ g O}} = 0.4\underline{5}00 \text{ mol O}$$

$$34.0 \text{ g F} \times \frac{1 \text{ mol F}}{18.99 \text{ g F}} = 1.7\underline{9}04 \text{ mol F}$$

Divide by 0.4478 :

$$\text{Xe: } \frac{0.4478}{0.4478} = 1; \qquad \text{O: } \frac{0.450}{0.4478} = 1.00; \qquad \text{F: } \frac{1.7904}{0.4478} = 3.9\underline{9}8 = 4.00$$

The simplest formula is $XeOF_4$. This is also the molecular formula. The Lewis formula is

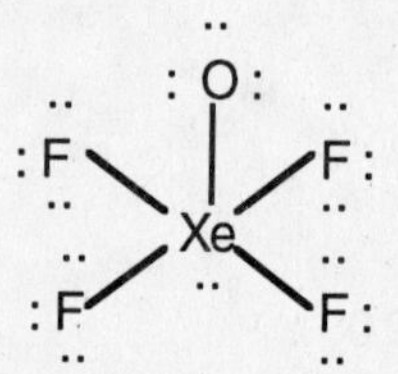

Number of electron pairs = 6, number of lone pairs = 1; hence, the geometry is square pyramidal. Because xenon has six electron pairs, it will require six orbitals to describe the bonding. This suggests sp^3d^2 hybridization.

10.65 $U(s) + ClF_n \rightarrow UF_6 + ClF(g)$

$$\text{mol } UF_6 = 3.53 \text{ g } UF_6 \times \frac{1 \text{ mol } UF_6}{352.07 \text{ g } UF_6} = 0.010\underline{0}3 \text{ mol } UF_6$$

$$\text{mol ClF} = n = \frac{PV}{RT} = \frac{2.50 \text{ atm} \times 0.343 \text{ L}}{0.082057 \text{ L}\bullet\text{atm}/(\text{K}\bullet\text{mol}) \times 348\text{K}} = 0.30\underline{0}2 \text{ mol ClF}$$

0.010 mol UF_6 = 0.060 mol F and 0.030 mol ClF = 0.030 mol F; therefore, the total moles of F from ClF_n = 0.090 mol F. Because mol ClF_n must equal mol ClF, mol ClF_n = 0.030 mol and n = 0.090 mol F ÷ 0.030 mol ClF_n = 3. The Lewis formula is

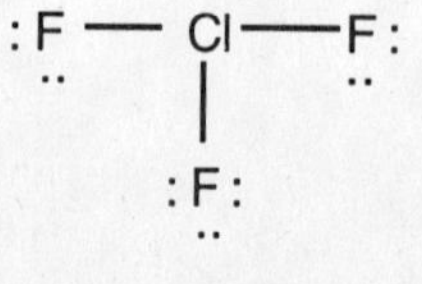

(continued)

Number of electron pairs = 5, number of lone pairs = 2; hence, the geometry is T-shaped. Because chlorine has five electron pairs, it will require five orbitals to describe the bonding. This suggests sp^3d hybridization.

10.66 $Br_2 + \text{excess } F_2 \rightarrow 2BrF_n$

$$\text{mol } Br_2 = n = \frac{PV}{RT} = \frac{(748/760) \text{ atm} \times 0.423 \text{ L}}{0.082057 \text{ L} \bullet \text{atm}/(\text{K} \bullet \text{mol}) \times 423\text{K}}$$

$$= 0.011\underline{9}9 = 0.0120 \text{ mol } Br_2$$

From the equation, 0.0120 mol Br_2 must produce 0.0240 mol BrF_n.

$$\text{Formula wt. } BrF_n = \frac{4.20 \text{ g}}{0.0240 \text{ mol}} = 175 \text{ g/mol}$$

$$175 \text{ g/mol} = 79.9 \text{ g Br/mol} + 19.00 \times n \text{ g F/mol}$$

$$n = \frac{175 - 79.9}{19.00} = 5.00$$

The Lewis formula is

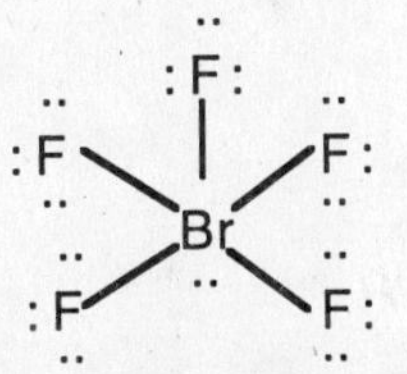

Number of electron pairs = 6, number of lone pairs = 1; hence, the geometry is square pyramidal. Because bromine has six electron pairs, it will require six orbitals to describe the bonding. This suggests sp^3d^2 hybridization.

10.67 N_2: Triple bond; bond length = 110 pm. Geometry is linear; sp hybrid orbitals are needed for one lone pair and one σ bond.

N_2F_2: Double bond; bond length = 122 pm. Geometry is trigonal planar; sp^2 hybrid orbitals are needed for one lone pair and two σ bonds.

N_2H_4: Single bond; bond length = 145 pm. Geometry is tetrahedral; sp^3 hybrid orbitals are needed for one lone pair and three σ bonds.

10.68 The bond length of C_2 is close to that of C_2H_4, suggesting the same bond order for C_2 as for C_2H_4, which has a C=C bond. Total electrons for C_2 = 2 x 6 = 12. The molecular orbital configuration for C_2 is $KK(\sigma_{2s})^2(\sigma_{2s}^*)^2(\pi_{2p})^4(\sigma_{2p})^0$. Bond order = 1/2(8 - 4) = 2. This agrees with the bond-length prediction.

10.69 HNO_3 resonance formulas:

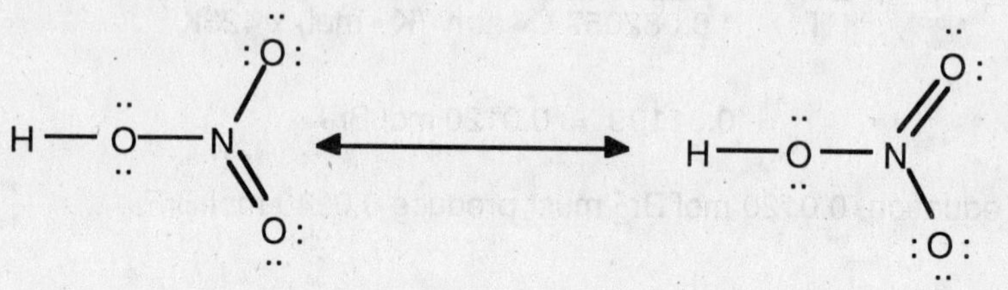

The geometry around the nitrogen is trigonal planar; therefore, the hybridization is sp^2.

Formation reaction: $H_2(g) + 3O_2(g) + N_2(g) \rightarrow 2HNO_3(g)$

$$2 \times \Delta H_f^\circ = [BE(H\text{-}H) + 3BE(O_2) + BE(N_2)] - [2BE(H\text{-}O) + 4BE(N\text{-}O) + 2BE(N{=}O)]$$
$$= [(432 + 3 \times 494 + 942) - (2 \times 459 + 4 \times 201 + 2 \times 607)\ kJ/2\ mol]$$
$$= -80\ kJ/2\ mol = -40\ kJ/mol$$

Resonance energy = -40 kJ - (-135 kJ) = 95 kJ

10.70 This is the other resonance formula of benzene.

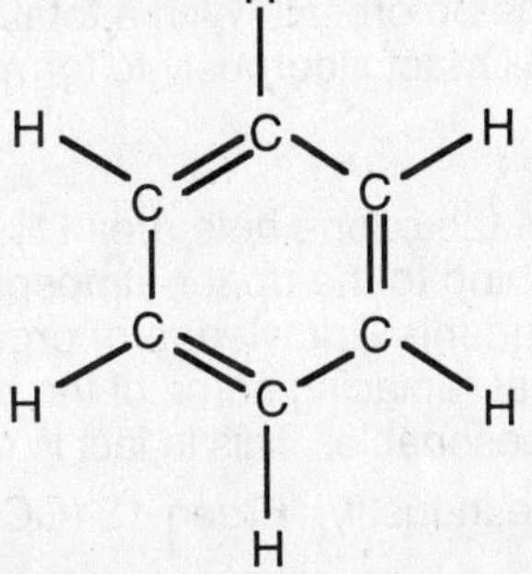

Geometry is trigonal planar; therefore, the hybridization is sp^2.

Formation reaction: $3H_2(g) + 6C(graphite) \rightarrow C_6H_6(g)$ [+ C(graphite) → C(g)]

$$\Delta H_f^\circ = [3BE(H\text{-}H) + 6 \times \Delta H\ C(g)] - [6BE(C\text{-}H) + 3BE(C\text{-}C) + 3BE(C{=}C)]$$

$$= [(3 \times 432 + 6 \times 715) - (6 \times 411 + 3 \times 346 + 3 \times 602)\ kJ/mol]$$

$$= +276\ kJ/mol$$

Resonance energy = +276 kJ - (-83 kJ) = 359 kJ

■ Solutions to Conceptual Problems

10.71 Chlorinated fluorocarbons (CFCs), such as CCl_2F_2, have been recently implicated in the formation of the ozone hole (see the essay at the end of the chapter). The problem develops from the generation of chlorine atoms formed on interaction with ultraviolet light in the stratosphere. The chlorine atoms react with ozone to form oxygen according to the equation:

$$Cl + O_3(g) \rightarrow ClO(g) + O_2(g)$$

Since ozone protects the earth's surface by absorbing harmful UV, its loss in the atmosphere is of great concern. Explain why we would expect Cl atoms to be reactive and suggest either a replacement for CFCs or an alternative approach to minimize the loss of ozone in the upper atmosphere.

(continued)

Discussion

The reactivity of chlorine atoms might be discussed as follows: The electronic structure of the chlorine atom has an unpaired electron in the 3p orbital. With a total of seven electrons in the valence shell, the chorine atom will react vigorously to form a closed-shell octet.

Two conceivable ways to approach the problem of the ozone hole are: (1) replace the CFCs with alternative compounds; (2) add something to the upper atmosphere to counteract the effect of the Cl atoms. Though students' knowledge of organic chemistry may be limited, they might rationalize that replacing some of the reactive Cl atoms with less reactive atoms such as H or C is reasonable. This in fact is one approach being implemented in the U.S. appliance industry. Freon-12 (CCl_2F_2) is being replaced by HFC-13a (CH_2FCH_3). A replacement being considered for use in chillers for commercial buildings is HFC-123 ($CHCl_2CF_3$). Based on periodicity, students may also suggest replacing the Cl atoms with larger members of the halogen group. CF_3I and CF_3Br are currently being considered in fire fighting applications.

Students might consider adding agents to the stratospheric ozone hole to repair the damage that has already been done. One might minimize the damage done by introducing either ozone or another UV absorber into the stratosphere. These would probably be expensive solutions and perhaps simplistic. Recently, it has been suggested that the introduction radical gases (gases having unpaired electrons) into the stratosphere to react with or "scavenge" chlorine atoms to form stable halides would be effective. Another possibility would be the addition of a compound to reform ozone from dioxygen. Ozone is produced in smog at ground level by the following reactions, among others:

$$2NO(g) + O_2(g) \rightarrow 2NO_2(g)$$

$$NO_2(g) + \text{Sunlight} \rightarrow NO(g) + O(g)$$

$$O(g) + O_2(g) \rightarrow O_3(g)$$

For a discussion of the ozone depletion problem, see "Ozone Depletion's Recurring Surprises Challenge Atmospheric Scientists," Chemical and Engineering News, May 24, 1993; "Chlorofluorocarbons and Stratospheric Ozone," J. Chem. Educ. Vol. 64, p. 387, 1987.

10.72 The octet rule is based on the observation that atoms tend to lose, gain, or share electrons until they are surrounded by eight valence electrons. As with any model, there are many exceptions to the octet rule, including molecules with odd electrons, too few electrons, and too many electrons. The Lewis structure of PCl_5 is an example of one of these exceptions. How would you account for the exception from the octet rule in this case? Given you answer, predict a model for the number of valence electrons in stable complexes of the transition metals and the lanthanides and actinides. What exceptions might you expect in these cases?

Discussion

The phosphorus atom in PCl_5 has ten electrons around it. Since s and p atomic orbitals are the basis for the octet rule, d orbitals are commonly invoked to account for these exceptions to the octet rule (although there is some controversy about this).

For the transition metals, we might consider a valence shell that includes s, p, and d orbitals for a total of 18 electrons. Similarly, for the lanthanides/actinides, we might consider a valence shell the includes s, p, d, and f orbitals for a total of 32 electrons. Any compounds exceeding these numbers of electrons in the valence shell would be exceptions.

11. STATES OF MATTER; LIQUIDS AND SOLIDS

■ Solutions to Exercises

Note on significant figures: The final answer to each mathematical solution is given first with one nonsignificant figure (the rightmost significant figure is underlined) and then is rounded to the correct number of figures. Intermediate answers usually also have at least one nonsignificant figure.

11.1 First calculate the heat required to vaporize 1.00 kg of ammonia:

$$1.00 \text{ kg } NH_3 \times \frac{1000 \text{ g}}{1 \text{ kg}} \times \frac{1 \text{ mol } NH_3}{17.03 \text{ g}} \times \frac{23.4 \text{ kJ}}{1 \text{ mol}} = 13\underline{7}4.04 \text{ kJ}$$

The amount of water at 0°C that can be frozen to ice at 0°C with this heat is

$$1374.04 \text{ kJ} \times \frac{1 \text{ mol } H_2O}{6.01 \text{ kJ}} \times \frac{18.01 \text{ g } H_2O}{1 \text{ mol } H_2O} = 41\underline{1}7.54 \text{ g} = 4.12 \text{ kg } H_2O$$

11.2 Use the two-point form of the Clausius-Clapeyron equation to calculate P_2:

$$\text{Log} \frac{P_2}{760 \text{ mmHg}} = \frac{26.8 \times 10^3 \text{ J/mol}}{2.303 \times 8.31 \text{ J/(K} \bullet \text{mol)}} \left[\frac{1}{319 \text{ K}} - \frac{1}{308 \text{ K}} \right]$$

$$1400.4 \text{ K} \left[\frac{-1.12 \times 10^{-4}}{\text{K}} \right] = -0.15\underline{6}8$$

(continued)

Converting to antilogs gives

$$\frac{P_2}{760 \text{ mmHg}} = 0.69\underline{6}8$$

$$P_2 = 0.6968 \times 760 \text{ mmHg} = 52\underline{9}.6 = 530 \text{ mmHg}$$

11.3 Use the two-point form of the Clausius-Clapeyron equation to solve for ΔH_{vap}:

$$\text{Log}\ \frac{757 \text{ mmHg}}{522 \text{ mmHg}} = \frac{\Delta H_{vap}}{2.303 \times 8.31 \text{ J/(K•mol)}} \left[\frac{1}{368 \text{ K}} - \frac{1}{378 \text{ K}}\right]$$

$$0.16\underline{1}4 = \frac{\Delta H_{vap}}{19.\underline{1}37 \text{J/(K•mol)}} \left[\frac{7.188 \times 10^{-5}}{\text{K}}\right]$$

$$\Delta H_{vap} = 4.2\underline{9}7 \times 10^4 \text{ J/mol} \ \ (43.0 \text{ kJ/mol})$$

11.4 a. Liquefy methyl chloride by a sufficient increase in pressure below 144°C.

b. Liquefy oxygen by compressing to 50 atm below -119°C.

11.5 a. Propanol has a hydrogen atom bonded to an oxygen atom. Therefore, hydrogen bonding is expected. Because propanol is polar (from the O-H bond), we also expect dipole-dipole forces. Weak London forces exist too because such forces exist between all molecules.

b. Linear carbon dioxide (Chapter 10) is not polar, so only London forces exist between CO_2 molecules.

c. Bent sulfur dioxide (Chapter 10) is polar, so we expect dipole-dipole forces; we also expect the usual London forces.

11.6 The order of increasing vapor pressure is butane (C_4H_{10}), propane (C_3H_8), and ethane (C_2H_6). Because London forces tend to increase with increasing molecular weight, we would expect the molecule with the highest molecular weight to have the lowest vapor pressure.

11.7 Because ethanol has an H atom bonded to an O atom, strong hydrogen bonding exists in ethanol but not in methyl chloride. Hydrogen bonding explains the lower vapor pressure of ethanol compared to methyl chloride.

11.8 a. Zinc, a metal, is a metallic solid.

b. Sodium iodide, an ionic substance, exists as an ionic solid.

c. Silicon carbide, a compound in which carbon and silicon might be expected to form covalent bonds to other carbon and silicon atoms, exists as a covalent network solid.

d. Methane, at room temperature a gaseous molecular compound with covalent bonds, freezes as a molecular solid.

11.9 Only $MgSO_4$ is an ionic solid; C_2H_5OH, CH_4, and CH_3Cl form molecular solids; thus, $MgSO_4$ should have the highest melting point. Of the molecular solids, CH_4 has the lowest molecular weight (16.0 amu) and would be expected to have the lowest melting point. Both C_2H_5OH and CH_3Cl have approximately the same molecular weights (46.0 amu vs. 50.5 amu), but C_2H_5OH exhibits strong hydrogen bonding and therefore would be expected to have the higher melting point. The order of increasing melting points is CH_4, CH_3Cl, C_2H_5OH, and $MgSO_4$.

11.10 The atom at the center of each unit cell belongs entirely to that cell. In addition, each of the four corners of the cell contains one atom, which is shared by a total of four unit cells. Therefore, the corners contribute one whole atom. This is summarized as follows:

$$\frac{\text{"Atoms"}}{\text{Unit cell}} = 1 \text{ central atom} + \left[4 \text{ corners} \times \frac{1/4 \text{ atom}}{1 \text{ corner}}\right] = 2 \text{ atoms}$$

11.11 Use the edge length to calculate the volume of the unit cell. Then use the density to determine the mass of one atom. Divide the molar mass by the mass of one atom.

$$V = (3.509 \times 10^{-10} \text{ m})^3 = 4.3\underline{2}1 \times 10^{-29} \text{ m}^3$$

$$d = \frac{0.534 \text{ g}}{\text{cm}^3} \times \left[\frac{100 \text{ cm}}{1 \text{ m}}\right]^3 = 5.34 \times 10^5 \text{ g/m}^3$$

$$\text{Mass of 1 unit} = d \times V$$

$$= (5.34 \times 10^5 \text{ g/m}^3) \times (4.321 \times 10^{-29} \text{ m}^3) = 2.3\underline{0}74 \times 10^{-23} \text{ g}$$

There are two atoms in a body-centered cubic unit cell; thus, the mass of one lithium atom is

$$1/2 \times 2.3074 \times 10^{-23} \text{ g} = 1.1\underline{5}37 \times 10^{-23} \text{ g}$$

(continued)

The known atomic weight of lithium is 6.941 amu, so Avogadro's number is

$$N_A = \frac{6.941 \text{ g/mol}}{1.1537 \times 10^{-23} \text{ g/atom}} = 6.0\underline{1}6 \times 10^{23} = 6.02 \times 10^{23} \text{ atoms/mol}$$

11.12 Use Avogadro's number to convert the molar mass of potassium to the mass per one atom.

$$\frac{39.0983 \text{ g K}}{1 \text{ mol K}} \times \frac{1 \text{ mol K}}{6.022 \times 10^{23} \text{ atom}} = \frac{6.49\underline{2}5 \times 10^{-23} \text{ g K}}{\text{atom}}$$

There are two K atoms per unit cell; therefore, the mass per unit cell is

$$\frac{6.4925 \times 10^{-23} \text{ g K}}{\text{atom}} \times \frac{2 \text{ atoms}}{1 \text{ unit cell}} = \frac{1.29\underline{8}5 \times 10^{-22} \text{ g}}{\text{unit cell}}$$

The density of 0.856 g/cm^3 is equal to the mass of one unit cell divided by its unknown volume, V. After solving for V, determine the edge length from the cube root of the volume.

$$0.856 \text{ g/cm}^3 = \frac{1.29\underline{8}5 \times 10^{-22} \text{ g}}{V}$$

$$V = \frac{1.2985 \times 10^{-22} \text{ g}}{0.856 \text{ g/cm}^3} = 1.5\underline{1}7 \times 10^{-22} \text{ cm}^3 \;\; (1.517 \times 10^{-28} \text{ m}^3)$$

$$\text{Edge length} = \sqrt[3]{1.517 \times 10^{-28} \text{ m}^3} = 5.3\underline{3}3 \times 10^{-10}$$

$$= 5.33 \times 10^{-10} \text{ m (533 pm)}$$

■ Answers to Review Questions

11.1 The six different phase transitions, with examples in parentheses, are melting (snow melting), sublimation (dry ice subliming directly to carbon dioxide gas), freezing (water freezing), vaporization (water evaporating), condensation (dew forming on the ground), and gas-solid condensation or deposition (frost forming on the ground).

11.2 Iodine can be purified by heating it in a beaker covered with a dish containing ice or ice water. Only pure iodine should sublime, crystallizing on the cold bottom surface of the dish above the iodine. The common impurities in iodine do not sublime, nor do they vaporize significantly.

11.3 The vapor pressure of a liquid is the partial pressure of the vapor over the liquid, measured at equilibrium. In molecular terms, vapor pressure involves molecules of a liquid vaporizing from the liquid phase, colliding with any surface above the liquid, and exerting pressure on it. The equilibrium is a dynamic one because molecules of the liquid are continually leaving the liquid phase and returning to it from the vapor phase.

11.4 Steam at 100°C will melt more ice than the same weight of water at 100°C because it contains much more energy in the form of its heat of vaporization. It will transfer this energy to the ice and condense in doing so; obviously, water at 100°C cannot condense because it is a liquid.

11.5 The heat of fusion is smaller than the heat of vaporization because melting requires only enough energy for molecules to escape from their sites in the crystal lattice, leaving other molecular attractions intact. In vaporization, sufficient energy must be added to break almost all molecular attractions.

11.6 Evaporation leads to cooling of a liquid because the gaseous molecules require heat to evaporate; as they leave the other liquid molecules, they remove the heat energy required to vaporize them. This leaves less energy in the liquid, whose temperature then drops.

11.7 As the temperature increases for a liquid and its vapor in the closed vessel, the two, which are separated by a meniscus, gradually become identical. The meniscus first becomes fuzzy and then disappears altogether as the temperature reaches the critical temperature. Above this temperature, only the vapor exists.

11.8 A permanent gas can be liquefied only by lowering the temperature below its critical temperature, in addition to compressing the gas.

11.9 The pressure in the cylinder of nitrogen at room temperature (*above* its critical temperature of -147°C) decreases continuously as gas is released because the number of molecules in the vapor phase, which governs the pressure, decreases continuously. The pressure in the cylinder of propane at room temperature (*below* its critical temperature) is constant because liquid propane and gaseous propane exist at equilibrium in the cylinder. The pressure will remain constant at the vapor pressure of propane until only gaseous propane remains. At that point, the pressure will decrease until all of the propane is gone.

11.10 The vapor pressure of a liquid depends on the intermolecular forces in the liquid phase, since the ease with which a molecule leaves the liquid phase depends on how strongly it is attracted to the other molecules. If such molecules attract each other strongly, the vapor pressure will be relatively low; if they attract each other weakly, the vapor pressure will be relatively high.

11.11 Surface tension makes a liquid act as though it had a skin because for an object to break through the surface, the surface area must increase. This requires energy, so there is some resistance to the object breaking through the surface.

11.12 London forces, also known as dispersion forces, originate between any two molecules that are weakly attracted to each other by means of small instantaneous dipoles that occur as a result of the varying positions of the electrons during their movement about their nuclei.

11.13 Hydrogen bonding is a weak to moderate attractive force that exists between a hydrogen atom covalently bonded to a very electronegative atom, X, and a lone pair of electrons on another small, electronegative atom, Y. (X and Y may be the same or different elements.) The hydrogen bonding in water involves a hydrogen atom of one water molecule bonding to a lone pair of electrons on the oxygen atom of another water molecule.

11.14 Molecular substances have relatively low melting points because the forces broken by melting are weak intermolecular attractions in the solid state, not strong bonding attractions.

11.15 A crystalline solid has a well-defined, orderly structure; a noncrystalline solid has a random arrangement of structural units.

11.16 In a face-centered cubic cell, there are atoms at the center of each face of the unit cell, in addition to those at the corners.

11.17 The structure of thallium(I) iodide is a simple cubic lattice for both the metal ions and the anions. Thus, the structure consists of two interpenetrating cubic lattices of cation and anion.

11.18 The coordination number of Cs^+ in CsCl is 8; the coordination number of Na^+ in NaCl is 6; and the coordination number of Zn^{2+} in ZnS is 4.

11.19 Starting with the edge length of a cubic crystal, we can calculate the volume of a unit cell by cubing the edge length. Then, knowing the density of the crystalline solid, we can calculate the mass of the atoms in the unit cell. Then the mass of the atoms in the unit cell is divided by the number of atoms in the unit cell, giving the mass of one atom. Dividing the mass of one mole of the crystal by the mass of one atom yields a value for Avogadro's number.

11.20 X rays can strike a crystal, and be reflected, at various angles; at most angles, the reflected waves will be out of phase and will interfere destructively. At certain angles, however, the reflected waves will be in phase and will interfere constructively, giving rise to a diffraction pattern.

■ Solutions to Practice Problems

Note on significant figures: The final answer to each problem is given first with one nonsignificant figure (the rightmost significant figure is underlined) and then is rounded to the correct number of figures. Intermediate answers usually also have at least one nonsignificant figure.

11.21 a. Vaporization
b. Freezing of eggs and sublimation of ice
c. Condensation
d. Gas-solid condensation, deposition
e. Freezing

11.22 a. Sublimation
b. Vaporization
c. Sublimation of the filament and gas-solid condensation of the vapor
d. Freezing
e. Melting, fusion

11.23 Dropping a line from the intersection of a 450-mmHg line with the diethyl ether curve in Figure 11.7 intersects the temperature axis about 0.8 of the distance between 0°C and 20°C, giving a boiling point for diethyl ether of about 16°C.

11.24 Dropping a line from the intersection of a 350-mmHg line with the carbon tetrachloride curve in Figure 11.7 intersects the temperature axis about 0.5 of the distance between 40°C and 60°C, giving a boiling point of carbon tetrachloride of about 50°C.

11.25 The total amount of energy provided by the heater in 4.54 min is

$$4.54 \text{ min} \times \frac{60 \text{ s}}{1 \text{ min}} \times \frac{3.48 \text{ J}}{\text{s}} = 94\underline{7}.95 \quad (0.94\underline{7}95 \text{ kJ})$$

The heat of fusion per mole of I_2 is

$$\frac{0.9479 \text{ kJ}}{15.5 \text{ g } I_2} \times \frac{2 \times 126.9 \text{ g } I_2}{\text{mol } I_2} = 15.\underline{5}2 = \frac{15.5 \text{ kJ}}{\text{mol } I_2}$$

11.26 The total amount of energy provided by the heater in 6.92 min is

$$6.92\text{ min} \times \frac{60\text{ s}}{1\text{ min}} \times \frac{4.66\text{ J}}{\text{s}} = 19\underline{3}4.8\text{ J}\ \ (1.93\underline{4}8\text{ kJ})$$

The heat of fusion per mole of Cd is

$$\frac{1.9348\text{ kJ}}{35.8\text{ g Cd}} \times \frac{112.4\text{ g Cd}}{\text{mol Cd}} = 6.0\underline{7}4 = \frac{6.07\text{ kJ}}{\text{mol Cd}}$$

11.27 The heat absorbed per 10.0 g of alcohol is

$$10.0\text{ g alcohol} \times \frac{\text{mol alcohol}}{60.1\text{ alcohol}} \times \frac{42.1\text{ kJ}}{\text{mol alcohol}} = 7.0\underline{0}4 = 7.00\text{ kJ}$$

11.28 For 31.4 g of butane, the heat energy needed is

$$31.4\text{ g } C_4H_{10} \times \frac{\text{mol } C_4H_{10}}{58.12\text{ g } C_4H_{10}} \times \frac{21.3\text{ kJ}}{\text{mol } C_4H_{10}} = 11.\underline{5}07 = 11.5\text{ kJ}$$

11.29 Because all of the heat released by freezing the water is used to evaporate the remaining water, you must first calculate the amount of heat released in the freezing:

$$9.31\text{ g } H_2O \times \frac{\text{mol } H_2O}{18.02\text{ g } H_2O} \times \frac{6.01\text{ kJ}}{\text{mol } H_2O} = 3.1\underline{0}505\text{ kJ}$$

Finally, calculate the mass of H_2O that was vaporized by the 3.10505 kJ of heat:

$$3.10505\text{ kJ} \times \frac{18.02\text{ g } H_2O}{\text{mol } H_2O} \times \frac{\text{mol } H_2O}{44.9\text{ kJ}} = 1.2\underline{4}6 = 1.25\text{ g } H_2O$$

11.30 Enough ice must have been added so that the heat consumed in melting the ice is equal to the heat released in cooling the water from 21.0°C to 0.0°C.

$$\text{Heat released by cooling} = [(36.2\text{ g})(0.0^\circ\text{C} - 21.0^\circ\text{C})(4.18\text{ J/g}\bullet{}^\circ\text{C})]$$

$$= -3.1\underline{7}7 \times 10^3\text{ J} = -3.177\text{ kJ}$$

Convert this heat (3.177 kJ) to the mass of ice melted by it:

$$3.177\text{ kJ} \times \frac{\text{mol } H_2O}{6.01\text{ kJ}} \times \frac{18.02\text{ g } H_2O}{\text{mol } H_2O} = 9.5\underline{2}7 = 9.53\text{ g } H_2O$$

11.31 Calculate how much heat is released by cooling 64.3 g of H_2O from 55°C to 15°C.

$$\text{Heat rel'd.} = (64.3\text{ g})(15\ ^\circ\text{C} - 55\ ^\circ\text{C})\left(\frac{4.18\text{ J}}{1\text{ g}\bullet^\circ\text{C}}\right)$$

$$= -1.\underline{0}7509 \times 10^4\text{ J} = -1\underline{0}.7509\text{ kJ}$$

The heat released is used first to melt the ice and then to warm the liquid from 0°C to 15°C. Let the mass of ice equal y grams. Then for fusion, and for warming, we have

$$\text{Fusion: } (y\text{ g } H_2O)\left(\frac{1\text{ mol } H_2O}{18.02\text{ g } H_2O}\right)\left(\frac{6.01\text{ kJ}}{1\text{ mol } H_2O}\right) = 0.33\underline{3}5y\text{ kJ}$$

$$\text{Warming: } (y\text{ g } H_2O)\ (15^\circ\text{C} - 0^\circ\text{C})\left(\frac{4.18\text{ J}}{\text{g}\bullet^\circ\text{C}}\right) = 6\underline{2}.70y\text{ J } (0.06\underline{2}70y\text{ kJ})$$

Because the total heat required for melting and warming must equal the heat released by cooling, equate the two and solve for y.

$$10.7509\text{ kJ} = 0.3335y\text{ kJ} + 0.0627y\text{ kJ} = y(0.3335 + 0.0627)\text{ kJ}$$

$$y = 1\underline{0}.7509\text{ kJ} \div 0.39\underline{6}2\text{ kJ} = 2\underline{7}.13\text{ (grams)}$$

Thus, 27 g of ice were added.

11.32 If all the steam condensed, the quantity of heat used to warm the water in the flask must equal the heat released by condensation and cooling of the steam. Thus, first find the quantity of heat used to warm the water in the flask.

$$\text{Heat required} = (275\text{ g})(76^\circ\text{C} - 21^\circ\text{C})(4.18\text{ J/(g}\bullet^\circ\text{C))}$$

$$= 6\underline{3}.22 \times 10^3\text{ J } (6\underline{3}.22\text{ kJ})$$

Now let y = the mass of the steam. Then find the quantity of heat released by condensation:

$$y\text{ g } H_2O \times \frac{1\text{ mol } H_2O}{18.02\text{ g } H_2O} \times \frac{-40.7\text{ kJ}}{1\text{ mol } H_2O} = -2.2\underline{5}9y\text{ kJ}$$

Now find the quantity of heat released by cooling y grams of H_2O from 100°C to 76°C:

(continued)

Heat rel'd. = y (76°C - 100°C) (4.18 J/(g•°C)) = -100.3y J (-0.1$\underline{0}$03y kJ)

The total quantity of heat released in condensing and cooling the steam is equal in magnitude, but opposite in sign, to the quantity of heat required to warm the water in the flask:

-6$\underline{3}$.22 kJ = -2.259y kJ + (-0.1003y) kJ

$$\text{y g steam} = \frac{63.22 \text{ kJ}}{(2.259 + 0.1003) \text{ kJ}} = 2\underline{6}.8$$

Thus, 27 g of steam condensed.

11.33 At the normal boiling point, the vapor pressure of a liquid is 760.0 mmHg. Use the Clausius-Clapeyron equation to find P_2 when P_1 = 760.0 mmHg, T_1 = 334.9 K, and T_2 = 298.2 K.

ΔH_{vap} = 31.4 kJ/mol

$$\text{Log} \frac{P_2}{P_1} = \frac{\Delta H_{vap}}{2.303 R} \times \frac{(T_2 - T_1)}{T_2 T_1} = \frac{31.4 \times 10^3 \text{ J/mol}}{(2.303)[8.31 \text{ J/(K•mol)}]}$$

$$\times \frac{(298.2 - 334.9)}{(298.2\text{K} - 334.9\text{K})}$$

$$\text{Log} \frac{P_2}{P_1} = -0.60\underline{2}9; \text{ log } P_2 = \text{log } P_1 - 0.6029 = \text{log } (760.0) - 0.6029$$

$$= 2.27\underline{7}9$$

P_2 = antilog (2.2779) = 18$\underline{9}$.6 = 190. mmHg

11.34 At methanol's normal boiling point, its vapor pressure is 760.0 mmHg. Use the Clausius-Clapeyron equation to calculate P_2 when P_1 = 760 mmHg, T_1 = 338.2 K, and T_2 = 295.2 K.

ΔH_{vap} = 37.4 kJ/mol

$$\text{Log}\,\frac{P_2}{P_1} = \frac{\Delta H_{vap}}{2.303\,R} \times \frac{(T_2 - T_1)}{T_2T_1} = \frac{37.4 \times 10^3\ \text{J/mol}}{(2.303)[8.31\ \text{J/(K•mol)}]}$$

$$\times \frac{(295.2 - 338.2)}{(295.2\text{K})(338.2\text{K})}$$

$$\text{Log}\,\frac{P_2}{P_1} = -0.84\underline{1}7;\ \log P_2 = \log P_1 - 0.8417 = \log(760.0) - 0.8417$$

$$= 2.03\underline{9}1$$

$$P_2 = \text{antilog}\,(2.0391) = 10\underline{9}.42 = 109\ \text{mmHg}$$

11.35 From the Clausius-Clapeyron equation,

$$\Delta H_{vap} = 2.303R\left[\frac{T_2T_1}{(T_2 - T_1)}\right]\left[\log\frac{P_2}{P_1}\right]$$

$$= 2.303[8.31\text{J/(K•mol)}]\left[\frac{(553.2\ \text{K})(524.2\ \text{K})}{(553.2 - 524.2)\text{K}}\right]\left[\log\frac{760.0\ \text{mmHg}}{400.0\ \text{mmHg}}\right]$$

$$= 5.3\underline{3}4 \times 10^4\ \text{J/mol} = 53.3\ \text{kJ/mol}$$

11.36 From the Clausius-Clapeyron equation,

$$\Delta H_{vap} = 2.303R\left[\frac{T_2T_1}{(T_2 - T_1)}\right]\left[\log\frac{P_2}{P_1}\right]$$

$$= 2.303[8.31\text{J/(K•mol)}]\left[\frac{(319.7\ \text{K})(301.2\ \text{K})}{(319.7 - 301.2)\text{K}}\right]\left[\log\frac{760.0\ \text{mmHg}}{400.0\ \text{mmHg}}\right]$$

$$= 2.7\underline{7}68 \times 10^4\ \text{J/mol} = 27.8\ \text{kJ/mol}$$

11.37 The phase diagram for oxygen is shown below. It is plotted from these points: triple point = -219°C, boiling point = -183°C, and critical point = -118°C.

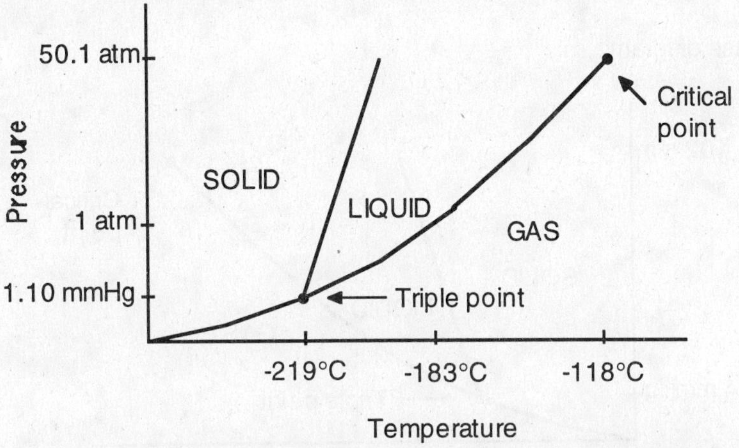

11.38 The phase diagram for argon below is plotted from these points: triple point = -189°C, boiling point = -186°C, and critical point = -122°C.

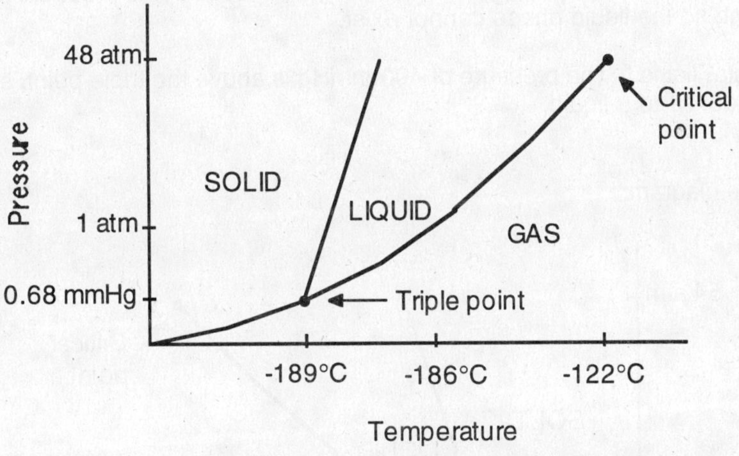

11.39 Liquefied at 25°C: SO_2 and C_2H_2. To liquefy CH_4, lower its temperature below -82°C and then compress it. To liquefy CO, lower its temperature below -140°C and then compress it.

11.40 a. If CF_4 is in the tank, it's not in liquid form because the liquid phase cannot exist above -46°C.

b. If C_4H_{10} is in the tank, it's not in liquid form because 21°C is above its boiling point (1.0 atm).

11.41 Br_2 phase diagram:

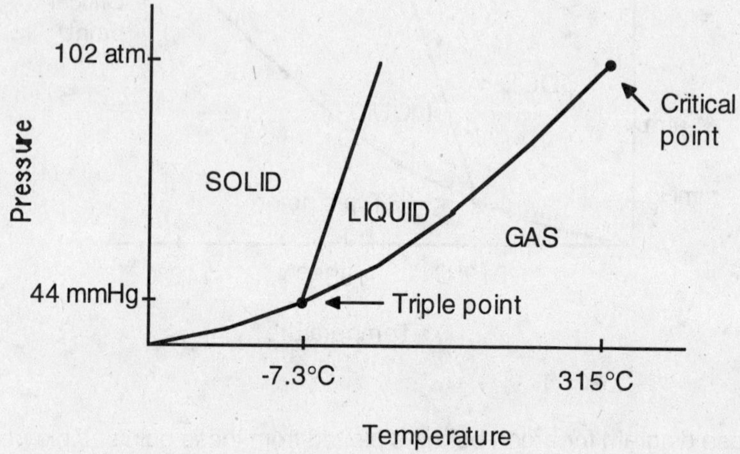

a. Circle "solid." The pressure of 40 mmHg is lower than the pressure at the triple point, so the liquid phase cannot exist.

b. Circle "liquid." The pressure of 400 mmHg is above the triple point, so the gas will condense to a liquid.

11.42 K_r phase diagram:

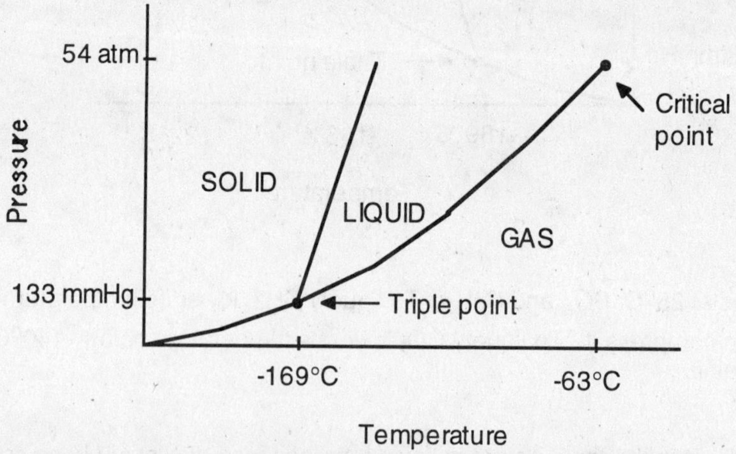

a. Circle "sublimes." The pressure of 130 mmHg is lower than the pressure at the triple point, so the liquid phase cannot exist.

b. Circle "melts." The pressure of 760 mmHg is higher than the pressure at the triple point, so the solid melts to the liquid phase.

11.43 Yes, the heats of vaporization of 0.9, 5.6, and 20.4 kJ/mol (for H_2, N_2, and Cl_2, respectively) increase in the order of the respective molecular weights of 2.016, 28.02, and 71.0. (London forces increase in order of increasing molecular weight.)

11.44 The heats of vaporization of 1.8 kJ/mol for Ne and 6.8 kJ/mol for O_2 increase in the order of the respective molecular weights of 20.1 and 32.0 because only London forces are involved. The heat of vaporization of 34.5 kJ/mol for methanol (molecular weight = 32.0) is higher than that of oxygen because of strong hydrogen bonding between H and O.

11.45 a. London forces
b. London and dipole-dipole forces, H-bonding
c. London and dipole-dipole forces
d. London forces

11.46 Both (a) and (d) exhibit only London forces. Both (b) and (c) have dipole moments.

11.47 The order is $CCl_4 < SiCl_4 < GeCl_4$ (in order of increasing molecular weight).

11.48 The order is He < Ar < Kr (in order of increasing atomic weight).

11.49 CCl_4 has the lowest vapor pressure because it has the largest molecular weight and thus the greatest intermolecular forces.

11.50 ClF has the smallest molecular weight and hence the smallest intermolecular forces and the highest vapor pressure. Thus, it should have the lowest boiling point.

11.51 The order of increasing vapor pressure is $HOCH_2CH_2OH$, FCH_2CH_2OH, FCH_2CH_2F. There is no hydrogen bonding in the third molecule; the second molecule can hydrogen-bond only at one end; and the first molecule can hydrogen-bond at both ends, for the strongest interaction.

11.52 The order is $CH_3CH_2OCH_2CH_3 > CH_3CH_2CH_2CH_2OH > HOCH_2CH_2CH_2OH$. There is no hydrogen bonding in the first molecule; the second molecule can hydrogen-bond only at one end; and the third molecule can hydrogen-bond at both ends, for the strongest interaction.

11.53 The order is $CH_4 < C_2H_6 < CH_3OH < CH_2OHCH_2OH$. The weakest forces are the London forces in CH_4 and C_2H_6, which increase with molecular weight. The next strongest interaction is in CH_3OH, which can hydrogen-bond at only one end of the molecule. The strongest interaction is in the last molecule, which can hydrogen-bond at both ends.

11.54 The order is $C_2H_6 < C_3H_8 < (CH_3)_3N < C_4H_9OH$. The weakest forces are the London forces in C_2H_6, C_3H_8, and $(CH_3)_3N$, which increase with molecular weight. The strongest interaction is in C_4H_9OH, which can hydrogen-bond from oxygen to hydrogen, a stronger force than in the other three molecules.

11.55 a. Metallic
b. Metallic
c. Covalent network
d. Covalent network
e. Molecular

11.56 a. Ionic
b. Ionic
c. Molecular
d. Molecular
e. Molecular

11.57 a. Metallic
b. Covalent network (like diamond)
c. Molecular
d. Molecular

11.58 a. Not molecular (ionic)
b. Molecular
c. Not molecular (metallic)
d. Molecular

11.59 The order is $(C_2H_5)_2O < C_4H_9OH < KCl < CaO$. Melting points increase in the order of attraction between molecules or ions in the solid state. Hydrogen bonding in C_4H_9OH causes it to melt at a higher temperature than $(C_2H_5)_2O$. Both KCl and CaO are ionic solids with much stronger attraction than the organic molecules. In CaO, the higher charges cause the lattice energy to be higher than in KCl.

11.60 The order is $C_2H_6 < CH_3OH < NaCl < Si$. Melting points increase in the order of attraction between molecules or atoms. Hydrogen bonding in CH_3OH causes it to melt at a higher temperature than C_2H_6. Because NaCl is an ionic solid, it melts at a higher temperature than either of the previous two. Silicon is a covalent network solid with the highest melting point.

11.61 a. Low-melting and brittle
b. High-melting, hard, and brittle
c. Malleable and electrically conducting
d. Hard and high-melting

11.62 a. Metallic (from conductivity and luster)
b. Covalent network (from high-melting, hard, and nonconducting liquid)
c. Ionic (from high melting point and conducting liquid)
d. Molecular (from low melting point and odor or vapor pressure at room temperature)

11.63 a. LiCl b. SiC c. CHI_3 d. Co

11.64 a. Pb b. $CaCl_2$ c. P_4S_3 d. BN

11.65 In a simple cubic lattice with one atom at each lattice point, there are atoms only at the corners of unit cells. Each corner is shared by eight unit cells, and there are eight corners per unit cell. Therefore, there is one atom per unit cell.

11.66 There are two atoms per unit cell, one from the corners (see Problem 11.65) and one atom at the center of the unit cell.

11.67 Calculate the volume of the unit cell, change density to g/m^3, and then convert volume to mass, using density:

$$\text{Volume} = (2.866 \times 10^{-10}\ \text{m})^3 = 2.354 \times 20^{-29}\ \text{m}^3$$

$$\left[\frac{7.87\ \text{g}}{\text{cm}^3}\right] \times \left[\frac{10^2\ \text{cm}}{\text{m}}\right]^3 = 7.87 \times 10^6\ \text{g/m}^3$$

$$\text{Mass of one cell} = (7.87 \times 10^6\ \text{g/m}^3) \times (2.354 \times 10^{-29}\ \text{m}^3) = 1.8\underline{5}26 \times 10^{-22}\ \text{g}$$

Because Fe is a body-centered cubic cell, there are two Fe atoms in the cell, and

$$\text{Mass of one Fe atom} = (1.8526 \times 10^{-22}\ \text{g}) \div 2 = 9.2\underline{6}3 \times 10^{-23}\ \text{g}$$

Using the molar mass to calculate the mass of one Fe atom, you find the agreement is good:

$$\frac{55.85\ \text{g Fe}}{1\ \text{mol Fe}} \times \frac{1\ \text{mol Fe}}{6.022 \times 10^{23}\ \text{Fe atoms}} = 9.27\underline{4}3 \times 10^{-23}\ \text{g/Fe atom}$$

11.68 Calculate the volume of the unit cell, change density to g/m^3, and then convert volume to mass, using density:

$$\text{Volume} = (3.524 \times 10^{-10}\ \text{m})^3 = 4.376 \times 20^{-29}\ \text{m}^3$$

$$\left[\frac{8.91\ \text{g}}{\text{cm}^3}\right] \times \left[\frac{10^2\ \text{cm}}{\text{m}}\right]^3 = 8.91 \times 10^6\ \text{g/m}^3$$

$$\text{Mass of one cell} = (8.91 \times 10^6\ \text{g/m}^3) \times (4.376 \times 10^{-29}\ \text{m}^3) = 3.8\underline{9}9 \times 10^{-22}\ \text{g}$$

Because Ni is a face-centered unit cell, there are four Ni atoms per cell, and

$$\text{Mass of one Ni atom} = (3.899 \times 10^{-22}\ \text{g}) \div 4 = 9.7\underline{4}7 \times 20^{-23}\ \text{g}$$

Using the molar mass to calculate Avogadro's number, you obtain

$$\frac{58.70\ \text{g Ni}}{1\ \text{mol Ni}} \times \frac{1\ \text{atom Ni}}{9.747 \times 10^{-23}\ \text{g Ni}} = 6.0\underline{2}2 \times 10^{23}\ \text{atoms Ni/mol Ni}$$

11.69 There are four Cu atoms in the face-centered cubic structure, so the mass of one cell is

$$4\ \text{Cu atoms} \times \frac{1\ \text{mol Cu}}{6.022 \times 10^{23}\ \text{Cu atoms}} \times \frac{63.5\ \text{g Cu}}{1\ \text{mol Cu}} = 4.2\underline{1}8 \times 10^{-22}\ \text{g}$$

$$\text{Cell volume} = \frac{4.2\underline{1}8 \times 10^{-22}\ \text{g}}{8.93\ \text{g/cm}^3} = 4.7\underline{2}3 \times 10^{-23}\ \text{cm}^3$$

All edges are the same length in a cubic cell, so the edge length, l, is

$$l = \sqrt[3]{V} = \sqrt[3]{4.7\underline{2}3 \times 10^{-23}\ \text{cm}^3} = 3.6\underline{1}4 \times 10^{-8}$$

$$= 3.61 \times 10^{-8}\ \text{cm}\ \ (361\ \text{pm})$$

11.70 There are two Ba atoms in the face-centered cubic structure, so the mass of one cell is

$$2\ \text{Ba atoms} \times \frac{1\ \text{mol Ba}}{6.022 \times 10^{23}\ \text{Ba atoms}} \times \frac{137.33\ \text{g Ba}}{1\ \text{mol Ba}} = 4.5\underline{6}1 \times 10^{-22}\ \text{g}$$

$$\text{Cell volume} = \frac{4.5\underline{6}1 \times 10^{-22}\ \text{g}}{3.51\ \text{g/cm}^3} = 1.2\underline{9}94 \times 10^{-22}\ \text{cm}^3$$

(continued)

All edges are the same length in a cubic cell, so the edge length, l, is

$$l = \sqrt[3]{V} = \sqrt[3]{1.2\underline{9}94 \times 10^{-22}\ \text{cm}^3} = 5.0\underline{6}51 \times 10^{-8}$$

$$= 5.07 \times 10^{-8}\ \text{cm}\ \ (507\ \text{pm})$$

11.71 Calculate the volume from the edge length of 407.9 pm (4.079×10^{-8} cm) and then use it to calculate the mass of the unit cell:

$$\text{Cell volume} = (4.079 \times 10^{-8}\ \text{cm})^3 = 6.787 \times 10^{-23}\ \text{cm}^3$$

$$\text{Cell mass} = (19.3\ \text{g/cm}^3)(6.878 \times 20^{-23}\ \text{cm}^3) = 1.3\underline{2}7 \times 10^{-21}\ \text{g}$$

Calculate the mass of one gold atom:

$$1\ \text{Au atom} \times \frac{1\ \text{mol Au}}{6.022 \times 10^{23}\ \text{Au atoms}} \times \frac{196.97\ \text{g Au}}{1\ \text{mol Au}} = 3.27\underline{1} \times 10^{-22}\ \text{g Au}$$

$$\frac{1.327 \times 10^{-21}\ \text{g}}{1\ \text{unit cell}} \times \frac{1\ \text{Au atom}}{3.271 \times 10^{-22}\ \text{g Au}} = \frac{4.05\underline{7}\ \text{Au atom}}{\text{cell}}$$

Since there are 4 atoms per unit cell, it is a face centered cubic.

11.72 Calculate the volume from the edge (288.5 pm = 2.885×10^{-8} cm). Use it to calculate the mass:

$$\text{Cell volume} = (2.885 \times 10^{-8}\ \text{cm})^3 = 2.401 \times 10^{-23}\ \text{cm}^3$$

$$\text{Cell mass} = (7.20\ \text{g/cm}^3)(2.401 \times 10^{-23}\ \text{cm}^3) = 1.7\underline{2}9 \times 10^{-22}\ \text{g}$$

Calculate the mass of one chromium atom:

$$1\ \text{Cr atom} \times \frac{1\ \text{mol Cr}}{6.022 \times 10^{23}\ \text{Cr atoms}} \times \frac{51.996\ \text{g Cr}}{1\ \text{mol Cr}}$$

$$= 8.6\underline{3}4 \times 10^{-23}\ \text{g Cr}$$

$$\frac{1.729 \times 10^{-22}\ \text{g}}{1\ \text{unit cell}} \times \frac{1\ \text{Cr atom}}{8.634 \times 10^{-23}\ \text{g Cr}} = \frac{2.002\ \text{Cr atom}}{\text{cell}}$$

Since there are 2 atoms per unit cell, it is a body centered cubic.

11.73 Calculate the volume from the edge (316.5 pm = 3.165×10^{-8} cm). Use it to calculate the mass:

$$\text{Cell volume} = (3.165 \times 10^{-8}\ \text{cm})^3 = 3.17\underline{0}5 \times 10^{-23}\ \text{cm}^3$$

For a body-centered cubic lattice, there are two atoms per cell, so their mass is

$$2\ \text{W atoms} \times \frac{1\ \text{mol W}}{6.022 \times 10^{23}\ \text{W atoms}} \times \frac{183.8\ \text{g W}}{1\ \text{mol W}} = 6.10\underline{4}3 \times 10^{-22}\ \text{g W}$$

$$\text{Density} = \frac{6.1043 \times 10^{-22}\ \text{g W}}{3.1705 \times 10^{-23}\ \text{cm}^3} = 19.2\underline{5}3 = 19.25\ \text{g/cm}^3$$

11.74 Calculate the volume from the edge (495.0 pm = 4.950×10^{-8} cm). Use it to calculate the mass:

$$\text{Cell volume} = (4.950 \times 10^{-8}\ \text{cm})^3 = 1.21\underline{2}9 \times 10^{-22}\ \text{cm}^3$$

For a face-centered cubic lattice, there are four atoms per cell, so their mass is

$$4\ \text{Pb atoms} \times \frac{1\ \text{mol Pb}}{6.022 \times 10^{23}\ \text{Pb atoms}} \times \frac{207.2\ \text{g Pb}}{1\ \text{mol Pb}} = 1.3763 \times 10^{-21}\ \text{g Pb}$$

$$\text{Density} = \frac{1.3763 \times 10^{-21}\ \text{g Pb}}{1.2129 \times 10^{-22}\ \text{cm}^3} = 11.3\underline{4}7 = 11.35\ \text{g/cm}^3$$

11.75 Use Avogadro's number to calculate the number of atoms in 1.74 g (= d x 1.000 cm^3):

$$1.74\ \text{g Mg} \times \frac{1\ \text{mol Mg}}{24.305\ \text{g Mg}} \times \frac{6.022 \times 10^{23}\ \text{Mg atoms}}{1\ \text{mol Mg}}$$

$$= 4.311 \times 10^{22}\ \text{Mg atoms}$$

Because the space occupied by the Mg atoms = 0.741 cm^3, each atom's volume is

$$\text{Volume 1 Mg atom} = \frac{0.741\ \text{cm}^3}{4.311 \times 10^{22}\ \text{Mg atoms}} = 1.7\underline{1}9 \times 10^{-23}\ \text{cm}^3$$

$$\text{Volume} = \frac{4\pi r^3}{3}$$

(continued)

so

$$r = \sqrt[3]{\frac{3V}{4\pi}}$$

$$r = \sqrt[3]{\frac{3}{4\pi}(1.719 \times 10^{-23}\ cm^3)} = 1.6\underline{0}1 \times 10^{-8} = 1.60 \times 10^{-8}\ cm$$

$$= 160.\ pm$$

11.76 Use Avogadro's number to calculate the number of atoms in 3.51 g (= d x 1.000 cm^3):

$$3.51\ g\ Ba \times \frac{1\ mol\ Ba}{137.33\ g\ Ba} \times \frac{6.022 \times 10^{23}\ Ba\ atoms}{1\ mol\ Ba}$$

$$= 1.539 \times 10^{22}\ Ba\ atoms$$

Because the space occupied by the Ba atoms = 0.680 cm^3, each atom's volume is

$$\text{Volume 1 Ba atom} = \frac{0.680\ cm^3}{1.539 \times 10^{22}\ Ba\ atoms} = 4.4\underline{1}8 \times 10^{-23}\ cm^3$$

$$\text{Volume} = \frac{4\pi r^3}{3}$$

so

$$r = \sqrt[3]{\frac{3V}{4\pi}}$$

$$r = \sqrt[3]{\frac{3}{4\pi}(4.418 \times 10^{-23}\ cm^3)} = 2.1\underline{9}3 \times 10^{-8} = 2.19 \times 10^{-8}\ cm$$

$$= 219\ pm$$

■ Solutions to Unclassified Problems

11.77 Water vapor condensed directly to solid water (frost) without forming liquid water (Section 11.2). After heating, most of the frost melted to liquid water, which then vaporized to water vapor. Some of the frost may have sublimed directly to water vapor (Section 11.2).

11.78 Water vapor condenses directly to solid water (snow) in the upper atmosphere (Section 11.2). After falling through the warm air mass, the snow melts to liquid water (rain). After falling on a sunny spot, the rain is vaporized to water vapor.

11.79 From Table 5.6, the vapor pressures are 21.1 mmHg at 23°C and 12.8 mmHg at 15°C. If the moisture did not begin to condense until the air had been cooled to 15°C, then the partial pressure of water in the air at 23°C must have been 12.8 mmHg. The relative humidity is

$$\%\text{ relative humidity} = \frac{12.8\text{ mmHg}}{21.1\text{ mmHg}} \times 100\% = 60.\underline{66} = 60.7\%$$

11.80 The vapor pressure of water at 21°C is 18.7 mmHg (Table 5.6). Therefore,

$$\%\text{ relative humidity} = 58\% = \frac{x\text{ mmHg}}{18.7\text{ mmHg}} \times 100\%;\ x = 1\underline{0}.8\text{ mmHg}$$

The partial pressure of water in the air at 21°C is thus 10.8 mmHg. The water will begin to condense when the temperature drops to the temperature at which the vapor pressure of water is 10.8 mmHg. This temperature is between 12°C and 13°C (Table 5.6).

11.81 After labeling the problem data as below, use the Clausius-Clapeyron equation to obtain ΔH_{vap}, which can then be used to calculate the boiling point.

At T_1 = 299.3 K, P_1 = 100.0 mmHg; at T_2 = 333.8 K, P_2 = 400.0 mmHg

$$\text{Log}\ \frac{400.0}{100.0} = \frac{\Delta H_{vap}}{2.303 \times 8.314\text{ J/(K}\bullet\text{mol)}} \times \frac{333.8\text{ K} - 299.3\text{ K}}{333.8\text{ K} \times 299.3\text{ K}}$$

$$0.6021 = \Delta H_{vap}\ (1.804 \times 10^{-5}\text{ mol/J})$$

$$\Delta H_{vap} = 33.4 \times 10^3\text{ J/mol (33.4 kJ/mol)}$$

Now use this value of ΔH_{vap} and the following data to calculate the boiling point:

At T_1 = 299.3 K, P_1 = 100.0 mmHg; at T_2 (boiling pt.), P_2 = 760 mmHg

$$\text{Log}\ \frac{760.0}{100.0} = \frac{33.4 \times 10^3\text{ J/mol}}{2.303 \times 8.314\text{ J/(K}\bullet\text{mol)}} \times \left[\frac{1}{299.3\text{ K}} - \frac{1}{T_2}\right]$$

$$0.8808 = 1.744 \times 10^3\text{ K}\left[\frac{1}{299.3\text{K}} - \frac{1}{T_2}\right]$$

(continued)

$$\frac{1}{T_2} = \frac{1}{299.3\text{K}} - \frac{0.8808}{1.744 \times 10^3\ \text{K}} = 2.8\underline{3}6 \times 10^{-3}/\text{K}$$

$$T_2 = 35\underline{2}.6 = 353\ \text{K}\ (80^\circ\text{C})$$

11.82 After labeling the problem data as below, use the Clausius-Clapeyron equation to obtain ΔH_{vap}, which can then be used to calculate the boiling point.

At $T_1 = 293.2$ K, $P_1 = 17.5$ mmHg; at $T_2 = 353.2$ K, $P_2 = 355.1$ mmHg

$$\text{Log}\ \frac{355.1}{17.5} = \frac{\Delta H_{vap}}{2.303 \times 8.314\ \text{J/(K•mol)}} \times \frac{353.2\ \text{K} - 293.2\ \text{K}}{353.2\ \text{K} \times 293.2\ \text{K}}$$

$$1.307 = \Delta H_{vap}\ (3.026 \times 10^{-5}\ \text{mol/J})$$

$$\Delta H_{vap} = 43.2 \times 10^3\ \text{J/mol}\ (43.2\ \text{kJ/mol})$$

Now use this value of ΔH_{vap} and the following data to calculate the boiling point:

At $T_1 = 293.32$ K, $P_1 = 17.5$ mmHg; at T_2 (boiling pt.), $P_2 = 760$ mmHg

$$\text{Log}\ \frac{760.0}{17.5} = \frac{43.2 \times 10^3\ \text{J/mol}}{2.303 \times 8.314\ \text{J/(K•mol)}} \times \left[\frac{1}{293.2\ \text{K}} - \frac{1}{T_2}\right]$$

$$1.638 = 2.256 \times 10^3\ \text{K} \left[\frac{1}{293.2\text{K}} - \frac{1}{T_2}\right]$$

$$\frac{1}{T_2} = \frac{1}{293.2\text{K}} - \frac{1.638}{2.256 \times 10^3\ \text{K}} = 2.6\underline{8}46 \times 10^{-3}/\text{K}$$

$$T_2 = 37\underline{2}.49 = 372\ \text{K}\ (99^\circ\text{C})$$

11.83 a. As this gas is compressed at 20°C, it will condense into a liquid because 20°C is above the triple point but below the critical point.

b. As this gas is compressed at -70°C, it will condense directly to the solid phase because the temperature of -70°C is below the triple point.

c. As this gas is compressed at 40°C, it will not condense because 40°C is above the critical point.

11.84 a. As I_2 vapor is cooled at 120 atm, no change to a distinct liquid will be observed, but the I_2 will condense to a solid phase at some definite temperature.

b. As I_2 vapor is cooled at 1 atm, the vapor will condense to a liquid and then freeze to the solid phase.

c. As I_2 vapor is cooled at 50 mmHg (below the triple point), the vapor will condense directly to the solid phase without going through the liquid phase.

11.85 In propanol, hydrogen bonding exists between the hydrogen of the OH group and the lone pair of electrons of oxygen of the OH group of an adjacent propanol molecule. For two adjacent propanol molecules, the hydrogen bond may be represented as follows:

C_3H_7-O-H•••O(H)C_3H_7

11.86 In hydrogen peroxide, hydrogen bonding exists between any hydrogen and the lone pair of electrons of oxygen of an adjacent hydrogen peroxide. For two adjacent hydrogen peroxide molecules, the hydrogen bond may be represented as follows:

HO-O-H•••O($\dot{O}H_2$)

11.87 Ethylene glycol molecules are capable of hydrogen bonding to each other, whereas pentane molecules are not. The greater intermolecular forces in ethylene glycol are reflected in greater resistance to flow (viscosity) and high boiling point.

11.88 Pentylamine molecules are capable of hydrogen bonding to each other, but triethylamine molecules are not. The greater intermolecular forces in pentylamine cause a higher boiling point and greater resistance to flow.

11.89 Aluminum (Group IIIA) forms a metallic solid. Silicon (Group IVA) forms a covalent network solid (Section 11.8). Phosphorus (Group VA) forms a molecular solid. Sulfur (Group VIA) forms a molecular (amorphous) solid.

11.90 AlF_3 forms an ionic solid. SiF_4, PF_3, and SF_4 form molecular solids.

11.91 a. Lower: KCl. The lattice energy should be lower for ions with a lower charge. A lower lattice energy implies a lower melting point.

b. Lower: CCl_4. Both are molecular solids, so the compound with the lower molecular weight should have weaker London forces and therefore the lower melting point.

c. Lower: Zn. Melting points for Group IIB metals are lower than for metals near the middle of the transition-metal series.

d. Lower: C_2H_5Cl. Ethyl chloride cannot hydrogen-bond, but acetic acid can. The compound with the weaker intermolecular forces has the lower melting point.

11.92 a. Lower: C_6H_{14}. A molecular solid has a lower melting point than an ionic solid.

b. Lower: 1-propanol. The 1-propanol can hydrogen-bond only at one end; ethylene glycol can hydrogen-bond at both ends, so its intermolecular forces are stronger.

c. Lower: Na. Sodium is a metallic solid, but Si is a covalent network solid (high melting point).

d. Lower: CH_4. Both form molecular solids, but CH_4 has a lower molecular weight with weaker London forces and therefore the lower melting point.

11.93 The face-centered cubic structure means that one atom is at each lattice point. All edges are the same length in such a structure, so the volume is

$$\text{Volume} = l^3 = (3.839 \times 10^{-8}\ \text{cm})^3 = 5.65\underline{7}9 \times 10^{-23}\ \text{cm}^3$$

$$\text{Mass of unit cell} = dV = (22.42\ \text{g/cm}^3)(5.6579 \times 10^{-23}\ \text{cm}^3) = 1.26\underline{8}5 \times 10^{-21}\ \text{g}$$

There are four atoms in a face-centered cubic cell, so

$$\text{Mass of 1 Ir atom} = \text{mass of unit cell} \div 4 = (1.2685 \times 10^{-21}\ \text{g}) \div 4$$

$$= 3.17\underline{1}2 \times 10^{-22}\ \text{g}$$

$$\text{Molar mass of Ir} = (3.1712 \times 10^{-22}\ \text{g/Ir atom}) \times (6.022 \times 10^{23}\ \text{Ir atoms/mol})$$

$$= 190.96 = 191.0\ \text{g/mol}\ \text{(The atomic weight} = 191.0\ \text{amu.)}$$

11.94 The body-centered cubic structure means that one atom is at each lattice point. All edges are the same length in such a structure, so the volume is

$$\text{Volume} = l^3 = (3.306 \times 10^{-8}\ \text{cm})^3 = 3.61\underline{3}3 \times 10^{-23}\ \text{cm}^3$$

$$\text{Mass of unit cell} = dV = (16.69\ \text{g/cm}^3)(3.6133 \times 10^{-23}\ \text{cm}^3) = (6.03\underline{0}6 \times 10^{-22}\ \text{g})$$

There are two atoms in a body-centered cubic cell, so

$$\text{Mass of 1 Ta atom} = \text{mass of unit cell} \div 2 = (6.0306 \times 10^{-22}\ \text{g}) \div 2$$

$$= 3.01\underline{5}3 \times 10^{-22}\ \text{g}$$

(continued)

$$\text{Molar mass of Ta} = (3.0153 \times 10^{-22} \text{ g/Ta atom}) \times (6.022 \times 10^{23} \text{ Ta atoms/ mol})$$

$$= 181.\underline{5}8 = 181.6 \text{ g/mol (The atomic weight} = 181.6 \text{ amu.)}$$

11.95 From Problem 11.69, the cell edge length (l) is 36$\underline{1}$.4 pm. There are four copper atom radii along the diagonal of a unit-cell face. Because the diagonal square = $l^2 + l^2$ (Pythagorean theorem),

$$4r = \sqrt{2l^2} = \sqrt{2}\, l, \text{ or } r = \frac{\sqrt{2}}{4}(36\underline{1}.4 \text{ pm}) = 12\underline{7}.7 = 128 \text{ pm}$$

11.96 A body-centered cubic cell has two atoms per cell. The mass of the unit cell is

$$2 \text{ Rb atoms} \times \frac{1 \text{ mol Rb}}{6.022 \times 10^{23} \text{ Rb atoms}} \times \frac{85.468 \text{ g Rb}}{1 \text{ mol Rb}} = 2.83\underline{8}5 \times 10^{-22} \text{ g}$$

The volume of the unit cell is

$$\text{Volume} = \frac{2.8385 \times 10^{-22} \text{ g}}{1.532 \text{ g/cm}^3} = 1.85\underline{2}8 \times 10^{-22} \text{ cm}^3$$

$$l = \sqrt[3]{\text{volume}} = \sqrt[3]{1.85\underline{2}8 \times 10^{-22} \text{ cm}^3} = 5.70\underline{0}9 \times 10^{-8} \text{ cm}$$

Because the corner spheres touch the body-centered sphere, the length of the body diagonal (diagonal passing through the center of the cell) must be four times the radius of the Rb atom. Also, from the geometry of a cube and the Pythagorean theorem, the square of the body diagonal equals $l^2 + d^2$ (l is the unit-cell edge length and d is the diagonal along a face of the unit cell). Because d^2 $l^2 + l^2$, or $d = \sqrt{2}\, l$, you can write

$$(\text{Body diagonal})^2 = l^2 + d^2 = l^2 + 2l^2 = 3l^2$$

$$\text{Body diagonal} = \sqrt{3}\,(l) = \sqrt{3}\,(5.7009 \times 10^{-8} \text{ cm}) = 9.87\underline{4}2 \times 10^{-8} \text{ cm}$$

$$\text{Radius of Rb atom} = (9.8742 \times 10^{-8} \text{ cm}) \div 4 = 2.46\underline{8}6 \times 10^{-8} \text{ cm (246.9 pm)}$$

11.97 The body diagonal (diagonal passing through the center of the cell) is four times the radius, r, of a sphere. Also, from the geometry of a cube and the Pythagorean theorem, the body diagonal equals $\sqrt{3}$ (l), where l is the edge length of the unit cell (see Problem 11.96). Thus

$$4r = \sqrt{3}\ (l)$$

or

$$l = 4r/(\sqrt{3})$$

Because the unit cell contains two spheres, the volume occupied by the spheres is

$$V_{spheres} = 2 \times \left[(4/3)\ \pi r^3\right]$$

and

$$V_{cell} = l^3 = \left[\frac{4r}{\sqrt{3}}\right]^3 = \frac{64r^3}{3\sqrt{3}}$$

Finally, to obtain the % volume of the cell occupied, divide $V_{spheres}$ by V_{cell}:

$$\% V = \frac{V_{spheres}}{V_{cell}} \times 100\% = \frac{2\left[\frac{4\pi r^3}{3}\right]}{\frac{64r^3}{3\sqrt{3}}} \times 100\% = \frac{\pi\sqrt{3}}{8} \times 100\%$$

$$= 68.01 = 68\%$$

11.98 Because the spheres touch along the diagonal of a face, d, the radius, r, of the spheres is

$$r = d/4 = l(\sqrt{2})/4$$

or

$$l = 4r/\sqrt{2}$$

$$V_{cell} = l^3 = \left[\frac{4r}{\sqrt{2}}\right]^3$$

For a face-centered cubic structure, there are four spheres per cell, so the volume, $V_{spheres}$, occupied by the spheres is

(continued)

$$V_{spheres} = 4\left[\frac{4\pi r^3}{3}\right]$$

$$\%V = \frac{V_{spheres}}{V_{cell}} \times 100\% = \frac{4\left[\frac{4\pi r^3}{3}\right]}{\left[\frac{4r}{\sqrt{2}}\right]^3} \times 100\% = \frac{\pi\sqrt{2}}{6} \times 100\%$$

$$= 74.04 = 74\%$$

11.99 a. The boiling point increases as the size (number of electrons in the atom or molecule) increases. The London forces or dispersion forces increase.

b. Hydrogen bonding occurs between the H-F molecules, which is much stronger than the London forces.

c. In addition to the dispersion forces, the hydrogen halides are polar (have dipole moments) so there are dipole-dipole interactions.

11.100 a. The boiling point increases as the size (number of electrons in the molecule) increases. The London forces or dispersion forces increase.

b. Hydrogen bonding occurs between the NH_3 molecules.

c. The nitrogen family compounds consist of polar molecules, whereas the carbon family molecules are nonpolar. So the nitrogen family has dipole-dipole interactions as well as the contributions from the dispersion forces.

11.101 a. Diamond and silicon carbide are giant molecules with all the bonds between the atoms being strong covalent bonds. Graphite is a layered structure and the forces holding the layers together are weak dispersion forces.

b. Silicon dioxide is a giant molecule with an infinite array of O-Si-O bonds. Each silicon is bonded to four oxygen atoms. Carbon dioxide is a discrete, nonpolar, molecule.

11.102. The first member atoms are very small and highly electronegative. Oxygen readily forms double bonds, whereas the larger atoms have difficulty getting close enough together to form multiple bonds. So we have O_2 vs. S_8.

Because oxygen has a higher attraction for hydrogen, the dipole moment is so large that it is possible to have hydrogen bonding in H_2O, but there is no hydrogen bonding with H_2S.

11.103 a. CO_2 consists of discrete non-polar molecules that are held together in the solid by weak dispersion forces. SiO_2 is a giant molecule with all the atoms held together by strong covalent bonds.

b. HF(l) has extensive hydrogen bonding between the molecules. HCl(l) boils much lower because it doesn't have H-bonding.

c. SiF_4 is a larger molecule (it has more electrons than CF_4), so it has stronger dispersion forces and a higher boiling point than CF_4. Both molecules are tetrahadrally symmetrical and therefore non-polar.

11.104 a. Li^+ $H:^-$ $:\overset{\overset{H}{|}}{\underset{\underset{H}{|}}{N}}-H$ $H-\overset{\overset{H}{|}}{\underset{\underset{H}{|}}{C}}-H$ $\overset{..}{\underset{..}{O}}=C=\overset{..}{\underset{..}{O}}$

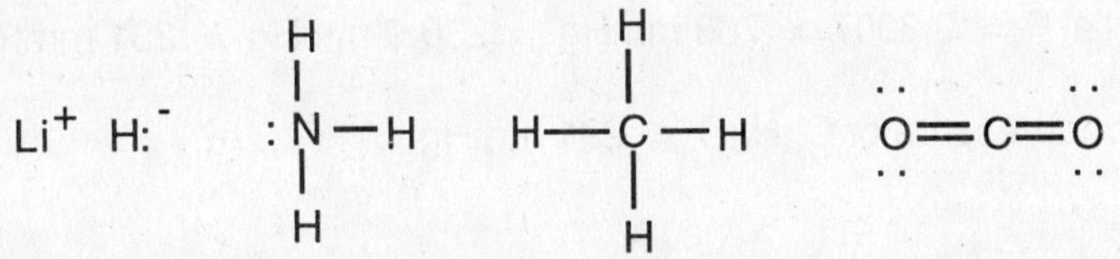

b. LiH has the highest boiling point because of the ionic bonding and crystalline lattice of LiH.

c. CH_4 has the lowest boiling point because it is a small nonpolar molecule. The only intermolecular forces are dispersion forces.

d. NH_3 in a polar molecule which also has H-bonding between the molecules.

■ Cumulative Skills Problems
(require skills from previous chapters)

11.105 Use the ideal gas law to calculate n, the number of moles of N_2:

N_2: $n = \dfrac{PV}{RT} = \dfrac{(745/760) \text{ atm} \times 5.40 \text{ L}}{[0.082057 \text{ L·atm/(K·mol)}] \times 293 \text{ K}} = 0.220\underline{1}$ mol

C_3H_8O: $n = 0.6149 \text{ g } C_3H_8O \times \dfrac{1 \text{ mol } C_3H_8O}{60.094 \text{ g } C_3H_8O} = 0.01023\underline{2}$ mol

$X_{C_3H_8O} = \dfrac{0.010232 \text{ mol } C_3H_8O}{(0.010232 \text{ mol} + 0.2201 \text{ mol})} = 0.0444\underline{2}$ mol fraction

Partial P = 0.04442 × 745 mmHg = 33.$\underline{0}$93 mmHg = 33.1 mmHg

Vapor pressure of C_3H_8O = 33.1 mmHg

11.106 Use the ideal gas law to calculate n, the number of moles of N_2:

$$N_2: \ n = \frac{PV}{RT} = \frac{(768/760) \text{ atm} \times 6.35 \text{ L}}{[0.082057 \text{ L•atm/(K•mol)}] \times 298 \text{ K}} = 0.26\underline{2}4 \text{ mol}$$

$$C_3H_6O: \ n = 6.550 \text{ g } C_3H_6O \times \frac{1 \text{ mol } C_3H_6O}{58.05 \text{ g } C_3H_6O} = 0.112\underline{8}3 \text{ mol}$$

$$X_{C_3H_6O} = \frac{0.11283 \text{ mol } C_3H_6O}{(0.11283 \text{ mol} + 0.2624 \text{ mol})} = 0.30\underline{0}7 \text{ mol fraction}$$

Partial P = 0.3007 x 768 mmHg = 23$\underline{0}$.9 mmHg = 231 mmHg

Vapor pressure of C_3H_6O = 231 mmHg

11.107 Calculate the moles of HCN in 10.0 mL of the solution (density = 0.687 g HCN/mL HCN):

$$10.0 \text{ mL HCN} \times \frac{0.687 \text{ g HCN}}{\text{mL HCN}} \times \frac{1 \text{ mol HCN}}{27.03 \text{ g HCN}} = 0.25\underline{4}1 \text{ mol HCN}$$

0.2541 mol HCN(l) → 0.2541 mol HCN(g)

(ΔH_f^o = 105 kJ/mol) (ΔH_f^o = 135 kJ/mol)

ΔH^o = 0.2541 mol x [135 kJ/mol - 105 kJ/mol] = 7.$\underline{6}$23 = 7.6 kJ

11.108 Calculate moles of CH_3OH in the 10.0 mL solution (density = 0.787 g CH_3OH/mL CH_3OH):

$$10.0 \text{ mL } CH_3OH \times \frac{0.787 \text{ g } CH_3OH}{\text{mL } CH_3OH} \times \frac{1 \text{ mol } CH_3OH}{32.05 \text{ g } CH_3OH}$$

= 0.24$\underline{5}$6 mol CH_3OH

0.2456 mol CH_3OH(l) → 0.2456 mol CH_3OH(g)

(ΔH_f^o = -238.6 kJ/mol) (ΔH_f^o = -201.2 kJ/mol)

ΔH^o = 0.2456 mol x [-201.2 kJ/mol - (-238.6 kJ/mol)] = 9.1$\underline{8}$54 = 9.19 kJ

11.109 First convert the mass to moles; then multiply by the standard heat of formation to obtain the heat absorbed in vaporizing this mass:

$$25.0 \text{ g } P_4 \times \frac{1 \text{ mol } P_4}{123.92 \text{ g } P_4} = 0.20\underline{1}7 \text{ mol } P_4$$

$$0.20\underline{1}7 \text{ mol } P_4 \times \frac{95.4 \text{ J}}{°\text{C}\bullet\text{mol } P_4} \times (44.1°\text{C} - 25.0°\text{C})$$

$$= 36\underline{7}.52 \text{ J} = 0.36\underline{7}52 \text{ kJ}$$

$$2.63 \text{ kJ/mol } P_4 \times 0.2017 \text{ mol } P_4 = 0.53\underline{0}47 \text{ kJ}$$

$$\text{Total heat} = 0.53047 \text{ kJ} + 0.36752 \text{ kJ} = 0.89\underline{7}99 = 0.898 \text{ kJ}$$

11.110 First convert the mass to moles; then multiply by the standard heat of formation to obtain the heat absorbed in vaporizing this mass:

$$25.0 \text{ g Na} \times \frac{1 \text{ mol Na}}{22.99 \text{ g Na}} = 1.0\underline{8}7 \text{ mol Na}$$

$$1.0\underline{8}7 \text{ mol Na} \times \frac{28.2 \text{ J}}{°\text{C}\bullet\text{mol Na}} \times (97.8°\text{C} - 25.0°\text{C}) = 22\underline{3}1 \text{ J } (2.2\underline{3}1 \text{ kJ})$$

$$2.60 \text{ kJ/mol Na} \times 1.087 \text{ mol Na} = 2.8\underline{2}6 \text{ kJ}$$

$$\text{Total heat} = 2.826 \text{ kJ} + 2.231 \text{ kJ} = 5.0\underline{5}7 = 5.06 \text{ kJ}$$

11.111 Use the ideal gas law to calculate the total number of moles of monomer and dimer:

$$n = \frac{PV}{RT} = \frac{(436/760) \text{ atm} \times 1.000 \text{ L}}{[0.082057 \text{ L}\bullet\text{atm/(K}\bullet\text{mol)}] \times 373.75 \text{ K}}$$

$$= 0.018\underline{7}057 \text{ mol monomer + dimer}$$

$$(0.0187057 \text{ monomer + dimer}) \times \frac{0.630 \text{ mol dimer}}{1 \text{ mol dimer + monomer}}$$

$$= 0.011\underline{7}8 \text{ mol dimer}$$

$$0.0187057 \text{ mol both} - 0.01178 \text{ mol dimer} = 0.006\underline{9}2 \text{ mol monomer}$$

(continued)

$$\text{Mass dimer} = 0.01\underline{1}78 \text{ mol dimer} \times \frac{120.1 \text{ g dimer}}{\text{mol dimer}} = 1.4\underline{1}4 \text{ g dimer}$$

$$\text{Mass monomer} = 0.00692 \text{ mol monomer} \times \frac{60.05 \text{ g monomer}}{\text{mol monomer}}$$

$$= 0.4\underline{1}55 \text{ g monomer}$$

$$\text{Density} = \frac{1.414 \text{ g} + 0.4155 \text{ g}}{1.000 \text{ L}} = 1.8\underline{2}9 = 1.83 \text{ g/L vapor}$$

11.112 Use the ideal gas law to calculate the total number of moles of monomer and dimer:

$$n = \frac{PV}{RT} = \frac{(146/760) \text{ atm} \times 1.000 \text{ L}}{[0.082057 \text{ L•atm/(K•mol)}] \times 344.45 \text{ K}}$$

$$= 0.0067\underline{9}6 \text{ mol monomer + dimer}$$

Let X_d = mole fraction of dimer; then write one equation in one unknown based on 1.000 L of gas:

$$0.702 \text{ g both} = 0.006796 \text{ mol } X_d \text{ (120.1 g/mol)} + 0.006796 \text{ (1 - } X_d\text{) (60.05 g/mol)}$$

$$0.702 \text{ g} = 0.8162\, X_d + 0.4081 - 0.4081 X_d$$

$$X_d = \frac{(0.702 - 0.4081)}{0.4081} = 0.72\underline{0}1 = 0.720 \text{ (= mol fraction dimer)}$$

■ Solutions to Conceptual Problems

11.113 The storage of water is critical in areas with minimal rainfall. One problem is the steady evaporation that occurs in open containers or reservoirs. The Clausius- Clapeyron equation provides a description of how decreasing the temperature of a solution will decrease its vapor pressure and thus evaporation. Chemical methods of minimizing evaporation exist as well. Suggest a chemical substance that could be used to decrease the evaporation of water from a storage site. Include in your answer discussion of the phenomenon involved.

(continued)

Discussion

Evaporation depends not only on vapor pressure, but also on wind, sunlight, and surface area. By circulating dry air over the water surface, wind increases the rate of evaporation. Sunlight provides the energy needed for evaporation, so increases the rate of evaporation. Increasing the surface area also increases the rate of evaporation. Control of any of these variables could be used to decrease the rate of evaporation. Adding a substance that forms an insoluble layer over the water might be used to reduce the surface area available for evaporation. In addition to being insoluble, the substance should have a density less than water (so it floats on the surface), have a low vapor pressure (so it doesn't evaporate readily), and be nontoxic. The student should consult one or more handbooks for substances with the appropriate properties. A possibility would be hexadecane, $C_{16}H_{34}$, a paraffin hydrocarbon with density of 0.773 and boiling point of 287°C. The vapor pressure could be lowered by adding a solute to the water by Raoult's law (discussed in Chapter 12). The addition of a solute to the water might be possible for certain uses, but it would not be desirable for drinking water, unless the solute could be removed later. The costs of adding the solute, then removing it, say by reverse osmosis, would likely be prohibitive.

11.114 A friend has two containers that contain 50 mL of water and 50 mL of ethanol (CH_3CH_2OH). When the contents of these two containers are mixed together, do you expect the total volume to be 100 mL? Is it possible to predict the total volume before two liquids are mixed? What is your prediction in this case and why?

Discussion

Volumes of liquids are not necessarily additive. When 50 mL of water is mixed with 50 mL of ethanol at 20°C, the total volume is 96.5 mL. A simple analogy can make this understandable. Imagine that you mixed a peck of potatoes and a peck of shelled peas. The peas would likely fill the spaces between the potatoes so that the total volume would be much less than two pecks. The situation with molecules is more complicated than this however. For example, the hydrogen bonding forces in a liquid such as water give a liquid in which there is considerable order and therefore a structure in which water molecules do not pack as tightly as they otherwise might. The addition of ethanol probably breaks up the ordering among water molecules, giving a more compact structure. It is also possible to find cases in which the addition of two liquids gives a total volume that is greater than their sum. In general, it is difficult to predict the volume of a solution of two liquids, because of the complexity of intermolecular forces. Fortunately, the total volume of two dilute aqueous solutions may be approximated by the sum of the individual volumes.

12. SOLUTIONS

Solutions to Exercises

Note on significant figures: The final answer to each mathematical solution is given first with one nonsignificant figure (the rightmost significant figure is underlined) and then is rounded to the correct number of figures. Intermediate answers usually also have at least one nonsignificant figure.

12.1 An example of a solid solution prepared from a liquid and a solid is a dental filling made of liquid mercury and solid silver.

12.2 The C_4H_9OH molecules will be more soluble in water because their -OH ends can form hydrogen bonds with water.

12.3 The Na^+ ion has a larger energy of hydration because its ionic radius is smaller, giving Na^+ a more concentrated electric field than K^+.

12.4 Write Henry's law ($S = k_HP$) for 159 mmHg (P_2) and divide it by Henry's law for 1 atm, or 760 mmHg (P_1). Then substitute the experimental values of P_1, P_2, and S_1 to solve for S_2.

$$\frac{S_2}{S_1} = \frac{k_HP_2}{k_HP_1} = \frac{P_2}{P_1}$$

$$S_2 = \frac{P_2S_1}{P_1} = \frac{(159\text{ mmHg})(0.0404\text{ g O}_2/\text{L})}{760\text{ mmHg}} = 8.4\underline{5}2 \times 10^{-3}$$

$$= 8.45 \times 10^{-3}\text{ g O}_2/\text{L}$$

12.5 The mass of HCl in 20.2% HCl (0.202 = fraction of HCl) is

$$0.202 \times 35.0 \text{ g} = 7.0\underline{7}0 = 7.07 \text{ g HCl}$$

The mass of H_2O in 20.2% HCl is

$$35.0 \text{ g solution} - 7.07 \text{ g HCl} = 27.\underline{9}3 = 27.9 \text{ g } H_2O$$

12.6 Calculate the moles of toluene, using its molar mass of 92.14 g/mol:

$$35.6 \text{ g toluene} \times \frac{1 \text{ mol toluene}}{92.14 \text{ g toluene}} = 0.38\underline{6}3 \text{ mol toluene}$$

To calculate molality, divide the moles of toluene by the mass in kg of the solvent (C_6H_6):

$$\text{Molality} \times \frac{0.3863 \text{ mol toluene}}{0.125 \text{ kg solvent}} = 3.0\underline{9}04 = 3.09 \text{ m toluene}$$

12.7 The number of moles of toluene = 0.3863 (previous exercise); the number of moles of benzene is

$$125 \text{ g benzene} \times \frac{1 \text{ mol benzene}}{78.11 \text{ g benzene}} = 1.6\underline{0}03 \text{ mol benzene}$$

The total number of moles is 1.6003 + 0.3863 = $1.9\underline{8}66$, and the mole fractions are

$$\text{Mol fraction benzene} \times \frac{1.6003 \text{ mol benzene}}{1.9866 \text{ mol}} = 0.80\underline{5}54 = 0.806$$

$$\text{Mol fraction toluene} = \frac{0.3863 \text{ mol toluene}}{1.9866 \text{ mol}} = 0.19\underline{4}4 = 0.194$$

The sum of the mole fractions = 1.000.

12.8 This solution contains 0.120 moles of methanol dissolved in 1.00 kg of ethanol. The number of moles in 1.00 kg of ethanol is

$$1.00 \times 10^3 \text{ g } C_2H_5OH \times \frac{1 \text{ mol } C_2H_5OH}{46.07 \text{ g } C_2H_5OH} = 21.\underline{7}06 \text{ mol } C_2H_5OH$$

The total number of moles is 21.706 + 0.120 = $21.\underline{8}26$, and the mole fractions are

(continued)

$$\text{Mol fraction } C_2H_5OH = \frac{21.\underline{7}06 \text{ mol } C_2H_5OH}{21.826 \text{ mol}} = 0.99\underline{4}501 = 0.995$$

$$\text{Mol fraction } CH_3OH = \frac{0.12\underline{0} \text{ mol } CH_3OH}{21.826 \text{ mol}} = 0.0054\underline{9}8 = 0.00550$$

The sum of the mole fractions is 1.000.

12.9 One mole of solution contains 0.250 moles methanol and 0.750 moles ethanol. The mass of this amount of ethanol, the solvent, is

$$0.750 \text{ mol } C_2H_5OH \times \frac{46.07 \text{ g } C_2H_5OH}{1 \text{ mol } C_2H_5OH} = 34.\underline{5}5 \text{ g } C_2H_5OH \ (0.03455 \text{ kg})$$

The molality of methanol in the ethanol solvent is

$$\frac{0.250 \text{ mol } CH_3OH}{0.03455 \text{ kg } C_2H_5OH} = 7.2\underline{3}58 = 7.24 \text{ m } CH_3OH$$

12.10 Assume an amount of solution contains one kilogram of water. The mass of urea in this mass is

$$3.42 \text{ mol urea} \times \frac{60.05 \text{ g urea}}{1 \text{ mol urea}} = 20\underline{5}.4 \text{ g urea}$$

The total mass of solution is 205.4 + 1000.0 g = 120$\underline{5}$.4 g. The volume and molarity are

$$\frac{1205.4 \text{ g}}{1.045 \text{ g/mL}} = 115\underline{3}.49 \text{ mL } (1.15\underline{3}49 \text{ L})$$

$$\text{Molarity} = \frac{3.42 \text{ mol urea}}{1.15349 \text{ L solution}} = 2.9\underline{6}49 \text{ mol/L} = 2.96 \text{ M}$$

12.11 Assume a volume equal to 1.000 L of solution. Then

$$\text{Mass of solution} = 1.029 \text{ g/mL} \times (1.000 \times 10^3 \text{ mL}) = 1029 \text{ g}$$

$$\text{Mass of urea} = 2.00 \text{ mol urea} \times \frac{60.05 \text{ g urea}}{1 \text{ mol urea}} = 120.1 \text{ g urea}$$

$$\text{Mass of water} = (1029 - 120.1) \text{ g} = 90\underline{8}.9 \text{ g water } (0.90\underline{8}9 \text{ kg})$$

$$\text{Molality} = \frac{2.00 \text{ mol urea}}{0.9089 \text{ kg solvent}} = 2.2\underline{0}04 = 2.20 \text{ m urea}$$

12.12 Calculate the moles of naphthalene and moles of chloroform:

$$0.515 \text{ g } C_{10}H_8 \times \frac{1 \text{ mol } C_{10}H_8}{128.17 \text{ g } C_{10}H_8} = 0.0040\underline{1}8 \text{ mol } C_{10}H_8$$

$$60.8 \text{ g } CHCl_3 \times \frac{1 \text{ mol } CHCl_3}{119.38 \text{ g } CHCl_3} = 0.50\underline{9}29 \text{ mol } CHCl_3$$

The total number of moles is 0.004018 + 0.50929 = 0.51$\underline{3}$3 mol, and the mole fraction of chloroform is

$$\text{Mol fraction } CHCl_3 = \frac{0.50\underline{9}29 \text{ mol } CHCl_3}{0.5133} = 0.99\underline{2}1$$

$$\text{Mol fraction } C_{10}H_8 = \frac{0.004018 \text{ mol } C_{10}H_8}{0.5133 \text{ mol}} = 0.0078\underline{2}8$$

The vapor pressure lowering is

$$\Delta P = P^o X_{C_{10}H_8} = (156 \text{ mmHg})(0.007828) = 1.2\underline{2}1 = 1.22 \text{ mmHg}$$

Use Raoult's law to calculate the vapor pressure of chloroform:

$$P = P^o X_{CHCl_3} = (156 \text{ mmHg})(0.9921) = 15\underline{4}.7 = 155 \text{ mmHg}$$

12.13 Solve for c_m in the freezing-point-depression equation ($\Delta T = K_f c_m$; K_f in Table 12.3):

$$c_m = \frac{\Delta T}{K_f} = \frac{0.150°C}{1.858°C/m} = 0.080\underline{7}3 \text{ m}$$

Use the molal concentration to solve for the mass of ethylene glycol:

$$\frac{0.08073 \text{ m glycol}}{1 \text{ kg solvent}} \times 0.0378 \text{ kg solvent} = 0.0030\underline{5}1 \text{ mol glycol}$$

$$0.003051 \text{ mol glycol} \times \frac{62.1 \text{ g glycol}}{1 \text{ mol glycol}} = 0.18\underline{9}4 = 0.189 \text{ g glycol}$$

12.14 Calculate the moles of ascorbic acid (vit. C) from the molality, and then divide the mass of 0.930 g by the number of moles to obtain the molar mass:

$$\frac{0.0555 \text{ mol vit. C}}{1 \text{ kg } H_2O} \times 0.0950 \text{ kg } H_2O = 0.0052\underline{7}2 \text{ mol vit. C}$$

$$\frac{0.930 \text{ g vit. C}}{0.005272 \text{ mol vit. C}} = 17\underline{6}.4 = 176 \text{ g/mol}$$

The molecular weight of ascorbic acid, or vitamin C, is 176 amu.

12.15 The molal concentration of white phosphorus is

$$c_m = \frac{\Delta T}{K_f} = \frac{0.159°C}{2.40°C/m} = 0.066\underline{2}5 \text{ m}$$

The number of moles of white phosphorus (P_x) present in this solution is

$$\frac{0.066\underline{2}5 \text{ mol } P_x}{1 \text{ kg } CS_2} \times 0.0250 \text{ kg } CS_2 = 0.0016\underline{5}6 \text{ mol } P_x$$

The molar mass of white phosphorus equals the mass divided by moles:

$$0.205 \text{ g} \div 0.001656 \text{ mol} = 12\underline{3}.7 = 124 \text{ g/mol}$$

Thus, the molecular weight of P_x is 124 amu. The number of P atoms in the molecule of white phosphorus is obtained by dividing the molecular weight by the atomic weight of P:

$$\frac{124 \text{ amu } P_x}{30.97 \text{ amu P}} = 4.0\underline{0}3 = 4.00$$

Hence, the molecular formula is P_4 (x = 4).

12.16 The number of moles of sucrose is

$$5.0 \text{ g sucrose} \times \frac{1 \text{ mol sucrose}}{342.3 \text{ g sucrose}} = 0.01\underline{4}6 \text{ mol sucrose}$$

The molarity of the solution is

$$\frac{0.0146 \text{ mol sucrose}}{0.100 \text{ L}} = 0.1\underline{4}6 \text{ M sucrose}$$

The osmotic pressure, π, is equal to MRT and is calculated as follows:

$$\frac{0.146 \text{ mol sucrose}}{\text{L}} \times \frac{0.0821 \text{ L•atm}}{\text{mol•K}} \times 293 \text{ K} = 3.\underline{5}1 = 3.5 \text{ atm}$$

12.17 The number of ions from each formula unit is i. Here,

$$i = 1 + 2 = 3$$

The boiling-point elevation is

$$\Delta T_b = K_b c_m = 3 \times \frac{0.512^\circ C}{m} \times 0.050\ m = 0.07\underline{6}81 = 0.077^\circ C$$

The boiling point of aqueous $MgCl_2$ is 100.077°C.

12.18 $AlCl_3$ would be most effective in coagulating colloidal sulfur because of the greater magnitude of charge on the Al ion (+3).

■ Answers to Review Questions

12.1 An example of a gaseous solution is air, in which nitrogen (78%) acts as a solvent for a gas such as oxygen (21%). Recall that the solvent is the component present in greater amount. An example of a liquid solution containing a gas is any carbonated beverage in which water acts as the solvent for carbon dioxide gas. Ethanol in water is an example of a liquid-liquid solution. An example of a solid solution is any gold-silver alloy.

12.2 Glycerol is miscible in water because its -OH group can form strong hydrogen bonds that can break up and replace the hydrogen bonds between the water molecules. Thus, "like dissolves like" because the -OH group in glycerol is like that in water. Benzene has very limited solubility in water because it can only replace the strong hydrogen bonds between water molecules with weaker London forces between water and benzene. Thus, this is not a case of two "like" molecules mixing together.

12.3 The two factors that explain differences in solubilities are (1) the natural tendency of substances to mix together, or the natural tendency of substances to become disordered, and (2) the *relative* forces of attraction between solute species (or solvent species) compared to that between the solute and solvent species. The strongest interactions are always achieved.

12.4 This is not a case of "like dissolves like." There are strong hydrogen-bonding forces between the water molecules. For the octane to mix with water, hydrogen bonds must be broken and replaced by the much weaker London forces between water and octane. Thus, octane does not dissolve in water, because the maximum forces of attraction among molecules are obtained if it does not dissolve.

12.5 In most cases, the wide differences in solubility can be explained in terms of the different energies of attraction between ions in the crystal and between ions and water. Hydration energy is used to measure the attraction of ions for water molecules, and lattice energy is used to measure the attraction of positive ions for negative ions in the crystal lattice. An ionic substance is soluble when the hydration energy is much larger than the lattice energy. An ionic substance is insoluble when the lattice energy is much larger than the hydration energy.

12.6 A sodium chloride crystal dissolves in water because of two factors. The positive Na^+ ion is strongly attracted to the oxygen (negative end of the water dipole) and dissolves as $Na^+(aq)$. The negative Cl^- ion is strongly attracted to the hydrogens (positive end of the water dipole) and dissolves as $Cl^-(aq)$.

12.7 When the temperature (energy of a solution) is increased, the solubility of an ionic compound usually increases. A number of salts are exceptions to this rule, particularly a number of calcium salts such as calcium acetate, calcium sulfate, and calcium hydroxide (although the solubilities of calcium bromide, calcium chloride, calcium fluoride, and calcium iodide all increase with temperature).

12.8 Calcium chloride is an example of a salt that releases heat when it dissolves (exothermic heat of solution). Ammonium nitrate is an example of a salt that absorbs heat when it dissolves (endothermic heat of solution).

12.9 Fewer fish can live in an aquarium of fixed volume in the summer than in the winter because fish breathe the air dissolved in the water. In the summer the solubility of air (including oxygen) in the water is lower because of the higher water temperature.

12.10 A carbonated beverage must be stored in a closed container because carbonated beverages must contain more carbon dioxide than is soluble in water at atmospheric pressure. It is possible to add carbon dioxide under pressure to a closed container before it is sealed and increase the solubility of carbon dioxide. This is an illustration of Le Chatelier's principle, which states that the equilibrium between gaseous and dissolved carbon dioxide is shifted in favor of the dissolved carbon dioxide by an increase in pressure.

12.11 According to Le Chatelier's principle, a gas is more soluble in a liquid at higher pressures because when the gas dissolves in the liquid, the system decreases in volume, tending to decrease the applied pressure. However, when a solid dissolves in a liquid, there is very little volume change. Thus, pressure has very little effect on the solubility of a solid in a liquid.

12.12 The four ways to express the concentration of a solute in a solution are (1) molarity, which is moles per liter; (2) mass percentage of solute, which is the percentage by mass of solute contained in a given mass of solution; (3) molality, which is the moles of solute per kilogram of solvent; and (4) mole fraction, which is the moles of the component substance divided by the total moles of solution.

12.13 The vapor pressure of the solvent over the more dilute solution is larger than the vapor pressure of solvent over the more concentrated solution. Thus, the more dilute solution loses more solvent by evaporation and becomes more concentrated. Because the cabinet is closed, the excess solvent molecules condense into the more concentrated solution, diluting it. The changes in solvent composition stop when the concentrations of each solution become equal.

12.14 In fractional distillation, the vapor that first appears over a solution will have a greater mole fraction of the more volatile component. If a portion of this is vaporized and condensed, the liquid will be still richer in the more volatile component. After successive distillation stages, eventually the more volatile component will be obtained in pure form (Figure 12.17).

12.15 The boiling point of the solution is higher because the nonvolatile solute lowers the vapor pressure of the solvent. Thus, the temperature must be increased to a value greater than the boiling point of the pure solvent to achieve a vapor pressure equal to atmospheric pressure.

12.16 It is possible to prepare drinking water from seawater by freezing the seawater, which then forms almost pure ice. The dissolved salts are left behind in a concentrated solution, which does not freeze because the salts lower the freezing point of the solution. After the concentrated solution is drained off, the "sea ice" can be melted and used for drinking.

12.17 One application is the use of ethylene glycol in automobile radiators as antifreeze; the glycol-water mixture usually has a freezing point well below the average temperature low during the winter. A second application is spreading sodium chloride on icy roads in the winter to melt the ice. The ice usually melts because at equilibrium a concentrated solution of NaCl usually freezes at a temperature below that of the roads.

12.18 The lettuce wilts if left too long in a salad dressing containing vinegar and salt because of osmosis. There is a higher salt concentration outside the lettuce membrane than inside, and the water flows through the membrane in order to lower the salt concentration outside to equal the salt concentration inside the lettuce.

12.19 If a pressure greater than the osmotic pressure of the ocean water is applied, the natural osmotic flow can be reversed. Then the water solvent flows from the ocean water through a membrane to a more dilute solution or to pure water, leaving behind the salt and other ionic compounds from the ocean in a more concentrated solution.

12.20 Part of the light from the sun is scattered in the direction of an observer by fine particles in the clouds (Tyndall effect), rather than being completely absorbed by the clouds. The scattered light becomes visible against the darker background of dense clouds.

12.21 The examples are an aerosol—a fog; a foam—whipped cream; an emulsion—mayonnaise; a sol—solid silver chloride dispersed in water; and a gel—fruit jelly.

12.22 Iron(III) hydroxide precipitates because electrons from the negative electrode neutralize the positive charge of the colloidal iron(III) hydroxide. This allows the colloidal particles to approach each other closely enough to aggregate and finally precipitate.

12.23 A Cottrell precipitator works by aggregating the fine particles of smoke (an aerosol type of colloid) into a solid that can be readily removed from the gas. The electrodes conduct the high voltage into the smoke; the high-voltage current apparently removes the excess of electrical charge on the particles that prevents them from aggregating, thus allowing aggregation and removal.

12.24 Each stearic acid molecule orients itself with its acid group (-COOH) into the heavier water below the benzene and with its hydrocarbon group into the lighter benzene layer on the surface of the water. As the benzene evaporates, the stearic acid molecules are left on the water surface. Each molecule keeps its acid group positioned in the water, so a single layer results.

12.25 Soap removes oil from a fabric by absorbing the oil into the hydrophobic centers of the soap micelles and off the surface of the fabric. Rinsing removes the micelles from contact with the fabric and leaves only water on the fabric, which can then be dried.

12.26 An example of an anionic detergent is sodium lauryl sulfate detergent ($-OSO_3^-$ group). An example of a cationic detergent is the positively charged nitrogen detergents (R_4N^+ group).

■ Solutions to Practice Problems

Note on significant figures: The final answer to each problem is given first with one nonsignificant figure (the rightmost significant figure is underlined) and then is rounded to the correct number of figures. Intermediate answers usually also have at least one nonsignificant figure. Atomic weights are rounded to two decimal places, except for that of hydrogen.

12.27 An example of a liquid solution prepared by dissolving a gas in a liquid is household ammonia, which consists of ammonia (NH_3) gas dissolved in water.

12.28 An example of a solid solution prepared from two solids is almost any alloy, such as 18 kt gold, which consists of 25% silver dissolved in 75% gold.

12.29 Boric acid would be more soluble in ethanol because this acid is polar and is more soluble in a more polar solvent. It can also hydrogen-bond to ethanol but not to benzene.

12.30 Naphthalene is more soluble in benzene because nonpolar naphthalene must break the strong hydrogen bonds between ethanol molecules and replace them with weaker London forces.

12.31 The order of increasing solubility is $H_2O < CH_2OHCH_2OH < C_{10}H_{22}$. The solubility in nonpolar hexane increases with the decreasing polarity of the solute.

12.32 Acetic acid is more soluble because it is polar like ethanol and can form hydrogen bonds to it.

12.33 The Ca^{2+} ion has both a greater charge and a smaller ionic radius than K^+, so Ca^{2+} should have a greater energy of hydration.

12.34 The F^- ion has a smaller radius than the Cl^- ion, so F^- should have a greater hydration energy.

12.35 The order is $Ba(IO_3)_2 < Sr(IO_3)_2 < Ca(IO_3)_2 < Mg(IO_3)_2$. The iodate ion is fairly large, so the lattice energy for all these iodates should change to a smaller degree than the hydration energy of the cations. Therefore, solubility should increase with decreasing cation radius.

12.36 Fluorides (smaller anion): As the cation radius increases, the lattice energy of the fluoride salts decreases to a greater degree than the cation hydration energy decreases. Permanganates (larger anion): As the cation radius increases, the lattice energy of the permanganate salts decreases to a smaller degree than the cation hydration energy decreases. Therefore, the solubility of the fluorides increases with increasing cation radius, whereas the solubility of the permanganates decreases with increasing cation radius.

12.37 Using Henry's law, let S_1 = the solubility at 1.00 atm (P_1), and S_2 = the solubility at 5.50 atm (P_2).

$$S_2 = S_1 \times \frac{P_2}{P_1} = (0.161 \text{ g/100 mL}) \times \frac{5.50 \text{ atm}}{1.00 \text{ atm}} = 0.88\underline{5}5$$

$$= 0.886 \text{ g/100 mL}$$

12.38 The partial pressure of N_2 in air at 4.79 atm is

$$4.79 \text{ atm} \times 0.781 = 3.7\underline{4}099 \text{ atm}$$

Substituting this for P_2 in $S_2 = S_1 \times P_2/P_1$ (Henry's law), you obtain

$$S_2 = (0.00175 \text{ g/100 mL}) \times \frac{3.74099 \text{ atm}}{1.00 \text{ atm}} = 0.0065\underline{4}6$$

$$= 0.00655 \text{ g/100 mL}$$

12.39 First calculate the mass of KI in the solution; then calculate the mass of water needed.

$$\text{Mass KI} = 145\text{ g} \times \frac{2.50\text{ g KI}}{100\text{ g soln}} = 3.6\underline{2}50\text{ g KI}$$

$$\text{Mass } H_2O = 145\text{ g soln} - 3.625\text{ g KI} = 14\underline{1}.375\text{ g } H_2O$$

Dissolve 3.6$\underline{2}$5 g KI in 14$\underline{1}$.3 g of water.

12.40 First calculate the mass of Na_2SO_4 in the solution; then calculate the mass of water needed.

$$\text{Mass of } Na_2SO_4 = 455\text{ g soln} \times \frac{6.50\text{ g } Na_2SO_4}{100\text{ g soln}} = 29.\underline{5}7\text{ g } Na_2SO_4$$

$$\text{Mass of water} = 455\text{ g soln} - 29.57\text{ g } Na_2SO_4 = 42\underline{5}.43\text{ g water}$$

Dissolve 29.$\underline{5}$7 g NaSO in 42$\underline{5}$.43 g of water.

12.41 Multiply the mass of KI by 100 g of soln per 2.50 g KI (reciprocal of percentage).

$$0.258\text{ g KI} \times \frac{100\text{ g soln}}{2.50\text{ g KI}} = 10.\underline{3}2 = 10.3\text{ g soln}$$

12.42 Multiply the mass of sodium sulfate by 100 g of soln per 6.50 g Na_2SO_4 (reciprocal of percentage).

$$1.50\text{ g } Na_2SO_4 \times \frac{100\text{ g soln}}{6.50\text{ g } Na_2SO_4} = 23.\underline{0}7 = 23.1\text{ g soln}$$

12.43 Convert mass of vanillin to moles, convert mg of ether to kg, and divide for molality.

$$0.0372\text{ g vanillin} \times \frac{1\text{ mol vanillin}}{152.2\text{ g vanillin}} = 2.4\underline{4}4 \times 10^{-4}\text{ mol vanillin}$$

$$168.5\text{ mg ether} \times 1\text{ kg}/10^6\text{ mg} = 168.5 \times 10^{-6}\text{ kg ether}$$

$$\text{Molality} = \frac{2.444 \times 10^{-4}\text{ mol vanillin}}{168.5 \times 10^{-6}\text{ kg ether}} = 1.4\underline{5}04 = 1.45\text{ } m\text{ vanillin}$$

12.44 Convert mass of lauryl alcohol (LA) to moles, convert g of ethanol to kg, and divide.

$$15.6 \text{ g LA} \times \frac{1 \text{ mol LA}}{186.3 \text{ g LA}} = 8.3\underline{7}4 \times 10^{-2} \text{ mol LA}$$

$$148 \text{ g ethanol} \times 1 \text{ kg}/10^3 \text{ g} = 148 \times 10^{-3} \text{ kg ethanol}$$

$$\text{Molality} = \frac{8.374 \times 10^{-2} \text{ mol LA}}{148 \times 10^{-3} \text{ kg ethanol}} = 0.56\underline{5}8 = 0.566 \text{ m LA}$$

12.45 Convert mass of fructose to moles and then multiply by 1 kg H_2O per 0.125 mol fructose (the reciprocal of molality).

$$1.75 \text{ g fruct.} \times \frac{1 \text{ mol fruct.}}{180.16 \text{ g fruct.}} \times \frac{1 \text{ kg } H_2O}{0.125 \text{ mol fruct.}}$$

$$= 0.077\underline{7}08 \text{ kg} \quad (77.7 \text{ g } H_2O)$$

12.46 Multiply the kg of $CHCl_3$ by the molality to get moles of caffeine; then convert moles to mass.

$$0.0450 \text{ kg } CHCl_3 \times \frac{0.0946 \text{ mol caff.}}{1 \text{ kg } CHCl_3} \times \frac{194.2 \text{ g caff.}}{1 \text{ mol caff.}}$$

$$= 0.82\underline{6}7 = 0.827 \text{ g caffeine}$$

12.47 Convert masses to moles and then calculate the mole fractions.

$$65.0 \text{ g alc.} \times \frac{1 \text{ mol alc.}}{60.09 \text{ g alc.}} = 1.0\underline{8}1 = 1.08 \text{ mol alc.}$$

$$35.0 \text{ g } H_20 \times \frac{1 \text{ mol } H_20}{18.02 \text{ g } H_20} = 1.9\underline{4}2 = 1.094 \text{ mol } H_20$$

$$\text{Mol fraction alc.} = \frac{\text{mol alc.}}{\text{total mol}} = \frac{1.081 \text{ mol}}{3.023 \text{ mol}} = 0.35\underline{7}59 = 0.358$$

$$\text{Mol fraction } H_2O = \frac{\text{mol } H_2O}{\text{total mol}} = \frac{1.942 \text{ mol}}{3.023 \text{ mol}} = 0.64\underline{2}4 = 0.642$$

12.48 Convert masses to moles and then calculate the mole fractions.

$$2.25 \times 10^3 \text{ g gly. } \times \frac{1 \text{ mol gly.}}{62.07 \text{ g gly.}} = 36.\underline{2}5 \text{ mol gly.}$$

$$2.00 \times 10^3 \text{ g } H_2O \times \frac{1 \text{ mol } H_2O}{18.02 \text{ g } H_2O} = 11\underline{0}.98 \text{ mol } H_2O$$

$$\text{Mol fraction } H_2O = \frac{\text{mol } H_2O}{\text{total mol}} = \frac{11\underline{0}.98 \text{ mol}}{147.23 \text{ mol}} = 0.75\underline{3}8 = 0.754$$

$$\text{Mol fraction gly.} = \frac{\text{mol gly.}}{\text{total mol}} = \frac{36.25 \text{ mol}}{147.23 \text{ mol}} = 0.24\underline{6}2 = 0.246$$

12.49 In the solution, for every 0.750 mol of NaClO there is 1.00 kg, or 1.00×10^3 g, H_2O, so

$$1.00 \times 10^3 \text{ g } H_2O \times \frac{1 \text{ mol } H_2O}{18.02 \text{ g } H_2O} = 55.\underline{4}9 \text{ mol } H_2O$$

$$\text{Total mol} = 55.49 \text{ mol } H_2O + 0.750 \text{ mol NaClO} = 56.24 \text{ mol}$$

$$\text{Mol fraction NaClO} = \frac{\text{mol NaClO}}{\text{mol soln}} = \frac{0.750 \text{ mol}}{56.24 \text{ mol}} = 0.013\underline{3}3 = 0.0133$$

12.50 In the solution, for every 0.655 mol of H_2O_2, there is 1.00 kg, or 1.00×10^3 g, of H_2O, so

$$1.00 \times 10^3 \text{ g } H_2O \times \frac{1 \text{ mol } H_2O}{18.02 \text{ g } H_2O} = 55.\underline{4}9 \text{ mol } H_2O$$

$$\text{Total mol} = 55.49 \text{ mol } H_2O + 0.655 \text{ mol } H_2O_2 = 56.\underline{1}45 \text{ mol}$$

$$\text{Mol fraction } H_2O_2 = \frac{\text{mol } H_2O_2}{\text{mol soln}} = \frac{0.655 \text{ mol}}{56.145 \text{ mol}} = 0.011\underline{6}6 = 0.0117$$

12.51 The total moles of solution = 3.31 mol H_2O + 1.00 mol HCl = 4.31 mol.

$$\text{Mol fraction HCl} = \frac{1 \text{ mol HCl}}{4.31 \text{ mol}} = 0.23\underline{2}01 = 0.232$$

$$3.31 \text{ mol } H_2O \times \frac{18.02 \text{ g } H_2O}{1 \text{ mol } H_2O} \times \frac{1 \text{ kg } H_2O}{10^3 \text{ g } H_2O} = 5.9\underline{6}5 \times 10^{-2} \text{ kg } H_2O$$

$$\text{Molality} = \frac{1.00 \text{ mol HCl}}{5.965 \times 10^{-2} \text{ kg } H_2O} = 16.\underline{7}6 = 16.8 \text{ m}$$

12.52 The total moles of solution = 1.00 mol NH_3 + 2.44 mol H_2O = 3.44 mol.

$$\text{Mol fraction } NH_3 = \frac{1 \text{ mol } NH_3}{3.44 \text{ mol}} = 0.29\underline{0}6 = 0.291$$

$$2.44 \text{ mol } H_2O \times \frac{18.02 \text{ g } H_2O}{1 \text{ mol } H_2O} \times \frac{1 \text{ kg } H_2O}{10^3 \text{ g } H_2O} = 4.3\underline{9}7 \times 10^{-2} \text{ kg } H_2O$$

$$\text{Molality} = \frac{1.00 \text{ mol } NH_3}{4.397 \times 10^{-2} \text{ kg } H_2O} = 22.\underline{7}4 = 22.7 \text{ m}$$

12.53 The mass of 1.000 L of solution is 1.022 kg. In the solution, there are 0.585 moles of $H_2C_2O_4$ (OA) for every 1.0000 kg of water. Convert this number of moles to mass.

$$0.585 \text{ mol OA} \times \frac{90.04 \text{ g OA}}{1 \text{ mol OA}} \times \frac{1 \text{ kg OA}}{10^3 \text{ g OA}} = 0.052\underline{6}7 \text{ kg OA}$$

The total mass of the solution containing 1.000 kg H_2O and 0.585 moles of OA is calculated as follows:

$$\text{Mass} = 1.0000 \text{ kg } H_2O + 0.05267 \text{ kg OA} = 1.052\underline{6}7 \text{ kg}$$

Use this to relate the mass of 1.000 L (1.022 kg) of solution to the amount of solute:

$$1.022 \text{ kg soln} \times \frac{0.585 \text{ mol OA}}{1.05267 \text{ kg soln}} = 0.56\underline{7}9 \text{ mol OA}$$

$$\text{Molarity} = \frac{0.5679 \text{ mol OA}}{1.000 \text{ L soln}} = 0.56\underline{7}9 = 0.568 \text{ M}$$

12.54 The mass of 1.000 L of solution is 1.049 kg. In the solution, there is 0.710 mole of citric acid (CA) for every 1.000 kg of water. Convert this number of moles to mass.

$$0.710 \text{ mol CA} \times \frac{192.1 \text{ g CA}}{1 \text{ mol CA}} \times \frac{1 \text{ kg CA}}{10^3 \text{ g CA}} = 0.13\underline{6}39 \text{ kg CA}$$

The total mass of solution containing 1.000 kg H_2O and 0.710 moles of CA is calculated as follows:

$$\text{Mass} = 1.0000 \text{ kg } H_2O + 0.13639 \text{ kg CA} = 1.13\underline{6}39 \text{ kg}$$

Use this to relate the mass of 1.000 L (1.049 kg) of solution to the amount of solute:

$$1.049 \text{ kg soln} \times \frac{0.710 \text{ mol CA}}{1.13639 \text{ kg soln}} = 0.65\underline{5}4 \text{ mol CA}$$

$$\text{Molarity} = \frac{0.6554 \text{ mol CA}}{1.000 \text{ L soln}} = 0.65\underline{5}4 = 0.655 \text{ M}$$

12.55 In 1.000 L of vinegar, there is 0.763 mole of acetic acid. The total mass of the 1.000-L solution is 1.004 kg. Start by calculating the mass of acetic acid (AA) in the solution.

$$0.763 \text{ mol AA} \times \frac{60.05 \text{ g AA}}{1 \text{ mol AA}} = 45.\underline{8}2 \text{ g AA } (0.045\underline{8}2 \text{ kg AA})$$

The mass of water may be found by difference:

$$\text{Mass } H_2O = 1.004 \text{ kg soln} - 0.04582 \text{ kg AA} = 0.95\underline{8}2 \text{ kg } H_2O$$

$$\text{Molality} = \frac{0.763 \text{ mol AA}}{0.9582 \text{ lg } H_2O} = 0.79\underline{6}2 = 0.796 \text{ m AA}$$

12.56 In 1.000 L of beverage, there is 0.271 mole of tartaric acid. The total mass of the 1.000-L beverage is 1.016 kg. Start by calculating the mass of tartaric acid (TA) in the beverage.

$$0.271 \text{ mol TA} \times \frac{150.1 \text{ g TA}}{1 \text{ mol TA}} = 40.\underline{6}8 \text{ g TA } (0.040\underline{6}8 \text{ kg TA})$$

The mass of water may be found by difference:

$$\text{Mass } H_2O = 1.016 \text{ kg soln} - 0.04068 \text{ kg TA} = 0.97\underline{5}3 \text{ kg } H_2O$$

$$\text{Molality} = \frac{0.271 \text{ mol TA}}{0.9753 \text{ kg } H_2O} = 0.27\underline{7}8 = 0.278 \text{ m TA}$$

12.57 To find the mole fraction of sucrose, first find the amounts of both sucrose (suc.) and water:

$$20.2 \text{ g suc.} \times \frac{1 \text{ mol suc.}}{342.3 \text{ g suc.}} = 0.059\underline{0}1 \text{ mol suc.}$$

$$60.5 \text{ g } H_2O \times \frac{1 \text{ mol } H_2O}{18.02 \text{ g } H_2O} = 3.3\underline{5}7 \text{ mol } H_2O$$

$$X_{suc.} = \frac{0.05901 \text{ mol suc.}}{(3.357 + 0.05901) \text{ mol}} = 0.017\underline{2}7$$

From Raoult's law, the vapor pressure (P) and lowering (ΔP) are

$$P = P^o_{H_2O} X_{H_2O} = P^o_{H_2O}(1 - X_{suc.}) = (42.2 \text{ mmHg})(1 - 0.01727)$$

$$= 41.\underline{4}7 = 41.5 \text{ mmHg}$$

$$\Delta P = P^o_{H_2O} X_{suc.} = (42.2 \text{ mmHg})(0.017\underline{2}7) = 0.72\underline{8}7 = 0.729 \text{ mmHg}$$

12.58 To find the mole fraction of benzene, first find the amounts of both benzene (ben.) and naphthalene (nap.):

$$1.20 \text{ g nap.} \times \frac{1 \text{ mol nap.}}{128.2 \text{ g nap.}} = 0.0093\underline{6}03 \text{ mol nap.}$$

$$25.6 \text{ g ben.} \times \frac{1 \text{ mol ben.}}{78.11 \text{ g ben.}} = 0.32\underline{7}7 \text{ mol ben.}$$

$$X_{nap.} = \frac{0.0093603 \text{ mol nap.}}{(0.3277 + 0.0093603)} = 0.027\underline{7}7$$

From Raoult's law, the vapor pressure (P) and lowering (ΔP) are

$$P = P^o_{ben.}(1 - X_{nap.}) = (86.0 \text{ mmHg})(1 - 0.027\underline{7}7) = 83.\underline{6}1$$

$$= 83.6 \text{ mmHg}$$

$$\Delta P = P^o_{ben.} X_{nap.} = (86.0 \text{ mmHg})(0.027\underline{7}7) = 2.3\underline{8}8 = 2.39 \text{ mmHg}$$

12.59 Find the molality of glycerol (gly.) in the solution first:

$$0.152 \text{ g gly.} \times \frac{1 \text{ mol gly.}}{92.095 \text{ g gly.}} = 0.0016\underline{5}04 \text{ mol gly.}$$

$$\text{Molality} = \frac{0.0016\underline{5}04 \text{ mol gly.}}{0.0200 \text{ kg solvent}} = 0.082\underline{5}2 \text{ m}$$

Substitute $K_b = 0.512^oC/m$ and $K_f = 1.858^oC/m$ (Table 12.3) into equations for ΔT_b and ΔT_f:

$$\frac{0.512°C}{\text{m gly.}} \times 0.082\underline{5}2 \text{ m gly.} = 0.042\underline{2}502°C$$

$$\frac{1.858°C}{\text{m gly.}} \times 0.08252 \text{ m gly.} = 0.15\underline{3}3°C$$

$$T_b = 100.000^oC + 0.0422502^oC = 100.042\underline{2}502 = 100.0423^oC$$

$$T_f = 0.000^oC - 0.1533^oC = -0.15\underline{3}3 = -0.153^oC$$

12.60 Find the molality of the solution first:

$$0.915 \text{ g } S_8 \times \frac{1 \text{ mol } S_8}{256.5 \text{ g } S_8} = 0.0035\underline{6}7 \text{ mol } S_8$$

$$\text{Molality} = \frac{0.0035\underline{6}7 \text{ mol } S_8}{0.1000 \text{ kg solvent}} = 0.035\underline{6}7 \text{ m}$$

Substitute $K_b = 3.08^oC/m$ and $K_f = 3.59^oC/m$ (Table 12.3) into equations for ΔT_b and ΔT_f:

$$\frac{3.08°C}{\text{m } S_8} \times 0.035\underline{6}7 \text{ m } S_8 = 0.10\underline{9}8°C$$

$$\frac{3.59°C}{\text{m } S_8} \times 0.03567 \text{ m } S_8 = 0.12\underline{8}05°$$

$$T_b = 118.5^oC + 0.1098^oC = 118.\underline{6}098 = 118.6^oC$$

$$T_f = 16.60^oC - 0.12805^oC = 16.4\underline{7}195 = 16.47^oC$$

12.61 Calculate ΔT_f, the freezing-point depression, and, using $K_f = 1.858^oC/m$ (Table 12.3), the molality, c_m.

$$\Delta T_f = 0.000^oC - (-0.086^oC) = 0.086^oC$$

$$c_m = \frac{\Delta T_f}{K_f} = \frac{0.086°C}{1.858°C/m} = 0.04\underline{6}28 = 0.046 \text{ m}$$

12.62 Calculate ΔT_f, the freezing-point depression, and, using $K_f = 1.858^oC/m$ (Table 12.3), the molality, c_m.

$$\Delta T_f = 0.000^oC - (-0.085^oC) = 0.085^oC$$

$$c_m = \frac{\Delta T_f}{K_f} = \frac{0.085°C}{1.858°C/m} = 0.04\underline{5}7 \text{ m}$$

Now find the amount of urea in the solution from the definition of molality:

$$Mol_{urea} = c_m \times (\text{kg } H_2O) = \frac{0.0457 \text{ mol urea}}{1 \text{ kg } H_2O} \times 0.1000 \text{ kg } H_2O$$

$$= 0.004\underline{5}7 \text{ mol}$$

$$0.00457 \text{ mol urea} \times \frac{60.06 \text{ g urea}}{1 \text{ mol urea}} = 0.2\underline{7}4 = 0.27 \text{ g urea}$$

12.63 Find the moles of unknown solute from the definition of molality:

$$Mol_{solute} = m \times \text{kg solvent} = \frac{0.0698 \text{ mol}}{1 \text{ kg solvent}} \times 0.002135 \text{ kg solvent}$$

$$= 1.4\underline{9}0 \times 10^{-4} \text{ mol}$$

$$\text{Molar mass} = \frac{0.0182 \text{ g}}{1.490 \times 10^{-4} \text{ mol}} = 12\underline{2}.1 = 122 \text{ g/mol}$$

The molecular weight is 122 amu.

12.64 Find the moles of the solute from the definition of molality:

$$\text{Mol}_{\text{solute}} = \text{m x kg solvent} = \frac{0.0368 \text{ mol}}{1 \text{ kg solvent}} \text{ x } 0.00975 \text{ kg solvent}$$

$$= 3.5\underline{8}8 \times 10^{-4} \text{ mol}$$

$$\text{Molar mass} = \frac{0.0653 \text{ g}}{3.588 \times 10^{-4} \text{ mol}} = 18\underline{2}.0 = 182 \text{ g/mol}$$

The molecular weight is 182 amu.

12.65 Calculate ΔT_f, the freezing-point depression, and, using $K_f = 1.858^oC/m$ (Table 12.3), the molality, c_m.

$$\Delta T_f = 26.84^oC - 25.70^oC = 1.14^oC$$

$$c_m = \frac{\Delta T_f}{K_f} = \frac{1.14°C}{8.00°C/m} = 0.14\underline{2}5 \text{ m}$$

Find the moles of solute by rearranging the definition of molality:

$$\text{Mol} = \text{m x kg solvent} = \frac{0.1425 \text{ mol}}{1 \text{ kg solvent}} \text{ x } 103 \times 10^{-6} \text{ kg solvent}$$

$$= 1.4\underline{6}7 \times 10^{-5} \text{ mol}$$

$$\text{Molar mass} = \frac{2.39 \times 10^{-3} \text{ g}}{1.467 \times 10^{-5} \text{ mol}} = 16\underline{2}.9 = 163 \text{ g/mol}$$

The molecular weight is 163 amu.

12.66 Calculate ΔT_f, the freezing-point depression, and, using $K_f = 1.858^oC/m$ (Table 12.3), the molality, c_m.

$$\Delta T_f = 5.455^oC - 4.880^oC = 0.575^oC$$

$$c_m = \frac{\Delta T_f}{K_f} = \frac{0.575°C}{5.065°C/m} = 0.11\underline{3}5 \text{ m}$$

(continued)

Find the moles of solute by rearranging the definition of molality:

$$\text{Mol} = m \times \text{kg solvent} = \frac{0.1135\ \text{mol}}{1\ \text{kg solvent}} \times 0.1000\ \text{kg solvent} = 0.011\underline{3}5\ \text{mol}$$

$$\text{Molar mass} = \frac{2.500\ \text{g}}{0.01135\ \text{mol}} = 22\underline{0}.2\ \text{g/mol}$$

The molecular weight is 220 amu.

12.67 Use the equation for osmotic pressure (π) to solve for the molarity of the solution.

$$M = \frac{\pi}{RT} = \frac{1.47\ \text{mmHg} \times \frac{1\ \text{atm}}{760\ \text{mmHg}}}{(0.0821\ \text{L}\bullet\text{atm/mol}\bullet\text{K})(21 + 273)} = 8.0\underline{1}3 \times 10^{-5}\ \text{mol/L}$$

Now find the number of moles in 106 mL (0.106 L) using the molarity.

$$0.106\ \text{L} \times \frac{8.013 \times 10^{-5}\ \text{mol}}{\text{L}} = 8.4\underline{9}4 \times 10^{-6}\ \text{mol}$$

$$\text{Molar mass} = \frac{0.582\ \text{g}}{8.494 \times 10^{-6}\ \text{mol}} = 6.8\underline{5}1 \times 10^{4}\ \text{g/mol}$$

The molecular weight is 6.85×10^4 amu.

12.68 Use the equation for osmotic pressure (π) to solve for the molarity of the solution.

$$M = \frac{\pi}{RT} = \frac{3.70\ \text{mmHg} \times \frac{1\ \text{atm}}{760\ \text{mmHg}}}{(0.0821\ \text{L}\bullet\text{atm/mol}\bullet\text{K})(25 + 273)} = 1.9\underline{8}9 \times 10^{-4}\ \text{mol/L}$$

Now find the number of moles in 100.0 mL (0.1000 L) using the molarity.

$$0.100\ \text{L} \times \frac{1.989 \times 10^{-4}\ \text{mol}}{\text{L}} = 1.9\underline{8}9 \times 10^{-5}\ \text{mol}$$

$$\text{Molar mass} = \frac{0.0216\ \text{g}}{1.989 \times 10^{-5}\ \text{mol}} = 1.0\underline{8}59 \times 10^{3}\ \text{g/mol}$$

The molecular weight is 1.09×10^3 amu.

12.69 Begin by noting that i = 3. Then calculate ΔT_f from the product of iK_fc_m:

$$3 \times \frac{1.858\,^{\circ}\mathrm{C}}{\mathrm{m}} \times 0.0085\ \mathrm{m} = -0.04\underline{7}4\,^{\circ}\mathrm{C}$$

The freezing point = 0.000°C - 0.0474°C = -0.04$\underline{7}$4 = -0.047°C.

12.70 Start by noting that i = 4. Then calculate ΔT_f from the product of iK_fc_m:

$$4 \times \frac{1.858\,^{\circ}\mathrm{C}}{\mathrm{m}} \times 0.0095\ \mathrm{m} = 0.07\underline{0}6\,^{\circ}\mathrm{C}$$

The freezing point = 0.000°C - 0.0706°C = -0.07$\underline{0}$6 = -0.071°C.

12.71 Begin by calculating the molarity of $Cr(NH_3)_5Cl_3$.

$$1.40 \times 10^{-2}\ \mathrm{g}\ Cr(NH_3)_5Cl_3 \times \frac{1\ \mathrm{mol}\ Cr(NH_3)_5Cl_3}{243.5\ \mathrm{g}\ Cr(NH_3)_5Cl_3}$$

$$= 5.7\underline{4}9 \times 10^{-5}\ \mathrm{mol}\ Cr(NH_3)_5Cl_3$$

$$\text{Molarity} = \frac{5.7\underline{4}9 \times 10^{-5}\ \mathrm{mol}\ Cr(NH_3)_5Cl_3}{0.0250\ \mathrm{L}} = 0.0022\underline{9}9\ \mathrm{M}$$

Now find the hypothetical osmotic pressure, assuming $Cr(NH_3)_5Cl_3$ does not ionize:

$$\pi = MRT = (2.30 \times 10^{-3}\ \mathrm{M}) \times \frac{0.0821\ \mathrm{L{\bullet}atm}}{\mathrm{K{\bullet}mol}} \times 298\ \mathrm{K} \times \frac{760\ \mathrm{mmHg}}{1\ \mathrm{atm}}$$

$$= 42.\underline{7}7\ \mathrm{mmHg}$$

The measured osmotic pressure is greater than the hypothetical osmotic pressure. The number of ions formed per formula unit = ratio of the measured pressure to the hypothetical pressure:

$$i = \frac{119\ \mathrm{mmHg}}{42.77\ \mathrm{mmHg}} = 2.7\underline{8}2 \cong 3\ \text{ions/ formula unit}$$

12.72 Begin by calculating the molality of the solution after rearranging $\Delta T_b = K_b c_m$:

$$c_m = \frac{\Delta T_b}{K_b} = \frac{5°C}{0.512°C/m} = \underline{9}.76 \text{ m}$$

The 9.76 m is the molality of the ions, not of the NaCl. The molality of the NaCl will be 9.76 m ÷ 2 = 4.88 m because i = 2. Thus 4.88 moles of NaCl must be dissolved in the 1 kg of water. This is converted to mass using the molar mass of NaCl:

$$\underline{4}.88 \text{ mol NaCl} \times \frac{58.5 \text{ g NaCl}}{1 \text{ mol NaCl}} = \underline{2}85 = 3 \times 10^2 \text{ g NaCl}$$

12.73 a. Aerosol (liquid water in air)
b. Sol [solid $Mg(OH)_2$ in liquid water]
c. Foam (air in liquid soap solution)
d. Sol (solid silt in liquid water)

12.74 a. Aerosol (liquid water in air)
b. Foam (air in liquid egg white)
c. Aerosol (solid dust particles in air)
d. Emulsion (liquid oil in liquid vinegar)

12.75 Because the As_2S_3 particles are negatively charged, the effective coagulation requires a highly charged cation, so $Al_2(SO_4)_3$ is the best choice.

12.76 The K_3PO_4 contains highly charged PO_4^{3-} ions that will be strongly attracted to the positively charged solution and will cause it to coagulate.

■ Solutions to Unclassified Problems

12.77 Using Henry's law [$S_2 = S_1 \times (P_2/P_1)$, where P_1 = 1.00 atm], find the solubility of each gas at P_2, its partial pressure. For N_2, P_2 = 0.800 x 1.00 atm = 0.800 atm; for O_2, P_2 = 0.200 atm.

$$N_2\text{: } S_2 \times \frac{P_2}{P_1} = (0.0175 \text{ g/L } H_2O) \times \frac{0.800 \text{ atm}}{1.00 \text{ atm}} = 0.014\underline{0}0 \text{ g/L } H_2O$$

$$O_2\text{: } S_2 \times \frac{P_2}{P_1} = (0.0393 \text{ g/L } H_2O) \times \frac{0.200 \text{ atm}}{1.00 \text{ atm}} = 0.0078\underline{6}0 \text{ g/L } H_2O$$

(continued)

In 1.00 L of the water, there are 0.0140 g of N_2 and 0.00786 g of O_2. If the water is heated to drive off both dissolved gases, the gas mixture that is expelled will contain 0.0140 g of N_2 and 0.00786 g of O_2. Convert both masses to moles using the molar masses:

$$0.0140 \text{ g } N_2 \times \frac{1 \text{ mol } N_2}{28.01 \text{ g } N_2} = 4.9\underline{9}8 \times 10^{-4} \text{ mol}$$

$$0.00786 \text{ g } O_2 \times \frac{1 \text{ mol } O_2}{32.00 \text{ g } O_2} = 2.4\underline{5}6 \times 10^{-4} \text{ mol}$$

Now calculate the mole fractions of each gas:

$$X_{N_2} = \frac{\text{mol } N_2}{\text{total mol}} = \frac{4.998 \times 10^{-4} \text{ mol}}{(4.998 + 2.456) \times 10^{-4} \text{ mol}} = 0.67\underline{0}51 = 0.671$$

$$X_{O_2} = 1 - X_{N_2} = 1 - 0.67\underline{0}51 = 0.32\underline{9}49 = 0.329$$

12.78 Using Henry's law [$S_2 = S_1 \times (P_2/P_1)$, where $P_1 = 1.00$ atm], find the solubility of each gas at P_2, its partial pressure. For CH_4, $P_2 = 0.900 \times 1.00 \text{ atm} = 0.900$ atm; for C_2H_6, $P_2 = 0.100 \times 1.00 \text{ atm} = 0.100$ atm.

$$CH_4: \; S_2 \times \frac{P_2}{P_1} = (0.023 \text{ g/ L } H_2O) \times \frac{0.900 \text{ atm}}{1.00 \text{ atm}} = 0.02\underline{0}7 \text{ g/ L } H_2O$$

$$C_2H_6: \; S_2 \times \frac{P_2}{P_1} = (0.059 \text{ g/ L } H_2O) \times \frac{0.100 \text{ atm}}{1.00 \text{ atm}} = 0.005\underline{9}0 \text{ g/ L } H_2O$$

When 1.00 L of the water saturated with the gas mixture is heated, 0.0207 g of CH_4 and 0.0059 g of C_2H_6 will be expelled. Convert these to moles, using molar masses:

$$0.0207 \text{ g } CH_4 \times \frac{1 \text{ mol } CH_4}{16.04 \text{ g } CH_4} = 1.\underline{2}91 \times 10^{-3} \text{ mol}$$

$$0.0059 \text{ g } C_2H_6 \times \frac{1 \text{ mol } C_2H_6}{30.07 \text{ g } C_2H_6} = 1.\underline{9}62 \times 10^{-4} \text{ mol}$$

Now find the mole fractions of each gas:

$$X_{CH_4} = \frac{\text{mol } CH_4}{\text{total mol}} = \frac{1.\underline{2}91 \times 10^{-3} \text{ mol}}{(1.291 + 0.196) \times 10^{-3} \text{ mol}} = 0.8\underline{6}819 = 0.87$$

$$X_{C_2H_6} = 1 - X_{CH_4} = 1 - 0.8\underline{6}819 = 0.1\underline{3}181 = 0.13$$

12.79 Assume a volume of 1.000 L, whose mass is then 1.024 kg. Use the percent composition given to find the mass of each of the components of the solution.

$$1.024 \text{ kg soln} \times \frac{8.50 \text{ kg } NH_4Cl}{100.00 \text{ kg soln}} = 0.08\underline{7}04 \text{ kg } NH_4Cl$$

$$\text{Mass of } H_2O = 1.024 \text{ kg soln} - 0.08704 \text{ kg } NH_4Cl = 0.9370 \text{ kg } H_2O$$

Convert mass of NH_4Cl and water to moles:

$$87.04 \text{ g } NH_4Cl \times \frac{1 \text{ mol } NH_4Cl}{53.49 \text{ g } NH_4Cl} = 1.6\underline{2}7 \text{ mol } NH_4Cl$$

$$937.0 \text{ g } H_2O \times \frac{1 \text{ mol } H_2O}{18.015 \text{ g } H_2O} = 52.\underline{0}1 \text{ mol } H_2O$$

$$\text{Molarity} = \frac{\text{mol } NH_4Cl}{\text{L soln}} = \frac{1.627 \text{ mol}}{1.00 \text{ L}} = 1.6\underline{2}7 = 1.63 \text{ M}$$

$$\text{Molality} = \frac{\text{mol } NH_4Cl}{\text{kg } H_2O} = \frac{1.627 \text{ mol}}{0.9370 \text{ kg } H_2O} = 1.7\underline{3}6 = 1.74 \text{ m}$$

$$X_{NH_4Cl} = \frac{\text{mol } NH_4Cl}{\text{total moles}} = \frac{1.627 \text{ mol}}{(52.01 + 1.627) \text{ mol}} = 0.030\underline{3}3 = 0.0303$$

12.80 Assume a volume of 1.000 L, whose mass is then 1.127 kg. Use the percent composition given to find the mass of each of the components of the solution.

$$1.127 \text{ kg soln} \times \frac{22.0 \text{ kg LiCl}}{100.00 \text{ kg soln}} = 0.24\underline{7}9 \text{ kg LiCl}$$

$$\text{Mass of } H_2O = 1.127 \text{ kg soln} - 0.2479 \text{ kg LiCl} = 0.87\underline{9}1 \text{ kg } H_2O$$

Convert mass of LiCl and water to moles:

$$247.9 \text{ g LiCl} \times \frac{1 \text{ mol LiCl}}{42.39 \text{ g LiCl}} = 5.8\underline{4}8 \text{ mol LiCl}$$

$$879.1 \text{ g } H_2O \times \frac{1 \text{ mol } H_2O}{18.015 \text{ g } H_2O} = 48.\underline{7}9 \text{ mol } H_2O$$

$$\text{Molarity} = \frac{\text{mol LiCl}}{\text{L soln}} = \frac{5.848 \text{ mol}}{1.00 \text{ L}} = 5.8\underline{4}8 = 5.85 \text{ M}$$

(continued)

$$\text{Molality} = \frac{\text{mol LiCl}}{\text{kg } H_2O} = \frac{5.848 \text{ mol}}{0.8791 \text{ kg } H_2O} = 6.6\underline{5}2 = 6.65 \text{ m}$$

$$X_{LiCl} = \frac{\text{mol LiCl}}{\text{total moles}} = \frac{5.848 \text{ mol}}{(5.848 + 48.79) \text{ mol}} = 0.10\underline{7}03 = 0.107$$

12.81 In 1.00 mol of gas mixture, there are 0.43 mol of propane (pro.) and 0.57 mol of butane (but.). Calculate the masses of these components first.

$$0.43 \text{ mol pro.} \times \frac{44.1 \text{ g pro.}}{1 \text{ mol pro.}} = 1\underline{9}.0 \text{ g pro.}$$

$$0.57 \text{ mol but.} \times \frac{58.12 \text{ g but.}}{1 \text{ mol but.}} = 3\underline{3}.1 \text{ g but.}$$

The mass of 1.00 mol of gas mixture is the sum of the masses of the two components:

$$19.0 \text{ g pro.} + 33.1 \text{ g but.} = 52.1 \text{ g mixture}$$

Therefore, in 52.1 g of the mixture there are 19.0 g of propane and 33.1 g of butane. For a sample with a mass of 58 g:

$$58 \text{ g mixture} \times \frac{19.0 \text{ g pro.}}{52.1 \text{ g mixture}} = 2\underline{1}.1 = 21 \text{ g pro.}$$

$$58 \text{ g mixture} \times \frac{33.1 \text{ g but.}}{52.1 \text{ g mixture}} = 3\underline{6}.8 = 37 \text{ g but.}$$

12.82 In 1.000 mol of gas mixture, there are 0.036 mol O_2, 0.056 mol N_2, and 0.908 mol He. Calculate the masses of these components first.

$$0.036 \text{ mol } O_2 \times \frac{32.00 \text{ g } O_2}{1 \text{ mol } O_2} = 1.\underline{1}5 \text{ g } O_2$$

$$0.056 \text{ mol } N_2 \times \frac{28.01 \text{ g } N_2}{1 \text{ mol } N_2} = 1.\underline{5}7 \text{ g } N_2$$

$$0.908 \text{ mol He} \times \frac{4.0026 \text{ g He}}{1 \text{ mol He}} = 3.6\underline{3}4 \text{ g He}$$

The mass of 1.000 mol of the gas mixture is the sum of the masses of the components:

$$1.15 \text{ g } O_2 + 1.57 \text{ g } N_2 + 3.634 \text{ g He} = 6.35 \text{ g mixture}$$

(continued)

For a sample with a mass of 7.84 g,

$$7.84 \text{ g mixture} \times \frac{1.15 \text{ g } O_2}{6.35 \text{ g mixture}} = 1.\underline{4}1 = 1.4 \text{ g } O_2$$

$$7.84 \text{ g mixture} \times \frac{1.57 \text{ g } N_2}{6.35 \text{ g mixture}} = 1.\underline{9}3 = 1.9 \text{ g } N_2$$

$$7.84 \text{ g mixture} \times \frac{3.634 \text{ g He}}{6.35 \text{ g mixture}} = 4.\underline{4}8 = 4.5 \text{ g He}$$

12.83 $P_{ED} = P^o_{ED}(X_{ED}) = 173 \text{ mmHg } (0.35) = 6\underline{0}.55 \text{ mmHg}$

$P_{PD} = P^o_{PD}(E_{PD}) = 127 \text{ mmHg } (0.65) = 8\underline{2}.55 \text{ mmHg}$

$P = P_{ED} + P_{PD} = 60.55 + 82.55 = 14\underline{3}.1 = 143 \text{ mmHg}$

12.84 $P_B = P^o_B(X_B) = 75 \text{ mmHg } (0.25) = 1\underline{8}.75 \text{ mmHg}$

$P_T = P^o_T(E_T) = 22 \text{ mmHg } (0.75) = 1\underline{6}.5 \text{ mmHg}$

$P = P_B + P_T = 18.75 + 16.5 = 3\underline{5}.25 = 35 \text{ mmHg}$

12.85 Calculate the moles of $KAl(SO_4)_2 \bullet 12H_2O$ using its molar mass of 474.4 g/mol, and use this to calculate the three concentrations.

a. The moles of $KAl(SO_4)_2 \bullet 12H_2O$ is calculated below, using the abbreviation of "Hyd" for the formula of $KAl(SO_4)_2 \bullet 12H_2O$.

$$\text{mol Hyd} = 0.1186 \text{ g Hyd} \times \frac{1 \text{ mol Hyd}}{474.4 \text{ g Hyd}} = 0.000250\underline{0}0 = 0.0002500 \text{ mol}$$

Note that one mol of $KAl(SO_4)_2 \bullet 12H_2O$ contains one mole of $KAl(SO_4)_2$ so calculating the molarity of $KAl(SO_4)_2$ can be performed using the moles of $KAl(SO_4)_2 \bullet 12H_2O$.

$$\frac{\text{mol}}{\text{L}} = \frac{0.0002500 \text{ mol Hyd}}{1.000 \text{ L soln.}} \times \frac{1 \text{ mol } KAl(SO_4)_2}{1 \text{ mol Hyd}} = 0.0002500 \text{ M } KAl(SO_4)_2$$

b. The molarity of the SO_4^{2-} ion will be twice that of the $KAl(SO_4)_2$.

$$\frac{\text{mol } SO_4^{2-}}{L} = 0.0002500 \text{ M } KAl(SO_4)_2 \times \frac{2 \text{ mol } SO_4^{2-}}{1 \text{ mol } KAl(SO_4)_2} = 0.0005000 \text{ M } SO_4^{2-}$$

c. Since the density of the solution is 1.00 g/mL, the mass of 1.000 L of solution is 1000 g, or 1.000 kg. Since molality is moles per 1.000 kg of solvent, the molality of $KAl(SO_4)_2$ equals 0.0002500 moles divided by 1.000 kg or 0.0002500 m (the same as the molarity).

12.86 Calculate the moles of $Al_2(SO_4)_3{\bullet}18H_2O$ using its molar mass of 342.3 g/mol, and use this to calculate the three concentrations.

a. The moles of $Al_2(SO_4)_3{\bullet}18H_2O$ is calculated below, using the abbreviation "Hyd" for the formula of $Al_2(SO_4)_3{\bullet}18H_2O$.

$$\text{mol Hyd} = 0.1369 \text{ g Hyd} \times \frac{1 \text{ mol Hyd}}{342.3 \text{ g Hyd}} = 0.00039999\underline{4} = 0.0003999 \text{ mol}$$

Note that one mole of $Al_2(SO_4)_3{\bullet}18H_2O$ contains one mole of $Al_2(SO_4)_3$ so calculating the molarity of $Al_2(SO_4)_3$ can be performed using the moles of $Al_2(SO_4)_3{\bullet}18H_2O$.

$$\frac{\text{mol}}{L} = \frac{0.00039994 \text{ mol Hyd}}{1.000 \text{ L soln.}} \times \frac{1 \text{ mol } Al_2(SO_4)_3}{1 \text{ mol Hyd}} = 0.0003999 \; Al_2(SO_4)_3$$

b. The molarity of the SO_4^{2-} ion will be three times that of the $Al_2(SO_4)_3$:

$$0.00039994 \text{ M } Al_2(SO_4)_3 \times \frac{3 \text{ mol } SO_4^{2-}}{1 \text{ mol } Al_2(SO_4)_3} = 0.00119\underline{9}8$$

$$= 0.001200 \text{ M } SO_4^{2-}$$

c. Since the density of the solution is 1.00 g/mL, the mass of 1.000 L of solution is 1000 g, or 1.000 kg. Since molality is moles per 1.000 kg of solvent, the molality of $Al_2(SO_4)_3$ equals 0.0003999 moles divided by 1.000 kg or 0.0003999 m (the same as the molarity).

12.87 In 1.00 kg of a saturated solution of urea, there are 0.44 kg of urea (a molecular solute) and 0.56 kg of water. First convert the mass of urea to moles.

$$0.44 \times 10^3 \text{ g urea} \times \frac{1 \text{ mol urea}}{60.06 \text{ g urea}} = 7.\underline{3}26 \text{ mol urea}$$

Then find the molality of the urea in the solution:

$$\text{Molality} = \frac{\text{mol urea}}{\text{kg } H_2O} \times \frac{7.326 \text{ mol } CaCl_2}{0.56 \text{ kg}} = 1\underline{3}.08 \text{ m}$$

$$\Delta T_f = K_f c_m = (1.858^oC/m)(13.08 \text{ m}) = 2\underline{4}.3^oC$$

$$T_f = 0.0^oC - 24.3^oC = -2\underline{4}.3 = -24^oC$$

12.88 In 1.00 kg of a saturated solution of $CaCl_2$, there are 0.32 kg of $CaCl_2$ (ionic) and 0.68 kg of water. First convert the mass of $CaCl_2$ to moles.

$$0.32 \times 10^3 \text{ g } CaCl_2 \times \frac{1 \text{ mol } CaCl_2}{111.0 \text{ g } CaCl_2} = 2.\underline{8}8 \text{ mol } CaCl_2$$

Then find the molality of the $CaCl_2$ in the solution:

$$\text{Molality} = \frac{\text{mol } CaCl_2}{\text{kg } H_2O} \times \frac{2.88 \text{ mol } CaCl_2}{0.68 \text{ kg}} = 4.\underline{2}4 \text{ m}$$

For ionic solutions,

$$\Delta T_f = K_f i \, c_m$$

Calcium chloride dissolves to give three ions; thus i = 3, and

$$\Delta T_f = (1.858^oC/m)(3)(4.24 \text{ m}) = 2\underline{3}.6^oC$$

$$T_f = 0.0^oC - 2\underline{3}.6^oC = -24^oC$$

12.89 $$M = \frac{\pi}{RT} = \frac{7.7 \text{ atm}}{(0.0821 \text{ L•atm/mol•K})(37 + 273)\text{K}} = 0.3\underline{0}2 = 0.30 \text{ mol/L}$$

12.90 $$M = \frac{\pi}{RT} = \frac{5.61 \text{ atm}}{(0.0821 \text{ L•atm/mol•K})(25 + 273)\text{K}} = 0.22\underline{9}2 = 0.229 \text{ mol/L}$$

12.91 Consider the equation $\Delta T_f = K_f\, i\, c_m$. For $CaCl_2$, i = 3; for glucose i = 1. Because c_m = 0.10 for both solutions, the product of i and c_m will be larger for $CaCl_2$, as will ΔT_f. The solution of $CaCl_2$ will thus have the lower freezing point.

12.92 Consider the equation $\Delta T_b = K_b\, i\, c_m$. For $CaCl_2$, i = 3; for KCl it is 2. The product of i and c_m will be larger for $CaCl_2$, as will ΔT_b. The solution of KCl will thus have the lower boiling point.

12.93 Assume that there is 1.000 L of the solution, which will contain 18 mol H_2SO_4, molar mass 98.09 g/mol. The mass of the solution is

$$\text{Mass solution} = \frac{18 \text{ mol} \times 98.09 \text{ g/mol}}{0.98} = 1\underline{8}02 \text{ g}$$

Thus, the density of the solution is

$$d = \frac{1802 \text{ g}}{1000 \text{ mL}} = 1.\underline{8}02 = 1.8 \text{ g/mL}$$

The mass of water in the solution is

$$\text{mass } H_2O = 1802\text{g} \times 0.02 = \underline{3}6.0\text{g} = 0.0\underline{3}60 \text{ kg}$$

Thus, the molality of the solution is

$$m = \frac{18 \text{ mol } H_2SO_4}{0.0360 \text{ Kg}} = \underline{5}00 = 5 \times 10^2 \text{ m}$$

12.94 Assume that there is 1.000 L of the solution, which will contain 15 mol H_3PO_4, molar mass 97.99 g/mol. The mass of the solution is

$$\text{Mass solution} = \frac{15 \text{ mol} \times 97.99 \text{ g/mol}}{0.85} = 1\underline{7}29 \text{ g}$$

Thus, the density of the solution is

$$d = \frac{1729 \text{ g}}{1000 \text{ mL}} = 1.\underline{7}29 = 1.7 \text{ g/mL}$$

The mass of water in the solution is

$$\text{mass } H_2O = 1729\text{g} \times 0.15 = 2\underline{5}9\text{g} = 0.2\underline{5}9 \text{ kg}$$

(continued)

Thus, the molality of the solution is

$$m = \frac{15 \text{ mol } H_3PO_4}{0.0259 \text{ Kg}} = 5\underline{7}.9 = 58 \text{ m}$$

12.95 Use the freezing point depression equation to find the molality of the solution. The freezing point of pure cyclohexane is 6.55°C and K_f = 20.2°C/m. Thus

$$m = \frac{\Delta T_f}{K_f} = \frac{(6.55 - 5.28)^{\circ}C}{20.2^{\circ}C/m} = 0.062\underline{8}7 \text{ m}$$

The moles of the compound are

$$\text{mol compound} = \frac{0.06287 \text{ mol}}{1 \text{ kg solvent}} \times 0.00538 \text{ kg} = 3.3\underline{8}2 \times 10^{-4} \text{ mol}$$

The molar mass of the compound is

$$\text{molar mass} = \frac{0.125 \text{ g}}{3.382 \times 10^{-4} \text{ mol}} = 36\underline{9}.6 \text{ g/mol}$$

The moles of the elements in 100g of the compound are

$$\text{mol Mn} = 28.17\text{g Mn} \times \frac{1 \text{ mol}}{54.94 \text{ g Mn}} = 0.512\underline{7}4 \text{ mol}$$

$$\text{mol C} = 30.80\text{g C} \times \frac{1 \text{ mol}}{12.01 \text{ g C}} = 2.56\underline{4}5 \text{ mol}$$

$$\text{mol O} = 41.03\text{g O} \times \frac{1 \text{ mol}}{16.00 \text{ g O}} = 2.56\underline{4}4 \text{ mol}$$

This gives mole ratios of 1 mol Mn to 5 mol C to 5 mol O. Therefore, the empirical formula of the compound is MnC_5O_5. The weight of this formula unit is approximately 195 amu. Since the molar mass of the compound is 370. g/mol, the value of n is

$$n = \frac{370 \text{ g/mol}}{195 \text{ g/unit}} = 2.00, \text{ or } 2$$

Therefore the formula of the compound is $Mn_2C_{10}O_{10}$.

12.96 Use the freezing point depression equation to find the molality of the solution. The freezing point of pure cyclohexane is 6.55°C and K_f = 20.2°C/m. Thus

$$m = \frac{\Delta T_f}{K_f} = \frac{(6.55 - 5.23)^{\circ}C}{20.2^{\circ}C/m} = 0.065\underline{3}4 \text{ m}$$

The moles of the compound are

$$\text{mol compound} = \frac{0.06534 \text{ mol}}{1 \text{ kg solvent}} \times 0.00672 \text{ kg} = 4.3\underline{9}1 \times 10^{-4} \text{ mol}$$

The molar mass of the compound is

$$\text{molar mass} = \frac{0.147 \text{ g}}{4.391 \times 10^{-4} \text{ mol}} = 33\underline{4}.8 \text{ g/mol}$$

The moles of the elements in 100g of the compound are

$$\text{mol Co} = 34.47\text{g Co} \times \frac{1 \text{ mol}}{58.93 \text{ g Co}} = 0.584\underline{9}3 \text{ mol}$$

$$\text{mol C} = 28.10\text{g C} \times \frac{1 \text{ mol}}{12.01 \text{ g C}} = 2.33\underline{9}7 \text{ mol}$$

$$\text{mol O} = 37.43\text{g O} \times \frac{1 \text{ mol}}{16.00 \text{ g O}} = 2.33\underline{9}4 \text{ mol}$$

This gives mole ratios of 1 mol Co to 4 mol C to 4 mol O. Therefore, the empirical formula of the compound is CoC_4O_4. The weight of this formula unit is approximately 171 amu. Since the molar mass of the compound is 335. g/mol, the value of n is

$$n = \frac{335 \text{ g/mol}}{171 \text{ g/unit}} = 1.96, \text{ or } 2$$

Therefore the formula of the compound is $Co_2C_8O_8$.

12.97 a. Use the freezing point depression equation to find the molality of the solution. The freezing point of pure water is 0°C and K_f = 1.86°C/m. Thus

$$m = \frac{\Delta T_f}{K_f} = \frac{(0 - (-2.2))^{o}C}{1.86^{o}C/m} = 1.1\underline{8}\ m$$

The moles of the compound are

$$\text{mol compound} = \frac{1.18\ \text{mol}}{1\ \text{kg solvent}} \times 0.100\ \text{kg} = 0.1\underline{1}8\ \text{mol}$$

The molar mass of the compound is

$$\text{molar mass} = \frac{18.0\ \text{g}}{0.118\ \text{mol}} = 1\underline{5}2.5\ \text{g/mol}$$

The moles of the elements in 100g of the compound are

$$\text{mol C} = 48.64\text{g C} \times \frac{1\ \text{mol}}{12.01\ \text{g C}} = 4.05\underline{0}0\ \text{mol}$$

$$\text{mol H} = 8.16\text{g H} \times \frac{1\ \text{mol}}{1.008\ \text{g H}} = 8.0\underline{9}5\ \text{mol}$$

$$\text{mol O} = 43.20\text{g O} \times \frac{1\ \text{mol}}{16.00\ \text{g O}} = 2.70\underline{0}0\ \text{mol}$$

This gives mole ratios of 1.5 mol C to 3 mol H to 1 mol O. Multiplying by two gives ratios of 3 mol C to 6 mol H to 2 mol O. Therefore, the empirical formula of the compound is $C_3H_6O_2$. The weight of this formula unit is approximately 74 amu. Since the molar mass of the compound is 153. g/mol, the value of n is

$$n = \frac{153\ \text{g/mol}}{74\ \text{g/unit}} = 2.06,\ \text{or}\ 2$$

Therefore the formula of the compound is $C_6H_{12}O_4$.

b. The molar mass of the compound is (to the nearest tenth of a gram):

$$6(12.01) + 12(1.008) + 4(16.00) = 148.\underline{1}56 = 148.2\ \text{g/mol}$$

12.98 a. Use the freezing point depression equation to find the molality of the solution. The freezing point of naphthalene is 80.0°C and K_f = 6.8°C/m. Thus

$$m = \frac{\Delta T_f}{K_f} = \frac{(80.0 - 78.0)^{o}C}{6.8^{o}C/m} = 0.29\underline{4}\ m$$

The moles of the compound are

$$\text{mol compound} = \frac{0.294\ \text{mol}}{1\ \text{kg solvent}} \times 0.00750\ \text{kg} = 2.2\underline{1} \times 10^{-3}\ \text{mol}$$

The molar mass of the compound is

$$\text{molar mass} = \frac{0.855\ \text{g}}{2.21 \times 10^{-3}\ \text{mol}} = 3\underline{8}7\ \text{g/mol}$$

The moles of the elements in 100g of the compound are

$$\text{mol C} = 39.50\,\text{g C} \times \frac{1\ \text{mol}}{12.01\ \text{g C}} = 3.28\underline{8}9\ \text{mol}$$

$$\text{mol H} = 2.21\text{g H} \times \frac{1\ \text{mol}}{1.008\ \text{g H}} = 2.1\underline{9}2\ \text{mol}$$

$$\text{mol Cl} = 58.30\,\text{g Cl} \times \frac{1\ \text{mol}}{35.45\ \text{g O}} = 1.644\underline{6}\ \text{mol}$$

This gives mole ratios of 2 mol C to 1.33 mol H to 1 mol Cl. Multiplying by three gives ratios of 6 mol C to 4 mol H to 3 mol Cl. Therefore, the empirical formula of the compound is $C_6H_4Cl_3$. The weight of this formula unit is approximately 182.5 amu. Since the molar mass of the compound is 387. g/mol, the value of n is

$$n = \frac{387\ \text{g/mol}}{182.5\ \text{g/unit}} = 2.12,\ \text{or } 2$$

Therefore the formula of the compound is $C_{12}H_8Cl_6$.

b. The molar mass of the compound is (to the nearest tenth of a gram):

$$12(12.01) + 8(1.008) + 6(35.45) = 364.\underline{8}84 = 364.9\ \text{g/mol}$$

12.99 Use the freezing point depression equation to find the molality of the solution. The freezing point of pure water is 0°C and K_f = 1.86°C/m. Thus

$$m = \frac{\Delta T_f}{K_f} = \frac{(0 - (-2.3))^{\circ}C}{1.86^{\circ}C/m} = 1.2\underline{4}\ m$$

Since this is a relatively dilute solution, we can assume that the molarity and molality of the fish blood are approximately equal. The calculated molarity is for the total number of particles, assuming that they behave ideally.

$$\pi = MRT = 1.2\underline{4}\ mol/L \times 0.08206\ L{\bullet}atm/K{\bullet}mol \times 298.2\ K$$

$$= 3\underline{0}.3 = 30.\ atm$$

12.100 Use the osmotic pressure equation to find the molarity of the solution.

$$M = \frac{\pi}{RT} = \frac{18\ atm}{0.08206\ L \bullet atm/K \bullet mol \times 295.2\ K} = 0.7\underline{4}3\ M$$

Since this is a relatively dilute solution, we can assume that the molality and molarity of the solution are approximately equal, and that the electrolyte(salt) is behaving ideally. The calculated temperature change is

$$\Delta T = K_f m = 1.86^{\circ}C/m \times 0.743\ m = 1.\underline{3}8^{\circ}C$$

Therefore, the solution will freeze at -1.4°C.

■ Cumulative-Skills Problems
(require skills from previous chapters)

12.101 Use the solubility rules in Table 3.1 to predict which ions will precipitate. Since only sodium, potassium, and ammonium carbonates are soluble, it follows that both $CaCO_3$ and Ag_2CO_3 will precipitate according to the following equations (moles written below each). Either equation could be written first; the precipitation of $CaCO_3$ is arbitrarily written first; the same results could be obtained by writing the precipitation of Ag_2CO_3 first and that of $CaCO_3$ second. From the ratios of moles given, it can be seen that the Ca^{2+} ion and the Ag^{+} ion will be precipitated completely, leaving excess CO_3^{2-} ion in solution.

(continued)

	$Ca(NO_3)_2(aq)$ +	$Na_2CO_3(aq)$ →	$2\ NaNO_3(aq)$ +	$CaCO_3(s)$
Before any pptn:	0.125 mol	0.375 mol	0 mol	0 mol
After $CaCO_3$ pptn:	0.000 mol	0.250 mol	0.250 mol	0.125 mol

	$2\ AgNO_3(aq)$ +	$Na_2CO_3(aq)$ →	$2\ NaNO_3(aq)$ +	$Ag_2CO_3(s)$
After $CaCO_3$ pptn:	0.200 mol	0.250 mol	0.250 mol	0 mol
After Ag_2CO_3 pptn:	0.000 mol	0.150 mol	0.450 mol	0.100 mol

The molarities of the various ions are calculated by dividing the moles from the second equation by the final volume of 2.000 L. Note that the moles of Na^+ ion comes from the moles of Na_2CO_3 left as well as the moles of $NaNO_3$ formed.

$$\text{M of } CO_3^{2-} \text{ left} = 0.150 \text{ mol} \div 2.000 \text{ L} = 0.0750 \text{ M}$$

$$\text{M of } NO_3^- \text{ left} = 0.450 \text{ mol} \div 2.000 \text{ L} = 0.225 \text{ M}$$

$$\text{mol } Na^+ \text{ left} = 0.450 \text{ mol (from } NaNO_3) + 2 \times 0.150 \text{ (from } Na_2CO_3) = 0.750 \text{ mol}$$

$$\text{M of } Na^+ \text{ left} = 0.750 \text{ mol} \div 2.000 \text{ L} = 0.375 \text{ M}$$

12.102 Use the solubility rules in Table 3.1 to predict which ions will precipitate. Since only sodium, potassium, and ammonium carbonates are soluble, it follows that both $Ca_3(PO_4)_2$ and Ag_3PO_4 will precipitate according to the following equations (moles written below each). Either equation could be written first; the precipitation of $Ca_3(PO_4)_2$ is arbitrarily written first; the same results could be obtained by writing the precipitation of Ag_3PO_4 first and that of $Ca_3(PO_4)_2$ second. From the ratios of moles given, it can be seen that the Ca^{2+} ion and the Ag^+ ion will be precipitated completely, leaving excess PO_4^{3-} ion in solution.

	$3\ Ca(NO_3)_2(aq)$ +	$2\ Na_3PO_4(aq)$ →	$6\ NaNO_3(aq)$ +	$Ca_3(PO_4)_2(s)$
Before any pptn:	0.100 mol	0.625 mol	0 mol	0 mol
After $Ca_3(PO_4)_2$ pptn:	0.000 mol	$0.55\underline{8}3$ mol	0.200 mol	$0.03\underline{3}3$ mol

(continued)

	$3\ AgNO_3(aq)$ +	$Na_3PO_4(aq)$ →	$3\ NaNO_3(aq)$ +	$Ag_3PO_4(s)$
After $Ca_3(PO_4)_2$ pptn:	0.150 mol	$0.55\underline{8}3$ mol	0.200 mol	0 mol
After Ag_2CO_3 pptn:	0.000 mol	$0.50\underline{8}3$ mol	0.350 mol	0.050 mol

The molarities of the various ions are calculated by dividing the moles from the second equation by the final volume of 4.000 L. Note that the moles of Na^+ ion comes from the moles of Na_3PO_4 left as well as the moles of $NaNO_3$ formed.

$$\text{M of } PO_4^{3-} \text{ left} = 0.50\underline{8}3 \text{ mol} \div 4.000 \text{ L} = 0.12\underline{7}1 = 0.127 \text{ M}$$

$$\text{M of } NO_3^{-} \text{ left} = 0.350 \text{ mol} \div 4.000 \text{ L} = 0.0875 \text{ M}$$

$$\text{mol } Na^{+} \text{ left} = 0.350 \text{ mol (from } NaNO_3) + 3 \times 0.50\underline{8}3 \text{ (from } Na_3PO_4) = 1.8\underline{7}5 \text{ mol}$$

$$\text{M of } Na^{+} \text{ left} = 1.8\underline{7}5 \text{ mol} \div 4.000 \text{ L} = 0.46\underline{8}75 = 0.469 \text{ M}$$

12.103

$Na^+(g) + Cl^-(g)$	→	$NaCl(s)$	$\Delta H = -787$ kJ/mol
$NaCl(s)$	→	$Na^+(aq) + Cl^-(aq)$	$\Delta H = +4$ kJ/mol
$Na^+(g) + Cl^-(g)$	→	$Na^+(aq) + Cl^-(aq)$	$\Delta H = -783$ kJ/mol

The heat of hydration of Na^+ is

$Na^+(g) + Cl^-(g)$	→	$Na^+(aq) + Cl^-(aq)$	$\Delta H = -783$ kJ/mol
$Cl^-(aq)$	→	$Cl^-(g)$	$\Delta H = +338$ kJ/mol
$Na^+(g)$	→	$Na^+(aq)$	$\Delta H = -445$ kJ/mol

12.104

$K^+(g) + Cl^-(g)$	→	$KCl(s)$	$\Delta H = -717$ kJ/mol
$KCl(s)$	→	$K^+(aq) + Cl^-(aq)$	$\Delta H = +18$ kJ/mol
$K^+(g) + Cl^-(g)$	→	$K^+(aq) + Cl^-(aq)$	$\Delta H = -699$ kJ/mol

(continued)

The heat of hydration of K^+ is

$K^+(g) + Cl^-(g)$	$\rightarrow$	$K^+(aq) + Cl^-(aq)$	$\Delta H = -699$ kJ/mol
$Cl^-(aq)$	$\rightarrow$	$Cl^-(g)$	$\Delta H = +338$ kJ/mol
$K^+(g)$	$\rightarrow$	$K^+(aq)$	$\Delta H = -361$ kJ/mol

Less heat is evolved in the hydration of $K^+(g)$ than in that for $Na^+(g)$ because smaller ions have larger hydration energies.

12.105 $$15.0 \text{ g } MgSO_4\bullet7H_2O \times \frac{1 \text{ mol}}{246.5 \text{ g}} = 0.060\underline{8}54 \text{ mol } MgSO_4\bullet7H_2O$$

$$0.060\underline{8}54 \text{ mol } MgSO_4\bullet7H_2O \times \frac{7 \text{ mol } H_2O}{1 \text{ mol hydrate}} \times \frac{18.0 \text{ g } H_2O}{1 \text{ mol } H_2O} = 7.6\underline{6}7 \text{ g } H_2O$$

$$\text{kg } H_2O = (100.0 \text{ g } H_2O + 7.667 \text{ g } H_2O) \times \frac{1 \text{ kg } H_2O}{1000 \text{ g } H_2O} = 0.107\underline{6}6 \text{ kg } H_2O$$

$$m = \frac{0.060854 \text{ mol } MgSO_4}{0.10766 \text{ kg } H_2O} = 0.56\underline{5}2 \text{ mol/ kg} = 0.565 \text{ m}$$

12.106 $$15.0 \text{ g } Na_2CO_3\bullet10H_2O \times \frac{1 \text{ mol}}{286.2 \text{ g}} = 0.052\underline{4}1 \text{ mol } Na_2CO_3\bullet10H_2O \text{ or } Na_2CO_3$$

$$0.052\underline{4}1 \text{ mol } Na_2CO_3\bullet10H_2O \times \frac{10 \text{ mol } H_2O}{1 \text{ mol hydrate}} \times \frac{18.0 \text{ g } H_2O}{1 \text{ mol } H_2O} = 9.4\underline{3}4 \text{ g } H_2O$$

$$\text{kg } H_2O = (100.0 \text{ g } H_2O + 9.434 \text{ g } H_2O) \times \frac{1 \text{ kg } H_2O}{1000 \text{ g } H_2O} = 0.109\underline{4}3 \text{ kg } H_2O$$

$$m = \frac{0.05241 \text{ mol } Na_2CO_3}{0.10943 \text{ kg } H_2O} = 0.47\underline{8}9 \text{ mol/ kg} = 0.479 \text{ m}$$

12.107 $$15.0 \text{ g } CuSO_4\bullet5H_2O \times \frac{1 \text{ mol}}{249.7 \text{ g}} = 0.060\underline{0}7 \text{ mol } CuSO_4\bullet5H_2O \text{ or } CuSO_4$$

$$100 \text{ g soln} \times \frac{1 \text{ mL soln}}{1.167 \text{ g soln}} \times \frac{1 \text{ L}}{1000 \text{ mL}} = 0.0856\underline{8}9 \text{ L}$$

$$M = \frac{0.060\underline{0}7 \text{ mol } CuSO_4}{0.085689 \text{ L}} = 0.70\underline{1}02 = 0.701 \text{ mol/L}$$

12.108 20.0 g Na$_2$S$_2$O$_3 \cdot$5H$_2$O $\times \dfrac{1 \text{ mol}}{248.2 \text{ g}}$ = 0.080$\underline{5}$8 mol Na$_2$S$_2$O$_3 \cdot$5H$_2$O or Na$_2$S$_2$O$_3$

100 g soln $\times \dfrac{1 \text{ mL soln}}{1.174 \text{ g soln}} \times \dfrac{1 \text{ L}}{1000 \text{ mL}}$ = 0.085$\underline{1}$7 L

M = $\dfrac{0.080\underline{5}8 \text{ mol Na}_2\text{S}_2\text{O}_3}{0.08517 \text{ L}}$ = 0.94$\underline{6}$1 = 0.946 mol/L

12.109 0.159°C $\times \dfrac{m}{1.858°C}$ = 0.085$\underline{5}$7 m = $\dfrac{0.08557 \text{ mol AA + H}^+}{1000 \text{ g (or 1000 mL) H}_2\text{O}}$

Note that 0.0830 mol AA + mol H$^+$ = 0.08557 mol/L (AA + H$^+$).

Mol H$^+$ = 0.08557 - 0.0830 = 0.002$\underline{5}$7 mol

% dissoc. = $\dfrac{\text{mol H}^+}{\text{mol AA}}$ (100) = $\dfrac{0.00257 \text{ mol}}{0.0830 \text{ mol}}$ (100) = 3.$\underline{0}$9 = 3.1%

12.110 0.210°C $\times \dfrac{m}{1.858°C}$ = 0.11$\underline{3}$0 m = $\dfrac{0.1130 \text{ mol FA + H}^+}{1000 \text{ g (or 1000 mL) H}_2\text{O}}$

Note that 0.109 mol FA + mol H$^+$ = 0.113 mol/L (FA + H$^+$).

mol H$^+$ = 0.113 - 0.109 = 0.00$\underline{4}$0 mol

% dissoc. = $\dfrac{\text{mol H}^+}{\text{mol FA}}$ (100) = $\dfrac{0.004 \text{ mol}}{0.109 \text{ mol}}$ (100) = $\underline{3}$.66 = 4%

12.111   Calculate the empirical formula first, using the mass of C, O, and H in 1.000 g:

1.434 g CO$_2$ $\times \dfrac{12.01 \text{ g C}}{44.01 \text{ g CO}_2}$ = 0.391$\underline{3}$2 g C

0.783 g H$_2$O $\times \dfrac{2.016 \text{ g H}}{18.016 \text{ g H}_2\text{O}}$ = 0.087$\underline{6}$1 g H

g O = 1.000 g - 0.39132 g - 0.08761 g = 0.52$\underline{1}$1 g O

Mol C = 0.39132 g C $\times$ 1 mol/12.01 g = 0.03258 mol C (lowest integer = 1)

(continued)

Mol H = 0.08761 g H x 1 mol/1.008 g = 0.08691 mol H (lowest integer = 8/3)

Mol O = 0.5211 g O x 1 mol/16.00 g = 0.03257 mol O (lowest integer = 1)

Therefore, the empirical formula is $C_3H_8O_3$. The formula weight from the freezing point is calculated by first finding the molality:

$0.0894^{o}C \times (m/1.858^{o}C) = 0.04811 = (0.048\underline{1}1 \text{ mol}/1000 \text{ g } H_2O)$

$(0.04811 \text{ mol}/1000 \text{ g } H_2O) \times 25.0 \text{ g } H_2O = 0.0012\underline{0}3 \text{ mol (in 25.0 g } H_2O)$

Molar mass = M_m = 0.1107 g/0.001203 mol = 92.02 g/mol

Because this is also the formula weight, the molecular formula is $C_3H_8O_3$ also.

12.112 Calculate the empirical formula first, using the mass of C, O, and H in 1.000 g:

$$1.418 \text{ g } CO_2 \times \frac{12.01 \text{ g C}}{44.01 \text{ g } CO_2} = 0.38\underline{7}0 \text{ g C}$$

$$0.871 \text{ g } H_2O \times \frac{2.016 \text{ g H}}{18.016 \text{ g } H_2O} = 0.097\underline{4}6 \text{ g H}$$

g O = 1.000 g - 0.3870 g - 0.09746 g = $0.51\underline{5}5$ g O

Mol C = 0.3870 g C x 1 mol/12.01 g = 0.0322 mol C (lowest integer = 1)

Mol H = 0.09746 g H x 1 mol/1.008 g = 0.09668 mol H (lowest integer = 3)

Mol O = 0.5155 g O x 1 mol/16.00 g = 0.03222 mol O (lowest integer = 1)

The empirical formula is CH_3O. For the freezing-point calculation, the molality is calculated as follows:

$0.0734^{o}C \times (m/1.858^{o}C) = 0.039505 = (0.39505 \text{ mol}/1000 \text{ g } H_2O)$

$(0.39504 \text{ mol}/1000 \text{ g } H_2O) \times 45.0 \text{ g } H_2O = 0.001778 \text{ mol}$

Molar mass = M_m = 0.1103 g/0.0017778 mol = 62.04 g/mol

Because the formula weight of CH_3O is 31.02, the molecular formula is $C_2H_6O_2$.

■ Solution to Conceptual Problem

12.113 When sunlight is incident on a 2-3 m swimming pool, the light penetrates the transparent water and is absorbed by the bottom of the pool. Under normal conditions, the absorbed solar energy is dissipated as heat through the pool by convection. By adding NaCl to the water, researchers have been able to create light harvesting ponds (one example is at En Bokek near the Dead Sea) that collect the sun's heat at the bottom of the pond. In this way, higher temperatures at the bottom of the pool since the heat is not being distributed through the pool evenly. The energy can be removed by pumping the salt solution through a heat exchanger.

How does the salt solution collect the heat at the bottom of the pool while minimizing convection? If a swimming pool 4 m x 8 m collects 100 kJ of solar power per minute, can solar pond generators rationally replace a 500 megawatt conventional power plant?

Discussion

The water in the pool has a temperature gradient, with cooler water at the top and hotter water at the bottom. When you dissolve salt in the water, a concentration gradient is established, because of the difference in solubility of the salt at different temperatures. Usually, the solubility is greater at the higher temperature. Moreover, the more concentrated solution is generally the more dense. Therefore, the more dense solution is at the bottom of the pool. Normally, when a body of water is heated is at the bottom, the hotter water is less dense and rises. The rising water causes convection currents, which mix the hotter water with the cooler water, so it maintains a uniform temperature. In the case of this salt pool (solar pond), however, the water at the bottom of the pool is more dense, so it does not rise, and normal convection is prevented. As a result, very high temperatures (near 100°C) are possible at the bottom of the pool. The heat energy can be recovered by pumping the hot salt solution through a heat exchanger.

To estimate the feasibility that solar ponds could replace conventional power plants, we first obtain the power obtained from our swimming pool.

$$(100 \text{ kJ/min})(1 \text{ min}/60 \text{ sec}) = 1.67 \text{ kJ/sec} = 1.67 \text{ kW}$$

To obtain 500 MW, we would need 5.0×10^5 kW/1.67 kW = 3.0×10^5 swimming pools. Now, let us convert the area covered by these swimming pools to acres, in order to give us conventional units of land area for comparison purposes. One acre equals 4.05×10^3 m^2 (this conversion factor is available in handbooks).

$$(1 \text{ acre}/4.05 \times 10^3 \text{ m}^2)(4 \text{ m} \times 8 \text{ m/pool})(3.0 \times 10^5 \text{ pools/power plant})$$

$$= 2365 \text{ acres/power plant}$$

(continued)

This estimate indicates that solar ponds are not likely to replace conventional power plants. However, there are perhaps areas available with sufficient sunlight to make solar ponds feasible. The 52 acre pond at En Bokek is a good example and produces 2.5 MW of power.

13. RATES OF REACTION

Solutions to Exercises

Note on significant figures: The final answer to each mathematical solution is given first with one nonsignificant figure (the rightmost significant figure is underlined) and then is rounded to the correct number of figures. Intermediate answers usually also have at least one nonsignificant figure.

13.1 Rate of formation of NO_2F = $\Delta[NO_2F]/\Delta t$. Rate of reaction of NO_2 = $-\Delta[NO_2]/\Delta t$. Divide each rate by the coefficient of the corresponding substance in the equation:

$$1/2\,\frac{\Delta[NO_2F]}{\Delta t} = -1/2\,\frac{\Delta[NO_2]}{\Delta t};\ \text{or}\ \frac{\Delta[NO_2F]}{\Delta t} = -\frac{\Delta[NO_2]}{\Delta t}$$

13.2 $$\text{Rate} = -\frac{\Delta[I^-]}{\Delta t} = -\frac{[0.00101\ M - 0.00169\ M]}{8.00\ s - 2.00\ s} = 1.\underline{1}3 \times 10^{-4} = 1.1 \times 10^{-4}\ M/s$$

13.3 The order with respect to CO is 0, and with respect to NO_2 is 2. The overall order = 2, the sum of the exponents in the rate law.

13.4 By comparing experiments 1 and 2, you see that the rate is quadrupled when the $[NO_2]$ is doubled. Thus, the reaction is second order in NO_2, and the rate law is

$$\text{Rate} = k[NO_2]^2$$

(continued)

The rate constant may be found by substituting experimental values into the rate-law expression. Based on values from Experiment 1,

$$k = \frac{\text{rate}}{[NO_2]^2} = \frac{7.1 \times 10^{-5}\ \text{mol/(L•s)}}{(0.010\ \text{mol/L})^2} = 0.7\underline{1}0 = 0.71\ \text{L/(mol•s)}$$

13.5 a. For $[N_2O_5]$ after 6.00×10^2 s, use the first-order rate law and solve for the concentration at time t:

$$\log \frac{[N_2O_5]_t}{[1.65 \times 10^{-2}\ M]} = \frac{-(4.80 \times 10^{-4}/\text{s})(6.00 \times 10^2\ \text{s})}{2.303}$$

$$\log\ [N_2O_5]_t = \log\ (1.65 \times 10^{-2}\ M) - \frac{(4.80 \times 10^{-4}/\text{s})(6.00 \times 10^2\ \text{s})}{2.303}$$

$$\log\ [N_2O_5]_t = -1.78\underline{2}5 - 0.12\underline{5}05 = -1.90\underline{7}6$$

$$[N_2O_5]_t = 1.2\underline{3}7 \times 10^{-2} = 0.0124\ \text{mol/L}$$

b. Use the first-order rate law and solve for time t. Let the initial concentration be 100.0%, and let the concentration at time t be 10.0%.

$$\log \frac{(10.0\%)}{(100.0\%)} = \frac{-(4.80 \times 10^{-4}\ /\text{s})t}{2.303}$$

$$t = \frac{-(2.303)(\log 0.100)}{4.80 \times 10^{-4}/\text{s}} = 4.7\underline{9}7 \times 10^3\ \text{s, or } 1.33\ \text{hr}$$

13.6 Use the expression relating the half-life and the rate constant to calculate the half-life, $t_{1/2}$.

$$t_{1/2} = \frac{0.693}{k} = \frac{0.693}{9.2/\text{s}} = 0.07\underline{5}3 = 0.075\ \text{s}$$

By definition, the half-life is the amount of time it takes to decrease the amount of substance present by one-half. Thus, it takes $0.07\underline{5}3$ s for concentration to decrease by 50%, and another $0.07\underline{5}3$ s for the concentration to decrease by 50% of the remaining 50% (to 25% left), for a total of $0.15\underline{0}6$, or 0.151, s.

13.7 Solve for E_a by substituting the given values into the two-temperature Arrhenius equation:

$$\log \frac{2.14 \times 10^{-2}}{1.05 \times 10^{-3}} = \frac{E_a}{2.303 \times 8.31\ \text{J/(K•mol)}} \left[\frac{1}{759\ \text{K}} - \frac{1}{836\ \text{K}}\right]$$

$$1.30\underline{9}20 = \frac{E_a}{19.\underline{1}38\ \text{J/(K•mol)}} \left[1.2\underline{1}35 \times 10^{-4}/\text{K}\right]$$

$$E_a = \frac{1.30920 \times 19.138\ \text{J/(K•mol)}}{1.2135 \times 10^{-4}/\text{K}} = 2.0\underline{6}5 \times 10^5 = 2.07 \times 10^5\ \text{J/mol}$$

Solve for the rate constant, k_2, at 865 K by using the same equation and using $E_a = 2.065 \times 10^5$ J/mol:

$$\log \frac{k_2}{2.14 \times 10^{-2}/(\text{M}^{1/2}\text{•s})} = \frac{2.065 \times 10^5\ \text{J/mol}}{2.303 \times 8.31\ \text{J/(K•mol)}} \left[\frac{1}{836\ \text{K}} - \frac{1}{865\ \text{K}}\right]$$

$$= 0.43\underline{2}71$$

Now calculate the antilog of 0.43271:

$$\frac{k_2}{2.14 \times 10^{-2}/(\text{M}^{1/2}\text{•s})} = 2.7\underline{0}84$$

$$k_2 = 5.7\underline{9}6 \times 10^{-2} = 5.80 \times 10^{-2}/(\text{M}^{1/2}\text{•s})$$

13.8 The net chemical equation is the overall sum of the two elementary reactions:

$$H_2O_2 + I^- \rightarrow H_2O + IO^-$$

$$H_2O_2 + IO^- \rightarrow H_2O + O_2 + I^-$$

$$2H_2O_2 \rightarrow 2H_2O + O_2$$

The IO^- is an intermediate; I^- is a catalyst. Neither of these appears in the net equation.

13.9 The reaction is bimolecular because it is an elementary reaction that involves two molecules.

13.10 For $NO_2 + NO_2 \rightarrow N_2O_4$, the rate law is

$$\text{Rate} = k[NO_2]^2$$

(The rate must be proportional to the concentration of both reactant molecules.)

13.11 The slow step is the rate-determining step. Therefore, the rate law predicted by the mechanism given is

$$\text{Rate} = k_1 [H_2O_2][I^-]$$

13.12 According to the rate-determining (slow) step, the rate law is

$$\text{Rate} = k_2[NO_3][NO]$$

Because NO_3 is an intermediate, it cannot appear in the overall equation. You can eliminate it by substituting a mathematical equality for it. Note that $[NO_3]$ appears in the first step, a fast equilibrium step. At equilibrium, the ratio k_1/k_{-1} is equal to the ratio of product over reactants:

$$\frac{k_1}{k_{-1}} = \frac{[NO_3]}{[NO][O_2]}$$

Rearranging to obtain an equality for $[NO_3]$ gives

$$[NO_3] = k_1/k_{-1}\ [NO][O_2]$$

Substituting the right-hand side into the rate law for $[NO_3]$ gives

$$\text{Rate} = k_2(k_1/k_{-1})\ [O_2][NO]^2$$

The $k_2(k_1/k_{-1})$ product is equal to the observed rate constant, k. Note that k_2 is a contribution from step 2, whereas the (k_1/k_{-1}) ratio is a contribution from step 1.

■ Answers to Review Questions

13.1 The four variables that can affect rate are (1) the concentrations of the reactants, although in some cases a particular reactant's concentration does not affect the rate; (2) the presence and concentration of a catalyst; (3) the temperature of the reaction; and (4) the surface area of any solid reactant or solid catalyst.

13.2 The rate of reaction of HBr can be defined as the decrease in HBr concentration (or the increase in Br_2 product formed) over the time interval Δt:

$$\text{Rate} = -1/4\,\frac{\Delta[\text{HBr}]}{\Delta t} = 1/2\,\frac{\Delta[\text{Br}_2]}{\Delta t} \quad \text{or} \quad -\frac{\Delta[\text{HBr}]}{\Delta t} = 2\,\frac{\Delta[\text{Br}_2]}{\Delta t}$$

13.3 Two physical properties used to determine the rate are color, or absorption of electromagnetic radiation, and pressure. If a reactant or product is colored, or absorbs a different type of electromagnetic radiation than the other species, then measurement of the change in color, or change in absorption of electromagnetic radiation, may be used to determine the rate. If a gas reaction involves a change in the number of gaseous molecules, then measurement of the pressure change may be used to determine the rate.

13.4 Use the general example of a rate law in Section 13.3, where A and B react to give D and E with C as a solid catalyst:

$$\text{Rate} = k\,[\text{A}]^m\,[\text{B}]^n\,[\text{C(s)}]^p$$

Note that the exponents m, n, and p are the orders of the individual reactants and catalyst. Assuming that m and n are positive numbers, the rate law predicts that increasing the concentrations of A and/or B will increase the rate. In addition, the rate will be increased by increasing the surface of the solid catalyst (making it as finely divided as possible, etc.). Finally, increasing the temperature will increase the rate constant, k, and increase the rate.

13.5 An example that illustrates that exponents have no relationship to coefficients is the reaction of nitric oxide and hydrogen from Example 13.12:

$$2\text{NO} + 2\text{H}_2 \rightarrow \text{N}_2 + 2\text{H}_2\text{O}.$$

The experimental rate law given there is

$$\text{Rate} = k[\text{NO}]^2\,[\text{H}_2]$$

Thus, the exponent for hydrogen is 1, not 2 like the coefficient, and the overall order is 3, not 4 like the sum of the coefficients.

13.6 The rate law for this reaction of iodide ion, arsenic acid, and hydrogen ion is

$$\text{Rate} = k[\text{I}^-]\,[\text{H}_3\text{AsO}_4]\,[\text{H}^+]$$

The overall order is 1 + 1 + 1 = 3 (third order).

13.7 Use m to symbolize the reaction order as is done in the text. Then from the table for m and the change in rate in the text, m = 2 when the rate is quadrupled (increased fourfold). Using the equation in the text gives the same result:

$$2^m = \text{new rate/old rate} = 4/1;\ \text{thus, } m = 2$$

13.8 Use m to symbolize the reaction order as is done in the text. The table for m and the change in rate in the text cannot be used in this case. When m = 0.5, the new rate should be found using the equation in the text:

$$2^{0.50} = \sqrt{2} = 1.41 = \text{new rate/old rate}$$

Thus, the new rate is 1.41 times the old rate.

13.9 Use the half-life concept to answer the question without an equation. If the half-life for the reaction of A(g) is 25 s, then the time for A(g) to decrease to 1/4 the initial value is 2 half-lives or 2 x 25 = 50 s. The time for A(g) to decrease to 1/8 the initial value is 3 half-lives, or 3 x 25 = 75 s.

13.10 The half-life of a first-order reaction is constant over the course of the reaction. The half-life of a second-order reaction depends on the initial concentration and becomes larger as time elapses. Thus, the reaction must be second order because the half-life increases from 20 s to 40 s after time has elapsed.

13.11 According to transition-state theory, the two factors that determine whether a collision results in reaction or not are (1) the molecules must collide with the proper orientation to form the activated complex, and (2) the activated complex formed must have a kinetic energy greater than the activation energy.

13.12 The potential-energy diagram for the exothermic reaction of A and B to give activated complex $AB^{‡}$ and products C and D is given below.

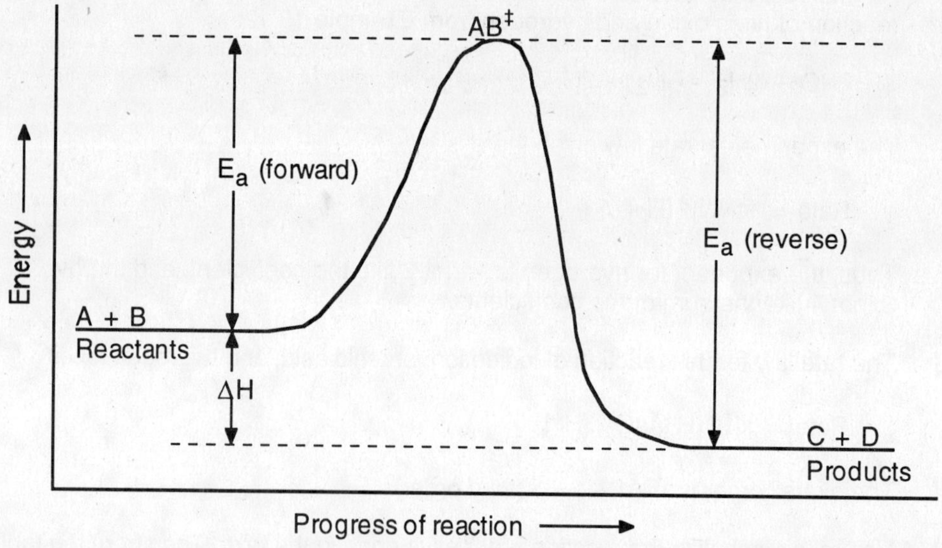

13.13 The activated complex for the reaction of NO_2 with NO_3 to give NO, NO_2, and O_2 has the structure below (dashed lines = bonds about to form or break):

O=N---O---O---N(=O)$_2$

13.14 The Arrhenius equation expressed with the base e is

$$k = Ae^{-Ea/RT}$$

With the base 10, the Arrhenius equation is

$$k = A\,10^{-Ea/(2.303RT)}$$

The A term is the frequency factor and is equal to the product of p and Z from collision theory. The term p is the fraction of collisions with reactant molecules properly oriented, and Z is the frequency of collisions. Thus, A is the number of collisions with the molecules properly oriented. The E_a term is the activation energy, the minimum energy of collision required for two molecules to react. The R term is the gas constant and T is the absolute temperature.

13.15 In the reaction of $NO_2(g)$ with CO(g), an example of an intermediate is the temporary formation of NO_3 from the reaction of two NO_2 molecules in the first step:

$$NO_2 + NO_2 \rightarrow NO_3 + NO$$

$$NO_3 + CO \rightarrow NO_2 + CO_2$$

13.16 It is generally impossible to predict the rate law from the equation alone because most reactions consist of several elementary steps whose combined result is summarized in the rate law. If these elementary steps are unknown, the rate law cannot be predicted.

13.17 The mechanism cannot be $2NO_2Cl \rightarrow 2NO_2 + Cl_2$ because the reaction is first order in NO_2Cl. The order in NO_2Cl would have to reflect the total number of molecules (two) for the proposed mechanism, but it does not.

13.18 The characteristic of the rate-determining step in a mechanism is that it is, relatively speaking, the slowest step of all of the elementary steps (even though it may occur in seconds). Thus, the rate of disappearance of reactant(s) is limited by the rate of this step.

13.19 For the rate of decomposition of N_2O_4,

$$\text{Rate} = k_1[N_2O_4]$$

For the rate of formation of N_2O_4,

$$\text{Rate} = k_{-1}[NO_2]^2$$

(continued)

At equilibrium the rates are equal, so

$$k_1[N_2O_4] = k_{-1}[NO_2]^2$$

$$[N_2O_4] = (k_{-1}/k_1)\,[NO_2]^2$$

13.20 A catalyst operates by providing a pathway (mechanism) that occurs faster than the uncatalyzed pathway (mechanism) of the reaction. The catalyst is not consumed because after reacting in an early step, it is regenerated in a later step.

13.21 In physical adsorption, molecules adhere to a surface through *weak* intermediate forces, whereas in chemisorption, the molecules adhere to the surface by *stronger* chemical bonding.

13.22 In the first step of catalytic hydrogenation of ethylene, the ethylene and hydrogen molecules diffuse to the catalyst surface and undergo chemisorption. Then the pi electrons of ethylene form temporary bonds to the metal catalyst and the hydrogen molecule breaks into two hydrogen atoms. The hydrogen atoms next migrate to an ethylene held in position on the metal catalyst surface, forming ethane. Finally, because it cannot bond to the catalyst, the ethane diffuses away from the surface.

■ Solutions to Practice Problems

Note on significant figures: The final answer to each problem is given first with one nonsignificant figure (the rightmost significant figure is underlined) and then is rounded to the correct number of figures. Intermediate answers usually also have at least one nonsignificant figure.

13.23 For the reaction $2NO_2 \rightarrow 2NO + O_2$, the rate of decomposition of NO_2 and the rate of formation of O_2 are, respectively,

$$\text{Rate} = -\Delta[NO_2]/\Delta t$$

$$\text{Rate} = \Delta[O_2]/\Delta t$$

To relate the two rates, divide each rate by the coefficient of the corresponding substance in the chemical equation and equate them.

$$-1/2\,\frac{\Delta[NO_2]}{\Delta t} = \frac{\Delta[O_2]}{\Delta t}$$

13.24 For the reaction $H_2 + I_2 \rightarrow 2HI$, the rate of formation of HI and the rate of decomposition of I_2 are, respectively,

$$\text{Rate} = \Delta[HI]/\Delta t$$

$$\text{Rate} = -\Delta[I_2]/\Delta t$$

To relate the two rates, divide each rate by the coefficient of the corresponding substance in the chemical equation and equate them.

$$1/2\,\frac{\Delta[HI]}{\Delta t} = -\frac{\Delta[I_2]}{\Delta t}$$

13.25 For the reaction $5Br^- + BrO_3^- + 6H^+ \rightarrow 3Br_2 + 3H_2O$, the rate of decomposition of Br^- and the rate of decomposition of BrO_3^- are, respectively,

$$\text{Rate} = -\Delta[Br^-]/\Delta t$$

$$\text{Rate} = -\Delta[BrO_3^-]/\Delta t$$

To relate the two rates, divide each rate by the coefficient of the corresponding substance in the chemical equation and equate them.

$$1/5\,\frac{\Delta[Br^-]}{\Delta t} = \frac{\Delta[BrO_3^-]}{\Delta t}$$

13.26 For the reaction $3I^- + H_3AsO_4 + 2H^+ \rightarrow I_3^- + H_3AsO_3 + H_2O$, the rate of decomposition of I^- and the rate of formation of I_3^- are, respectively,

$$\text{Rate} = -\Delta[I^-]/\Delta t$$

$$\text{Rate} = \Delta[I_3^-]/\Delta t$$

To relate the two rates, divide each rate by the coefficient of the corresponding substance in the chemical equation and equate them.

$$-1/3\,\frac{\Delta[I^-]}{\Delta t} = \frac{\Delta[I_3^-]}{\Delta t}$$

13.27 $$\text{Rate} = -\frac{\Delta[NH_4NO_2]}{\Delta t} = -\frac{[0.432\text{ M} - 0.500\text{ M}]}{3.00\text{ hr} - 0.00\text{ hr}} = 0.02\underline{2}7 = 0.023\text{ M/hr}$$

13.28 $\text{Rate} = -\dfrac{\Delta[FeCl_3]}{\Delta t} = -\dfrac{[0.02638\text{ M} - 0.03586\text{ M}]}{4.00\text{ m} - 0.00\text{ m}} = 0.0023\underline{7}0$

$= 0.00237\text{ M/min}$

13.29 $\text{Rate} = -\dfrac{\Delta[\text{Azo.}]}{\Delta t} = -\dfrac{[0.0129\text{ M} - 0.0150\text{ M}]}{10.00\text{ min} - 0.00\text{ min}} \times \dfrac{1\text{ min}}{60\text{ s}} = 3.\underline{5}0 \times 10^{-6}$

$= 3.5 \times 10^{-6}\text{ M/s}$

13.30 $\text{Rate} = -\dfrac{\Delta[NO_2]}{\Delta t} = -\dfrac{[0.1076\text{ M} - 0.1103\text{ M}]}{60.0\text{ s} - 0.0\text{ s}} = 4.\underline{5}0 \times 10^{-5} = 4.5 \times 10^{-5}\text{ M/s}$

13.31 If the rate law is rate = $k[H_2S][Cl_2]$, the order with respect to H_2S is 1 (first order), and the order with respect to Cl_2 is also 1 (first order). The overall order is 1 + 1 = 2, second order.

13.32 If the rate law is rate = $k[NO]^2[Cl_2]$, the order with respect to NO is 2 (second order), and the order with respect to Cl_2 is 1 (first order). The overall order is 2 + 1 = 3, third order.

13.33 If the rate law is rate = $k[MnO_4^-][H_2C_2O_4]$, the order with respect to MnO_4^- is 1 (first order), the order with respect to $H_2C_2O_4$ is 1 (first order), and the order with respect to H^+ is 0. The overall order is 2, second order.

13.34 If the rate law is rate = $k[H_2O_2][Fe^{2+}]$, the order with respect to H_2O_2 is 1 (first order), the order with respect to Fe^{2+} is 1 (first order), and the order with respect to H^+ is 0. The overall order is 2, second order.

13.35 The reaction rate doubles when the concentration of CH_3NNCH_3 is doubled, so the reaction is first order in azomethane. The rate equation should have the form

$$\text{Rate} = k[CH_3NNCH_3]$$

Substituting values for the rate and concentration yields a value for k:

$$k = \frac{\text{rate}}{[\text{Azo.}]} = \frac{2.8 \times 10^{-6}\text{ M/s}}{1.13 \times 10^{-2}\text{ M}} = 2.\underline{4}7 \times 10^{-4} = 2.5 \times 10^{-4}/\text{s}$$

13.36 The reaction rate doubles when the concentration of ethylene oxide is doubled, so the reaction is first order in ethylene oxide. The rate equation should have the form

$$\text{Rate} = k[C_2H_4O]$$

Substituting values for the rate and concentration yields a value for k:

$$k = \frac{\text{rate}}{[\text{Et. Ox.}]} = \frac{5.57 \times 10^{-7}\ M/s}{2.72 \times 10^{-3}\ M} = 2.0\underline{4}7 \times 10^{-4} = 2.05 \times 10^{-4}/s$$

13.37 Doubling [NO] quadruples the rate, so the reaction is second order in NO. Doubling $[H_2]$ doubles the rate, so the reaction is first order in H_2. The rate law should have the form

$$\text{Rate} = k[NO]^2[H_2]$$

Substituting values for the rate and concentrations yields a value for k:

$$k = \frac{\text{rate}}{[NO]^2[H_2]} = \frac{2.6 \times 10^{-5}\ M/s}{[6.4 \times 10^{-3}\ M]^2[2.2 \times 10^{-3}]} = 2.\underline{8}8 \times 10^2 = 2.9 \times 10^2/M^2s$$

13.38 Doubling [NO] quadruples the rate, so the reaction is second order in NO. Doubling $[O_2]$ doubles the rate, so the reaction is first order in O_2. The rate law should have the form

$$\text{Rate} = k[NO]^2[O_2]$$

Substituting values for the rate and concentrations yields a value for k:

$$k = \frac{\text{rate}}{[NO]^2[O_2]} = \frac{0.0281\ M/s}{[0.0125\ M]^2[0.0253\ M]} = 7.1\underline{0}8 \times 10^3 = 7.11 \times 10^3/M^2s$$

13.39 By comparing experiments 1 and 2, you see that tripling $[ClO_2]$ increases the rate ninefold; that is, $3^m = 9$, so $m = 2$ (and the reaction is second order in ClO_2). From experiments 2 and 3, you see that tripling $[OH^-]$ triples the rate, so the reaction is first order in OH^-. The rate law is

$$\text{Rate} = k[ClO_2]^2[OH^-]$$

Substituting values for the rate and concentrations yields a value for k:

$$k = \frac{\text{rate}}{[ClO_2]^2[OH^-]} = \frac{0.0248\ M/s}{[0.060\ M]^2[0.030\ M]} = 2.\underline{2}9 \times 10^2 = 2.3 \times 10^2/M^2s$$

13.40 By comparing experiments 2 and 3, you see that doubling $[I^-]$ doubles the rate, so the reaction is first order in I^-. From experiments 1 and 3, you see that doubling $[ClO^-]$ also doubles the rate, so the reaction is first order in ClO^-. From experiments 3 and 4, you see that doubling $[OH^-]$ halves the rate; that is, $2^m = 1/2$. Hence, m = -1, and the rate is inversely proportional to the first power of OH^-. The rate law is

$$\text{Rate} = k[I^-][ClO^-]/[OH^-]$$

Substituting values for the rate and concentrations yields a value for k:

$$k = \frac{\text{rate }[OH^-]}{[ClO^-][I^-]} = \frac{6.1 \times 10^{-2}\ M/s[0.010\ M]}{[0.010\ M]\ [0.010\ M]} = 6.\underline{1} = 6.1/s$$

13.41 Let $[SO_2Cl_2]_o$ = 0.0248 M and $[SO_2Cl_2]_t$ = the concentration after 4.5 hr. Substituting these and $k = 2.2 \times 10^{-5}/s$ into the first-order rate equation gives

$$\log \frac{[SO_2Cl_2]_t}{[0.0248\ M]} = \frac{-(2.2 \times 10^{-5}/s)\left(4.5\ hr \times \frac{3600\ s}{1\ hr}\right)}{2.303} = -0.1\underline{5}47$$

Taking the antilogs of both sides gives

$$\frac{[SO_2Cl_2]_t}{[0.0248\ M]} = 0.7\underline{0}03$$

$$[SO_2Cl_2] = 0.7003 \times [0.0248\ M] = 1.\underline{7}3 \times 10^{-2} = 1.7 \times 10^{-2}\ M$$

13.42 Let $[C_3H_6]_o$ = 0.0226 M and $[C_3H_6]_t$ = the concentration after 955 s. Substituting these concentrations and $k = 6.0 \times 10^{-4}/s$ into the first-order rate equation gives

$$\log \frac{[C_3H_6]_t}{[0.0226\ M]} = \frac{-(6.0 \times 10^{-4}/s)(955\ s)}{2.303} = -0.2\underline{4}88$$

Taking the antilogs of both sides gives

$$\frac{[C_3H_6]_t}{[0.0226\ M]} = 0.5\underline{6}38$$

$$[C_3H_6]_t = 0.5638 \times [0.0226\ M] = 1.\underline{2}7 \times 10^{-2} = 1.3 \times 10^{-2}\ M$$

13.43 First find the rate constant, k, by substituting experimental values into the first-order rate equation. Let $[Et.\ Cl.]_o = 0.00100$ M, $[Et.\ Cl.]_t = 0.00067$ M, and t = 155 s. Solving for k yields

$$k = \frac{-2.303 \log \dfrac{[0.00067\ M]_t}{[0.00100\ M]_o}}{155\ s} = 2.5\underline{8}4 \times 10^{-3}/s$$

Now let $[Et.\ Cl.]_t$ = the concentration after 256 s, $[Et.\ Cl.]_o$ again = 0.00100 M, and use the value of k of $1.5\underline{6}4 \times 10^{-3}/s$ to calculate $[Et.\ Cl.]_t$.

$$\log \frac{[Et.\ Cl.]_t}{[0.00100\ M]} = \frac{-(2.584 \times 10^{-3}/s)(256\ s)}{2.303} = -0.2\underline{8}72$$

Converting both sides to antilogs gives

$$\frac{[Et.\ Cl.]_t}{[0.00100\ M]} = 0.5\underline{1}61$$

$$[Et.\ Cl.]_t = 0.5161 \times [0.00100\ M] = 5.\underline{1}6 \times 10^{-4} = 5.2 \times 10^{-4}\ M$$

13.44 First find the rate constant, k, by substituting experimental values into the first-order rate equation. Let $[Cyb.]_o = 0.00150$ M, $[Cyb.]_t = 0.00119$ M, and t = 455 s. Solving for k yields

$$k = \frac{-2.303 \log \dfrac{[0.00119\ M]_t}{[0.00150\ M]_o}}{455\ s} = 5.0\underline{8}9 \times 10^{-4}/s$$

Now let $[Cyb.]_t$ = the concentration after 968 s, $[Cyb.]_o$ again = 0.00150 M, and use the value of k of $5.089 \times 10^{-4}/s$ to calculate $[Cyb.]_t$.

$$\log \frac{[Cyb.]_t}{[0.00150\ M]} = \frac{-(5.089 \times 10^{-4}/s)(968\ s)}{2.303} = -0.21\underline{3}9$$

Taking antilogs of both sides yields

$$\frac{[Cyb.]_t}{[0.00150\ M]} = 0.61\underline{1}08$$

$$[Cyb.]_t = 0.61108 \times [0.00150\ M] = 9.1\underline{6}6 \times 10^{-4}\ M = 9.17 \times 10^{-4}\ M$$

13.45 For a first-order reaction, divide 0.693 by the rate constant to find the half-life:

$t_{1/2} = 0.693/(6.3 \times 10^{-4}/s) = 1.\underline{1}0 \times 10^3 = 1.1 \times 10^3$ s (1$\underline{8}$.3 min)

$t_{25\%\text{ left}} = t_{1/4\text{ left}} = 2 \times t_{1/2} = 2 \times (1.1 \times 10^3\text{ s}) = 2.2 \times 10^3$ s (3$\underline{6}$.6 min)

$t_{12.5\%\text{ left}} = t_{1/8\text{ left}} = 3 \times t_{1/2} = 3 \times (1.1 \times 10^3\text{ s}) = 3.3 \times 10^3$ s (55 min)

13.46 For a first-order reaction, divide 0.693 by the rate constant to find the half-life:

$t_{1/2} = 0.693/(6.2 \times 10^{-4}/\text{min}) = 1.\underline{1}17 \times 10^3 = 1.1 \times 10^3$ min (1$\underline{8}$.6 hr)

$t_{25\%\text{ left}} = t_{1/4\text{ left}} = 2 \times t_{1/2} = 2 \times (1.117 \times 10^3\text{ min})$

$= 2.\underline{2}35 \times 10^3 = 2.2 \times 10^3$ min

$t_{12.5\%\text{ left}} = t_{1/8\text{ left}} = 3 \times t_{1/2} = 3 \times (1.117 \times 10^3\text{ min})$

$= 3.\underline{3}53 \times 10^3 = 3.4 \times 10^3$ min

13.47 For a first-order reaction, divide 0.693 by the rate constant to find the half-life:

$t_{1/2} = 0.693/(2.0 \times 10^{-6}/s) = 3.\underline{4}65 \times 10^5$ s (9$\underline{6}$.25 or 96 hr)

$t_{25\%\text{ left}} = t_{1/4\text{ left}} = 2 \times t_{1/2} = 2 \times 96.25\text{ hr} = 1\underline{9}2.5 = 1.9 \times 10^2$ hr

$t_{12.5\%\text{ left}} = t_{1/8\text{ left}} = 3 \times t_{1/2} = 3 \times 96.25\text{ hr} = 2\underline{8}8.75 = 2.9 \times 10^2$ hr

$t_{6.25\%\text{ left}} = t_{1/16\text{ left}} = 4 \times t_{1/2} = 4 \times 96.25\text{ hr} = 3\underline{8}5.0 = 3.9 \times 10^2$ hr

$t_{3.125\%\text{ left}} = t_{1/32\text{ left}} = 5 \times t_{1/2} = 5 \times 96.25\text{ hr} = 4\underline{8}1.25 = 4.8 \times 10^2$ hr

13.48 For a first-order reaction, divide 0.693 by the rate constant to find the half-life:

$t_{1/2} = 0.693/(1.27/s) = 0.54\underline{5}6$ s

$t_{25\%\text{ left}} = t_{1/4\text{ left}} = 2 \times t_{1/2} = 2 \times 0.5456\text{ s} = 1.0\underline{9}1 = 1.09$ s

$t_{12.5\%\text{ left}} = t_{1/8\text{ left}} = 3 \times t_{1/2} = 3 \times 0.5456\text{ s} = 1.6\underline{3}6 = 1.64$ s

$t_{6.25\%\text{ left}} = t_{1/16\text{ left}} = 4 \times t_{1/2} = 4 \times 0.5456\text{ s} = 2.1\underline{8}2 = 2.18$ s

$t_{3.125\%\text{ left}} = t_{1/32\text{ left}} = 5 \times t_{1/2} = 5 \times 0.5456\text{ s} = 2.7\underline{2}8 = 2.73$ s

13.49 Use the first-order rate equation and solve for time t. Let $[Cr^{3+}]_o$ = 100.0%; then the concentration at time t, $[Cr^{3+}]_t$, = (100.0% - 90.0%, or 10.0%). Use k = 2.0 x 10^{-6}/s.

$$\log \frac{[10.0\%]}{[100.0\%]} = \frac{-(2.0 \times 10^{-6}/\text{s})\,t}{2.303}$$

$$t = \frac{-(2.303)(\log 0.100)}{2.0 \times 10^{-6}/\text{s}} = 1.\underline{1}51 \times 10^{6}\text{ s, or } 3.2 \times 10^{2}\text{ hr}$$

13.50 Use the first-order rate equation and solve for time t. Let $[Fe^{3+}]_o$ = 100.0%; then the concentration at time t, $[Fe^{3+}]_t$, = (100.0% - 90.0%, or 10.0%). Use k = 1.27/s.

$$\log \frac{[10.0\%]}{[100.0\%]} = \frac{-(1.27/\text{s})\,t}{2.303}$$

$$t = \frac{-(2.303)(\log 0.100)}{1.27/\text{s}} = 1.8\underline{1}3 = 1.81\text{ s}$$

13.51 For the first-order plot, follow Figure 13.9 and plot log $[ClO_2]$ versus the time in seconds. The data used for plotting are

t. min	log $[ClO_2]$
0.00	-3.321
1.00	-3.365
2.00	-3.408
3.00	-3.452

The plot requires a graph with too many lines to be reproduced here, but it yields a straight line, demonstrating that the reaction is first order in $[ClO_2]$. The slope of the line may be calculated from the difference between the last point and the first point:

$$\text{Slope} = \frac{[-3.452 - (-3.321)]}{[3.00 - 0.00\text{ s}]} = \frac{-0.043666}{\text{s}}$$

Just as the slope, m, was obtained for the plot in Figure 13.9, you can also equate m to -k/2.303 and calculate k as follows:

$$k = -2.303\text{ m} = -2.303 \times (-0.04366/\text{s}) = 0.10\underline{0}56 = 0.101/\text{s}$$

13.52 For the first-order plot, follow Figure 13.9 and plot log [MA], the methyl acetate concentration, versus time in minutes. This plot does not yield a straight line, so the reaction is not first order. For the second-order plot, follow Figure 13.10 and plot 1/[MA] versus time in minutes. The data used for plotting are

Time, min	1/[MA]
0.00	100.0
3.00	135.1
4.00	146.4
5.00	157.7

The plot requires a graph with too many lines to be reproduced here, but it yields a straight line, demonstrating that the reaction is second order in [MA]. The slope of the line may be calculated from the difference between the last point and the first point:

$$\text{Slope} = \frac{[157.7 - 100.0]/M}{[5.00 - 0.00]\ \text{min}} = \frac{11.54}{M \bullet \text{min}}, \text{ or } \frac{0.192}{M \bullet s}$$

In this case, the slope equals the rate constant, so k = 11.5/(M•m), or 0.192/(M•s).

13.53 The potential-energy diagram is below. Because the activation energy for the forward reaction is +10 kJ and ΔH^o = -200 kJ, the activation energy for the reverse reaction is +210 kJ.

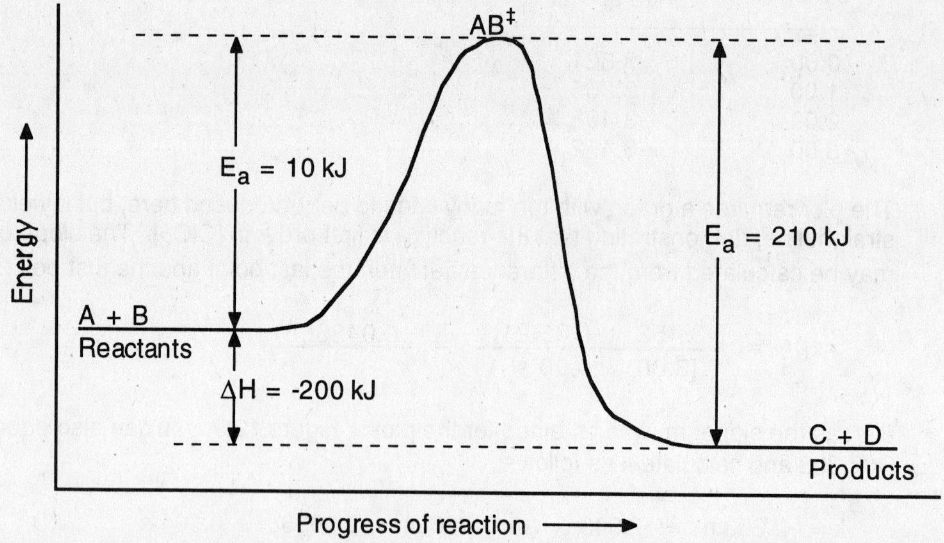

13.54 The potential-energy diagram is below. Because the activation energy for the forward reaction is +251 kJ and ΔH^o is +167 kJ, the activation energy for the reverse reaction is +84 kJ.

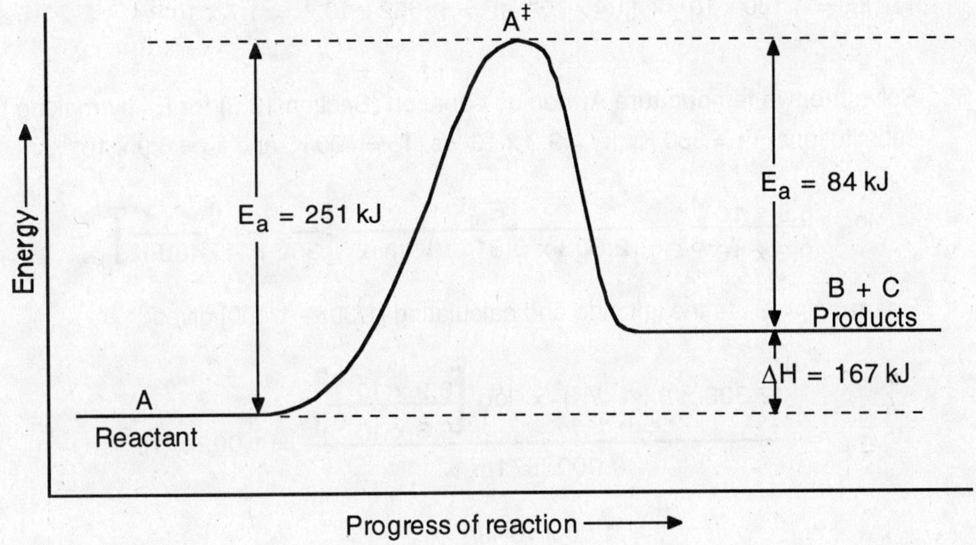

13.55 Solve the two-temperature Arrhenius equation (Section 13.6) for E_a by substituting T_1 = 308 K (from 35°C), $k_1 = 1.4 \times 10^{-4}$/s, T_2 = 318 K (from 45°C), and $k_2 = 5.0 \times 10^{-4}$/s:

$$\log \frac{5.0 \times 10^{-4}}{1.4 \times 10^{-4}} = \frac{E_a}{2.303 \times 8.31\ \text{J/(K•mol)}} \left[\frac{1}{308\ \text{K}} - \frac{1}{318\ \text{K}}\right]$$

Rearranging E_a to the left side and calculating [1/308 - 1/318] gives

$$E_a = \frac{(2.303 \times 8.31\ \text{J/K}) \times \log \left[\frac{5.0 \times 10^{-4}}{1.4 \times 10^{-4}}\right]}{0.00010209/\text{K}} = 1.\underline{0}37 \times 10^5$$

$$= 1.0 \times 10^5\ \text{J/mol}$$

To find the rate at 55°C (328 K), use the first equation, but let k_2 in the numerator be unknown and solve for k_2:

$$\log \frac{k_2}{1.4 \times 10^{-4}/\text{s}} = \frac{1.037 \times 10^5\ \text{J/mol}}{2.303 \times 8.31\ \text{J/(K•mol)}} \left[\frac{1}{308\ \text{K}} - \frac{1}{328\ \text{K}}\right] = 1.072$$

(continued)

$$\frac{k_2}{1.4 \times 10^{-4}/s} = 1.\underline{1}80 \times 10^1$$

$$k_2 = 1.180 \times 10^1 \times (1.4 \times 10^{-4}/s) = 1.\underline{6}52 \times 10^{-3} = 1.7 \times 10^{-3}/s$$

13.56 Solve the two-temperature Arrhenius equation (Section 13.6) for E_a by making these substitutions: $T_1 = 350$ K, $k_1 = 9.3 \times 10^{-6}$/s, $T_2 = 400$ K, and $k_2 = 6.9 \times 10^{-4}$/s.

$$\log \frac{6.9 \times 10^{-4}}{9.3 \times 10^{-6}} = \frac{E_a}{2.303 \times 8.31\ J/(K\bullet mol)} \left[\frac{1}{350\ K} - \frac{1}{400\ K}\right]$$

Rearranging E_a to the left side and calculating [1/305 - 1/400] gives

$$E_a = \frac{(2.303 \times 8.31\ J/K) \times \log\left[\frac{6.9 \times 10^{-4}}{9.3 \times 10^{-6}}\right]}{0.0003571/K} = 1.0\underline{0}2 \times 10^5$$

$$= 1.00 \times 10^5\ J/mol$$

To find the rate constant at 450 K, use the first equation with k_2 as the symbol for the rate constant at 450K and $E_a = 1.002 \times 10^5$ J/mol:

$$\log \frac{k_2}{9.3 \times 10^{-6}/s} = \frac{1.002 \times 10^5\ J/mol}{2.303 \times 8.31\ J/(K\bullet mol)} \left[\frac{1}{350\ K} - \frac{1}{450\ K}\right] = 3.324$$

$$\frac{k_2}{9.3 \times 10^{-6}/s} = 2.\underline{1}08 \times 10^3$$

$$k_2 = 2.108 \times 10^3 \times (9.3 \times 10^{-6}/s) = 1.\underline{9}6 \times 10^{-2} = 2.0 \times 10^{-2}/s$$

13.57 Because the rate constant is proportional to the rate of a reaction, tripling the rate at 25°C also means that the rate constant at 25°C is tripled. Thus, $k_{35} = 3k_{25}$, and the latter can be substituted for k_{35} in the Arrhenius equation:

$$\log \frac{3k_{25}}{k_{25}} = \frac{E_a}{2.303 \times 8.31\ J/(K\bullet mol)} \left[\frac{1}{298\ K} - \frac{1}{308\ K}\right]$$

$$0.4771 = (5.693 \times 10^{-6}\ mol/J)\ E_a$$

$$E_a = 0.4771 \div (5.693 \times 10^{-6}\ mol/J) = 8.3\underline{8}04 \times 10^4\ J/mol, \text{ or } 83.8\ kJ/mol$$

13.58 Because the rate constant is proportional to the rate of a reaction, quadrupling the rate at 25°C also means that the rate constant at 25°C is quadrupled. Thus, $k_{35} = 4k_{25}$, and the latter can be substituted for k_{35} in the Arrhenius equation:

$$\log \frac{4k_{25}}{k_{25}} = \frac{E_a}{2.303 \times 8.31\ \text{J/(K•mol)}} \left[\frac{1}{298\ \text{K}} - \frac{1}{308\ \text{K}}\right]$$

$$0.6021 = (5.693 \times 10^{-6}\ \text{mol/J})\ E_a$$

$$E_a = 0.6021 \div (5.693 \times 10^{-6}\ \text{mol/J}) = 1.0\underline{5}7 \times 10^5\ \text{J/mol, or 106 kJ/mol}$$

13.59 For plotting log k versus 1/T, the data below are used:

k	log k	1/T (1/K)
0.527	-0.2781	1.686×10^{-3}
0.776	-0.1101	1.658×10^{-3}
1.121	0.0496	1.631×10^{-3}
1.607	0.2060	1.605×10^{-3}

The plot requires a graph with too many lines to be reproduced here, but it yields a straight line. The slope of the line is calculated from the difference between the last and the first points:

$$\text{Slope} = \frac{[0.2060 - (-0.2781)]}{[1.605 \times 10^{-3} - 1.686 \times 10^{-3}]/\text{K}} = -5\underline{9}76\ \text{K}$$

Because the slope = $-E_a/(2.303\ R)$, you can solve for E_a using R = 8.31 J/K:

$$\frac{-E_a}{2.303 \times 8.31\ \text{J/ (K • mol)}} = -5976\ \text{K}$$

$$E_a = 5976\ \text{K} \times 2.303 \times 8.31\ \text{J/K} = 1.\underline{1}43 \times 10^5\ \text{J/mol, or } 1.1 \times 10^2\ \text{kJ/mol}$$

13.60 For plotting log k versus 1/T, the data below are used:

k	log k	1/T (K)
2.69×10^{-3}	-2.570	1.402×10^{-3}
6.21×10^{-3}	-2.207	1.364×10^{-3}
1.40×10^{-2}	-1.854	1.328×10^{-3}
3.93×10^{-2}	-1.406	1.294×10^{-3}

(continued)

The plot requires a graph with too many lines to be reproduced here, but it yields a straight line. The slope of the line may be calculated from the difference between the last point and the first point:

$$\text{Slope} = \frac{[-1.406 - (-2.570)]}{[1.294 \times 10^{-3} - 1.402 \times 10^{-3}]/\text{K}} = -1.0\underline{7}7 \times 10^4 \text{ K}$$

Because the slope = $-E_a/(2.303\ R)$, you can solve for E_a using R = 8.31 J/K • mol:

$$\frac{-E_a}{2.303 \times 8.31 \text{ J/(K} \bullet \text{mol)}} = -1.0\underline{7}7 \times 10^4 \text{ K}$$

$$E_a = 1.077 \times 10^4 \text{ K} \times 2.303 \times 8.31 \text{ J/K} \bullet \text{mol} = 2.\underline{0}61 \times 10^5 \text{ J/mol},$$

or 2.1×10^2 kJ/mol

13.61 The $NOCl_2$ is a reaction intermediate, being produced in the first reaction and consumed in the second. The overall reaction is the sum of the two elementary reactions:

$$NO + Cl_2 \rightarrow NOCl_2$$

$$NOCl_2 + NO \rightarrow 2NOCl$$

$$2NO + Cl_2 \rightarrow 2NOCl$$

13.62 The O atom is a reaction intermediate, being produced in the first reaction and consumed in the second. The overall reaction is the sum of the two elementary reactions:

$$O_3 \rightarrow O_2 + O$$

$$O_3 + O \rightarrow 2O_2$$

$$2O_3 \rightarrow 3O_2$$

13.63 a. Bimolecular b. Bimolecular c. Unimolecular d. Termolecular

13.64 a. Termolecular b. Bimolecular c. Bimolecular d. Unimolecular

13.65 a. Only O_3 occurs on the left side of the equation, so the rate law is

$$\text{Rate} = k[O_3]$$

b. Both $NOCl_2$ and NO occur on the left side of the equation, so the rate law is

$$\text{Rate} = k[NOCl_2][NO]$$

13.66 a. Only CS_2 occurs on the left side of the equation, so the rate law is

$$\text{Rate} = k[CS_2]$$

b. Both CH_3Br and OH^- occur on the left side of the equation, so the rate law is

$$\text{Rate} = k[CH_3Br][OH^-]$$

13.67 Step 1 of the isomerization of cyclopropane, C_3H_6, is slow, so the rate law for the overall reaction will be the rate law for this step, with $k_1 = k$, the overall rate constant:

$$\text{Rate} = k[C_3H_6]^2$$

13.68 Step 1 of the decomposition of NO_2Cl is slow, so the rate law for the overall reaction will be the rate law for this step, with $k_1 = k$, the overall rate constant:

$$\text{Rate} = k[NO_2Cl]$$

13.69 Step 2 of this reaction is slow, so the rate law for the overall reaction would appear to be the rate law for this step:

$$\text{Rate} = k_2[I\ ]^2[H_2]$$

However, the rate law includes an intermediate, the I atom, and cannot be used unless the intermediate is eliminated. This can only be done using an equation for step 1. At equilibrium, you can write the following equality for step 1:

$$k_1[I_2] = k_{-1}[I\ ]^2$$

Rearranging and then substituting for the $[I\ ]^2$ term yields

$$[I\ ]^2 = [I_2]\frac{k_1}{k_{-1}}$$

$$\text{Rate} = k_2\left[k_1/k_{-1}\right][I_2][H_2] = k[I_2][H_2] \quad (k = \text{the overall rate constant})$$

13.70 Step 2 of this reaction is slow, so the rate law for the overall reaction would appear to be the rate law for this step:

$$\text{Rate} = k_2[O_3][O]$$

However, the rate law includes an intermediate, the O atom, and cannot be used unless the intermediate is eliminated. This can only be done using an equation for step 1. At equilibrium, you can write the following equality for step 1:

$$k_1[O_3] = k_{-1}[O_2][O]$$

Rearranging and then substituting for the [O] term yields

$$[O] = \frac{[O_3]}{[O_2]}\left[\frac{k_1}{k_{-1}}\right]$$

$$\text{Rate} = k_2\left[\frac{k_1}{k_{-1}}\right]\frac{[O_3]^2}{[O_2]} = k\frac{[O_3]^2}{[O_2]} \quad (k = \text{the overall rate constant})$$

13.71 The Br^- ion is the catalyst. It is consumed in the first step and regenerated in the second step. It speeds up the reaction by providing a pathway with a lower activation energy than that of a reaction pathway involving no Br^-. The overall reaction is obtained by adding the two steps together:

$$2H_2O_2 \rightarrow 2H_2O + O_2$$

Bromide ion is added to the mixture to give the catalytic activity, and BrO^- is an intermediate.

13.72 The OH^- ion is the catalyst. It is consumed in the first step and regenerated in the second step. It speeds up the reaction by providing a pathway with a lower activation energy than that of a reaction pathway involving no OH^- ion. The overall reaction is obtained by adding the two steps together:

$$NH_2NO_2 \rightarrow N_2O + H_2O$$

You could add NaOH to give the catalytic activity.

■ Solutions to Unclassified Problems

13.73 All rates of reaction are calculated by dividing the decrease in concentration by the difference in times; hence, only the setup for the first rate (after 10 minutes) is given below. This setup is

$$\text{Rate}_{10\text{ min}} = -\frac{(1.29 - 1.50) \times 10^{-2}\text{ M}}{(10 - 0)\text{ min}} \times \frac{1\text{ min}}{60\text{ s}} = 3.\underline{5}0 \times 10^{-6}$$

$$= 3.5 \times 10^{-6}\text{ M/s}$$

A summary of the times and rates is given in the table.

Time, min	Rate
10	$3.\underline{50} \times 10^{-6} = 3.5 \times 10^{-6}$ M/s
20	$3.\underline{1}7 \times 10^{-6} = 3.2 \times 10^{-6}$ M/s
30	$2.\underline{50} \times 10^{-6} = 2.5 \times 10^{-6}$ M/s

13.74 All rates of reaction are calculated by dividing the decrease in concentration by the difference in times; only the setup for the first rate after 1.0 min is given below:

$$\text{Rate}_{1.0\text{ min}} = -\frac{(0.1076 - 0.1103\text{ M})}{(1.0 - 0.0)\text{ min}} \times \frac{1\text{ min}}{60\text{ s}} = 4.\underline{5}0 \times 10^{-5}$$

$$= 4.5 \times 10^{-5}\text{ M/s}$$

A summary of the times and rates is given in the table.

Time, min	Rate
1.0	$4.\underline{5}0 \times 10^{-5} = 4.5 \times 10^{-5}$ M/s
2.0	$4.\underline{33} \times 10^{-5} = 4.3 \times 10^{-5}$ M/s
3.0	$4.\underline{0}0 \times 10^{-5} = 4.0 \times 10^{-5}$ M/s

13.75 The calculation of the average concentration, and the division of the rate by this average concentration, are the same for all three time intervals. Thus, only the setup for the first interval is given:

$$k_{10\ min} = \frac{\text{rate}}{\text{avg. conc.}} = \frac{3.50 \times 10^{-6}\ M/s}{\left[\dfrac{(1.50 + 1.29) \times 10^{-2}\ M}{2}\right]} = 2.5\underline{0}8 \times 10^{-4}/s$$

A summary of the times, rate constants, and average rate constant is given in the table.

Time	Rate	k
10 min	3.50×10^{-6} M/s	$2.\underline{5}08 \times 10^{-4}$/s
20 min	3.17×10^{-6} M/s	$2.\underline{6}52 \times 10^{-4}$/s
30 min	2.50×10^{-6} M/s	$2.\underline{4}39 \times 10^{-4}$/s
-----	---- average k =	$2.\underline{5}33 \times 10^{-4} = 2.5 \times 10^{-4}$/s

13.76 The calculation of the average concentration, and the division of the rate by this average concentration, are the same for all three time intervals. Thus, only the setup for the first interval is given:

$$k_{1.0\ min} = \frac{\text{rate}}{\text{avg. conc.}} = \frac{4.50 \times 10^{-5}\ M/s}{\left[\dfrac{0.1103M + 0.1076\ M}{2}\right]^2} = 3.\underline{7}9 \times 10^{-3}/(M \bullet s)$$

A summary of the times, rate constants, and average rate constant is given in the table.

Time	Rate	k
1.0 min	4.50×10^{-5} M/s	$3.\underline{7}9 \times 10^{-3}/(M \bullet s)$
2.0 min	4.33×10^{-5} M/s	$3.\underline{8}3 \times 10^{-3}/(M \bullet s)$
3.0 min	4.00×10^{-5} M/s	$3.\underline{7}1 \times 10^{-3}/(M \bullet s)$
-----	---- average k =	$3.\underline{7}8 \times 10^{-3} = 3.8 \times 10^{-3}/(M \bullet s)$

13.77 Use the first-order rate equation: $k = 1.26 \times 10^{-4}$/s, the initial methyl acetate $[MA]_o$ = 100%, and $[MA]_t$ = (100% - 85%, or 15%).

$$t = \frac{-(2.303)\left(\log \dfrac{15\%}{100\%}\right)}{1.26 \times 10^{-4}/s} = 1.5\underline{0}59 \times 10^{4} = 1.51 \times 10^{4}\ s$$

13.78 Use the first-order rate equation: substitute k = 4.3 x 10^{-5}/s, the initial benzene diazonium chloride $[BC]_o$ = 100%, and $[BC]_t$ = (100% - 85%, or 15%).

$$t = \frac{-(2.303)\left(\log \frac{15\%}{100\%}\right)}{4.3 \times 10^{-5}/s} = 4.\underline{4}1 \times 10^4 = 4.4 \times 10^4 \text{ s}$$

13.79 Use k = 1.26 x 10^{-4}/s and substitute into the $t_{1/2}$ equation:

$$t_{1/2} = \frac{0.693}{k} = \frac{0.693}{1.26 \times 10^{-4}/s} = 5.5\underline{0}0 \times 10^3 = 5.50 \times 10^3 \text{ s} \ (1.5\underline{2}7 \text{ hr})$$

13.80 Use k = 4.3 x 20^{-5}/s and substitute into the $t_{1/2}$ equation:

$$t_{1/2} = \frac{0.693}{k} = \frac{0.693}{4.3 \times 10^{-5}/s} = 1.\underline{6}1 \times 10^4 = 1.6 \times 10^4 \text{ s} \ (4.\underline{4}7 \text{ hr})$$

13.81 First find the rate constant from the rearranged first-order rate equation, substituting the initial concentration of $[\text{comp.}]_o$ = 0.0350 M, and the $[\text{comp.}]_t$ = 0.0250 M.

$$k = \frac{-(2.303)\left(\log \frac{0.0250 \text{ M}}{0.0350 \text{ M}}\right)}{65 \text{ s}} = 5.\underline{1}8 \times 10^{-3}/s$$

Now arrange the first-order rate equation to solve for $[\text{comp.}]_t$; substitute the above value of k, again using $[\text{comp.}]_o$ = 0.0350 M.

$$\log \frac{[\text{comp.}]_t}{[0.0350 \text{ M}]_o} = \frac{-(5.18 \times 10^{-3}/s)(98 \text{ s})}{2.303} = -0.2\underline{2}04$$

Taking antilogarithms of both sides gives

$$\frac{[\text{comp.}]_t}{[0.0350 \text{ M}]_o} = 0.6\underline{0}2$$

$$[\text{comp}]_t = 0.602 \times [0.0350 \text{ M}]_o = 0.02\underline{1}08 = 0.021 \text{ M}$$

13.82 First find the rate constant from the rearranged first-order rate equation, substituting the initial concentration of $[\text{comp.}]_o = 0.1180$ M, and the $[\text{comp.}]_t = 0.0950$ M. Use k to find the fraction.

$$k = \frac{-(2.303)\left(\log \frac{0.0950\ M}{0.1180\ M}\right)}{5.2\ \text{min}} = 4.\underline{1}7 \times 10^{-2}/\text{min}$$

$$\log \frac{[\text{comp.}]_t}{[0.1180\ M]_o} = \frac{-(4.17 \times 10^{-2}/\text{min})(6.7\ \text{min})}{2.303} = -0.1\underline{2}1$$

Taking the antilogarithm of both sides yields

$$\frac{[\text{comp.}]_t}{[0.1180\ M]_o} = 0.7\underline{5}6 = 0.76 = \text{fraction remaining}$$

13.83 The log $[CH_3NNCH_3]$ and time data for the plot are tabulated below.

t. min	log $[CH_3NNCH_3]$
0	-1.824
10	-1.889
20	-1.959
30	-2.022

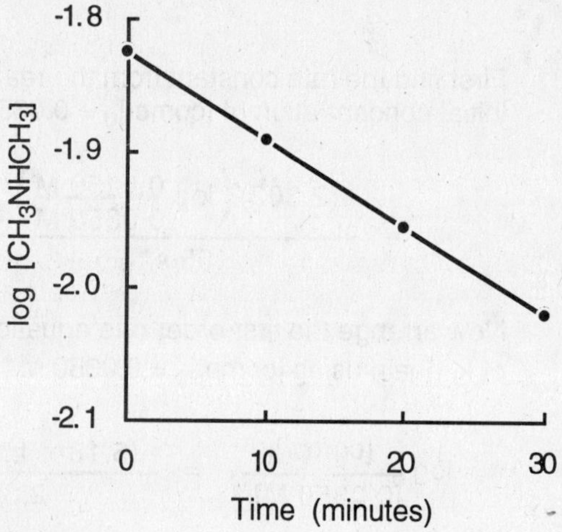

From the graph the slope, m, is calculated:

$$m = \frac{-2.022 - (-1.824)}{(30 - 0)\ \text{min}}$$

$$= -6.60 \times 10^{3}\ /\text{min}$$

Because the slope also = -k/2.303, solve for k by equating the two right-hand terms:

$$k = -(-6.60 \times 10^{-3}/\text{min})(2.303)(1\ \text{min}/60\ \text{s})$$

$$k = 2.\underline{5}3 \times 10^{-4} = 2.5 \times 10^{-4}/\text{s}$$

13.84 The log $[H_2O_2]$ and time data for the plot are tabulated below.

t. min	log $[H_2O_2]$
0.0	-1.000
5.0	-1.095
10.0	-1.188
15.0	-1.285

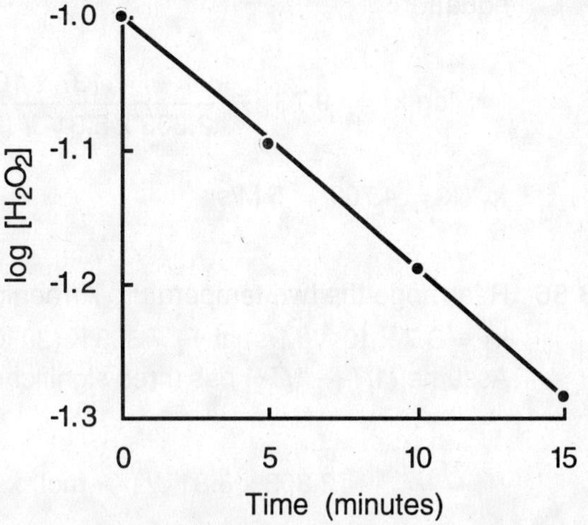

From the graph the slope, m, is calculated:

$$m = \frac{-1.285 - (-1.000)}{(15.0 - 0.0)\ \text{min}}$$

$$= -1.90 \times 10^{-2}\ /\text{min}$$

Because the slope also = -k/2.303, solve for k by equating the two right-hand terms:

$$k = -(-1.90 \times 10^{-2}/\text{min})(2.303)(1\ \text{min}/60\ \text{s})$$

$$k = 7.2\underline{9}2 \times 10^{-4} = 7.29 \times 10^{-4}/\text{s}$$

13.85 Rearrange the two-temperature Arrhenius equation to solve for E_a in J, using $k_1 = 0.498$ M/s at $T_1 = 592$ K (319°C) and $k_2 = 1.81$ M/s at 627 K (354°C). Assume $(1/T_1 - 1/T_2)$ has three significant figures.

$$E_a = \frac{[2.303 \times 8.31\ \text{J}/(\text{K}\bullet\text{mol}) \times \log\left[\frac{1.81}{0.498}\right]}{\left[\frac{1}{592\text{K}} - \frac{1}{627\text{K}}\right]} = 1.1\underline{3}7 \times 10^{5}$$

$$= 1.14 \times 10^{5}\ \text{J/mol} = 114\ \text{kJ/mol}$$

To obtain A, rearrange the log form of the one-temperature Arrhenius equation; substitute the value of E_a obtained above and use $k_1 = 0.498$ M/s at $T_1 = 592$ K.

$$\log A = \log 0.498 + \frac{1.137 \times 10^{5}\ \text{J/mol}}{2.303 \times 8.31\ \text{J}/(\text{K}\bullet\text{mol}) \times 592\text{K}} = 9.\underline{7}3$$

$$A = \underline{5}.37 \times 10^{9} = 5 \times 10^{9}$$

(continued)

To obtain k at 383°C (656 K), also use the log form of the one-temperature Arrhenius equation:

$$\log k \;=\; 9.73 \;-\; \frac{1.137 \times 10^{5}\ \text{J/ mol}}{2.303 \times 8.31\ \text{J/ (K} \bullet \text{mol)} \times 656\text{K}} \;=\; 0.\underline{6}73$$

$$k \;=\; 4.709 \;=\; 5\ \text{M/s}$$

13.86 Rearrange the two-temperature Arrhenius equation; solve for E_a using $k_1 = 8.7 \times 10^{-4}/(\text{M}\bullet\text{s})$ at $T_1 = 303$ K (30°C) and $k_2 = 1.5 \times 10^{-3}$ (M•s) at 313 K (40°C). Assume $(1/T_1 - 1/T_2)$ has three significant figures.

$$E_a \;=\; \frac{[2.303 \times 8.31\ \text{J/ (K} \bullet \text{mol)} \times \log\left[\dfrac{1.5 \times 10^{-3}}{8.7 \times 10^{-4}}\right]}{\left[\dfrac{1}{303\text{K}} - \dfrac{1}{313\text{K}}\right]} \;=\; 4.\underline{2}94 \times 10^{4}\quad (43\ \text{kJ/ mol})$$

To obtain A, rearrange the log form of the one-temperature Arrhenius equation; substitute the value of E_a obtained above and use $k_1 = 8.7 \times 10^{-4}$ M/s at $T_1 = 303$ K.

$$\log A \;=\; \log\,(8.7 \times 10^{-4}) \;+\; \frac{4.294 \times 10^{4}\ \text{J/ mol}}{2.303 \times 8.31\ \text{J/ (K} \bullet \text{mol)} \times 303\text{K}} \;=\; 4.\underline{3}44$$

$$A \;=\; 2.20 \times 10^{4} \;=\; 2 \times 10^{4}/(\text{M}\bullet\text{s})$$

To obtain k at 45°C (318 K), also use the log form of the one-temperature Arrhenius equation:

$$\log k \;=\; 4.34 \;-\; \frac{4.294 \times 10^{4}\ \text{J/ mol}}{2.303 \times 8.31\ \text{J/ (K} \bullet \text{mol)} \times 318\text{K}} \;=\; -\,2.\underline{7}12$$

$$k \;=\; \underline{1}.9 \times 10^{-3} \;=\; 2 \times 10^{-3}/(\text{M}\bullet\text{s})$$

13.87 If the reaction occurs in one step, the coefficients of NO_2 and CO in this elementary reaction are each 1, so the rate law should be

$$\text{Rate} \;=\; k[NO_2][CO]$$

13.88 If the reaction occurs in one step, the coefficients of CH_3Cl and OH^- in this elementary reaction are each 1, so the rate law should be

$$\text{Rate} = k[CH_3Cl][OH^-]$$

13.89 The slow first step determines the observed rate, so the overall rate constant, k, should be equal to k_1 for the first step and the rate law should be

$$\text{Rate} = k_1[NO_2Br] = k[NO_2Br]$$

13.90 The slow first step determines the observed rate, so the overall rate constant, k, should be equal to k_1 for the first step and the rate law should be:

$$\text{Rate} = k_1[(CH_3)_3CCl] = k[(CH_3)_3CCl]$$

13.91 The slow step determines the observed rate; assuming k_2 is the rate constant for the second step, the rate law would appear to be

$$\text{Rate} = k_2[NH_3][HCNO]$$

However, this rate law includes two intermediate substances that are neither reactants nor products. The rate law cannot be used unless both are eliminated. This can only be done using an equation from step 1. At equilibrium in step 1, you can write the following equality, assuming k_1 and k_{-1} are the rate constants for the forward and back reactions, respectively:

$$k_1[NH_4^+][CNO^-] = k_{-1}[NH_3][HCNO]$$

Rearranging and then substituting for the $[NH_3][HCNO]$ product gives

$$[NH_3][HCNO] = (k_1/k_{-1})[NH_4^+][CNO^-]$$

$$\text{Rate} = k_2(k_1/k_{-1})[NH_4^+][CNO^-] = k[NH_4^+][CNO^-] \quad (k = \text{overall rate constant})$$

13.92 The slow step determines the observed rate; assuming k_2 is the rate constant for the second step, and using $[HA^+]$ for $[CH_3C(OH^+)CH_3]$, the rate law would appear to be

$$\text{Rate} = k_2[HA^+][H_2O]$$

(continued)

However, this rate law includes the intermediate HA^+ and H_2O that are neither reactants nor products. The rate law cannot be used unless both are eliminated. This can only be done using an equation from step 1. At equilibrium in step 1, you can write the following equality, assuming k_1 and k_{-1} are the rate constants for the forward and back reactions, respectively:

$$k_1[A][H_3O^+] = k_{-1}[HA^+][H_2O]$$

Rearranging and then substituting for the $[HA^+][H_2O]$ product gives

$$\text{Rate} = k_2(k_1/k_{-1})[A][H_3O^+] = k[A][H_3O^+] \quad (k = \text{overall rate constant})$$

13.93. a. The reaction is first order in O_2 because the rate doubled with a doubling of the oxygen concentration. The reaction is second order in NO because the rate increased by a factor of 8 when both the NO and O_2 concentrations were doubled.

$$\text{Rate} = k[NO]^2[O_2]$$

b. The initial rate of the reaction for Experiment 4 can be determined by first calculating the value of the rate constant using Experiment 1 for the data.

$$0.80 \times 10^{-2}\ M/s = k[4.5 \times 10^{-2}\ M]^2[2.2 \times 10^{-2}\ M]$$

Solving for the rate constant gives

$$k = 1.\underline{7}96 \times 10^2\ M^{-2}s^{-1}$$

Now, use the data in Experiment 4 and the rate constant to determine the initial rate of the reaction.

$$\text{Rate} = 1.796 \times 10^2\ M^{-2}s^{-1}\ [3.8 \times 10^{-1}\ M]^2[4.6 \times 10^{-3}\ M]$$

$$\text{Rate} = 0.1\underline{1}9 = 0.12\ \text{mol/L}\bullet\text{s}$$

13.94 a. From the information given, the rate law is

$$\text{Rate} = k[CH_3Cl][H_2O]^2$$

Using just the units for each term in the rate law will give the units for the rate constant.

$$M/s = k(M)(M)^2$$

(continued)

Thus, the units for K must be

$$k = M^{-2}s^{-1}$$

b. Plugging into the rate equation gives

$$1.50\ M/s = k[0.20\ M][0.20\ M]^2$$

$$k = 1.\underline{8}75 \times 10^2 = 1.9 \times 10^2\ M^{-2}s^{-1}$$

13.95 a. i) Rate will decrease because OH^- will react with H_3O^+ and will lower its concentration.

ii) Rate will decrease because the dilution with water will decrease both the H_3O^+ and CH_3CSNH_2 concentrations.

iii) The rate will decrease because the H_3O^+ concentration will decrease since acetic acid is a weak acid.

b. i) The catalyst will provide another pathway and k will increase because E_a will be smaller.

ii) The rate constant changes with temperature and it will decrease with a decrease in temperature as fewer molecules will have enough energy to react.

13.96 a. i) The $[H_3O^+]$ is increased so the rate will increase.

ii) The $[H_3O^+]$ and $[CH_3COOCH_3]$ concentrations will be decreased so the rate of reaction will decrease.

iii) The CH_3COO^- will react with H_3O^+ to form CH_3COOH, a weak acid so the $[H_3O^+]$ will decrease and the rate of reaction will decrease.

b. i) The rate of reaction will increase, but the rate constant will remain the same.

ii) The rate constant will increase as more molecules have enough energy to react.

13.97 a. The diagram:

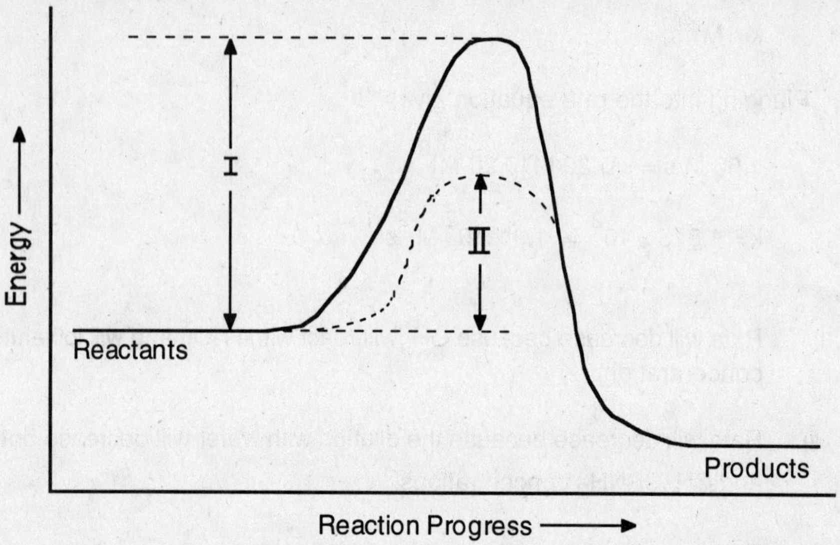

b In the diagram I represents the activation energy for the uncatalyzed reaction and II represents the activation energy for the catalyzed reaction. A catalyst provides another pathway for a chemical reaction and with a lower activation energy, more molecules have enough energy to react, so the reaction will be faster.

13.98 a. The diagram:

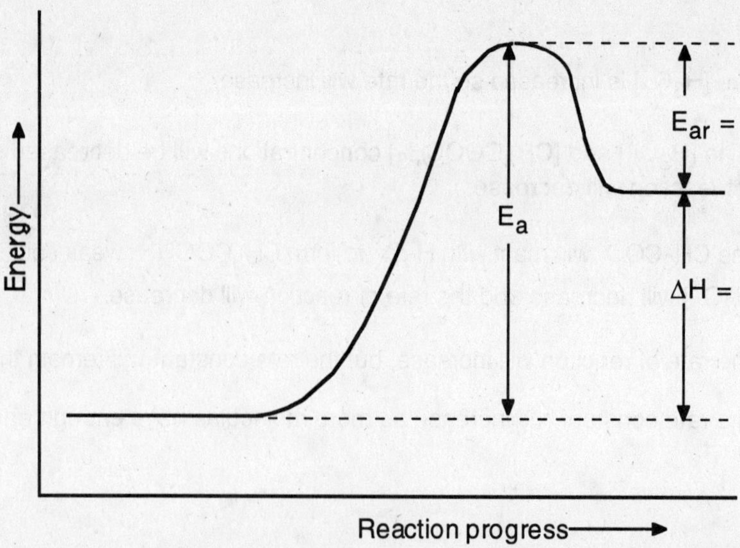

b. $E_a = E_{ar} + \Delta H$

$E_a = 23 + 57 = 80$ kJ

c. The forward reaction will be more sensitive to a temperature change as it has the largest activation energy. When E_a is large, the ratio of the number of molecules that have enough energy to react is larger than when E_a is smaller.

13.99. a. The rate of a chemical reaction is the change in the concentration of a reactant or product with time. For a reactant,

$-\Delta c/\Delta t$ or $-d[c]/dt$.

b. The rate changes because the concentration of the reactant has changed.

$\text{Rate} = k[A]^m$

c. $\text{Rate} = k[A]^m[B]^n$

The rate will equal k when the reactants all have 1.00 M concentrations.

13.100. Factors which affect the rates of reactions:

i) Concentrations of the reactants
ii) Temperature
iii) Catalysts

At higher concentrations, more molecules can undergo effective collisions and give more product per unit of time. The temperature affects the number of molecules that have enough energy to react. The higher the temperature the larger the fraction of molecules that have enough energy to react. Catalysts provide another pathway for reaction which has a lower activation energy so that more molecules have the minimum energy to react.

The value of the rate constant for a particular reaction depends upon the temperature and the activation energy. The concentration of catalyst and the solvent, if the reaction occurs in solution, can affect E_a.

■ Cumulative-Skills Problems
(require skills from previous chapters)

13.101 The balanced equation is: $2\ N_2O_5 \rightarrow 4\ NO_2 + O_2$.
Note that the moles of O_2 formed will be one half that of the moles of N_2O_5 decomposed. Now use the integrated form of the first order rate law to calculate the fraction of the 1.00 mol N_2O_5 decomposing in 20.0 hr or 1,200 min.

$$\log \frac{[N_2O_5]_t}{[N_2O_5]_o} = \frac{-kt}{2.303} = \frac{-(6.2 \times 10^{-4})(1{,}200)}{2.303} = -0.3\underline{2}305$$

$$\frac{[N_2O_5]_t}{[N_2O_5]_o} = 0.4\underline{7}527$$

Fraction of N_2O_5 decomposed = 1.000 - 0.4$\underline{7}$527 = 0.5$\underline{2}$473

Since there is 1.00 mol of N_2O_5 present at the start, the moles of N_2O_5 decomposed is 0.5$\underline{2}$473 mol. Thus,

$$\text{Mol } O_2 \text{ formed} = (1\ O_2/2\ N_2O_5) \times 0.52473 \text{ mol } N_2O_5 = 0.2\underline{6}237 \text{ mol } O_2$$

Now use ideal gas law to calculate the volume of this number of moles of O_2 gas at 45 °C and 780 mmHg.

$$V = \frac{nRT}{P} = \frac{(0.26237 \text{ mol})(0.08206 \text{ L} \bullet \text{atm}/(\text{K} \bullet \text{mol})(318 \text{ K})}{(780/760) \text{ atm}}$$

$$= 6.\underline{6}7 = 6.8 \text{ L}$$

13.102 The balanced equation is: $2\ H_2O_2 \rightarrow 2\ H_2O + O_2$.
Note that the moles of O_2 formed will be equal to the moles of H_2O_2 decomposed. Now use the integrated form of the first order rate law to calculate the fraction of the 1.00 mol H_2O_2 decomposing in 20.0 min or 1,200 sec.

$$\log \frac{[H_2O_2]_t}{[H_2O_2]_o} = \frac{-kt}{2.303} = \frac{-(7.40 \times 10^{-4})(1{,}200)}{2.303} = -0.385\underline{5}8$$

$$\frac{[H_2O2]_t}{[H_2O_2]_o} = 0.41\underline{1}54$$

(continued)

Fraction of H_2O_2 decomposed = 1.00 - 0.41154 = 0.58846

Since there is 1.00 mol of H_2O_2 present at the start, the moles of H_2O_2 decomposed is 0.58846 mol. Thus,

Mol O_2 formed = (1 O_2/ 1 H_2O_2) x 0.58846 mol H_2O_2 = 0.58846 mol O_2

Now use ideal gas law to calculate the volume of this number of moles of O_2 gas at 25 °C and 740 mmHg.

$$V = \frac{nRT}{P} = \frac{(0.58846 \text{ mol})(0.08206 \text{ L}\bullet\text{atm/ (K}\bullet\text{mol)}(298 \text{ K})}{(740/\ 760) \text{ atm}}$$

$$= 14.7 = 15 \text{ L}$$

13.103 Using the first-order rate law, the initial rate of decomposition is given by

Rate = $k[H_2O_2]$ = (7.40 x 10^{-4}/s) x (1.50 M H_2O_2) = 1.110 x 10^{-3} M H_2O_2/s

The heat liberated per second per mol of H_2O_2 can be found by first calculating the standard enthalpy of the decomposition of 1 mol of H_2O_2:

	$H_2O_2(aq)$	$\rightarrow$	$H_2O(l)$	+	$1/2O_2(g)$
ΔH^o_f =	-191.2 kJ/mol		-285.84 kJ/mol		0 kJ/mol

For the reaction, the standard enthalpy change is

ΔH^o = -285.84 kJ/mol - (-191.2 kJ/mol) = -94.64 kJ/mol H_2O_2

The heat liberated per second is:

$$\frac{94.64 \text{ kJ}}{\text{mol } H_2O_2} \times \frac{1.110 \times 10^{-3} \text{ mol } H_2O_2}{\text{L}\bullet\text{s}} \times 2.00 \text{ L} = 0.21010 = 0.210 \text{ kJ/s}$$

13.104 Using the second-order rate law, the initial rate of reaction of NO_2 is

$$\text{Rate} = k[NO_2]^2 = 0.515\ \text{L/(mol•s)} \times [0.0250\ M\ NO_2]^2 = 3.2\underline{1}8 \times 10^{-4}\ M\ NO_2/\text{s}$$

The heat liberated per second per mol of NO_2 can be found by first calculating the standard enthalpy for the reaction of 1 mol of NO_2:

	$NO_2(g)$	+	$CO(g)$	$\rightarrow$	$NO(g)$	+	$CO_2(g)$
ΔH^o_f =	33.2 kJ/mol		-110.5 kJ/mol		90.29 kJ/mol		-393.5 kJ/mol

For the reaction, the change in standard enthalpy is

$$\Delta H^o = 90.29 - 393.5 - (33.2) - (-110.5) = -22\underline{5}.9\ \text{kJ/mol}\ NO_2$$

The heat liberated per second is

$$\frac{225.9\ \text{kJ}}{\text{mol}\ NO_2} \times \frac{3.218 \times 10^{-4}\ \text{mol}\ NO_2}{\text{L•s}} \times 3.50\ \text{L} = 0.25\underline{4}4 = 0.254\ \text{kJ/s}$$

13.105 Use the ideal gas law ($P/RT = n/V$) to calculate the mol/L of each gas:

$$[O_2] = \frac{(345/760)\ \text{atm}}{0.082057\ \text{L•atm/(K•mol)} \times 612\ \text{K}} = 0.0090\underline{3}9\ \text{mol}\ O_2/\text{L}$$

$$[NO] = \frac{(155/760)\ \text{atm}}{0.082057\ \text{L•atm/(K•mol)} \times 612\text{K}} = 0.0040\underline{6}1\ \text{mol NO/L}$$

The rate of decrease of NO is

$$\frac{1.16 \times 10^{-5}\ \text{L}^2}{\text{mol}^2 \bullet \text{s}} \times \left(\frac{4.061 \times 10^{-3}\ \text{mol}}{1\ \text{L}}\right)^2 \times \frac{9.039 \times 10^{-3}\ \text{mol}}{1\ \text{L}}$$

$$= 1.7\underline{2}9 \times 10^{-12}\ \text{mol/(L•s)}$$

The rate of decrease in atm/s is found by multiplying by RT:

$$\frac{1.729 \times 10^{-12}\ \text{mol}}{\text{L•s}} \times \frac{0.082057\ \text{L•atm}}{\text{K•mol}} \times 612\ \text{K} = 8.6\underline{8}2 \times 10^{-11}\ \text{atm/s}$$

The rate of decrease in mmHg/s is

$$8.682 \times 10^{-11}\ \text{atm/s} \times (760\ \text{mmHg/atm}) = 6.598 \times 10^{-8} = 6.60 \times 10^{-8}\ \text{mmHg/s}$$

13.106 Use the ideal gas law (P/RT = n/V) to calculate the mol/L of each gas:

$$[H_2] = \frac{(324/760)\text{ atm}}{0.082057\text{ L} \bullet \text{atm/(K/ mol)} \times 1099\text{K}} = 0.0047\underline{2}7\text{ mol } H_2/\text{ L}$$

$$[NO] = \frac{(144/760)\text{ atm}}{0.082057\text{ L} \bullet \text{atm/(K/ mol)} \times 1099\text{K}} = 0.0021\underline{0}1\text{ mol NO/L}$$

The rate of NO decrease is twice the rate of H_2 decrease, so

$$2 \times \frac{1.10 \times 10^{-7}\text{ L}^2}{\text{mol}^2 \bullet \text{s}} \times \left(\frac{2.101 \times 10^{-3}\text{ mol}}{1\text{ L}}\right)^2 \times \frac{4.727 \times 10^{-3}\text{ mol}}{1\text{ L}}$$

$$= 4.5905 \times 10^{-15}\text{ mol/(L}\bullet\text{s)}$$

The rate of decrease in atm/s is found by multiplying by RT:

$$\frac{4.5905 \times 10^{-15}\text{ mol}}{\text{L}\bullet\text{s}} \times \frac{0.082057\text{ L}\bullet\text{atm}}{\text{K}\bullet\text{mol}} \times 1099\text{ K} = 4.1\underline{3}9 \times 10^{-13}\text{ atm/s}$$

The rate of decrease in mmHg/s is

$$4.139 \times 10^{-13}\text{ atm/s} \times (760\text{ mmHg/atm}) = 3.146 \times 10^{-10} = 3.15 \times 10^{-10}\text{ mmHg/s}$$

■ Solution to Conceptual Problem

13.107 The hydrogenation of aromatic hydrocarbons is often achieved using a solid platinum catalyst. The catalyst substantially increases the rate of the reaction. However, the limited surface area of the platinum typically results in no change in rate upon changing either the hydrogen concentration or the hydrocarbon concentration. What is the order of this reaction? Create a rate vs. concentration plot for this kind of reaction. Suggest how you would go about controlling the rate of a reaction that is independent of concentration.

Discussion

No change in reaction rate with change in concentrations corresponds to a zero-order rate law. The rate law has the form Rate = $k[A]^0[B]^0$, and a plot of [A] or [B] versus t is a straight line with a slope of k. Thus, the reaction is zero order. You cannot vary the rate by changing the concentrations of reactants; but, you can control the rate by varying the temperature and quantity of catalyst surface area. Heterogeneous reaction on a surface catalyst are common examples of zero-order reactions.

(Continued)

Assuming that the surface area of the catalyst is occupied with reactant molecules, increasing the reactant concentrations has no effect on rate. The enzyme-catalyzed reactions that occur in a biological cell may also be zero order. These are homogeneous reactions involving substrate molecules and enzyme molecules in solution. Once all of the active sites on the enzyme molecules are filled with substrate, the rate of reaction depends only the activation-energy barrier of the enzyme-substrate transition state.

14. CHEMICAL EQUILIBRIUM

Solutions to Exercises

Note on significant figures: The final answer to each mathematical solution is given first with one nonsignificant figure (the rightmost significant figure is underlined) and then is rounded to the correct number of figures. Intermediate answers usually also have at least one nonsignificant figure.

14.1 Use the "table" approach, giving the starting, change, and equilibrium number of moles of each.

Amt. (mol)	$CO(g)$	+	$H_2O(g)$	$\rightleftharpoons$	$CO_2(g)$	+	$H_2(g)$
Starting	1.00		1.00		0		0
Change	-x		-x		+x		+x
Equilibrium	(1.00 - x)		(1.00 - x)		x		x = 0.43

Because we are given that x = 0.43 in the statement of the problem, we can use that to calculate the equilibrium amounts of the reactants and products:

Equilibrium amount CO = 1.00 - 0.43 = 0.57 mol

Equilibrium amount H_2O = 1.00 - 0.43 = 0.57 mol

Equilibrium amount CO_2 = x = 0.43 mol

Equilibrium amount H_2 = x = 0.43 mol

14.2 For the equation $2NO_2 + 7H_2 \rightarrow 2NH_3 + 4H_2O$, the expression for the equilibrium constant, K_c, is

$$K_c = \frac{[NH_3]^2[H_2O]^4}{[NO_2]^2[H_2]^7}$$

Notice that each concentration term is raised to a power equal to that of its coefficient in the chemical equation.

For the equation $NO_2 + 7/2H_2 \rightarrow NH_3 + 2H_2O$, the expression for the equilibrium constant, K_c, is

$$K_c = \frac{[NH_3][H_2O]^2}{[NO_2][H_2]^{7/2}}$$

Note the correspondence between the power and coefficient for each molecule.

14.3 The chemical equation for the reaction is

$$CO(g) + H_2O(g) \rightleftharpoons H_2(g) + CO_2(g)$$

The expression for the equilibrium constant for this reaction is

$$K_c = \frac{[CO_2][H_2]}{[CO][H_2O]}$$

We obtain the concentration of each substance by dividing the moles of substance by its volume. The equilibrium concentrations are as follows: $[CO] = 0.057$ M, $[H_2O] = 0.057$ M, $[CO_2] = 0.043$ M, and $[H_2] = 0.043$ M. Substituting these values into the equation for the equilibrium constant gives

$$K_c = \frac{(0.043)(0.043)}{(0.057)(0.057)} = 5.\underline{6}9 \times 10^{-1} = 5.7 \times 10^{-1}$$

14.4 Use the "table" approach, giving the starting, change, and equilibrium concentrations of each by dividing moles by volume in liters.

Conc. (M)	$2H_2S(g)$	$\rightleftharpoons$	$2H_2(g)$	+	$S_2(g)$
Starting	0.0100		0		0
Change	-2x		+2x (= 0.00285)		+x
Equilibrium	0.0100 - 2x		2x		x

(continued)

Because the problem states that 0.00285 M H_2 was formed, we can use the 0.00285 M to calculate the other concentrations. The S_2 molarity should be one-half that, or $0.0014\underline{2}5$ M, and the H_2S molarity should be 0.0100 - 0.00285, or $0.007\underline{1}5$ M. Substituting into the equilibrium expression gives

$$K_c = \frac{[H_2]^2[S_2]}{[H_2S]^2} = \frac{(0.00285)^2(0.001425)}{(0.00715)^2} = 2.\underline{2}6 \times 10^{-4} = 2.3 \times 10^{-4}$$

14.5 Use the expression that relates K_c to K_p:

$$K_p = K_c(RT)^{\Delta n}$$

The Δn term is the sum of the coefficients of the gaseous products minus the sum of the coefficients of the gaseous reactants. In this case, $\Delta n = (2 - 1) = 1$, and K_p is

$$K_p = (3.26 \times 10^{-2})(0.0821 \times 464)^1 = 1.2\underline{4}1 = 1.24$$

14.6 For a heterogeneous equilibrium, the concentration terms for liquids and solids are omitted because such concentrations are constant at a given temperature and are incorporated into the measured value of K_c. For this case, K_c is defined

$$K_c = \frac{[Ni(CO)_4]}{[CO]^4}$$

14.7 Because the equilibrium constant is very large (> 10^4), the equilibrium mixture will contain mostly products. Rearrange the K_c expression to solve for $[NO_2]$.

$$[NO_2] = \sqrt{K_c[O_2][NO]^2} = \sqrt{4.0 \times 10^{13}(2.0 \times 10^{-6})(2.0 \times 10^{-6})^2} = 1.\underline{7}8 \times 10^{-2}$$

$$= 1.8 \times 10^{-2}$$

14.8 First divide moles by volume in liters to convert to molar concentrations, giving 0.00015 M CO_2 and 0.010 M CO. Substitute these values into the reaction quotient and calculate Q.

$$Q = \frac{[CO]^2}{[CO_2]} = \frac{(0.010)^2}{(0.00015)} = 6.\underline{6}6 \times 10^{-1} = 6.7 \times 10^{-1}$$

Because Q = 0.67 and is less than K_c, the reaction will go to the right, forming more CO.

14.9 Rearrange the K_c expression, and substitute for K_c (= 0.0415) and the given moles per 1.00 L to solve for moles per 1.00 L of PCl_5.

$$[PCl_5] = \frac{[PCl_3][Cl_2]}{K_c} = \frac{(0.020)(0.020)}{0.0415} = 9.\underline{6}3 \times 10^{-3}$$

$$= 9.6 \times 10^{-3} \text{ mol/ } 1.00 \text{ L}$$

Because the volume is 1.00 L, the moles of PCl_5 = 0.0096 M.

14.10 Use the "table" approach, giving the starting, change, and equilibrium number of moles of each.

Amt. (mol)	H_2(g	+	I_2(g)	$\rightleftharpoons$	2HI(g)
Starting	0.500		0.500		0
Change	-x		-x		+2x
Equilibrium	0.500 - x		0.500 - x		2x

Substitute the equilibrium concentrations into the expression for K_c (= 49.7).

$$K_c = \frac{[HI]^2}{[H_2][I_2]};\quad 49.7 = \frac{(2x)^2}{(0.500 - x)(0.500 - x)} = \frac{(2x)^2}{(0.500 - x)^2}$$

Taking the square root of both sides of the right-hand equation and solving for x gives:

$$\pm 7.05 = \frac{2x}{(0.500 - x)}, \text{ or } \pm 7.05(0.500 - x) = 2x$$

Using the positive root, x = 0.3$\underline{9}$0.

Using the negative root, x = 0.6$\underline{9}$8 (this must be rejected because 0.698 is greater than the 0.500 starting number of moles).

Substituting x = 0.3$\underline{9}$0 mol into the last line of the table to solve for equilibrium concentrations gives these amounts: 0.11 mol H_2, 0.11 mol I_2, and 0.78 mol HI.

14.11 Use the "table" approach for starting, change, and equilibrium concentrations of each species.

Conc. (M)	$PCl_5(g)$	$\rightleftharpoons$	$PCl_3(g)$	+	$Cl_2(g)$
Starting	1.00		0		0
Change	-x		+x		+x
Equilibrium	1.00 - x		x		x

Substitute the equilibrium concentration expressions from the table into the equilibrium equation and solve for x, using the quadratic formula.

$$[PCl_3][Cl_2] = K_c \times [PCl_5] = 0.0211(1.00 - x) = x^2$$

$$x^2 + 0.0211x - 0.0211 = 0$$

$$x = \frac{-0.0211 \pm \sqrt{(0.0211)^2 - 4(-0.0211)}}{2} = \frac{-0.0211 \pm 0.2913}{2}$$

x = -0.1562 (impossible; reject), or x = 0.13$\underline{5}$09 = 0.135 M (logical)

Solve for the equilibrium concentrations using x = 0.135 M: $[PCl_5]$ = 0.86 M, $[Cl_2]$ = 0.135 M, and $[PCl_3]$ = 0.135 M.

14.12 a. Increasing the pressure will cause a net reaction to occur from right to left, and more $CaCO_3$ will form.

b. Increasing the concentration of hydrogen will cause a net reaction to occur from right to left, forming more Fe and H_2O.

14.13 a. Because there are equal numbers of moles of gas on each side of the equation, increasing the pressure will not increase the amount of product.

b. Because the reaction increases the number of moles of gas, increasing the pressure will decrease the amount of product.

c. Because the reaction decreases the number of moles of gas, increasing the pressure will increase the amount of product.

14.14 Because this is an endothermic reaction and absorbs heat, high temperatures will be more favorable to the production of carbon monoxide.

14.15 Because this is an endothermic reaction and absorbs heat, high temperatures will give the best yield of carbon monoxide. Because the reaction increases the number of moles of gas, decreasing the pressure will increase the yield also.

■ Answers to Review Questions

14.1 A reasonable graph showing the decrease in concentration $N_2O_4(g)$ and the increase in concentration of $NO_2(g)$ is shown below:

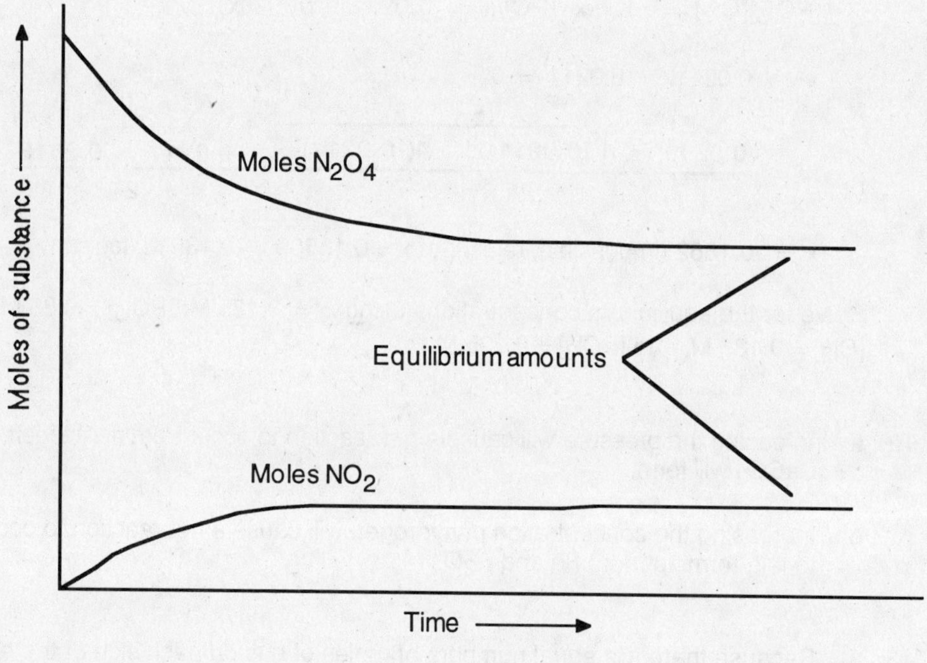

At first the concentration of N_2O_4 is large and the rate of the forward reaction is large, but then as the concentration of N_2O_4 decreases, the rate of the forward reaction decreases. In contrast, the concentration of NO_2 builds up from zero to a low concentration. Thus, the initial rate of the reverse reaction is zero, but it steadily increases as the concentration of NO_2 increases. Eventually the two rates become equal when the reaction reaches equilibrium. This is a dynamic equilibrium because both the forward and reverse reactions are occurring at all times even though there is no net change in concentration at equilibrium.

14.2 The 1.0 mol of $H_2(g)$ and 1.0 mol of $I_2(g)$ in the *first* mixture reaches equilibrium when the amounts of reactants decrease to 0.50 mol each and when the amount of product increases to 1.0 mol. The total number of moles of the reactants at the start is 2.0 mol, which is the same number of moles as in the *second* mixture, the 2.0 mol of HI that is to be allowed to come to equilibrium. The second mixture should produce the same number of moles of H_2, I_2, and HI at equilibrium because if the total number of moles is constant, it should not matter from which direction an equilibrium is approached.

14.3 The equilibrium constant for a gaseous reaction can be written using partial pressures instead of concentrations because all the reactants and products are in the same vessel. Therefore, at constant temperature, the pressure, P, is proportional to the concentration, n/V. (The ideal gas law says that $P = [n/V]RT$.)

14.4 The addition of reactions 1 and 2 yields reaction 3:

$$\text{(Reaction 1)} \quad HCN + OH^- \rightleftharpoons CN^- + H_2O$$

$$\text{(Reaction 2)} \quad H_2O \rightleftharpoons H^+ + OH^-$$

$$\text{(Reaction 3)} \quad HCN \rightleftharpoons H^+ + CN^-$$

The rule states that if a given equation can be obtained from the sum of other equations, the equilibrium constant for the given equation equals the product of the other equilibrium constants. Thus, K for reaction 3 is

$$K = K_1 \times K_2 = (4.9 \times 10^4) \times (1.0 \times 10^{-14}) = 4.9 \times 10^{-10}$$

14.5 a. Homogeneous equilibrium. All substances are gases and thus exist in one phase, a mixture of gases.

b. Heterogeneous equilibrium. The two copper compounds are solids, but the other substances are gases. This fulfills the definition of a heterogeneous equilibrium.

c. Homogeneous equilibrium. All substances are gases and thus exist in one phase, a mixture of gases.

d. Heterogeneous equilibrium. The two copper-containing substances are solids, but the other substances are gases. This fulfills the definition of a heterogeneous equilibrium.

14.6 Pure liquids and solids can be ignored in an equilibrium-constant expression because their concentrations do not change. (If a solid is present, it has not even dissolved in any gas or solution present.) In effect, concentrations of liquids and solids are incorporated into the value of K_c, as discussed in the chapter.

14.7 A qualitative interpretation of the equilibrium constant involves using the magnitude of the equilibrium constant to predict the relative amounts of reactants and products at equilibrium. If K_c is around 1, the equilibrium mixture contains appreciable amounts (same order of magnitude) of reactants and products. If K_c is large, the equilibrium mixture is mostly products. If K_c is small, the equilibrium mixture is mostly reactants. The type of reaction governs what "large" and "small" values are; but for some types of reactions, "large" might be no less than 10^2 to 10^4, whereas "small" might be no more than 10^{-4} to 10^{-2}.

14.8 The reaction quotient, Q_c, is an expression that has the same form as the equilibrium-constant expression but whose concentrations are not necessarily equilibrium concentrations. It is useful in determining whether a reaction mixture is at equilibrium or, if not, what direction the reaction will go as it approaches equilibrium.

14.9 The total number of moles of products (and reactants) in this problem is 2.00 mol, the same as the total number of moles of reactants (and products) in the example. Because the total number of moles is constant, this mixture of products should produce the same number of moles of both reactants and both products at equilibrium. (It does not matter from which direction an equilibrium is approached because the ratio of products to reactants is constant.) Thus, the concentration of H_2 is 0.0086 M; the other concentrations would also be the same.

14.10 The ways in which the equilibrium composition of a mixture can be altered are (1) changing concentrations by removing some of the products and/or some of the reactants, (2) changing the partial pressure of a gaseous reactant and/or a gaseous product, and (3) changing the temperature of the reaction mixture. (Adding a catalyst cannot alter the equilibrium concentration, but affects only the rate of reaction.)

14.11 The role of the platinum is that of a catalyst; it provides conditions (a finely divided surface) suitable for speeding up the attainment of equilibrium. The platinum has no effect on the equilibrium composition of the mixture even though it greatly increases the rate of reaction.

14.12 In some cases, a catalyst can affect the product in a reaction because it affects only the rate of one reaction out of several reactions that are possible. If two reactions are possible and the uncatalyzed rate of one is much slower but is the only reaction that is catalyzed, then the products with and without a catalyst will be different. The Ostwald process is a good example. In the absence of a catalyst, NH_3 burns in O_2 to form only N_2 and H_2O, even though it is possible for NH_3 to react to form NO and H_2O. Ostwald found that adding a platinum catalyst favors the formation of the NO and H_2O almost to the exclusion of the N_2 and H_2O.

14.13 Evidently, the equilibrium constant is of such a magnitude that the reaction can form significant amounts of substances on either side of the arrows under the right conditions. Let us write the reaction using iron oxide and hydrogen as reactants:

$$Fe_3O_4(s) + 4H_2(g) \rightleftharpoons 3Fe(s) + 4H_2O(g)$$

In the first case, the stream of H_2 gas furnishes an excess of H_2 and also removes the gaseous H_2O along with that stream. Thus, increasing the amount of one reactant and decreasing the amount of one product causes the reaction to shift in the direction of forming more products. In the second case, the stream of water vapor furnishes an excess of H_2O and also removes the gaseous H_2 along with that stream. Again, increasing the amount of one reactant and decreasing the amount of one product causes the reaction to shift in the direction of forming more products.

14.14 Four ways in which the yield of ammonia can be improved are (1) removing the gaseous NH_3 from the equilibrium by liquefying it, (2) increasing the nitrogen or hydrogen concentration, (3) increasing the total pressure on the mixture (the moles of gas decrease), and (4) lowering the temperature (ΔH^o is negative, so heat is evolved). Each causes a shift to the right in accordance with Le Chatelier's principle.

■ Solutions to Practice Problems

Note on significant figures: The final answer to each mathematical solution is given first with one nonsignificant figure (the rightmost significant figure is underlined) and then is rounded to the correct number of figures. Intermediate answers usually also have at least one nonsignificant figure.

14.15 Because the amount of NO_2 at the start is zero, the 0.30 mol of NO_2 at equilibrium must also equal the change, x, in the moles of N_2O_3. Use the "table" approach and insert the starting, change, and equilibrium number of moles:

Amt. (mol)	$N_2O_3(g)$	$\rightleftharpoons$	$NO_2(g)$	+	$NO(g)$
Starting	1.00		0		0
Change	-x (0.30)		+x (0.30)		+x (0.30)
Equilibrium	0.70		0.30		0.30

Hence, the equilibrium amounts are 0.70 mol N_2O_3, 0.30 mol NO_2, and 0.30 mol NO.

14.16 Use the "table" approach, giving the starting, change, and equilibrium number of moles of each.

Amt. (mol)	$PCl_5(g)$	⇌	$PCl_3(g)$	+	$Cl_2(g)$
Starting	1.500		0		0
Change	-x		+x		+x
Equilibrium	1.500 - x		x (= 0.203)		x

Hence, the equilibrium amounts are 1.297 mol PCl_5, 0.203 mol PCl_3, and 0.203 mol Cl_2.

14.17 Use the "table" approach, giving the starting, change, and equilibrium number of moles of each.

Amt. (mol)	2NO(g)	+	$Br_2(g)$	⇌	2NOBr(g)
Starting	0.0873		0.0437		0
Change	-2x		-x		+2x
Equilibrium	0.0873 - 2x		0.0437 - x		2x (= 0.0518)

Hence, x = 0.0259, and the equilibrium amounts are 0.0355 mol NO, 0.0178 mol Br_2, and 0.0518 mol NOBr.

14.18 Let the change in the moles of N_2 equal x. Because the amount of NH_3 at the start is zero, the 0.080 mol of NH_3 at equilibrium must also equal twice the change, 2x, in the moles of N_2. Thus, the change, x, in the moles of N_2 is 0.040 mol. Use the "table" approach and insert the starting, change, and equilibrium number of moles:

Amt. (mol)	$N_2(g)$	+	$3H_2(g)$	⇌	$2NH_3(g)$
Starting	1.000		3.000		0
Change	-x (0.040)		-3x (0.120)		+2x (0.080)
Equilibrium	0.960		2.880		0.080

Hence, the equilibrium amounts are 0.960 mol N_2, 2.880 mol H_2, and 0.080 mol NH_3.

14.19 Use the "table" approach, giving the starting, change, and equilibrium number of moles of each.

Amt. (mol)	$CO(g)$	+	$2H_2(g)$	⇌	$CH_3OH(g)$
Starting	0.1500		0.3000		0
Change	-x		-2x		+x
Equilibrium	0.1500 - x (= 0.1187)		0.3000 - 2x		x

Because 0.1500 - x = 0.1187, x = 0.0313. Therefore, the amounts of substances at equilibrium are 0.1187 mol CO, 0.2374 mol H_2, and 0.0313 mol CH_3OH.

14.20 Use the "table" approach, giving the starting, change, and equilibrium number of moles of each.

Amt. (mol)	$2SO_2(g)$	+	$O_2(g)$	⇌	$2SO_3(g)$
Starting	0.0400		0.0200		0
Change	-2x		-x		+2x
Equilibrium	0.0400 - 2x		0.0200 - x		2x (= 0.0296)

Therefore, x = 0.0148, and the amounts of substances at equilibrium are 0.0104 mol SO_2, 0.0052 mol O_2, and 0.0296 mol SO_3.

14.21 a. $K_c = \dfrac{[POCl][Cl_2]}{[POCl_3]}$ c. $K_c = \dfrac{[CH_3CHO]^2}{[C_2H_4]^2[O_2]}$

b. $K_c = \dfrac{[H_2O]^2}{[H_2]^2[O_2]}$ d. $K_c = \dfrac{[H_2O]^2[SO_2]^2}{[H_2S]^2[O_2]^3}$

14.22 a. $K_c = \dfrac{[PCl_5]}{[PCl_3][Cl_2]}$ c. $K_c = \dfrac{[NO]^2[Cl_2]}{[NOCl]^2}$

b. $K_c = \dfrac{[O_3]^2}{[O_2]^3}$ d. $K_c = \dfrac{[N_2]^2[H_2O]^6}{[NH_3]^4[O_2]^3}$

14.23 The reaction is $4NO(g) + 6H_2O(g) \rightleftharpoons 4NH_3(g) + 5O_2(g)$.

14.24 The reaction is $CH_4(g) + 2H_2S(g) \rightleftharpoons CS_2(g) + 4H_2(g)$.

14.25 The equilibrium-constant expressions when the reaction is halved and then reversed are

$$\text{Halved: } K_c = \frac{[N_2]^{1/2}[H_2O]}{[NO][H_2]} \qquad \text{Reversed: } K_c = \frac{[NO][H_2]}{[N_2]^{1/2}[H_2O]}$$

14.26 The equilibrium-constant expressions when the reaction is halved and then reversed are

$$\text{Halved: } K_c = \frac{[H_2O][Cl_2]}{[HCl]^2[O_2]^{1/2}} \qquad \text{Reversed: } K_c = \frac{[HCl]^2[O_2]^{1/2}}{[H_2O][Cl_2]}$$

14.27 Because $K_c = 1.84$ for $2HI \rightleftharpoons H_2 + I_2$, the value of K_c for $H_2 + I_2 \rightleftharpoons 2HI$ must be the reciprocal of K_c for the first reaction. Mathematically, this can be shown as follows:

$$\text{Forward: } K_c = \frac{[H_2][I_2]}{[HI]^2} = 1.84$$

$$\text{Reverse: } K_c = \frac{[HI]^2}{[H_2][I_2]} = \frac{1}{\dfrac{[H_2][I_2]}{[HI]^2}} = \frac{1}{K_c(f)}$$

Thus, for the reverse reaction K_c is calculated as follows:

$$K_c = 1 \div 1.84 = 5.4\underline{3}4 \times 10^{-1} = 5.43 \times 10^{-1}$$

14.28 Because $K_c = 27.8$ for $CS_2 + 4H_2 \rightleftharpoons CH_4 + 2H_2S$, the value of K_c for $1/2CS_2 + 2H_2 \rightleftharpoons 1/2CH_4 + H_2S$, the second (**2**) reaction, must be the square root of K_c for the first (**1**) reaction. Mathematically, this can be shown as follows:

$$\mathbf{1}\text{: } K_c(\mathbf{1}) = \frac{[CH_4][H_2S]^2}{[CS_2][H_2]^4}$$

$$\mathbf{2}\text{: } K_c = \frac{[CH_4]^{1/2}[H_2S]}{[CS_2]^{1/2}[H_2]^2} = \left[\frac{[CH_4][H_2S]^2}{[CS_2][H_2]^4}\right]^{1/2} = [K_c(\mathbf{1})]^{1/2}$$

Thus, for the second (**2**) reaction K_c is calculated as follows:

$$K_c = \sqrt{27.8} = 5.2\underline{7}2 = 5.27$$

14.29 First calculate the molar concentrations of each of the compounds in the equilibrium:

$[PCl_5] = 0.0158 \text{ mol } PCl_5 \div 5.00 \text{ L} = 0.0031\underline{6}0 \text{ M}$

$[PCl_3] = 0.0185 \text{ mol } PCl_3 \div 5.00 \text{ L} = 0.0037\underline{0}0 \text{ M}$

$[Cl_2] = 0.0870 \text{ mol } Cl_2 \div 5.00 \text{ L} = 0.017\underline{4}0 \text{ M}$

Now substitute these into the expression for K_c:

$$K_c = \frac{0.003160}{(0.003700)(0.01740)} = 49.\underline{0}8 = 49.1/\text{M}$$

14.30 First calculate the molar concentrations of each of the compounds in the equations:

$[H_2] = 0.650 \text{ mol } H_2 \div 8.00 \text{ L} = 0.081\underline{2}5 \text{ M}$

$[I_2] = 0.275 \text{ mol } I_2 \div 8.00 \text{ L} = 0.034\underline{3}8 \text{ M}$

$[HI] = 3.000 \text{ mol HI} \div 8.00 \text{ L} = 0.37\underline{5}0 \text{ M}$

Now substitute into the K_c expression:

$$K_c = \frac{(0.3750)^2}{(0.08125)(0.03438)} = 50.\underline{3}423 = 50.3$$

14.31 Substitute the following concentrations into the K_c expression:

$[CH_3OH] = 0.0313 \text{ mol} \div 1.50 \text{ L} = 0.020\underline{8}6 \text{ M}$

$[CO] = 0.1187 \text{ mol} \div 1.50 \text{ L} = 0.079\underline{1}3 \text{ M}$

$[H_2] = 0.2374 \text{ mol} \div 1.50 \text{ L} = 0.15\underline{8}2$

$$K_c = \frac{(0.02086)}{(0.07913)(0.1582)^2} = 1.0\underline{5}3 \times 10^1 = 1.05 \times 10^1 /\text{M}^2$$

14.32 Substitute the following concentrations into the K_c expression:

$$[SO_3] = 0.0296 \text{ mol} \div 2.000 \text{ L} = 0.014\underline{8}0 \text{ M}$$

$$[SO_2] = 0.0104 \text{ mol} \div 2.000 \text{ L} = 0.0052\underline{0}0 \text{ M}$$

$$[O_2] = 0.0052 \text{ mol} \div 2.000 \text{ L} = 0.002\underline{6}0 \text{ M}$$

$$K_c = \frac{(0.01480)^2}{(0.005200)^2\,(0.00260)} = 3.1\underline{1} \times 10^3 = 3.1 \times 10^3 \text{ /M}$$

14.33 For each mole of NOBr that reacts, (1.000 - 0.094 = 0.906) mol remains. Starting with 2.00 mol NOBr, 2 x 0.906 mol NOBr, or 1.812 mol NOBr, remains. Because the volume is 1.00 L, the concentration of NOBr at equilibrium is 1.812 M. Assemble a table of starting, change, and equilibrium concentrations:

Conc. (M)	$2NOBr(g)$	$\rightleftharpoons$	$2NO(g)$	+	$Br_2(g)$
Starting	2.00		0		0
Change	-2x		+2x		+x
Equilibrium	2.00 - 2x (= 1.812)		2x		x

Because 2.00 - 2x = 1.812, x = 0.094 M. Therefore, the equilibrium concentrations are [NOBr] = 1.812 M, [NO] = 0.188 M, and $[Br_2]$ = 0.094 M.

$$K_c = \frac{[NO]^2[Br_2]}{[NOBr]^2} = \frac{(0.188 \text{ M})^2(0.094 \text{ M})}{(1.812 \text{ M})^2} = 1.\underline{0}1 \times 10^{-3} = 1.0 \times 10^{-3} \text{ M}$$

14.34 A decomposition of 6.0% of NO_2 means that 0.060 x 2.00 mol = 0.12 mol was decomposed, leaving 1.88 mol at equilibrium. Because the concentration at the start was 2.00 mol/80.0 L, or 0.0250 M, the concentration at equilibrium is 1.88 mol/80.0 L, or 0.0235 M. Assemble a table of starting, change, and equilibrium concentrations:

Conc. (M)	$2NO_2(g)$	$\rightleftharpoons$	$2NO(g)$	+	$O_2(g)$
Starting	0.0250		0		0
Change	-2x		+2x		+x
Equilibrium	0.0250 - 2x (= 0.0235)		2x		x

(continued)

Because 0.0250 - 2x = 0.0235 M, x = 0.00075 M; thus, $[NO_2]$ = 0.0235 M, [NO] = 0.0015 M, and $[O_2]$ = 0.00075 M. The value of K_c is

$$K_c = \frac{(0.0015)^2(0.00075)}{(0.0235)^2} = 3.\underline{0}55 \times 10^{-6} = 3.1 \times 10^{-6} \text{ M}$$

14.35 a. $K_p = \dfrac{P_{HBr}^2}{P_{H_2}P_{Br_2}}$ c. $K_p = \dfrac{P_{H_2O}^2 P_{Cl_2}^2}{P_{HCl}^4 P_{O_2}}$

b. $K_p = \dfrac{P_{CH_4}P_{H_2S}^2}{P_{CS_2}P_{H_2}^4}$ d. $K_p = \dfrac{P_{CH_3OH}}{P_{CO}P_{H_2}^2}$

14.36 a. $K_p = \dfrac{P_{NO_2}^2}{P_{N_2O_4}}$ c. $K_p = \dfrac{P_{SO_3}^2}{P_{SO_2}^2 P_{O_2}}$

b. $K_p = \dfrac{P_{NOBr}^2}{P_{NO}^2 P_{Br_2}}$ d. $K_p = \dfrac{P_{NO}^4 P_{H_2O}^6}{P_{NH_3}^4 P_{O_2}^5}$

14.37 There are 3 mol of gaseous product for every 5 mol of gaseous reactant, so $\Delta n = 3 - 5$ $= -2$. Using this, calculate K_p from K_c:

$$K_p = K_c(RT)^{\Delta n} = 0.28\ (0.0821 \times 1173)^{-2} = 3.\underline{0}19 \times 10^{-5} = 3.0 \times 10^{-5}$$

14.38 There is 1 mol of gaseous product for every 3 mol of gaseous reactant, so $\Delta n = 1 - 3 = -2$. Using this, calculate K_p from K_c:

$$K_p = K_c(RT)^{\Delta n} = 10.5\ (0.0821 \times 500)^{-2} = 6.2\underline{3}1 \times 10^{-3} = 6.23 \times 10^{-3}$$

14.39 For each 1 mol of gaseous product, there are 1.5 mol of gaseous reactants; thus, $\Delta n = 1 - 1.5 = -0.5$. Using this, calculate K_c from K_p:

$$K_c = \frac{K_p}{(RT)^{\Delta n}} = \frac{6.55}{(0.0821 \times 900)^{-0.5}} = 6.55 \times (0.0821 \times 900)^{0.5}$$

$$= 56.\underline{3}03 = 56.3$$

14.40 For each 1 mol of gaseous product, there is 0.5 mol of gaseous reactant; thus, $\Delta n = 1 - 0.5 = 0.5$. Using this, calculate K_c from K_p:

$$K_c = \frac{K_p}{(RT)^{\Delta n}} = \frac{7.55 \times 10^{-2}}{(0.0821 \times 1115)^{0.5}} = 7.8\underline{9}1 \times 10^{-3} = 7.89 \times 10^{-3}$$

14.41 a. $K_c = \frac{[CO]^2}{[CO_2]}$ c. $K_c = \frac{[CO_2]}{[SO_2][O_2]^{1/2}}$

b. $K_c = \frac{[CO_2]}{[CO]}$ d. $K_c = [Pb^{2+}][I^-]^2$

14.42 a. $K_c = [NH_3][HCl]$ c. $K_c = [H_2O][CO_2]$

b. $K_c = \frac{[CO_2][N_2]^2}{[N_2O]^2}$ d. $K_c = \frac{1}{[Fe^{3+}][OH^-]^3}$

14.43 a. Nearly complete; K_c is very large (> 10^2 to 10^4), indicating nearly complete reaction.

b. Not complete; K_c is very small (< 10^{-2} to 10^{-4}), indicating very little reaction.

14.44 a. Not complete; K_c is very small (< 10^{-2} to 10^{-4}), indicating very little reaction.

b. Nearly complete; K_c is very large (> 10^2 to 10^4), indicating nearly complete reaction.

14.45 K_c is extremely small, indicating very little reaction at room temperature. Because the decomposition of HF yields equal amounts of H_2 and F_2, at equilibrium $[H_2] = [F_2]$. So for the decomposition of HF,

$$K_c = \frac{[H_2][F_2]}{[HF]^2} = \frac{[H_2]^2}{[HF]^2}$$

$$[H_2] = (K_c)^{1/2}\,[HF] = (1.0 \times 10^{-95})^{1/2}\,(1.0\ M) = 3.\underline{1}6 \times 10^{-48} = 3.2 \times 10^{-48}\ mol/L$$

This result does agree with what is expected from the very small magnitude of K_c.

14.46 K_c is extremely large, indicating nearly complete reaction at room temperature. The problem states that at equilibrium $[SO_2] = [O_2]$. So for the reaction of SO_2,

$$K_c = \frac{[SO_3]^2}{[SO_2]^2[O_2]} = \frac{[SO_3]^2}{[SO_2]^3}$$

Solving for $[SO_2]$

$$[SO_2] = \left[\frac{[SO_3]^2}{K_c}\right]^{1/3} = \left[\frac{(1.0)^2}{8.0 \times 10^{35}}\right]^{1/3} = 1.\underline{0}7 \times 10^{-12} = 1.1 \times 10^{-12}\ M$$

This agrees with what we expect from the magnitude of K_c.

14.47 Calculate Q, the reaction quotient, and compare it to the equilibrium constant. If Q is larger, the reaction will go to the left, and vice versa. In all cases, Q is found by combining these terms:

$$Q = \frac{[NO]^2[Br_2]}{[NOBr]^2}$$

a. $$Q = \frac{(0.0151)^2(0.0108)}{(0.0610)^2} = 6.6\underline{1}7 \times 10^{-4}\ (> 3.07 \times 10^{-4})$$

The reaction goes to the left.

b. $$Q = \frac{(0.0169)^2(0.0142)}{(0.115)^2} = 3.0\underline{6}6 \times 10^{-4}\ (= 3.07 \times 10^{-4})$$

The reaction is at equilibrium.

c. $$Q = \frac{(0.0123)^2(0.0201)}{(0.181)^2} = 9.2\underline{8}2 \times 10^{-5}\ (< 3.07 \times 10^{-4})$$

The reaction goes to the right.

d. $$Q = \frac{(0.0105)^2(0.0100)}{(0.0450)^2} = 5.4\underline{4}4 \times 10^{-4}\ (> 3.07 \times 10^{-4})$$

The reaction goes to the left.

14.48 Calculate Q, the reaction quotient, and compare it to the equilibrium constant. If Q is larger, the reaction goes to the left, and vice versa. For all, Q is found by combining these terms:

$$Q = \frac{[CS_2][H_2]^4}{[CH_4][H_2S]^2}$$

a. $Q = \frac{(1.51)(1.08)^4}{(1.15)(1.20)^2} = 1.2\underline{4}05$ (< 3.59; reaction goes to right)

b. $Q = \frac{(0.90)(1.78)^4}{(1.07)(1.20)^2} = 5.8\underline{6}3$ (> 3.59; reaction goes to left)

c. $Q = \frac{(1.10)(1.68)^4}{(1.10)(1.49)^2} = 3.5\underline{8}8$ (= 3.59; reaction is at equilibrium)

d. $Q = \frac{(1.25)(1.75)^4}{(1.45)(1.29)^2} = 4.8\underline{5}8$ (> 3.59; reaction goes to left)

14.49 Calculate Q, the reaction quotient, and compare it to the equilibrium constant. If Q is larger, the reaction will go to the left, and vice versa. Q is found by combining these terms:

$$Q = \frac{[CH_3OH]}{[CO][H_2]^2} = \frac{(0.020)}{(0.10)(0.10)^2} = 2.\underline{0}0 \times 10^1$$ (> 10.5; reaction goes to left)

14.50 Calculate Q, the reaction quotient, and compare it to the equilibrium constant. If Q is larger, the reaction will go to the left, and vice versa. Q is found by combining these terms:

$$Q = \frac{[SO_3]^2}{[SO_2]^2[O_2]} = \frac{(0.40)^2}{(0.20)^2(0.10)} = 4.\underline{0}0 \times 10^1$$ (> 0.0417)

The reaction goes to the left.

14.51 Substitute into the expression for K_c and solve for $[COCl_2]$:

$$K_c = 1.23 \times 10^3 = \frac{[COCl_2]}{[CO][Cl_2]} = \frac{(COCl_2)}{(0.012)(0.025)}$$

$$[COCl_2] = (1.23 \times 10^3)(0.012)(0.025) = 0.3\underline{6}9 = 0.37\ M$$

14.52 Substitute into the expression for K_c and solve for [NO]:

$$K_c = 0.0025 = \frac{[NO]^2}{[O_2][N_2]} = \frac{(NO)^2}{(0.031)(0.023)}$$

$$[NO] = [(0.0025)(0.031)(0.023)]^{1/2} = 1.\underline{3}3 \times 10^{-3} = 1.3 \times 10^{-3}\ M$$

14.53 Divide moles of substance by the volume of 5.0 L to obtain concentration. The starting concentrations are 3.0×10^{-4} M for both [I_2] and [Br_2]. Assemble a table of starting, change, and equilibrium concentrations.

Conc. (M)	$I_2(g)$	+	$Br_2(g)$	$\rightleftharpoons$	2 IBr(g)
Starting	3.0×10^{-4}		3.0×10^{-4}		0
Change	-x		-x		+2x
Equilibrium	(3.0×10^{-4}) - x		(3.0×10^{-4}) - x		2x

Substituting into the equilibrium-constant expression gives

$$K_c = 1.2 \times 10^2 = \frac{[IBr]^2}{[I_2][Br_2]} = \frac{(2x)^2}{(3.0 \times 10^{4} - x)(3.0 \times 10^{4} - x)}$$

Taking the square root of both sides yields

$$1\underline{0}.95 = \frac{(2x)}{(3.0 \times 10^{-4} - x)}$$

Rearranging and simplifying the right side gives

$$(3.0 \times 10^{-4} - x) = \frac{(2x)}{10.95} = (0.1\underline{8}2\ x)$$

$$x = 2.\underline{5}3 \times 10^{-4}\ M$$

Thus, [I_2] = [Br_2] = $4.\underline{7}0 \times 10^{-5} = 4.7 \times 10^{-5}$ M, and [IBr] = $5.\underline{0}6 \times 10^{-4} = 5.1 \times 10^{-4}$ M.

14.54 Divide moles of substance by the volume of 8.00 L to obtain concentration. The starting concentrations are 0.10$\underline{6}$25 M for both [N_2] and [O_2]. Assemble a table of starting, change, and equilibrium concentrations.

Conc. (M)	$N_2(g)$	+	$O_2(g)$	$\rightleftharpoons$	2NO(g)
Starting	0.10$\underline{6}$25		0.10$\underline{6}$25		0
Change	-x		-x		+2x
Equilibrium	0.10$\underline{6}$25 - x		0.10$\underline{6}$25 - x		2x

$$K_c = 0.0123 = \frac{[NO]^2}{[N_2][O_2]} = \frac{(2x)^2}{(0.10625 - x)(0.10625 - x)}$$

Taking the square root of both sides gives

$$0.11\underline{0}9 = \frac{(2x)}{(0.10625 - x)}$$

Rearranging and simplifying the right side yields

$$(0.10625 - x) = \frac{(2x)}{0.1109} = (18.03\,x)$$

$$x = 5.5\underline{8}3 \times 10^{-3}\,M$$

Thus [N_2] = [O_2] = 0.10$\underline{0}$6 = 0.101 M, and [NO] = 1.1$\underline{1}$6 x 10^{-2} = 1.12 x 10^{-2} M.

14.55 Divide moles of substance by the volume of 5.0 L to obtain concentration. The starting concentrations are 0.10 M for both [PCl_3] and [Cl_2]. Assemble a table of starting, change, and equilibrium concentrations.

Conc. (M)	$PCl_3(g)$	+	$Cl_2(g)$	$\rightleftharpoons$	$PCl_5(g)$
Starting	0.10		0.10		0
Change	-x		-x		+x
Equilibrium	0.10 - x		0.10 - x		x

Substituting into the equilibrium expression for K_c gives

$$K_c = 49 = \frac{[PCl_5]}{[PCl_3][Cl_2]} = \frac{x}{(0.10 - x)(0.10 - x)}$$

(continued)

Rearranging and solving for x yields

$$49(0.10 - x)^2 = 49(x^2 - 0.20x + 0.010) = x$$

$$49x^2 - 10.8x + 0.49 = 0 \text{ (quadratic equation)}$$

Using the solution to the quadratic gives

$$x = \frac{10.8 \pm \sqrt{(10.8)^2 - 4(49)(0.49)}}{2(49)}$$

$x = 0.157$ (impossible; reject), or $x = 0.06\underline{3}89 = 0.064$ M (logical)

Thus, $[PCl_3] = [Cl_2] = 0.036$ M, and $[PCl_5] = 0.064$ M; the vessel contains 0.18 mol PCl_3, 0.18 mol Cl_2, and 0.32 mol PCl_5.

14.56 Divide moles of substance by the volume of 1.00 L to obtain concentration. The starting concentration is 1.00 M for $[CO_2]$, but the concentration of carbon, a solid, is omitted. Assemble a table of starting, change, and equilibrium concentrations.

Conc. (M)	$CO_2(g)$	+	C(s)	$\rightleftharpoons$	2CO(g)
Starting	1.00				0
Change	-x				+2x
Equilibrium	1.00 - x				2x

Substituting into the equilibrium expression for K_c gives

$$K_c = 14.0 = \frac{[CO]^2}{[CO_2]} = \frac{(2x)^2}{(1.00 - x)}$$

Rearranging and solving for x yields

$$14.0 - 14.0x = 4x^2$$

$$4x^2 + 14.0x - 14.0 = 0 \text{ (quadratic equation)}$$

Using the solution to the quadratic gives

$$x = \frac{-14.0 \pm \sqrt{(14.0)^2 - 4(4)(-14.0)}}{2(4)}$$

$x = -4.31$ (impossible; reject), or $x = 0.81\underline{1}7 = 0.812$ M (logical)

Thus, $[CO_2] = 0.19$ M, and $[CO] = 1.62$ M.

14.57 Divide moles of substance by the volume of 10.00 L to obtain concentration. The starting concentrations are 0.1000 M for [CO] and 0.3000 M for [H_2]. Assemble a table of starting, change, and equilibrium concentrations.

Conc. (M)	CO(g)	+	$3H_2(g)$	$\rightleftharpoons$	$CH_4(g)$	+	$H_2O(g)$
Starting	0.1000		0.3000		0		0
Change	-x		-3x		+x		+x
Equilibrium	0.1000 - x		0.3000 - 3x		x		x

Substituting into the equilibrium expression for K_c gives

$$K_c = 3.92 = \frac{[CH_4][H_2O]}{[CO][H_2]^3} = \frac{x^2}{(0.1000 - x)[3(0.1000 - x)]^3}$$

$$= \frac{x^2}{27(0.1000 - x)^4}$$

Multiplying both sides by 27 and taking the square root of both sides gives

$$10.29 = \frac{x}{(0.1000 - x)^2}, \text{ or } 10.29x^2 - 3.058x + 0.1029 = 0$$

Using the solution to the quadratic equation yields

$$x = \frac{3.058 \pm \sqrt{(-3.058)^2 - 4(10.29)(0.1029)}}{2(10.29)}$$

x = 0.2585 (can't be > 0.1000, so reject), or x = 0.03868 = 0.0387 M (use)

Thus, [CO] = 0.0613 M, [H_2] = 0.1839 M, [CH_4] = 0.0387 M, and [H_2O] = 0.0387 M.

14.58 Divide moles of substance by the volume of 2.00 L to obtain concentration. The starting concentrations are 0.500 M for [N_2] and 1.50 M for [H_2]. Assemble a table of starting, change, and equilibrium concentrations.

Conc. (M)	$N_2(g)$	+	$3H_2(g)$	$\rightleftharpoons$	$2NH_3(g)$
Starting	0.500		1.50		0
Change	-x		-3x		+2x
Equilibrium	0.500 - x		1.50 - 3x		2x

(continued)

Substituting into the equilibrium expression for K_c gives

$$K_c = 0.159 = \frac{[NH_3]^3}{[N_2][H_2]^3} = \frac{(2x)^2}{(0.500 - x)(1.50 - 3x)^3}$$

$$= \frac{(2x)^2}{(0.500 - x)\,3^3\,(0.500 - x)^3}$$

Combining the value of K_c and 3^3 from the right-hand term gives

$$(3^3)0.159 = \frac{(2x)^2}{0.500 - x}$$

Taking the square root of both sides gives

$$2.0\underline{7}2 = \frac{2x}{(0.5000 - x)^2}$$

$$0.5180 - 2.072x + 2.072x^2 = 2x$$

$$2.072x^2 - 4.072x + 0.5180 = 0 \quad \text{(quadratic equation)}$$

Using the solution to the quadratic equation yields

$$x = \frac{4.072 \pm \sqrt{(-4.072)^2 - 4(2.072)(0.5180)}}{2(2.072)}$$

$x = 1.829$ (larger than 0.500, so reject), or $x = 0.13\underline{6}7$ M (logical; use)

Thus, $[N_2] = 0.36\underline{3}3 = 0.363$ M, $[H_2] = 1.0\underline{8}99 = 1.09$ M, and $[NH_3] = 0.27\underline{3}4$

$= 0.273$ M.

14.59 Forward direction

14.60 a. Forward direction b. Reverse direction

14.61 a. A pressure increase has no effect because number of moles of reactants equals that of products.

b. A pressure increase has no effect because number of moles of reactants equals that of products.

c. A pressure increase causes the reaction to go to the left because number of moles of reactants is less than that of products.

14.62 a. Pressure increase b. Pressure increase c. Pressure decrease

14.63 The fraction would not increase, because an increase in temperature decreases the amounts of products of an exothermic reaction.

14.64 The decomposition would be favorable at high temperatures because an increase in temperature increases the amounts of products of an endothermic reaction.

14.65 The value of ΔH^o is calculated from the $\Delta H^o{}_f$ values below each substance in the reaction:

$2NO_2(g)$	+	$7H_2(g)$	$\rightleftharpoons$	$2NH_3(g)$	+	$4H_2O(g)$
2(33.2)		7(0)		2(-45.9)		4(-241.8)

ΔH^o = -967.2 + (-91.8) - 66.4 = -1125.4 kJ/2 mol NO_2

The equilibrium constant will decrease with temperature because raising the temperature of an exothermic reaction will cause the reaction to go farther to the left.

14.66 The value of ΔH^o is calculated from the $\Delta H^o{}_f$ values below each substance in the reaction:

$CH_4(g)$	+	$2H_2S(g)$	$\rightleftharpoons$	$CS_2(g)$	+	$4H_2(g)$
-74.9		2(-20)		117		4(0)

ΔH^o = 117 - (-40) - (-74.9) = $23\underline{1}.9$ = 232 kJ/mol CH_4

The equilibrium constant will increase with temperature because raising the temperature of an endothermic reaction will cause the reaction to go farther to the right.

14.67 Because the reaction is exothermic, the formation of products will be favored by low temperatures. Because there are more molecules of gaseous products than of gaseous reactants, the formation of products will be favored by low pressures.

14.68 Because the reaction is exothermic, the formation of products will be favored by low temperatures. Because there are more molecules of gaseous reactants than of gaseous products, the formation of products will be favored by high pressures.

■ Solutions to Unclassified Problems

14.69 Substitute the concentrations into the equilibrium expression to calculate K_c.

$$K_c = \frac{[CH_3OH]}{[CO][H_2]^2} = \frac{(0.015)}{(0.096)(0.191)^2} = 4.\underline{2}8 = 4.3$$

14.70 Substitute the concentrations into the equilibrium expression to calculate K_c.

$$K_c = \frac{[SO_3]}{[SO_2][O_2]^2} = \frac{(0.0160)}{(0.0056)(0.0021)^{1/2}} = 6\underline{2}.3 = 62$$

14.71 Assume 100.00 g of gas: 90.55 g are CO, and 9.45 g are CO_2. The moles of each are

$$90.55 \text{ g CO} \times \frac{1 \text{ mol CO}}{28.01 \text{ g CO}} = 3.23\underline{2}8 \text{ mol;}$$

$$9.45 \text{ g } CO_2 \times \frac{1 \text{ mol } CO_2}{44.01 \text{ g } CO_2} = 0.21\underline{4}7 \text{ mol } CO_2$$

Total moles of gas = (3.2328 + 0.2147) mol = 3.44$\underline{7}$5 mol. Use the ideal gas law to convert to the volume of gaseous solution:

$$V = \frac{nRT}{P} = \frac{(3.4475 \text{ mol})[0.082057 \text{ L}\bullet\text{atm/(K}\bullet\text{mol)}](850 + 273 \text{ K})}{1.000 \text{ atm}} = 317.\underline{6}9 \text{ L}$$

The concentrations are

$$[CO] = \frac{3.2328 \text{ mol CO}}{317.69 \text{ L}} = 0.0101\underline{7}6 \text{ M;}$$

$$[CO_2] = \frac{0.2147 \text{ mol } CO_2}{317.69 \text{ L}} = 6.7\underline{5}8 \times 10^{-4} \text{ M}$$

Find K_c by substituting into the equilibrium expression:

$$K_c = \frac{[CO]^2}{[CO_2]} = \frac{(0.0101\underline{7}6)^2}{(6.758 \times 10^{-4})} = 0.15\underline{3}2 = 0.153 \text{ M}$$

14.72 Assume 100.0 g of gas: 65.8 g are NO_2 and 34.2 g are N_2O_4. The moles of each are

$$65.8 \text{ g } NO_2 \times \frac{1 \text{ mol } NO_2}{46.01 \text{ g } NO_2} = 1.4\underline{3}0 \text{ mol};$$

$$34.2 \text{ g } N_2O_4 \times \frac{1 \text{ mol } N_2O_4}{92.01 \text{ g } N_2O_4} = 0.37\underline{1}7 \text{ mol}$$

Total moles of gas = (1.430 + 0.3717) mol = 1.80$\underline{2}$ mol. Use the ideal gas law to convert to the volume of gaseous solution:

$$V = \frac{nRT}{P} = \frac{(1.802 \text{ mol})[0.082057 \text{ L•atm/(K•mol)}](25 + 273 \text{ K})}{1.00 \text{ atm}} = 44.\underline{0}6 \text{ L}$$

The concentrations are

$$[NO_2] = \frac{1.430 \text{ mol } NO_2}{44.06 \text{ L}} = 0.032\underline{4}6 \text{ M};$$

$$[N_2O_4] = \frac{0.3717 \text{ mol } N_2O_4}{44.06 \text{ L}} = 0.0084\underline{3}6 \text{ M}$$

Find K_c by substituting into the equilibrium expression:

$$K_c = \frac{[NO_2]^2}{[N_2O_4]} = \frac{(0.032\underline{4}6)^2}{(0.008436)} = 0.12\underline{4}8 = 0.125 \text{ M}$$

14.73 After calculating the concentrations after mixing, calculate Q, the reaction quotient, and compare it with K_c.

$$[N_2] = [H_2] = 1.00 \text{ mol} \div 2.00 \text{ L} = 0.500 \text{ M}$$

$$[NH_3] = 2.00 \text{ mol} \div 2.00 \text{ L} = 1.00 \text{ M}$$

$$Q = \frac{[NH_3]^2}{[N_2][H_2]^3} = \frac{(1.00)^2}{(0.500)(0.500)^3} = 16.\underline{0}0 = 16.0/M^2$$

Because Q is greater than K_c, the reaction will go in the reverse direction (to the left) to reach equilibrium.

14.74 To calculate the concentrations after mixing, the volume can be assumed to be 1.00 L, or symbolized as V. Because the volumes in the numerator and denominator cancel each other, it does not matter. Assume a 1.00-L volume and calculate Q, the reaction quotient, and compare it with K_c.

$$[I_2] = [Br_2] = 0.0010 \text{ mol}/1.00 \text{ L}$$

$$[IBr] = 0.200 \div 1.00 \text{ L} = 0.200 \text{ mol}/1.00 \text{ L}$$

$$Q = \frac{[IBr]^2}{[I_2][Br_2]} = \frac{(0.200 \text{ mol}/1.00 \text{ L})^2}{(0.0010 \text{ mol}/1.00 \text{ L})(0.0010 \text{ mol}/1.00 \text{ L})} = 4.00 \times 10^4$$

Because Q is greater than K_c, the reaction will go in the reverse direction (to the left) to reach equilibrium.

14.75 To calculate the concentrations after mixing, assume the volume to be 1.00 L, symbolized as V. Because the volumes in the numerator and denominator cancel each other, they do not matter. Assume a 1.00-L volume and calculate Q, the reaction quotient, and compare it with K_c.

$$[CO] = [H_2O] = [CO_2] = [H_2] = 1.00 \text{ mol}/1.00 \text{ L}$$

$$Q = \frac{[CO_2][H_2]}{[CO][H_2O]} = \frac{(1.00 \text{ mol}/1.00 \text{ L})(1.00 \text{ mol}/1.00 \text{ L})}{(1.000 \text{ mol}/1.00 \text{ L})(1.00 \text{ mol}/1.00 \text{ L})} = 1.00$$

Because Q is greater than K_c, the reaction will go in the reverse direction (left) to reach equilibrium.

14.76 After calculating the concentrations after mixing, calculate Q, the reaction quotient, and compare it with K_c.

$$[CO] = 0.10 \text{ mol} \div 2.0 \text{ L} = 0.050 \text{ M}$$

$$[H_2] = 0.20 \text{ mol} \div 2.0 \text{ L} = 0.10 \text{ M}$$

$$[CH_3OH] = 0.50 \text{ mol} \div 2.0 \text{ L} = 0.25 \text{ M}$$

$$Q = \frac{[CH_3OH]}{[CO][H_2]^2} = \frac{(0.25)}{(0.050)(0.10)^2} = 5.00 \times 10^2 = 5.0 \times 10^2/\text{M}^2$$

Because Q is greater than K_c, the reaction will go in the reverse direction (left) to reach equilibrium.

14.77 Assemble a table of starting, change, and equilibrium concentrations, letting 2x = the change in [HBr].

Conc. (M)	$2HBr(g)$	$\rightleftharpoons$	$H_2(g)$	+	$Br_2(g)$
Starting	0.010		0		0
Change	-2x		+x		+x
Equilibrium	0.010 - 2x		x		x

$$K_c = 0.016 = \frac{[H_2][Br_2]}{[HBr]^2} = \frac{(x)(x)}{(0.010 - 2x)^2}$$

$$0.1\underline{2}6 = \frac{(x)}{(0.010 - 2x)}$$

$$1.26 \times 10^{-3} - (2.52 \times 10^{-1})x = x$$

$$x = (1.26 \times 10^{-3}) \div 1.252 = 1.\underline{0}06 \times 10^{-3} = 1.0 \times 10^{-3}\ M$$

Therefore, [HBr] = 0.008 M, or 0.008 mol; $[H_2]$ = 0.0010 M, or 0.0010 mol; and $[Br_2]$ = 0.0010 M, or 0.0010 mol.

14.78 The starting concentration of [IBr] = 0.010 mol ÷ 1.0 L = 0.010 M. Assemble a table of starting, change, and equilibrium concentrations.

Conc. (M)	$2IBr(g)$	$\rightleftharpoons$	$I_2(g)$	+	$Br_2(g)$
Starting	0.010		0		0
Change	-2x		+x		+x
Equilibrium	0.010 - 2x		x		x

$$K_c = 0.026 = \frac{[I_2][Br_2]}{[IBr]^2} = \frac{(x)(x)}{(0.010 - 2x)^2}$$

$$0.1\underline{6}1 = \frac{(x)}{(0.010 - 2x)}$$

$$1.61 \times 10^{-3} - (3.22 \times 10^{-1})x = x$$

$$x = (1.61 \times 10^{-3}) \div 1.322 = 1.\underline{2}17 \times 10^{-3} = 1.2 \times 10^{-3}\ M$$

Therefore, [IBr] = 0.00$\underline{7}$6 M; or 0.008 mol; $[I_2]$ = 0.001$\underline{2}$1 M, or 0.0012 mol; and $[Br_2]$ = 0.001$\underline{2}$1 M, or 0.0012 mol.

14.79 The starting concentration of $COCl_2$ = 1.00 mol ÷ 25.00 L = 0.0400 M. Assemble a table of starting, change, and equilibrium concentrations.

Conc. (M)	$COCl_2(g)$	⇌	$Cl_2(g)$	+	$CO(g)$
Starting	0.0400		0		0
Change	-x		+x		+x
Equilibrium	0.0400 - x		x		x

Substituting into the equilibrium expression for K_c gives

$$K_c = 8.05 \times 10^{-4} = \frac{[CO][Cl_2]}{[COCl_2]} = \frac{(x)(x)}{(0.0400 - x)}$$

Rearranging and solving for x yields

$$3.22 \times 10^{-5} - (8.05 \times 10^{-4})x - x^2 = 0$$

$$x^2 + (8.05 \times 10^{-4})x - 3.22 \times 10^{-5} = 0 \text{ (quadratic equation)}$$

Using the solution to the quadratic gives

$$x = \frac{-8.05 \times 10^{-4} \pm \sqrt{(8.05 \times 10^{-4})^2 - 4(1)(-3.22 \times 10^{-5})}}{2}$$

x = -6.09×10^{-3} (impossible; reject), or x = $5.2\underline{8}6 \times 10^{-3}$ (logical; use)

% dissoc. = (change ÷ starting) x 100% = (0.005286 ÷ 0.0400) x 100%

= $13.\underline{2}1$ = 13.2%

14.80 The starting concentration of N_2O_4 = 0.0300 mol ÷ 1.00 L = 0.0300 M. Assemble a table of starting, change, and equilibrium concentrations.

Conc. (M)	$N_2O_4(g)$	⇌	$2NO_2(g)$
Starting	0.0300		0
Change	-x		+2x
Equilibrium	0.0300 - x		2x

Substituting into the equilibrium expression for K_c gives

(continued)

$$K_c = 0.125 = \frac{[NO_2]^2}{[N_2O_4]} = \frac{(2x)^2}{(0.0300 - x)}$$

Rearranging and solving for x yields

$$4x^2 + 0.125x - 3.75 \times 10^{-3} = 0 \text{ (quadratic equation)}$$

Using the solution to the quadratic gives

$$x = \frac{-0.125 \pm \sqrt{(0.125)^2 - 4(4)(-3.75 \times 10^{-3})}}{2(4)}$$

$x = -5.00 \times 10^{-2}$ (impossible; reject), or $x = 1.8\underline{7}5 \times 10^{-2}$ M (logical)

% dissoc. = (change ÷ starting) x 100% = (0.01875 ÷ 0.0300) x 100%
= 62.$\underline{5}$0 = 62.5%

14.81 Using 1.00 mol/10.00 L, or 0.100 M, and 4.00 mol/10.00 L, or 0.400 M, for the respective starting concentrations for CO and H_2, assemble a table of starting, change, and equilibrium concentrations.

Conc. (M)	CO(g)	+	$3H_2$(g)	⇌	CH_4(g)	+	H_2O(g)
Starting	0.100		0.400		0		0
Change	-x		-3x		+x		+x
Equilibrium	0.100 - x		0.400 -3 x		x		x

Substituting into the equilibrium expression for K_c gives

$$K_c = 3.92 = \frac{[CH_4][H_2O]}{[CO][H_2]^3} = \frac{x^2}{(0.100 - x)(0.400 - 3x)^3}$$

$$f(x) = \frac{x^2}{(0.100 - x)(0.400 - 3x)^3}$$

Because K_c is > 1 (> 50% reaction), choose x = 0.05 (about half of CO reacting), and use that for the first entry in the table of x, f(x), and interpretations:

(continued)

x	f(x)	Interpretation
0.05	3.20	x > 0.05
0.06	8.45	x < 0.06
0.055	5.18	x < 0.055
0.0525	4.07	x < 0.0525 (but close)
0.052	3.87	f(x) of 3.87 ≅ 3.92

At equilibrium, concentrations and moles are CO: 0.048 M and 0.48 mol; H_2: 0.244 M and 2.44 mol; CH_4: 0.052 M and 0.52 mol; and H_2O: 0.052 M and 0.52 mol.

14.82 Using 1.00 mol/10.00 L, or 0.100 M, and 4.00 mol/10.00 L, or 0.400 M, for the respective starting concentrations for N_2 and H_2, assemble a table of starting, change, and equilibrium concentrations.

Conc. (M)	$N_2(g)$	+	$3H_2(g)$	$\rightleftharpoons$	$2NH_3(g)$
Starting	0.100		0.400		0
Change	-x		-3x		+2x
Equilibrium	0.100 - x		0.400 - 3x		2x

Substituting into the equilibrium expression for K_c gives

$$K_c = 0.153 = \frac{[NH_3]^2}{[N_2][H_2]^3} = \frac{(2x)^2}{(0.100 - x)(0.400 - 3x)^3}$$

$$f(x) = \frac{4x^2}{(0.100 - x)(0.400 - 3x)^3}$$

Because K_c is < 1 (< 50% reaction), choose x = 0.010 (about 10% of N_2 reacting), and use that for the first entry in the table of x, f(x), and interpretations:

x	f(x)	Interpretation
0.010	0.0877	x too small
0.020	0.5089	x too big
0.015	0.237	x still too big
0.012	0.136	x < 0.153 (but close)
0.013	0.1651	f(x) of .1651 ≅ 0.153
0.0125	0.150	best f(x)

At equilibrium, concentrations and moles are N_2: 0.087 M and 0.87 mol; H_2: 0.362 M and 3.62 mol; and NH_3: 0.025 M and 0.25 mol.

14.83 The dissociation is endothermic.

14.84 The value of K_c and the ratio of products to reactants decrease with temperature, so the reaction must be exothermic.

14.85 For $N_2 + 3H_2 \rightleftharpoons 2NH_3$, K_p is defined in terms of pressures as

$$K_p = \frac{P_{NH_3}{}^2}{P_{N_2}P_{H_2}{}^3}$$

But by the ideal gas law, where [i] = mol/L,

$P_i = (n_iRT)/V$, or $P_i = [\,i\,]RT$

Substituting the right-hand equality into the K_p expression gives

$$K_p = \frac{[NH_3]^2 (RT)^2}{[N_2](RT)\ [H_2]^3 (RT)^3} = \frac{[NH_3]^2}{[N_2]\,[H_2]^3}(RT)^{-2}$$

$K_p = K_c\,(RT)^{-2}$, or $K_c = K_p\,(RT)^2$

14.86 For $COCl_2 \rightarrow CO + Cl_2$, K_p is defined in terms of pressure as

$$K_p = \frac{P_{CO}P_{Cl_2}}{P_{COCl_2}}$$

Substituting [i]RT = P for each term as in the above problem yields

$$K_p = \frac{[CO](RT)[Cl_2](RT)}{[COCl_2](RT)} = \frac{[CO][Cl_2]}{[COCl_2]}(RT)$$

$K_p = K_cRT$, or $K_c = K_p/(RT)$

14.87 a. The change in the number of moles of gas for the reaction is $\Delta n = 2 - 1 = 1$. Using this, calculate K_p from K_c:

$$K_p = K_c(RT)^{\Delta n} = 0.153\ (0.08206 \times 1123)^1 = 14.\underline{0}99 = 14.1$$

b. Use the table approach, giving the starting, change, and equilibrium pressures, in atm.

Press. (atm)	C(s)	+	$CO_2(g)$	⇌	2CO(g)
Starting			1.50		0
Change			-x		+2x
Equilibrium			1.50 - x		2x

Substituting into the equilibrium-constant expression gives

$$K_p = 14.10 = \frac{P^2_{CO}}{P_{CO_2}} = \frac{(2x)^2}{(1.50 - x)}$$

Rearranging and solving for x gives a quadratic equation.

$$4x^2 + 14.10x - 21.15 = 0$$

Using the quadratic formula gives

$$x = \frac{-14.10 \pm \sqrt{(14.10)^2 - (4)(4)(-21.15)}}{2(4)}$$

$$x = 1.1\underline{3}5 \text{ (positive root)}$$

Thus, at equilibrium, the pressures of CO and CO_2 are

$$P_{CO_2} = 1.50 - 1.1\underline{3}5 = 0.3\underline{6}5 = 0.37 \text{ atm}$$

$$P_{CO} = 2x = 2(1.1\underline{3}5) = 2.2\underline{7}0 = 2.27 \text{ atm}$$

c. Because the reaction is endothermic, the equilibrium will shift to the left and the pressure of CO will decrease.

14.88 a. The change in the number of moles of gas for the reaction is $\Delta n = 2 - 1 = 1$. Using this, calculate K_p from K_c:

$$K_p = K_c(RT)^{\Delta n} = 0.238\,(0.08206 \times 1173)^1 = 22.\underline{9}09 = 22.9$$

b. The total pressure of the system at equilibrium is 6.40 atm. This is equal to the sum of the pressures of CO and CO_2. Thus, the pressure of CO_2 can be expressed as

$$P_{CO_2} = 6.40 - P_{CO}$$

Substituting into the equilibrium-constant expression gives

$$K_p = 22.91 = \frac{P^2_{CO}}{P_{CO_2}} = \frac{P^2_{CO}}{6.40 - P_{CO}}$$

Rearranging and solving for x (= P_{CO}) gives a quadratic equation.

$$x^2 + 22.91x - 146.6 = 0$$

Using the quadratic formula gives

$$x = \frac{-22.91 \pm \sqrt{(22.91)^2 - (4)(1)(-146.6)}}{2(1)}$$

$$x = 5.2\underline{1}3 \text{ (positive root)}$$

Thus, at equilibrium, the pressures of CO and CO_2 are

$$P_{CO_2} = 6.40 - 5.2\underline{1}3 = 1.1\underline{8}7 = 1.19 \text{ atm}$$

$$P_{CO} = x = 5.2\underline{1}3 = 5.21 \text{ atm}$$

c. Carbon doesn't appear in the equilibrium expression so it doesn't have any effect on the position of the equilibrium.

14.89 a. The molar mass of PCl_5 is 208.22 g/mol. Thus, the initial concentration of PCl_5 is

$$\frac{35.8 \text{ g } PCl_5 \times \dfrac{1 \text{ mol } PCl_5}{208.22 \text{ g } PCl_5}}{5.0 \text{ L}} = 0.03\underline{4}4 \text{ M}$$

Use the table approach, giving the starting, change, and equilibrium concentrations.

(continued)

Conc. (M)	$PCl_3(g)$	+	$Cl_2(g)$	$\rightleftharpoons$	$PCl_5(g)$
Starting	0		0		0.0344
Change	x		x		-x
Equilibrium	x		x		0.0344 - x

Substituting into the equilibrium-constant expression gives

$$K_c = 4.1 = \frac{[PCl_5]}{[PCl_3][Cl_2]} = \frac{0.0344 - x}{x^2}$$

Rearranging and solving for x gives a quadratic equation.

$$4.1x^2 + x - 0.0344 = 0$$

Using the quadratic formula gives

$$x = \frac{-1 \pm \sqrt{(1)^2 - (4)(4.1)(-0.0344)}}{2(4.1)}$$

$$x = 0.0306 \text{ (positive root)}$$

Thus, at equilibrium, $[PCl_3] = [Cl_2] = x = 0.031$ M. The concentration of PCl_5 is

$$[PCl_5] = .0344 - x = 0.0344 - 0.0306 = 0.0038 = 0.004 \text{ M.}$$

b. The fraction of PCl_5 decomposed is

$$\text{fraction decomposed} = \frac{0.0306 \text{ M}}{0.0344 \text{ M}} = 0.889 = 0.89$$

c. There would be a greater pressure so less PCl_5 would decompose in order to minimize the increase in pressure.

14.90 a. The total pressure at equilibrium is

$$P_{NH_3} + P_{H_2S} = 0.660 \text{ atm}$$

Since the pressures of NH_3 and H_2S are equal, this gives

$$P_{NH_3} = P_{H_2S} = \frac{0.660 \text{ atm}}{2} = 0.330 \text{ atm}$$

(continued)

The equilibrium-constant expression is

$$K_p = P_{NH_3} \cdot P_{H_2S} = (0.330)^2 = 0.10\underline{8}9 = 0.109$$

b. Since $P_{H_2S} = 3P_{NH_3}$, the equilibrium-constant expression becomes

$$K_p = P_{NH_3} \cdot P_{H_2S} = 3P^2_{NH_3} = 0.10\underline{8}9$$

$$P_{NH_3} = \sqrt{\frac{0.1089}{3}} = 0.19\underline{0}5 = 0.191 \text{ atm}$$

The partial pressure of H_2S is

$$P_{H_2S} = 3(0.1905) = 0.57\underline{1}5 = 0.572 \text{ atm}$$

c. Use the table approach, giving the starting, change, and equilibrium pressures, in atm.

Press. (atm)	$NH_4HS(s)$	$\rightleftharpoons$	$NH_3(g)$	+	$H_2S(g)$
Starting			0.750		0.500
Change			-x		-x
Equilibrium			0.750 - x		0.500 - x

Substituting into the equilibrium-constant expression gives

$$K_p = P_{NH_3} \cdot P_{H_2S} = (0.750 - x)(0.500 - x) = 0.10\underline{8}9$$

Rearranging and solving for x gives a quadratic equation.

$$x^2 - 1.250x + 0.2661 = 0$$

Using the quadratic formula gives

$$x = \frac{-(-1.250) \pm \sqrt{(-1.250)^2 - (4)(1)(0.2661)}}{2(1)}$$

$$x = 0.27\underline{2}1 \text{ (negative root)}$$

(continued)

Thus, at equilibrium, the partial pressures of NH_3 and H_2S are

$$P_{NH_3} = 0.750 - 0.2721 = 0.47\underline{7}9 = 0.478 \text{ atm}$$

$$P_{H_2S} = 0.500 - 0.2721 = 0.22\underline{7}9 = 0.228 \text{ atm}$$

For NH_4HS, use the ideal gas equation to convert atm (= x) to moles.

$$n = \frac{PV}{RT} = \frac{(0.2721 \text{ atm})(1.00 \text{ L})}{(0.08206 \text{ L} \bullet \text{atm/K} \bullet \text{mol})(298 \text{ K})} = 0.011\underline{1}3 = 0.0111 \text{ mol}$$

14.91 The initial moles of $SbCl_5$ (molar mass 299.01 g/mol) are

$$65.4 \text{ g } SbCl_5 \times \frac{1 \text{ mol } SbCl_5}{299.01 \text{ g } SbCl_5} = 0.21\underline{8}7 \text{ mol}$$

The initial pressure of $SbCl_5$ is

$$P = \frac{nRT}{V} = \frac{(0.2187 \text{ mol})(0.08206 \text{ L} \bullet \text{atm/K} \bullet \text{mol})(468 \text{K})}{5.00 \text{ L}} = 1.6\underline{8}0 \text{ atm}$$

Use the table approach, giving the starting, change, and equilibrium pressures, in atm.

Press. (atm)	$SbCl_5(g)$	$\rightleftharpoons$	$SbCl_3(g)$	+	$Cl_2(g)$
Starting	1.680		0		0
Change	-x		x		x
Equilibrium	(1.680)(0.642)		(1.680)(0.358)		(1.680)(0.358)

At equilibrium, 35.8% of the $SbCl_5$ is decomposed, so x = (1.680)(0.358) in this table. The equilibrium-constant expression is

$$K_p = \frac{P_{SbCl_3} \bullet P_{Cl_2}}{P_{SbCl_5}} = \frac{(1.680 \times 0.358)^2}{1.680 \times 0.642} = 0.33\underline{5}4 = 0.335$$

14.92 Use the table approach, giving the starting, change, and equilibrium concentrations. The volume of the system is 1.00 L.

Conc. (M)	$SO_2(g)$	+	$1/2O_2(g)$	$\rightleftharpoons$	$SO_3(g)$
Starting	0.0216		0.0148		0
Change	-x		-1/2x		+x (= 0.0175)
Equilibrium	0.0216 - 0.0175 = 0.00410		0.0148 - 1/2(0.0175) = 0.00605		0.0175

Substituting into the equilibrium-constant expression gives

$$K_c = \frac{[SO_3]}{[SO_2][O_2]^{1/2}} = \frac{0.0175}{(0.00410)(0.00605)^{1/2}} = 54.\underline{8}7 = 54.9$$

14.93 a. The initial concentration of SO_2Cl_2 (molar mass 134.97 g/mol) is

$$\frac{8.25 \text{ g } SO_2Cl_2 \times \dfrac{1 \text{ mol } SO_2Cl_2}{134.97 \text{ g } SO_2Cl_2}}{1.00 \text{ L}} = 0.059\underline{8}0 \text{ M}$$

Use the table approach, giving the starting, change, and equilibrium concentrations.

Conc. (M)	$SO_2Cl_2(g)$	$\rightleftharpoons$	$SO_2(g)$	+	$Cl_2(g)$
Starting	0.05980		0		0
Change	-x		x		x
Equilibrium	0.05980 - x		x		x

Substituting into the equilibrium-constant expression gives

$$K_c = \frac{[SO_2][Cl_2]}{[SO_2Cl_2]} = \frac{x^2}{(0.05980 - x)} = 0.045$$

Rearranging and solving for x gives a quadratic equation.

$$x^2 + 0.045x - 0.002691 = 0$$

(continued)

Using the quadratic formula gives

$$x = \frac{-(0.045) \pm \sqrt{(0.045)^2 - (4)(1)(-0.002691)}}{2(1)}$$

$x = 0.03\underline{4}0$ (positive root)

The concentrations at equilibrium are $[SO_2] = [Cl_2] = x = 0.034$ M. For SO_2Cl_2,

$[SO_2Cl_2] = 0.05980 - x = 0.05980 - 0.03\underline{4}0 = 0.02\underline{5}8 = 0.026$M.

b. The fraction of SO_2Cl_2 decomposed is

$$\text{Fraction decomposed} = \frac{0.034 \text{ M}}{0.05980 \text{ M}} = 0.5\underline{6}9 = 0.57$$

c. This would shift the equilibrium to the left and decrease the fraction of SO_2Cl_2 that has decomposed.

14.94 a. The initial concentration of $COCl_2$ (molar mass 98.91 g/mol) is

$$\frac{6.55 \text{ g } COCl_2 \times \dfrac{1 \text{ mol } COCl_2}{98.91 \text{ g } COCl_2}}{1.00 \text{ L}} = 0.066\underline{2}2 \text{ M}$$

Use the table approach, giving the starting, change, and equilibrium concentrations.

Conc. (M)	$COCl_2(g)$	$\rightleftharpoons$	$CO(g)$	+	$Cl_2(g)$
Starting	0.06622		0		0
Change	-x		x		x
Equilibrium	0.06622 - x		x		x

Substituting into the equilibrium-constant expression gives

$$K_c = \frac{[CO][Cl_2]}{[COCl_2]} = \frac{x^2}{(0.06622 - x)} = 0.0046$$

Rearranging and solving for x gives a quadratic equation.

$$x^2 + 0.0046x - 3.046 \times 10^{-4} = 0$$

(continued)

Using the quadratic formula gives

$$x = \frac{-(0.0046) \pm \sqrt{(0.0046)^2 - (4)(1)(-3.046 \times 10^{-4})}}{2(1)}$$

$x = 0.01\underline{5}3$ (positive root)

The concentrations at equilibrium are $[CO] = [Cl_2] = x = 0.015$ M. For $COCl_2$,

$$[COCl_2] = 0.06622 - x = 0.06622 - 0.01\underline{5}3 = 0.05\underline{0}9 = 0.051 \text{ M.}$$

b. The fraction of $COCl_2$ decomposed is

$$\text{Fraction decomposed} = \frac{0.0153 \text{ M}}{0.06622 \text{ M}} = 0.2\underline{3}1 = 0.23$$

c. This would shift the equilibrium to the left and decrease the fraction of $COCl_2$ that has decomposed.

14.95 a. First, determine the initial concentration of the dimer, assuming complete reaction. The reaction can be described as $2A \rightarrow D$. Therefore, the initial concentration of dimer is one-half of the concentration of monomer, or 2.0×10^{-4} M. Next, allow the dimer to dissociate into the monomer in equilibrium. Use the table approach, giving the starting, change, and equilibrium concentrations.

Conc. (M)	D(g)	$\rightleftharpoons$	2A(g)
Starting	2.0×10^{-4}		0
Change	$-x$		$2x$
Equilibrium	$2.0 \times 10^{-4} - x$		$2x$

Substituting into the equilibrium-constant expression gives

$$K_c = \frac{[A]^2}{[D]} = \frac{(2x)^2}{(2.0 \times 10^{-4} - x)} = \frac{1}{3.2 \times 10^4} = 3.125 \times 10^{-5}$$

Rearranging and solving for x gives a quadratic equation.

$$4x^2 + (3.125 \times 10^{-5})x - (6.250 \times 10^{-9}) = 0$$

(continued)

Using the quadratic formula gives

$$x = \frac{-(3.125 \times 10^{-5}) \pm \sqrt{(3.125 \times 10^{-5})^2 - (4)(4)(-6.250 \times 10^{-9})}}{2(4)}$$

$x = 3.\underline{5}8 \times 10^{-5}$ (positive root)

Thus, the concentrations at equilibrium are

$[CH_3COOH] = 2x = 2(3.\underline{5}8 \times 10^{-5}) = 7.\underline{1}6 \times 10^{-5} = 7.2 \times 10^{-5}$ M

$[Dimer] = 2.0 \times 10^{-4} - 3.\underline{5}8 \times 10^{-5} = 1.\underline{6}4 \times 10^{-4} = 1.6 \times 10^{-4}$ M.

b. Some hydrogen bonding can occur that results in a more stable system. The proposed structure of the dimer is

$$\begin{array}{c} \text{O} \cdots \text{H}-\text{O} \\ \parallel \qquad \qquad \backslash \\ CH_3-C \qquad \qquad C-CH_3 \\ \backslash \qquad \qquad \parallel \\ \text{O}-\text{H} \cdots \text{O} \end{array}$$

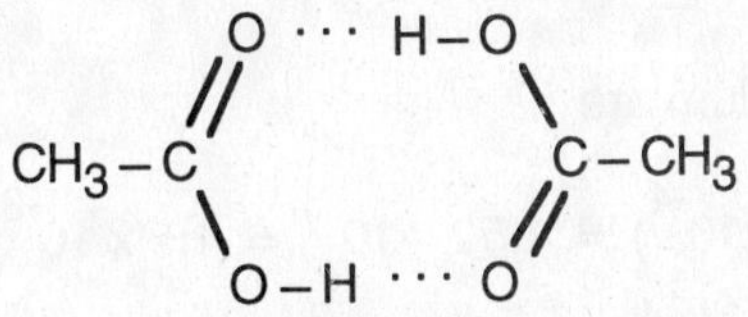

c. An increase in the temperature would help facilitate bond breaking and would decrease the amount of dimer. We could also use a Le Chatelier type of argument.

14.96 a. First, determine the initial pressure of the dimer, assuming complete reaction. The reaction can be described as $2A \rightarrow D$. Therefore, the initial pressure of the dimer is one-half of the pressure of monomer, or $3.\underline{7}5 \times 10^{-3}$ atm. Next, allow the dimer to dissociate into the monomer in equilibrium. Use the table approach, giving the starting, change, and equilibrium pressures.

Press. (atm)	D(g)	⇌	2A(g)
Starting	3.75×10^{-3}		0
Change	$-x$		$2x$
Equilibrium	$3.75 \times 10^{-3} - x$		$2x$

(continued)

Substituting into the equilibrium-constant expression gives

$$K_c = \frac{P_A^2}{P_D} = \frac{(2x)^2}{(3.75 \times 10^{-3} - x)} = \frac{1}{1.3 \times 10^3} = 7.692 \times 10^{-4}$$

Rearranging and solving for x gives a quadratic equation.

$$4x^2 + (7.692 \times 10^{-4})\,x - (2.885 \times 10^{-6}) = 0$$

Using the quadratic formula gives

$$x = \frac{-(7.692 \times 10^{-4}) \pm \sqrt{(7.692 \times 10^{-4})^2 - (4)(4)(-2.885 \times 10^{-6})}}{2(4)}$$

$x = 7.\underline{5}9 \times 10^{-4}$ (positive root)

Thus, the concentrations at equilibrium are

$$[CH_3COOH] = 2x = 2\,(7.\underline{5}9 \times 10^{-4}) = 1.\underline{5}2 \times 10^{-3} = 1.5 \times 10^{-3}\text{ atm}$$

$$[\text{Dimer}] = 3.75 \times 10^{-3} - 7.\underline{5}9 \times 10^{-4} = 2.\underline{9}9 \times 10^{-3} = 3.0 \times 10^{-3}\text{ atm}$$

b. Some hydrogen bonding can occur that results in a more stable system. The proposed structure of the dimer is

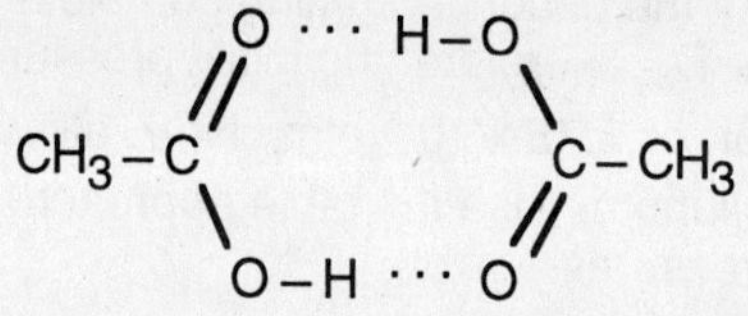

c. An decrease in the temperature would decrease bond breaking in the dimer so this would decrease the amount of dimer.

14.97 The molar mass of Br_2 is 159.82 g/mol. Thus, the initial concentration of Br_2 is

$$\frac{18.22\text{ g }Br_2 \times \dfrac{1\text{ mol }Br_2}{159.82\text{ g }Br_2}}{1.00\text{ L}} = 0.11\underline{4}0\text{ M}$$

(continued)

Use the table approach, giving the starting, change, and equilibrium concentrations.

Conc. (M)	$2NO(g)$ +	$Br_2(g)$ $\rightleftharpoons$	$2NOBr(g)$
Starting	0.112	$0.11\underline{4}0$	0
Change	-2x	-x	2x
Equilibrium	0.112 - 2x	0.1140 - x	2x (= 0.0824 M)
	$= 0.02\underline{9}60$	$= 0.07\underline{2}80$	= 0.0824

Substituting into the equilibrium-constant expression gives

$$K_c = \frac{[NOBr]^2}{[NO]^2[Br_2]} = \frac{(0.0824)^2}{(0.02960)^2(0.07280)} = 1\underline{0}6.4 = 1.1 \times 10^2$$

14.98 The molar mass of Br_2 is 159.82 g/mol. Thus, the initial moles of Br_2 are

$$1.52 \text{ g } Br_2 \times \frac{1 \text{ mol } Br_2}{159.82 \text{ g } Br_2} = 9.5\underline{1}1 \times 10^{-3} \text{ mol}$$

The initial pressures of NO and Br_2 are

$$P_{NO} = \frac{nRT}{V} = \frac{(0.0322 \text{ mol})(0.08206 \text{ L} \bullet \text{atm/K} \bullet \text{mol})(298\text{K})}{1.00 \text{ L}} = 0.78\underline{7}4 \text{ atm}$$

$$P_{Br_2} = \frac{nRT}{V} = \frac{(9.511 \times 10^{-3} \text{ mol})(0.08206 \text{ L} \bullet \text{atm/K} \bullet \text{mol})(298\text{K})}{1.00 \text{ L}}$$

$$= 0.23\underline{2}6 \text{ atm}$$

Use the table approach, giving the starting, change, and equilibrium pressures.

Press. (atm)	$2NO(g)$ +	$Br_2(g)$ $\rightleftharpoons$	$2NOBr(g)$
Starting	$0.78\underline{7}4$	$0.23\underline{2}6$	0
Change	-2x	-x	2x
Equilibrium	$0.78\underline{7}4$ - 2x	$0.23\underline{2}6$ - x	2x (= 0.438 atm)
	$= 0.34\underline{9}4$	$= 0.01\underline{3}6$	= 0.438

(continued)

Substituting into the equilibrium-constant expression gives

$$K_p = \frac{P^2_{NOBr}}{P^2_{NO}P_{Br_2}} = \frac{(0.438)^2}{(0.3494)^2\,(0.01\underline{3}6)} = 1\underline{1}5.5 = 1.2 \times 10^2$$

■ Cumulative-Skills Problems

(require skills from previous chapters)

14.99 For $Sb_2S_3(s) + 3H_2(g) \rightleftharpoons 2Sb(s) + 3H_2S(g)$ [$+3Pb^{2+} \rightarrow 3PbS(s) + 6H^+$]:

Starting M of $H_2(g)$ = 0.0100 mol ÷ 2.50 L = 0.00400 M H_2

1.029 g PbS ÷ 239.26 g PbS/mol H_2S = $4.30\underline{0}7 \times 10^{-3}$ mol H_2S

4.3007×10^{-3} mol H_2S ÷ 2.50 L = [1.7203×10^{-3}] = M of H_2S

Conc. (M)	$3H_2(g)$	+	$Sb_2S_3(s)$	$\rightleftharpoons$	$3H_2S(g)$	+	$2Sb(s)$
Starting	0.00400				0		
Change	-0.0017203				+0.0017203		
Equilibrium	0.0022797				0.0017203		

Substituting into the equilibrium expression for K_c gives

$$K_c = \frac{[H_2S]^3}{[H_2]^3} = \frac{(0.0017203)^3}{(0.0022797)^3} = 4.2\underline{9}7 \times 10^{-1} = 4.30 \times 10^{-1}$$

14.100 For $LaCl_3 + H_2O \rightleftharpoons LaOCl(s) + 2HCl(g)$ [$+2Ag^+ \rightarrow 2AgCl(s) + 2H^+$]:

M of H_2O = 0.0200, mol HCl = 0.025048, and M of HCl = 0.020039

Conc. (M)	$H_2O(g)$ + $LaCl_3(s)$	$\rightleftharpoons$	$2HCl(g)$ + $LaOCl(s)$
Starting	0.0200		0
Change	-0.0100195		+0.020039
Equilibrium	0.0099805		0.020039

$$K_c = \frac{[HCl]^2}{[H_2O]} = \frac{(0.020039)^2}{(0.0099805)} = 4.023 \times 10^{-2}$$

$$= 4.02 \times 10^{-2} \quad \text{(0.0099805 approx. three sig. figs)}$$

14.101 For $PCl_5(g) \rightleftharpoons PCl_3(g) + Cl_2(g)$:

Starting M of PCl_5 = 0.0100 mol ÷ 2.00 L = 0.00500 M

Conc. (M)	$PCl_5(g)$	$\rightleftharpoons$	$PCl_3(g)$ +	$Cl_2(g)$
Starting	0.00500		0	0
Change	-x		+x	+x
Equilibrium	0.00500 - x		x	x

Substituting into the equilibrium expression for K_c gives

$$K_c = 4.15 \times 10^{-2} = \frac{[PCl_3][Cl_2]}{[PCl_5]} = \frac{(x)(x)}{(0.00500 - x)}$$

$$x^2 + (4.15 \times 10^{-2})x - 2.075 \times 10^{-4} = 0 \quad \text{(quadratic)}$$

Solving the quadratic gives $x = -4.60 \times 10^{-2}$ (impossible) and $x = 4.51 \times 10^{-3}$ M (use).

Total M of gas = 0.00451 + 0.00451 + 0.00049 = 0.009510 M

$$P = (n/V)RT = (0.009510\ M)[0.082057\ L{\bullet}atm/(K{\bullet}mol)](523\ K)$$

$$= 0.4081 = 0.408\ atm$$

14.102 For $SbCl_5(g) \rightleftharpoons SbCl_3(g) + Cl_2(g)$:

Starting M of $SbCl_5$ = 0.0125 mol ÷ 3.50 L = 0.003571 M

Conc. (M)	$SbCl_5(g)$	$\rightleftharpoons$	$SbCl_3(g)$	+	$Cl_2(g)$
Starting	0.003571		0		0
Change	-x		+x		+x
Equilibrium	0.003571 - x		x		x

Substituting into the equilibrium expression for K_c gives

$$K_c = 2.50 \times 10^{-2} = \frac{[SbCl_3][Cl_2]}{[SbCl_5]} = \frac{(x)(x)}{(0.003751 - x)}$$

$$x^2 + (2.50 \times 10^{-2})x - 8.927 \times 10^{-5} = 0 \text{ (quadratic)}$$

Solving the quadratic gives $x = -2.817 \times 10^{-2}$ (impossible) and $x = 3.17 \times 10^{-3}$ M (use).

Total M of gas = 0.00317 + 0.00317 + 0.00040 = 0.006740 M

$$P = (n/V)RT = (0.00674 \text{ M})[0.082057 \text{ L•atm/(K•mol)}](521 \text{ K})$$

$$= 0.2881 = 0.288 \text{ atm}$$

15. ACIDS AND BASES

Solutions to Exercises

Note on significant figures: The final answer to each mathematical solution is given first with one nonsignificant figure (the rightmost significant figure is underlined) and then is rounded to the correct number of figures. Intermediate answers usually also have at least one nonsignificant figure.

15.1 See labels below reaction:

$$\underset{\text{acid}}{H_2CO_3(aq)} + \underset{\text{base}}{CN^-(aq)} \rightleftharpoons \underset{\text{acid}}{HCN(aq)} + \underset{\text{base}}{HCO_3^-(aq)}$$

H_2CO_3 is the proton donor (Bronsted-Lowry acid) on the left and HCN is the proton donor (Bronsted-Lowry acid) on the right. The CN^- and HCO_3^- ions are proton acceptors (Bronsted-Lowry bases). HCN is the conjugate acid of CN^-.

15.2 Part (a) involves molecules with all single bonds; part (b) does not, so bonds are drawn in.

a. 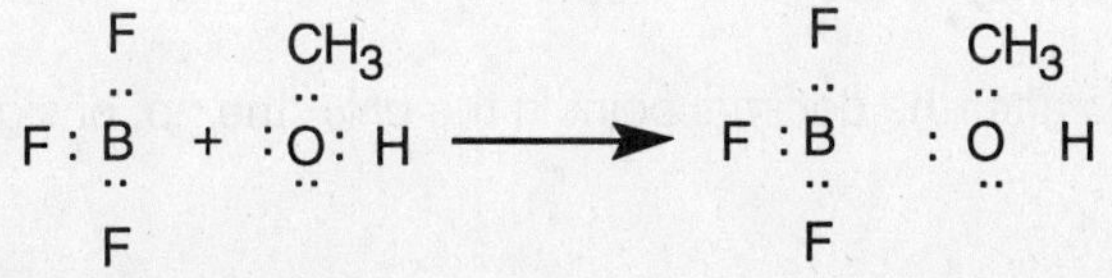

Lewis acid   Lewis base

b.

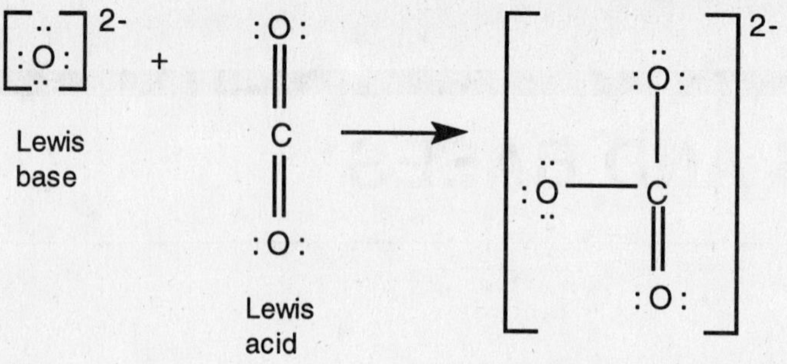

15.3 The $HC_2H_3O_2$ is a stronger acid than H_2S, and HS^- is a stronger base than the $C_2H_3O_2^-$ ion. The equilibrium favors the weaker acid and weaker base; therefore, the reactants are favored.

15.4 a. PH_3 b. HI c. H_2SO_3 d. H_3AsO_4 e. HSO_4^-

15.5 A 0.125 M solution of $Ba(OH)_2$, a strong base, ionizes completely to yield 0.125 M Ba^{2+} ion and 2 x 0.125 M, or 0.250 M, OH^- ion. Use the K_w equation to calculate the $[H^+]$.

$$[H^+] = \frac{K_w}{[OH^-]} = \frac{1.0 \times 10^{-14}}{(0.250)} = 4.\underline{00} \times 10^{-14} = 4.0 \times 10^{-14}\ M$$

15.6 The K_w equation gives 1.0×10^{-9} M $[H^+]$, which is basic (< 1.00×10^{-7} M of a neutral solution):

$$[H^+] = \frac{K_w}{[OH^-]} = \frac{1.0 \times 10^{-14}}{(1.0 \times 10^{-5})} = 1.\underline{0}0 \times 10^{-9} = 1.0 \times 10^{-9}\ M\ \text{(basic)}$$

15.7 Calculate the negative log of the $[H^+]$:

$$pH = -\log[H^+] = -\log(0.045) = 1.3\underline{4}6 = 1.35$$

Note: The number of places after the decimal point in the pH = the no. of sig. figs. of $[H^+]$.

15.8 Calculate the pOH of 0.025 M OH^- and then subtract from 14.00 to find pH:

$$pOH = -\log[OH^-] = -\log(0.025) = 1.6\underline{0}2$$

$$pH = 14.00 - 1.602 = 12.3\underline{9}7 = 12.40$$

15.9 Because pH = 3.16, by definition the log $[H^+]$ = -3.16. Enter this on the calculator and convert to the antilog (number) of -3.16.

$$\text{antilog}(-3.16) = 10^{-3.16} = 6.\underline{9}1 \times 10^{-4} = 6.9 \times 10^{-4}\ M$$

15.10 Find the pOH by subtracting the pOH from 14.00. Then enter -3.40 on the calculator to convert to the antilog (number) corresponding to -3.40.

$$\text{pOH} = 14.00 - 10.6 = 3.\underline{4}0$$

$$\text{antilog}(-3.40) = 10^{-3.40} = \underline{3}.98 \times 10^{-4} = 4 \times 10^{-4}\ M$$

■ Answers to Review Questions

15.1 An Arrhenius acid in aqueous solution is a substance that increases the concentration of the hydrogen ion, $H^+(aq)$. An example of such an acid is $HClO_4$, perchloric acid. One mole of this acid supplies 1 mole of H^+ to an aqueous solution. An Arrhenius base in aqueous solution is a substance that increases the concentration of the hydroxide ion, $OH^-(aq)$. An example of such a base is NaOH, sodium hydroxide. One mole of this base supplies 1 mole of OH^- to an aqueous solution.

15.2 You can classify these acids using the information in Section 15.1. Also recall that all diatomic acids of Group VIIA halides are strong except for HF.

a. Weak	c. Strong	e. Weak
b. Weak	d. Strong	f. Weak

15.3 In Section 15.1 we are told that all neutralizations involving strong acids and bases evolve 55.90 kJ of heat per 1 mol H^+. Thus, the thermochemical evidence for the Arrhenius concept is based on the fact that when 1 mole of any strong acid (1 mol H^+) is neutralizedby 1 mole of any strong base (1 mol OH^-), the heat of neutralization is always the same (ΔH^o = -55.90 kJ/mol).

15.4 A Bronsted-Lowry acid is a molecule or ion that donates an H^+ ion (proton donor) to a base in a proton-transfer reaction. A Bronsted-Lowry base is a molecule or ion that accepts an H^+ ion (proton acceptor) from an acid in a proton-transfer reaction. An example of an acid-base equation:

$$\underset{\text{acid}}{HF(aq)} + \underset{\text{base}}{NH_3(aq)} \rightarrow \underset{\text{acid}}{NH_4^+(aq)} + \underset{\text{base}}{F^-(aq)}$$

15.5 The conjugate acid of a base is a species that differs from the base by only one H^+. Consider the base HSO_3^-. Its conjugate acid would be H_2SO_3 but not H_2SO_4. H_2SO_4 differs from HSO_3^- by one H and one O.

15.6 You can write the equations by considering that $H_2PO_3^-$ is both a Bronsted-Lowry acid and a Bronsted-Lowry base. The $H_2PO_3^-$ acts as a Bronsted-Lowry acid when it reacts with a base such as OH^- :

$$H_2PO_3^-(aq) \quad + \quad OH^-(aq) \quad \rightarrow \quad HPO_3^{2-}(aq) \quad + \quad H_2O(l)$$

The $H_2PO_3^-$ acts as a Bronsted-Lowry base when it reacts with an acid such as HF:

$$H_2PO_3^-(aq) \quad + \quad HCl(aq) \quad \rightarrow \quad H_3PO_3(aq) \quad + \quad Cl^-(aq)$$

15.7 The Bronsted-Lowry concept enlarges on the Arrhenius concept in the following ways: (1) It expands the concept of a base to include any species that accepts protons, not just the OH^- ion or compounds containing the OH^- ion. (2) It enlarges the concepts of acids and bases to include ions as well as molecules. (3) It enables us to write acid-base reactions in nonaqueous solutions as well as aqueous solutions, whereas the Arrhenius concept applies only to aqueous solutions. (4) It allows some species to be considered as acids or bases, depending on the other reactant with which they are mixed.

15.8 According to the Lewis concept, an acid is an electron-pair acceptor and a base is an electron-pair donor. An example is

$$\underset{\text{acid}}{Ag^+(aq)} \quad + \quad \underset{\text{base}}{2(:NH_3)} \quad \rightarrow \quad Ag(NH_3)_2^+(aq)$$

15.9 Recall that the weaker the acid, the stronger it holds on to its proton(s). Thus, if a reaction mixture consists of a stronger acid and base and a weaker acid and base, the weaker-acid side will always be favored because the proton(s) will bond more strongly to the weaker acid.

15.10 The fact that acetic acid forms a weaker acid than formic acid means that acetic acid holds on to its proton more strongly than does formic acid. Because a stronger base picks up a proton more readily than does a weaker base, the acetate ion is a stronger base than the formate ion.

15.11 The two factors that determine the strength of an acid are (1) the polarity of the bond to which the H atom is attached and (2) the strength of the bond—that is, how tightly the proton is held by the atom to which it is bonded. An *increase in the polarity* of the bond makes it easier to remove the proton, increasing the strength of the acid. An *increase in the strength* of the bond makes it more difficult to remove the proton, decreasing the strength of the acid. The strength of the bond depends in turn on the size of the atom, so that larger atoms have weaker bonds, whereas smaller atoms have stronger bonds.

15.12 The self-ionization of water is the reaction of two water molecules in which a proton is transferred from one molecule to the other to form H_3O^+ and OH^- ions. At 25°C, the K_w expression is $K_w = [H_3O^+][OH^-] = 1.0 \times 10^{-14}$.

15.13 The pH = -log [H^+] of an aqueous solution. Measure pH by using electrodes and a pH meter, or by interpolating the pH from the color changes of a series of acid–base indicators.

15.14 a solution of pH 4 has a $[H^+] = 10^{-4}$ M and is more acidic than a solution of pH = 5, which has a $[H^+] = 10^{-5}$ M.

15.15 For a neutral solution, $[H^+] = [OH^-]$; thus, the $[H^+]$ of a neutral solution at 37°C is the square root of K_w at 37°C:

$$[H^+] = \sqrt{2.5 \times 10^{-14}} = 1.\underline{5}8 \times 10^{-7}\text{ M; pH} = -\log(1.58 \times 10^{-7}) = 6.8\underline{0}1 = 6.80$$

15.16 Because pH + pOH = pK_w at any temperature,

$$\text{pH} + \text{pOH} = -\log(2.5 \times 10^{-14}) = 13.60$$

■ Solutions to Practice Problems

Note on significant figures: The final answer to each problem is given first with one nonsignificant figure (the rightmost significant figure is underlined) and then is rounded to the correct number of figures. Intermediate answers usually also have at least one nonsignificant figure.

15.17 The reaction with labels of "acid" or "base" written below is as follows:

$$\underset{\text{base}}{CN^-(aq)} + \underset{\text{acid}}{H_2O(l)} \rightleftharpoons \underset{\text{acid}}{HCN(aq)} + \underset{\text{base}}{OH^-(aq)}$$

15.18 The reaction with labels of "acid" or "base" written below is as follows:

$$\underset{\text{base}}{OH^-(aq)} + \underset{\text{acid}}{H_2PO_4^-(l)} \rightleftharpoons \underset{\text{base}}{HPO_4^{2-}(aq)} + \underset{\text{acid}}{H_2O(l)}$$

15.19 a. SO_4^{2-} b. HS^- c. HPO_4^{2-} d. NH_3

15.20 a. ClO⁻ b. PO₄³⁻ c. CN⁻ d. CH₃NH₂

15.21 a. HCN b. H₂CO₃ c. HSeO₄⁻ d. HPO₄²⁻

15.22 a. HIO b. HSO₃⁻ c. H₂SO₃ d. HCHO₂

15.23 Each equation is given below with the labels for acid or base and conjugate pairs:

a. HSO_4^- (aq) + NH_3 (aq) ⇌ SO_4^{2-} (aq) + NH_4^+ (aq)
   acid   base   base   acid
   conjugate ←——————————→ conjugate
                  conjugate ←——————————→ conjugate

b. HPO_4^- (aq) + NH_4^+ (aq) ⇌ $H_2PO_4^-$ (aq) + NH_3 (aq)
   base   acid   acid   base
   conjugate ←——————————→ conjugate
                  conjugate ←——————————→ conjugate

c. $Al(H_2O)_6^{3+}$ (aq) + H_2O(l) ⇌ $Al(H_2O)_5(OH)^{2+}$ (aq) + H_3O^+ (aq)
   acid   base   base   acid
   conjugate ←——————————→ conjugate
                  conjugate ←——————————→ conjugate

d. SO_3^{2-} (aq) + NH_4^+ (aq) ⇌ HSO_3^- (aq) + NH_3 (aq)
   base   acid   acid   base
   conjugate ←——————————→ conjugate
                  conjugate ←——————————→ conjugate

15.24 Each equation is given below with the labels for acid or base and conjugate pairs:

a. $H_2PO_4^-$ (aq) + HCO_3^- (aq) ⇌ HPO_4^{2-} (aq) + H_2CO_3 (aq)
   acid   base   base   acid
   conjugate ←——————————→ conjugate
                  conjugate ←——————————→ conjugate

b. F⁻ (aq) + HSO_4^- (aq) ⇌ HF (aq) + SO_4^{2-} (aq)
   base   acid   acid   base
   conjugate ←——————————→ conjugate
                  conjugate ←——————————→ conjugate

c. HSO_4^- (aq) + H_2O(l) ⇌ SO_4^{2-} (aq) + H_3O^+ (aq)
   acid   base   base   acid
   conjugate ←——————————→ conjugate
                  conjugate ←——————————→ conjugate

d. $H_2S\,(aq) + CN^-\,(aq) \rightleftharpoons HS^-\,(aq) + HCN\,(aq)$

H_2S (aq)	+	CN^- (aq)	⇌	HS^- (aq)	+	HCN (aq)
acid		base		base		acid

conjugate ⟷ conjugate (H_2S acid and HS^- base)

conjugate ⟷ conjugate (CN^- base and HCN acid)

15.25 For (a), the completed equation is $CO_2 + OH^- \rightarrow HCO_3^-$. For (b), the completed equation is $AlCl_3 + Cl^- \rightarrow AlCl_4^-$. For (a), the CO_2 is a Lewis acid and the OH^- ion is a Lewis base. For (b), the $AlCl_3$ is a Lewis acid and the Cl^- ion is a Lewis base. The Lewis formulas are shown below and on the next page.

a.

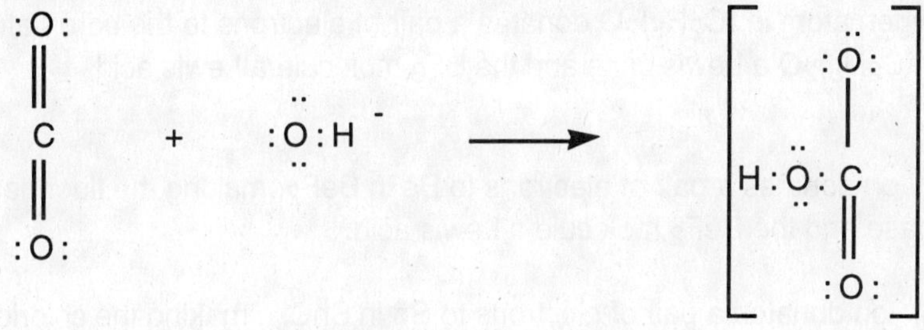

b.

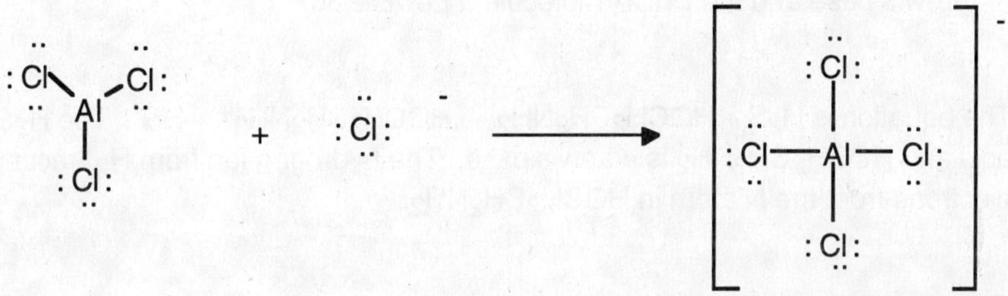

15.26 For (a), the completed equation is $CN^- + H_2S \rightarrow HCN + HS^-$. For (b), the completed equation is $BF_3 + PH_3 \rightarrow F_3B{:}PH_3$. For (a), the H_2S is a Lewis acid and the CN^- ion is a Lewis base. For (b), the BF_3 is a Lewis acid and the PH_3 is a Lewis base. The Lewis formulas are shown below.

a.

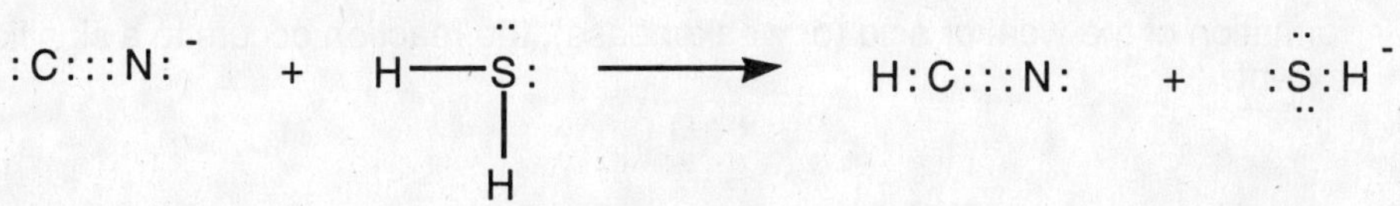

b. 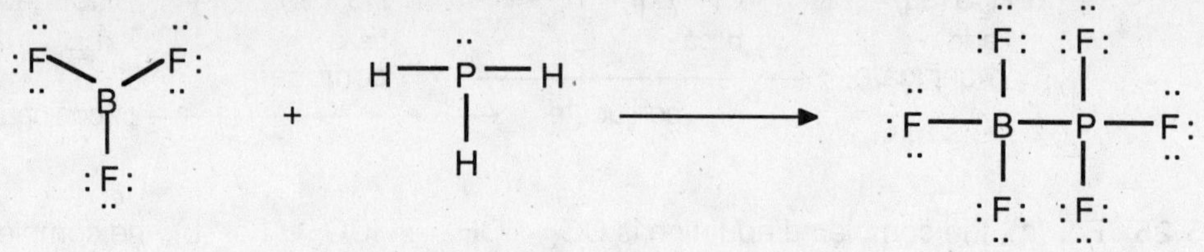

15.27 a. Each water molecule donates a pair of electrons to chromium(III), making the water molecule a Lewis base and the Cr^{3+} ion a Lewis acid.

b. The oxygen atom in $(C_2H_5)_2O$ donates a pair of electrons to the boron atom in BF_3, making $(C_2H_5)_2O$ a Lewis base and the BF_3 molecule a Lewis acid.

15.28 a. Each F^- ion donates a pair of electrons to Be in BeF_2, making the fluoride ion a Lewis base and the BeF_2 molecule a Lewis acid.

b. Each Cl^- ion donates a pair of electrons to Sn in $SnCl_4$, making the chloride ion a Lewis base and the $SnCl_4$ molecule a Lewis acid.

15.29 The equation is $H_2S + HOCH_2CH_2NH_2 \rightarrow HOCH_2CH_2NH_3^+ + HS^-$. The H_2S is a Lewis acid and $HOCH_2CH_2NH_2$ is a Lewis base. The hydrogen ion from H_2S accepts a pair of electrons from the N atom in $HOCH_2CH_2NH_2$.

15.30 The equation is $CaO + SO_2 \rightarrow CaSO_3$. The CaO is a Lewis base and SO_2 is a Lewis acid. The oxide ion from CaO donates a pair of electrons to the sulfur of SO_2, forming the SO_3^{2-} ion in $CaSO_3$.

15.31 The reaction is $HSO_4^- + ClO^- \rightarrow HClO + SO_4^{2-}$. According to Table 15.2, HClO is a weaker acid than HSO_4^-. Because the equilibrium for this type of reaction favors formation of the weaker acid (or weaker base), the reaction occurs to a significant extent.

15.32 The reaction is $HCN + SO_4^{2-} \rightarrow CN^- + HSO_4^-$. According to Table 15.2, HCN is a weaker acid than HSO_4^-. Because the equilibrium for this type of reaction favors formation of the weaker acid (or weaker base), the reaction occurs in the opposite direction.

15.33 a. NH_4^+ is a weaker acid than H_3PO_4, so the left-hand species are favored at equilibrium.

b. HCN is a weaker acid than H_2S, so the left-hand species are favored at equilibrium.

c. H_2O is a weaker acid than HCO_3^-, so the right-hand species are favored at equilibrium.

d. H_2O is a weaker acid than $Al(H_2O)_6^{3+}$, so the right-hand species are favored at equilibrium.

15.34 a. HCO_3^- is a weaker acid than NH_4^+, so the right-hand species are favored at equilibrium.

b. H_2S is a weaker acid than H_2CO_3, so the left-hand species are favored at equilibrium.

c. H_2O is a weaker acid than HCN, so the left-hand species are favored at equilibrium.

d. HCN is a weaker acid than H_2CO_3, so the right-hand species are favored at equilibrium.

15.35 Trichloroacetic acid is the stronger acid because in general the equilibrium favors the formation of the weaker acid, which is formic acid in this case.

15.36 The BF_4^- ion is the weaker base because the equilibrium favors the formation of the weaker base.

15.37 a. H_2S is stronger because acid strength decreases with increasing anion charge for polyprotic acid species.

b. H_2SO_3 is stronger because, for a series of oxoacids, acid strength increases with increasing electronegativity.

c. HBr is stronger because Br is more electronegative than Se. Within a period, acid strength increases as electronegativity increases.

d. HIO_4 is stronger because acid strength increases with the number of oxygen atoms bonded to the central atom.

e. H_2S is stronger because, within a group, acid strength increases with the increasing size of the central atom in binary acids.

15.38 a. $HNO_2 < HNO_3$: Acid strength increases with the number of oxygen atoms bonded to the central atom.

b. $HCO_3^- < H_2CO_3$: Acid strength decreases with increasing anion charge for polyprotic acid species.

c. $H_2S < H_2Te$: Within a group, acid strength increases with increasing size of the central atom for binary acids.

d. $H_2S < HCl$: Within a period, acid strength increases with increasing electronegativity of the central atom for binary acids.

e. $H_3AsO_4 < H_3PO_4$: For a series of oxoacids, acid strength increases with increasing electronegativity.

15.39 a. $[H^+] = 1.25$ M; $[OH^-] = K_w \div [H^+] = 8.\underline{0}0 \times 10^{-15}$ M

b. $[OH^-] = 0.25$ M; $[H^+] = K_w \div [OH^-] = 4.\underline{0}0 \times 10^{-14}$ M

c. $[OH^-] = 0.070$ M; $[H^+] = K_w \div [OH^-] = 1.\underline{4}2 \times 10^{-13}$ M

d. $[H^+] = 0.50$ M; $[OH^-] = K_w \div [H^+] = 2.\underline{0}0 \times 10^{-14}$ M

15.40 a. $[OH^-] = 1.25$ M; $[H^+] = K_w \div [OH^-] = 8.\underline{0}0 \times 10^{-15}$ M

b. $[OH^-] = 0.30$ M; $[H^+] = K_w \div [OH^-] = 3.\underline{3}3 \times 10^{-14}$ M

c. $[H^+] = 0.030$ M; $[OH^-] = K_w \div [H^+] = 3.\underline{3}3 \times 10^{-13}$ M

d. $[H^+] = 0.200$ M; $[OH^-] = K_w \div [H^+] = 5.\underline{0}0 \times 10^{-14}$ M

15.41 The $[H^+] = 0.050$ M (HCl is a strong acid); using K_w, the $[OH^-] = 2.0 \times 10^{-13}$ M.

15.42 The $[H^+] = 0.020$ M (HNO_3 is a strong acid); using K_w, the $[OH^-] = 5.0 \times 10^{-13}$ M.

15.43 Because the $Sr(OH)_2$ forms $2OH^-$ per formula unit, the $[OH^-] = 2 \times 0.0050 = 0.010$ M.

$$[H^+] = K_w \div [OH^-] = (1.0 \times 10^{-14}) \div 0.010 \text{ M} = 1.\underline{0}0 \times 10^{-12} = 1.0 \times 10^{-12} \text{ M}$$

15.44 Because $Mg(OH)_2$ forms 2 OH^- per formula unit, the $[OH^-] = 2 \times 3.2 \times 10^{-4} = 6.4 \times 10^{-4}$ M.

$$[H^+] = K_w \div [OH^-] = (1.0 \times 10^{-14}) \div (6.4 \times 10^{-4}) = 1.\underline{5}6 \times 10^{-11} = 1.6 \times 10^{-11} \text{ M}$$

15.45 a. Basic; 5×10^{-9} M $H^+ < 1.0 \times 10^{-7}$ M H^+

b. Acidic; 2×10^{-6} M $H^+ > 1.0 \times 10^{-7}$ M H^+

c. Neutral; 1×10^{-7} M H^+ is the same as 1.0×10^{-7} M H^+ of neutral water

d. Acidic; 2×10^{-6} M $H^+ > 1.0 \times 10^{-7}$ M H^+

15.46 a. Basic; 5×10^{-11} M $H^+ < 1.0 \times 10^{-7}$ M H^+

b. Acidic; 2×10^{-6} M $H^+ > 1.0 \times 10^{-7}$ M H^+

c. Acidic; 1.7×10^{-5} M $H^+ > 1.0 \times 10^{-7}$ M H^+

d. Basic; 6×10^{-10} M $H^+ < 1.0 \times 10^{-7}$ M H^+

15.47 The $[H^+]$ calculated below is $> 1.0 \times 10^{-7}$ M, so the solution is acidic.

$$[H^+] = K_w \div [OH^-] = (1.0 \times 10^{-14}) \div (1.5 \times 10^{-9} \text{ M}) = 6.\underline{6}6 \times 10^{-6} = 6.7 \times 10^{-6} \text{ M}$$

15.48 The $[H^+]$ calculated below is $< 1.0 \times 10^{-7}$ M, so the solution is basic.

$$[H^+] = K_w \div [OH^-] = (1.0 \times 10^{-14}) \div (8.4 \times 10^{-5} \text{ M}) = 1.\underline{1}9 \times 10^{-10} = 1.2 \times 10^{-10} \text{ M}$$

15.49 a. pH 5.8 = acidic
b. pH 1.9 = acidic
c. pH 7.0 = neutral
d. pH 12.3 = basic

15.50 a. pH 9.5 = basic
b. pH 4.3 = acidic
c. pH 2.8 = acidic
d. pH 13.8 = basic

15.51 a. Acidic ($3.5 < 7.0$)
b. Neutral ($7.0 = 7.0$)
c. Basic ($9.0 > 7.0$)
d. Acidic ($5.5 < 7.0$)

15.52 a. Basic ($12.0 > 7.0$)
b. Neutral ($7.0 = 7.0$)
c. Acidic ($4.0 < 7.0$)
d. Acidic ($5.7 < 7.0$)

15.53 Record the same number of places after the decimal point in the pH as the number of significant figures in the $[H^+]$.

a. $-\log(1.0 \times 10^{-8}) = 8.00$

b. $-\log(5.0 \times 10^{-12}) = 11.3\underline{0}1 = 11.30$

c. $-\log(7.5 \times 10^{-3}) = 2.1\underline{2}4 = 2.12$

d. $-\log(6.35 \times 10^{-9}) = 8.19\underline{7}2 = 8.197$

15.54 Record the same number of places after the decimal point in the pH as the number of significant figures in the $[H^+]$.

a. $-\log(1.0 \times 10^{-4}) = 4.00$

b. $-\log(3.2 \times 10^{-10}) = 9.4\underline{9}4 = 9.49$

c. $-\log(2.3 \times 10^{-5}) = 4.6\underline{3}8 = 4.64$

d. $-\log(2.91 \times 10^{-11}) = 10.53\underline{6}1 = 10.536$

15.55 Record the same number of places after the decimal point in the pH as the number of significant figures in the $[H^+]$.

$-\log(7.5 \times 10^{-3}) = 2.1\underline{2}49 = 2.12$

15.56 Record the same number of places after the decimal point in the pH as the number of significant figures in the $[H^+]$.

$-\log(5.0 \times 10^{-3}) = 2.3\underline{0}1 = 2.30$

15.57 a. $pOH = -\log(5.25 \times 10^{-9}) = 8.27\underline{9}8$; $pH = 14.00 - 8.2798 = 5.7\underline{2}02 = 5.72$

b. $pOH = -\log(8.3 \times 10^{-3}) = 2.0\underline{8}09$; $pH = 14.00 - 2.0809 = 11.9\underline{1}908 = 11.92$

c. $pOH = -\log(3.6 \times 10^{-12}) = 11.44\underline{3}6$; $pH = 14.00 - 11.4436 = 2.5\underline{5}63 = 2.56$

d. $pOH = -\log(2.1 \times 10^{-8}) = 7.67\underline{7}7$; $pH = 14.00 - 7.6777 = 6.3\underline{2}2 = 6.32$

15.58 a. $pOH = -\log(4.83 \times 10^{-11}) = 10.31\underline{6}05$; $pH = 14.00 - 10.31605 = 3.6\underline{8}39 = 3.68$

b. $pOH = -\log(3.2 \times 10^{-5}) = 4.4\underline{9}4$; $pH = 14.00 - 4.494 = 9.5\underline{0}51 = 9.51$

c. $pOH = -\log(2.7 \times 10^{-10}) = 9.5\underline{6}8$; $pH = 14.00 - 9.568 = 4.4\underline{3}1 = 4.43$

d. $pOH = -\log(5.0 \times 10^{-4}) = 3.3\underline{0}1$; $pH = 14.00 - 3.301 = 10.6\underline{9}8 = 10.70$

15.59 First convert the $[OH^-]$ to $[H^+]$ using the K_w equation. Then find the pH, recording the same number of places after the decimal point in the pH as the number of significant figures in the $[H^+]$.

$$[H^+] = K_w \div [OH^-] = (1.0 \times 10^{-14}) \div (0.0040) = 2.\underline{5}0 \times 10^{-12}\ M$$

$$-\log (2.50 \times 10^{-12}) = 11.6\underline{0}2 = 11.60$$

15.60 First convert the $[OH^-]$ to $[H^+]$ using the K_w equation. Then find the pH, recording the same number of places after the decimal point in the pH as the number of significant figures in the $[H^+]$.

$$[H^+] = K_w \div [OH^-] = (1.0 \times 10^{-14}) \div (0.050) = 2.\underline{0}0 \times 10^{-13}\ M$$

$$-\log (2.00 \times 10^{-13}) = 12.6\underline{9}8 = 12.70$$

15.61 From the definition pH = $-\log [H^+]$, $-pH = \log [H^+]$, so enter the negative value of the pH on the calculator and use the inverse and log keys (or 10^x) key to find the antilog of -pH:

$$\log [H^+] = -pH = -5.12$$

$$[H^+] = \text{antilog} (-5.12) = 10^{-5.12} = 7.\underline{5}8 \times 10^{-6} = 7.6 \times 10^{-6}\ M$$

15.62 From the definition pH = $-\log [H^+]$, $-pH = \log [H^+]$, so enter the negative value of the pH on the calculator and use the inverse and log keys (or 10^x) key to find the antilog of -pH:

$$\log [H^+] = -pH = -3.85$$

$$[H^+] = \text{antilog} (-3.85) = 10^{-3.85} = 1.\underline{4}1 \times 10^{-4} = 1.4 \times 10^{-4}\ M$$

15.63 From the definition pH = $-\log [H^+]$, $-pH = \log [H^+]$, so enter the negative value of the pH on the calculator and use the inverse and log keys (or 10^x) key to find the antilog of -pH. Then use the K_w equation to calculate $[OH^-]$ from $[H^+]$.

$$\log [H^+] = -pH = -11.63$$

$$[H^+] = \text{antilog} (-11.63) = 10^{-11.63} = 2.\underline{3}4 \times 10^{-12}\ M$$

$$[OH^-] = K_w \div [H^+] = (1.0 \times 10^{-14}) \div (2.34 \times 10^{-12}) = 4.\underline{2}7 \times 10^{-3} = 4.3 \times 10^{-3}\ M$$

15.64 From the definition pH = -log [H^+], -pH = log [H^+], so enter the negative value of the pH on the calculator and use the inverse and log keys (or 10^x) key to find the antilog of -pH. Then use the K_w equation to calculate [OH^-] from [H^+].

$$\log [H^+] = -pH = -9.61$$

$$[H^+] = \text{antilog}(-9.61) = 10^{-9.61} = 2.\underline{4}5 \times 10^{-10}\ M$$

$$[OH^-] = K_w \div [H^+] = (1.0 \times 10^{-14}) \div (2.45 \times 10^{-10}) = 4.\underline{0}8 \times 10^{-5} = 4.1 \times 10^{-5}\ M$$

15.65 First calculate the molarity of the OH^- ion from the mass of NaOH. Then convert the [OH^-] to [H^+] using the K_w equation. Then find the pH, recording the same number of places after the decimal point in the pH as the number of significant figures in the [H^+].

$$\frac{5.80\ g\ NaOH}{1.00\ L} \times \frac{1\ mol\ NaOH}{40.01\ g\ NaOH} = \frac{0.1450\ mol\ NaOH}{L} = 0.14\underline{5}0\ M\ OH^-$$

$$[H^+] = K_w \div [OH^-] = (1.0 \times 10^{-14}) \div (0.1450) = 6.\underline{8}96 \times 10^{-14}\ M$$

$$pH = -\log [H^+] = -\log (6.896 \times 10^{-14}) = 13.1\underline{6}14 = 13.16$$

15.66 First calculate the molarity of the $Ba(OH)_2$ from the mass of $Ba(OH)_2$, and multiply the molarity of $Ba(OH)_2$ by 2 to obtain [OH^-]. Then convert the [OH^-] to [H^+] using the K_w equation. Then find the pH, recording the same number of places after the decimal point in the pH as the number of significant figures in the [H^+].

$$\frac{6.78\ g\ Ba(OH)_2}{1.00\ L} \times \frac{1\ mol\ Ba(OH)_2}{171.4\ g\ Ba(OH)_2} = \frac{0.039\underline{5}56\ mol\ Ba(OH)_2}{L}$$

$$[OH^-] = 2\ mol\ OH^-/1\ mol\ Ba(OH)_2 = 2 \times 0.039556\ M\ Ba(OH)_2 = 0.079\underline{1}1\ M$$

$$[H^+] = K_w \div [OH^-] = (1.0 \times 10^{-14}) \div (0.07912) = 1.\underline{2}64 \times 10^{-13}\ M$$

$$pH = -\log [H^+] = -\log (1.264 \times 10^{-13}) = 12.8\underline{9}82 = 12.90$$

15.67 Figure 16.13 shows that the methyl-red indicator is yellow at pH values above about 5.5 (slightly past the midpoint of the range for methyl red). Bromthymol blue is yellow at pH values up to about 6.5 (slightly below the midpoint of the range for bromthymol blue). Therefore, the pH of the solution is between 5.5 and 6.5, and the solution is acidic. (*Note:* Normal rain has a pH of 5.6, as indicated in the "A Chemist Looks At" discussion of acid rain.)

15.68 Thymol blue is yellow at pH values above about 2.5. Bromphenol blue is yellow at pH values below about 3.5. Therefore, the pH of the aspirin solution must be in the range of 2.5 to 3.5. The solution is, of course, acidic.

■ Solutions to Unclassified Problems

15.69 a. BaO is a base; $BaO + H_2O \rightarrow Ba^{2+} + 2OH^-$

b. H_2S is an acid; $H_2S + H_2O \rightarrow H_3O^+ + HS^-$

c. CH_3NH_2 is a base; $CH_3NH_2 + H_2O \rightarrow CH_3NH_3^+ + OH^-$

d. SO_2 is an acid; $SO_2 + 2H_2O \rightarrow H_3O^+ + HSO_3^-$

15.70 a. P_4O_{10} is an acid; $P_4O_{10} + 10H_2O \rightarrow 4H_3O^+ + 4H_2PO_4^-$

b. K_2O is a base; $K_2O + H_2O \rightarrow 2K^+ + 2OH^-$

c. N_2H_4 is a base; $N_2H_4 + H_2O \rightarrow N_2H_5^+ + OH^-$

d. H_2Se is an acid; $H_2Se + H_2O \rightarrow H_3O^+ + HSe^-$

15.71 a. $H_2O_2(aq) + S^{2-}(aq) \rightarrow HO_2^-(aq) + HS^-(aq)$

b. $HCO_3^-(aq) + OH^-(aq) \rightarrow CO_3^{2-}(aq) + H_2O(l)$

c. $NH_4^+(aq) + CN^-(aq) \rightarrow NH_3(aq) + HCN(aq)$

d. $H_2PO_4^-(aq) + OH^-(aq) \rightarrow HPO_4^{2-}(aq) + H_2O(l)$

15.72 a. $H_2O(l) + HCl(aq) \rightarrow H_3O^+(aq) + Cl^-(aq)$

b. $HCO_3^-(aq) + HF(aq) \rightarrow H_2CO_3(aq) + F^-(aq)$

c. $NH_3(aq) + HBrO(aq) \rightarrow NH_4^+(aq) + BrO^-(aq)$

d. $H_2PO_4^-(aq) + H_2SO_3(aq) \rightarrow H_3PO_4(aq) + HSO_3^-(aq)$

15.73 a. The ClO⁻ ion is a Bronsted base and water is a Bronsted acid. The complete chemical equation is $ClO^-(aq) + H_2O(l) \rightleftharpoons HClO(aq) + OH^-(aq)$. The equilibrium does not favor the products because ClO⁻ is a weaker base than OH⁻. In Lewis language, a proton from H_2O acts as a Lewis acid by accepting a pair of electrons on the oxygen of ClO⁻.

$$H^+ + [:\ddot{\underset{..}{Cl}}:\ddot{\underset{..}{O}}:]^- \longrightarrow :\ddot{\underset{..}{Cl}}:\ddot{\underset{..}{O}}:H$$

b. The NH_2^- ion is a Bronsted base and NH_4^+ is a Bronsted acid. The complete chemical equation is $NH_4^+ + NH_2^- \rightarrow 2NH_3$. The equilibrium favors the products because the reactants form the solvent, a weakly ionized molecule. In Lewis language, the proton from NH_4^+ acts as a Lewis acid by accepting a pair of electrons on the nitrogen of NH_2^-.

$$H^+ + \begin{bmatrix} :\ddot{N}:H \\ \underset{H}{} \end{bmatrix}^- \rightarrow H:\ddot{\underset{H}{N}}:H$$

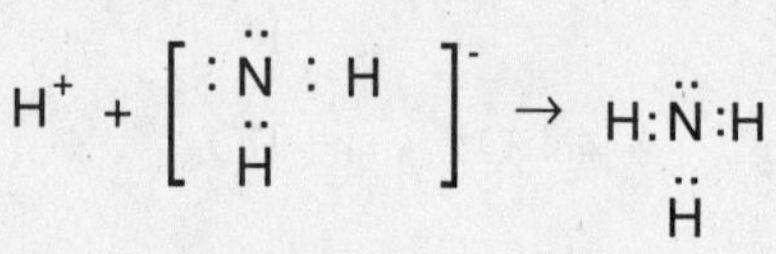

15.74 a. The HS⁻ ion is a Bronsted base and water is a Bronsted acid. The complete chemical equation is $HS^-(aq) + H_2O(l) \rightleftharpoons H_2S(aq) + OH^-(aq)$. The equilibrium does not favor the products because the HS⁻ ion is a weaker base than OH⁻. In Lewis language, a proton from H_2O acts as a Lewis acid by accepting a pair of electrons on the sulfur of HS⁻.

$$H^+ + [:\ddot{\underset{..}{S}}:H]^- \longrightarrow H:\ddot{\underset{..}{S}}:H$$

b. The complete chemical equation is $Fe^{3+}(aq) + CN^-(aq) \rightarrow Fe(CN)^{2+}$. The reaction cannot be described in Bronsted language because no proton transfer occurs. In Lewis language, Fe^{3+} acts as a Lewis acid by sharing a pair of electrons on the carbon of CN⁻.

$$Fe^{3+} + [:C \equiv N:]^- \longrightarrow [Fe:C \equiv N:]^{2+}$$

15.75 Table 15.2 shows that HNO_2 is a stronger acid than HF. Because an acid-base reaction normally goes in the direction of the weaker acid, the reaction is more likely to go in the direction written:

$$HNO_2 + F^- \rightleftharpoons HF + NO_2^-$$

15.76 The HS^- ion is acting as a base and will form its conjugate acid H_2S. Table 15.2 shows that H_2S is a stronger acid than HCN. Because an acid-base reaction normally goes in the direction of the weaker acid, the reaction is more likely to go in the opposite direction:

$$H_2S + CN^- \rightleftharpoons HS^- + HCN.$$

15.77 The order is $H_2S < H_2Se < HBr$. H_2Se is stronger than H_2S because, within a group, acid strength increases with increasing size of the central atom in binary acids. HBr is a strong acid, whereas the others are weak acids.

15.78 The order is $HBrO < HBrO_2 < HClO_2$. $HClO_2$ is stronger than $HBrO_2$ because, for oxoacids, acid strength increases with increasing electronegativity. $HBrO_2$ is stronger than HBrO because acid strength increases with increasing oxidation number of Br.

15.79 The KOH is a strong base and is fully ionized in solution, so you can use its formula and molar concentration to determine the $[OH^-]$ of the solution. Therefore the 0.25 M KOH contains 0.25 M OH^-. The $[H^+]$ is obtained from the K_w expression:

$$K_w = 1.0 \times 10^{-14} = [H^+] \times 0.25 \text{ M } OH^-$$

$$[H^+] = \frac{1.0 \times 10^{-14}}{0.25} = 4.\underline{0}0 \times 10^{-14} = 4.0 \times 10^{-14} \text{ M}$$

15.80 The $Sr(OH)_2$ is a strong base and is fully ionized in solution, so you can use its formula and molar concentration to determine the $[OH^-]$ of the solution. Therefore the 0.35 M $Sr(OH)_2$ contains 2 x 0.35 = 0.70 M OH^-. The $[H^+]$ is obtained from the K_w expression:

$$K_w = 1.0 \times 10^{-14} = [H^+] \times 0.70 \text{ M } OH^-$$

$$[H^+] = \frac{1.0 \times 10^{-14}}{0.70} = 1.\underline{4}2 \times 10^{-14} = 1.4 \times 10^{-14} \text{ M}$$

15.81 Enter the H^+ concentration of 1.5×10^{-3} into the calculator, press the log key, and press the sign key to change the negative log to a positive log. This follows the negative log definition of pH. The number of decimal places of the pH should equal the significant figures in the H^+.

$$pH = -\log [H^+] = -\log (1.5 \times 10^{-3}) = 2.8\underline{2}3 = 2.82$$

15.82 Enter the H^+ concentration of 2.5×10^{-2} into the calculator, press the log key, and press the sign key to change the negative log to a positive log. This follows the negative log definition of pH. The number of decimal places of the pH should equal the significant figures in the H^+.

$$pH = -\log [H^+] = -\log (2.5 \times 10^{-2}) = 1.6\underline{0}2 = 1.60$$

15.83 Find the pOH from the pH using pH + pOH = 14.00. Then calculate the $[OH^-]$ from the pOH by entering the pOH into the calculator, pressing the sign key to change the positive log to a negative log, and finding the antilog. On some calculators the antilog is found by using the inverse of the log; on other calculators the antilog is found using the 10^x key. The number of significant figures in the $[OH^-]$ should equal the number of decimal places in the pOH.

$$pOH = 14.00 - 3.15 = 10.85$$

$$[OH^-] = \text{antilog}(-10.85) = 10^{-10.85} = 1.\underline{4}2 \times 10^{-11} = 1.4 \times 10^{-11}\ M$$

15.84 Find the pOH from the pH using pH + pOH = 14.00. Then calculate the $[OH^-]$ from the pOH by entering the pOH into the calculator, pressing the sign key to change the positive log to a negative log, and finding the antilog. On some calculators the antilog is found by using the inverse of the log; on other calculators the antilog is found using the 10^x key. The number of significant figures in the $[OH^-]$ should equal the number of decimal places in the pOH.

$$pOH = 14.00 - 4.05 = 9.95$$

$$[OH^-] = \text{antilog}(-9.95) = 10^{-9.95} = 1.\underline{1}2 \times 10^{-10} = 1.1 \times 10^{-10}\ M$$

15.85 a. $H_2SO_4(aq) + 2NaHCO_3(aq) \rightarrow Na_2SO_4(aq) + CO_2(g) + H_2O(l)$ (molecular)

$H^+(aq) + HCO_3^-(aq) \rightarrow CO_2(g) + H_2O(l)$ (total ionic)

b. The total moles of H^+ from the H_2SO_4 is

$$\text{mol } H^+ = \frac{0.437 \text{ mol } H_2SO_4}{1 \text{ L}} \times 0.02500 \text{ L} \times \frac{2 \text{ mol } H^+}{1 \text{ mol } H_2SO_4}$$

$$= 0.021\underline{8}5 \text{ mol } H^+$$

(continued)

The moles of H^+ that reacted with the NaOH is given by

$$\text{mol } H^+ = \frac{0.108 \text{ mol NaOH}}{1 \text{ L}} \times 0.0354 \text{ L} = 0.0038\underline{2}3 \text{ mol } H^+$$

The moles of $NaHCO_3$ present in the original sample is equal to the moles of H^+ that reacted with the HCO_3^-, which is given by

$$\text{Total moles } H^+ - \text{moles } H^+ \text{ reacted with the NaOH} = \text{moles } HCO_3^-$$

$$0.021\underline{8}5 \text{ mol} - 0.003823 \text{ mol} = 0.018\underline{0}3 = 0.0180 \text{ mol } NaHCO_3$$

c. The mass of $NaHCO_3$ (molar mass 84.01 g/mol) present in the original sample is

$$0.01803 \text{ mol } NaHCO_3 \times \frac{84.01 \text{ g } NaHCO_3}{1 \text{ mol } NaHCO_3} = 1.5\underline{1}4 \text{ g}$$

Thus, the percent $NaHCO_3$ in the original sample is given by

$$\% \, NaHCO_3 = \frac{1.514 \text{ g}}{2.500 \text{ g}} \times 100\% = 60.\underline{5}6 = 60.6\%$$

The percent KCl in the original sample is

$$\% \, KCl = 100 - 60.56 = 39.\underline{4}4 = 39.4\%$$

15.86 a. $CO_3^{2-}(aq) + 2H^+(aq) \rightarrow H_2O(l) + CO_2(g)$

b. The total moles of H^+ from the HCl is

$$\text{mol } H^+ = \frac{0.798 \text{ mol HCl}}{1 \text{ L}} \times 0.02500 \text{ L} = 0.019\underline{9}5 \text{ mol } H^+$$

The moles of H^+ that reacted with the NaOH is given by

$$\text{mol } H^+ = \frac{0.108 \text{ mol NaOH}}{1 \text{ L}} \times 0.0287 \text{ L} = 0.0031\underline{0}0 \text{ mol } H^+$$

The moles of Na_2CO_3 present in the original sample is equal to the moles of H^+ that reacted with the CO_3^{2-}, which is given by

(continued)

Total moles H^+ - moles H^+ reacted with the NaOH = moles CO_3^{2-}

0.01995 mol - 0.003100 mol = 0.01685 = 0.0169 mol Na_2CO_3

c. The mass of Na_2CO_3 (molar mass 105.99 g/mol) present in the original sample is

$$0.01685 \text{ mol } Na_2CO_3 \times \frac{105.99 \text{ g } Na_2CO_3}{1 \text{ mol } Na_2CO_3} = 1.785 \text{ g}$$

Thus, the percent Na_2CO_3 in the original sample is given by

$$\% \, Na_2CO_3 = \frac{1.785 \text{ g}}{2.500 \text{ g}} \times 100\% = 71.40 = 71.4\%$$

The percent NaCl in the original sample is

$$\% \, NaCl = 100 - 71.40 = 28.60 = 28.6\%$$

15.87 $HCO_3^- (aq) + H_2O (l) \rightleftharpoons H_3O^+ (aq) + CO_3^{2-} (aq)$

$HCO_3^- (aq) + H_2O (l) \rightleftharpoons H_2CO_3 (aq) + OH^- (aq)$

$HCO_3^- (aq) + Na^+ (aq) + OH^- (aq) \rightleftharpoons Na^+ (aq) + CO_3^{2-} (aq) + H_2O (l)$

$HCO_3^- (aq) + H^+ (aq) + Cl^- (aq) \rightleftharpoons H_2O (l) + CO_2 (g) + Cl^- (aq)$

15.88 $H_2PO_4^-$ is amphiprotic or amphoteric.

$H_2PO_4^- (aq) + H_2O (l) \rightleftharpoons H_3O^+ (aq) + HPO_4^{2-} (aq)$

$H_2PO_4^- (aq) + H_2O (l) \rightleftharpoons H_3PO_4 (aq) + OH^- (aq)$

$H_2PO_4^- (aq) + K^+ (aq) + OH^- (aq) \rightleftharpoons K^+ (aq) + HPO_4^{2-} (aq) + H_2O (l)$

$H_2PO_4^- (aq) + H^+ (aq) + NO_3^- (aq) \rightleftharpoons H_3PO_4 (aq) + NO_3^- (aq)$

15.89 $CaH_2\,(s) + 2H_2O\,(l) \rightarrow Ca(OH)_2\,(s) + H_2\,(g)$

The hydride ion is a stronger base because it took an H^+ from water leaving the OH^- ion. Every time a strong base is added to water it will react with the water leaving the OH^- as the product so the strong base cannot exist in water.

15.90 $Ca_3N_2\,(s) + 6H_2O\,(l) \rightarrow 3Ca(OH)_2\,(s) + 2NH_3\,(g)$

$NaNH_2\,(s) + H_2O\,(l) \rightarrow NaOH\,(aq) + NH_3\,(g)$

N^{3-} is a stronger base than NH_2^- because it has more negative charge so that it will have a greater attraction for H^+. Also we could consider that NH_2^- is a N^{3-} that has already reacted with two H^+'s. Stronger bases than OH^- will produce OH^- in aqueous solution.

15.91 a. $2HF\,(l) \rightleftharpoons H_2F^+ + F^-$

b. NaF will be a base because F^- is a conjugate base of HF.

c. $HClO_4 + HF \rightarrow H_2F^+ + ClO_4^-$

The conjugate acid is H_2F^+.

15.92 a. $2NH_3\,(l) \rightleftharpoons NH_4^+ + NH_2^-$

b. NH_2^- is the conjugate base of NH_3 so $NaNH_2$ will be a base.

c. $NaNH_2 + NH_4Cl \rightarrow 2NH_3 + NaCl$

15.93 The reaction of ammonia with water is given by

$$NH_3\,(aq) + H_2O\,(l) \rightleftharpoons NH_4^+\,(aq) + OH^-\,(aq)$$

The initial concentration of NH_3 (molar mass 17.03 g/mol) is

$$\text{Molarity} = \frac{4.25\text{ g }NH_3 \times \dfrac{1\text{ mol }NH_3}{17.03\text{ g }NH_3}}{0.2500\text{ L}} = 0.99\underline{8}2\text{ M}$$

(continued)

Since the NH_3 is 0.42% reacted, the concentration of OH^- is

$$[OH^-] = 0.9982\ M \times 0.0042 = 0.004\underline{1}9\ M$$

$$pOH = -\log[OH^-] = -\log(0.00419) = 2.3\underline{7}8$$

$$pH = 14 - pOH = 14 - 2.3\underline{7}8 = 11.6\underline{2}2 = 11.62$$

15.94 The reaction of $C_2H_5NH_2$ with water is given by

$$C_2H_5NH_2\,(aq) + H_2O\,(l) \rightleftharpoons C_2H_5NH_3^+\,(aq) + OH^-\,(aq)$$

The initial concentration of $C_2H_5NH_2$ (molar mass 45.09 g/mol) is given by

$$\text{Molarity} = \frac{0.675\ g\ C_2H_5NH_2 \times \dfrac{1\ mol\ C_2H_5NH_2}{45.09\ g\ C_2H_5NH_2}}{0.1000\ L} = 0.14\underline{9}7\ M$$

Since the $C_2H_5NH_2$ is 0.98% reacted, the concentration of OH^- is

$$[OH^-] = 0.1497\ M \times 0.0098 = 0.001\underline{4}7\ M$$

$$pOH = -\log[OH^-] = -\log(0.00147) = 2.8\underline{3}3$$

$$pH = 14 - pOH = 14 - 2.8\underline{3}3 = 11.1\underline{6}7 = 11.17$$

■ Cumulative-Skills Problems
(require skills from previous chapters)

15.95 For $(HO)_mYO_n$ acids, acid strength increases with n, regardless of the number of OH's. The structure of H_3PO_4 is $(HO)_3PO$; because H_3PO_3 and H_3PO_4 have about the same acidity, H_3PO_3 must also have n = 1; thus, m = 2. This leaves one H, which must bond to phosphorus, giving a structure of $(HO)_2(O)PH$. Assuming that only 2H's react with NaOH, the mass of NaOH that reacts with 1.00 g of H_3PO_3 (PA) is calculated as follows:

$$1.00\ g\ PA \times \frac{1\ mol\ PA}{81.994\ g\ PA} \times \frac{2\ mol\ NaOH}{1\ mol\ PA} \times \frac{40.00\ g\ NaOH}{1\ mol\ NaOH}$$

$$= 0.97\underline{5}6 = 0.976\ g\ NaOH$$

15.96 For $(HO)_mYO_n$ acids, acid strength increases with n, regardless of the number of OH's. The structure of H_3PO_4 is $(HO)_3PO$; because H_3PO_2 and H_3PO_4 have the same acidity, H_3PO_2 must also have n = 1; thus, m = 1. This leaves two H's, which must bond to phosphorus, giving a structure of $(HO)(O)PH_2$. Assuming that only 1H reacts with NaOH, the mass of NaOH that reacts with 1.00 g of H_3PO_2 (HA) is calculated as follows:

$$1.00 \text{ g HA} \times \frac{1 \text{ mol HA}}{65.994 \text{ g HA}} \times \frac{1 \text{ mol NaOH}}{1 \text{ mol HA}} \times \frac{40.00 \text{ g NaOH}}{1 \text{ mol NaOH}}$$

$$= 0.60\underline{6}1 = 0.606 \text{ g NaOH}$$

15.97 BF_3 acts as a Lewis acid, accepting an electron pair from NH_3:

$$BF_3 + :NH_3 \rightarrow F_3B:NH_3$$

The NH_3 acts as a Lewis base in donating an electron pair to BF_3. When 10.0 g of each are mixed, the BF_3 is the limiting reagent because it has the higher formula weight. The mass of $BF_3:NH_3$ formed is

$$10.0 \text{ g } BF_3 \times \frac{1 \text{ mol } BF_3}{67.81 \text{ g } BF_3} \times \frac{1 \text{ mol } BF_3:NH_3}{1 \text{ mol } BF_3} \times \frac{84.84 \text{ g } BF_3:NH_3}{1 \text{ mol } BF_3:NH_3}$$

$$= 12.\underline{5}1 = 12.5 \text{ g } BF_3:NH_3$$

15.98 BF_3 is the Lewis acid, accepting an electron pair from ether:

$$BF_3 + :OR_2 \rightarrow F_3B:OR_2$$

The ether ($:OR_2$) acts as a Lewis base in donating an electron pair to BF_3. When 10.0 g of BF_3 and 20.0 g of ether are mixed, the BF_3 is the limiting reagent because the formula weights are nearly equal. The mass of $BF_3:OR_2$ formed is

$$10.0 \text{ g } BF_3 \times \frac{1 \text{ mol } BF_3}{67.81 \text{ g } BF_3} \times \frac{1 \text{ mol } BF_3:OR_2}{1 \text{ mol } BF_3} \times \frac{141.93 \text{ g } BF_3:OR_2}{1 \text{ mol } BF_3:OR_2}$$

$$= 20.\underline{9}3 = 20.9 \text{ g } BF_3:OR_2$$

■ Solution to Conceptual Problem

15.99 The pH of natural rain is approximately 5.7, due to the reaction with CO_2 in the atmosphere to form carbonic acid (see essay on acid rain in Ch. 16). Through similar mechanisms, gaseous NO_2 and SO_2, generated as by-products of industrial energy production, can form acids on reaction with water in the atmosphere according to the equations shown below:

$$2SO_2(g) + O_2(g) \rightarrow 2SO_3(g)$$

$$SO_3(g) + H_2O(l) \rightarrow H_2SO_4(aq)$$

$$2NO_2(g) + H_2O(l) \rightarrow HNO_3(aq) + HNO_2(aq)$$

As a result, some rainwater has a pH as low as 1.0. This has led to devastating effects on the environment, including the decay of marble and stone buildings and sculpture. Since the production of NO_2 and SO_2 gases occur to some extent at industrial power plant smoke stacks, design a system to remove these pollutants from these waste gases utilizing what you know about acid-base chemistry. If the typical output of NO_2 or SO_2 is 100 lb/day, estimate the feasibility of your method bases on the consumption of chemical you propose.

Discussion

Several industrial processes already exist or have been proposed, and the student may find these in a library search. One example involves the reacting of the acid pollutants in stack gas with an aqueous suspension of slaked lime, $Ca(OH)_2$. Sulfurous acid ($H_2O + SO_2$) reacts with calcium hydroxide to produce insoluble calcium sulfite.

$$Ca(OH)_2 + H_2SO_3 \rightarrow CaSO_3 + 2H_2O$$

Thus, one mole of SO_2 (64.1 g) reacts with one mole of $Ca(OH)_2$ (74.1 g); or, 64.1 lb SO_2 reacts with 74.1 lb $Ca(OH)_2$. Calculation shows that 100 lb SO_2 would require 116 lb $Ca(OH)_2$. The cost of the slaked lime would be not be great. However, one would have to add to this the cost of disposal of the product $CaSO_3$. A better method would involve a reversible scheme in which the SO_2 could be recovered and sold to manufacture sulfuric acid.

16. ACID-BASE EQUILIBRIA

Solutions to Exercises

Note on answers to equilibrium calculations: The rounded answer (with correct number of significant figures) is always given first. After the stepwise solution, the numerical answer is given again, but with one nonsignificant figure. (The rightmost significant figure is underlined.)

16.1 Rounded answer: $K_a = 1.4 \times 10^{-4}$ (correct no. of sig. figs.). To solve, assemble a table of starting, change, and equilibrium concentrations. Use HL as the symbol for lactic acid.

Conc. (M)	HL	$\rightleftharpoons$	H^+	+	L^-
Starting	0.025		0		0
Change	-x		+x		+x
Equilibrium	0.025 - x		x		x

The value of x equals the value of the molarity of the H^+ ion, which can be obtained from the pH: $[H^+]$ = antilog (-pH) = antilog (-2.75) = $1.\underline{7}8 \times 10^{-3}$ M. After using x to substitute into the equilibrium-constant expression, substitute this value for x and solve for K_a:

$$K_a = \frac{[H^+][L^-]}{[HL]} = \frac{[x]^2}{(0.025 - x)} = \frac{(1.78 \times 10^{-3})^2}{0.0232} = 1.\underline{3}6 \times 10^{-4} = 1.4 \times 10^{-4}$$

$$\text{Degree of ionization} = \frac{1.78 \times 10^{-3}}{0.025} = 0.071\ (7.1\%)$$

16.2 Rounded answer: $[H^+] = [C_2H_3O_2^-] = 1.3 \times 10^{-3}$ M; pH = 2.89. To solve, assemble a table of starting, change, and equilibrium concentrations. Use HAc as the symbol for acetic acid.

Conc. (M)	HAc	⇌	H^+	+	Ac^-
Starting	0.10		0		0
Change	-x		+x		+x
Equilibrium	0.10 - x		x		x

Write the equilibrium-constant expression in terms of chemical symbols and then substitute x and (0.10 - x):

$$\frac{[H^+][Ac^-]}{[HAc]} = K_a = \frac{[x]^2}{(0.10 - x)} = 1.7 \times 10^{-5}$$

Solve the equation for x, assuming that x is much smaller than 0.10, so that (0.10 - x) ≅ 0.10. Solve by taking the square root of $1.7 \times 10^{-5}(0.10)$.

$$\frac{[x]^2}{(0.10)} \cong 1.7 \times 10^{-5};\ x^2 \cong (1.7 \times 10^{-5})(0.10) = 1.7 \times 10^{-6}$$

$$x = 1.\underline{3}03 \times 10^{-3}\ M = [H^+] = [Ac^-]$$

Check to make sure that the assumption that (0.10 - x) ≅ 0.10 is valid:

$$0.10 - 1.3 \times 10^{-3} = 0.0987, \text{ or} \cong 0.10 \text{ to two sig. figs.}$$

The pH of the solution = - log (1.303×10^{-3}) = 2.8$\underline{8}$505 (rounded answer above).

$$\text{Degree of ionization} = \frac{1.303 \times 10^{-3}}{0.10} = 0.013\ (1.3\%)$$

16.3 Rounded answer: $[H^+] = 5.8 \times 10^{-4}$ M; pH = 3.24 (correct no. of sig. figs.). To solve, assemble a table of starting, change, and equilibrium concentrations. Let HPy symbolize pyruvic acid.

Conc. (M)	HPy	⇌	H^+	+	Py^-
Starting	0.0030		0		0
Change	-x		+x		+x
Equilibrium	0.0030 - x		x		x

(continued)

Write the equilibrium-constant expression in terms of chemical symbols and then substitute the terms x and (0.0030 - x):

$$\frac{[H^+][Py^-]}{[HPy]} = K_a = \frac{[x]^2}{(0.0030 - x)} = 1.4 \times 10^{-4}$$

In this case, x cannot be neglected compared to 0.0030 M. (If it is neglected, the calculated $[H^+]$ is 6.48×10^{-4}, which when subtracted from 0.0030 yields 0.00235, a significant change.) The *quadratic formula* must be used. Reorganize the above equilibrium-constant expression into the form $ax^2 + bx + c = 0$, and substitute for a, b, and c in the quadratic formula.

$$x^2 + 1.4 \times 10^{-4} - 4.20 \times 10^{-7} = 0$$

$$x = \frac{-1.4 \times 10^{-4} \pm \sqrt{1.96 \times 10^{-8} + 1.68 \times 10^{-6}}}{2} = -7.00 \times 10^{-5} \pm 6.52 \times 10^{-4}$$

Using the + root: $x = [H^+] = 5.\underline{8}2 \times 10^{-4}$ M.

$$pH = -\log(5.82 \times 10^{-4}) = 3.2\underline{3}507 \text{ (rounded answer above)}$$

16.4 Rounded answer: $[H^+] = 0.051$ M; pH = 1.29; $[SO_3^{2-}] = 6.3 \times 10^{-8}$ M. To solve, note that $K_{a1} = 1.3 \times 10^{-2} > K_{a2} = 6.3 \times 10^{-8}$, and hence the second ionization and K_{a2} can be neglected. Assemble a table of starting, change, and equilibrium concentrations:

Conc. (M)	H_2SO_3 $\rightleftharpoons$	H^+	+	HSO_3^-
Starting	0.25	0		0
Change	-x	+x		+x
Equilibrium	0.25 - x	x		x

Write the equilibrium-constant expression in terms of chemical symbols and then substitute x and (0.25 - x):

$$\frac{[H^+][HSO_3^-]}{[H_2SO_3]} = K_{a1} = \frac{[x]^2}{(0.25 - x)} = 1.3 \times 10^{-2}$$

This gives $x^2 + 0.013x - 0.00325 = 0$

(continued)

In this case, x cannot be neglected compared to 0.25 M. (If it is neglected, the calculated [H^+] is 5.7 x 10^{-2}, which when subtracted from 0.25 yields 0.193, a significant change.) The *quadratic formula* must be used. Reorganize the above equilibrium-constant expression into the form $ax^2 + bx + c = 0$, and substitute for a, b, and c in the quadratic formula.

$$x = \frac{-1.3 \times 10^{-2} \pm \sqrt{1.69 \times 10^{-4} + 1.30 \times 10^{-2}}}{2} = -6.50 \times 10^{-3} \pm 5.738 \times 10^{-2}$$

Using the + root, x = [H^+] = 5.0̲88 x 10^{-2} M.

$$pH = -\log(5.088 \times 10^{-2}) = 1.2\underline{9}3$$

To calculate [SO_3^{2-}], note from the equilibrium concentrations in the table above that [H^+] = [HSO_3^-] (= 5.088 x 10^{-2} M). In exact terms, [H^+] = (0.051 + y) from the ionization of HSO_3, and [HSO_3^-] = (0.051 - y) from ionization. Substituting into the K_{a2} expression gives

$$\frac{[H^+][SO_3^-]}{[HSO_3^-]} = K_{a2} = \frac{(0.051 + y)(SO_3^{2-})}{(0.051 - y)} = 6.3 \times 10^{-8}$$

Assuming that y is much smaller than 0.051, note that the (0.051 + y) cancels the (0.051 - y) term, leaving

$$[SO_3^{2-}] \cong 6.3 \times 10^{-8}\ M$$

16.5 Rounded answer: K_b = 3.3 x 10^{-6} (correct no. of sig. figs.). To solve, convert the pH to [OH^-]:

$$pOH = 14.00 - pH = 14.00 - 9.84 = 4.16$$

$$[OH^-] = \text{antilog}(-4.16) = 6.\underline{9}2 \times 10^{-5}\ M$$

Using the symbol Qu for quinine, assemble a table of starting, change, and equilibrium concentrations.

Conc. (M)	Qu	+	H_2O	⇌	HQu^+	+	OH^-
Starting	0.0015				0		0
Change	-x				+x		+x
Equilibrium	0.0015 - 6.92 x 10^{-5}				6.92 x 10^{-5}		6.92 x 10^{-5}

(continued)

Write the equilibrium-constant expression in terms of chemical symbols; then substitute the terms after subtracting in the denominator, and solve for K_b:

$$K_b = \frac{[HQu^+][OH^-]}{[Qu]} = \frac{(6.92 \times 10^{-5})^2}{(0.00143)} = 3.\underline{3}48 \times 10^{-6} \text{ (rounded answer above)}$$

16.6 Rounded answer: $[H^+] = 5.3 \times 10^{-12}$ M. To solve, assemble a table of starting, change, and equilibrium concentrations:

Conc. (M)	NH_3	+	H_2O	$\rightleftharpoons$	NH_4^+	+	OH^-
Starting	0.20				0		0
Change	-x				+x		+x
Equilibrium	0.20 - x				x		x

Write the equilibrium-constant expression in terms of chemical symbols, then substitute the terms and solve for $[OH^-]$ and then $[H^+]$:

$$\frac{[NH_4^+][OH^-]}{[NH_3]} \cong \frac{(x)^2}{(0.20)} \cong 1.8 \times 10^{-5}$$

$x = [OH^-] \cong 1.89 \times 10^{-3}$ M (Note that x is negligible compared to 0.20.)

$[H^+] = (1.0 \times 10^{-14}) \div (1.89 \times 10^{-3}) = 5.\underline{2}9 \times 10^{-12}$ M (rounded answer above)

16.7 a. NH_4NO_3 yields an acidic solution because NH_4^+ hydrolyzes in water to form the H_3O^+ ion (and NH_3); the NO_3^- ion does not hydrolyze.

b. KNO_3 does not change the pH (7.00) of neutral water because neither ion hydrolyzes.

c. $Al(NO_3)_3$ yields an acidic solution because Al^{3+} hydrolyzes in water to form the H_3O^+ ion; the NO_3^- ion does not hydrolyze.

16.8 a. Calculate K_b of the F^- ion from the K_a of its conjugate acid, HF:

$$K_b = K_w \div K_a = (1.0 \times 10^{-14}) \div (6.8 \times 10^{-4}) = 1.\underline{4}7 \times 10^{-11} = 1.5 \times 10^{-11}$$

b. Calculate K_a of $C_6H_5NH_3^+$ from K_b of its conjugate base, $C_6H_5NH_2$:

$$K_a = K_w \div K_b = (1.0 \times 10^{-14}) \div (4.2 \times 10^{-10}) = 2.\underline{3}8 \times 10^{-5} = 2.4 \times 10^{-5}$$

16.9 Rounded answer: $[OH^-] = 1.5 \times 10^{-6}$ M; pH = 8.19. Assemble the usual table, letting $[Ben^-]$ equal the equilibrium concentration of the benzoate anion (the only ion that hydrolyzes). Then calculate K_b of the Ben^- ion from K_a of its conjugate acid, HBen. Assume x is much smaller than the 0.015 M concentration in the denominator and solve for x in the numerator of the equilibrium-constant expression. Finally, calculate pOH from the $[OH^-]$ and pH from the pOH.

Conc. (M)	Ben^-	+	H_2O	⇌	HBen	+	OH^-
Starting	0.015				0		0
Change	-x				+x		+x
Equilibrium	0.015 - x				x		x

$$K_b = K_w \div K_a = (1.0 \times 10^{-14}) \div (6.3 \times 10^{-5}) = 1.\underline{5}8 \times 10^{-10}$$

$$\frac{[HBen][OH^-]}{[Ben^-]} \cong \frac{(x)^2}{(0.015)} \cong 1.58 \times 10^{-10}$$

$x = [OH^-] \cong 1.\underline{5}39 \times 10^{-6}$ M (Note that x is negligible compared to 0.015.)

Thus, the concentration of benzoic acid in the solution is 1.5×10^{-6} M.

$$pOH = -\log[OH^-] = -\log(1.539 \times 10^{-6}) = 5.8\underline{1}2$$

$$pH = 14.00 - 5.812 = 8.1\underline{8}8 = 8.19$$

16.10 Rounded answer: $[CHO_2^-] = 8.5 \times 10^{-5}$ M; degree of ionization $= 8.5 \times 10^{-4}$ (0.085%). Assemble the usual table, using starting $[H^+] = 0.20$ M from 0.20 M HCl, and letting HFo symbolize $HCHO_2$. Assume x is negligible compared to 0.10 M and 0.20 M, and solve for x in the numerator. Calculate the degree of ionization from $[CHO_2^-]$.

Conc. (M)	HFo	⇌	H^+	+	Fo^-
Starting	0.10		0.20		0
Change	-x		+x		+x
Equilibrium	0.10 - x		0.20 + x		x

$$\frac{[H^+][Fo^-]}{[HFo]} = K_a = \frac{(0.20 + x)(x)}{(0.10 - x)} = 1.7 \times 10^{-4}$$

(continued)

$$\frac{(0.20)(x)}{(0.10)} \cong 1.7 \times 10^{-4}; \; x = [Fo^-] \cong 8.\underline{5}0 \times 10^{-5} \; (0.085\%)$$

Degree of ionization = $(8.50 \times 10^{-5}) \div 0.10 = 8.\underline{5}0 \times 10^{-4}$ (0.085%)

16.11 Rounded answer: $[H^+] = 2.4 \times 10^{-4}$ M; pH = 3.63. Assemble the usual table, using a starting $[CHO_2^-]$ of 0.018 M from 0.018 M $NaCHO_2$, and symbolizing $HCHO_2$ as HFo and the CHO_2^- anion as Fo^-. Assume x is negligible compared to 0.025 M and 0.018 M, and solve for the x in the numerator.

Conc. (M)	HFo	$\rightleftharpoons$	H^+	+	Fo^-
Starting	0.025		0		0.018
Change	-x		+x		+x
Equilibrium	0.025 - x		x		0.018 + x

$$\frac{[H^+][Fo^-]}{[HFo]} = K_a = \frac{(0.018 + x)(x)}{(0.025 - x)} = 1.7 \times 10^{-4}$$

$$\frac{(x)(0.018)}{(0.025)} \cong 1.7 \times 10^{-4}; \; x = [Fo^-] \cong 2.\underline{3}6 \times 10^{-4} \text{ M}$$

$$pH = -\log[H^+] = -\log(2.36 \times 10^{-4}) = 3.6\underline{2}7$$

16.12 Rounded answer: $[H^+] = 5.5 \times 10^{-6}$ M; pH = 5.26. Find the mol/L of HAc ($HC_2H_3O_2$) and the mol/L of Ac^- ($C_2H_3O_2^-$), and assemble the usual table. Substitute the equilibrium concentrations into the equilibrium-constant expression; then assume x is negligible compared to the starting concentrations of both HAc and Ac^-. Solve for x in the numerator of the equilibrium-constant expression, and calculate the pH from this value.

Total volume = 0.030 L + 0.070 L = 0.100 L

(0.15 mol HAc/L) x 0.030 L = 0.0045 mol HAc (÷ 0.100 L total volume = 0.045 M)

(0.20 mol Ac^-/L) x 0.070 L = 0.0140 mol Ac^- (÷ 0.100 L total volume = 0.14 M)

Now substitute these starting concentrations into the table.

(continued)

Conc. (M)	HAc ⇌	H^+	+	Ac^-
Starting	0.045	0		0.14
Change	-x	+x		+x
Equilibrium	0.045 - x	x		0.14 + x

$$\frac{[H^+][Ac^-]}{[HAc]} = K_a = \frac{(x)(0.14 + x)}{(0.045 - x)} = 1.7 \times 10^{-5}$$

$$\frac{(x)(0.14)}{(0.045)} \cong 1.7 \times 10^{-5};\ x = [H^+] \cong 5.\underline{4}6 \times 10^{-6}\ M$$

$$pH = -\log[H^+] = -\log(5.46 \times 10^{-6}) = 5.2\underline{6}2$$

16.13 Rounded answer: $[H^+] = 1.5 \times 10^{-4}$; pH = 3.83. The OH^- ion reacts with the $HCHO_2$ (HFo) to form additional CHO_2^- (Fo^-) plus H_2O. Calculate the stoichiometric amounts of NaOH; then subtract the moles of NaOH from the moles of HFo. Add the resulting moles of Fo^- to the 0.018 starting moles of Fo^- already present.

(0.10 mol NaOH/L) x 0.0500 L = 0.0050 mol NaOH (reacts with 0.0050 mol HFo)

Mol HFo left = (0.025 - 0.0050) mol = 0.020 mol HFo

Total mol Fo^- now present = 0.018 orig. + 0.0050 formed from NaOH
= 0.023 mol Fo^-

$$[HFo] = \frac{0.020\ \text{mol HFo}}{1.050\ \text{L soln}} = 0.01\underline{9}0\ M;\ [Fo^-] = \frac{0.023\ \text{mol}\ Fo^-}{1.050\ \text{L soln}} = 0.02\underline{1}9\ M$$

Now account for the ionization of HFo to Fo^- at equilibrium by assembling the usual table. Assume x is negligible compared to 0.0190 M and 0.0219 M, and solve the equilibrium-constant expression for x in the numerator. Calculate the pH from x, the $[H^+]$.

Conc. (M)	HFo ⇌	H^+	+	Fo^-
Starting	0.0190	0		0.0219
Change	-x	+x		+x
Equilibrium	0.0190 - x	x		0.0219 + x

(continued)

$$\frac{[H^+][Fo^-]}{[HFo]} = K_a = \frac{(0.0219 + x)(x)}{(0.0190 - x)} = 1.7 \times 10^{-4}$$

$$\frac{(x)(0.0219)}{(0.0190)} \cong 1.7 \times 10^{-4};\ x = [H^+] \cong 1.\underline{4}7 \times 10^{-4}\ M$$

$$pH = -\log [H^+] = -\log (1.47 \times 10^{-4}) = 3.8\underline{3}1$$

16.14 Rounded answer: $[H^+]$ = 0.025 M; pH = 1.60. All the OH^- (from the NaOH) reacts with the H^+ from HCl. Calculate the stoichiometric amounts of OH^- and H^+ and subtract the mol of OH^- from the mol of H^+. Then divide the remaining H^+ by the total volume of 0.025 L + 0.015 L, or 0.040 L, to find the $[H^+]$. Then calculate the pH.

$$\text{Mol } H^+ = (0.10 \text{ mol HCl/L}) \times 0.025 \text{ L HCl} = 0.0025 \text{ mol } H^+$$

$$\text{Mol } OH^- = (0.10 \text{ mol NaOH/L}) \times 0.015 \text{ L NaOH} = 0.0015 \text{ mol } OH^-$$

$$\text{Mol } H^+ \text{ left} = (0.0025 - 0.0015) \text{ mol } H^+ = 0.0010 \text{ mol } H^+$$

$$[H^+] = 0.0010 \text{ mol } H^+ \div 0.040 \text{ L total volume} = 0.025 \text{ M}$$

$$pH = -\log [H^+] = -\log (0.025) = 1.6\underline{0}2$$

16.15 Rounded answer: $[OH^-]$ = 9.4×10^{-7} M; pH = 7.97. At the equivalence point, equal molar amounts of HF and NaOH react to form a solution of NaF. Start by calculating the moles of HF. Use this to calculate the volume of NaOH needed to neutralize all of the HF (and use the moles of HF as the moles of F^- formed at the equivalence point). Add the volume of NaOH to the original 0.025 L to find the total volume of solution.

$$(0.10 \text{ mol HF/L}) \times 0.025 \text{ L} = 0.0025 \text{ mol HF}$$

$$\text{Volume NaOH} = 0.0025 \text{ mol HF} \div (0.15 \text{ mol NaOH/L}) = 0.01\underline{6}6 \text{ L}$$

$$\text{Total volume} = 0.0166 \text{ L} + 0.025 \text{ L HF soln} = 0.0416 \text{ L}$$

$$[F^-] = (0.0025 \text{ mol } F^- \text{ from HF}) \div 0.0416 \text{ L} = 0.06\underline{0}09 \text{ M}$$

Because the F^- hydrolyzes to OH^- and HF, use this to calculate the $[OH^-]$. Start by calculating the hydrolysis constant of F^- from the K_a of its conjugate acid, HF. Then assemble the usual table of concentrations, assume x is negligible, and calculate $[OH^-]$ and pH.

$$K_b = K_w \div K_a = (1.0 \times 10^{-14}) \div (6.8 \times 10^{-4}) = 1.\underline{4}7 \times 10^{-11}$$

(continued)

Conc. (M)	F^-	+	H_2O	$\rightleftharpoons$	HF	+	OH^-
Starting	0.06009				0		0
Change	-x				+x		+x
Equilibrium	0.06009 - x				x		x

$$\frac{[HF][OH^-]}{[F^-]} = K_b = \frac{(x)^2}{(0.06009 - x)} = 1.47 \times 10^{-11}$$

$$\frac{(x)^2}{(0.06009)} \cong 1.47 \times 10^{-11};\ x = [OH^-] \cong 9.\underline{3}9 \times 10^{-7}\ M$$

$pOH = -\log [OH^-] = -\log (9.39 \times 10^{-7}) = 6.0\underline{2}6$

$pH = 14.00 - 6.026 = 7.9\underline{7}4$

16.16 Rounded answer: $[H^+] = 6.5 \times 10^{-6}$ M; pH = 5.19. At the equivalence point, equal molar amounts of NH_3 and HCl react to form a solution of NH_4Cl. Start by calculating the moles of NH_3. Use this to calculate the volume of HCl needed to neutralize all of the NH_3 (and use the moles of NH_3 as the moles of NH_4^+ formed at the equivalence point). Add the volume of HCl to the original 0.035 L to find the total volume of solution.

$(0.20\ \text{mol}\ NH_3/L) \times 0.035\ L = 0.007\underline{0}0\ \text{mol}\ NH_3$

$\text{Volume HCl} = 0.00700\ \text{mol}\ NH_3 \div (0.12\ \text{mol HCl/L}) = 0.05\underline{8}3\ L$

$\text{Total volume} = 0.035\ L + 0.0583\ L\ NH_3\ \text{soln} = 0.0933\ L$

$[NH_4^+] = (0.00700\ \text{mol}\ NH_4^+\ \text{from}\ NH_3) \div 0.0933\ L = 0.0750\ M$

Because the NH_4^+ hydrolyzes to H_3O^+ and NH_3, use this to calculate the $[H^+]$. Start by calculating the hydrolysis constant of NH_4^+ from the K_b of its conjugate base, NH_3. Then assemble the usual table of concentrations, assume x is negligible, and calculate $[H^+]$ and pH.

$K_a = K_w \div K_b = (1.0 \times 10^{-14}) \div (1.8 \times 10^{-5}) = 5.\underline{5}6 \times 10^{-10}$

(continued)

Conc. (M)	NH_4^+	+	H_2O	$\rightleftharpoons$	NH_3	+	H_3O^+
Starting	0.0750				0		0
Change	-x				+x		+x
Equilibrium	0.0750 - x				x		x

$$\frac{[NH_3][H_3O^+]}{[NH_4^+]} = K_a = \frac{(x)^2}{(0.0750 - x)} = 5.56 \times 10^{-10}$$

$$\frac{(x)^2}{(0.0750)} \cong 5.56 \times 10^{-10};\ [H^+] \cong 6.\underline{4}6 \times 10^{-6}\ M$$

$$pH = -\log [H^+] = -\log (6.46 \times 10^{-6}) = 5.1\underline{8}9$$

■ Answers to Review Questions

16.1 The equation is

$$HCN(aq) \rightleftharpoons H^+(aq) + CN^-(aq)$$

The equilibrium-constant expression is

$$K_a = \frac{[H^+][CN^-]}{[HCN]}$$

16.2 HCN is the weakest acid. Its K_a of 4.9×10^{-10} is < K_a of 1.7×10^{-5} of $HC_2H_3O_2$; $HClO_4$ is a strong acid, of course.

16.3 Both methods involve direct measurement of the concentrations of the hydrogen ion and the anion of the weak acid and calculation of the concentration of the un-ionized acid. All concentrations are substituted into the K_a expression to obtain a value for K_a. In the first method, the electrical conductivity of a solution of the weak acid is measured. The conductivity is proportional to the concentration of the hydrogen ion and anion. In the second method, the pH of a known starting concentration of weak acid is measured. The pH is converted to $[H^+]$, which will be equal to the [anion].

16.4 The degree of ionization of a weak acid decreases as the concentration of the acid added to the solution increases. Compared to low concentrations, at high concentrations there is less water for each weak acid molecule to react with the weak acid as it ionizes:

$$HA(aq) + H_2O(l) \rightleftharpoons H_3O^+(aq) + A^-(aq)$$

16.5 You can neglect x if C_a/K is ≥ 100. In this case, C_a/K = [(0.0010 M ÷ 6.8 x 10^{-4}) = 1.47], which is significantly less than 100 and x cannot be neglected in the (0.010 - x) term. This says the degree of ionization is significant.

16.6 The ionization of the first H^+ is $H_2PHO_3(aq) \rightleftharpoons H^+(aq) + HPHO_3^-(aq)$. The ionization of the second H^+ is $HPHO_3^-(aq) \rightleftharpoons H^+(aq) + PHO_3^{2-}(aq)$.

$$K_{a1} = \frac{[H^+][HPHO_3^-]}{[H_2PHO_3]};\quad K_{a2} = \frac{[H^+][PHO_3^{2-}]}{[HPHO_3^-]}$$

16.7 As shown in Example 16.4, the concentration of a -2 anion of a polyprotic acid in a solution of the diprotic acid alone is approximately equal to the value of K_{a2}. For oxalic acid, begin by noting that the $[H^+] \cong [HC_2O_4^-]$. Then substitute $[H^+]$ for the $[HC_2O_4^-]$ term in the equilibrium-constant equation for K_{a2}:

$$K_{a2} = \frac{[H^+][C_2O_4^{2-}]}{[HC_2O_4^-]} = \frac{[H^+][C_2O_4^{2-}]}{[H^+]} \cong [C_2O_4^{2-}]$$

16.8 The balanced chemical equation for the ionization of aniline is

$$C_6H_5NH_2(aq) + H_2O(l) \rightleftharpoons C_6H_5NH_3^+(aq) + OH^-(aq)$$

The equilibrium-constant equation or expression for K_b is defined without an $[H_2O]$ term; this term is included in the value for K_b, as discussed in Section 16.3. The expression is

$$K_b = \frac{[C_6H_5NH_3^+][OH^-]}{[C_6H_5NH_2]}$$

16.9 First decide whether any of the three is a strong base or not. Because all of the molecules are among the nitrogen-containing weak bases listed in Table 16.2, none is a strong base. Next, recognize that the greater the $[OH^-]$, the stronger the weak base. Because $[OH^-]$ can be calculated from the square root of the product of K_b and concentration, the larger the K_b, the greater the $[OH^-]$ and the stronger the weak base. Thus, CH_3NH_2 is the strongest of these three weak bases because its K_b is the largest.

16.10 Anilinium chloride is not a weak base nor a weak acid but a salt that contains the anilinium ion, $C_6H_5NH_3^+$, and the Cl^- ion. The chloride ion does not hydrolyze because it could only form HCl, a strong acid. The anilinium ion does hydrolyze as follows:

$$C_6H_5NH_3^+ + H_2O(l) \rightleftharpoons C_6H_5NH_2 + H_3O^+$$

(continued)

The equilibrium-constant expression for this reaction is

$$K_a = \frac{[C_6H_5NH_2][H_3O^+]}{[C_6H_5NH_3^+]}$$

Obtain the value for K_a by calculating the value of K_w/K_b, where K_b is the ionization constant for aniline, $C_6H_5NH_2$.

16.11 The common-ion effect is the shift in an ionic equilibrium caused by the addition of a solute that furnishes an ion that is common to, or takes part in, the equilibrium. If the equilibrium involves the ionization of a weak acid, then the common ion is usually the anion formed by the ionization of the weak acid. If the equilibrium involves the ionization of a weak base, then the common ion is usually the cation formed by the ionization of the weak base. An example is the addition of F^- ion (as NaF) to a solution of the weak acid HF, which ionizes as shown below:

$$HF(aq) \rightleftharpoons H^+(aq) + F^-(aq)$$

The effect of adding F^- to this equilibrium is that it causes a shift in the equilibrium composition to the left. The additional F^- reacts with H^+ and lowers its concentration and raises the concentration of the HF.

16.12 The addition of CH_3NH_3Cl to 0.10 M CH_3NH_2 exerts a common-ion effect, causing the equilibrium below to exhibit a shift in composition to the left:

$$CH_3NH_2(aq) + H_2O(l) \rightleftharpoons CH_3NH_3^+(aq) + OH^-(aq)$$

This shift lowers the equilibrium concentration of the OH^- ion, which increases the $[H^+]$. An increase in $[H^+]$ lowers the pH below 11.8. The shift in composition to the left occurs according to Le Chatelier's principle, which states that a system shifts so as to counteract any change in composition.

16.13 A buffer is most often a solution of a mixture of two substances that is able to resist pH changes when limited amounts of acid or base are added to it. A buffer must contain a *weak acid* and its *conjugate (weak) base.* Strong acids and/or bases cannot form effective buffers because a buffer acts by converting H^+ (strong acid) to the un-ionized (weak) buffer acid and by converting OH^- (strong base) to the un-ionized (weak) buffer base. An example of a buffer pair is a mixture of H_2CO_3 and HCO_3^-, the principal buffer in the blood.

16.14 The capacity of a buffer is the amount of acid or base with which the buffer can react before exhibiting a significant pH change. (A significant change in blood pH might mean 0.01–0.02 pH units; for other systems a significant change might mean 0.5 pH units.) A high-capacity buffer might be of the type discussed for Figure 16.10: 1 mol of buffer acid and 1 mol of buffer base. A low-capacity buffer might involve quite a bit less than these amounts: 0.01–0.05 mol of buffer acid and buffer base.

16.15 The pH of a weak base before titration is relatively high—around pH 10 for a 0.1 M solution of a typical weak base. As a strong acid titrant is added, the $[OH^-]$ decreases and the pH decreases. At 50% neutralization, a buffer of equal amounts of buffer acid and base is formed. The $[OH^-]$ equals the K_b, or the pOH equals the pK_b. At the equivalence point, the pH is governed by the hydrolysis of the salt of the weak base formed and is usually in the pH 4–6 region. After the equivalence point, the pH decreases to a level just greater than the pH of the strong acid titrant.

16.16 If the pH is 8.0, an indicator that changes color in the basic region would be needed. Of the indicators mentioned in the text, phenolphthalein (pH 8.2–10.0) and thymol blue would work. In actual practice, cresol red (pH 7.2–8.8) would be the best choice because the pH should be closer to the middle of the range than to one end.

■ Solutions to Practice Problems

Note on answers to equilibrium calculations: The rounded answer (with correct number of significant figures) is always given first. After the stepwise solution, the numerical answer is given again, but with one nonsignificant figure. (The rightmost significant figure is underlined.)

16.17 a. $HOCN (aq) + H_2O (l) \rightleftharpoons H_3O^+ (aq) + OCN^- (aq)$

b. $HIO_3 (aq) + H_2O (l) \rightleftharpoons H_3O^+ (aq) + IO_3^- (aq)$

c. $HBrO (aq) + H_2O (l) \rightleftharpoons H_3O^+ (aq) + BrO^- (aq)$

d. $HCO_2H (aq) + H_2O (l) \rightleftharpoons H_3O^+ (aq) + CO_2H^- (aq)$

16.18 a. $HN_3 (aq) + H_2O (l) \rightleftharpoons H_3O^+ (aq) + N_3^- (aq)$

b. $HClO_2 (aq) + H_2O (l) \rightleftharpoons H_3O^+ (aq) + ClO_2^- (aq)$

c. $HNO_2 (aq) + H_2O (l) \rightleftharpoons H_3O^+ (aq) + NO_2^- (aq)$

d. $HCN (aq) + H_2O (l) \rightleftharpoons H_3O^+ (aq) + CN^- (aq)$

16.19 HAc will be used throughout as an abbreviation for acrylic acid and Ac^- for the acrylate ion. At the start, the H^+ from the self-ionization of water is so small that it is approximately zero. Once the acrylic acid solution is prepared, some of the 0.10 M HAc ionizes to H^+ and Ac^-. Then let x = the mol/ L of HAc that ionizes, forming x mol/ L of H^+ and x mol/ L of Ac^-, and leaving (0.10 - x) M HAc in solution. We can summarize the situation in tabular form:

Conc. (M)	HAc	$\rightleftharpoons$	H^+	+	Ac^-
Starting	0.10		0		0
Change	-x		+x		+x
Equilibrium	0.10 - x		x		x

The equilibrium-constant equation is:

$$K_a = \frac{[H^+][Ac^-]}{[HAc]} = \frac{x^2}{(0.10 - x)}$$

The value of x can be obtained from the pH of the solution:

$$x = [H^+] = \text{antilog}(-pH) = \text{antilog}(-2.63) = 2.\underline{3}4 \times 10^{-3} = 0.00\underline{2}34 \text{ M}$$

Note that (0.10 - x) = (0.10 - 0.00234) = 0.09766, which is significantly different from 0.10, so that x cannot be neglected in the calculation. Thus we substitute for x in both the numerator and denominator to obtain the value of K_a:

$$K_a = \frac{x^2}{(0.10 - x)} = \frac{(0.00234)^2}{(0.10 - 0.00234)} = 5.\underline{6}06 \times 10^{-5} = 5.6 \times 10^{-5}$$

16.20 HN_3 will be used throughout as an abbreviation for hydrazoic acid and N_3^- for the hydrazoate ion. At the start, the H^+ from the self-ionization of water is so small that it is approximately zero. Once the hydrazoic acid solution is prepared, some of the 0.20 M HN_3 ionizes to H^+ and N_3^-. Then let x = the mol/ L of HN_3 that ionizes, forming x mol/ L of H^+ and x mol/ L of N_3^-, and leaving (0.20 - x) M HN_3 in solution. We can summarize the situation in tabular form:

(continued)

Conc. (M)	HN_3	$\rightleftharpoons$	H^+	+	N_3^-
Starting	0.20		0		0
Change	-x		+x		+x
Equilibrium	0.20 - x		x		x

The equilibrium-constant equation is:

$$K_a = \frac{[H^+][N_3^-]}{[HN_3]} = \frac{x^2}{(0.20 - x)}$$

The value of x can be obtained from the pH of the solution:

$$x = [H^+] = \text{antilog}(\text{-pH}) = \text{antilog}(-3.21) = 6.\underline{1}6 \times 10^{-4} = 0.0006\underline{1}6 \text{ M}$$

Note that (0.20 - x) = (0.20 - 0.000616) = 0.1993, which is not significantly different from 0.20, so that x can be neglected in the calculation. Thus we substitute for x in only the numerator, not the denominator, to obtain the value of K_a:

$$K_a = \frac{x^2}{(0.20 - x)} \cong \frac{(0.000616)^2}{(0.20)} \cong 1.\underline{8}9 \times 10^{-6} = 1.9 \times 10^{-6}$$

16.21 Rounded answer: degree of ionization = 0.00015; pH = 5.42. To solve, assemble a table of starting, change, and equilibrium concentrations. Use HBo as the symbol for boric acid and Bo- as the symbol for $B(OH)_4^-$.

Conc. (M)	HBo	$\rightleftharpoons$	H^+	+	Bo^-
Starting	0.025		0		0
Change	-x		+x		+x
Equilibrium	0.025 - x		x		x

The value of x equals the value of the molarity of the H^+ ion, which can be obtained from the equilibrium-constant expression. Substitute into the equilibrium-constant expression and solve for x.

$$\frac{[H^+][Bo^-]}{[HBo]} = K_a = \frac{[x]^2}{(0.025 - x)} = 5.9 \times 10^{-10}$$

Solve the equation for x, assuming that x is much smaller than 0.025, so that (0.025 - x) $\cong$ 0.025. Solve by taking the square root of $(5.9 \times 10^{-10}) \times (0.025)$.

(continued)

$$\frac{[x]^2}{(0.025)} \cong 5.9 \times 10^{-10};\ x^2 \cong 5.9 \times 10^{-10}\,(0.025) = 1.\underline{4}75 \times 10^{-11}$$

$x = 3.\underline{8}4 \times 10^{-6}$ M = [H^+]; degree of ionization = $(3.84 \times 10^{-6})/0.025 = 0.0001\underline{5}3$

Check to make sure that the assumption that $(0.025 - x) \cong 0.025$ is valid:

$0.025 - (3.84 \times 10^{-6}) = 0.02499$, or $\cong 0.025$ to two sig. figs.

pH = - log [H^+] = - log (3.84×10^{-6}) = $5.4\underline{1}56$ (rounded answer above)

16.22 Rounded answer: [H^+] = 4.5×10^{-3}; pH = 2.35; degree of ionization = 0.038. To solve, assemble a table of starting, change, and equilibrium concentrations. Use HFo as the symbol for formic acid and Fo- as the symbol for CHO_2^-.

Conc. (M)	HFo	⇌	H^+	+	Fo^-
Starting	0.12		0		0
Change	-x		+x		+x
Equilibrium	0.12 - x		x		x

The value of x equals the value of the molarity of the H^+ ion, which can be obtained from the equilibrium-constant expression. Substitute into the equilibrium-constant expression and solve for x.

$$\frac{[H^+][Fo^-]}{[HFo]} = K_a = \frac{[x]^2}{(0.12 - x)} = 1.7 \times 10^{-4}$$

Solve the equation for x, assuming that x is much smaller than 0.12, so that $(0.12 - x) \cong 0.12$. Solve by taking the square root of $(1.7 \times 10^{-4}) \times (0.12)$.

$$\frac{(x)^2}{(0.12)} \cong 1.7 \times 10^{-4};\ x^2 \cong 1.7 \times 10^{-4}\,(0.12) = 2.\underline{0}4 \times 10^{-5}$$

$x \cong 4.\underline{5}16 \times 10^{-3}$ M = [H^+] (rounded answer above)

Check to make sure that the assumption that $(0.12 - x) \cong 0.12$ is valid:

$0.12 - (4.516 \times 10^{-3}) = 0.11548$, or $\cong 0.12$ to two sig. figs. (This is a borderline case; the quadratic gives [H^+] = 4.43×10^{-3} M, which is not much different.)

pH = - log [H^+] = - log (4.516×10^{-3}) = $2.3\underline{4}52$ (rounded answer above)

Degree of ionization = $(4.5 \times 10^{-3}) \div 0.12 = 0.03\underline{7}58$ (3.8%)

16.23 Rounded answer: $[H^+] = [C_6H_4NH_2COO^-] = 1.0 \times 10^{-3}$ M. To solve, assemble a table of starting, change, and equilibrium concentrations. Use HPaba as a symbol for p-aminobenzoic acid (PABA), and use $Paba^-$ as the symbol for the -1 anion.

Conc. (M)	HPaba	⇌	H^+	+	$Paba^-$
Starting	0.050		0		0
Change	-x		+x		+x
Equilibrium	0.050 - x		x		x

Write the equilibrium-constant expression in terms of chemical symbols and then substitute the terms x and (0.050 - x):

$$\frac{[H^+][Paba^-]}{[HPaba]} = K_a = \frac{[x]^2}{(0.050 - x)} = 2.2 \times 10^{-5}$$

Technically, x should not be neglected compared to 0.050 M. If it is neglected, the calculated $[H^+]$ is 1.04×10^{-3}, which when subtracted from 0.050 yields 0.0489, a significant change. However, the *quadratic formula* gives the same value of x within three significant figures, 1.04×10^{-3} M. Therefore, neglect x and avoid the quadratic:

$$x^2 = (2.2 \times 10^{-5}) \times 0.050 = 1.1 \times 10^{-6}$$

$$x = [H^+] = [Paba^-] \cong 1.\underline{0}4 \times 10^{-3} \text{ M}$$ (rounded answer above)

16.24 Rounded answer: $[H^+] = [C_4H_3N_2O_3^-] = 4.4 \times 10^{-3}$ M. To solve, assemble a table of starting, change, and equilibrium concentrations. Use HBar as a symbol for barbituric acid and Bar^- as the symbol for the -1 anion.

Conc. (M)	HBar	⇌	H^+	+	Bar^-
Starting	0.20		0		0
Change	-x		+x		+x
Equilibrium	0.20 - x		x		x

Write the equilibrium-constant expression in terms of chemical symbols and then substitute the terms x and (0.20 - x):

$$\frac{[H^+][Bar^-]}{[HBar]} = K_a = \frac{[x]^2}{(0.20 - x)} = 9.8 \times 10^{-5}$$

(continued)

Assuming x is negligible compared to 0.20, so that (0.20 - x) ≅ (0.20), solve for x:

$$x^2 \cong (9.8 \times 10^{-5}) \times 0.20 = 1.96 \times 10^{-5}$$

$$x = [H^+] = [Bar^-] \cong 4.\underline{4}2 \times 10^{-3} \text{ M}$$

16.25 Rounded answer: $[HC_2H_3O_2]$ = 0.26 M. To solve, first convert the pH to $[H^+]$, which also equals the $[C_2H_3O_2^-]$, here symbolized as $[Ac^-]$. Then assemble the usual table and substitute into the equilibrium-constant expression to solve for the $[HC_2H_3O_2]$, here symbolized as [HAc].

$$[H^+] = \text{antilog}(-2.68) = 2.\underline{0}89 \times 10^{-3} \text{ M}$$

Conc. (M)	HAc	⇌	H^+	+	Ac^-
Starting	x		0		0
Change	-2.089×10^{-3}		$+2.089 \times 10^{-3}$		$+2.089 \times 10^{-3}$
Equilibrium	$x - (2.089 \times 10^{-3})$		2.089×10^{-3}		2.089×10^{-3}

Write the equilibrium-constant expression in terms of chemical symbols, and then substitute the x and the $x - (2.089 \times 10^{-3})$ terms into the expression:

$$\frac{[H^+][Ac^-]}{[HAc]} = K_a = \frac{(2.089 \times 10^{-3})^2}{(x - 0.002089)} = 1.7 \times 10^{-5}$$

Solve the equation for x, assuming that 0.002089 is much smaller than x, so that you can say that (x - 0.002089) ≅ (x). Solve by rearranging to

$$(x) = [HAc] \cong \frac{(2.089 \times 10^{-3})^2}{1.7 \times 10^{-5}} \cong 0.2\underline{5}6 \text{ M}$$

Subtracting 0.00289 from 0.26 indeed does not change the two significant figures (0.26).

16.26 Rounded answer: $[HC_3H_5O_3]$ = 0.071 M. To solve, first convert the pH to $[H^+]$, which also equals the $[C_3H_5O_3^-]$, here symbolized as $[Lac^-]$. Then assemble the usual table and substitute into the equilibrium-constant expression to solve for the $[HC_3H_5O_3]$, here symbolized as [HLac].

$$[H^+] = \text{antilog}(-2.51) = 3.09 \times 10^{-3} \text{ M}$$

(continued)

Conc. (M)	HLac	$\rightleftharpoons$	H^+	+	Lac^-
Starting	x		0		0
Change	-3.09×10^{-3}		$+3.09 \times 10^{-3}$		$+3.09 \times 10^{-3}$
Equilibrium	$x - (3.09 \times 10^{-3})$		3.09×10^{-3}		3.09×10^{-3}

Write the equilibrium-constant expression in terms of chemical symbols, and then substitute the x and the x - (3.09×10^{-3}) terms into the expression:

$$\frac{[H^+][Lac^-]}{[HLac]} = K_a = \frac{(3.09 \times 10^{-3})^2}{(x - 0.00309)} = 1.4 \times 10^{-4}$$

Solve the equation for x. It appears that 0.00309 is not much smaller than x, so do not neglect x in the denominator.

$$\frac{(3.09 \times 10^{-3})^2}{1.4 \times 10^{-4}} = 6.82 \times 10^{-2} = (x - 0.00309)$$

$$x = (6.82 \times 10^{-2}) + 0.00309 = 7.\underline{1}2 \times 10^{-2} = [HLac]$$

16.27 Rounded answer: $[H^+] = 4.9 \times 10^{-3}$ M; pH = 2.31. To solve, assemble the usual table of starting, change, and equilibrium concentrations of HF and F^- ions.

Conc. (M)	HF	$\rightleftharpoons$	H^+	+	F^-
Starting	0.040		0		0
Change	-x		+x		+x
Equilibrium	0.040 - x		x		x

Write the equilibrium-constant expression in terms of chemical symbols and then substitute the terms x and (0.040 - x):

$$\frac{[H^+][F^-]}{[HF]} = K_a = \frac{[x]^2}{(0.040 - x)} = 6.8 \times 10^{-4}$$

In this case, x cannot be neglected compared to 0.040 M. (If it is neglected, subtracting the calculated $[H^+]$ from 0.040 yields a significant change.) The *quadratic formula* must be used. Reorganize the equilibrium-constant expression into the form $ax^2 + bx + c = 0$, and substitute for a, b, and c in the quadratic formula.

$$x^2 + (6.8 \times 10^{-4}) - (2.72 \times 10^{-5}) = 0$$

(continued)

$$x = \frac{-6.8 \times 10^{-4} \pm \sqrt{(6.8 \times 10^{-4})^2 + (4)(2.72 \times 10^{-5})}}{2}$$

$$x = \frac{-6.8 \times 10^{-4} \pm \sqrt{1.092 \times 10^{-4}}}{2}$$

$$x = 9.768 \times 10^{-3} \div 2 = 4.\underline{8}8 \times 10^{-3}$$

$$pH = -\log(4.88 \times 10^{-3}) = 2.3\underline{1}1 \text{ (rounded answer above)}$$

16.28 Rounded answer: $[H^+] = 3.8 \times 10^{-3}$ M; pH = 2.42. To solve, assemble the usual table of starting, change, and equilibrium concentrations of $HC_2H_2ClO_2$, symbolized as HChl, and the $C_2H_2ClO_2^-$ ion, symbolized as Chl^-.

Conc. (M)	HChl	⇌	H^+	+	Chl^-
Starting	0.015		0		0
Change	-x		+x		+x
Equilibrium	0.015 - x		x		x

Write the equilibrium-constant expression in terms of chemical symbols and then substitute the terms x and (0.015 - x):

$$\frac{[H^+][Chl^-]}{[HChl]} = K_a = \frac{[x]^2}{(0.015 - x)} = 1.3 \times 10^{-3}$$

In this case, x cannot be neglected compared to 0.015 M. (If it is neglected, subtracting the calculated $[H^+]$ from 0.015 yields a significant change.) The *quadratic formula* must be used. Reorganize the equilibrium-constant expression into the form $ax^2 + bx + c = 0$, and substitute for a, b, and c in the quadratic formula.

$$x^2 + (1.3 \times 10^{-3}) - 1.95 \times 10^{-5} = 0$$

$$x = \frac{-1.3 \times 10^{-3} \pm \sqrt{(1.3 \times 10^{-3})^2 + (4)(1.95 \times 10^{-5})}}{2}$$

$$x = \frac{-1.3 \times 10^{-3} \pm \sqrt{7.969 \times 10^{-5}}}{2}$$

$$x = (7.626 \times 10^{-3}) \div 2 = 3.\underline{8}13 \times 10^{-3}$$

$$pH = -\log(3.813 \times 10^{-3}) = 2.4\underline{1}87 \text{ (rounded answer above)}$$

16.29 Rounded answer: $[H^+] = 3.61 \times 10^{-1}$ M. To solve, assemble the usual table of starting, change, and equilibrium concentrations of $(NO_2)_2C_6H_3CO_2H$, symbolized as HDin, and the $(NO_2)_2C_6H_3CO_2^-$ ion, symbolized as Din^-.

Conc. (M)	HDin	⇌	H^+	+	Din^-
Starting	2.00		0		0
Change	-x		+x		+x
Equilibrium	2.00 - x		x		x

Write the equilibrium-constant expression in terms of chemical symbols and then substitute the terms x and (2.00 - x):

$$\frac{[H^+][Din^-]}{[HDin]} = K_a = \frac{[x]^2}{(2.00 - x)} = 7.94 \times 10^{-2}$$

In this case, x cannot be neglected compared to 2.00 M. (If it is neglected, subtracting the calculated $[H^+]$ from 2.00 yields a significant change.) The *quadratic formula* must be used. Reorganize the equilibrium-constant expression into the form $ax^2 + bx + c = 0$, and substitute for a, b, and c in the quadratic formula.

$$x^2 + (7.94 \times 10^{-2}) - 1.588 \times 10^{-1} = 0$$

$$x = \frac{-7.94 \times 10^{-2} \pm \sqrt{(7.94 \times 10^{-2})^2 + (4)(1.588 \times 10^{-1})}}{2}$$

$$x = \frac{-7.94 \times 10^{-2} \pm \sqrt{6.41504 \times 10^{-1}}}{2}$$

$$x = (7.215 \times 10^{-1}) \div 2 = 3.6\underline{0}76 \times 10^{-1}$$

(The pH is not required, but the value of pH = $0.44\underline{2}7$.)

16.30 Rounded answer: $[H^+] = 1.00 \times 10^{-4}$ M; pH = 4.000. To solve, assemble the usual table of starting, change, and equilibrium concentrations of $BrC_6H_4CO_2H$, symbolized as HpBr, and the $BrC_6H_4CO_2^-$ ion, symbolized as pBr^-.

(continued)

Conc. (M)	HpBr	$\rightleftharpoons$	H^+	+	pBr^-
Starting	2.00×10^{-4}		0		0
Change	-x		+x		+x
Equilibrium	(2.00×10^{-4}) - x		x		x

Write the equilibrium-constant expression in terms of chemical symbols and then substitute the terms x and (2.00×10^{-4} - x):

$$\frac{[H^+][pBr^-]}{[HpBr]} = K_a = \frac{[x]^2}{(2.00 \times 10^{-4}) - x} = 1.00 \times 10^{-4}$$

In this case, x cannot be neglected compared to 2.00×10^{-4} M. (If it is neglected, subtracting the calculated [H^+] from 2.00 yields a significant change.) The *quadratic formula* must be used. Reorganize the equilibrium-constant expression into the form $ax^2 + bx + c = 0$, and substitute for a, b, and c in the quadratic formula.

$$x^2 + (1.00 \times 10^{-4}) - 2.00 \times 10^{-8} = 0$$

$$x = \frac{-1.00 \times 10^{-4} \pm \sqrt{(1.00 \times 10^{-4})^2 + 4(2.00 \times 10^{-8})}}{2}$$

$$x = \frac{-1.00 \times 10^{-4} \pm \sqrt{9.00 \times 10^{-8}}}{2}$$

$$x = (2.000 \times 10^{-4}) \div 2 = 1.0\underline{0}0 \times 10^{-4}$$

(The pH is not required, but the value of pH = 4.000.)

16.31 Rounded answers: (a) [H^+] = 3.7×10^{-3} M; (b) [$C_8H_4O_4^{2-}$] = K_{a2} = 3.9×10^{-6} M. The $C_8H_4O_4^{2-}$ ion will be symbolized as the [Ph^{2-}] ion. To solve, note that $K_{a1} = 1.2 \times 10^{-3} > K_{a2} = 3.9 \times 10^{-6}$, and hence the second ionization and K_{a2} can be neglected. Assemble a table of starting, change, and equilibrium concentrations. Let $H_2Ph = H_2C_8H_4O_4$ and $HPh^- = HC_8H_4O_4^-$.

Conc. (M)	H_2Ph	$\rightleftharpoons$	H^+	+	HPh^-
Starting	0.015		0		0
Change	-x		+x		+x
Equilibrium	0.015 - x		x		x

(continued)

Write the equilibrium-constant expression in terms of chemical symbols and then substitute x and (0.015 - x):

$$\frac{[H^+][HPh^-]}{[H_2Ph]} = K_{a1} = \frac{[x]^2}{(0.015 - x)} = 1.2 \times 10^{-3}$$

In this case, x cannot be neglected in the (0.015 M - x) term. Solving the quadratic gives

$$x = (7.3697 \times 10^{-3}) \div 2$$

$$x = [H^+] \cong 3.\underline{6}84 \times 10^{-3}\ M \quad \text{(rounded answer above)}$$

Because $[HPh^-]$ is $\cong [H^+]$, these terms cancel in the K_{a2} expression. This reduces to the equality that the concentration of the -2 phthalate ion, $[Ph^{2-}]$, = K_{a2} = 3.9×10^{-6} M.

16.32 Rounded answer: $[H^+] = 1.5 \times 10^{-5}$ M; $[CO_3^{2-}] = 4.8 \times 10^{-11}$ M. To solve, note that $K_{a1} = 4.3 \times 10^{-7} > K_{a2} = 4.8 \times 10^{-11}$, and hence the second ionization and K_{a2} can be neglected. Assemble a table of starting, change, and equilibrium concentrations.

Conc. (M)	H_2CO_3	$\rightleftharpoons$	H^+	+	HCO_3^-
Starting	5.45×10^{-4}		0		0
Change	-x		+x		+x
Equilibrium	(5.45×10^{-4}) - x		x		x

Write the equilibrium-constant expression in terms of chemical symbols and then substitute x and $(5.45 \times 10^{-4} - x)$:

$$\frac{[H^+][HCO_3^-]}{[H_2CO_3]} = K_{a1} = \frac{[x]^2}{(0.000545 - x)} = 4.3 \times 10^{-7}$$

In this case, x cannot be neglected compared to 5.45×10^{-4} M. (If it is neglected, the calculated $[H^+]$ when subtracted from 5.45×10^{-4} M yields a significant change.) The *quadratic formula* must be used. Reorganize the equilibrium-constant expression into the form $ax^2 + bx + c = 0$, and substitute for a, b, and c in the quadratic formula.

$$x = \frac{-4.3 \times 10^{-7} \pm \sqrt{(4.3 \times 10^{-7})^2 + 4(2.3435 \times 10^{-10})}}{2}$$

$$x = \frac{-4.3 \times 10^{-7} \pm \sqrt{(4.3 \times 10^{-7})^2 + 9.375 \times 10^{-10}}}{2}$$

(continued)

$x = [H^+] = (3.0192 \times 10^{-5}) \div 2 = 1.\underline{5}10 \times 10^{-5}$ M (rounded answer above)

To calculate $[CO_3^{2-}]$, note from the equilibrium concentrations in the table above that $[HCO_3^-] \cong [H^+]$ (= 1.5×10^{-5} M). In exact terms, $[H^+] = (1.5 \times 10^{-5} + y)$ from the ionization of HCO_3, and $[HCO_3^-] = (1.5 \times 10^{-5} - y)$ from ionization. Substituting into the K_{a2} expression gives

$$\frac{[H^+][CO_3^{2-}]}{[HCO_3^-]} = K_{a2} = \frac{[(1.5 \times 10^{-5} + y)(CO_3^{2-})]}{[1.5 \times 10^{-5} - y]} = 4.8 \times 10^{-11}$$

Assuming that y is much smaller than 1.5×10^{-5}, note that the $(1.5 \times 10^{-5}) + y$ term cancels the $(1.5 \times 10^{-5}) - y$ term, leaving

$$[CO_3^{2-}] \cong 4.8 \times 10^{-11} \text{ M}$$

16.33 The equation is

$$C_2H_5NH_2(aq) + H_2O(l) \rightleftharpoons C_2H_5NH_3^+(aq) + OH^-(aq)$$

The K_b expression is

$$K_b = \frac{[C_2H_5NH_3^+][OH^-]}{[C_2H_5NH_2]}$$

16.34 The equation is

$$C_5H_5N(aq) + H_2O(l) \rightleftharpoons C_5H_5NH^+(aq) + OH^-(aq)$$

The K_b expression is

$$K_b = \frac{[C_5H_5NH^+][OH^-]}{[C_5H_5N]}$$

16.35 Rounded answer: $K_b = 3.2 \times 10^{-5}$ (correct no. of sig. figs.). To solve, convert the pH to $[OH^-]$:

$$pOH = 14.00 - pH = 14.00 - 11.34 = 2.66$$

$$[OH^-] = \text{antilog}(-2.66) = 2.\underline{1}88 \times 10^{-3} \text{ M}$$

(continued)

Using the symbol EtN for ethanolamine, assemble a table of starting, change, and equilibrium concentrations.

Conc. (M)	EtN	+	H_2O	⇌	$HEtN^+$	+	OH^-
Starting	0.15				0		0
Change	-x				+x		+x
Equilibrium	0.15 - (2.188 x 10^{-3})				2.188 x 10^{-3}		2.188 x 10^{-3}

Write the equilibrium-constant expression in terms of chemical symbols and then substitute the terms and solve for K_b:

$$K_b = \frac{[HEtN^+][OH^-]}{[EtN]} = \frac{(2.188 \times 10^{-3})^2}{(0.15 - 0.002188)} = 3.\underline{2}3 \times 10^{-5}$$

16.36 Rounded answer: $K_b = 7.4 \times 10^{-5}$ (correct no. of sig. figs.). To solve, convert the pH to $[OH^-]$:

$$pOH = 14.00 - pH = 14.00 - 11.63 = 2.37$$

$$[OH^-] = \text{antilog}(-2.37) = 4.\underline{2}66 \times 10^{-3}\ M$$

Using the symbol TMeN for trimethylamine, assemble a table of starting, change, and equilibrium concentrations.

Conc. (M)	TMeN	+	H_2O	⇌	$HTMeN^+$	+	OH^-
Starting	0.25				0		0
Change	-x				+x		+x
Equilibrium	0.25 - (4.266 x 10^{-3})				4.266 x 10^{-3}		4.266 x 10^{-3}

Write the equilibrium-constant expression in terms of chemical symbols and then substitute the terms and solve for K_b:

$$K_b = \frac{[HTMeN^+][OH^-]}{[TMeN]} = \frac{(4.266 \times 10^{-3})^2}{(0.25 - 0.004266)} = 7.\underline{4}058 \times 10^{-5}$$

16.37 Rounded answer: $[OH^-] = 5.7 \times 10^{-3}$ M; pH = 11.76. To solve, assemble a table of starting, change, and equilibrium concentrations:

Conc. (M)	CH_3NH_2	+	H_2O	$\rightleftharpoons$	$CH_3NH_3^+$	+	OH^-
Starting	0.080				0		0
Change	-x				+x		+x
Equilibrium	0.080 - x				x		x

Write the equilibrium-constant expression in terms of chemical symbols and then substitute the terms and the value of K_b:

$$\frac{[CH_3NH_3^+][OH^-]}{[CH_3NH_2]} = K_b = \frac{(x)^2}{(0.080 - x)} = 4.4 \times 10^{-4}$$

In this case, x cannot be neglected compared to 0.080 M. (If it is neglected, subtracting the calculated $[H^+]$ from 0.080 yields a significant change.) The *quadratic formula* must be used. Reorganize the equilibrium-constant expression into the form $ax^2 + bx + c = 0$, and substitute for a, b, and c in the quadratic formula.

$$x^2 + (4.4 \times 10^{-4})x - 3.52 \times 10^{-5} = 0$$

$$x = \frac{-4.4 \times 10^{-4} \pm \sqrt{(4.4 \times 10^{-4})^2 + (4)(3.52 \times 10^{-5})}}{2}$$

$$x = \frac{-4.4 \times 10^{-4} \pm \sqrt{1.4099 \times 10^{-4}}}{2}$$

$x = (1.143 \times 10^{-2}) \div 2 = 5.\underline{7}17 \times 10^{-3} = [OH^-]$ (rounded answer above)

$pOH = -\log(5.717 \times 10^{-3}) = 2.2\underline{4}28$

$pH = 14.00 - 2.2428 = 11.7\underline{5}72$ (rounded answer above)

16.38 Rounded answer: $[OH^-]$ = 4.1×10^{-5} M; pH = 9.61. To solve, assemble a table of starting, change, and equilibrium concentrations:

Conc. (M)	$HONH_2$ + H_2O ⇌	$HONH_3^+$	+ OH^-
Starting	0.15	0	0
Change	-x	+x	+x
Equilibrium	0.15 - x	x	x

Write the equilibrium-constant expression in terms of chemical symbols and then substitute the terms and the value of K_b:

$$\frac{[HONH_3^+][OH^-]}{[HONH_2]} = K_b = \frac{(x)^2}{(0.15\ -\ x)} = 1.1 \times 10^{-8}$$

Solve the equation for x, assuming that x can be neglected compared to 0.15 M, so that you can say that (0.15 - x) ≅ (0.15). Solve by taking the square root of $(1.1 \times 10^{-8})(0.15)$.

$$\frac{(x)^2}{(0.15)} \cong 1.1 \times 10^{-8};\ x^2 \cong 1.1 \times 10^{-8}\,(0.15) = 1.\underline{6}5 \times 10^{-9}$$

$x = [OH^-] \cong 4.\underline{0}6 \times 10^{-5}$ M (rounded answer above)

$pOH = -\log(4.06 \times 10^{-5}) = 4.3\underline{9}1$

$pH = 14.00 - 4.391 = 9.6\underline{0}9$ (rounded answer above)

16.39 a. No hydrolysis occurs, because the iodide ion is the anion of a strong acid.

b. Hydrolysis occurs. Equation:

$$CHO_2^- + H_2O \rightleftharpoons HCHO_2 + OH^-$$

Equilibrium-constant expression:

$$K_b = \frac{K_w}{K_a} = \frac{[HCHO_2][OH^-]}{[CHO_2^-]}$$

c. Hydrolysis occurs. Equation:

$$CH_3NH_3^+ + H_2O \rightleftharpoons H_3O^+ + CH_3NH_2$$

Equilibrium-constant expression:

$$K_a = \frac{K_w}{K_b} = \frac{[H_3O^+][CH_3NH_2]}{[CH_3NH_3^+]}$$

d. Hydrolysis occurs. Equation:

$$IO^- + H_2O \rightleftharpoons HIO + OH^-$$

Equilibrium-constant expression:

$$K_b = \frac{K_w}{K_a} = \frac{[HIO][OH^-]}{[IO^-]}$$

16.40 a. Hydrolysis occurs. Equation:

$$NO_2^- + H_2O \rightleftharpoons HNO_2 + OH^-$$

Equilibrium-constant expression:

$$K_b = \frac{K_w}{K_a} = \frac{[HNO_2][OH^-]}{[NO_2^-]}$$

b. No hydrolysis occurs because the bromide ion is the anion of a strong acid.

c. No hydrolysis occurs. The only reaction that can tentatively be written is

$NO_3^- + H_2O \rightleftharpoons HNO_3 + OH^-$ (Impossible because HNO_3 is a strong acid.)

d. Hydrolysis occurs. Equation:

$$NH_2NH_3^+ + H_2O \rightleftharpoons H_3O^+ + NH_2NH_2$$

Equilibrium-constant expression:

$$K_a = \frac{K_w}{K_b} = \frac{[NH_2NH_2][H_3O^+]}{[NH_2NH_3^+]}$$

16.41 Acid ionization is $Zn(H_2O)_6^{2+}(aq) + H_2O(l) \rightleftharpoons Zn(H_2O)_5(OH)^+(aq) + H_3O^+(aq)$.

16.42 Acid ionization is $Cu(H_2O)_6^{2+}(aq) + H_2O(l) \rightleftharpoons Cu(H_2O)_5(OH)^+(aq) + H_3O^+(aq)$.

16.43 a. $Fe(NO_3)_3$ is a salt of a weak base, $Fe(OH)_3$, and a strong acid, HNO_3, so it would be expected to be acidic. Fe^{3+} is not in Group IA or IIA, so it would be expected to form a metal hydrate ion that hydrolyzes to form an acidic solution.

b. Na_2CO_3 is a salt of a strong base, NaOH, and the anion of a weak acid, HCO_3^-, so it would be expected to be basic.

c. $Ca(CN)_2$ is a salt of a strong base, $Ca(OH)_2$, and a weak acid, HCN, so it would be expected to be basic.

d. NH_4ClO_4 is a salt of a weak base, NH_3, and a strong acid, $HClO_4$, so it would be expected to be acidic.

16.44 a. Na_2S is a salt of a strong base, NaOH, and the anion of a weak acid, HS^-, so it would be expected to be basic.

b. $Cu(NO_3)_2$ is a salt of a weak base, $Cu(OH)_2$, and a strong acid, HNO_3, so it would be expected to be acidic. Cu^{2+} is not in Group IA or IIA, so it would be expected to form a metal hydrate ion that hydrolyzes to form an acidic solution.

c. $KClO_4$ is a salt of a strong base, KOH, and a strong acid, $HClO_4$, so no reaction would occur to change the neutral pH (7.00) of water.

d. CH_3NH_3Cl is a salt of a weak base, CH_3NH_2, and a strong acid, HCl, so it would be expected to be acidic.

16.45 a. Both ions hydrolyze:

$$NH_4^+ + H_2O \rightleftharpoons NH_3 + H_3O^+;$$

$$C_2H_3O_2^- + H_2O \rightleftharpoons HC_2H_3O_2 + OH^-$$

Calculate the K_a and K_b constants of each, to compare them:

$$NH_4^+ \text{ as an acid: } K_a = \frac{K_w}{K_b} = \frac{1.0 \times 10^{-14}}{1.8 \times 10^{-5}} = 5.\underline{5}5 \times 10^{-10}$$

$$C_2H_3O_2^- \text{ as a base: } K_b = \frac{K_w}{K_a} = \frac{1.0 \times 10^{-14}}{1.7 \times 10^{-5}} = 5.\underline{8}8 \times 10^{-10}$$

Because the K_b for the hydrolysis of $C_2H_3O_2^-$ is slightly larger than the constant K_a for the hydrolysis of NH_4^+, the solution will be slightly basic but close to pH 7.0.

b. Both ions hydrolyze:

$$C_6H_5NH_3^+ + H_2O \rightleftharpoons C_6H_5NH_2 + H_3O^+;$$

$$C_2H_3O_2^- + H_2O \rightleftharpoons HC_2H_3O_2 + OH^-$$

Calculate the K_a and K_b constants of each, to compare them:

$$C_6H_5NH_3^+ \text{ as an acid: } K_a = \frac{K_w}{K_b} = \frac{1.0 \times 10^{-14}}{4.2 \times 10^{-10}} = 2.\underline{3}8 \times 10^{-5}$$

$$C_2H_3O_2^- \text{ as a base: } K_b = \frac{K_w}{K_a} = \frac{1.0 \times 10^{-14}}{1.7 \times 10^{-5}} = 5.\underline{8}8 \times 10^{-10}$$

Because the constant K_a for the hydrolysis of $C_6H_5NH_3^+$ is larger than the hydrolysis constant (K_b) for the hydrolysis of $C_2H_3O_2^-$, the solution will be acidic and significantly less than pH 7.0.

16.46 a. Both ions hydrolyze:

$$NH_4^+ + H_2O \rightleftharpoons NH_3 + H_3O^+; CNO^- + H_2O \rightleftharpoons HCNO + OH^-$$

Calculate the K_a and K_b constants of each, to compare them:

$$NH_4^+ \text{ as an acid: } K_a = \frac{K_w}{K_b} = \frac{1.0 \times 10^{-14}}{1.8 \times 10^{-5}} = 5.\underline{5}5 \times 10^{-10}$$

$$CNO^- \text{ as a base: } K_b = \frac{K_w}{K_a} = \frac{1.0 \times 10^{-14}}{3.5 \times 10^{-4}} = 2.\underline{8}6 \times 10^{-11}$$

Because the constant K_a for the hydrolysis of NH_4^+ is larger than the hydrolysis constant (K_b) for the hydrolysis of CNO^-, the solution will be acidic and significantly less than pH 7.0.

b. Both ions hydrolyze:

$$C_5H_5NH^+ + H_2O \rightleftharpoons C_5H_5N + H_3O^+; \quad CNO^- + H_2O \rightleftharpoons HCNO + OH^-$$

Calculate the K_a and K_b constants of each, to compare them:

$$C_5H_5NH^+ \text{ as an acid: } K_a = \frac{K_w}{K_b} = \frac{1.0 \times 10^{-14}}{1.4 \times 10^{-9}} = 7.\underline{1}4 \times 10^{-6}$$

(continued)

$$CNO^- \text{ as a base: } K_b = \frac{K_w}{K_a} = \frac{1.0 \times 10^{-14}}{3.5 \times 10^{-4}} = 2.\underline{86} \times 10^{-11}$$

Because the constant K_a for the hydrolysis of $C_5H_5NH^+$ is larger than the hydrolysis constant (K_b) for the hydrolysis of CNO^-, the solution will be acidic and significantly less than pH 7.0.

16.47 a. The reaction is $NO_2^- + H_2O \rightleftharpoons HNO_2 + OH^-$, and the constant K_b is obtained by dividing K_w by the K_a of the conjugate acid, HNO_2:

$$K_b = K_w \div K_a = (1.0 \times 10^{-14}) \div (4.5 \times 10^{-4}) = 2.\underline{2}2 \times 10^{-11} = 2.2 \times 10^{-11}$$

b. The reaction is $C_5H_5NH^+ + H_2O \rightleftharpoons H_3O^+ + C_5H_5N$, and the constant K_a is obtained by dividing K_w by the K_b of the conjugate base, C_5H_5N:

$$K_a = K_w \div K_b = (1.0 \times 10^{-14}) \div (1.4 \times 10^{-9}) = 7.\underline{1}4 \times 10^{-6} = 7.1 \times 10^{-6}$$

16.48 a. The reaction is $ClO^- + H_2O \rightleftharpoons HClO + OH^-$, and the constant K_b is obtained by dividing K_w by the K_a of the conjugate acid, HClO:

$$K_b = K_w \div K_a = (1.0 \times 10^{-14}) \div (3.5 \times 10^{-8}) = 2.\underline{8}57 \times 10^{-7} = 2.9 \times 10^{-7}$$

b. The reaction is $NH_3OH^+ + H_2O \rightleftharpoons H_3O^+ + NH_2OH$, and the constant K_a is obtained by dividing K_w by the K_b of the conjugate base, NH_2OH:

$$K_a = K_w \div K_b = (1.0 \times 10^{-14}) \div (1.1 \times 10^{-8}) = 9.\underline{0}9 \times 10^{-7} = 9.1 \times 10^{-7}$$

16.49 Rounded answer: $[OH^-] = [CH_3CH_2CO_2H] \cong 4.4 \times 10^{-6}$ M; pH = 8.64. Assemble the usual table, letting $[Pr^-]$ equal the equilibrium concentration of the propionate anion (the only ion that hydrolyzes). Then calculate K_b of the Pr^- ion from K_a of its conjugate acid, HPr. Assume x is much smaller than the 0.025 M concentration in the denominator and solve for x in the numerator of the equilibrium-constant expression. Finally, calculate pOH from the $[OH^-]$ and pH from the pOH.

Conc. (M)	Pr^-	+	H_2O	$\rightleftharpoons$	HPr	+	OH^-
Starting	0.025				0		0
Change	-x				+x		+x
Equilibrium	0.025 - x				x		x

(continued)

$K_b = K_w \div K_a = (1.0 \times 10^{-14}) \div (1.3 \times 10^{-5}) = 7.\underline{6}9 \times 10^{-10}$

$$\frac{[HPr][OH^-]}{[Pr^-]} \cong \frac{(x)^2}{(0.025)} \cong 7.\underline{6}9 \times 10^{-10}$$

$x = [OH^-] = [Pr^-] \cong 4.\underline{3}8 \times 10^{-6}$ M (Note that x is negligible compared to 0.025.)

$pOH = -\log [OH^-] = -\log (4.38 \times 10^{-6}) = 5.3\underline{5}8$

$pH = 14.00 - 5.358 = 8.6\underline{4}2$ (rounded answer above)

16.50 Rounded answer: $[OH^-] = [HCN] \cong 4.5 \times 10^{-4}$ M; $[H^+] = 2.2 \times 10^{-11}$; $[CN^-] = 9.5 \times 10^{-3}$ M; pH = 10.66. Assemble the usual table, letting $[CN^-]$ equal the equilibrium concentration of the cyanide anion (the only ion that hydrolyzes). Then calculate K_b of the CN^- ion from K_a of its conjugate acid, HCN. Assume x is much smaller than the 0.010 M concentration in the denominator and solve for x in the numerator of the equilibrium-constant expression. (This is a borderline case, but the quadratic gives a value of 4.415×10^{-4} M for x, not much different than the approximation.) Finally, calculate pOH from the $[OH^-]$ and pH from the pOH.

Conc. (M)	CN^-	+	H_2O	⇌	HCN	+	OH^-
Starting	0.010				0		0
Change	-x				+x		+x
Equilibrium	0.010 - x				x		x

$K_b = K_w \div K_a = (1.00 \times 10^{-14}) \div (4.9 \times 10^{-10}) = 2.\underline{0}4 \times 10^{-5}$

$$\frac{[HCN][OH^-]}{[CN^-]} \cong \frac{(x)^2}{(0.010)} \cong 2.\underline{0}4 \times 10^{-5}$$

$x = [OH^-] = [HCN] \cong 4.\underline{5}2 \times 10^{-4}$M (Note that x is negligible compared to 0.010.)

$pOH = -\log [OH^-] = -\log (4.52 \times 10^{-4}) = 3.3\underline{4}48$

$pH = 14.00 - 3.3448 = 10.6\underline{5}52$ (rounded answer above)

16.51 Rounded answer: $[H^+] = [C_5H_5N] \cong 1.0 \times 10^{-3}$ M; pH = 2.99. Assemble the usual table, letting $[PyNH^+]$ equal the equilibrium concentration of the pryridinium cation (the only ion that hydrolyzes). Then calculate K_a of the $PyNH^+$ ion from K_b of its conjugate base, PyN. Assume x is much smaller than the 0.15 M concentration in the denominator and solve for x in the numerator of the equilibrium-constant expression. Finally, calculate pH from the $[H^+]$.

Conc. (M)	$PyNH^+$ + H_2O	⇌	H_3O^+	+	PyN
Starting	0.15		0		0
Change	-x		+x		+x
Equilibrium	0.15 - x		x		x

$$K_a = K_w \div K_b = (1.0 \times 10^{-14}) \div (1.4 \times 10^{-9}) = 7.\underline{1}4 \times 10^{-6}$$

Write the equilibrium-constant expression in terms of chemical symbols and then substitute the terms and solve for K_b:

$$\frac{[PyN][H_3O^+]}{[PyNH^+]} \cong \frac{(x)^2}{(0.15)} \cong 7.\underline{1}4 \times 10^{-6}$$

$x = [H^+] \cong 1.\underline{0}3 \times 10^{-3}$ M (Note that x is negligible compared to 0.15.)

$\text{pH} = -\log [H^+] = -\log (1.03 \times 10^{-3}) = 2.9\underline{8}7$ (rounded answer above)

16.52 Rounded answer: $[H^+] = [CH_3NH_2] \cong 2.8 \times 10^{-6}$ M; pH = 5.55. Assemble the usual table, letting $[MeNH_3^+]$ equal the equilibrium concentration of the methylammonium cation (the only ion that hydrolyzes). Then calculate K_a of the $MeNH_3^+$ ion from K_b of its conjugate base, $MeNH_2$. Assume x is much smaller than the 0.35 M concentration in the denominator and solve for x in the numerator of the equilibrium-constant expression. Finally, calculate pH from the $[H^+]$.

Conc. (M)	$MeNH_3^+$ + H_2O	⇌	H_3O^+	+	$MeNH_2$
Starting	0.35		0		0
Change	-x		+x		+x
Equilibrium	0.35 - x		x		x

$$K_a = K_w \div K_b = (1.0 \times 10^{-14}) \div (4.4 \times 10^{-4}) = 2.\underline{2}7 \times 10^{-11}$$

(continued)

Write the equilibrium-constant expression in terms of chemical symbols and then substitute the terms and solve for K_b:

$$\frac{[MeNH_2][H_3O^+]}{[MeNH_3^+]} \cong \frac{(x)^2}{(0.35)} \cong 2.\underline{2}7 \times 10^{-11}$$

$x = [H^+] = [MeNH_2] \cong 2.\underline{8}2 \times 10^{-6}$ M (Note that x is negligible compared to 0.35.)

pH = - log $[H^+]$ = - log (2.82×10^{-6}) = 5.5$\underline{4}$9 (rounded answer above)

16.53 Degree of ionization—rounded answers: (a) 0.029 (2.9%); (b) 0.0065 (0.65%). To solve, assemble a table of starting, change, and equilibrium concentrations for each part. For each part, assume x is much smaller than the 0.80 M starting concentration of HF. Then solve for x in the numerator of each equilibrium-constant expression by using the product of 6.8×10^{-4} and other terms.

a. Conc. (M)	HF	$\rightleftharpoons$	H^+	+	F^-
Starting	0.80		0		0
Change	-x		+x		+x
Equilibrium	0.80- x		x		x

$$\frac{[H^+][F^-]}{[HF]} = K_a \cong \frac{(x)^2}{(0.80)} \cong 6.8 \times 10^{-4}$$

$x^2 = 6.8 \times 10^{-4} \times 0.80$ M

$x \cong 2.\underline{3}3 \times 10^{-2}$ M = $[H^+]$ (rounded answer above)

$0.80 - (2.33 \times 10^{-2}) = 0.7767$, or $\cong 0.78$ to two sig. figs. (This is a borderline case; the quadratic gives $[H^+] = 2.298$ or 2.30×10^{-2} M, not much different.)

Degree of ionization = $(2.3 \times 10^{-2}) \div 0.80 = 0.0287$ (2.87%)

b. Conc. (M)	HF	$\rightleftharpoons$	H^+	+	F^-
Starting	0.80		0.10		0
Change	-x		+x		+x
Equilibrium	0.80- x		0.10 + x		x

(continued)

Assuming x is negligible compared to 0.10 and to 0.80, substitute into the equilibrium-constant expression (0.10) for $[H^+]$ from 0.10 M HCl and (0.80) from the HF:

$$\frac{[H^+][F^-]}{[HF]} = K_a \cong \frac{(0.10)(x)}{(0.80)} \cong 6.8 \times 10^{-4}$$

$$x \cong \frac{(0.80)6.8 \times 10^{-4}}{(0.10)} \cong 5.\underline{44} \times 10^{-3}$$

$0.80 - (5.44 \times 10^{-3}) = 0.794$, or $\cong 0.79$ to two sig. figs. (Using just [0.10 + x] and the quadratic gives $x = 5.17 \times 10^{-3}$.)

Degree of ionization (quadratic) = $(5.17 \times 10^{-3}) \div 0.80 = 0.006\underline{46}$ (0.6$\underline{46}$%)

16.54 Degree of ionization—rounded answers: (a) 0.029 (2.9%); (b) 0.0017 (0.17%). To solve, assemble a table of starting, change, and equilibrium concentrations for each part. For each part, assume x is much smaller than the 0.20 M starting concentration of $HCHO_2$. Then solve for x in each equilibrium-constant expression by using the product of 1.7×10^{-4} and each starting molarity of formic acid.

a.

Conc. (M)	$HCHO_2$ $\rightleftharpoons$	H^+	+	CHO_2^-
Starting	0.20	0		0
Change	-x	+x		+x
Equilibrium	0.20 - x	x		x

$$\frac{[H^+][CHO_2^-]}{[HCHO_2]} = K_a \cong \frac{(x)^2}{(0.20)} \cong 1.7 \times 10^{-4}$$

$x^2 = (1.7 \times 10^{-4})(0.20\ M)$

$x \cong 5.\underline{83} \times 10^{-3}\ M = [H^+]$ (rounded answer above)

$0.20 - (5.83 \times 10^{-3}) = 0.194$, or $\cong 0.20$ to two sig. figs.

Degree of ionization = $(5.83 \times 10^{-3}) \div 0.20 = 0.02\underline{9}1$ (2.9%)

b.

Conc. (M)	$HCHO_2$ ⇌	H^+	+	CHO_2^-
Starting	0.20	0.10		0
Change	-x	+x		+x
Equilibrium	0.20 - x	0.10 + x		x

Assuming x is negligible compared to 0.10 and to 0.20, substitute into the equilibrium-constant expression (0.10) for $[H^+]$ from 0.10 M HCl and (0.20) from the $HCHO_2$:

$$\frac{[H^+][CHO_2^-]}{[HCHO_2]} = K_a \cong \frac{(0.10)(x)}{(0.20)} \cong 1.7 \times 10^{-4}$$

$$x \cong \frac{(0.20)1.7 \times 10^{-4}}{(0.10)} \cong 3.\underline{4}0 \times 10^{-4}$$

$0.20 - (3.40 \times 10^{-4}) = 0.199$, or $\cong 0.20$ to two sig. figs.

Degree of ionization $= (3.40 \times 10^{-4}) \div 0.20 = 0.001\underline{7}$ (0.17%)

16.55 Rounded answer: pH = 3.17 ($[H^+] = 6.8 \times 10^{-4}$ M). Assemble the usual table, using a starting NO_2^- of 0.10 M, from 0.10 M KNO_2, and a starting HNO_2 of 0.15 M. Assume x is negligible compared to 0.10 M and 0.15 M, and solve for the x in the numerator.

Conc. (M)	HNO_2 ⇌	H^+	+	NO_2^-
Starting	0.15	0		0.10
Change	-x	+x		+x
Equilibrium	0.15 - x	x		0.10 + x

$$\frac{[H^+][NO_2^-]}{[HNO_2]} = K_a = \frac{(x)(0.10 + x)}{(0.15 - x)} = 4.5 \times 10^{-4}$$

$$\frac{(x)(0.10)}{(0.15)} \cong 4.5 \times 10^{-4};\ x = [H^+] \cong 6.\underline{7}5 \times 10^{-4}\ M$$

$\text{pH} = -\log [H^+] = -\log (6.75 \times 10^{-4}) = 3.1\underline{7}06$ (rounded answer above)

16.56 Rounded answer: pH = 3.76 ([H^+] = 1.8×10^{-4} M). Use a starting OCN^- of 0.20 M and a starting HOCN of 0.10 M. Assume x is negligible compared to 0.20 M and 0.10 M.

Conc. (M)	HOCN $\rightleftharpoons$	H^+	+	OCN^-
Starting	0.10	0		0.20
Change	-x	+x		+x
Equilibrium	0.10 - x	x		0.20 + x

$$\frac{[H^+][OCN^-]}{[HOCN]} = K_a = \frac{(x)(0.20 + x)}{(0.10 - x)} = 3.5 \times 10^{-4}$$

$$\frac{(x)(0.20)}{(0.10)} \cong 3.5 \times 10^{-4};\; x = [H^+] \cong 1.\underline{7}5 \times 10^{-4}\text{ M}$$

$$pH = -\log[H^+] = -\log(1.75 \times 10^{-4}) = 3.7\underline{5}69 \text{ (rounded answer above)}$$

16.57 Rounded answer: pH = 10.47 ([OH^-] = 2.9×10^{-4} M). Assemble the usual table, using a starting $CH_3NH_3^+$ of 0.15 M, from 0.15 M CH_3NH_3Cl, and a starting CH_3NH_2 of 0.10 M. Assume x is negligible compared to 0.15 M and 0.10 M, and solve for the x in the numerator.

Conc. (M)	CH_3NH_2	+ H_2O $\rightleftharpoons$	OH^-	+	$CH_3NH_3^+$
Starting	0.10		0		0.15
Change	-x		+x		+x
Equilibrium	0.10 - x		x		0.15 + x

$$\frac{[CH_3NH_3^+][OH^-]}{[CH_3NH_2]} = K_b = \frac{(0.15 + x)(x)}{(0.10 - x)} = 4.4 \times 10^{-4}$$

$$\frac{(0.15)(x)}{(0.10)} \cong 4.4 \times 10^{-4};\; x = [OH^-] \cong 2.\underline{9}3 \times 10^{-4}\text{ M}$$

$$pOH = -\log[OH^-] = -\log(2.93 \times 10^{-4}) = 3.5\underline{3}2$$

$$pH = 14.00 - pOH = 14.00 - 3.532 = 10.4\underline{6}8 \text{ (rounded answer above)}$$

16.58 Rounded answer: pH = 10.85 ($[OH^-]$ = 7.0×10^{-4} M). Assemble the usual table, using a starting $C_2H_5NH_3^+$ of 0.10 M, from 0.10 M $C_2H_5NH_3Br$, and a starting $C_2H_5NH_2$ of 0.15 M. Assume x is negligible compared to 0.10 M and 0.15 M, and solve for the x in the numerator.

Conc. (M)	$C_2H_5NH_2$ + H_2O ⇌	OH^-	+ $C_2H_5NH_3^+$
Starting	0.15	0	0.10
Change	-x	+x	+x
Equilibrium	0.15 - x	x	0.10 + x

$$\frac{[C_2H_5NH_3^+][OH^-]}{[C_2H_5NH_2]} = K_b = \frac{(0.10 + x)(x)}{(0.15 - x)} = 4.7 \times 10^{-4}$$

$$\frac{(0.10)(x)}{(0.15)} \cong 4.7 \times 10^{-4};\ x = [OH^-] \cong 7.\underline{0}5 \times 10^{-4}\ M$$

$$pOH = -\log [OH^-] = -\log (7.05 \times 10^{-4}) = 3.1\underline{5}18$$

$$pH = 14.00 - pOH = 14.00 - 3.1518 = 10.8\underline{4}82 \text{ (rounded answer above)}$$

16.59 Rounded answer: pH = 3.45 ($[H^+]$ = 3.5×10^{-4} M). Find the mol/L of HF and the mol/L of F^-, and assemble the usual table. Substitute the equilibrium concentrations into the equilibrium-constant expression; then assume x is negligible compared to the starting concentrations of both HF and F^-. Solve for x in the numerator of the equilibrium-constant expression, and calculate the pH from this value.

Total volume = 0.045 L + 0.035 L = 0.080 L

(0.10 mol HF/L) x 0.035 L = 0.0035 mol HF (÷ 0.080 L total volume = 0.04375 M)

(0.15 mol F^-/L) x 0.045 L = 0.00675 mol F^- (÷ 0.080 L total volume = 0.084375 M)

Now substitute these starting concentrations into the usual table:

Conc. (M)	HF ⇌	H^+	+ F^-
Starting	0.04375	0	0.084375
Change	-x	+x	+x
Equilibrium	0.04375 - x	x	0.084375 + x

(continued)

$$\frac{[H^+][F^-]}{[HF]} = K_a = \frac{(x)(0.084375 + x)}{(0.04375 - x)} = 6.8 \times 10^{-4}$$

$$\frac{(x)(0.084375)}{(0.04375)} \cong 6.8 \times 10^{-4};\ x = [H^+] \cong 3.\underline{5}2 \times 10^{-4}\ M$$

$$pH = -\log [H^+] = -\log (3.52 \times 10^{-4}) = 3.4\underline{5}3$$

16.60 Rounded answer: pH = 9.46 ($[OH^-]$ = 2.9×10^{-5} M). Find the mol/L of NH_3 and the mol/L of NH_4^+, and assemble the usual table. Substitute the equilibrium concentrations into the equilibrium-constant expression; then assume x is negligible compared to the starting concentrations of both NH_3 and NH_4^+. Solve for x in the numerator of the equilibrium-constant expression, and calculate the pH from this value.

Total volume = 0.115 L + 0.145 L = 0.260 L

(0.30 mol NH_3/L) x 0.115 L = 0.0345 mol NH_3 (÷ 0.260 L total vol = 0.1327 M)

(0.15 mol NH_4^+/L) x 0.145 L = 0.02175 mol NH_4^+ (÷ 0.260 L total vol = 0.08365 M)

Now substitute these starting concentrations into the usual table:

Conc. (M)	NH_3 + H_2O	$\rightleftharpoons$	NH_4^+ +	OH^-
Starting	0.1327		0.08365	0
Change	-x		+x	+x
Equilibrium	0.1327 - x		0.08365 + x	x

$$\frac{[NH_4^+][OH^-]}{[NH_3]} = K_b = \frac{(0.08365 + x)(x)}{(0.1327 - x)} = 1.8 \times 10^{-5}$$

$$\frac{(0.08365)(x)}{(0.1327)} \cong 1.8 \times 10^{-5};\ x \cong 2.\underline{8}55 \times 10^{-5}\ M$$

$$pOH = -\log [OH^-] = -\log (2.855 \times 10^{-5}) = 4.5\underline{4}4$$

pH = 14.00 - 4.544 = 9.4$\underline{5}$6 (rounded answer above)

16.61 Rounded answer: pH before HCl = 9.26; pH after HCl = 9.09. First use the 0.10 M NH_3 and 0.10 M NH_4^+ to calculate the [OH^-] and pH before HCl is added. Assemble a table of starting, change, and equilibrium concentrations. Assume x is negligible compared to 0.10 M and substitute the approximate concentrations into the equilibrium-constant expression.

Conc. (M)	NH_3	+	H_2O	$\rightleftharpoons$	NH_4^+	+	OH^-
Starting	0.10				0.10		0
Change	-x				+x		+x
Equilibrium	0.10 - x				0.10 + x		x

$$\frac{[NH_4^+][OH^-]}{[NH_3]} = K_b \cong \frac{(0.10)(x)}{(0.10)} \cong 1.8 \times 10^{-5}$$

x = [OH^-] $\cong$ 1.8×10^{-5} M (Note that x is negligible compared to 0.10.)

[H^+] = $(1.00 \times 10^{-14}) \div (1.8 \times 10^{-5})$ = $5.\underline{5}55 \times 10^{-10}$ M

pH before HCl added = $9.2\underline{5}53$ (rounded answer above)

Now calculate the pH after the 0.012 L (12 mL) of 0.20 M HCl is added by noting that the H^+ ion reacts with the NH_3 to form additional NH_4^+. Calculate the stoichiometric amount of HCl; then subtract the moles of HCl from the moles of NH_3. Add the resulting moles of NH_4^+ to the 0.0125 starting moles of NH_4^+ in the 0.125 L of buffer.

(0.20 mol HCl/L) x 0.012 L = 0.0024 mol HCl (reacts with 0.0024 mol NH_3)

Mol NH_3 left = ($0.01\underline{2}5$ - 0.0024) mol = $0.01\underline{0}1$ mol NH_3

Total mol NH_4^+ now present = 0.0125 orig. + 0.0024 formed from HCl = 0.0149 mol NH_4^+

$$[NH_3] = \frac{0.0101 \text{ mol } NH_3}{0.137 \text{ L soln}} = 0.07\underline{3}7 \text{ M}; \quad [NH_4^+] = \frac{0.0149 \text{ mol } NH_4^+}{0.137 \text{ L soln}} = 0.1\underline{0}8 \text{ M}$$

Now account for the ionization of NH_3 to NH_4^+ and OH^- at equilibrium by assembling the usual table. Assume x is negligible compared to 0.0737 M and 0.108 M, and solve the equilibrium-constant expression for x in the numerator. Calculate the pH from x, the [OH^-].

(continued)

Conc. (M)	NH_3 + H_2O ⇌	NH_4^+ +	OH^-
Starting	0.0737	0.108	0
Change	-x	+x	+x
Equilibrium	0.0737 - x	0.108 + x	x

$$\frac{[NH_4^+][OH^-]}{[NH_3]} = K_b \cong \frac{(0.108)(x)}{(0.0737)} \cong 1.8 \times 10^{-5}$$

$x = [OH^-] \cong 1.\underline{2}2 \times 10^{-5}$ M (Note that x is negligible compared to 0.10.)

$[H^+] = (1.00 \times 10^{-14}) \div (1.22 \times 10^{-5}) = 8.\underline{1}9 \times 10^{-10}$ M

pH after HCl added = 9.0$\underline{8}$6 (rounded answer above)

16.62 Rounded answer: pH before HCl = 3.73; pH after HCl = 3.68. First use the 0.525 L of 0.50 M $HCHO_2$ (HFo) and 0.475 L of 0.50 M $NaCHO_2$ (Na^+Fo^-) to calculate the $[H^+]$ and pH before HCl is added. Assemble a table of starting, change, and equilibrium concentrations. Assume x is negligible compared to both concentrations; substitute into the K_a expression.

0.50 M HFo x 0.525 L = 0.263 mol HFo (÷ 1.00 L = 0.2$\underline{6}$3 M)

0.50 M Fo^- x 0.475 L = 0.238 mol Fo^- (÷ 1.00 L = 0.2$\underline{3}$8 M)

Conc. (M)	HFo ⇌	H^+ +	Fo^-
Starting	0.263	0	0.238
Change	-x	+x	+x
Equilibrium	0.263 - x	x	0.238 + x

$$\frac{[H^+][Fo^-]}{[HFo]} = K_a \cong \frac{(x)(0.238)}{(0.263)} \cong 1.7 \times 10^{-4}$$

$x = [H^+] \cong 1.\underline{8}8 \times 10^{-4}$ M

pH = - log $[H^+]$ = - log (1.88×10^{-4}) = 3.7$\underline{2}$6 (rounded answer above)

Now calculate the pH after the 0.0085 L (8.5 mL) of 0.15 M HCl is added by noting that the H^+ ion reacts with the Fo^- to form additional HFo. Calculate the stoichiometric amount of HCl; then subtract the moles of HCl from the moles of Fo^-. Add the resulting moles of HFo to the 0.02227 starting moles of HFo in the 0.085 L of buffer.

(continued)

(0.15 mol HCl/L) x 0.0085 L = 0.00128 mol HCl (reacts with 0.00128 mol Fo^-)

Mol Fo^- left = (0.02023 - 0.00128) mol = 0.01895 mol Fo^-

Total mol HFo present = 0.02227 orig. + 0.00128 formed from HCl = 0.02355 mol HFo

$$[HFo] = \frac{0.02355 \text{ mol HFo}}{0.0935 \text{ L soln}} = 0.2\underline{5}18 \text{ M}; \quad [Fo^-] = \frac{0.01895 \text{ mol } Fo^-}{0.0935 \text{ L soln}} = 0.2\underline{0}26 \text{ M}$$

Now account for the ionization of HFo to H^+ and Fo^- at equilibrium by assembling the usual table. Assume x is negligible compared to 0.2518 M and 0.2026M, and solve the equilibrium-constant expression for x in the numerator. Calculate the pH from x, the $[H^+]$.

Conc. (M)	HFo	⇌	H^+	+	Fo^-
Starting	0.2518		0		0.2026
Change	-x		+x		+x
Equilibrium	0.2518 - x		x		0.2026 + x

$$\frac{[H^+][F^-]}{[HFo]} = K_a \cong \frac{(x)(0.2026)}{(0.2518)} \cong 1.7 \times 10^{-4}$$

$x = [H^+] \cong 2.\underline{1}1 \times 10^{-4}$ M

pH = - log $[H^+]$ = - log (2.11×10^{-4}) = 3.6$\underline{7}$57 (rounded answer above)

16.63 Use the Henderson-Hasselbalch equation, where $pK_a = -\log K_a$.

$$pH = -\log K_a + \log \frac{[\text{buff. base}]}{[\text{buff. acid}]} = -\log (1.4 \times 10^{-3}) + \log \frac{[0.15 \text{ M}]}{[0.10 \text{ M}]}$$

$$= 3.0\underline{2}99 = 3.03$$

16.64 Use the Henderson-Hasselbalch equation ($pK_a = -\log K_a$) and do the calculations in one step.

$$pH = -\log K_a + \log \frac{[\text{buff. base}]}{[\text{buff. acid}]} = -\log (1.3 \times 10^{-5}) + \log \frac{[0.10 \text{ M}]}{[0.20 \text{ M}]}$$

$$= 4.5\underline{8}502 = 4.59$$

16.65 Calculate the K_a of the pyridinium ion from the K_b of pyridine and then use the Henderson-Hasselbalch equation, where $pK_a = -\log K_a$.

$$K_a = K_w \div K_b = (1.0 \times 10^{-14}) \div (1.4 \times 10^{-9}) = 7.\underline{1}4 \times 10^{-6}$$

$$pH = -\log K_a + \log \frac{[\text{buff. base}]}{[\text{buff. acid}]} = -\log (7.14 \times 10^{-6}) + \log \frac{[0.15\ M]}{[0.10\ M]}$$

$$= 5.3\underline{2}2 = 5.32$$

16.66 Calculate the K_a of the methylammonium ion from the K_b of methylamine and then use the Henderson-Hasselbalch equation, where $pK_a = -\log K_a$.

$$K_a = K_w \div K_b = (1.0 \times 10^{-14}) \div (4.4 \times 10^{-4}) = 2.\underline{2}7 \times 10^{-11}$$

$$pH = -\log K_a + \log \frac{[\text{buff. base}]}{[\text{buff. acid}]} = -\log (2.27 \times 10^{-11}) + \log \frac{[0.20\ M]}{[0.15\ M]}$$

$$= 10.7\underline{6}8 = 10.77$$

16.67 Symbolize acetic acid as HAc and sodium acetate as Na^+Ac^-. Use the Henderson-Hasselbalch equation to find the log of $[Ac^-]/[HAc]$. Then solve for $[Ac^-]$ and for moles of NaAc in the 2.0 L of solution.

$$pH = -\log K_a + \log \frac{[Ac^-]}{[HAc]} = 5.00 = -\log (1.7 \times 10^{-5}) + \log \frac{[Ac^-]}{[0.10\ M]}$$

$$5.00 = 4.770 + \log [Ac^-] + \log 1/[0.10]$$

$$\log[Ac^-] = 5.00 - 4.770 - 1.00 = -0.770$$

$$[Ac^-] = 0.1\underline{6}98\ M$$

$$\text{Mol NaAc} = 0.1698\ \text{mol/L} \times 2.0\ \text{L} = 0.3\underline{3}96 = 0.34\ \text{mol}$$

16.68 Use the Henderson-Hasselbalch equation to find the log of $[F^-]/[HF]$. Then solve for $[HF]$ and for moles of HF in the 0.5000 L of solution.

$$pH = -\log K_a + \log \frac{[F^-]}{[HF]} = 3.50 = -\log (6.8 \times 10^{-4}) + \log \frac{[0.30\ M]}{[HF]}$$

(continued)

$3.50 = 3.167 + \log [0.30] + \log 1/[HF]$

$\log [HF] = 3.167 - 0.523 - 3.50 = -0.856$

$[HF] = 0.1\underline{3}93$ M

Mol HF = 0.1393 mol/L x 0.5000 L = $0.06\underline{9}65$ = 0.070 mol HF

16.69 Rounded answer: pH = 2.0 ($[H^+]$ = 0.01 M). All the OH^- (from the NaOH) reacts with the H^+ from HCl. Calculate the stoichiometric amounts of OH^- and H^+ and subtract the mol of OH^- from the mol of H^+. Then divide the remaining H^+ by the total volume of 0.020 L + 0.025 L, or 0.045 L, to find the $[H^+]$. Then calculate the pH.

Mol H^+ = (0.10 mol HCl/L) x 0.025 L HCl = 0.0025 mol H^+

Mol OH^- = (0.10 mol NaOH/L) x 0.020 L NaOH = 0.0020 mol OH^-

Mol H^+ left = (0.0025 - 0.0020) mol H^+ = 0.0005 mol H^+

$[H^+]$ = 0.0005 mol H^+ ÷ 0.045 L total volume = $0.0\underline{1}1$ M

$pH = -\log [H^+] = -\log (0.011) = 1.\underline{9}54$

16.70 Rounded answer: pH = 12.22 ($[OH^-]$ = 0.017 M). All the H^+ (from the HCl) reacts with the OH^- from NaOH. Calculate the stoichiometric amounts of OH^- and H^+ and subtract the mol of H^+ from the mol of OH^-. Then divide the remaining OH^- by the total volume of 0.035 L + 0.025 L, or 0.060 L, to find the $[OH^-]$. Then calculate the pOH and pH.

Mol OH^- = (0.10 mol NaOH/L) x 0.035 L NaOH = 0.0035 mol OH^-

Mol H^+ = (0.10 mol HCl/L) x 0.025 L HCl = 0.0025 mol H^+

Mol OH^- left = (0.0035 - 0.0025) mol OH^- = 0.0010 mol OH^-

$[OH^-]$ = 0.0010 mol OH^- ÷ 0.060 L total volume = $0.01\underline{6}66$M

$pOH = -\log [OH^-] = -\log (0.01666) = 1.7\underline{7}8$

$pH = 14.00 - pH = 14.00 - 1.778 = 12.2\underline{2}2$

16.71 Rounded answer: pH = 8.59 ([OH^-] = 3.9×10^{-6} M). Use HBen to symbolize benzoic acid and Ben^- to symbolize the benzoate anion. At the equivalence point, equal molar amounts of HBen and NaOH react to form a solution of NaBen. Start by calculating the moles of HBen. Use this to calculate the volume of NaOH needed to neutralize all of the HBen (and use the moles of HBen as the moles of Ben^- formed at the equivalence point). Add the volume of NaOH to the original 0.050 L to find the total volume of solution.

Mol HBen = 1.24 g HBen ÷ (122.1 g HBen/mol HBen) = 0.01016 mol HBen

Volume NaOH = 0.01016 mol NaOH ÷ (0.180 mol NaOH/L) = $0.056\underline{4}4$ L

Total volume = 0.05644 L + 0.050 L HBen soln = 0.10644 L

[Ben^-] = (0.01016 mol Ben^- from HBen) ÷ 0.10644 L = 0.09545 M

Because the Ben^- hydrolyzes to OH^- and HBen, use this to calculate the [OH^-]. Start by calculating the K_b constant of Ben^- from the K_a of its conjugate acid, HBen. Then assemble the usual table of concentrations, assume x is negligible, and calculate [OH^-] and pH.

$$K_b = K_w \div K_a = (1.0 \times 10^{-14}) \div (6.3 \times 10^{-5}) = 1.\underline{5}9 \times 10^{-10}$$

Conc. (M)	Ben^- + H_2O	⇌	HBen	+	OH^-
Starting	0.09545		0		0
Change	-x		+x		+x
Equilibrium	0.09545 - x		x		x

$$\frac{[HBen][OH^-]}{[Ben^-]} = K_b = \frac{(x)^2}{(0.09545 - x)} = 1.59 \times 10^{-10}$$

$$\frac{(x)^2}{(0.09545)} \cong 1.59 \times 10^{-10};\ x = [OH^-] \cong 3.\underline{8}95 \times 10^{-6}$$

$$pOH = -\log [OH^-] = -\log (3.895 \times 10^{-6}) = 5.4\underline{0}94$$

$$pH = 14.00 - 5.4094 = 8.5\underline{9}05$$

16.72 Rounded answer: pH = 8.84 ([OH^-] = 6.9 x 10^{-6} M). Use HPr to symbolize propionic acid and Pr^- to symbolize the propionate anion. At the equivalence point, equal molar amounts of HPr and NaOH react to form a solution of NaPr. Start by calculating the moles of HPr. Use this to calculate the volume of NaOH needed to neutralize all of the HPr (and use the moles of HPr as the moles of Pr^- formed at the equivalence point). Add the volume of NaOH to the original 0.050 L to find the total volume of solution.

Mol HPr = 0.400 g HPr ÷ (74.08 g HPr/mol HPr) = 0.00540 mol HPr

Volume NaOH = 0.00540 mol NaOH ÷ (0.150 mol NaOH/L) = 0.036$\underline{0}$0 L

Total volume = 0.03600 L + 0.050 L HPr soln = 0.08600 L

[Pr^-] = (0.00540 mol Pr^- from HPr) ÷ 0.08600 L = 0.06279 M

Because the Pr^- hydrolyzes to OH^- and HPr, use this to calculate the [OH^-]. Start by calculating the K_b constant of Pr^- from the K_a of its conjugate acid, HPr. Then assemble the usual table of concentrations, assume x is negligible, and calculate [OH^-] and pH.

$$K_b = K_w \div K_a = (1.0 \times 10^{-14}) \div (1.3 \times 10^{-5}) = 7.\underline{6}9 \times 10^{-10}$$

Conc. (M)	Pr^- + H_2O	⇌	HPr	+	OH^-
Starting	0.06279		0		0
Change	-x		+x		+x
Equilibrium	0.06279 - x		x		x

$$\frac{[HPr][OH^-]}{[Pr^-]} = K_b = \frac{(x)^2}{(0.06279 - x)} = 7.69 \times 10^{-10}$$

$$\frac{(x)^2}{(0.06279)} \cong 7.69 \times 10^{-10};\ x = [OH^-] \cong 6.\underline{9}48 \times 10^{-6} \text{ M}$$

$$pOH = -\log[OH^-] = -\log(6.948 \times 10^{-6}) = 5.1\underline{5}8$$

$$pH = 14.00 - 5.158 = 8.8\underline{4}1$$

16.73 Rounded answer: pH = 5.97 ([H^+] = 1.1 x 10^{-6} M). Use EtN to symbolize ethylamine and $EtNH^+$ to symbolize the ethylammonium cation. At the equivalence point, equal molar amounts of EtN and HCl react to form a solution of EtNHCl. Start by calculating the moles of EtN. Use this to calculate the volume of HCl needed to neutralize all of the EtN (and use the moles of EtN as the moles of $EtNH^+$ formed at the equivalence point). Add the volume of HCl to the original 0.032 L to find the total volume of solution.

(continued)

(0.087 mol EtN/L) x 0.032 L = 0.00278 mol EtN

Volume HCl = 0.00278 mol HCl ÷ (0.15 mol HCl/L) = 0.0185 L

Total volume = 0.0185 L + 0.032 L EtN soln = 0.0505 L

$[EtNH^+]$ = (0.00278 mol $EtNH^+$ from EtN) ÷ 0.0505 L = 0.0550 M

Because the $EtNH^+$ hydrolyzes to H_3O^+ and EtN, use this to calculate the $[H^+]$. Start by calculating the K_a constant of $EtNH^+$ from the K_b of its conjugate base, EtN. Then assemble the usual table of concentrations, assume x is negligible, and calculate $[H^+]$ and pH.

$K_a = K_w \div K_b = (1.0 \times 10^{-14}) \div (4.7 \times 10^{-4}) = 2.\underline{1}3 \times 10^{-11}$

Conc. (M)	$EtNH^+$	+	H_2O	$\rightleftharpoons$	EtN	+	H_3O^+
Starting	0.0550				0		0
Change	-x				+x		+x
Equilibrium	0.0550 - x				x		x

$$\frac{[EtN][H^+]}{[EtNH^+]} = K_a = \frac{(x)^2}{(0.0550 - x)} = 2.13 \times 10^{-11}$$

$$\frac{(x)^2}{(0.0550)} \cong 2.13 \times 10^{-11};\ x = [H^+] \cong 1.08 \times 10^{-6} \text{ M}$$

$pH = -\log [H^+] = -\log (1.08 \times 10^{-6}) = 5.9\underline{6}56$

16.74 Rounded answer: $[H^+] = 2.8 \times 10^{-4}$ M; pH = 3.55. Use HyN to symbolize hydroxylamine and $HyNH^+$ to symbolize the hydroxylammonium cation. At the equivalence point, equal molar amounts of HyN and HCl react to form a solution of HyNHCl. Start by calculating the moles of HyN. Use this to calculate the volume of HCl needed to neutralize all of the HyN (and use the moles of HyN as the moles of $HyNH^+$ formed at the equivalence point). Add volumes.

(0.020 mol HyN/L) x 0.022 L = 0.00440 mol HyN

Volume HCl = 0.00440 mol HCl ÷ (0.15 mol HCl/L) = 0.0293 L

Total volume = 0.0293 L + 0.022 L HyN soln = 0.0513 L

$[HyNH^+]$ = (0.00440 mol $HyNH^+$ from HyN) ÷ 0.0513 L = 0.0858 M

(continued)

Use $HyNH^+$ hydrolysis to calculate the $[H^+]$. Calculate the K_a constant of $HyNH^+$ from the K_b of HyN. Assemble the concentration table; assume x is negligible and calculate $[H^+]$.

$$K_a = K_w \div K_b = (1.0 \times 10^{-14}) \div (1.1 \times 10^{-8}) = 9.\underline{0}9 \times 10^{-7}$$

Conc. (M)	$HyNH^+$	+	H_2O	⇌	HyN	+	H_3O^+
Starting	0.0858				0		0
Change	-x				+x		+x
Equilibrium	0.0858 - x				x		x

$$\frac{[HyN][H^+]}{[HynH^+]} = K_a = \frac{(x)^2}{(0.0858 - x)} = 9.09 \times 10^{-7}$$

$$\frac{(x)^2}{(0.0858)} \cong 9.09 \times 10^{-7};\ x = [H^+] \cong 2.\underline{7}9 \times 10^{-4}\ M$$

$$pH = -\log [H^+] = -\log (2.79 \times 10^{-4}) = 3.5\underline{5}4$$

16.75 Rounded answer: pH = 9.08; ($[OH^-]$ = 1.2×10^{-5} M). Calculate the stoichiometric amounts of NH_3 and HCl, which forms NH_4^+. Then divide NH_3 and NH_4^+ by the total volume of 0.500 L + 0.200 L, or 0.700 L, to find the starting concentrations. Calculate the $[OH^-]$, the pOH, and the pH.

Mol NH_3 = (0.10 mol NH_3/L) x 0.500 L = 0.0500 mol NH_3

Mol HCl = (0.15 mol HCl/L) x 0.200 L = 0.0300 mol HCl (" 0.0300 mol NH_4^+)

Mol NH_3 left = 0.0500 mol - 0.0300 mol HCl = 0.0200 mol NH_3

$0.02\underline{0}0$ mol NH_3 ÷ 0.700 L = 0.0286 M NH_3

$0.03\underline{0}0$ mol NH_4^+ ÷ 0.700 L = 0.0429 M NH_4^+

(continued)

Conc. (M)	NH_3	+	H_2O	$\rightleftharpoons$	NH_4^+	+	OH^-
Starting	0.0286				0.0429		0
Change	-x				+x		+x
Equilibrium	0.0286 - x				0.0429 + x		x

$$\frac{[NH_4^+][OH^-]}{[NH_3]} = K_b \cong \frac{(0.0429)(x)}{(0.0286)} \cong 1.8 \times 10^{-5}$$

$x = [OH^-] \cong 1.\underline{2}0 \times 10^{-5}$ M (Note that x is negligible compared to 0.286.)

$[H^+] = (1.0 \times 10^{-14}) \div (1.20 \times 10^{-5}) = 8.\underline{3}3 \times 10^{-10}$ M

pH = 9.0$\underline{7}$9 (rounded answer above)

16.76 Rounded answer: pH = 4.45 ($[H^+] = 3.6 \times 10^{-5}$ M). Use the symbols HAc for acetic acid and Ac^- for the acetate ion from sodium acetate. Calculate the stoichiometric amounts of HAc and Ac^-. Then divide the HAc and Ac^- by the total volume of 0.035 L + 0.025 L, or 0.060 L, to find the starting concentrations. Calculate the $[H^+]$ and pH.

Mol HAc = (0.15 mol HAc/L) x 0.035 L HAc = 0.00525 mol HAc

Mol Ac^- = (0.10 mol NaAc/L) x 0.025 L NaAc = 0.00250 mol Ac^-

0.00525 mol HAc ÷ 0.060 L = 0.0875 M HAc

0.00250 mol Ac^- ÷ 0.060 L = 0.0417 M Ac^-

Conc. (M)	HAc	$\rightleftharpoons$	H^+	+	Ac^-
Starting	0.0875		0		0.0417
Change	-x		+x		+x
Equilibrium	0.0875 - x		x		0.0417 + x

$$\frac{[H^+][Ac^-]}{[HAc]} = K_a \cong \frac{(0.0417)(x)}{(0.0875)} \cong 1.7 \times 10^{-5}$$

$[H^+] \cong 3.\underline{5}7 \times 10^{-5}$ M

pH = - log $[H^+]$ = - log (3.57×10^{-5}) = 4.4$\underline{4}$7

■ Solutions to Unclassified Problems

16.77 Rounded answer: $K_a = 1.1 \times 10^{-3}$ (correct no. of sig. figs.). To solve, assemble a table of starting, change, and equilibrium concentrations. Use HSal to symbolize salicylic acid and use Sal^- for the anion. Start by converting pH to $[H^+]$:

$$[H^+] = \text{antilog}(-pH) = \text{antilog}(-2.43) = 3.72 \times 10^{-3} \text{ M}$$

$$\text{Starting M of HSal} = (2.2 \text{ g} \div 138 \text{ g/mol}) \div 1.00 \text{ L} = 0.01\underline{5}9 \text{ M}$$

Conc. (M)	HSal	⇌	H^+	+	Sal^-
Starting	0.0159		0		0
Change	-x		+x		+x
Equilibrium	0.0159 - x		x		x

The value of x equals the value of the molarity of the H^+ ion, which is 3.72×10^{-3} M. Substitute into the equilibrium-constant expression to find K_a:

$$K_a = \frac{[H^+][Sal^-]}{[HSal]} = \frac{[x]^2}{(0.0159 - x)} = \frac{(3.72 \times 10^{-3})^2}{(0.0159 - 0.00372)} = 1.\underline{1}3 \times 10^{-3}$$

16.78 Rounded answer: $K_a = 3.6 \times 10^{-3}$ (correct no. of sig. figs.). To solve, assemble a table of starting, change, and equilibrium concentrations. Use HCyn to symbolize cyanoacetic acid and use Cyn^- for the anion. Start by converting pH to $[H^+]$:

$$[H^+] = \text{antilog}(-pH) = \text{antilog}(-1.89) = 1.\underline{2}9 \times 10^{-2} \text{ M}$$

$$\text{Starting M of HCyn} = (5.0\text{g} \div 85.1 \text{ g/mol}) \div 1.00 \text{ L} = 0.0588 \text{ M}$$

Conc. (M)	HCyn	⇌	H^+	+	Cyn^-
Starting	0.0588		0		0
Change	-x		+x		+x
Equilibrium	0.0588 - x		x		x

The value of x equals the value of the molarity of the H^+ ion, which is 1.29×10^{-2} M. Substitute into the equilibrium-constant expression to find K_a:

$$K_a = \frac{[H^+][Cyn^-]}{[HCyn]} = \frac{[x]^2}{(0.0588 - x)} = \frac{(1.29 \times 10^{-2})^2}{(0.0588 - 0.00129)} = 3.\underline{6}2 \times 10^{-3}$$

16.79 Rounded answer: $K_a = 1.1 \times 10^{-2}$ (correct no. of sig. figs.). To solve, assemble a table of starting, change, and equilibrium concentrations. Start by converting pH to $[H^+]$:

$$[H_+] = \text{antilog}(\text{-pH}) = \text{antilog}(-1.73) = 1.86 \times 10^{-2}\ M$$

Starting M of HSO_4^- = 0.050 M

Conc. (M)	HSO_4^-	$\rightleftharpoons$	H^+	+	SO_4^{2-}
Starting	0.050		0		0
Change	-x		+x		+x
Equilibrium	0.050 - x		x		x

The value of x equals the value of the molarity of the H^+ ion, which is 1.86×10^{-2} M. Substitute into the equilibrium-constant expression to find K_a:

$$K_{a2} = \frac{[H^+][SO_4^{2-}]}{[HSO_4^-]} = \frac{[x]^2}{(0.050 - x)} = \frac{(1.86 \times 10^{-2})^2}{(0.050 - 0.0186)} = 1.\underline{1}01 \times 10^{-2}$$

16.80 Rounded answer: $K_a = 6.3 \times 10^{-8}$ (correct no. of sig. figs.). To solve, assemble a table of starting, change, and equilibrium concentrations. Start by converting pH to $[H^+]$:

$$[H_+] = \text{antilog}(\text{-pH}) = \text{antilog}(-4.10) = 7.94 \times 10^{-5}\ M$$

Starting M of $H_2PO_4^-$ = 0.10 M

Conc. (M)	$H_2PO_4^-$	$\rightleftharpoons$	H^+	+	HPO_4^{2-}
Starting	0.10		0		0
Change	-x		+x		+x
Equilibrium	0.10 - x		x		x

The value of x equals the value of the molarity of the H^+ ion, which is 7.94×10^{-5} M. Substitute into the equilibrium-constant expression to find K_{a2}:

$$K_{a2} = \frac{[H^+][HPO_4^{2-}]}{[H_2PO_4^-]} = \frac{[x]^2}{(0.10 - x)} = \frac{(7.94 \times 10^{-5})^2}{(0.10 - 7.94 \times 10^{-5})} = 6.\underline{3}04 \times 10^{-8}$$

16.81 For the base ionization (hydrolysis) of CN^- to $HCN + OH^-$, the base-ionization constant is

$$K_w \div K_a = (1.0 \times 10^{-14}) \div (4.9 \times 10^{-10}) = 2.\underline{0}4 \times 10^{-5} = 2.0 \times 10^{-5}$$

For the base ionization (hydrolysis) of CO_3^{2-} to $HCO_3^- + OH^-$, the base-ionization constant is calculated from the ionization constant (K_{a2}) of HCO_3^-, the conjugate acid of CO_3^{2-}:

$$K_w \div K_a = (1.00 \times 10^{-14}) \div (4.8 \times 10^{-11}) = 2.\underline{0}8 \times 10^{-4} = 2.1 \times 10^{-4}$$

Because the constant of CO_3^{2-} is larger, it is the stronger base.

16.82 For the base ionization (hydrolysis) of PO_4^{3-} to $HPO_4^{2-} + OH^-$, the base-ionization constant is calculated from the ionization constant (K_{a3}) of HPO_4^{2-}, the conjugate acid of PO_4^{3-}:

$$K_w \div K_a = (1.0 \times 10^{-14}) \div (4.8 \times 10^{-13}) = 2.\underline{0}8 \times 10^{-2} = 2.1 \times 10^{-2}$$

For the base ionization (hydrolysis) of SO_4^{2-} to $HSO_4^- + OH^-$, the base-ionization constant is calculated from the ionization constant (K_{a2}) of HSO_4^-, the conjugate acid of SO_4^{2-}:

$$K_w \div K_a = (1.0 \times 10^{-14}) \div (1.1 \times 10^{-2}) = 9.\underline{0}9 \times 10^{-13} = 9.1 \times 10^{-13}$$

Because the constant of PO_4^{3-} is larger, it is the stronger base.

16.83 Rounded answer: $[H^+] = 1.4 \times 10^{-3}$ M; pH = 2.84. Assume Al^{3+} is $Al(H_2O)_6^{3+}$. Assemble the usual table to calculate the $[H^+]$ and pH. Use the usual equilibrium-constant expression.

Conc. (M)	$Al(H_2O)_6^{3+}$	+	H_2O	$\rightleftharpoons$	H_3O^+	+	$Al(H_2O)_5(OH)^{2+}$
Starting	0.15				0		0
Change	-x				+x		+x
Equilibrium	0.15 - x				x		x

$$K_a = 1.4 \times 10^{-5}$$

(continued)

$$\frac{[Al(H_2O)_5(OH)^{2+}][H_3O^+]}{[Al(H_2O)_6^{3+}]} \cong \frac{(x)^2}{(0.15)} \cong 1.\underline{4} \times 10^{-5}$$

$x = [H^+] \cong 1.\underline{4}49 \times 10^{-3}$ M (Note that x is negligible compared to 0.15.)

pH = - log [H^+] = - log (1.449×10^{-3}) = 2.8$\underline{3}$8 (rounded answer above)

16.84 Rounded answer: [H^+] = 6.1×10^{-6} M; pH = 5.21. Assume Zn^{2+} is $Zn(H_2O)_6^{2+}$. Assemble the usual table to calculate the [H^+] and pH.

Conc. (M)	$Zn(H_2O)_6^{2+}$	+	H_2O	$\rightleftharpoons$	H_3O^+	+	$Zn(H_2O)_5(OH)^+$
Starting	0.15				0		0
Change	-x				+x		+x
Equilibrium	0.15 - x				x		x

$K_a = 2.5 \times 10^{-10}$

Write the equilibrium-constant expression in terms of chemical symbols and substitute.

$$\frac{[Zn(H_2O)_5(OH)^+][H_3O^+]}{[Zn(H_2O)_6^{2+}]} \cong \frac{(x)^2}{(0.15)} \cong 2.\underline{5} \times 10^{-10}$$

$x = [H^+] \cong 6.\underline{1}2 \times 10^{-6}$ M (Note that x is negligible compared to 0.15.)

pH = - log [H^+] = - log (6.12×10^{-6}) = 5.2$\underline{1}$3 (rounded answer above)

16.85 Rounded answer: [H^+] = 6.9×10^{-4} M; pH = 3.16. Calculate the concentrations of the tartaric acid (H_2Tar) and the hydrogen tartrate ion ($HTar^-$).

(11.0 g H_2Tar ÷ 150.1 g/mol) ÷ 1.00 L = 0.07328 M H_2Tar

(20.0 g $KHTar^-$ ÷ 188.2 g/mol) ÷ 1.00 L = 0.1063 M [$HTar^-$]

Conc. (M)	H_2Tar	$\rightleftharpoons$	H^+	+	$HTar^-$
Starting	0.07328		0		0.1063
Change	-x		+x		+x
Equilibrium	0.07328 - x		x		0.1063 + x

(continued)

Neglecting x compared to 0.07328 and 0.1063 and substituting into the K_{a1} expression gives

$$\frac{(x)(0.1063)}{(0.07328)} \cong 1.0 \times 10^{-3};\ x = [H^+] \cong 6.\underline{8}9 \times 10^{-4}\ M$$

$$pH = -\log [H^+] = -\log (6.89 \times 10^{-4}) = 3.1\underline{6}1$$

16.86 Rounded answer: $[H^+] = 6.4 \times 10^{-8}$ M; pH = 7.20. Calculate the concentrations of the dihydrogen phosphate ion ($H_2PO_4^-$) and the hydrogen phosphate ion (HPO_4^{2-}).

$$(13.0\ g\ NaH_2PO_4 \div 120.0\ g/mol) \div 1.00\ L = 0.1082\ M\ H_2PO_4^-$$

$$(15.0\ g\ Na_2HPO_4 \div 142.0\ g/mol) \div 1.00\ L = 0.1056\ M\ [HPO_4^{2-}]$$

Conc. (M)	$H_2PO_4^-$	$\rightleftharpoons$	H^+	+	HPO_4^{2-}
Starting	0.1082		0		0.1056
Change	-x		+x		+x
Equilibrium	0.1082 - x		x		0.1056 + x

Neglecting x compared to 0.1082 and 0.1056 and substituting into the K_{a1} expression gives

$$\frac{(x)(0.1056)}{(0.1082)} \cong 6.2 \times 10^{-8};\ x = [H^+] \cong 6.\underline{3}53 \times 10^{-8}\ M$$

$$pH = -\log [H^+] = -\log (6.353 \times 10^{-8}) = 7.1\underline{9}7$$

16.87 Use the Henderson-Hasselbalch equation, where $[H_2CO_3]$ = the buffer acid, $[HCO_3^-]$ = the buffer base, and K_{a1} of carbonic acid is the ionization constant.

$$pH = -\log K_a + \log \frac{[HCO_3^-]}{[H_2CO_3]} = 7.40 = -\log (4.3 \times 10^{-7}) + \log \frac{[HCO_3^-]}{[H_2CO_3]}$$

$$\log \frac{[HCO_3^-]}{[H_2CO_3]} = 7.40 - 6.366 = 1.0\underline{3}4$$

$$[HCO_3^-]/[H_2CO_3] = 1\underline{0}.81/1 = 11/1$$

16.88 Use the Henderson-Hasselbalch equation, where $[C_{18}H_{21}NO_3]$ = [Cod] = the buffer base and $[HC_{18}H_{21}NO_3^+]$ = $[HCod^+]$ = the buffer acid. The ionization constant is calculated as follows:

$$K_a = K_w \div K_b = (1.0 \times 10^{-14}) \div (6.2 \times 10^{-9}) = 1.\underline{6}1 \times 10^{-6}$$

$$pH = -\log K_a + \log \frac{[Cod]}{[HCod^+]} = 4.50 = -\log (1.61 \times 10^{-6}) + \log \frac{[Cod]}{[HCod^+]}$$

$$\log \frac{[Cod]}{[HCod^+]} = 4.50 - 5.793 = -1.2\underline{9}3$$

$$\log [HCod^+]/[Cod] = 1.293$$

$$[HCod^+]/[Cod] = 1\underline{9}.6/1 = 20/1$$

16.89 Rounded answer: $[H^+] = 4 \times 10^{-3}$ M; pH = 2.4. All the OH^- (from the NaOH) reacts with the H^+ from HCl. Calculate the stoichiometric amounts of OH^- and H^+ and subtract the mol of OH^- from the mol of H^+. Then divide the remaining H^+ by the total volume of 0.456 L + 0.285 L, or 0.741 L, to find the $[H^+]$. Then calculate the pH.

Mol H^+ = (0.10 mol HCl/L) x 0.456 L HCl = 0.0456 mol H^+

Mol OH^- = (0.15 mol NaOH/L) x 0.285 L NaOH = 0.0428 mol OH^-

Mol H^+ left = (0.0456 - 0.0428) mol H^+ = 0.0028 mol H^+

$[H^+]$ = 0.0028 mol H^+ ÷ 0.741 L total volume = $0.00\underline{3}8$ M

pH = - log $[H^+]$ = - log (0.0038) = $2.\underline{4}2$

16.90 Rounded answer: $[OH^-]$ = 0.12 M; pH = 13.08. All the H^+ (from the $HClO_4$) reacts with the OH^- from KOH. Calculate the stoichiometric amounts of OH^- and H^+ and subtract the mol of H^+ from the mol of OH^-. Then divide the remaining OH^- by the volume of 0.115 L to find the $[OH^-]$. Then calculate the pOH and pH.

Mol OH^- = 2.0 g KOH ÷ (56.1 g KOH/mol KOH) = $0.03\underline{5}7$ mol OH^-

Mol H^+ = (0.19 mol $HClO_4$/L) x 0.115 L HCl = $0.02\underline{1}9$ mol H^+

Mol OH^- left = (0.0357 - 0.0219) mol OH^- = $0.01\underline{3}8$ mol OH^-

(continued)

$[OH^-]$ = 0.0138 mol OH^- ÷ 0.115 L volume = $0.1\underline{2}0$ M

pOH = - log $[OH^-]$ = - log (0.120) = 0.9208

pH = 14.00 - pOH = 14.00 - 0.9208 = $13.0\underline{7}9$

16.91 Rounded answer: $[H^+]$ = 7.8×10^{-4} M; pH = 3.11. Use BzN to symbolize benzylamine and $BzNH^+$ to symbolize the benzylammonium cation. At the equivalence point, equal molar amounts of BzN and HCl react to form a solution of BzNHCl. Start by calculating the moles of BzN. Use this to calculate the volume of HCl needed to neutralize all of the BzN (and use the moles of BzN as the moles of $BzNH^+$ formed at the equivalence point). Add the volume of HCl to the original 0.025 L to find the total volume of solution.

(0.025 mol BzN/L) x 0.065 L = 0.00162 mol BzN

Volume HCl = 0.00162 mol HCl ÷ (0.050 mol HCl/L) = 0.0324 L

Total volume = 0.025 L + 0.0324 L = $0.05\underline{7}4$ L

$[BzNH^+]$ = (0.00162 mol $BzNH^+$ from BzN) ÷ 0.0574 L = $0.02\underline{8}2$ M

Because the $BzNH^+$ hydrolyzes to H_3O^+ and BzN, use this to calculate the $[H^+]$. Start by calculating the K_a constant of $BzNH^+$ from the K_b of its conjugate base, BzN. Then assemble the usual table of concentrations, assume x is negligible, and calculate $[H^+]$ and pH.

$K_a = K_w \div K_b = (1.00 \times 10^{-14}) \div (4.7 \times 10^{-10}) = 2.\underline{1}3 \times 10^{-5}$

Conc. (M)	$BzNH^+$	+	H_2O	$\rightleftharpoons$	BzN	+	H_3O^+
Starting	0.0282				0		0
Change	-x				+x		+x
Equilibrium	0.0282 - x				x		x

$$\frac{[BzN][H^+]}{[BzNH^+]} = K_a = \frac{(x)^2}{(0.0282 - x)} = 2.13 \times 10^{-5}$$

$$\frac{(x)^2}{(0.0282)} \cong 2.13 \times 10^{-5};\ x = [H^+] \cong 7.\underline{7}502 \times 10^{-4}\ M$$

pH = - log $[H^+]$ = - log (7.7502×10^{-4}) = $3.1\underline{1}06$

16.92 Rounded answer: $[OH^-]$ = 3.0×10^{-7} M; pH = 7.47. At the equivalence point, equal molar amounts of HSO_4^- and NaOH react to form a solution of Na_2SO_4. Start by calculating the moles of HSO_4^-. Use this to calculate the volume of NaOH needed to neutralize all of the HSO_4^- (and use the moles of HSO_4^- as the moles of SO_4^{2-} formed at the equivalence point). Add the volume of NaOH to the original 0.050 L to find the total volume of solution.

$$\text{Mol } HSO_4^- = 1.24 \text{ g } NaHSO_4 \div (120.1 \text{ g } NaHSO_4/\text{mol } NaHSO_4)$$

$$= 0.01032 \text{ mol } HSO_4^-$$

$$\text{Volume NaOH} = 0.01032 \text{ mol NaOH} \div (0.180 \text{ mol NaOH/L}) = 0.057\underline{3}3 \text{ L}$$

$$\text{Total volume} = 0.05733 \text{ L} + 0.0500 \text{ L } HSO_4^- \text{ soln} = 0.1073 \text{ L}$$

$$[SO_4^{2-}] = (0.01032 \text{ mol } SO_4^{2-} \text{ from } HSO_4^-) \div 0.1073 \text{ L} = 0.0962 \text{ M}$$

Because the SO_4^{2-} hydrolyzes to OH^- and HSO_4^-, use this to calculate the $[OH^-]$. Start by calculating the K_a constant of SO_4^{2-} from the K_a of its conjugate acid, HSO_4^-. Assemble the usual table of concentrations, assume x is negligible, and calculate $[OH^-]$ and pH.

$$K_b = K_w \div K_a = (1.0 \times 10^{-14}) \div (1.1 \times 10^{-2}) = 9.09 \times 10^{-13}$$

Conc. (M)	SO_4^{2-}	+	H_2O	$\rightleftharpoons$	HSO_4^-	+	OH^-
Starting	0.0962				0		0
Change	-x				+x		+x
Equilibrium	0.0962 - x				x		x

$$\frac{[HSO_4^-][OH^-]}{[SO_4^{2-}]} = K_b = \frac{(x)^2}{(0.0962 - x)} = 9.09 \times 10^{-13}$$

$$\frac{(x)^2}{(0.0962)} \cong 9.09 \times 10^{-13}; \; x = [OH^-] \cong 2.\underline{9}6 \times 10^{-7} \text{ M}$$

$$\text{pOH} = -\log [OH^-] = -\log (2.96 \times 10^{-7}) = 6.5\underline{2}8$$

$$\text{pH} = 14.00 - 6.528 = 7.4\underline{7}2 \text{ (ignoring ionization of water to } OH^-)$$

16.93 a. $[H^+] \cong 0.100$ M

b. Rounded answer: $[H^+] = 0.11$ M. The 0.100 M H_2SO_4 ionizes to 0.100 M H^+ and 0.100 M HSO_4^-. Assemble the usual table and substitute into the K_{a2} equilibrium-constant expression for H_2SO_4. Solve the resulting quadratic.

Conc. (M)	HSO_4^- + H_2O	⇌	SO_4^{2-}	+	H^+
Starting	0.100		0		0.100
Change	-x		+x		+x
Equilibrium	0.100 - x		x		0.100 + x

$x^2 + 0.111x - (1.10 \times 10^{-3}) = 0$

$$x = \frac{-0.111 \pm \sqrt{(0.111)^2 + (4)(1.10 \times 10^{-3})}}{2}$$

$$x = \frac{-0.111 \pm \sqrt{1.6721 \times 10^{-2}}}{2}$$

$x = (1.8308 \times 10^{-2}) \div 2 = 9.\underline{1}54 \times 10^{-3}$; $[H^+] = 0.10\underline{0} + 0.009\underline{1}54 = 0.1\underline{0}9$

16.94 a. $[H^+] \cong 0.150$ M

b. Rounded answer: $[H^+] = 0.16$ M. The 0.150 M H_2SeO_4 ionizes to 0.150 M H^+ and 0.150 M $HSeO_4^-$. Assemble the usual table and substitute into the K_{a2} equilibrium-constant expression for H_2SeO_4. Solve the resulting quadratic.

Conc. (M)	$HSeO_4^-$ + H_2O	⇌	SeO_4^{2-}	+	H^+
Starting	0.150		0		0.150
Change	-x		+x		+x
Equilibrium	0.150 - x		x		0.150 + x

$x^2 + 0.162x - (1.80 \times 10^{-3}) = 0$

$$x = \frac{-0.162 \pm \sqrt{(0.162)^2 + 4(1.8 \times 10^{-3})}}{2}$$

$$x = \frac{-0.162 \pm \sqrt{3.344 \times 10^{-2}}}{2}$$

$x = (2.08 \times 10^{-2}) \div 2 = 1.\underline{0}4 \times 10^{-2}$; $[H^+] = 0.150 + 0.0104 = 0.1\underline{6}04$

16.95 a. From the pH, calculate the H^+ ion concentration:

$$[H_3O^+] = 10^{-pH} = 10^{-5.82} = 1.\underline{5}1 \times 10^{-6} \text{ M}$$

Use the table approach, giving the starting, change, and equilibrium concentrations.

Conc. (M)	$CH_3NH_3^+$	+	H_2O	$\rightleftharpoons$	CH_3NH_2	+	H_3O^+
Starting	0.10				0		0
Change	-x				x		x
Equilibrium	0.10 - x				x		x
	≈ 0.10				= 1.51 x 10^{-6}		= 1.51 x 10^{-6}

Substituting into the equilibrium-constant expression gives

$$K_a = \frac{[CH_3NH_2][H_3O^+]}{[CH_3NH_3^+]} = \frac{x^2}{0.10 - x} \approx \frac{(1.51 \times 10^{-6})^2}{0.10} = 2.\underline{2}8 \times 10^{-11}$$

$$= 2.3 \times 10^{-11}$$

b. Now use K_w to calculate the value of K_b:

$$K_b = \frac{K_w}{K_a} = \frac{1.0 \times 10^{-14}}{2.28 \times 10^{-11}} = 4.\underline{3}8 \times 10^{-4} = 4.4 \times 10^{-4}$$

c. The equilibrium concentration of $CH_3NH_3^+$ is approximately 0.450 mol/1.00 L = 0.450 M. Use $[CH_3NH_2] \approx$ 0.250 M. From K_a we get

$$K_a = \frac{[CH_3NH_2][H_3O^+]}{[CH_3NH_3^+]} \approx \frac{(0.250)\,[H_3O^+]}{(0.450)} = 2.\underline{2}8 \times 10^{-11}$$

Solving for $[H_3O^+]$ gives

$$[H_3O^+] \approx \frac{(2.28 \times 10^{-11})(0.450)}{(0.250)} \approx 4.\underline{1}0 \times 10^{-11}$$

Thus, the pH is

$$pH = -\log(4.10 \times 10^{-11}) = 10.3\underline{8}7 = 10.39$$

16.96 a. From the pOH, calculate the OH^- ion concentration:

$$[OH^-] = 10^{-pOH} = 10^{-5.31} = 4.\underline{9}0 \times 10^{-6}\ M$$

Use the table approach, giving the starting, change, and equilibrium concentrations.

Conc. (M)	$C_6H_5COO^-$ + H_2O	⇌	C_6H_5COOH	+ OH^-
Starting	0.15		0	0
Change	-x		x	x
Equilibrium	0.15 - x		x	x
	≈ 0.15		= 4.90×10^{-6}	= 4.90×10^{-6}

Substituting into the equilibrium-constant expression gives

$$K_b = \frac{[C_6H_5COOH][OH^-]}{[C_6H_5COO^-]} = \frac{x^2}{0.15 - x} \approx \frac{(4.90 \times 10^{-6})^2}{0.15} \approx 1.\underline{6}0 \times 10^{-10}$$

$$= 1.6 \times 10^{-10}$$

b. Now use K_w to calculate the value of K_a:

$$K_a = \frac{K_w}{K_b} = \frac{1.0 \times 10^{-14}}{1.60 \times 10^{-10}} = 6.\underline{2}5 \times 10^{-5} = 6.3 \times 10^{-5}$$

c. The reaction is: $C_6H_5COOH + H_2O \rightleftharpoons C_6H_5COO^- + H_3O^+$

From the pH, calculate the H^+ and $C_6H_5COO^-$ ion concentrations:

$$[C_6H_5COO^-] = [H_3O^+] = 10^{-pH} = 10^{-2.83} = 1.\underline{4}8 \times 10^{-3}\ M$$

Use the value of K_a to calculate the solubility of C_6H_5COOH = y.

$$K_a = \frac{[C_6H_5COO^-][H_3O^+]}{[C_6H_5COOH]} = \frac{(1.48 \times 10^{-3})^2}{y} = 6.\underline{2}5 \times 10^{-5}$$

Solving for y gives the molar solubility

$$y = \frac{(1.48 \times 10^{-3})^2}{(6.25 \times 10^{-5})} = 0.03\underline{5}0 = 0.035\ M$$

16.97 a. True. Weak acids have small K_a values so most of the solute is present as undissociated molecules.

b. True. Weak acids have small K_a values so most of the solute is present as the molecule.

c. False. The hydroxide concentration only equals the hydronium concentration in neutral solutions.

d. False. If HA were a strong acid the pH would be equal to 2.

e. False. The H_3O^+ would be 0.010 if HA were a strong acid.

f. True. For every HA molecule that dissociates one H_3O^+ is generated along with one A^-.

16.98 a. True. Approximately 1% or less of most weak bases react with water.

b. True. Very little base has reacted with water. For weak bases most of the base is present as the molecule.

c. False. This is a basic solution so there will be very little H^+.

d. False. If B were a strong base, the $[OH^-]$ would be equal to 0.10 M.

e. True. $B + H_2O \rightleftharpoons BH^+ + OH^-$

f. False. Most B is present as B so little OH^- is produced upon addition of the base to water.

16.99 a. At the equivalence point, moles acid = moles base. Thus

$$\text{moles acid} = 0.115 \text{ mol/L} \times 0.03383 \text{ L} = 3.8\underline{9}0 \times 10^{-3}$$

The molar mass is

$$\text{Molar mass} = \frac{0.288 \text{ g}}{3.890 \times 10^{-3} \text{ mol}} = 74.\underline{0}4 = 74.0 \text{ g/mol}$$

b. At the 50% titration point, pH = pK_a. Since the pH measurement wasn't made, it is necessary to obtain a $[A^-]/[HA]$ ratio. The ratio can be in terms of moles, %, mL, M or ... because the units end up canceling out. First, use the pH to obtain the H_3O^+ ion concentration. Express the other concentrations in mL for convenience.

$[H_3O^+] = 10^{-4.92} = 1.2\underline{0} \times 10^{-5}$ M

$HA_o \approx 33.83$ mL

$[HA] \approx 33.83 - 17.54 = 16.29$ mL

$[A^-] \approx OH^- = 17.54$ mL

$$K_a = [H_3O^+] \times \frac{[A^-]}{[HA]} = 1.20 \times 10^{-5}\ M \times \frac{17.54\ mL}{16.29\ mL} = 1.\underline{2}9 \times 10^{-5}$$

$$= 1.3 \times 10^{-5}$$

16.100 a. At the equivalence point, moles base = moles acid. Thus

moles base = 0.135 mol/L x 0.03924 L = $5.2\underline{9}7 \times 10^{-3}$

The molar mass is

$$\text{Molar mass} = \frac{0.239\ g}{5.297 \times 10^{-3}\ mol} = 45.\underline{1}1 = 45.1\ g/mol$$

b. At the 50% titration point pK_b = pOH. Since the pH measurement wasn't made, it is necessary to obtain a $[HB^+]/[B]$ ratio. The ratio can be in terms of moles, %, mL, M or ... because the units end up canceling out. First, use the pH to obtain the OH^- ion concentration. Express the other concentrations in mL for convenience.

pOH = 14.00 - 10.73 = 3.27

$[OH^-] = 10^{-3.27} = 5.3\underline{7}0 \times 10^{-4}$

$B_o \approx 39.24$ mL

$[B] \approx 39.24 - 18.35 = 20.89$ mL

$[HB^+] \approx H^+ = 18.35$ mL

(continued)

$$K_b = [OH^-] \times \frac{[HB][OH^-]}{[B]} = 5.370 \times 10^{-4}\ M \times \frac{18.35\ mL}{20.89\ mL} = 4.7\underline{2} \times 10^{-4}$$

$$= 4.7 \times 10^{-4}$$

16.101 a. Initial pH: Use the table approach, giving the starting, change, and equilibrium concentrations.

Conc. (M)	NH_3	+	H_2O	$\rightleftharpoons$	NH_4^+	+	OH^-
Starting	0.10				0		0
Change	-x				x		x
Equilibrium	0.10 - x ≈ 0.10				x		x

Substituting into the equilibrium-constant expression gives

$$K_b = \frac{[NH_4^+][OH^-]}{[NH_3]} = \frac{x^2}{0.10 - x} \approx \frac{x^2}{0.10} = 1.8 \times 10^{-5}$$

Rearranging and solving for x gives

$$x = [OH^-] = \sqrt{(0.10)(1.8 \times 10^{-5})} = 1.\underline{3}4 \times 10^{-3}\ M$$

$$pOH = -\log(1.34 \times 10^{-3}) = 2.8\underline{7}3$$

$$pH = 14 - pOH = 14 - 2.873 = 11.1\underline{2}7 = 11.13$$

30% titration point: Express the concentrations as percents for convenience.

$[NH_3] \approx 70\%$ and $[NH_4^+] \approx 30\%$

Plug into the equilibrium-constant expression.

$$K_b = \frac{[NH_4^+][OH^-]}{[NH_3]} = \frac{(30\%)(x)}{70\%} = 1.8 \times 10^{-5}$$

Solving for x gives

$$x = [OH^-] = 1.8 \times 10^{-5} \times \frac{70\%}{30\%} = 4.\underline{2}0 \times 10^{-5}\ M$$

$$pOH = -\log(4.20 \times 10^{-5}) = 4.3\underline{7}7$$

$$pH = 14 - pOH = 14 - 4.377 = 9.6\underline{2}3 = 9.62$$

(continued)

50% titration point:

$[NH_4^+] = [NH_3]$

$\therefore\ K_b = [OH^-] = 1.8 \times 10^{-5}\ M$

$pOH = -\log(1.8 \times 10^{-5}) = 4.7\underline{4}4$

$pH = 14 - pOH = 14 - 4.744 = 9.2\underline{5}6 = 9.26$

100% titration point:

The NH_4Cl that is produced has undergone a two-fold dilution.

$\therefore\ [NH_4^+] = 0.05\underline{0}0\ M$

Use the table approach, giving the starting, change, and equilibrium concentrations.

Conc. (M)	NH_4^+ + H_2O	⇌	NH_3	+	H_3O^+
Starting	0.0500		0		0
Change	-x		x		x
Equilibrium	0.0500 - x ≈ 0.0500		x		x

Plug into the equilibrium-constant expression.

$$K_a = \frac{[NH_3][H_3O^+]}{[NH_4^+]} = \frac{x^2}{0.0500 - x} \approx \frac{x^2}{0.0500} = \frac{1.0 \times 10^{-14}}{1.8 \times 10^{-5}}$$

Solving for x gives

$$x = [H_3O^+] = \sqrt{0.0500 \times \frac{1.0 \times 10^{-14}}{1.8 \times 10^{-5}}} = 5.\underline{2}7 \times 10^{-6}\ M$$

$pH = -\log(5.27 \times 10^{-6}) = 5.2\underline{7}8 = 5.28$

b. The solution is acidic because the NH_4^+ ion reacts with water to produce acid. Ammonium chloride is the salt of a weak base and a strong acid.

16.102 a. Initial pH: Use the table approach, giving the starting, change, and equilibrium concentrations. Here, A = $CH_3CH_2COO^-$.

Conc. (M)	HA	+	H_2O	$\rightleftharpoons$	H_3O^+	+	A^-
Starting	0.15				0		0
Change	-x				x		x
Equilibrium	0.15 - x ≈ 0.15				x		x

Substituting into the equilibrium-constant expression gives

$$K_a = \frac{[H_3O^+][A^-]}{[HA]} = \frac{x^2}{0.15 - x} \approx \frac{x^2}{0.15} = 1.3 \times 10^{-5}$$

Rearranging and solving for x gives

$$x = [H_3O^+] = \sqrt{(0.15)(1.3 \times 10^{-5})} = 1.\underline{4}0 \times 10^{-3} \text{ M}$$

$$pH = -\log(1.40 \times 10^{-3}) = 2.8\underline{5}3 = 2.85$$

50% titration point:

$$[H_3O^+] = [A^-] = K_a = 1.3 \times 10^{-5} \text{ M}$$

$$pH = -\log(1.3 \times 10^{-5}) = 4.8\underline{8}6 = 4.89$$

60% titration point: Express the concentrations as percents for convenience.

$$[HA] \approx 40\% \text{ and } [A^-] \approx 60\%$$

Plug into the equilibrium-constant expression.

$$K_a = \frac{[H_3O^+][A^-]}{[HA]} = \frac{(60\%)(x)}{40\%} = 1.3 \times 10^{-5}$$

Solving for x gives

$$x = [H_3O^+] = 1.3 \times 10^{-5} \times \frac{40\%}{60\%} = 8.\underline{6}7 \times 10^{-6} \text{ M}$$

$$pH = -\log(8.67 \times 10^{-6}) = 5.0\underline{6}2 = 5.06$$

(continued)

100% titration point:

The salt produced has undergone a 2-fold dilution.

∴ $[A^-] = 0.07\underline{5}0$ M

Use the table approach, giving the starting, change, and equilibrium concentrations.

Conc. (M)	A^-	+	H_2O	⇌	HA	+	OH^-
Starting	0.0750				0		0
Change	-x				x		x
Equilibrium	0.0750 - x ≈ 0.0750				x		x

Plug into the equilibrium-constant expression.

$$K_b = \frac{[HA][OH^-]}{[A^-]} = \frac{x^2}{0.0750 - x} \approx \frac{x^2}{0.0750} = \frac{1.0 \times 10^{-14}}{1.3 \times 10^{-5}}$$

Solving for x gives

$$x = [OH^-] = \sqrt{0.0750 \times \frac{1.0 \times 10^{-14}}{1.3 \times 10^{-5}}} = 7.\underline{5}9 \times 10^{-6} \text{ M}$$

$$\text{pOH} = -\log(7.59 \times 10^{-6}) = 5.1\underline{2}0$$

$$\text{pH} = 14 - \text{pOH} = 14 - 5.1\underline{2}0 = 8.8\underline{8}0 = 8.88$$

b. The solution is basic because the propionate ion reacts with water to produce OH^-. Potassium propionate is the salt of a weak acid and a strong base.

16.103 a. Select the conjugate pair that has a pK_a value closest to a pH of 2.88.

$pK_a\ (H_2C_2O_4) = 1.25$

$pK_a\ (H_3PO_4) = 2.16$

pK_a (HCOOH) = 3.77

Therefore, the best pair is H_3PO_4 and $H_2PO_4^-$.

b. Using the Henderson-Hasselbalch equation,

$$pH = pK_a + \log [A^-]/[HA]$$

$$2.88 = 2.16 + \log [A^-]/[HA]$$

$$\log [A^-]/[HA] = 0.72; \quad [A^-]/[HA] = 10^{0.72} = 5.\underline{2}4$$

Therefore, 5.2 times as much conjugate base is needed than acid. Since the concentrations of H_3PO_4 and $H_2PO_4^-$ are equal, we could use 30.0 mL $H_2PO_4^-$.

$$\text{mL } H_3PO_4 = 30.0 \text{ mL} \times \frac{1 \text{ mL } H_3PO_4}{5.\underline{2}4 \text{ mL } H_2PO_4^-} = 5.\underline{7}7 = 5.8 \text{ mL}$$

Combine these two volumes together and then add enough water to make 50 mL of solution. Of course it is arbitrary as to how much $H_2PO_4^-$ that we will start with.

16.104 a. Select the conjugate pair that has a pK_a value closest to a pH of 6.96.

$$pK_a\ (NH_4^+) = 9.25$$

$$pK_a\ (H_2CO_3) = 6.37$$

$$pK_a\ (H_2PO_4^-) = 7.21$$

Therefore, the best pair is $H_2PO_4^-$ and HPO_4^{2-}.

b. Using the Henderson-Hasselbalch equation,

$$pH = pK_a + \log [A^-]/[HA]$$

$$6.96 = 7.21 + \log [A^-]/[HA]$$

$$\log [A^-]/[HA] = -0.25; \quad [A^-]/[HA] = 10^{0.25} = 0.5\underline{6}2$$

Therefore, 0.56 times as much conjugate base is needed than acid. Since the concentrations of $H_2PO_4^-$ and HPO_4^{2-} are equal, we could use 50.0 mL of the $H_2PO_4^-$ solution.

$$\text{mL } HPO_4^{2-} = 50.0 \text{ mL} \times \frac{0.5\underline{6}2 \text{ mL } HPO_4^{2-}}{1 \text{ mL } H_2PO_4^-} = 2\underline{8}.1 = 28 \text{ mL}$$

Combine these two volumes and then add enough water to make 100 mL of solution. Of course it is arbitrary as to what volume of $H_2PO_4^-$ we will start with.

16.105 a. At the equivalence point, moles base = moles acid. Therefore

$$M_{NH_2OH} \times 25.0 \text{ mL} = 0.150 \text{ M} \times 35.8 \text{ mL}$$

$$M_{NH_2OH} = \frac{0.150 \text{ M} \times 35.8 \text{ mL}}{25.0 \text{ mL}} = 0.21\underline{4}8 = 0.215 \text{ M}$$

b. Concentration of NH_3OH^+ at the equivalence point is

$$[NH_3OH^+] = \frac{(25.0 \text{ mL})(0.215 \text{ M})}{25.0 \text{ mL} + 35.8 \text{ mL}} = 0.088\underline{4}0 \text{ M}$$

$$K_a = \frac{K_w}{K_b} = \frac{1.0 \times 10^{-14}}{1.1 \times 10^{-8}} = 9.\underline{0}9 \times 10^{-7}$$

Use the table approach, giving the starting, change, and equilibrium concentrations.

Conc. (M)	NH_3OH^+ + H_2O ⇌	NH_2OH +	H_3O^+
Starting	$0.088\underline{4}0$	0	0
Change	$-x$	x	x
Equilibrium	$0.088\underline{4}0 - x \approx 0.088\underline{4}0$	x	x

Substituting into the equilibrium-constant expression gives

$$K_a = \frac{[NH_2OH][H_3O^+]}{[NH_3OH^+]} = \frac{x^2}{0.0.08840 - x} \approx \frac{x^2}{0.08840} = 9.\underline{0}9 \times 10^{-7}$$

Rearranging and solving for x gives

$$x = [H_3O^+] = \sqrt{(0.08840)(9.\underline{0}9 \times 10^{-7})} = 2.\underline{8}3 \times 10^{-4} \text{ M}$$

$$pH = -\log(2.\underline{8}4 \times 10^{-4}) = 3.5\underline{4}7 = 3.55$$

c. Need an indicator to change pH around pH of 3-4. Therefore, the appropriate indicator is bromophenol blue. Select an indicator that changes color around the equivalence point.

16.106 a. At the equivalence point, moles base = moles acid. The molar mass of $NaHCO_3$ is 84.01 g/mol. Therefore

$$\text{mol base} = \text{mol acid} = 0.562 \text{ g } NaHCO_3 \times \frac{1 \text{ mol } NaHCO_3}{84.01 \text{ g } NaHCO_3}$$

$$= 6.6\underline{9}0 \times 10^{-3} \text{ mol}$$

The molarity of the base is

$$M_{NaOH} = \frac{6.6\underline{9}0 \times 10^{-3} \text{ mol}}{0.04236 \text{ L}} = 0.15\underline{7}9 = 0.158 \text{ M}$$

b. The concentration of CO_3^{2-} at the equivalence point is

$$[CO_3^{2-}] = \frac{6.6\underline{9}0 \times 10^{-3} \text{ mol}}{0.02500 \text{ L} + 0.04236 \text{ L}} = 0.099\underline{3}2 \text{ M}$$

$$K_b = \frac{K_w}{K_a} = \frac{1.0 \times 10^{-14}}{4.8 \times 10^{-11}} = 2.\underline{0}8 \times 10^{-4}$$

Use the table approach, giving the starting, change, and equilibrium concentrations.

Conc. (M)	CO_3^{2-} + H_2O	$\rightleftharpoons$	HCO_3^-	+	OH^-
Starting	0.099$\underline{3}$2		0		0
Change	-x		x		x
Equilibrium	0.099$\underline{3}$2 - x ≈ 0.099$\underline{3}$2		x		x

Substituting into the equilibrium-constant expression gives

$$K_b = \frac{[HCO_3^-][OH^-]}{[CO_3^{2-}]} = \frac{x^2}{0.0.09932 - x} \approx \frac{x^2}{0.09932} = 2.\underline{0}8 \times 10^{-4}$$

Rearranging and solving for x gives

$$x = [OH^-] = \sqrt{(0.09932)(2.\underline{0}8 \times 10^{-4})} = 4.\underline{5}5 \times 10^{-3} \text{ M}$$

$$pOH = -\log(4.\underline{5}5 \times 10^{-3}) = 2.3\underline{4}2$$

$$pH = 14 - pOH = 14 - 2.3\underline{4}2 = 11.6\underline{5}8 = 11.66$$

c. Alizarin yellow R. It changes color around the equivalence point.

16.107 a. $H_3O^+(aq) + NH_3(aq) \rightarrow NH_4^+(aq) + H_2O(l)$

b. $NH_4^+(aq) + OH^-(aq) \rightarrow NH_3(aq) + H_2O(l)$

c. Initial:

$$\text{mol } NH_3 = 1.0 \text{ mol/L} \times 0.100 \text{ L} = 0.1\underline{0}0 = 0.10 \text{ mol}$$

$$\text{mol } NH_4^+ = 0.50 \text{ mol/L} \times 0.100 \text{ L} \times \frac{2 \text{ mol } NH_4^+}{1 \text{ mol } (NH_4)_2SO_4} = 0.1\underline{0}0 = 0.10 \text{ mol}$$

After addition of HCl: The HCl is completely consumed by producing 0.01$\underline{0}$0 mol of NH_4^+. The amount of NH_3 is decreased by 0.01$\underline{0}$0 mol.

$$\text{mol HCl added} = 1.00 \text{ mol/L} \times 0.0100 \text{ L} = 0.010\underline{0}0 \text{ mol}$$

$$\text{mol } NH_3 = 0.1\underline{0}0 - 0.010\underline{0}0 = 0.09\underline{0}0 \text{ mol}$$

$$\text{mol } NH_4^+ = 0.1\underline{0}0 + 0.010\underline{0}0 = 0.1\underline{1}0 \text{ mol}$$

The reaction is: $NH_3 + H_2O \rightleftharpoons NH_4^+ + OH^-$

Substituting into the equilibrium-constant expression gives

$$K_b = \frac{[NH_4^+][OH^-]}{[NH_3]} = \frac{(0.110)[OH^-]}{0.0\underline{9}0} = 1.8 \times 10^{-5}$$

Rearranging and solving for $[OH^-]$ gives

$$[OH^-] = 1.8 \times 10^{-5} \times \frac{0.0\underline{9}0}{0.110} = \underline{1}.47 \times 10^{-5} \text{ M}$$

$$\text{pOH} = -\log(\underline{1}.47 \times 10^{-5}) = 4.\underline{8}3$$

$$\text{pH} = 14 - \text{pOH} = 14 - 4.\underline{8}3 = 9.\underline{1}7 = 9.2$$

d. This is a buffer system, so the ratio of NH_4^+/NH_3 has not changed very much.

16.108 a. Use the table approach, giving the starting, change, and equilibrium concentrations. The initial concentration of formic acid is
0.150 mol/0.425 L = 0.35<u>2</u>9 M

Conc. (M)	HCOOH + H_2O $\rightleftharpoons$	H_3O^+ +	$HCOO^-$
Starting	0.35<u>2</u>9	0	0
Change	-x	x	x
Equilibrium	0.35<u>2</u>9 - x ≈ 0.35<u>2</u>9	x	x

Substituting into the equilibrium-constant expression gives

$$K_a = \frac{[HCOO^-][H_3O^+]}{[HCOOH]} \approx \frac{x^2}{0.3529} = 1.7 \times 10^{-4}$$

Rearranging and solving for x gives

$$x = [H_3O^+] = [HCOO^-] = \sqrt{(0.3529)(1.7 \times 10^{-4})} = 7.\underline{7}4 \times 10^{-3} = 7.7 \times 10^{-3}\ M$$

b. To decrease the $HCOO^-$ concentration by a factor of ten, the concentration must be reduced to 7.74×10^{-4} M. The change in equilibrium is

$$HCOOH(aq) + H_2O(l) \rightleftharpoons H_3O^+(aq) + HCOO^-(aq)$$

$$0.3529 - (y + 7.74 \times 10^{-4}) \approx 0.3529 \quad y + 7.74 \times 10^{-4} \approx y \quad 7.74 \times 10^{-4}$$

$$[H_3O^+] = \frac{[HCOOH]K_a}{[HCOO^-]} = \frac{(0.3529)(1.7 \times 10^{-4})}{7.74 \times 10^{-4}} = 0.07\underline{7}5\ M$$

0.07<u>7</u>5 M x 425 mL = 2.00 M x V

V = 16.4 = 16 mL

c. This is a common ion effect. An increase in the H_3O^+ concentration shifts the equilibrium so that the percent dissociation of HCOOH is decreased.

16.109 a. From the pH $[H_3O^+] = 10^{-7.44} = 3.\underline{6}3 \times 10^{-8}$ M. The reaction is

$$H_2PO_4^- (aq) + H_2O \rightleftharpoons H_3O^+ (aq) + HPO_4^{2-} (aq) \qquad K_2 = 6.2 \times 10^{-8}$$

$$K_2 = \frac{[HPO_4^{2-}][H_3O^+]}{[H_2PO_4^-]}; \quad \frac{[H_2PO_4^-]}{[HPO_4^{2-}]} = \frac{[H_3O^+]}{K_2} = \frac{3.63 \times 10^{-8}}{6.2 \times 10^{-8}}$$

$$= \frac{0.5\underline{8}55}{1} = 0.59 \text{ to } 1$$

Or using the Henderson-Hasselbalch equation

$pH = pK_a + \log \text{base/acid}$

$7.44 = 7.21 + \log \text{base/acid}$

$\log \text{base/acid} = 0.23$; $\text{base/acid} = 1.\underline{7}1$

$\therefore$ $\text{acid/base} = 1/1.71 = 0.5\underline{8}57/1$

b. $$[H_3O^+] = K_2 \times \frac{[HPO_4^{2-}]}{[H_2PO_4^-]}$$

If HPO_4^{2-} = 1.00 mol, then when 25% of it is converted into $H_2PO_4^-$, 0.75 mol remain, and $H_2PO_4^-$ becomes $0.5\underline{8}6 + 0.25 = 0.8\underline{3}6$ mol. Therefore,

$$[H_3O^+] = 6.2 \times 10^{-8} \times \frac{0.836}{0.75} = 6.\underline{9}1 \times 10^{-8}$$

$$pH = -\log (6.\underline{9}1 \times 10^{-8}) = 7.1\underline{6}1 = 7.16$$

c. Assume that $H_2PO_4^-$ = $0.5\underline{8}6$ mol. Then $H_2PO_4^- = 0.5\underline{8}6 - 0.15 \times 0.5\underline{8}6$

$$= 0.85 \times 0.5\underline{8}6 = 0.4\underline{9}8 \text{ mol left}$$

The moles of HPO_4^{2-} = $1.00 + 0.15 \times 0.5\underline{8}6 = 1.0\underline{8}8$ mol.

$$[H_3O^+] = 6.2 \times 10^{-8} \times \frac{0.498}{1.088} = 2.\underline{8}4 \times 10^{-8} \text{ M}$$

$$pH = -\log (2.84 \times 10^{-8}) = 7.5\underline{4}7 = 7.55$$

16.110 a. From the pH $[H_3O^+] = 10^{-7.42} = 3.\underline{8}0 \times 10^{-8}$ M. The reaction is

$$H_2CO_3\,(aq) + H_2O \rightleftharpoons H_3O^+\,(aq) + HCO_3^-\,(aq) \qquad K_1 = 4.3 \times 10^{-7}$$

$$K_1 = \frac{[HCO_3^-][H_3O^+]}{[H_2CO_3]};\quad \frac{[H_2CO_3]}{[HCO_3^-]} = \frac{[H_3O^+]}{K_1} = \frac{3.80 \times 10^{-8}}{4.3 \times 10^{-7}}$$

$$= \frac{0.08\underline{8}4}{1} = 0.088 \text{ to } 1$$

Or using the Henderson-Hasselbalch equation

$pH = pK_a + \log \text{base/acid}$

$7.42 = 6.3\underline{6}7 + \log \text{base/acid}$

$\log \text{base/acid} = 1.0\underline{5}3$; $\text{base/acid} = 1\underline{1}.30$

$\text{acid/base} = 1/1\underline{1}.30 = 0.08\underline{8}5$

b. If $HCO_3^- = 1.00$ mol, then when 15% of it is converted into H_2CO_3, 0.85 mol remain, and H_2CO_3 becomes $0.08\underline{8}4 + 0.15 = 0.2\underline{3}8$ mol. Therefore,

$$[H_3O^+] = 4.3 \times 10^{-7} \times \frac{0.238}{0.85} = 1.\underline{2}0 \times 10^{-7}$$

$$pH = -\log(1.\underline{2}0 \times 10^{-7}) = 6.9\underline{2}0 = 6.92$$

c. Assume that $H_2CO_3 = 0.08\underline{8}4$ mol. Then $HCO_3^- = 0.08\underline{8}4 - 0.25 \times 0.08\underline{8}4$

$$= 0.75 \times 0.08\underline{8}4 = 0.06\underline{6}3 \text{ mol left}$$

The moles of $HCO_3^- = 1.000 + 0.25 \times 0.08\underline{8}4 = 1.02\underline{2}1$ mol.

$$[H_3O^+] = 4.3 \times 10^{-7} \times \frac{0.0663}{1.0221} = 2.\underline{7}9 \times 10^{-8}$$

$$pH = -\log(2.\underline{7}9 \times 10^{-8}) = 7.5\underline{4}7 = 7.55$$

16.111 a.

$$H_2A + H_2O \rightleftharpoons H_3O^+ + HA^- \qquad K_1$$

$$HA^- + H_2O \rightleftharpoons H_3O^+ + A^{2-} \qquad K_2$$

b. $H_2A >> H_3O^+ = HA^- >> A^{2-}$

c. $H_2A + H_2O \rightleftharpoons H_3O^+ + HA^-$ $\quad K_1 = 1.0 \times 10^{-3}$

$0.0250 - x \qquad x \qquad x$

$$K_1 = \frac{[HA^-][H_3O^+]}{[H_2A]} = \frac{x^2}{0.0250 - x} = 1.0 \times 10^{-3}$$

Rearranging into a quadratic and solving for x gives

$$x^2 + (1.0 \times 10^{-3})x + (-2.50 \times 10^{-5}) = 0$$

$$x = \frac{-(1.0 \times 10^{-3}) \pm \sqrt{(1.0 \times 10^{-3})^2 - (4)(1)(-2.50 \times 10^{-5})}}{2}$$

$$x = [H_3O^+] = [HA^-] = 4.\underline{5}2 \times 10^{-3} = 4.5 \times 10^{-3} \text{ M}$$

$$pH = -\log(4.\underline{5}2 \times 10^{-3}) = 2.3\underline{4}4 = 2.34$$

$$[H_2A] = 0.0250 - x = 0.0250 - 4.52 \times 10^{-3} = 0.020\underline{4}8 = 0.0205 \text{ M}$$

d. $HA^- + H_2O \rightleftharpoons H_3O^+ + A^{2-}$ $\quad K_2 = 4.6 \times 10^{-5}$

$4.52 \times 10^{-3} - y \approx 4.52 \times 10^{-3} \qquad 4.52 \times 10^{-3} + y \approx 4.52 \times 10^{-3} \qquad y$

$$[A^{2-}] = \frac{[HA^-]K_2}{[H_3O^+]} = \frac{(4.52 \times 10^{-5})(4.6 \times 10^{-5})}{(4.52 \times 10^{-5})} = 4.6 \times 10^{-5} \text{ M}$$

16.112 a. $H_2A + H_2O \rightleftharpoons H_3O^+ + HA^-$, $\quad K_1$

$HA^- + H_2O \rightleftharpoons H_3O^+ + A^{2-}$, $\quad K_2$

$H_2A + 2H_2O \rightleftharpoons 2H_3O^+ + A^{2-}$, $\quad K_1 \times K_2$

b. $H_2A >> H_3O^+ = HA^- >> A^{2-}$

c. $H_2A + H_2O \rightleftharpoons H_3O^+ + HA^-$ $K_1 = 4.0 \times 10^{-4}$

$0.0100 - x$ x x

$$K_1 = \frac{[HA^-][H_3O^+]}{[H_2A]} = \frac{x^2}{0.0100 - x} = 4.0 \times 10^{-4}$$

Rearranging into a quadratic and solving for x gives

$$x^2 + (4.0 \times 10^{-4})x + (-4.00 \times 10^{-6}) = 0$$

$$x = \frac{-(4.0 \times 10^{-4}) \pm \sqrt{(4.0 \times 10^{-4})^2 - (4)(1)(-4.00 \times 10^{-6})}}{2}$$

$x = [H_3O^+] = [HA^-] = 1.\underline{8}1 \times 10^{-3} = 1.8 \times 10^{-3}$ M

$pH = -\log(1.\underline{8}1 \times 10^{-3}) = 2.7\underline{4}2 = 2.74$

$[H_2A] = 0.0100 - x = 0.0100 - 1.\underline{8}1 \times 10^{-3} = 0.008\underline{1}9 = 0.0082$ M

d. $HA^- + H_2O \rightleftharpoons H_3O^+ + A^{2-}$ $K_2 = 4.6 \times 10^{-5}$

$1.81 \times 10^{-3} - y$ $1.81 \times 10^{-3} + y$ y

$\approx 1.81 \times 10^{-3}$ $\approx 1.81 \times 10^{-3}$

$$[A^{2-}] = \frac{[HA^-]K_2}{[H_3O^+]} = \frac{(1.81 \times 10^{-3})(9.0 \times 10^{-6})}{(1.81 \times 10^{-3})} = 9.0 \times 10^{-6} \text{ M}$$

16.113 a. From the pH, $[H_3O^+] = 10^{-4.45} = 3.\underline{5}5 \times 10^{-5}$ M. The $NaCH_3COO$ is generated by the acid-base reaction resulting in the formation of a buffer solution. The reaction is

$CH_3COOH + H_2O \rightleftharpoons H_3O^+ + CH_3COO^-$ $K_a = 1.7 \times 10^{-5}$

$0.15 - x$ x x

$$K_a = \frac{[H_3O^+][CH_3COO^-]}{[CH_3COOH]}; \quad \frac{[CH_3COO^-]}{[CH_3COOH]} = \frac{K_a}{[H_3O^+]}$$

(continued)

$$\frac{x}{0.15 - x} = \frac{1.7 \times 10^{-5}}{3.55 \times 10^{-5}}; \; x = \text{mol } CH_3COO^-$$

Solving for x gives x = 0.04$\underline{8}$6 mol. Therefore, the molar concentration is

$$[CH_3COO^-] = 0.04\underline{8}6 \text{ mol}/0.375 \text{ L} = 0.1\underline{2}96 = 0.13 \text{ M}$$

b. moles OH^- = moles CH_3COO^-

0.04$\underline{8}$6 mol = 0.25 M x V_{NaOH}

V_{NaOH} = 0.0486 mol/0.25 M = 0.1$\underline{9}$4 = 0.19 L

c. The volume of the original acid is V = 0.375 L - 0.1$\underline{9}$4 L = 0.1$\underline{8}$1 L. Therefore, the concentration of the original acid is

$$[CH_3COOH] = 0.15 \text{ mol}/0.181 \text{ L} = 0.8\underline{2}9 = 0.83 \text{ M}$$

16.114 a. From the pH, $[H_3O^+] = 10^{-4.56} = 2.\underline{7}5 \times 10^{-5}$ M. The CH_3COOH is generated by the acid-base reaction resulting in the formation of a buffer solution. The reaction is

$$\begin{array}{lllll} CH_3COO^- & + \; H_3O^+ & \rightleftharpoons & H_2O & + \; CH_3COOH \quad K_a = 1.7 \times 10^{-5} \\ 0.10 - x & & & x & \quad x \end{array}$$

$$K_a = \frac{[H_3O^+][CH_3COO^-]}{[CH_3COOH]}; \; \frac{[CH_3COO^-]}{[CH_3COOH]} = \frac{K_a}{[H_3O^+]}$$

$$\frac{0.10 - x}{x} = \frac{1.7 \times 10^{-5}}{2.75 \times 10^{-5}}; \; x = \text{mol } CH_3COOH$$

Solving for x gives x = 0.06$\underline{1}$8 mol. Therefore, the molar concentration is

$$[CH_3COO^-] = 0.06\underline{1}8 \text{ mol}/0.650 \text{ L} = 0.09\underline{5}1 = 0.095 \text{ M}$$

b. moles HCl = moles CH_3COOH

0.06$\underline{1}$8 mol = 0.15 M x V_{NaOH}

V_{NaOH} = 0.0618 mol/0.15 M = 0.4$\underline{1}$2 = 0.41 L

c. The volume of the original acid is V = 0.650 L - 0.4<u>1</u>2 L = 0.2<u>3</u>8 L. Therefore, the concentration of the original acid is

$$[CH_3COOH] = 0.10 \text{ mol}/0.238 \text{ L} = 0.4\underline{2}0 = 0.42 \text{ M}$$

■ Cumulative-Skills Problems
(require skills from previous chapters)

16.115 Use the pH to calculate $[H^+]$, and then use the K_a of 1.7×10^{-5} and the K_a expression to calculate the molarity of acetic acid (HAc), assuming ionization is negligible. Convert molarity to mass percentage using the formula weight of 60.05 g/mol of HAc.

$$[H^+] = \text{antilog}(-2.45) = 3.\underline{5}48 \times 10^{-3} \text{ M}$$

Write the equilibrium-constant expression in terms of chemical symbols and then substitute the x and the (0.003548 M) terms into the expression:

$$\frac{[H^+][Ac^-]}{[HAc]} = K_a \cong \frac{(0.003548)^2}{(x)} \cong 1.7 \times 10^{-5}$$

Solve the equation for x, assuming that 0.003548 is much smaller than x.

$$x = (0.003548)^2 \div (1.7 \times 10^{-5}) \cong [HAc] \cong 0.7\underline{4}04 \text{ M}$$

$$(0.7404 \text{ mol HAc/L}) \times (60.05 \text{ g/mol}) \times (1 \text{ L}/1090 \text{ g})(100\%) = 4.\underline{0}78 = 4.1\% \text{ HAc}$$

16.116 Use the pH to calculate $[OH^-]$, and then use the K_b of 1.8×10^{-5} and the K_b expression to calculate the molarity of ammonia (NH_3), assuming ionization is negligible. Convert molarity to mass percentage using the formula weight of 17.03 g/mol of NH_3.

$$[OH^-] = \text{antilog}[-(14.00 - 11.87)] = 7.\underline{4}13 \times 10^{-3} \text{ M}$$

Write the equilibrium-constant expression in terms of chemical symbols, and then substitute the x and the (0.007413 M) terms into the expression:

$$\frac{[NH_4^+][OH^-]}{[NH_3]} = K_b \cong \frac{(0.007413)^2}{(x)} \cong 1.8 \times 10^{-5}$$

(continued)

Solve the equation for x, assuming that 0.007413 is much smaller than x.

$$x = (0.007413)^2 \div (1.8 \times 10^{-5}) \cong [NH_3] \cong 3.053\ M$$

$$(3.053\ \text{mol}\ NH_3/L) \times (17.03\ g/\text{mol}) \times (1\ L/1000\ g) \times 100\% = 5.\underline{1}99 = 5.2\%\ NH_3$$

16.117 Find the $[H^+]$ and $[C_2H_3O_2^-]$ by solving for the approximate $[H^+]$, noting that x is much smaller than the starting 0.92 M of acetic acid. The usual table is used but is not shown; only the final setup for $[H^+]$ is shown. Use $K_f = 1.858^oC/m$ for the constant for water.

$$[H^+] = [C_2H_3O_2^-] = (1.7 \times 10^{-5} \times 0.92)^{1/2} \cong 0.003\underline{9}54\ M$$

$$\text{Total molarity of acid + ions} = 0.92 + 0.003954 = 0.9\underline{2}39$$

$$\text{Molality} = m = 0.9239\ \text{mol} \div (1.000\ L \times 0.953\ kg\ H_2O/L) = 0.9695\ m$$

$$\text{Freezing point} = -\Delta T_f = -K_f c_m = (-1.858^oC/m) \times 0.9695\ m = -1.801 = -1.8^oC$$

16.118 Find the $[NH_4^+]$ and $[OH^-]$ by solving for the approximate $[OH^-]$, noting that x is much smaller than the starting 0.87 M of ammonia. The usual table is used but is not shown; only the final setup for $[OH^-]$ is shown. Use $K_f = 1.858^oC/m$ for the constant for water.

$$[NH_4^+] = [OH^-] = (1.8 \times 10^{-5} \times 0.87)^{1/2} \cong 0.003\underline{9}57\ M$$

$$\text{Total molarity of base and ions} = 0.87 + 0.003957 = 0.8739\ M$$

$$kg\ NH_3/L = 0.87\ \text{mol}/L \times 0.017\ kg/\text{mol} = 0.0148\ kg\ NH_3/L$$

$$kg\ H_2O/L = 0.992\ kg\ \text{soln}/L - 0.0148\ kg\ NH_3/L = 0.977\ kg\ H_2O/L$$

$$\text{Molality} = m = 0.8739\ \text{mol} \div (1.000\ L \times 0.977\ kg\ \text{solv}/L) = 0.8944\ m$$

$$\text{Freezing point} = -\Delta T_f = -K_f c_m = (-1.858^oC/m) \times 0.8944\ m = -1.\underline{6}62 = -1.7^oC$$

16.119 The $[H^+]$ = -antilog (-4.35) = $4.\underline{4}6 \times 10^{-5}$ M. Note that 0.465 L of 0.0941 M NaOH will produce 0.043756 mol of acetate, Ac^-, ion. Rearranging the K_a expression for acetic acid (HAc) and Ac^-, and canceling the volume in the mol/L of each, you obtain:

$$\frac{[HAc]}{[Ac^-]} = \frac{[H^+]}{K_a} = \frac{4.46 \times 10^{-5}}{1.7 \times 10^{-5}} = \frac{2.63}{1.00} \cong \frac{x \text{ mol HAc}}{0.043756 \text{ mol Ac}^-}$$

$$x = 0.1\underline{1}49 \text{ mol HAc}$$

Total mol HAc added = 0.1149 + 0.043756 = 0.1587 mol HAc

Mol/L of pure HAc = 1049 g HAc/L x (1 mol HAc/60.05 g) = $17.4\underline{6}7$ mol/L

L of pure HAc needed = 0.1587 mol HAc x (L/17.467 mol) = $0.009\underline{0}86$ L (9.1 mL)

16.120 The $[H^+]$ = -antilog (-3.50) = $3.\underline{1}6 \times 10^{-4}$ M. Note that 0.325 L of 0.0857 M NaOH will produce 0.027852 mol of formate, Fo^-, ion. Rearranging the K_a expression for formic acid (HFo) and Fo^-, and canceling the volume in the mol/L of each, you obtain:

$$\frac{[HFo]}{[Fo^-]} = \frac{[H^+]}{K_a} = \frac{3.16 \times 10^{-4}}{1.7 \times 10^{-4}} = \frac{1.858}{1.00} \cong \frac{x \text{ mol HFo}}{0.027852 \text{ mol Fo}^-}$$

$$x = 0.05175 \text{ mol HFo}$$

Total mol HFo added = 0.05175 + 0.027852 = 0.079602 mol HFo

Mol/L of pure HFo = 1220 g HFo/L x (1 mol HFo/46.027 g) = $26.5\underline{0}6$ mol/L

L of pure HFo needed = $0.07\underline{9}602$ mol HAc x (L/26.506 mol) = 0.003003 L (3.0 mL)

■ Solution to Conceptual Problem

16.121 There are many equilibria involved in blood chemistry. The major system that controls pH buffering is the carbonic/bicarbonate buffer system show below:

$$H^+(aq) + HCO_3^-(aq) \rightleftharpoons H_2CO_3(aq) \rightleftharpoons H_2O(l) + CO_2(g) \quad (1)$$

The result of this equilibrium is to maintain the pH of the body at ~7.4. A complementary equilibrium involves oxygen transport through the body. In this case, hemoglobin (Hb) reversibly binds both O2 and H+ according to the following equilibrium:

$$HbH^+ + O_2 \rightleftharpoons HbO_2 + H^+ \quad (2)$$

When the body undergoes strenuous exertion, many factors are involved adjusting the equilibrium above to ensure sufficient deliver of oxygen to relevant tissues. Based on your knowledge of Le Chatelier's principle and acid-base equilibria, what changes occur in these equilibria and how does the body further aid oxygen delivery?

Discussion

Students should have little difficulty with reasoning from Le Chatelier's principle if they have some basic knowledge of respiration. Oxygen gas combines with hemoglobin in the lungs. The high partial pressure of oxygen gas in the lungs shifts the hemoglobin equilibrium to the right, increasing the hydrogen-ion concentration. This increase in hydrogen-ion concentration affects the bicarbonate buffer equilibrium, shifting it to the right. Thus, carbon dioxide gas is released. Overall, in the lungs, oxygen gas is taken up and carbon dioxide is released. If the student has done some reading on the subject, he or she may find out that the body senses an increase of carbon dioxide concentration (from physical exertion) in the lung tissue and induces fast, deep breathing (hyperventilation). This lowers the carbon dioxide concentration and increases the partial pressure of oxygen, thus aiding respiration.

In the blood capillaries of muscle tissue, the shifts in equilibria are the opposite. An increase in carbon dioxide concentration (caused by physical exertion) shifts the bicarbonate buffer equilibrium to the left, and therefore increases the hydrogen-ion concentration. This increase in hydrogen-ion concentration shifts the hemoglobin equilibrium to the left. Thus, the oxyhemoglobin, HbO_2, releases oxygen to the muscle tissue. (This oxygen is picked up by myoglobin in the muscle tissue.)
References: "Acid-base control and acid-base disorders." in Clinical Chemistry, edited by Lawrence A. Kaplan and Amadeo J. Pesce, Mosby, St. Louis, p. 388, 1984; Biochemistry, 3rd edition, by Lubert Stryer, Freeman, San Francisco, p. 156, 1988.

17. SOLUBILITY AND COMPLEX-ION EQUILIBRIA

■ Solutions to Exercises

Note on answers to equilibrium calculations: The rounded answer (with correct number of significant figures) is always given first. After the stepwise solution, the numerical answer is given again, but with one nonsignificant figure. (The rightmost significant figure is underlined.)

17.1 a. $BaSO_4(s) \rightleftharpoons Ba^{2+}(aq) + SO_4^{2-}(aq)$; $K_{sp} = [Ba^{2+}][SO_4^{2-}]$

b. $Fe(OH)_3(s) \rightleftharpoons Fe^{3+}(aq) + 3OH^-(aq)$; $K_{sp} = [Fe^{3+}][OH^-]^3$

c. $Ca_3(PO_4)_2(s) \rightleftharpoons 3Ca^{2+}(aq) + 2PO_4^{3-}(aq)$; $K_{sp} = [Ca^{2+}]^3[PO_4^{3-}]^2$

17.2 Rounded answer: $K_{sp} = 1.8 \times 10^{-10}$ (correct no. of sig. figs.). Calculate the molar solubility. Then assemble the usual concentration table and substitute from it the equilibrium concentrations into the equilibrium-constant expression. (Because no concentrations can be given for solid AgCl, dashes are written; in later problems, similar spaces will be left blank.)

$$\frac{1.9 \times 10^{-3}\ g}{L} \times \frac{1\ mol}{143\ g} = 1.\underline{3}3 \times 10^{-5}\ M$$

(continued)

Conc. (M)	$AgCl(s)$	$\rightleftharpoons$	Ag^+	+	Cl^-
Starting	—		0		0
Change	—		$+1.33 \times 10^{-5}$		$+1.33 \times 10^{-5}$
Equilibrium	—		1.33×10^{-5}		1.33×10^{-5}

$K_{sp} = [Ag^+][Cl^-] = (1.33 \times 10^{-5})(1.33 \times 10^{-5}) = 1.\underline{7}68 \times 10^{-10}$

(rounded answer above)

17.3 Rounded answer: $K_{sp} = 4.5 \times 10^{-36}$. Calculate the molar solubility. Then assemble the usual concentration table and substitute from it into the equilibrium-constant expression. (Because no concentrations can be given for solid $Pb_3(AsO_4)_2$, spaces are left blank.)

$$\frac{3.0 \times 10^{-5}\ \text{g}}{\text{L}} \times \frac{1\ \text{mol}}{899\ \text{g}} = 3.\underline{3}4 \times 10^{-8}\ \text{M}$$

Conc. (M)	$Pb_3(AsO_4)_2(s)$	$\rightleftharpoons$	$3Pb^{2+}$	+	$2AsO_4^{3-}$
Starting			0		0
Change			$+3(3.34 \times 10^{-8})$		$+2(3.34 \times 10^{-8})$
Equilibrium			$3(3.34 \times 10^{-8})$		$2(3.34 \times 10^{-8})$

$K_{sp} = [Pb^{2+}]^3[AsO_4^{3-}]^2 = [3 \times (3.34 \times 10^{-8})]^3(2 \times 3.34 \times 10^{-8})^2 = 4.\underline{4}89 \times 10^{-36}$

(rounded answer above)

17.4 Rounded answer: 0.67 g $CaSO_4$/L. Assemble the usual concentration table. Let x = the molar solubility of $CaSO_4$. When x mol $CaSO_4$ dissolves in 1 L of solution, x mol Ca^{2+} and x mol SO_4^{2-} form.

Conc. (M)	$CaSO_4(s)$	$\rightleftharpoons$	Ca^{2+}	+	SO_4^{2-}
Starting			0		0
Change			+x		+x
Equilibrium			x		x

Substitute the equilibrium concentrations into the equilibrium-constant expression and solve for x. Then convert to g $CaSO_4$ per L.

(continued)

$[Ca^{2+}][SO_4^{2-}] = K_{sp}$

$(x)(x) = x^2 = 2.4 \times 10^{-5}$

$x = \sqrt{(2.4 \times 10^{-5})} = 4.\underline{8}9 \times 10^{-3}$ M (rounded answer above)

$\frac{4.9 \times 10^{-3}\ \text{mol}}{\text{L}} \times \frac{136\ \text{g}}{1\ \text{mol}} = 0.6\underline{6}64$ g/L

17.5 Rounded answers: (a) 6.3×10^{-3} M; (b) 4.4×10^{-5} M.

a. Let x = the molar solubility of BaF_2. Assemble the usual concentration table and substitute from the table into the equilibrium-constant expression.

Conc. (M)	$BaF_2(s)$ ⇌	Ba^{2+}	+	$2F^-$
Starting		0		0
Change		+x		+2x
Equilibrium		x		2x

$[Ba^{2+}][F^-]^2 = K_{sp}$

$(x)(2x)^2 = 4x^3 = 1.0 \times 10^{-6}$

$\sqrt[3]{\frac{1.0 \times 10^{-6}}{4}} = 6.\underline{2}99 \times 10^{-3}$ M

b. At the start, before any BaF_2 dissolves, the solution contains 0.15 M F^-. At equilibrium, x mol of solid BaF_2 dissolves to yield x mol Ba^{2+} and 2x mol F^-. Assemble the usual concentration table, and substitute the equilibrium concentrations into the equilibrium-constant expression. As an approximation, assume x is negligible compared to 0.15 M F^-.

Conc. (M)	$BaF_2(s)$ ⇌	Ba^{2+}	+	$2F^-$
Starting		0		0.15
Change		+x		+2x
Equilibrium		x		0.15 + 2x

$[Ba^{2+}][F^-]^2 = K_{sp}$

(continued)

$$(x)(0.15 + 2x)^2 \cong (x)(0.15)^2 \cong 1.0 \times 10^{-6}$$

$$x \cong \frac{1.0 \times 10^{-6}}{(0.15)^2} \cong 4.\underline{4}44 \times 10^{-5}\ M$$

Note that adding 2x to 0.15 M will not change it (to two significant figures), so 2x is negligible compared to 0.15 M. The solubility of 4.4×10^{-5} M in 0.15 M NaF is lower than the solubility of 6.3×10^{-3} M in pure water.

17.6 Rounded answer: $Q_c = 8.5 \times 10^{-5}$ (> K_{sp} of 2.4×10^{-5}, so precipitation occurs at equilibrium). Calculate the ion product, Q_c, after evaporation, assuming no precipitation has occurred. Compare it with the K_{sp}.

$$Q_c = [Ca^{2+}][SO_4^{2-}]$$

$$Q_c = (2 \times 0.0052)(2 \times 0.0041) = 8.528 \times 10^{-5}\ M^2 \ \ (> K_{sp}, \text{ so precipitation occurs})$$

The solution is supersaturated before equilibrium is reached. At equilibrium, precipitation occurs and the solution is saturated.

17.7 Rounded answer: $Q_c = 8.5 \times 10^{-9}$ (< K_{sp}, so no precipitation occurs). Calculate the concentrations of Pb^{2+} and SO_4^{2-}, assuming no precipitation. Use a total volume of 0.456 L + 0.255 L, or 0.711 L.

$$[Pb^{2+}] = \frac{\frac{0.00016\ mol}{L} \times 0.255\ L}{0.711\ L} = 5.\underline{7}4 \times 10^{-5}\ M$$

$$[SO_4^{2-}] = \frac{\frac{0.00023\ mol}{L} \times 0.456\ L}{0.711\ L} = 1.\underline{4}8 \times 10^{-4}\ M$$

Calculate the ion product and compare it to K_{sp}.

$$Q_c = [Pb^{2+}][SO_4^{2-}] = (5.74 \times 10^{-5})(1.48 \times 10^{-4}) = 8.\underline{4}9 \times 10^{-9}\ M^2$$

Because Q_c is less than the K_{sp} of 1.7×10^{-8}, no precipitation occurs and the solution is unsaturated.

17.8 Rounded answers: 3.6×10^{-11} M; 7.3×10^{-4}% remaining. Because K_{sp} is small, almost all the Pb^{2+} and CrO_4^{2-} react until one ion, the *limiting reactant,* is essentially completely precipitated. It will be seen below that this is Pb^{2+}. Assuming complete precipitation of the Pb^{2+} ion, calculate the amount of CrO_4^{2-} left as follows:

$$(0.0010 \text{ mol } CrO_4^{2-}/L) \times 0.50 \text{ L} = 5.0 \times 10^{-4} \text{ mol } CrO_4^{2-} \quad (= 5.0 \times 10^{-4} \text{ M in 1.00 L})$$

$$- (0.00001 \text{ mol } Pb^{2+}/L) \times 0.50 \text{ L} = 5.0 \times 10^{-6} \text{ mol } Pb^{2+} \quad (= 5.0 \times 10^{-6} \text{ M in 1.00 L})$$

$$= 4.\underline{95} \times 10^{-4} \text{ mol } CrO_4^{2-} \quad (4.\underline{95} \times 10^{-4} \text{ M in 1.00 L})$$

At equilibrium, a small amount of $PbCrO_4$ will dissolve, producing an unknown concentration, x, of Pb^{2+} ion. Assemble the usual table for exact equilibrium concentrations. Then assume that x is negligible compared to 4.95×10^{-4} M, and do an approximate calculation of the Pb^{2+} concentration.

Conc. (M)	$PbCrO_4(s)$	$\rightleftharpoons$	Pb^{2+}	+	CrO_4^{2-}
Starting			0		4.95×10^{-4}
Change			+x		+x
Equilibrium			x		$(4.95 \times 10^{-4}) + x$

$$[Pb^{2+}][CrO_4^{2-}] = K_{sp}$$

$$(x)[(4.95 \times 10^{-4}) + x] \cong (x)(4.95 \times 10^{-4}) \cong 1.8 \times 10^{-14}$$

$$x \cong \frac{1.8 \times 10^{-14}}{4.95 \times 10^{-4}} \cong 3.\underline{6}36 \times 10^{-11} \text{ M} \cong [Pb^{2+}]$$

The percentage of Pb^{2+} remaining in solution is

$$[(3.636 \times 10^{-11}) \div (5.0 \times 10^{-6})] \times 100\% = 7.\underline{2}7 \times 10^{-4}\%$$

17.9 The solubility of AgCN would increase as the pH decreases because the increasing concentration of H^+ would react with the CN^- to form the weakly ionized acid HCN. As CN^- is removed, more AgCN dissolves to replace the cyanide:

$$AgCN(s) \rightleftharpoons Ag^+(aq) + CN^-(aq) \quad [+ H^+ \rightarrow HCN]$$

In the case of AgCl, the chloride ion is the conjugate base of a strong acid and would therefore not be affected by any amount of hydrogen ion.

17.10 Rounded answer: pH range = 0.0 to 2.51. Find the *higher end* of the pH range by calculating the minimum concentration of S^{2-} required to prevent precipitation of the most soluble metal sulfide. In this case, this is FeS because its K_{sp} of 6×10^{-18} is much larger than the K_{sp} of 6×10^{-36} for CuS. For FeS, this is

$$[Fe^{2+}][S^{2-}] = K_{sp}$$

$$(0.050)[S^{2-}] = 6 \times 10^{-18}$$

$$[S^{2-}] = \underline{1}.2 \times 10^{-16}\ M$$

Substituting into the overall equilibrium-constant ($K_{a1}K_{a2}$) expression for H_2S, and assuming 0.10 M for saturated H_2S, solve for $[H^+]$:

$$\frac{[H^+]^2[S^{2-}]}{[H_2S]} = 1.1 \times 10^{-20}$$

$$[H^+]^2 = (1.1 \times 10^{-20})(0.10\ M) \div (1.2 \times 10^{-16}) = 9.16 \times 10^{-6};\ [H^+] = \underline{3}.02 \times 10^{-3}\ M$$

Higher end of pH range = - log (3.02×10^{-3}) = 2.$\underline{5}$18 = 2.5

Find the *lower end* of the pH range by calculating the sulfide ion concentration required to just begin precipitation of CuS, using the same calculations.

$$[Cu^{2+}][S^{2-}] = K_{sp}$$

$$(0.050)[S^{2-}] = 6 \times 10^{-36}$$

$$[S^{2-}] = 1.2 \times 10^{-34}\ M$$

$$\frac{[H^+]^2[S^{2-}]}{[H_2S]} = 1.1 \times 10^{-20}$$

$$[H^+]^2 = [(1.1 \times 10^{-20})(0.10\ M)] \div (1.2 \times 10^{-34}) = 9.16 \times 10^{12};\ [H^+] = \underline{3}.02 \times 10^{6}\ M$$

Theoretical lower end of pH range = - log (3.02×10^6) = -6.4 (practical end: pH = 0.0)

17.11 Rounded answer: 1.2×10^{-9} M. Because $K_f = 4.8 \times 10^{12}$, and because the starting concentration of NH_3 is much larger than that of the Cu^{2+} ion, you can make a rough assumption that most of the copper(II) is converted to $Cu(NH_3)_4{}^{2+}$ ion. This ion then dissociates slightly to give a small concentration of Cu^{2+} and additional NH_3. The amount of NH_3 remaining at the start after reacting with 0.015 M Cu^{2+} is

(continued)

[0.100 M - (4 x 0.015 M)] = 0.040 M starting NH_3

Assemble the usual concentration table, using this starting concentration for NH_3 and assuming that the starting concentration of Cu^{2+} is zero.

Conc. (M)	$Cu(NH_3)_4^{2+}$	$\rightleftharpoons$ Cu^{2+}	+	$4NH_3$
Starting	0.015	0		0.040
Change	-x	+x		+4x
Equilibrium	0.015 - x	x		0.040 + 4x

Even though this reaction is the opposite of the equation for the formation constant, the formation-constant expression can be used. Simply substitute all exact equilibrium concentrations into the formation-constant expression; then simplify the exact equation by assuming that x is negligible compared to 0.015 and 4x is negligible compared to 0.040.

$$K_f = \frac{[Cu(NH_3)_4^{2+}]}{[Cu^{2+}][NH_3]^4} = \frac{(0.015 - x)}{(x)(0.040 + 4x)^4} \cong \frac{(0.015)}{(x)(0.040)^4} \cong 4.8 \times 10^{12}$$

Rearrange and solve for x:

$$x \cong (0.015) \div [(4.8 \times 10^{12})(0.040)^4] \cong 1.\underline{22} \times 10^{-9}\ M \cong [Cu^{2+}]$$

17.12 Rounded answers: $Q_c = 3.3 \times 10^{-21}$; no precipitation. Start by calculating the $[Ag^+]$ in equilibrium with the $Ag(CN)_2^-$ formed from Ag^+ and CN^-. Then use the $[Ag^+]$ to decide whether or not AgI will precipitate, by calculating the ion product and comparing it with the K_{sp} of 8.3×10^{-17} for AgI. Assume that all of the 0.0045 M Ag^+ reacts with CN^- to form 0.0045 M $Ag(CN)_2^-$, and calculate the remaining CN^-. Use these as starting concentrations for the usual concentration table.

[0.20 M KCN - (2 x 0.0045 M)] = 0.191 M starting CN^-

Conc. (M)	$Ag(CN)_2^-$	$\rightleftharpoons$ Ag^+	+	$2CN^-$
Starting	0.0045	0		0.191
Change	-x	+x		+2x
Equilibrium	0.0045 - x	x		0.191 + 2x

(continued)

Even though this reaction is the opposite of the equation for the formation constant, the formation-constant expression can be used. Simply substitute all exact equilibrium concentrations into the formation-constant expression; then simplify the exact equation by assuming that x is negligible compared to 0.0045 and 2x is negligible compared to 0.191.

$$K_f = \frac{[Ag(CN)_2^-]}{[Ag^+][CN^-]^2} = \frac{(0.0045 - x)}{(x)(0.191 + 2x)^2} \cong \frac{(0.0045)}{(x)(0.191)^2} \cong 5.6 \times 10^{18}$$

Rearrange and solve for x:

$$x \cong (0.0045) \div [(5.6 \times 10^{18})(0.191)^2] \cong 2.\underline{2}02 \times 10^{-20}\ M \cong [Ag^+]$$

Now calculate the ion product for AgI:

$$Q_c = [Ag^+][I^-] = (2.20 \times 10^{-20})(0.15) = 3.\underline{3}0 \times 10^{-21}\ M^2$$

Because Q_c is less than the K_{sp} of 8.3×10^{-17}, no precipitate will form and the solution is unsaturated.

17.13 Rounded answer: 0.44 M. Using the rule from Chapter 14, obtain the overall equilibrium constant for this reaction from the product of the individual equilibrium constants of the two individual equations whose sum gives this equation:

$$AgBr(s) \rightleftharpoons Ag^+(aq) + Br^-(aq)$$

$$K_{sp} = 5.0 \times 10^{-13}$$

$$Ag^+(aq) + 2S_2O_3^{2-}(aq) \rightleftharpoons Ag(S_2O_3)_2^{3-}(aq)$$

$$K_f = 2.9 \times 10^{13}$$

$$AgBr(s) + 2S_2O_3^{2-}(aq) \rightleftharpoons Ag(S_2O_3)_2^{3-}(aq) + Br^-(aq)$$

$$K_c = K_{sp} \times K_f = 14.5$$

Assemble the usual table, using 1.0 M as the starting concentration of $S_2O_3^{2-}$ and x as the unknown concentration of $Ag(S_2O_3)_2^{3-}$ formed.

(continued)

Conc. (M)	$AgBr(s)$ +	$2S_2O_3^{2-}$ ⇌	$Ag(S_2O_3)_2^{3-}$ +	Br^-
Starting		1.0	0	0
Change		-2x	+x	+x
Equilibrium		1.0 - 2x	x	x

The equilibrium-constant expression can now be used. Simply substitute all exact equilibrium concentrations into the equilibrium-constant expression. The solution can be obtained without using the quadratic equation.

$$K_c = \frac{[Ag(S_2O_3)_2^{3-}][Br^-]}{[S_2O_3^{2-}]^2} = \frac{(x)^2}{(1.0 - 2x)^2} = 14.5$$

Take the square root of both sides of the two right-hand terms, and solve for x:

$$\frac{x}{(1.0 - 2x)} = 3.\underline{8}08$$

$$x = 3.808\,(1.0 - 2x)$$

$$7.62x + x = 3.808$$

$x = 0.4\underline{4}17$ = molar solubility of AgBr in 1.0 M $Na_2S_2O_3$

■ Answers to Review Questions

17.1 The solubility equation is $Ni(OH)_2(s) \rightleftharpoons Ni^{2+}(aq) + 2OH^-(aq)$. If the molar solubility of $Ni(OH)_2$ = x molar, the concentrations of the ions in the solution must be x M Ni^{2+} and 2x M OH^-. Substituting into the equilibrium-constant expression gives

$$K_{sp} = [Ni^{2+}][OH^-]^2 = (x)(2x)^2 = 4x^3$$

17.2 Calcium sulfate is less soluble in a solution containing sodium sulfate because the increase in sulfate from the sodium sulfate causes the equilibrium composition in the equation below to shift to the left:

$$CaSO_4(s) \rightleftharpoons Ca^{2+}(aq) + SO_4^{2-}(aq)$$

The result is a decrease in both the calcium ion and the calcium sulfate concentrations.

(continued)

17.3 Substitute the 0.10 M concentration of chloride into the solubility product expression and solve for $[Ag^+]$:

$$[Ag^+] = \frac{K_{sp}}{[Cl^-]} = \frac{1.8 \times 10^{-10}}{0.10} = 1.8\underline{0} \times 10^{-9} = 1.8 \times 10^{-9}\ M$$

17.4 In order to predict whether or not PbI_2 will precipitate when lead nitrate and potassium iodide are mixed, the concentrations of Pb^{2+} and I^- after mixing would have to be calculated first (if the concentrations are not known or are not given). Then the value of Q_c, the ion product, would have to be calculated for PbI_2. Finally, Q_c would have to be compared with the value of K_{sp}. If $Q_c > K_{sp}$, then a precipitate will form at equilibrium. If Q_c is $\leq$ than K_{sp}, no precipitate will form.

17.5 Barium fluoride, normally insoluble in water, dissolves in dilute hydrochloric acid because the fluoride ion, once it forms, reacts with the hydrogen ion to form weakly ionized HF:

$$BaF_2(s) \rightleftharpoons Ba^{2+}(aq) + 2F^-(aq)\ [+ 2H^+ \rightarrow 2HF(aq)]$$

17.6 Metal ions such as Pb^{2+} and Zn^{2+} are separated by controlling the $[S^{2-}]$ in a solution of saturated H_2S by means of adjusting the pH correctly. Because the K_{sp} of 2.5×10^{-27} for PbS is smaller than the K_{sp} of 1.1×10^{-21} for ZnS, the pH can be adjusted to make the $[S^{2-}]$ just high enough to precipitate PbS, without precipitating ZnS.

17.7 When NaCl is first added to a solution of $Pb(NO_3)_2$, a precipitate of $PbCl_2$ forms. As more NaCl is added, the excess chloride reacts further with the insoluble $PbCl_2$, forming soluble complex ions of $PbCl_3^-$ and $PbCl_4^{2-}$:

$$Pb^{2+}(aq) + 2Cl^-(aq) \rightleftharpoons PbCl_2(s)$$

$$PbCl_2(s) + Cl^-(aq) \rightleftharpoons PbCl_3^-(aq)$$

$$PbCl_3^-(aq) + Cl^-(aq) \rightleftharpoons PbCl_4^{2-}(aq)$$

17.8 When a small amount of NaOH is added to a solution of $Al_2(SO_4)_3$, a precipitate of $Al(OH)_3$ forms at first. As more NaOH is added, the excess hydroxide ion reacts further with the insoluble $Al(OH)_3$, forming a soluble complex ion of $Al(OH)_4^-$.

17.9 The Ag^+, Cu^{2+}, and Ni^{2+} ions can be separated in two steps: (1) Add HCl to precipitate just the Ag^+ as AgCl, leaving the others in solution. (2) After pouring the solution away from the precipitate, add 0.3 M HCl and H_2S to precipitate only the CuS away from the Ni^{2+} ion, whose sulfide is soluble under these conditions.

17.10 By controlling the pH through the appropriate buffer, one can control the $[CO_3^{2-}]$ using the equilibrium reaction:

$$H^+(aq) + CO_3^{2-}(aq) \rightleftharpoons HCO_3^-(aq)$$

Calcium carbonate is much more insoluble than magnesium carbonate and thus can precipitate in weakly basic solution whereas magnesium carbonate will not. Magnesium carbonate will precipitate only in highly basic solution. (There is the possibility that $Mg(OH)_2$ might precipitate, but the $[OH^-]$ is too low for this to occur.)

■ Solutions to Practice Problems

Note on answers to equilibrium calculations: The rounded answer (with correct number of significant figures) is always given first. After the stepwise solution, the numerical answer is given again, but with one nonsignificant figure. (The rightmost significant figure is underlined.)

17.11 a. Soluble (Group IA salts are soluble.)

b. Insoluble (Carbonates are generally insoluble.)

c. Insoluble ($PbSO_4$ is an insoluble sulfate.)

d. Soluble (All ammonium salts are soluble.)

17.12 a. Insoluble ($PbBr_2$ is an insoluble bromide.)

b. Soluble (All Group IA salts are soluble.)

c. Soluble (Bromides are generally soluble.)

d. Insoluble ($BaSO_4$ is an insoluble sulfate.)

17.13 a. $K_{sp} = [Ba^{2+}][CO_3^{2-}]$

b. $K_{sp} = [Pb^{2+}][OH^-]^2$

c. $K_{sp} = [Mg^{2+}]^3[PO_4^{3-}]^2$

d. $K_{sp} = [Ag^+]^2[CrO_4^{2-}]$

17.14 a. $K_{sp} = [Ba^{2+}][CrO_4^{2-}]$

b. $K_{sp} = [Pb^{2+}][Cl^-]^2$

c. $K_{sp} = [Mg^{2+}]^3[AsO_4^{3-}]^2$

d. $K_{sp} = [Ag^+]^2[S^{2-}]$

17.15 Rounded answer: K_{sp} = 9.3×10^{-10} (correct no. of sig. figs.). Calculate molar solubility, assemble the usual table, and substitute from it the equilibrium concentrations into the equilibrium-constant expression. (Dashes are given for $AgBrO_3(s)$; in later problems, the dashes are omitted.)

$$\frac{7.2 \times 10^{-3}\ g}{L} \times \frac{1\ mol}{236\ g} = 3.\underline{0}5 \times 10^{-5}\ M$$

Conc. (M)	$AgBrO_3(s)$ $\rightleftharpoons$	Ag^+	+	BrO_3^-
Starting	—	0		0
Change	—	$+3.05 \times 10^{-5}$		$+3.05 \times 10^{-5}$
Equilibrium	—	3.05×10^{-5}		3.05×10^{-5}

K_{sp} = $[Ag^+][BrO_3^-]$ = $(3.05 \times 10^{-5})(3.05 \times 10^{-5})$ = $9.\underline{3}02 \times 10^{-10}$

(rounded answer above)

17.16 Rounded answer: K_{sp} = 8.6×10^{-5} (correct no. of sig. figs.). Assemble the usual concentration table and substitute from it the equilibrium concentrations into the equilibrium-constant expression.

Conc. (M)	$MgC_2O_4(s)$ $\rightleftharpoons$	Mg^{2+}	+	$C_2O_4^{2-}$
Starting		0		0
Change		$+9.3 \times 10^{-3}$		$+9.3 \times 10^{-3}$
Equilibrium		9.3×10^{-3}		9.3×10^{-3}

K_{sp} = $[Mg^{2+}][C_2O_4^{2-}]$ = $(9.3 \times 10^{-3})(9.3 \times 10^{-3})$ = $8.\underline{6}49 \times 10^{-5}$

(rounded answer above)

17.17 Rounded answer: K_{sp} = 1.3×10^{-7} (correct no. of sig. figs.). Calculate the molar solubility. Then assemble the usual concentration table and substitute from it the equilibrium concentrations into the equilibrium-constant expression.

$$\frac{0.13\ g}{0.100\ L} \times \frac{1\ mol}{413\ g} = 3.\underline{1}5 \times 10^{-3}\ M$$

(continued)

Conc. (M)	$Cu(IO_3)_2(s)$ $\rightleftharpoons$	Cu^{2+}	+	$2IO_3^-$
Starting		0		0
Change		$+3.15 \times 10^{-3}$		$+2 \times 3.15 \times 10^{-3}$
Equilibrium		3.15×10^{-3}		$2 \times 3.15 \times 10^{-3}$

$K_{sp} = [Cu^{2+}][IO_3^-]^2 = (3.15 \times 10^{-3})[2(3.15 \times 10^{-3})]^2 = 1.\underline{2}502 \times 10^{-7}$

17.18 Rounded answer: $K_{sp} = 1.2 \times 10^{-12}$ (correct no. of sig. figs.). Calculate the molar solubility. Then assemble the usual concentration table and substitute from it the equilibrium concentrations into the equilibrium-constant expression.

$$\frac{0.022\ g}{L} \times \frac{1\ mol}{332\ g} = 6.\underline{6}2 \times 10^{-5}\ M$$

Conc. (M)	$Ag_2CrO_4(s)$ $\rightleftharpoons$	$2Ag^+$	+	CrO_4^{2-}
Starting		0		0
Change		$+2(6.62 \times 10^{-5})$		6.62×10^{-5}
Equilibrium		$2(6.62 \times 10^{-5})$		6.62×10^{-5}

$K_{sp} = [Ag^+]^2[CrO_4^{2-}] = [2(6.62 \times 10^{-5})]^2(6.62 \times 10^{-5}) = 1.\underline{1}61 \times 10^{-12}$

17.19 Rounded answer: $K_{sp} = 1.8 \times 10^{-11}$ (correct no. of sig. figs.). Calculate the pOH from pH and then convert pOH to $[OH^-]$. Then assemble the usual concentration table and substitute from it the equilibrium concentrations into the equilibrium-constant expression.

$pOH = 14.00 - 10.52 = 3.48$

$[OH^-] = \text{antilog}(-pOH) = \text{antilog}(-3.48) = 3.\underline{3}11 \times 10^{-4}\ M$

$[Mg^{2+}] = [OH^-] \div 2 = (3.311 \times 10^{-4}) \div 2 = 1.655 \times 10^{-4}\ M$

Conc. (M)	$Mg(OH)_2(s)$ $\rightleftharpoons$	Mg^{2+}	+	$2OH^-$
Starting		0		0
Change		$+1.655 \times 10^{-4}$		$+3.311 \times 10^{-4}$
Equilibrium		1.655×10^{-4}		3.311×10^{-4}

$K_{sp} = [Mg^{2+}][OH^-]^2 = (1.655 \times 10^{-4})(3.311 \times 10^{-4})^2 = 1.\underline{8}14 \times 10^{-11}$

17.20 Rounded answer: K_{sp} = 5.6 x 10^{-6} (correct no. of sig. figs.). Calculate the pOH from pH and then convert pOH to [OH^-]. Then assemble the usual concentration table and substitute from it the equilibrium concentrations into the equilibrium-constant expression.

pOH = 14.00 - 12.35 = 1.65

[OH^-] = antilog (-pOH) = antilog (-1.65) = 2.$\underline{2}$38 x 10^{-2} M

[Ca^{2+}] = [OH^-] ÷ 2 = (2.238 x 10^{-2}) ÷ 2 = 1.119 x 10^{-2}M

Conc. (M)	$Ca(OH)_2(s)$	$\rightleftharpoons$	Ca^{2+}	+	$2OH^-$
Starting			0		0
Change			+1.119 x 10^{-2}		+2.238 x 10^{-2}
Equilibrium			1.119 x 10^{-2}		2.238 x 10^{-2}

K_{sp} = [Ca^{2+}][OH^-]^2 = (1.119 x 10^{-2})(2.238 x 10^{-2})^2 = 5.$\underline{6}$04 x 10^{-6}

17.21 Rounded answers: 5.0 x 10^{-4} M and 0.092 g $SrSO_4$/L. Assemble the usual concentration table. Let x = the molar solubility of $SrSO_4$. When x mol $SrSO_4$ dissolves in 1 L of solution, x mol Sr^{2+} and x mol SO_4^{2-} form.

Conc. (M)	$SrSO_4(s)$	$\rightleftharpoons$	Sr^{2+}	+	SO_4^{2-}
Starting			0		0
Change			+x		+x
Equilibrium			x		x

Substitute the equilibrium concentrations into the equilibrium-constant expression and solve for x. Then convert to g $SrSO_4$ per L.

[Sr^{2+}][SO_4^{2-}] = K_{sp}

(x)(x) = x^2 = 2.5 x 10^{-7}

x = $\sqrt{(2.5 \times 10^{-7})}$ = 5.$\underline{0}$0 x 10^{-4} M (rounded answer above)

$$\frac{5.0 \times 10^{-4}\ \text{mol}}{\text{L}} \times \frac{184\ \text{g}}{1\ \text{mol}} = 0.09\underline{2}0\ \text{g/L}$$

17.22 Rounded answer: 2.4×10^{-3} g $BaSO_4$/L. Assemble the usual concentration table. Let x = the molar solubility of $BaSO_4$. When x mol $BaSO_4$ dissolves in 1 L of solution, x mol Ba^{2+} and x mol SO_4^{2-} form.

Conc. (M)	$BaSO_4(s)$	$\rightleftharpoons$	Ba^{2+}	+	SO_4^{2-}
Starting			0		0
Change			+x		+x
Equilibrium			x		x

Substitute the equilibrium concentrations into the equilibrium-constant expression and solve for x. Then convert to g $BaSO_4$ per L.

$$[Ba^{2+}][SO_4^{2-}] = K_{sp}$$

$$(x)(x) = x^2 = 1.1 \times 10^{-10}$$

$$x = \sqrt{(1.1 \times 10^{-10})} = 1.\underline{0}48 \times 10^{-5}\ M$$

$$\frac{1.048 \times 10^{-5}\ mol}{L} \times \frac{233\ g}{1\ mol} = 2.\underline{4}4 \times 10^{-3}\ g/L\ \text{(rounded answer above)}$$

17.23 Rounded answer: 1.9×10^{-3} M. Let x = the molar solubility of PbF_2. Assemble the usual concentration table and substitute from the table into the equilibrium-constant expression.

Conc. (M)	$PbF_2(s)$	$\rightleftharpoons$	Pb^{2+}	+	$2F^-$
Starting			0		0
Change			+x		+2x
Equilibrium			x		2x

$$[Pb^{2+}][F^-]^2 = K_{sp}$$

$$(x)(2x)^2 = 4x^3 = 2.7 \times 10^{-8}$$

$$x = \sqrt[3]{\frac{2.7 \times 10^{-8}}{4}} = 1.\underline{8}8 \times 10^{-3}\ M$$

17.24 Rounded answer: 1.2×10^{-3} M. Let x = the molar solubility of MgF_2. Assemble the usual concentration table and substitute from the table into the equilibrium-constant expression.

Conc. (M)	$MgF_2(s)$	$\rightleftharpoons$	Mg^{2+}	+	$2F^-$
Starting			0		0
Change			+x		+2x
Equilibrium			x		2x

$$[Mg^{2+}][F^-]^2 = K_{sp}$$

$$(x)(2x)^2 = 4x^3 = 7.1 \times 10^{-9}$$

$$x = \sqrt[3]{\frac{7.1 \times 10^{-9}}{4}} = 1.\underline{2}1 \times 10^{-3}\ M$$

17.25 Rounded answer: 3.1×10^{-4} g/L. Let x = the molar solubility of $SrSO_4$. At the start, before any $SrSO_4$ dissolves, the solution contains 0.15 M SO_4^{2-}. At equilibrium, x mol of solid $SrSO_4$ dissolves to yield x mol Sr^{2+} and x mol SO_4^{2-}. Assemble the usual concentration table, and substitute the equilibrium concentrations into the equilibrium-constant expression. As an approximation, assume x is negligible compared to 0.15 M SO_4^{2-}.

Conc. (M)	$SrSO_4(s)$	$\rightleftharpoons$	Sr^{2+}	+	SO_4^{2-}
Starting			0		0.15
Change			+x		+x
Equilibrium			x		0.15 + x

$$[Sr^{2+}][SO_4^{2-}] = K_{sp}$$

$$(x)(0.15 + x) \cong (x)(0.15) \cong 2.5 \times 10^{-7}$$

$$x \cong \frac{2.5 \times 10^{-7}}{0.15} \cong 1.\underline{6}6 \times 10^{-6}\ M$$

$$\frac{1.66 \times 10^{-6}\ mol}{L} \times \frac{184\ g}{1\ mol} = 3.\underline{0}54 \times 10^{-4}\ g/L$$

Note that adding x to 0.15 M will not change it (to two significant figures), so x is negligible compared to 0.15 M.

17.26 Rounded answer: 2.9×10^{-11} g/L. Let x = the molar solubility of $PbCrO_4$. At the start, before any $PbCrO_4$ dissolves, the solution contains 0.20 M CrO_4^{2-}. At equilibrium, x mol of solid $PbCrO_4$ dissolves to yield x mol Pb^{2+} and x mol CrO_4^{2-}. Assemble the usual concentration table, and substitute the equilibrium concentrations into the equilibrium-constant expression. As an approximation, assume x is negligible compared to 0.20 M CrO_4^{2-}.

Conc. (M)	$PbCrO_4(s)$ $\rightleftharpoons$	Pb^{2+}	+	CrO_4^{2-}
Starting		0		0.20
Change		+x		+x
Equilibrium		x		0.20 + x

$$[Pb^{2+}][CrO_4^{2-}] = K_{sp}$$

$$(x)(0.20 + x) \cong (x)(0.20) \cong 1.8 \times 10^{-14}$$

$$x \cong \frac{1.8 \times 10^{-14}}{0.20} \cong 9.\underline{0}0 \times 10^{-14} \text{ M}$$

$$\frac{9.00 \times 10^{-14} \text{ mol}}{\text{L}} \times \frac{323 \text{ g}}{1 \text{ mol}} = 2.\underline{9}07 \times 10^{-11} \text{ g/L}$$

Note that adding x to 0.20 M will not change it (to two significant figures), so x is negligible compared to 0.20 M.

17.27 Rounded answer: 1.1×10^{-6} g/L. Calculate the value of K_{sp} from the solubility, using the concentration table. Then, using the common-ion calculation, assemble another concentration table. Use 0.020 M NaF as the starting concentration of F^- ion. Substitute the equilibrium concentrations from the table into the equilibrium-constant expression. As an approximation, assume that 2x is negligible compared to 0.020 M F^- ion.

$$\frac{0.0076 \text{ g}}{\text{L}} \times \frac{1 \text{ mol}}{62.3 \text{ g}} = 1.\underline{2}2 \times 10^{-4} \text{ M}$$

Conc. (M)	$MgF_2(s)$ $\rightleftharpoons$	Mg^{2+}	+	$2F^-$
Starting		0		0
Change		$+1.22 \times 10^{-4}$		$+2 \times 1.22 \times 10^{-4}$
Equilibrium		1.22×10^{-4}		$2 \times 1.22 \times 10^{-4}$

(continued)

$K_{sp} = [Mg^{2+}][F^-]^2 = (1.22 \times 10^{-4})[2(1.22 \times 10^{-4})]^2 = 7.26 \times 10^{-12}$

Now use K_{sp} to calculate the molar solubility of MgF_2.

Conc. (M)	$MgF_2(s)$ ⇌	Mg^{2+}	+	$2F^-$
Starting		0		0.020
Change		+x		+2x
Equilibrium		x		0.020 + 2x

$[Mg^{2+}][F^-]^2 = K_{sp}$

$(x)(0.020 + 2x)^2 \cong (x)(0.020)^2 = 7.\underline{2}6 \times 10^{-12}$

$$x \cong \frac{7.26 \times 10^{-12}}{(0.020)^2} \cong 1.\underline{8}15 \times 10^{-8}\ M$$

$$\frac{1.815 \times 10^{-8}\ mol}{L} \times \frac{62.3\ g}{1\ mol} = 1.\underline{1}3 \times 10^{-6}\ g/L$$

17.28 Rounded answer: 1.5 g/L. Calculate the value of K_{sp} from the solubility, using the concentration table. Then, using the common-ion calculation, assemble another concentration table. Use 0.75 M Na_2SO_4 as the starting concentration of SO_4^{2-} ion. Substitute the equilibrium concentrations from the table into the equilibrium-constant expression. As an approximation, assume that x is negligible compared to 0.75 M SO_4^{2-} ion.

$$\frac{8.0\ g}{1\ L} \times \frac{1\ mol}{312\ g} = 0.02\underline{5}6\ M\ Ag_2SO_4$$

Conc. (M)	$Ag_2SO_4(s)$ ⇌	$2Ag^+$	+	SO_4^{2-}
Starting		0		0
Change		+2 x 0.0256		+0.0256
Equilibrium		2 x 0.0256		0.0256

$K_{sp} = [Ag^+]^2[SO_4^{2-}] = (2 \times 0.0256)^2(0.0256) = 6.\underline{7}1 \times 10^{-5}$

Now use K_{sp} to calculate the molar solubility of Ag_2SO_4.

(continued)

Conc. (M)	$Ag_2SO_4(s)$ $\rightleftharpoons$	$2Ag^+$	+	SO_4^{2-}
Starting		0		0.75
Change		+2x		+x
Equilibrium		2x		0.75 + x

$[Ag^+]^2[SO_4^{2-}] = K_{sp}$

$(2x)^2(0.75 + x) \cong (2x)^2(0.75) = 6.71 \times 10^{-5}$

$$x \cong \sqrt{\frac{6.71 \times 10^{-5}}{4 \times 0.75}} \cong 4.\underline{7}29 \times 10^{-3}\text{ M}$$

$$\frac{4.729 \times 10^{-3}\text{ mol}}{\text{L}} \times \frac{312\text{ g}}{1\text{ mol}} = 1.\underline{4}75\text{ g/L}$$

17.29 Rounded answer: 0.4 g/L. The concentration table follows.

Conc. (M)	$MgC_2O_4(s)$ $\rightleftharpoons$	Mg^{2+}	+	$C_2O_4^{2-}$
Starting		0		0.020
Change		+x		+x
Equilibrium		x		0.020 + x

The equilibrium-constant equation is

$K_{sp} = [Mg^{2+}][C_2O_4^{2-}]$

$8.5 \times 10^{-5} = x(0.020 + x)$

$x^2 + 0.020x - (8.5 \times 10^{-5}) = 0$

Solving the quadratic equation gives

$$x = \frac{-0.020 \pm \sqrt{(0.020)^2 + 4(8.5 \times 10^{-5})}}{2} = 0.00\underline{3}6$$

The solubility in grams per liter is

$$0.0036\ \frac{\text{mol}}{\text{L}} \times 112\ \frac{\text{g}}{\text{mol}} = 0.\underline{4}03\text{ g/L} = 0.4\text{ g/L}$$

17.30 Rounded answer: 0.03 g/L. The concentration table follows.

Conc. (M)	$SrSO_4(s)$	$\rightleftharpoons$	Sr^{2+}	+	SO_4^{2-}
Starting			0		0.0015
Change			+x		+x
Equilibrium			x		0.0015 + x

$$K_{sp} = 2.5 \times 10^{-7} = x(0.0015 + x)$$

$$x^2 + 0.0015x - (2.5 \times 10^{-7}) = 0$$

Solving the quadratic equation gives

$$x = \frac{-0.0015 \pm \sqrt{(0.0015)^2 + 4(2.5 \times 10^{-7})}}{2} = \underline{1}.51 \times 10^{-4}\ M$$

17.31 a. Rounded answer: $Q_c = 2.5 \times 10^{-6}$ (> K_{sp} of 1.0×10^{-6}; precipitation occurs at equilibrium). Calculate Q_c, the ion product, of the solution using the concentrations in the problem as the concentrations present after mixing and assuming no precipitation. Then compare Q_c with K_{sp} to determine whether precipitation has occurred. Start by defining the ion product with brackets as used for the definition of K_{sp}; then use parentheses for the concentrations.

$$Q_c = [Ba^{2+}][F^-]^2$$

$$Q_c = (0.025)(0.010)^2 = 2.\underline{5}0 \times 10^{-6}\ (> 1.0 \times 10^{-6})$$

If no precipitation occurs immediately, the solution will be supersaturated; but at equilibrium, precipitation will occur and the solution will be saturated.

b. Rounded answer: $Q_c = 6.5 \times 10^{-5}$ (> K_{sp} of 1.0×10^{-5}; precipitation occurs at equilibrium). Calculate Q_c, the ion product, of the solution using the concentrations in the problem as the concentrations present after mixing and assuming no precipitation. Then compare Q_c with K_{sp} to determine whether precipitation has occurred. Start by defining the ion product with brackets as used for the definition of K_{sp}; then use parentheses for the concentrations.

$$Q_c = [Mg^{2+}][CO_3^{2-}]$$

$$Q_c = (0.0017)(0.038) = 6.\underline{4}6 \times 10^{-5}\ (> 1.0 \times 10^{-5})$$

If no precipitation occurs immediately, the solution will be supersaturated; but at equilibrium, precipitation will occur and the solution will be saturated.

17.32 a. Rounded answer: $Q_c = 1.9 \times 10^{-5}$ (< K_{sp} of 3.5×10^{-5}; no precipitation at equilibrium). Calculate Q_c, the ion product, of the solution using the concentrations in the problem as the concentrations present after mixing and assuming no precipitation. Then compare Q_c with K_{sp} to determine whether precipitation has occurred. Start by defining the ion product with brackets as used for the definition of K_{sp}; then use parentheses for the concentrations.

$$Q_c = [Sr^{2+}][CrO_4^{2-}]$$

$$Q_c = (0.016)(0.0012) = 1.\underline{9}2 \times 10^{-5} \quad (< 3.5 \times 10^{-5})$$

No precipitation occurs and the solution will be unsaturated.

b. Rounded answer: $Q_c = 6.9 \times 10^{-5}$ (> K_{sp} of 1.6×10^{-5}; precipitation occurs at equilibrium). Calculate Q_c, the ion product, of the solution using the concentrations in the problem as the concentrations present after mixing and assuming no precipitation. Then compare Q_c with K_{sp} to determine whether precipitation has occurred. Start by defining the ion product with brackets as used for the definition of K_{sp}; then use parentheses for the concentrations.

$$Q_c = [Pb^{2+}][Cl^-]^2$$

$$Q_c = (0.0048)(0.12)^2 = 6.\underline{9}12 \times 10^{-5} \quad (> 1.6 \times 10^{-5})$$

If no precipitation occurs immediately, the solution will be supersaturated; but at equilibrium, precipitation will occur and the solution will be saturated.

17.33 Rounded answer: $Q_c = 2.5 \times 10^{-8}$ (> K_{sp} of 1.8×10^{-14}, so precipitation occurs at equilibrium). Calculate the ion product, Q_c, after preparation of the solution, assuming no precipitation has occurred. Compare it with the K_{sp}.

$$Q_c = [Pb^{2+}][CrO_4^{2-}]$$

$$Q_c = (5.0 \times 10^{-4})(5.0 \times 10^{-5}) = 2.\underline{5}0 \times 10^{-8}\ M^2 \quad (> K_{sp} \text{ of } 1.8 \times 10^{-14})$$

The solution is supersaturated before equilibrium is reached. At equilibrium, precipitation occurs and the solution is saturated.

17.34 Rounded answer: $Q_c = 5.0 \times 10^{-9}$ (< K_{sp} of 1.7×10^{-8}, so no precipitation occurs at equilibrium). Calculate the ion product, Q_c, after preparation of the solution and assuming no precipitation has occurred. Compare it with the K_{sp}.

$$Q_c = [Pb^{2+}][SO_4^{2-}]$$

$$Q_c = (5.0 \times 10^{-4})(1.0 \times 10^{-5}) = 5.\underline{0}0 \times 10^{-9}\ M^2 \quad (< K_{sp} \text{ of } 1.7 \times 10^{-8})$$

At equilibrium, no precipitation occurs and the solution is unsaturated.

17.35 Rounded answer: $Q_c = 2.5 \times 10^{-12}$ (< K_{sp}, so no precipitation occurs). Calculate the concentrations of Mg^{2+} and OH^-, assuming no precipitation. Use a total volume of 1.0 L + 1.0 L, or 2.0 L. (Note that the concentrations are halved when the volume is doubled.)

$$[Mg^{2+}] = \frac{\frac{0.0020\ \text{mol}}{\text{L}} \times 1.0\ \text{L}}{2.0\ \text{L}} = 1.\underline{0}0 \times 10^{-3}\ M$$

$$[OH^-] = \frac{\frac{0.00010\ \text{mol}}{\text{L}} \times 1.0\ \text{L}}{2.0\ \text{L}} = 5.\underline{0}0 \times 10^{-5}\ M$$

Calculate the ion product and compare it to K_{sp}.

$$Q_c = (Mg^{2+})(OH^-)^2 = (1.00 \times 10^{-3})(5.00 \times 10^{-5})^2 = 2.\underline{5}0 \times 10^{-12}\ M^3$$

Because Q_c is less than the K_{sp} of 1.8×10^{-11}, no precipitation occurs and the solution is unsaturated.

17.36 Rounded answer: $Q_c = 3.7 \times 10^{-5}$ (> K_{sp}, so precipitation occurs). Calculate the concentrations of Ca^{2+} and SO_4^{2-}, assuming no precipitation. Use a total volume of 0.045 L + 0.055 L, or 0.100 L.

$$[Ca^{2+}] = \frac{\frac{0.015\ \text{mol}}{\text{L}} \times 0.045\ \text{L}}{0.100\ \text{L}} = 0.006\underline{7}5\ M$$

$$[SO_4^{2-}] = \frac{\frac{0.010\ \text{mol}}{\text{L}} \times 0.055\ \text{L}}{0.100\ \text{L}} = 0.005\underline{5}0\ M$$

Calculate the ion product and compare it to K_{sp}.

(continued)

$$Q_c = [Ca^{2+}][SO_4^{2-}] = (0.00675)(0.00550) = 3.\underline{7}1 \times 10^{-5}\ M^2$$

Because Q_c is greater than the K_{sp} of 2.4×10^{-5}, the solution is supersaturated before equilibrium is reached. At equilibrium, precipitation occurs and the solution is saturated.

17.37 Rounded answer: $Q_c = 1.4 \times 10^{-9}$ (< K_{sp}, so no precipitation occurs). Calculate the concentrations of Ba^{2+} and F^-, assuming no precipitation. Use a total volume of 0.045 L + 0.075 L, or 0.120 L.

$$[Ba^{2+}] = \frac{\frac{0.0015\ mol}{L} \times 0.045\ L}{0.120\ L} = 5.\underline{6}25 \times 10^{-4}\ M$$

$$[F^-] = \frac{\frac{0.0025\ mol}{L} \times 0.075\ L}{0.120\ L} = 1.\underline{5}6 \times 10^{-3}\ M$$

Calculate the ion product and compare it to K_{sp}.

$$Q_c = [Ba^{2+}][F^-]^2 = (5.625 \times 10^{-4})(1.56 \times 10^{-3})^2 = 1.\underline{3}6 \times 10^{-9}\ M^3$$

Because Q_c is less than the K_{sp} of 1.0×10^{-6}, no precipitation occurs and the solution is unsaturated.

17.38 Rounded answer: $Q_c = 1.1 \times 10^{-6}$ (< K_{sp}, so no precipitation occurs). Calculate the concentrations of Pb^{2+} and Cl^-, assuming no precipitation. Use a total volume of 0.040 L + 0.065 L, or 0.105 L.

$$[Pb^{2+}] = \frac{\frac{0.010\ mol}{L} \times 0.065\ L}{0.105\ L} = 6.\underline{1}9 \times 10^{-3}\ M$$

$$[Cl^-] = \frac{\frac{0.035\ mol}{L} \times 0.040\ L}{0.105\ L} = 1.\underline{3}3 \times 10^{-2}\ M$$

Calculate the ion product and compare it to K_{sp}.

$$Q_c = [Pb^{2+}][Cl^-]^2 = (6.19 \times 10^{-3})(1.33 \times 10^{-2})^2 = 1.\underline{0}9 \times 10^{-6}\ M^3$$

Because Q is less than the K_{sp} of 1.6×10^{-5}, no precipitation occurs and the solution is unsaturated.

17.39 Rounded answer: 1.8×10^{-3} mol of $CaCl_2$. A mixture of $CaCl_2$ and K_2SO_4 can only precipitate $CaSO_4$ because KCl is soluble. Use the K_{sp} expression to calculate the $[Ca^{2+}]$ needed to just begin precipitating the 0.020 M SO_4^{2-} (in essentially a saturated solution). Then convert to moles.

$$[Ca^{2+}][SO_4^{2-}] = K_{sp} = 2.4 \times 10^{-5}$$

$$[Ca^{2+}] = \frac{K_{sp}}{[SO_4^{2-}]} = \frac{2.4 \times 10^{-5}}{2.0 \times 10^{-2}} = 1.\underline{2}0 \times 10^{-3}\ M$$

The number of moles in 1.5 L of this calcium-containing solution is

$$1.5\ L \times (1.20 \times 10^{-3})\ mol/L = 1.\underline{8}0 \times 10^{-3}\ mol\ Ca^{2+} = 1.\underline{8}0 \times 10^{-3}\ mol\ CaCl_2$$

17.40 Rounded answer: 6.2×10^{-7} g of $MgSO_4$. A mixture of $MgSO_4$ and NaOH can only precipitate $Mg(OH)_2$ because Na_2SO_4 is soluble. Use the K_{sp} expression to calculate the $[Mg^{2+}]$ needed to just begin precipitating the 0.040 M OH^- (in essentially a saturated solution). Then convert to moles and finally to grams.

$$[Mg^{2+}][OH^-]^2 = K_{sp} = 1.8 \times 10^{-11}$$

$$[Mg^{2+}] = \frac{K_{sp}}{[OH^-]} = \frac{1.8 \times 10^{-11}}{(0.040)^2} = 1.\underline{1}25 \times 10^{-8}\ M$$

The number of moles, and grams, in 0.456 L (456 mL) of this magnesium-containing solution is

$$0.456\ L \times (1.125 \times 10^{-8}\ mol/L) = 5.\underline{1}3 \times 10^{-9}\ mol\ Mg^{2+} = 5.\underline{1}3 \times 10^{-9}\ mol\ MgSO_4$$

$$(5.13 \times 10^{-9}\ mol\ MgSO_4) \times (120\ g\ MgSO_4/1\ mol\ MgSO_4) = 6.\underline{1}56 \times 10^{-7}\ g\ MgSO_4$$

17.41 Rounded answers: 6.6×10^{-6} M; 0.013% remaining. Because K_{sp} is small, almost all the Ag^+ and CrO_4^{2-} react until one ion, the *limiting reactant,* is essentially completely precipitated. It will be seen below that this is Ag^+. Assuming complete precipitation of the Ag^+ ion, the amount of CrO_4^{2-} left can be calculated: Subtract half the moles of Ag^+ from the moles of CrO_4^{2-} to find the moles of CrO_4^{2-} unprecipitated (2 Ag^+ are used to form Ag_2CrO_4).

(continued)

$(0.10 \text{ mol } CrO_4^{2-}/L) \times 0.025 \text{ L} = 2.\underline{5}0 \times 10^{-3} \text{ mol } CrO_4^{2-}$ (= $5.\underline{0} \times 10^{-2}$ M in 0.050 L)

$- (0.5 \times 0.10 \text{ mol } Ag^{+}/L) \times 0.025 \text{ L} = 1.\underline{2}5 \times 10^{-3} \text{ mol } Ag_2CrO_4$ (= $2.\underline{5} \times 10^{-2}$ M in 0.050 L)

= Unprecipitated CrO_4^{2-} = $1.\underline{2}5 \times 10^{-3}$ mol CrO_4^{2-} (= $2.\underline{5} \times 10^{-2}$ M in 0.050 L)

At equilibrium, a small amount of Ag_2CrO_4 will dissolve, producing an unknown concentration, x, of Ag^{+} ion. Assemble the usual table for exact equilibrium concentrations. Then assume that x is negligible compared to 2.5×10^{-2} M, and do an approximate calculation of the Ag^{+} concentration.

Conc. (M)	$Ag_2CrO_4(s)$	$\rightleftharpoons$	$2Ag^{+}$	+	CrO_4^{2-}
Starting			0		0.0250
Change			+2x		+x
Equilibrium			2x		0.0250 + x

$$[Ag^{+}]^2[CrO_4^{2-}] = K_{sp}$$

$$(2x)^2(0.0250 + x) \cong (2x)^2(0.0250) \cong 1.1 \times 10^{-12}$$

$$[Ag^{+}] = 2x \cong \sqrt{\frac{1.1 \times 10^{-12}}{0.0250}} \cong 6.\underline{6}3 \times 10^{-6} \text{ M}$$

To calculate the percentage remaining of the initial Ag^{+}, first calculate the initial concentration of Ag^{+}.

$$[Ag^{+}] = \frac{2.50 \times 10^{-3} \text{ mol}}{0.0500 \text{ L}} = 5.\underline{0}0 \times 10^{-2} \text{ M}$$

$$\% \, Ag^{+} \text{ remaining} = [(6.63 \times 10^{-6} \text{ M}) \div (5.00 \times 10^{-2})] \times 100\% = 0.01\underline{3}2\%$$

17.42 Rounded answers: 6.2×10^{-5} M; 0.12% remaining. Because K_{sp} is small, almost all the Ca^{2+} and CO_3^{2-} react completely. The stoichiometric calculation is as follows:

$$\begin{array}{lcl} (0.10 \text{ mol } Ca^{2+}/L) \times 0.025 \text{ L} & = & 2.\underline{5}0 \times 10^{-3} \text{ mol } Ca^{2+} \\ -\ (0.10 \text{ mol } CO_3^{2-}/L) \times 0.025 \text{ L} & = & 2.\underline{5}0 \times 10^{-3} \text{ mol } CO_3^{2-} \\ \hline & = & 0 \text{ mol } Ca^{2+} \text{ and } CO_3^{2-} \end{array}$$

At equilibrium, a small amount of $CaCO_3$ will dissolve, producing an unknown concentration, x, of Ca^{2+} and CO_3^{2-} ions. Assemble the usual table for exact equilibrium concentrations. Then substitute into the K_{sp} equilibrium expression.

Conc. (M)	$CaCO_3(s)$	$\rightleftharpoons$	Ca^{2+}	+	CO_3^{2-}
Starting			0		0
Change			+x		+x
Equilibrium			x		x

$[Ca^{+}][CO_3^{2-}] = K_{sp}$

$(x)(x) = 3.8 \times 10^{-9}$

$x = \sqrt{3.8 \times 10^{-9}} = 6.16 \times 10^{-5} \text{ M} = [Ca^{2+}]$

To calculate the percentage remaining of the initial Ca^{2+}, first calculate the initial concentration of Ca^{2+}.

$$[Ca^{2+}] = \frac{2.50 \times 10^{-3} \text{ mol}}{0.0500 \text{ L}} = 5.\underline{0}0 \times 10^{-2} \text{ M}$$

$\% \; Ca^{2+} \text{ remaining} = [(6.16 \times 10^{-5} \text{ M}) \div (5.00 \times 10^{-2})] \times 100\% = 0.1\underline{2}3\%$

17.43 Rounded answer: 6.9×10^{-9} M. Because the $AgNO_3$ solution is relatively concentrated, ignore the dilution of the solution of Cl^- and I^- from the addition of $AgNO_3$. The $[Ag^+]$ just as the AgCl begins to precipitate can be calculated from the K_{sp} expression for AgCl, using the $[Cl^-] = 0.015$ M. Therefore, for AgCl

$[Ag^+][Cl^-] = K_{sp}$

(continued)

$[Ag^+][0.015] = 1.8 \times 10^{-10}$

$[Ag^+] = \frac{1.8 \times 10^{-10}}{0.015} = 1.\underline{2}0 \times 10^{-8}$ M

The $[I^-]$ at this point can be obtained by substituting the $[Ag^+]$ into the K_{sp} expression for AgI. Therefore, for AgI

$[Ag^+][I^-] = K_{sp}$

$[1.20 \times 10^{-8}][I^-] = 8.3 \times 10^{-17}$

$[I^-] = \frac{8.3 \times 10^{-17}}{1.2 \times 10^{-8}} = 6.\underline{9}1 \times 10^{-9}$ M

17.44 Rounded answer: 2.1×10^{-5} M. Because the problem asks you to find $[Cl^-]$ at the point when Ag_2CrO_4 just begins to precipitate, we start with the K_{sp} expression for Ag_2CrO_4 and ignore temporarily what may have happened to the Cl^- ion. Also, because the $AgNO_3$ solution is relatively concentrated, ignore the dilution of the solution of Cl^- and CrO_4^{2-} from the addition of $AgNO_3$. The $[Ag^+]$ just as the Ag_2CrO_4 begins to precipitate can be calculated from the K_{sp} expression for Ag_2CrO_4, using the $[CrO_4^{2-}] = 0.015$ M. Thus, for Ag_2CrO_4

$[Ag^+]^2[CrO_4^{2-}] = K_{sp}$

$[Ag^+]^2[0.015] = 1.1 \times 10^{-12}$

$[Ag^+] = \sqrt{\frac{1.1 \times 10^{-12}}{0.015}} = 8.\underline{5}6 \times 10^{-6}$ M

The $[Cl^-]$ at this point can be obtained by substituting the $[Ag^+]$ into the K_{sp} expression for AgCl. Therefore, for AgCl

$[Ag^+][Cl^-] = K_{sp}$

$[8.56 \times 10^{-6}][Cl^-] = 1.8 \times 10^{-10}$

$[Cl^-] = \frac{1.8 \times 10^{-10}}{8.56 \times 10^{-6}} = 2.\underline{1}02 \times 10^{-5}$ M

Because this Cl^- concentration is extremely small compared to the initial 0.015 M concentration, you can also deduce that essentially all of the chloride ion has precipitated before the Ag_2CrO_4 begins to precipitate.

17.45 The net ionic equation is $MgC_2O_4(s) + 2H^+(aq) \rightarrow Mg^{2+}(aq) + H_2C_2O_4(aq)$.

17.46 The net ionic equation is $CaF_2(s) + 2H^+(aq) \rightarrow Ca^{2+}(aq) + 2HF(aq)$.

17.47 Calculate the value of K for the reaction of H^+ with both the SO_4^{2-} and F^- anions as they form by the slight dissolving of the insoluble salts. These constants are the reciprocals of the K_a values of the conjugate acids of these anions.

$$F^- + H^+ \rightarrow HF\text{: } K = \frac{1}{K_a} = \frac{1}{6.8 \times 10^{-4}} = 1.\underline{4}7 \times 10^3$$

$$SO_4^{2-} + H^+ \rightarrow HSO_4^-\text{: } K = \frac{1}{K_{a2}} = \frac{1}{1.1 \times 10^{-2}} = 9.\underline{0}9 \times 10^1$$

Because K for the fluoride ion is relatively larger, more of BaF_2 will dissolve in acid than $BaSO_4$.

17.48 Calculate the value of K for the reaction of H^+ with both the SO_4^{2-} and PO_4^{3-} anions as they form by the slight dissolving of the insoluble salts. These constants are the reciprocals of the K_a values of the conjugate acids of these anions.

$$PO_4^{3-} + H^+ \rightarrow HPO_4^{2-}\text{: } K = \frac{1}{K_{a3}} = \frac{1}{4.8 \times 10^{-13}} = 2.\underline{0}8 \times 10^{12}$$

$$SO_4^{2-} + H^+ \text{ “ } HSO_4^-\text{: } K = \frac{1}{K_{a2}} = \frac{1}{1.1 \times 10^{-2}} = 9.\underline{0}9 \times 10^1$$

Because K for the phosphate ion is relatively larger, more of $Ca_3(PO_4)_2$ will dissolve in acid than $CaSO_4$.

17.49 Rounded answer: pH range = 0 to 0.8. Find the *higher end* of the pH range by calculating the minimum concentration of S^{2-} required to prevent precipitation of the more soluble metal sulfide. In this case, this is CoS because its K_{sp} of 4×10^{-21} is much larger than the K_{sp} of 1.6×10^{-52} for HgS. For CoS, this is

$[Co^{2+}][S^{2-}] = K_{sp}$

$(0.10)[S^{2-}] = 4 \times 10^{-21}$

$[S^{2-}] = \underline{4}.0 \times 10^{-20}$ M

(continued)

Substituting into the overall equilibrium-constant ($K_{a1}K_{a2}$) expression for H_2S, and assuming 0.10 M for saturated H_2S, solve for $[H^+]$:

$$\frac{[H^+]^2[S^{2-}]}{[H_2S]} = 1.1 \times 10^{-20}$$

$[H^+]^2 = (1.1 \times 10^{-20})(0.10\ M) \div (4.0 \times 10^{-20}) = 2.75 \times 10^{-2}$; $[H^+] = \underline{1}.658 \times 10^{-1}\ M$

Higher end of pH range $= -\log(1.658 \times 10^{-1}) = 0.\underline{7}803$

Find the *lower end* of the pH range by calculating the sulfide ion concentration required to just begin precipitation of HgS, using the same calculations.

$[Hg^{2+}][S^{2-}] = K_{sp}$

$(0.10)[S^{2-}] = 1.6 \times 10^{-52}$

$[S^{2-}] = 1.6 \times 10^{-51}\ M$

$$\frac{[H^+]^2[S^{2-}]}{[H_2S]} = 1.1 \times 10^{-20}$$

$[H^+]^2 = (1.1 \times 10^{-20})(0.10\ M) \div (1.6 \times 10^{-51}) = 6.875 \times 10^{29}$; $[H^+] = 8.\underline{2}9 \times 10^{14}\ M$

Theoretical lower end of pH $= -\log(8.29 \times 10^{14}) = -14.91$ (practical end: pH = 0.0)

17.50 Rounded answer: pH range = 0.0 to 1.7. Find the *higher end* of the pH range by calculating the minimum concentration of S^{2-} required to prevent precipitation of the most soluble metal sulfide. In this case, this is NiS because its K_{sp} of 3×10^{-19} is much larger than the K_{sp} of 8×10^{-27} for CdS. For NiS, this is

$[Ni^{2+}][S^{2-}] = K_{sp}$

$(0.10)[S^{2-}] = 3 \times 10^{-19}$

$[S^{2-}] = \underline{3}.0 \times 10^{-18}\ M$

Substituting into the overall equilibrium-constant ($K_{a1}K_{a2}$) expression for H_2S, and assuming 0.10 M for saturated H_2S, solve for $[H^+]$:

(continued)

$$\frac{[H^+]^2[S^{2-}]}{[H_2S]} = 1.1 \times 10^{-20}$$

$[H^+]^2 = (1.1 \times 10^{-20})(0.10\ M) \div (3.0 \times 10^{-18}) = \underline{3}.66 \times 10^{-4}$; $[H^+] = \underline{1}.91 \times 10^{-2}\ M$

Higher end of pH range $= -\log(1.91 \times 10^{-2}) = 1.\underline{7}17$

Find the *lower end* of the pH range by calculating the sulfide ion concentration required to just begin precipitation of CdS, using the same calculations.

$[Cd^{2+}][S^{2-}] = K_{sp}$

$(0.10)[S^{2-}] = 8 \times 10^{-27}$

$[S^{2-}] = \underline{8}.0 \times 10^{-26}\ M$

$$\frac{[H^+]^2[S^{2-}]}{[H_2S]} = 1.1 \times 10^{-20}$$

$[H^+]^2 = (1.1 \times 10^{-20})(0.10\ M) \div (8.0 \times 10^{-26}) = \underline{1}.375 \times 10^4$; $[H^+] = \underline{1}.17 \times 10^2\ M$

Theoretical lower end of pH $= -\log(1.17 \times 10^2) = -2.06$ (practical end: pH = 0.0)

17.51 The equation is

$Ag^+(aq) + 2CN^-(aq) \rightarrow Ag(CN)_2^-(aq)$

The K_f expression is

$$K_f = \frac{[Ag(CN)_2^-]}{[Ag^+][CN^-]^2} = 5.6 \times 10^{18}$$

17.52 The equation is

$Ag^+(aq) + 2S_2O_3^{2-}(aq) \rightarrow Ag(S_2O_3)_2^{3-}(aq)$

The K_f expression is

$$K_f = \frac{[Ag(S_2O_3)_2^{3-}]}{[Ag^+][S_2O_3^{2-}]^2} = 2.9 \times 10^{13}$$

17.53 Rounded answer: 5.5×10^{-19} M. Assume that the only $[Ag^+]$ is that in equilibrium with the $Ag(CN)_2^-$ formed from Ag^+ and CN^-. (In other words, assume that all of the 0.015 M Ag^+ reacts with CN^- to form 0.015 M $Ag(CN)_2^-$.) Subtract the CN^- that forms 0.015 M $Ag(CN)_2^-$ from the initial 0.100 M CN^-. Use this as starting concentration of CN^- for the usual concentration table.

[0.100 M NaCN - (2 x 0.015 M)] = 0.070 M starting CN^-

Conc. (M)	$Ag(CN)_2^-$	$\rightleftharpoons$	Ag^+	+	$2CN^-$
Starting	0.015		0		0.070
Change	-x		+x		+2x
Equilibrium	0.015 - x		x		0.070 + 2x

Even though this reaction is the opposite of the equation for the formation constant, the formation-constant expression can be used. Simply substitute all exact equilibrium concentrations into the formation-constant expression; then simplify the exact equation by assuming that x is negligible compared to 0.015 and 2x is negligible compared to 0.070.

$$K_f = \frac{[Ag(CN)_2^-]}{[Ag^+][CN^-]^2} = \frac{(0.015 - x)}{(x)(0.070 + 2x)^2} \cong \frac{(0.015)}{(x)(0.070)^2} \cong 5.6 \times 10^{18}$$

Rearrange and solve for x:

$$x \cong (0.015) \div [(5.6 \times 10^{18})(0.070)^2] \cong 5.\underline{4}6 \times 10^{-19} \text{ M} \cong [Ag^+]$$

17.54 Rounded answer: 1.9×10^{-4} M. You can make a rough assumption that most of the $Zn(OH)_4^{2-}$ that dissociates forms Zn^{2+} and OH^- ions. Assemble the usual concentration table, using 0.20 M as the starting concentration for $Zn(OH)_4^{2-}$ and assuming that the starting concentration of Zn^{2+} and OH^- ions is zero.

Conc. (M)	$Zn(OH)_4^{2-}$	$\rightleftharpoons$	Zn^{2+}	+	$4OH^-$
Starting	0.20		0		0
Change	-x		+x		+4x
Equilibrium	0.20 - x		x		4x

(continued)

Even though this reaction is the opposite of the equation for the formation constant, the formation-constant expression can be used. Simply substitute all exact equilibrium concentrations into the formation-constant expression; then simplify the exact equation by assuming that x is negligible compared to 0.20.

$$K_f = \frac{[Zn(OH)_4^{2-}]}{[Zn^{2+}][OH^-]^4} = \frac{(0.20 - x)}{(x)(4x)^4} \cong \frac{(0.20)}{(x)(4x)^4} \cong 2.8 \times 10^{15}$$

Rearrange and solve for x:

$$x \cong \sqrt[5]{\frac{(0.20)}{(256 \times 2.8 \times 10^{15})}} \cong 1.\underline{9}459 \times 10^{-4}\ M \cong [Zn^{2+}]$$

17.55 Rounded answers: $Q_c = 3 \times 10^{-8}$; precipitation occurs at equilibrium. Start by calculating the $[Cd^{2+}]$ in equilibrium with the $Cd(NH_3)_4^{2+}$ formed from Cd^{2+} and NH_3. Then use the $[Cd^{2+}]$ to decide whether or not CdC_2O_4 will precipitate, by calculating the ion product and comparing it with the K_{sp} of 1.5×10^{-8} for CdC_2O_4. Assume that all of the 0.0020 M Cd^{2+} reacts with NH_3 to form 0.0020 M $Cd(NH_3)_4^{2+}$, and calculate the remaining NH_3. Use these as starting concentrations for the usual concentration table.

$[0.10\ M\ NH_3 - (4 \times 0.0020\ M)] = 0.0\underline{9}2\ M$ starting NH_3

Conc. (M)	$Cd(NH_3)_4^{2+}$ ⇌	Cd^{2+}	+	$4NH_3$
Starting	0.0020	0		0.092
Change	-x	+x		+4x
Equilibrium	0.0020 - x	x		0.092 + 4x

Even though this reaction is the opposite of the equation for the formation constant, the formation-constant expression can be used. Simply substitute all exact equilibrium concentrations into the formation-constant expression; then simplify the exact equation by assuming that x is negligible compared to 0.0020 and 4x is negligible compared to 0.092.

$$K_f = \frac{[Cd(NH_3)_4^{2+}]}{[Cd^{2+}][NH_3]^4} = \frac{(0.0020 - x)}{(x)(0.092 + 4x)^4} \cong \frac{(0.0020)}{(x)(0.092)^4} \cong 1.0 \times 10^7$$

Rearrange and solve for x:

(continued)

$x \cong (0.0020) \div [(1.0 \times 10^7)(0.092)^4] \cong \underline{2}.79 \times 10^{-6}\ M \cong [Cd^{2+}]$

Now calculate the ion product for CdC_2O_4:

$$Q_c = [Cd^{2+}][C_2O_4^{2-}] = (2.79 \times 10^{-6})(0.010) = 2.\underline{7}9 \times 10^{-8}\ M^2$$

Because Q_c is greater than the K_{sp} of 1.5×10^{-8}, the solution is supersaturated before equilibrium. At equilibrium, a precipitate will form and the solution will be saturated.

17.56 Rounded answers: $Q_c = 8 \times 10^{-10}$; precipitation occurs at equilibrium. Start by calculating the $[Ni^{2+}]$ in equilibrium with the $Ni(NH_3)_6^{2+}$ formed from Ni^{2+} and NH_3. Then use the $[Ni^{2+}]$ to decide whether or not $Ni(OH)_2$ will precipitate, by calculating the ion product and comparing it with the K_{sp} of 2.0×10^{-15} for $Ni(OH)_2$. Assume that all of the 0.0020 M Ni^{2+} reacts with NH_3 to form 0.0020 M $Ni(NH_3)_6^{2+}$, and calculate the remaining NH_3. Use these as starting concentrations for the usual concentration table.

$$[0.10\ M\ NH_3 - (6 \times 0.0020\ M)] = 0.0\underline{8}8\ M \text{ starting } NH_3$$

Conc. (M)	$Ni(NH_3)_6^{2+}$ $\rightleftharpoons$	Ni^{2+}	+	$6NH_3$
Starting	0.0020	0		0.088
Change	-x	+x		+6x
Equilibrium	0.0020 - x	x		0.088 + 6x

Even though this reaction is the opposite of the equation for the formation constant, the formation-constant expression can be used. Simply substitute all exact equilibrium concentrations into the formation-constant expression; then simplify the exact equation by assuming that x is negligible compared to 0.0020 and 6x is negligible compared to 0.088.

$$K_f = \frac{[Ni(NH_3)_6^{2+}]}{[Ni^{2+}][NH_3]^6} = \frac{(0.0020 - x)}{(x)(0.088 + 6x)^6} \cong \frac{(0.0020)}{(x)(0.088)^6} \cong 5.6 \times 10^8$$

Rearrange and solve for x:

$$x \cong (0.0020) \div [5.6 \times 10^8 \times (0.088)^6] \cong \underline{7}.69 \times 10^{-6}\ M \cong [Ni^{2+}]$$

Now calculate the ion product for $Ni(OH)_2$:

$$Q_c = [Ni^{2+}][OH^-]^2 = (7.69 \times 10^{-6})(0.010)^2 = \underline{7}.69 \times 10^{-10}\ M^3$$

Because Q_c is greater than the K_{sp} of 2.0×10^{-15}, the solution is supersaturated before equilibrium. At equilibrium, a precipitate will form and the solution will be saturated.

17.57 Rounded answer: 3.0×10^{-3} M. Using the rule from Chapter 14, obtain the overall equilibrium constant for this reaction from the product of the individual equilibrium constants of the two individual equations whose sum gives this equation.

$$CdC_2O_4(s) \rightleftharpoons Cd^{2+}(aq) + C_2O_4^{2-}(aq) \qquad K_{sp} = 1.5 \times 10^{-8}$$

$$Cd^{2+}(aq) + 4NH_3(aq) \rightleftharpoons Cd(NH_3)_4^{2+}(aq) \qquad K_f = 1.0 \times 10^{7}$$

$$CdC_2O_4(s) + 4NH_3(aq) \rightleftharpoons Cd(NH_3)_4^{2+}(aq) + C_2O_4^{2-}(aq)$$

$$K_c = K_{sp} \times K_f = 0.15$$

Assemble the usual table, using 0.10 M as the starting concentration of NH_3 and x as the unknown concentration of $Cd(NH_3)_4^{2+}$ formed.

Conc. (M)	$CdC_2O_4(s)$	+	$4NH_3$	$\rightleftharpoons$	$Cd(NH_3)_4^{2+}$	+	$C_2O_4^{2-}$
Starting			0.10		0		0
Change			-4x		+x		+x
Equilibrium			0.10 - 4x		x		x

The equilibrium-constant expression can now be used. Simply substitute all exact equilibrium concentrations into the equilibrium-constant expression.

$$K_c = \frac{[Cd(NH_3)_4^{2+}][C_2O_4^{2-}]}{[NH_3]^4} = \frac{(x)^2}{(0.10 - 4x)^4} = 0.15$$

Take the square root of both sides of the two right-hand terms, rearrange into a quadratic equation, and solve for x:

$$\frac{x}{(0.10 - 4x)^2} = 0.38\underline{7}$$

$$16x^2 - 3.38x + 0.010 = 0$$

$$x = \frac{3.38 \pm \sqrt{(-3.38)^2 - 4(16)(0.010)}}{2(16)} = 0.208 \text{ (too large) and } 3.\underline{0}01 \times 10^{-3}$$

Using the smaller root, $x = 3.\underline{0}01 \times 10^{-3}$ M = theoretical molar solubility of CdC_2O_4.

17.58 Rounded answer: 1×10^{-8} M. Using the rule from Chapter 14, obtain the overall equilibrium constant for this reaction from the product of the individual equilibrium constants of the two individual equations whose sum gives this equation.

$$NiS(s) \rightleftharpoons Ni^{2+}(aq) + S^{2-}(aq) \quad K_{sp} = 3 \times 10^{-19}$$

$$Ni^{2+}(aq) + 6NH_3(aq) \rightleftharpoons Ni(NH_3)_6^{2+}(aq) \quad K_f = 5.6 \times 10^{8}$$

$$NiS(s) + 6NH_3(aq) \rightleftharpoons Ni(NH_3)_6^{2+}(aq) + S^{2-}(aq)$$

$$K_c = K_{sp} \times K_f = \underline{1}.7 \times 10^{-10}$$

Assemble the usual table, using 0.10 M as the starting concentration of NH_3 and x as the unknown concentration of $Ni(NH_3)_6^{2+}$ formed.

Conc (M)	$NiS(s)$ +	$6NH_3$	$\rightleftharpoons$	$Ni(NH_3)_6^{2+}$	+	S^{2-}
Starting		0.10		0		0
Change		-6x		+x		+x
Equilibrium		0.10 - 6x		x		x

The equilibrium-constant expression can now be used. Simply substitute all exact equilibrium concentrations into the equilibrium-constant expression. It will be necessary to simplify the exact equation to obtain a solution without using a higher-order equation.

$$K_c = \frac{[Ni(NH_3)_6^{2+}][S^{2-}]}{[NH_3]^6} = \frac{(x)^2}{(0.10 - 6x)^6} = 1.7 \times 10^{-10}$$

Take the square root of both sides of the two right-hand terms. Note that the resulting equation contains an x^3 term and an x term.

$$\frac{x}{(0.10 - 6x)^3} = \underline{1}.3 \times 10^{-5}$$

Assume that 6x is negligible compared to 0.10 M, and solve the above equation for x:

$$\frac{x}{(0.10)^3} \cong \underline{1}.3 \times 10^{-5}$$

$x \cong \underline{1}.3 \times 10^{-8}$ = theoretical molar solubility of NiS in 0.10 M NH_3

17.59 The Pb^{2+}, Cd^{2+}, and Sr^{2+} ions can be separated in two steps: (1) Add HCl to precipitate only the Pb^{2+} as $PbCl_2$, leaving the others in solution. (2) After pouring the solution away from the precipitate, add 0.3 M HCl and H_2S to precipitate only the CdS away from the Sr^{2+} ion, whose sulfide is soluble under these conditions.

17.60 The Hg^{2+}, Ca^{2+}, and Na^+ ions can be separated in two steps: (1) Add 0.3 M HCl and H_2S to precipitate only the HgS, leaving the others in solution. (2) After pouring the solution away from the precipitate, add NH_3 and $(NH_4)_2HPO_4$ to precipitate only the $Ca_3(PO_4)_2$ away from the Na^+ ion, whose phosphate is soluble under these conditions.

17.61 a. Ag^+ is not possible because no precipitate formed with HCl.

b. Ca^{2+} is not possible if the compound contains only one cation because Mn^{2+} is indicated by the evidence. However, if the compound consists of two or more cations, Ca^{2+} is possible because no reactions were described involving Ca^{2+}.

c. Mn^{2+} is possible because a precipitate was obtained with basic sulfide ion.

d. Cd^{2+} is not possible because no precipitate was obtained with acidic sulfide solution.

17.62 $PbCrO_4$ is not the compound formed from H_2S because it doesn't contain sulfide ion. CdS is possible because it contains sulfide ion and precipitates in acidic sulfide (Analytical Group II). Neither MnS nor Ag_2S is possible because neither falls into Analytical Group II.

■ Solutions to Unclassified Problems

17.63 Rounded answer: 1.3×10^{-4} M. Assemble the usual concentration table. Let x = the molar solubility of $PbSO_4$. When x mol $PbSO_4$ dissolves in 1 L of solution, x mol Pb^{2+} and x mol SO_4^{2-} form.

Conc. (M)	$PbSO_4(s)$ $\rightleftharpoons$	Pb^{2+}	+	SO_4^{2-}
Starting		0		0
Change		+x		+x
Equilibrium		x		x

(continued)

Substitute the equilibrium concentrations into the equilibrium-constant expression and solve for x.

$[Pb^{2+}][SO_4^{2-}] = K_{sp}$

$(x)(x) = x^2 = 1.7 \times 10^{-8}$

$x = \sqrt{(1.7 \times 10^{-8})} = 1.\underline{3}03 \times 10^{-4}$ M (rounded answer above)

17.64 Rounded answer: 1.3×10^{-26} M. Assemble the usual concentration table. Let x = the molar solubility of HgS. When x mol HgS dissolves in 1 L of solution, x mol Hg^{2+} and x mol S^{2-} form.

Conc. (M)	HgS(s)	$\rightleftharpoons$	Hg^{2+}	+	S^{2-}
Starting			0		0
Change			+x		+x
Equilibrium			x		x

Substitute the equilibrium concentrations into the equilibrium-constant expression and solve for x.

$[Hg^{2+}][S^{2-}] = K_{sp}$

$(x)(x) = x^2 = 1.6 \times 10^{-52}$

$x = \sqrt{(1.6 \times 10^{-52})} = 1.\underline{2}6 \times 10^{-26}$ M (rounded answer above)

17.65 Rounded answers: (a) 6.9×10^{-7} M; (b) 3.2×10^{-4} g/L. Let x = the molar solubility of Hg_2Cl_2. Assemble the usual concentration table and substitute from the table into the equilibrium-constant expression.

Conc. (M)	$Hg_2Cl_2(s)$	$\rightleftharpoons$	Hg_2^{2+}	+	$2Cl^-$
Starting			0		0
Change			+x		+2x
Equilibrium			x		2x

$[Hg_2^{2+}][Cl^-]^2 = K_{sp}$

(continued)

$(x)(2x)^2 = 4x^3 = 1.3 \times 10^{-18}$

$$x = \sqrt[3]{\frac{1.3 \times 10^{-18}}{4}} = 6.\underline{8}7 \times 10^{-7} \text{ M}$$

a. Molar solubility $= 6.\underline{8}7 \times 10^{-7}$ M

b. $$\frac{6.\underline{8}7 \times 10^{-7} \text{ mol}}{\text{L}} \times \frac{472.1 \text{ g } Hg_2Cl_2}{1 \text{ mol}} = 3.\underline{2}4 \text{ g} \times 10^{-4} \ Hg_2Cl_2/\text{L}$$

17.66 Rounded answers: (a) 6.3×10^{-5} M; (b) 8.6×10^{-3} g/L. Let x = the molar solubility of $MgNH_4PO_4$. Assemble the usual concentration table and substitute from the table into the equilibrium-constant expression.

Conc. (M)	$MgNH_4PO_4(s)$	$\rightleftharpoons$	Mg^{2+}	+	NH_4^+	+	PO_4^{3-}
Starting			0		0		0
Change			+x		+x		+x
Equilibrium			x		x		x

$[Mg^{2+}][NH_4^+][PO_4^{3-}] = K_{sp}$

$(x)(x)(x) = x^3 = 2.5 \times 10^{-13}$

$$x = \sqrt[3]{2.5 \times 10^{-13}} = 6.\underline{2}99 \times 10^{-5} \text{ M}$$

a. Molar solubility $= 6.\underline{2}99 \times 10^{-5}$ M

b. $$\frac{6.299 \times 10^{-5} \text{ mol}}{\text{L}} \times \frac{137.3 \text{ g } MgNH_4PO_4}{1 \text{ mol}} = 8.\underline{6}48 \text{ g} \times 10^{-3} \ MgNH_4PO_4 \text{ /L}$$

17.67 Rounded answers: (a) 5.2×10^{-6} M; (b) pOH = 4.81. Let x = the molar solubility of $Ce(OH)_3$. Assemble the usual concentration table and substitute from the table into the equilibrium-constant expression.

Conc. (M)	$Ce(OH)_3(s)$	$\rightleftharpoons$	Ce^{3+}	+	$3OH^-$
Starting			0		0
Change			+x		+3x
Equilibrium			x		3x

(continued)

$[Ce^{3+}][OH^-]^3 = K_{sp}$

$(x)(3x)^3 = 27x^4 = 2.0 \times 10^{-20}$

$$x = \sqrt[4]{\frac{2.0 \times 10^{-20}}{27}} = 5.\underline{2}16 \times 10^{-6}\ M$$

a. Molar solubility = $5.\underline{2}16 \times 10^{-6}$ M

b. $[OH^-] = 3x = 1.\underline{5}6 \times 10^{-5}$ M; pOH = $4.8\underline{0}54$

17.68 Rounded answers: (a) 3.2×10^{-6} M; (b) 1.1×10^{-3} g/L. Let x = the molar solubility of $Cu_2Fe(CN)_6$. Assemble the usual concentration table and substitute from the table into the equilibrium-constant expression.

Conc. (M)	$Cu_2Fe(CN)_6(s)$	$\rightleftharpoons$	$2Cu^{2+}$	+	$Fe(CN)_6^{4-}$
Starting			0		0
Change			+2x		+x
Equilibrium			2x		x

$[Cu^{2+}]^2[Fe(CN)_6^{4-}] = K_{sp}$

$(2x)^2(x) = 4x^3 = 1.3 \times 10^{-16}$

$$x = \sqrt[3]{\frac{1.3 \times 10^{-16}}{4}} = 3.\underline{1}9 \times 10^{-6}\ M$$

a. Molar solubility = $3.\underline{1}9 \times 10^{-6}$ M

b. $$\frac{3.19 \times 10^{-6}\ mol}{L} \times \frac{339.0\ g\ Cu_2Fe(CN)_6}{1\ mol} = 1.\underline{0}8 \times 10^{-3}\ g\ Cu_2Fe(CN)_6/L$$

17.69 Rounded answer: 0.45 M; 26 g/L. Calculate the pOH from pH, and then convert pOH to $[OH^-]$. Then assemble the usual concentration table and substitute from it the equilibrium concentrations into the equilibrium-constant expression.

pOH = 14.00 - 8.80 = 5.20

$[OH^-]$ = antilog (-pOH) = antilog (-5.20) = $6.\underline{3}09 \times 10^{-6}$ M

(continued)

Conc. (M)	$Mg(OH)_2(s)$	$\rightleftharpoons$	Mg^{2+}	+	$2OH^-$
Starting			0		0
Change			+x		—
Equilibrium			x		6.309×10^{-6}

$$K_{sp} = [Mg^{2+}][OH^-]^2 = (x)(6.309 \times 10^{-6})^2 = 1.8 \times 10^{-11}$$

$$x = \frac{1.8 \times 10^{-11}}{(6.309 \times 10^{-6})^2} = 0.4\underline{5}2 \text{ M}$$

$$\frac{0.452 \text{ mol}}{\text{L}} \times \frac{58.3 \text{ g } Mg(OH)_2}{1 \text{ mol}} = 2\underline{6}.3 \text{ g } Mg(OH)_2/\text{L}$$

17.70 Rounded answers: 0.22 M; 52 g/L. Calculate the pOH from pH, and then convert pOH to $[OH^-]$. Then assemble the usual concentration table and substitute from it the equilibrium concentrations into the equilibrium-constant expression. Let 2x equal the molar solubility of the Ag^+ ion because you are seeking x, the molar solubility of Ag_2O.

$$pOH = 14.00 - 10.50 = 3.50$$

$$[OH^-] = \text{antilog}(-pOH) = \text{antilog}(-3.50) = 3.\underline{1}6 \times 10^{-4} \text{ M}$$

Conc. (M)	$Ag_2O(s)$ + $2H_2O$	$\rightleftharpoons$	$2Ag^+$	+	$2OH^-$
Starting			0		0
Change	x		+2x		———
Equilibrium			2x		3.16×10^{-4}

$$K_c = [Ag^+]^2[OH^-]^2 = (2x)^2(3.16 \times 10^{-4})^2 = 2.0 \times 10^{-8}$$

$$x = \sqrt{\frac{2.0 \times 10^{-8}}{4 \times (3.16 \times 10^{-4})^2}} = 0.2\underline{2}4 \text{ M } Ag_2O$$

$$\frac{0.224 \text{ mol}}{1\text{L}} \times \frac{231.7 \text{ g } Ag_2O}{1 \text{ mol}} = 5\underline{1}.9 \text{ g } Ag_2O/\text{L}$$

17.71 Rounded answer: 1.8×10^{-9} M. Let x = the change in M of Mg^{2+}, and 0.10 M = the starting OH^- concentration. Then assemble the usual concentration table and substitute from it the equilibrium concentrations into the equilibrium-constant expression. Assume 2x is negligible compared to 0.10 M and perform an approximate calculation.

Conc. (M)	$Mg(OH)_2(s)$	$\rightleftharpoons$	Mg^{2+}	+	$2OH^-$
Starting			0		0.10
Change			+x		+2x
Equilibrium			x		0.10 + 2x

$$K_{sp} = [Mg^{2+}][OH^-]^2 = (x)(0.10 + 2x)^2 \cong (x)(0.10)^2 \cong 1.8 \times 10^{-11}$$

$$x \cong \frac{1.8 \times 10^{-11}}{(0.10)^2} \cong 1.\underline{8}0 \times 10^{-9} \text{ M } Mg(OH)_2$$

17.72 Rounded answer: 4.6×10^{-24} M. Let x = the change in M of Al^{3+}, and 0.0010 M = the starting OH^- concentration. Then assemble the usual concentration table and substitute from it the equilibrium concentrations into the equilibrium-constant expression. Assume 3x is negligible compared to 0.0010 M and perform an approximate calculation.

Conc. (M)	$Al(OH)_3(s)$	$\rightleftharpoons$	Al^{3+}	+	$3OH^-$
Starting			0		0.0010
Change			+x		+3x
Equilibrium			x		0.0010 + 3x

$$K_{sp} = [Al^{3+}][OH^-]^3 = (x)(0.0010 + 3x)^3 \cong (x)(0.0010)^3 \cong 4.6 \times 10^{-33}$$

$$x \cong \frac{4.6 \times 10^{-33}}{(0.0010)^3} \cong 4.\underline{6}0 \times 10^{-24} \text{ M } Al(OH)_3$$

17.73 Rounded answer: just slightly greater than 8.0×10^{-3} M. To begin precipitation, you must add just slightly more sulfate ion than that required to give a saturated solution. Use the K_{sp} expression to calculate the $[SO_4^{2-}]$ needed to just begin precipitating the 0.0030 M Ca^{2+}.

$$[Ca^{2+}][SO_4^{2-}] = K_{sp} = 2.4 \times 10^{-5}$$

$$[SO_4^{2-}] = \frac{K_{sp}}{[Ca^{2+}]} = \frac{2.4 \times 10^{-5}}{3.0 \times 10^{-3}} = 8.\underline{0}0 \times 10^{-3} \text{ M}$$

When the sulfate ion concentration slightly exceeds 8.0×10^{-3} M, precipitation begins.

17.74 Rounded answer: just slightly greater than 1.4×10^{-2} M. To begin precipitation, you must add just slightly more chromate ion than that required to give a saturated solution. Use the K_{sp} expression to calculate the $[CrO_4^{2-}]$ needed to just begin precipitating the 0.0025 M Sr^{2+}.

$$[Sr^{2+}][CrO_4^{2-}] = K_{sp} = 3.5 \times 10^{-5}$$

$$[CrO_4^{2-}] = \frac{K_{sp}}{[Sr^{2+}]} = \frac{3.5 \times 10^{-5}}{2.5 \times 10^{-3}} = 1.\underline{4}0 \times 10^{-2}\ M$$

When the chromate ion concentration slightly exceeds 1.4×10^{-2} M, precipitation begins.

17.75 Rounded answer: $Q_c = 1.0 \times 10^{-5}$ (< K_{sp}, so the solution is unsaturated). Calculate the concentrations of Pb^{2+} and Cl^-. Use a total volume of 3.20 L + 0.80 L, or 4.00 L.

$$[Pb^{2+}] = \frac{\dfrac{1.25 \times 10^{-3}\ mol}{L} \times 3.20\ L}{4.00\ L} = 1.0\underline{0} \times 10^{-3}\ M$$

$$[Cl^-] = \frac{\dfrac{5.0 \times 10^{-1}\ mol}{L} \times 0.80\ L}{4.00\ L} = 1.\underline{0}0 \times 10^{-1}\ M$$

Calculate the ion product and compare it to K_{sp}.

$$Q_c = [Pb^{2+}][Cl^-]^2 = (1.00 \times 10^{-3})(1.00 \times 10^{-1})^2 = 1.\underline{0}0 \times 10^{-5}\ M^3$$

Because Q_c is less than the K_{sp} of 1.6×10^{-5}, no precipitation occurs and the solution is not saturated.

17.76 Rounded answer: $Q_c = 1.8 \times 10^{-11}$ (= K_{sp}, so solution is saturated). Calculate the concentrations of Mg^{2+} and OH^-. Use a total volume of 0.150 L + 0.050 L, or 0.200 L.

$$[Mg^{2+}] = \frac{\dfrac{2.4 \times 10^{-5}\ mol}{L} \times 0.150\ L}{0.200\ L} = 1.\underline{8}0 \times 10^{-5}\ M$$

$$[OH^-] = \frac{\dfrac{4.0 \times 10^{-3}\ mol}{L} \times 0.050\ L}{0.200\ L} = 1.\underline{0}0 \times 10^{-3}\ M$$

(continued)

Calculate the ion product and compare it to K_{sp}.

$$Q_c = (Mg^{2+})(OH^-)^2 = (1.80 \times 10^{-5})(1.00 \times 10^{-3})^2 = 1.\underline{8}0 \times 10^{-11}\ M^3$$

Because Q_c equals the K_{sp} of 1.8×10^{-11}, no precipitation occurs and the solution is saturated.

17.77 Rounded answer: 5.5×10^{-6} g of NaCl. A mixture of $AgNO_3$ and NaCl can only precipitate AgCl because $NaNO_3$ is soluble. Use the K_{sp} expression to calculate the $[Cl^-]$ needed to prepare a saturated solution (just before precipitating the 0.0015 M Ag^+). Then convert to moles and finally to grams.

$$[Ag^+][Cl^-] = K_{sp} = 1.8 \times 10^{-10}$$

$$[Cl^-] = \frac{K_{sp}}{[Ag^+]} = \frac{1.8 \times 10^{-10}}{(1.5 \times 10^{-3})} = 1.\underline{2}0 \times 10^{-7}\ M$$

The number of moles, and grams, in 0.785 L (785 mL) of this chloride-containing solution is

$$0.785\ L \times (1.20 \times 10^{-7}\ mol/L) = 9.\underline{4}2 \times 10^{-8}\ mol\ Cl^- = 9.\underline{4}2 \times 10^{-8}\ mol\ NaCl$$

$$(9.42 \times 10^{-7}\ mol\ NaCl) \times (58.5\ g\ NaCl/1\ mol\ NaCl) = 5.\underline{5}1 \times 10^{-6}\ g\ NaCl$$

17.78 Rounded answer: 2.4×10^{-6} g of Na_2SO_4. A mixture of $BaCl_2$ and Na_2SO_4 can only precipitate $BaSO_4$ because NaCl is soluble. Use the K_{sp} expression to calculate the $[SO_4^{2-}]$ needed to prepare a saturated solution (just before precipitating the 0.0028 M Ba^{2+}). Then convert to moles and finally to grams.

$$[Ba^{2+}][SO_4^{2-}] = K_{sp} = 1.1 \times 10^{-10}$$

$$[SO_4^{2-}] = \frac{K_{sp}}{[Ba^{2+}]} = \frac{1.1 \times 10^{-10}}{(2.8 \times 10^{-3})} = 3.\underline{9}3 \times 10^{-8}\ M$$

The number of moles and grams in 0.435 L of this sulfate-containing solution is

$$(0.435\ L \times 3.93 \times 10^{-8}\ mol/L) = 1.\underline{7}09 \times 10^{-8}\ mol\ SO_4^{2-} = 1.\underline{7}09 \times 10^{-8}\ mol\ Na_2SO_4$$

$$(1.709 \times 10^{-8}\ mol\ Na_2SO_4) \times (142\ g\ Na_2SO_4/1\ mol\ Na_2SO_4) = 2.\underline{4}2 \times 10^{-6}\ g\ Na_2SO_4$$

17.79 Rounded answers: 2.1×10^{-4}; 0.021% remaining. Because K_{sp} is small, almost all the Ba^{2+} and SO_4^{2-} react completely. The stoichiometric calculation follows:

$$(0.10 \text{ mol } Ba^{2+}/L) \times 0.050 \text{ L} = 5.\underline{0}0 \times 10^{-3} \text{ mol } Ba^{2+}$$

$$- (0.10 \text{ mol } SO_4^{2-}/L) \times 0.050 \text{ L} = 5.\underline{0}0 \times 10^{-3} \text{ mol } SO_4^{2-}$$

$$= 0 \text{ mol } Ba^{2+} \text{ and } SO_4^{2-}$$

At equilibrium, a small amount of $BaSO_4$ will dissolve, producing an unknown concentration, x, of Ba^{2+} and SO_4^{2-} ions. Assemble the usual table for exact equilibrium concentrations. Then substitute into the K_{sp} expression.

Conc. (M)	$BaSO_4(s)$	$\rightleftharpoons$	Ba^{2+}	+	SO_4^{2-}
Starting			0		0
Change			+x		+x
Equilibrium			x		x

$$[Ba^{2+}][SO_4^{2-}] = K_{sp}$$

$$(x)(x) = 1.1 \times 10^{-10}$$

$$x = \sqrt{(1.1 \times 10^{-10})} = 1.\underline{0}4 \times 10^{-5} \text{ M} = [SO_4^{2-}]$$

To calculate the percentage remaining of the initial SO_4^{2-}, first calculate the initial concentration of SO_4^{2-} in the 0.100 L after mixing, assuming precipitation has not started.

$$[SO_4^{2-}] = \frac{\frac{0.10 \text{ mol}}{1 \text{ L}} \times 0.0500 \text{ L}}{0.100 \text{ L}} = 5.\underline{0}0 \times 10^{-2} \text{ M}$$

$$\% \; SO_4^{2-} \text{ not precipitated} = [(1.04 \times 10^{-5} \text{ M}) \div (5.00 \times 10^{-2} \text{ M})] \times 100\% = 0.02\underline{0}8\%$$

17.80 Rounded answers: 2.7×10^{-4}; 0.027 % remaining. Because K_{sp} is small, almost all the Ag^+ and Cl^- react completely. The stoichiometric calculation follows:

$$(0.10 \text{ mol } Ag^+/L) \times 0.050 \text{ L} = 5.\underline{0}0 \times 10^{-3} \text{ mol } Ag^+$$

$$- (0.10 \text{ mol } Cl^-/L) \times 0.050 \text{ L} = 5.\underline{0}0 \times 10^{-3} \text{ mol } Cl^-$$

$$= 0 \text{ mol } Ag^+ \text{ and } Cl^-$$

At equilibrium, a small amount of AgCl will dissolve, producing an unknown concentration, x, of Ag^+ and Cl^- ions. Assemble the usual table for exact equilibrium concentrations. Then substitute into the K_{sp} expression.

Conc. (M)	AgCl(s)	$\rightleftharpoons$	Ag^+	+	Cl^-
Starting			0		0
Change			+x		+x
Equilibrium			x		x

$$[Ag^+][Cl^-] = K_{sp}$$

$$(x)(x) = 1.8 \times 10^{-10}$$

$$x = \sqrt{(1.8 \times 10^{-10})} = 1.\underline{3}4 \times 10^{-5} \text{ M} = [Cl^-]$$

To calculate the percentage remaining of the initial Cl^-, first calculate the initial concentration of Cl^- in the 0.100 L after mixing, assuming precipitation has not started.

$$[Cl^-] = \frac{\frac{0.10 \text{ mol}}{1 \text{ L}} \times 0.0500 \text{ L}}{0.100 \text{ L}} = 5.\underline{0}0 \times 10^{-2} \text{ M}$$

$$\% \; Cl^- \text{ not precipitated} = [(1.34 \times 10^{-5} \text{ M}) \div (5.00 \times 10^{-2} \text{ M})] \times 100\% = 0.02\underline{6}8\%$$

17.81 Rounded answer: 4.7×10^{-2} M. From the magnitude of K_f, assume that Fe^{3+} and SCN^- react essentially completely to form 2.00 M $Fe(SCN)^{2+}$ at equilibrium. Use 2.00 M as the starting concentration of $Fe(SCN)^{2+}$ for the usual concentration table.

Conc. (M)	Fe^{3+}	+	SCN^-	⇌	$FeSCN^{2+}$
Starting	0		0		2.00
Change	+x		+x		-x
Equilibrium	x		x		2.00 - x

Substitute all exact equilibrium concentrations into the formation-constant expression; then simplify the exact equation by assuming that x is negligible compared to 2.00 M.

$$K_f = \frac{[FeSCN^{2+}]}{[Fe^{3+}][SCN^-]} = \frac{(2.00 - x)}{(x)(x)} \cong \frac{(2.00)}{x^2} \cong 9.0 \times 10^2$$

Rearrange and solve for x:

$$x \cong \sqrt{\frac{2.00}{9.0 \times 10^2}} \cong 4.\underline{7}1 \times 10^{-2} \text{ M} \cong [Fe^{3+}]$$

Fraction dissociated = $[(4.\underline{7}1 \times 10^{-2}) \div (2.00)] \times 100\%$

= $0.02\underline{3}5$ (< 0.03, so acceptable)

17.82 Rounded answer: 0.14 M. From the magnitude of K_f, assume that Co^{2+} and SCN^- react essentially completely to form 2.00 M $Co(SCN)^+$ at equilibrium. Use 2.00 M as the starting concentration of $Co(SCN)^+$ for the usual concentration table.

Conc. (M)	Co^{2+}	+	SCN^-	⇌	$CoSCN^+$
Starting	0		0		2.00
Change	+x		+x		-x
Equilibrium	x		x		2.00 - x

Substitute all exact equilibrium concentrations into the formation-constant expression; then simplify the exact equation by assuming that x is negligible compared to 2.00 M.

$$K_f = \frac{[CoSCN^+]}{[Co^{2+}][SCN^-]} = \frac{(2.00 - x)}{(x)(x)} \cong \frac{(2.00)}{x^2} \cong 1.0 \times 10^2$$

(continued)

Rearrange and solve for x:

$$x \cong \sqrt{\frac{2.00}{1.0 \times 10^2}} \cong 0.1\underline{4}1 \text{ M} \cong [Co^{2+}]$$

The value of x = 0.14 affects the term 2.00 - x slightly in the second significant figure. From the quadratic formula, you obtain x = 0.1$\underline{3}$7, which also rounds to 0.14.

17.83 Rounded answer: 1.4×10^{-2} M. Using the rule from Chapter 14, obtain the overall equilibrium constant for this reaction from the product of the individual equilibrium constants of the two individual equations whose sum gives this equation:

$$AgBr(s) \rightleftharpoons Ag^+(aq) + Br^-(aq) \qquad K_{sp} = 5.0 \times 10^{-13}$$

$$Ag^+(aq) + 2NH_3(aq) \rightleftharpoons Ag(NH_3)_2{}^+(aq) \qquad K_f = 1.7 \times 10^7$$

$$AgBr(s) + 2NH_3(aq) \rightleftharpoons Ag(NH_3)_2{}^{2+}(aq) + Br^-(aq)$$

$$K_c = K_{sp} \times K_f = 8.\underline{5}0 \times 10^{-6}$$

Assemble the usual table, using 5.0 M as the starting concentration of NH_3 and x as the unknown concentration of $Ag(NH_3)_2{}^+$ formed.

Conc. (M)	AgBr(s)	+	$2NH_3$	$\rightleftharpoons$	$Ag(NH_3)_2{}^+$	+	Br^-
Starting			5.0		0		0
Change			-2x		+x		+x
Equilibrium			5.0 - 2x		x		x

The equilibrium-constant expression can now be used. Simply substitute all exact equilibrium concentrations into the equilibrium-constant expression; it will not be necessary to simplify the equation because taking the square root of both sides removes the x^2 term.

$$K_c = \frac{[Ag(NH_3)_2{}^+][Br^-]}{[NH_3]^2} = \frac{x^2}{(5.0 - 2x)^2} = 8.\underline{5}0 \times 10^{-6}$$

Take the square root of both sides of the two right-hand terms and solve for x.

$$\frac{x}{(5.0 - 2x)} = 2.\underline{9}15 \times 10^{-3}$$

$$x + (5.8 \times 10^{-3})\,x = 1.\underline{4}575 \times 10^{-2}$$

$x = 1.\underline{4}49 \times 10^{-2}$ M = the molar solubility of AgBr in 5.0 M NH_3

17.84 Rounded answer: 7.5×10^{-5} M. Using the rule from Chapter 14, obtain the overall equilibrium constant for this reaction from the product of the individual equilibrium constants of the two individual equations whose sum gives this equation:

$$AgI(s) \rightleftharpoons Ag^+(aq) + I^-(aq) \quad K_{sp} = 8.3 \times 10^{-17}$$

$$Ag^+(aq) + 2NH_3(aq) \rightleftharpoons Ag(NH_3)_2^+(aq) \quad K_f = 1.7 \times 10^7$$

$$AgI(s) + 2NH_3(aq) \rightleftharpoons Ag(NH_3)_2^{2+}(aq) + I^-(aq)$$

$$K_c = K_{sp} \times K_f = 1.\underline{4}1 \times 10^{-9}$$

Assemble the usual table, using 2.0 M as the starting concentration of NH_3 and x as the unknown concentration of $Ag(NH_3)_2^+$ formed.

Conc. (M)	AgI(s) +	$2NH_3$	$\rightleftharpoons$	$Ag(NH_3)_2^+$	+ I^-
Starting		2.0		0	0
Change		-2x		+x	+x
Equilibrium		2.0 - 2x		x	x

The equilibrium-constant expression can now be used. Simply substitute all exact equilibrium concentrations into the equilibrium-constant expression; it will not be necessary to simplify the equation because taking the square root of both sides removes the x^2 term.

$$K_c = \frac{[Ag(NH_3)_2^+][I^-]}{[NH_3]^2} = \frac{x^2}{(2.0 - 2x)^2} = 1.\underline{4}1 \times 10^{-9}$$

Take the square root of both sides of the two right-hand terms and solve for x.

$$\frac{x}{(2.0 - 2x)} = 3.\underline{7}5 \times 10^{-5}$$

$$x + (7.5 \times 10^{-5})x = 7.\underline{5}0 \times 10^{-5}$$

$x = 7.\underline{4}99 \times 10^{-5}$ M = the molar solubility of AgI in 2.0 M NH_3

17.85 Rounded answers: $[C_2O_4^{2-}] = 3.6 \times 10^{-4}$ M; $[Zn^{2+}] = 4.2 \times 10^{-6}$ M; $K_f = 2.5 \times 10^9$. Start by recognizing that because each zinc oxalate produces one oxalate ion, the solubility of zinc oxalate equals the oxalate concentration:

$$ZnC_2O_4(s) + 4NH_3 \rightleftharpoons Zn(NH_3)_4^{2+} + C_2O_4^{2-}$$

Thus, the $[C_2O_4^{2-}] = 3.6 \times 10^{-4}$ M. Now calculate the $[Zn^{2+}]$ in equilibrium with the oxalate ion, using the K_{sp} expression for zinc oxalate.

$$K_{sp} = [Zn^{2+}][C_2O_4^{2-}] = 1.5 \times 10^{-9}$$

$$[Zn^{2+}] = \frac{1.5 \times 10^{-9}}{(3.6 \times 10^{-4})} = 4.\underline{1}6 \times 10^{-6} \text{ M}$$

To calculate K_f, the $[Zn(NH_3)_4^{2+}]$ term must be calculated. This can be done by recognizing that the molar solubility of ZnC_2O_4 is the sum of the concentration of Zn^{2+} and $Zn(NH_3)_4^{2+}$ ions:

$$\text{Molar solubility of } ZnC_2O_4 = [Zn^{2+}] + [Zn(NH_3)_4^{2+}]$$

$$3.6 \times 10^{-4} = (4.16 \times 10^{-6}) + [Zn(NH_3)_4^{2+}]$$

$$[Zn(NH_3)_4^{2+}] = (3.6 \times 10^{-4}) - (4.16 \times 10^{-6}) = 3.\underline{5}58 \times 10^{-4} \text{ M}$$

Now the $[NH_3]$ term must be calculated by subtracting the ammonia in $[Zn(NH_3)_4^{2+}]$ from the starting NH_3 of 0.0150 M:

$$[NH_3] = 0.0150 - 4\,[Zn(NH_3)_4^{2+}] = 0.0150 - 4(3.558 \times 10^{-4}) = 0.013\underline{5}7 \text{ M}$$

Solve for K_f by substituting the known concentrations into the K_f expression:

$$K_f = \frac{[Zn(NH_3)_4^{2+}]}{[Zn^{2+}][NH_3]^4} = \frac{3.558 \times 10^{-4}}{(4.\underline{1}6 \times 10^{-6})(0.01357)^4} = 2.\underline{5}2 \times 10^9$$

17.86 Rounded answers: $[C_2O_4^{2-}] = 6.1 \times 10^{-3}$ M; $[Cd^{2+}] = 2.5 \times 10^{-6}$ M; $K_f = 1.0 \times 10^7$. Start by recognizing that because each cadmium oxalate produces one oxalate ion, the solubility of cadmium oxalate equals the oxalate concentration:

$$CdC_2O_4(s) + 4NH_3 \rightleftharpoons Cd(NH_3)_4^{2+} + C_2O_4^{2-}$$

Thus, the $[C_2O_4^{2-}] = 6.1 \times 10^{-3}$ M. Now calculate the $[Cd^{2+}]$ in equilibrium with the oxalate ion, using the K_{sp} expression for cadmium oxalate.

$$K_{sp} = [Cd^{2+}][C_2O_4^{2-}] = 1.5 \times 10^{-8}$$

$$[Cd^{2+}] = \frac{1.5 \times 10^{-8}}{(6.1 \times 10^{-3})} = 2.\underline{4}6 \times 10^{-6}\ M$$

To calculate K_f, the $[Cd(NH_3)_4^{2+}]$ term must be calculated. This can be done by recognizing that the molar solubility of CdC_2O_4 is the sum of the concentration of Cd^{2+} and $Cd(NH_3)_4^{2+}$ ions:

$$\text{Molar solubility of } CdC_2O_4 = [Cd^{2+}] + [Cd(NH_3)_4^{2+}]$$

$$6.1 \times 10^{-3} = (2.46 \times 10^{-6}) + [Cd(NH_3)_4^{2+}]$$

$$[Cd(NH_3)_4^{2+}] = (6.1 \times 10^{-3}) - (2.46 \times 10^{-6}) \cong 6.1 \times 10^{-3}\ M$$

Now the $[NH_3]$ term must be calculated by subtracting the ammonia in $[Cd(NH_3)_4^{2+}]$ from the starting NH_3 of 0.0150 M:

$$[NH_3] = 0.150 - 4[Cd(NH_3)_4^{2+}] = 0.150 - 4(6.1 \times 10^{-3}) = 0.1\underline{25}6\ M$$

Solve for K_f by substituting the known concentrations into the K_f expression:

$$K_f = \frac{[Cd(NH_3)_4^{2+}]}{[Cd^{2+}][NH_3]^4} = \frac{6.1 \times 10^{-3}}{(2.46 \times 10^{-6})(0.1256)^4} = 9.\underline{9}6 \times 10^6$$

17.87 The OH^- formed by ionization of NH_3 (to NH_4^+ and OH^-) is a common ion that will precipitate Mg^{2+} as the slightly soluble $Mg(OH)_2$ salt. The simplest way to treat the problem is to calculate the $[OH^-]$ of 0.10 M NH_3 before the soluble Mg^{2+} salt is added. As the soluble Mg^{2+} salt is added, the $[Mg^{2+}]$ will increase until the solution will be saturated with respect to $Mg(OH)_2$ (the next Mg^{2+} ions to be added will precipitate). Calculate the $[Mg^{2+}]$ at the point at which precipitation begins.

(continued)

To calculate the $[OH^-]$ of 0.10 M NH_3, let x = the mol/ L of NH_3 that ionizes, forming x mol/ L of NH_4^+ and x mol/ L of OH^-, and leaving (0.10 - x) M NH_3 in solution. We can summarize the situation in tabular form:

Conc. (M)	NH_3	+	H_2O	$\rightleftharpoons$	NH_4^+	+	OH^-
Starting	0.10				~0		0
Change	-x				+x		+x
Equilibrium	0.10 - x				x		x

The equilibrium-constant equation is:

$$K_c = \frac{[NH_4^+][OH^-]}{[NH_3]} = \frac{x^2}{(0.10 - x)} \cong \frac{x^2}{(0.10)}$$

The value of x can be obtained by rearranging and taking the square root:

$$[OH^-] \cong \sqrt{1.8 \times 10^{-5} \times 0.10} = 1.3\underline{4} \times 10^{-3}\text{ M}$$

Note that (0.10 - x) is not significantly different from 0.10 so x can be neglected in the (0.10 - x) term.

Now use the K_{sp} of $Mg(OH)_2$ to calculate the $[Mg^{2+}]$ of a saturated solution of $Mg(OH)_2$, which essentially will be the Mg^{2+} ion concentration when $Mg(OH)_2$ begins to precipitate.

$$[Mg^{2+}] = \frac{K_{sp}}{[OH^-]^2} = \frac{1.8 \times 10^{-11}}{(1.34 \times 10^{-3})^2} = 1.\underline{0}0 \times 10^{-5} = 1.0 \times 10^{-5}\text{ M}$$

17.88 The OH^- formed by ionization of N_2H_4 (to $N_2H_5^+$ and OH^-) is a common ion that will precipitate Mg^{2+} as the slightly soluble $Mg(OH)_2$ salt. The simplest way to treat the problem is to calculate the $[OH^-]$ of 0.20 M N_2H_4 before the soluble Mg^{2+} salt is added. As the soluble Mg^{2+} salt is added, the $[Mg^{2+}]$ will increase until the solution will be saturated with respect to $Mg(OH)_2$ (the next Mg^{2+} ions to be added will precipitate). Calculate the $[Mg^{2+}]$ at the point at which precipitation begins.

(continued)

To calculate the $[OH^-]$ of 0.20 M N_2H_4, let x = the mol/ L of N_2H_4 that ionizes, forming x mol/ L of $N_2H_5^+$ and x mol/ L of OH^-, and leaving (0.20 - x) M N_2H_4 in solution. We can summarize the situation in tabular form:

Conc. (M)	N_2H_4	+	H_2O	$\rightleftharpoons$	$N_2H_5^+$	+	OH^-
Starting	0.20				~0		0
Change	-x				+x		+x
Equilibrium	0.20 - x				x		x

The equilibrium-constant equation is:

$$K_c = \frac{[N_2H_5^+][OH^-]}{[N_2H_4]} = \frac{x^2}{(0.20 - x)} \cong \frac{x^2}{(0.20)}$$

The value of x can be obtained by rearranging and taking the square root:

$$[OH^-] \cong \sqrt{1.7 \times 10^{-6} \times 0.20} = 5.\underline{8}3 \times 10^{-4}\ M$$

Note that (0.20 - x) is not significantly different from 0.10 so x can be neglected in the (0.20 - x) term.

Now use the K_{sp} of $Mg(OH)_2$ to calculate the $[Mg^{2+}]$ of a saturated solution of $Mg(OH)_2$, which essentially will be the Mg^{2+} ion concentration when $Mg(OH)_2$ begins to precipitate.

$$[Mg^{2+}] = \frac{K_{sp}}{[OH^-]^2} = \frac{1.8 \times 10^{-11}}{(5.83 \times 10^{-4})^2} = 5.\underline{2}9 \times 10^{-5} = 5.3 \times 10^{-5}\ M$$

17.89 a. Use the solubility information to calculate K_{sp}. The reaction is

$$Cu(IO_3)_2(s) \rightleftharpoons Cu^{2+}(aq) + 2IO_3^-(aq)$$

$$2.7 \times 10^{-3} \qquad 2 \times (2.7 \times 10^{-3})$$

$$K_{sp} = [Cu^{2+}][IO_3^-]^2 = [2.7 \times 10^{-3}][2 \times (2.7 \times 10^{-3})]^2 = 7.\underline{8}7 \times 10^{-8}$$

Set up an equilibrium. The reaction is

$$Cu(IO_3)_2(s) \rightleftharpoons Cu^{2+}(aq) + 2IO_3^-(aq)$$

$$y \qquad 0.35 + 2y \approx 0.35$$

(continued)

$K_{sp} = [y][0.35]^2 = 7.87 \times 10^{-8}$

Molar solubility $= y = 6.\underline{4}2 \times 10^{-7} = 6.4 \times 10^{-7}$ M

b. Set up an equilibrium. The reaction is

$$Cu(IO_3)_2(s) \rightleftharpoons Cu^{2+}(aq) + 2IO_3^-(aq)$$

$$y + 0.35 \approx 0.35 \qquad 2y$$

$K_{sp} = [0.35][2y]^2 = 7.87 \times 10^{-8}$ M

Molar solubility $= y = 2.\underline{3}7 \times 10^{-4} = 2.\underline{4} \times 10^{-4}$ M

c. Yes, $Cu(IO_3)_2$ is a 1:2 electrolyte. It takes two IO_3^- ions to combine with one Cu^{2+} ion. The IO_3^- ion is involved as a square term in the K_{sp} expression.

17.90 a. Use the solubility information to calculate K_{sp}. The reaction is

$$Pb(IO_3)_2(s) \rightleftharpoons Pb^{2+}(aq) + 2IO_3^-(aq)$$

$$4.0 \times 10^{-5}\text{ M} \qquad 2 \times (4.0 \times 10^{-5})\text{ M}$$

$K_{sp} = [Pb^{2+}][IO_3^-]^2 = [4.0 \times 10^{-5}][2 \times (4.0 \times 10^{-5})]^2 = 2.\underline{5}6 \times 10^{-13}$

Set up an equilibrium. The reaction is

$$Pb(IO_3)_2(s) \rightleftharpoons Pb^{2+}(aq) + 2IO_3^-(aq)$$

$$y + 0.15 \approx 0.15 \qquad 2y$$

$K_{sp} = [0.15][2y]^2 = 2.56 \times 10^{-13}$ M

Molar solubility $= y = 6.\underline{5}3 \times 10^{-7} = 6.5 \times 10^{-7}$ M

b. Because of a common ion effect or a consideration of Le Chatelier's principle, the molar solubility would be less.

17.91 a. Set up an equilibrium. The reaction and equilibrium-constant expression are

$$PbI_2(s) \rightleftharpoons Pb^{2+}(aq) + 2I^-(aq) \qquad K_{sp} = [Pb^{2+}][I^-]^2 = 6.5 \times 10^{-9}$$

The Pb^{2+} ion concentration is 0.0150 M. Plug this in and solve for the iodine ion concentration.

$$[I^-] = \sqrt{\frac{6.5 \times 10^{-9}}{0.0150}} = 6.\underline{5}8 \times 10^{-4} = 6.6 \times 10^{-4} M$$

b. Solve the equilibrium-constant expression for the lead ion concentration.

$$[Pb^{2+}] = \frac{6.5 \times 10^{-9}}{(2.0 \times 10^{-3})^2} = 1.\underline{6}3 \times 10^{-3} M$$

The percent of the lead(II) ion remaining in solution is

$$\% \, Pb^{2+} \text{ remaining} = \frac{1.63 \times 10^{-3}}{0.0150} \times 100\% = 1\underline{0}.9 = 11\%$$

17.92 a. Set up an equilibrium. The reaction and equilibrium-constant expression are

$$CaF_2(s) \rightleftharpoons Ca^{2+}(aq) + 2F^-(aq) \qquad K_{sp} = [Ca^{2+}][F^-]^2 = 3.4 \times 10^{-11}$$

Solve the equilibrium-constant expression for the fluoride ion concentration. The calcium ion concentration is 0.00750 M.

$$[F^-] = \sqrt{\frac{3.4 \times 10^{-11}}{0.00750}} = 6.\underline{7}3 \times 10^{-5} = 6.7 \times 10^{-5} M$$

b. Solve the equilibrium-constant expression for the calcium ion concentration. The fluoride ion concentration is 9.5×10^{-4} M.

$$[Ca^{2+}] = \frac{3.4 \times 10^{-11}}{(9.5 \times 10^{-4})^2} = 3.\underline{7}7 \times 10^{-5} M$$

The percent of the calcium ion precipitated is

$$\% \, Ca^{2+} \text{ precipitated} = \frac{0.00750 - 3.77 \times 10^{-5}}{0.00750} \times 100\% = 9\underline{9}.4 = 99\%$$

17.93 a. The reaction and equilibrium-constant expression are

$$Co(OH)_2(s) \rightleftharpoons Co^{2+}(aq) + 2OH^-(aq) \qquad K_{sp} = [Co^{2+}][OH^-]^2$$

From the molar solubility, $[Co^{2+}] = 5.4 \times 10^{-6}$ M and $[OH^-] = 2 \times (5.4 \times 10^{-6})$ M. Therefore,

$$K_{sp} = [5.4 \times 10^{-6}][2 \times (5.4 \times 10^{-6})]^2 = 6.\underline{3}0 \times 10^{-16} = 6.3 \times 10^{-16}$$

b. From the pOH (14 - pH), the $[OH^-] = 10^{-3.57} = 2.\underline{6}9 \times 10^{-4}$ M. The molar solubility is equal to the cobalt ion concentration at equilibrium.

$$[Co^{2+}] = \frac{K_{sp}}{[OH^-]^2} = \frac{6.3 \times 10^{-16}}{(2.69 \times 10^{-4})^2} = 8.\underline{7}1 \times 10^{-9} = 8.7 \times 10^{-9}\ M$$

c. A common ion effect (OH^-) in part (b) decreases the solubility of Co^{2+}.

17.94 a. The reaction and equilibrium-constant expression are

$$Be(OH)_2(s) \rightleftharpoons Be^{2+}(aq) + 2OH^-(aq) \qquad K_{sp} = [Be^{2+}][OH^-]^2$$

From the molar solubility, $[Be^{2+}] = 8.6 \times 10^{-7}$ M and $[OH^-] = 2 \times (8.6 \times 10^{-7})$ M. Therefore,

$$K_{sp} = [8.6 \times 10^{-7}][2 \times (8.6 \times 10^{-7})]^2 = 2.\underline{5}4 \times 10^{-18} = 2.5 \times 10^{-18}$$

b. This is a buffer system. The reaction is

$$NH_3 + H_2O \rightleftharpoons NH_4^+ + OH^- \qquad K_b = 1.8 \times 10^{-5}$$

Set up the equilibrium and calculate the OH^- ion concentration.

$$K_b = \frac{[NH_4^+][OH^-]}{[NH_3]} = \frac{(0.25)[OH^-]}{(1.50)} = 1.8 \times 10^{-5}$$

$$[OH^-] = 1.\underline{0}8 \times 10^{-4}\ M$$

The molar solubility is equal to the beryllium ion concentration at equilibrium.

$$[Be^{2+}] = \frac{K_{sp}}{[OH^-]^2} = \frac{2.5 \times 10^{-18}}{(1.08 \times 10^{-4})^2} = 2.\underline{1}4 \times 10^{-10} = 2.1 \times 10^{-10}\ M$$

c. A common ion effect (a Le Chatelier stress) decreases the solubility of Be^{2+}.

17.95 a. $AgCl(s) \rightleftharpoons Ag^{+}(aq) + Cl^{-}(aq)$ $K_{sp} = 1.8 \times 10^{-10}$

$Ag^{+}(aq) + 2NH_3(aq) \rightleftharpoons Ag(NH_3)_2^{+}(aq)$ $K_f = 1.7 \times 10^{7}$

$AgCl(s) + 2NH_3(aq) \rightleftharpoons Ag(NH_3)_2^{+}(aq) + Cl^{-}(aq)$

$$K = K_{sp} \times K_f = (1.8 \times 10^{-10})(1.7 \times 10^{7}) = 3.\underline{0}6 \times 10^{-3} = 3.1 \times 10^{-3}$$

b. The equilibrium-constant expression is

$$\frac{[Ag(NH_3)_2^{+}][Cl^{-}]}{[NH_3]^2} = \frac{y^2}{(0.80)^2} = 3.06 \times 10^{-3}$$

$$y = [Ag(NH_3)_2^{+}] = 0.04\underline{4}3 = 0.044 \text{ M}$$

$$\text{mol AgCl dissolved} = \text{mol } Ag(NH_3)_2^{+} = 0.044 \text{ mol/L} \times 1.00 \text{ L} = 0.044 \text{ mol}$$

$$\text{mol } NH_3 \text{ reacted} = 0.0443 \text{ mol } Ag(NH_3)_2^{+} \times \frac{2 \text{ mol } NH_3}{1 \text{ mol } Ag(NH_3)_2^{+}} = 0.08\underline{8}6 \text{ mol}$$

The total moles of NH_3 added $= 0.80 \text{ M} \times 1.00 \text{ L} + 0.08\underline{8}6 = 0.8\underline{8}9 = 0.89$ mol

17.96 a. $AgBr(s) \rightleftharpoons Ag^{+}(aq) + Br^{-}(aq)$ $K_{sp} = 5.0 \times 10^{-13}$

$Ag^{+}(aq) + 2S_2O_3^{2-}(aq) \rightleftharpoons Ag(S_2O_3)_2^{3-}$ $K_f = 2.9 \times 10^{13}$

$AgBr(s) + 2S_2O_3^{2-}(aq) \rightleftharpoons Ag(S_2O_3)^{3-}(aq) + Br^{-}(aq)$

$$K = K_{sp} \times K_f = (5.0 \times 10^{-13})(2.9 \times 10^{13}) = 1\underline{4}.5 = 15$$

$$\text{mol AgBr} = 2.5 \text{ g AgBr} \times \frac{1 \text{ mol AgBr}}{187.77 \text{ g AgBr}} = 0.01\underline{3}3 \text{ mol}$$

$$\text{mol } Ag(S_2O_3)_2^{3-} = \text{mol AgBr} = 0.0133 \text{ mol}$$

$$\frac{[Ag(S_2O_3)_2^{2-}][Br^{-}]}{[S_2O_3^{2-}]^2} = \frac{(0.0133)^2}{[S_2O_3^{2-}]^2} = 14.5$$

(continued)

Take the square root of both sides to obtain the moles of $S_2O_3^{2-}$ at equilibrium.

$$\frac{0.0133}{[S_2O_3^{2-}]} = \sqrt{14.5}$$

$[S_2O_3^{2-}] = 0.003\underline{4}9$ mol/L

The moles of $S_2O_3^{2-}$ in $Ag(S_2O_3)_2^{3-}$ = 2 x 0.0133 = $0.02\underline{6}6$ mol

Total moles of $S_2O_3^{2-}$ = $0.02\underline{6}6$ + 0.00349 = $0.03\underline{0}1$ = 0.030 mol

The moles of $Na_2S_2O_3$ added = moles of $S_2O_3^{2-}$ = 0.030 mol

17.97 a. The reaction and equilibrium-constant expression are

$$H_2S(aq) + 2H_2O \rightleftharpoons 2H_3O^+(aq) + S^{2-}(aq)$$

$$K = K_1 \times K_2 = \frac{[H_3O^+]^2[S^{2-}]}{[H_2S]} = 1.1 \times 10^{-20}$$

From the pH, $[H_3O^+] = 10^{-pH} = 10^{-3.25} = 5.\underline{6}2 \times 10^{-4}$ M. Also, assume $[H_2S] \approx 0.10$ M. Solve the equilibrium-constant expression for $[S^{2-}]$.

$$[S^{2-}] = \frac{(1.1 \times 10^{-20})(0.10)}{(5.62 \times 10^{-4})^2} = 3.\underline{4}8 \times 10^{-15} = 3.5 \times 10^{-15}\ \text{M}$$

b. The reaction and equilibrium-constant expression are

$$FeS(s) \rightleftharpoons Fe^{2+}(aq) + S^{2-}(aq) \qquad K_{sp} = [Fe^{2+}][S^{2-}] = 6 \times 10^{-18}$$

Solving for $[Fe^{2+}]$ gives

$$[Fe^{2+}] = \frac{(6 \times 10^{-18})}{(3.48 \times 10^{-15})} = \underline{1}.72 \times 10^{-3}\ \text{M}$$

The percent Fe^{2+} that has precipitated is given by

$$\%\ Fe^{2+}\ \text{precipitated} = \frac{0.10 - 1.72 \times 10^{-3}}{0.10} \times 100\% = 9\underline{8}.2 = 98\ \%$$

17.98 a. The reaction and equilibrium-constant expression are

$$H_2S(aq) + 2H_2O \rightleftharpoons 2H_3O^+(aq) + S^{2-}(aq)$$

$$K = K_1 \times K_2 = \frac{[H_3O^+]^2[S^{2-}]}{[H_2S]} = 1.1 \times 10^{-20}$$

From the pH, $[H_3O^+] = 10^{-pH} = 10^{-3.10} = 7.\underline{9}4 \times 10^{-4}$ M. Also, assume $[H_2S] \approx 0.10$ M. Solve the equilibrium-constant expression for $[S^{2-}]$.

$$[S^{2-}] = \frac{(1.1 \times 10^{-20})(0.10)}{(7.94 \times 10^{-4})^2} = 1.\underline{7}4 \times 10^{-15} = 1.7 \times 10^{-15} \text{ M}$$

b. The reaction and equilibrium-constant expression are

$$NiS(s) \rightleftharpoons Ni^{2+}(aq) + S^{2-}(aq) \qquad K_{sp} = [Ni^{2+}][S^{2-}] = 3 \times 10^{-19}$$

Solving for $[Ni^{2+}]$ gives

$$[Ni^{2+}] = \frac{(3 \times 10^{-19})}{(1.74 \times 10^{-15})} = \underline{1}.72 \times 10^{-4} \text{ M}$$

The percent Ni^{2+} that has precipitated is given by

$$\% \ Fe^{2+} \text{ precipitated} = \frac{0.15 - 1.72 \times 10^{-4}}{0.15} \times 100\% = 9\underline{9}.9 = 1.0 \times 10^{2} \%$$

17.99 a. The moles of NH_4Cl (molar mass 53.49 g/mol) are given by

$$\text{mol } NH_4Cl = 26.7 \text{ g} \times \frac{1 \text{ mol } NH_4Cl}{53.49 \text{ g } NH_4Cl} = 0.49\underline{9}2 \text{ mol}$$

The reaction and equilibrium-constant expression are

$$NH_3 + H_2O \rightleftharpoons NH_4^+ + OH^- \qquad K_b = \frac{[NH_4^+][OH^-]}{[NH_3]} = 1.8 \times 10^{-5}$$

Solve the equilibrium-constant expression for $[OH^-]$. The molarity of the NH_4Cl is $0.49\underline{9}2$ mol/1.0 L = $0.4\underline{9}92$ M.

$$[OH^-] = \frac{(1.8 \times 10^{-5})(4.2)}{(0.4992)} = 1.\underline{5}1 \times 10^{-4} = 1.5 \times 10^{-4} \text{ M}$$

b. The reaction and equilibrium-constant expression are

$$Mg(OH)_2 \rightleftharpoons Mg^{2+} + 2OH^- \qquad K_{sp} = [Mg^{2+}][OH^-]^2 = 1.8 \times 10^{-11}$$

Solving for $[Mg^{2+}]$ gives

$$[Mg^{2+}] = \frac{(1.8 \times 10^{-11})}{(1.51 \times 10^{-4})^2} = 7.\underline{8}9 \times 10^{-4} = 7.9 \times 10^{-4} \text{ M}$$

The percent Mg^{2+} that has been removed is given by

$$\% \, Mg^{2+} \text{ removed} = \frac{0.075 - 7.89 \times 10^{-4}}{0.075} \times 100\% = 9\underline{8}.9 = 99\,\%$$

17.100 a. The moles of $(NH_4)_2SO_4$ (molar mass 132.15 g/mol) are given by

$$\text{mol } (NH_4)_2SO_4 = 75.8 \text{ g} \times \frac{1 \text{ mol } (NH_4)_2SO_4}{132.15 \text{ g } (NH_4)_2SO_4} = 0.57\underline{3}6 \text{ mol}$$

The moles of NH_4^+ = 2 x 0.5736 = 1.1$\underline{4}$7 mol. The reaction and equilibrium-constant expression are

$$NH_3 + H_2O \rightleftharpoons NH_4^+ + OH^- \qquad K_b = \frac{[NH_4^+][OH^-]}{[NH_3]} = 1.8 \times 10^{-5}$$

Solve the equilibrium-constant expression for $[OH^-]$. The molarity of the NH_4Cl is 1.1$\underline{4}$7 mol/1.0 L = 1.1$\underline{4}$7 M.

$$[OH^-] = \frac{(1.8 \times 10^{-5})(1.6)}{(1.147)} = 2.\underline{5}1 \times 10^{-5} = 2.5 \times 10^{-5} \text{ M}$$

b. The reaction and equilibrium-constant expression are

$$Mn(OH)_2 \rightleftharpoons Mn^{2+} + 2OH^- \qquad K_{sp} = [Mn^{2+}][OH^-]^2 = 4.6 \times 10^{-14}$$

Solving for $[Mn^{2+}]$ gives

$$[Mn^{2+}] = \frac{(4.6 \times 10^{-14})}{(2.51 \times 10^{-5})^2} = 7.\underline{3}0 \times 10^{-5} = 7.3 \times 10^{-5} \text{ M}$$

The percent Mn^{2+} that has been removed is given by

$$\% \, Mn^{2+} \text{ removed} = \frac{0.058 - 7.30 \times 10^{-5}}{0.058} \times 100\% = 9\underline{9}.9 = 1.0 \times 10^{2}\,\%$$

■ Cumulative-Skills Problems

(require skills from previous chapters)

17.101 Rounded answer: 0.18 M SO_4^{2-}. Using the K_{sp} of 1.1×10^{-21} for ZnS, calculate the $[S^{2-}]$ needed to maintain a saturated solution (without precipitation):

$$[S^{2-}] = \frac{1.1 \times 10^{-21}}{1.5 \times 10^{-4}\ M\ Zn^{2+}} = 7.\underline{3}3 \times 10^{-18}$$

Next use the overall H_2S ionization expression to calculate the $[H^+]$ needed to achieve this $[S^{2-}]$ level:

$$[H^+] = \sqrt{\frac{1.1 \times 10^{-20}\ (0.10\ M\ H_2S)}{7.33 \times 10^{-18}\ M\ S^{2-}}} = 1.\underline{2}25 \times 10^{-2}\ M$$

Finally, calculate the buffer ratio of $[SO_4^{2-}]/[HSO_4^-]$ from the H_2SO_4 K_{a2} expression, where K_{a2} has the value 1.1×10^{-2}:

$$\frac{[SO_4^{2-}]}{[HSO_4^-]} = \frac{K_{a2}}{[H^+]} = \frac{1.1 \times 10^{-2}}{1.\underline{2}25 \times 10^{-2}} = \frac{0.8979}{1.000}$$

If $[HSO_4^-] = 0.20$ M, then

$$[SO_4^{2-}] = 0.8979 \times 0.20\ M = 0.1\underline{7}95\ M$$

17.102 Rounded answer: 0.3 M. Using the K_{sp} of 4×10^{-21} for CoS, calculate the $[S^{2-}]$ needed to maintain a saturated solution (without precipitation):

$$[S^{2-}] = \frac{4 \times 10^{-21}}{1.8 \times 10^{-4}\ M\ Co^{2+}} = \underline{2}.22 \times 10^{-17}$$

Next use the overall H_2S ionization expression to calculate the $[H^+]$ needed to achieve this $[S^{2-}]$ level:

$$[H^+] = \sqrt{\frac{1.1 \times 10^{-20}\ (0.10\ M\ H_2S)}{2.22 \times 10^{-17}\ M\ S^{2-}}} = \underline{7}.04 \times 10^{-3}\ M$$

Finally, calculate the buffer ratio of $[SO_4^{2-}]/[HSO_4^-]$ from the H_2SO_4 K_{a2} expression, where K_{a2} has the value 1.1×10^{-2}:

(continued)

$$\frac{[SO_4^{2-}]}{[HSO_4^-]} = \frac{K_{a2}}{[H^+]} = \frac{1.1 \times 10^{-2}}{7.04 \times 10^{-3}} = \frac{1.56}{1.000}$$

If $[HSO_4^-] = 0.20$ M, then

$$[SO_4^{2-}] = \underline{1}.56 \times 0.20 \text{ M} = 0.\underline{3}12 \text{ M}$$

17.103 Rounded answer: 2.7×10^{-4} M. Begin by solving for $[H^+]$ in the buffer. Ignoring changes in $[HCHO_2]$ as a result of ionization in the buffer, you obtain

$$[H^+] \cong 1.7 \times 10^{-4} \times \frac{0.45 \text{ M}}{0.20 \text{ M}} \cong 3.\underline{8}25 \times 10^{-4}$$

You should verify that this approximation is valid (you obtain this same result from the Henderson-Hasselbalch equation). The equilibrium for the dissolution of CaF_2 in acidic solution is obtained by subtracting twice the acid ionization of HF from the solubility equilibrium of CaF_2:

$$CaF_2(s) \rightleftharpoons Ca^{2+}(aq) + 2F^-(aq) \qquad K_{sp}$$

$$2H^+(aq) + 2F^-(aq) \rightleftharpoons 2HF(aq) \qquad 1/(K_a)^2$$

$$2H^+(aq) + CaF_2(s) \rightleftharpoons Ca^{2+}(aq) + 2HF(aq) \qquad K_c = K_{sp}/(K_a)^2$$

Therefore, $K_c = (3.4 \times 10^{-11})/(6.8 \times 10^{-4})^2 = 7.\underline{35} \times 10^{-5}$. In order to solve the equilibrium-constant equation, you require the concentration of HF, which you obtain from the acid-ionization constant for HF.

$$K_a = \frac{[H^+][F^-]}{[HF]}, \text{ or } 6.8 \times 10^{-4} = (3.825 \times 10^{-4})[F^-]/[HF]$$

$$[F^-]/[HF] = 1.778, \text{ or } [F^-] = 1.\underline{7}78\,[HF]$$

Let x be the solubility of CaF_2 in the buffer. Then $[Ca^{2+}] = x$, and $[F^-] + [HF] = 2x$. Substituting from the previous equation, you obtain

$$2x = 1.778[HF] + [HF] = 2.778[HF], \text{ or } [HF] = 2x/2.\underline{7}78$$

You can now substitute for $[H^+]$ and $[HF]$ into the equation for K_c.

$$K_c = \frac{[Ca^{2+}][HF]^2}{[H^+]^2} = \frac{x(2x/2.778)^2}{(3.825 \times 10^{-4})^2} = 7.35 \times 10^{-5}$$

(continued)

$7.35 \times 10^{-5} = (3.543 \times 10^{6})x^3$

$x^3 = 2.075 \times 10^{-11}$

$x = 2.\underline{7}48 \times 10^{-4}$

17.104 Rounded answer: 1.2×10^{-3} M. Begin by solving for $[H^+]$ in the buffer. Ignoring changes in $[HC_2H_3O_2]$ as a result of ionization in the buffer, you obtain

$$[H^+] \cong 1.7 \times 10^{-5} \times \frac{0.45\ M}{0.20\ M} \cong 3.\underline{8}25 \times 10^{-5}$$

You should verify that this approximation is valid (you obtain this same result from the Henderson-Hasselbalch equation). The equilibrium for the dissolution of MgF_2 in acidic solution is obtained by subtracting twice the acid ionization of HF from the solubility equilibrium of MgF_2:

$$MgF_2(s) \rightleftharpoons Mg^{2+}(aq) + 2F^-(aq) \quad K_{sp}$$

$$2H^+(aq) + 2F^-(aq) \rightleftharpoons 2HF(aq) \quad 1/(K_a)^2$$

$$2H^+(aq) + MgF_2(s) \rightleftharpoons Mg^{2+}(aq) + 2HF(aq) \quad K_c = K_{sp}/(K_a)^2$$

Therefore, $K_c = (6.5 \times 10^{-9}) / (6.8 \times 10^{-4})^2 = 1.\underline{4}06 \times 10^{-2}$. In order to solve the equilibrium-constant equation, you require the concentration of HF, which you obtain from the acid-ionization constant for HF.

$$K_a = \frac{[H^+][F^-]}{[HF]}, \text{ or } 6.8 \times 10^{-4} = (3.825 \times 10^{-5})[F^-]/[HF]$$

$[F^-]/[HF] = 17.78$, or $[F^-] = 1\underline{7}.78\ [HF]$

Let x be the solubility of MgF_2 in the buffer. Then $[Mg^{2+}] = x$, and $[F^-] + [HF] = 2x$. Substituting from the previous equation, you obtain

$2x = 17.78[HF] + [HF] = 18.78[HF]$, or $[HF] = 2x/1\underline{8}.78$

You can now substitute for $[H^+]$ and $[HF]$ into the equation for K_c.

$$K_c = \frac{[Mg^{2+}][HF]^2}{[H^+]^2} = \frac{x(2x/18.78)^2}{(3.825 \times 10^{-5})^2} = 1.406 \times 10^{-2}$$

$1.406 \times 10^{-2} = (7.752 \times 10^{6})x^3$

(continued)

$x^3 = 1.812 \times 10^{-9}$

$x = 1.\underline{2}19 \times 10^{-3}$

17.105 Rounded answers: $[SO_4^{2-}] = [Mg^{2+}] = 0.109$ M; $[Ba^{2+}] = 1.0 \times 10^{-9}$ M; and $[OH^-] = 1.3 \times 10^{-5}$ M. The net ionic equation is

$$Ba^{2+}(aq) + 2OH^-(aq) + Mg^{2+}(aq) + SO_4^{2-}(aq) \rightleftharpoons BaSO_4(s) + Mg(OH)_2(s)$$

Start by calculating the mol/L of each ion after mixing and before precipitation. Use a total volume of 0.0450 + 0.0670 L = 0.112 L.

$$\text{M of } SO_4^{2-} \text{ and } Mg^{2+} = (0.350 \text{ mol/L} \times 0.0670 \text{ L}) \div 0.112 \text{ L} = 0.20\underline{9}4 \text{ M}$$

$$\text{M of } Ba^{2+} = (0.250 \text{ mol/L} \times 0.0450 \text{ L}) \div 0.112 \text{ L} = 0.10\underline{0}4 \text{ M}$$

$$\text{M of } OH^- = (2 \times 0.250 \text{ mol/L} \times 0.0450 \text{ L}) \div 0.112 \text{ L} = 0.20\underline{0}89 \text{ M}$$

Assemble a table showing the precipitation of $BaSO_4$ and $Mg(OH)_2$.

Conc. (M)	Ba^{2+} +	SO_4^{2-} +	Mg^{2+} +	$2OH^-$ $\rightleftharpoons BaSO_4(s) + Mg(OH)_2(s)$
Starting	0.1004	0.2094	0.2094	0.20089
Change	-0.1004	-0.1004	-0.1004	-0.20089
Equilibrium	0.0000	0.1090	0.1090	0.0000

To calculate $[Ba^{2+}]$, use the K_{sp} expression for $BaSO_4$.

$$[Ba^{2+}] = \frac{1.1 \times 10^{-10}}{0.1090 \text{ M } SO_4^{2-}} = 1.\underline{0}09 \times 10^{-9} \text{ M}$$

$[SO_4^{2-}] = 0.10\underline{9}0$ M

Calculate the $[OH^-]$ using the K_{sp} expression for $Mg(OH)_2$:

$$[OH^-] = \sqrt{\frac{1.8 \times 10^{-11}}{0.1090 \text{ M } Mg^{2+}}} = 1.\underline{2}8 \times 10^{-5} \text{ M}$$

$[Mg^{2+}] = 0.10\underline{9}0$ M

17.106 Rounded answers: $[SO_4^{2-}]$ = 0.0067 M; $[Pb^{2+}]$ = 2.5×10^{-6} M; $[Ag^+]$ = 0.0133 M; and $[Cl^-]$ = 1.4×10^{-8} M. The net ionic equation is

$$Pb^{2+}(aq) + 2Cl^-(aq) + 2Ag^+(aq) + SO_4^{2-}(aq) \rightleftharpoons PbSO_4(s) + 2AgCl(s)$$

Start by calculating the mol/L of each ion after mixing and before precipitation. Use a total volume of 0.0250 + 0.0500 L = 0.0750 L.

$$\text{M of } SO_4^{2-} = (0.0150 \text{ mol/L} \times 0.0500 \text{ L}) \div 0.0750 \text{ L} = 0.010\underline{0}0 \text{ M}$$

$$\text{M of } Pb^{2+} = (0.0100 \text{ mol/L} \times 0.0250 \text{ L}) \div 0.0750 \text{ L} = 0.0033\underline{3}3 \text{ M}$$

$$\text{M of } Ag^{+} = (2 \times 0.0150 \text{ mol/L} \times 0.0500 \text{ L}) \div 0.0750 \text{ L} = 0.020\underline{0}0 \text{ M}$$

$$\text{M of } Cl^{-} = (2 \times 0.0100 \text{ mol/L} \times 0.0250 \text{ L}) \div 0.07500 = 0.0066\underline{6}6 \text{ M}$$

Assemble a table showing the precipitation of $PbSO_4$ and AgCl.

Conc. (M)	Pb^{2+} +	SO_4^{2-} +	Ag^+ +	Cl^- $\rightleftharpoons$ $PbSO_4$ (s) + AgCl(s)
Starting	0.003333	0.01000	0.02000	0.006667
Change	-0.003333	-0.003333	-0.006667	-0.006667
Equilibrium	0.000000	$0.006\underline{6}67$	$0.013\underline{3}33$	0.000000

To calculate $[Pb^{2+}]$, use the K_{sp} expression for $PbSO_4$.

$$[Pb^{2+}] = \frac{1.7 \times 10^{-8}}{0.006667 \text{ M } SO_4^{2-}} = 2.\underline{5}49 \times 10^{-6} \text{ M}$$

$$[SO_4^{2-}] = 0.006\underline{6}67 \text{ M}$$

Calculate the $[Cl^-]$ using the K_{sp} expression for AgCl:

$$[Cl^-] = \frac{1.8 \times 10^{-10}}{0.01333 \text{ M } Ag^+} = 1.\underline{3}503 \times 10^{-8} \text{ M}$$

$$[Ag^+] = 0.013\underline{3}3 \text{ M}$$

Solution to Conceptual Problem

17.107 The oceans contain many of the elements in the form of ions, and various schemes have been put forth to recover some of these (such as gold). Presently, seawater is the principal source of magnesium metal (the process is described on p. 909-910 of the text). Seawater contains the following quantities of the alkaline earth elements (in grams per kilogram of seawater): Mg, 1.272; Ca, 0.400; Sr, 0.013; the density of seawater is 1.035 g/mL. Devise a scheme using a precipitation reaction to obtain a pure strontium compound that could then be used as the starting material to produce other compounds and the metal. Note that any precipitation you use would have to account for the presence of the other alkaline earth elements, particularly calcium, which is similar to strontium. Do any necessary calculations to show the feasibility of the scheme you propose. (Note that any precipitating anion you use might already be present in seawater at such concentration that the strontium ion present does not precipitate. How would this affect your calculation?)

Discussion

Unlike magnesium, strontium hydroxide is soluble, so a different precipitating agent would be needed. Looking at Table 17.1 would give the student some possibilities. For example, magnesium, calcium, and strontium carbonates are all insoluble, and a scheme based on fractional precipitation with CO_2 is possible. The sulfates $MgSO_4$, $CaSO_4$, and $SrSO_4$ differ significantly in solubilities, and a scheme based on sulfate precipitation (say with sodium sulfate) would seem quite feasible and not too expensive. Magnesium sulfate is soluble, whereas calcium sulfate is slightly soluble, and strontium sulfate quite insoluble. The student would need to calculate the quantity of sodium sulfate needed to precipitate strontium sulfate, but not calcium sulfate (a fractional precipitation problem). Sulfate ion is present in seawater (2.649 g/ kg seawater; students can find this value in various sources, including Encyclopedia Britannica). This would need to be known to obtain the quantity of sodium sulfate to use per liter of seawater, but the feasibility of using a sulfate precipitation is possible without this information.

Optionally, you might ask students to suggest how to prepare the metal from the compound they obtain in the precipitation. To obtain the metal, the student might suggest the electrolysis of strontium chloride, in analogy with the process used for magnesium. The problem then would be to obtain the chloride from the precipitated compound. If this is a carbonate, this presents no problem; they would dissolve the carbonate with hydrochloric acid. If the compound is a sulfate, the student will have some difficulty knowing how to proceed. A useful source of information on chemical technology is Kirk-Othmer's Encyclopedia of Chemical Technology. Presently, strontium is obtained by reducing the oxide with aluminum.

18. THERMODYNAMICS AND EQUILIBRIUM

Solutions to Exercises

Note on units and significant figures: The mol unit is omitted from all thermodynamic parameters such as S^o, ΔS^o, etc. The final answer to each mathematical solution is given first with one nonsignificant figure (the rightmost significant figure is underlined) and then is rounded to the correct number of figures. Intermediate answers usually also have at least one nonsignificant figure.

18.1 Calculate the work, w, done using w = F x d = (mg) x d. Then use w to calculate ΔE.

$$w = (mg) \times d = (2.20 \text{ kg} \times 9.80 \text{ m/s}^2) \times 0.250 \text{ m} = 5.3\underline{9}0 \text{ kg}\bullet\text{m}^2/\text{s}^2 = 5.39 \text{ J}$$

$$\Delta U = q + w = (-1.50 \text{ J}) + (5.3\underline{9}0 \text{ J}) = 3.8\underline{9}0 = 3.89 \text{ J}$$

18.2 At 1.00 atm and 25°C, the volume occupied by 1.00 mol of any of the gases in the equation is 22.41 L x (298/273) = 24.46 L. Find the change in volume:

$$CH_4(24.46 \text{ L}) + 2O_2(2 \times 24.46 \text{ L}) \rightarrow CO_2(24.46 \text{ L}) + 2H_2O(l)$$

$$\Delta V = 24.46 \text{ L} - (3 \times 24.46 \text{ L}) = -48.\underline{9}2 \text{ L}$$

Next calculate the work, w, done on the system (its value decreases), using 1.00 atm = 1.013×10^5 Pa, and 1.00 L = 1.00×10^{-3} m^3. Add this to the heat, q_p, at constant P:

$$w = -P\Delta V = -(1.013 \times 10^5 \text{ Pa}) \times (-48.92 \times 10^{-3} \text{ m}^3) = 4.9\underline{5}56 \times 10^3 \text{ J} = 4.96 \text{ kJ}$$

$$\Delta U = q_p + w = (-890.\underline{2} \text{ kJ}) + (+4.9\underline{5}56 \text{ kJ}) = -885.\underline{2}4 = -885.2 \text{ kJ}$$

18.3 When the liquid evaporates, it absorbs heat: ΔH_{vap} = 42.3 kJ/mol at 25°C. The entropy change is

$$\Delta S = \frac{\Delta H_{vap}}{T} = \frac{42.3 \times 10^3 \text{ J/mol}}{298 \text{ K}} = 14\underline{1}.9 \text{ J/K·mol}$$

The entropy of 1 mole of vapor is calculated using the entropy of 1 mol of liquid (161 J/K):

$$S^o = 161 \text{ J/K} + 141.9 \text{ J/K} = 302.9 = 303 \text{ J/K}$$

18.4 a. ΔS^o is positive because there is an increase in moles of gas (Δn_{gas} = +1) from a solid reactant forming a mole of gas. (Entropy increases.)

b. ΔS^o is positive because there is an increase in moles of gas (Δn_{gas} = +1) from a liquid reactant forming a mole of gas. (Entropy increases.)

c. ΔS^o is negative because there is a decrease in moles of gas (Δn_{gas} = -1) from liquid and gaseous reactants forming two moles of solid. (Entropy decreases.)

d. ΔS^o is positive because there is an increase in moles of gas (Δn_{gas} = +1) from solid and liquid reactants forming a mole of gas and four moles of an ionic compound. (Entropy increases.)

18.5 The reaction and standard entropies are given below. Multiply the S^o values by their stoichiometric coefficients and subtract the entropy of the reactant from the sum of the product entropies.

$C_6H_{12}O_6$	$\rightarrow$ $2C_2H_5OH(l)$	+	$2CO_2(g)$
S^o: 212	2 x 161		2 x 213.7

$$\Delta S^o = \Sigma S^o(\text{products}) - \Sigma S^o(\text{reactants}) = (2 \times 161) + (2 \times 213.7) - 212 = 53\underline{7}.4$$

$$\Delta S^o = 537 \text{ J/K}$$

18.6 The reaction, standard enthalpy changes, and standard entropies are as follows:

	$CH_4(g)$	+ $2O_2(g)$	$\rightarrow$ $CO_2(g)$	+ $2H_2O(g)$
ΔH_f^o:	-74.9	2 x 0	-393.5	2 x (-241.8)
S^o:	186.1	2 x 205.0	213.7	2 x 188.7

(continued)

Calculate ΔH^o and ΔS^o for the reaction.

$$\Delta H^o = [-393.5 + 2 \times (-241.8)]kJ - [-74.9 + 0]kJ = -802.2 \text{ kJ}$$

$$\Delta S^o = [213.7 + 2 \times 188.7]J/K - [186.1 + 2 \times (205.0)]J/K = -5.0 \text{ J/K}$$

Now substitute into the equation for ΔG^o in terms of ΔH^o and ΔS^o (as -5.0×10^{-3} kJ/K):

$$\Delta G^o = \Delta H^o - T\Delta S^o = -802.2 \text{ kJ} - (298 \text{ K})(-5.0 \times 10^{-3} \text{ kJ/K}) = -800.\underline{7}1 = -800.7 \text{ kJ}$$

18.7 Write the values of ΔG_f^o multiplied by their stoichiometric coefficients below each formula:

	$CaCO_3(s)$	$\rightarrow$	$CaO(s)$	+	$CO_2(g)$
ΔG_f^o:	-1128.8		-603.5		-394.4

Subtract ΔG_f^o of reactant from that of the products:

$$\Delta G_f^o = [(-603.5) + (-394.4) - (-1128.8)]kJ = 130.\underline{9} \text{ kJ}$$

18.8 a.

	C(graphite)	+	$2H_2(g)$	$\rightarrow$	$CH_4(g)$
ΔG_f^o:	0		0		-50.8

$$\Delta G^o = [(-50.8) - (0)]kJ = -50.8 \text{ kJ (spontaneous reaction)}$$

b.

	$2H_2(g)$	+	$O_2(g)$	$\rightarrow$	$2H_2O(l)$
ΔG_f^o:	0		0		2 x (-237.2)

$$\Delta G^o = [(2 \times -237.2) - (0)]kJ = -474.\underline{4} \text{ kJ (spontaneous reaction)}$$

c.

	$4HCN(g)$	+	$5O_2(g)$	$\rightarrow$	$2H_2O(l)$	+	$4CO_2(g)$	+	$2N_2(g)$
ΔG_f^o:	4 x 125		0		2 x (-237.2)		4 x (-394.4)		0

$$\Delta G^o = [(2 \times -237.2) + 4 \times (-394.4) - (4 \times 125)]kJ$$
$$= -255\underline{2}.0 \text{ kJ (spontaneous reaction)}$$

d.

	$Ag^+(aq)$	+	$I^-(aq)$	$\rightarrow$	$AgI(s)$
ΔG_f^o:	77.1		-51.7		-66.3

$$\Delta G^o = [(-66.3) - (77.1 - 51.7)]kJ = -91.\underline{7} \text{ kJ (spontaneous reaction)}$$

18.9 a. $K = P_{CO_2}$ $(= K_p)$

b. $K = [Pb^{2+}][I^-]^2$ $(= K_{sp})$

c. $K = \frac{P_{CO_2}}{[H^+][HCO_3^-]}$

18.10 First calculate ΔG^o using the ΔG_f^o values from Table 18.2.

$$CaCO_3(s) \rightleftharpoons CaO(s) + CO_2(g)$$

ΔG_f^o: -1128.8 -603.5 -394.4

Subtract ΔG_f^o of reactant from that of the products:

$$\Delta G_f^o = [(-603.5) + (-394.4) - (-1128.8)]\text{kJ} = 130.\underline{9} \text{ kJ}$$

Use the equation $\Delta G^o = -2.303 RT \log K$ to calculate K_p, equating K to K_p. Use 298 as an exact number and use the four significant figures for ΔG^o to assign four significant figures to the log K_p:

$$\log K_p = \frac{\Delta G^o}{-2.303 RT} = \frac{130.9 \text{ kJ}}{(-2.303)(0.008314 \text{ kJ/K})(298 \text{ K})} = -22.9\underline{4}1$$

$K_p = 1.\underline{1}4 \times 10^{-23} = 1.1 \times 10^{-23}$ (only two sig. figs. to the right of the decimal in log K_p)

18.11 First calculate ΔG^o using the ΔG_f^o values in the exercise.

$$Mg(OH)_2(s) \rightarrow Mg^{2+}(aq) + 2OH^-(aq)$$

ΔG_f^o: -933.9 -456.0 2 x (-157.3)

Subtract ΔG_f^o of reactant from that of the products:

$$\Delta G_f^o = [2 \times (-157.3) + (-456.0) - (-933.9)]\text{kJ} = 163.\underline{3} \text{ kJ}$$

Use the equation $\Delta G^o = -2.303 RT \log K$ to calculate K_{sp}, equating K to K_{sp}. Use 298 as an exact number and use the four significant figures for ΔG^o to assign four significant figures to the log K_{sp}:

continued)

$$\log K_{sp} = \frac{\Delta G^o}{-2.303\ RT} = \frac{163.3\ kJ}{(-2.303)(0.008314\ kJ/K)(298\ K)} = -28.6\underline{1}9$$

$K_{sp} = 2.\underline{4}04 \times 10^{-29} = 2.4 \times 10^{-29}$ (two sig. figs. to the right of the decimal in $\log K_{sp}$)

18.12 From Table 6.2 and 18.1, you have

	$H_2O(l)$	$\rightarrow$ $H_2O(g)$
ΔH_f^o:	-285.8	-241.8
S^o:	69.9	188.7

Calculate ΔH^o and ΔS^o from these values.

$$\Delta H^o = [-241.8 - (-285.8)]\ kJ = 44.\underline{0}\ kJ$$

$$\Delta S^o = [188.7 - 69.9]\ J/K = 118.8\ J/K$$

Substitute ΔH^o and ΔS^o into the Gibbs equation:

$$\Delta G^o = \Delta H^o - T\Delta S^o = 44.0\ kJ - (318\ K)(0.1188\ kJ/K) = 6.\underline{2}2\ kJ$$

For this reaction, $K = K_p = P_{H_2O}$. Substitute ΔG^o into the K equation at 318 K:

$$\log K_p = \frac{\Delta G^o}{-2.303\ RT} = \frac{6.22\ kJ}{(-2.303)(0.008314\ kJ/K)(318\ K)} = -1.\underline{0}22$$

$K = \underline{9}.506 \times 10^{-2} = 1 \times 10^{-1}$ = vapor pressure of H_2O in atm (0.0946 in Table 5.6)

18.13 First calculate ΔH^o and ΔS^o, using the given ΔH_f^o and S^o values.

	$MgCO_3(s)$	$\rightleftharpoons$	$MgO(s)$ +	$CO_2(g)$
ΔH_f^o:	-1112		-601.2	-393.5
ΔS^o:	65.9		26.9	213.7

$$\Delta H^o = [-601.2 + (-393.5 - (-1112)]\ kJ = 11\underline{7}.3\ kJ$$

$$\Delta S^o = [(26.9 + 213.7) - 65.9]\ J/K = 174.7\ J/K$$

Substitute these values into the expression relating T, ΔH^o, and ΔS^o.

$$T = \frac{\Delta H^o}{\Delta S^o} = \frac{117.3\ kJ}{0.1747\ kJ/K} = 67\underline{1}.4\ K$$ (lower than that for $CaCO_3$)

■ Answers to Review Questions

18.1 A spontaneous process is a chemical and/or a physical change that occurs by itself without the continuing intervention of an outside agency. Three examples are (1) a rock on a hilltop rolls down, (2) heat flows from a hot object to a cold one, and (3) iron rusts in moist air. Three examples of nonspontaneous processes are (1) a rock rolls uphill by itself, (2) heat flows from a cold object to a hot one, and (3) rust is converted to iron and oxygen.

18.2 Because liquids are more random or disordered than are solids, liquid benzene contains more entropy than does the same quantity of frozen benzene.

18.3 The second law of thermodynamics states that for a spontaneous process, the total entropy of a system and its surroundings always increases. As stated in Section 18.2, a spontaneous process actually creates disorder, or entropy.

18.4 The relationship between entropy and enthalpy can be expressed in terms of the following equation

$$\Delta S = \frac{\Delta H - \Delta G}{T}$$

At equilibrium, $\Delta G = 0$, so the equation reduces to $\Delta H/T$ whereas, when not at equilibrium, $\Delta G \neq 0$, so this is not the case. In contrast to a phase change at equilibrium, the entropy change for a spontaneous chemical reaction (at constant pressure) does not equal $\Delta H/T$ because entropy is created by the spontaneous reaction. This can be an increase in the entropy of the surroundings or of the system. An example of the latter is the reaction $N_2O_4(g) \rightarrow 2NO_2(g)$, where one reactant molecule forms two product molecules, thus increasing the randomness.

18.5 The standard entropy of hydrogen gas at 25°C can be obtained by starting at 0 K as a reference point, where the entropy of perfect crystals of hydrogen is zero. Then warm to room temperature in small increments and calculate ΔS^o for each incremental temperature change (say, 2 K) by dividing the heat absorbed by the average temperature (1 K is used as the average for 0 K to 2 K), and also take into account the entropy increases that accompany a phase change.

18.6 To predict the sign of ΔS^o, look for a change Δn_{gas} in the number of moles of gas. If there is an increase in moles of gas in the products (Δn_{gas} is positive), then ΔS^o should be positive. A decrease in moles of gas in the products suggests that ΔS^o should be negative.

18.7 Free energy, G, equals H - TS; that is, it is the difference between the enthalpy of a system and the product of temperature and entropy. The free-energy change, ΔG, equals $\Delta H - T\Delta S$.

18.8 The standard free-energy change, ΔG^o, equals $\Delta H^o - T\Delta S^o$; that is, it is the difference between the standard enthalpy change of a system and the product of temperature and the standard entropy change of a system. The standard free-energy change of formation is the free-energy change when one mole of a substance is formed from its elements in their stable states at 1 atm and at a standard temperature, usually 25°C.

18.9 If ΔG^o for a reaction is negative, the equation for the reaction is spontaneous in the direction written; that is, the reactants form the products as written. If it is positive, then the equation as written is nonspontaneous.

18.10 In principle, if a reaction is carried out so that no entropy is produced, the useful work obtained is the maximum useful work, w_{max}, and is equal to ΔG of the reaction.

18.11 When gasoline burns in an automobile engine, the change in free energy shows up as useful work. Gasoline, a mixture of hydrocarbons such as C_8H_{18} or octane, burns to yield energy, gaseous CO_2, and gaseous H_2O.

18.12 A nonspontaneous reaction can be made to occur by coupling it with a spontaneous reaction having a sufficiently negative ΔG^o to furnish the required energy. (The net ΔG^o of the coupled reactions must be negative.)

18.13 As a spontaneous reaction proceeds, the free energy decreases until equilibrium is reached at a minimum ΔG. See the diagram below.

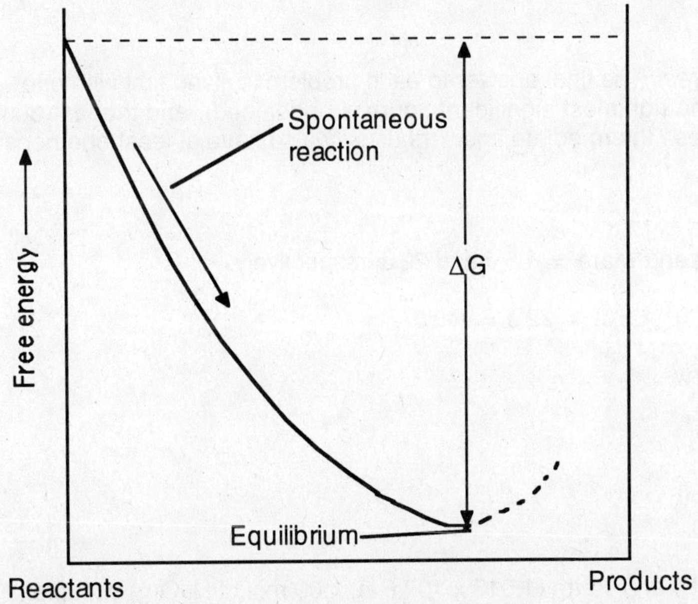

18.14 Because the equilibrium constant is related to ΔH^o and ΔS^o by $-RT \ln K = \Delta H^o - T\Delta S^o$, heat measurements alone can be used to obtain it. The standard enthalpy, ΔH^o, is the heat of reaction measured at constant pressure. The standard entropy change, ΔS^o, can be calculated from standard entropies, which are obtained from heat-capacity data.

18.15 The four combinations are as follows: (1) A negative ΔH^o and a positive ΔS^o always give a negative ΔG^o and a spontaneous reaction. (2) A positive ΔH^o and a negative ΔS^o always give a positive ΔG^o and a nonspontaneous reaction. (3) A negative ΔH^o and a negative ΔS^o may give a negative or a positive ΔG^o. At low temperatures, ΔG^o will usually be negative and the reaction spontaneous; at high temperatures, ΔG^o will usually be positive and the reaction nonspontaneous. (4) A positive ΔH^o and a positive ΔS^o may give a negative or a positive ΔG^o. At low temperatures, ΔG^o will usually be positive and the reaction nonspontaneous; at high temperatures, ΔG^o will usually be negative and the reaction spontaneous.

18.16 You can estimate the temperature at which a nonspontaneous reaction becomes spontaneous by substituting 0 for ΔG^o into the equation $\Delta G^o = \Delta H^o - T\Delta S^o$ and then solving for T using the form $T = \Delta H^o/\Delta S^o$.

■ Solutions to Practice Problems

Note on significant figures: The final answer to each problem is given first with one nonsignificant figure (the rightmost significant figure is underlined), and then is rounded to the correct number of figures. Intermediate answers usually also have at least one nonsignificant figure.

18.17 The values of q and w are = -65 J and 22 J, respectively.

$$\Delta U = q + w = (-65\text{ J}) + 22\text{ J} = -43\text{ J}$$

18.18 $\Delta U = 0 = q + w$

$0 = (-65\text{ J}) + w$

$w = +65\text{ J}$

18.19 At 100°C (373 K) and 1 atm (1.013×10^5 Pa), 1.00 mol of $H_2O(g)$ occupies

$$22.41\text{ L} \times \frac{373\text{ K}}{273\text{ K}} = 30.\underline{6}19\text{ L} \quad (30.619 \times 10^{-3}\text{ m}^3)$$

(continued)

The work done by the chemical system in pushing back the atmosphere is

$$w = -P\Delta V = -(1.013 \times 10^5 \text{ Pa}) \times (30.619 \times 10^{-3} \text{ m}^3) = -3.10\underline{1}7 \times 10^3 \text{ J}$$

$$= -3.10\underline{1}7 \text{ kJ}$$

$$\Delta U = q_p + w = (40.6\underline{6} \text{ kJ}) + (-3.1017 \text{ kJ}) = 37.5\underline{8}3 = 37.58 \text{ kJ}$$

18.20 At 25°C (298 K) and 1 atm (1.013×10^5 Pa), the decrease in volume in going from 4 mol to 2 mol of gas is

$$-2 \times 22.41 \text{ L} \times \frac{298 \text{ K}}{273 \text{ K}} = -48.9\underline{2}4 \text{ L} \ (-48.9\underline{2}4 \times 10^{-3} \text{ m3})$$

The work done on the chemical system by the atmosphere is

$$w = -P\Delta V = -(1.013 \times 10^5 \text{ Pa}) \times (-48.924 \times 10^{-3} \text{ m}^3) = 4.95\underline{6}0 \times 10^3 \text{ J} = 4.95\underline{6}0 \text{ kJ}$$

$$\Delta U = q_p + w = (-91.\underline{8} \text{ kJ}) + 4.956 \text{ kJ} = -86.\underline{8}44 = -8.68 \text{ kJ}$$

18.21 Use the equilibrium relation between ΔS and ΔH_{vap}, as 29.6×10^3 J, at the boiling point:

$$\Delta S = \frac{\Delta H_{vap}}{T} = \frac{29.6 \times 10^3 \text{ J}}{334.4 \text{ K}} = 88.\underline{5}2 = 88.5 \text{ J/ K}$$

18.22 Use the equilibrium relation between ΔS and ΔH_{vap}, as 26.7×10^3 J, at the boiling point:

$$\Delta S = \frac{\Delta H_{vap}}{T} = \frac{26.7 \times 10^3 \text{ J}}{308.8 \text{ K}} = 86.\underline{4}6 = 86.5 \text{ J/ K}$$

18.23 When the liquid condenses, it releases heat: ΔH_{cond} = -37.4 kJ/mol, or -37.4×10^3 J/mol, at 25°C. The entropy change is

$$\Delta S = \frac{\Delta H_{cond}}{T} = \frac{-37.4 \times 10^3 \text{ J}}{298 \text{ K}} = -12\underline{5}.503 \text{ J/ K}$$

The entropy of 1 mole of liquid is calculated using the entropy of 1 mol of vapor (252 J/K):

$$S_{liq} = S_{vap} + \Delta S_{cond} = 252 \text{ J/K} + (-125.503 \text{ J/K}) = 12\underline{6}.497 = 126 \text{ J/K}$$

18.24 When the liquid condenses, it releases heat: ΔH_{cond} = -29 kJ/mol, or -29 x 10^3 J/mol, at 25°C. The entropy change is

$$\Delta S = \frac{\Delta H_{cond}}{T} = \frac{-29 \times 10^3 \text{ J}}{298 \text{ K}} = -9\underline{7}.3 \text{ J/K}$$

The entropy of 1 mole of liquid is calculated using the entropy of 1 mol of vapor (248 J/K):

$$S_{liq} = S_{vap} + \Delta S_{cond} = 248 \text{ J/K} + (-97.3 \text{ J/K}) = 15\underline{0}.7 = 151 \text{ J/K}$$

18.25 a. ΔS^o is negative because there is a decrease in moles of gas (Δn_{gas} = -2) from 3 moles of gaseous reactants forming 1 mole of gaseous product. (Entropy decreases.)

b. ΔS^o is not predictable because there is no change in moles of gas (Δn_{gas} = 0) from 2 moles of gaseous reactants forming 2 moles of gaseous products.

c. ΔS^o is positive because there is an increase in moles of gas (Δn_{gas} = +1) from 5 moles of gaseous reactants forming 6 moles of gaseous products. (Entropy increases.)

d. ΔS^o is positive because there is an increase in moles of gas (Δn_{gas} = +1) from a solid reactant and 1 mole of gaseous reactant forming 2 moles of gaseous products. (Entropy increases.)

18.26 a. ΔS^o is positive because there is an increase in moles of gas (Δn_{gas} = +1) from 2 moles of gaseous reactant forming 3 moles of gaseous products. (Entropy increases.)

b. ΔS^o is negative because there is a decrease in moles of gas (Δn_{gas} = -1) from a liquid reactant and 3 moles of gaseous reactant forming 2 moles of gaseous product and liquid product. (Entropy decreases.)

c. ΔS^o is negative because there is a decrease in moles of gas (Δn_{gas} = -1) from 1 mole of gaseous reactant forming 1 mole of solid product. (Entropy decreases.)

d. ΔS^o is positive because there is an increase in moles of gas (Δn_{gas} = +2) from a solid reactant forming 2 moles of gaseous product plus solid product. (Entropy increases.)

18.27 The reaction and standard entropies are given below. Multiply the S^o values by their stoichiometric coefficients and subtract the entropy of the reactant from the sum of the product entropies. In part (b), note that aqueous NaCl exists as $Na^+(aq)$ and $Cl^-(aq)$ ions.

a. $2Na(s) \quad + \quad Cl_2(g) \quad \rightarrow \quad 2NaCl(s)$

S^o: 2 x 51.4   223.0   2 x 72.1

$\Delta S^o = \Sigma S^o(\text{products}) - \Sigma S^o(\text{reactants}) = [(2 \times 72.1) - (2 \times 51.4) + 223.0] = -181.\underline{6}$

$\Delta S^o = -181.6 \text{ J/K}$

b. $NaCl(s) \quad \rightarrow \quad Na^+(aq) \quad + \quad Cl^-(aq)$

S^o: 72.1   60.2   55.1

$\Delta S^o = \Sigma S^o(\text{products}) - \Sigma S^o(\text{reactants}) = [(60.2 + 55.1) - (72.1)] = 43.\underline{2}$

$\Delta S^o = 43.2 \text{ J/K}$

c. $CS_2(l) \quad + \quad 3O_2(g) \quad \rightarrow \quad CO_2(g) \quad + \quad 2SO_2(g)$

S^o: 151.0   3 x 205.0   213.7   2 x 248.1

$\Delta S^o = \Sigma S^o(\text{products}) - \Sigma S^o(\text{reactants})$

$\Delta S^o = [213.7 + (2 \times 248.1) - (151.0 + 3 \times 205.0)] = -56.\underline{1} = -56.1 \text{ J/K}$

d. $2CH_3OH(l) \quad + \quad 3O_2(g) \quad \rightarrow \quad 2CO_2(g) \quad + \quad 4H_2O(g)$

S^o: 2 x 127   3 x 205.0   2 x 213.7   4 x 188.7

$\Delta S^o = \Sigma S^o(\text{products}) - \Sigma S^o(\text{reactants})$

$\Delta S^o = [(2 \times 213.7) + (4 \times 188.7) - (2 \times 127 + 3 \times 205.0)] = 31\underline{3}.2 = 313 \text{ J/K}$

18.28 The reaction and standard entropies are given below. Multiply the S^o values by their stoichiometric coefficients and subtract the entropy of the reactant from the sum of the product entropies. In part (b), note that aqueous $CaCl_2$ exists as $Ca^+(aq)$ and $2Cl^-(aq)$ ions.

a. $2Ca(s) \quad + \quad O_2(g) \quad \rightarrow \quad 2CaO(s)$

S^o: 2 x 41.6   205.0   2 x 38.2

$\Delta S^o = \Sigma S^o(\text{products}) - \Sigma S^o(\text{reactants}) = [(2 \times 38.2) - (2 \times 41.6 + 205.0)] = -211.\underline{8}$

$\Delta S^o = -211.8 \text{ J/K}$

b. $CaCl_2(s) \rightarrow Ca^{2+}(aq) + 2Cl^-(aq)$

S^o: 114   -55.2   2 x 55.1

$\Delta S^o = \Sigma S^o(\text{products}) - \Sigma S^o(\text{reactants}) = [-55.2 + (2 \times 55.1) - (114)]$

$= -5\underline{9}.0 = -59$ J/K

c. $CS_2(g) + 4H_2(g) \rightarrow CH_4(g) + 2H_2S(g)$

S^o: 237.8   4 x 130.6   186.1   2 x 205.6

$\Delta S^o = \Sigma S^o(\text{products}) - \Sigma S^o(\text{reactants})$

$\Delta S^o = [(186.1) + (2 \times 205.6) - (237.8 + 4 \times 130.6)] = -162.\underline{9} = -162.9$ J/K

d. $C_2H_4(g) + 3O_2(g) \rightarrow 2CO_2(g) + 2H_2O(g)$

S^o: 219.2   3 x 205.0   2 x 213.7   2 x 188.7

$\Delta S^o = \Sigma S^o(\text{products}) - \Sigma S^o(\text{reactants})$

$\Delta S^o = [(2 \times 213.7) + (2 \times 188.7) - (219.2 + 3 \times 205.0)] = -29.\underline{4} = -29.4$ J/K

18.29 $CH_4(g) + 2O_2(g) \rightarrow CO_2(g) + 2H_2O(l)$

S^o: 186.1   2 x 205.0   213.7   2 x 69.9

$\Delta S^o = \Sigma S^o(\text{products}) - \Sigma S^o(\text{reactants})$

$\Delta S^o = [(213.7) + (2 \times 69.9) - (186.1 + 2 \times 205.0)] = -242.\underline{6} = -242.6$ J/K

S decreases as expected from decrease in moles of gas.

18.30 $CaCO_3(s) + 2H^+(aq) \rightarrow Ca^{2+}(aq) + H_2O(l) + CO_2(g)$

S^o: 92.9   2 x 0   -55.2   69.9   213.7

$\Delta S^o = \Sigma S^o(\text{products}) - \Sigma S^o(\text{reactants}) = [(-55.2 + 69.9 + 213.7) - (92.9 + 2 \times 0)]$

$\Delta S^o = 135.5$ J/K

S increases as expected from increase in moles of gas.

18.31 The reaction, with standard enthalpies of formation and standard entropies written underneath, is

	$2CH_3OH(l)$	+	$3O_2(g)$	$\rightarrow$	$2CO_2(g)$	+	$4H_2O(l)$
ΔH_f^o:	2 x (-238.6) 3 x 0		2 x (-393.5)				4 x (-285.840)
ΔS^o:	2 x 127		3 x 205.0		2 x 213.7		4 x 69.940

Calculate ΔH^o and ΔS^o for the reaction.

$$\Delta H^o = [2 \times (-393.5) + 4 \times (-285.840)] \text{ kJ} - [2 \times (-238.6) + 3 \times 0] \text{ kJ}$$

$$= -1453.\underline{1}60 \text{ kJ} = -1453.2 \text{ kJ}$$

$$\Delta S^o = [2 \times 213.7 + 4 \times 69.940] \text{ J/K} - [2 \times 127 + (3 \times 205.0)] \text{ J/K} = -16\underline{1}.84 = -162 \text{ J/K}$$

Now substitute into the equation for ΔG^o in terms of ΔH^o and ΔS^o (as -0.16184 kJ/K).

$$\Delta G^o = \Delta H^o - T\Delta S^o = -1453.160 \text{ kJ} - (298.2 \text{ K}) \times (-0.16184 \text{ kJ/K})$$

$$= -1404.\underline{8}99 = -1404.9 \text{ kJ}$$

18.32 The reaction, standard enthalpy changes, and standard entropies are as follows:

	$4HCN(l)$	+	$5O_2(g)$	$\rightarrow$	$4CO_2(g)$	+	$2H_2O(g)$	+	$2N_2(g)$
ΔH_f^o:	4 x 105		2 x 0		4 x (-393.5)		2 x (-241.826)		2 x 0
S^o:	4 x 112.8		5 x 205.0		4 x 213.7		2 x 188.72		2 x 191.5

Calculate ΔH^o and ΔS^o for the reaction; then substitute into the equation for ΔG^o:

$$\Delta H^o = [4 \times -393.5 + 2 \times (-241.826 + 0)] \text{ kJ} - [4 \times 105 + 0] \text{ kJ} = -247\underline{7}.652 \text{ kJ}$$

$$\Delta S^o = [4 \times 213.7 + 2 \times 188.72 + 2 \times 191.5] \text{ J/K} - [4 \times 112.8 + 5 \times 205.0] \text{ J/K}$$

$$\Delta S^o = 139.\underline{0}4 \text{ J/K} \ (0.139\underline{0}4 \text{ kJ/K})$$

$$\Delta G^o = \Delta H^o - T\Delta S^o = -2477.652 \text{ kJ} - (298 \text{ K})(0.13904 \text{ kJ/K}) = -251\underline{9}.08 = -2519 \text{ kJ}$$

18.33 a. $Na(s) + 1/2Cl_2(g) \rightarrow NaCl(s)$

b. $1/2H_2(g) + C(graphite) + 1/2N_2(g) \rightarrow HCN(l)$

c. $S(rhombic) + O_2(g) \rightarrow SO_2(g)$

d. $P(red) + 3/2H_2(g) \rightarrow PH_3(g)$

18.34 a. $Ca(s) + 1/2O_2(g) \rightarrow CaO(s)$

b. $C(graphite) + 5/2H_2(g) + 1/2N_2(g) \rightarrow CH_3NH_2(g)$

c. $C(graphite) + 2S(rhombic) \rightarrow CS_2(l)$

d. $4P(red) + 5O_2(g) \rightarrow P_4O_{10}(s)$

18.35 Write the values of ΔG_f^o multiplied by their stoichiometric coefficients below each formula; then subtract ΔG_f^o of reactant from that of the products.

a.

	$CH_4(g)$	+	$2O_2(g)$	$\rightarrow$	$CO_2(g)$	+	$2H_2O(g)$
ΔG_f^o:	-50.8		2 x 0		-394.4		2 x (-228.6)

$\Delta G^o = [(-394.4) + 2(-228.6) - (-50.8 + 0)]kJ = -800.8\ kJ$

b.

	$CaCO_3(s)$	+	$2H^+(aq)$	$\rightarrow$	$Ca^{2+}(aq)$	+	$H_2O(l)$	+	$CO_2(g)$
ΔG_f^o:	-1128.8		2 x 0		-553.04		-237.192		-394.4

$\Delta G^o = [(-553.04) + (-237.192) + (-394.4) - (-1128.8 + 0)]kJ = -55.\underline{8}32\ kJ$

$\Delta G^o = -55.8\ kJ$

18.36 Write the values of ΔG_f^o below each formula; then subtract ΔG_f^o values.

a.

	$C_2H_4(g)$	+	$3O_2(g)$	$\rightarrow$	$2CO_2(g)$	+	$2H_2O(g)$
ΔG_f^o:	68.36		3 x 0		2 x (-394.4)		2 x (-228.6)

$\Delta G^o = [2(-394.4) + 2(-228.6) - (68.36 + 0)]kJ = -1314.\underline{36} = -1314.4\ kJ$

b.

	$Na_2CO_3(s)$	+	$H^+(aq)$	$\rightarrow$	$2Na^+(aq)$	+	$HCO_3^-(aq)$
ΔG_f^o:	-1048.1		0		2 x (-261.87)		-587.06

$\Delta G^o = [2(-261.87) + (-587.06) - (-1048.1 + 0)]kJ = -62.7\ kJ$

18.37 a. Spontaneous reaction
b. Spontaneous reaction
c. Nonspontaneous reaction
d. Equilibrium mixture—significant amounts of both
e. Nonspontaneous reaction

18.38 a. Spontaneous reaction
b. Nonspontaneous reaction
c. Nonspontaneous reaction
d. Spontaneous reaction
e. Equilibrium mixture—significant amounts of both

18.39 Calculate ΔH^o and ΔG^o, using the given ΔH_f^o and ΔG_f^o values.

a.	$Al_2O_3(s)$	+ 2Fe(s)	$\rightarrow$	$Fe_2O_3(s)$	+	2Al(s)
ΔH_f^o:	-1676	2 x 0		-825.5		2 x 0
ΔG_f^o:	-1582	2 x 0		-743.6		2 x 0

ΔH^o = [(-825.5) + 0 - (-1676) - 0] kJ = $8\underline{5}0.5$ = 851 kJ

ΔG^o = [(-743.6) + 0 - (-1582) - 0] kJ = $83\underline{8}.4$ = 838 kJ

The reaction is endothermic, absorbing 850 kJ of heat. The large positive value for ΔG^o indicates that the equilibrium composition is mainly reactants.

b.	$COCl_2(g)$	+ $H_2O(l)$	$\rightarrow$	$CO_2(g)$	+	2HCl(g)
ΔH_f^o:	-220	-285.840		-393.5		2 x (-92.31)
ΔG_f^o:	-206	-237.192		-394.4		2 x (-95.30)

ΔH^o = [-393.5 + (2)(-92.31) - (-220) - (-285.840)] kJ = -72.28 = -72 kJ

ΔG^o = [-394.4 + (2)(-95.30) - (-206) - (-237.192)] kJ = $-14\underline{1}.808$ = -142 kJ

The reaction is exothermic; the ΔG^o value indicates mainly products at equilibrium.

18.40 Calculate ΔH^o and ΔG^o, using the given ΔH_f^o and ΔG_f^o values.

a.	2PbO(s)	+ $N_2(g)$	$\rightarrow$	2Pb(s)	+	2NO(g)
ΔH_f^o:	2(-219.0)	0		2 x 0		2 x 90.29
ΔG_f^o:	2(-189.2)	0		2 x 0		2 x 86.60

(continued)

$$\Delta H^o = [2(90.29) + 0 - 2(-219.0) - 0] \text{ kJ} = 618.\underline{5}8 = 618.6 \text{ kJ}$$

$$\Delta G^o = [2(86.60) + 0 - 2(-189.2) - 0] \text{ kJ} = 55\underline{1}.60 = 551.6 \text{ kJ}$$

The reaction is endothermic, absorbing 617 kJ of heat. The large positive value for ΔG^o indicates that the equilibrium composition is mainly reactants.

b.	$CS_2(l)$ +	$2H_2O(l)$	→	$CO_2(g)$ +	$2H_2S(g)$
ΔH_f^o:	87.9	2 x (-285.840)		-393.5	2 x (-20.2)
ΔG_f^o:	63.6	2 x (-237.192)		-394.4	2 x (-33)

$$\Delta H^o = [-393.5 + (2)(-20.2) - (87.9) - 2(-285.840)] \text{ kJ} = 49.\underline{8}8 = 49.9 \text{ kJ}$$

$$\Delta G^o = [(-394.4) + (2)(-33) - (63.6) - 2(-237.192)] \text{ kJ} = -4\underline{9}.616 = -50. \text{ kJ}$$

The reaction is endothermic; ΔG^o indicates that the equilibrium composition is mainly products.

18.41 Calculate ΔG^o, using the given ΔG_f^o values.

	$2H_2(g)$ +	$O_2(g)$	→	$2H_2O(l)$
ΔG_f^o:	0	0		2 x (-237.2)

$$\Delta G^o = [(2 \times -237.2) - (0)]\text{kJ} = -474.\underline{4} \text{ kJ}$$

Maximum work = ΔG^o = -474.4 kJ. Because the maximum work is stipulated, no entropy is produced.

18.42 Calculate ΔG^o, using the given ΔG_f^o values.

	$H_2(g)$ +	$Cl_2(g)$	→	$2HCl(g)$
ΔG_f^o:	0	0		2 x (-95.3)

$$\Delta G^o = [(2 \times -95.3) - (0)]\text{kJ} = -190.\underline{6} \text{ kJ}$$

Maximum work = ΔG^o = -190.6 kJ. No entropy is produced.

18.43 Calculate ΔG^o per 1 mol Zn(s), using the given ΔG_f^o values.

	$Zn(s)$	+	$Cu^{2+}(aq)$	→	$Zn^{2+}(aq)$	+	$Cu(s)$
ΔG_f^o:	0		64.98		-147.21		0

ΔG^o = [(-147.21) + (0) - (64.98) - 0]kJ = -212.19 kJ/mol Zn

-212.19 kJ/mol Zn x (5.00 g ÷ 65.39 g/mol Zn) = -16.$\underline{2}$24 = -16.2 kJ

Maximum work = ΔG^o = -16.2 kJ. Because maximum work is stipulated, no entropy is produced.

18.44 Calculate ΔG^o per 1 mol Zn(s), using the given ΔG_f^o values.

	$Zn(s)$	+	$2H^+(aq)$	→	$Zn^{2+}(aq)$	+	$H_2(g)$
ΔG_f^o:	0		0		-147.21		0

ΔG^o = [(-147.21) + (0) - 0 - 0]kJ = -147.21 kJ/mol Zn

-147.21 kJ/mol Zn x (5.00 g ÷ 65.39 g/mol Zn) = -11.$\underline{2}$56 = -11.3 kJ

Maximum work = ΔG^o = (-11.3 kJ) = -11.3 kJ. No entropy is produced.

18.45 a. $K = K_p = \dfrac{P_{CO_2}P_{H_2}}{P_{CO}P_{H_2O}}$

b. $K = [Mg^{2+}][OH^-]^2 = K_{sp}$

c. $K = [Li^+][OH^-]^2\ P_{H_2}$

18.46 a. $K = K_p = \dfrac{P_{CH_3OH}}{P_{CO}P_{H_2}^2}$

b. $K = \dfrac{1}{[Ag^+]^2[CrO_4^{2-}]} = \dfrac{1}{K_{sp}}$

c. $K = \dfrac{[Ca^{2+}]P_{CO_2}}{[H^+]^2}$

18.47 First calculate ΔG^o, using the ΔG_f^o values from Table 18.2.

	$H_2(g)$	+	$Cl_2(g)$	$\rightarrow$	$2HCl(g)$
ΔG_f^o:	0		0		2 x (-95.3)

ΔG^o = 2(-95.3) kJ = -190.6 kJ

Use the equation ΔG^o = -2.303 RT log K to calculate K. Use 298 as an exact number, use R = 0.008314 kJ/K, and use the four significant figures for ΔG^o to assign four significant figures to the log K:

$$\log K = \frac{\Delta G^\circ}{-2.303\ RT} = \frac{-190.6\ kJ}{(-2.303)(0.008314\ kJ/K)(298\ K)} = 33.40\underline{4}$$

K = $2.\underline{5}3 \times 10^{33}$ = 2.5×10^{33} (only two sig. figs. to the right of the decimal in log K)

18.48 First calculate ΔG^o, using the ΔG_f^o values from Table 18.2.

	C(graphite)	+	$O_2(g)$	$\rightarrow$	$CO_2(g)$
ΔG_f^o:	0		0		-394.4

Subtract ΔG_f^o of reactants from that of the product:

ΔG^o = (-394.4) kJ = -394.4 kJ

Use the equation ΔG^o = -2.303 RT log K to calculate K. Use 298 as an exact number, use R = 0.008314 kJ/K, and use the four significant figures for ΔG^o to assign four significant figures to the log K:

$$\log K = \frac{\Delta G^\circ}{-2.303\ RT} = \frac{-394.4\ kJ}{(-2.303)(0.008314\ kJ/K)(298\ K)} = 69.1\underline{2}2$$

K = $1.\underline{3}2 \times 10^{69}$ = 1.3×10^{69} (only two sig. figs. to the right of the decimal in log K)

18.49 First calculate ΔG^o, using the ΔG_f^o values from Table 18.2.

	$CO(g)$	+	$3H_2(g)$	$\rightarrow$	$CH_4(g)$	+	$H_2O(g)$
ΔG_f^o:	-137.2		0		-50.8		-228.6

Subtract ΔG_f^o of reactants from that of the products:

$$\Delta G^o = [(-50.8 + -228.6) - (-137.2 + 0)]\text{kJ} = -142.2 \text{ kJ}$$

Use the equation $\Delta G^o = -2.303 \text{ RT} \log K_p$ to calculate K_p, which is equal to K. Use 298 as an exact number, use R = 0.008314 kJ/K, and use the four significant figures for ΔG^o to assign four significant figures to the log K_p:

$$\log K_p = \frac{\Delta G^o}{-2.303 \text{ RT}} = \frac{-142.2 \text{ kJ}}{(-2.303)(0.008314 \text{ kJ/K})(298 \text{ K})} = 24.9\underline{2}1$$

$K_p = 8.\underline{33} \times 10^{24} = 8.3 \times 10^{24}$ (only two sig. figs. to the right of the decimal in log K)

18.50 First calculate ΔG^o, using the ΔG_f^o values from Appendix C.

	$CO(g)$	+	$2H_2(g)$	$\rightarrow$	$CH_3OH(g)$
ΔG_f^o:	-137.2		0		-161.9

Subtract ΔG_f^o of reactants from that of the product:

$$\Delta G^o = [(-161.9) - (-137.2 + 0)] \text{ kJ} = -24.7 \text{ kJ}$$

Use the equation $\Delta G^o = -2.303 \text{ RT} \log K_p$ to calculate K_p, which is equal to K.

$$\log K_p = \frac{\Delta G^o}{-2.303 \text{ RT}} = \frac{-24.7 \text{ kJ}}{(-2.303)(0.008314 \text{ kJ/K})(298 \text{ K})} = 4.3\underline{2}8$$

$K_p = 2.\underline{1}2 \times 10^4 = 2.1 \times 10^4$ (only two sig. figs. to the right of the decimal in log K)

18.51 First calculate ΔG^o, using the ΔG_f^o values from Appendix C.

	Mg(s)	+	Cu^{2+}(aq)	→	Mg^{2+}(aq)	+	Cu(s)
ΔG_f^o:	0		64.98		-456.01		0

Subtract ΔG_f^o of reactants from that of the products:

$$\Delta G^o = [(-456.01 + 0) - (64.98 + 0)]\text{kJ} = -520.99 \text{ kJ}$$

Use the equation $\Delta G^o = -2.3026 \, RT \log K_c$ to calculate K_c, which is equal to K. Use 298.15 as the temperature, use R = 0.0083144 kJ/K, and use the five significant figures for ΔG^o to assign five significant figures to the log K_c:

$$\log K = \frac{\Delta G^\circ}{-2.3026 \, RT} = \frac{-520.99 \text{ kJ}}{(-2.3026)(0.0083144 \text{ kJ/ K})(298.15 \text{ K})} = 91.27\underline{3}6$$

$K_c = 1.8\underline{7}8 \times 10^{91} = 1.88 \times 10^{91}$

(only three sig. figs. to the right of the decimal in log K)

18.52 First calculate ΔG^o, using the ΔG_f^o values from Appendix C.

	Zn(s)	+	Cu^{2+}(aq)	→	Zn^{2+}(aq)	+	Cu(s)
ΔG_f^o:	0		64.98		-147.21		0

Subtract ΔG_f^o of reactants from that of the products:

$$\Delta G^o = [(-147.21 + 0) - (64.98 + 0)]\text{kJ} = -212.19 \text{ kJ}$$

Use the equation $\Delta G^o = -2.3026 \, RT \log K_c$ to calculate K_c, which is equal to K. Use 298.15 as the temperature, use R = 0.0083144 kJ/K, and use the five significant figures for ΔG^o to assign five significant figures to the log K_c:

$$\log K = \frac{\Delta G^\circ}{-2.3026 \, RT} = \frac{-212.19 \text{ kJ}}{(-2.3026)(0.0083144 \text{ kJ/ K})(298.15 \text{ K})} = 37.17\underline{4}1$$

$K_c = 1.4\underline{9}3 \times 10^{37} = 1.49 \times 10^{37}$

(only three sig. figs. to the right of the decimal in log K)

18.53 From Tables 6.2 and 18.1, you have

	C(graphite)	+	$CO_2(g)$	⇌	2CO(g)
ΔH_f^o:	0		-393.5		2(-110.5)
S^o:	5.7		213.7		2(197.5)

Calculate ΔH^o and ΔS^o from these values.

$$\Delta H^o = [2(-110.5) - (0 + -393.5)]\text{ kJ} = 172.\underline{5}\text{ kJ}$$

$$\Delta S^o = [2(197.5) - (5.7 + 213.7)]\text{ J/K} = 175.\underline{6}\text{ J/K}\ \ (0.1756\text{ kJ/K})$$

$$\Delta G^o = \Delta H^o - T\Delta S^o = 172.5\text{ kJ} - (1273\text{ K})(0.1756\text{ kJ/K}) = -51.\underline{0}4\text{ kJ}$$

Substitute ΔG^o into the K_p equation at 1273 K, using R = 0.008314 kJ/K:

$$\log K_p = \frac{\Delta G^o}{-2.303\ RT} = \frac{-51.04\text{ kJ}}{(-2.303)(0.008314\text{ kJ/K})(1273\text{ K})} = 2.0\underline{9}40$$

$$K_p = 1.\underline{2}41 \times 10^2 = 1.2 \times 10^2$$

(only two sig. figs. to the right of the decimal in log K_p)

Because K_p is greater than 1, the data predict that combustion of carbon should form significant amounts of CO product at equilibrium.

18.54 From Tables 6.2 and 18.1, you have

	$N_2(g)$	+	$O_2(g)$	⇌	2NO(g)
ΔH_f^o:	0		0		2(90.3)
S^o:	191.5		205.0		2(210.6)

Calculate ΔH^o and ΔS^o from these values.

$$\Delta H^o = [2(90.3) - (0 + 0)]\text{ kJ} = 180.\underline{6}\text{ kJ}$$

$$\Delta S^o = [2(210.6) - (191.5 + 205.0)]\text{ J/K} = 24.\underline{7}\text{ J/K}\ \ (0.0247\text{ kJ/K})$$

Substitute ΔH^o and ΔS^o into the Gibbs equation:

$$\Delta G^o = \Delta H^o - T\Delta S^o = 180.6\text{ kJ} - (2273\text{ K})(0.0247\text{ kJ/K}) = 124.\underline{4}57\text{ kJ}$$

(continued)

Substitute ΔG^o into the K_p equation at 2273 K, using R = 0.008314 kJ/K:

$$\log K_p = \frac{\Delta G^o}{-2.303\ RT} = \frac{124.457\ kJ}{(-2.303)(0.008314\ kJ/K)(2273\ K)} = -2.85\underline{9}67$$

$K_p = 1.3\underline{8}1 \times 10^{-3} = 1.38 \times 10^{-3}$

(only three sig. figs. to the right of the decimal in log K_p)

Because K_p is slightly less than 1 but greater than 10^{-4}, the data predict that combustion of nitrogen and oxygen should form small (but not significant) amounts of NO product at equilibrium.

18.55 First calculate ΔH^o and ΔS^o, using the given ΔH_f^o and S^o values.

	$2NaHCO_3(s)$	$\rightarrow$	$Na_2CO_3(s)$	+	$H_2O(g)$	+	$CO_2(g)$
ΔH_f^o:	2 x (-947.7)		-1130.8		-241.826		-393.5
ΔS^o:	2 x 102		139		188.72		213.7

$\Delta H^o = [(-1130.8) + (-241.826) + (-393.5) - 2(-947.7)]\ kJ = 129.\underline{2}74\ kJ$

$\Delta S^o = [(139 + 188.7 + 213.7) - 2(102)]\ J/K = 33\underline{7}.4\ J/K\ \ (0.3374\ kJ/K)$

Substitute these values into $\Delta G^o = \Delta H^o - T\Delta S^o$, let $\Delta G^o = 0$, and rearrange to solve for T.

$$T = \frac{\Delta H^o}{\Delta S^o} = \frac{129.274\ kJ}{0.3374\ kJ/K} = 38\underline{3}.14 = 383\ K$$

18.56 First calculate ΔH^o and ΔS^o, using the given ΔH_f^o and S^o values.

	$2HgO(s)$	$\rightarrow$	$2Hg(g)$	+	$O_2(g)$
ΔH_f^o:	2 x (-90.8)		2 x 61.3		0
ΔS^o:	2 x 70.3		2 x 174.9		205.0

$\Delta H^o = [(2 \times 61.3 + 0) - 2(-90.8)]\ kJ = 304.\underline{2}\ kJ$

$\Delta S^o = [(2 \times 174.9 + 205.0) - 2(70.3)]\ J/K = 414.\underline{2}\ J/K\ \ (0.414\underline{2}\ kJ/K)$

(continued)

Substitute these values into $\Delta G^o = \Delta H^o - T\Delta S^o$, let $\Delta G^o = 0$, and rearrange to solve for T.

$$T = \frac{\Delta H^o}{\Delta S^o} = \frac{304.2 \text{ kJ}}{0.4142 \text{ kJ/K}} = 734.\underline{4}2 = 734.4 \text{ K}$$

■ Solutions to Unclassified Problems

18.57 The sign of ΔS^o should be positive because there is an increase in moles of gas ($\Delta n_{gas} = +5$) as the solid reactant forms 5 moles of gas. The reaction is endothermic, denoting a positive ΔH^o. The fact that the reaction is spontaneous implies that the product $T\Delta S^o$ is larger than ΔH^o, so that ΔG^o is negative, as required for a spontaneous reaction.

18.58 The sign of ΔS^o should be negative because there is a decrease in moles of gas ($\Delta n_{gas} = -3$) as the 7 moles of gaseous reactants form 4 moles of gaseous product plus liquid product. The reaction is exothermic, denoting a negative ΔH^o. The fact that the reaction is spontaneous implies that the negative ΔH^o is larger than the positive value of $-T\Delta S^o$, so that ΔG^o is negative, as required for a spontaneous reaction.

18.59 The ΔH value ~ BE(H-H) + BE(Cl-Cl) - BE(H-Cl) ~ [432 + 240 - 2(428)] kJ ~ -184 kJ, and thus the reaction is exothermic. ΔS^o should be positive because there is an increase in disorder with the formation of unsymmetrical molecules from symmetrical H_2 and Cl_2. The reaction should be spontaneous because the contributions of both the ΔH term and the $-T\Delta S$ term are negative.

18.60 The ΔH value ≅ BE(HC) + BE(CN) + 2BE(HH) - 3BE(CH) - 3BE(CN) - 2BE(NH) ≅ [411 + 887 + 2(432) - 3(411) - 305 - 2(386)] kJ ~ -148 kJ. Thus, the reaction is exothermic. ΔS^o should be negative because there is a decrease in moles of gas ($\Delta n_{gas} = -2$) with the formation of 1 mole of gas from 3 moles of gas. The reaction is spontaneous because the contributions of the ΔH term dominate the $-T\Delta S$ term.

18.61 When the liquid freezes, it releases heat: $\Delta H_{fus} = -69.0$ J/g at 16.6°C (289.6 K). The entropy change is

$$\Delta S = \frac{\Delta H_{fus}}{T} = \frac{-69.0 \text{ J/g}}{289.8 \text{ K}} \times \frac{60.05 \text{ g}}{1 \text{ mol}} = -14.\underline{3}0 = -14.3 \text{ J/(K•mol)}$$

18.62 When the liquid evaporates, it absorbs heat: ΔH_{vap} = 29.1 kJ/mol, or 2.91 x 10^4 J/mol, at 56°C (329 K). The entropy change is:

$$\Delta S = \frac{\Delta H_{vap}}{T} = \frac{2.91 \times 10^4 \text{ J/mol}}{329 \text{ K}} = 88.\underline{4}5 = 88.4 \text{ J/(K•mol)}$$

18.63 a. ΔS^o is negative because there is a decrease in moles of gas (Δn_{gas} = -1) from 1 mole of gaseous reactant forming aqueous and liquid products. (Entropy decreases.)

b. ΔS^o is positive because there is an increase in moles of gas (Δn_{gas} = +5) from a solid reactant forming 5 moles of gas. (Entropy increases.)

c. ΔS^o is positive because there is an increase in moles of gas (Δn_{gas} = +3) from 2 moles of gaseous reactant forming 5 moles of gaseous products. (Entropy increases.)

d. ΔS^o is negative because there is a decrease in moles of gas (Δn_{gas} = -1) from 3 moles of gaseous reactants forming 2 moles of gaseous products. (Entropy decreases.)

18.64 a. Entropy decreases; ΔS^o is negative because there is a decrease in the moles of gas (Δn_{gas} = -2) from 4 moles of gaseous reactant forming 2 moles of gaseous products.

b. Entropy increases; ΔS^o is certainly positive because there is an increase in the moles of gas (Δn_{gas} = +2) from a solid reactant forming 2 moles of gas.

c. Entropy decreases; ΔS^o is negative because there is a decrease in the moles of gas (Δn_{gas} = -3) from 3 moles of gaseous reactant forming a liquid product.

d. Entropy increases; ΔS^o is certainly positive because there is an increase in moles of gas (Δn_{gas} = +1) from solid and liquid reactants forming 1 mole of gaseous product.

18.65 ΔS^o is negative because there is a decrease in the moles of gas (Δn_{gas} = -2) from 3 moles of gaseous reactant forming 1 mole of gaseous product plus liquid product.

18.66 ΔS^o is positive because there is an increase in moles of gas (Δn_{gas} = +2) from 3 moles of gaseous reactant forming 5 moles of gaseous products.

18.67 Calculate ΔS^o from the individual S^o values:

	$C_2H_5OH(l)$	+	$O_2(g)$	→	$CH_3COOH(l)$	+	$H_2O(l)$
S^o:	161		205.0		160		69.9

$\Delta S^o = \Sigma S^o(\text{products}) - \Sigma S^o(\text{reactants})$

$\Delta S^o = [(160 + 69.9) - (161 + 205.0)] = -13\underline{6}.1 = -136 \text{ J/K}$

18.68 Calculate ΔS^o from the individual S^o values:

	$CO(g)$	+	$2H_2(g)$	→	$CH_3OH(l)$
S^o:	197.5		2 x 130.6		127

$\Delta S^o = \Sigma S^o(\text{products}) - \Sigma S^o(\text{reactants})$

$\Delta S^o = [(127) - (197.5 + 2 \times 130.6)] = -33\underline{1}.7 = -332 \text{ J/K}$

18.69 Calculate ΔG^o, using the ΔG_f^o values from Table 18.2.

	$H_2(g)$	+	$SO_2(g)$	→	$H_2S(g)$	+	$O_2(g)$
ΔG_f^o:	0		-300.2		-33		0

$\Delta G^o = (-33 + 0) - (-300.2 + 0) \text{ kJ} = 26\underline{7}.2 = 267 \text{ kJ}$

Because ΔG^o is positive, the reaction is nonspontaneous as written, at 25°C.

18.70 Calculate ΔG^o, using the ΔG_f^o values from Table 18.2.

	$N_2(g)$	+	$CH_4(g)$	→	$HCN(g)$	+	$NH_3(g)$
ΔG_f^o:	0		-50.8		125		-16

$\Delta G^o = (-16 + 125) - (-50.8 + 0) \text{ kJ} = 15\underline{9}.8 = 160 \text{ kJ}$

Because ΔG^o is positive, the reaction is nonspontaneous as written, at 25°C.

18.71 At low (room) temperature, ΔG^o or $(\Delta H^o - T\Delta S^o)$ must be positive; but at higher temperatures, ΔG^o or $(\Delta H^o - T\Delta S^o)$ must be negative. Thus, at the higher temperatures the $-T\Delta S^o$ term must become more negative than ΔH^o. Thus, ΔS^o must be positive and so must ΔH^o be positive. If either was negative, ΔG^o would not become negative at higher temperatures.

18.72 At low (room) temperature, ΔG^o or $(\Delta H^o - T\Delta S^o)$ must be negative; but at higher temperatures, ΔG^oor $(\Delta H^o - T\Delta S^o)$ must be positive. Thus, at the higher temperatures the $-T\Delta S^o$ term must become more positive than ΔH^o. Thus, ΔS^o must be negative and so must ΔH^o be negative. If either was positive, ΔG^o would not become negative at lower temperatures.

18.73 First calculate ΔG^o, using the ΔG_f^o values in the exercise.

	$CaF_2(s)$	$\rightarrow$	$Ca^{2+}(aq)$	+	$2F^-(aq)$
ΔG_f^o:	-1162		-553		2 x (-276.5)

Subtract ΔG_f^o of reactant from that of the products:

$$\Delta G_f^o = [-553 + 2 \times (-276.5) - (-1162)]\text{kJ} = 5\underline{6}.0 \text{ kJ}$$

Use the equation $\Delta G^o = -2.303 RT \log K$ to calculate K_{sp}, equating K to K_{sp}.

$$\log K_{sp} = \frac{\Delta G^o}{-2.303 RT} = \frac{56.0 \text{ kJ}}{(-2.303)(0.008314 \text{ kJ/K})(298 \text{ K})} = -9.\underline{8}145$$

$K_{sp} = \underline{1}.53 \times 10^{-10} = 2 \times 10^{-10}$ (one sig. fig. to the right of the decimal in $\log K_{sp}$)

18.74 First calculate ΔG^o, using the ΔG_f^o values in the exercise.

	$BaSO_4(s)$	$\rightarrow$	$Ba^{2+}(aq)$	+	$SO_4^{2-}(aq)$
ΔG_f^o:	-1353		-561		-742

Subtract ΔG_f^o of reactant from that of the products:

$$\Delta G^o = [-561 + (-742) - (-1353)]\text{kJ} = 5\underline{0}.0 \text{ kJ}$$

(continued)

Use the equation $\Delta G^o = -2.303\ RT \log K$ to calculate K_{sp}, equating K to K_{sp}.

$$\log K_{sp} = \frac{\Delta G^o}{-2.303\ RT} = \frac{50.0\ kJ}{(-2.303)(0.008314\ kJ/K)(298\ K)} = -8.\underline{7}63$$

$K_{sp} = \underline{1}.7 \times 10^{-9} = 2 \times 10^{-9}$ (one sig. fig. to the right of the decimal in log K_{sp})

18.75 From Appendix C, you have

	$COCl_2(g)$	$\rightarrow$	$CO(g)$	+	$Cl_2(g)$
ΔH_f^o:	-220		-110.5		0
S^o:	283.74		197.5		223.0

Calculate ΔH^o and ΔS^o from these values.

$\Delta H^o = [(-110.5) + 0 - (-220)]\ kJ = 10\underline{9}.5\ kJ$

$\Delta S^o = [197.5 + 223.0 - 283.74]\ J/K = 136.\underline{7}6\ J/K\ (0.136\underline{7}6\ kJ/K)$

At 25°C: $\Delta G^o = \Delta H^o - T\Delta S^o = 109.5\ kJ - (298\ K)(0.13676\ kJ/K) = 6\underline{8}.7 = 69\ kJ$

At 800°C: $\Delta G^o = \Delta H^o - T\Delta S^o = 109.5\ kJ - (1073\ K)(0.13676\ kJ/K) = -3\underline{7}.2$

$= -37\ kJ$

Thus, ΔG^o changes from a positive value and a nonspontaneous reaction at 25°C to a negative value and a spontaneous reaction at 800°C.

18.76 From Appendix C, you have

	$CS_2(g)$	+	$4H_2(g)$	$\rightleftharpoons$	$CH_4(g)$	+	$2H_2S(g)$
ΔH_f^o:	117		4 x 0		-74.9		2 x (-20)
S^o:	237.8		4 x 130.6		186.1		2 x 205.6

Calculate ΔH^o and ΔS^o from these values.

$\Delta H^o = [(-74.9) + 2 \times (-20) - (117 + 0)]\ kJ = -23\underline{1}.9\ kJ$

$\Delta S^o = [(186.1 + 2 \times 205.6) - (237.8 + 4 \times 130.6)]\ J/K$

$= -162.\underline{9}\ J/K\ (-0.162\underline{9}\ kJ/K)$

(continued)

At 25°C: $\Delta G^o = \Delta H^o - T\Delta S^o$ = -231.9 kJ - (298 K)(-0.1629 kJ/K) = -18<u>3</u>.3

= -183 kJ

At 650°C: $\Delta G^o = \Delta H^o - T\Delta S^o$ = -231.9 kJ - (923 K)(-0.1629 kJ/K) = -8<u>1</u>.54

= -82 kJ

Thus, ΔG^o has a negative value at both temperatures, with the reaction being spontaneous at both temperatures. However, ΔG^o is more negative at 25°C than at 650°C, and so the relative proportion of the products is greater at 25°C.

18.77 a.

			$\Delta H°$, kJ
$CO_2(g) + 2H_2(g)$	$\rightarrow$	$CH_2O(g) + H_2O(g)$	35.5
$CH_2O(g) + 2H_2(g)$	$\rightarrow$	$CH_4(g) + H_2O(g)$	-201
$C(s) + O_2(g)$	$\rightarrow$	$CO_2(g)$	-393
$2H_2O(g)$	$\rightarrow$	$2H_2(g) + O_2(g)$	484
$C(s) + 2H_2(g)$	$\rightarrow$	$CH_4(g)$	-74.5 = 75 kJ

b.

	$C(s)$	+ $2H_2(g)$	$\rightarrow$ $CH_4(g)$
$S°$, J/mol•K	5.7	2(130.6)	186.1

$S°_{rxn}$ = 186.1 - [5.7 + 2(130.6)] = -80.8 J/mol•K

c. $\Delta G° = \Delta H° - T\Delta S°$

= -74.5 kJ/mol - (298 K) (-80.8×10^{-3} kJ/mol•K)

= -50.42 = -50.4 kJ/mol

11.78 a.

			$\Delta H°$, kJ
$CH_3CHO(g) + 2H_2(g)$	$\rightarrow$	$C_2H_6(g) + H_2O(l)$	-205
$C_2H_5OH(l) + 1/2\ O_2(g)$	$\rightarrow$	$CH_3CHO(g) + H_2O(l)$	-348/2
$2C(s) + 3H_2(g) + 1/2\ O_2(g)$	$\rightarrow$	$C_2H_5OH(l)$	-556/2
$2H_2O(g)$	$\rightarrow$	$2H_2(g) + O_2(g)$	484
$2H_2O(l)$	$\rightarrow$	$2H_2O(g)$	2(44)
$2C(s) + 3H_2(g)$	$\rightarrow$	$C_2H_6(g)$	-8<u>5</u>.0 = 85 kJ

b. $2C(s) + 3H_2(g) \rightarrow C_2H_6(g)$

$S°$, J/mol•K 2(5.7) 3(130.6) 229.5

$S°_{rxn} = 229.5 - [2(5.7) + 3(130.6)] = -173.7$ J/mol•K

c. $\Delta G° = \Delta H° - T\Delta S°$

$= -85.0 \text{ kJ/mol} - (298 \text{ K})(-173.7 \times 10^{-3} \text{ kJ/mol•K})$

$= -33$ kJ/mol

18.79 $\Delta H°_{rxn} = -393.5 + 2(-285.8) - (-238.6) = -726.5$ kJ

$\Delta G° = \Delta H° - T\Delta S°$

$-702.6 \times 10^3 \text{ J} = -726.5 \times 10^3 \text{ J} - (298 \text{ K})(\Delta S°_{rxn})$

$$\Delta S°_{rxn} = -\frac{(-702.6 \times 10^3 \text{ J}) - (-726.5 \times 10^3 \text{ J})}{298 \text{ K}} = 80.\underline{2}0 = 80.2 \text{ J/K}$$

$\Delta S°_{rxn} = -80.2 \text{ J/K} = 2(70) + 214 - [127 + 3/2\, S°(O_2)]$

$S°(O_2) = 20\underline{4}.8 = 205$ J/mol•K

18.80 $\Delta H°_{rxn} = -241.8 + (-393.5) - [-116.0 + 2/3(143)] = -614.6$ kJ

$\Delta G° = \Delta H° - T\Delta S°$

$-621.7 \times 10^3 \text{ J} = -614.6 \times 10^3 \text{ J} - (298 \text{ K})(\Delta S°_{rxn})$

$$\Delta S°_{rxn} = -\frac{(-621.7 \times 10^3 \text{ J}) - (-614.6 \times 10^3 \text{ J})}{298 \text{ K}} = 23.\underline{8}3 = 23.8 \text{ J/K}$$

$\Delta S°_{rxn} = 23.8 \text{ J/K} = 189 + 214 - [219 + 2/3\, S°(O_3)]$

$S°(O_3) = -24\underline{0}.3 = 240.$ J/mol•K

18.81 a. The first reaction is

	$SnO_2(s)$	+ $2H_2(g)$	→	$Sn(s)$	+ $2H_2O(g)$
ΔH_f°	- 580.7	0		0	-241.8
S°	52.3	130.6		51.5	205

$\Delta H^\circ_{rxn} = 2(-241.8) - (-580.7) = 97.1 \text{ kJ}$

$\Delta S^\circ_{rxn} = 2(205) + 51.5 - [(2)(130.6) + 52.3] = 14\underline{8}.0 = 148 \text{ J/K}$

The second reaction is

	$SnO_2(s)$	+ $C(s)$	→	$Sn(s)$	+ $CO_2(g)$
ΔH_f°	-580.7	0		0	-393.5
S°	52.3	5.7		51.5	197.5

$\Delta H^\circ_{rxn} = -393.5 - (-580.7) = 187.2 \text{ kJ}$

$\Delta S^\circ_{rxn} = 197.5 + 51.5 - (5.7 + 52.3) = 19\underline{1}.0 = 191 \text{ J/K}$

b. For H_2, at what temperature does $\Delta G = 0$?

$0 = \Delta H^\circ - T\Delta S^\circ$

$$T = \frac{\Delta H}{\Delta S} = \frac{97.1 \times 10^3 \text{ J}}{148.0 \text{ J/K}} = 656.1 = 656 \text{ K}$$

At temperatures greater than 837 K, the reaction will be spontaneous.

For C, at what temperature does $\Delta G = 0$?

$$T = \frac{\Delta H}{\Delta S} = \frac{187.2 \times 10^3 \text{ J}}{191.0 \text{ J/K}} = 98\underline{0}.1 = 980. \text{ K}$$

At temperatures greater than 900 K, the reaction will be spontaneous.

c. From a consideration of temperature, the process with hydrogen would be preferred. But hydrogen is very expensive and carbon is cheap. On this basis, carbon would be preferred. Tin is produced commercially using carbon as the reducing agent.

18.82 a. $\Delta G^\circ_{rxn} = 3(-228.6) - (-763.1) = 77.3 \text{ kJ}$

Since ΔG° is positive, K will be less than 1.

b. $\Delta H^\circ_{rxn} = 3(-241.8) - (-839.9) = 114.5 \text{ kJ}$

$\Delta G^\circ_{rxn} = \Delta H^\circ - T\Delta S^\circ$

$77.3 \times 10^3 \text{ J} = 114.5 \times 10^3 \text{ J} - (298 \text{ K}) \Delta S^\circ$

$\Delta S^\circ_{rxn} = 12\underline{4}.8 = 125 \text{ J/mol}\bullet\text{K}$

c. $T = \dfrac{\Delta H}{\Delta S} = \dfrac{114.5 \times 10^3 \text{ J}}{124.8 \text{ J/K}} = 917.4 = 917 \text{ K}$

d. The driving force is the change in entropy. At high temperatures, the term $T\Delta S$ becomes very important.

18.83 a. Formic acid is favored as it is of lower energy than CO and H_2O.

b. The change in entropy for the decomposition of formic acid is positive as a mole of gas is produced. So the change in entropy would be the driving force for this reaction.

18.84 a. It depends upon whether the products are of lower energy than the reactants (exothermic) or if the products are of higher energy than the reactants (endothermic). If the products have stronger bonds than the reactants, the reaction will be exothermic. If the reactants have stronger bonds than the products, the reaction will be endothermic.

b. i) For a highly exothermic reaction, the driving force is the change in enthalpy. The $T\Delta S$ term under these conditions won't be of great significance.

ii) The driving force will be the entropy. The change in entropy must be positive and the $T\Delta S$ must be greater than ΔH.

18.85 a. If ΔG° is negative than K must be greater than 1. Consequently, the products will predominate.

b. The molecules must have enough energy to react when they collide with each other. So it depends upon the activation energy for the reaction. Usually it is necessary to heat solids for a reaction to occur as it is difficult to have effective collisions.

18.86 a. Since a gaseous product is produced, $\Delta S°$ for the reaction would be positive.

b. Since $\Delta H°$ is negative and $S°$ is positive, this would give a negative value for $\Delta G°$ and K would be greater than 1. Products would predominate.

c. Evidently the reaction doesn't take place at room temperature. Since the reaction requires a catalyst, we must conclude that there is a high activation energy without the catalyst. Also it is necessary to heat the mixture so that a greater fraction of the molecules would have enough energy to react.

18.87 a. $C_4H_{10}(g) + 13/2\ O_2(g) \rightarrow 4CO_2(g) + 5H_2O(l)$

$\Delta H°_{rxn} = -49.57\ kJ/g \times 58.12\ g/mol = -288\underline{1}.0\ kJ/mol$

$-2881.0 = 4(-393.5) + 5(-285.8) - [13/2\ (0) + \Delta H_f°\ (C_4H_{10})\ ]$

$\Delta H_f°\ (C_4H_{10}) = -12\underline{2}.0 = 122\ kJ/mol$

b. $\Delta G°_{rxn} = 4(-394.4) + 5(-237.2) - [13/2\ (0) + (-15.7)\ ]$

$\Delta G°_{rxn} = -2747.9\ kJ/mol$

c. $\Delta G° = \Delta H° - T\Delta S°$

$-2747.9\ kJ/mol = -2881.0\ kJ/mol - (298\ K)\ \Delta S°$

$\Delta S°_{rxn} = -0.44\underline{6}6\ kJ/mol \bullet K = -447\ J/mol{\bullet}K$

18.88 a. $C_2H_6O_2(l) + 5/2\ O_2(g) \rightarrow 2CO_2(g) + 3H_2O(l)$

$\Delta H°_{rxn} = -19.18\ kJ/g \times 62.07\ g/mol = -119\underline{0}.4\ kJ/mol$

$-1190.4 = 2(-393.5) + 3(-285.8) - [5/2\ (0) + \Delta H_f°\ (C_2H_6O_2)\ ]$

$\Delta H_f°\ [C_2H_6O_2] = -454.0\ kJ/mol$

b. $\Delta G°_{rxn} = 2(-394.4) + 3(-237.2) - [5/2\ (0) + (-322.5)]$

$\Delta G°_{rxn} = -1177.9\ kJ/mol$

c. $\Delta G^\circ = \Delta H^\circ - T\Delta S^\circ$

$-1177.9 \times 10^3\ J = -1190.4.0 \times 10^3\ J - (298\ K)(\Delta S^\circ_{rxn})$

$$\Delta S^\circ_{rxn} = -\frac{(-1177.9 \times 10^3\ J) - (-1190.4 \times 10^3\ J)}{298\ K} = 41.\underline{9}4 = 41.9\ J/K$$

18.89 a. $\Delta H^\circ_{rxn} = -1302 - (-1289) = -13\ kJ/mol$

$\Delta S^\circ_{rxn} = 89 - 176 = -87\ J/mol{\bullet}K$

$\Delta G^\circ = -13 \times 10^3\ J/mol - (298\ K)(-87\ J/mol{\bullet}K) = 1\underline{2}.93 \times 10^3\ J/mol$

$$\Delta G^\circ = -RT \ln K; \quad \ln K = -\frac{12.93 \times 10^3\ J/mol}{8.314\ J/K \bullet mol \times 298\ K} = -5.\underline{2}2$$

$K = e^{-5.22} = 5.\underline{4}1 \times 10^{-3} = 5.4 \times 10^{-3}$

b. The change in entropy is negative, greater order, so this causes H_3PO_4 to be a weak acid. The energy change enhances the acid strength of H_3PO_4, but the entropy is a very important term.

18.90 a. $\Delta H^\circ_{rxn} = -635 - (-608) = -27\ kJ/mol$

$\Delta S^\circ_{rxn} = 109 - 234 = -125\ J/mol{\bullet}K$

$\Delta G^\circ = -27 \times 10^3\ J - (298\ K)(-125\ kJ/mol{\bullet}K) = 10.25 \times 10^3\ J/mol$

$$\Delta G^\circ = -RT \ln K; \quad \ln K = -\frac{10.25 \times 10^3\ J/mol}{8.314\ J/K \bullet mol \times 298\ K} = -4.\underline{1}4$$

$K = 1.\underline{5}9 \times 10^{-2} = 1.6 \times 10^{-2}$

b. The change in entropy is negative, greater order, so this causes H_2SO_3 to be a weak acid. The energy change enhances the acid strength of H_2SO_3, but the entropy is a very important term.

■ Cumulative-Skills Problems
(require skills from previous chapters)

18.91 For the dissociation of HBr, assume that ΔH and ΔS are constant over the temperature range from 25°C to 375°C, and calculate the value of each to use to calculate K at 375°C. Start by calculating ΔH° and ΔS° at 25°C, using $\Delta H_f°$ and S° values.

	$2HBr(g)$	→	$H_2(g)$	+	$Br_2(g)$
$\Delta H_f°$:	2 x (-36)		0		30.91
ΔS°:	2 x 198.59		130.6		245.38

$\Delta H° = [(30.91 + 0) - 2(-36)]\ kJ = 10\underline{2}.91\ kJ\ (\cong \Delta H \text{ at } 375°C)$

$\Delta S° = [(245.38 + 130.6) - 2(198.59)]\ J/K = -21.\underline{2}0\ J/K$

$= -0.02120\ kJ/K\ (\cong \Delta S \text{ at } 375°C)$

Substitute ΔH° and ΔS° into the Gibbs equation at 375°C (648 K):

$\Delta G° \cong \Delta H - T\Delta S \cong 102.91\ kJ - (648\ K)(-0.02120\ kJ/K) \cong 11\underline{6}.6\ kJ$

Substitute ΔG° into the K equation at 648 K, using R = 0.008314 kJ/K:

$$\log K = \frac{\Delta G°}{-2.303\ RT} = \frac{116.6\ kJ}{(-2.303)(0.008314\ kJ/K)(648\ K)} = -9.3\underline{9}76$$

$K = 4.\underline{0}0 \times 10^{-10}$ (only two sig. figs. to the right of the decimal in log K)

Assuming $x = [H_2] = [Br_2]$, and assuming $[HBr] = (1.00 - 2x) \cong 1.00$ atm, substitute into the equilibrium expression:

$$\frac{[H_2][Br_2]}{[HBr]^2} \cong \frac{(x)(x)}{(1.00)^2} \cong 4.\underline{0}0 \times 10^{-10}$$

Solve for the approximate pressure of x:

$$x \cong \sqrt{(4.00 \times 10^{-10})(1.00)^2} \cong 2.\underline{0}0 \times 10^{-5}\ atm$$

(continued)

The percent dissociation at 1.00 atm is

$$\% \text{ dissociation} = \frac{2(2.00 \times 10^{-5} \text{ atm})}{1.00 \text{ atm}} \times 100\% = 0.0040\% \text{ (at 1.00 atm)}$$

Based on Le Chatelier's principle, pressure has no effect on equilibrium, so the % dissociation is $4.\underline{0} \times 10^{-3} = 0.0040\%$.

18.92 For the formation of HI, assume that ΔH and ΔS are constant over the temperature range from 25°C to 205°C, and calculate the value of each to use to calculate K at 205°C. Start by calculating ΔH^o and ΔS^o at 25°C, using ΔH_f^o and S^o values. From Appendix C, we have

	$H_2(g)$	+	$I_2(g)$	⇌	$2HI(g)$
ΔH_f^o:	0		62.442		2 x 25.9
S^o:	130.6		260.58		2 x 206.33

Calculate ΔH^o and ΔS^o from these values.

$$\Delta H^o = [2(25.9) - (0 + 62.442)] \text{ kJ} = -10.\underline{6}42 \text{ kJ} \quad (\cong \Delta H^o \text{ at } 205^oC)$$

$$\Delta S^o = [2(206.33) - (130.6 + 260.58)] \text{ J/K}$$

$$= 21.\underline{4}8 \text{ J/K} \; (0.02148 \text{ kJ/K}) \quad (\cong \Delta S \text{ at } 205^oC)$$

Substitute ΔH^o and ΔS^o into the Gibbs equation at 205°C (478 K):

$$\Delta G^o \cong \Delta H - T\Delta S \cong -10.642 \text{ kJ} - (478 \text{ K})(0.02148 \text{ kJ/K}) \cong -20.\underline{9}09 \text{ kJ}$$

Substitute ΔG^o into the K equation at 478 K, using R = 0.008314 kJ/K:

$$\log K = \frac{\Delta G^\circ}{-2.303 \, RT} = \frac{-20.909 \text{ kJ}}{(-2.303)(0.008314 \text{ kJ/ K})(478 \text{ K})} = 2.2\underline{8}46$$

$K = 1.\underline{9}3 \times 10^2$ (only two sig. figs. to the right of the decimal in log K)

Letting [HI] = 2x, $[H_2]$ = 0.500 mol - x, and $[I_2]$ = 1.00 mol - x, substitute into the equilibrium expression:

$$\frac{[HI]}{[H_2][I_2]} = \frac{(2x)^2}{(0.500 - x)(1.00 - x)} = 1.\underline{9}3 \times 10^2$$

(continued)

Rearranging gives a quadratic: $189x^2 - 289.5x + 96.5 = 0$. Using the quadratic formula, solve for x:

$$x = \frac{-(-289.5) \pm \sqrt{(-289.5)^2 - 4 \times 189 \times 96.5}}{2 \times 189} = \frac{289.5 \pm 104.19}{378}$$

$= 0.4902$ mol (negative root)

$[HI] = 2x = 0.98\underline{0}4$ mol

The mol fraction of HI is

$$\frac{0.9804 \text{ mol HI}}{(0.500 - 0.4902) \text{ mol } H_2 + (1.00 - 0.4902) \text{ mol } I_2 + 0.9804 \text{ mol HI}}$$

$$= 0.65\underline{3}6 = 0.654$$

18.93 For the dissociation of NH_3, assume that ΔH and ΔS are constant over the temperature range from 25°C to 345°C, and calculate values of each to calculate K at 345°C. First calculate ΔH^o and ΔS^o, using ΔH_f^o and S^o values from Appendix C.

	$2NH_3(g)$	$\rightarrow$	$3H_2(g)$	+	$N_2(g)$
ΔH_f^o:	2 x (-45.9)		0		0
S^o:	2 x 193		3 x 130.6		191.5

$\Delta H^o = [0 + 0 - 2(-45.9)]$ kJ $= 91.\underline{8}$ kJ ~ ΔH at 345°C

$\Delta S^o = [(3 \times 130.6 + 191.5) - 2(193)]$ J/K $= 19\underline{7}.3$ J/K

$= 0.19\underline{7}3$ kJ/K ($\cong \Delta S$ at 345°C)

Substitute ΔH^o and ΔS^o into the Gibbs equation at 345°C (618 K):

$\Delta G^o \cong \Delta H - T\Delta S \cong 91.8 \text{ kJ} - (618 \text{ K})(0.1973 \text{ kJ/K}) \cong -30.\underline{1}3$ kJ

Substitute ΔG^o into the K equation at 618 K, using R = 0.008314 kJ/K:

$$\log K = \frac{\Delta G^\circ}{-2.303\,RT} = \frac{-30.13 \text{ kJ}}{(-2.303)(0.008314 \text{ kJ/K})(618 \text{ K})} = 2.5\underline{4}62$$

$K = 3.\underline{5}17 \times 10^2 = K_p$

(continued)

Now obtain K_c.

$$K_c = K_p(RT)^{-2}$$

$$= (3.517 \times 10^2)(0.0821 \times 618)^{-2} = 0.1\underline{3}66$$

The starting concentration of NH_3 is 1.00 mol/20.0L = 0.0500M. You obtain the following table:

	$2NH_3(g)$	$\rightarrow$	$3H_2(g)$	+	$N_2(g)$
Starting	0.0500		0		0
Change	-2x		+3x		+x
Equilibrium	0.0500 - 2x		3x		x

The equilibrium equation is

$$\frac{[H_2]^3[N_2]}{[NH_3]^2} = K_c$$

or

$$\frac{(3x)^3x}{(0.0500 - 2x)^2} = 0.1\underline{3}66$$

$$\frac{x^4}{(0.0500 - 2x)^2} = \frac{0.1366}{27} = 5.060 \times 10^{-3}$$

Taking the square root of both sides of this equation gives

$$\frac{x^2}{0.0500 - 2x} = 7.\underline{1}13 \times 10^{-2}$$

which rearranges to

$$x^2 + (1.423 \times 10^{-1})x - (3.556 \times 10^{-3}) = 0$$

From the quadratic formula, you obtain

$$x = \frac{-0.1423 \pm \sqrt{(0.1423)^2 + 4 \times 3.556 \times 10^{-3}}}{2}$$

The positive root is

$$x = 0.02168$$

(continued)

Hence,

$$[NH_3] = 0.0500 - 2(0.02168) = 6.64 \times 10^{-3}$$

$$\% \ NH_3 \text{ dissociated} = \left(1 - \frac{6.64 \times 10^{-3}}{0.0500}\right) \times 100\%$$

$$= 8\underline{6}.7\% = 87\%$$

18.94 The reaction, with ΔH_f^o and S^o values underneath the equation, is

	$CO(g)$	+	$3H_2(g)$	→	$CH_4(g)$	+	$H_2O(g)$
ΔH_f^o:	-110.5		0		-74.87		-241.826
S^o:	197.5		3 x 130.6		186.1		188.72

Assume that ΔH^o and ΔS^o are constant over the temperature range from 25°C to 785°C.

$$\Delta H^o = [(-74.87 - 241.826) - (-110.5)] \text{ kJ} = -206.\underline{1}96 \text{ kJ}$$

$$\Delta S^o = [(186.1 + 188.72) - (197.5 + 3 \times 130.6)] \text{ J/K} = -214.\underline{4}8 \text{ J/K}$$

Now calculate ΔG^o at 785°C (1058 K).

$$\Delta G^o = \Delta H^o - T\Delta S^o = -206.196 \text{ kJ} - (1058\text{K})(-0.21448 \text{ kJ/K}) = 20.\underline{7}24 \text{ kJ}$$

The equilibrium constant at 785°C is obtained from the equation

$$\log K = \frac{\Delta G^\circ}{-2.303 \ RT} = \frac{20.724 \text{ kJ}}{(-2.303)(0.008314 \text{ kJ/ K})(1058 \text{ K})} = -1.0\underline{2}301$$

$$K = K_p = 0.09\underline{4}84$$

$$K_c = K_p(RT)^2 = 0.09484(0.0821 \times 1058)^2 = 715.6$$

The starting concentrations of CO and H_2 are

$$[CO] = \frac{0.0100 \text{ mol}}{25.0 \text{ L}} = 4.00 \times 10^{-4} \text{ M}$$

(continued)

$$[H_2] = \frac{0.0300 \text{ mol}}{25.0 \text{ L}} = 1.20 \times 10^{-3} \text{ M}$$

You obtain the following table:

	$CO(g)$	+	$3H_2(g)$	→	$CH_4(g)$	+	$H_2O(g)$
Starting	4.00×10^{-4}		1.20×10^{-3}		0		0
Change	$-x$		$-3x$		$+x$		$+x$
Equilibrium	$(4.00 \times 10^{-4}) - x$		$(1.20 \times 10^{-3}) - 3x$		x		x

The equilibrium equation is

$$\frac{[CH_4][H_2O]}{[CO][H_2]^3} = K_c$$

or

$$\frac{x^2}{[(4.00 \times 10^{-4}) - x]\,[(1.20 \times 10^{-3}) - 3x]^3} = 7.156 \times 10^2$$

The denominator can be written as $3^3\,[(4.00 \times 10^{-4}) - x]^4$ so that this equation can be reduced to a quadratic and solved exactly. Multiply both sides of the equation by 3^3 and then take the square root. This gives

$$\frac{x}{(4.00 \times 10^{-4} - x)^2} = 139.00$$

This can be rearranged into the following quadratic:

$$x^2 + (-7.99 \times 10^{-3})x + (1.60 \times 10^{-7}) = 0$$

From the quadratic formula you obtain

$$x = \frac{-(-7.99 \times 10^{-3}) \pm \sqrt{(-7.99 \times 10^{-3})^2 - 4 \times 1 \times (1.60 \times 10^{-7})}}{2 \times 1}$$

$$= \frac{7.99 \times 10^{-3} \pm 7.95 \times 10^{-3}}{2} = 2.00 \times 10^{-5} \text{ (negative root)}$$

$$x = [CH_4] = [H_2O] = 2.\underline{0}0 \times 10^{-5}$$

Mol CH_4 = mol H_2O = 25.0 L × (2.00×10^{-5} mol/L) = $5.\underline{0}0 \times 10^{-4}$ = 5.0×10^{-4} mol

18.95 First calculate ΔG^o at each temperature, using $\Delta G^o = -2.303RT \log K_a$:

$$25.0^oC: \Delta G^o = -(2.303)(0.008314 \text{ kJ/K})(298.2 \text{ K})(\log 1.754 \times 10^{-5}) = 27.15\underline{5}1 \text{ kJ}$$

$$50.0^oC: \Delta G^o = -(2.303)(0.008314 \text{ kJ/K})(323.2 \text{ K})(\log 1.633 \times 10^{-5}) = 29.62\underline{3}7 \text{ kJ}$$

Next solve two equations in two unknowns, assuming that ΔH^o and ΔS^o are constant over the range of 25°C to 50°C. Use 0.2982 K(kJ/J) and 0.3232 K(kJ/J) to convert ΔS^o in J to $T\Delta S^o$ in kJ.

$$1.\ 27.1551 \text{ kJ} = \Delta H^o - [0.2982 \text{ K(kJ/J)} \Delta S^o]$$

$$2.\ 29.6237 \text{ kJ} = \Delta H^o - [0.3232 \text{ K(kJ/J)} \Delta S^o]$$

Then rearrange equation 2 and substitute for ΔH^o into equation 2:

$$3a.\ \Delta H^o = [(0.3232\ \Delta S^o) \text{ K(kJ/J)} + 29.6237]$$

$$3b.\ 27.1551 \text{ kJ} = [(0.3232\ \Delta S^o) \text{ K(kJ/J)} + 29.6237] - [(0.2982\ \Delta S^o) \text{ K(kJ/J)}]$$

Solve for ΔS^o:

$$\Delta S^o = \frac{(29.6237 - 27.1551) \text{ kJ}}{(0.2982 - 0.3232) \text{ K(kJ/J)}} = -98.\underline{7}4 = -98.7 \text{ J/K}$$

Substitute this value into equation 3a and solve for ΔH^o:

$$\Delta H^o = [(0.3232) \text{ K(kJ/J)} \times (-98.74 \text{ J/K})] + 29.6237 \text{ kJ} = -2.2\underline{8}9 = -2.29 \text{ kJ}$$

18.96 First calculate ΔG^o at each temperature, using $\Delta G^o = -2.303RT \log K_{sp}$:

$$25.0^oC: \Delta G^o = -(2.303)(0.008314 \text{ kJ/K})(298.2 \text{ K})(\log 1.782 \times 10^{-10}) = 55.66\underline{4}2 \text{ kJ}$$

$$35.0^oC: \Delta G^o = -(2.303)(0.008314 \text{ kJ/K})(308.2 \text{ K})(\log 4.159 \times 10^{-10}) = 55.35\underline{8}7 \text{ kJ}$$

Next solve two equations in two unknowns, assuming that ΔH^o and ΔS^o are constant over the range of 25°C to 35°C. Use 0.2982 K(kJ/J) and 0.3082 K(kJ/J) to convert ΔS^o in J to $T\Delta S^o$ in kJ.

$$1.\ 55.6642 \text{ kJ} = \Delta H^o - [0.2982 \text{ K(kJ/J)} \Delta S^o]$$

$$2.\ 55.3587 \text{ kJ} = \Delta H^o - [0.3082 \text{ K(kJ/J)} \Delta S^o]$$

(continued)

Then rearrange equation 2 and substitute for ΔH^o into equation 2:

3a. $\Delta H^o = [(0.3082\ \Delta S^o)\ K(kJ/J) + 55.3587]$

3b. $55.6642\ kJ = [(0.3082\ \Delta S^o)\ K(kJ/J) + 55.3587] - [(0.2982\ \Delta S^o)\ K(kJ/J)]$

Solve for ΔS^o:

$$\Delta S^o = \frac{(55.6642 - 55.3587)\ kJ}{(0.3082 - 0.2982)\ K(kJ/J)} = 3\underline{0}.55 = 31\ J/K$$

Substitute this value into equation 3a and solve for ΔH^o:

$$\Delta H^o = [(0.3082)\ K(kJ/J) \times (30.55\ J/K)] + 55.3587\ kJ = 64.\underline{7}7 = 64.8\ kJ$$

■ Solution to Conceptual Problem

18.97 The overall reaction for the corrosion of iron to form rust is shown below.

$$4Fe(s) + 3O_2(g) \rightarrow 2Fe_2O_3(s)$$

This process is an ongoing problem for many of the structures and materials used today and has a tremendous economic impact. Using thermodynamic data in Appendix C, perform some appropriate calculations. What does this suggest about controlling rust by trying to shift the equilibrium concentrations of reactants and products? Consider the various methods of controlling rust formation. What principles are involved?

Discussion

Changes of concentration and temperature all affect equilibrium composition. However, equilibrium favors the product overwhelmingly ($K = 10^{261}$ at 25°C). Alloying the iron with other substances might reduce the effective concentration (activity) or iron sufficiently to shift the equilibrium composition to the reactant side. Changes of oxygen partial pressure and temperature would affect the equilibrium concentrations, but are not normally available to us.

Looking at the conventional methods of rust control, we see two basic schemes. The simplest is to keep the iron from coming into contact with oxygen. This is the purpose of protective coatings like paint. Tin plate protects iron in a similar way. Slowing the rate at which oxygen gets to iron to react might be considered as an extension of the idea of preventing oxygen from coming into contact with iron. Chromium is often added to iron to give corrosion-resistant alloys, such as stainless steel. These steels form a chromium oxide coating that impedes further oxidation.

(continued)

The second basic scheme for controlling rust is "cathodic protection." Because the explanation of cathodic protection depends on electrochemistry, which is not discussed until the next chapter, students might not come up with this solution. (Rusting and cathodic protection are discussed on pages 813 to 815.) If this solution does come up, it presents the instructor with the opportunity to discuss the importance of mechanism, as well as thermodynamics, to the outcome of a reaction. Zinc plating, as in tin plating, protects by excluding oxygen from the iron surface. However, if there is a break in the zinc surface, the conditions for cathodic protection become available. Zinc becomes the anode, whereas iron becomes the cathode. The half-reactions are

$$Zn(s) \rightarrow Zn^{2+}(aq) + 2e^- \qquad \text{(anode)}$$

$$O_2(g) + 2H_2O(l) + 4e^- \rightarrow 4OH^-(aq) \qquad \text{(cathode)}$$

Thus, zinc oxidizes instead of iron. The protection lasts as long as the zinc coating.

19. ELECTROCHEMISTRY

■ Solutions to Exercises

Note on significant figures: The final answer to each mathematical solution is given first with one nonsignificant figure (the rightmost significant figure is underlined) and then is rounded to the correct number of figures. Intermediate answers usually also have at least one nonsignificant figure.

19.1 The electrode reactions are

Cathode: $Ag^{+}(aq) + e^{-} \rightleftharpoons Ag(s)$

Anode: $Ni(s) \rightleftharpoons Ni^{2+}(aq) + 2e^{-}$

A sketch of the cell is given below:

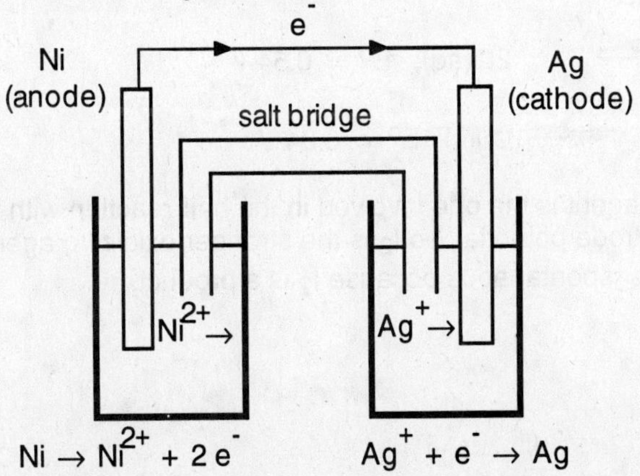

19.2 The notation for the cell is $Zn(s)|Zn^{2+}(aq)||H^{+}(aq)|H_2(g)|Pt(s)$.

19.3 Given below are the half-cell reactions and their sum, the overall cell reaction:

$$Cd(s) \rightleftharpoons Cd^{2+}(aq) + 2e^-$$
$$2H^+(aq) + 2e^- \rightleftharpoons H_2(g)$$
$$\overline{Cd(s) + 2H^+(aq) \rightleftharpoons Cd^{2+}(aq) + H_2(g)}$$

19.4 Because maximum work, w_{max}, = $-nFE_{cell}$, write out the half-reactions to determine n:

$$Zn(s) \rightleftharpoons Zn^{2+}(aq) + 2e^-; \quad Cu^{2+}(aq) + 2e^- \rightleftharpoons Cu(s)$$

Because both are 2-electron half-reactions, n = 2. The maximum work per 1 mole of Zn(s) is thus

$$w_{max} = -nFE_{cell} = -(2)(9.65 \times 10^4 \text{ C})(1.10 \text{ V}) = -2.1\underline{2}3 \times 10^5 \text{ V}\bullet\text{C} = -2.12 \times 10^5 \text{ J}$$

The maximum work for 6.54 g of Zn(s) is

$$6.54 \text{ g Zn} \times \frac{1 \text{ mol Zn}}{65.39 \text{ g Zn}} \times \frac{-2.123 \times 10^5 \text{ J}}{1 \text{ mol Zn}} = -2.1\underline{2}3 \times 10^4 = -2.12 \times 10^4 \text{ J}$$

19.5 The E^o of the $NO_3^- \rightleftharpoons NO(g)$ half-reaction is +0.96 V; the E^o of the $Ag^+ \rightleftharpoons Ag(s)$ half-reaction is +0.80 V. Thus, the NO_3^- is the stronger oxidizing agent because it has the more positive E^o.

19.6 The half-reactions and the corresponding E^o values are

$$I_2(s) + 2e^- \rightleftharpoons 2I^-(aq); \quad E^o = 0.54 \text{ V}$$

$$Cu^{2+}(aq) + 2e^- \rightleftharpoons Cu(s); \quad E^o = 0.34 \text{ V}$$

The stronger oxidizing agent is the one involved in the half-reaction with the larger (more positive) standard electrode potential, so I_2 is the stronger oxidizing agent. Thus, the reaction as written is nonspontaneous because I_2 is a product.

19.7 The half-reactions and the corresponding E^o values are

$$Zn^{2+}(aq) + 2e^- \rightleftharpoons Zn(s); \quad E^o = -0.76\ V$$

$$Cu^{2+}(aq) + 2e^- \rightleftharpoons Cu(s); \quad E^o = 0.34\ V$$

To obtain the standard cell emf, reverse the zinc half-reaction and reverse its half cell potential, and then add the half-cell potentials:

$$Zn(s) \rightleftharpoons Zn^{2+}(aq) + 2e^- \qquad -E^o_{Zn} = 0.76\ V$$

$$Cu^{2+}(aq) + 2e- \rightleftharpoons Cu(s) \qquad E^o_{Cu} = 0.34\ V$$

$$Zn(s) + Cu^{2+}(aq) \rightleftharpoons Zn^{2+}(aq) + Cu(s) \qquad E^o_{cell} = 1.10\ V$$

19.8 The half-cell reactions, the corresponding E^o values, and the addition of these to obtain the overall cell reaction and standard-cell emf are as follows:

$$Sn^{2+}(aq) \rightleftharpoons Sn^{4+}(aq) + 2e^- \qquad -E^o_{Sn} = -0.15\ V$$

$$2Hg^{2+} + 2e- \rightleftharpoons Hg_2^{2+}(aq) \qquad E^o_{Hg} = 0.90\ V$$

$$Sn^{2+}(aq) + 2Hg^{2+}(aq) \rightleftharpoons Sn^{4+}(aq) + Hg_2^{2+}(aq) \quad E^o_{cell} = 0.75\ V$$

Noting that n = 2 in the overall cell reaction, calculate ΔG^o:

$$\Delta G^o = -nFE^o_{cell} = -(2)(9.65 \times 10^4\ C)(0.75\ V) = -1.\underline{4}475 \times 10^5 = -1.4 \times 10^5\ J$$

$$\Delta G^o = -1.4 \times 10^2\ kJ$$

19.9 Write the equation with ΔG_f^o's beneath each substance:

	Mg(s) +	$Cu^{2+}(aq)$ ⇌	$Mg^{2+}(aq)$ +	Cu(s); n = 2
ΔG_f^o:	0	65.0	-456.0	0

$$\Delta G^o = -456.0 - (65.0) = -521.\underline{0}\ kJ\ (-521.0 \times 10^3\ C{\bullet}V)$$

$$\Delta G^o = -nFE^o_{cell}$$

(continued)

$$-521.0 \times 10^3 \text{ C}\bullet\text{V} = -(2)(9.65 \times 10^4 \text{ C})(E^o_{cell})$$

Rearrange and solve for E^o_{cell}. Decide the significant figures from the four significant figures of ΔG^o, assuming the F of 9.65×10^4 is an exact number and not three significant figures:

$$E^o_{cell} = \frac{-521.0 \times 10^3 \text{ C}\bullet\text{V}}{-(2)(9.65 \times 10^4 \text{ C})} = 2.69\underline{9}4 = 2.699 \text{ V}$$

19.10 Note that $K = K_c$; then rearrange the equation $E^o_{cell} = (0.0592 \text{ V})/n \log K_c$ to solve for $\log K_c$. Begin by noting that the standard emf is 0.56 V. This is calculated from the $-E^o$ value of 0.41 V for $Fe(s) \rightleftharpoons Fe^{2+}(aq)$ and the E^o value of 0.15 V for $Sn^{4+}(aq) \rightleftharpoons Sn^{2+}(aq)$.

$$0.56 \text{ V} = \frac{0.0592 \text{ V}}{2} \log K_c$$

$\log K_c = 1\underline{8}.91$ (two sig. figs. from the two sig. figs. of E^o_{cell})

$K_c = 8.1 \times 10^{1\underline{8}} = 10^{19}$ (Neither of the digits in 8.1 is significant.)

19.11 The cell reaction is $Zn(s) + 2Ag^+(aq) \rightleftharpoons Zn^{2+}(aq) + 2Ag(s)$; $n = 2$. The reaction quotient, Q, is

$$\frac{[Zn^{2+}]}{[Ag^+]^2} = \frac{0.200}{(0.00200)^2} = 5.00 \times 10^4$$

E^o_{cell} is the sum of $-(-0.76 \text{ V}) + 0.80 \text{ V}$, or 1.56 V. Substitute this and Q into the Nernst equation to calculate E_{cell}:

$$E_{cell} = E^\circ_{cell} - \frac{0.0592}{n} \log Q = 1.56 - \frac{0.0592}{2} \log (5.00 \times 10^4)$$

$$E_{cell} = 1.5\underline{6} \text{ V} - 0.139\underline{0}9 \text{ V} = 1.4\underline{2}09 = 1.42 \text{ V}$$

19.12 First calculate E^o_{cell}:

$$E^o_{cell} = -0.23 \text{ V} - (-0.76 \text{ V}) = 0.53 \text{ V}$$

Substitute this value and E = 0.34 V into the Nernst equation and solve for log Q:

$$E_{cell} = E^\circ_{cell} - \frac{0.0592}{n} \log Q; \quad 0.34 \text{ V} = 0.53 \text{ V} - \frac{0.0592}{2} \log Q$$

$$\log Q = (0.53 \text{ V} - 0.34 \text{ V}) \times \frac{2}{0.0592 \text{ V}} = 6.\underline{4}19$$

$$Q = \frac{[Zn^{2+}]}{[Ni^{2+}]} = \frac{1.00}{[Ni^{2+}]} = \text{antilog } 6.\underline{4}19 = \underline{2}.624 \times 10^6$$

$$[Ni^{2+}] = \underline{3}.81 \times 10^{-7} = 4 \times 10^{-7} \text{ M}$$

19.13 a. The cathode reaction is $K^+(l) + e^- \rightarrow K(l)$.

The anode reaction is $Cl^-(l) \rightarrow 1/2Cl_2(g) + e^-$.

b. The cathode reaction is $K^+(l) + e^- \rightarrow K(l)$.

The anode reaction is $4OH^-(l) \rightarrow O_2(g) + 2H_2O(g) + 4e^-$.

19.14 Two possible cathode reactions are

$Ag^+(aq) + e^- \rightarrow Ag(s)$; $E^o = 0.80$ V

$2H_2O(l) + 2e^- \rightarrow H_2(g) + 2OH^-(aq)$; $E^o = -0.41$ V (at pH 7.00)

Because the electrode potential for silver is larger (more positive), it is easier to reduce; thus, the cathode reaction is the first half-reaction above. The only possible anode reaction is

$2H_2O(l) \rightarrow O_2(g) + 4H^+(aq) + 4e^-$

19.15 From the silver electrode equation ($Ag^+ + e^- \rightarrow Ag(s)$), you can write

$$1 \text{ mol Ag} = 1 \text{ mol } e^-$$

Because 1 mol of e^- is equivalent to 9.65×10^4 C, the charge equivalent to 365 mg silver is

$$0.365 \text{ g Ag} \times \frac{1 \text{ mol Ag}}{107.9 \text{ g}} \times \frac{1 \text{ mol } e^-}{1 \text{ mol Ag}} \times \frac{9.65 \times 10^4 \text{ C}}{1 \text{ mol } e^-} = 32\underline{6}.4 \text{ C}$$

Using the time of 216 min as 1.296×10^4 s, calculate the current in amps:

$$\text{Current} = \frac{\text{charge}}{\text{time}} = \frac{326.4 \text{ C}}{1.296 \times 10^4 \text{ s}} = 2.5\underline{1}88 \times 10^{-2} = 2.52 \times 10^{-2} \text{ A}$$

19.16 When the current flows for 185 s, the charge is

$$0.0565 \text{ A} \times 185 \text{ s} = 10.\underline{4}52 \text{ C}$$

The electrode reaction is

$$2H_2O \rightarrow 4H^+ + O_2 + 4e^-$$

Thus, 4 moles of electrons are equivalent to 1 mol of O_2. The mass of O_2 liberated is

$$10.45 \text{ C} \times \frac{1 \text{ mol } e^-}{9.65 \times 10^4 \text{ C}} \times \frac{1 \text{ mol } O_2}{4 \text{ mol } e^-} \times \frac{32.00 \text{ g } O_2}{1 \text{ mol } O_2} = 8.6\underline{6}3 \times 10^{-4}$$

$$= 8.66 \times 10^{-4} \text{ g } O_2$$

■ Answers to Review Questions

19.1 A voltaic cell is an electrochemical cell in which a spontaneous reaction generates an electric current (energy). An electrolytic cell is an electrochemical cell that requires electrical current (energy) to drive a nonspontaneous reaction to the right.

19.2 In both the voltaic and electrolytic cells, the cathode is the electrode at which reduction occurs, and the anode is the electrode at which oxidation occurs. In an electrochemical cell, cations move toward the cathode, and anions move toward the anode. In a voltaic cell it is reversed.

19.3 The SI unit of electrical potential is the volt (V).

19.4 The faraday (F) is the magnitude of charge on one mole of electrons; it equals 9.65×10^4 C, or 9.65×10^4 J/V.

19.5 It is necessary to measure the voltage of a voltaic cell when no current is flowing because the cell voltage exhibits its maximum value only when no current flows. Even if the current flows just for the time of measurement, the voltage drops enough so that what is measured is significantly less than the maximum.

19.6 Standard electrode potentials are defined relative to a standard electrode potential of zero volts (0.00, 0.000 V, etc.) for the $H^+/H_2(g)$ electrode. Because the cell emf is measured using the hydrogen electrode at standard conditions and a second electrode at standard conditions, the cell emf equals the E^o of the half-reaction at the second electrode.

19.7 The SI unit of energy = joules = coulombs x volts.

19.8 The mathematical relationships are as follows:

$$\Delta G^o = -nFE^o_{cell}$$

$$\Delta G^o = -2.303\ RT \log K$$

Combining these two equations gives

$$\log K = nFE^o_{cell}/2.303\ RT$$

19.9 The first step in the corrosion of iron is

$$2Fe(s) + O_2(g) + 2H_2O(l) \rightarrow 4OH^- + 2Fe^{2+}$$

The Nernst equation for this reaction is

$$E_{cell} = E^o_{cell} - (0.0592/4) \log [OH^-]^4[Fe^{2+}]^2$$

If the pH increases, the $[OH^-]$ increases, and thus E_{cell} becomes more negative (this predicts that the reaction becomes less spontaneous). If the pH decreases, the $[OH^-]$ decreases, and thus E_{cell} becomes more positive (this predicts that the reaction becomes more spontaneous).

19.10 The zinc-carbon cell has a zinc can as the anode; the cathode is a graphite rod surrounded by a paste of manganese dioxide and carbon black. Around this is a second paste of ammonium and zinc chlorides. The electrode reactions involve oxidation of zinc metal to zinc(II) ion, and reduction at the cathode of $MnO_2(s)$ to $Mn_2O_3(s)$. The lead storage battery consists of a spongy lead anode and a lead dioxide cathode, both immersed in aqueous sulfuric acid. At the anode, the lead is oxidized to lead sulfate; at the cathode, lead dioxide is reduced to lead sulfate.

19.11 A fuel cell is essentially a battery that does not use up its electrodes. Instead, it operates with a continuous supply of reactants (fuel). An example is the hydrogen-oxygen fuel cell, in which oxygen is reduced at one electrode to the hydroxide ion and hydrogen is oxidized at the other electrode to water (H in the +1 oxidation state). Such a cell produces electrical energy in a spacecraft for long periods of time.

19.12 During the rusting of iron, one end of a drop of water exposed to air acts as one electrode of a voltaic cell; at this electrode, an oxygen molecule is reduced by 4 electrons to 4 hydroxide ions. Oxidation of metallic iron to iron(II) ion at the center of the drop of water supplies the electrons, and the center serves as the other electrode of the voltaic cell. Thus, electrons flow from the center of the drop through the iron to the end of the drop.

19.13 When iron or steel is connected to an active metal such as zinc, a voltaic cell is formed with zinc as the anode and iron as the cathode. Any type of moisture forms the electrolyte solution, and the zinc metal is then oxidized to zinc(II) ion in preference to the oxidation of iron metal. Oxygen is reduced at the cathode to hydroxide ions. If iron or steel is exposed to oxygen while connected to a less active metal such as tin, a voltaic cell is formed with iron as the anode and tin as the cathode, and iron is oxidized to iron(II) ion rather than tin being oxidized to tin(II) ion. Thus, exposed iron corrodes rapidly in a tin can. Fortunately, as long as the iron is covered by the tin, it cannot corrode.

19.14 The addition of ionic species such as strongly ionized sulfuric acid facilitates the passage of current through the solution.

19.15 Sodium metal can be prepared by electrolysis of molten sodium chloride.

19.16 The anode reaction in the electrolysis of molten potassium hydroxide is

$$4OH^- \rightarrow O_2(g) + 2H_2O(g) + 4e^-$$

19.17 The reason different products are obtained is that water, instead of Na^+, is reduced at the cathode during the electrolysis of aqueous NaCl. This is because the water has a more positive E^o (smaller decomposition voltage). At the anode, water, instead of chloride ion, is oxidized because water has a less positive E^o (smaller decomposition voltage).

19.18 The Nernst equation for the electrode reaction of $2Cl^-(aq) \rightarrow Cl_2(g) + 2e^-$ is

$$E = -1.36\text{ V} - (0.0592/2) \log (1\text{ atm}/[Cl^-]^2) = -1.36\text{ V} + 0.0592 \log [Cl^-]$$

This equation implies that E increases as $[Cl^-]$ increases. For a sufficiently large $[Cl^-]$, Cl^- will be more readily oxidized than the water solvent.

■ Solutions to Practice Problems

Note on significant figures: The final answer to each problem is given first with one nonsignificant figure (the rightmost significant figure is underlined) and then is rounded to the correct number of figures. Intermediate answers usually also have at least one nonsignificant figure.

19.19 Sketch of the cell:

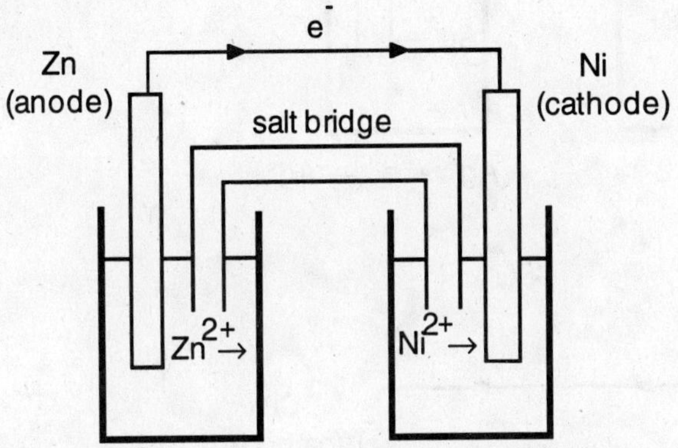

$Zn \rightarrow Zn^{2+} + 2\,e^-$ $\qquad Ni^{2+} + 2e^- \rightarrow Ni$

19.20 Sketch of the cell:

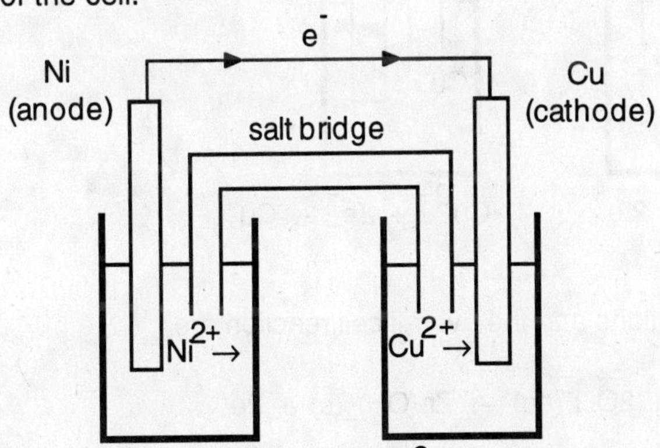

$Ni \rightarrow Ni^{2+} + 2\,e^-$ $\qquad Cu^{2+} + 2e^- \rightarrow Cu$

19.21 Sketch of the cell:

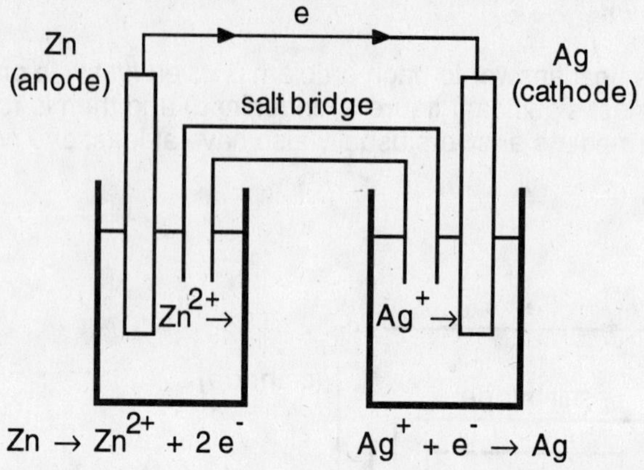

19.22 Sketch of the cell:

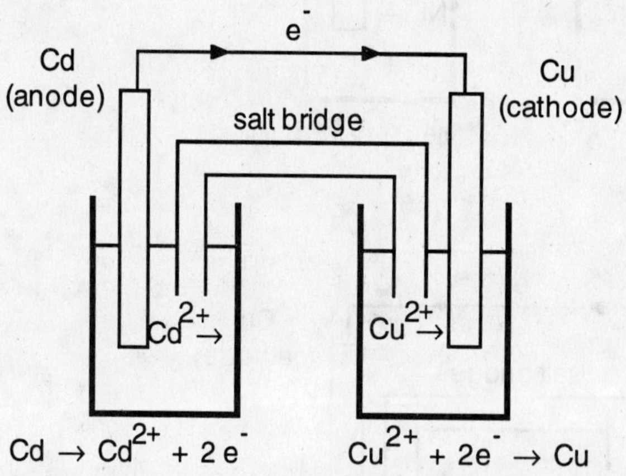

19.23 The electrode half-reactions and the overall cell reaction are

Anode: $Zn(s) + 2OH^-(aq) \rightarrow Zn(OH)_2(s) + 2e^-$

Cathode: $Ag_2O(s) + H_2O\,(l) + 2e^- \rightarrow 2Ag(s) + 2OH^-(aq)$

Overall: $Zn(s) + Ag_2O(s) + H_2O(l) \rightarrow Zn(OH)_2(s) + 2Ag(s)$

19.24 The electrode half-reactions and the overall cell reaction are

Anode: $Zn(s) + 2OH^-(aq) \rightarrow Zn(OH)_2(s) + 2e^-$

Cathode: $HgO(s) + H_2O(l) + 2e^- \rightarrow Hg(l) + 2OH^-(aq)$

Overall: $Zn(s) + HgO(s) + H_2O(l) \rightarrow Zn(OH)_2(s) + Hg(l)$

19.25 Because of its less negative E°, Pb^{2+} is reduced at the cathode and is written on the right; Cd(s) is oxidized at the anode and is written first, at the left, in the cell notation. The notation is $Cd(s)|Cd^{2+}(aq)||Pb^{2+}(aq)|Pb(s)$.

19.26 Because of its less negative E°, H^+ is reduced at the cathode and is written on the right; Al(s) is oxidized at the anode and is written first, at the left, in the cell notation. The notation is $Al(s)|Al^{3+}(aq)||H^+(aq)|H_2(g)|Pt$.

19.27 Because of its less negative E°, H^+ is reduced at the cathode and is written on the right; Ni(s) is oxidized at the anode and is written first, at the left, in the cell notation. The notation is $Ni(s)|Ni^{2+}(1\ M)||H^+(1\ M)|H_2(g)|Pt$.

19.28 Because of its less negative E°, Fe^{2+} is reduced at the cathode and is written on the right; Zn(s) is oxidized at the anode and is written first, at the left, in the cell notation. The notation is $Zn(s)|Zn^{2+}(0.20\ M)||Fe^{3+}(0.30\ M)|Fe(s)$.

19.29 The Fe(s), on the left, is the reducing agent. The Ag^+, on the right, is the oxidizing agent, gaining just one electron. Multiplying its half-reaction by 2 to equalize the numbers of electrons and writing both half-reactions gives the overall cell reaction:

$Fe(s) \rightarrow Fe^{2+}(aq) + 2e^-$

$2Ag^+(aq) + 2e^- \rightarrow 2Ag(s)$

$Fe(s) + 2Ag^+(aq) \rightarrow Fe^{2+}(aq) + 2Ag(s)$

19.30 The $H_2(g)$, on the left, is the reducing agent. The $Br_2(l)$, on the right, is the oxidizing agent. The half-reactions and the overall cell reaction are

$$H_2(g) \rightarrow 2H^+(aq) + 2e^-$$

$$Br_2(l) + 2e^- \rightarrow 2Br^-(aq)$$

$$H_2(g) + Br_2(l) \rightarrow 2H^+(aq) + 2Br^-(aq)$$

19.31 The half-cell reactions, the overall cell reaction, and the sketch are

$$Cd(s) \rightarrow Cd^{2+}(aq) + 2e^-$$

$$Ni^{2+}(aq) + 2e^- \rightarrow Ni(s)$$

$$Cd(s) + Ni^{2+}(aq) \rightarrow Cd^{2+}(aq) + Ni(s)$$

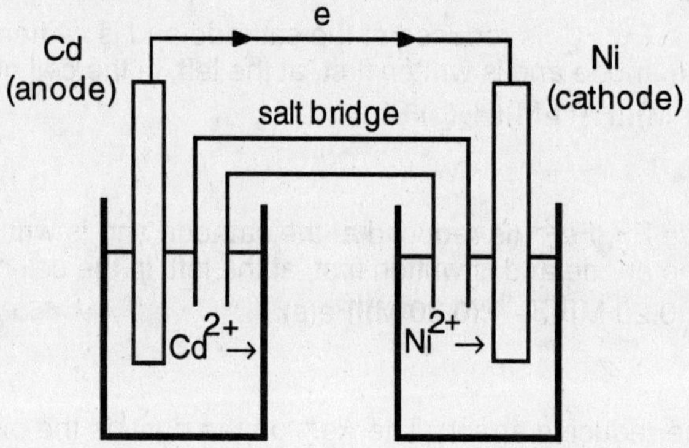

19.32 The half-cell reactions, the overall cell reaction, and the sketch are

$Zn(s) \rightarrow Zn^{2+}(aq) + 2e^-$

$2Ag^+(aq) + 2e^- \rightarrow 2Ag(s)$

$Zn(s) + 2Ag^+(aq) \rightarrow Zn^{2+}(aq) + 2Ag(s)$

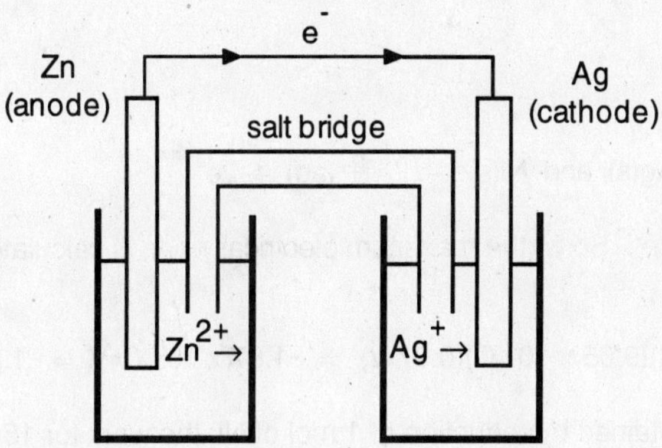

19.33 The half-cell reactions are

$2Fe^{3+}(aq) + 2e^- \rightarrow 2Fe^{2+}(aq)$ and $Zn(s) \rightarrow Zn^{2+}(aq) + 2e^-$

Therefore, for the cell, n = 2e⁻. So w, the maximum electrical work, is calculated as follows:

$$w_{max} = -nFE_{cell} = -(2)(9.65 \times 10^4\ C)(0.72\ V) = -1.\underline{3}89 \times 10^5\ C{\bullet}V$$

$$= -1.\underline{3}89 \times 10^5\ J$$

Because this is the work obtained by reduction of 2 mol of Fe^{3+}, the work for 1 mol is

$$w_{max} = (-1.389 \times 10^5\ J)/2\ mol = -6.\underline{9}45 \times 10^4 = -6.9 \times 10^4\ J$$

19.34 The half-cell reactions are

$Cl_2(aq) + 2e^- \rightarrow 2Cl^-(aq)$ and $Zn(s) \rightarrow Zn^{2+}(aq) + 2e^-$

Therefore, for the cell, n = 2e⁻. So w, the maximum electrical work, is calculated as follows:

(continued)

$$w_{max} = -nFE_{cell} = -(2)(9.65 \times 10^4\ C)(0.853\ V) = -1.6\underline{4}6 \times 10^5\ C{\bullet}V = -1.6\underline{4}6 \times 10^5\ J$$

Because this is the work obtained by reduction of 1 mol of Zn, the work for 20.0 g is

$$20.0\ g\ Zn \times \frac{1\ mol\ Zn}{65.39\ g\ Zn} \times \frac{-1.646 \times 10^5\ J}{1\ mol\ Zn} = -5.0\underline{3}44 \times 10^4$$

$$= -5.03 \times 10^4\ J$$

19.35 The half-cell reactions are

$$2Ag^+(aq) + 2e^- \rightarrow 2Ag(s) \text{ and } Ni(s) \rightarrow Ni^{2+}(aq) + 2e^-$$

Therefore, for the cell, $n = 2e^-$. So w, the maximum electrical work, is calculated as follows:

$$w_{max} = -nFE_{cell} = -(2)(9.65 \times 10^4\ C)(0.97\ V) = -1.\underline{8}7 \times 10^5\ C{\bullet}V = -1.\underline{8}7 \times 10^5\ J$$

Because this is the work obtained by reduction of 1 mol of Ni, the work for 15.0 g is

$$15.0\ g\ Ni \times \frac{1\ mol\ Ni}{58.69\ g\ Ni} \times \frac{-1.87 \times 10^5\ J}{1\ mol\ Ni} = -4.\underline{7}7 \times 10^4 = -4.8 \times 10^4\ J$$

19.36 The half-cell reactions are

$$3O_2(g) + 12H^+ + 12e^- \rightarrow 6H_2O(l) \text{ and } 4Al(s) \rightarrow 4Al^{3+}(aq) + 12e^-$$

Therefore, for the cell, $n = 12e^-$. So w, the maximum electrical work, is calculated as follows:

$$w_{max} = -nFE_{cell} = -(12)(9.65 \times 10^4\ C)(1.15\ V) = -1.3\underline{3}1 \times 10^6\ C{\bullet}V = -1.3\underline{3}1 \times 10^6\ J$$

Because this is the work obtained by reduction of 4 mol of Al, the work for 25.0 g is

$$25.0\ g\ Al \times \frac{1\ mol\ Al}{26.98\ g\ Al} \times \frac{-1.331 \times 10^6\ J}{4\ mol\ Al} = -3.0\underline{8}3 \times 10^5 = -3.08 \times 10^5\ J$$

19.37 The species of interest all occur as reactants in Table 19.1. The order from top to bottom in which these species occur is the order of increasing oxidizing power of these species. The half-reactions, E^o values, and the order of increasing oxidizing power (increasing value of E^o) are

$NO_3^-(aq) + 4H^+(aq) + 3e^- \rightleftharpoons NO(g) + 2H_2O(l)$ 0.96 V

$O_2(g) + 4H^+(aq) + 4e^- \rightleftharpoons 2H_2O(l)$ 1.23 V

$H_2O_2(aq) + 2H^+(aq) + 2e^- \rightleftharpoons 2H_2O(l)$ 1.78 V

In summary, the increasing order of strength is $NO_3^-(aq)$, $O_2(g)$, and $H_2O_2(aq)$.

19.38 The species of interest all occur as reactants in Table 19.1. The order from top to bottom in which these species occur is the order of increasing oxidizing power of these species. The half-reactions, their E^o values, and the order of increasing oxidizing power (increasing value of E^o) are

$I_2(s) + 2e^- \rightleftharpoons 2I^-(aq)$ 0.54 V (weakest oxidizing agent)

$Ag^+(aq) + e^- \rightleftharpoons Ag(s)$ 0.80 V

$MnO_4^-(aq) + 8H^+(aq) + 5e^- \rightleftharpoons$
$Mn^{2+}(aq) + 4H_2O(l)$ 1.49 V (strongest oxidizing agent)

19.39 The species of interest all occur as products in Table 19.1. The order from bottom to top in which these species occur is the order of increasing reducing power of these species. The half-reactions, their E^o values, and the order of increasing reducing power (decreasing value of E^o) are

$Cu^{2+}(aq) + e^- \rightleftharpoons Cu^+(aq)$ 0.16 V (weakest reducing agent)

$Fe^{2+}(aq) + 2e^- \rightleftharpoons Fe(s)$ -0.41 V

$Zn^{2+}(aq) + 2e^- \rightleftharpoons Zn(s)$ -0.76 V (strongest reducing agent)

19.40 The species of interest all occur as products in Table 19.1. The order from bottom to top in which these species occur is the order of increasing reducing power of these species. The half-reactions, their E^o values, and the order of increasing reducing power (decreasing value of E^o) are

$I_2(s) + 2e^- \rightleftharpoons 2I^-(aq)$ 0.54 V (weakest reducing agent)

$Sn^{4+}(aq) + 2e^- \rightleftharpoons Sn^{2+}(aq)$ 0.15 V

$Al^{3+}(aq) + 3e^- \rightleftharpoons Al(s)$ -1.66 V (strongest reducing agent)

19.41 a. The reduction half-reactions and standard potentials are

$Fe^{3+}(aq) + e^- \rightleftharpoons Fe^{2+}(aq)$ $E^o = 0.77$ V

$Sn^{4+}(aq) + 2e^- \rightleftharpoons Sn^{2+}(aq)$ $E^o = 0.15$ V

To calculate E^o_{cell}, reverse the sign of E^o for the iron half-reaction, giving $E^o_{cell} = -0.62$ V. Thus, the reaction is not spontaneous.

b. The reduction half-reactions and standard potentials are

$O_2(g) + 4H^+(aq) + 4e^- \rightleftharpoons 2H_2O(l)$ $E^o = 1.23$ V

$MnO_4^-(aq) + 8H^+ + 5e^- \rightleftharpoons Mn^{2+}(aq)$ $E^o = 1.49$ V

To calculate E^o_{cell}, reverse the sign of E^o for the oxygen half-reaction, giving $E^o_{cell} = 0.26$ V. Thus, the reaction is spontaneous.

19.42 a. The reduction half-reactions and standard potentials are

$Fe^{3+}(aq) + e^- \rightleftharpoons Fe^{2+}(aq)$ $E^o = 0.77$ V

$Cr_2O_7^{2-}(aq) + 14H^+(aq) + 6e^- \rightleftharpoons 2Cr^{2+}(aq) + 7H_2O(l)$ $E^o = 1.33$ V

To calculate E^o_{cell}, reverse the sign of E^o for the iron half-reaction, giving $E^o_{cell} = 0.56$ V. Thus, dichromate ion will spontaneously oxidize iron(II) ion.

b. The reduction half-reactions and standard potentials are

$$Cu^{2+}(aq) + 2e^- \rightleftharpoons Cu(s) \qquad E^o = 0.34\ V$$

$$Ni^{2+}(aq) + 2e^- \rightleftharpoons Ni(s) \qquad E^o = -0.23\ V$$

To calculate E^o_{cell}, reverse the sign of E^o for the copper half-reaction, giving $E^o_{cell} = -0.57$ V. Thus, copper will not reduce Ni(II) ion spontaneously.

19.43 The pertinent reduction half-reactions and their corresponding E^o values are

$$Br_2(l) + 2e^- \rightleftharpoons 2Br^-(aq) \qquad E^o = 1.07\ V$$

$$Cl_2(g) + 2e^- \rightleftharpoons 2Cl^-(aq) \qquad E^o = 1.36\ V$$

$$F_2(g) + 2e^- \rightleftharpoons 2F^-(aq) \qquad E^o = 2.87\ V$$

From these, you see that Cl_2 is a stronger oxidizing agent under standard conditions than is Br_2. To calculate E^o_{cell}, reverse the sign of the bromine half-reaction, giving $E^o_{cell} = 0.29$ V, a spontaneous reaction. The E^o_{cell} for the Cl_2 oxidation of F^- ion would be -1.51 V, a nonspontaneous reaction.

The possible reactions involving the oxidation of the anions by Cl_2 can be summarized as follows:

$Cl_2(g) + 2Br^-(aq) \rightarrow 2Cl^-(aq) + Br_2(l)$; spontaneous reaction

$Cl_2(g) + 2F^-(aq) \rightarrow 2Cl^-(aq) + F_2(g)$; nonspontaneous reaction

19.44 The possible reactions involving the oxidation of the ions by $Cr_2O_7^{2-}$ can be summarized as follows:

$Cr_2O_7^{2-}(aq) + 6Br^-(aq) + 14H^+(aq) \rightarrow 2Cr^{3+}(aq) + 3Br_2(l) + 7H_2O(l)$;

spontaneous reaction

$5Cr_2O_7^{2-}(aq) + 6Mn^{2+}(aq) + 22H^+(aq) \rightarrow 10Cr^{3+}(aq) + 6MnO_4^-(aq) + 11H_2O(l)$;

nonspontaneous reaction

To understand, consider the pertinent reduction half-reactions and their corresponding E^o's:

(continued)

$Br_2(l) + 2e^- \rightleftharpoons 2Br^-(aq)$ $E^o = 1.07\ V$

$Cr_2O_7^{2-}(aq) + 14H^+(aq) + 6e^- \rightleftharpoons 2Cr^{3+}(aq) + 7H_2O(l)$ $E^o = 1.33\ V$

$MnO_4^-(aq) + 8H^+(aq) + 5e^- \rightleftharpoons Mn^{2+}(aq) + 4H_2O(l)$ $E^o = 1.49\ V$

From these, you see that $Cr_2O_7^{2-}$ is a stronger oxidizing agent under standard conditions than is Br_2. To calculate E^o_{cell}, reverse the sign of the bromine half-reaction, giving $E^o_{cell} = 0.26\ V$, a spontaneous reaction. The E^o_{cell} for the $Cr_2O_7^{2-}$ oxidation of Mn^{2+} ion would be -0.16 V, a nonspontaneous reaction.

19.45 The pertinent reduction half-reactions and their corresponding E^o values are

$Cr^{3+}(aq) + 3e^- \rightleftharpoons Cr(s)$ $E^o = -0.74\ V$

$Hg_2^{2+}(aq) + 2e^- \rightleftharpoons 2Hg(l)$ $E^o = 0.80\ V$

Reverse the first half-reaction, multiply by 2, and reverse the sign of its E^o. Then multiply the second equation by 3; this allows you to combine both half-reactions:

$2Cr(s) \rightleftharpoons 2Cr^{3+}(aq) + 6e^-$ $-E^o = 0.74\ V$

$3Hg_2^{2+}(aq) + 6e^- \rightleftharpoons 6Hg(l)$ $E^o = 0.80\ V$

$2Cr(s) + 3Hg_2^{2+}(aq) \rightarrow 2Cr^{3+}(aq) + 6Hg(l)$ $E^o_{cell} = 1.54\ V$

19.46 The pertinent reduction half-reactions and their corresponding E^o values are

$Sn^{2+}(aq) + 2e^- \rightleftharpoons Sn(s)$ $E^o = -0.14\ V$

$Cu^{2+}(aq) + 2e^- \rightleftharpoons Cu(s)$ $E^o = 0.34\ V$

Reversing the first half-reaction and reversing the sign of its E^o allows you to combine both half-reactions:

$Sn(s) \rightleftharpoons Sn^{2+}(aq) + 2e^-$ $-E^o = 0.14\ V$

$Cu^{2+}(aq) + 2e^- \rightleftharpoons Cu(s)$ $E^o = 0.34\ V$

$Sn(s) + Cu^{2+}(aq) \rightarrow Sn^{2+}(aq) + Cu(s)$ $E^o_{cell} = 0.48\ V$

19.47 The pertinent reduction half-reactions and their corresponding E^o values are

$Cr^{3+}(aq) + 3e^- \rightleftharpoons Cr(s)$ $E^o = -0.74$ V

$I_2(s) + 2e^- \rightleftharpoons 2I^-(aq)$ $E^o = 0.54$ V

Reverse the first half-reaction, multiply by 2, and reverse the sign of its E^o. Then multiply the second equation by 3; this allows you to combine both half-reactions:

$2Cr(s) \rightleftharpoons 2Cr^{3+}(aq)$ $-E^o = 0.74$ V

$3I_2(s) + 6e^- \rightleftharpoons 6I^-(aq)$ $E^o =$ 0.54 V

$2Cr(s) + 3I_2(s) \rightarrow 2Cr^{3+}(aq) + 6I^-(aq)$ $E^o_{cell} = 1.28$ V

19.48 The pertinent reduction half-reactions and their corresponding E^o values are

$Al^{3+}(aq) + 3e^- \rightleftharpoons Al(s)$ $E^o = -1.66$ V

$Hg_2^{2+}(aq) + 2e^- \rightleftharpoons 2Hg(l)$ $E^o = 0.80$ V

Reverse the first half-reaction, multiply by 2, and reverse the sign of its E^o. Then multiply the second equation by 3; this allows you to combine both half-reactions:

$2Al(s) \rightleftharpoons 2Al^{3+}(aq) + 6e^-$ $-E^o = 1.66$ V

$3Hg_2^{2+}(aq) + 6e^- \rightleftharpoons 6Hg(l)$ $E^o = 0.80$ V

$2Al(s) + 3Hg_2^{2+}(aq) \rightarrow 2Al^{3+}(aq) + 6Hg(l)$ $E^o_{cell} = 2.46$ V

19.49 First calculate E^o_{cell}; then find ΔG^o by multiplying E^o_{cell} by the electron change and the value of the faraday. The half-reactions and their respective E^o's are

$Cu^{2+}(aq) + 2e^- \rightleftharpoons Cu(s)$ $E^o = 0.34$ V

$NO_3^-(aq) + 4H^+(aq) + 3e^- \rightleftharpoons NO(g) + 2H_2O(l)$ $E^o = 0.96$ V

Reverse the first half-reaction, multiply by 3, and reverse the sign of its E^o. Then multiply the second equation by 2; this allows you to combine both half-reactions:

(continued)

$3Cu(s)$	$\rightleftharpoons$	$3Cu^{2+}(aq) + 6e^-$	$-E^o = -0.34$ V
$2NO_3^-(aq) + 8H^+(aq) + 6e^-$	$\rightleftharpoons$	$2NO(g) + 4H_2O(l)$	$E^o = 0.96$ V
$3Cu(s) + 2NO_3^-(aq) + 8H^+(aq)$	$\rightarrow$	$3Cu^{2+}(aq) + 2NO(g) + 4H_2O(l)$	$E^o_{cell} = 0.62$ V

$\Delta G^o = -nFE^o_{cell} = -(6)(9.65 \times 10^4 \text{ C})(0.62 \text{ V}) = -3.\underline{5}8 \times 10^5 \text{ C} \bullet \text{V} = -3.6 \times 10^5 \text{ J}$

19.50 First calculate E^o_{cell}; then find ΔG^o by multiplying E^o_{cell} by the electron change and the value of the faraday. The half-reactions and their respective E^os are

$Al^{3+}(aq) + 3e^-$	$\rightleftharpoons$	$Al(s)$	$E^o = -1.66$ V
$O_2(g) + 4H^+(aq) + 4e^-$	$\rightleftharpoons$	$2H_2O(l)$	$E^o = 1.23$ V

Reverse the first half-reaction, multiply by 4, and reverse the sign of its E^o. Then multiply the second equation by 3; this allows you to combine both half-reactions:

$4Al(s)$	$\rightleftharpoons$	$4Al^{3+}(aq) + 12e^-$	$-E^o = 1.66$ V
$3O_2(g) + 12H^+(aq) + 12e^-$	$\rightleftharpoons$	$6H_2O(l)$	$E^o = 1.23$ V
$4Al(s) + 3O_2(g) + 12H^+(aq)$	$\rightarrow$	$4Al^{3+}(aq) + 6H_2O(l)$	$E^o_{cell} = 2.89$ V

$\Delta G^o = -nFE^o_{cell} = -(12)(9.65 \times 10^4 \text{ C})(2.89 \text{ V}) = -3.3\underline{4}6 \times 10^6 \text{ C} \bullet \text{V} = -3.35 \times 10^6 \text{ J}$

19.51 First calculate E^o_{cell}; then multiply by the electron change and the value of the faraday to find ΔG^o. The half-reactions and their respective E^o's are

$I_2(s) + 2e^-$	$\rightleftharpoons$	$2I^-(aq)$	$E^o = 0.54$ V
$Cl_2(g) + 2e^-$	$\rightleftharpoons$	$2Cl^-(aq)$	$E^o = 1.36$ V

Reverse the first half-reaction, and reverse the sign of its E^o. Then combine both half-reactions:

$2I^-(aq)$	$\rightleftharpoons$	$I_2(s) + 2e^-$	$-E^o = -0.54$ V
$Cl_2(g) + 2e^-$	$\rightleftharpoons$	$2Cl^-(aq)$	$E^o = 1.36$ V
$2I^-(aq) + Cl_2(g)$	$\rightarrow$	$I_2(s) + 2Cl^-(aq)$	$E^o_{cell} = 0.82$ V

$\Delta G^o = -nFE^o_{cell} = -(2)(9.65 \times 10^4 \text{ C})(0.82 \text{ V}) = -1.\underline{5}8 \times 10^5 \text{ C} \bullet \text{V} = -1.6 \times 10^5 \text{ J}$

19.52 First calculate E^o_{cell}; then use n = 1 and the value of the faraday to find ΔG^o.

$$Na^+(aq) + e^- \rightleftharpoons Na(s) \qquad E^o = -2.71\ V$$

$$1/2Cl_2(g) + e^- \rightleftharpoons Cl^-(aq) \qquad E^o = 1.36\ V$$

Reverse the first half-reaction, and reverse the sign of its E^o. Then combine both:

$$Na(s) \rightleftharpoons Na^+(aq) + e^- \qquad -E^o = 2.71\ V$$

$$1/2Cl_2(g) + e^- \rightleftharpoons Cl^-(aq) \qquad E^o = 1.36\ V$$

$$Na(s) + 1/2Cl_2(g) \rightarrow Na^+(aq) + Cl^-(aq) \qquad E^o_{cell} = 4.07\ V$$

$$\Delta G^o = -nFE^o_{cell} = -(1)(9.65 \times 10^4\ C)(4.07\ V) = -3.9\underline{2}7 \times 10^5\ C{\bullet}V = -3.93 \times 10^5\ J$$

19.53 Write the equation with ΔG_f^o's beneath each substance. Use ΔG^o to solve for E^o_{cell}.

	Al(s)	+	3Ag+(aq)	→	Al3+(aq)	+	3Ag(s); n = 3
ΔG_f^o:	0		3(77.111)		-481.2		3(0)

$$\Delta G^o = [(-481.2) - 3(77.111)] = -712.\underline{5}3\ kJ\ (-712.\underline{5}3 \times 10^3\ C{\bullet}V)$$

$$\Delta G^o = -nFE^o_{cell}$$

$$-712.53 \times 10^3\ C{\bullet}V = -(3)(9.65 \times 10^4\ C)(E^o_{cell})$$

$$E^o_{cell} = \frac{-712.53 \times 10^3\ C{\bullet}V}{-(3)(9.65 \times 10^4\ C)} = 2.4\underline{6}1 = 2.46\ V$$

19.54 Write the equation with ΔG_f^o's beneath each substance. Use ΔG^o to solve for E^o_{cell}.

	2Al(s)	+	3Cu2+(aq)	→	2Al3+(aq)	+	3Cu(s); n = 6
ΔG_f^o:	2(0)		3(64.98)		(2)-481.2		3(0)

$$\Delta G^o = [2(-481.2) - 3(64.98)] = -1157.\underline{3}4\ kJ\ (-1157.\underline{3}4 \times 10^3\ C{\bullet}V)$$

$$\Delta G^o = -nFE^o_{cell}$$

(continued)

$$-1157.34 \times 10^3 \text{ C}\bullet\text{V} = -(6)(9.65 \times 10^4 \text{ C})(E^o_{cell})$$

$$E^o_{cell} = \frac{-1157.34 \times 10^3 \text{ C}\bullet\text{V}}{-(6)(9.65 \times 10^4 \text{ C})} = 1.99\underline{88} = 1.999 \text{ V}$$

19.55 Write the equation with ΔG_f^o's beneath each substance:

$$PbO_2(s) + 2HSO_4^-(aq) + 2H^+(aq) + Pb(s) \rightarrow 2PbSO_4(s) + 2H_2O(l); \quad n = 2$$

ΔG_f^o: -189.2   2(-752.87)   0   0   2(-811.24)   2(-237.192)

$$\Delta G^o = [2(-811.24) + 2(-237.192) - (-189.2) - 2(-752.87)]$$

$$= -401.\underline{9}24 \text{ kJ} \quad (-401.924 \times 10^3 \text{ C}\bullet\text{V})$$

$$\Delta G^o = -nFE^o_{cell} = -401.\underline{9}24 \times 10^3 \text{ C}\bullet\text{V} = -(2)(9.65 \times 10^4 \text{ C})(E^o_{cell})$$

Rearrange and solve for E^o_{cell}. Determine the significant figures from the three significant figures of F.

$$E^o_{cell} = \frac{-401.924 \times 10^3 \text{ C}\bullet\text{V}}{-(2)(9.65 \times 10^4 \text{ C})} = 2.0\underline{8}25 = 2.08 \text{ V}$$

19.56 Write the equation with ΔG_f^o's beneath each substance:

$$5H_2C_2O_4(aq) + 2MnO_4^-(aq) + 6H^+(aq) \rightarrow 10CO_2(g) + 8H_2O(l) + 2Mn^{2+}(aq); \quad n = 10$$

ΔG_f^o: 5(-698)   2(-425.1)   6(0)   10(-394.4)   8(-237.192)   2(-223)

$$\Delta G^o = [10(-394.4) + 8(-237.192) + 2(-223) - 5(-698) - 2(-425.1)] = -194\underline{7}.336 \text{ kJ}$$

$$\Delta G^o = -194\underline{7}.336 \times 10^3 \text{ C}\bullet\text{V} = -nFE^o_{cell}$$

(continued)

$$-1947.336 \times 10^3 \text{ C•V} = -(10)(9.65 \times 10^4 \text{ C})(E°_{cell})$$

Rearrange and solve for $E°_{cell}$. Determine the significant figures from the three significant figures of F.

$$E°_{cell} = \frac{-1947.6 \times 10^3 \text{ C•V}}{-(10)(9.65 \times 10^4 \text{ C})} = 2.0\underline{1}79 = 2.02 \text{ V}$$

19.57 Rearrange the equation $E°_{cell} = (0.0592 \text{ V})/n \log K$ to solve for $\log K_c$. Begin by calculating $E°_{cell}$. Then substitute into the $\log K_c$ expression.

$$Sn^{4+}(aq) + 2e^- \rightleftharpoons Sn^{2+}(aq) \qquad E° = 0.15 \text{ V}$$

$$Fe^{3+}(aq) + e^- \rightleftharpoons Fe^{2+}(aq) \qquad E° = 0.77 \text{ V}$$

Reverse the first half-reaction and reverse the sign of its E°. Then multiply the second equation by 2; this allows you to combine both half-reactions:

$$Sn^{2+}(aq) \rightleftharpoons Sn^{4+}(aq) + 2e^- \qquad -E° = -0.15 \text{ V}$$

$$2Fe^{3+}(aq) + 2e^- \rightleftharpoons 2Fe^{2+}(aq) \qquad E° = 0.77 \text{ V}$$

$$Sn^{2+}(aq) + 2Fe^{3+}(aq) \rightleftharpoons Sn^{4+}(aq) + 2Fe^{2+}(aq) \qquad E°_{cell} = 0.62 \text{ V}$$

Now calculate log K.

$$0.62 \text{ V} = \frac{0.0592 \text{ V}}{2} \log K$$

$\log K = 2\underline{0}.94$ (two sig. figs. from the two sig. figs. of $E°_{cell}$)

$K = 8.7 \times 10^{2\underline{0}}$, or rounding, $K = 10^{21}$ (Neither of the digits in 8.7 is significant.)

19.58 Rearrange the equation E^o_{cell} = (0.0592 V)/n log K to solve for log K_c. Begin by calculating E^o_{cell}. Then substitute into the log K_c expression.

$Hg_2^{2+}(aq) + 2e^- \rightleftharpoons 2Hg(l)$ $E^o = 0.80$ V

$Fe^{3+}(aq) + e^- \rightleftharpoons Fe^{2+}(aq)$ $E^o = 0.77$ V

Reverse the first half-reaction and reverse the sign of its E^o. Then multiply the second equation by 2; this allows you to combine both half-reactions:

$2Hg(l) \rightleftharpoons Hg_2^{2+}(aq) + 2e^-$ $-E^o = -0.80$ V

$2Fe^{3+}(aq) + 2e^- \rightleftharpoons 2Fe^{2+}(aq)$ $E^o = 0.77$ V

$2Hg(l) + 2Fe^{3+}(aq) \rightarrow Hg_2^{2+}(aq) + 2Fe^{2+}(aq)$ $E^o_{cell} = -0.03$ V

Now calculate log K_c.

$$-0.03\text{ V} = \frac{0.0592\text{ V}}{2}\log K_c$$

log K_c = $-\underline{1}.013$ (one sig. fig. from the one sig. fig. of E^o_{cell})

K_c = 9.7 x $10^{-\underline{2}}$, or rounding, $K_c = 10^{-1}$ (Neither of the digits in 9.7 is significant.)

19.59 Rearrange the equation E^o_{cell} = (0.0592 V)/n log K to solve for log K_c. Begin by calculating E^o_{cell}. Then substitute into the log K_c expression.

$Cu^{2+}(aq) + e^- \rightleftharpoons Cu^+(aq)$ $E^o = 0.16$ V

$Cu^+(aq) + e^- \rightleftharpoons Cu(s)$ $E^o = 0.52$ V

Reverse the first half-reaction and reverse the sign of its E^o. Then combine with the second equation:

$Cu^+(aq) \rightleftharpoons Cu^{2+}(aq) + e^-$ $-E^o = -0.16$ V

$Cu^+(aq) + e^- \rightleftharpoons Cu(s)$ $E^o = 0.52$ V

$2Cu^+(aq) \rightleftharpoons Cu(s) + Cu^{2+}(aq)$ $E^o_{cell} = 0.36$ V

(continued)

Now calculate log K_c.

$$0.36\ V = \frac{0.0592\ V}{1} \log K_c$$

log K_c = 6.081 (two sig. figs. from the two sig. figs. of E^o_{cell})

K_c = 1.2 x 10^6 (one sig. fig. because only one sig. fig. to the right of the decimal point)

19.60 Rearrange the equation E^o_{cell} = (0.0592 V)/n log K_c to solve for log K_c. Begin by calculating E^o_{cell}. Then substitute into the log K_c expression.

$$ClO_4^-(aq) + H_2O(l) + 2e^- \rightleftharpoons ClO_3^-(aq) + 2OH^-(aq) \qquad E^o = 0.17\ V$$

$$ClO_3^-(aq) + H_2O(l) + 2e^- \rightleftharpoons ClO_2^-(aq) + 2OH^-(aq) \qquad E^o = 0.35\ V$$

Reverse the first half-reaction and reverse the sign of its E^o. Then combine with the second equation:

$$ClO_3^-(aq) + 2OH^-(aq) \rightleftharpoons ClO_4^-(aq) + H_2O(l) + 2e^- \qquad -E^o = -0.17\ V$$

$$ClO_3^-(aq) + H_2O(l) + 2e^- \rightleftharpoons ClO_2^-(aq) + 2OH^-(aq) \qquad E^o = 0.35\ V$$

$$2ClO_3^-(aq) \rightleftharpoons ClO_4^-(aq) + ClO_2^-(aq) \qquad E^o_{cell} = 0.18\ V$$

Now calculate log K_c.

$$0.18\ V = \frac{0.0592\ V}{2} \log K_c$$

log K_c = 6.081 (two sig. figs. from the two sig. figs. of E^o_{cell})

K_c = 1.2 x 10^6 (one sig. fig. because only one sig. fig. to the right of the decimal point)

19.61 First calculate E^o_{cell}:

$$E^o_{cell} = -(-0.74\ V) + (-0.23\ V) = 0.51\ V$$

The cell reaction is $2Cr(s) + 3Ni^{2+}(aq) \rightarrow 2Cr^{3+}(aq) + 3Ni(s)$; n = 6.

Substitute E^o_{cell} = 0.51 V, 1.0 x 10^{-2} M Cr^{3+}, and 2.0 M Ni^{2+} into the Nernst equation and solve for E_{cell}:

$$E_{cell} = E^\circ cell - \frac{0.0592}{n} \log Q = 0.51\ V - \frac{0.0592}{6} \log \frac{(1.0 \times 10^{-2})^2}{(2.0)^3}$$

$$E_{cell} = 0.51\ V - (-0.048\underline{3}7\ V) = 0.5\underline{5}837 = 0.56\ V$$

19.62 First calculate E^o_{cell}:

$$E^o_{cell} = -(-0.23\ V) + (-0.14\ V) = 0.09\ V$$

The cell reaction is $Ni(s) + Sn^{2+}(aq) \rightarrow Ni^{2+}(aq) + Sn(s)$; n = 2.

Substitute E^o_{cell} = 0.09 V, 1.0 x 10^{-4} M Sn^{2+}, and 1.0 M Ni^{2+} into the Nernst equation and solve for E_{cell}:

$$E_{cell} = E^\circ cell - \frac{0.0592}{n} \log Q = 0.09\ V - \frac{0.0592}{2} \log \frac{(1.0)}{(1.0 \times 10^{-4})}$$

$$E_{cell} = 0.09\ V - (0.11\underline{8}4\ V) = -0.0\underline{2}84 = -0.03\ V$$

19.63 First calculate E^o_{cell}:

$$E^o_{cell} = -(1.07\ V) + (1.49\ V) = 0.42\ V$$

The cell reaction is $2MnO_4^-(aq) + 10Br^-(aq) + 16H^+(aq)$

$$\rightarrow 2Mn^{2+}(aq) + 5Br_2(l) + 8H_2O(l);\ n = 10.$$

Substitute E^o_{cell} = 0.42 V, 0.010 M MnO_4^-, 0.010 M Br^-, 0.15 M Mn^{2+}, and 1.0 M H^+ into the Nernst equation and solve for E_{cell}:

(continued)

$$E_{cell} = E^\circ_{cell} - \frac{0.0592}{n} \log Q$$

$$= 0.42 \text{ V} - \frac{0.0592}{10} \log \frac{(0.15)^2}{(0.010)^2 (0.010)^{10} (1.0)^{16}}$$

$$E_{cell} = 0.42 \text{ V} - (0.132\underline{3} \text{ V}) = 0.2\underline{8}77 = 0.29 \text{ V}$$

19.64 First calculate E^o_{cell}:

$$E^o_{cell} = -(0.54 \text{ V}) + (1.33 \text{ V}) = 0.79 \text{ V}$$

The cell reaction is $Cr_2O_7^{2-}(aq) + 6I^-(aq) + 14H^+(aq)$

$$\rightarrow 2Cr^{3+}(aq) + 3I_2(s) + 7H_2O(l); \quad n = 6.$$

Substitute E^o_{cell} = 0.79 V, 0.020 M $Cr_2O_7^{2-}$, 0.015 M I^-, 0.20 M Cr^{3+}, and 1.0 M H^+ into the Nernst equation and solve for E_{cell}:

$$E_{cell} = E^\circ_{cell} - \frac{0.0592}{n} \log Q$$

$$= 0.79 \text{ V} - \frac{0.0592}{6} \log \frac{(0.20)^2}{(0.020)(0.015)^6 (1.0)^{14}}$$

$$E_{cell} = 0.79 \text{ V} - (0.11\underline{0}9 \text{ V}) = 0.6\underline{7}9 = 0.68 \text{ V}$$

19.65 Note that E^o_{cell} = 0.170 V. Substitute this value and E = 0.240 V into the Nernst equation and solve for log Q:

$$E_{cell} = E^\circ_{cell} - \frac{0.0592}{n} \log Q; \quad 0.240 \text{ V} = 0.170 \text{ V} - \frac{0.0592}{2} \log Q$$

$$\log Q = (0.240 \text{ V} - 0.170 \text{ V}) \times \left(- \frac{2}{0.0592 \text{ V}} \right) = -2.\underline{3}648$$

$$Q = \frac{[Cd^{2+}]}{[Ni^{2+}]} = \frac{[Cd^{2+}]}{1.0} = \text{antilog}(-2.\underline{3}648) = \underline{4}.31 \times 10^{-3}$$

$$[Cd^{2+}] = \underline{4}.31 \times 10^{-3} = 4 \times 10^{-3} \text{ M}$$

19.66 The overall cell reaction for this cell is

$$Zn(s) + 2H^+(aq) \rightarrow Zn^{2+}(aq) + H_2(g)$$

Use the usual methods to calculate that $E°_{cell}$ for this reaction is +0.76 V. Write the Nernst equation for any reaction at 25°C using the unit of volts after the 0.0592 constant:

$$E_{cell} = E°_{cell} - \frac{0.0592\ V}{n} \log Q$$

Into the numerator of Q substitute 1.0 M for $[Zn^{2+}(aq)]$ and 1.0 atm pressure for $[H_2(g)]$. Write the unknown $[H^+]^2$ in the denominator of Q (solid zinc has no concentration). Using n = 2, obtain the setup for the Nernst equation:

$$0.475\ V = 0.76\ V - \frac{0.0592\ V}{2} \log \frac{(1.0)(1.0)}{[H^+]^2}$$

Collecting the emf terms and rearranging yields

$$\log \frac{(1.0)(1.0)}{[H^+]^2} = 0.2\underline{8}5\ V \times \frac{2}{0.0592\ V} = 9.\underline{6}28$$

This can be solved for $[H^+]$ and then $[H^+]$ can be used to calculate pH, or pH can be found directly by recognizing that the log of $1/[H^+]$ = pH. The latter approach is used here:

$$\log \frac{1}{[H^+]^2} = -\log [H^+]^2 = 2 \times pH = 9.\underline{6}28$$

$$pH = 9.628 \div 2 = 4.\underline{8}14 = 4.8$$

19.67 a. The cathode reaction is $Ca^{2+}(l) + 2e^- \rightarrow Ca(l)$.

The anode reaction is $2Cl^-(l) \rightarrow Cl_2(g) + 2e^-$.

b. The cathode reaction is $Cs^+(l) + e^- \rightarrow Cs(l)$.

The anode reaction is $4OH^-(l) \rightarrow O_2(g) + 2H_2O(g) + 4e^-$.

19.68 a. The cathode reaction is $Mg^{2+}(l) + 2e^- \rightarrow Mg(l)$.

The anode reaction is $2Br^-(l) \rightarrow Br_2(l) + 2e^-$.

b. The cathode reaction is $Ca^{2+}(l) + 2e^- \rightarrow Ca(l)$.

The anode reaction is $4OH^-(l) \rightarrow O_2(g) + 2H_2O(g) + 4e^-$.

19.69 a. Two possible cathode reactions are

$Na^+(aq) + e^- \rightarrow Na(s)$; $E^o = -2.71$ V

$2H_2O(l) + 2e^- \rightarrow H_2(g) + 2OH^-(aq)$; $E^o = -0.83$ V

Because the electrode potential for water is larger (less negative), it is easier to reduce; thus, the cathode reaction is the second half-reaction above. The possible anode reactions are

$2H_2O(l) \rightarrow O_2(g) + 4H^+(aq) + 4e^-$; $E^o = -1.23$ V

$2SO_4^{2-}(aq) \rightarrow S_2O_8^{2-}(aq) + 2e^-$; $E^o = -2.01$ V

Because the electrode potential for water is less negative, it is easier to oxidize; thus, the anode reaction is the first half-reaction above.

The overall reaction is

$2H_2O(l) \rightarrow 2H_2(g) + O_2(g)$

b. Two possible cathode reactions are

$K^+(aq) + e^- \rightarrow K(s)$; $E^o = -2.92$ V

$2H_2O(l) + 2e^- \rightarrow H_2(g) + 2OH^-(aq)$; $E^o = -0.83$ V

Because the electrode potential for water is larger (less negative), it is easier to reduce; thus, the cathode reaction is the second half-reaction above. Next consider the possible anode reactions:

$2H_2O(l) \rightarrow O_2(g) + 4H^+(aq) + 4e^-$; $E^o = -1.23$ V

$2Br^-(aq) \rightarrow Br_2(l) + 2e^-$; $E^o = -1.07$ V

Because the electrode potential for bromide is less negative, it is easier to oxidize; thus, the anode reaction is the second half-reaction above.

The overall reaction is

$2Br^-(aq) + 2H_2O(l) \rightarrow Br_2(l) + H_2(g) + 2OH^-$

19.70 a. Two possible cathode reactions are

$$Cu^{2+}(aq) + 2e^- \rightarrow Cu(s); \; E^o = 0.34 \text{ V}$$

$$2H_2O(l) + 2e^- \rightarrow H_2(g) + 2OH^-(aq); \; E = -0.41 \text{ V (at pH 7.00)}$$

Because the electrode potential for Cu^{2+} is larger (less negative), it is easier to reduce; thus, the cathode reaction is the first half-reaction above. The possible anode reactions are

$$2H_2O(l) \rightarrow O_2(g) + 4H^+(aq) + 4e^-; \; E^o = -1.23 \text{ V}$$

$$2Cl^-(aq) \rightarrow Cl_2(g) + 2e^-; \; E^o = -1.36 \text{ V}$$

The anode reaction is the first half-reaction above. The overall reaction is

$$2Cu^{2+}(aq) + 2H_2O(l) \rightarrow 2Cu(s) + 4H^+(aq) + O_2(g)$$

b. The half-reactions that occur at each electrode, and the overall reaction for $Cu(NO_3)_2$, are the same as for part (a).

19.71 From the aluminum electrode equation ($Al^{3+}(l) + 3e^- \rightarrow Al(s)$): 1 mol Al = 3 mol e^-. Because 1 mol of e^- is equivalent to 9.65×10^4 C, the charge equivalent to 5.12 kg aluminum is

$$5.12 \times 10^3 \text{ g} \times \frac{1 \text{ mol Al}}{26.98 \text{ g}} \times \frac{3 \text{ mol e}^-}{1 \text{ mol Al}} \times \frac{9.65 \times 10^4 \text{ C}}{1 \text{ mol e}^-} = 5.4\underline{9}38 \times 10^7$$

$$= 5.49 \times 10^7 \text{ C}$$

19.72 From the chloride electrode equation ($2Cl^-(aq) \rightarrow Cl_2(g) + 2e^-$): 1 mol Cl_2 = 2 mol e^-. Because 1 mol of e^- is equivalent to 9.65×10^4 C, the charge equivalent to 1.18 kg of Cl_2 is

$$1.18 \times 10^3 \text{ g} \times \frac{1 \text{ mol Cl}_2}{70.905 \text{ g}} \times \frac{2 \text{ mol e}^-}{1 \text{ mol Cl}_2} \times \frac{9.65 \times 10^4 \text{ C}}{1 \text{ mol e}^-} = 3.2\underline{1}2 \times 10^6 \text{ C}$$

Using the current of 5.00×10^2 A, calculate the time in s:

$$\text{Time} = \frac{\text{charge}}{\text{current}} = \frac{3.212 \times 10^6 \text{ C}}{5.00 \times 10^2 \text{ A}} = 6.4\underline{2}4 \times 10^3 = 6.42 \times 10^3 \text{ s}$$

19.73 From the lithium electrode equation ($Li^+(l) + e^- \rightarrow Li(s)$): 1 mol Li = 1 mol e^-. Because 1 mol of e^- is equivalent to 9.65×10^4 C, the mass equivalent to 5.00×10^3 C is

$$5.00 \times 10^3 \text{ C} \times \frac{1 \text{ mol e}^-}{9.65 \times 10^4 \text{ C}} \times \frac{1 \text{ mol Li}}{1 \text{ mol e}^-} \times \frac{6.941 \text{ g Li}}{1 \text{ mol Li}} = 0.35\underline{9}6$$

$$= 0.360 \text{ g Li}$$

19.74 From the cadmium electrode equation ($Cd^{2+}(aq) + 2e^- \rightarrow Cd(s)$): 1 mol Cd = 2 mol e^-. Because 1 mol of e^- is equivalent to 9.65×10^4 C, the mass equivalent to a current of 1.51 A for 156 min is

$$\frac{1.51 \text{ C}}{1 \text{ s}} \times (156 \times 60) \text{ s} \times \frac{1 \text{ mol e}^-}{9.65 \times 10^4 \text{ C}} \times \frac{1 \text{ mol Cd}}{2 \text{ mol e}^-} \times \frac{112.4 \text{ g Cd}}{1 \text{ mol Cd}} = 8.2\underline{3}1$$

$$= 8.23 \text{ g Cd}$$

■ Solutions to Unclassified Problems

19.75 The cell notation is $Mg(s)|Mg^{2+}(aq)||Cl^-(aq)|Cl_2(g)|Pt(s)$. The reactions are

Anode: $Mg(s) \rightarrow Mg^{2+}(aq) + 2e^-$

Cathode: $Cl_2(g) + 2e^- \rightarrow 2Cl^-(aq)$

$E^o_{cell} = 1.36 \text{ V} - (-2.38 \text{ V}) = 3.74 \text{ V}$

19.76 The cell notation is $Mg(s)|Mg^{2+}(aq)||Ag^+(aq)|Ag(s)$. The reactions are

Anode: $Mg(s) \rightarrow Mg^{2+}(aq) + 2e^-$

Cathode: $Ag^+(aq) + e^- \rightarrow Ag(s)$

$E^o_{cell} = 0.80 \text{ V} - (-2.38 \text{ V}) = 3.18 \text{ V}$

19.77 In each case, calculate the standard cell emf; if it is positive, the reaction is spontaneous.

a. The half-reactions and the corresponding E° values are

$Ni^{2+}(aq) + 2e^- \rightleftharpoons Ni(s) \qquad E^o = -0.23\ V$

$Fe^{3+}(aq) + e^- \rightleftharpoons Fe^{2+}(aq) \qquad E^o = 0.77\ V$

To obtain the standard cell emf, reverse the nickel half-reaction and reverse its half-cell potential, double the iron reaction, and then add the half-cell potentials:

$Ni(s) \rightleftharpoons Ni^{2+}(aq) + 2e^- \qquad -E^o_{Ni} = 0.23\ V$

$2Fe^{3+} + 2e- \rightleftharpoons 2Fe^{2+}(aq) \qquad E^o_{Fe} = 0.77\ V$

$Ni(s) + 2Fe^{3+}(aq) \rightarrow Ni^{2+}(aq) + 2Fe^{2+}(aq) \qquad E^o_{cell} = 1.00\ V$ (spontaneous)

b. The half-reactions and the corresponding E° values are

$Sn^{4+}(aq) + 2e^- \rightleftharpoons Sn^{2+}(aq) \qquad E^o = 0.15\ V$

$Fe^{3+}(aq) + e^- \rightleftharpoons Fe^{2+}(aq) \qquad E^o = 0.77\ V$

To obtain the standard cell emf, reverse the tin half-reaction and reverse its half-cell potential, double the iron reaction, and then add the half-cell potentials:

$Sn^{2+}(aq) \rightleftharpoons Sn^{4+}(aq) + 2e^- \qquad -E^o_{Sn} = -0.15\ V$

$2Fe^{3+} + 2e- \rightleftharpoons 2Fe^{2+}(aq) \qquad E^o_{Fe} = 0.77\ V$

$Sn^{2+}(aq) + 2Fe^{3+}(aq) \rightarrow Sn^{4+}(aq) + 2Fe^{2+}(aq) \qquad E^o_{cell} = 0.62\ V$

The reaction is spontaneous.

19.78 In each case, calculate the standard cell emf.

a. The half-reactions and the corresponding E° values are

$Cl_2(g) + 2e^- \rightleftharpoons 2Cl^-(aq) \qquad E^o = 1.36\ V$

$MnO_4^-(aq) + 8H^+(aq) + 5e^- \rightleftharpoons Mn^{2+}(aq) + 4H_2O(l) \qquad E^o = 1.49\ V$

(continued)

To obtain the standard cell emf, reverse the chloride half-reaction and reverse its half-cell potential, cross-multiply by 5 and 2, and then add half-cell potentials:

$$10Cl^-(aq) \rightleftharpoons 5Cl_2(g) + 10e^- \qquad -E^o_{Cl} = -1.36\ V$$

$$2MnO_4^-(aq) + 16H^+(aq) + 10e^- \rightleftharpoons 2Mn^{2+}(aq) + 8H_2O(l) \qquad E^o_M = 1.49\ V$$

$$10Cl^-(aq) + 2MnO_4^-(aq) + 16H^+(aq)$$

$$\rightarrow 5Cl_2(g) + 2Mn^{2+}(aq) + 8H_2O(l) \qquad E^o_{cell} = 0.13\ V$$

The reaction is spontaneous.

b. The half-reactions and the corresponding E^o values are

$$Cl_2(g) + 2e^- \rightleftharpoons 2Cl^-(aq) \qquad E^o = 1.36\ V$$

$$Cr_2O_7^{2-}(aq) + 14H^+(aq) + 6e^- \rightleftharpoons 2Cr^{3+}(aq) + 7H_2O(l) \qquad E^o = 1.33\ V$$

To obtain the standard cell emf, reverse the chloride half-reaction and reverse its half-cell potential, multiply it by 3, and then add half-cell potentials:

$$6Cl^-(aq) \rightleftharpoons 3Cl_2(g) + 6e^- \qquad -E^o = -1.36\ V$$

$$Cr_2O_7^{2-}(aq) + 14H^+(aq) + 6e^- \rightleftharpoons 2\ Cr^{3+}(aq) + 7H_2O(l) \qquad E^o = 1.33\ V$$

$$6Cl^-(aq) + Cr_2O_7^{2-}(aq) + 14H^+(aq)$$

$$\rightarrow 3Cl_2(g) + 2Cr^{3+}(aq) + 7H_2O(l) \qquad E^o_{cell} = -0.03\ V$$

The reaction is not spontaneous.

19.79 First use K_{sp} to calculate $[Pb^{2+}]$:

$$[Pb^{2+}] = \frac{K_{sp}}{[SO_4^{2-}]} = \frac{1.7 \times 10^{-8}}{1.0} = 1.7 \times 10^{-8}\ M$$

Then calculate E^o_{cell}:

$$E^o_{cell} = -(-0.13\ V) + 0.00\ V = 0.13\ V$$

(continued)

The cell reaction is $Pb(s) + 2H^+(aq) \rightarrow Pb^{2+}(aq) + H_2(g)$; n = 2.

Substitute E^o_{cell} = 0.13 V, 1.7×10^{-8} M Pb^{2+}, and 1.0 M H^+ into the Nernst equation and solve for E_{cell}:

$$E_{cell} = E^\circ_{cell} - \frac{0.0592}{n} \log Q = 0.13 \text{ V} - \frac{0.0592}{2} \log \frac{(1.7 \times 10^{-8})}{(1.0)^2}$$

$$E_{cell} = 0.13 \text{ V} + (0.22\underline{9}9 \text{ V}) = 0.3\underline{5}99 = 0.36 \text{ V}$$

19.80 First use K_{sp} to calculate $[Ag^+]$:

$$[Ag^+] = \frac{K_{sp}}{[Cl^-]} = \frac{1.8 \times 10^{-10}}{1.0} = 1.8 \times 10^{-10} \text{ M}$$

Then calculate E^o_{cell}:

$$E^o_{cell} = -0.00 \text{ V} + 0.80 \text{ V} = 0.80 \text{ V}$$

The cell reaction is $H_2(g) + 2Ag^+(aq) \rightarrow 2Ag(s) + 2H^+(aq)$; n = 2.

Substitute E^o_{cell} = 0.80 V, 1.8×10^{-10} M Ag^+, and 1.0 M H^+ into the Nernst equation and solve for E_{cell}:

$$E_{cell} = E^\circ_{cell} - \frac{0.0592}{n} \log Q = 0.80 \text{ V} - \frac{0.0592}{2} \log \frac{(1.0)^2}{(1.8 \times 10^{-10})^2}$$

$$E_{cell} = 0.80 \text{ V} - (0.57\underline{6}8 \text{ V}) = 0.2\underline{2}32 = 0.22 \text{ V}$$

19.81 a. Rearrange the equation E^o_{cell} = (0.0592 V)/n log K_c to solve for log K_c. Note that E^o_{cell} = 0.010 V and that n = 2. Now calculate log K_c.

$$0.010 \text{ V} = \frac{0.0592 \text{ V}}{2} \log K_c$$

log K_c = 0.3$\underline{3}$78 (two sig. figs. from the two sig. figs. of E^o_{cell})

K_c = 2.$\underline{1}$76 = 2.2 (two sig. figs. from the two sig. figs. to the right of the decimal)

b. Now let $x = [Sn^{2+}]$ and $1.0\ M - x = [Pb^{2+}]$ and substitute into the equilibrium expression.

$$K_c = \frac{[Sn^{2+}]}{[Pb^{2+}]} = \frac{[x]}{[1.0 - x]} = 2.176$$

$$x = 0.6\underline{8}51$$

$$[Pb^{2+}] = 1.0 - x = 0.\underline{3}149 = 0.3\ M$$

19.82 a. Rearrange the equation $E°_{cell} = (0.0592\ V)/n \log K_c$ to solve for $\log K_c$. Note that $E°_{cell} = 0.030$ V and that $n = 1$. Now calculate $\log K_c$.

$$0.030\ V = \frac{0.0592\ V}{1} \log K_c$$

$\log K_c = 0.5\underline{0}67$ (two sig. figs. from the two sig. figs. of $E°_{cell}$)

$K_c = 3.\underline{2}11 = 3.2$ (two sig. figs. from the two sig. figs. to the right of the decimal)

b. After mixing equal volumes, the starting amounts = 0.50 M. Now let $x = [Fe^{3+}]$ and $0.50\ M - x = [Ag^+] = [Fe^{2+}]$. Now substitute into the equilibrium expression.

$$K_c = \frac{[Fe^{3+}]}{[Ag^+][Fe^{2+}]} = \frac{[x]}{[0.50 - x]^2} = 3.21$$

This equation can be rearranged to give

$$3.21(0.25 - x + x^2) = x$$

$$3.21x^2 - 4.21x + 0.8025 = 0$$

$$x = \frac{-(-4.21) \pm \sqrt{(-4.21)^2 - (4)(3.21)(0.8025)}}{(2)(3.21)} = 0.2\underline{3}146\ M$$

$$[Fe^{2+}] = 0.50 - x = 0.2\underline{6}8 = 0.27\ M$$

19.83 The number of faradays is (a) 1 F; (b) 2 F; (c) 0.11 F; (d) 0.028 F. The number of coulombs is calculated as follows:

a. $1.0 \text{ mol e} \times \frac{1\text{F}}{1 \text{ mol e}} \times \frac{9.65 \times 10^4 \text{ C}}{1 \text{ F}} = 9.\underline{6}5 \times 10^4 = 9.6 \times 10^4 \text{ C}$

b. $2.0 \text{ mol e} \times \frac{1\text{F}}{1 \text{ mol e}} \times \frac{9.65 \times 10^4 \text{ C}}{1 \text{ F}} = 1.\underline{9}3 \times 10^5 = 1.9 \times 10^5 \text{ C}$

c. $1.0 \text{ g} \times \frac{1 \text{ mol}}{18.0 \text{ g}} \times \frac{2.0 \text{ mol e}}{1 \text{ mol}} \times \frac{1\text{F}}{1 \text{ mol e}} \times \frac{9.65 \times 10^4 \text{ C}}{1 \text{ F}} = 1.\underline{0}7 \times 10^4$

$= 1.1 \times 10^4 \text{ C}$

d. $1.0 \text{ g} \times \frac{1 \text{ mol}}{35.5 \text{ g}} \times \frac{2.0 \text{ mol e}}{2 \text{ mol}} \times \frac{1\text{F}}{1 \text{ mol e}} \times \frac{9.65 \times 10^4 \text{ C}}{1 \text{ F}} = 2.\underline{7}2 \times 10^3$

$= 2.7 \times 10^3 \text{ C}$

19.84 The number of faradays is (a) 1.0 F; (b) 3.0 F; (c) 0.017 F; (d) 0.015 F. The number of coulombs is calculated as follows:

a. $1.0 \text{ mol e} \times \frac{1\text{F}}{1 \text{ mol e}} \times \frac{9.65 \times 10^4 \text{ C}}{1 \text{ F}} = 9.\underline{6}5 \times 10^4 = 9.6 \times 10^4 \text{ C}$

b. $3.0 \text{ mol e} \times \frac{1\text{F}}{1 \text{ mol e}} \times \frac{9.65 \times 10^4 \text{ C}}{1 \text{ F}} = 2.\underline{8}95 \times 10^5 = 2.9 \times 10^5 \text{ C}$

c. $1.0 \text{ g} \times \frac{1 \text{ mol}}{119 \text{ g}} \times \frac{2.0 \text{ mol e}}{1 \text{ mol}} \times \frac{1\text{F}}{1 \text{ mol e}} \times \frac{9.65 \times 10^4 \text{ C}}{1 \text{ F}} = 1.\underline{6}2 \times 10^3$

$= 1.6 \times 10^3 \text{ C}$

d. $1.0 \text{ g} \times \frac{1 \text{ mol}}{197 \text{ g}} \times \frac{3.0 \text{ mol e}}{1 \text{ mol}} \times \frac{1\text{F}}{1 \text{ mol e}} \times \frac{9.65 \times 10^4 \text{ C}}{1 \text{ F}} = 1.\underline{4}6 \times 10^3$

$= 1.5 \times 10^3 \text{ C}$

19.85 Find the moles of I_2 produced from the reaction $2I^- \rightarrow I_2 + 2e^-$ and the current and time.

$$65.3 \text{ s} \times (10.5 \times 10^{-3} \text{ A}) \times \frac{1 \text{ C}}{1 \text{ A}\bullet\text{s}} \times \frac{1 \text{ mol e}}{9.65 \times 10^4 \text{ C}} \times \frac{1 \text{ mol } I_2}{2 \text{ mol e}}$$

$$= 3.5\underline{5}3 \times 10^{-6} \text{ mol } I_2$$

From the information in the problem, 1 mol I_2 reacts with 1 mol of H_3AsO_3 or 1 mol As, so

$$3.5\underline{5}3 \times 10^{-6} \text{ mol } I_2 \times \frac{1 \text{ mol } H_3AsO_3}{1 \text{ mol } I_2} \times \frac{1 \text{ mol As}}{1 \text{ mol } H_3AsO_3} \times \frac{74.92 \text{ g As}}{1 \text{ mol As}}$$

$$= 2.6\underline{6}1 \times 10^{-4} \text{ g As}$$

19.86 Find the moles of OH^- produced from the reaction $2H_2O(l) + 2e^- \rightarrow H_2(g) + 2OH^-$ and the current and time.

$$115 \text{ s} \times \frac{15.6 \times 10^{-3} \text{ C}}{\text{s}} \times \frac{1.0 \text{ mol e}}{9.65 \times 10^4 \text{ C}} \times \frac{2 \text{ mol } OH^-}{2 \text{ mol e}}$$

$$= 1.8\underline{5}9 \times 10^{-5} \text{ mol } OH^-$$

From the information in the problem, 1 mol lactic acid (LA) reacts with 1 mol of OH^-, so

$$1.859 \times 10^{-5} \text{ mol } OH^- \times \frac{1 \text{ mol LA}}{1 \text{ mol } OH^-} \times \frac{90.08 \text{ g LA}}{1 \text{ mol LA}} = 1.6\underline{7}45 \times 10^{-3} \text{ g LA}$$

19.87 a.

	$E°$
$Cd(s) \rightleftharpoons Cd^{2+}(aq) + 2e^-$	0.40 V
$2Ag^+(aq) + 2e^- \rightleftharpoons 2Ag(s)$	0.80 V
$Cd(s) + 2Ag^+(aq) \rightleftharpoons 2Ag(s) + Cd^{2+}(aq)$	1.20 V

b. Addition of S^{2-} would greatly decrease the $Cd^{2+}(aq)$ and help shift the equilibrium to the right and thus create a greater driving force. From an analysis of the Nernst equation we get the same answer.

c. No effect. The size of the electrode makes no difference in the potential. There are large batteries and small ones which have the same potential.

			$\underline{E°}$
19.88 a.	$Fe(s) \rightleftharpoons Fe^{2+}(aq) + 2e^-$		0.41 V
	$Cu^{2+} + 2e^- \rightleftharpoons Cu(s)$		0.34 V
	$Fe(s) + Cu^{2+}(aq) \rightleftharpoons Cu(s) + Fe^{2+}(aq)$		0.75 V

b. Since the $Cu^{2+}(aq)$ concentration is greatly decreased and it is a reactant, there would be less of a tendency for the cell reaction to occur so the potential would be less. We get a similar answer from a consideration of the Nernst equation.

c. No effect. The size of the electrode affects the amount of current that can be delivered, but it doesn't have any effect on the potential difference between the two half cells. This depends upon the nature of the materials.

19.89 Anode: $2H_2O(l) \rightarrow O_2(g) + 4H^+(aq) + 4e^-$

Cathode: $Cu^{2+}(aq) + 2e^- \rightarrow Cu(s)$

19.90 Anode: $2Cl^-(aq) \rightarrow Cl_2(g) + 2e^-$

Cathode: $2H_2O(aq) + 2e^- \rightarrow H_2(g) + 2OH^-(aq)$

19.91 a. First, find the moles of silver (molar mass 107.87 g/mol).

$$\text{mol Ag} = 2.48 \text{ g Ag} \times \frac{1 \text{ mol Ag}}{107.87 \text{ g Ag}} = 0.022\underline{9}9 \text{ mol Ag}$$

Since it takes 1 mol of e^- per mol of Ag^+, the number of coulombs is

$$\text{coul} = 0.02299 \text{ mol } e^- \times \frac{96{,}500 \text{ coul}}{1 \text{ mol } e^-} = 22\underline{1}9 \text{ coul}$$

Since amp x s = coul, the number of seconds is

$$s = \frac{2219 \text{ coul}}{1.50 \text{ amp}} = 1.4\underline{7}9 \times 10^3 = 1.48 \times 10^3 \text{ s}$$

b. Compared to Ag^{+}, it would take three times as many moles of e^{-} per mole of Cr^{3+}. Therefore, the mass of Cr^{3+} (molar mass 52.00 g/mol) that can be deposited is

$$\text{mol } Cr^{3+} = 0.02299 \text{ mol } e^{-} \times \frac{1 \text{ mol } Cr^{3+}}{3 \text{ mol } e^{-}} = 0.0076\underline{6}3 \text{ mol } Cr^{3+}$$

$$\text{mass } Cr^{3+} = 0.007663 \text{ mol } Cr^{3+} \times \frac{52.00 \text{ g } Cr^{3+}}{1 \text{ mol } Cr^{3+}} = 0.39\underline{8}4 = 0.398 \text{ g}$$

19.92 a. First, find the moles of copper (molar mass 63.55 g/mol).

$$\text{mol Cu} = 1.58 \text{ g Ag} \times \frac{1 \text{ mol Cu}}{63.55 \text{ g Cu}} = 0.024\underline{8}6 \text{ mol Cu}$$

Since it takes 2 mol of e^{-} per mol of Cu^{2+}, the number of coulombs is

$$\text{coul} = 0.02486 \text{ mol Cu} \times \frac{2 \text{ mol } e^{-}}{1 \text{ mol Cu}} \times \frac{96{,}500 \text{ coul}}{1 \text{ mol } e^{-}} = 47\underline{9}8 \text{ coul}$$

Since amp x s = coul, the number of seconds is

$$\text{s} = \frac{4798 \text{ coul}}{1.25 \text{ amp}} = 3.8\underline{3}8 \times 10^{3} = 3.84 \times 10^{3} \text{ s}$$

b. Compared to Cu^{2+}, it would take one mole of Cu^{2+} per two moles of Ag^{+}. Therefore, the mass of Ag^{+} (molar mass 107.87 g/mol) that can be deposited is

$$\text{mol } Ag^{+} = 0.02486 \text{ mol } Cu^{2+} \times \frac{2 \text{ mol } Ag^{+}}{1 \text{ mol } Cu^{2+}} = 0.049\underline{7}2 \text{ mol } Ag^{+}$$

$$\text{mass } Ag^{+} = 0.04972 \text{ mol } Ag^{+} \times \frac{107.87 \text{ g } Ag^{+}}{1 \text{ mol } Ag^{+}} = 5.3\underline{6}3 = 5.36 \text{ g}$$

19.93 a. First, calculate the number of coulombs required to deposit the Au. Since the efficiency is 90.0% you get

$$\text{coul} = 2.75 \text{ amp} \times 3.50 \text{ h} \times \frac{3600 \text{ s}}{1 \text{ h}} \times 0.900 = 31\underline{1}85 \text{ coul}$$

(continued)

The Faradays are

$$\text{Faradays} = 31185 \text{ coul} \times \frac{1 \text{ Faraday}}{96{,}500 \text{ coul}} = 0.32\underline{3}2 = 0.323 \text{ F}$$

b. The moles of Au (molar mass 196.97g/mol) that was deposited is

$$\text{mol Au} = 21.221 \text{ g Au} \times \frac{1 \text{ mol Au}}{196.97 \text{ g Au}} = 0.1077\underline{3}7 \text{ mol Au}$$

Since 1 F = 1 mol e^-, the ratio of moles of e^- to moles Au is

$$\frac{\text{mol } e^-}{\text{mol Au}} = \frac{0.3232}{0.107737} = \frac{3.00}{1}$$

Thus the ion must be Au^{3+}. The reaction is

$$Au^{3+} + 3e^- \rightarrow Au$$

19.94 a. First, calculate the number of coulombs required to deposit the V. Since the efficiency is 95.0% you get

$$\text{coul} = 2.50 \text{ amp} \times 1.90 \text{ h} \times \frac{3600 \text{ s}}{1 \text{ h}} \times 0.950 = 16\underline{2}45 \text{ coul}$$

The Faradays are

$$\text{Faradays} = 16245 \text{ coul} \times \frac{1 \text{ Faraday}}{96{,}500 \text{ coul}} = 0.16\underline{8}3 = 0.168 \text{ F}$$

b. The moles of V (molar mass 50.94g/mol) that was deposited is

$$\text{mol V} = 2.850 \text{ g V} \times \frac{1 \text{ mol V}}{50.94 \text{ g V}} = 0.0559\underline{4}8 \text{ mol V}$$

Since 1 F = 1 mol e^-, the ratio of moles of e^- to moles V is

$$\frac{\text{mol } e^-}{\text{mol V}} = \frac{0.1683}{0.055948} = \frac{3.00}{1}$$

Thus the ion must be $^{3+}$. The reaction is

$$V^{3+} + 3e^- \rightarrow V$$

19.95 a.

	E°
$Zn(s) \rightleftharpoons Zn^{2+}(aq) + 2e^-$	0.76 V
$Cu^{2+}(aq) + 2e^- \rightleftharpoons Cu(s)$	0.34 V
$Cu^{2+}(aq) + Zn(s) \rightleftharpoons Zn^{2+}(aq) + Cu(s)$	1.10 V

Use the Nernst equation to calculate the voltage of the cell.

$$E = E^o - \frac{0.0592\ V}{n} \log \frac{[Zn^{2+}]}{[Cu^{2+}]}$$

$$E = 1.10\ V - \frac{0.0592\ V}{2} \log \frac{[0.20]}{[0.015]} = 1.0\underline{6}7 = 1.07\ V$$

b. First, calculate the moles of electrons passing through the cell.

$$\text{mol } e^- = 1.00 \text{ amp } \times 225 \text{ s } \times \frac{1 \text{ mol } e^-}{96{,}500 \text{ coul}} = 0.0023\underline{3}2 \text{ mol } e^-$$

The moles of Cu deposited is

$$\text{mol Cu} = 0.002332 \text{ mol } e^- \times \frac{1 \text{ mol Cu}}{2 \text{ mol } e^-} = 0.0011\underline{6}6 \text{ mol}$$

The moles of Cu remaining in the 1.00 L of solution are

mol Cu remaining = 0.015 - 0.001166 = $0.01\underline{3}8$ mol

Since the volume of the solution is 1.00 L, the molarity of Cu^{2+} is 0.014 M.

19.96 a.

	E°
$Fe(s) \rightleftharpoons Fe^{2+}(aq) + 2e^-$	0.41 V
$Cu^{2+}(aq) + 2e^- \rightleftharpoons Cu(s)$	0.34 V
$Fe(s) + Cu^{2+}(aq) \rightleftharpoons Fe^{2+}(aq) + Cu(s)$	0.75 V

Use the Nernst equation to calculate the voltage of the cell.

$$E = E^o - \frac{0.0592\ V}{n} \log \frac{[Fe^{2+}]}{[Cu^{2+}]}$$

(continued)

$$E = 0.75\ V - \frac{0.0592\ V}{2} \log \frac{[0.15]}{[0.028]} = 0.7\underline{2}8 = 0.73\ V$$

b. First, calculate the moles of electrons passing through the cell.

$$\text{mol } e^- = 1.25 \text{ amp} \times 335 \text{ s} \times \frac{1 \text{ mol } e^-}{96{,}500 \text{ coul}} = 0.0043\underline{3}9 \text{ mol } e^-$$

The moles of Cu deposited is

$$\text{mol Cu} = 0.004339 \text{ mol } e^- \times \frac{1 \text{ mol Cu}}{2 \text{ mol } e^-} = 0.0021\underline{7}0 \text{ mol}$$

The moles of Cu remaining in the 1.00 L of solution are

$$\text{mol Cu remaining} = 0.028 - 0.002170 = 0.02\underline{5}8 \text{ mol}$$

Since the volume of the solution is 1.00 L, the molarity of Cu^{2+} is 0.016 M.

19.97 a.

	Cd(s)	+ Co^{2+}(aq)	→ Cd^{2+}(aq)	+ Co(s)
ΔG° (kJ/mol)	0	-51.5	-77.7	0

$$\Delta G^\circ_{cell} = -77.7 - (-51.5) = -26.2 \text{ kJ}$$

b. $\Delta G^\circ = -nFE^\circ$

$$-26.2 \times 10^3 \text{ J} = -2(96{,}500 \text{ coul})\ E^\circ$$

$$E^\circ = 0.13\underline{5}8 \text{ V}$$

The half-reactions and voltages are

$$Cd(s) \rightarrow Cd^{2+}(aq) + 2e^- \qquad 0.40\ V$$

$$Co^{2+}(aq) + 2e^- \rightarrow Co(s) \qquad ?$$

$$Cd(s) + Co^{2+}(aq) \rightarrow Cd^{2+}(aq) + Co(s) \qquad 0.136\ V$$

$$E^\circ_{ox} + E^\circ_{red} = E^\circ_{cell}$$

$$0.40\ V + E^\circ_{red} = 0.136\ V$$

$$E^\circ_{red} = -0.26\ V\ (Co^{2+} + 2e^- \rightarrow Co(s))$$

19.98 a.

	$2Tl(s)$	+	$Pb^{2+}(aq)$	$\rightarrow$	$2Tl^{+}(aq)$	+	$Pb(s)$
$\Delta G°$ (kJ/mol)	0		-24.3		2(-32.4)		0

$\Delta G°_{cell} = 2(-32.4) - (-24.3) = -40.5$ kJ

b. $\Delta G° = -nFE°$

-40.5×10^{3} J $= -2(96{,}500$ coul$)$ E°

$E° = 0.20\underline{9}8$ V

The half-reactions and voltages are

$2Tl(s) \rightarrow 2Tl^{+}(aq) + 2e^{-}$?

$Pb^{2+}(aq) + 2e^{-} \rightarrow Pb(s)$ -0.13 V

$Cd(s) + Co^{2+}(aq) \rightarrow Cd^{2+}(aq) + Co(s)$ 0.2098 V

$E°_{ox} + E°_{red} = E°_{cell}$

$E°_{ox} + (-0.13\ V) = 0.2098$ V

$E°_{red} = -0.3\underline{4}0 = 0.34$ V $(Tl^{+}(aq) + e^{-} \rightarrow Tl(s))$

■ Cumulative-Skills Problems
(require skills from previous chapters)

19.99 The standard emf of this cell is calculated by summing the electrode potentials of the oxidizing and reducing agents. Then use $E°_{cell}$ to calculate $\Delta G°$ from $-nFE°_{cell}$.

$3\ Zn(s) \rightarrow 3\ Zn^{2+}(aq) + 6e^{-}$ $-E°_{Zn} = 0.76$ V

$2\ Cr^{3+}(aq) + 6e^{-} \rightarrow 2\ Cr(s)$ $E°_{Cr} = -0.74$ V

$2\ Cr^{3+}(aq) + 3\ Zn(s) \rightarrow 3\ Zn^{2+}(aq) + 2\ Cr(s)$ $E°_{cell} = 0.02$ V

$\Delta G° = -nFE°_{cell} = -(6 \times 9.65 \times 10^{4}\ C \times 0.02\ V) = -\underline{1}.1580 \times 10^{4}\ J = -\underline{1}1.58$ kJ

(continued)

Use the data in Appendix C to calculate ΔH^{o} for the overall cell reaction:

$2\,Cr^{3+}(aq)$	+	$3\,Zn(s)$	→	$3\,Zn^{2+}(aq)$	+	$2\,Cr(s)$
2(-1971)		3(0)		3(-152.4)		2(0)

$\Delta H^{o} = [\,3\,\Delta H^{o}_{f}(Zn^{2+}) + 2\,\Delta H^{o}_{f}(Cr)\,] - [\,2\,\Delta H^{o}_{f}(Cr^{3+}) + 3\,\Delta H^{o}_{f}(Zn)\,]$

$\Delta H^{o} = [\,3(-152.4) + 2(0)\,] - [\,2(-1971) + 3(0)\,]$

$\Delta H^{o} = 338\underline{4}.8 = 3485\ kJ$

Now calculate ΔS^{o} from the $\Delta G^{o} = \Delta H^{o} - T\Delta S^{o}$ relation:

$T\Delta S^{o} = \Delta H^{o} - \Delta G^{o} = 348\underline{4}.8\ kJ - (-\underline{1}1.58\ kJ) = 34\underline{9}6.38\ kJ$

$\Delta S^{o} = 34\underline{9}6.38\ kJ \div 298\ K = 11.\underline{7}3 = 11.7\ kJ/K$

19.100 The standard emf of this cell is calculated by summing the electrode potentials of the oxidizing and reducing agents. Then use E^{o}_{cell} to calculate ΔG^{o} from $-nF\,E^{o}_{cell}$.

$2\,Cr(s) \rightarrow 2\,Cr^{3+}(aq) + 6e^{-}$		$-E^{o}_{Cr} = 0.74\ V$
$3\,Fe^{2+}(aq) + 6e^{-} \rightarrow 3\,Fe(s)$		$E^{o}_{Cr} = -0.41\ V$
$2Cr(s) + 3\,Fe^{2+}(aq) \rightarrow 2\,Cr^{3+}(aq) + 3\,Fe(s)$		$E^{o}_{cell} = 0.33\ V$

$\Delta G^{o} = -nFE^{o}_{cell} = -(6 \times 9.65 \times 10^{4}\ C \times 0.33\ V) = -1.\underline{9}107 \times 10^{5}\ J = -1\underline{9}1.07\ kJ$

Use the data in Appendix C to calculate ΔH^{o} for the overall cell reaction:

$2Cr(s)$	+	$3\,Fe^{2+}(aq)$	→	$2\,Cr^{3+}(aq)$	+	$3\,Fe(s)$
2(0)		3(-87.9)		2(-1971)		2(0)

$\Delta H^{o} = [\,2\,\Delta H^{o}_{f}(Cr^{3+}) + 3\,\Delta H^{o}_{f}(Fe)\,] - [\,2\,\Delta H^{o}_{f}(Cr) + 3\,\Delta H^{o}_{f}(Fe^{2+})\,]$

$\Delta H^{o} = [\,2(-1971) + 3(0)\,] - [\,2(0) + 3(-87.9)\,]$

$\Delta H^{o} = 367\underline{8}.3 = 3678\ kJ$

Now calculate ΔS^{o} from the $\Delta G^{o} = \Delta H^{o} - T\Delta S^{o}$ relation:

$T\Delta S^{o} = \Delta H^{o} - \Delta G^{o} = 367\underline{8}.3\ kJ - (-1\underline{9}1.07\ kJ) = -386\underline{9}.37\ kJ$

$\Delta S^{o} = -386\underline{9}.37\ kJ \div 298\ K = -12.\underline{9}8 = -13.0\ kJ/K$

19.101 First calculate E^o_{cell}:

$$E^o_{cell} = -(1.07\ V) + (1.23\ V) = 0.16\ V$$

The cell reaction is $O_2(g) + 4H^+(aq) + 4Br^-(aq) \rightarrow 2H_2O(l) + 2Br_2(l)$; n = 4.

Convert pH to $[H^+]$: $[H^+]$ = antilog (-3.60) = $2.\underline{5}1 \times 10^{-4}$ M.

Substitute E^o_{cell} = 0.16 V, 1 M Br^-, 1 M Br_2, and 2.51×10^{-4} M H^+ into the Nernst equation and solve for E_{cell}:

$$E_{cell} = E^\circ_{cell} - \frac{0.0592}{n} \log Q = 0.16\ V - \frac{0.0592}{4} \log \frac{1}{(2.51 \times 10^{-4})^4}$$

$$E_{cell} = E^\circ_{cell} - \frac{0.0592}{n} \log Q = 0.16\ V - \frac{0.0592}{4} \log (2.519 \times 10^{14})$$

$$E_{cell} = E^\circ_{cell} - \frac{0.0592}{n} \log Q = 0.16\ V - \frac{0.0592}{4} (14.4\underline{0}1)$$

E_{cell} = 0.16 V - 0.213$\underline{1}$ V (using 0.0592 as exact no.) = -0.0$\underline{5}$31 = -0.05 V

(nonspontaneous reaction)

19.102 First calculate E^o_{cell}:

$$E^o_{cell} = -(1.82\ V) + (2.07\ V) = 0.25\ V$$

The cell reaction is $O_3(g) + 2H^+(aq) + 2Co^+(aq) \rightarrow 2Co^{3+}(aq) + O_2(g) + H_2O(l)$; n = 2.

Convert pH to $[H^+]$: $[H^+]$ = antilog (-8.40) = $3.\underline{9}81 \times 10^{-9}$ M.

Substitute E^o_{cell} = 0.25 V, 1 M Br^-, 1 M Co^{2+}, 1 M Co^{3+}, and 3.981×10^{-9} M H^+ into the Nernst equation and solve for E_{cell}:

$$E_{cell} = E^\circ_{cell} - \frac{0.0592}{n} \log Q = 0.25\ V - \frac{0.0592}{2} \log \frac{1}{(3.981 \times 10^{-9})^2}$$

$$E_{cell} = E^\circ_{cell} - \frac{0.0592}{n} \log Q = 0.25\ V - \frac{0.0592}{2} \log (6.\underline{3}09 \times 10^{16})$$

$$E_{cell} = E^\circ_{cell} - \frac{0.0592}{n} \log Q = 0.25\ V - \frac{0.0592}{2} (16.8\underline{0}0)$$

E_{cell} = 0.25 V - 0.497$\underline{2}$8 V (using 0.0592 as exact no.) = -0.2$\underline{4}$728 = -0.25 V

(nonspontaneous reaction)

19.103 In Problem 19.89, E^o_{cell} was found to be 0.16 V. Neglecting ionization of HOCN, the $[H^+]$ of the buffer is

$$[H^+] \cong 3.5 \times 10^{-4} \times \frac{(0.10\ M\ HOCN)}{(0.10\ M\ OCN^-)} \cong 3.5 \times 10^{-4}\ M$$

From the Nernst equation for the reaction of O_2 with $4Br^-$ and $4H^+$:

$$E_{cell} = E^\circ_{cell} - \frac{0.0592}{n} \log Q = 0.16\ V - \frac{0.0592}{4} \log \frac{1}{(3.5 \times 10^{-4})^4}$$

$$E_{cell} = E^\circ_{cell} - \frac{0.0592}{n} \log Q = 0.16\ V - \frac{0.0592}{4} \log (6.6639 \times 10^{13})$$

$$E_{cell} = E^\circ_{cell} - \frac{0.0592}{n} \log Q = 0.16\ V - \frac{0.0592}{4} (13.8\underline{2}37)$$

E_{cell} = 0.16 V - 0.204$\underline{5}$9 V (using 0.0592 as exact no.) = -0.0$\underline{4}$459 = -0.04 V

(nonspontaneous rection)

19.104 In Problem 19.90, E^o_{cell} was found to be 0.25 V. Neglecting ionization of HClO, the $[H^+]$ of the buffer is

$$[H^+] \cong 3.5 \times 10^{-8} \times \frac{(0.10\ M\ HClO)}{(0.10\ M\ ClO^-)} \cong 3.5 \times 10^{-8}\ M$$

$$E_{cell} = E^\circ_{cell} - \frac{0.0592}{n} \log Q = 0.25\ V - \frac{0.0592}{2} \log \frac{1}{(3.5 \times 10^{-8})^2}$$

$$E_{cell} = E^\circ_{cell} - \frac{0.0592}{n} \log Q = 0.25\ V - \frac{0.0592}{2} \log (8.\underline{1}63 \times 10^{14})$$

$$E_{cell} = E^\circ_{cell} - \frac{0.0592}{n} \log Q = 0.25\ V - \frac{0.0592}{2} (14.9\underline{1}1)$$

E_{cell} = 0.25 V - 0.441$\underline{3}$9 V (using 0.0592 as exact no.) = -0.1$\underline{9}$139 = -0.19 V

(nonspontaneous reaction)

19.105 The cell reaction is $1/2H_2(g) + Ag^+(aq) \rightarrow Ag(s) + H^+(aq)$; n = 1. Calculate E^o_{cell}:

$$E^o_{cell} = -0.00\ V + 0.80\ V = 0.80\ V$$

Substitute $E_{cell} = 0.45$ V, $E^o_{cell} = 0.80$ V, and 1.0 M H^+ into the Nernst equation and solve for log $[Ag^+]$:

$$E_{cell} = E^\circ cell - \frac{0.0592}{n} \log Q; \quad 0.45\ V = 0.80\ V - \frac{0.0592}{1} \log \frac{(1.0)}{[Ag^+]}$$

$$\log [Ag^+] = \frac{1(0.45 - 0.80)}{0.0592} = -5.\underline{9}12\ V$$

$$[Ag^+] = 10^{-5.912} = \underline{1}.224 \times 10^{-6}\ M$$

$$K_{sp} = [Ag^+][SCN^-] = (1.224 \times 10^{-6})(0.10) = \underline{1}.224 \times 10^{-7} = 1 \times 10^{-7}$$

19.106 The overall cell reaction is: $Hg_2{}^{2+}(aq) + H_2(g) \rightarrow 2Hg(l) + 2H^+(aq)$, where n = 2. E^o_{cell} is:

$$E^o_{cell} = 0.80\ V - 0.00\ V = 0.80\ V$$

Substitute $E_{cell} = 0.33$ V, $E^o_{cell} = 0.80$ V, $P_{H2} = 1.0$ atm, and $Q = 1.0/[Hg_2{}^{2+}][1.0]$ into the Nernst equation; then solve for log $[Hg_2{}^{2+}]$:

$$E_{cell} = E^o_{cell} - \frac{0.0592\ V}{n} \log Q$$

$$0.33\ V = 0.80\ V - \frac{0.0592\ V}{n} \log \frac{1.0}{\left[Hg_2^{2+}\right]}$$

$$\log \left[Hg_2^{2+}\right] = \frac{2(0.33 - 0.80)}{0.0592\ V} = -1\underline{5}.87$$

$$\left[Hg_2^{2+}\right] = 1.3 \times 10^{-1\underline{6}}$$ (neither of the two digits in 1.3 is significant)

$$K_{sp} = [H_2{}^{2+}][Cl^-]^2 = (1.3 \times 10^{-16})(0.10)^2 = 1.3 \times 10^{-1\underline{8}} = 10^{-18}$$

Note that the answer does not have any significant figures; the eight in the exponent is the last significant figure.

■ Solution to Conceptual Problem

19.107 The development of lightweight batteries is an ongoing research effort combining many of the physical sciences. You are a member of an engineering team trying to develop a lightweight battery that will effectively react with $O_2(g)$ from the atmosphere as the oxidizing agent. A reducing agent must be chosen for this battery that will be lightweight, have nontoxic products, and reacts spontaneously with oxygen. Using data from Appendix I, suggest a likely reducing agent, being sure that the above conditions are met. Are there any drawbacks to your selection?

Discussion

The most important consideration in battery design is the spontaneity of the cell reaction, since this determines the cell voltage. The half-reactions for the reduction of oxygen gas given in Appendix I are:

$$O_2 + 2H_2O + 4e^- \rightarrow 4OH^- \qquad E^\circ = +0.40\text{ V}$$

$$O_2 + 4H^+ + 4e^- \rightarrow 2H_2O \qquad E^\circ = +1.23\text{ V}$$

These represent the reduction of oxygen under basic and under acidic conditions. Thus, in basic solution, we would require an oxidation half-reaction (anode half-reaction) with a potential greater than -0.40 V to obtain a spontaneous reaction.

The density of the reducing agent is the next most important consideration. Hydrogen, the element with the lowest density, should certainly be considered. It does have some drawbacks, however. Because it is a gas, some method of storage needs to be considered. Liquid storage and metal hydride storage have been considered. Storage of liquid hydrogen present problems of safety and weight. Storage of hydrogen as the hydride is used in nickel-hydride cells presently available for portable computers, cellular phones, etc. (These batteries are rechargeable, but do not use oxygen.) The metal hydride obviously adds weight to the battery. Other elements the student might consider are Na, Li, Al, K, Ca, and Zn, which have favorable power-to-mass ratios. Lithium and sodium might present some disposal problems, since they are very reactive metals. Batteries using aluminum with atmospheric oxygen are available.

20. NUCLEAR CHEMISTRY

Solutions to Exercises

Note on significant figures: The final answer to each mathematical solution is given first with one nonsignificant figure (the rightmost significant figure is underlined) and then is rounded to the correct number of figures. Intermediate answers usually also have at least one nonsignificant figure.

20.1 The nuclide symbol for potassium-40 is $^{40}_{19}K$. Similarly, the nuclide symbol for calcium-40 is $^{40}_{20}Ca$. The equation for beta emission is

$$^{40}_{19}K \rightarrow ^{40}_{20}Ca + ^{0}_{-1}e$$

20.2 Write the nuclear equation for the decay of plutonium-239. Plutonium has atomic number 94. Thus, the nuclide symbol is $^{239}_{94}Pu$. An alpha particle has the symbol $^{4}_{2}He$. For the unknown product nucleus, write $^{A}_{Z}X$, where A is the mass number and Z is the atomic number. The nuclear equation is

$$^{239}_{94}Pu \rightarrow ^{A}_{Z}X + ^{4}_{2}He$$

From the superscripts, you can write

$$239 = A + 4; \quad A = 235$$

(continued)

Similarly, from the subscripts you can write

$$94 = Z + 2; \quad Z = 92$$

Hence, A = 235 and Z = 92, so the product is ${}^{235}_{92}X$. Because element 92 is uranium, you write the product nucleus as ${}^{235}_{92}U$.

20.3 a. ${}^{118}_{50}Sn$ has atomic number 50. It has 50 protons and 68 neutrons. Because its atomic number is less than 83 and it has an even number of protons and neutrons, it is expected to be stable.

b. ${}^{76}_{33}As$ has atomic number 33. It has 33 protons and 43 neutrons. Because stable odd-odd nuclei are rare, you would expect ${}^{76}_{33}As$ to be one of the radioactive isotopes.

c. ${}^{227}_{89}Ac$ has atomic number 89. Because its atomic number is greater than 83, ${}^{227}_{89}Ac$ is radioactive.

20.4 a. ${}^{13}_{7}N$ has 7 protons and 6 neutrons (fewer neutrons than protons). Its mass no. is < than that of N = 14, so it is expected to decay by electron capture or positron emission (more likely, because this is a light isotope).

b. ${}^{26}_{11}Na$ has 11 protons and 15 neutrons. It is expected to decay by beta emission.

20.5 a. The abbreviated notation is ${}^{40}_{20}Ca(d,p){}^{41}_{20}Ca$.

b. The nuclear equation is ${}^{12}_{6}C + {}^{2}_{1}H \rightarrow {}^{13}_{6}C + {}^{1}_{1}H$.

20.6 You can write the nuclear equation as follows:

$${}^{A}_{Z}X + {}^{1}_{0}n \rightarrow {}^{14}_{6}C + {}^{1}_{1}H$$

To balance this equation in charge (subscripts) and mass number (superscripts), write the equations

(continued)

$A + 1 = 14 + 1$ (from superscripts)

$Z + 0 = 6 + 1$ (from subscripts)

Hence, A = 14 and Z = 7. Therefore, the nucleus that produces carbon-14 by this reaction is $^{14}_{7}N$.

20.7 Because an activity of 1.0 Ci is 3.7 x 10^{10} nuclei/s, the rate of decay in this sample is

$$\text{Rate} = 13\ \text{Ci} \times \frac{3.7 \times 10^{10}\ \text{nuclei/s}}{1.0\ \text{Ci}} = 4.\underline{8}1 \times 10^{11}\ \text{nuclei/s}$$

The number of nuclei in this 2.5-μg sample of $^{99m}_{43}Tc$ can be calculated by noting that the molar mass in grams is approximately equal to the mass number.

$$2.5 \times 10^{-6}\ \text{g Tc–99m} \times \frac{1\ \text{mol Tc–99m}}{99\ \text{g Tc–99m}} \times \frac{6.02 \times 10^{23}\ \text{Tc–99m nuclei}}{1\ \text{mol Tc–99m}}$$

$$= 1.\underline{5}2 \times 10^{16}\ \text{Tc–99m nuclei}$$

Finally, solve the rate equation for k:

$$k = \frac{\text{rate}}{N_t} = \frac{4.81 \times 10^{11}\ \text{nuclei/s}}{1.52 \times 10^{16}\ \text{nuclei}} = 3.\underline{1}6 \times 10^{-5} = 3.2 \times 10^{-5}/\text{s}$$

20.8 First, substitute the value of k into the equation relating half-life to the decay constant:

$$t_{1/2} = \frac{0.693}{k} = \frac{0.693}{4.18 \times 10^{-9}/\text{s}} = 1.6\underline{5}8 \times 10^{8}\ \text{s}$$

Convert the half-life from seconds to years:

$$1.658 \times 10^{8}\ \text{s} \times \frac{1\ \text{min}}{60\ \text{s}} \times \frac{1\ \text{hour}}{60\ \text{min}} \times \frac{1\ \text{day}}{24\ \text{hours}} \times \frac{1\ \text{year}}{365\ \text{days}} = 5.2\underline{5}7 = 5.26\ \text{years}$$

20.9 Convert the half-life from years to seconds. Then calculate k. Next use the rate equation to find the decay constant of the sample and finally the activity. The conversion of the half-life to seconds gives

$$28.1\ \text{years} \times \frac{365\ \text{days}}{1\ \text{year}} \times \frac{24\ \text{hours}}{1\ \text{day}} \times \frac{60\ \text{min}}{1\ \text{hour}} \times \frac{60\ \text{s}}{1\ \text{min}} = 8.8\underline{6}1 \times 10^{8}\ \text{s}$$

(continued)

Because $t_{1/2} = 0.693/k$, solve this for k and substitute the half-life in seconds.

$$k = \frac{0.693}{t_{1/2}} = \frac{0.693}{8.861 \times 10^{8}\ s} = 7.8\underline{2}08 \times 10^{-10}\ /s$$

Before substituting into the rate equation, you need to know the number of nuclei in a sample containing 5.2×10^{-9} of strontium-90.

$$5.2 \times 10^{-9} g\ Sr\text{–}90 \times \frac{1\ mol\ Sr\text{–}90}{90\ g} \times \frac{6.02 \times 10^{23}\ Sr\text{–}90\ nuclei}{1\ mol\ Sr\text{–}90}$$

$$= 3.\underline{4}78 \times 10^{13}\ Sr\text{-}90\ nuclei$$

Now substitute into the rate equation:

$$Rate = kN_t = (7.8208 \times 10^{-10}/s) \times (3.478 \times 10^{13}\ nuclei)$$

$$= 2.\underline{7}2 \times 10^{4}\ nuclei/s$$

Calculate the activity by dividing the rate (disintegrations of nuclei per second) by 3.70×10^{10} disintegrations of nuclei per second per curie.

$$Activity = \frac{2.72 \times 10^{4}\ nuclei/s}{3.7 \times 10^{10}\ nuclei/(s \cdot Ci)} = 7.\underline{3}51 \times 10^{-7} = 7.4 \times 10^{-7}\ Ci$$

20.10 The decay constant, k, is $0.693/t_{1/2}$. If you substitute this into the equation

$$\log \frac{N_t}{N_0} = \frac{-kt}{2.303}$$

you get

$$\log \frac{N_0}{N_t} = \frac{0.693\ t}{2.303\ t_{1/2}}$$

Substitute the values $t = 25.0$ y and $t_{1/2} = 10.76$ y into the equation and solve for N_o/N_t.

$$\log \frac{N_0}{N_t} = \frac{0.693 \times 25.0\ y}{2.303 \times 10.76\ y} = 0.69\underline{9}1$$

$$\frac{N_0}{N_t} = 5.0\underline{0}1$$

The fraction of krypton-85 remaining after 25.0 years is N_t/N_0.

$$\frac{N_t}{N_0} = \frac{1}{5.001} = 0.19\underline{9}9 = 0.200$$

20.11 Substitute $k = 0.693/t_{1/2}$ into the equation for the number of nuclei in a sample after time t:

$$\log \frac{N_t}{N_o} = \frac{-kt}{2.303} = \frac{2.303\ t_{1/2}}{0.693\ t}$$

Hence,

$$t = \frac{2.303\ t_{1/2}}{0.693} \times \log \frac{N_o}{N_t}$$

To obtain N_o/N_t, assume that the ratio of ${}^{14}_{6}C$ to ${}^{12}_{6}C$ in the atmosphere has remained constant. Then you can say that 1.00 gram of total carbon from the jawbone gave 15.3 disintegrations per minute. The ratio of the number of ${}^{14}_{6}C$ originally present to the number that existed at the time of dating equals the ratio of rates of disintegration.

$$\frac{N_o}{N_t} = \frac{15.3}{4.5} = 3.\underline{4}00$$

Therefore, substituting this value of N_o/N_t and $t_{1/2}$ = 5730 y into the previous equation gives

$$t = \frac{2.303\ t_{1/2}}{0.693} \log \frac{N_o}{N_t} = \frac{2.303 \times 5730\ y}{0.693} \log 3.4 = 1.\underline{0}1 \times 10^4 = 1.0 \times 10^4 \text{ years}$$

20.12 Write the nuclear masses below each nuclide symbol. Then calculate Δm.

$$ {}^{234}_{90}Th \rightarrow {}^{234}_{91}Pa + {}^{0}_{-1}e $$

233.9942 233.9934 0.000549 amu

$$\Delta m = (233.9934 + 0.000549 - 233.9942) \text{ amu} = -0.000\underline{2}51 \text{ amu}$$

a. The mass change for molar amounts in this reaction is -0.000251 g, or -2.51×10^{-7} kg. The energy change is

$$\Delta E = (\Delta m)c^2 = (-2.51 \times 10^{-7} \text{ kg})(3.00 \times 10^8 \text{ m/s})^2 = -\underline{2}.259 \times 10^{10} \text{ J/mol}$$

For 1.00 g Th-234, the energy change is

$$1.00 \text{ g Th-234} \times \frac{1 \text{ mol Th-234}}{234 \text{ g Th-234}} \times \frac{-2.259 \times 10^{10} \text{ J}}{1 \text{ mol Th-234}} = -\underline{9}.653 \times 10^7 = -1 \times 10^8 \text{ J}$$

b. Convert the mass change for the reaction from amu to grams.

$$\Delta m = -2.51 \times 10^{-4} \times \frac{1\text{ g}}{1\text{ amu} \times 6.02 \times 10^{23}} = -\underline{4}.169 \times 10^{-28}\text{ g}$$

$$= -4.169 \times 10^{-31}\text{ kg}$$

$$\Delta E = (\Delta m)c^2 = (-4.169 \times 10^{-31}\text{ kg})(3.00 \times 10^8\text{ m/s})^2 = -\underline{3}.752 \times 10^{-14}\text{ J}$$

Convert this to MeV:

$$\Delta E = -3.752 \times 10^{-14}\text{ J} \times \frac{1\text{ MeV}}{1.602 \times 10^{-13}\text{ J}} = -0.\underline{2}342 = -0.2\text{ MeV}$$

■ Answers to Review Questions

20.1 The two types of nuclear reactions and their equations are

Radioactive decay: $^{238}_{92}\text{U} \rightarrow {}^{234}_{90}\text{Th} + {}^{4}_{2}\text{He}$

Nuclear bombardment reactions: $^{27}_{13}\text{Al} + {}^{4}_{2}\text{He} \rightarrow {}^{30}_{15}\text{P} + {}^{1}_{0}\text{n}$

20.2 Magic numbers are the numbers of nuclear particles in completed shells of protons or neutrons. Examples of nuclei with magic numbers of protons are

$^{4}_{2}\text{He}$, $^{16}_{8}\text{O}$, and $^{40}_{20}\text{Ca}$.

20.3 To predict whether a nucleus will be stable, look for nuclei that have one of the magic numbers of protons and neutrons. Also look for nuclei that have an even number of protons and an even number of neutrons. Nuclei that fall in the band of stability are also very stable. There are no stable nuclei above atomic number 83.

20.4 The five common types of radioactive decay and the usual condition that leads to each type are listed below.

Alpha emission:	$Z > 83$.
Beta emission:	N/Z is too large.
Positron emission:	N/Z is too small.
Electron capture:	N/Z is too small.
Gamma emission:	The nucleus is in an excited state.

20.5 The isotopes that begin each of the natural radioactive decay series are uranium-238, uranium-235, and thorium-232.

20.6 The equations are as follows:

a. $^{14}_{7}N + ^{4}_{2}He \rightarrow ^{17}_{8}O + ^{1}_{1}H$

b. $^{27}_{13}Al + ^{4}_{2}He \rightarrow ^{30}_{15}P + ^{1}_{0}n$

20.7 Particle accelerators are devices used to accelerate electrons, protons, and alpha particles and other ions to very high speeds. They operate by accelerating the charged particle toward a plate with charge opposite to that of the particle. Particle accelerators are required to accelerate alpha particles to speeds high enough to penetrate nuclei of large positive charge, which normally scatter alpha particles.

20.8 Before the discovery of transuranium elements, it was thought that americium (Z = 95) and curium (Z = 96) should be placed after actinium (Z = 89) in the periodic table as d-block transition elements. However, Seaborg and others discovered that these elements had properties similar to the lanthanides and placed them in a second series under the lanthanides.

20.9 The Geiger counter measures alpha particles by means of a tube filled with gas. When particles pass through the tube, they ionize the gas, freeing electrons, which creates a pulse of current that is detected by electronic equipment and is counted. A scintillation counter consists of phosphor, a substance that emits photons when struck by radiation; zinc sulfide is used for the detection of alpha particles; and sodium iodide containing thallium(I) iodide is used for gamma rays. The photons travel from the phosphor to a photoelectric detector, such as a photomultiplier, which magnifies the effect and gives a pulse of electric current that is measured.

20.10 A curie (Ci) equals 3.700×10^{10} nuclear disintegrations per second. A rad is the dosage of radiation that deposits 1×10^{-2} J of energy per kilogram of tissue. A rem is a unit of radiation dosage for biological destruction; it equals the rad multiplied by the relative biological effectiveness (RBE).

20.11 It will take cesium-137 three times its half-life of 30.2 y, or 90.6 y, to decay to 1/8 its original mass: 1 to 1/2 to 1/4 to 1/8 = 3 half-lives.

20.12 Because the $^{40}_{18}Ar$ was produced by radioactive decay of $^{40}_{19}K$, half of the initial amount of $^{40}_{19}K$ has decomposed. The age equals the half-life of 1.28×10^{9} y.

20.13 A radioactive tracer is a radioactive isotope added to a chemical, biological, or physical system to study it. For instance, $^{131}I^{-}$ is used as a tracer in the study of the dissolving of lead(II) iodide and its equilibrium in a saturated solution.

20.14 Isotope dilution is a technique designed to determine the quantity of a substance in a mixture, or to determine the total volume of a solution, by adding a known amount of an isotope to it. After removing a portion of the mixture, the fraction by which the isotope has been diluted provides a way of determining the quantity of substance or the total volume of solution.

20.15 Neutron activation analysis is an analysis based on the conversion of stable isotopes to radioactive isotopes by bombarding a sample with neutrons. An unstable nucleus results, which then emits gamma rays or radioactive particles (such as beta particles). The amount of stable isotope is proportional to the measured emission.

20.16 The reason the deuteron, ${}^{2}_{1}H$, has a mass smaller than the sum of the masses of its constituents is that when nucleons come together to form a nucleus, energy is released. There must therefore be an equivalent decrease in mass because mass and energy are equivalent.

20.17 Iron-56 has a binding energy per nucleon that is near the maximum value. Two light nuclei, such as two C-12 nuclei, will undergo fusion (with the release of energy) as long as the product nuclei are lighter than iron-56 (which is the case with Na-23 and H-1).

20.18 The nuclear fission reactor operates by means of a chain reaction of nuclear fissions, controlled to produce energy without explosion. A tokamak nuclear fusion reactor controls nuclear fusion by using a doughnut-shaped magnetic field to hold plasma away from material.

■ Solutions to Practice Problems

Note on significant figures: The final answer to each problem is given first with one nonsignificant figure (the rightmost significant figure is underlined) and then is rounded to the correct number of figures. Intermediate answers usually also have at least one nonsignificant figure.

20.19 ${}^{87}_{37}Rb \rightarrow {}^{87}_{38}Sr + {}^{0}_{-1}e$

20.20 ${}^{32}_{15}P \rightarrow {}^{32}_{16}S + {}^{0}_{-1}e$

20.21 ${}^{232}_{90}Th \rightarrow {}^{228}_{88}Ra + {}^{4}_{2}He$

20.22 ${}^{226}_{88}Ra \rightarrow {}^{222}_{86}Rn + {}^{4}_{2}He$

20.23 Let X be the product nucleus. The nuclear equation is

$$^{18}_{9}\text{F} \rightarrow {}^{A}_{Z}\text{X} + {}^{0}_{1}\text{e}$$

From the superscripts: 18 = A + 0, or A = 18; from the subscripts: 9 = Z + 1, or Z = 8. Thus, the product of the reaction is $^{18}_{8}$X, and because element 8 is oxygen, symbol O, the nuclear equation is

$$^{18}_{9}\text{F} \rightarrow {}^{18}_{8}\text{O} + {}^{0}_{1}\text{e}$$

20.24 Let X be the product nucleus. The nuclear equation is

$$^{22}_{11}\text{Na} \rightarrow {}^{A}_{Z}\text{X} + {}^{0}_{1}\text{e}$$

From the superscripts: 22 = A + 0, or A = 22; from the subscripts: 11 = Z + 1, or Z = 10. Thus, the product nucleus is $^{22}_{10}$X, and because element 10 is neon, symbol Ne, the nuclear equation is

$$^{22}_{11}\text{Na} \rightarrow {}^{22}_{10}\text{Ne} + {}^{0}_{1}\text{e}$$

20.25 Let X be the product nucleus. The nuclear equation is

$$^{210}_{84}\text{Po} \rightarrow {}^{A}_{Z}\text{X} + {}^{4}_{2}\text{He}$$

From the superscripts: 210 = A + 4, or A = 206; from the subscripts: 84 = Z + 2, or Z = 82. Thus, the product nucleus is $^{206}_{82}$X, and because element 82 is lead, symbol Pb, the nuclear equation is

$$^{210}_{84}\text{Po} \rightarrow {}^{206}_{82}\text{Pb} + {}^{4}_{2}\text{He}$$

20.26 Let X be the product nucleus. The nuclear equation is

$$^{227}_{89}\text{Ac} \rightarrow {}^{A}_{Z}\text{X} + {}^{4}_{2}\text{He}$$

From the superscripts: 227 = A + 4, or A = 223; from the subscripts: 89 = Z + 2, or Z = 87. Thus, the product nucleus is $^{223}_{87}\text{X}$, and because element 87 is francium, symbol Fr, the nuclear equation is

$$^{227}_{89}\text{Ac} \rightarrow {}^{223}_{87}\text{Fr} + {}^{4}_{2}\text{He}$$

20.27 a. Neither nucleus has an atomic number that is a magic number of protons. Find how many neutrons are in each nucleus.

Sb: No. of neutrons = A - Z = 122 - 51 = 71

Xe: No. of neutrons = A - Z = 136 - 54 = 82

Because 82 is a magic number for neutrons (implying stability of nucleus), you predict that $^{136}_{54}\text{Xe}$ is stable and $^{122}_{51}\text{Sb}$ is radioactive.

b. $^{204}_{82}\text{Pb}$ has a magic number of protons (82), so it is expected to be the stable nucleus and $^{204}_{85}\text{At}$ is radioactive (atomic number greater than 83).

c. Rb does not have an atomic number that is a magic number of protons. Find the numbers of neutrons in the two isotopes.

$^{87}_{37}\text{Rb}$: No. of neutrons = 87 - 37 = 50

$^{80}_{37}\text{Rb}$: No. of neutrons = 80 - 37 = 43

Because 50 is a magic number for neutrons, you predict that $^{87}_{37}\text{Rb}$ is stable and $^{80}_{37}\text{Rb}$ is radioactive.

20.28 a. Ag does not have an atomic number that is a magic number of protons. Find the numbers of neutrons in the two isotopes.

$^{102}_{47}$Ag: No. of neutrons = 102 - 47 = 55

$^{109}_{47}$Ag: No. of neutrons = 109 - 47 = 62

Neither isotope has a magic number of neutrons, but $^{109}_{47}$Ag has an even number of neutrons, whereas $^{102}_{47}$Ag has an odd number of neutrons. Because stable odd-odd nuclei are rare, $^{109}_{47}$Ag is expected to be stable and $^{102}_{47}$Ag radioactive.

b. Neither nucleus has a magic number of protons or neutrons, but Ne has an even number of both protons and neutrons, whereas $^{25}_{12}$Mg has an odd number of neutrons. Therefore, $^{24}_{10}$Ne is expected to be stable and $^{25}_{12}$Mg radioactive.

c. No element with atomic number greater than 83 has a stable isotope. Thus, $^{223}_{90}$Th is radioactive and $^{203}_{81}$Tl is stable.

20.29 a. α-emission is most likely for nuclei with $Z > 83$

b. positron emission (more likely; $Z < 20$) or electron capture because mass no. < that of Cu = 63.5

c. β-emission (mass no. > He = 4)

20.30 a. positron emission or electron capture ($N/Z = 30/30 = 1$)

b. β-emission ($N/Z = 4/2 > 1$)

c. α-emission ($Z > 83$)

20.31 α-emission decreases the mass number by 4; β-emission does not affect the mass number. $^{219}_{86}$Rn belongs to the $^{235}_{92}$U decay series because the difference in mass numbers is 16, which is divisible by 4. $^{220}_{86}$Rn belongs to the $^{232}_{90}$Th decay series because the difference in mass numbers is 12, which is divisible by 4.

20.32 α-emission decreases the mass number by 4; β-emission does not affect the mass number. $^{227}_{89}$Ac belongs to the $^{235}_{92}$U decay series because the difference in mass numbers is 8, which is divisible by 4. $^{225}_{89}$Ac belongs to the $^{241}_{94}$Pu decay series because the difference in mass numbers is 16, which is divisible by 4.

20.33 a. $^{26}_{12}$Mg(d, α) $^{24}_{11}$Na b. $^{16}_{8}$O(n,p) $^{16}_{7}$N

20.34 a. $^{14}_{7}$N(p, α) $^{11}_{6}$C b. $^{63}_{29}$Cu(α, n) $^{66}_{31}$Ga

20.35 a. $^{27}_{13}\text{Al} + ^{2}_{1}\text{H} \rightarrow ^{25}_{12}\text{Mg} + ^{4}_{2}\text{He}$ b. $^{10}_{5}\text{B} + ^{4}_{2}\text{He} \rightarrow ^{13}_{6}\text{C} + ^{1}_{1}\text{H}$

20.36 a. $^{45}_{21}\text{Sc} + ^{1}_{0}\text{n} \rightarrow ^{42}_{19}\text{K} + ^{4}_{2}\text{He}$ b. $^{63}_{29}\text{Cu} + ^{1}_{1}\text{H} \rightarrow ^{63}_{30}\text{Zn} + ^{1}_{0}\text{n}$

20.37 $$\frac{13.8\text{ MeV}}{1\text{ proton}} = \frac{13.8 \times 10^6\text{ eV}}{\text{proton}} \times \frac{1.602 \times 10^{-19}\text{ J}}{1\text{ eV}} \times \frac{1\text{ kJ}}{10^3\text{ J}} \times \frac{6.02 \times 10^{23}\text{ protons}}{1\text{ mol}}$$

$$= 1.3\underline{3}09 \times 10^9 = 1.33 \times 10^9\text{ kJ/mol}$$

20.38 $$\frac{23.1\text{ MeV}}{1\alpha} = \frac{23.1 \times 10^6\text{ eV}}{1\alpha} \times \frac{1.602 \times 10^{-19}\text{ J}}{1\text{ eV}} \times \frac{1\text{ kJ}}{10^3\text{ J}} \times \frac{6.02 \times 10^{23}\ \alpha\text{'s}}{1\text{ mol}}$$

$$= 2.2\underline{2}78 \times 10^9 = 2.23 \times 10^9\text{ kJ/mol}$$

20.39 a. $^{6}_{3}\text{Li} + ^{1}_{0}\text{n} \rightarrow ^{A}_{Z}\text{X} + ^{3}_{1}\text{H}$

From the superscripts: 6 + 1 = A + 3, or A = 6 + 1 - 3 = 4; from the subscripts: 3 + 0 = Z + 1, or Z = 3 - 1 = 2. The product nucleus is $^{4}_{2}$X. The element with Z = 2 is helium (He), so the missing nuclide is $^{4}_{2}$He.

b. The reaction may be written

$$^{232}_{90}\text{Th} + ^{A}_{Z}\text{X} \rightarrow ^{235}_{90}\text{U} + ^{1}_{0}\text{n}$$

From the superscripts: 232 + A = 235 + 1, or A = 235 + 1 - 232 = 4; from the subscripts: 90 + Z = 92 + 0, or Z = 92 - 90 = 2. The projectile nucleus is $^{4}_{2}\text{X}$. The element with Z = 2 is helium, so the missing nuclide is $^{4}_{2}\text{He}$,or α. The reaction is then written

$$^{232}_{90}\text{Th}(\alpha,\text{n})\,^{235}_{92}\text{U}.$$

20.40 a. $^{27}_{13}\text{Al} + ^{3}_{1}\text{H} \rightarrow ^{27}_{12}\text{Mg} + ^{A}_{Z}\text{X}$

From the superscripts: 27 + 3 = 27 + A, or A = 27 + 3 - 27 = 3; from the subscripts: 13 + 1 = 12 + Z, or Z = 13 + 1 - 12 = 2. The ejected particle is $^{3}_{2}\text{X}$. The element with Z = 2 is helium, so the missing nuclide is $^{3}_{2}\text{He}$.

b. The reaction may be written

$$^{12}_{6}\text{C} + ^{3}_{1}\text{H} \rightarrow ^{14}_{6}\text{C} + ^{A}_{Z}\text{X}$$

From the superscripts: 12 + 3 = 14 + A, or A = 12 + 3 - 14 = 1; from the subscripts: 6 + 1 = 6 + Z, or Z = 6 + 1 - 6 = 1. The ejected particle is $^{1}_{1}\text{X}$. The element with Z = 1 is hydrogen, so the missing nuclide is $^{1}_{1}\text{H}$ or p. The reaction is then written

$$^{12}_{6}\text{C}(^{3}\text{H, p})\,^{14}_{6}\text{C}$$

20.41 The reaction may be written

$$^{A}_{Z}\text{X} + ^{4}_{2}\text{He} \rightarrow ^{242}_{96}\text{Cm} + ^{1}_{0}\text{n}$$

From the superscripts: A + 4 = 242 + 1, or A = 242 + 1 - 4 = 239; from the subscripts: Z + 2 = 96 + 0, or Z = 96 - 2 = 94. The element with Z = 94 is plutonium (the target nucleus was $^{239}_{94}\text{Pu}$).

20.42 The reaction may be written

$$^{A}_{Z}X + ^{4}_{2}He \rightarrow ^{245}_{98}Cf + ^{1}_{0}n$$

From the superscripts: A + 4 = 245 + 1, or A = 245 + 1 - 4 = 242; from the subscripts: Z + 2 = 98 + 0, or Z = 98 - 2 = 96. The element with Z = 96 is curium (the target nucleus was $^{242}_{96}Cm$).

20.43 The rate of decay is 8.94 x 10^{10} nuclei/s. The number of nuclei in the sample is

$$0.250 \times 10^{-3} \text{ g H-3} \times \frac{1 \text{ mol H-3}}{3.02 \text{ g H-3}} \times \frac{6.02 \times 10^{23} \text{ nuclei}}{1 \text{ mol H-3}} = 4.983 \times 10^{19} \text{ nuclei}$$

The rate equation is rate = kN_t. Solve for k.

$$k = \frac{\text{rate}}{N_t} = \frac{8.94 \times 10^{10} \text{ nuclei/s}}{4.98 \times 10^{19} \text{ nuclei}} = 1.7\underline{9}4 \times 10^{-9}/\text{s} = 1.79 \times 10^{-9}/\text{s}$$

20.44 Convert the mass of Pu-238 to the number of nuclei. The molar mass in grams is approximately equal to the mass number.

$$2.8 \times 10^{-6} \text{ g Pu-238} \times \frac{1 \text{ mol Pu-238}}{238 \text{ g Pu-238}} \times \frac{6.02 \times 10^{23} \text{ nuclei}}{1 \text{ mol Pu-238}} = 7.\underline{0}8 \times 10^{15} \text{ nuclei}$$

Solve the rate equation for k and substitute.

$$k = \frac{\text{rate}}{N_t} \times \frac{1.8 \times 10^{6} \text{ nuclei/s}}{7.08 \times 10^{15} \text{ nuclei}} = 2.\underline{5}42 \times 10^{-10}/\text{s} = 2.5 \times 10^{-10}/\text{s}$$

20.45 Find the rate of decay from the activity.

$$\text{Rate} = 20.4 \text{ Ci} \times \frac{3.700 \times 10^{10} \text{ nuclei/s}}{1 \text{ Ci}} = 7.5\underline{4}8 \times 10^{11} \text{ nuclei/s}$$

Convert the mass of S-35 to the number of nuclei. The molar mass in grams is approximately equal to the mass number.

$$0.48 \times 10^{-3} \text{ g S-35} \times \frac{1 \text{ mol S-35}}{35 \text{ g S-35}} \times \frac{6.02 \times 10^{23} \text{ nuclei}}{1 \text{ mol S-35}} = 8.\underline{2}56 \times 10^{18} \text{ nuclei}$$

(continued)

Solve the rate equation for k and substitute.

$$k = \frac{\text{rate}}{N_t} = \frac{7.548 \times 10^{11}\ \text{nuclei/s}}{8.256 \times 10^{18}\ \text{nuclei}} = 9.\underline{1}4 \times 10^{-8}\ \text{/s} = 9.1 \times 10^{-8}\text{/s}$$

20.46 Find the rate of decay from the activity.

$$\text{Rate} = 45.3\ \text{Ci} \times \frac{3.700 \times 10^{10}\ \text{nuclei/s}}{1\ \text{Ci}} = 1.6\underline{7}61 \times 10^{12}\ \text{nuclei/s}$$

Convert the mass of the Na-24 to the number of nuclei. The molar mass in grams is approximately equal to the mass number.

$$5.2 \times 10^{-6}\ \text{g} \times \frac{1\ \text{mol Na-24}}{24\ \text{g Na-24}} \times \frac{6.02 \times 10^{23}\ \text{nuclei}}{1\ \text{mol Na-24}} = 1.\underline{3}04 \times 10^{17}\ \text{nuclei}$$

Solve the rate equation for k and substitute.

$$k = \frac{\text{rate}}{N_t} = \frac{1.6\underline{7}61 \times 10^{12}\ \text{nuclei/s}}{1.304 \times 10^{17}\ \text{nuclei}} = 1.\underline{2}85 \times 10^{-5}\ \text{/s} = 1.3 \times 10^{-5}\text{/s}$$

20.47 $$t_{1/2} = \frac{0.693}{k} = \frac{0.693}{4.6 \times 10^{-19}\text{/s}} \times \frac{1\ \text{min}}{60\ \text{s}} \times \frac{1\ \text{h}}{60\ \text{min}} \times \frac{1\ \text{d}}{24\ \text{h}} \times \frac{1\ \text{y}}{365\ \text{d}} = 4.\underline{7}77 \times 10^{10}\text{y}$$

$$= 4.8 \times 10^{10}\ \text{y}$$

20.48 $$t_{1/2} = \frac{0.693}{k} = \frac{0.693}{1.03 \times 10^{-14}\text{/s}} \times \frac{1\ \text{min}}{60\ \text{s}} \times \frac{1\ \text{h}}{60\ \text{min}} \times \frac{1\ \text{d}}{24\ \text{h}} \times \frac{1\ \text{y}}{365\ \text{d}} = 2.1\underline{3}3 \times 10^{6}\text{y}$$

$$= 2.13 \times 10^{6}\ \text{y}$$

20.49 $$k = \frac{0.693}{t_{1/2}} = \frac{0.693}{5.73 \times 10^{3}\text{y}} \times \frac{1\ \text{y}}{365\ \text{d}} \times \frac{1\ \text{d}}{24\ \text{h}} \times \frac{1\ \text{h}}{60\ \text{min}} \times \frac{1\ \text{min}}{60\ \text{s}} = 3.8\underline{3}505 \times 10^{-12}\text{/s}$$

$$3.84 \times 10^{-12}\ \text{/s}$$

20.50 $$k = \frac{0.693}{t_{1/2}} = \frac{0.693}{2.5\ \text{y}} \times \frac{1\ \text{y}}{365\ \text{d}} \times \frac{1\ \text{d}}{24\ \text{h}} \times \frac{1\ \text{h}}{60\ \text{min}} \times \frac{1\ \text{min}}{60\ \text{s}} = 8.\underline{7}89 \times 10^{-9}\text{/s} = 8.8 \times 10^{-9}\text{/s}$$

20.51 $k = \frac{0.693}{t_{1/2}} = \frac{0.693}{2.69\ d} \times \frac{1\ d}{24\ h} \times \frac{1\ h}{60\ min} \times \frac{1\ min}{60\ s} = 2.9\underline{8}2 \times 10^{-6}/s$

Before substituting into the rate equation, find the number of gold nuclei from the mass.

$$0.43 \times 10^{-3}\ g\ Au\text{-}198 \times \frac{1\ mol\ Au\text{-}198}{198\ g\ Au\text{-}198} \times \frac{6.02 \times 10^{23}\ Au\text{-}198\ nuclei}{1\ mol\ Au\text{-}198}$$

$$= 1.\underline{3}1 \times 10^{18}\ Au\text{-}198\ nuclei$$

Now find the rate.

$$Rate = kN_t = (2.982 \times 10^{-6}/s)(1.31 \times 10^{18}\ nuclei) = 3.\underline{9}06 \times 10^{12}\ nuclei/s$$

$$Activity = \frac{3.906 \times 10^{12}\ nuclei/s}{3.70 \times 10^{10}\ nuclei/s{\cdot}Ci} = 1.\underline{0}55 \times 10^{2}\ Ci = 1.1 \times 10^{2}\ Ci$$

20.52 $k = \frac{0.693}{t_{1/2}} = \frac{0.693}{2.05\ y} \times \frac{1 y}{365\ d} \times \frac{1\ d}{24\ h} \times \frac{1\ h}{60\ min} \times \frac{1\ min}{60\ s} = 1.0\underline{7}2 \times 10^{-8}/s$

Find the number of cesium nuclei from the mass.

$$0.75 \times 10^{-3}\ g\ Cs\text{-}134 \times \frac{1\ mol\ Cs\text{-}134}{134\ g\ Cs\text{-}134} \times \frac{6.02 \times 10^{23}\ Cs\text{-}134\ nuclei}{1\ mol\ Cs\text{-}134}$$

$$= 3.37 \times 10^{18}\ nuclei$$

Now find the rate.

$$Rate = kN_t = (1.072 \times 10^{-8}/s)(3.37 \times 10^{18}\ nuclei) = 3.\underline{6}1 \times 10^{10}\ nuclei/s$$

$$Activity = \frac{3.61 \times 10^{10}\ nuclei/s}{3.7 \times 10^{10}\ nuclei/(s \cdot Ci)} = 0.9\underline{7}56\ Ci = 0.98\ Ci$$

20.53 Find k from the half-life.

$$k = \frac{0.693}{t_{1/2}} = \frac{0.693}{14.3\ d} \times \frac{1\ d}{24\ h} \times \frac{1\ h}{60\ min} \times \frac{1\ min}{60\ s} = 5.6\underline{0}9 \times 10^{-7}/s$$

Solve the rate equation for N_t.

$$N_t = \frac{rate}{k} = \frac{6.0 \times 10^{12}\ nuclei/s}{5.609 \times 10^{-7}/s} = 1.\underline{0}7 \times 10^{19}\ nuclei$$

(continued)

Convert N_t to the mass of P-32.

$$1.\underline{0}7 \times 10^{19} \text{ P-32 nuclei} \times \frac{1 \text{ mol P-32}}{6.02 \times 10^{23} \text{ P-32 nuclei}} \times \frac{32 \text{ g P-32}}{1 \text{ mol P-32}}$$

$$= 5.\underline{6}8 \times 10^{-4} \text{ g} = 5.7 \times 10^{-4} \text{ g}$$

20.54 $k = \frac{0.693}{t_{1/2}} = \frac{0.693}{88 \text{ d}} \times \frac{1 \text{ d}}{24 \text{ h}} \times \frac{1 \text{ h}}{60 \text{ min}} \times \frac{1 \text{ min}}{60 \text{ s}} = 9.\underline{1}1 \times 10^{-8}/\text{s}$

Solve the rate equation for N_t.

$$N_t = \frac{\text{rate}}{k} = \frac{7.7 \times 10^{11} \text{ nuclei/s}}{9.11 \times 10^{-8}/\text{s}} = 8.\underline{4}5 \times 10^{18} \text{ nuclei}$$

Convert N_t to the mass of S-35.

$$8.45 \times 10^{18} \text{ S-35 nuclei} \times \frac{1 \text{ mol S-35}}{6.02 \times 10^{23} \text{ S-35 nuclei}} \times \frac{35 \text{ g S-35}}{1 \text{ mol S-35}}$$

$$= 4.\underline{9}1 \times 10^{-4} \text{ g} = 4.9 \times 10^{-4} \text{ g}$$

20.55 The decay constant, k, is equal to $0.693/t_{1/2}$. For a first-order reaction (like radioactive decay)

$$\log \frac{N_t}{N_o} = \frac{-kt}{2.303}$$

Substituting for k in the previous equation and removing the negative sign gives

$$\log \frac{N_o}{N_t} = \frac{0.693 \, t}{2.303 t_{1/2}} = \frac{0.693(24.0 \text{ h})}{2.303(15.0 \text{ h})} = 0.48\underline{1}5$$

Taking the antilog of both sides of this equation gives

$$\frac{N_o}{N_t} = \text{antilog } (0.4815) = 3.0\underline{3}0$$

The fraction of Na-24 remaining at time t is N_t/N_o, so

$$\frac{N_t}{N_o} = \frac{1}{3.030} = 0.33\underline{0}03$$

After 24.0 h, 33.0% of the Na-24 remains.

$$5.0 \ \mu\text{g} \times 0.33003 = 1.\underline{6}5015 \ \mu\text{g} = 1.7 \ \mu\text{g}$$

20.56 $\log\left(\frac{N_0}{N_t}\right) = \frac{0.693\ t}{2.303\ t_{1/2}}$

$$\log\left(\frac{N_0}{N_t}\right) = \frac{0.693\ (7.0\ d)}{2.303\ (8.07\ d)} = 0.2\underline{6}10$$

$$\frac{N_0}{N_t} = 1.\underline{8}24$$

The fraction of I-131 remaining at time t is N_t/N_0, so

$$\frac{N_t}{N_0} = \frac{1}{1.824} = 0.5\underline{4}34$$

After 7.0 days, 54.3% of the I-131 remains; therefore, the mass of I-131 remaining is

$$5.0\ \mu g\ \text{I-131} \times 0.5\underline{4}34 = 2.\underline{7}4\ \mu g\ \text{I-131} = 2.7\ \mu g\ \text{I-131}$$

20.57 After 1.52 h, the amount of Ag-112 is $(1 - 0.280)N_0 = 0.720\ N_0$.

$$\log\left(\frac{N_0}{N_t}\right) = \frac{kt}{2.303} = \frac{0.693\ t}{2.303\ t_{1/2}}$$

Solve for $t_{1/2}$.

$$t_{1/2} = \frac{0.693\ t}{2.303\log\left(\frac{N_0}{N_t}\right)} = \frac{(0.693)\ (1.52\ h)}{2.303\log\left(\frac{N_0}{0.720\ N_0}\right)} = 3.2\underline{0}59\ h = 3.21\ h$$

20.58 After 69.9 d, the amount of Zn-65 is $(1 - 0.180)\ N_0 = 0.820\ N_0$.

$$\log\left(\frac{N_0}{N_t}\right) = \frac{0.693\ t}{2.303 t_{1/2}}$$

Solve for $t_{1/2}$.

$$t_{1/2} = \frac{0.693\ t}{2.303\log\left(\frac{N_0}{N_t}\right)} = \frac{(0.693)\ (69.9\ d)}{2.303\log\left(\frac{N_0}{0.820\ N_0}\right)} = 24\underline{4}.05\ d = 244\ d$$

20.59 The rate equation for radioactive decay is

$$\text{Rate} = kN_t$$

Rearrange this to give an expression for N_t.

$$N_t = \frac{\text{rate}}{k}$$

At time t_0, rate = 125 nuclei/s, so N_0 = 125/k nuclei. At time t_t (= 10.0 d), rate = 107 nuclei/s, so N_t = 107/k nuclei.

For radioactive decay,

$$\log\left(\frac{N_0}{N_t}\right) = \frac{0.693\ t}{2.303\ t_{1/2}}$$

Solve for $t_{1/2}$.

$$t_{1/2} = \frac{0.693\ t}{2.303 \log\left(\frac{N_0}{N_t}\right)} = \frac{0.693(10.0\ \text{d})}{2.303 \log\left(\frac{125/k}{107/k}\right)} = 44.\underline{5}6\ \text{d} = 44.6\ \text{d}$$

20.60 The rate equation for radioactive decay is rate = kN_t, or

$$N_t = \frac{\text{rate}}{k}$$

At time 0, $N_0 = \frac{88\ \text{nuclei/s}}{k}$. At time t, $N_t = \frac{53\ \text{nuclei/s}}{k}$.

For radioactive decay,

$$\log\left(\frac{N_0}{N_t}\right) = \frac{0.693\ t}{2.303\ t_{1/2}}$$

Therefore,

$$t_{1/2} = \frac{0.693\ t}{2.303 \log\left(\frac{N_0}{N_t}\right)} = \frac{0.693(9.5\ \text{h})}{2.303 \log\left(\frac{88/k}{53/k}\right)} = 1\underline{2}.98\ \text{h} = 13\ \text{h}$$

20.61 $\log\left(\frac{N_0}{N_t}\right) = \frac{0.693\ t}{2.303\ t_{1/2}}$

Rearrange this to give an expression for t.

$$t = \frac{2.303\ t_{1/2} \log\left(\frac{N_0}{N_t}\right)}{0.693}$$

Because the number of C-14 nuclei present is directly proportional to the rate of decay (rate = kN_t), the ratio of the number of nuclei present at time 0 to the number of nuclei present at time t is equal to the ratio of the rates of disintegrations at those times.

$$\frac{N_0}{N_t} = \frac{15.3 \text{ nuclei/s}}{8.1 \text{ nuclei/s}} = 1.\underline{8}9$$

For C-14, $t_{1/2}$ = 5730 y, so

$$t = \frac{2.303(5730 \text{ y})\ (\log 1.89)}{0.693} = 5.\underline{2}6 \times 10^3 \text{ y} = 5.3 \times 10^3 \text{ y}$$

20.62 $\log\left(\frac{N_0}{N_t}\right) = \frac{0.693\ t}{2.303\ t_{1/2}}$

Hence,

$$t = \frac{2.303\ t_{1/2} \log\left(\frac{N_0}{N_t}\right)}{0.693}$$

The ratio N_0/N_t may be found from the ratio of the rates of decay.

$$\frac{N_0}{N_t} = \frac{\text{rate}_0}{\text{rate}_t} = \frac{15.3 \text{ nuclei/s}}{12.1 \text{ nuclei/s}} = 1.2\underline{6}4$$

$t_{1/2}$ for C-14 is 5730 y, so

$$t = \frac{2.303(5730 \text{ y})\ (\log 1.264)}{0.693} = 1.9\underline{3}7 \times 10^3 = 1.94 \times 10^3 \text{y}$$

20.63 The half-life of C-14 is 5730 y, and the age of the sandals is 9.0 x 10^3 y. Substitute into the equation.

$$\log \frac{N_t}{N_0} = \frac{-0.693\ t}{2.303\ t_{1/2}} = \frac{-0.693 \times (9.0 \times 10^3\ y)}{2.303 \times (5730\ y)} = -0.47\underline{2}64$$

Taking the antilogarithm of both sides, you obtain

$$\frac{N_t}{N_0} = \frac{(\text{activity})_t}{(\text{activity})_0} = 0.3\underline{3}679$$

where $(\text{activity})_0$ is the original activity per gram sample, which was 15.3 disintegrations/min, and $(\text{activity})_t$ is the activity 9.0 x 10^3 y later.

$$(\text{activity})_t = 15.3 \text{ disintegrations/(min} \bullet \text{g) x } 0.3\underline{3}679$$

$$= 5.\underline{1}529 \text{ disintegrations/(min} \bullet \text{g)}$$

$$= 5.2 \text{ disintegrations/(min} \bullet \text{g)}$$

20.64 Substitute the half-life of C-14 (5730 y) and the age of the mammoth bones (1.13 x 10^4 y) into the equation.

$$\log \frac{N_t}{N_0} = \frac{-0.693\ t}{2.303\ t_{1/2}} = \frac{-0.693 \times (1.13 \times 10^4\ y)}{2.303 \times (5730\ y)} = -0.59\underline{3}42$$

Thus,

$$\frac{N_t}{N_0} = \frac{(\text{activity})_t}{(\text{activity})_0} = 0.25\underline{5}02$$

so the activity for a gram sample is

$$(\text{activity})_t = 15.3 \text{ disintegrations/(min} \bullet \text{g) x } 0.25502$$

$$= 3.9\underline{0}18 \text{ disintegrations/(min} \bullet \text{g)}$$

$$= 3.90 \text{ disintegrations/(min} \bullet \text{g)}$$

20.65 $$\Delta m = \frac{\Delta E}{c^2} = \frac{-185 \times 10^3\ \text{J}}{(3.00 \times 10^8\ \text{m/s})^2} = \frac{-1.85 \times 10^5\ \text{kg m}^2/\text{s}^2}{(3.00 \times 10^8\ \text{m/s})^2}$$

$$= -2.0\underline{5}55 \times 10^{-12} \text{ kg} = -2.06 \times 10^{-12} \text{ kg } (-2.06 \times 10^{-9} \text{ g})$$

20.66 $\Delta m = \frac{\Delta E}{c^2} = \frac{-297 \times 10^3 \text{ J}}{(3.00 \times 10^8 \text{ m/s})^2} = \frac{-2.97 \times 10^5 \text{ kg m}^2/\text{s}^2}{(3.00 \times 10^8 \text{ m/s})^2}$

$$= -3.3\underline{0}0 \times 10^{-12} \text{ kg} = -3.30 \times 10^{-12} \text{ kg } (-3.30 \times 10^{-9} \text{ g})$$

20.67 $\quad {}^{2}_{1}\text{H} \quad + \quad {}^{3}_{1}\text{H} \quad \rightarrow \quad {}^{4}_{2}\text{He} \quad + \quad {}^{1}_{0}\text{n}$

Masses: 2.01345 3.01550 4.00150 1.00867 amu

$$\Delta m = 4.00150 + 1.00867 - 2.01345 - 3.01550 = -0.018\underline{7}8$$

The mass change is –0.01878 amu for one ${}^{2}_{1}$H nucleus, or -0.01878 g for 1 mol ${}^{2}_{1}$H.

$$\Delta E = (\Delta m)c^2 = (-0.01878 \times 10^{-3} \text{ kg})(2.998 \times 10^8 \text{ m/s})^2$$
$$= (-1.878 \times 10^{-5} \text{ kg})(2.998 \times 10^8 \text{ m/s})^2$$
$$= -1.68\underline{7}9 \times 10^{12} \text{ kg} \bullet \text{m}^2/\text{s}^2$$
$$\Delta E = -1.68\underline{7}9 \times 10^{12} \text{ J} = -1.688 \times 10^{12} \text{ J (for 1 mol)}$$

Convert the mass change for one ${}^{2}_{1}$H nucleus to kilograms.

$$\Delta m = -0.01878 \text{ amu} \times \frac{1 \text{ g}}{6.022 \times 10^{23} \times 1 \text{ amu}}$$

$$= -3.1186 \times 10^{-26} \text{ g } (-3.1186 \times 10^{-29} \text{ kg})$$

$$\Delta E = (-3.1186 \times 10^{-29} \text{ kg})(2.998 \times 10^8 \text{ m/s})^2 \left(\frac{1 \text{ MeV}}{1.602 \times 10^{-13} \text{ kg} \cdot \text{m}^2/\text{s}^2}\right)$$

$$= -17.4\underline{9}7 \text{ MeV} = -17.50 \text{ MeV}$$

20.68 $\quad {}^{1}_{1}\text{H} \quad + \quad {}^{1}_{1}\text{H} \quad \rightarrow \quad {}^{2}_{1}\text{H} \quad + \quad {}^{0}_{1}\text{e}$

Masses: 1.00728 1.00728 2.01345 0.000549 amu

$$\Delta m = [2.01345 + 0.000549 - (2 \times 1.00728)] \text{ amu} = -0.0005\underline{6}1 \text{ amu}$$

The mass change for reaction of 2 mol ${}^{1}_{1}$H is -0.000561 g, or -0.0002$\underline{8}$05 g per mol ${}^{1}_{1}$H .

(continued)

$$\Delta E = (\Delta m)c^2 = (-0.0002\underline{8}05 \times 10^{-3}\ \text{kg})\ (3.00 \times 10^8\ \text{m/s})^2$$

$$= (-2.\underline{8}05 \times 10^{-7}\ \text{kg})\ (3.00 \times 10^8\ \text{m/s})^2$$

$$= -2.\underline{5}24 \times 10^{10}\ \text{kg} \bullet \text{m}^2/\text{s}^2 = -2.5 \times 10^{10}\ \text{J}$$

Convert the mass change for one $^{1}_{1}\text{H}$ nucleus to kilograms.

$$\Delta m \ = \ -0.0002\underline{8}05\ \text{amu} \times \frac{1\ \text{g}}{6.02 \times 10^{23} \times 1\ \text{amu}}$$

$$= -4.\underline{6}6 \times 10^{-28}\ \text{g} \ \ (-4.\underline{6}6 \times 10^{-31}\ \text{kg})$$

$$\Delta E = (-4.\underline{6}6 \times 10^{-31}\ \text{kg})\ (3.00 \times 10^8\ \text{m/s})^2 \left(\frac{1\ \text{MeV}}{1.602 \times 10^{-13}\ \text{kg}{\cdot}\text{m}^2/\text{s}^2}\right)$$

$$= -0.2\underline{6}2\ \text{MeV} = -0.26\ \text{MeV}$$

20.69 Mass of 3 protons $= 3 \times 1.00728\ \text{amu} = 3.02184\ \text{amu}$

Mass of 3 neutrons $= 3 \times 1.00867\ \text{amu} = \underline{3.02601\ \text{amu}}$

Total mass of nucleons $= 6.04785\ \text{amu}$

Mass defect $=$ total nucleon mass - nuclear mass $= (6.04785 - 6.01348)$ amu

$= 0.03437$ amu

$$\Delta E = (\Delta mc)^2 = 0.03437\ \text{amu} \times \frac{1\text{g}}{1\ \text{amu} \times 6.022 \times 10^{23}} \times \frac{1\ \text{kg}}{10^3\ \text{g}} \times (2.998 \times 10^8\ \text{m/s})^2$$

$$\times \frac{1\ \text{MeV}}{1.602 \times 10^{-13}\ \text{J}} = 32.0\underline{2}1\ \text{MeV}$$

$$\text{Binding energy per nucleon} = \frac{32.0\underline{2}1\ \text{MeV}}{6\ \text{nucleons}} = 5.33\underline{6}8\ \text{MeV/nucleon}$$

$$= 5.337\ \text{MeV/nucleon}$$

20.70 Mass of 26 protons $= 26 \times 1.00728\ \text{amu} = 26.18928\ \text{amu}$

Mass of 30 neutrons $= 30 \times 1.00867\ \text{amu} = \underline{30.26010\ \text{amu}}$

Total mass of nucleons $= 56.449\underline{3}8\ \text{amu}$

Mass defect $=$ total nucleon mass - nuclear mass $= (56.44938 - 55.9207)$ amu

$= 0.528\underline{6}8$ amu

(continued)

$$\Delta E = (\Delta m)c^2 = 0.528\underline{6}8 \text{ amu} \times \frac{1 \text{ g}}{1 \text{ amu} \times 6.0220 \times 10^{23}} \times \frac{1 \text{ kg}}{10^3 \text{ g}} \times (2.9979 \times 10^8 \text{ m/s})^2$$

$$\times \frac{1 \text{ MeV}}{1.6022 \times 10^{-13} \text{ J}} = 492.\underline{4}58 \text{ MeV}$$

$$\text{Binding energy per nucleon} = \frac{492.458 \text{ MeV}}{56 \text{ nucleons}} = 8.79\underline{3}89 \text{ MeV/nucleon}$$

$$= 8.794 \text{ MeV/nucleon}$$

■ Solutions to Unclassified Problems

20.71 Na-20, having fewer neutrons than the stable Na-23, is expected to decay to a nucleus with a lower atomic number (and hence a higher N/Z ratio) by electron capture or positron emission. Na-26, having more neutrons than the stable isotope, is expected to decay by beta emission to give a nucleus with a higher atomic number (and hence a lower N/Z ratio).

20.72 Al-24 has fewer neutrons than the stable Al-27, so it is expected to decay to a nucleus with a lower atomic number and higher N/Z value. Al-24 is expected to decay by positron emission or electron capture. Al-30 has more neutrons than the stable Al-27, so it is expected to decay to a nucleus with a higher atomic number and lower N/Z value. Al-30 is expected to decay by beta emission.

20.73 The overall reaction may be written

$$^{235}_{92}\text{U} \rightarrow \, ^{207}_{82}\text{Pb} + n\,^{4}_{2}\text{He} + m\,^{0}_{-1}\text{e}$$

From the superscripts:

$$235 = 207 + (n \times 4) + [m \times 0], \text{ or } n = \frac{235 - 207}{4} = 7$$

From the subscripts:

$$92 = 82 + (n \times 2) + m \times (-1), \text{ or } m = (-92 - 82 - 7 \times 2) = 4$$

Therefore, there are 7 a emissions and 4 b emissions.

20.74 The overall reaction is

$$^{232}_{90}\text{Th} \rightarrow ^{208}_{82}\text{Pb} + n\,^{4}_{2}\text{He} + m\,^{0}_{-1}\text{e}$$

From the superscripts

$$232 = 208 + (n \times 4) + [m \times 0], \text{ or } n = \frac{232 - 208}{4} = 6$$

From the subscripts

$$90 = 82 + (n \times 2) + [m \times (-1)], \text{ or } m = \frac{90 - 82 - 6 \times 2}{-1} = 4$$

Therefore, there are 6 a emissions and 4 b emissions.

20.75 $^{209}_{83}\text{Bi} + ^{4}_{2}\text{He} \rightarrow ^{A}_{85}\text{At} + 2\,^{1}_{0}\text{n}$

From the superscripts: 209 + 4 = A + (2 x 1), or A = 213 - 2 = 211.

The reaction is $^{209}_{83}\text{Bi} + ^{4}_{2}\text{He} \rightarrow ^{211}_{85}\text{At} + 2\,^{1}_{0}\text{n}$.

20.76 $^{209}_{83}\text{Bi} + ^{2}_{1}\text{H} \rightarrow ^{A}_{84}\text{Po} + ^{1}_{0}\text{n}$

From the superscripts: 209 + 2 = A + 1, or A = 211 - 1 = 210.

The reaction is $^{209}_{83}\text{Bi} + ^{2}_{1}\text{H} \rightarrow ^{210}_{84}\text{Po} + ^{1}_{0}\text{n}$.

20.77 $^{238}_{92}\text{U} + ^{12}_{6}\text{C} \rightarrow ^{A}_{Z}\text{X} + 4\,^{1}_{0}\text{n}$

From the superscripts: 238 + 12 = A + (4 x 1), or A = 250 - 4 = 246; from the subscripts: 92 + 6 = Z + (4 x 0), or Z = 98.

The element with z = 98 is californium (Cf), so the equation is

$$^{238}_{92}\text{U} + ^{12}_{6}\text{C} \rightarrow ^{246}_{98}\text{Cf} + 4\,^{1}_{0}\text{n}$$

20.78 $^{246}_{96}\text{Cm} + ^{12}_{6}\text{C} \rightarrow ^{A}_{Z}\text{X} + 4\,^{1}_{0}\text{n}$

From the superscripts: 246 + 12 = A + 4 x 1, or A = 258 - 4 = 254; from the subscripts: 96 + 6 = Z + (4 x 0), or Z = 102.

The element with z = 102 is nobelium (No), so the equation is

$$^{246}_{96}\text{Cm} + ^{12}_{6}\text{C} \rightarrow ^{254}_{102}\text{No} + 4\,^{1}_{0}\text{n}.$$

20.79 $\log\left(\frac{N_0}{N_t}\right) = \frac{0.693\ t}{2.303\ t_{1/2}}$

Hence,

$$t = \frac{2.303\ t_{1/2} \log\left(\frac{N_0}{N_t}\right)}{0.693}$$

The ratio N_0/N_t may be found from the ratio of the rates of decay.

$$\frac{N_0}{N_t} = \frac{\text{rate}_0}{\text{rate}_t} = \frac{\text{rate}_0}{0.78\ \text{rate}_0} = 1.\underline{28}$$

$$t = \frac{2.303\ (12.3\ y)\ (\log 1.28)}{0.693} = 4.\underline{3}8\ y = 4.4\ y$$

20.80 Because Rb-87 and Sr-87 have the same mass number, equal masses of the nuclides will contain equal numbers of nuclei. Thus, the mass of Rb-87 required to produce 5.3 µg of Sr-87 is 5.3 µg. The amount of Rb-87 initially present is the amount remaining plus the amount converted to Sr-87. Therefore,

$$\frac{N_0}{N_t} = \frac{102.1\ \mu g + 5.3\ \mu g}{102.1\ \mu g} = 1.052$$

$$\log\left(\frac{N_0}{N_t}\right) = \frac{0.693\ t}{2.303\ t_{1/2}}$$

or

$$t = \frac{2.303\ t_{1/2} \log\left(\frac{N_0}{N_t}\right)}{0.693} = \frac{2.303(4.8 \times 10^{10}\ y)\ (\log 1.052)}{0.693} = 3.\underline{5}1 \times 10^9\ y = 3.5 \times 10^9\ y$$

20.81 In the annihilation, the final mass is 0; that is, all mass is converted to energy. Therefore,

$$\Delta m = 0 - \text{mass of positron} - \text{mass of electron}$$

$$= -2(0.000549)\ \text{amu} \times \frac{1\ \text{g}}{1\ \text{amu} \times 6.02 \times 10^{23}} \times \frac{1\ \text{kg}}{10^3\ \text{g}} = -1.824 \times 10^{-30}\ \text{kg}$$

The energy of each photon is $-\Delta E/2$.

$$E_{photon} = \frac{-\Delta E}{2} = \frac{-(\Delta m)c^2}{2} = \frac{-(-1.824 \times 10^{-30}\ \text{kg})\ (3.00 \times 10^8\ \text{m/s})^2}{2}$$

$$E = 8.2\underline{0}8 \times 10^{-14}\ \text{J}$$

The energy of a photon is related to its wavelength by the equation

$$E = \frac{hc}{\lambda}$$

where h is Planck's constant. Therefore,

$$\lambda = \frac{hc}{E} = \frac{(6.626 \times 10^{-34}\ \text{J} \cdot \text{s})\ (3.00 \times 10^8\ \text{m/s})}{(8.208 \times 10^{-14}\ \text{J})} = 2.4\underline{2}1 \times 10^{-12}\ \text{m}$$

The wavelength of the photons is 2.42 pm.

20.82 Because $E = \frac{hc}{\lambda}$,

$$\lambda = \frac{hc}{E} = \frac{(6.626 \times 10^{-34}\ \text{J} \cdot \text{s})\ (3.00 \times 10^8\ \text{m/s})}{0.143\ \text{MeV}} \times \frac{1\ \text{MeV}}{1.602 \times 10^{-13}\ \text{J}}$$

$$= 8.6\underline{7}7 \times 10^{-12}\ \text{m} = 8.68 \times 10^{-12}\ \text{m}\ (8.68\ \text{pm})$$

To obtain the difference in mass, note that

$$\Delta E = -E = (\Delta m)c^2$$

Therefore,

$$\Delta m = \frac{-E}{c^2} = \frac{-0.143\ \text{MeV}}{(3.00 \times 10^8\ \text{m/s})^2} \times \frac{1.602 \times 10^{-13}\ \text{J}}{1\ \text{MeV}}$$

$$= -2.5\underline{4}54 \times 10^{-31} = -2.55 \times 10^{-31}\ \text{kg}\ (-2.55 \times 10^{-28}\text{g})$$

The difference in mass between a Tc-99m nucleus and a Tc-99 nucleus is 2.55×10^{-28} g.

20.83 $\;{}^{1}_{0}n \; + \; {}^{235}_{92}U \; \rightarrow \; {}^{136}_{53}I \; + \; {}^{96}_{39}Y \; + \; 4\,{}^{1}_{0}n$

Masses: 1.00867 234.9935 135.8401 95.8629 1.00867 amu

$$\Delta m = [135.8401 + 95.8629 + 4 \times 1.00867 - (1.00867 + 234.9935)] \text{ amu}$$
$$= -0.264\underline{4}9 \text{ amu}$$

When 1 mol of U-235 decays, the change in mass is -0.26449 g. Therefore, for 1.00 kg of U-235, the change in mass is

$$1.00 \times 10^3 \text{ g U-235} \times \frac{1 \text{ mol U-235}}{234.9935 \text{ g U-235}} \times \frac{-0.26449 \text{ g}}{1 \text{ mol U-235}} = -1.1255 \text{ g}$$

Converting mass to energy gives:

$$E = (\Delta m)c^2 = (-1.1255 \times 10^{-3} \text{ kg})\,(3.00 \times 10^8 \text{ m/s})^2$$
$$= -1.0\underline{1}29 \times 10^{14} \text{ J (rounds to } -1.01 \times 10^{11} \text{ kJ)}$$

For the combustion of carbon:

	C(graphite)	+	O_2(g)	→	CO_2(g)
ΔH_f°:	0		0		-394 kJ/mol

For the reaction, $\Delta H = \Delta E = -394$ kJ/mol.

$$1.00 \times 10^3 \text{ g C} \times \frac{1 \text{ mol C}}{12.01 \text{ g C}} \times \frac{(-394 \text{ kJ})}{1 \text{ mol C}} = -3.2\underline{8}0 \times 10^4 \text{ kJ} = 3.28 \times 10^4 \text{ kJ}$$

The energy released in the fission of 1.00 kg of U-235 (1.01×10^{11} kJ) is larger by several orders of magnitude than the energy released by burning 1.00 kg of C (3.28×10^4 kJ).

20.84 $\;4\,{}^{1}_{1}H \; \rightarrow \; {}^{4}_{2}He \; + \; 2\,{}^{0}_{1}e$

Masses: 1.00728 4.00150 0.000549 amu

$$\Delta m = [4.00150 + (2 \times 0.000549) - (4 \times 1.00728) \text{ amu}] = -0.02652 \text{ amu}$$

When 4 mol ${}^{1}_{1}H$ are fused, the change in mass is -0.02652 g. Therefore, for 1.00 kg ${}^{1}_{1}H$, the mass change is

(continued)

$$1.00 \times 10^3\ g \times \frac{1\ mol\ ^1_1H}{1.00728\ g\ ^1_1H} \times \frac{(-0.02652\ g)}{4\ mol\ ^1_1H} = -6.582\ g$$

$$\Delta E = (\Delta m)c^2 = (-6.582 \times 10^{-3}\ kg)\ (3.00 \times 10^8\ m/s)^2$$

$$= -5.9\underline{2}38 \times 10^{14}\ J = -5.92 \times 10^{14}\ J\ (-5.92 \times 10^{11}\ kJ)$$

The energy released in the fusion of 1.00 kg of ^1_1H is 5.92 x 10^{11} kJ, which is several orders of magnitude larger than the energy released in the combustion of 1.00 kg of C (3.28 x 10^4 kJ; see Problem 20.83).

20.85 a. $^{47}_{20}Ca \rightarrow\ ^{47}_{21}Sc +\ ^{0}_{-1}\beta$

b. Use the half-life to calculate the rate constant.

$$k = \frac{0.693}{t_{1/2}} = \frac{0.693}{4.536\ d} = 0.15\underline{2}7\ d^{-1}$$

Now use the rate law to find the initial amount. The time is 1 day.

$$\ln A_o = \ln A + kt$$

$$\ln A_o = \ln 10.0 + 0.1527\ d^{-1}\ (1\ d)$$

$$\ln A_o = 2.4\underline{5}5$$

$$A_o = 11.\underline{6}4\ \mu g \text{ of Ca-47}$$

Convert this to mass of $CaSO_4$ (molar mass 136.15 g/mol).

$$\mu g\ CaSO_4 = 11.64\ \mu g\ ^{47}Ca \times \frac{136.15\ g\ CaSO_4}{40.08\ g\ Ca}$$

$$= 39.\underline{5}4\ \mu g \text{ or } 3.95 \times 10^{-5}\ g\ CaSO_4$$

20.86 a. $^{32}_{15}P \rightarrow ^{32}_{16}S + ^{0}_{-1}\beta$

b. Use the half-life to calculate the rate constant.

$$k = \frac{0.693}{t_{1/2}} = \frac{0.693}{14.28\ d} = 0.048\underline{5}3\ d^{-1}$$

Now use the rate law to find the amount remaining after 45.0 days.

$$\ln A = \ln A_o - kt$$

$$\ln A = \ln 225 - 0.04853\ d^{-1}\ (45.0\ d)$$

$$\ln A = 3.2\underline{3}2$$

$$A = 25.\underline{3}3 \text{ mg remaining}$$

The percent decayed is.

$$\% \text{ decayed} = \frac{225 \text{ mg} - 25.33 \text{ mg}}{225 \text{ mg}} \times 100\% = 88.\underline{7}4 = 88.7\ \%$$

20.87 a. $^{82}_{35}Br \rightarrow ^{82}_{36}Kr + ^{0}_{-1}\beta$

$$2H^{82}Br(g) \rightarrow H_2 + 2Kr$$

b. Use the half-life to calculate the rate constant.

$$k = \frac{0.693}{t_{1/2}} = \frac{0.693}{1.471\ d} = 0.47\underline{1}1\ d^{-1}$$

Now use the rate law to find the amount remaining after 48.0 hours (2.00 days).

$$\ln A = \ln A_o - kt$$

$$\ln A = \ln 0.0150 - 0.4711\ d^{-1}\ (2.00\ d)$$

$$\ln A = -5.1\underline{4}2$$

$$A = 5.8\underline{4}6 \times 10^{-3} = 5.85 \times 10^{-3} \text{ mol remaining}$$

(continued)

The reaction can be summarized as follows:

$$2HBr \rightarrow H_2 + 2Kr$$

$$0.0150 - y \quad y \quad 2y$$

$$y = 0.0150 - 0.005846 = 0.00\underline{9}15 \text{ mol HBr reacted}$$

To calculate the pressure in the flask, you need the total moles of gas. This is

$$\text{total mol} = \text{mol HBr} + \text{mol } H_2 + \text{mol Kr} = (0.0150 - y) + y + 2y$$

$$= 0.0150 + 2y = 0.0150 + 2(0.00915) = 0.03\underline{3}3 \text{ mol}$$

$$P = \frac{nRT}{V} = \frac{0.0333 \text{ mol} \times 0.08206 \text{ L} \bullet \text{atm/K} \bullet \text{mol} \times 295 \text{ K}}{1.00 \text{ L}}$$

$$= 0.8\underline{0}6 = 0.81 \text{ atm}$$

20.88 a. $${}^{132}_{52}Te \rightarrow {}^{132}_{54}Xe + 2\,{}^{0}_{-1}\beta$$

b. $$H_2{}^{132}Te(g) \rightarrow H_2 + Xe$$

c. Use the half-life to calculate the rate constant.

$$k = \frac{0.693}{t_{1/2}} = \frac{0.693}{3.26 \text{ d}} = 0.21\underline{2}6 \text{ d}^{-1}$$

Now use the rate law to find the amount remaining after 100. hours (100./24 days).

$$\ln A = \ln A_o - kt$$

$$\ln A = \ln 0.0125 - 0.2126 \text{ d}^{-1} (100./24 \text{ d})$$

$$\ln A = -5.2\underline{6}8$$

$$A = 5.1\underline{5}4 \times 10^{-3} = 5.15 \times 10^{-3} \text{ mol remaining}$$

The reaction can be summarized as follows:

$$H_2Te \rightarrow H_2 + Xe$$

$$0.0125 - y \quad y \quad y$$

(continued)

$$y = 0.0125 - 0.005154 = 0.00\underline{7}35 \text{ mol } H_2Te \text{ reacted}$$

To calculate the pressure in the flask, you need the total moles of gas. This is

$$\text{total mol} = \text{mol } H_2Te + \text{mol } H_2 + \text{mol Xe} = (0.0125 - y) + y + y$$

$$= 0.0125 + y = 0.0125 + 0.00735 = 0.01\underline{9}9 \text{ mol}$$

$$P = \frac{nRT}{V} = \frac{0.0199 \text{ mol} \times 0.08206 \text{ L} \bullet \text{atm/K} \bullet \text{mol} \times 295 \text{ K}}{1.50 \text{ L}}$$

$$= 0.3\underline{2}1 = 0.32 \text{ atm}$$

■ Cumulative-Skills Problems

(require skills from previous chapters)

20.89 Because the rate = kN_t, first find the value of k in reciprocal seconds:

$$k = \frac{0.693}{t_{1/2}} = \frac{0.693}{14.3 \text{ d}} = 4.8\underline{4}6 \times 10^{-2}/\text{d}$$

$$\frac{4.846 \times 10^{-2}}{\text{d}} \times \frac{1 \text{ d}}{24 \text{ h}} \times \frac{1 \text{ h}}{60 \text{ min}} \times \frac{1 \text{ min}}{60 \text{ s}} = 5.6\underline{0}9 \times 10^{-7}/\text{s}$$

Now calculate the number of P-32 nuclei in the sample (N_t). For this, you need the formula weight of Na_3PO_4 containing 15.6% of P-32 and (100.0 - 15.6)% = 84.4% naturally occurring P (you will see that this value differs only slightly from the formula weight of the naturally occurring isotopic mixture). The formula weight of Na_3PO_4 containing naturally occurring P is 163.9 amu; the formula weight of Na_3PO_4 when the P is 100% P-32 is 165.0 amu (you can assume that the atomic mass of P-32 is 32.0 amu). The formula weight of Na_3PO_4 containing 15.6% P-32 is obtained from the weighted average of the formula weights:

$$(163.9 \text{ amu} \times 0.844) + (165.0 \text{ amu} \times 0.156) = 16\underline{4}.1 \text{ amu}$$

The moles of P in the sample equal

$$0.0545 \text{ g } Na_3PO_4 \times \frac{1 \text{ mol } Na_3PO_4}{164.1 \text{ g } Na_3PO_4} \times \frac{1 \text{ mol P}}{1 \text{ mol } Na_3PO_4} = 3.3\underline{2}1 \times 10^{-4} \text{ mol P}$$

and the moles of P-32 equal

$$(3.321 \times 10^{-4} \text{ mol P}) \times 0.156 = 5.1\underline{8}1 \times 10^{-5} \text{ mol P-32}$$

(continued)

Then, the number of P-32 nuclei is

$$(5.181 \times 10^{-5} \text{ mol P-32}) \times \frac{6.022 \times 10^{23} \text{ P-32 nuclei}}{1 \text{ mol P-32}} = 3.1\underline{2}0 \times 10^{19} \text{ P-32 nuclei}$$

Finally, calculate the rate of disintegrations:

$$\text{Rate} = kN_t = (5.609 \times 10^{-7}/\text{s}) \times (3.120 \times 10^{19} \text{ P-32 nuclei})$$

$$= 1.7\underline{5}0 \times 10^{13} = 1.75 \times 10^{13} \text{ P-32 nuclei/s}$$

20.90 Because the rate = kN_t, first find the value of k in reciprocal seconds:

$$k = \frac{0.693}{t_{1/2}} = \frac{0.693}{87.9 \text{ d}} = 7.8\underline{8}4 \times 10^{-3}/\text{d}$$

$$\frac{7.884 \times 10^{-3}}{\text{d}} \times \frac{1 \text{ d}}{24 \text{ h}} \times \frac{1 \text{ h}}{60 \text{ min}} \times \frac{1 \text{ min}}{60 \text{ s}} = 9.1\underline{2}5 \times 10^{-8}/\text{s}$$

Now calculate N_t. The formula weight of $Na_2S_2O_3$ containing 22.3% S-35 is the weighted average of the formula weight containing naturally occurring sulfur (158.1 amu) and the formula weight containing S-35 (164.0 amu):

$$(158.1 \text{ amu} \times 0.777) + (164.0 \times 0.223) = 15\underline{9}.4 \text{ amu}$$

The moles of S in the sample equal

$$0.0381 \text{ g } Na_2S_2O_3 \times \frac{1 \text{ mol } Na_2S_2O_3}{159.4 \text{ g } Na_2S_2O_3} \times \frac{2 \text{ mol S}}{1 \text{ mol } Na_2S_2O_3} = 4.7\underline{8}0 \times 10^{-4} \text{ mol S}$$

and the moles of S-35 equal

$$(4.780 \times 10^{-4} \text{ mol S}) \times 0.223 = 1.0\underline{6}6 \times 10^{-4} \text{ mol S-35}$$

The number of S-35 nuclei is

$$(1.0\underline{6}6 \times 10^{-5} \text{ mol S-35}) \times \frac{6.022 \times 10^{23} \text{ S-35 nuclei}}{1 \text{ mol S-35}} = 6.4\underline{2}0 \times 10^{19} \text{ S-35 nuclei}$$

The rate of disintegrations is

$$\text{Rate} = kN_t = (9.1\underline{2}5 \times 10^{-8}/\text{s}) \times (6.4\underline{2}0 \times 10^{19} \text{ S-35 nuclei})$$

$$= 5.8\underline{5}8 \times 10^{12} = 5.86 \times 10^{12} \text{ S-35 nuclei/s}$$

20.91 First calculate the value of the decay constant, k, for Po-210 to four significant figures using 2.303 log 2 = 0.6933:

$$k = \frac{0.6933}{t_{1/2}} = \frac{0.6933}{138.4\ d} = (5.00\underline{9}3 \times 10^{-3})/d$$

To calculate g PoO_2 decomposed, first calculate g PoO_2 left after 2 days; use the first-order relationship between time and concentration:

$$-\log \frac{[A]_t}{[A]_0} = -\log \frac{[g\ PoO_2]_t}{[g\ PoO_2]_0} = -\log \frac{[g\ PoO_2]_{2\ d}}{[1.0000\ g]} = \frac{kt}{2.303}$$

$$-\log \frac{[g\ PoO_2]_{2\ d}}{[1.0000\ g]} = \frac{(5.0093 \times 10^{-3})/d \times 2.00\ d}{2.303}$$

$$-\log\ [g\ PoO_2]_{2\ d} = (4.3\underline{5}02 \times 10)^{-3} - \log\ [1.0000\ g] = 4.3\underline{5}02 \times 10^{-3}$$

$$\log\ [g\ PoO_2]_{2\ d} = -(4.3\underline{5}02 \times 10^{-3})$$

$$[g\ PoO_2]_{2\ d} = 0.99\underline{0}0333 \text{ g left after 2 days}$$

$$g\ PoO_2 \text{ decomposed after 2 d} = 1.000\underline{0}\ g - 0.99\underline{0}0333 = 0.00\underline{9}9667\ g$$

$$\text{Mol He} = \text{mol } PoO_2 \text{ decomposed} = 0.00\underline{9}9667\ g\ Po \times \frac{1\ mol\ PoO_2}{242\ g\ PoO_2} \times \frac{1\ mol\ He}{1\ mol\ PoO_2}$$

$$= 0.0000\underline{4}11 \text{ mol He}$$

Now calculate the volume of He at 25°C and 735 mmHg using R = 0.08206.

$$V = \frac{nRT}{P} = \frac{(0.0000411\ mol)\ [0.08206\ L \cdot atm/(K \cdot mol)]\ (298\ K)}{(735/760)\ atm}$$

$$= 0.001039\ L = 0.001\ L \quad (1\ mL)$$

20.92 First calculate the value of the decay constant, k, for Ra-226 to four significant figures using 2.303 log 2 = 0.6933:

$$k = \frac{0.6933}{t_{1/2}} = \frac{0.6933}{1602\ y} = (4.32\underline{7}7 \times 10^{-4})/y$$

To calculate g $RaBr_2$ decomposed, first calculate g $RaBr_2$ left after 20.2 y; use the first-order relationship between time and concentration:

$$-\log \frac{[A]_t}{[A]_0} = -\log \frac{[g\ RaBr_2]_t}{[g\ RaBr_2]_0} = -\log \frac{[g\ RaBr_2]_{20\ y}}{[0.5430\ g]} = \frac{kt}{2.303}$$

(continued)

$$-\log \frac{[\text{g RaBr}_2]_{20\text{ y}}}{[0.5430\text{ g}]} = \frac{(4.3277 \times 10^{-4})/\text{y} \times 20.2\text{y}}{2.303}$$

$$\log \frac{[\text{g RaBr}_2]_{20\text{ y}}}{[0.5430\text{ g}]} = -(3.7\underline{9}59 \times 10^{-3})$$

$$\frac{[\text{g RaBr}_2]_{20\text{ y}}}{[0.5430\text{ g}]} = 0.9\underline{9}129$$

$$[\text{g RaBr}_2]_{20\text{ y}} = 0.99129 \times 0.5430\text{ g} = 0.53\underline{8}2\text{ g}$$

g $RaBr_2$ decomposed after 20.2 y = 0.543$\underline{0}$ g -0.53$\underline{8}$2 g = 0.00$\underline{4}$8 g

$$\text{Mol Rn} = \text{mol RaBr}_2\text{ decomposed} = 0.00\underline{4}8\text{ g RaBr}_2 \times \frac{1\text{ mol}}{385.8\text{ g RaBr}_2} \times \frac{1\text{ mol Rn}}{1\text{ mol RaBr}_2}$$

$$= \underline{1}.244 \times 10^{-5}\text{ mol Rn}$$

Now calculate the volume of Rn at 23°C and 785 mmHg using R = 0.08206.

$$V = \frac{nRT}{P} = \frac{(1.244 \times 10^{-5}\text{ mol})\,[0.08206\text{ L}\cdot\text{atm}/(\text{K}\cdot\text{mol})]\,(296\text{ K})}{(785/760)\text{ atm}}$$

$$= \underline{2}.92 \times 10^{-4} = 3 \times 10^{-4}\text{ L (0.3 mL)}$$

20.93 $2p + 2n \rightarrow$ He-4

On a mole basis, the mass difference, Δm, is

-2.01456	g/2 mol protons
-2.01734	g/2 mol neutrons
+4.00150	g/mol He

Δm = -0.03040 g/mol He = -3.04$\underline{0}$ x 10^{-5} kg/mol He

$$\Delta E = (\Delta m)c^2 = -3.040 \times 10^{-5}\text{ kg/mol He} \times (2.998 \times 10^8\text{ m/s})^2 = -2.73\underline{2}3 \times 10^{12}\text{ kg}\cdot\text{m}^2/\text{s}^2$$

$$= -2.73\underline{2}3 \times 10^{12}\text{ J} = -2.73\underline{2}3 \times 10^9\text{ kJ}$$

Next calculate ΔH° for burning of ethane:

	$C_2H_6(g)$	+ $7/2\ O_2(g)$	→ $2CO_2(g)$	+ $3\ H_2O(g)$
ΔH_f° =	-84.667 kJ	0	2(-393.5 kJ)	3(-241.826 kJ)

$$\Delta H_{\infty} = [2(-393.\underline{5}) + 3(-241.826) - (-84.667) + 0]\text{ kJ} = 1427.\underline{8}1$$

$$= -1.427\underline{8}1 \times 10^3\text{ kJ/mol ethane}$$

(continued)

Now calculate the mol of ethane needed to obtain 2.736 x 10^9 kJ heat:

$$2.73\underline{2}3 \times 10^9 \text{ kJ} \times \frac{1 \text{ mol ethane}}{1.427\underline{8}1 \times 10^3 \text{ kJ}} = 1.91\underline{3}6 \times 10^6 \text{ mol ethane}$$

Finally, convert moles to liters at 25°C and 725 mmHg using R = 0.08205.

$$V = \frac{nRT}{P} = \frac{(1.9136 \times 10^6 \text{ mol})\,[0.08206 \text{ L} \cdot \text{atm/(K} \cdot \text{mol)}]\,(298 \text{ K})}{(725/760) \text{ atm}}$$

$$= 4.9\underline{0}48 \times 10^7 = 4.90 \times 10^7 \text{ L}$$

20.94 $^{239}_{94}\text{Pu} \rightarrow {}^{4}_{2}\text{He} + {}^{235}_{92}\text{U}$

On a mole basis, the mass difference, Δm, is

-239.0006	g/mol Pu-239
+234.9935	g/mol U-235
+4.00150	g/mol He-4

Δm = -0.0056 g/mol Pu-239 = 5.$\underline{6}$0 x 10^{-6} kg/mol Pu-239

$\Delta E = (\Delta m)c^2$ = -5.60 x 10^{-6} kg/mol Pu x (3.00 x 10^8 m/s$)^2$ = -5.$\underline{0}$4 x 10^{11} kg • m^2/s^2

= -5.$\underline{0}$4 x 10^{11} J/mol Pu

For 215 mg of plutonium-239, the energy released is

$$\Delta E = 215 \times 10^{-3} \text{ g Pu} \times \frac{1 \text{ mol Pu}}{239.0006 \text{ g}} \times \frac{-5.04 \times 10^{11} \text{ J}}{1 \text{ mol Pu}} = -4.53 \times 10^8 \text{ J}$$

Next calculate 25.0% of this energy:

4.53 x 10^8 J x 0.250 = 1.$\underline{1}$325 x 10^8 J

Next calculate the E°_{cell} for the standard voltaic Zn/Cu^{2+} cell:

E°_{cell} = 0.34 V – (–0.76 V) = +1.10 V = 1.10 V/mol Zn(s)

Now calculate J using F = 9.65 × 10^4 C and remembering that C × V = J (Chapter 19):

$$\Delta G^\circ = -nFE^\circ_{cell} = -\frac{2 \text{ mol e}}{\text{mol Zn(s)}} \times \frac{9.65 \times 10^4 \text{ C}}{\text{mol e}} \times 1.10 \text{ V} = -2.1\underline{2}3 \times 10^5 \text{ J/mol Zn(s)}$$

(continued)

Moles of Zn(s) needed to release $1.\underline{1}325 \times 10^6$ J:

$$1.\underline{1}325 \times 10^8 \text{ J} \times \frac{\text{mol Zn(s)}}{-2.1\underline{2}3 \times 10^5 \text{ J}} = 5\underline{3}3.4 \text{ mol Zn(s)}$$

Mass of Zn(s) needed:

$$5.334 \text{ mol Zn(s)} \times \frac{65.39 \text{ g Zn(s)}}{\text{mol Zn(s)}} = 3.\underline{4}88 \times 10^4 = 3.5 \times 10^4 \text{ g Zn(s)}$$

21. METALLURGY AND CHEMISTRY OF THE MAIN-GROUP METALS

■ Answers to Review Questions

21.1 Seven free metals have been found: gold, silver, copper, iron, lead, mercury, and tin.

21.2 Four characteristics of a metal are luster, heat conductivity, electrical conductivity, and ductility or malleability.

21.3 A rock is a naturally occurring solid material composed of one or more minerals. A mineral is a naturally occurring inorganic solid substance or solid solution with a definite crystalline form. An ore is a rock or mineral from which a metal or nonmetal can be economically produced. Bauxite is a rock.

21.4 The kinds of metal compounds found most commonly in ores are hydroxides, oxides, and sulfides (Table 21.2).

21.5 The basic steps in producing a pure metal are (1) preliminary treatment to separate the metal-containing mineral from the rest of the ore (bauxite is treated with NaOH to dissolve the aluminum and precipitate the rest); (2) reduction to the metal (the aluminum oxide from the bauxite is reduced to aluminum by electrolysis); and (3) refining to increase the purity (the aluminum might be further refined).

21.6 Flotation separates the ore mineral by coating the valuable chemical substance with a flotation agent to float it, allowing it to be removed at the top of the vessel. The gangue settles to the bottom and is left behind.

21.7 Roasting a zinc ore burns the sulfur away by converting it to gaseous sulfur dioxide.

21.8 Metals obtained by electrolysis from their compounds include lithium sodium, magnesium, and aluminum.

21.9 Hydrogen is preferred over carbon for the reducing of tungsten oxide to tungsten metal.

21.10 The three different methods for refining a metal are (1) the distillation of zinc to remove it from lead, cadmium, and iron impurities; (2) the Mond process for purifying nickel by forming volatile nickel tetracarbonyl, which is distilled away from impurities; and (3) the electrolytic purification of copper by deposition at the cathode of an electrolysis cell.

21.11 When Avogadro's number, N, of metal atoms have formed a crystal, N molecular orbitals are also formed, encompassing the entire crystal. The number of individual energy levels is so large that they merge into a band of continuous energies. The result is a regular array of positive atomic cores surrounded by a "sea" of electrons from the valence shells. These electrons are free to move throughout the crystal, so that when even a low voltage is applied to a metal, the highest-energy electrons in the valence shell are easily excited into the unoccupied orbitals, creating an electrical conductor.

21.12 An energy band is formed in a metal as metal atoms are brought together. When two metal atoms approach each other, their outer orbitals overlap to form two molecular orbitals. When a third atom is brought to this diatomic metal molecule, the three outer orbitals all overlap, forming a larger delocalized outer orbital. When a large number, N (close to Avogadro's number), of atoms are brought together, the atoms will form N molecular orbitals delocalized over the entire crystalline metal molecule. This is a band of electrons of continuous energies.

21.13 The mineral source is given after the name of the metal: lithium—lithium aluminum silicate; sodium—sodium chloride; magnesium—seawater (Dow process) and dolomite or magnesite; calcium—calcium oxide; aluminum—bauxite; tin—cassiterite; and lead—galena.

21.14 The reactions are

$$2Li(s) + 2H_2O(l) \rightarrow 2LiOH(aq) + H_2(g)$$

$$4Li(s) + O_2(g) \rightarrow 2Li_2O(s)$$

$$2Na(s) + 2H_2O(l) \rightarrow 2NaOH(aq) + H_2(g)$$

$$2Na(s) + O_2(g) \rightarrow Na_2O_2(s)$$

21.15 The reaction is similar to that of lithium carbonate with calcium hydroxide:

$$Li_2CO_3(aq) + Ba(OH)_2(aq) \rightarrow 2LiOH(aq) + BaCO_3(s)$$

21.16 Lithium hydroxide is contaminated by reaction with carbon dioxide absorbed from the air:

$$2LiOH(s) + CO_2(g) \rightarrow Li_2CO_3(s) + H_2O(l)$$

$$Li_2CO_3(s) + H_2O(l) + CO_2(g) \rightarrow 2LiHCO_3(s)$$

21.17 Potassium is expected to be more reactive than lithium because metals become more reactive going down Group IA. This is partly because potassium is much larger, so its 4s electron is lost more readily than the 2s electron of lithium.

21.18 The reaction is $2Na(s) + 2C_2H_5OH(l) \rightarrow H_2(g) + 2NaOC_2H_5(aq)$.

21.19 a. Cathode reaction: $Na(l) + e^- \rightarrow Na(l)$
Anode reaction: $2Cl^-(l) \rightarrow Cl_2(g) + 2e^-$

b. Cathode reaction: $Na(l) + e^- \rightarrow Na(l)$
Anode reaction: $4OH^-(l) \rightarrow O_2(g) + 2H_2O(g) + 4e^-$

21.20 Sodium hydroxide is manufactured by the electrolysis of aqueous sodium chloride, which also produces chlorine gas as a major product.

21.21 The uses are given after each compound: sodium chloride—used for making sodium hydroxide and in seasoning; sodium hydroxide—used in aluminum production and in producing sodium compounds such as soap; and sodium carbonate—used to make glass and as washing soda with many detergent preparations.

21.22 The main step in the Solvay process involves the reaction of carbon dioxide with ammonia and sodium chloride to form sodium bicarbonate:

$$NH_3(g) + H_2O(l) + CO_2(g) + NaCl(aq) \rightarrow NaHCO_3(s) + NH_4Cl(aq)$$

21.23 Limestone is heated to form CaO and gaseous CO_2. CaO then reacts with seawater to form calcium hydroxide, which precipitates the Mg^{2+} in seawater as magnesium hydroxide. The overall equation for the two reactions (given in the text) of CaO with seawater and Mg^{2+} is

$$CaO(s) + H_2O(l) + Mg^{2+} \rightarrow Mg(OH)_2(s) + Ca^{2+}(aq)$$

The magnesium hydroxide is then neutralized to magnesium chloride:

$$Mg(OH)_2(s) + 2HCl(aq) \rightarrow MgCl_2(aq) + 2H_2O(l)$$

The magnesium chloride is then dried by evaporating the water and electrolyzed at 700°C:

$$MgCl_2(l) \rightarrow Mg(l) + Cl_2(g)$$

21.24 $2Mg(s) + O_2(g) \rightarrow 2MgO(s)$

$Mg(s) + H_2O(g) \rightarrow MgO(s) + H_2(g)$

$2Mg(s) + CO_2(g) \rightarrow 2MgO(s) + C(s)$

21.25 a. Calcium oxide is prepared industrially from calcium carbonate:

$$CaCO_3(s) \rightarrow CaO(s) + CO_2(g)$$

b. Calcium hydroxide is prepared from the reaction of calcium and water:

$$CaO(s) + H_2O(l) \rightarrow Ca(OH)_2(aq)$$

21.26 $CaCO_3(s) + 2HCl(aq) \rightarrow CO_2(g) + CaCl_2(aq)$

21.27 $Ca(OH)_2 + CO_2(g) \rightarrow CaCO_3(s) + H_2O(l)$

$Ca(OH)_2 + Na_2CO_3(aq) \rightarrow CaCO_3(s) + 2NaOH(aq)$

21.28 The metallic character of the elements decreases from left to right across a row of elements and increases going down a column of elements.

21.29 No, aluminum ores are not widespread because most of the aluminum is found in aluminum-containing clays and not in a convenient mine. Also, much bauxite occurs in tropical and subtropical regions, where mining is not easy.

21.30 A brief flow chart is

$$\text{Bauxite} + NaOH \rightarrow Al(OH)_3 + \text{heat} \rightarrow Al_2O_3 + C + \text{electrolysis in } Na_3AlF_6$$
$$\rightarrow Al + CO_2(g)$$

21.31 $Fe_2O_3(s) + 2Al(s) \rightarrow 2Fe(l) + Al_2O_3(s)$

21.32 Some major uses of aluminum oxide are making abrasives for grinding tools, fusing with small amounts of other metal oxides to make synthetic sapphires and rubies, and making industrial ceramics.

21.33 This means that aluminum hydroxide reacts with both acids and bases. For example,

$$Al(OH)_3(s) + 3HCl(aq) \rightarrow 3H_2O(l) + AlCl_3(aq)$$

$$Al(OH)_3(s) + NaOH(aq) \rightarrow Na^+(aq) + Al(OH)_4^-(aq)$$

21.34 To purify municipal water, aluminum sulfate and calcium hydroxide are added to waste water, forming a gelatinous precipitate of aluminum hydroxide. Colloidal particles of clay (usually present in the waste water) and other substances adhere to the aluminum hydroxide, whose particles are large enough to be filtered from the water to purify it.

21.35 Tin metal would not make a good structural metal in cold climates because below 13°C the stable white metallic allotrope undergoes a transition to the brittle powder allotrope called gray tin.

21.36 Lead(IV) oxide, PbO_2, is formed by first packing a paste of PbO into the lead metal grids of the storage battery. When the battery is charged at the factory, the PbO is oxidized by electrolysis to PbO_2. This gives the proper cathode for a new battery.

21.37 Lead pigments are no longer used for house paints because of the possibility of lead poisoning. If chips of lead paint are eaten by children, or if the dust from lead paint that has been removed is breathed by adults or children, lead(II) ion can enter the blood stream. The lead(II) ion can ultimately inhibit the production of red blood cells, causing anemia. (The Pb^{2+} ion can also be absorbed in the brains of children, causing irreversible brain damage.)

21.38 $Pb(NO_3)_2(aq) + CrO_4^{2-}(aq) \rightarrow PbCrO_4(s) + 2NO_3^-(aq)$

■ Solutions to Practice Problems

Note on significant figures: The final answer to each problem is given first with one nonsignificant figure (the rightmost significant figure is underlined), and then is rounded to the correct number of figures. Intermediate answers usually also have at least one nonsignificant figure.

21.39 As for copper, the impure lead metal serves as the anode and the pure lead serves as the cathode. During the electrolysis, lead(II) ions leave the anode and deposit on the cathode; the electrolyte of $PbSiF_6$ may be considered the form of lead(II) that reacts at the cathode.

Anode reaction: $Pb(s) \rightarrow Pb^{2+}(aq) + 2e^-$

Cathode reaction: $PbSiF_6(aq) + 2e^- \rightarrow Pb(s) + SiF_6^{2-}(aq)$

21.40 As for copper, the impure nickel metal serves as the anode and the pure nickel serves as the cathode. During the electrolysis, nickel(II) ions leave the anode and deposit on the cathode; the electrolyte of $NiSO_4$ may be considered the form of nickel(II) that reacts at the cathode.

Anode reaction: $Ni(s) \rightarrow Ni^{2+}(aq) + 2e^-$

Cathode reaction: $NiSO_4(aq) + 2e^- \rightarrow Ni(s) + SO_4^{2-}$

21.41 $Fe_2O_3(s) + 3H_2(g) \rightarrow 2Fe(s) + 3H_2O(g)$

21.42 $3MnO_2(s) + 4Al(s) \rightarrow 3Mn(s) + 2Al_2O_3(s)$

21.43 In the reaction (Problem 21.41), 3 mol of H_2 are used to form 2 mol of iron. Using the respective atomic masses of 55.85 g/mol for Fe and 2.016 g/mol for H_2, you can calculate the mass of Fe as follows:

$$10\underline{0}0 \text{ g } H_2 \times \frac{1 \text{ mol } H_2}{2.016 \text{ g } H_2} \times \frac{2 \text{ mol Fe}}{3 \text{ mol } H_2} \times \frac{55.85 \text{ g Fe}}{1 \text{ mol Fe}} = 1.8\underline{4}6 \times 10^4 \text{ g}$$

$$= 18.5 \text{ kg Fe}$$

21.44 In the reaction (Problem 21.42), 4 mol of Al are used to form 3 mol of manganese. Using the respective atomic masses of 54.94 g/mol for Mn and 26.98 g/mol for Al, you can calculate the mass of Mn as follows:

$$1\underline{0}00 \text{ g Al} \times \frac{1 \text{ mol Al}}{26.98 \text{ g Al}} \times \frac{3 \text{ mol Mn}}{4 \text{ mol Al}} \times \frac{54.94 \text{ g Mn}}{1 \text{ mol Mn}} = 1.5\underline{2}7 \times 10^3 \text{ g}$$

$$= 1.53 \text{ kg Mn}$$

21.45 Each mole of aluminum is formed from Al^{3+} by reacting with 3 mol of electrons. Using Faraday's law of electrolysis, and 3600 s (= 1 hr) to convert time to seconds, you can calculate the mass of aluminum. Using 26.98 g/mol of Al as its atomic mass, the calculation is

$$94.0 \text{ A} \times 3600 \text{ s} \times \frac{1 \text{ mol e}^-}{9.65 \times 10^4 \text{ A•s}} \times \frac{1 \text{ mol Al}}{3 \text{ mol e}^-} \times \frac{26.98 \text{ g Al}}{1 \text{ mol Al}} = 31.\underline{5}3$$

$$= 31.5 \text{ g Al}$$

21.46 Each mole of copper is formed from Cu^{2+} by reacting with 2 mol of electrons. Using Faraday's law of electrolysis, and 4500 s (= 1.25 hr) to convert time to seconds, you can calculate the mass of copper. Using 63.55 g/mol of Cu as its atomic mass, the calculation is

$$56.7 \text{ A} \times 4500 \text{ s} \times \frac{1 \text{ mol e}^-}{9.65 \times 10^4 \text{ A•s}} \times \frac{1 \text{ mol Cu}}{2 \text{ mol e}^-} \times \frac{63.55 \text{ g Cu}}{1 \text{ mol Cu}} = 84.\underline{0}1$$

$$= 84.0 \text{ g Cu}$$

21.47 The equation with ΔH°_f's recorded beneath each substance is

$PbS(s)$	+ $3/2O_2(g)$	$\rightarrow$ $PbO(s)$	+ $SO_2(g)$
-98.3	0	-219.0	-296.8 (kJ)

$$\Delta H^\circ = -219.0 + (-296.8) - (-98.3 + 0) = -417.5 \text{ kJ (exothermic)}$$

21.48 The equation with ΔH°_f's recorded beneath each substance is

$3MnO_2(s)$	+ $4Al(s)$	$\rightarrow$ $3Mn(s)$	+ $2Al_2O_3(s)$
3(-520.9)	0	0	2(-1676) (kJ)

$$\Delta H^\circ = 2(-1676) + 0 - 3(-520.9) - 0 = -178\underline{9}.3 = -1789 \text{ kJ (exothermic)}$$

21.49 The balanced *molecular* equation is given first, followed by the net *ionic* equation:

$K_2SO_4(aq) + Ba(OH)_2(aq) \rightarrow 2KOH(aq) + BaSO_4(s)$

$SO_4^{2-}(aq) + Ba^{2+}(aq) \rightarrow BaSO_4(s)$

21.50 $Na_2CO_3(aq) + CO_2(g) + H_2O(l) \rightarrow 2NaHCO_3(s)$

21.51 $CO_2(g) + NH_3(g) + NaCl(aq) + H_2O(l) \xrightarrow{\Delta} NaHCO_3(s) + NH_4Cl(aq)$

$2NaHCO_3(s) \rightarrow Na_2CO_3(s) + CO_2(g) + H_2O(g)$

$Na_2CO_3(s) + Ca(OH)_2(aq) \rightarrow 2NaOH(aq) + CaCO_3(s)$

21.52 $P_4 + 5O_2(g) \rightarrow P_4O_{10}(s)$

$P_4O_{10}(s) + 6H_2O(l) \rightarrow 4H_3PO_4(aq)$

$2NaCl(aq) + 2H_2O(l)$ (electrolysis) $\rightarrow 2NaOH(aq) + H_2(g) + Cl_2(g)$

$H_3PO_4 + 3NaOH(aq) \rightarrow Na_3PO_4(aq) + 3H_2O(l)$

21.53 a. $2K(s) + Br_2(l) \rightarrow 2KBr(s)$

b. $2K(s) + 2H_2O(l) \rightarrow 2KOH(aq) + H_2(g)$

c. $2NaOH(s) + CO_2(g) \rightarrow Na_2CO_3(s) + H_2O(l)$

d. $Li_2CO_3(aq) + 2HNO_3(aq) \rightarrow H_2O(l) + 2LiNO_3(aq) + CO_2(g)$

e. $K_2SO_4(aq) + Pb(NO_3)_2(aq) \rightarrow PbSO_4(s) + 2KNO_3(aq)$

21.54 a. $2LiHCO_3(s)$ (heat) $\rightarrow Li_2CO_3(s) + CO_2(g) + H_2O(g)$

b. $Na_2SO_4(aq) + BaCl_2(aq) \rightarrow BaSO_4(s) + 2NaCl(aq)$

c. $K_2CO_3(aq) + Ca(OH)_2(aq) \rightarrow CaCO_3(s) + 2KOH(aq)$

d. $2Li(s) + 2HCl(aq) \rightarrow 2LiCl(aq) + H_2(g)$

e. $4Na(s) + ZrCl_4(g) \rightarrow 4NaCl(s) + Zr(s)$

21.55 $^{227}_{89}Ac \rightarrow {}^{223}_{87}Fr + {}^{4}_{2}He$

21.56 $^{223}_{87}Fr \rightarrow {}^{223}_{88}Ra + {}^{0}_{-1}e$

21.57 $Ca(OH)_2$ can be identified directly by adding an anion that will precipitate the Ca^{2+} (and not the Na^+). For example, adding CO_3^{2-} will precipitate $CaCO_3$ but will not precipitate Na^+ ion.

21.58 $Ba(OH)_2$ can be identified directly by adding SO_4^{2-} ion, precipitating $BaSO_4$ but not K_2SO_4, which is soluble (see Table 3.1). Thus, the presence of a solid will indicate $Ba(OH)_2$.

21.59 $BaCl_2$ can be separated by adding SO_4^{2-} ion, precipitating $BaSO_4$ and leaving $MgCl_2$ in solution because it is soluble (see Table 3.1). Pouring off the solution of $MgCl_2$ will separate it from the $BaSO_4$ at the bottom of the container.

21.60 $MgCl_2$ can be separated by adding OH^- ion, precipitating $Mg(OH)_2$ and leaving NaCl in solution because it is soluble (see Table 3.1). Pouring off the solution of NaCl will separate it from the $Mg(OH)_2$ at the bottom of the container.

21.61 $^{230}_{90}Th \rightarrow {}^{226}_{88}Ra + {}^{4}_{2}He$

21.62 $^{223}_{88}Ra \rightarrow {}^{219}_{86}Rn + {}^{4}_{2}He$

21.63 a. $BaCO_3(s) \xrightarrow{\Delta} BaO(s) + CO_2(g)$

b. $Ba(s) + 2H_2O(l) \rightarrow Ba(OH)_2(aq) + H_2(g)$

c. $Mg(OH)_2(s) + 2HNO_3(aq) \rightarrow 2H_2O(l) + Mg(NO_3)_2(aq)$

d. $Mg(s) + NiCl_2(aq) \rightarrow Ni(s) + MgCl_2(aq)$

e. $2NaOH(aq) + MgSO_4(aq) \rightarrow Mg(OH)_2(s) + Na_2SO_4(aq)$

21.64 a. $2KOH(aq) + MgCl_2(aq) \rightarrow Mg(OH)_2(s) + 2KCl(aq)$

b. $Mg(s) + CuSO_4(aq) \rightarrow Cu(s) + MgSO_4(aq)$

c. $Sr(s) + 2H_2O(l) \rightarrow Sr(OH)_2(aq) + H_2(g)$

d. $SrCO_3(s) + 2HCl(aq) \rightarrow CO_2(g) + H_2O(l) + SrCl_2(aq)$

e. $Ba(OH)_2(aq) + CO_2(g) \rightarrow BaCO_3(s) + H_2O(l)$

21.65 The equation is $Ca(HCO_3)_2(aq) + Ca(OH)_2(aq) \rightarrow 2CaCO_3(s) + 2H_2O(l)$. Using a 1:1 ratio of $Ca(OH)_2$ to Ca^{2+} ion in the reaction, calculate the mass of $Ca(OH)_2$:

$$0.0500 \text{ L } Ca^{2+} \times \frac{0.12 \text{ mol } Ca^{2+}}{\text{L } Ca^{2+}} \times \frac{1\ Ca(OH)_2}{1\ Ca^{2+}} \times \frac{74.09 \text{ g } Ca(OH)_2}{1 \text{ mol } Ca(OH)_2}$$

$$= 0.4\underline{4}4 \text{ g } Ca(OH)_2 = 0.44 \text{ g } Ca(OH)_2$$

21.66 The equation is $Ca(OH)_2(aq) + Mg^{2+}(aq) \rightarrow Mg(OH)_2(s) + Ca^{2+}(aq)$. Using a 1:1 ratio of $Ca(OH)_2$ to Mg^{2+} ion in the reaction, calculate the mass of $Ca(OH)_2$:

$$0.050 \text{ L } Mg^{2+} \times \frac{0.21 \text{ mol } Mg^{2+}}{\text{L } Mg^{2+}} \times \frac{1\ Ca(OH)_2}{1\ Mg^{2+}} \times \frac{74.09 \text{ g } Ca(OH)_2}{1 \text{ mol } Ca(OH)_2}$$

$$= 0.7\underline{7}7 \text{ g } Ca(OH)_2 = 0.78 \text{ g } Ca(OH)_2$$

21.67 $Al(H_2O)_6{}^{3+}(aq) + HCO_3{}^-(aq) \rightarrow Al(H_2O)_5OH^{2+}(aq) + H_2O(l) + CO_2(g)$

21.68 a. As the pH begins to rise, a precipitate of $Al(OH)_3$ forms. As the pH gets even higher, the precipitate redissolves.

b. $Al^{3+}(aq) + 3OH^-(aq) \rightarrow Al(OH)_3(s)$

$Al(OH)_3(s) + OH^-(aq) \rightarrow Al(OH)_4{}^-(aq)$

21.69 Test portions of solutions of each compound with the others; the results can differentiate the compounds. For example, if one solution is poured into one of the other solutions and gives no precipitate, then that means $BaCl_2$ was mixed with KOH and the third solution is $Al_2(SO_4)_3$. Adding the third solution of $Al_2(SO_4)_3$ to both of the first two solutions will form a precipitate with only $BaCl_2$. [$Al_2(SO_4)_3$ and excess KOH form soluble $Al(OH)_4{}^-$.] If instead one solution is poured into one of the other solutions and a precipitate forms, then that means that $BaCl_2$ was mixed with $Al_2(SO_4)_3 \rightarrow (BaSO_4)$, and the third solution is KOH, etc. Thus, all three are identified.

21.70 Test portions of solutions of each compound with the others; the results can differentiate the compounds. For example, if one solution is poured into both of the other solutions and both give a precipitate [→ $Al(OH)_3$] and → [$Mg(OH)_2$ + $BaSO_4$], that solution must be $Ba(OH)_2$. Continuing the addition of the $Ba(OH)_2$ will dissolve the $Al(OH)_3$ as $Al(OH)_4^-$ but will not dissolve the $Mg(OH)_2$ + $BaSO_4$. Thus, all three will be identified. If instead $MgSO_4$ is poured into $AlCl_3$, no reaction will occur and that will mean the third solution is $Ba(OH)_2$.

21.71 $AlCl_3$ has a trigonal planar geometry.

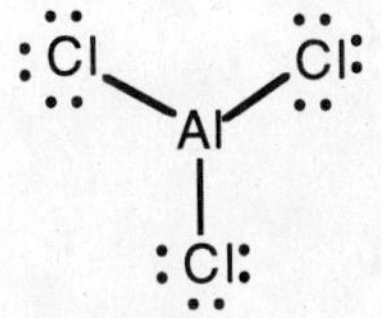

21.72 The electron-dot formula is

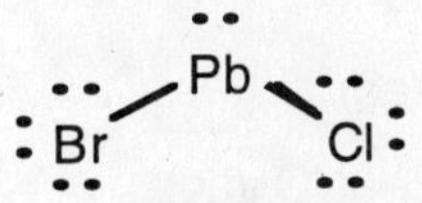

The VSEPR model predicts an angular, or bent, geometry.

21.73 $Sn(H_2O)_6^{2+}(aq) + H_2O(l) \rightarrow Sn(H_2O)_5(OH)^+(aq) + H_3O^+(aq)$

21.74 $Pb(H_2O)_6^{2+}(aq) + H_2O(l) \rightarrow Pb(H_2O)_5(OH)^+(aq) + H_3O^+(aq)$

21.75 The half-reactions and their sum are as follows:

$$PbO_2 + 4H^+ + 2e^- \rightarrow Pb^{2+} + 2H_2O$$

$$2Cl^- \rightarrow Cl_2 + 2e^-$$

$$PbO_2 + 4H^+ + 2Cl^- \rightarrow Pb^{2+} + Cl_2 + 2H_2O$$

Adding $2Cl^-$ to both sides gives

$$PbO_2 + 4HCl \rightarrow PbCl_2 + Cl_2 + 2H_2O$$

21.76 $ClO^- + H_2O + 2e^- \rightarrow Cl^- + 2OH^-$

$Pb(OH)_3^- + OH^- \rightarrow PbO_2(s) + 2H_2O + 2e^-$

$Pb(OH)_3^- + ClO^- + H_2O + OH^- \rightarrow PbO_2(s) + Cl^- + 2H_2O + 2OH^-$

Eliminating one H_2O and one OH from each side simplifies this to the final (ultimate) equation:

$$Pb(OH)_3^- + ClO^- \rightarrow PbO_2(s) + Cl^- + H_2O + OH^-$$

21.77 a. $Al_2O_3(s) + 3H_2SO_4(aq) \rightarrow Al_2(SO_4)_3(aq) + 3H_2O(l)$

b. $Al(s) + 3AgNO_3(aq) \rightarrow 3Ag(s) + Al(NO_3)_3(aq)$

c. $Pb(NO_3)_2(aq) + 2NaI(aq) \rightarrow PbI_2(s) + 2NaNO_3(aq)$

d. $8Al(s) + 3Mn_3O_4(s) \rightarrow 9Mn(s) + 4Al_2O_3(s)$

e. $2Ga(OH)_3(s) \rightarrow Ga_2O_3(s) + 3H_2O(g)$

21.78 a. $3Pb(NO_3)_2(aq) + 2Al(s) \rightarrow 3Pb(s) + 2Al(NO_3)_3(aq)$

b. $Pb(NO_3)_2(aq) + Na_2CrO_4(aq) \rightarrow PbCrO_4(s) + 2NaNO_3(aq)$

c. $Al_2(SO_4)_3(aq) + 6LiOH(aq) \rightarrow 2Al(OH)_3(s) + 3Li_2SO_4(aq)$

d. $2Al(s) + 6HCl(aq) \rightarrow 3H_2(g) + 2AlCl_3(aq)$

e. $Sn(s) + 2HBr(aq) \rightarrow H_2(g) + SnBr_2(aq)$

■ Solutions to Unclassified Problems

21.79 The equation is $6Fe^{2+} + K_2Cr_2O_7 + 14H^+ \rightarrow 6Fe^{3+} + 2Cr^{3+} + 7H_2O + 2K^+$. Using a 6:1 ratio of Fe to $K_2Cr_2O_7$ (= dich.), calculate the mass percentage of iron in the sample:

$$0.04114 \text{ L dich.} \times \frac{0.01618 \text{ mol dich.}}{\text{L dich.}} \times \frac{6 \text{ Fe}}{1 \text{ dich.}} \times \frac{55.85 \text{ g Fe}}{1 \text{ mol Fe}}$$

$$= 0.223\underline{0}57 \text{ g Fe}$$

$$(0.223057 \text{ g Fe} \div 0.2886 \text{ g}) \times 100\% = 77.2\underline{8}9 = 77.29\% \text{ Fe}$$

21.80 The equation is $6Fe^{2+} + K_2Cr_2O_7 + 14H^+ \rightarrow 6Fe^{3+} + 2Cr^{3+} + 7H_2O + 2K^+$. Using a 6:1 ratio of Fe to $K_2Cr_2O_7$ (= dich.), calculate the mass percentage of iron in the sample:

$$0.04548 \text{ L dich.} \times \frac{0.01599 \text{ mol dich.}}{\text{L dich.}} \times \frac{6 \text{ Fe}}{1 \text{ dich.}} \times \frac{55.85 \text{ g Fe}}{1 \text{ mol Fe}}$$

$$= 0.243\underline{6}9 \text{ g Fe}$$

$$(0.24369 \text{ g Fe} \div 0.4834 \text{ g}) \times 100\% = 50.4\underline{1}2 = 50.41\% \text{ Fe}$$

21.81 The equations are $2HCl + Mg(OH)_2(s) \rightarrow MgCl_2 + 2H_2O$ and $HCl + NaOH \rightarrow NaCl + H_2O$. Using a 1:1 ratio of HCl to NaOH, calculate the mol of HCl reacting with NaOH:

$$\left[\frac{0.4987 \text{ mol HCl}}{\text{L}} \times 0.05000 \text{ L}\right] - \left[\frac{0.2456 \text{ mol NaOH}}{\text{L}} \times 0.03923 \text{ L}\right]$$

$$= 0.0153\underline{0}01 \text{ mol HCl}$$

Using a 2:1 ratio of HCl to $Mg(OH)_2$, calculate the mass percentage of $Mg(OH)_2$:

$$0.0153001 \text{ mol HCl} \times \frac{1 \text{ Mg(OH)}_2}{2 \text{ HCl}} \times \frac{58.33 \text{ g Mg(OH)}_2}{1 \text{ mol Mg(OH)}_2} \times \frac{1}{5.436 \text{ g}} \times 100\%$$

$$= 8.209\%$$

21.82 The equations are $2HCl + CaCO_3(s) \rightarrow CaCl_2 + H_2O + CO_2$ and $HCl + NaOH \rightarrow NaCl + H_2O$. Using a 1:1 ratio of HCl to NaOH, calculate the mol of HCl reacting with NaOH:

$$\left[\frac{0.5068 \text{ mol HCl}}{\text{L}} \times 0.05000 \text{ L}\right] - \left[\frac{0.2601 \text{ mol NaOH}}{\text{L}} \times 0.04123 \text{ L}\right]$$

$$= 0.0146\underline{1}6 \text{ mol HCl}$$

Using a 2:1 ratio of HCl to $CaCO_3$, calculate the mass percentage of $CaCO_3$:

$$0.014616 \text{ mol HCl} \times \frac{1 \text{ CaCO}_3}{2 \text{ HCl}} \times \frac{100.1 \text{ g CaCO}_3}{1 \text{ mol CaCO}_3} \times \frac{1}{0.9863 \text{ g}} \times 100\%$$

$$= 74.17\%$$

21.83 The equation with $\Delta H^o{}_f$'s recorded beneath each substance is

$$\begin{array}{lllllll} Fe_2O_3(s) & + & 2Al(s & \rightarrow & 2Fe(s) & + & Al_2O_3(s) \\ -825.5 & & 0 & & 0 & & -1676 \ (kJ) \end{array}$$

$\Delta H^o = -1676 - (-825.5) = -85\underline{0}.5 = -851$ kJ/mol Fe_2O_3, or -425 kJ mol/Fe

21.84 The equation with $\Delta H^o{}_f$'s recorded beneath each substance is

$$\begin{array}{lllllll} 3CaO(s) & + & 2Al(s) & \rightarrow & 3Ca(s) & + & Al_2O_3(s) \\ 3(-635.1) & & 0 & & 0 & & -1676 \ (kJ) \end{array}$$

$\Delta H^o = -1676 - 3(-635.1) = 22\underline{9}.3 = 229$ kJ/3 mol Ca, or 76.4 kJ mol/Ca

21.85 The equation is $CaCO_3(s) + 2HCl \rightarrow CO_2(g) + CaCl_2 + H_2O(l)$. Using a 1:1 ratio of CO_2 to $CaCO_3$, calculate the mol of CO_2:

$$\frac{(745/760 \text{ atm})(0.03456 \text{ L})}{(294 \text{ K})(0.0821 \text{ L}\bullet\text{atm/K}\bullet\text{mol})} = 0.0014\underline{0}35 \text{ mol } CO_2$$

Finally, convert mol CO_2 to mass percentage of $CaCO_3$:

$$\frac{0.0014035 \text{ mol } CO_2 \times \frac{1 \ CaCO_3}{1 \ CO_2} \times \frac{100.1 \text{ g } CaCO_3}{1 \text{ mol } CaCO_3}}{0.1662 \text{ g}} \times 100\% = 84.\underline{5}3$$

$$= 84.5\% \ CaCO_3$$

21.86 The equation is $MgCO_3(s) + 2HCl \rightarrow CO_2(g) + MgCl_2 + H_2O(l)$. Using a 1:1 ratio of CO_2 to $MgCO_3$, calculate the mol of CO_2:

$$\frac{(758/760 \text{ atm})(0.03771 \text{ L})}{(295 \text{ K})(0.0821 \text{ L}\bullet\text{atm/K}\bullet\text{mol})} = 0.0015\underline{5}29 \text{ mol } CO_2$$

Finally, convert mol CO_2 to mass percentage of $MgCO_3$:

$$\frac{0.0015529 \text{ mol } CO_2 \times \frac{1 \ MgCO_3}{1 \ CO_2} \times \frac{84.3 \text{ g } MgCO_3}{1 \text{ mol } MgCO_3}}{0.1504 \text{ g}} \times 100\% = 87.\underline{0}4$$

$$= 87.0\% \ MgCO_3$$

21.87 The equation is $NaCl + NH_3 + H_2O + CO_2 \rightarrow NaHCO_3 + NH_4Cl$. Using a 1:1 ratio of $NaCl$ to $NaHCO_3$, calculate the mass of NaCl in grams as follows:

$$10.00 \text{ g NaHCO}_3 \times \frac{1 \text{ mol NaHCO}_3}{84.00 \text{ g NaHCO}_3} \times \frac{1 \text{ NaCl}}{1 \text{ NaHCO}_3} \times \frac{58.44 \text{ g NaCl}}{1 \text{ mol NaCl}} = 6.95\underline{7}1$$

$$= 6.957 \text{ g NaCl}$$

21.88 The equation is $Li_2CO_3 + Ca(OH)_2 \rightarrow 2LiOH + CaCO_3$. Using a 1:2 ratio of Li_2CO_3 to LiOH, calculate the mass of Li_2CO_3 in grams as follows:

$$10.00 \text{ g LiOH} \times \frac{1 \text{ mol LiOH}}{23.95 \text{ g LiOH}} \times \frac{1 \text{ Li}_2\text{CO}_3}{2 \text{ LiOH}} \times \frac{73.89 \text{ g Li}_2\text{CO}_3}{1 \text{ mol Li}_2\text{CO}_3} = 15.4\underline{2}58$$

$$= 15.43 \text{ g Li}_2\text{CO}_3$$

21.89 The equation is $Cr_2O_3 + 2Al \rightarrow 2Cr + Al_2O_3$. Using a 1:2 ratio of Cr_2O_3 to Al, calculate the mass of Al in grams as follows:

$$10.00 \text{ g Cr}_2\text{O}_3 \times \frac{1 \text{ mol Cr}_2\text{O}_3}{152.0 \text{ g Cr}_2\text{O}_3} \times \frac{2 \text{ Al}}{1 \text{ Cr}_2\text{O}_3} \times \frac{26.98 \text{ g Al}}{1 \text{ mol Al}} = 3.55\underline{0}0$$

$$= 3.550 \text{ g Al}$$

21.90 The equation is $3MnO_2 + 4Al \rightarrow 3Mn + 2Al_2O_3$. Using a 3:4 ratio of MnO_2 to Al, calculate the mass of Al in grams as follows:

$$10.00 \text{ g MnO}_2 \times \frac{1 \text{ mol MnO}_2}{86.94 \text{ g MnO}_2} \times \frac{4 \text{ Al}}{3 \text{ MnO}_2} \times \frac{26.98 \text{ g Al}}{1 \text{ mol Al}} = 4.13\underline{7}7 = 4.138 \text{ g Al}$$

21.91

	$SrCO_3(s)$	$\rightarrow$ $SrO(s)$	+ $CO_2(g)$	
ΔH_f°:	-1218	-592.0	-393.5	kJ
S°:	97.1	55.5	13.7	J/K

$$\Delta H^\circ = [-592.0 - 393.5 - (-1218)] \text{ kJ} = 23\underline{2}.5 \text{ kJ}$$

$$\Delta S^\circ = [55.5 + 213.7 - 97.1] \text{ J/K} = 172.1 \text{ J/K } (172.1 \times 10^{-3} \text{ kJ/K})$$

$$\Delta G^\circ = \Delta H^\circ - T\Delta S^\circ$$

(continued)

$$0 = 232.5\ \text{kJ} - T(172.1 \times 10^{-3}\text{kJ/K})$$

$$T = \frac{232.5}{172.1 \times 10^{-3}}\ \text{K} = 13\underline{5}0.9 = 1.35 \times 10^{3}\ \text{K}$$

21.92

	$BaCO_3(s)$ $\rightarrow$	$BaO(s)$ +	$CO_2(g)$	
ΔH°_f:	-1219	-548.1	-393.5	kJ
S°:	112	72.07	213.7	J/K

$$\Delta H^\circ = [-548.1 - 393.5 - (-1219)]\ \text{kJ} = 27\underline{7}.4\ \text{kJ}$$

$$\Delta S^\circ = [72.07 + 213.7 - 112]\ \text{J/K} = 17\underline{3}.77\ \text{J/K}\ (0.17\underline{3}77\ \text{kJ/K})$$

$$\Delta G^\circ = \Delta H^\circ - T\Delta S^\circ$$

$$0 = 277.4\ \text{kJ} - T(0.17377\ \text{kJ/K})$$

$$T = \frac{277.4}{0.17377}\ \text{K} = 15\underline{9}6 = 1.60 \times 10^{3}\ \text{K}$$

21.93 The overall disproportionation reaction can be considered as the sum of the following reactions:

$$2e^- + 2In^+(aq) \rightarrow 2In(s)$$
$$In^+(aq) \rightarrow In^{3+}(aq) + 2e^-$$

$$3In^+(aq) \rightarrow 2In(s) + In^{3+}$$

$$E^\circ_{cell} = E^\circ_{cathode} - E^\circ_{anode} = [-0.21 - (-0.40)]\ \text{V} = 0.19\ \text{V}$$

$$\Delta G^\circ = -nFE^\circ_{cell} = -(2)\ (9.65 \times 10^4\ \text{C})\ (0.19\ \text{J/C}) = -3.\underline{6}67 \times 10^4\ \text{J}\ (-37\ \text{kJ})$$

ΔG° for the reaction as written is negative, so the disproportionation does occur spontaneously.

21.94 The overall disproportionation can be considered as the sum of the following reactions:

$$2Tl^+(aq) + 2e^- \rightarrow 2Tl(s)$$
$$Tl^+(aq) \rightarrow Tl^{3+} + 2e^-$$
$$3Tl^+(aq) \rightarrow 2Tl(s) + Tl^{3+}(aq)$$

$$E^\circ_{cell} = E^\circ_{cathode} - E^\circ_{anode} = [-0.34 - (1.25)]\text{ V} = -1.59\text{ V}$$

$$\Delta G^\circ = -nFE^\circ_{cell} = -(2)(9.65 \times 10^4\text{ C})(-1.59\text{ J/C}) = +3.0\underline{6}87 \times 10^5\text{ J (307 kJ)}$$

ΔG° is positive for the reaction, so the disproportionation does not occur spontaneously.

21.95 The body-centered cubic cell contains 2 atoms of Na (1 atom in the center and 1/8 atom at each of eight corners). Let a be the cell dimension of the cubic cell. The cell diagonal has length $\sqrt{3}$ a. If the spheres touch along the diagonal, then the length of the diagonal is 4 times the radius of the spheres.

$$4r = \sqrt{3}\,a, \text{ or } a = \frac{4r}{\sqrt{3}}$$

Find the volume occupied by 2 atoms of Na.

$$2\text{ atoms Na} \times \frac{1\text{ mol Na}}{6.02 \times 10^{23}\text{ atoms Na}} \times \frac{22.99\text{ g Na}}{1\text{ mol Na}} \times \frac{1\text{ cm}^3}{0.97\text{ g Na}} = 7.\underline{8}7 \times 10^{-23}\text{ cm}^3$$

This volume is equal to the volume of the cubic cell, a^3.

$$a = \left(\frac{4r}{\sqrt{3}}\right)^3 = 7.87 \times 10^{-23}\text{ cm}$$

Solve for r.

$$\frac{4r}{\sqrt{3}} = \sqrt[3]{7.87 \times 10^{-23}\text{ cm}^3} = 4.\underline{2}9 \times 10^{-8}\text{ cm}$$

$$r = \frac{4.29 \times 10^{-8}(\sqrt{3})}{4}\text{cm} = 1.\underline{8}57 \times 10^{-8} = 1.9 \times 10^{-8}\text{ cm } (1.9 \times 10^2\text{ pm})$$

21.96 The body-centered cubic lattice contains 2 atoms of Li. If a is the length of the side of the unit cell, then the cell diagonal has length $\sqrt{3}$ a. If the spheres are touching along the diagonal, then the length of the diagonal is 4 times the radius of the spheres.

$$4r = \sqrt{3}a, \text{ or } a = \frac{4r}{\sqrt{3}}$$

Find the volume occupied by 2 atoms of Li.

$$2 \text{ atoms Li} \times \frac{1 \text{ mol Li}}{6.02 \times 10^{23} \text{ atoms}} \times \frac{6.94 \text{ g Li}}{1 \text{ mol Li}} \times \frac{1 \text{ cm}^3}{0.53 \text{ g Li}} = 4.\underline{3}5 \times 10^{-23} \text{ cm}^3$$

This volume is equal to the volume of the cubic cell, a^3.

$$a^3 = \left(\frac{4r}{\sqrt{3}}\right)^3 = 4.\underline{3}5 \times 10^{-23} \text{ cm}^3$$

Solve for r.

$$r = \left[\sqrt[3]{4.35 \times 10^{-23} \text{ cm}^3}\right]\left(\frac{\sqrt{3}}{4}\right) = 1.\underline{5}2 \times 10^{-8} = 1.5 \times 10^{-8} \text{ cm } (1.5 \times 10^2 \text{ pm})$$

21.97 $Mg^{2+} + Ca(OH)_2 \rightarrow Mg(OH)_2 + Ca^{2+}$

1 mol Mg^{2+} requires 1 mol $Ca(OH)_2$, so

$$1272 \text{ g } Mg^{2+} \times \frac{1 \text{ mol } Mg^{2+}}{24.305 \text{ g } Mg^{2+}} \times \frac{1 \text{ mol } Ca(OH)_2}{1 \text{ mol } Mg^{2+}} \times \frac{74.10 \text{ g } Ca(OH)_2}{1 \text{ mol } Ca(OH)_2} = 387\underline{8}.02$$

$$= 3878 \text{ g}$$

At least 3878 g of $Ca(OH)_2$ would be required to precipitate the Mg^{2+}.

21.98 1 mol Al_2O_3 ≡ 101.96 g Al_2O_3

1 mol Al_2O_3 2 mol Al ≡ 53.96 g Al

Therefore, 101.96 g Al_2O_3 ≡ 53.96 g Al.

$$1.00 \times 10^6 \text{ g bauxite} \times \frac{52 \text{ g } Al_2O_3}{100.0 \text{ g bauxite}} \times \frac{53.96 \text{ g Al}}{101.96 \text{ g } Al_2O_3} = 2.\underline{7}52 \times 10^5$$

$$= 2.8 \times 10^5 \text{ g Al}$$

2.8×10^5 g Al can be obtained from 1.00 metric ton of bauxite.

21.99 $2LiOH(s) + CO_2(g) \rightarrow Li_2CO_3(s) + H_2O(g)$

Using the ideal gas law, find the amount of CO_2.

$$n = \frac{PV}{RT} = \frac{(30.0\ \text{mmHg})\ (1\ \text{atm}/760\ \text{mmHg})\ (1.00\ \text{L})}{0.08206\ \text{L} \cdot \text{atm/K} \cdot \text{mol}\ (298\ \text{K})} = 1.6\underline{1}4 \times 10^{-3}\ \text{mol}$$

From the equation above,

$1\ \text{mol}\ CO_2 \equiv 2\ \text{mol LiOH}$

$1\ \text{mol LiOH} \equiv 23.95\ \text{g LiOH}$

So

$$1.614 \times 10^{-3}\ \text{mol}\ CO_2 \times \frac{2\ \text{mol LiOH}}{1\ \text{mol}\ CO_2} \times \frac{23.95\ \text{g LiOH}}{1\ \text{mol LiOH}} = 7.7\underline{3}1 \times 10^{-2}$$

$$= 7.73 \times 10^{-2}\ \text{g LiOH}$$

21.100 $6KOH(aq) + 3Cl_2(g) \rightarrow KClO_3(s) + 5KCl(aq) + 3H_2O(l)$

$3\ \text{mol}\ Cl_2 \equiv 1\ \text{mol}\ KClO_3$

$1\ \text{mol}\ KClO_3 \equiv 122.55\ \text{g}\ KClO_3$

Using the ideal gas law, find the amount of Cl_2.

$$n = \frac{PV}{RT} = \frac{(784\ \text{mmHg})\left(\frac{1\ \text{atm}}{760\ \text{mmHg}}\right)(156\ \text{L}\ Cl_2)}{\left(0.08206\ \frac{\text{L} \cdot \text{atm}}{\text{K} \cdot \text{mol}}\right)(298\ \text{K})} = 6.5\underline{8}08\ \text{mol}\ Cl_2$$

$$6.5808\ \text{mol}\ Cl_2 \times \frac{1\ \text{mol}\ KClO_3}{3\ \text{mol}\ Cl_2} \times \frac{122.55\ \text{g}\ KClO_3}{1\ \text{mol}\ KClO_3} = 26\underline{8}.8 = 269\ \text{g}\ KClO_3$$

22. CHEMISTRY OF THE NONMETALS

Answers to Review Questions

22.1 Silicon forms SiF_6^{2-} because it can utilize sp^3d^2 hybridization, but carbon cannot.

22.2 Catenation is the ability of an atom to bond covalently to like atoms, as in ethylene, $H_2C{=}CH_2$.

22.3 In the diamond allotrope, each carbon atom is tetrahedrally covalently bonded to four other carbon atoms. Moving one plane of atoms in the diamond crystal relative to another requires the breaking of many strong carbon-carbon bonds, yielding one of the hardest substances known. In the graphite allotrope, each layer consists of carbon atoms bonded to three other carbon atoms to give a hexagonal pattern of carbon atoms in a plane. One layer of carbon atoms in graphite is held to another layer only by van der Waals forces, yielding a substance that is soft and slippery. The buckminsterfullerene allotrope has a stable "soccer-ball" structure, yielding an allotrope that is different from the other two.

22.4 Carbon black is used in rubber tires and in black inks. Graphite is used in pencil "leads" and in electrodes. Carbon fibers are used in textiles such as rayon. Diamonds are used as jewels and as industrial abrasives.

22.5 According to Le Chatelier's principle, high pressure would be expected to transform graphite to diamond (which is more dense) because this would reduce the volume and relieve the pressure.

22.6 Carbon monoxide attaches to the iron in the hemoglobin of red blood cells and blocks the combination of hemoglobin with oxygen molecules normally carried by the hemoglobin.

22.7 $CO_2(g) + H_2O(l) \rightleftharpoons H_2CO_3(aq)$

$H_2CO_3(aq) \rightleftharpoons H^+(aq) + HCO_3^-(aq)$

$HCO_3^-(aq) \rightleftharpoons H^+(aq) + CO_3^{2-}(aq)$

22.8 To prepare ultrapure silicon, quartz sand (SiO_2) is reduced with coke at 3000°C to silicon. Then the impure silicon is converted to $SiCl_4$, which is purified before being reduced back to pure silicon.

22.9 The quartz crystal used is cut to the precise dimension that will respond most strongly to the nominal frequency of the alternating current. When the alternating current frequency deviates from its nominal frequency (and thus from the frequency of the quartz crystal), a feedback mechanism adjusts the alternating current frequency back to its nominal frequency.

22.10
$$\begin{array}{c}OH\\|\\HO-Si-OH\\|\\OH\end{array} + \begin{array}{c}OH\\|\\HO-Si-OH\\|\\OH\end{array} \rightarrow \begin{array}{c}OH\quad\ OH\\|\qquad\ |\\HO-Si-O-Si-OH\\|\qquad\ |\\OH\quad\ OH\end{array} + H_2O$$

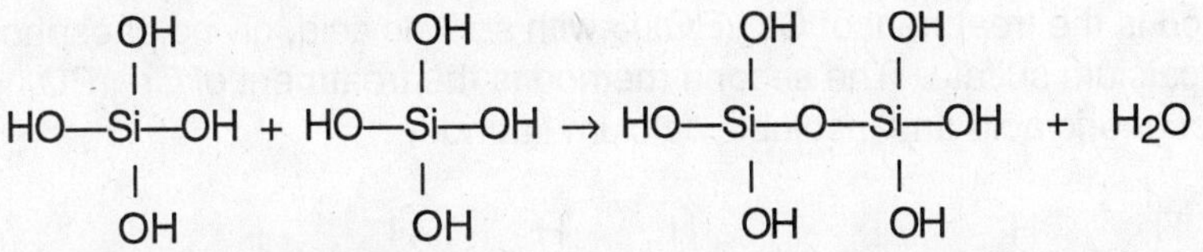

22.11 Spodumene, $LiAl(SiO_3)_2$, is a long-chain silicate mineral. Such a mineral consists of silicate chains in which SiO_4 tetrahedra are linked to other SiO_4 tetrahedra.

22.12 Polymers are materials of very high molecular weight that are made by linking together repeating units of low molecular weight. In most polymers, the units are linked by bonds such as C—C—C, C—O—C, etc. In a silicone polymer, the units are linked by Si—O—Si bonds with hydrocarbon groups such as —CH_3 attached to the silicon atoms.

22.13 Certain bacteria in the soil and in the roots of plants convert N_2 to ammonium and nitrate compounds. The plants then use these nitrogen compounds to make proteins and other complex nitrogen compounds. Animals eat the plants. Ultimately the animals die and bacteria in the decaying organic matter convert the nitrogen compounds back to N_2.

22.14 Rutherford removed oxygen from the air by burning a substance that would combine with the oxygen. He also removed any carbon dioxide formed by the burning by reacting it with KOH. The gas left contained primarily nitrogen, as well as small amounts (< 1%) of noble gases.

22.15 These oxides are nitrous oxide, N_2O, nitric oxide, NO, N_2O_3, nitrogen dioxide, NO_2, dinitrogen tetroxide, N_2O_4, and N_2O_5. The oxidation numbers in each are +1, +2, +3, +4, +4, and +5, respectively.

22.16 Natural gas, or CH_4, plus steam reacts to form CO and H_2. The CO is reacted with steam to form CO_2 plus additional H_2. The CO_2 is removed from the H_2 by dissolving it in liquid water; the H_2 is then reacted with N_2 in the Haber process to form NH_3.

22.17 Ammonia is burned in the presence of a platinum catalyst to form NO, which is then reacted with O_2 to form NO_2. The NO_2 is dissolved in water to form HNO_3 and NO. The NO is recycled back to the second step to react with more O_2 to form more NO_2.

22.18 White phosphorus is a molecular solid with the formula P_4. The phosphorus atoms are arranged at the corners of a regular tetrahedron with a 60° P—P—P bond angle, an angle considerably smaller than the normal tetrahedral bond angle. This accounts for the weakness of the P—P bonds and their high reactivity, as stronger bonds such as P—O can replace them.

22.19 $P_4(s) + 5O_2(g) \rightarrow P_4O_{10}(s)$

$P_4O_{10}(s) + 6H_2O(l) \rightarrow 4H_3PO_4(aq)$

22.20 The first method is the treatment of $Ca_3(PO_4)_2$ with sulfuric acid, giving phosphoric acid and insoluble calcium sulfate. The second method is the treatment of $Ca_3(PO_4)_2$ with HF, giving phosphoric acid and insoluble calcium fluoride.

22.21
$$H_3PO_4 + \begin{matrix} H & H \\ O & O \\ | & | \\ HO-P-O-P-OH \\ | & | \\ O & O \end{matrix} \rightarrow \begin{matrix} H & H & H \\ O & O & O \\ | & | & | \\ HO-P-O-P-O-P-OH \\ | & | & | \\ O & O & O \end{matrix} + H_2O$$

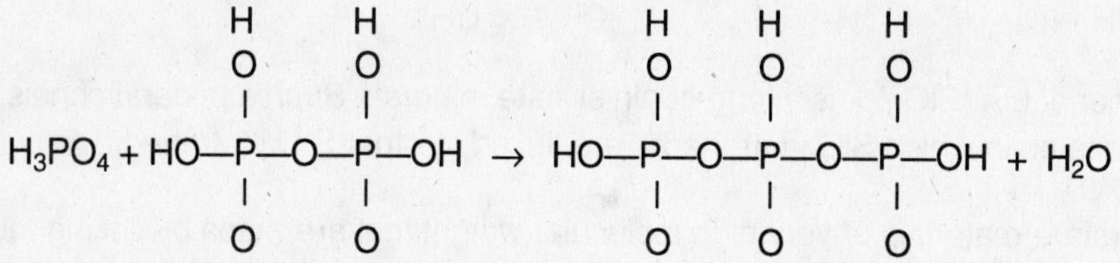

22.22 Polyphosphates are added to detergents to form complexes with metal ions and thus prevent their precipitation onto clothes.

22.23 Oxygen is a very electronegative element, and its bonding involves only the s and p orbitals, in contrast to bonding using the d orbitals in sulfur, etc. Molecular oxygen is a reactive gas but forms mainly compounds in which its oxidation state is -2, compared to compounds of sulfur, etc., which exhibit positive oxidation states as well as the -2 state.

22.24 Priestley prepared oxygen by heating mercury(II) oxide:

$2HgO(s) \xrightarrow{\Delta} 2Hg(l) + O_2(g)$

22.25 The most important commercial means of producing oxygen is by distillation of liquid air. Air is filtered from dust particles, cooled to freeze out water and carbon dioxide, liquefied, and finally warmed until nitrogen and argon distill, leaving liquid oxygen behind.

22.26 Oxides are binary oxygen compounds where oxygen is in the -2 oxidation state, whereas for peroxides the oxidation number of oxygen is -1 and the anion is O_2^{2-}. In the superoxides, the oxidation number of oxygen is -1/2 and the anion is O_2^-. An example of each is H_2O (oxide), H_2O_2 (peroxide), and KO_2 (superoxide).

22.27 CrO_3 is an example of an acidic oxide; Cr_2O_3, MgO, and Fe_3O_4 are basic oxides.

22.28 Three natural sources of sulfur or sulfur compounds are sulfate minerals, sulfide minerals, and coal or petroleum products.

22.29 Rhombic sulfur is a yellow crystalline solid with a lattice consisting of crown-shaped S_8 molecules; that is, eight sulfur atoms are arranged in a crown-shaped ring.

22.30 The monoclinic sulfur allotrope can be prepared from rhombic sulfur by first melting the rhombic sulfur and then cooling it to crystals of monoclinic sulfur. A liquid sulfur allotrope of long spiral chains of sulfur atoms can be prepared by heating rhombic sulfur above 160°C but keeping it below 200°C. Plastic sulfur allotrope can be prepared by pouring this liquid sulfur allotrope into water. Finally, gaseous allotropes of S_8, S_6, S_4, and S_2 molecules can be formed by boiling sulfur at 445°C.

22.31 The Frasch process involves melting sulfur deposits with superheated water, using air to force the melted sulfur upward to the surface, and cooling it to form solid sulfur.

22.32 The initial burning of hydrogen sulfide produces some sulfur as well as sulfur dioxide:

$$8H_2S(g) + 4O_2(g) \rightarrow S_8(s) + 8H_2O(g)$$

$$2H_2S(g) + 3O_2(g) \rightarrow 2SO_2(g) + 2H_2O(g)$$

The sulfur dioxide reacts with hydrogen sulfide to form more sulfur:

$$16H_2S(g) + 8SO_2(g) \rightarrow 3S_8(s) + 16H_2O(g)$$

22.33 a. $2HCl(aq) + ZnS(s) \rightarrow ZnCl_2(aq) + H_2S(g)$

b. $S_8(s) + 8O_2(g) \rightarrow 8SO_2(g)$

22.34 First step: $S_8(s) + 8O_2(g) \rightarrow 8SO_2(g)$

Second step: $2SO_2(g) + O_2(g) \rightarrow 2SO_3(g)$

Third step: $SO_3(g) + H_2O(l) \rightarrow H_2SO_4(aq)$

22.35

H_2S	Reducing agent	$S_8(s)$ is usual product.
$SO_2(g)$	Reducing agent	SO_4^{2-} or sulfuric acid is usual product.
H_2SO_4	Oxidizing agent	SO_2 is usual product.

22.36 a. $16\,H_2S(g) + 8SO_2(g) \rightarrow 3S_8(s) + 16H_2O(g)$

b. $Cr_2O_7^{2-}(aq) + 3SO_2(g) + 2H^+(aq) \rightarrow 2Cr^{3+}(aq) + 3SO_4^{2-}(aq) + H_2O(l)$

c. $Cu(s) + 2H_2SO_4(l) \rightarrow CuSO_4(aq) + 2H_2O(l) + SO_2(g)$

d. $8Na_2SO_3(aq) + S_8(s) \rightarrow 8Na_2S_2O_3(aq)$

22.37 $4HCl + MnO_2(s) \rightarrow MnCl_2(aq) + Cl_2(g) + 2H_2O(l)$

22.38 a. $I_2(aq) + Cl^-(aq) \rightarrow$ NR

b. $Cl_2(aq) + 2Br^-(aq) \rightarrow Br_2(aq) + 2Cl^-(aq)$

c. $Br_2(aq) + 2I^-(aq) \rightarrow I_2(s) + 2Br^-(aq)$

d. $Br_2(aq) + Cl^-(aq) \rightarrow$ NR

22.39 Add chlorine water and methylene chloride. For NaCl, there will be no reaction; for NaBr, an orange organic layer will appear; and for NaI, there will be a violet organic layer.

22.40 Chlorine is used in preparing chlorinated hydrocarbons, as a bleaching agent, and as a disinfectant.

22.41 HBr cannot be prepared by adding sulfuric acid to NaBr because the hot concentrated acid will oxidize the bromide ion to bromine.

22.42 An aqueous solution of sodium hypochlorite should be basic because HClO is a weak acid. A solution of sodium perchlorate should be neutral because $HClO_4$ is a strong acid and NaOH is a strong base.

22.43 Sodium hypochlorite is prepared by reaction of chlorine with NaOH:

$$Cl_2(g) + 2NaOH(aq) \rightarrow NaClO(aq) + NaCl(aq) + H_2O(l)$$

22.44 Industrially, sodium chloride is electrolyzed to form chlorine gas. The chlorine gas is then heated with sodium hydroxide solution, forming $NaClO_3$ (and NaCl). The ClO_3^- is electrolyzed at the anode to form the ClO_4^- anion. This is mixed with sulfuric acid, and the $HClO_4$ is distilled at reduced pressure (below 92°C) to isolate $HClO_4$.

22.45 Ramsay and Rayleigh passed a high-voltage electrical discharge through what they thought was a tube of pure nitrogen. They observed a series of red and green emission lines and concluded they had discovered another element in the tube, which they called argon.

22.46 Bartlett found that PtF_6 reacted with molecular oxygen. Because the first ionization energy of xenon was slightly less than that of molecular oxygen, he reasoned that PtF_6 ought to react with xenon also.

■ Solutions to Practice Problems

Note on significant figures: The final answer to each problem is given first with one nonsignificant figure (the rightmost significant figure is underlined) and then is rounded to the correct number of figures. Intermediate answers usually also have at least one nonsignificant figure.

22.47 a. Carbon has four valence electrons. These are directed tetrahedrally, and the orbitals should be sp^3 hybrid orbitals. Each C—H bond is formed by the overlap of a 1s orbital of a hydrogen atom with one of the singly occupied sp^3 hybrid orbitals of the carbon atom.

b. Silicon has four valence electrons. (Two more electrons are added for the -2 charge.) These are directed octahedrally, and the orbitals should be sp^3d^2 hybrid orbitals. Each Si—F bond is formed by the overlap of a 2p orbital of a fluorine atom with one of the singly occupied sp^3d^2 hybrid orbitals of the silicon atom.

c. Carbon has four valence electrons. The single C—C bond is a sigma bond formed by the overlap of the sp^2 orbital of the middle carbon and the sp^3 orbital of the CH_3 carbon. The other three orbitals of the CH carbon are used to form the C—H bonds by overlapping with the orbital of the hydrogen atom. The C=C double bond is a sigma bond formed by the overlap of the sp^2 hybrid orbitals of those carbon atoms, and a pi bond formed by the overlap of the unhybridized p orbitals.

d. Silicon has four valence electrons. These are directed tetrahedrally, and the orbitals should be sp^3 hybrid orbitals. Each Si—H bond is formed by the overlap of a 1s orbital of a hydrogen atom with one of the singly occupied sp^3 hybrid orbitals of the silicon atom.

22.48 a. Carbon has four valence electrons. These are directed tetrahedrally, and the orbitals should be sp^3 hybrid orbitals. Each C—Cl bond is formed by the overlap of a 3p orbital of a chlorine atom with one of the singly occupied sp^3 hybrid orbitals of the carbon atom.

b. Carbon has four valence electrons, and nitrogen has five valence electrons. In the carbon-nitrogen triple bond, these are linear, and the orbitals should be sp hybrid orbitals. The C—H bond is formed by the overlap of a 1s orbital of a hydrogen atom with one of the singly occupied sp hybrid orbitals of the carbon atom.

c. Silicon has four valence electrons. These are directed tetrahedrally, and the orbitals should be sp^3 hybrid orbitals. Each Si—F bond is formed by the overlap of a 2p orbital of a fluorine atom with one of the singly occupied sp^3 hybrid orbitals of the silicon atom.

d. Carbon has four valence electrons. In the CH_3 group, these are directed tetrahedrally, and the orbitals should be sp^3 hybrid orbitals. Each C—H bond is formed by the overlap of a 1s orbital of a hydrogen atom with one of the singly occupied sp^3 hybrid orbitals of the carbon atom. The C—C bond between the CH_3 and the COOH groups is formed by the overlap of the sp^2 orbital of the middle carbon and the sp^3 orbital of the CH_3 carbon. In the COOH group, there is a C=O double bond and a C—O single bond. The C=O double bond is a sigma bond formed by the overlap of the sp^2 hybrid orbitals, and a pi bond formed by the overlap of the unhybridized p orbitalsof the C and O atoms. The C—O bond is a sigma bond formed by the overlap of the C sp^2 hybrid orbital and the O sp^3 hybrid orbital. The O—H bond is similar to the C—H bond.

22.49 a. The equation with ΔH^o_f's recorded beneath each substance is

$$CH_4(g) \rightarrow C(graphite) + 2H_2(g)$$

-74.87 0 0 (kJ)

$\Delta H^o = 0 - (-74.87) = 74.87$ kJ

b. The equation with ΔH^o_f's recorded beneath each substance is

$$C_2H_6(g) \rightarrow C_2H_4(g) + H_2(g)$$

-84.667 52.47 0 (kJ)

$\Delta H^o = 0 + 52.47 - (-84.667) = 137.1\underline{37} = 137.14$ kJ

22.50 a. The equation with ΔH^o_f's recorded beneath each substance is

$$CO(g) + 2H_2(g) \rightarrow CH_3OH(g)$$

-110.5 0 -201.2 (kJ)

$\Delta H^o = -\ 201.2 + 0 - (-110.5) = -90.7$ kJ

b. The equation with ΔH^o_f's recorded beneath each substance is

$$CO(g) + 3H_2(g) \rightarrow CH_4(g) + H_2O(g)$$

-110.5 0 -74.87 -241.826 (kJ)

$\Delta H^o = -74.87 + -(241.826) - (-110.5 + 0) = -206.\underline{1}96 = -206.2$ kJ

22.51 a. $CO_2(g) + Ba(OH)_2(aq) \rightarrow BaCO_3(s) + H_2O(l)$

b. $MgCO_3(s) + 2HBr(aq) \rightarrow CO_2(g) + H_2O(l) + MgBr_2(aq)$

22.52 a. $NaHCO_3(aq) + HC_2H_3O_2(aq) \rightarrow CO_2(g) + H_2O(l) + NaC_2H_3O_2(aq)$

b. $Ca(HCO_3)_2(aq) + Ca(OH)_2(aq) \rightarrow 2CaCO_3(s) + 2H_2O(l)$

22.53 $C(s) + O_2(air) \rightarrow CO_2(g)$

$CO_2(g) + 2NaOH(aq) \rightarrow Na_2CO_3(aq) + H_2O(l)$

$Na_2CO_3(aq)$ + (evaporation) $\rightarrow Na_2CO_3(s)$

22.54 $C_2H_6(g) + 2H_2O(g) \xrightarrow{Ni} 2CO(g) + 5H_2(g)$

$CO(g) + 2H_2(g)$ - (catalyst) $\rightarrow CH_3OH(g)$

22.55 The equation is $Mg_3N_2(s) + 6H_2O(l) \rightarrow 3Mg(OH)_2(aq) + 2NH_3(g)$. Using a 1:2 ratio of Mg_3N_2 to NH_3, calculate the mass of NH_3 formed as follows:

$$5.00 \text{ g } Mg_3N_2 \times \frac{1 \text{ mol } Mg_3N_2}{100.915 \text{ g } Mg_3N_2} \times \frac{2\ NH_3}{1 Mg_3N_2} \times \frac{17.03 \text{ g } NH_3}{1 \text{ mol } NH_3} = 1.6\underline{8}7$$

$$= 1.69 \text{ g } NH_3$$

22.56 The equation is $4NH_3(g) + 5O_2(g) \rightarrow 4NO(g) + 6H_2O(l)$. Using a 5:4 ratio of O_2 to NO, calculate the mass of O_2 required as follows:

$$5.00 \text{ g NO} \times \frac{1 \text{ mol NO}}{30.0 \text{ g NO}} \times \frac{5O_2}{4NO} \times \frac{32.0 \text{ g } O_2}{1 \text{ mol } O_2} = 6.6\underline{6}6 = 6.67 \text{ g } O_2$$

22.57 Prepare HNO_3 from NH_3:

$$4NH_3(g) + 5O_2(g) \xrightarrow{Pt} 4NO(g) + 6H_2O(g)$$

$$2NO(g) + O_2(g) \rightarrow 2NO_2(g)$$

$$3NO_2(g) + H_2O(l) \rightarrow 2HNO_3(aq) + NO(g)$$

(continued)

To prepare N_2O, use HNO_3 just prepared:

$$NH_3(g) + HNO_3(aq) \rightarrow NH_4NO_3(aq)$$

$$NH_4NO_3(s) \xrightarrow{\Delta} N_2O(g) + 2H_2O(g)$$

22.58 Prepare NH_3 and HNO_3 separately:

$$2NaOH(aq) + (NH_4)_2SO_4(aq) \rightarrow Na_2SO_4(aq) + 2NH_3(g) + H_2O(l)$$

$$H_2SO_4(l) + NaNO_3(s) \rightarrow NaHSO_4(s) + HNO_3(g)$$

React NH_3 and HNO_3 to give NH_4NO_3, which can be decomposed to N_2O:

$$NH_3(g) + HNO_3(aq) \rightarrow NH_4NO_3(aq)$$

$$NH_4NO_3(s) \xrightarrow{\Delta} N_2O(g) + 2H_2O(g)$$

22.59 The reduction of NO_3^- ion to NH_4^+ ion is an eight-electron reduction, and the oxidation of zinc to zinc(II) ion is a two-electron oxidation. Balancing the equation involves multiplying the zinc half-reaction by 4 to achieve an eight-electron oxidation. The final equation is

$$4Zn(s) + NO_3^-(aq) + 10H^+(aq) \rightarrow 4Zn^{2+}(aq) + NH_4^+(aq) + 3H_2O(l)$$

22.60 The reduction of NO_3^- ion to NO is a three-electron reduction, and the oxidation of silver to silver(I) ion is a one-electron oxidation. Balancing the equation involves multiplying the silver half-reaction by 3 to achieve a three-electron oxidation. The final equation is

$$3Ag(s) + NO_3^-(aq) + 4H^+(aq) \rightarrow 3Ag^+(aq) + NO(g) + 2H_2O(l)$$

22.61 PBr_4^+ has four pairs of electrons around the P atom, arranged in a tetrahedral fashion. The hybridization of P is sp^3. Each sp^3 hybrid orbital is used in the formation of a P—Br bond.

22.62 The PCl_6^- ion should have an octahedral geometry. Phosphorus has five valence electrons, which, together with the electron added for the charge, accounts for the sp^3d^2 hybridization of P. Each sp^3d^2 hybrid orbital is used in the formation of a P—Cl bond.

22.63 $H_3PO_3 + H_2SO_4 \rightarrow H_3PO_4 + SO_2 + H_2O$

22.64

$+3e^-$

$$\overset{+3}{H_3PO_3} + 2\,H\overset{+5}{N}O_3 \rightarrow H_3\overset{+5}{P}O_4 + 2\,\overset{+2}{N}O + H_2O$$

$-2e^-$

$$3H_3PO_3 + 2HNO_3 \rightarrow 3H_3PO_4 + 2NO + H_2O$$

22.65 In the equation, there is a 2:1 ratio of H_3PO_4 to $Ca_3(PO_4)_2$, which can be used to calculate the mass of H_3PO_4 from the mass of $Ca_3(PO_4)_2$. The mass of $Ca_3(PO_4)_2$ in the 10.0 g of rock is

$$0.746 \times 10.0 \text{ g rock} = 7.46 \text{ g } Ca_3(PO_4)_2 \text{ (= CaP)}$$

The mass of H_3PO_4 can be calculated from this mass as follows:

$$7.46 \text{ g CaP} \times \frac{1 \text{ mol CaP}}{310.2 \text{ g CaP}} \times \frac{2H_3PO_4}{1CaP} \times \frac{98.0 \text{ g } H_3PO_4}{1 \text{ mol } H_3PO_4} = 4.7\underline{1}3$$

$$= 4.71 \text{ g } H_3PO_4$$

22.66 In the equation, there is a 3:1 ratio of $Ca(H_2PO_4)_2$ to $Ca_3(PO_4)_2$, which can be used to calculate the mass of $Ca(H_2PO_4)_2$ from the mass of $Ca_3(PO_4)_2$. The mass of $Ca_3(PO_4)_2$ in the 10.0 g of rock is

$$0.712 \times 10 \text{ g rock} = 7.12 \text{ g } Ca_3(PO_4)_2 \text{ (= CaP)}$$

The mass of $Ca(H_2PO_4)_2$ can be calculated from this mass as follows:

$$7.12 \text{ g CaP} \times \frac{1 \text{ mol CaP}}{310.2 \text{ g CaP}} \times \frac{3Ca(H_2PO_4)_2}{1CaP} \times \frac{234.0 \text{ g } Ca(H_2PO_4)_2}{1 \text{ mol } Ca(H_2PO_4)_2}$$

$$= 16.\underline{1}1 = 16.1 \text{ g } Ca(H_2PO_4)_2$$

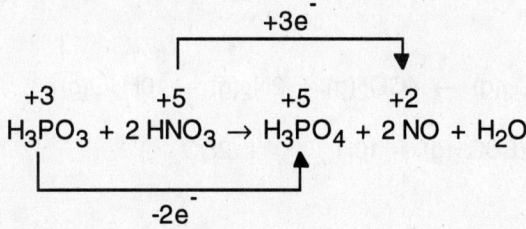

22.67 a. $4Li(s) + O_2(g) \rightarrow 2Li_2O(s)$

b. Organic materials burn in excess O_2 to give CO_2 and H_2O. The nitrogen becomes N_2:

$$4CH_3NH_2(g) + 9O_2(g) \rightarrow 4CO_2(g) + 2N_2(g) + 10H_2O(g)$$

c. $2(C_2H_5)_2S + 15O_2 \rightarrow 8CO_2(g) + 10H_2O(g) + 2SO_2(g)$

22.68 a. $2Ca(s) + O_2(g) \rightarrow 2CaO(s)$

b. $4PH_3(g) + 8O_2(g) \rightarrow P_4O_{10}(s) + 6H_2O(g)$

c. $4HOCH_2CH_2NH_2(l) + 13O_2(g) \rightarrow 8CO_2(g) + 14H_2O(g) + 2N_2(g)$

22.69 In the equation, there is a 2:1 ratio of Na_2O_2 to O_2, which can be used to calculate the mass of Na_2O_2 needed to form 15.0 g of O_2. The mass of Na_2O_2 is calculated as follows:

$$15.0 \text{ g } O_2 \times \frac{1 \text{ mol } O_2}{32.0 \text{ g } O_2} \times \frac{2Na_2O_2}{1O_2} \times \frac{77.98 \text{ g } Na_2O_2}{1 \text{ mol } Na_2O_2} = 73.\underline{1}06$$

$$= 73.1 \text{ g } Na_2O_2$$

22.70 In the equation, there is a 2:3 ratio of $KClO_3$ to O_2, which can be used to calculate the mass of $KClO_3$ needed to form 15.0 g of O_2. The mass of $KClO_3$ is calculated as follows:

$$15.0 \text{ g } O_2 \times \frac{1 \text{ mol } O_2}{32.0 \text{ g } O_2} \times \frac{2KClO_3}{3O_2} \times \frac{122.55 \text{ g } KClO_3}{1 \text{ mol } KClO_3} = 38.\underline{2}9$$

$$= 38.3 \text{ g } KClO_3$$

22.71 a. $x_S + 6x_F = 0$

The oxidation number of F in compounds is always -1.

$$x_S = -6x_F = -6(-1) = +6$$

b. $x_S + 3x_O = 0$

The oxidation number of O in most compounds is -2.

$$x_S = -3x_O = -3(-2) = +6$$

c. $x_S + 2x_H = 0$

The oxidation number of H in most compounds is +1.

$$x_S = -2x_H = -2(+1) = -2$$

d. $x_{Ca} + x_S + 3x_O = 0$

The oxidation number of Ca in compounds is +2; the oxidation number of O in most compounds is -2.

$$x_S = -x_{Ca} - 3x_O = -(+2) - 3(-2) = +4$$

22.72 a. $x_S = 0$

The oxidation number in any element is 0.

b. $x_{Ca} + x_S = 0$

The oxidation number of Ca in compounds is +2.

$$x_S = -x_{Ca} = -(+2) = -2$$

c. $x_{Ca} + x_S + 4x_O = 0$

The oxidation number of Ca in compounds is +2; the oxidation number of O in most compounds is -2.

$$x_S = -x_{Ca} - 4x_O = -(+2) - 4(-2) = +6$$

d. $x_S + 4x_{Cl} = 0$

The oxidation number of Cl in a binary compound with a less electronegative element is -1.

$$x_S = -4x_{Cl} = -4(-1) = +4$$

22.73 The reduction of $8H_2SeO_3$ to Se_8 is a 32-electron reduction, and the oxidation of $8H_2S$ to S_8 is a 16-electron oxidation. Balancing the equation requires multiplying the H_2S half-reaction by 2 to achieve a 32-electron oxidation. The final equation is

$$8H_2SeO_3(aq) + 16H_2S(g) \rightarrow Se_8(s) + 2S_8(s) + 24H_2O(l)$$

22.74 $H_2SO_4(aq) + 2I^-(aq) + 2H^+(aq) \rightarrow SO_2(g) + I_2(s) + 2H_2O(l)$

22.75 In the equation, there is a 2:1 ratio of $NaHSO_3$ to Na_2CO_3, which can be used to calculate the mass of $NaHSO_3$ from 25.0 g of Na_2CO_3. The mass of $NaHSO_3$ is calculated as follows:

$$25.0 \text{ g } Na_2CO_3 \times \frac{1 \text{ mol } Na_2CO_3}{106.0 \text{ g } Na_2CO_3} \times \frac{2NaHSO_3}{1Na_2CO_3} \times \frac{104.0 \text{ g } NaHSO_3}{1 \text{ mol } NaHSO_3}$$

$$= 49.\underline{0}56 = 49.1 \text{ g } NaHSO_3$$

22.76 In the equation, there is an 8:1 ratio of $Na_2S_2O_3$ to S_8, which can be used to calculate the mass of $Na_2S_2O_3$ from 25.0 g of S_8. The mass of $Na_2S_2O_3$ is calculated as follows:

$$25.0 \text{ g } S_8 \times \frac{1 \text{ mol } S_8}{256.5 \text{ g } S_8} \times \frac{8\ Na_2S_2O_3}{1\ S_8} \times \frac{158.1 \text{ g } Na_2S_2O_3}{1 \text{ mol } Na_2S_2O_3} = 12\underline{3}.2$$

$$= 123 \text{ g } Na_2S_2O_3$$

22.77 $Ba(ClO_3)_2(aq) + H_2SO_4(aq) \rightarrow 2HClO_3(aq) + BaSO_4(s)$

22.78 $Ba(ClO_2)_2(aq) + H_2SO_4(aq) \rightarrow 2HClO_2(aq) + BaSO_4(s)$

22.79

2 x ($-1e^-$)

$$\overset{-1}{2\ HCl} + K_2\overset{+6}{Cr_2}O_7 \rightarrow \overset{0}{Cl_2} + 2\ \overset{+3}{Cr^{+3}}$$

2 x ($+3e^-$)

$$6HCl + K_2Cr_2O_7 \rightarrow 3Cl_2 + 2Cr^{3+}$$

Balance O: $6HCl + K_2Cr_2O_7 \rightarrow 3Cl_2 + 2Cr^{3+} + 7H_2O$

Balance H: $8H^+ + 6HCl + K_2Cr_2O_7 \rightarrow 3Cl_2 + 2Cr^{3+} + 7H_2O$

Balance K: $8H^+ + 6HCl + K_2Cr_2O_7 \rightarrow 3Cl_2 + 2Cr^{3+} + 7H_2O + 2K^+$

(Note that charge is balanced.)

22.80

$$2 \times (-5e^-)$$

$$\overset{0}{I_2} + \overset{+5}{HNO_3} \rightarrow 2\overset{+5}{HIO_3} + \overset{+4}{NO_2}$$

$$+1e^-$$

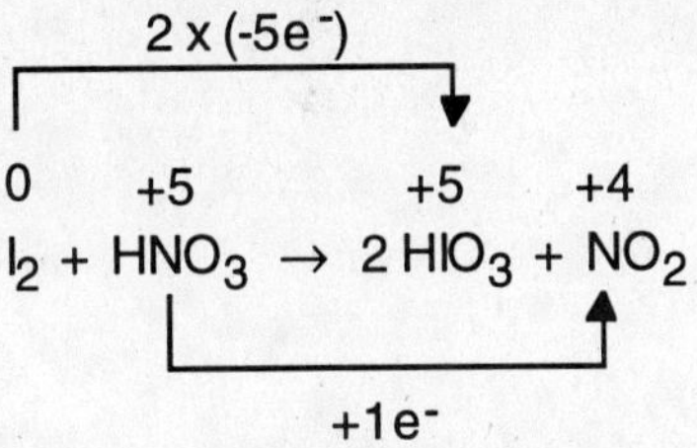

$$I_2 + 10HNO_3 \rightarrow 2HIO_3 + 10NO_2$$

Balance O: $I_2 + 10HNO_3 \rightarrow 2HIO_3 + 10NO_2 + 4H_2O$

22.81 a. The electron-dot formula of Cl_2O is

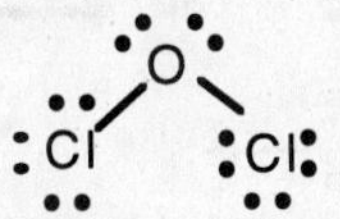

The VSEPR model predicts a bent (angular) molecular geometry. You can describe the four electron pairs on O using sp^3 hybrid orbitals. The diagramming for the bond formation follows:

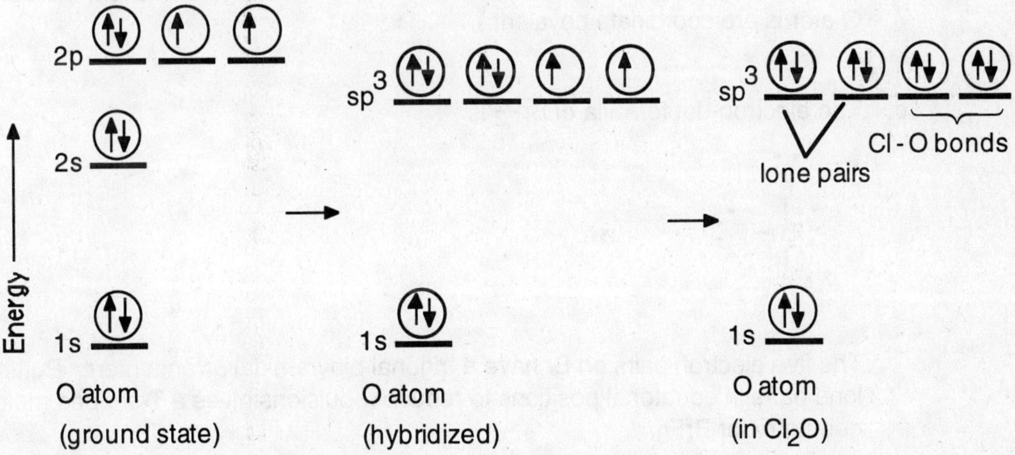

b. An electron-dot formula of BrO_3^- is

$$\left[\begin{array}{c} :\ddot{O}: \\ :\ddot{B}r:\ddot{O}: \\ :\ddot{O}: \end{array} \right]^-$$

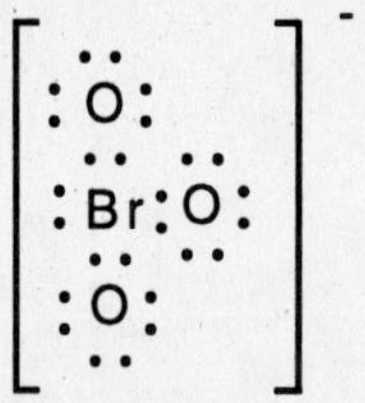

The VSEPR model predicts a trigonal pyramidal geometry. You can describe the four electron pairs on Br using sp^3 hybrid orbitals. The diagramming for the bond formation follows:

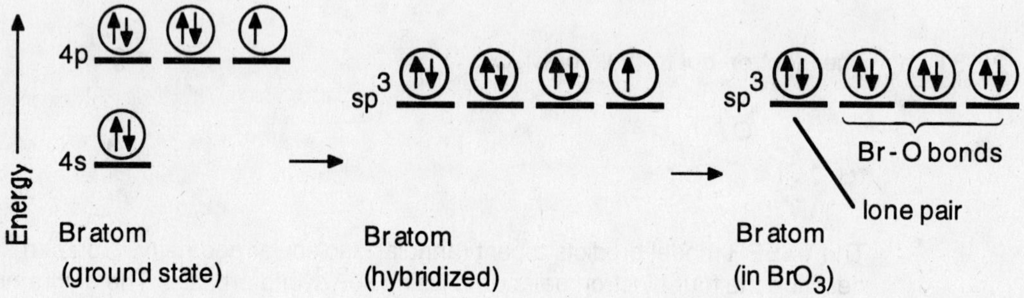

(*Note:* The additional electron accounts for the -1 charge of the ion; the bonds to O atoms are coordinate covalent.)

c. The electron-dot formula of BrF_3 is

$$\begin{array}{c} :\ddot{F}: \\ | \\ :\ddot{B}r-\ddot{F}: \\ | \\ :\ddot{F}: \end{array}$$

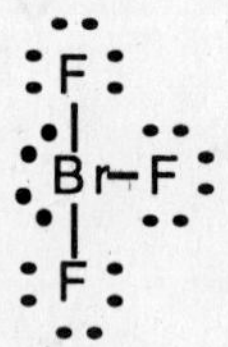

The five electron pairs on Br have a trigonal bipyramidal arrangement. Putting the lone pairs in equatorial positions to reduce repulsions gives a T-shaped molecular geometry for BrF_3.

(continued)

You can describe the five electron pairs on Br in terms of sp^3d hybrid orbitals. The diagramming for the bond formation follows:

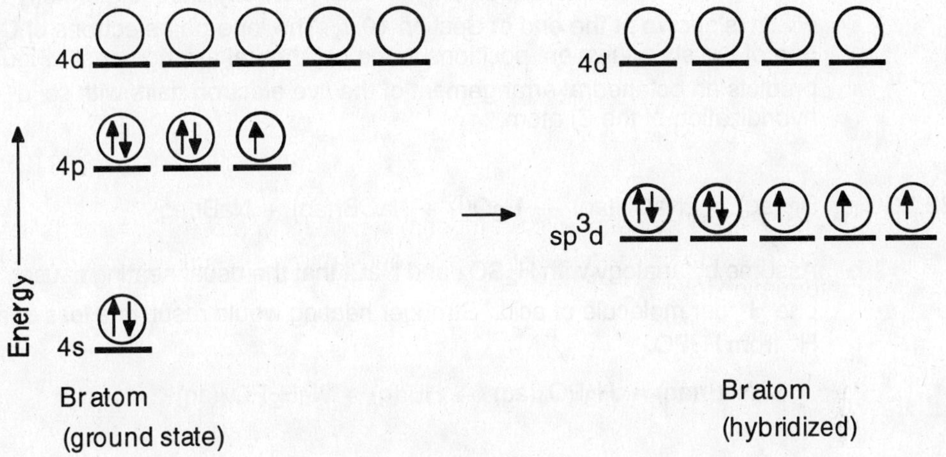

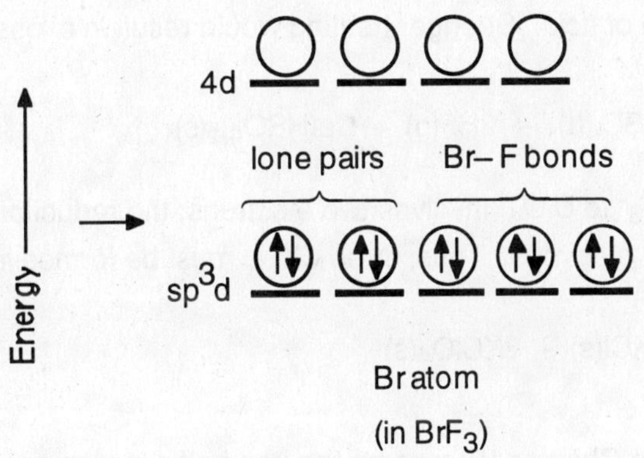

22.82 a. The electron-dot formula of HClO involves a pair of bonding electrons between H and O, and a pair of bonding electrons between Cl and O. Chlorine has three lone pairs of electrons. The VSEPR model predicts a bent (angular) molecular geometry like the geometry of H_2O discussed in Section 10.1 and shown in Figure 10.7. There is a total of four electron pairs around oxygen; two are bonding and two are lone pairs. Figure 10.4 predicts a tetrahedral arrangement of the four electron pairs with sp^3 hybridization of the O atom and a Cl—O—H structure.

b. The electron-dot formula of ClO_4^- involves a pair of bonding electrons between each oxygen and chlorine. Thus, chlorine has four pairs of electrons. The VSEPR model predicts a tetrahedral molecular geometry like the geometry of $SiCl_4$ discussed in Example 10.1 and shown for CH_4 in Figure 10.7. Figure 10.4 predicts a tetrahedral arrangement of the four electron pairs with sp^3 hybridization of the Cl atom.

c. The electron-dot formula of ClF_5 involves a pair of bonding electrons between each fluorine and chlorine. In addition, chlorine has a lone pair of electrons. The VSEPR model predicts a square pyramidal geometry like the geometry of IF_5, which is shown at the end of Section 10.1. The lone-pair electrons of Cl occupy one of the six equivalent positions in the octahedral arrangement. Figure 10.9 predicts an octahedral arrangement of the five electron pairs with sp^3d^2 hybridization of the Cl atom.

22.83 a. $Br_2(aq) + 2NaOH(aq) \rightarrow H_2O(l) + NaOBr(aq) + NaBr(aq)$

b. Assume by analogy with H_2SO_4 and NaCl that the usual heating causes a loss of one H^+ per molecule of acid. Stronger heating would result in a loss of additional H^+ from $H_2PO_4^-$.

$$NaBr(aq) + H_3PO_4(aq) \rightarrow HBr(g) + NaH_2PO_4(aq)$$

22.84 a. Assume by analogy with H_2SO_4 and NaCl that the usual heating causes a loss of one H^+ per molecule of acid. Stronger heating would result in a loss of the remaining H^+ of HSO_4^-.

$$CaF_2(s) + 2H_2SO_4(l) \rightarrow 2HF(g) + Ca(HSO_4)_2(s)$$

b. The oxidation of ClO_3^- to ClO_4^- involves two electrons; the reduction of ClO_3^- to Cl^- involves a gain of six electrons. Thus, three ClO_4^- must be formed in the final equation:

$$4KClO_3(s) \rightarrow KCl(s) + 3KClO_4(s)$$

22.85 Following the approach in Chapter 19, reverse the iron half-reaction and the sign of its E^o, and double the half-reaction to obtain the same number of electrons as the OCl^- half-reaction:

$$2Fe^{2+}(aq) \rightarrow 2Fe^{3+}(aq) + 2e^- \qquad -E^o = -0.77\ V$$

$$OCl^-(aq) + H_2O(l) + 2e^- \rightarrow Cl^-(aq) + 2OH^-(aq) \qquad E^o = 0.90\ V$$

$$OCl^-(aq) + H_2O(l) + 2Fe^{2+}(aq) \rightarrow 2Fe^{3+}(aq) + Cl^-(aq) + 2OH^-(aq) \qquad E^o_{cell} = 0.13\ V$$

Because E^o_{cell} is positive, the reaction is spontaneous at standard conditions.

22.86 Following the approach of Chapter 19, reverse the Br^- half-reaction and the sign of its E^o. Each half-reaction has the same number of electrons, so the half-reactions are simply added:

$2Br^-(aq)$	$\rightarrow Br_2(l) + 2e^-$	$-E^o$	$= -1.07$ V
$OCl^-(aq) + H_2O(l) + 2e^-$	$\rightarrow Cl^-(aq) + 2OH^-(aq)$	E^o	$= 0.90$ V
$OCl^-(aq) + H_2O(l) + 2Br^-(aq)$	$\rightarrow Br_2(l) + Cl^-(aq) + 2OH^-(aq)$	E^o_{cell}	$= -0.17$ V

Because E^o_{cell} is negative, the reaction is nonspontaneous at standard conditions.

22.87 The total number of valence electrons is 8 + (4 x 7) = 36. These are distributed to give the following Lewis formula:

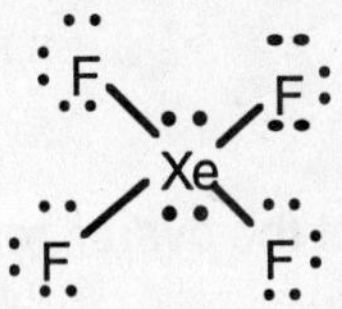

The six electron pairs on Xe would have an octahedral arrangement, suggesting sp^3d^2 hybridization. The lone pairs on Xe would be directed above and below the molecule, which has a square planar geometry.

22.88 The total number of valence electrons is 8 + (4 x 6) = 32. This gives the following Lewis formula:

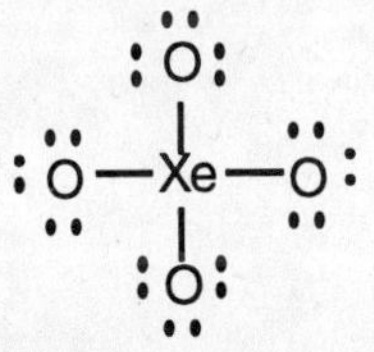

Because there are four electron pairs on Xe, the arrangement of pairs is tetrahedral, suggesting sp^3 hybridization. The molecular geometry would be predicted to be tetrahedral.

22.89 From the information given,

$XeF_2 \rightarrow Xe + O_2 + F^-$ (basic solution; not balanced)

The oxygen (and hydrogen) will be balanced by OH^- and H_2O from the basic solution. Balance half-reactions.

Reduction: $XeF_2 \rightarrow Xe + F^-$

Balance F atoms: $XeF_2 \rightarrow Xe + 2F^-$

Balance charge: $2e^- + XeF_2 \rightarrow Xe + 2F^-$

Oxidation: $H_2O \rightarrow O_2$

Balance O atoms: $2H_2O \rightarrow O_2$

Balance H atoms: $2H_2O \rightarrow 4H^+ + O_2$

Balance charge: $2H_2O \rightarrow 4H^+ + O_2 + 4e^-$

Convert to base: $2H_2O + 4OH^- \rightarrow 4H_2O + O_2 + 4e^-$

Double the Xe half-reaction and add half-reactions to get the overall reaction:

$4e^- + 2XeF_2 \rightarrow 2Xe + 4F^-$

$4OH^- \rightarrow O_2 + 2H_2O + 4e^-$

$2XeF_2 + 4OH^- \rightarrow 2Xe + 4F^- + O_2 + 2H_2O$

22.90 From the information given,

$XeO_3 + I^- \rightarrow Xe + I_2$ (acidic solution; not balanced)

Balance by the half-reaction method.

Reduction: $XeO_3 \rightarrow Xe$

Balance O atoms: $XeO_3 \rightarrow Xe + 3H_2O$

Balance H atoms: $6H^+ + XeO_3 \rightarrow Xe + 3H_2O$

Balance charge: $6e^- + 6H^+ + XeO_3 \rightarrow Xe + 3H_2O$

Oxidation: $I^- \rightarrow I_2$

Balance I atoms: $2I^- \rightarrow I_2$

Balance charge: $2I^- \rightarrow I_2 + 2e^-$

(continued)

Triple the Iodine half-reaction and add the half-reactions:

$$6e^- + 6H^+ + XeO_3 \rightarrow Xe + 3H_2O$$

$$6I^- \rightarrow 3I_2 + 6e^-$$

$$XeO_3 + 6H^+ + 6I^- \rightarrow Xe + 3H_2O + 3I_2$$

■ Solutions to Unclassified Problems

22.91 a. The yellow solid is sulfur, S_8, which burns to form SO_2 gas.

b. The white solid is phosphorus(V) oxide, P_4O_{10}, which reacts with water to form H_3PO_4.

c. The colorless gas is carbon monoxide, CO, which bonds to iron in hemoglobin.

d. The hard, lustrous gray solid is elemental silicon, S, which is used in semiconductors.

22.92 a. The white, waxy solid is phosphorus, P_4, which normally inflames in air.

b. The viscous liquid that reacts with table sugar is concentrated sulfuric acid, H_2SO_4.

c. The acid that reacts with copper to form brown fumes of NO_2 is nitric acid, HNO_3.

d. The greenish-yellow gas that dissolves in NaOH is chlorine, Cl_2.

22.93 Test the pH of each solution. Sodium sulfate, Na_2SO_4, is neutral, and both $NaHSO_4$ and $NaHSO_3$ are acidic. To differentiate the latter two, add HCl to acidify solutions of each. Only the $NaHSO_3$ will evolve SO_2 with its characteristic suffocating odor.

22.94 Add aqueous silver nitrate to a sample of each solution. The solution containing F^- does not form a precipitate, whereas the Cl^- and I^- solutions do. To differentiate the latter two, observe the colors of the silver(I) salts: AgCl is white and AgI is yellow. You can also identify the I^- ion by adding an aqueous solution of Cl_2 plus the solvent CH_2Cl_2. The I^- reacts with Cl_2 to give I_2, which has a violet color in the CH_2Cl_2. The Cl^- does not react.

22.95

$-2 \times (1e^-)$

$$\overset{+7}{KMnO_4} + 2\,\overset{-1}{HCl} \rightarrow \overset{0}{Cl_2} + \overset{+2}{Mn^{2+}}$$

$+5e^-$

$$2KMnO_4 + 10HCl \rightarrow 5Cl_2 + 2Mn^{2+}$$

Balance O: $2KMnO_4 + 10HCl \rightarrow 5Cl_2 + 2Mn^{2+} + 8H_2O$

Balance H: $6H^+ + 2KMnO_4 + 10HCl \rightarrow 5Cl_2 + 2Mn^{2+} + 8H_2O$

Balance K: $6H^+ + 2KMnO_4 + 10HCl \rightarrow 5Cl_2 + 2Mn^{2+} + 2K^+ + 8H_2O$

The source of H^+ is HCl, so the overall equation is

$$2KMnO_4 + 16HCl \rightarrow 5Cl_2 + 2MnCl_2 + 2KCl + 8H_2O$$

$$12.0 \text{ g } KMnO_4 \times \frac{1 \text{ mol } KMnO_4}{158.0 \text{ g } KMnO_4} \times \frac{16 \text{ mol HCl}}{2 \text{ mol } KMnO_4} \times \frac{1 \text{ L}}{1.50 \text{ mol HCl}} = 0.40\underline{5}06 \text{ L}$$

$$= 0.405 \text{ L}$$

0.405 L of 1.50 M HCl(aq) would be required.

22.96 First write the balanced equation.

$2 \times (-5e^-)$

$$\overset{0}{I_2} + \overset{+5}{HNO_3} \rightarrow 2\,\overset{+5}{HIO_3} + \overset{+4}{NO_2}$$

$+1e^-$

$$I_2 + 10HNO_3 \rightarrow 2HIO_3 + 10NO_2$$

Balance O: $I_2 + 10HNO_3 \rightarrow 2HIO_3 + 10NO_2 + 4H_2O$

$$15.0 \text{ g } HIO_3 \times \frac{1 \text{ mol } HIO_3}{175.9 \text{ g } HIO_3} \times \frac{10 \text{ mol } HNO_3}{2 \text{ mol } HIO_3} \times \frac{1 \text{ L}}{15.8 \text{ mol } HNO_3}$$

$$= 0.026\underline{9}8 = 0.0270 \text{ L}$$

0.0270 L of 15.8 M HNO_3 would be required.

22.97 The molecular weight of $Ca(H_2PO_4)_2 \bullet H_2O$ is 252.07 amu.

$$\text{Mass \% } Ca(H_2PO_4)_2 \bullet H_2O = \frac{\text{mass } Ca(H_2PO_4)_2 \bullet H_2O}{\text{mass fertilizer}} \times 100\%$$

Assume a sample of 100.0 g fertilizer. This contains 15.5 g P. Convert this to a mass of $Ca(H_2PO_4)_2 \bullet H_2O$:

$$15.5 \text{ g P} \times \underbrace{\frac{100.00 \text{ g } Ca(H_2PO_4)_2 \bullet H_2O}{24.57 \text{ g P}}}_{\text{from mass P in 100.00 g } Ca(H_2PO_4)_2 \bullet H_2O} = 63.085 \text{ g } Ca(H_2PO_4)_2 \bullet H_2O$$

$$\text{Mass \% } Ca(H_2PO_4)_2 \bullet H_2O = \frac{63.085 \text{ g } Ca(H_2PO_4)_2 \bullet H_2O}{100 \text{ g fertilizer}} \times 100\%$$

$$= 63.\underline{0}85 = 63.1\%$$

22.98 In 100.0 g of the fertilizer there are 30.0 g of $Ca(H_2PO_4)_2 \bullet H_2O$ and 10.0 g $CaHPO_4$. Calculate the amount of P in these masses.

$$30.0 \text{ g } Ca(H_2PO_4)_2 \bullet H_2O \times \frac{1 \text{ mol } Ca(H_2PO_4)_2 \bullet H_2O}{252.1 \text{ g } Ca(H_2PO_4)_2 \bullet H_2O} \times \frac{2 \text{ mol P}}{1 \text{ mol } Ca(H_2PO_4)_2 \bullet H_2O}$$

$$\times \frac{30.97 \text{ g}}{1 \text{ mol P}} = 7.3\underline{7}1 \text{ g P}$$

$$10.0 \text{ g } CaHPO_4 \times \frac{1 \text{ mol } CaHPO_4}{136.1 \text{ g } CaHPO_4} \times \frac{1 \text{ mol P}}{1 \text{ mol } CaHPO_4} \times \frac{30.97 \text{ g P}}{1 \text{ mol P}} = 2.2\underline{7}6 \text{ g P}$$

The total mass of P in the 100.0 = g sample is (7.371 + 2.276) g = 9.6$\underline{4}$7 g. Therefore, the mass % P = 9.6$\underline{4}$7% = 9.65%.

22.99 $2NaCl + 2H_2O \rightarrow 2NaOH + H_2 + Cl_2$ (electrolysis)

$2NaOH + Cl_2 \rightarrow NaClO + NaCl + H_2O$ (spontaneous)

One mol of NaClO is produced for every two mol of NaCl. Thus the reaction requires two electrons per mol of NaClO produced. Convert the 1.0$\underline{0}$ x 10^3 L of NaOCl to time:

(continued)

$$1.00 \times 10^3 \text{ L} \times \frac{1 \text{ mL}}{10^{-3} \text{L}} \times \frac{1.00 \text{ g soln}}{1 \text{ mL}} \times \frac{5.25 \text{ g NaOCl}}{100 \text{ g soln}} \times \frac{1 \text{ mol NaOCl}}{74.44 \text{ g NaOCl}} \times \frac{2 \text{ mol e}^-}{1 \text{ mol NaOCl}}$$

$$\times \frac{9.65 \times 10^4 \text{ C}}{1 \text{ mol e}^-} \times \frac{1 \text{ s}}{2.50 \times 10^3 \text{C}} = 5.4\underline{4}4 \times 10^4 \text{ s} \times \frac{1 \text{ h}}{3.6 \times 10^3 \text{ s}} = 15.\underline{1}2 \text{ h} = 15.1 \text{ h}$$

22.100 The electrode reaction is

$$ClO_3^-(aq) + H_2O(l) \rightarrow ClO_4^-(aq) + 2H^+(aq) + 2e^-$$

$$\frac{2.50 \times 10^3 \text{ C}}{\text{s}} \times \frac{3.6 \times 10^3 \text{s}}{1 \text{ h}} \times \frac{1 \text{ mol e}^-}{9.65 \times 10^4 \text{ C}} \times \frac{1 \text{ mol NaClO}_4}{2 \text{ mol e}^-} \times \frac{122.4 \text{ g NaClO}_4}{1 \text{ mol NaClO}_4}$$

$$= 5.7\underline{0}7 \times 10^3 \text{ g/h} = 5.71 \times 10^3 \text{ g/h} \ (5.71 \text{ kg/h})$$

22.101 $H_2O + NaOCl + 2I^- \rightarrow I_2 + NaCl + 2OH^-$

$I_2 + 2Na_2S_2O_3 \rightarrow 2NaI + Na_2S_4O_6$

$$34.6 \text{ mL Na}_2\text{S}_2\text{O}_3 \text{ soln} \times \frac{10^{-3} \text{L}}{1 \text{ mL}} \times \frac{0.100 \text{ mol Na}_2\text{S}_2\text{O}_3}{1 \text{ L soln}} \times \frac{1 \text{ mol I}_2}{2 \text{ mol Na}_2\text{S}_2\text{O}_3}$$

$$\times \frac{1 \text{ mol NaOCl}}{1 \text{ mol I}_2} \times \frac{74.44 \text{ g NaOCl}}{1 \text{ mol NaOCl}} = 0.1288 \text{ g NaOCl}$$

If the density of bleach is taken as 1.00 g/mL, then 5.00 mL = 5.00 g.

$$\text{Mass \% NaOCl} = \frac{\text{mass NaOCl}}{\text{mass bleach}} \times 100\% = \frac{0.1288 \text{ g}}{5.00 \text{ g}} \times 100\% = 2.5\underline{7}56\% = 2.58\%$$

22.102 In acid,

$$6H^+(aq) + IO_3^-(aq) + 5I^-(aq) \rightarrow 3I_2(aq) + 3H_2O(l)$$

Find the amount of I_2 initially present.

$$10.00 \text{ mL IO}_3^- \equiv 1.000 \times 10^{-2} \text{ L IO}_3^- \times \frac{0.0500 \text{ mol IO}_3^-}{1 \text{ L IO}_3^-} \times \frac{3 \text{ mol I}_2}{1 \text{ mol IO}_3^-}$$

$$= 1.5\underline{00} \times 10^{-3} \text{ mol I}_2$$

(continued)

The reaction for thiosulfate with I_2 is

$$I_2(aq) + 2S_2O_3^{2-}(aq) \rightarrow 2I^-(aq) + S_4O_6^{2-}(aq)$$

Find the amount of excess I_2:

$$29.5 \text{ mL } S_2O_3^{2-} \equiv 2.95 \times 10^{-2} \text{ L } S_2O_3^{2-} \times \frac{0.0300 \text{ mol } S_2O_3^{2-}}{1 \text{ L } S_2O_3^{2-}} \times \frac{1 \text{ mol } I_2}{2 \text{ mol } S_2O_3^{2-}}$$

$$= 4.4\underline{2}5 \times 10^{-4} \text{ mol } I_2$$

The amount of I_2 consumed by the ascorbic acid is

$$1.5\underline{0}0 \times 10^{-3} \text{ mol} - 4.4\underline{2}5 \times 10^{-4} \text{ mol} = (1.0\underline{5}75 \times 10^{-3}) \text{ mol } I_2$$

From the stoichiometry of the equation, there must have been 1.0575×10^{-3} mol of ascorbic acid present in the original 30.0 = g sample. Convert moles to mass:

$$1.0\underline{5}75 \times 10^{-3} \text{ mol ascorbic acid} \times \frac{176.1 \text{ g ascorbic acid}}{1 \text{ mol ascorbic acid}}$$

$$= 0.18\underline{6}22 \text{ g ascorbic acid}$$

$$\text{Mass \% ascorbic acid} = \frac{0.18622 \text{ g ascorbic acid}}{30.0 \text{ g drink mix}} \times 100\% = 0.62\underline{0}7 = 0.621\%$$

In 100.0 g of drink mix, there would be 0.621 g of ascorbic acid.

23. THE TRANSITION ELEMENTS

Solutions to Exercises

23.1 Because silver nitrate did not give a precipitate of AgCl, all of the chlorine atoms must be attached to the platinum. There are two potassium ions left over. These correspond to the second and third ions in the three-ion formula unit. This complex has the formula

$K_2[PtCl_6]$

23.2 a. Pentaamminechlorocobalt(III) chloride

b. Potassium aquapentacyanocobaltate(III)

c. Pentaaquahydroxoiron(III) ion

23.3 a. $K_4[Fe(CN)_6]$ b. $[Co(NH_3)_4Cl_2]Cl$ c. $PtCl_4^{2-}$

23.4 a. These two compounds display coordination isomerism because they differ in how the ligands are distributed to two metal atoms.

b. These are linkage isomers because they differ in that the SCN^- ligand is attached to Mn by the S atom in one case and the N atom in the other.

c. These are ionization isomers because they differ in the anion that is coordinated to the metal.

d. These two compounds are hydrate isomers because they differ in the placement of water molecules in the complex.

23.5 The structural formula of the compound is

$[Co(NH_3)_4(H_2O)Cl]Cl_2$

A possible structural isomer of this compound is

$[Co(NH_3)_4Cl_2]Cl \bullet H_2O$

This would be an example of a hydrate isomer.

23.6 a. No geometric isomers.

b.

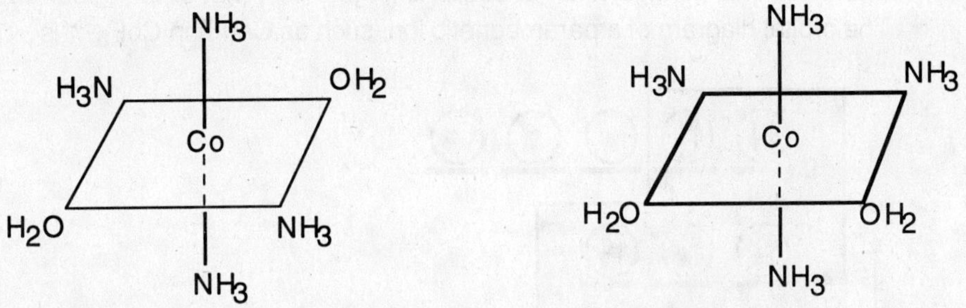

c.

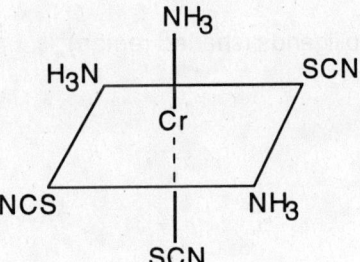

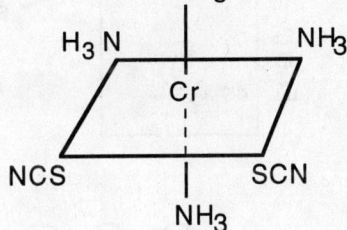

d. No geometric isomers.

23.7 a. No optical isomers.

b.

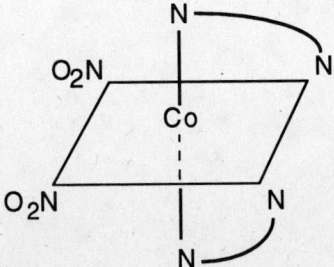

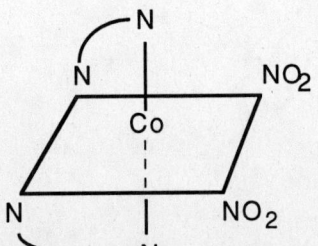

c.

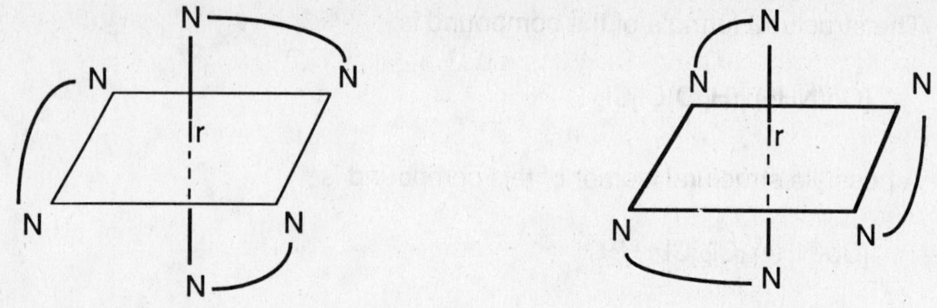

d. No optical isomers.

23.8 The electronic configuration of cobalt is [Ar]$3d^7 4s^2$; that of the cobalt(III) ion is [Ar]$3d^6$. The orbital diagram of a paramagnetic ion, such as Co(III) in CoF_6^{3-}, is

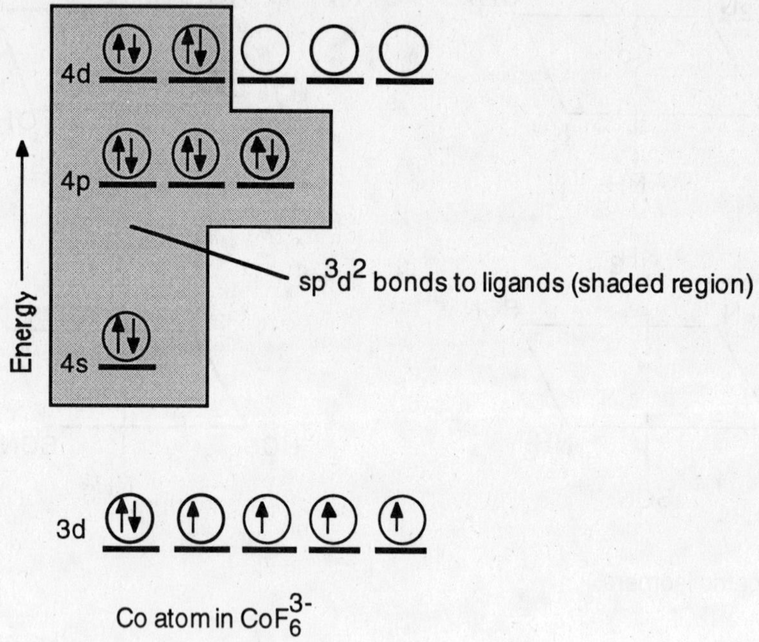

(continued)

Bonding uses sp^3d^2 hybrid orbitals on Co; there are four unpaired electrons. The orbital diagram of a diamagnetic ion, such as Co(III) in $Co(NH_3)_6^{3+}$, is

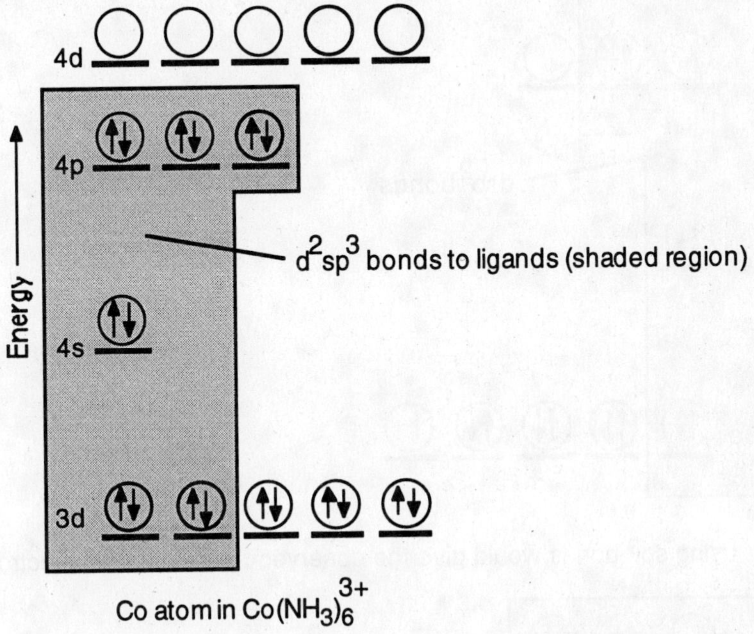

Bonding uses d^2sp^3 hybrid orbitals on Co; there are no unpaired electrons.

23.9 The Co^{2+} ion has the configuration $[Ar]3d^7$. The orbital diagram using dsp^2 bonds would give only one unpaired electron.

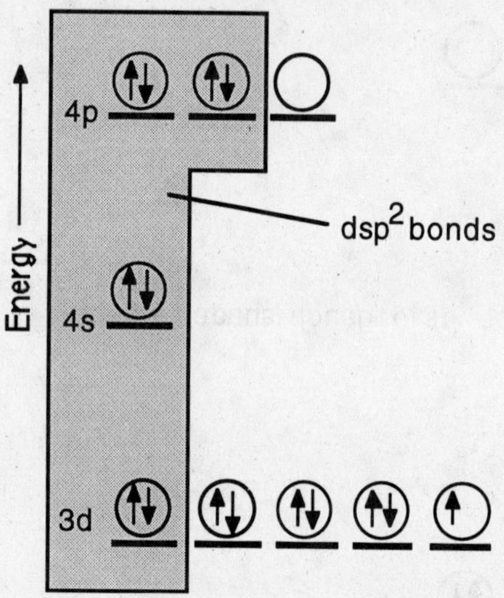

However, using sp^3 bonds would give the observed three unpaired electrons:

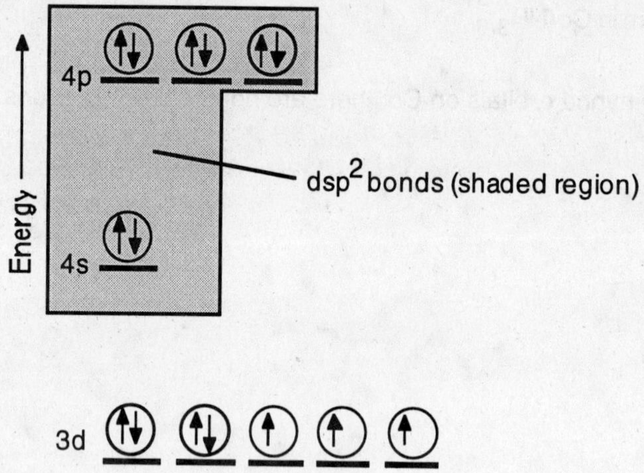

Therefore, the geometry is expected to be tetrahedral.

23.10 The electron configuration of Ni^{2+} is $[Ar]3d^8$. The distribution of electrons among the d orbitals of Ni in $Ni(H_2O)_6^{2+}$ is as follows:

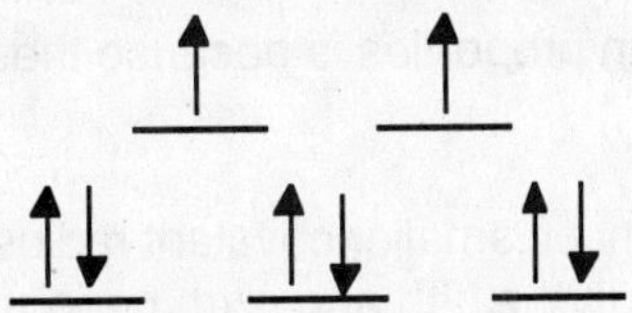

Note that there is only one possible distribution of electrons, giving two unpaired electrons.

23.11 The electronic configuration of the Co^{2+} ion is $[Ar]3d^7$. The distribution of the d electrons in $CoCl_4^{2-}$ is as follows:

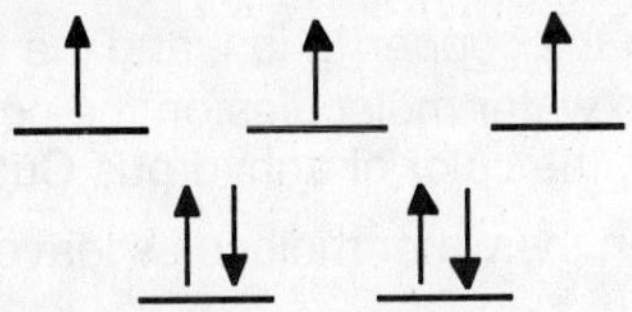

23.12 The approximate wavelength of the maximum absorption for $Fe(H_2O)_6^{3+}$, which is pale purple, is 530 nm. The approximate wavelength of the maximum absorption for $Fe(CN)_6^{3-}$, which is red, is 500 nm. The shift is in the expected direction because CN^- is a more strongly bonding ligand than H_2O. As a result, Δ should increase and the wavelength of the absorption should decrease when H_2O is replaced by CN^-.

■ Answers to Review Questions

23.1 Characteristics of the transition elements that set them apart from the main-group elements are the following: (1) The transition elements are metals with high melting points (only the IIB elements have low melting points). Most main-group elements have low melting points. (2) Each of the transition metals has several oxidation states (except for the IIIB and IIB elements). Most main-group metals have only one oxidation state in addition to 0. (3) Transition-metal compounds are often colored and many are paramagnetic. Most main-group compounds are colorless and diamagnetic.

23.2 Technetium has the electron configuration $[Kr]\,4d^55s^2$.

23.3 Molybdenum has the highest melting point of any element in the fifth period because it has the maximum number of unpaired electrons, which contribute to the strength of the metal bonding.

23.4 One reason iron, cobalt, and nickel are similar in properties is because these elements have similar covalent radii.

23.5 Nickel falls in the fourth period; as such, it has much smaller covalent radius than the corresponding metals, palladium and platinum, in the fifth and sixth periods. However, palladium and platinum have very similar covalent radii.

23.6 $Cr(s) + 2HCl(aq) + 6H_2O(l) \rightarrow Cr(H_2O)_6^{2+} + 2Cl^-(aq) + H_2(g)$

$Cu(s) + HCl(aq) \rightarrow NR$

23.7 $Cr_2O_3(s) + 6HCl(aq) + 9H_2O(l) \rightarrow 2Cr(H_2O)_6^{3+} + 6Cl^-(aq)$

23.8 Four of the water molecules are associated with the copper(II) ion, and the fifth is hydrogen bonded to the sulfate ion as well as to water molecules on the copper(II). Heating changes the blue color to a white color, the color of anhydrous $CuSO_4$. The blue color is associated with $Cu(H_2O)_4^{2+}$. When the water molecules leave the copper ion, the blue color is lost.

23.9 Two Cu^{2+} ions gain a total of two electrons in forming one Cu_2O. The HCHO loses two electrons in forming $HCOO^-$ ion. Thus, the final equation is

$$2Cu^{2+}(aq) + HCHO(aq) + 5OH^-(aq) \rightarrow Cu_2O(s) + HCOO^-(aq) + 3H_2O(l)$$

23.10 Werner showed that the electrical conductance of a solution of $[Pt(NH_3)_4Cl_2]Cl_2$ corresponded to that of three ions in solution, and that two of the chloride ions could be precipitated as AgCl whereas the other two could not.

23.11 A complex ion is a metal atom or ion with Lewis bases attached to it through coordinate covalent bonds. A ligand is a Lewis base attached to a metal ion in a complex; it may be either a molecule or an anion, rarely a cation. The coordination number of a metal atom in a complex ion is the total number of bonds the metal forms with ligands. An example of a complex ion is $Fe(CN)_6^{4-}$; an example of a ligand is CN^-; and the coordination number of the preceding complex ion is 6.

23.12 A bidentate ligand is a ligand that bonds to a metal ion through two atoms. Two examples are ethylenediamine, $H_2N—C_2H_4—NH_2$, and the oxalate ion, $^-O_2C—CO_2^-$.

23.13

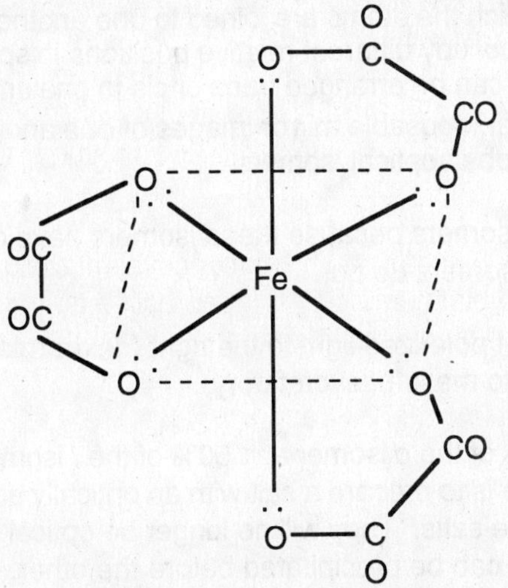

23.14 "Hexacyano" means that there are six CN^- ligands bonded to the iron cation. The Roman numeral II means that the oxidation state of the iron cation is +2, so that the overall charge of the complex ion is -4. This requires four potassium ions to counterbalance the -4 charge.

23.15 The three properties are isomerism, paramagnetism, and color (or absorption of visible and ultraviolet radiation).

23.16 a. Ionization isomerism involves isomers that are alike in that the same anions are present in the formula but different anions are coordinated to the metal ion. For example, the sulfate ion is coordinated to cobalt in $[Co(NH_3)_5(SO_4)]Br$, but the bromide ion is coordinated to cobalt in $[Co(NH_3)_5Br]SO_4$.

b. Hydrate isomerism involves differences in the placement of water molecules in the complex ion. For example, $CrCl_3 \bullet 6H_2O$ exists as $[Cr(H_2O)_6]Cl_3$, $[Cr(H_2O)_4Cl_2]Cl \cdot 2H_2O$, and one other isomer.

c. Coordination isomers are those in which both the cation and anion are complex and the ligands are distributed between the two metal atoms in different ways. For example,

$$[Cu(NH_3)_4]\,[PtCl_4] \text{ and } [Pt(NH_3)_4]\,[CuCl_4]$$

d. Linkage isomers are those in which two different donor atoms on the ligand may bond to the metal ion. For example, the SCN^- can bond to a metal ion through the sulfur atom or through the nitrogen atom.

23.17 Geometric isomers are isomers in which the atoms are joined to one another in the same way but differ because some atoms occupy different relative positions in space. In $[Pt(NH_3)_2Cl_2]$, the two NH_3's (or Cl's) can be arranged trans or cis to one another. Optical isomers are isomers that are nonsuperimposable mirror images of one another. See Figure 23.15 for an example of two cobalt optical isomers.

23.18 Compounds A and B are geometric isomers because these isomers have different physical properties whereas optical isomers do not.

23.19 A *d* optical isomer rotates the plane of polarized light to the right (dextrorotatory), and an *l* ioptical isomer rotates the plane to the left (levorotatory).

23.20 A racemic mixture is a mixture of 50% of the *d* isomer and 50% of the *l* isomer. One method of resolving a racemic mixture is to prepare a salt with an optically active ion of the opposite charge and crystallize the salts. They will no longer be optical isomers and will have different solubilities, so one can be precipitated before the other.

23.21 According to valence bond theory, a ligand orbital containing two electrons overlaps an unoccupied orbital on the metal ion.

23.22 a. In the high-spin complex ion, all 3d orbitals of Fe^{2+} are occupied (four of the 3d orbitals each contain only one electron). Because they are occupied, those orbitals cannot be used for ligand bonding. Instead, d^2sp^3 hybrid orbitals form from the 4s, the three 4p, and two of the 4d orbitals. Each of six ligands donates a pair of electrons to one of these sp^3d^2 hybrid orbitals.

b. In the low-spin complex ion, the six electrons in the 3d orbitals are paired, so they occupy only three of the 3d orbitals. Then sp^3d^2 hybrid orbitals form from two of the 3d, the 4s, and the three 4p orbitals. Each of six ligands donates a pair of electrons to one of these d^2sp^3 hybrid orbitals.

23.23 The d orbitals of a transition-metal atom may have different energies in the octahedral field of six negative charges because the electron pairs of the ligands point directly at the d_{z2} and $d_{x2 - y2}$ orbitals. These orbitals are raised much more in energy than the other three d orbitals because they occupy space between the ligands. Thus, there is a crystal field splitting between the first two d orbitals mentioned and the d_{xy}, d_{xz}, and d_{yz} orbitals.

23.24 Crystal field splitting is the difference in energy between the two sets of d orbitals for a given structure (such as octahedral) in complex ions. It is determined experimentally by measuring the energy of light absorbed by complex ions.

23.25 Pairing energy, P, is the energy required to place two electrons in the same orbital. If the crystal field splitting (Δ) is small because of weak-bonding ligands, then the pairing energy will be larger and the complex will be high-spin. If the crystal field splitting (Δ) is large because of strong-bonding ligands, then the pairing energy will be smaller and the complex will be low-spin.

23.26 a. A high-spin Fe(II) octahedral complex is

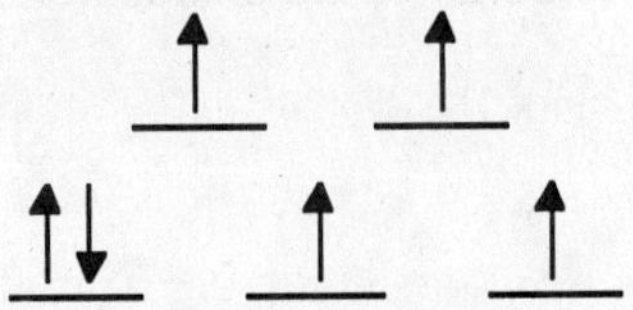

b. A low-spin Fe(II) octahedral complex is

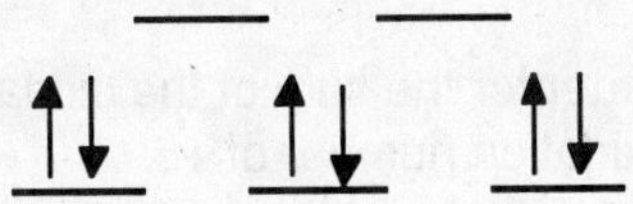

23.27 The spectrochemical series is the arrangement of ligands in order of the relative size of the crystal field splittings (Δ) they induce in the d orbitals of a given oxidation state of a given metal ion. The order is the same, no matter what metal or oxidation state is involved. For Cl^-, H_2O, NH_3, and CN^-, the order of increasing crystal field splitting is

$$Cl^- < H_2O < NH_3 < CN^-$$

where CN- always acts as a strong-bonding ligand.

23.28 The complex absorbing red light would appear as a mixture of blue and green (approximately).

■ Solutions to Practice Problems

Note on significant figures: The final answer to each problem is given first with one nonsignificant figure (the rightmost significant figure is underlined) and then is rounded to the correct number of figures. Intermediate answers usually also have at least one nonsignificant figure.

23.29 a. The charge on carbonate ion is -2. In order for $FeCO_3$ to be neutral, the oxidation number of iron must be +2.

b. The oxidation number of oxygen is -2. In order for the sum of the oxidation numbers of all atoms to be 0, manganese must be in the +4 oxidation state.

c. The oxidation number of the chlorine is -1, so the oxidation number of copper must be +1.

d. The oxidation number of oxygen is -2. The oxidation number of chlorine is -1. In order for the sum of the oxidation numbers to be zero, the oxidation number of chromium must be +6.

$$+6 + 2(-2) + 2(-1) = 0$$

23.30 a. The charge on sulfate ion is -2. In order for $CoSO_4$ to be neutral, the oxidation number of Co must be +2.

b. The oxidation number of oxygen is -2. In order for the sum of the oxidation numbers to be 0, tantalum must have an oxidation number of +5.

$$2(+5) + 5(-2) = 0$$

c. The oxidation number of H is +1; the oxidation number of O is -2; the oxidation number of Cl is -1. Therefore, the oxidation number of Cu is +2.

$$2(+2) + 3(+1) + 3(-2) + 1(-1) = 0$$

23.31 The half-reactions are

$$Fe^{2+} \rightarrow Fe^{3+} + e^-$$

$$4H^+ + NO_3^- + 3e^- \rightarrow NO + 2H_2O$$

The balanced equation is

$$3Fe^{2+} + NO_3^- + 4H^+ \rightarrow 3Fe^{3+} + NO + 2H_2O$$

23.32

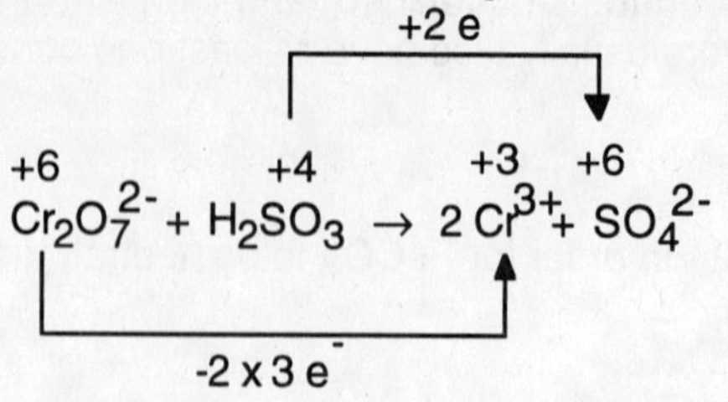

Balance electron charges: $Cr_2O_7^{2-} + 3H_2SO_3 \rightarrow 2Cr^{3+} + 3SO_4^{2-}$

Balance oxygen with water: $Cr_2O_7^{2-} + 3H_2SO_3 \rightarrow 2Cr^{3+} + 3SO_4^{2-} + 4H_2O$

Balance hydrogen: $Cr_2O_7^{2-} + 3H_2SO_3 + 2H^+ \rightarrow 2Cr^{3+} + 3SO_4^{2-} + 4H_2O$

23.33 Because one mole of chloride ion is precipitated per formula unit of the complex, the chlorine atoms must be present as chloride ion. All the other ligands are coordinated to the cobalt. An appropriate formula is

$$[Co(NH_3)_4(NO_2)_2]Cl$$

23.34 Because there is no precipitate with silver nitrate, both chlorine atoms must be coordinated to the cobalt. An appropriate formula is

$$[Co(NH_3)_4Cl_2]NO_3$$

23.35 a. 4. There are four cyanide groups coordinated to the gold atom.

b. 6. There are four ammonia molecules and two water molecules coordinated to the cobalt.

c. 4. Each of the ethylenediamine molecules bonds to the gold atom through two nitrogen atoms.

d. 6. Each ethylenediamine molecule bonds to the chromium atom through two nitrogen atoms, and the oxalate ion bonds to the chromium through two oxygen atoms.

23.36 a. 6. There are six ammonia molecules around the nickel atom.

b. 4. There are four ammonia molecules around the copper atom.

c. 6. Each ethylenediamine molecule bonds to the chromium atom through two nitrogen atoms.

d. 4. There are four cyanide ions around the nickel atom.

23.37 a. The charge on the $Ni(CN)_4^{2-}$ ion is -2, to balance the charge of +2 from the $2K^+$ ions. Each cyanide ion has a charge of -1. The sum of the oxidation number of nickel and the charge on the cyanide ions must equal the charge, so

$$-2 = [4 \times (-1)] + 1 \times [\text{ox. no. (Ni)}]$$

$$\text{Ox. no. (Ni)} = -2 - (-4) = +2$$

b. The charge on ethylenediamine is 0, so the oxidation number of Mo is equal to the charge on the complex ion.

Ox. no. (Mo) = +3

c. Oxalate ion has a charge of -2, so

Ox. no. of Cr = charge of complex ion -3 x (charge of oxalate ion)

= (-3) - [3 × (-2)] = +3

d. Chloride ion has a charge of -1, so the charge on the complex ion is +2. The NH_3 ligands are neutral and contribute nothing to the charge of the complex ion.

Ox. no. of Co = charge of complex ion - charge of nitrite ligand

= +2 - (-1) = +3

23.38 a. Nitrite ion has a charge of -1, so the charge on the complex ion is +1. The en ligands are neutral. Both the chloride and nitrite ligands have a charge of -1, so

Ox. no. of Co = charge of complex ion - charge of chloride - charge of nitrite

= +1 - (-1) - (-1) = +3

b. Each chloride ligand has a charge of -1.

Ox. no. of Pt = charge of complex ion - 4 x (charge of chloride ion)

= -2 - [4 x (-1)] = +2

c. Potassium has a charge of +1, so the complex ion has a charge of -3. Each cyanide ligand has a charge of -1.

Ox. no. of Cr = charge of complex ion - 6 x (charge of cyanide)

= -3 - [6 × (-1)] = +3

d. The water ligands are neutral and the hydroxide ligand has a charge of -1, so

Ox. no. of Fe = charge of complex ion - charge of hydroxide

= +2 - (-1) = +3

23.39 a. The charge of each chloride ligand is -1; the charge on the oxalate ligand is -2. The ammonia ligands are neutral.

Ox. no. of Cr = charge of complex ion - charge of oxalate - 2 × (charge of chloride)

$$= -1 - (-2) - 2(-1) = +3$$

b.

Formula	Name
NH_3	Ammine
Cl^-	Chloro
$C_2O_4^{2-}$	Oxalato

c. 6. The chromium atom has one bond to each of the NH_3 ligands and Cl^- ligands, and two bonds are to the $C_2O_4^{2-}$ ligand.

d. If each NH_3 was replaced by one Cl^- and the $C_2O_4^{2-}$ ligand was replaced by two Cl^- ligands, there would be a total of six Cl^- ligands bonded to a chromium atom in the +3 oxidation state.

Charge on complex ion = ox. no. of Cr + 6 × (charge on chloride ligand)

$$= +3 + [6 \times (-1)] = -3$$

23.40 a. The only charged ligands are the hydroxo ligands, with a charge of -1.

Ox. no. of Mn = charge of complex ion - charge of hydroxo ligand

$$= +2 - (-1) = +3$$

b.

Formula	Name
H_2O	Aqua
NH_3	Ammine
OH^-	Hydroxo

c. 6. There are two ammonia molecules, three water molecules, and a hydroxide ion, each singly bonded to the manganese atom.

d. If each of the ligands was replaced by chloride ions, there would be six chloride ions bonded to a manganese atom in the +3 oxidation state.

Charge of complex ion = ox. no. of Mn + 6 x (charge of chloride ion)

$$= +3 + [6 \times (-1)] = -3$$

23.41 a. Potassium hexafluoroferrate(III)
b. Diamminediaquacopper(II) ion
c. Ammonium aquapentafluoroferrate(III)
d. Dicyanoargentate(I) ion

23.42 a. Potassium octacyanomolybdate(IV)
b. Hexafluorochromate(III) ion
c. Trioxalatovanadate(IV) ion
d. Potassium tetrachloroferrate(II)

23.43 a. Pentacarbonyliron(0)
b. Dicyanobis(ethylenediamine)rhodium(III) ion
c. Tetraamminesulfatochromium(III) chloride
d. Tetraoxomanganate(VII) ion (Permanganate is the usual name.)

23.44 a. Octacarbonyltungsten(0)
b. Diaquabis(ethylenediamine)cobalt(VI) sulfate
c. Potassium octacyanomolybdate(VII)
d. Tetraoxochromate(VI) ion (Chromate is the usual name.)

23.45 a. The charge on the complex ion equals

$$\text{Ox. no. of Mn} + 6 \times (\text{charge on cyanide ion}) = +3 + 6(-1) = -3$$

Hence, the formula is $K_3[Mn(CN)_6]$.

b. The charge on the complex ion equals

$$\text{Ox. no. of Zn} + 4 \times (\text{charge on cyanide ion}) = 2 + 4(-1) = -2$$

Hence, the formula is $Na_2[Zn(CN)_4]$.

c. The charge on the complex ion equals

$$\text{Ox. no. of Co} + 2 \times (\text{charge on chloride ion})$$

Note that the ammine ligand (NH_3) is neutral. You get $+3 + 2(-1) = +1$. The formula is $[Co(NH_3)_4Cl_2]NO_3$.

d. The charge on the cation equals the ox. no. of Cr (+3). The charge on the anion equals

$$\text{Ox. no. of Cu} + 4 \times (\text{charge on chloride ion}) = +2 + 4(-1) = -2$$

The formula is $[Cr(NH_3)_6]_2[CuCl_4]_3$.

23.46 a. The charge on the complex ion equals

$$\text{Ox. no. of Cu} + 2 \times (\text{charge on cyanide ion}) = +2 + 2(-1) = 0$$

The formula is $[Cu(H_2O)_2(CN)_2]$.

b. The charge on the complex ion is

$$\text{Ox. no. of Pt} + 6 \times (\text{charge on chloride ion}) = +4 + 6(-1) = -2$$

The formula is $K_2[PtCl_6]$.

c. The charge on the complex ion equals the

Ox. no. of Ni (+2)

The formula is $[Ni(NH_3)_4](ClO_4)_2$.

d. The charge on the cation equals the ox. no. of Pt(+2). The charge on the anion equals

$$\text{Ox. no. of Cu} + 4 \times (\text{charge on chloride ion}) = +2 + 4(-1) = -2$$

Therefore, the formula is $[Pt(NH_3)_4]\,[CuCl_4]$.

23.47 a. Linkage isomerism. In one case, the SCN^- ligand is attached to Co at the S; in the other, it is attached at the N.

b. Coordination isomerism. In one case, the NH_3 ligands are attached to Co; in the other, they are attached to Cr.

c. Hydrate isomerism. In the first compound, both H_2O molecules are coordinated to the Co atom; in the other, only one H_2O is directly attached to Co.

d. Ionization isomerism. In the first compound, a chloride ion is produced upon ionization; in the other, a nitrite ion is produced.

23.48 a. Ionization isomerism; as in Problem 23.47(d).

b. Coordination isomerism; same as Problem 23.47(b).

c. Linkage isomerism. In the first compound, the nitrite ion is bonded to the cobalt through an O atom; in the other, the nitrogen atom is bonded to the Co.

d. Hydrate isomerism. In the first compound, the water is attached directly to the chromium atom. In the second, the last water is not coordinated to the chromium.

23.49 The given compound must consist of a chloride ion (which can be precipitated with $AgNO_3$ solution) and a $[Co(NH_3)_4Br_2]^+$ ion, giving $[Co(NH_3)_4Br_2]Cl$. An ionization isomer is $[Co(NH_3)_4BrCl]Br$.

23.50 The complex must consist of a bromide ion, which can be precipitated with $AgNO_3$ solution, and a $[Co(NH_3)_4C_2O_4]^+$ ion. Therefore, the formula is $[Co(NH_3)_4C_2O_4]Br$. The isomer would be $[Co(NH_3)_4Br]C_2O_4$.

23.51 a.

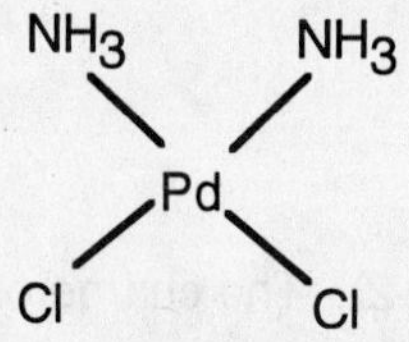

cis

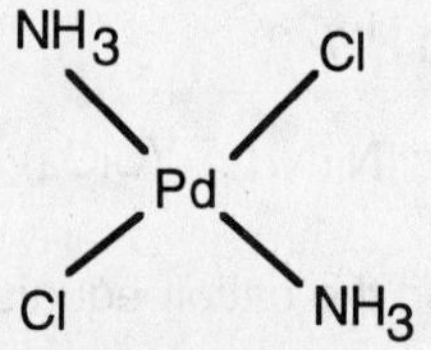

trans

b. No geometric isomerism.

c. No geometric isomerism.

d.

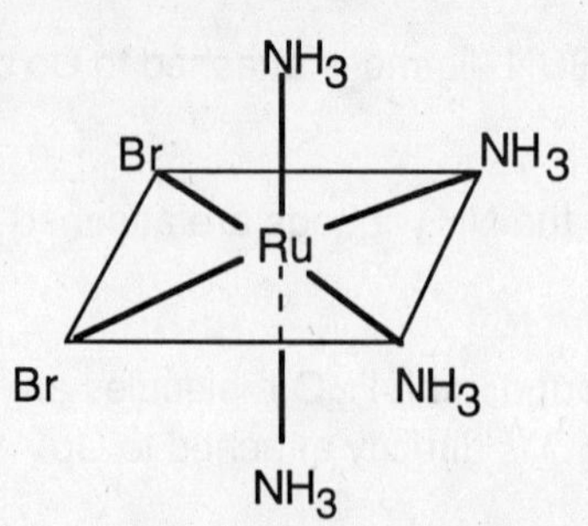

cis

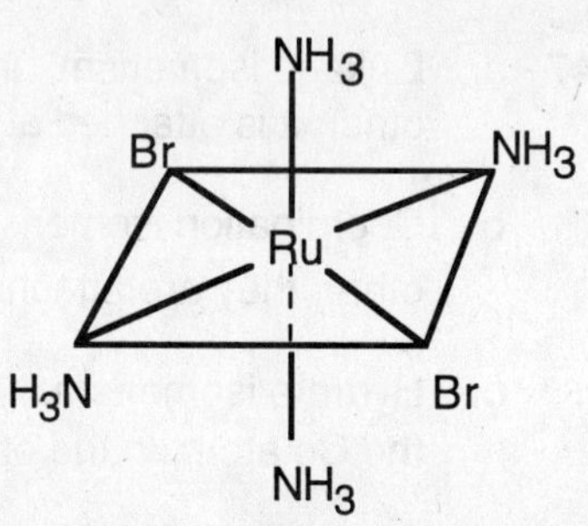

trans

23.52 a.

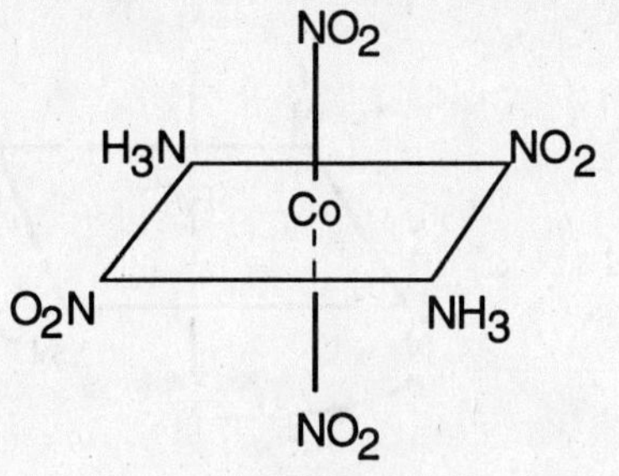

trans

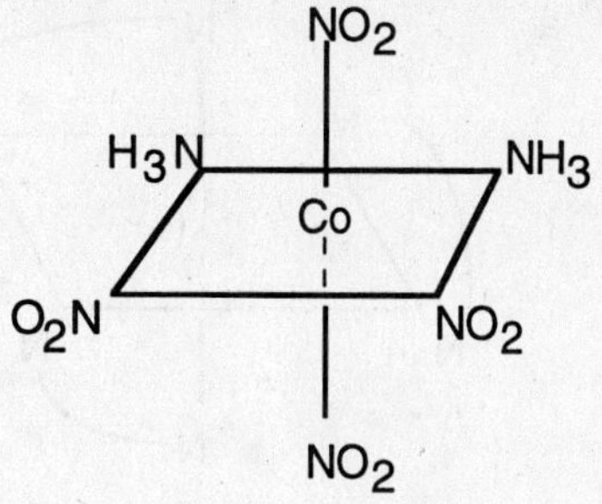

cis

b. No geometric isomerism.

c.

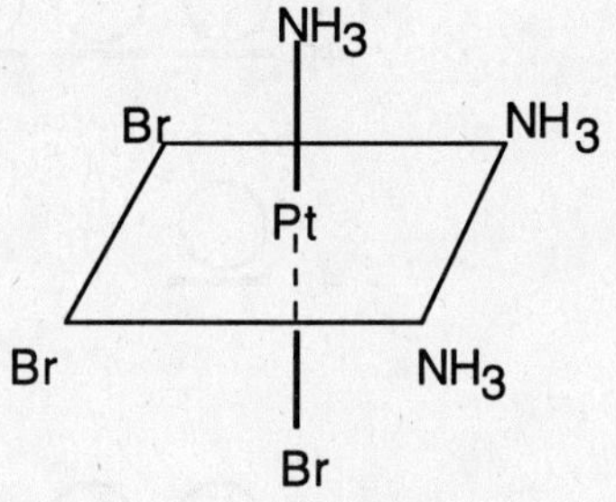

cis

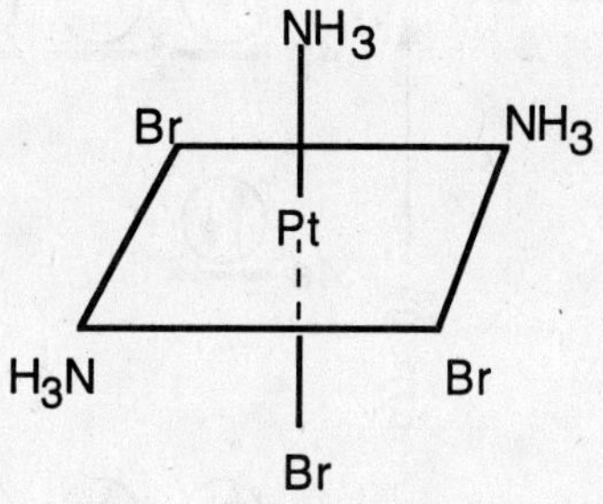

trans

d. No geometric isomerism.

23.53 a.

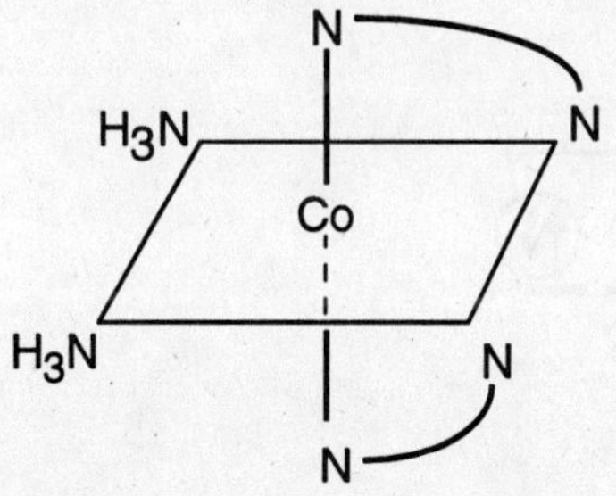

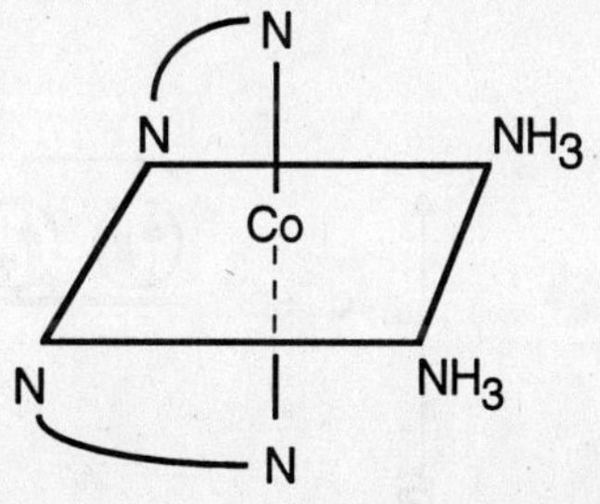

b. No optical isomers.

23.54 a.

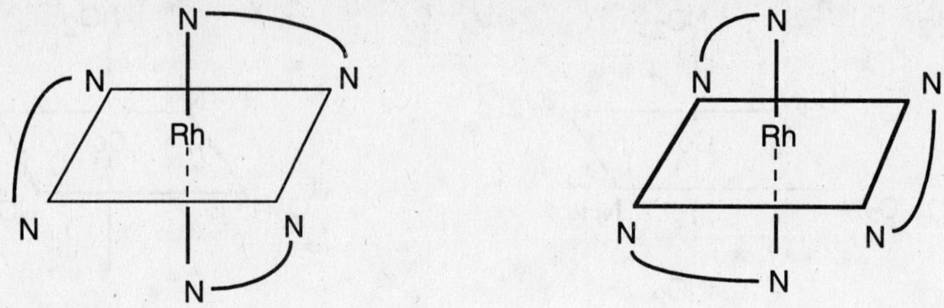

b. No optical isomers.

23.55 a.

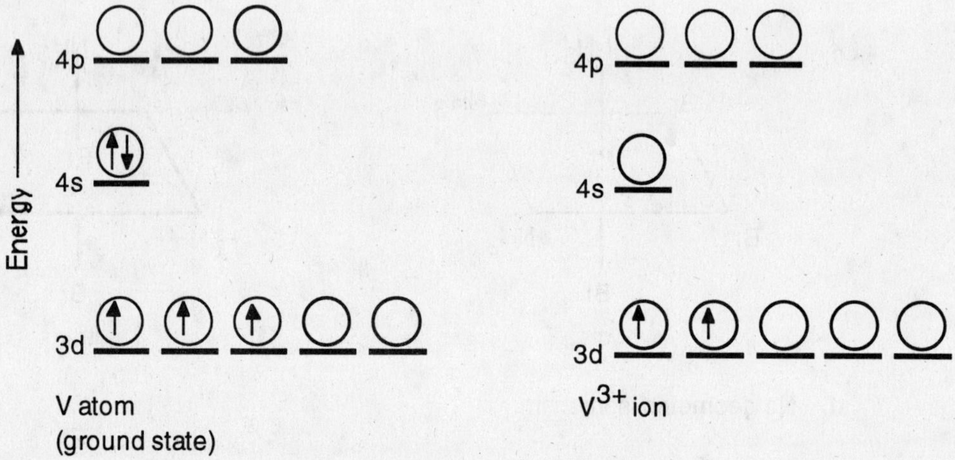

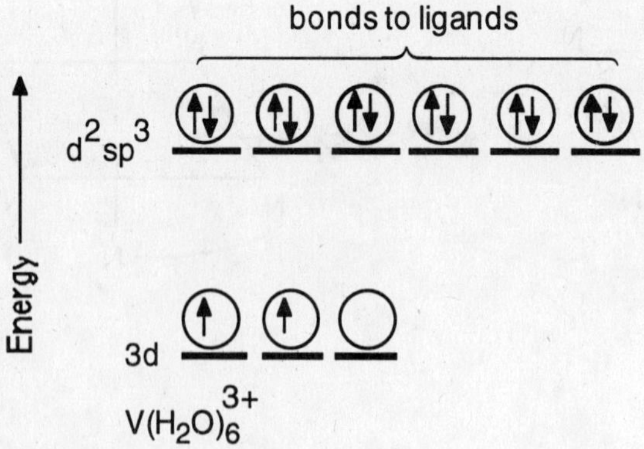

b.

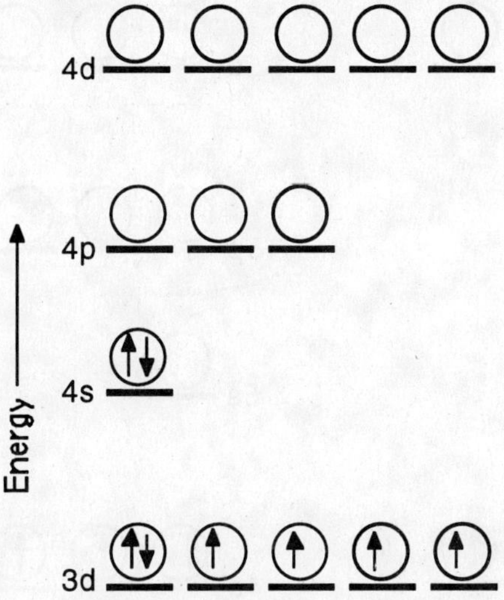

Fe atom
(ground state)

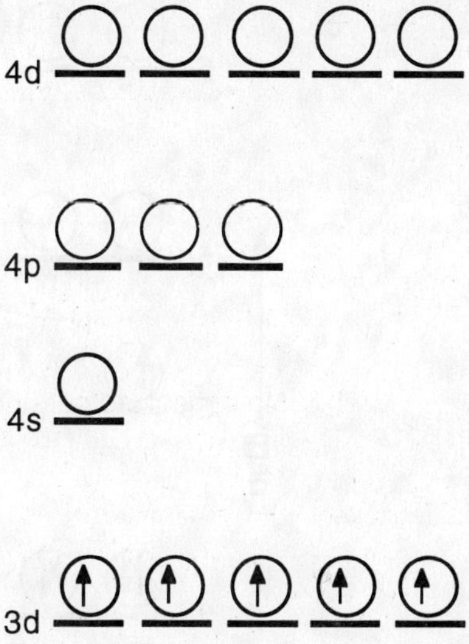

Fe^{3+} ion

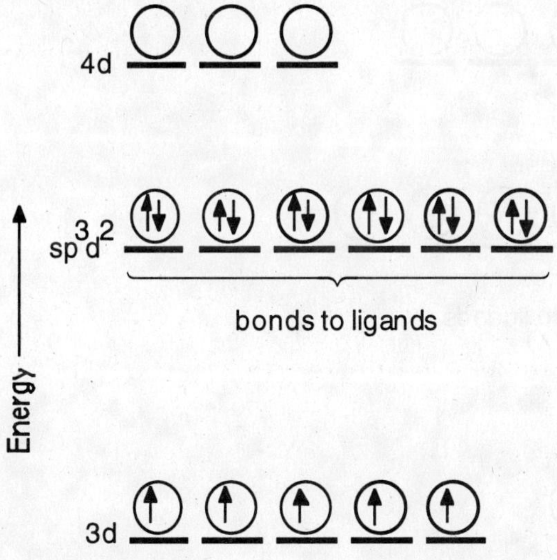

$Fe(en)_3^{3+}$

c.

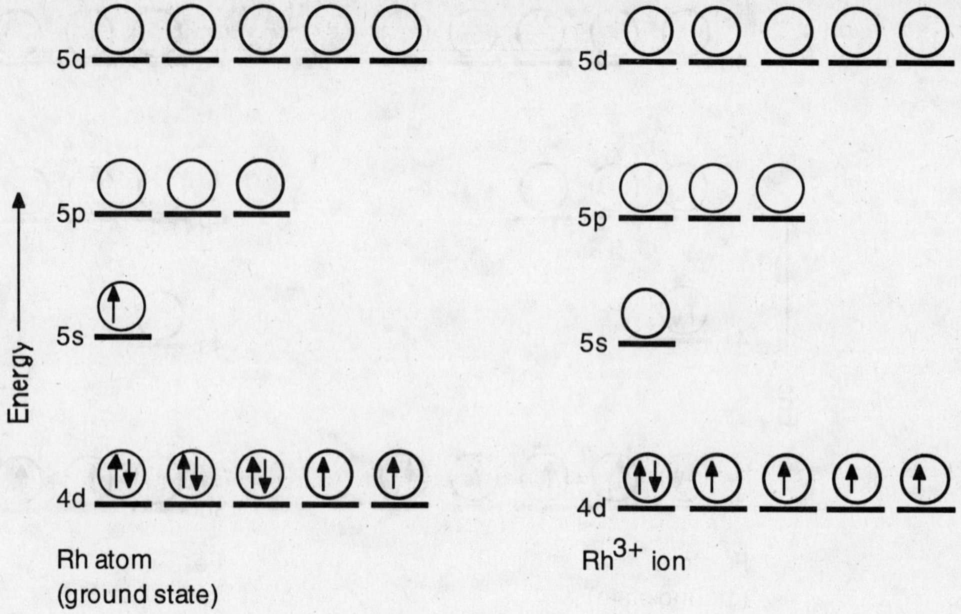

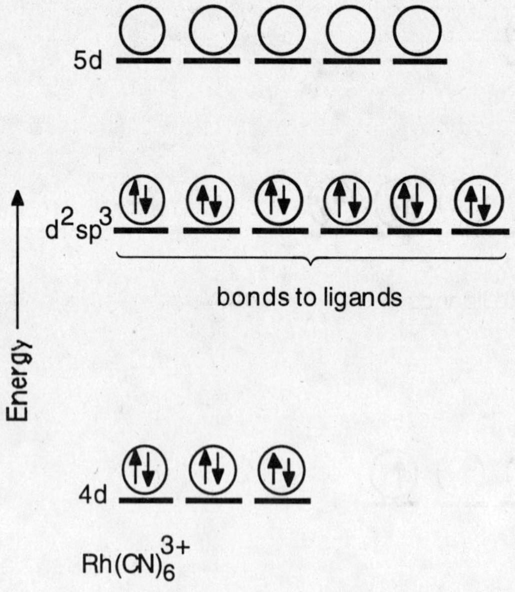

In the low-spin complex, two of the unpaired electrons in 4d orbitals are paired with two others to provide two empty d orbitals for cyanide ligands.

23.56 a.

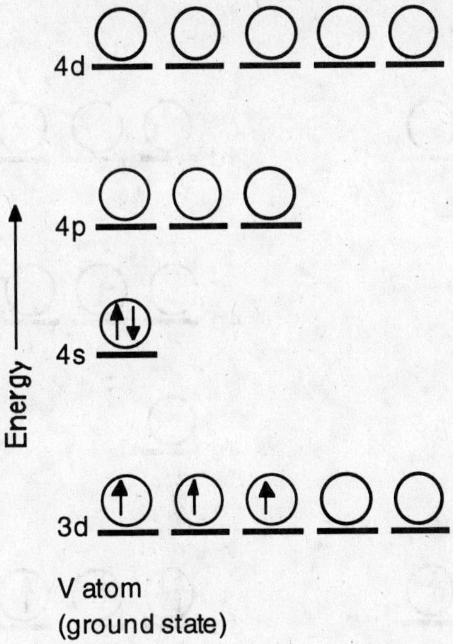
4d
4p
4s
3d
Energy
V atom
(ground state)

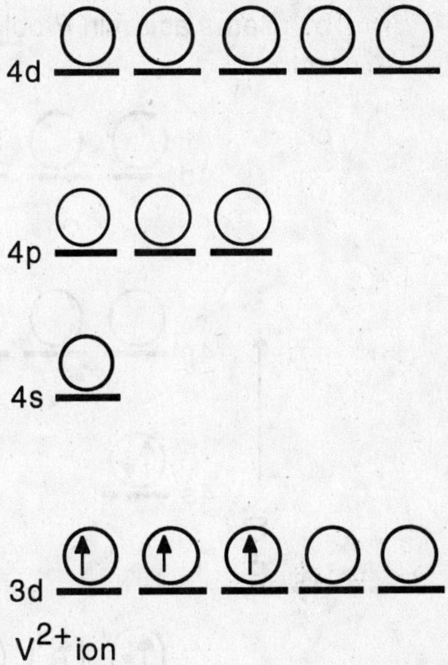
4d
4p
4s
3d
V^2+ ion

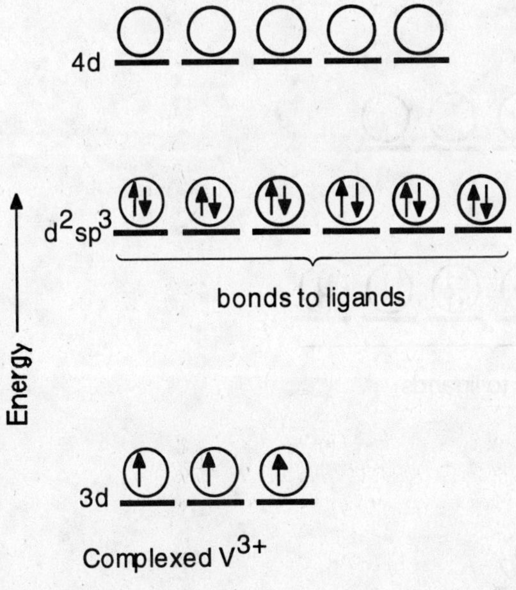
4d
d^2sp^3
bonds to ligands
Energy
3d
Complexed V^3+

b. Same as (b) in Problem 23.55.

c.

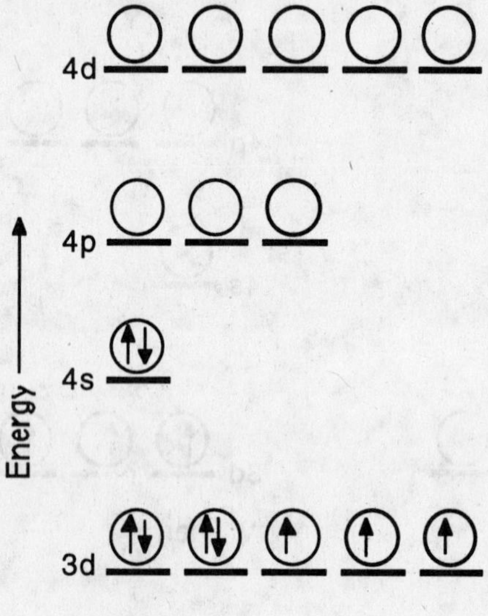

Co atom
(ground state)

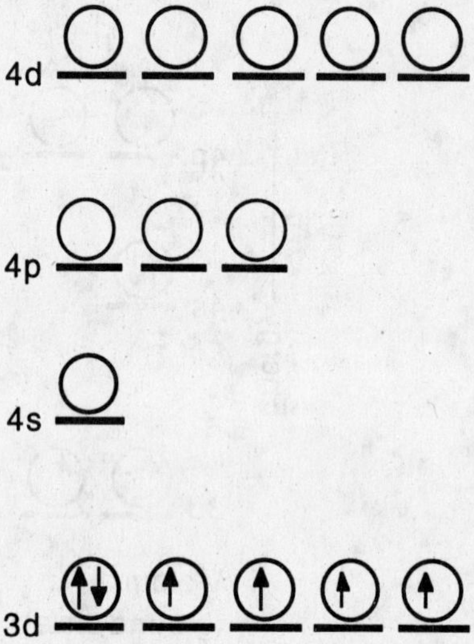

Co^{3+} ion

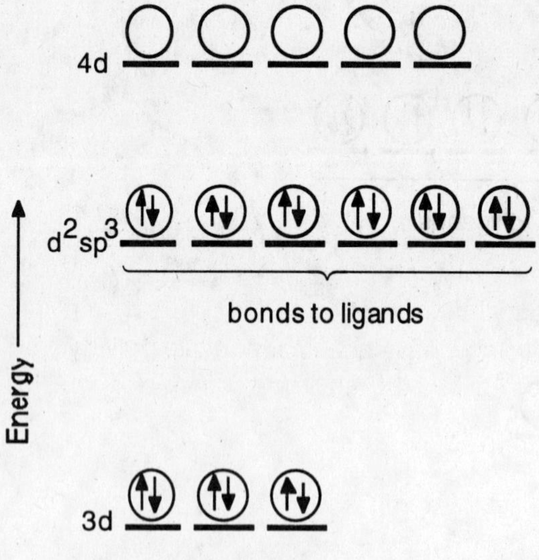

Complexed Co^{3+}

23.57 a. Pt: $[Xe]4f^{14}5d^{9}6s^{1}$

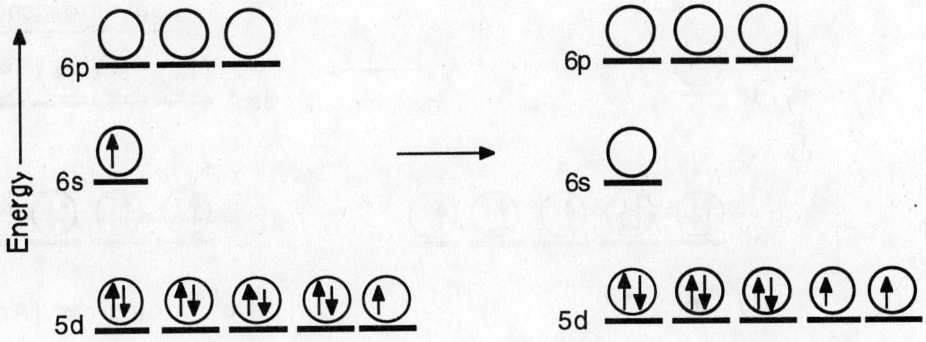

Pt atom: $[Xe]4f^{14}$ Pt^{2+} ion: $[Xe]4f^{14}$

If there are no unpaired electrons in the complex, the configuration of the complexed Pt must be

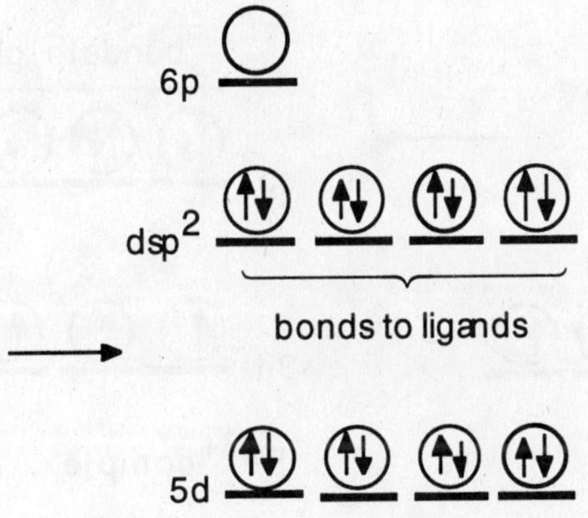

Complexed Pt^{2+}: $[Xe]4f^{14}$

in which the unpaired d electrons have been paired up to provide a 5d orbital for bonding to a ligand in a dsp^2 hybrid orbital. The geometry for dsp^2 hybridization is square planar.

b.

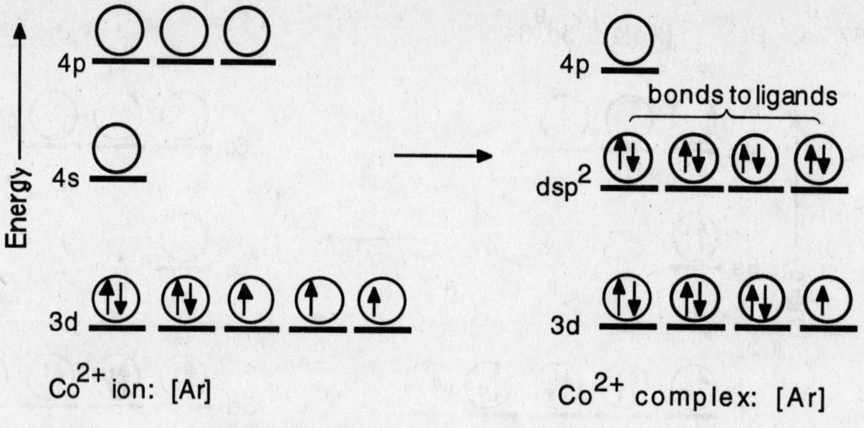

If there is only one unpaired electron in the complex, the configuration of the complexed Co must be dsp^2 hybridization with square planar geometry.

c.

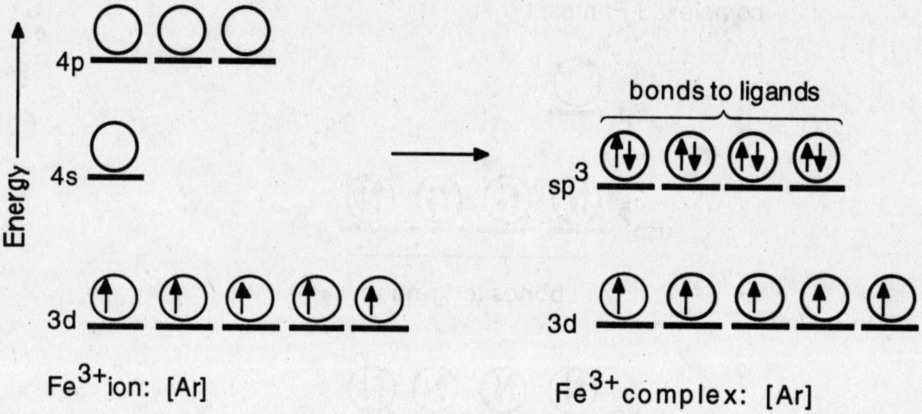

If there are five unpaired electrons in the complex, the configuration of the Fe must be sp^3 hybridization with tetrahedral geometry.

d.

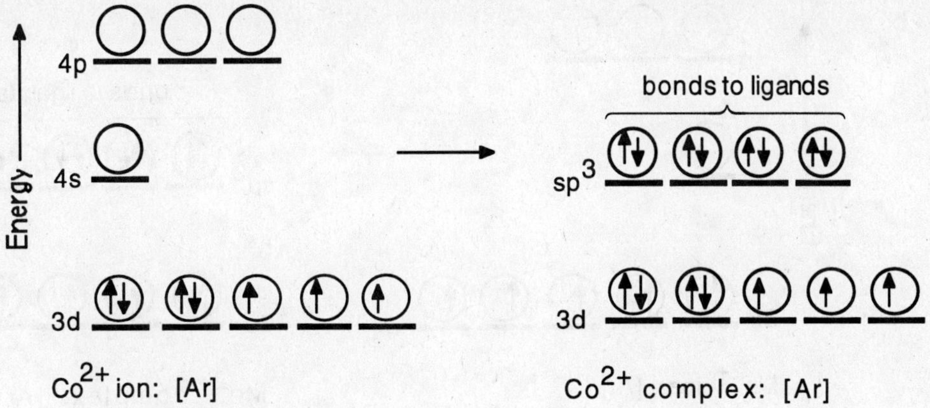

If there are three unpaired electrons in the complex, the configuration of the cobalt must be sp^3 hybrid with tetrahedral geometry.

23.58 a

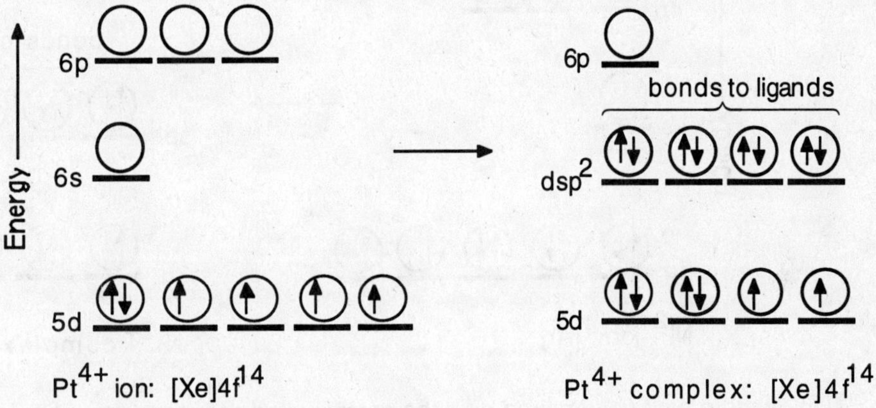

If there are two unpaired e- in the complex, the configuration of the Pt must be dsp^2 hybrid with square planar geometry.

b.

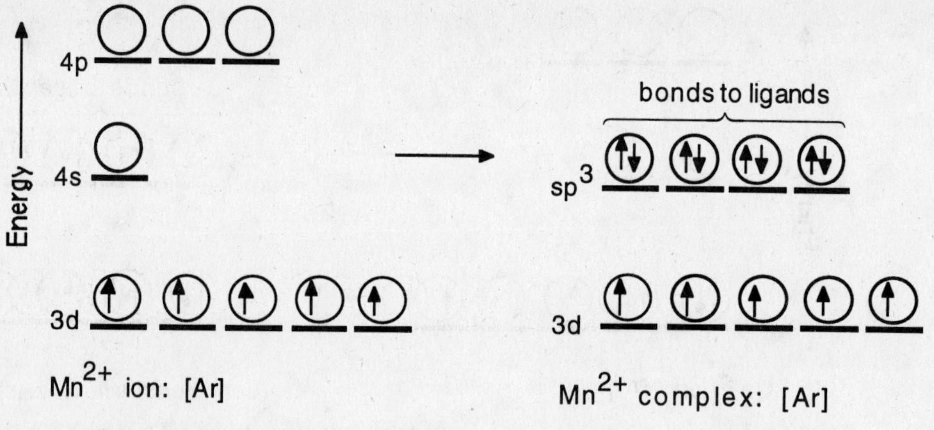

If there are five unpaired e- in the complex, the configuration of the Mn must be sp^3 hybrid with tetrahedral geometry.

c.

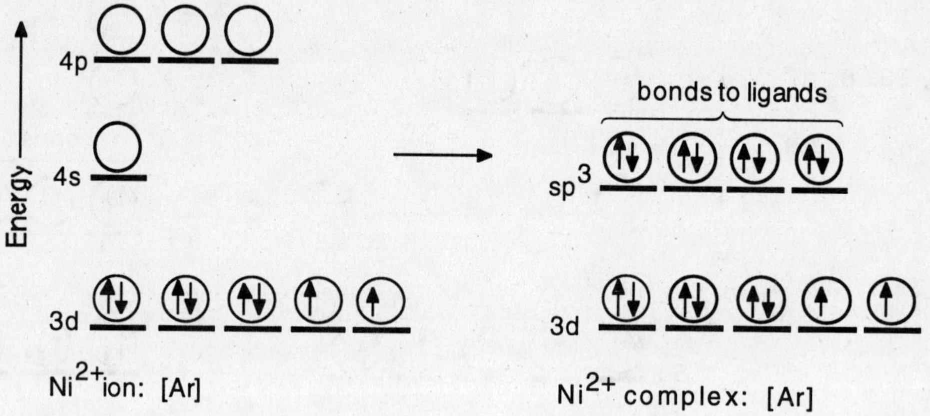

If there are two unpaired e- in the complex, the configuration of Ni must be sp^3 hybrid with tetrahedral geometry.

d.

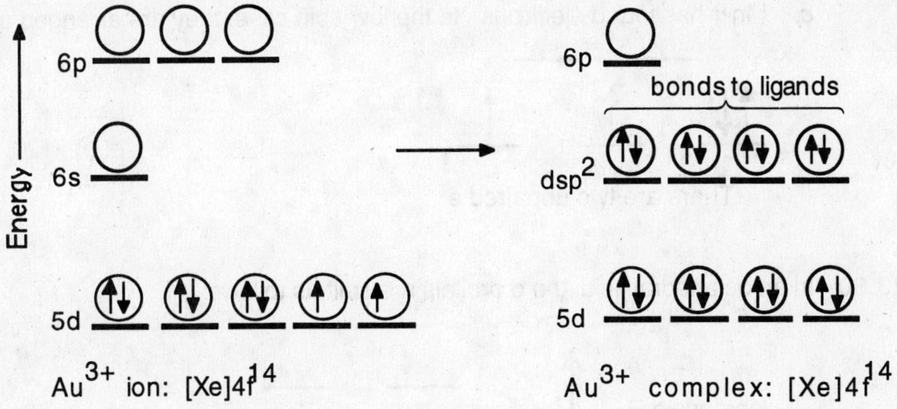

If there are no unpaired e- in the complex, the configuration of the gold atom must be dsp^2 hybrid with square planar geometry.

23.59 In an octahedral field, the d orbitals are split so that three of them are of a lower energy than the remaining two.

a. V^{3+} has two d electrons arranged as shown:

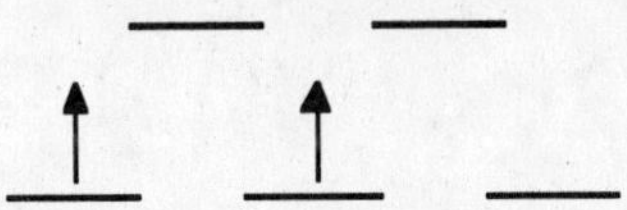

There are two unpaired electrons.

b. Co^{2+} has seven d electrons. In the high-spin case, they are arranged as follows:

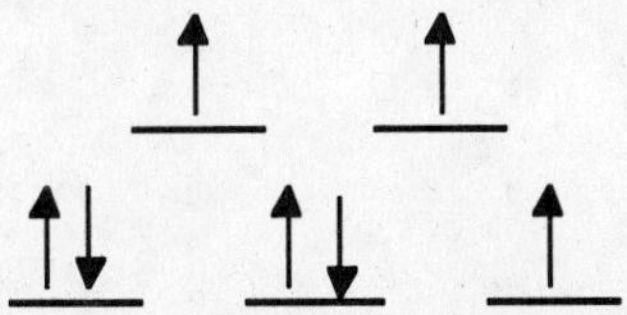

There are three unpaired e-.

c. Mn^{3+} has four d electrons. In the low-spin case, they are arranged as follows:

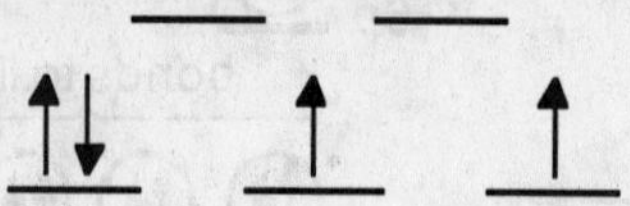

There are two unpaired e-.

23.60 In an octahedral field, the d orbitals are split as follows:

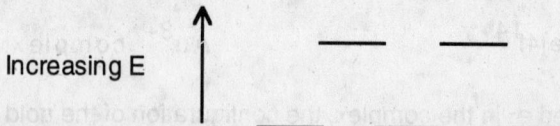

a. Zr^{2+} has two electrons in the 4d subshell. In the complex, they are arranged as follows:

There are two unpaired e-.

b. Os^{4+} has four electrons in the 5d subshell. In the low-spin case, they are arranged as follows:

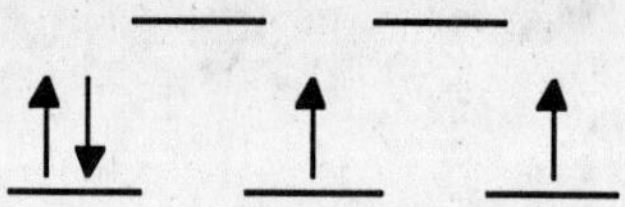

There are two unpaired e-.

c. Mn^{2+} has five d electrons. In the high-spin case, they are arranged as follows:

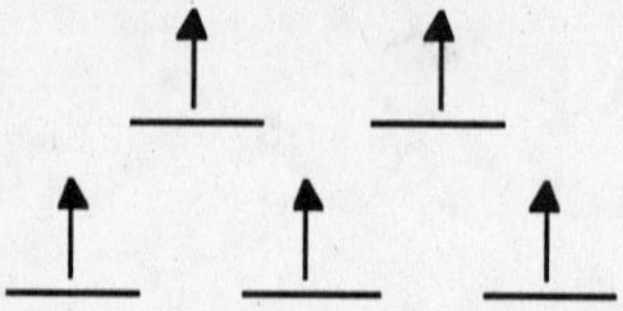

There are five unpaired electrons.

23.61 a. Pt^{2+} has eight electrons in the 5d subshell. Because the complex is diamagnetic, the crystal field felt by the d orbitals is most likely square planar (no low-spin tetrahedral complexes are known). The arrangement of the d electrons is

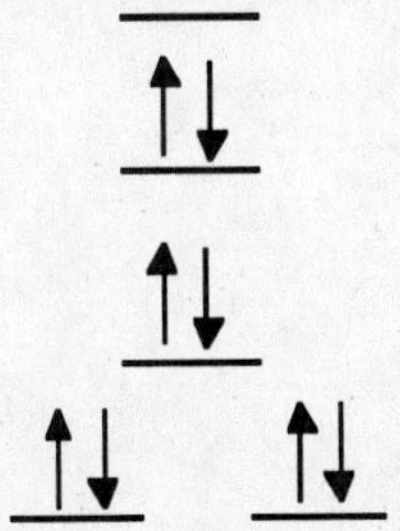

b. Co^{2+} has seven d electrons. One unpaired electron implies a low-spin complex. A square planar field will lead to low-spin complexes, so the geometry is probably square planar, with the d electrons arranged as follows:

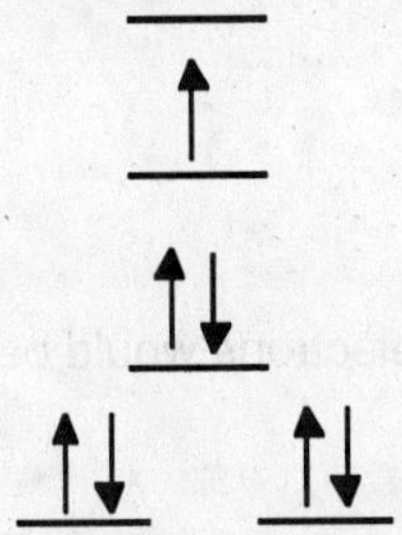

c. Fe^{3+} has five d electrons. If they are all unpaired, that is, a high-spin complex, the field felt by the metal ion is most likely tetrahedral with the d electrons arranged as follows:

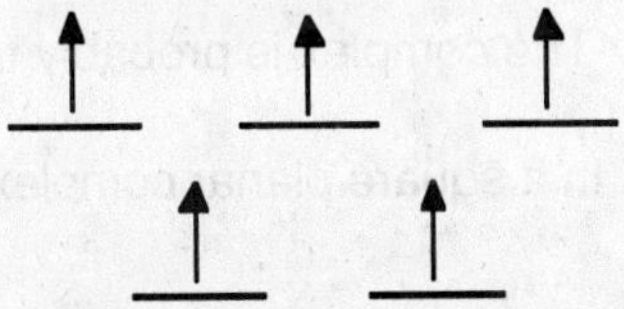

d. Co^{2+} has seven d electrons. Three unpaired electrons imply a high-spin complex. A tetrahedral field will lead to high-spin complexes, so the geometry is probably tetrahedral with the d electrons arranged as follows:

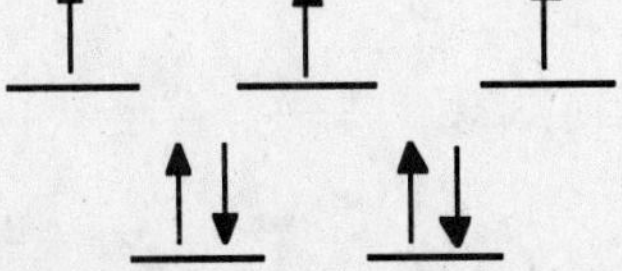

23.62 a. Pt^{4+} has six electrons in the 5d subshell. The presence of only two unpaired electrons can be explained by a square planar field, with the d electrons arranged as follows:

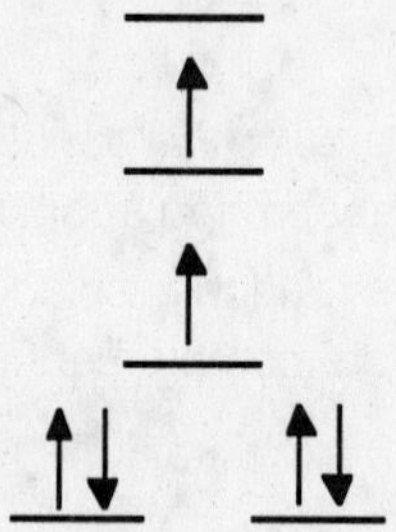

b. Mn^{2+} has five d electrons. Because these are all unpaired, it is a high-spin complex and probably has tetrahedral geometry.

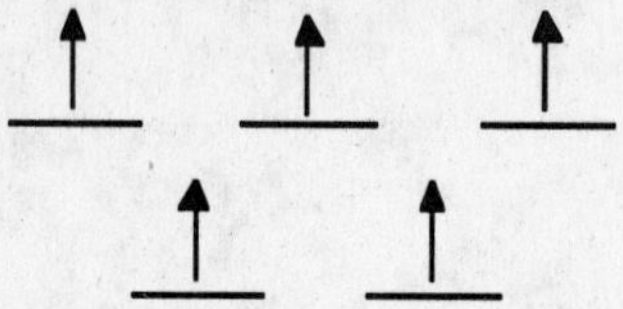

c. Ni^{2+} has eight d e^-. In a tetrahedral complex, eight d electrons would be arranged as follows:

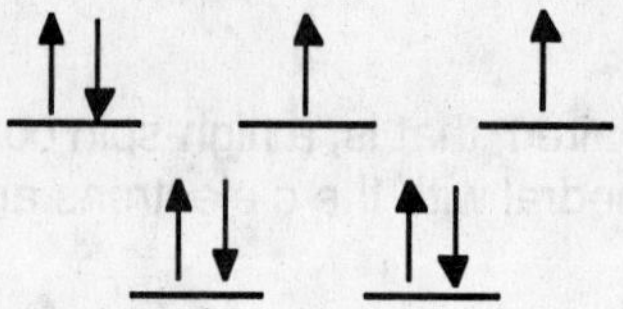

giving two unpaired electrons, as observed. The complex is probably tetrahedral.

d. Au^{3+} has eight electrons in the 5d subshell. In a square planar complex, these would be arranged as follows:

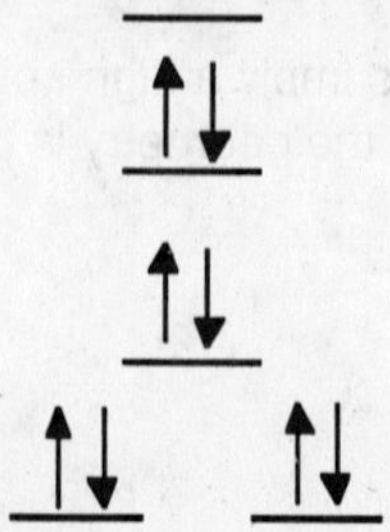

with no unpaired electrons. The complex is probably square planar.

23.63 Purple (from Table 23.8)

23.64 Orange (from Table 23.8)

23.65 Yes. According to the spectrochemical series, H_2O is a more weakly bonding ligand than NH_3, so Δ should decrease. The wavelength of the absorption should increase ($\lambda = hc/\Delta$). The light absorbed by $Co(NH_3)_6^{3+}$ is violet-blue, and the replacement of one NH_3 by H_2O shifts this toward blue. Thus, the observed (complementary) color is shifted toward red, as is observed.

23.66 Yes. According to the spectrochemical series, F^- is a more weakly bonding ligand than ethylenediamine, so Δ should decrease and the wavelength of the absorbed light should increase. The wavelength absorbed by $Co(en)_3^{3+}$ is violet-blue, and replacement of all the ligands by F^- shifts the absorption toward the red end of the spectrum. The color of the complex (the complement of the absorbed light) should shift toward blue, as is observed.

23.67 $$\Delta = \frac{hc}{\lambda} = \frac{(6.626 \times 10^{-34}\ J \cdot s)(2.998 \times 10^{8}\ m/s)}{(500 \times 10^{-9}\ m)} \times 6.02 \times 10^{23}/mol$$

$$= 2.3\underline{9}17 \times 10^5 = 2.39 \times 10^5\ J/mol\ \ (239\ kJ/mol)$$

23.68 $$\Delta = \frac{hc}{\lambda} = \frac{(6.626 \times 10^{-34}\ J \cdot s)(2.998 \times 10^{8}\ m/s)}{(680 \times 10^{-9}\ m)} \times 6.02 \times 10^{23}/mol$$

$$= 1.7\underline{5}86 \times 10^5 = 1.76 \times 10^5\ J/mol\ \ (176\ kJ/mol)$$

■ Solutions to Unclassified Problems

23.69 The color in transition-metal complexes is due to absorption of light when a d electron moves to a higher energy level. Because Sc^{3+} has no d electrons, it is expected to be colorless.

23.70 Zn^{2+} ion has a completely filled d subshell, so there are no openings for d electrons to be promoted to a higher energy level for which the difference in energy would correspond to the energy of visible light.

23.71 If $[Co(NH_3)_4Cl_2]^+$ had a regular planar hexagonal geometry, three geometric isomers would be expected.

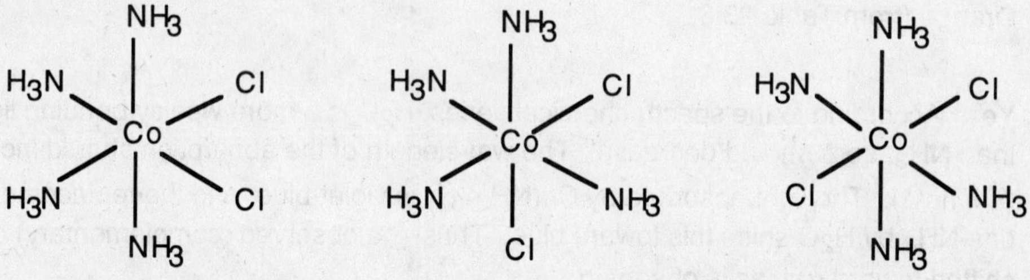

The known existence of only two isomers of the complex, by itself, is not sufficient to rule out hexagonal geometry. There is the possibility that the third isomer has simply not been made. Other information is required to eliminate hexagonal geometry as a possibility.

23.72 If $[Pt(NH_3)_3Cl_3]^+$ had a regular planar hexagonal geometry, three geometric isomers would be expected.

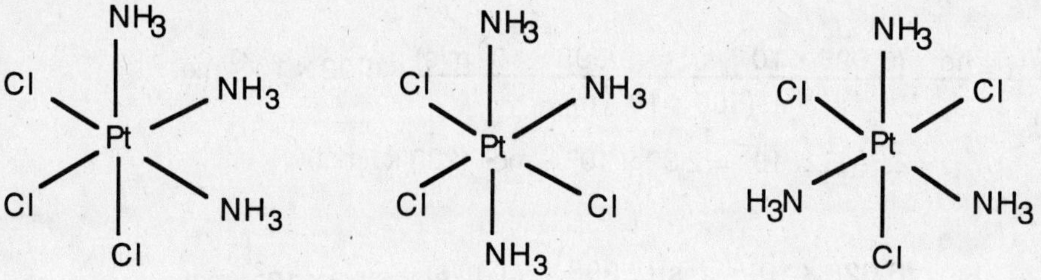

No. Same explanation as for Problem 23.71.

23.73 $K_c = 2.1 \times 10^{-13} = \dfrac{[Cu^{2+}][NH_3]^4}{[Cu(NH_3)_4{}^{2+}]}$ for $Cu(NH_3)_4{}^{2+} \rightleftharpoons Cu^{2+}(aq) + 4\,NH_3(aq)$

Concentration (M)	$Cu^{2+}(aq)$	+ $4NH_3(aq)$	$\rightleftharpoons$ $Cu(NH_3)_4{}^{2+}(aq)$
Start	0.10	0.40	0
Change	-x	-4x	+x
Equilibrium	0.10 - x	0.40 - 4x	x

(continued)

Substituting into the equilibrium expression, you obtain

$$2.1 \times 10^{-13} = \frac{(0.10 - x)[4(0.10 - x)]^4}{x}$$

To simplify, let y = 0.10 - x. Then x = 0.10 - y. The above equation becomes

$$2.1 \times 10^{-13} \cong \frac{y\,(4y)^4}{(0.10\ -\ y)}$$

Assume that y is small compared to 0.1. Then 0.10 - y ≅ 0.10. Then

$$2.1 \times 10^{-13} \cong \frac{y\,(4y)^4}{(0.10)}$$

$$y^5 \cong \frac{(2.1 \times 10^{-13})\,(0.10)}{4^4} \cong 8.\underline{2}03 \times 10^{-17}$$

$$y \cong \sqrt[5]{8.203 \times 10^{-17}} \cong 6.\underline{0}64 \times 10^{-4}$$

Going back to the assumption above, you find that to two significant figures, 0.10 - 0.0006$\underline{1}$ = 0.10. The assumption was valid.

At equilibrium:

$[Cu^{2+}]$ = (0.10 - x) M = y = 6.$\underline{0}$64 × 10^{-4} = 6.1 × 10^{-4} M

$[NH_3]$ = 4(0.10 - x) M = 4y = 4 × 6.$\underline{0}$64 × 10^{-4} = 2.$\underline{4}$3 × 10^{-3} = 2.4 × 10^{-3} M

$[Cu(NH_3)_4{}^{2+}]$ = x = (0.10 - y) = 0.10 M

23.74 $K_d = 5.9 \times 10^{-8}\ M^2$ for $Ag(NH_3)_2^+ \rightleftharpoons Ag^+ + 2NH_3$

$$K_d = \frac{[Ag^+][NH_3]^2}{[Ag(NH_3)_2^+]}$$

Concentration (M)	$Ag^+(aq)$	+	$2NH_3\ (aq)$	$\rightleftharpoons$	$Ag(NH_3)_2^+(aq)$
Start	0.10		0.20		0
Change	-x		-2x		+x
Equilibrium	0.10 - x		0.20 - 2x		x

$$K_d = 5.9 \times 10^{-8} = \frac{(0.10 - x)[2(0.10 - x)]^2}{x}$$

Let y = 0.10 - x; then x = 0.10 - y.

$$5.9 \times 10^{-8} = \frac{(y)(2y)^2}{0.10 - y}$$

Assume y < 0.10, so 0.10 - y ≅ 0.10.

$$5.9 \times 10^{-8} \cong \frac{4y^3}{0.10}$$

$$y^3 \cong \frac{(5.9 \times 10^{-8})(0.10)}{4} \cong 1.\underline{4}75 \times 10^{-9}$$

$$y \cong \sqrt[3]{1.475 \times 10^{-9}} \cong 1.138 \times 10^{-3}$$

(Checking the assumption above, you note that 0.10 - 0.0011 = 0.10. The assumption was valid.)

At equilibrium:

$[Ag^+] = (0.10 - x)\ M = y\ M = 1.\underline{1}38 \times 10^{-3} = 1.1 \times 10^{-3}\ M$

$[NH_3] = 2(0.10 - x)\ M = 2y\ M = 2.\underline{2}76 \times 10^{-3} = 2.3 \times 10^{-3}\ M$

$[Ag(NH_3)_2^+] = x\ M = (0.10 - y)\ M = 0.10\ M$

■ Solution to Conceptual Problem

23.75 The transition elements have a unique reactivity due to the many different oxidation states that can exist. The 18-electron rule (similar to the octet rule) has been suggested to describe stable transition metal compounds. What is the aufbau principle basis for this suggestion? Give an example of a stable transition metal compound that is consistent with this rule. What about the lanthanides and actinides?

Discussion

The aufbau principle states that you add electrons to the outer orbitals of an atom as you add protons to the nucleus (to increase the atomic number). With each filling of a shell, you expect some stability, because there will be a jump in energy with the addition of an electron into a new shell. The octet rule follows from the filling of a shell of s and p orbitals, for a total of eight electrons.

With the transition metals comes a d subshell of ten more electrons, for a total of 18. The $FeCl_6^{4-}$ ion provides an example of a stable transition-metal complex that obeys the 18-electron rule. The complex ion consists of Fe^{2+} (with configuration $[Ar]4d^6$) complexed to six Cl^- ions. Fe^{2+} contains six electrons in its outer shell, and each Cl^- ion contributes a pair of electrons to the complex. Thus, the total electrons in the outer shell of iron is 6 + 6(2) = 18.

For the lanthanides and actinides, many complexes obey this 18-electron rule, but a few complexes follow an analogous 32-electron rule in which there is an additional f subshell. The f subshell can hold a total of 14 electrons, which together with the s, p, and d subshells (18 electrons) gives a total of 32 electrons. An example of a complex following the 32-electron rule is $Lu(H_2O)_9^{3+}$. The complex consists of Lu^{3+} (configuration $[Xe]4f^{14}$) complexed to nine H_2O molecules. Lutetium(III) ion contains 14 electrons in its outer shell, and each H_2O molecule contributes a pair of electrons to the complex. Thus, the total electrons in the outer shell of iron is 14 + 9(2) = 32.

24. ORGANIC CHEMISTRY

■ Solutions to Exercises

24.1 a. The longest continuous chain is numbered as follows:

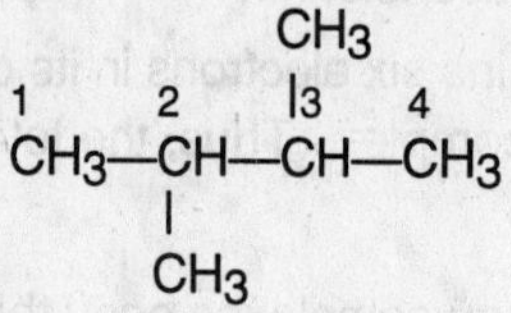

The name of the compound is 2,3-dimethylbutane.

b. The longest continuous chain is numbered as follows:

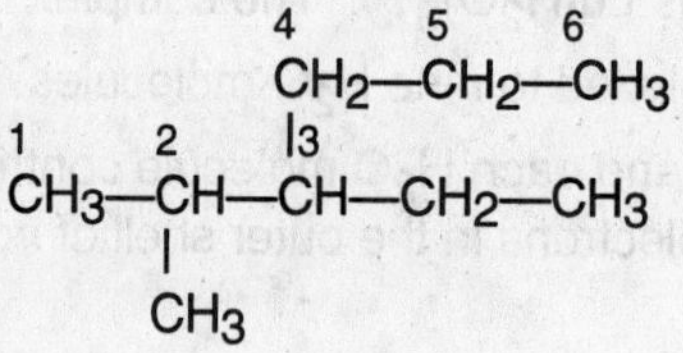

The name of the compound is 3-ethyl-2-methylhexane.

24.2 First write out the carbon skeleton for octane.

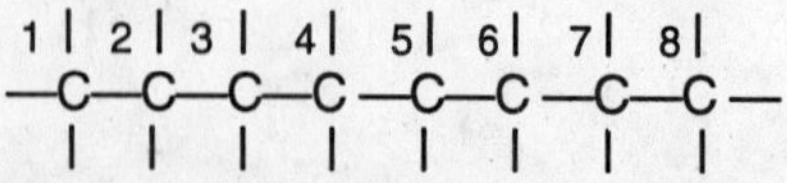

Then attach the alkyl groups.

(continued)

```
              CH3
   1   2  3   |  4    5   6   7   8
  —C——C——C————C————C———C———C———C—
   |   |  |   |    |   |   |   |
              CH3
```

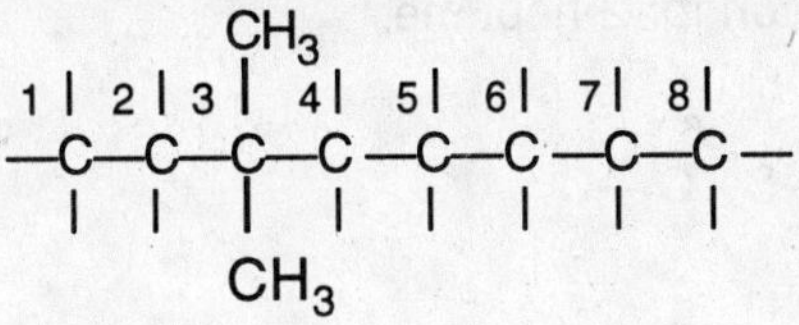

Finally, fill out the structure with H atoms.

```
                  CH3
                   |
  CH3—CH2—C—CH2—CH2—CH2—CH2—CH3
                   |
                  CH3
```

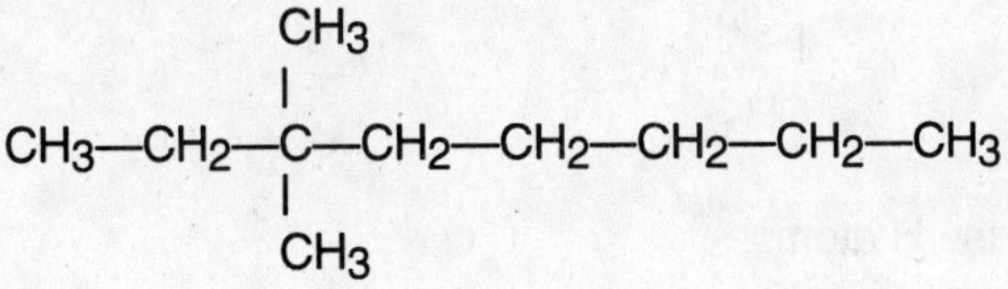

24.3   a.   The numbering of the carbon chain is

```
       1    2   3    4
      CH3—C═CH—CH—CH3
           |       5 |
          CH3       CH2
                   6 |
                   CH3
```

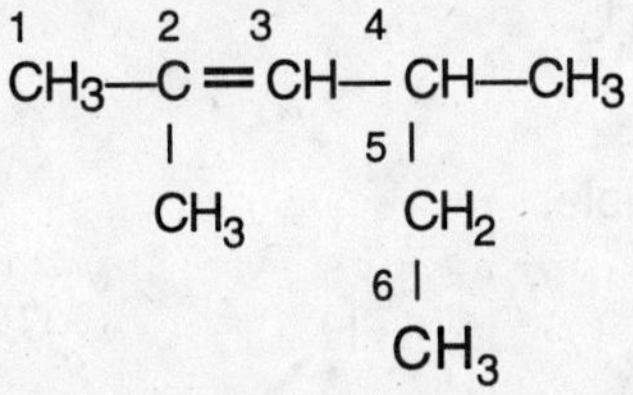

Because the longest chain containing a double bond has six carbons, this is a hexene. It is a 2-hexene because the double bond is between carbons 2 and 3. The name of the compound is 2,4-dimethyl-2-hexene.

b.   The numbering of the longest chain with a double bond is

```
       6    5    4    3
      CH3—CH2—CH2—CH—CH2—CH2—CH3
                    2 |
                     CH
                    1 ‖
                     CH2
```

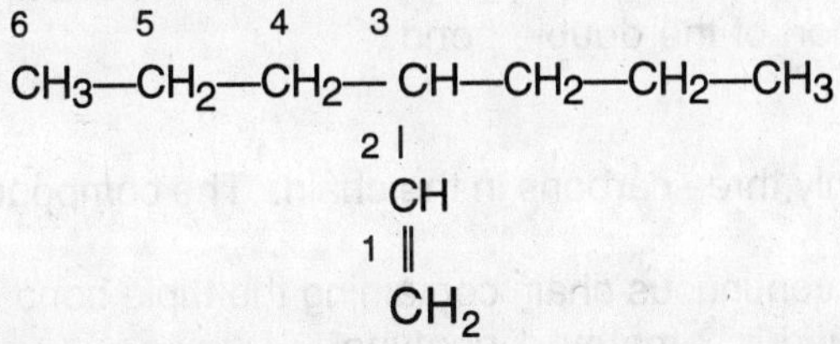

The longest chain containing the double bond has six carbon atoms; therefore, this is a hexene. It is a 1-hexene because the double bond is between carbons 1 and 2. The name of the compound is 3-propyl-1-hexene.

24.4 First write out the carbon skeleton for 2-heptene.

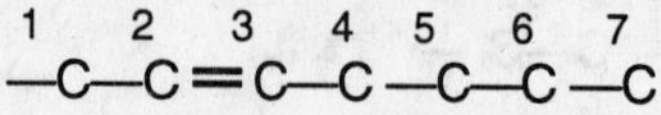

Then add the alkyl groups.

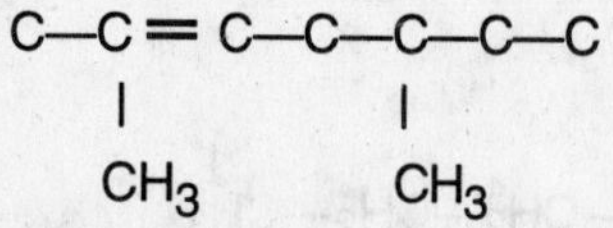

Finally, add the H atoms.

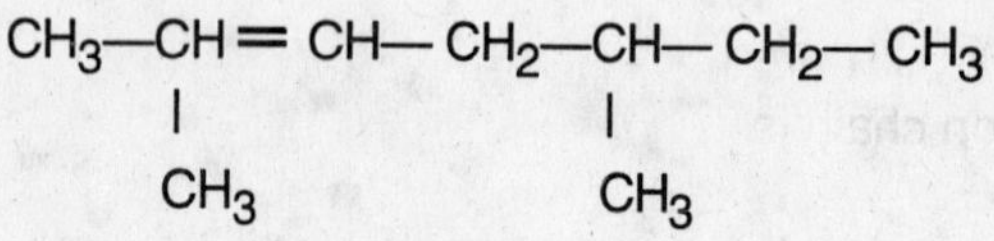

24.5 a. Geometric isomers are possible.

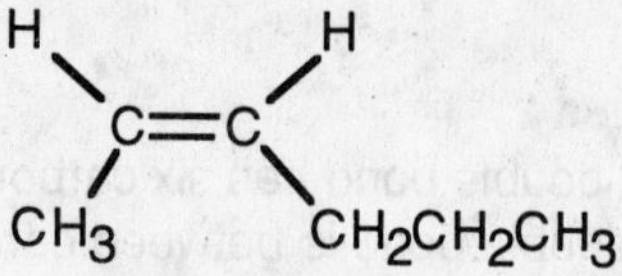

cis-2-hexene

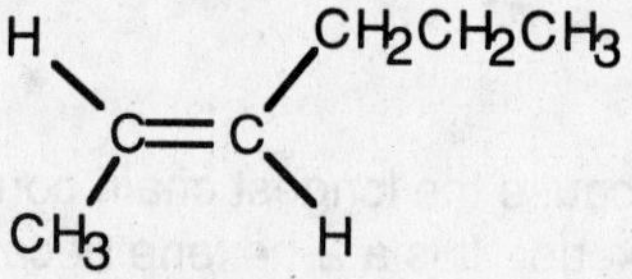

trans-2-hexene

b. No geometric isomers are possible because there are two H atoms attached to the second carbon of the double bond.

24.6 a. There are only three carbons in the chain. The compound is propyne.

b. The longest continuous chain containing the triple bond has five carbon atoms. The compound is 3-methyl-1-pentyne.

24.7 a. This compound has an ethyl group attached to a benzene ring.

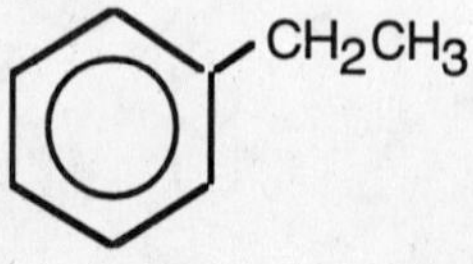

b. This compound has one phenyl group attached to each carbon atom in the ethane molecule.

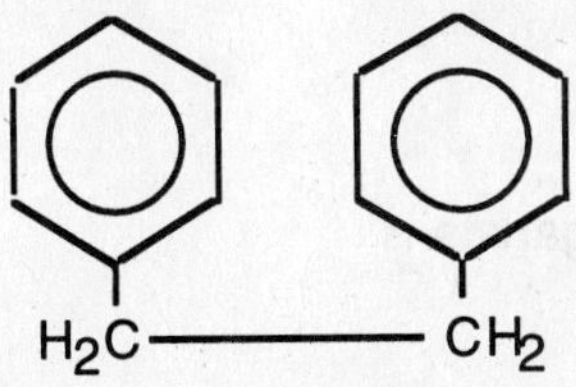

24.8 According to Markownikoff's rule, when HBr is added across the double bond in 1-butene, the H will add to carbon 1 (the C atom with the most bonds to H atoms) and the Br will add to carbon 2.

$$H_2C{=}CH-CH_2-CH_3 + HBr \rightarrow H_3C-\underset{\underset{Br}{|}}{CH}-CH_2-CH_3$$

The product is 2-bromobutane.

24.9 Because the compound has an —OH group, it is an alcohol. The longest carbon chain in the molecule has six carbons.

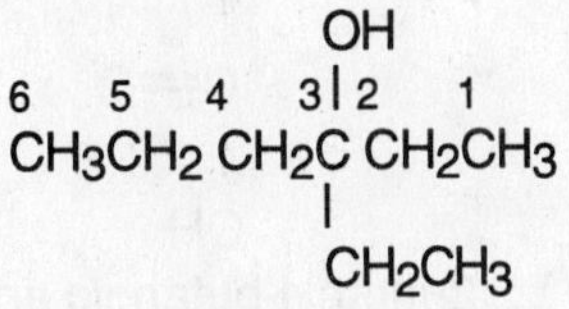

$$\underset{CH_2CH_3}{\underset{|}{\overset{6\quad 5\quad 4\quad\overset{OH}{\underset{|}{3|2}}\quad 1}{CH_3CH_2\ CH_2C\ CH_2CH_3}}}$$

The name of the compound is 3-ethyl-3-hexanol.

24.10 a. Dimethyl ether   b. Methyl ethyl ether

24.11 a. There are two alkyl groups attached to the carbonyl; therefore, the compound is a ketone. There are five carbon atoms in the chain. The name of the compound is 2-pentanone.

b. There is a hydrogen atom attached to the carbonyl, so the compound is an aldehyde. The numbering of the stem carbon chain is

$$\overset{1\ \overset{O}{\overset{\|}{\ }}\quad 2\quad 3\quad 4}{H-C-CH_2C\ CH_3}$$

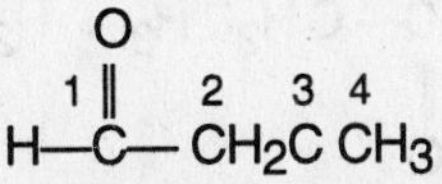

The name of the compound is butanal.

24.12 The oxidation half-reaction is

$$CH_3-\underset{\underset{\displaystyle OH}{|}}{CH_2} \rightarrow CH_3-\underset{\underset{\displaystyle O}{\|}}{C}-H + 2H^+ + 2e^-$$

The half-reaction for the reduction of permanganate is

$$MnO_4^- + 8H^+ + 5e^- \rightarrow Mn^{2+} + 4H_2O$$

Multiplying the first reaction by 5 and the second by 2 and adding them gives the final balanced equation for the oxidation of ethanol to acetaldehyde.

$$5\,CH_3-\underset{\underset{\displaystyle OH}{|}}{CH_2} + 2\,MnO_4^- + 6\,H^+ \rightarrow 5\,CH_3-\underset{\underset{\displaystyle O}{\|}}{C}-H + 2\,Mn^{2+} + 8\,H_2O$$

24.13 a. This is a tertiary alcohol, which is unreactive with most oxidizing agents.

b. Aldehydes are easily oxidized to carboxylic acids.

$$CH_3-\underset{\underset{\displaystyle CHO}{|}}{CH}-CH_2-CH_3 + (O) \rightarrow CH_3-\underset{\underset{\underset{\underset{\displaystyle OH}{|}}{C=O}}{|}}{CH}-CH_2-CH_3$$

2-methyl-butanoic acid

c. Aldehydes can also be reduced to alcohols.

$$CH_3\underset{\underset{\displaystyle CHO}{|}}{C}HCH_2CH_3 + 2\,(H) \rightarrow CH_3\underset{\underset{\underset{\underset{\displaystyle OH}{|}}{CH_2}}{|}}{C}HCH_2CH_3$$

2-methyl-1-butanol

d. Secondary alcohols can be oxidized to form ketones.

$$CH_3\underset{\underset{\displaystyle OH}{|}}{C}HCH_2CH_3 + (O) \rightarrow CH_3-\underset{\underset{\displaystyle O}{\|}}{C}-CH_2CH_3 + H_2O$$

2-butanone

24.14 Esters are often prepared by heating an alcohol and a carboxylic acid in the presence of an inorganic acid. The choice of acid and alcohol depend on the ester that is desired. The R groups on the acid and alcohol must correspond to those in the ester.

$$CH_3-CH_2-\overset{\overset{O}{\|}}{C}-OH + HO-CH_3 \xrightarrow{H^+} CH_3CH_2-\overset{\overset{O}{\|}}{C}-O-CH_3 + H_2O$$

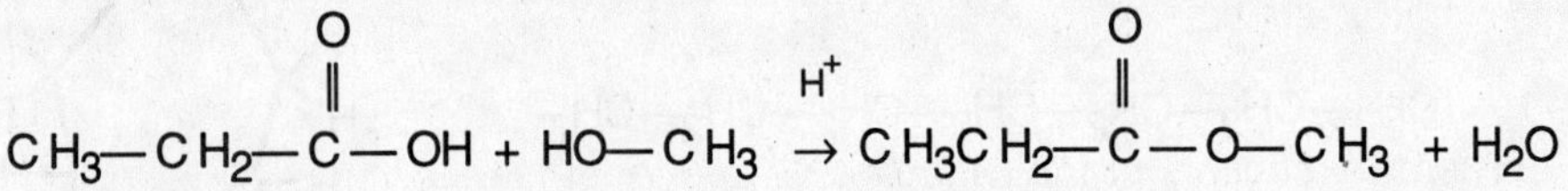

24.15 This addition polymer is formed when vinylidene chloride adds to itself across the double bond.

$$\cdots + CH_2=CCl_2 + CH_2=CCl_2 + CH_2=CCl_2 \rightarrow$$
$$-CH_2-CCl_2-CH_2-CCl_2-CH_2-CCl_2-$$

■ Answers to Review Questions

Note: For simplicity, all carbon atoms have been omitted in ring structures.

24.1 The formula of an alkane with 30 carbon atoms is $C_{30}H_{62}$.

24.2 The molecules increase regularly in molecular weight. Therefore, you expect their intermolecular forces and thus their melting points to increase.

24.3
$H_3C-CH_2-CH_2-CH_2-CH_2-CH_3$

$H_3C-\underset{\underset{CH_3}{|}}{CH}-CH_2-CH_2-CH_3$

$H_3C-CH_2-\underset{\underset{CH_3}{|}}{CH}-CH_2-CH_3$

$H_3C-\underset{\underset{CH_3}{|}}{CH}-\underset{\underset{CH_3}{|}}{CH}-CH_3$

$CH_3-\underset{\underset{CH_3}{|}}{\overset{\overset{CH_3}{|}}{C}}-CH_2-CH_3$

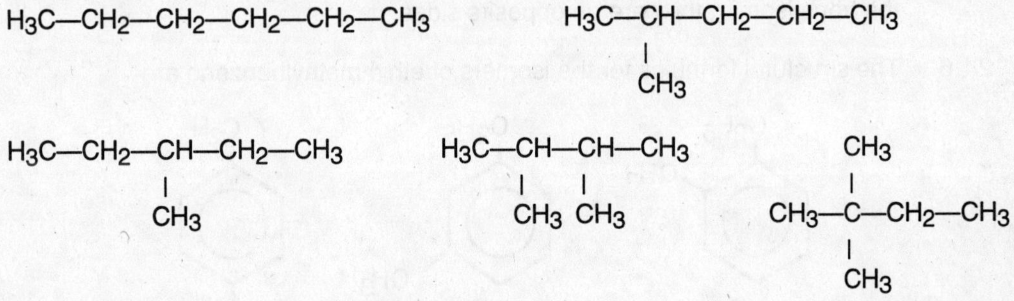

24.4 The structures of a seven-carbon alkane, cycloalkane, alkene, and aromatic hydrocarbon are

$CH_3—CH_2—CH_2—CH_2—CH_2—CH_2—CH_3$

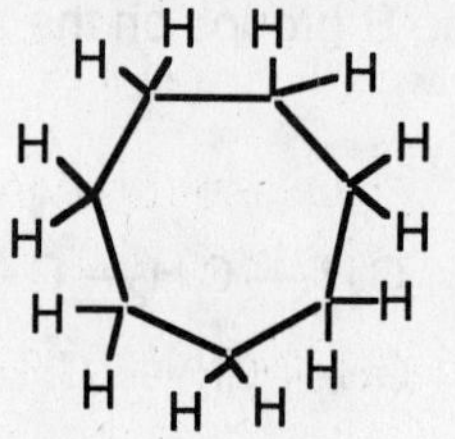

$CH_3═CH—CH_2—CH_2—CH_2—CH_2—CH_3$

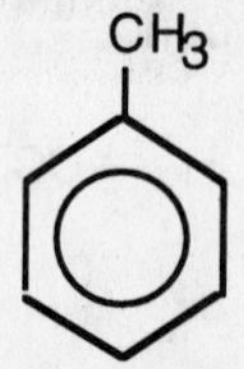

(toluene)

24.5 The two isomers of 2-butene are the cis and trans geometric isomers:

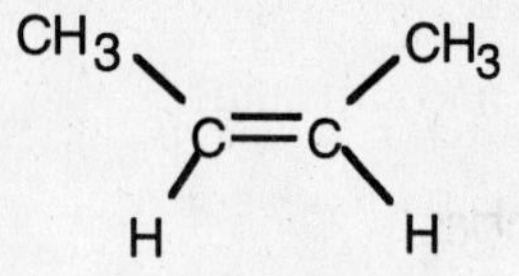

cis-2-butene

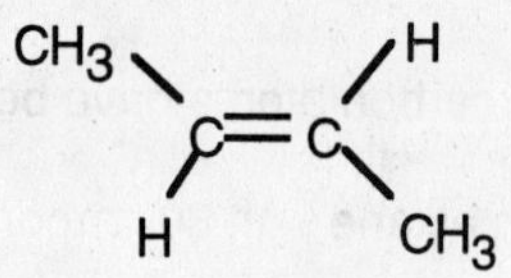

trans-2-butene

In the cis-2-butene, the two methyl groups are on the same side of the double bond; in the trans isomer, they are on opposite sides.

24.6 The structural formulas for the isomers of ethyl-methylbenzene are

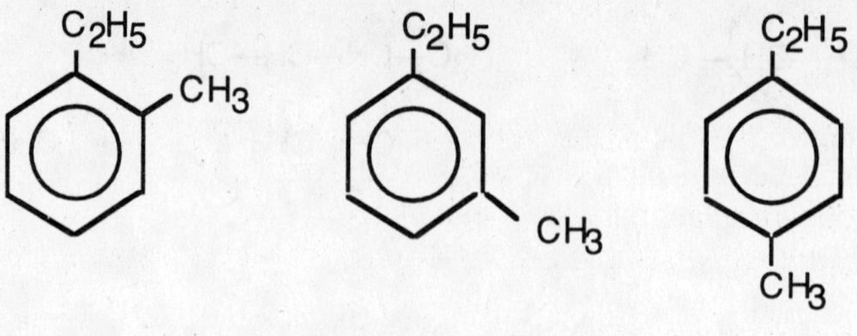

24.7 Methane: source—natural gas; use—home fuel.
Octane: source—petroleum; use—auto fuel.
Ethylene: source—petroleum refining; use—chemical industry raw material.
Acetylene: source—methane; use—acetylene torch.

24.8 CH_3CH_2Cl CH_2ClCH_2Cl $CHCl_2CHCl_2$ CCl_3CCl_3
CH_3CHCl_2 $CH_2ClCHCl_2$ $CHCl_2CCl_3$
CH_3CCl_3 CH_2ClCCl_3

24.9 A substitution reaction is a reaction in which part of the reagent molecule is substituted for a hydrogen atom on a hydrocarbon or hydrocarbon group. For example,

$$CH_4 + Cl_2 \rightarrow CH_3Cl + HCl$$

An addition reaction is a reaction in which parts of the reagent are added to each carbon atom of a carbon-carbon multiple bond, which then becomes a C—C single bond. For example,

$$CH_2{=}CH_2 + Br_2 \rightarrow CH_2Br{-}CH_2Br$$

24.10 The major product of HCl plus acetylene should be $Cl_2HC{-}CH_3$ because Markownikoff's rule predicts this.

24.11 The octane-number scale gives the "antiknock" characteristics of a gasoline. If a gasoline begins to burn before the spark plug fires, the engine "knocks." The octane-number scale is based on heptane and 2,2,4-trimethylpentane, given octane numbers of 0 and 100, respectively. The higher the octane number, the better the antiknock characteristics of the gasoline.

24.12 A functional group is a reactive portion of a molecule that undergoes predictable reactions no matter what the rest of the molecule is like. An example is a C=C bond, which always reacts with bromine or other addition reagents to add part of each reagent to each carbon atom.

24.13 An aldehyde is different from a ketone, carboxylic acid, and ester in that a hydrogen atom is always attached to the carbonyl group in addition to a hydrocarbon group.

24.14 Methanol: source—$CO + H_2$; use—solvent.
Ethanol: source—fermentation of glucose; use—solvent.
Ethylene glycol: source—ethylene; use—antifreeze.
Glycerol: source—from soap making; use—foods.
Formaldehyde: source—oxidation of methanol; use—plastics and resins.

24.15 a. CO is a carbonyl group (ketone).

b. $CH_3O{-}C$ is an ether group.

c. C=C is a double bond.

d. COOH is a carboxylic acid group.

e. CHO is an aldehyde (carbonyl).

f. CH_2OH, or -OH, is a hydroxyl group (primary alcohol).

24.16 A primary alcohol is oxidized to an acid in two steps. An overall example reaction is

$$CH_3CH_2OH + 2[O] \rightarrow CH_3CO_2H + H_2O$$

A secondary alcohol is oxidized to a ketone. A reaction is

$$CH_3CHOHCH_3 + [O] + H^+ \rightarrow CH_3COCH_3 + H_2O$$

Tertiary alcohols are not oxidized. An aldehyde is oxidized to an acid. The reaction is

$$CH_3CHO + [O] \rightarrow CH_3CO_2H$$

A ketone is not oxidized.

24.17 Ethyl ethanoate (acetate) is $CH_3CH_2OOCCH_3$; methyl propanoate is $CH_3CH_2CO_2CH_3$.

24.18 $CH_3CO_2C_2H_5(l) + NaOH(aq) \rightarrow CH_3CO_2^-(aq) + Na^+(aq) + C_2H_5OH(l)$

24.19 The source of basicity of an amine is the pair of electrons on the nitrogen. An example of amine-acid reaction is

$$(CH_3CH_2)_3N\!: + CH_3CO_2H \rightarrow CH_3CO_2^- + (CH_3CH_2)_3NH^+$$

24.20 A condensation polymer is formed by splitting out a small molecule such as water between two molecules (or monomers), whereas addition polymers form when molecules add to one another, giving a chain.

Example of addition:

$$2n\,H_2C{=}CH_2 \rightarrow [-H_2C-CH_2-CH_2-CH_2-]_n$$

Example of condensation:

$$nHOROH + nHOOC-R-COOH \rightarrow [-O-R-O-CO-R-CO-]_n + 2nH_2O$$

■ Solutions to Practice Problems

Note: For simplicity, all carbon atoms have been omitted in ring structures.

24.21 a.

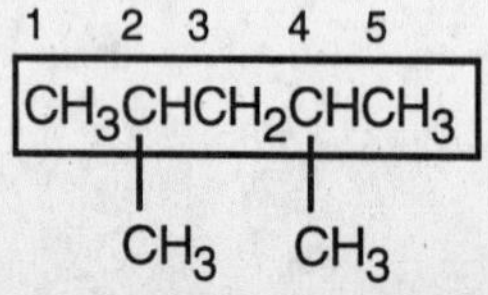

longest chain 2,4-dimethylpentane

b. 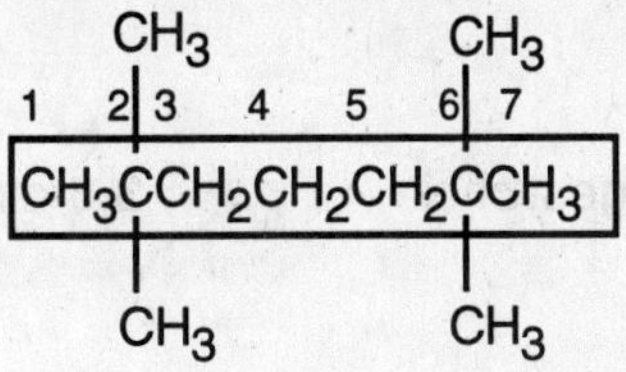 longest chain 2,2,6,6-tetramethylheptane

c. 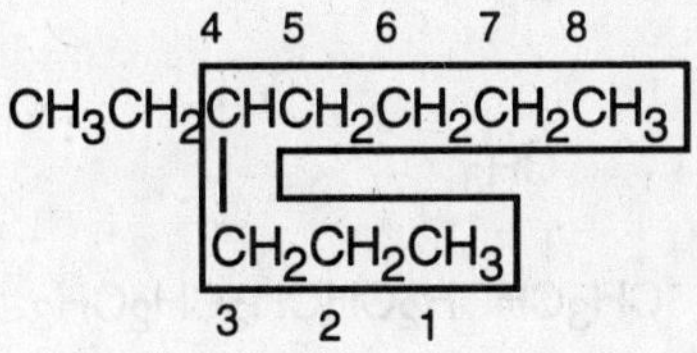 longest chain 4-ethyloctane

d. 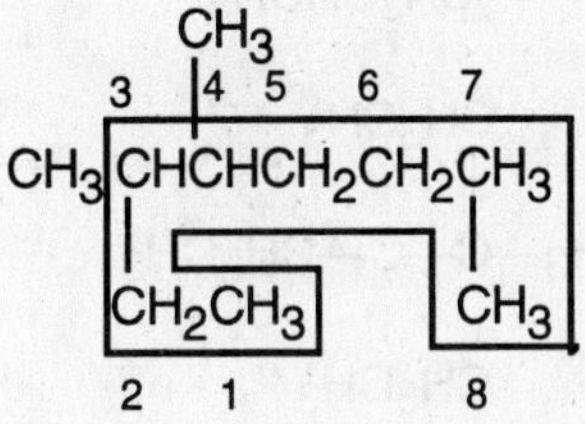 longest chain 3,4-dimethyloctane

24.22 a. 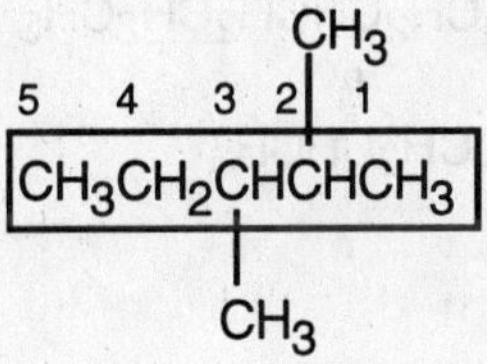 longest chain 2,3-dimethylpentane

b. 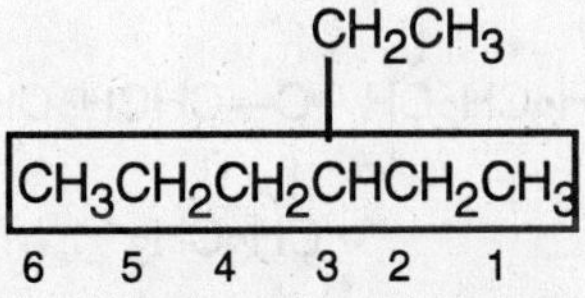 longest chain 3 ethyl-hexane

c. 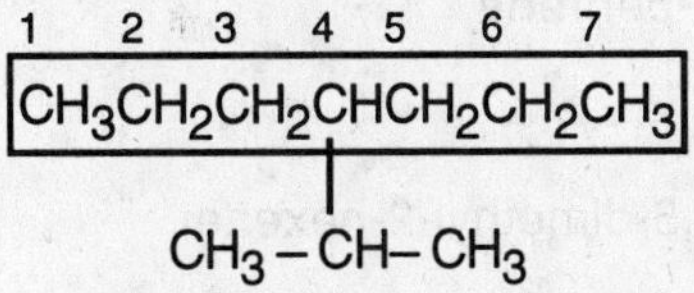 longest chain 4-isopropylheptane

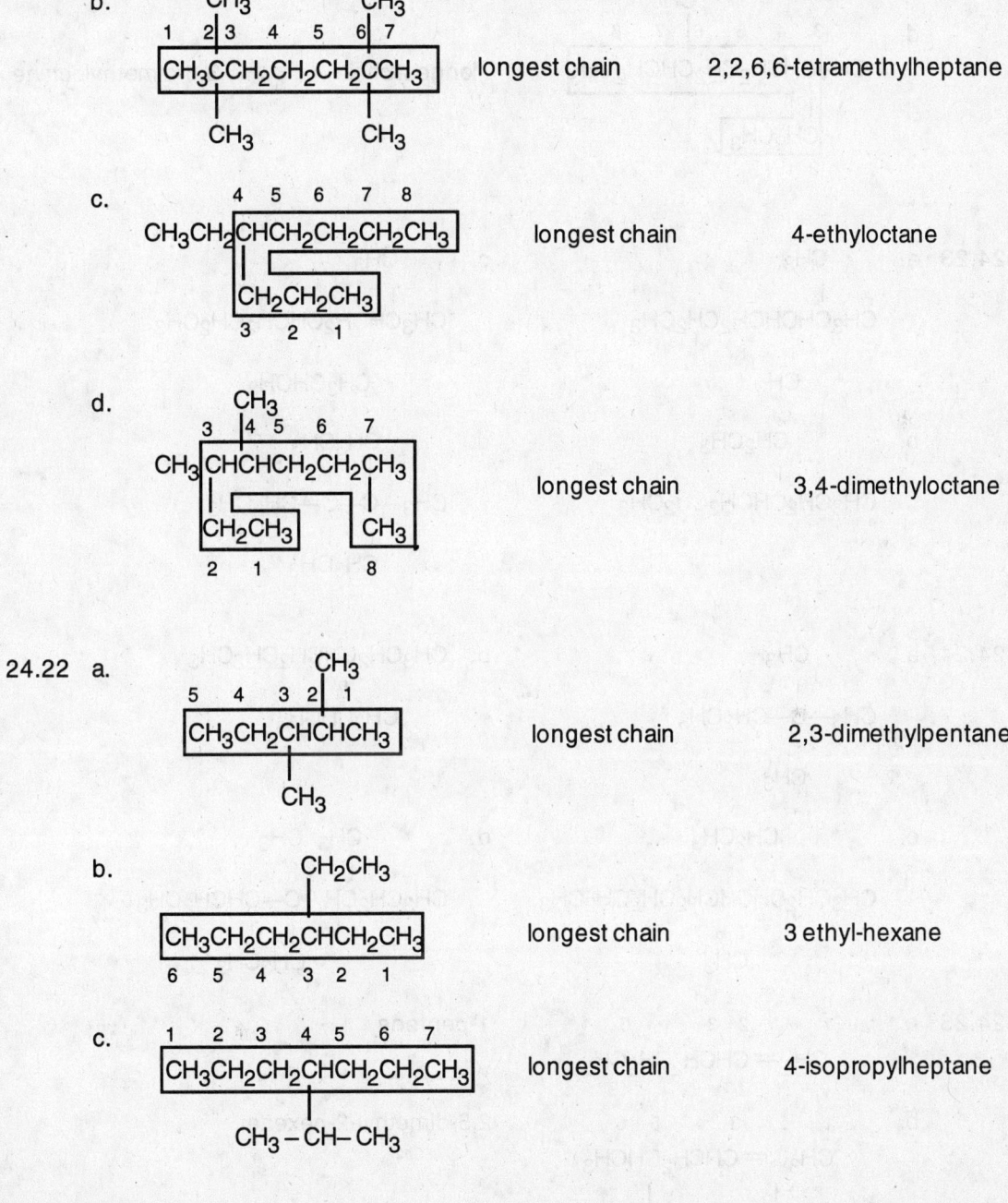

d. 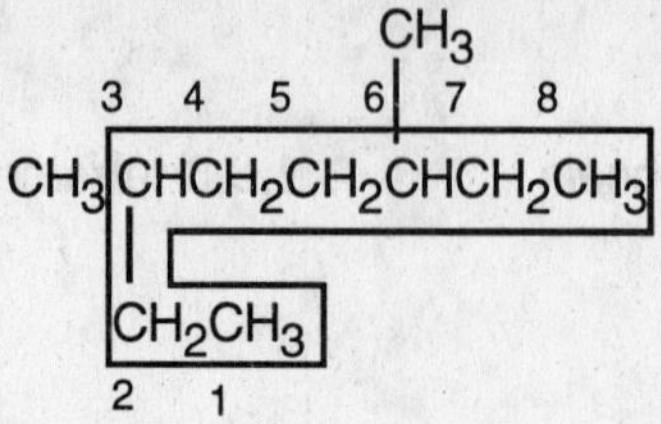 longest chain 3,6-dimethyloctane

24.23 a. 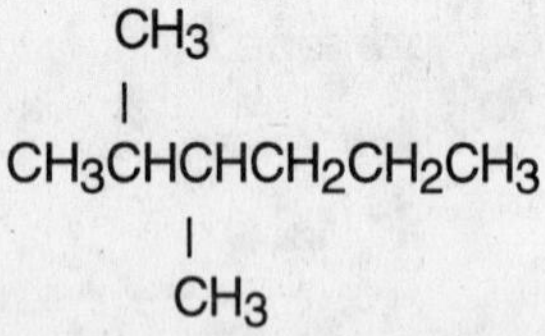

c. 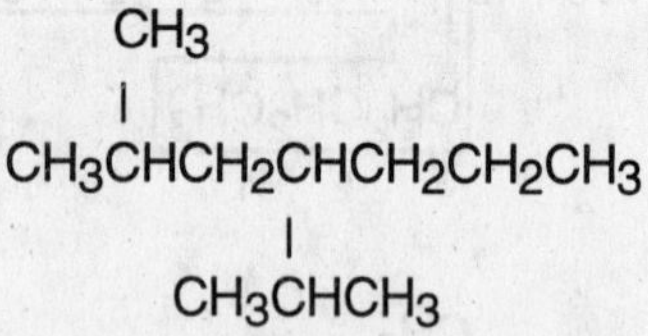

b. 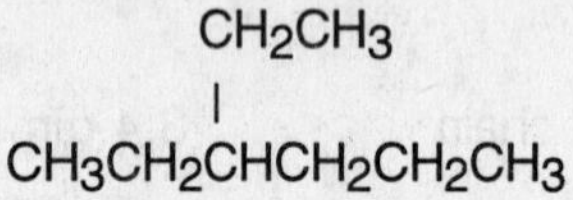

d. 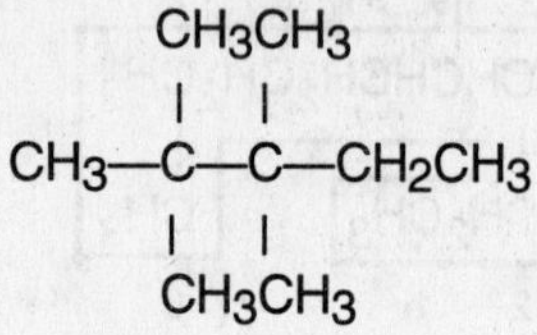

24.24 a. 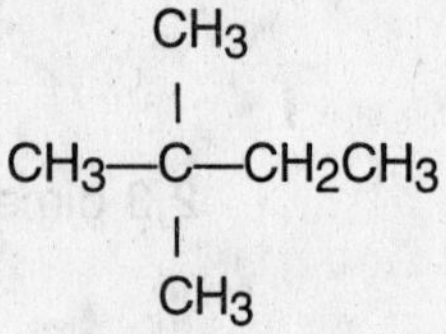

b. 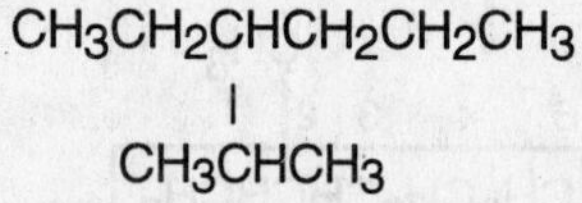

c. 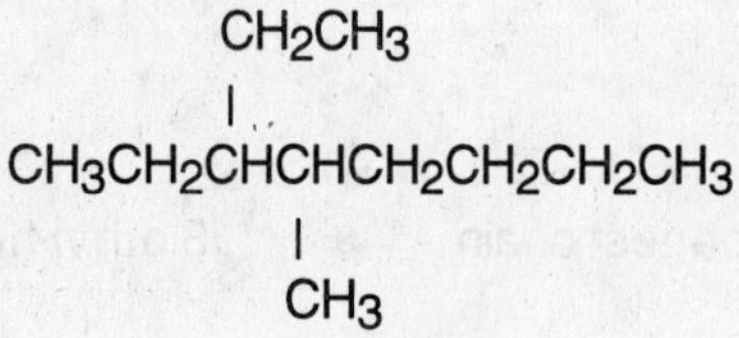

d. 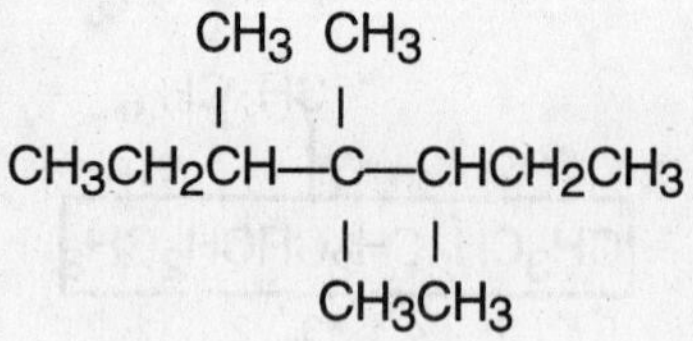

24.25 a. 1 2 3 4 5 $CH_2{=}CHCH_2CH_2CH_3$ 1-pentene

b. 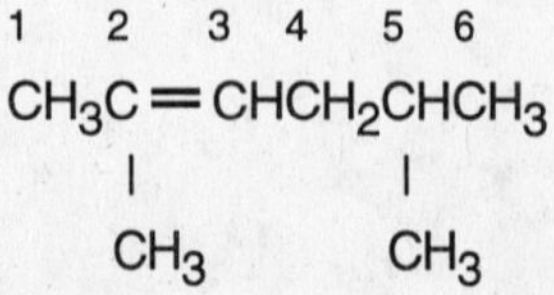 2,5-dimethyl-2-hexene
```
   1   2   3   4  5  6
   CH3C = CHCH2CHCH3
    |           |
    CH3        CH3
```

24.26 a. 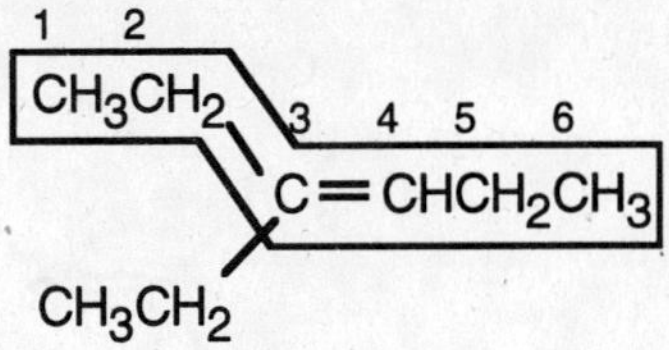

longest chain 3-ethyl-3-hexene

b. 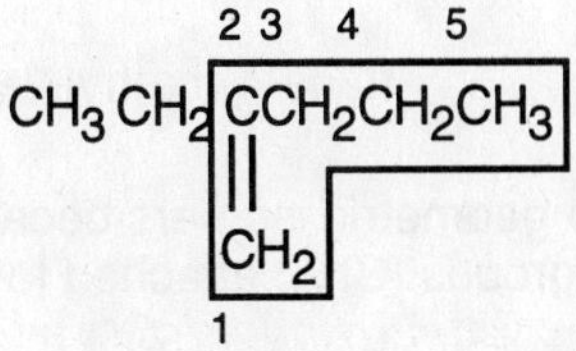

longest chain that includes double bond:

2-ethyl-1-pentene

24.27 a. 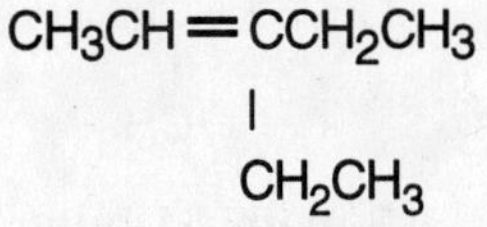

b. 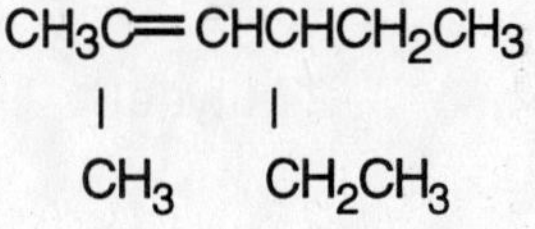

24.28 a. 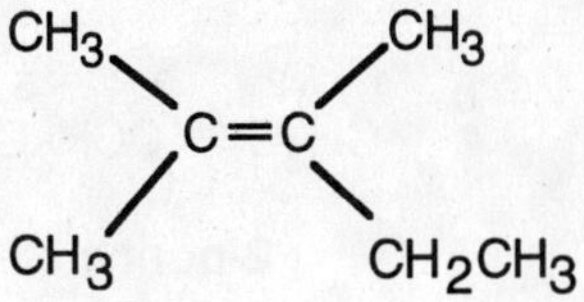

b. 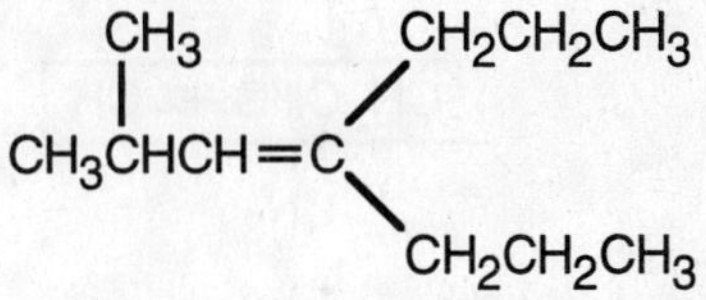

24.29 a. 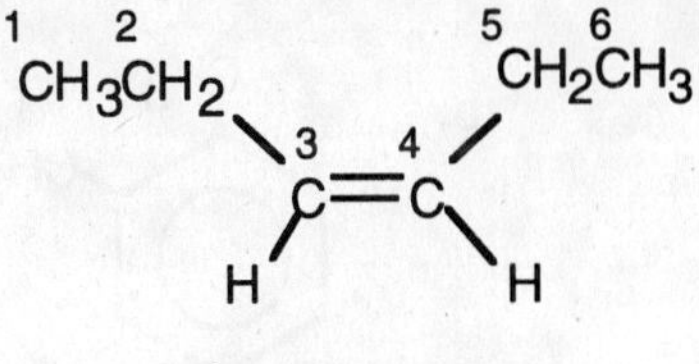

cis-3-hexene

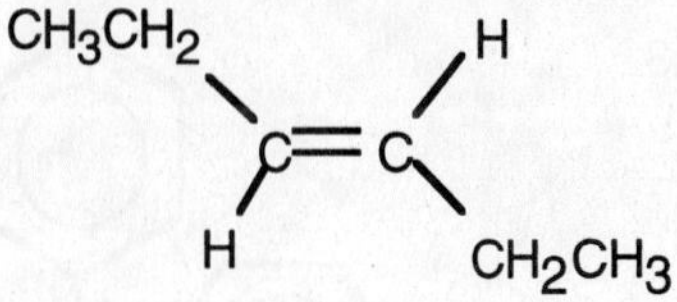

trans-3-hexene

b. 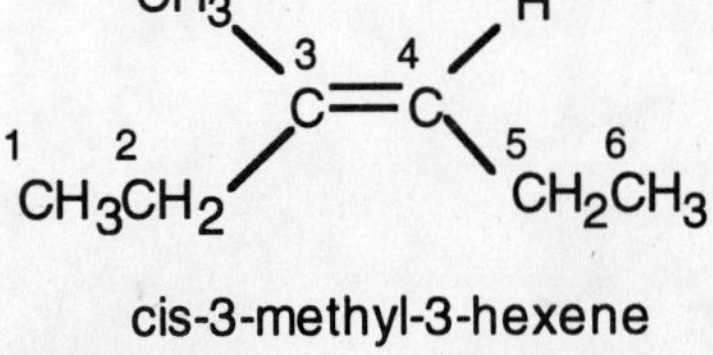

cis-3-methyl-3-hexene

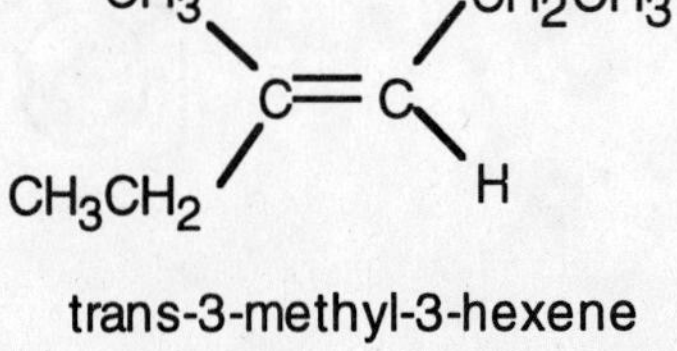

trans-3-methyl-3-hexene

24.30 a.

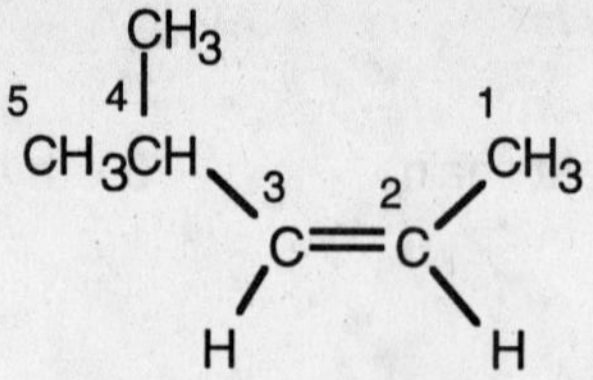

cis-4-methyl-2-pentene

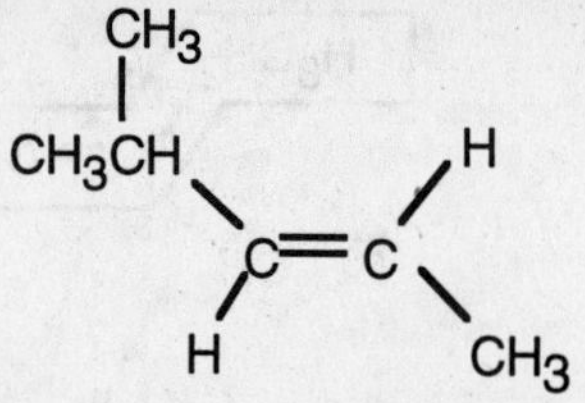

trans-4-methyl-2-pentene

b. This compound (2-methyl-2-pentene) has no geometric isomers because the second carbon in the chain has two identical groups (CH_3) attached to it.

24.31 a. 1 2 3 4

$CH_3C\equiv CCH_3$

2-butyne

b.

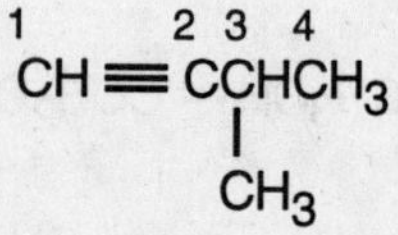

3-methyl-1-butyne

24.32 a.

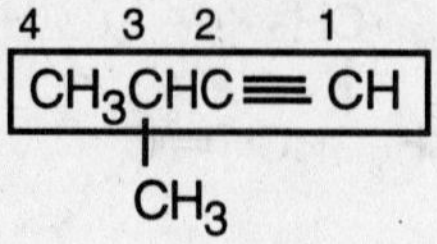

longest chain

3-methyl-1-butyne

b. 1 2 3 4 5

$CH_3C\equiv CCH_2CH_3$

2-pentyne

24.33 a.

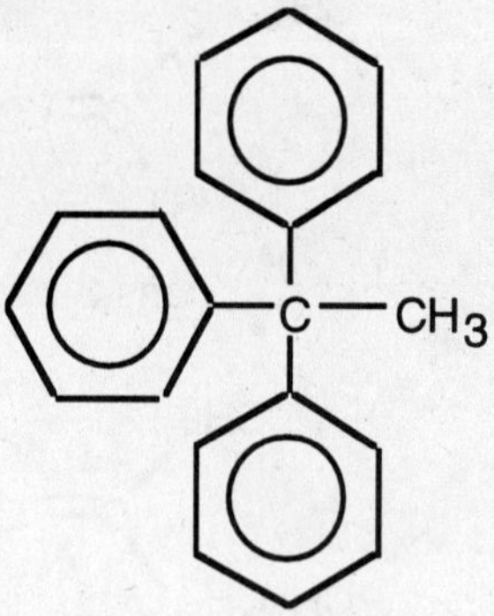

b.

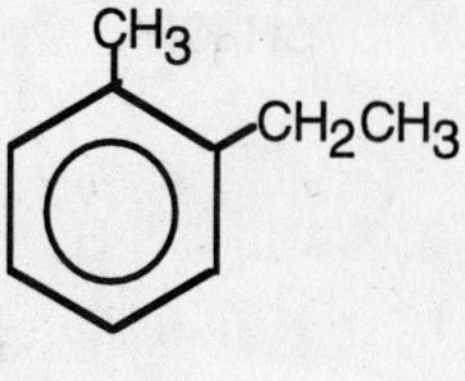

24.34 a. 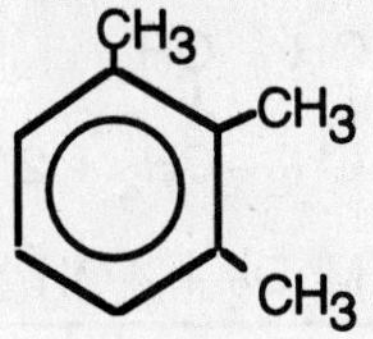

b. 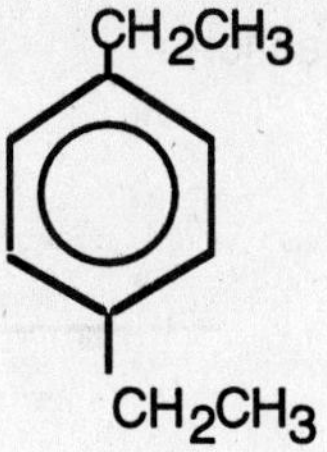

24.35 a. $C_3H_6 + \frac{9}{2} O_2 \rightarrow 3CO_2 + 3H_2O$

Remove the fraction:

$$2C_3H_6 + 9O_2 \rightarrow 6CO_2 + 6H_2O$$

b.

$$CH_2{=}CH_2 + MnO_4^- + H_2O \rightarrow \overset{OH}{\overset{|}{CH_2}}-\overset{OH}{\overset{|}{CH_2}} + MnO_2$$

Oxidation: $CH_2{=}CH_2 \rightarrow \overset{OH}{\overset{|}{CH_2}}-\overset{OH}{\overset{|}{CH_2}}$

Balance O: $2\,OH^- + CH_2{=}CH_2 \rightarrow \overset{OH}{\overset{|}{CH_2}}-\overset{OH}{\overset{|}{CH_2}}$

Balance e^-: $2\,OH^- + CH_2{=}CH_2 \rightarrow \overset{OH}{\overset{|}{CH_2}}-\overset{OH}{\overset{|}{CH_2}} + 2\,e^-$

Reduction: $MnO_4^- \rightarrow MnO_2$

Balance O: $MnO_4^- + 2\,H_2O \rightarrow MnO_2 + 4\,OH^-$

Balance e^-: $MnO_4^- + 2\,H_2O + 3\,e^- \rightarrow MnO_2 + 4\,OH^-$

(continued)

Half-reactions

$$3\left(2\,OH^- + CH_2=CH_2 \rightarrow \underset{\underset{OH}{|}}{CH_2}-\underset{\underset{OH}{|}}{CH_2} + 2\right)$$

$$2\,(MnO_4^- + 2\,H_2O + 3\,e^- \rightarrow MnO_2 + 4\,OH^-)$$

$$\cancel{6}\,OH^- + 3\,CH_2=CH_2 + 2\,MnO_4^- + 4\,H_2O \rightarrow 3\,\underset{\underset{OH}{|}}{CH_2}-\underset{\underset{OH}{|}}{CH_2} + 2\,MnO_2 + \overset{2}{\cancel{8}}\,OH^-$$

$$3\,CH_2=CH_2 + 2\,MnO_4^- + 4\,H_2O \rightarrow 3\,\underset{\underset{OH}{|}}{CH_2}-\underset{\underset{OH}{|}}{CH_2} + 2\,MnO_2 + 2\,OH^-$$

c. $CH_2=CH_2 + Br_2 \rightarrow \underset{\underset{Br}{|}}{CH_2}-\underset{\underset{Br}{|}}{CH_2}$

d. C$_6$H$_5$CH$_3$ + Br$_2$ $\xrightarrow{FeBr_3}$ p-BrC$_6$H$_4$CH$_3$ + HBr

e. C$_6$H$_5$CH$_3$ + HNO$_3$ $\xrightarrow{H_2SO_4}$ p-O$_2$NC$_6$H$_4$CH$_3$ + H$_2$O

24.36 a. $C_4H_{10} + \dfrac{13}{2}O_2 \rightarrow 4\,CO_2 + 5\,H_2O$

Remove the fraction:

$2\,C_4H_{10} + 13\,O_2 \rightarrow 8\,CO_2 + 10\,H_2O$

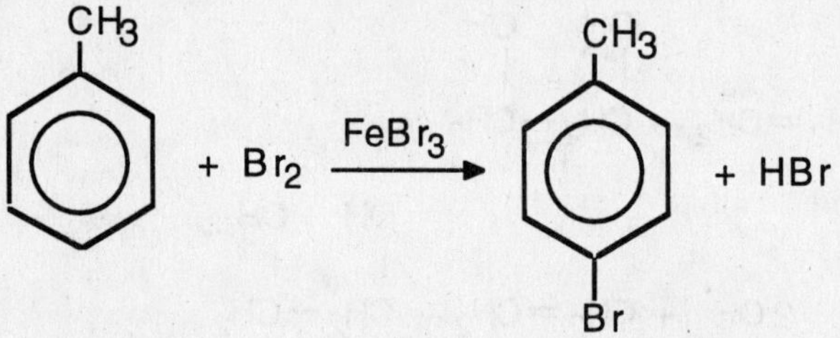

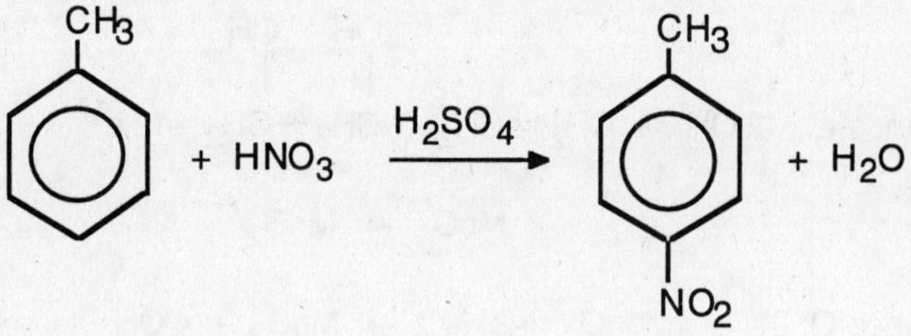

b.

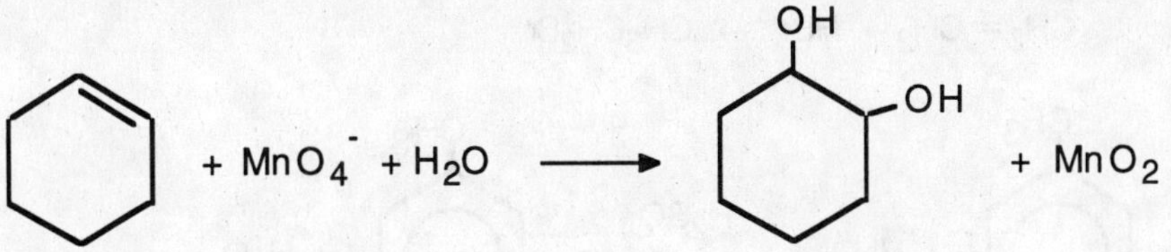

cyclohexene + MnO$_4^-$ + H$_2$O ⟶ cyclohexane-1,2-diol + MnO$_2$

Oxidation:

cyclohexene + 2 OH$^-$ ⟶ cyclohexane-1,2-diol   balance O

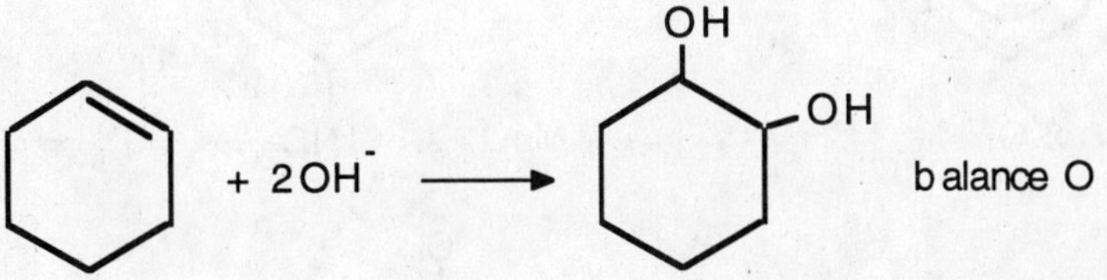

Oxidation: balance charge:

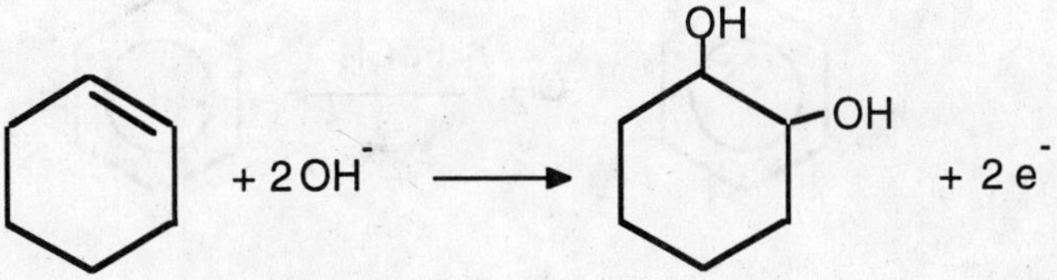

cyclohexene + 2 OH$^-$ ⟶ cyclohexane-1,2-diol + 2 e$^-$

Reduction: MnO$_4^-$ ⟶ MnO$_2$

MnO$_4^-$ + 2 H$_2$O ⟶ MnO$_2$ + 4 OH$^-$ (Balance O.)

MnO$_4^-$ + 2 H$_2$O + 3 e$^-$ ⟶ MnO$_2$ + 4 OH$^-$ (Balance charge.)

Add half-reactions:

3 (cyclohexene + 2 OH$^-$ ⟶ cyclohexane-1,2-diol + 2 e$^-$)

2 (MnO$_4^-$ + 2 H$_2$O + 3 e$^-$ ⟶ MnO$_2$ + 4 OH$^-$)

─────────────────────────────────

3 cyclohexene + 2 MnO$_4^-$ + 4 H$_2$O ⟶ 3 cyclohexane-1,2-diol + 2 MnO$_2$ + 2 OH$^-$

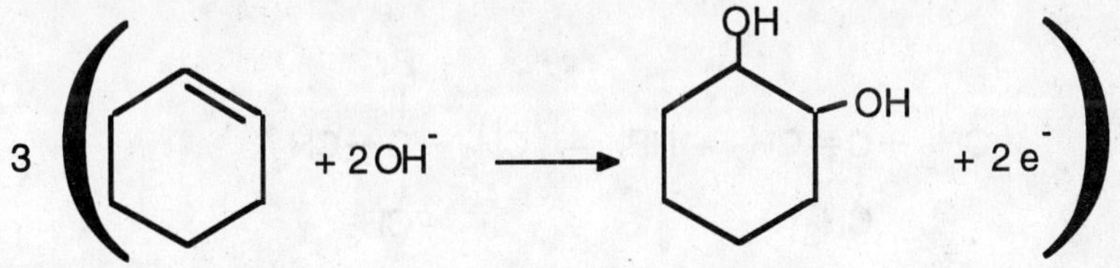

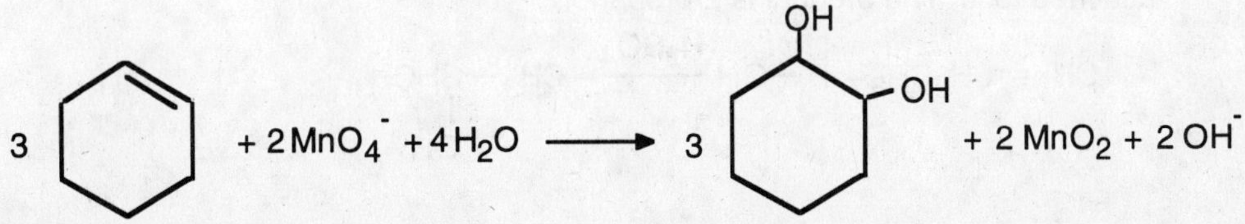

c. $CH_2=CH_2 + HBr \rightarrow CH_3CH_2Br$

d. [toluene] + HNO_3 $\xrightarrow{H_2SO_4}$ [p-nitrotoluene] + H_2O

e. [benzene] + Cl_2 $\xrightarrow{FeCl_3}$ [chlorobenzene] + HCl

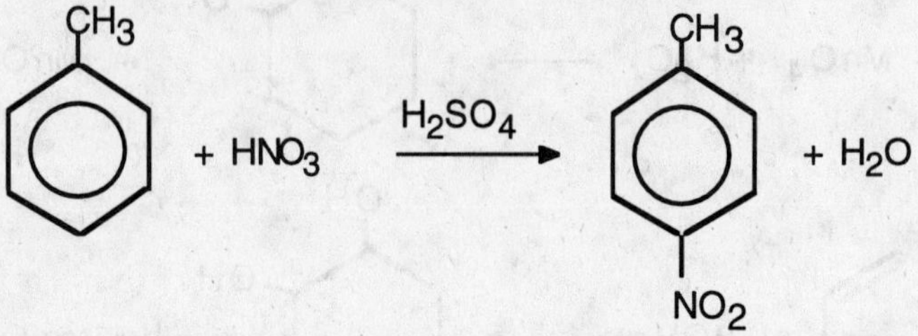

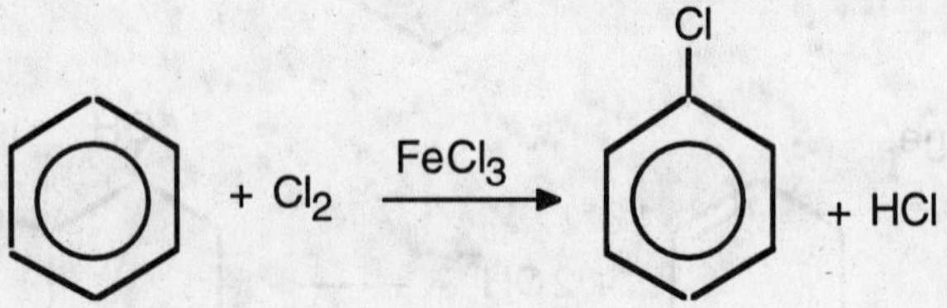

24.37 $CH_3CH_3 + Cl_2 \rightarrow CH_3CH_2Cl + HCl$

24.38 $CH_3CH_2CH_3 + Br_2 \rightarrow CH_3CHBrCH_3 + HBr$

24.39 According to Markownikoff's rule, the major product is the one obtained when the H atom adds to the carbon atom of the double bond that already has more hydrogen atoms attached to it. Therefore, 3-bromo-2-methylpropane is the major product.

$$CH_3-\underset{\underset{CH_3}{|}}{C}=CH_2 + HBr \rightarrow CH_3-\underset{\underset{CH_3}{|}}{\overset{\overset{Br}{|}}{C}}-CH_3$$

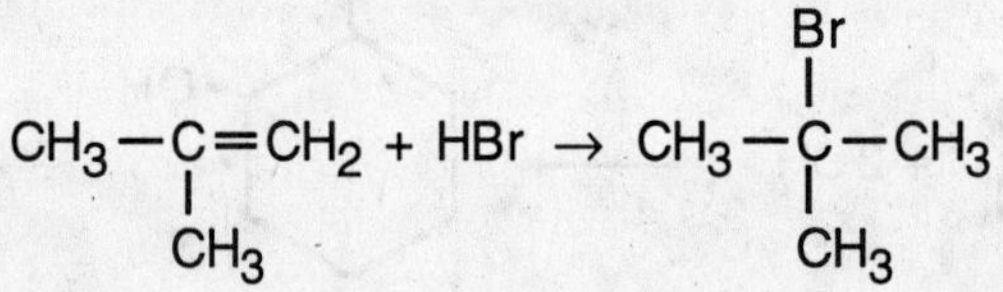

24.40 According to Markownikoff's rule, the major product is the one obtained when the H atom adds to the carbon atom of the double bond that already has more hydrogen atoms attached to it. The product is 2-propanol.

$$CH_2=CH_2CH_3 + H-OH \xrightarrow{H_2SO_4} CH_3-\underset{\underset{OH}{|}}{C}HCH_3$$

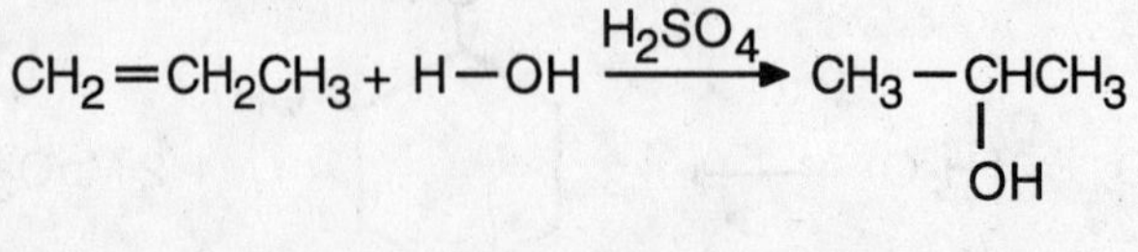

24.41 a.

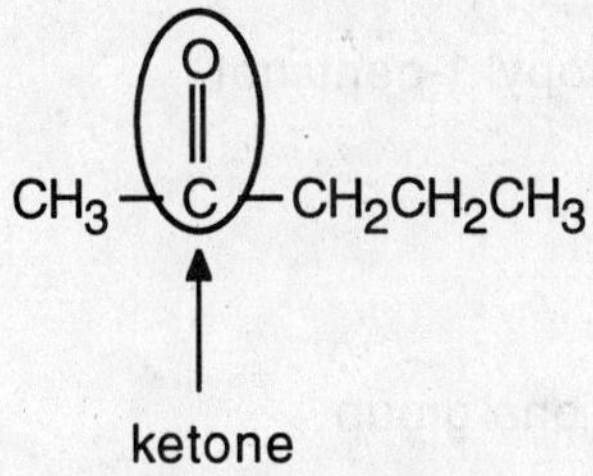

b.

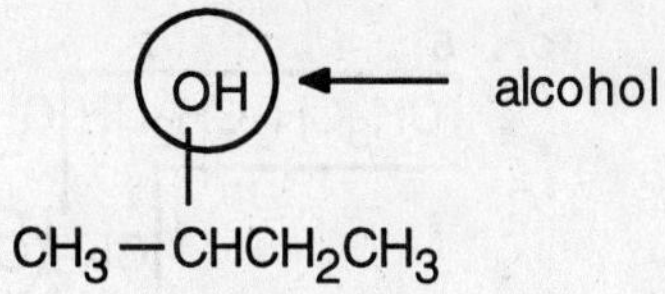

c.

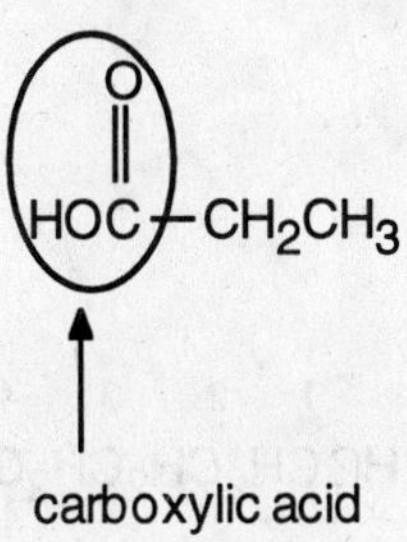

d.

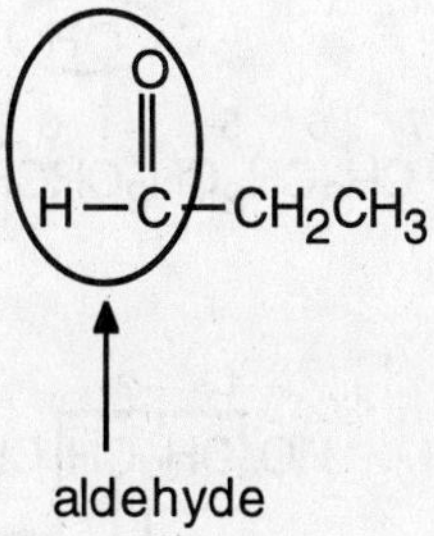

24.42 a.

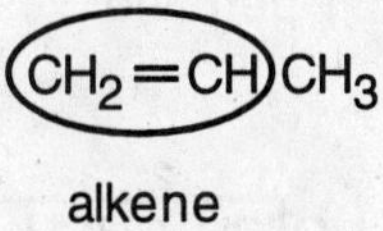

b.

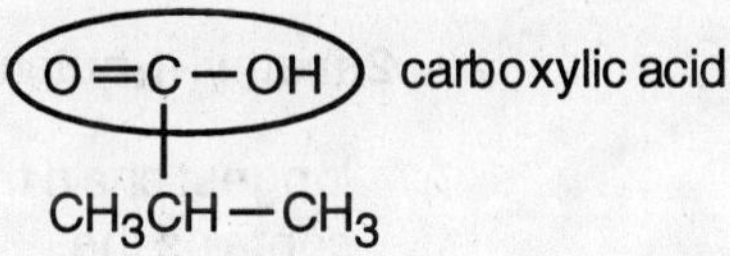

c.

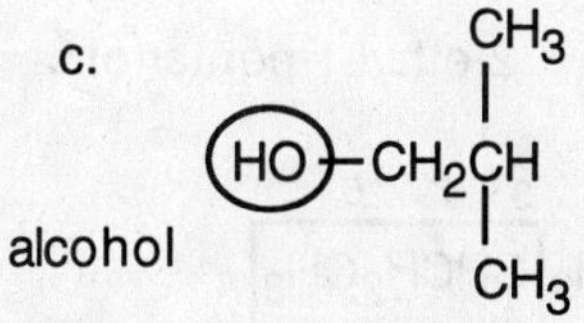

d.

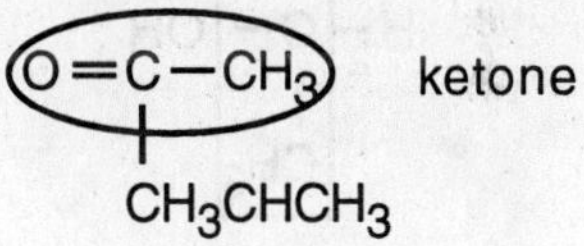

24.43 a.

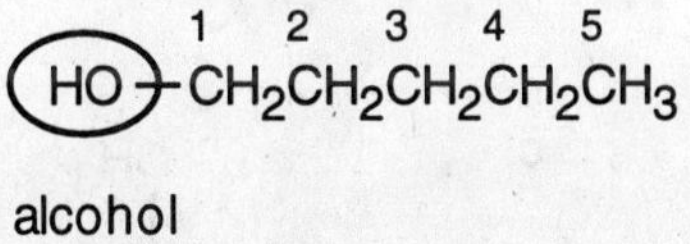

1-pentanol

b.

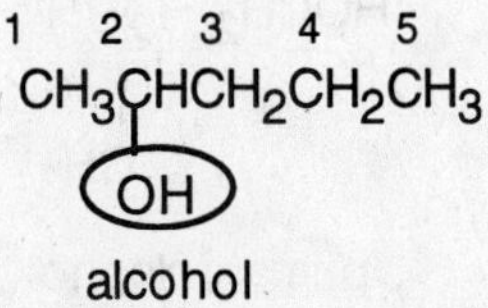

2-pentanol

c.
$\overset{5}{CH_3}\overset{4}{CH_2}\overset{3}{CH_2}\overset{2}{CH}\,\overset{}{CH_2CH_2CH_3}$
$\overset{1}{CH_2}$—(OH) ← alcohol
↑ longest chain containing functional group
2-propyl-1-pentanol

d. (OH) alcohol
$\overset{7}{CH_3}\overset{6}{CH_2}\overset{5}{CH_2}\overset{4}{CH}\overset{3}{CH_2}\overset{2}{CH_2}\overset{1}{CH_3}$ 4-heptanol

24.44 a. HO—$\overset{1}{CH_2}\overset{2}{CH}$—$CH_2CH_3$
       $\overset{3}{CH_2}\overset{4}{CH_2}\overset{5}{CH_3}$
2-ethyl-1-pentanol

b. $\overset{1}{HOCH_2}\overset{2}{CH_2}\overset{3}{CH_2}\overset{4}{CH_2}$
   $\overset{5}{CH_3}$
1-pentanol

c. $\overset{3}{CH_3}\,\overset{4}{CH}\overset{5}{CH_2CH_3}$
   $H\overset{2}{-}C—OH$
   $\overset{1}{CH_3}$
3-methyl-2-pentanol

d. HO—$\overset{1}{CH_2}\overset{2}{CH}\overset{3}{CH_2}\overset{4}{CH_3}$
      CH_3
2-methyl-1-butanol
longest chain that contains the functional group

24.45 a. ⒸH_3—CH—ⒸH_2CH_3
              OH
secondary alcohol

c. $HOCH_2$—ⒸH
         CH_3
         CH_3
primary alcohol

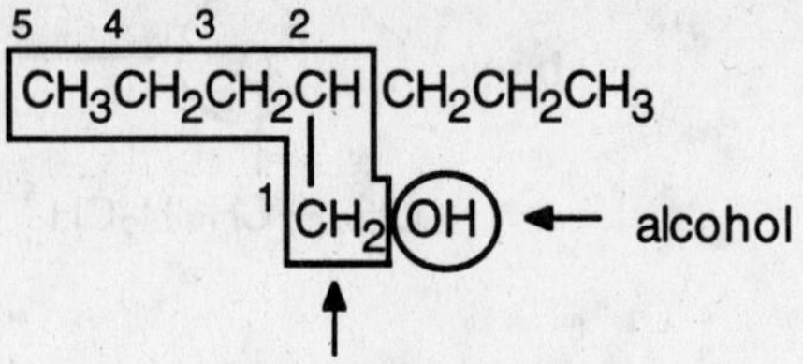

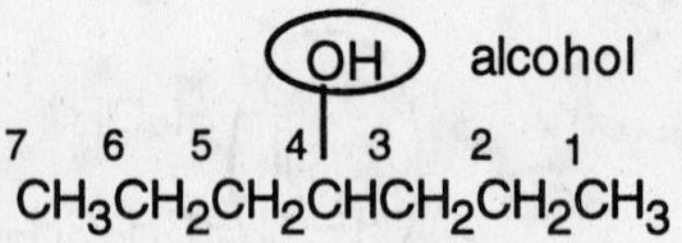

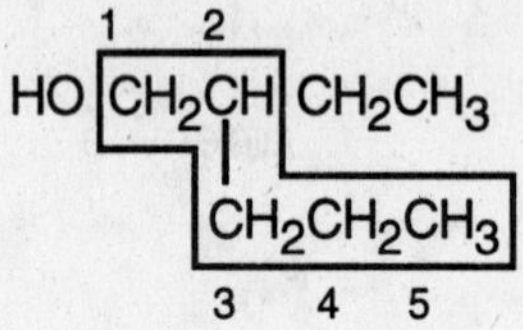

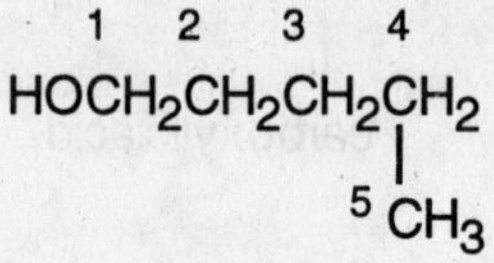

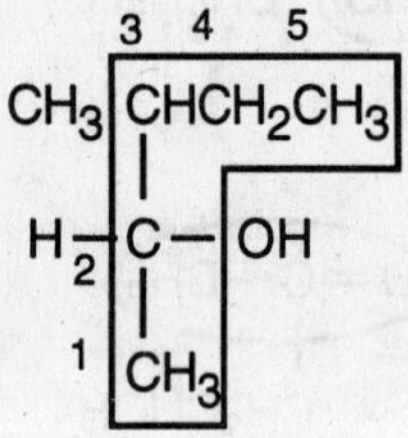

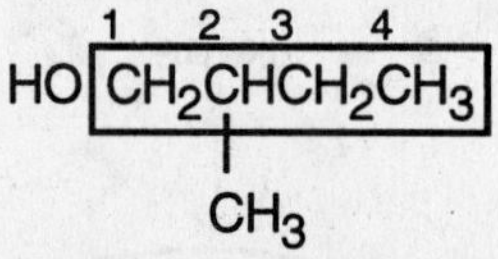

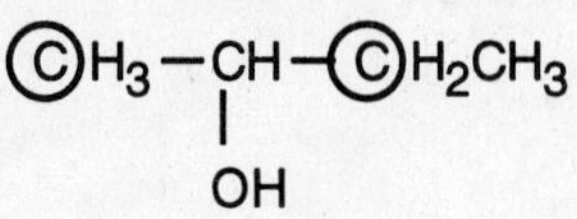

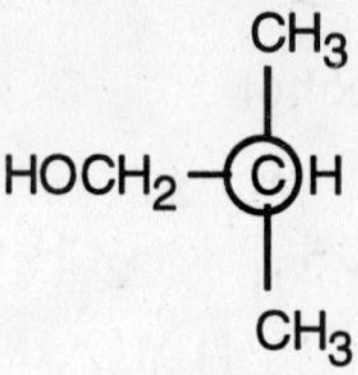

b.

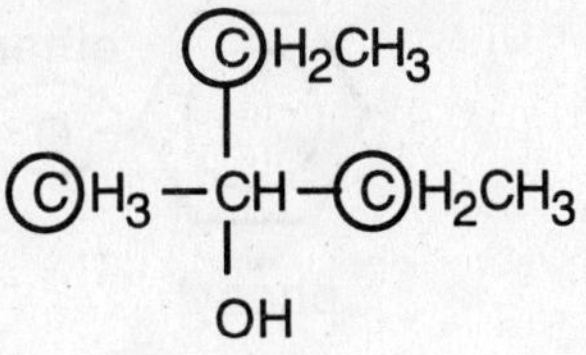

Ⓒ H₂CH₃
|
ⒸH₃—CH—ⒸH₂CH₃
|
OH

tertiary alcohol

d.

CH₃
|
HO—CH₂—Ⓒ—CH₃
|
CH₃

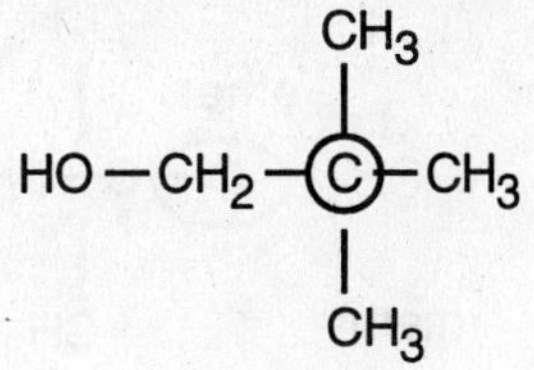

primary alcohol

24.46 a.

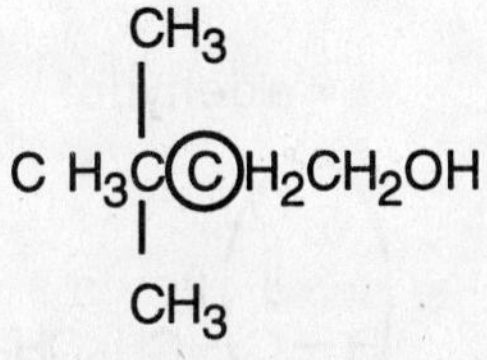

CH₃
|
CH₃ⒸH₂CH₂OH
|
CH₃

primary alcohol

c.

CH₂OH
|
CH₃CH₂ⒸCH₃
|
CH₃

primary alcohol

b.

ⒸH₃
|
CH₃CH₂ⒸH₂COH
|
ⒸH₃

tertiary alcohol

d.

ⒸH₂CH₃
|
CH₃ⒸH₂—C—OH
|
ⒸH₂CH₃

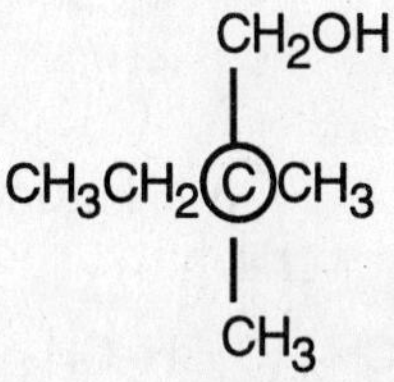

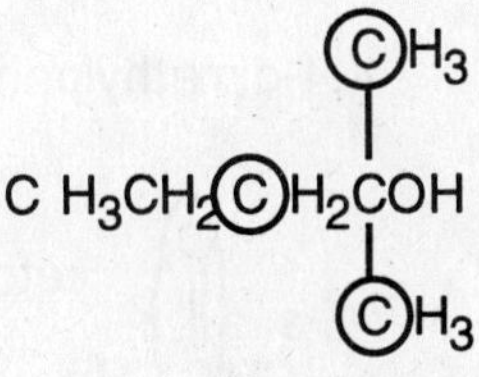

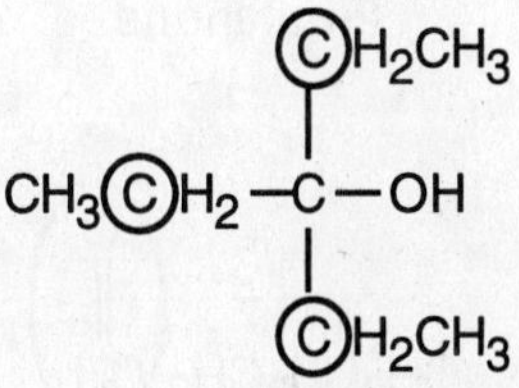

tertiary alcohol

24.47 a.   CH₃CH₂—Ⓞ—CH₂CH₂CH₃   ethyl propyl ether
           ethyl   propyl
                    ↑
                  ether

b.
     CH₃
      |
   HC—Ⓞ—CH₃   methyl isopropyl ether
      | ↑
     CH₃   methyl
      ↑ ether
   isopropyl

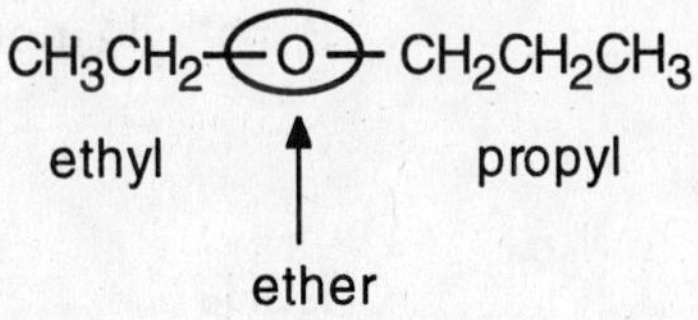

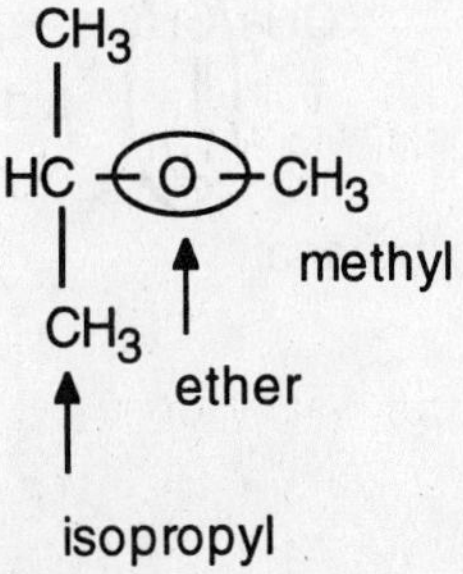

24.48 a.

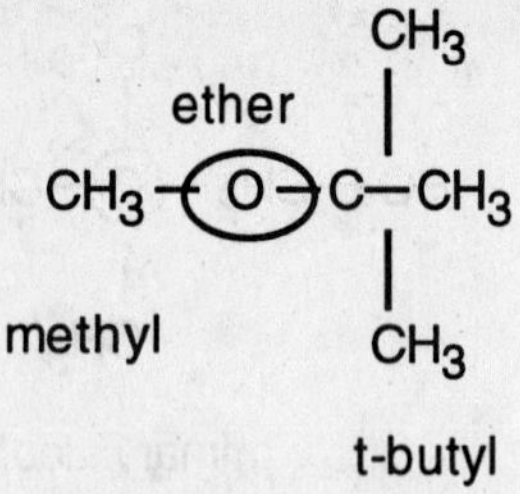

methyl t-butyl ether

b.

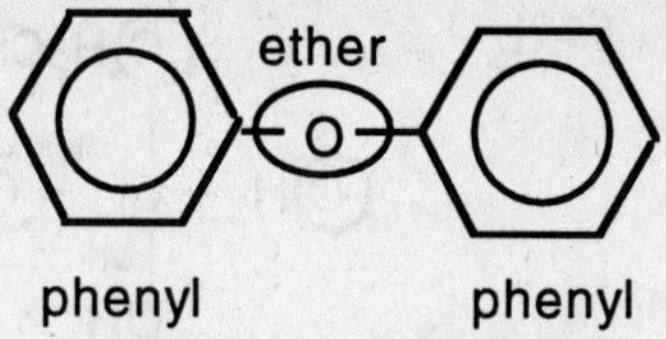

diphenyl ether

24.49 a.

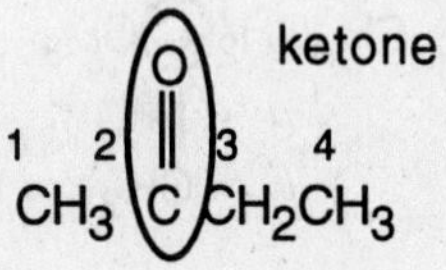

2-butanone

b.

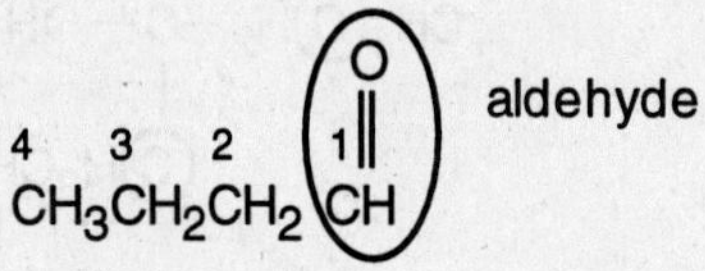

butanal

c.

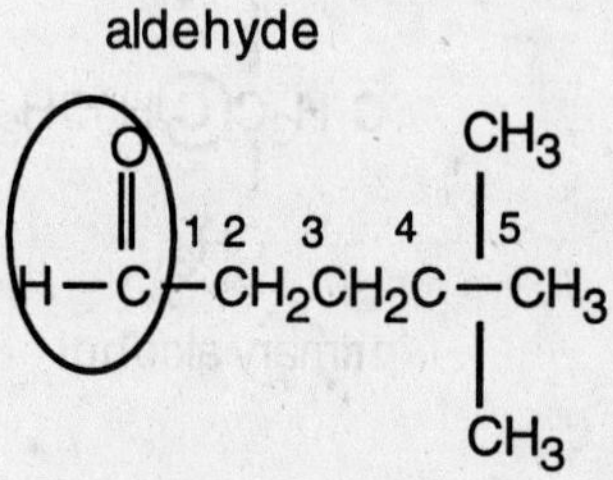

4,4-dimethylpentanal

d.

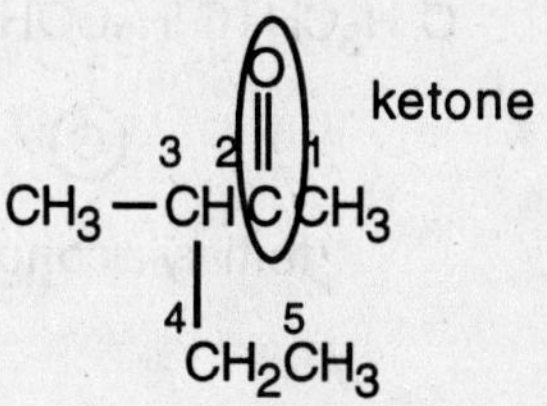

3-methyl-2-pentanone

24.50 a.

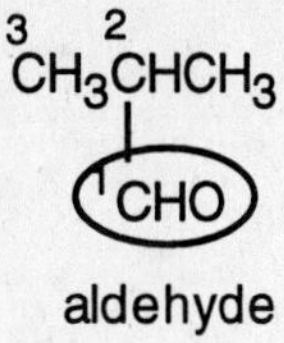

2-methylpropanal

c.

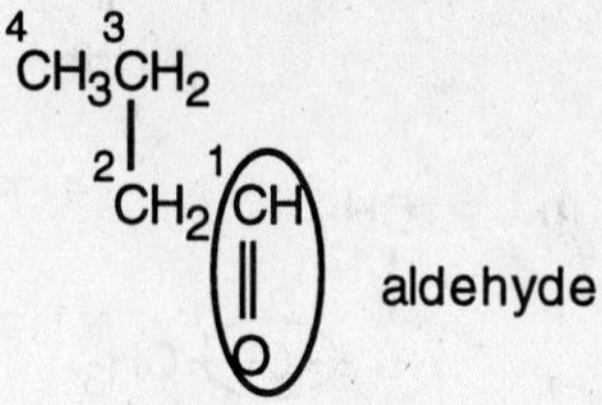

butanal

b. $\overset{4}{C}H_3\overset{3}{C}HCH_3$
    $\overset{2}{C}=O$ ... $\overset{1}{C}H_3$ (ketone circled around C2=O)
    ketone
    3-methyl-2-butanone

d. $\overset{6}{C}H_3\overset{5}{C}HCH_2$
    $\overset{4}{C}H_2\overset{3}{C}\overset{2}{C}H_2\overset{1}{C}H_3$, C3=O (ketone)
    ketone
    5-methyl-3-hexanone

24.51

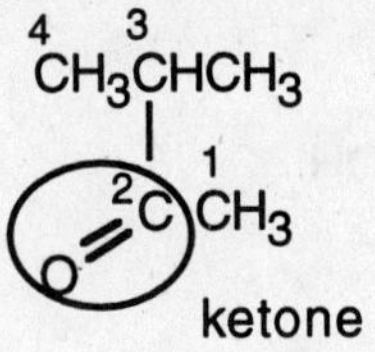

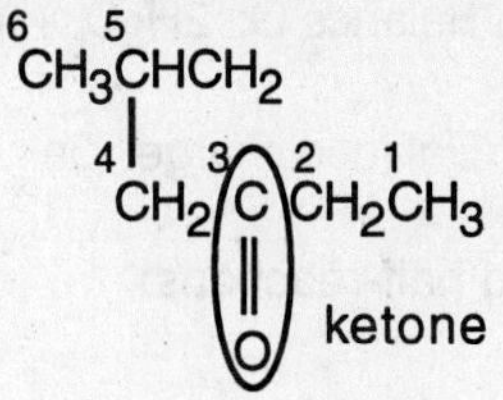

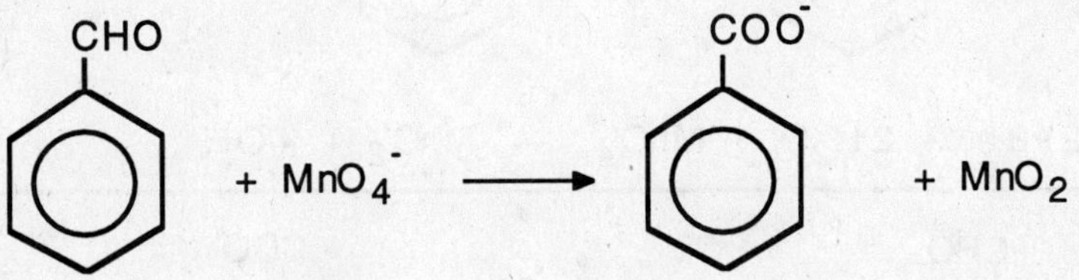

C$_6$H$_5$CHO + MnO$_4^-$ → C$_6$H$_5$COO$^-$ + MnO$_2$

Oxidation: C$_6$H$_5$CHO → C$_6$H$_5$COO$^-$

Balance O: 2 OH$^-$ + C$_6$H$_5$CHO → C$_6$H$_5$COO$^-$ + H$_2$O

Balance H: 3 OH$^-$ + C$_6$H$_5$CHO → C$_6$H$_5$COO$^-$ + 2 H$_2$O

Balance charge: 3 OH$^-$ + C$_6$H$_5$CHO → C$_6$H$_5$COO$^-$ + 2 H$_2$O + 2 e$^-$

Reduction: MnO$_4^-$ → MnO$_2$

(continued)

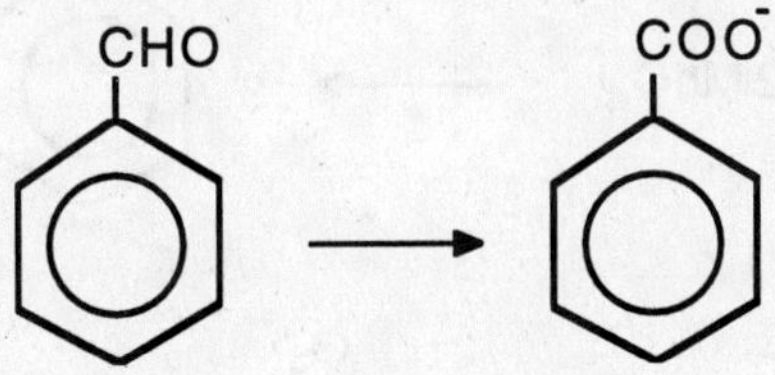

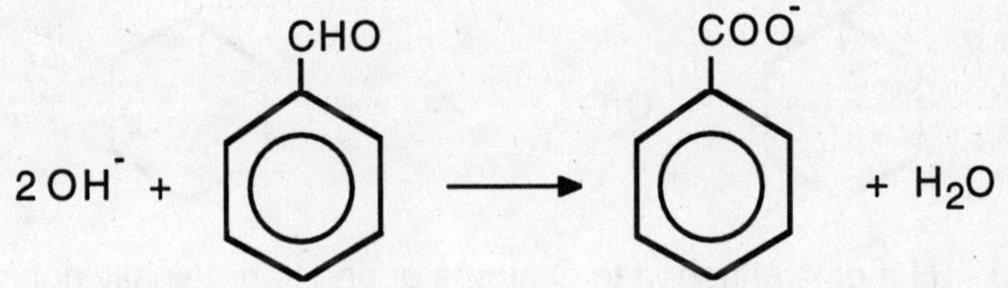

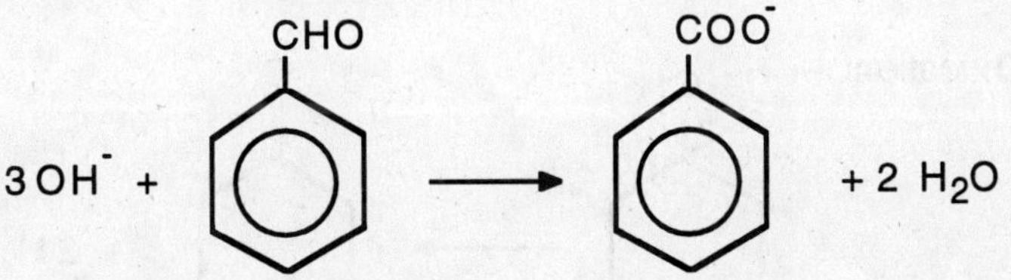

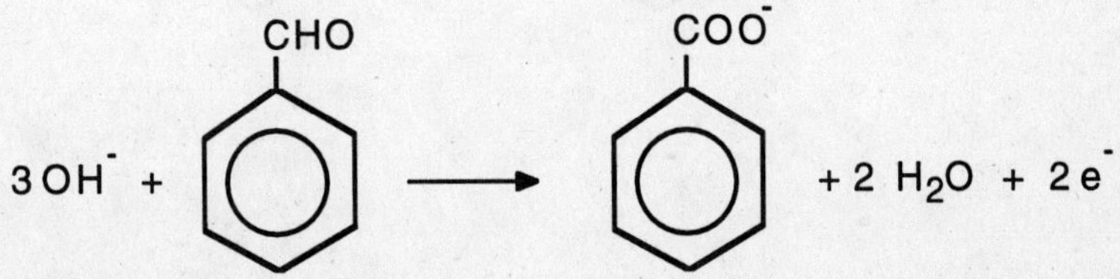

Balance O: $2H_2O + MnO_4^- \rightarrow MnO_2 + 4OH^-$

Balance charge: $3e^- + 2H_2O + MnO_4^- \rightarrow MnO_2 + 4OH^-$

Add half-reactions:

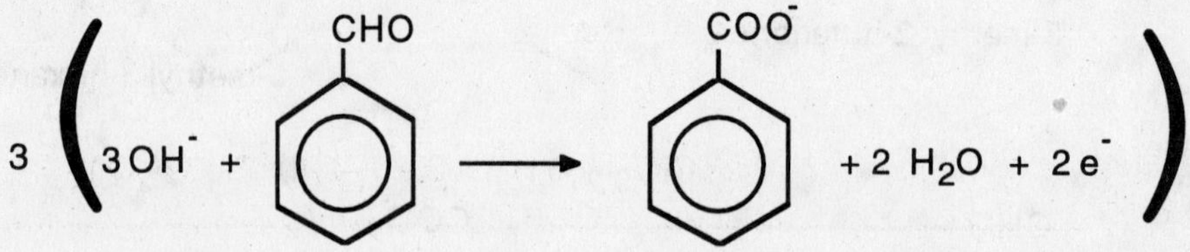

$$2(3e^- + 2H_2O + MnO_4^- \rightarrow MnO_2 + 4OH^-)$$

$$OH^- + 3\,C_6H_5\text{-CHO} + 2MnO_4^- \rightarrow 3\,C_6H_5\text{-COO}^- + 2H_2O + 2MnO_2$$

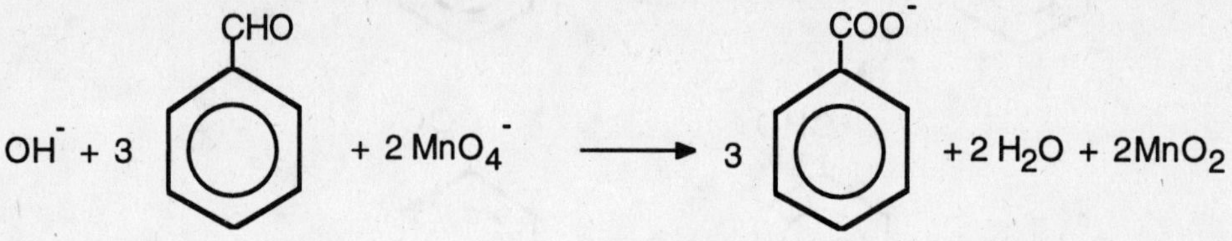

24.52

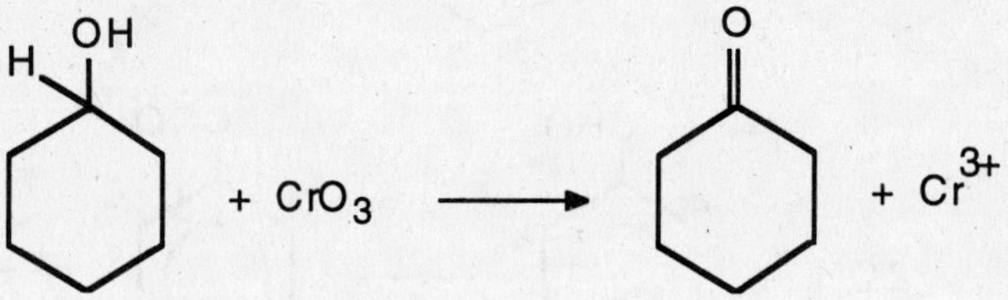

(*Note:* H atoms attached to C atoms of rings are usually not shown; here one is shown to clarify the balancing of H atoms.)

Oxidation:

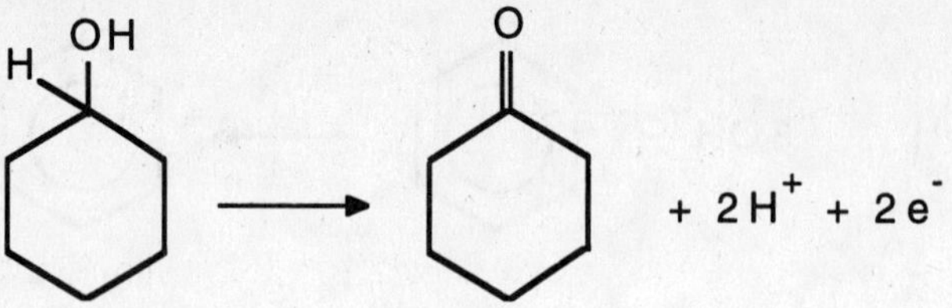

(continued)

Reduction: $CrO_3 + 6H^+ + 3e^- \rightarrow Cr^{3+} + 3H_2O$

Add half-reactions:

$$3\left(\text{cyclohexanol} \longrightarrow \text{cyclohexanone} + 2H^+ + 2e^-\right)$$

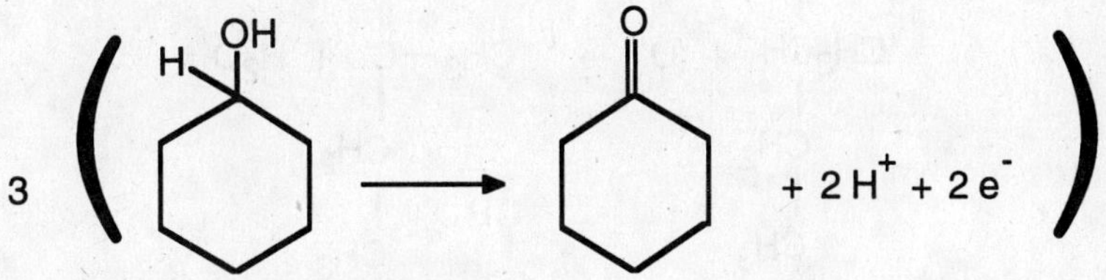

$$2(CrO_3 + 6H^+ + 3e^- \rightarrow Cr^{3+} + 3H_2O)$$

---

3 cyclohexanol $+ 2\,CrO_3 + 6\,H^+ \longrightarrow 3$ cyclohexanone $+ 2\,Cr^{3+} + 6\,H_2O$

24.53 a. $\underset{\underset{CH_2CHO}{|}}{CH_3CHCH_3} + (O) \rightarrow \underset{\underset{CH_2COOH}{2|\ \ 1}}{\overset{4\ \ \ 3}{CH_3CHCH_3}}$

aldehyde an acid
(3-methylbutanoic acid)

b. No reaction. Tertiary alcohols are not easily oxidized.

c. PhCOOH $+ 4(H) \longrightarrow$ PhCH$_2$OH $+ H_2O$

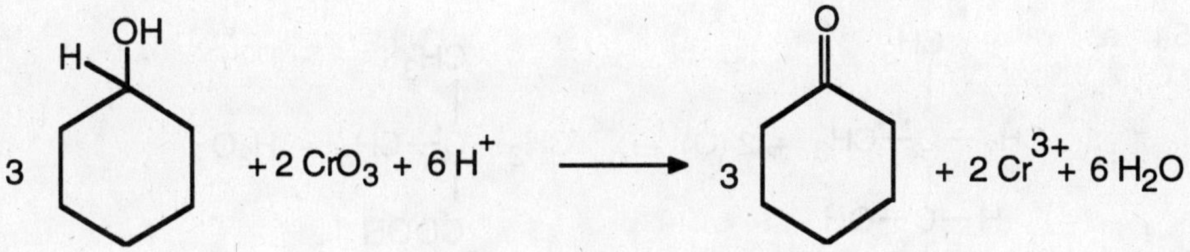

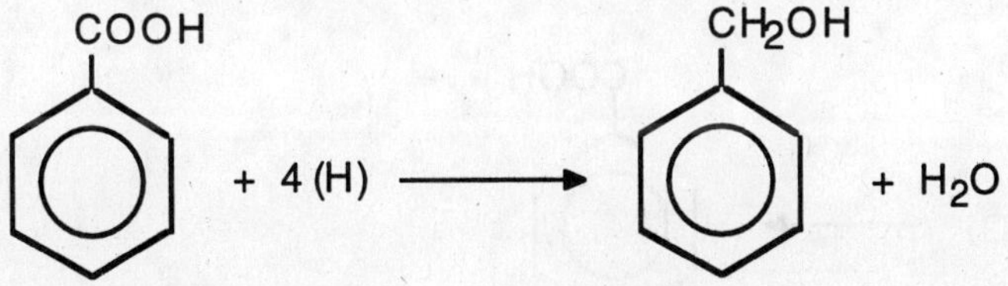

acid an alcohol (phenylmethanol)

d.

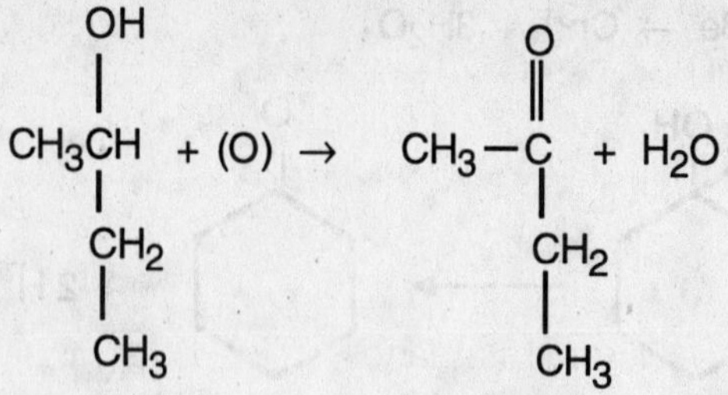

secondary alcohol

a ketone (2-butanone)

24.54 a.

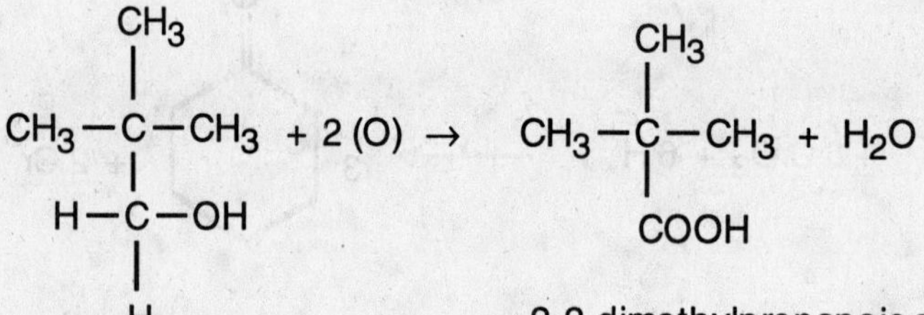

2,2-dimethylpropanoic acid

Primary alcohols are oxidized to acids.

b.
$$CH_3\underset{\underset{OH}{|}}{\overset{\overset{CH_2CH_3}{|}}{C}}CH_2CH_3 + (O) \rightarrow \text{no reaction.}$$

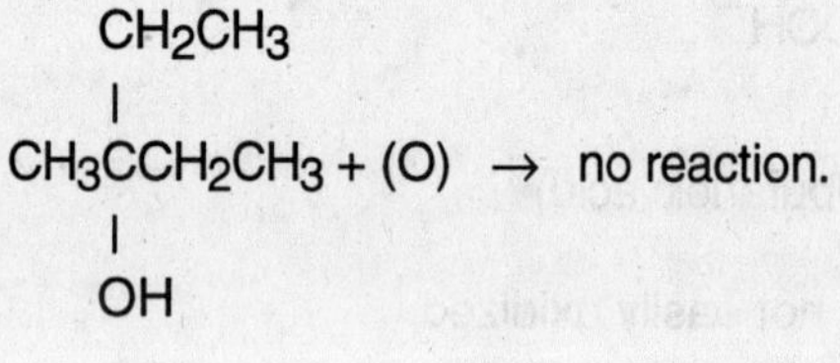

Tertiary alcohols are not easily oxidized.

c.

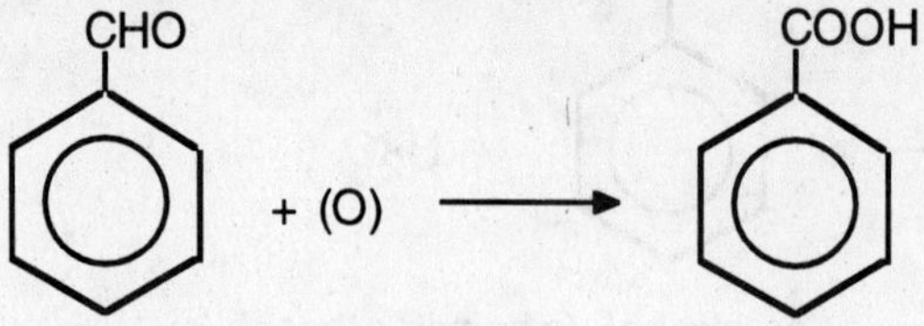

aldehydes are oxidized to carboxylic acids

benzoic acid

d.

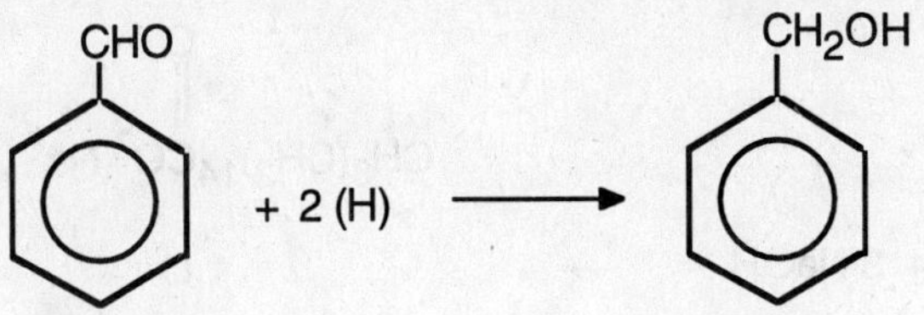

aldehydes are reduced to primary alcohols — phenyl methanol

24.55 a.

$$CH_3CH_2CH_2\overset{\overset{\large O}{\|}}{C}OH + CH_3CH_2OH \xrightleftharpoons{H^+} CH_3CH_2CH_2\overset{\overset{\large O}{\|}}{C}OCH_2CH_3 + H_2O$$

butyric acid — ethyl butyrate

b.

$$H\overset{\overset{\large O}{\|}}{C}OCH_3 + NaOH \rightarrow H\overset{\overset{\large O}{\|}}{C}O^- \; Na^+ + CH_3OH$$

methyl formate — sodium formate — methanol

24.56 a.

$$CH_3\overset{\overset{\large O}{\|}}{C}OH + CH_3\overset{\overset{\large OH}{|}}{C}HCH_3 \xrightarrow{H^+} CH_3\overset{\overset{\large O}{\|}}{C}O\overset{\overset{\large CH_3}{|}}{C}HCH_3$$

acetic acid — isopropyl alcohol — isopropyl acetate

b.

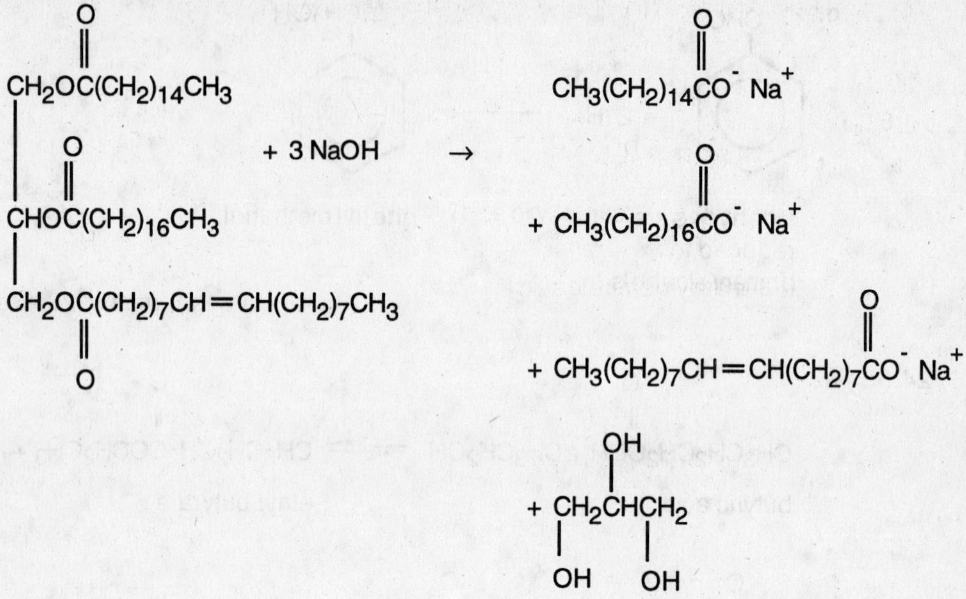

CH₂OC(CH₂)₁₄CH₃ (=O)
|
CHOC(CH₂)₁₆CH₃ (=O)   + 3 NaOH →   CH₃(CH₂)₁₄COO⁻ Na⁺
| + CH₃(CH₂)₁₆COO⁻ Na⁺
CH₂OC(CH₂)₇CH=CH(CH₂)₇CH₃ (=O)
                                       + CH₃(CH₂)₇CH=CH(CH₂)₇COO⁻ Na⁺

                                       + CH₂CHCH₂
                                         | | |
                                         OH OH OH

24.57 a. C₆H₅–NH₂ primary amine   b. CH₃CH₂–NH–CH₂CH₃ secondary amine

24.58 a. C₆H₅–C(=O)NH₂ amide   b. CH₃CH₂CH₂–NH₂ primary amine

24.59 nCF₂=CF₂ → —CF₂—CF₂—CF₂—CF₂—CF₂—CF₂—

24.60 nCH₂=CHCl → —CH₂—CH—CH₂—CH—CH₂—CH—
                          | | |
                          Cl   Cl   Cl

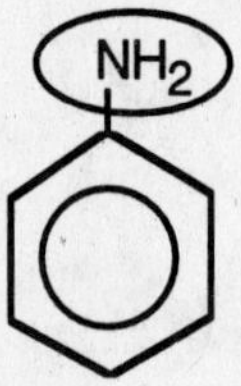

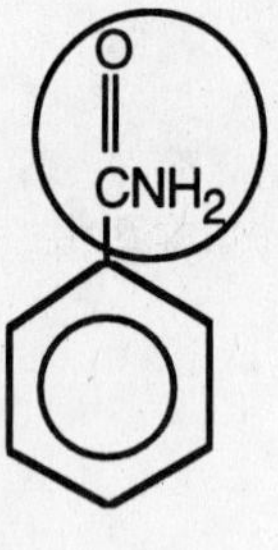

■ Solutions to Unclassified Problems

24.61 a.

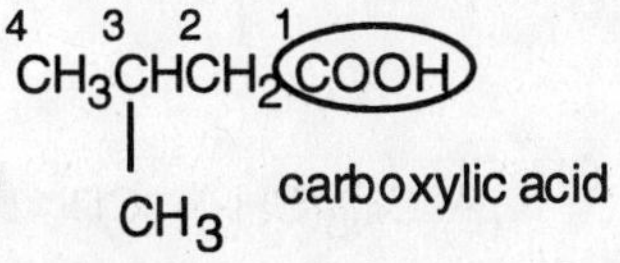

3-methylbutanoic acid

b.

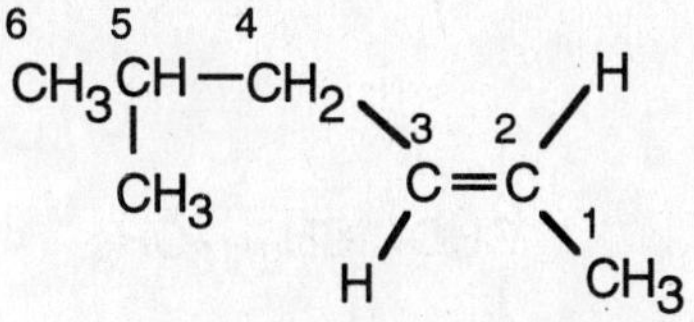

trans-5-methyl-2-hexene

c.

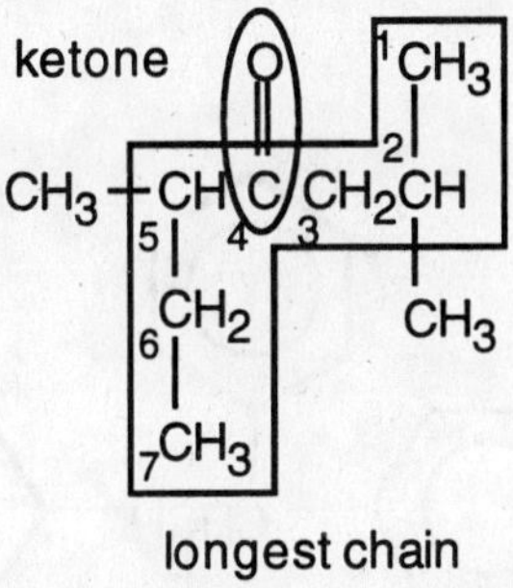

2,5-dimethyl-4-heptanone

d.

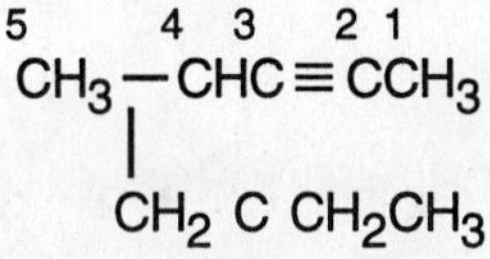

4-methyl-2-pentyne

24.62 a.

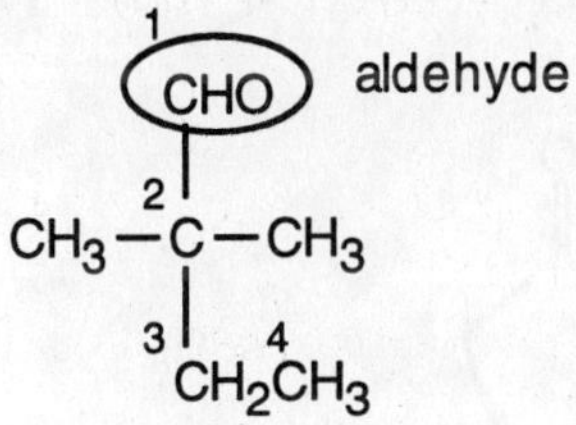

2,2-dimethylbutanal

c.

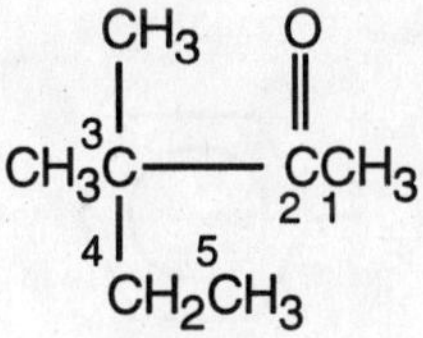

3,3-dimethyl-2-pentanone

b.

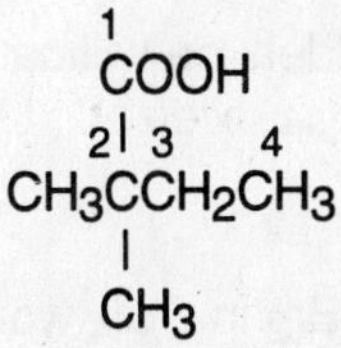

2,2-dimethylbutanoic acid

d.

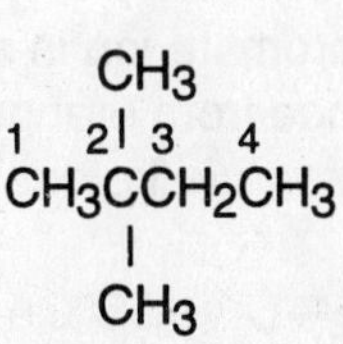

2,2-dimethylbutane

24.63 a.

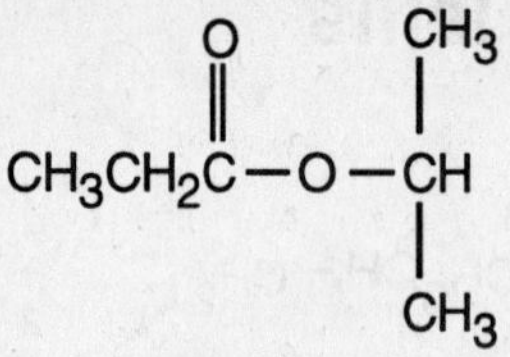

c.

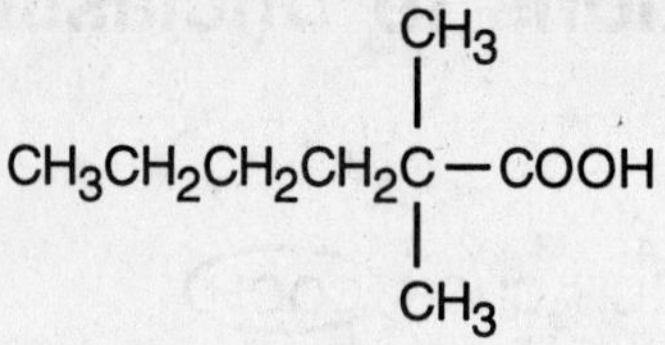

b.

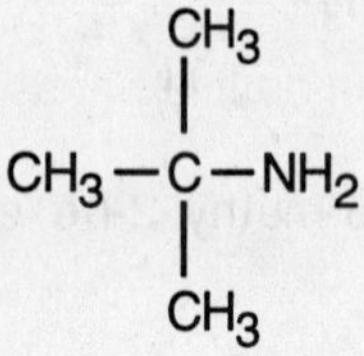

d.

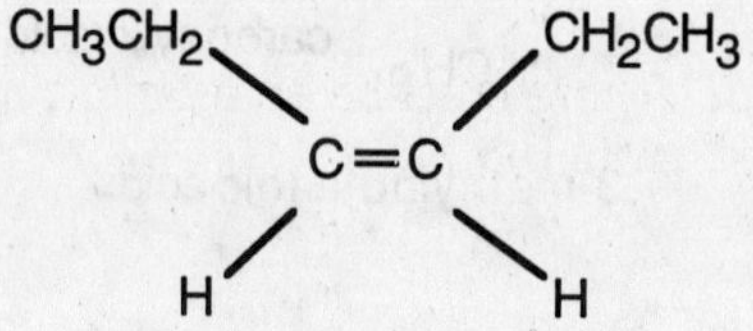

22.64 a.

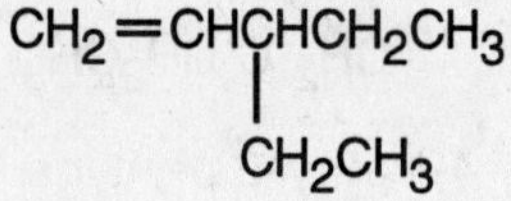

b.

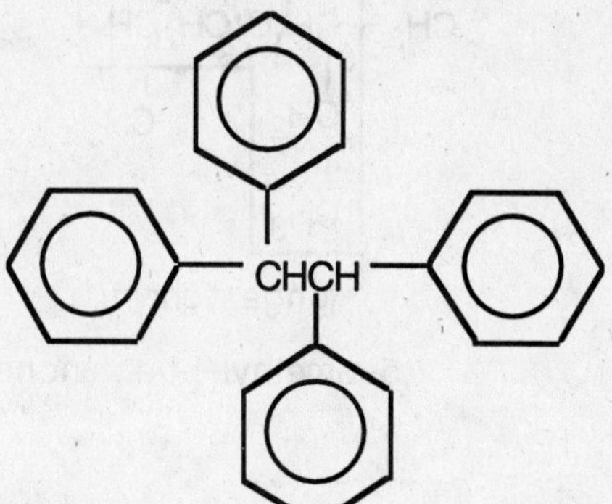

c.

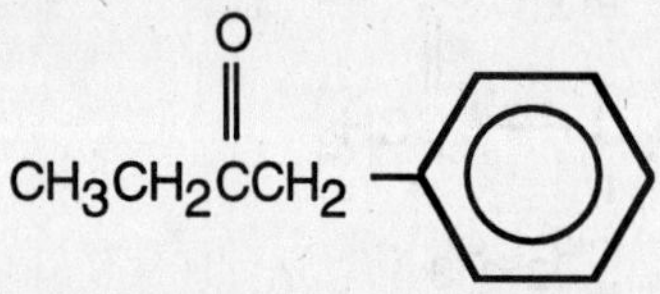

d.

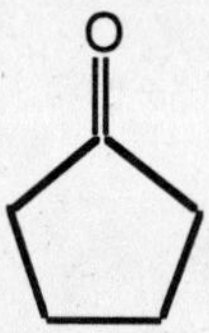

24.65 a. Addition of dichromate ion in acidic solution to propionaldehyde will cause the reagent to change from orange to green as the aldehyde is oxidized. Under similar conditions, acetone (a ketone) would not react.

b. Addition of $CH_2{=}CH{-}C{\equiv}C{-}CH{=}CH_2$ to a solution of Br_2 in CCl_4 would cause the bromine color to disappear as the Br_2 was added to the multiple bonds. Addition of benzene to Br_2 in CCl_4 results in no reaction. Aromatic rings are not susceptible to attack by Br_2 in the absence of a catalyst.

24.66 a. Addition of an acidic solution of dichromate ion to acetaldehyde will cause the reagent to change from orange to green as the aldehyde is oxidized. Under similar conditions, acetic acid will not be oxidized.

b. Addition of a solution of Br_2 in CCl_4 to 2-methyl-cyclohexene will cause the bromine color to disappear as the dibromomethylcyclohexene is formed. Toluene will not react instantaneously with Br_2 in CCl_4, as does the alkene.

24.67 a. $CH_2{=}CH_2$, ethylene

b. There must be an aromatic ring in the compound or the double bonds would react with Br_2.

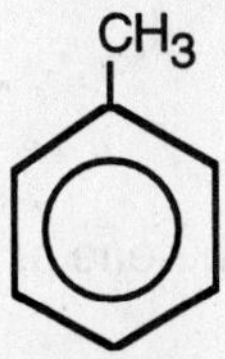

is the correct formula

c. CH_3NH_2

d. CH_3OH

24.68 a.

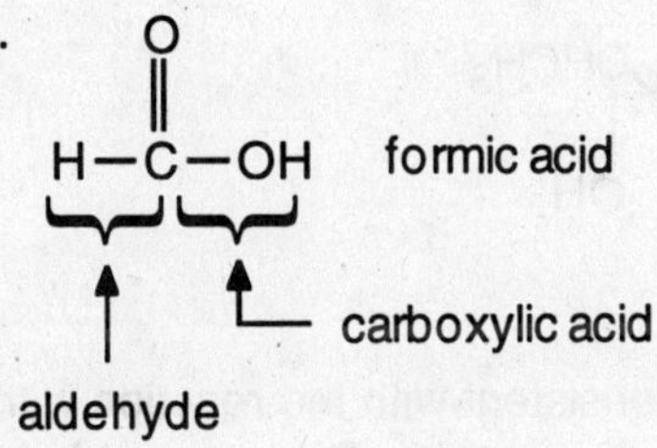

formic acid

d.

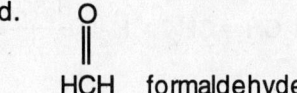

formaldehyde

c. $CH_3CH_2CH_2CH_3$, butane d. CH_4, methane

24.69 Assume 100.0 g of the unknown. This contains 85.6 g C and 14.4 g H. Convert these amounts to moles.

$$85.6\ \text{g C} \times \frac{1\ \text{mol C}}{12.01\ \text{g C}} = 7.1\underline{2}7\ \text{mol C}$$

$$14.4\ \text{g H} \times \frac{1\ \text{mol H}}{1.008\ \text{g H}} = 14.\underline{2}857\ \text{mol H}$$

The molar ratio of H to C is 14.2857:7.12, or 2.00:1. The empirical formula is therefore CH_2. This formula unit has a mass of [12.011 + 2 x (1.008)] amu = 14.027 amu.

(continued)

$$\frac{56.1 \text{ amu}}{1 \text{ molecule}} \times \frac{1 \text{ formula unit}}{14.03 \text{ amu}} = \frac{4.00 \text{ formula units}}{1 \text{ molecule}}$$

The molecular formula is $(CH_2)_4$, or C_4H_8. The formula (C_nH_{2n}) indicates that the compound is either an alkene or a cycloalkane. Because it reacts with water and H_2SO_4, it must be an alkene. The product of the addition of H_2O to a double bond is an alcohol. Because the alcohol produced can be oxidized to a ketone, it must be a secondary alcohol. The only secondary alcohol with four carbon atoms is 2-butanol.

$$CH_3CH_2\underset{\underset{OH}{|}}{C}HCH_3$$

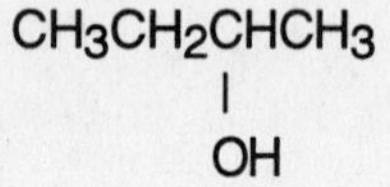

The original hydrocarbon from which it was produced is either 1-butene or 2-butene.

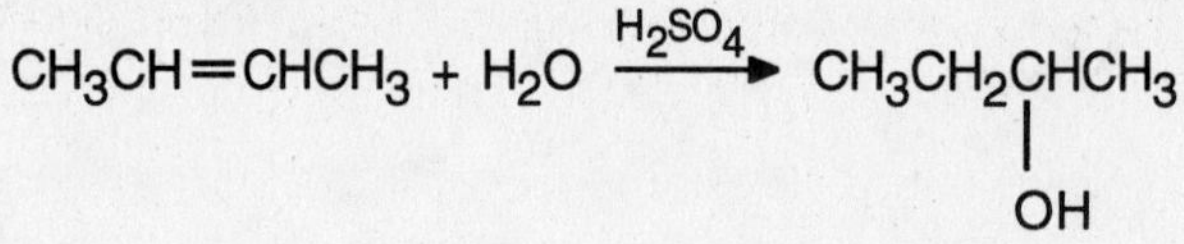

$$CH_3CH=CHCH_3 + H_2O \xrightarrow{H_2SO_4} CH_3CH_2\underset{\underset{OH}{|}}{C}HCH_3$$

or

$$CH_3CH_2CH=CH_2 + H_2O \xrightarrow{H_2SO_4} CH_3CH_2\underset{\underset{OH}{|}}{C}HCH_3$$

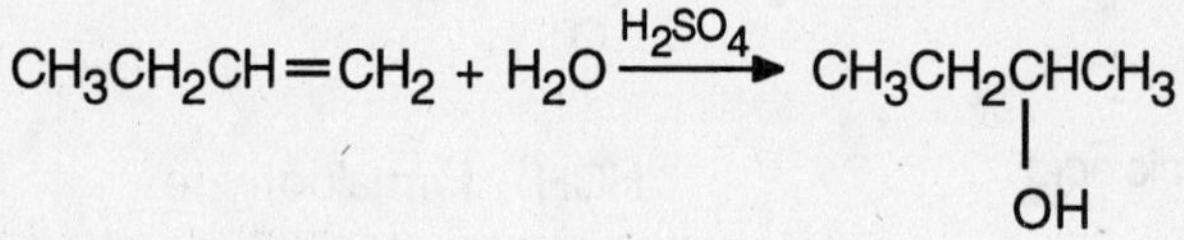

24.70 The fragrant odor suggests an ester, which is consistent with the reaction with acid. Esters are hydrolyzed in acid to give an alcohol and an acid. The only alcohol with a molecular weight of 32.0 amu is methanol, CH_3OH. Carboxylic acids can be reduced to give primary alcohols. In a 100.0-g sample of the alcohol, there are 60.0 g C, 13.4 g H, and 26.6 g O. Convert these masses to moles.

$$60.0 \text{ g C} \times \frac{1 \text{ mol C}}{12.011 \text{ g C}} = 4.995 \text{ mol C}$$

$$13.4 \text{ g H} \times \frac{1 \text{ mol H}}{1.008 \text{ g H}} = 13.29 \text{ mol H}$$

$$26.6 \text{ g O} \times \frac{1 \text{ mol O}}{16.00 \text{ g O}} = 1.662 \text{ mol O}$$

(continued)

Divide each by 1.662 (smallest number of moles).

$$\frac{4.995}{1.662}\text{ mol C} = 3.005\text{ mol C} = 3\text{ mol C}$$

$$\frac{13.29}{1.662}\text{ mol H} = 7.996\text{ mol H} = 8\text{ mol H}$$

$$\frac{1.662}{1.662}\text{ mol O} = 1.00\text{ mol O} = 1\text{ mol O}$$

The empirical formula of the alcohol is C_3H_8O, which has a formula weight of 60.1 amu, equal to the molecular weight of the alcohol. The only primary alcohol with this formula is 1-propanol, $CH_3CH_2CH_2OH$. The acid that gives 1-propanol when it is reduced is propanoic acid. Therefore, the original compound was the ester of methanol and propanoic acid:

```
        O
        ||
CH3CH2COCH3
```

methyl propanoate

(The common name for propanoic acid is propionic acid, so the ester could also be named methyl propionate.)

■ Solution to Conceptual Problem

24.71 Many organic reactions involve the reaction of two different molecules to form large macromolecules. The formation of the polymer nylon is a good example with industrial relevance. For example, 1.3 mullion tons were produced in 1993. Section 24.8 describes the preparation of nylon-6,6. Modifications of the polymer chain can change the physical properties of the polymer. Suggest modifications to the reactants in the preparation of nylon-6,6 that would yield a more rigid polymer system. (Hint: Table 24.8 lists several other polymers and their uses.)

Discussion

Polymerization reactions can be modified in many ways to enhance rigidity. One possibility would be to modify the number of monomer units in the chain. As you increase the chain length, there are more intermolecular interactions available among chains, and this can lead to enhanced crystalline behavior.

(continued)

For example, high density polyethylene (with a molecular weight on the order of a million) is used for liquid containers such as milk jugs. In contrast, low density polyethylene (with a molecular weight on the order of ten thousand) is used to make sandwich bags.

A monomer having substituents such as OH groups can also increase the intermolecular forces between chains and therefore give a more rigid polymer. Also, a monomer that gives a polymer containing double bonds in the polymer chains can give a more rigid product.

Another means of giving a more rigid polymer would be to increase the number of sites on the monomer that can polymerize. Such a monomer would form a polymer with cross links (that is, with links between chains), whose effect would be to give a more rigid product. Thus, instead of forming nylon with the amine $H_2N(CH_2)_6NH_2$, you could use an amine having an additional NH_2 substituent on the carbon chain, for example $H_2N(CH_2)_3CHNH_2(CH_2)_2NH_2$. Crosslinking is the basis for the vulcanization of rubber, a process developed by Charles Goodyear. Unvulcanized rubber becomes tacky in the summer. Heating it with sulfur yields crosslinks that give a product with a stiffer, but still rubberlike character.

25. BIOCHEMISTRY

Answers to Review Questions

25.1 Biological systems with positive free-energy changes proceed by using body energy supplied through coupling with a chemical reaction having negative free-energy change.

25.2 The primary structure of a protein refers to the order, or sequence, of the amino-acid units in the protein polymer. What makes one protein different from another of the same size is the arrangement of the various possible amino acids in the sequence. For example, there are 120 different sequences possible for a polypeptide with just five different amino acids. The basis of the unique conformation of a protein is the folding and coiling into a three-dimensional conformation in aqueous solution on the basis of the different side-chain amino-acid units. In stable conformations, the nonpolar side chains are buried within the structure away from water and the polar groups are on the surface of the conformation, where they can hydrogen bond with water.

25.3 The secondary structure of a protein is the simpler coiled or parallel arrangement of the protein chain. The tertiary structure refers to the folded nature of the structure.

25.4 An enzyme is a specific body catalyst, usually a globular protein, that possesses active sites. The specificity of an enzyme is explained by the active sites at which substrates will bind. Substrates fit into the active sites as a key fits into a lock.

25.5 A monosaccharide is a molecule with three to nine carbon atoms, all of which but one bear an -OH group. The remaining carbon is always a carbonyl carbon, either an aldehyde or a ketone. The four major monosaccharides are D-glucose, D-fructose, D-ribose, and 2-deoxy-D-ribose. D-glucose is blood sugar, an important energy source for cell function. D-fructose is a common sugar in fruits and a food source. D-ribose and 2-deoxy-D-ribose are parts of nucleic acids, carriers of species inheritance.

(continued)

An oligosaccharide is a molecule formed from monosaccharides in which the hemiacetal carbon atom of one is attached to the alcohol oxygen atom of another by a condensation reaction. The most common oligosaccharide is sucrose, a disaccharide. A polysaccharide is a long-chain molecular carbohydrate built of only one type of monomer or sometimes two alternating monomers. Some are made of glucose molecules only.

25.6 Cellulose forms the plant cell walls and is a linear polymer of β-D-glucopyranose units. Amylose is the major energy-storage substance of plants and is a linear polymer of α-D-glucopyranose units.

25.7 It is possible for glucose to exist in three forms in any solution such as blood: the straight-chain form, the α-D-glucopyranose form, and the β-D-glucopyranose form.

25.8 The complementary base pairs are the nucleotide bases that form strong hydrogen bonds with one another: adenine and thymine, adenine and uracil, and guanine and cytosine. A DNA molecule consists of two polynucleotide chains with base pairing along their entire lengths. The two chains are coiled about each other to form a double helix.

25.9 The only difference between ribonucleotides and deoxyribonucleotides is the sugar: Ribonucleotides contain β-D-ribose, and deoxyribonucleotides contain 2-deoxy-β-D-ribose, with both sugars having furanose rings. Both form polymers by condensation (loss of H_2O).

25.10 The genetic code is the relationship between the nucleotide sequence in DNA and the amino-acid sequence in proteins. Genetic information is coded into the linear sequence of nucleotides in the DNA molecule, and this coding then directs the synthesis of the specific proteins that make a cell unique.

25.11 A codon is a code structure in messenger RNA; each such structure has a particular sequence of three nucleotides, usually denoted simply by their bases. An anticodon is a triplet sequence in transfer RNA complementary to the codon in DNA; the transfer RNA uses the anticodon to carry an amino acid to a ribosome. The messenger RNA and transfer RNA bond to each other through the codon and anticodon, respectively.

25.12 There are four RNA bases, and three of these are arranged in a specific order in each codon. Since there are 4 different possibilities for each base in the codon, there are $4 \bullet 4 \bullet 4 = 64$ different triplet codons.

25.13 A polypeptide is produced as follows: Imagine that you have a ribosome with messenger RNA attached in the proper way for translation and the first codon is in position to be read. (The messenger RNA has been synthesized with a sequence of bases complementary to that of the gene before attachment.) Transfer RNAs bring up various amino acids to be bonded to each other until a termination codon appears, to signal the end of the chain, which is then released from the ribosome.

25.14 Fats and oils have the same basic structure, but fats are solids; oils are liquids at room temperature due to the presence of unsaturated fatty acids.

25.15 A triacylglycerol is a triester formed from glycerol and three fatty acids bonded to the oxygens of the glycerol.

25.16 A biological membrane is composed of proteins inserted into a phospholipid matrix. A phospholipid resembles a triacylglycerol, but only two fatty acids form ester bonds; the third —OH group of glycerol is bonded to a phosphate group that in turn is bonded to an alcohol through an oxygen attached to phosphorus. Because the phospholipids form a bilayer, with the interior of the lipid bilayer being hydrophobic hydrocarbon chains of the fatty acids of the phospholipids, they present a barrier to charged or polar substances (hydrophilic). The proteins in the membrane do catalyze the transport of particular substances across the membrane, some of which may be hydrophilic.

■ Solutions to Practice Problems

25.17 The free-energy change for coupled reactions is the sum of the free-energy changes for the individual reactions.

$$\Delta G_{overall} = +7.32 \text{ kcal/mol} + (-10.07) \text{ kcal/mol}$$

$$= -2.75 \text{ kcal/mol}$$

25.18 $\Delta G_{overall} = \Delta G_{energy\text{-}releasing} + \Delta G_{energy\text{-}requiring}$

so

$$\Delta G_{energy\text{-}requiring} = \Delta G_{overall} - \Delta G_{energy\text{-}releasing}$$

$$= -1.5 \text{ kcal/mol} - (-5.2 \text{ kcal/mol})$$

$$= +3.7 \text{ kcal/mol}$$

25.19 In general, reactions that break complex molecules into simpler molecules are energy-releasing, so the breakdown of a fat to give CO_2 and H_2O is expected to release free energy.

25.20 The overall change yields a more complex, or more highly organized, molecule. Thus, ΔS will be negative and $-T\Delta S$ will be positive. Unless ΔH is sufficiently large and negative (not usually the case), $\Delta G = \Delta H - T\Delta S$ will be positive. That is, free energy will be required.

25.21

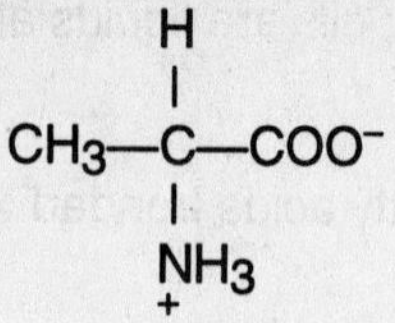

25.22

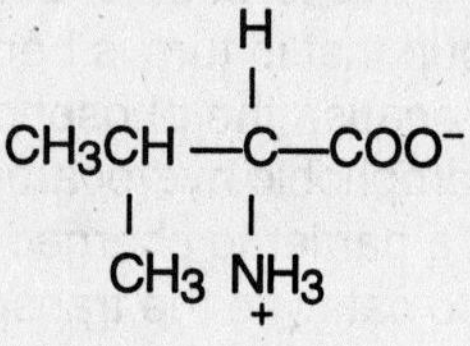

25.23 There are two possible dipeptides containing one molecule each of L-alanine and L-histidine:

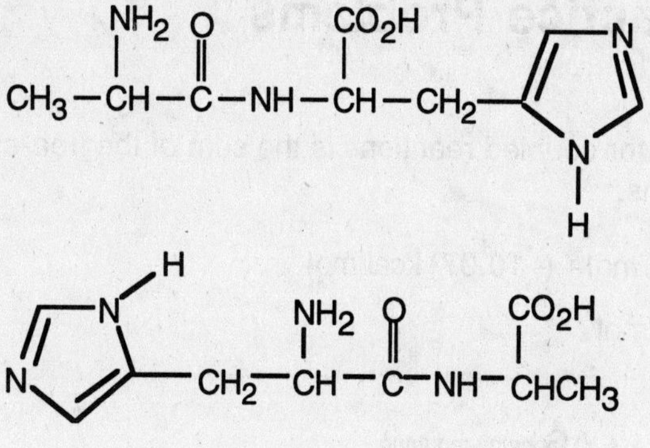

25.24 The number of possible tripeptides is 6 (3 • 2 • 1). Two possibilities are

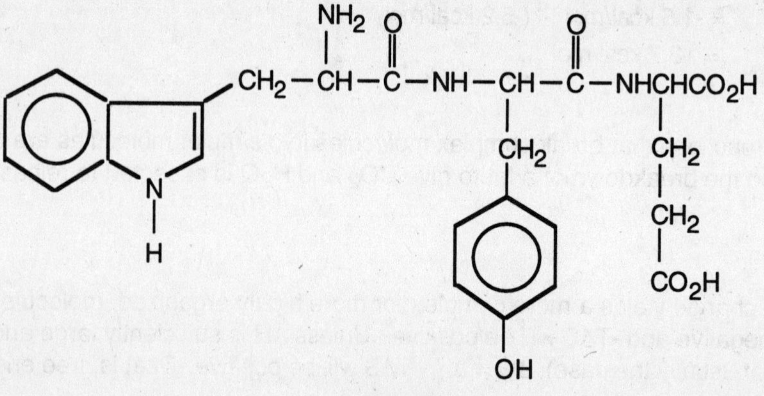

trp-tyr-glu

(continued)

trp-glu-tyr

25.25

$$\begin{array}{c} \text{O} \\ \parallel \\ \text{C}-\text{H} \\ \text{H}-\text{C}-\text{OH} \\ \text{CH}_2\text{OH} \end{array}$$

25.26

$$\begin{array}{c} \text{CH}_2\text{OH} \\ \text{C}=\text{O} \\ \text{H}-\text{C}-\text{OH} \\ \text{CH}_2\text{OH} \end{array}$$

25.27

$$\begin{array}{cc} \begin{array}{c} \text{O} \\ \parallel \\ \text{C}-\text{H} \\ \text{H}-\text{C}-\text{OH} \\ \text{CH}_2\text{OH} \end{array} & \begin{array}{c} \text{O} \\ \parallel \\ \text{C}-\text{H} \\ \text{HO}-\text{C}-\text{H} \\ \text{CH}_2\text{OH} \end{array} \end{array}$$

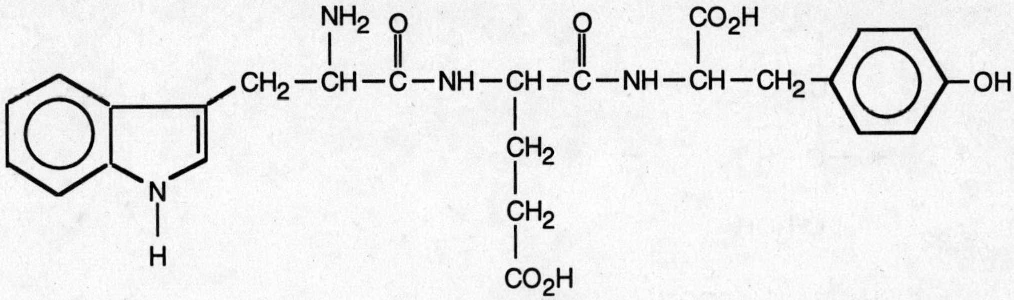

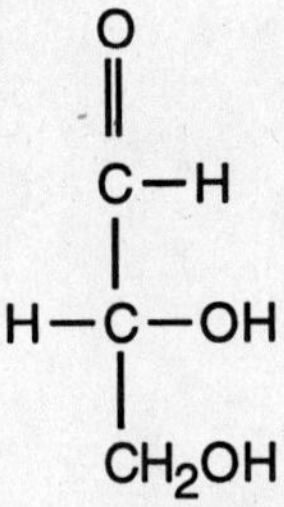

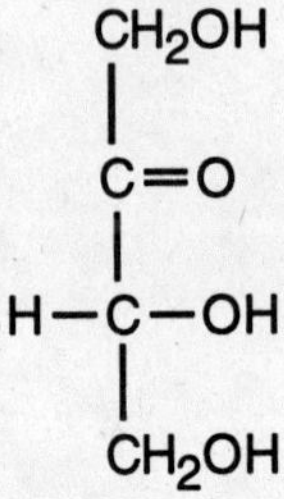

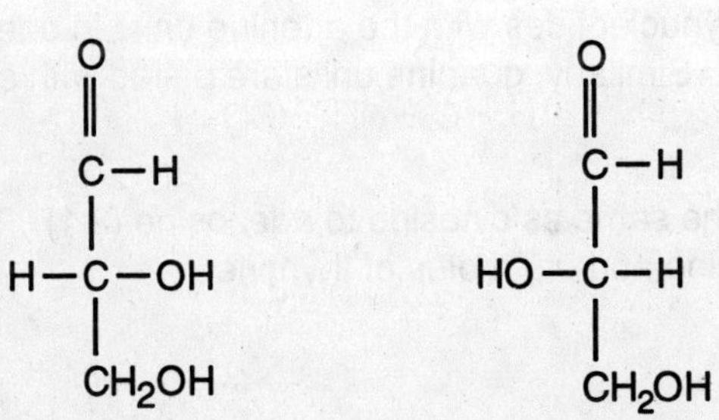

25.28

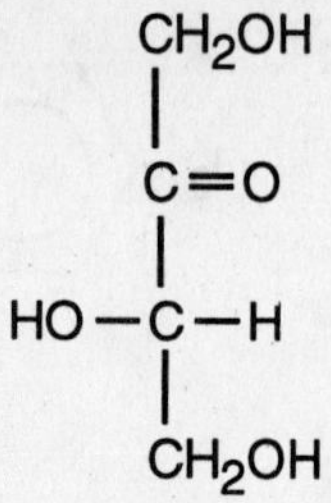

25.29 cyclic form of D-ribose

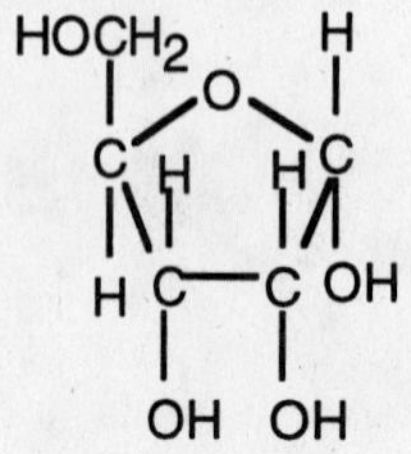

25.30 cyclic form of D-fructose

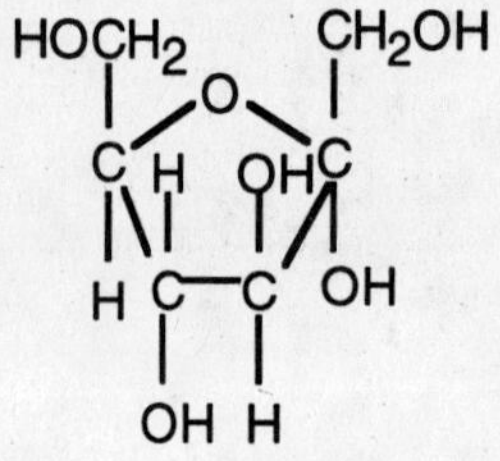

25.31 DNA consists of two strands of polynucleotides with the adenine units in one strand paired to thymine units in the other. Similarly, guanine units are paired with cytosine.

25.32 The ratio of guanine to thymine is the same as cytosine to adenosine (3:1). Thus, the sample contains 1.5 moles of guanine and 0.5 moles of thymine.

25.33 Adenosine consists of ribose and adenine.

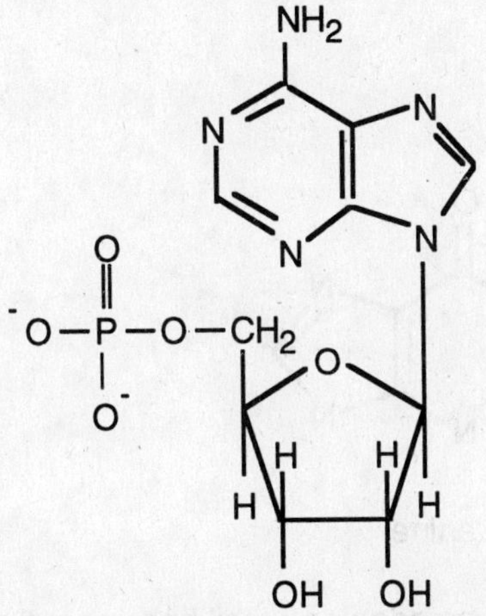

25.34 Deoxyadenosine consists of deoxyribose and adenine.

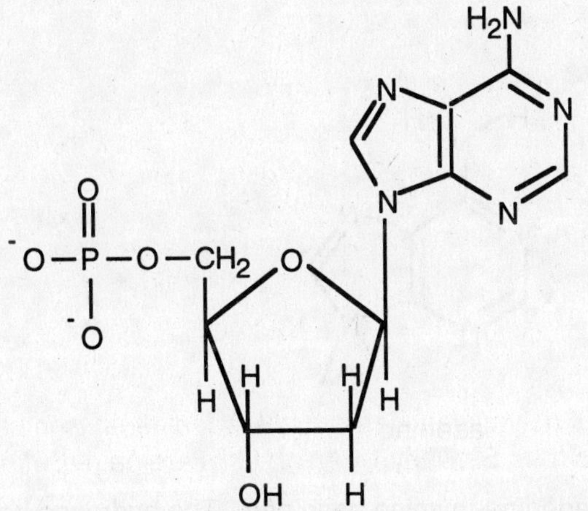

25.35 Three hydrogen bonds link a guanine-cytosine base pair.

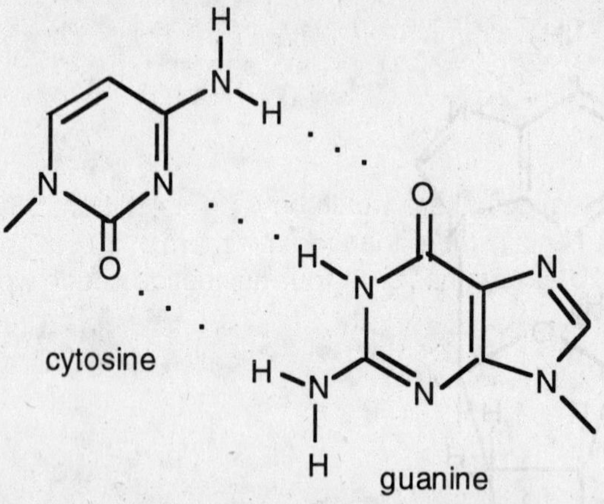

Because only two hydrogen bonds link an adenine-uracil base pair, the bonding would be expected to be stronger in the guanine-cytosine pair.

25.36

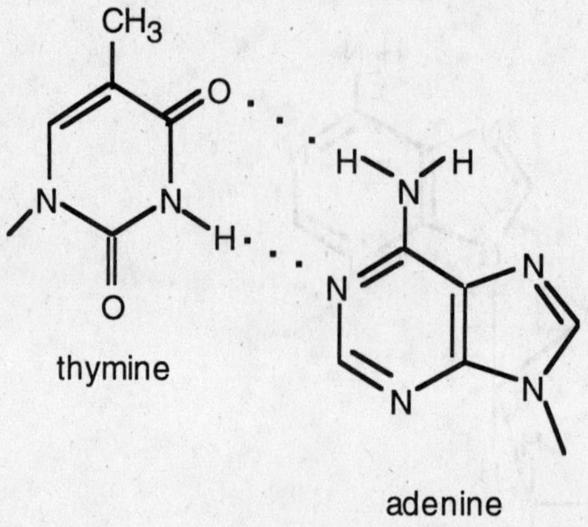

Two hydrogen bonds link an adenine-thymine base pair. The hydrogen bonding in a uracil-adenine base pair is the same as in a thymine-adenine base pair, so the strength of the bonding should be the same.

There are three hydrogen bonds in a guanine-cytosine base pair, so the bonding should be stronger between guanine and cytosine than between adenine and thymine.

25.37 If a codon consisted of two nucleotides, there would be 4 x 4 or 16 possible codons using the four nucleotides. Because there are 20 amino acids that must be represented uniquely by a codon for protein synthesis, the codon must be longer than two nucleotides. A two-nucleotide codon would not be workable as an amino-acid code.

25.38 If the codon were four nucleotides, there would be 4 x 4 x 4 x 4 or 256 possible codons using the four nucleotides. This is more than the minimum of 20 codons required for all the amino acids used in protein synthesis. A four-nucleotide codon would be workable as an amino-acid code.

25.39 When DNA is denatured, the hydrogen bonds between base pairs are broken. DNA with a greater percent composition of guanine and cytosine would be denatured less readily than DNA with a greater percent composition of adenine and thymine because there are more hydrogen bonds between the former than between the latter.

25.40 RNA of greater percent composition of guanine and cytosine would be denatured less readily than RNA with a lower such composition because denaturation involves disruption of hydrogen bonds and the higher the percent composition of guanine and cytosine, the more hydrogen bonds there are in the RNA.

25.41 Mark off the message into triplets, beginning at the left.

GGA | UCC | CGC | UUU | GGG | CUG | AAA | UAG
Gly-Ser-Arg-Phe-Gly-Leu-Lys

Note that the UAG at the right codes for the end of the sequence.

25.42 Mark the message off into triplets, beginning at the left.

AUU | GGC | GCG | AGA | UCG | AAU | GAG | CCC | AGU
Ile-Gly-Ala-Arg-Ser-Asn-Glu-Pro-Ser

25.43 The codons have bases that are complementary to those in the anticodon.

Anticodons	GAC	UGA	GGG	ACC
Codons	CUG	ACU	CCC	UGG

25.44

Codons	UUG	CAC	ACU	GAA
Anticodons	AAC	GUG	UGA	CUU

25.45 Consult Table 25.2 to find what nucleotides correspond to the amino acids in the sequence.

leu-ala-val-glu-asp-cys-met-trp-lys

CUU GCU GUU GAA GAU UGU AUG UGG AAA

25.46 tyr-ile-pro-his-leu-his-thr-ser-phe-met

UAU AUU CCU CAU CUU CAU ACU UCU UUU AUG

25.47

$$\begin{array}{c}CH_2OH\\|\\CHOH\\|\\CH_2OH\end{array} + 3\ CH_3(CH_2)_{16}COOH \rightarrow \begin{array}{c}CH_2OC(CH_2)_{16}CH_3\\\|\\O\\|\\CHOC(CH_2)_{16}CH_3\\\|\\O\\|\\CH_2OC(CH_2)_{16}CH_3\\\|\\O\end{array} + 3\ H_2O$$

glycerol stearic acid triacylglycerol

25.48

$$\begin{array}{c}CH_2OH\\|\\CHOH\\|\\CH_2OH\end{array} + 2\ CH_3(CH_2)_7CH=CH(CH_2)_7COOH + CH_3(CH_2)_{12}COOH \rightarrow$$

glycerol oleic acid myristic acid

$$3\ H_2O + \begin{array}{c}CH_2OC(CH_2)_7CH=CH(CH_2)_7CH_3\\\|\\O\\|\\CHOC(CH_2)_7CH=CH(CH_2)_7CH_3\\\|\\O\\|\\CH_2OC(CH_2)_{12}CH_3\\\|\\O\end{array}$$

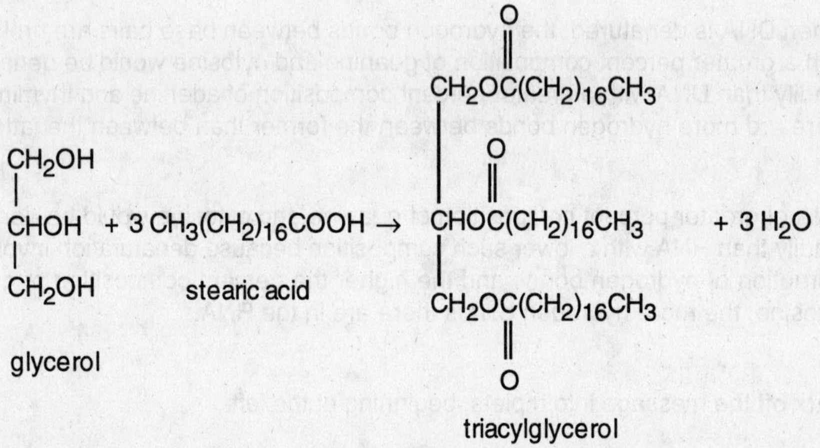

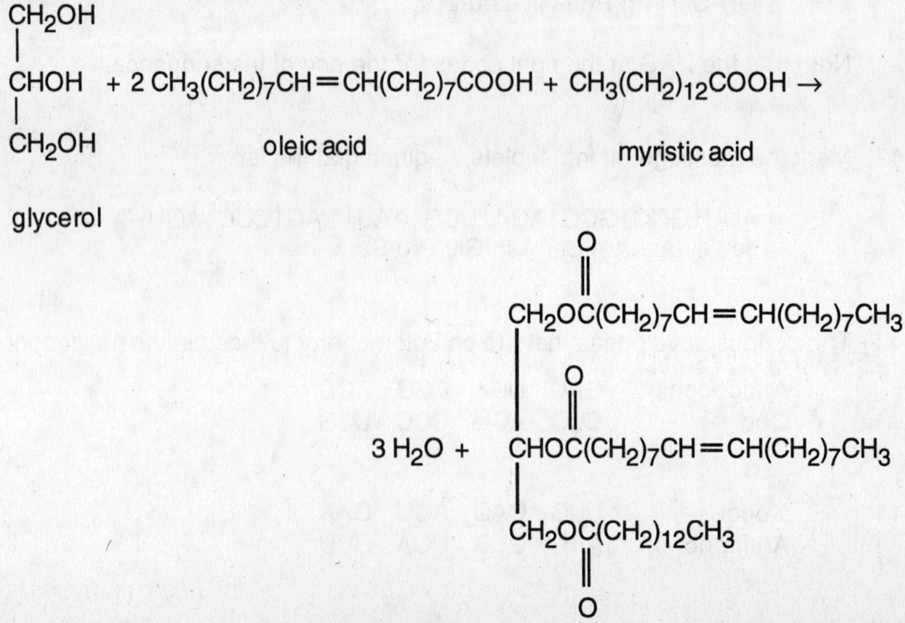

25.49

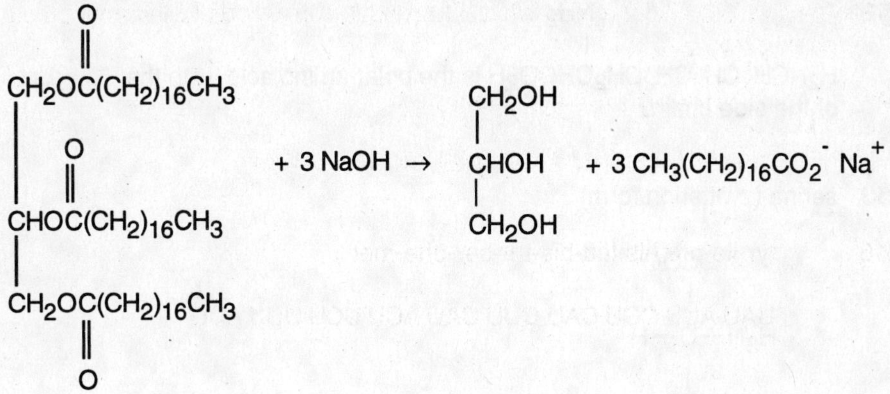

$$\begin{array}{c} CH_2OC(CH_2)_{16}CH_3 \\ \|\phantom{OC(CH_2)_{16}CH_3} \\ O \\ CHOC(CH_2)_{16}CH_3 \\ \|\phantom{OC(CH_2)_{16}CH_3} \\ O \\ CH_2OC(CH_2)_{16}CH_3 \\ \|\phantom{OC(CH_2)_{16}CH_3} \\ O \end{array} + 3\,NaOH \rightarrow \begin{array}{c} CH_2OH \\ | \\ CHOH \\ | \\ CH_2OH \end{array} + 3\,CH_3(CH_2)_{16}CO_2^-\,Na^+$$

25.50

$$\begin{array}{c} CH_2OC(CH_2)_7CH=CH(CH_2)_7CH_3 \\ \| \\ O \\ CHOC(CH_2)_7CH=CH(CH_2)_7CH_3 \\ \| \\ O \\ CH_2OC(CH_2)_{12}CH_3 \\ \|\phantom{OC(CH_2)_{12}CH_3} \\ O \end{array} + 3\,NaOH \rightarrow$$

$$\begin{array}{c} CH_2OH \\ | \\ CHOH \\ | \\ CH_2OH \end{array}$$

$$+\ 2\,CH_3(CH_2)_7CH=CH(CH_2)_7CO_2^-\,Na^+$$

$$+\ CH_3(CH_2)_{12}CO_2^-\,Na^+$$

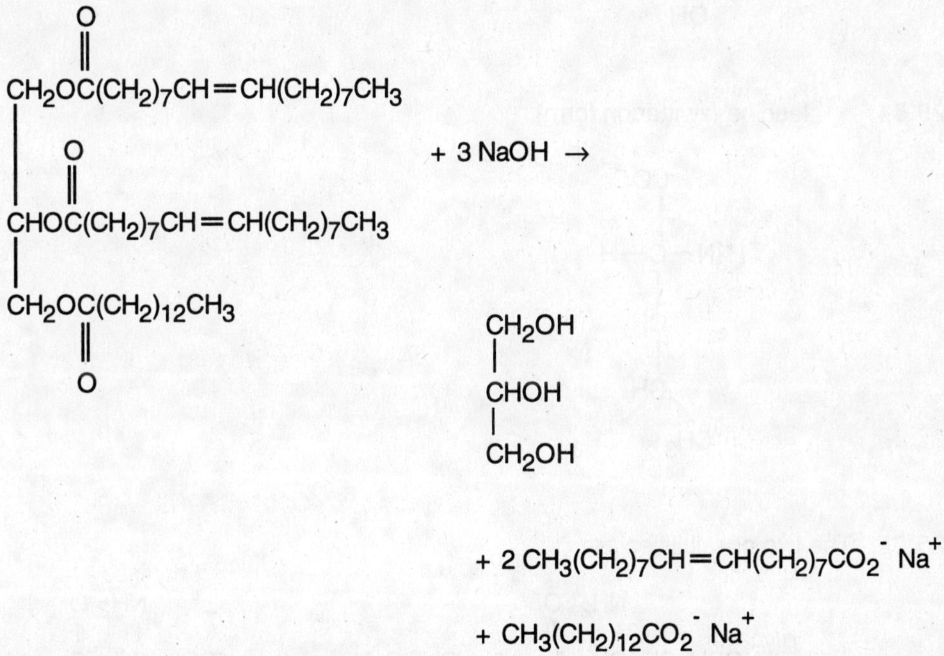

■ Solutions to Unclassified Problems

25.51

$$CH_3SCH_2CH_2\overset{\displaystyle NH_2}{\underset{\displaystyle |}{CH}}-CO_2H$$ has a nonpolar side chain. The other amino acid has a polar SH group in the side chain.

$H_2NCH_2CH_2CH_2CH_2CHCO_2H$ is the polar amino acid, with the -NH_2 group at the end of the side chain. (with NH_2 on the carbon bearing CO_2H)

25.53 serine (zwitterion form)

$$^+H_3N-\underset{\underset{OH}{\underset{|}{CH_2}}}{\overset{\overset{COO^-}{|}}{\underset{|}{C}}}-H$$

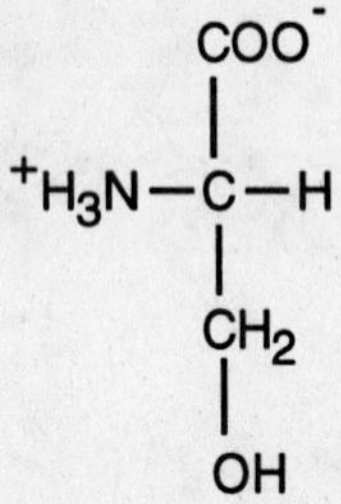

25.54 leucine (zwitterion form)

$$^+H_3N-\underset{\underset{CH_3CH_3}{\underset{\diagup\;\diagdown}{CH}}}{\underset{|}{\underset{CH_2}{\overset{\overset{COO^-}{|}}{\underset{|}{C}}}}}-H$$

25.55 The two possibilities are

$$CH_3SCH_2CH_2\underset{\underset{NH_2}{|}}{CH}-\underset{\overset{O}{\|}}{C}-NH-\underset{\underset{CH_2SH}{|}}{CH}CO_2H \quad \text{and} \quad HSCH_2\underset{\underset{NH_2}{|}}{CH}-\underset{\overset{O}{\|}}{C}-NH-\underset{\underset{CH_2SCH_3}{\underset{|}{CH_2}}}{\underset{|}{CH}}CO_2H$$

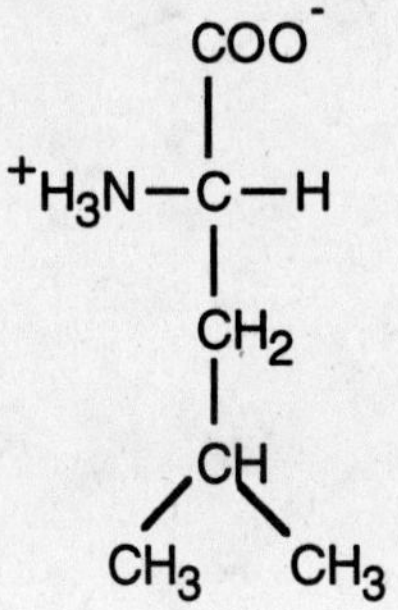

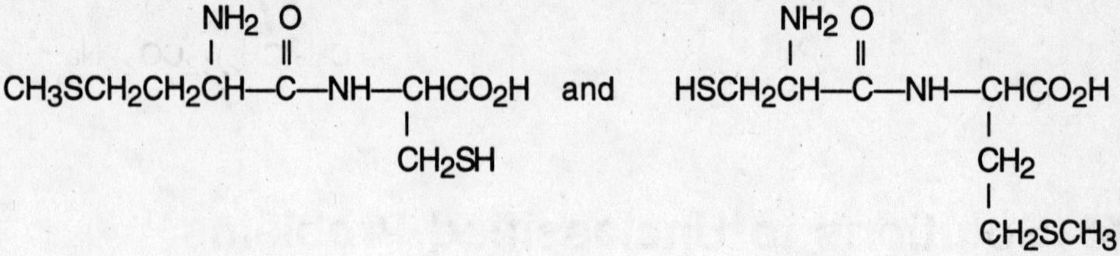

25.56 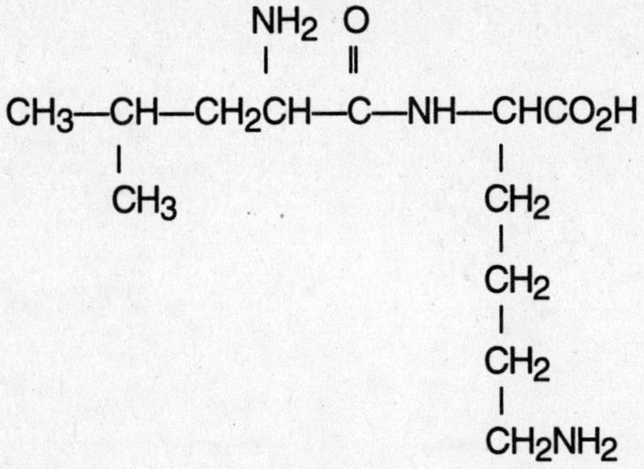 or 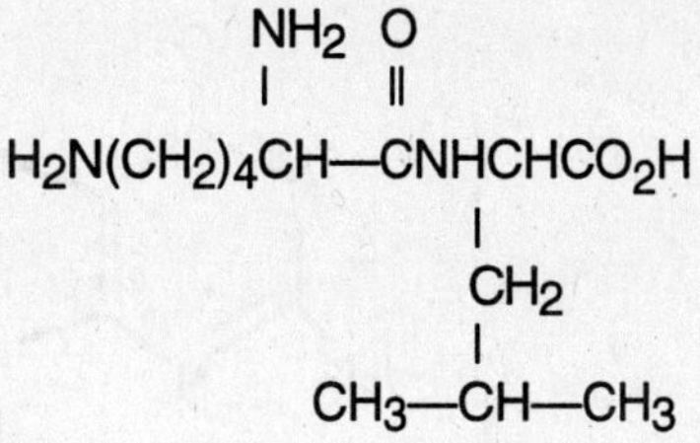

25.57 The number of possible sequences is 6 • 5 • 4 • 3 • 2 • 1, or 720.

25.58 arg-pro-glu-gly-asn-gln
arg-pro-glu-gly-gln-asn
arg-pro-glu-gln-gly-asn

25.59

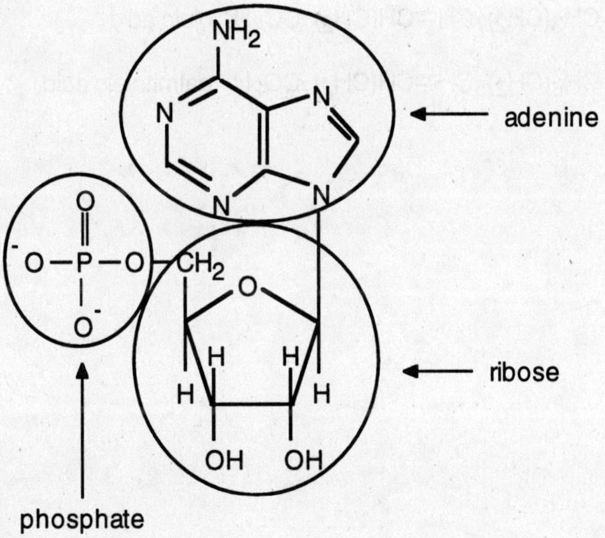

25.60

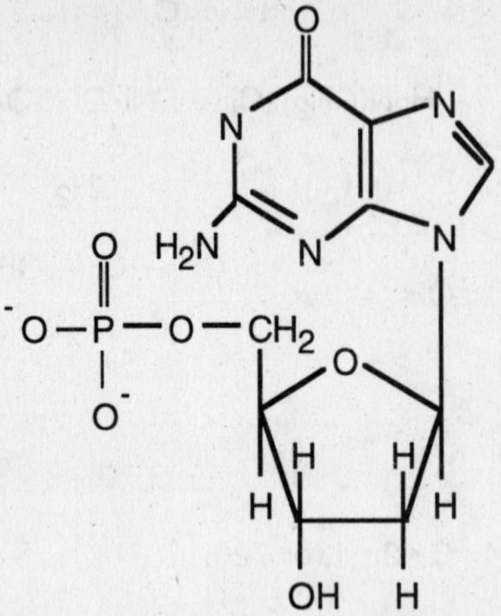

25.61 The triacylglycerol is made up of glycerol and three fatty acids.

CH_2OH	$CH_3(CH_2)_{14}CO_2H$ palmitic acid
|	
$CHOH$	$CH_3(CH_2)_7CH{=}CH(CH_2)_7CO_2H$ oleic acid
|	
CH_2OH	$CH_3(CH_2)_5CH{=}CH(CH_2)_7CO_2H$ palmitoleic acid
glycerol	

25.62

$$\begin{array}{l}\text{CH}_2\text{OC(CH}_2)_{14}\text{CH}_3 \\ \quad \| \\ \quad \text{O} \\ | \\ \text{CHOC(CH}_2)_7\text{CH}=\text{CH(CH}_2)_7\text{CH}_3 \quad + \ 3\ \text{NaOH} \rightarrow \\ \quad \| \\ \quad \text{O} \\ | \\ \text{CH}_2\text{OC(CH}_2)_7\text{CH}=\text{CH(CH}_2)_5\text{CH}_3 \\ \quad \| \\ \quad \text{O}\end{array}$$

$$\begin{array}{l}\text{CH}_2\text{OH} \\ | \\ \text{CHOH} \\ | \\ \text{CH}_2\text{OH}\end{array}$$

$+ \ CH_3(CH_2)_{14}CO_2^- \ Na^+$

$+ \ CH_3(CH_2)_7CH=CH(CH_2)_7CO_2^- \ Na^+$

$+ \ CH_3(CH_2)_5CH=CH(CH_2)_7CO_2^- \ Na^+$

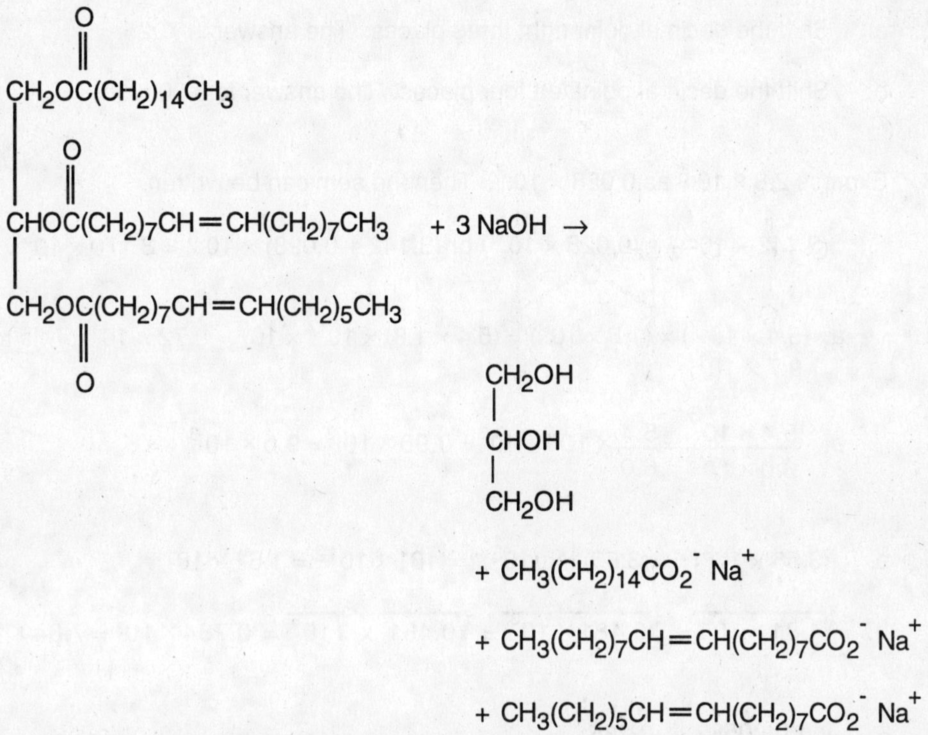

APPENDIX A. MATHEMATICAL SKILLS

■ Solutions to Exercises

1. a. Either leave it as 4.38 or write it as 4.38×10^0.

   b. Shift the decimal point left and count the number of positions shifted (3). The answer is 4.380×10^3 assuming the terminal zero is significant.

   c. Shift the decimal point right and count the number of positions shifted (4). The answer is 4.83×10^{-4}.

2. a. Shift the decimal point right three places. The answer is 7025.

   b. Shift the decimal point left four places. The answer is 0.000897.

3. Express 2.8×10^{-6} as 0.028×10^{-4}. Then the sum can be written

$$(3.142 \times 10^{-4}) + (0.028 \times 10^{-4}) \text{ or } (3.142 + 0.028) \times 10^{-4} = 3.170 \times 10^{-4}.$$

4. a. $(5.4 \times 10^{-7}) \times (1.8 \times 10^{8}) = (5.4 \times 1.8) \times 10^{-7} \times 10^{8} = 9.72 \times 10^{1}$. This rounds to 9.7×10^{1}.

   b. $\dfrac{5.4 \times 10^{-7}}{6.0 \times 10^{-5}} = \dfrac{5.4}{6.0} \times 10^{-7} \times 10^{5} = 0.90 \times 10^{-2} = 9.0 \times 10^{-3}$

5. a. $(3.56 \times 10^{3})^{4} = (3.56)^{4} \times (10^{3})^{4} = 161 \times 10^{12} = 1.61 \times 10^{14}$

   b. $\sqrt[3]{4.81 \times 10^{2}} = \sqrt[3]{0.481 \times 10^{3}} = \sqrt[3]{0.481} \times \sqrt[3]{10^{3}} = 0.784 \times 10^{1} = 7.84 \times 10^{0}$

6. a. $\log 0.00582 = -2.235$

   b. $\log 689 = 2.838$

7. a. $\text{antilog } 5.728 = 5.35 \times 10^{5}$

   b. $\text{antilog } (-5.728) = 1.87 \times 10^{-6}$

8. $x = \dfrac{-0.850 \pm \sqrt{(0.850)^2 - 4(1.80)(-9.50)}}{2(1.80)} = \dfrac{-0.850 \pm \sqrt{69.12}}{3.60} = \dfrac{-0.850 \pm 8.314}{3.60}$

   The positive root is

$$\frac{7.46}{3.60} = 2.07$$